Continental Shelves of the World
Their Evolution During the Last Glacio-Eustatic Cycle

IUGS/GSL publishing agreement

This volume is published under an agreement between the International Union of Geological Sciences and the Geological Society of London and arises from IGCP projects 396 'Continental Shelves in the Quaternary', 464 'Continental Shelves during the Last Glacial Cycle: Knowledge and Applications' and 526 'Risks, Resources and Record of the Past on the Continental Shelf'.

GSL is the publisher of choice for books related to IUGS activities, and the IUGS receives a fee for all books published under this agreement.

Books published under this agreement are subject to the Society's standard rigorous proposal and manuscript review procedures.

It is recommended that reference to all or part of this book should be made in one of the following ways:

CHIOCCI, F. L. & CHIVAS, A. R. (eds) 2014. *Continental Shelves of the World: Their Evolution During the Last Glacio-Eustatic Cycle*. Geological Society, London, Memoirs, **41**.

BARRIE, J. V., HETHERINGTON, R. & MACLEOD, R. 2014. Pacific margin, Canada shelf physiography: a complex history of glaciation, tectonism, oceanography and sea-level change. *In*: CHIOCCI, F. L. & CHIVAS, A. R. (eds) *Continental Shelves of the World: Their Evolution During the Last Glacio-Eustatic Cycle*. Geological Society, London, Memoirs, **41**, 305–313, http://dx.doi.org/10.1144/M41.22

GEOLOGICAL SOCIETY MEMOIR NO. 41

Continental Shelves of the World
Their Evolution During the Last Glacio-Eustatic Cycle

EDITED BY

F. L. CHIOCCI
Sapienza University of Rome, Italy

and

A. R. CHIVAS
University of Wollongong, Australia

2014
Published by
The Geological Society
London

THE GEOLOGICAL SOCIETY

The Geological Society of London (GSL) was founded in 1807. It is the oldest national geological society in the world and the largest in Europe. It was incorporated under Royal Charter in 1825 and is Registered Charity 210161.

The Society is the UK national learned and professional society for geology with a worldwide Fellowship (FGS) of over 10 000. The Society has the power to confer Chartered status on suitably qualified Fellows, and about 2000 of the Fellowship carry the title (CGeol). Chartered Geologists may also obtain the equivalent European title, European Geologist (EurGeol). One fifth of the Society's fellowship resides outside the UK. To find out more about the Society, log on to www.geolsoc.org.uk.

The Geological Society Publishing House (Bath, UK) produces the Society's international journals and books, and acts as European distributor for selected publications of the American Association of Petroleum Geologists (AAPG), the Indonesian Petroleum Association (IPA), the Geological Society of America (GSA), the Society for Sedimentary Geology (SEPM) and the Geologists' Association (GA). Joint marketing agreements ensure that GSL Fellows may purchase these societies' publications at a discount. The Society's online bookshop (accessible from www.geolsoc.org.uk) offers secure book purchasing with your credit or debit card.

To find out about joining the Society and benefiting from substantial discounts on publications of GSL and other societies worldwide, consult www.geolsoc.org.uk, or contact the Fellowship Department at: The Geological Society, Burlington House, Piccadilly, London W1J 0BG: Tel. + 44 (0)20 7434 9944; Fax + 44 (0)20 7439 8975; E-mail: enquiries@geolsoc.org.uk.

For information about the Society's meetings, consult *Events* on www.geolsoc.org.uk. To find out more about the Society's Corporate Affiliates Scheme, write to enquiries@geolsoc.org.uk.

Published by The Geological Society from:
The Geological Society Publishing House, Unit 7, Brassmill Enterprise Centre, Brassmill Lane, Bath BA1 3JN, UK

The Lyell Collection: www.lyellcollection.org
Online bookshop: www.geolsoc.org.uk/bookshop
Orders: Tel. + 44 (0)1225 445046, Fax + 44 (0)1225 442836

The publishers make no representation, express or implied, with regard to the accuracy of the information contained in this book and cannot accept any legal responsibility for any errors or omissions that may be made.

British Library Cataloguing in Publication Data

A catalogue record for this book is available from the British Library.
ISBN 978-1-86239-686-9
ISSN 0435-4052

Distributors

For details of international agents and distributors see:
www.geolsoc.org.uk/agentsdistributors

Typeset by Techset Composition India (P) Ltd, Bangalore and Chennai, India
Printed by Berforts Information Press Ltd, Oxford, UK

Contents

Acknowledgements

We wish to thank the many participants in several IGCP shelf projects over more than eighteen years for their scientific contributions, discussion and friendship.

IGGP 396, Continental Shelves in the Quaternary, leader Wyss W.-S. Yim (Hong Kong, China)

IGCP 464, Continental Shelves during the Last Glacial Cycle: Knowledge and Applications, leaders Francesco L. Chiocci (Italy) and Allan R. Chivas (Australia)

IGCP 526, Risks Resources and Record of the Past on the Continental Shelf, leaders Francesco Latino Chiocci (Italy), Lindsay Collins (Australia), Michael Mahiques (Brazil) and Renée Hetherington (Canada)

The Memoir would not have reached completion without the diligence of many referees, the professional encouragement, skill and patience of Angharad Hills from the Geological Society of London and the editorial assistance of Eleonora Morelli in Rome.

Chapter 1

An overview of the continental shelves of the world

FRANCESCO L. CHIOCCI[1] & ALLAN R. CHIVAS[2]*

[1]*Dipartimento Scienze della Terra, Università La Sapienza di Roma, Piazzale Aldo Moro 5, Rome, Italy*

[2]*GeoQuEST Research Centre, School of Earth & Environmental Sciences, University of Wollongong, Northfields Avenue, Wollongong, NSW 2522, Australia*

*Corresponding author (e-mail: toschi@uow.edu.au)

This Memoir explores the variability and controlling processes of sedimentation, morphology and tectonics on the world's continental shelves, with emphasis on their evolution during the last glacio-eustatic cycle. This work builds on some earlier volumes on continental shelves, notably those by Trumbul *et al.* (1958), Boillot (1978), De Batist & Jacobs (1996), Nittrouer *et al.* (2007) and Li *et al.* (2012) among others, although there has not been a previous comprehensive global-scale synthesis.

The subject material presented here was developed in several International Geoscience Programmes (formerly the International Geological Correlation Programme, IGCP) concerned with shelves. Project 396 'Continental Shelves in the Quaternary' (1996–2000) was followed by IGCP 464 'Continental Shelves during the Last Glacial Cycle: Knowledge and Applications' (2001–2007) and finally by IGCP 526 'Risks, Resources and Record of the Past on the Continental Shelf' (2007–2011). In this Memoir, 23 papers are devoted to the description of different aspects of shelves from all seven continents and which are representative of the world's shelves. However, there remain significant gaps in our compilation. A glance at Figure 1.1 indicates the sparse coverage of data, particularly using modern techniques of investigation, for Africa, Russia and the Polar regions. However, new data are presented from some frontier provinces, such as Myanmar, Morocco and Ecuador.

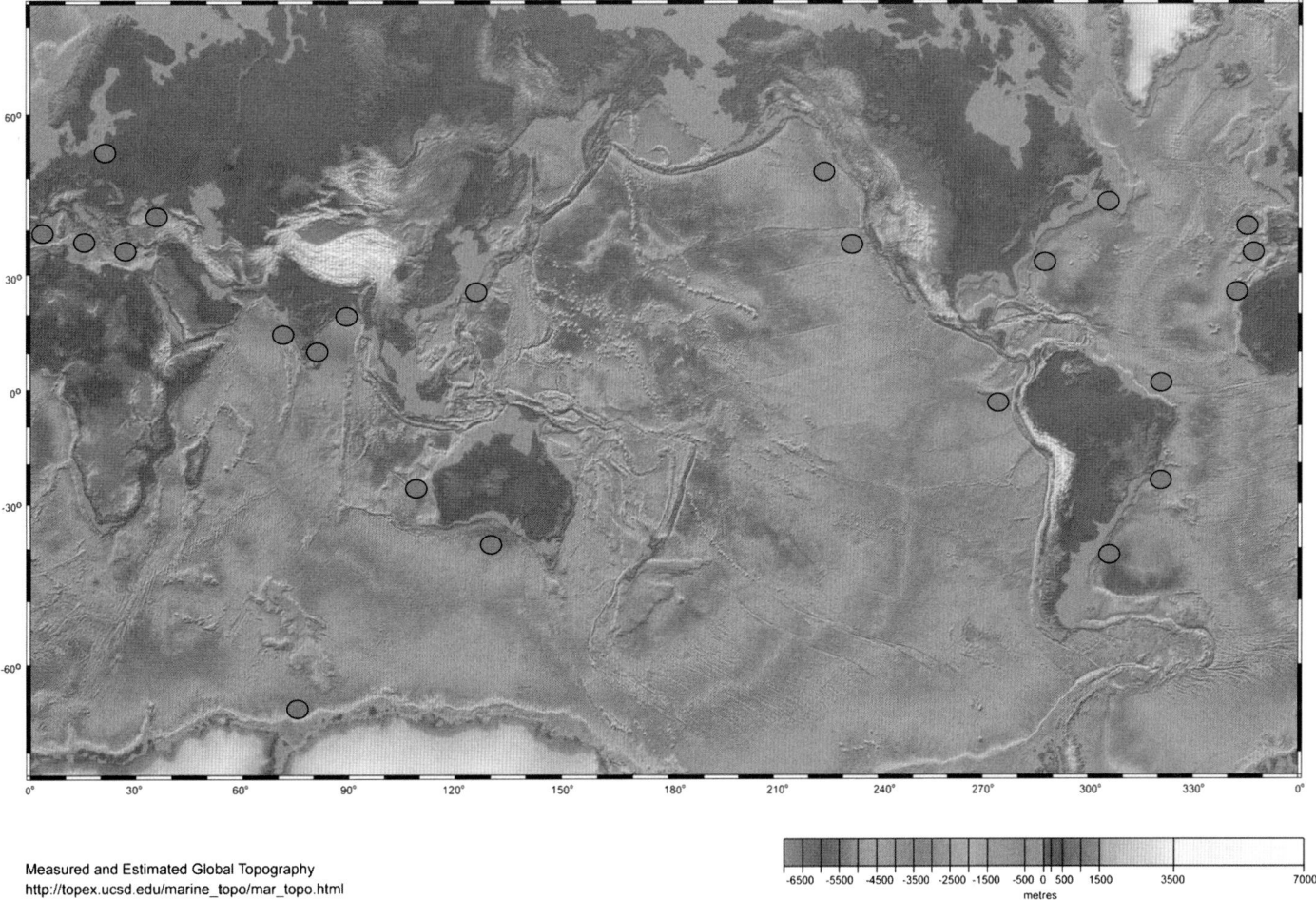

Measured and Estimated Global Topography
http://topex.ucsd.edu/marine_topo/mar_topo.html

metres

Fig. 1.1. The location of the shelf areas presented in this volume. Base map from Smith, W. H. F. & Sandwell, D. T. 1997. Global sea floor topography from satellite altimetry and ship depth soundings. *Science*, **277**, 1956–1962. Reprinted with permission from AAAS.

From: CHIOCCI, F. L. & CHIVAS, A. R. (eds) 2014. *Continental Shelves of the World: Their Evolution During the Last Glacio-Eustatic Cycle.*
Geological Society, London, Memoirs, **41**, 1–5. http://dx.doi.org/10.1144/M41.1

Why study continental shelves?

The continental shelf is the part of the seafloor most used by society. Although shelf areas account for little more than 8% of the marine areas of the world, they are the part of the sea most used for navigation, recreation, fishing and aquaculture, mineral exploration, waste disposal and, increasingly in the future, in the production of renewable energy from wave, tidal currents and wind (Barrie & Conway 2014). It is worth noting that the continental shelf in geological terms (and, therefore, in this volume) differs from the 'continental shelf' as legally defined in international laws and treaties (Emery 1981). The latter encompasses the whole continental margin (shelf, slope and rise).

Mineral resources (Rona 1972; Archer 1973; United Nations 2004) include ore deposits: that is, relict shelf sandbodies or erosional lags made up of heavy minerals of direct economic interest (e.g. ilmenite, magnetite, cassiterite, chromite, rutile, zircon, rutile, gold and diamonds). Shelf sandbodies can also be exploited for aggregates (i.e. quartz-rich sand and gravels) to be used for concrete aggregate, beach nourishment and specific industrial manufacture. Hydrocarbons (oil and gas) are the most relevant economic resources of the shelf (Earney 1990). Oil production from continental shelves grew from a few per cent in the 1960s to 25% of global production by the beginning of this century. Since then, the shelf proportion has remained constant, with a decrease in onshore production offset by increasing deep-water production (US Energy Information Administration, http://www.eia.gov). Finally, phosphate occurs on some shelves as nodules, crusts, hardgrounds or placers in regions that experienced specific nutrient-rich bio-oceanographic conditions or received detritus from eroded phosphate-bearing formations (Filippelli 2011).

Not only do mineral resources make the shelf the most relevant part of the marine realm but shelf waters are also, by far, the richest part of the ocean (Wei *et al.* 2010) in terms of biomass, ecological system services and source of food for humans. The richness of life on the shelf is due to the concurrence of the large amount of nutrients provided by landmasses, and by the light penetration in shallow water that favours pelagic and benthic communities (in contrast to the darkness of abyssal waters). As a consequence, coasts and shelves furnish society with a large proportion of ecosystem services (Costanza *et al.* 1997), comparable to that provided by all terrestrial habitats (Tsunogai *et al.* 1999).

Shelves also preserve detailed and commonly unique archives of environmental changes in the complex interaction among climate, sea level and sediment input from landmasses, recorded in the depositional and erosional features of the subsurface (Gao & Collins 2014; Lobo & Ridente 2014). In detail, changes in

rainfall and hydrology, and possibly storm energy and frequency, may leave their signature in shelf stratigraphy and morphology. Sediment cores collected on the shelf, because of the high deposition rate and proximity to sedimentary sources, are very sensitive to even minor changes in climate. They, therefore, can be used to distinguish between local and global effects, and possibly to assist in differentiating between human-induced and natural environmental changes (Asioli *et al.* 2001). The shift in the limits of ice sheets at high latitudes or the transition to and from carbonate sedimentation at low latitudes can be defined with precision on the shelves (Bailey & Flemming 2008). During the last sea-level lowstand, some shelf areas were flat coastal plains suitable for early human settlement. These experienced dramatic changes in sea level during deglaciation when an average rate of 1 m per century of sea-level rise occurred. The constant rise in base level and the damming of incised valleys by transgressive littoral barriers favoured river flooding and the formation of coastal marshes and swamps. On low-gradient shelves or in land-locked areas, fast to dramatic drowning of emerged land occurred. We can speculate that this process accelerated the dispersal of the Neolithic population, for instance into the interior of Europe at that time. Possible migration routes and civilization trends (compartmentalization of cultures during the Upper Palaeolithic for instance) may have a link with palaeoenvironmental changes on the shelf (Evans *et al.* 2014).

Controlling factors

The processes controlling Quaternary sedimentation on continental shelves are complex and varied. Continental shelves are the result of interactions between endogenous (e.g. geodynamic) and exogenous (e.g. climate-related) processes (Fig. 1.2). Plate tectonics and geodynamics determine the overall physiography of the margin, its width and slope, the deep structure of the subsurface, and the relevance of mass wasting and steady-state erosion. By contrast, climate determines the type of sediment (carbonate in tropical areas, siliciclastic in temperate climes and glacial in high latitudes). Furthermore, sediment distribution as well as bedforms and sedimentary structures on the seafloor are the result of a complex interplay between hydrodynamic processes (waves, tides and currents), the position of sediment sources (rivers, estuaries, wind) and the amount of sediment delivered to each shelf. Shelf deposits make up the bulk of the continental margin with thicknesses of several kilometres if subsidence permits. Their stratigraphic architecture is fundamentally controlled by sea-level changes that migrate the shoreline and related depositional

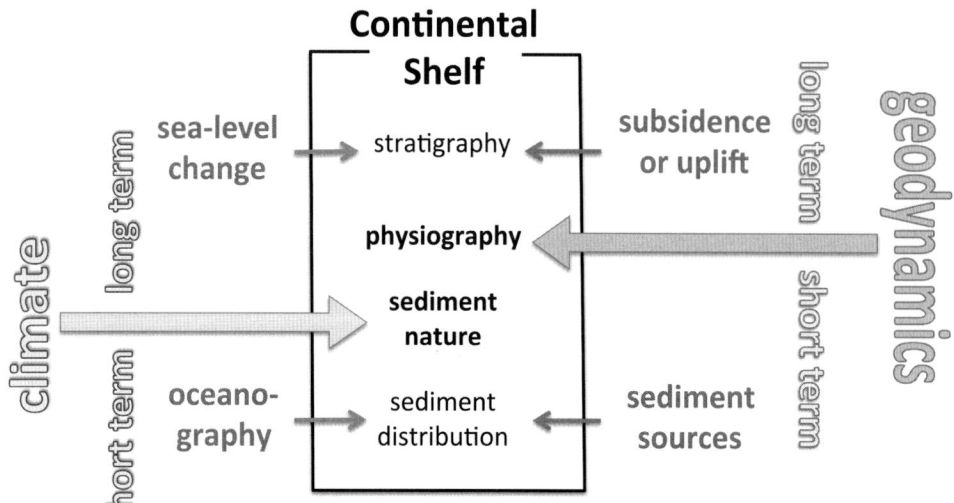

Fig. 1.2. Factors that control shelf character, on both long- and short-term geological timescales.

systems across the shelf. The resulting deposits (systems tracts in sequence stratigraphy) may, in fact, be preserved or destroyed by the vertical mobility of the continental margin that may create the space to accommodate and, thus, preserve shelf sediments. Depth of palaeo-shelf-breaks may also be used to reconstruct such vertical movement (Fraccascia *et al.* 2013).

Climate-driven eustatic changes played a particularly significant role in the last 2 Ma, when the combination of astronomically forced insolation cycles and the position of land masses (especially in the Northern Hemisphere) caused glaciation over large parts of North America and Eurasia, and forced sea level to fluctuate between its present position and a depth of 120–130 m at a fast rate. The shelves therefore underwent dramatic environmental changes, with repetitive cycles of emergence and submersion, displacement of the shoreline from the shelf break to the inner edge of the continental margin, with consequent alternation of sedimentation and erosion, whose traces can be found in the shelf shallow stratigraphy, inherited morphology and the presence of relict sediment.

This volume

The chapters in this Memoir are presented in an order starting from the western North Atlantic continental margin and progressing eastwards to the Mediterranean, Indian and Pacific Ocean shelves.

The methods used to study shelf sediments focus on stratigraphy, particularly seismic stratigraphy and coring, geochemistry and palaeobiology. The fact that no single chapter in this Memoir fully integrates all such aspects indicates that large teams of investigators (and resources) are required and emphasizes that much work remains to be carried out on shelf environments. In the first part of each chapter authors provide basic data for their shelf, such as size, geometry, climate, tides, geodynamic setting and sediment types, to facilitate intercomparison and to provide first-level information for readers unfamiliar with a given shelf. Thereafter, the chapter presents more specific aspects of the studied shelf, while maintaining a geographically broad perspective.

The wide continental shelves of Atlantic Canada are characterized by a series of banks separated by transverse troughs. **Shaw et al. (2014)** present a detailed description of the shelves, and focus on the glacial history in the last glacial cycle, the glacial land systems, the geographical changes caused by glacio-isostasy, the processes on the upper continental slopes and the sediment mobility on the offshore banks.

Miller et al. (2014) discuss sedimentation on the storm-dominated, microtidal, continental shelf and slope of the eastern US passive continental margin. The sediments record sea-level changes providing a classic example of the interplay between eustasy, tectonism and sedimentation.

The Brazilian shelf is described in two chapters by **Vital (2014)** (northern shelf) and **Nagai et al. (2014)** (southern shelf). In the former, a review of the shelf physiography and the spatial distribution of sediments are used to relate them to the sedimentological processes and physical characteristics of the environment, and to the sedimentary history of this shelf system. A key factor in the development of the southern Brazilian shelf is the influence of the northerly flowing sediment plume from the Rio de La Plata, the activity of which has varied throughout the Holocene. Other major factors include variability in sea level and the intensity of the south-flowing Brazil Current.

The Argentine continental shelf is one of the largest and smoothest siliciclastic shelves in the world, mostly emplaced on a passive continental margin. **Violante et al. (2014)** provide a full description of the shelf where sea-level fluctuations, sediment dynamics and climatic/oceanographic processes are the most important conditioning factors in the modelling of the shelf.

The Baltic Sea is a semi-enclosed intracontinental sea where thermohaline stratification of the waters occurs. **Uścinowicz (2014)** describes the evolution of the Baltic Sea caused by the evolution of the last Scandinavian ice sheet during the Pleistocene and the sediment distribution of the Quaternary cover of its seabed.

The NW Iberian continental shelf is a narrow, gently dipping, geomorphological structure with a well-defined shelf break. **Rey et al. (2014)** present a detailed overview of the most distinct features and processes on the shelf, among which waves and tides, seasonal upwelling, and coast-parallel currents are recognized.

In the southern part of Atlantic Iberia, **Lobo et al. (2014b)** describe the outer part of the Gibraltar Strait (Gulf of Cadiz), where the influence of northward-flowing intermediate water pouring out of the Mediterranean interacts with shelf morphology and structural features.

Mhammdi et al. (2014) present a summary of the state of knowledge of the recent sedimentation and processes in the NW African shelf in the Atlantic Sea. The sedimentary processes along the shelf are driven both by long-term (Quaternary glacial–interglacial periods) and short-term factors (fluvial and aeolian sediment supply, local climate and hydrodynamic conditions).

The Iberian Mediterranean shelves are divided by **Lobo et al. (2014a)** into three different geographical segments (the Northeastern Shelf, the Southeastern Shelf and the Northern Alboran Sea Shelf), where the Quaternary stratigraphic architecture is controlled, respectively, by regressive–transgressive cycles, a declining fluvial influence and short mountain rivers.

Martorelli et al. (2014) describe the structure of the shelves surrounding the Italian peninsula. Despite its small extent, Italy shows a variety of geodynamic domains (back-arc to foredeep to foreland), so that the striking differences in shelf morphology and stratigraphy can be used to depict the controlling factors on continental shelf physiography and sedimentology.

The Hellenic shelf, located within one of the most seismically active areas of the world, is presented by **Ferentinos et al. (2014)**. The tectonic activity – coupled with eustatic sea-level changes and water-circulation patterns – controls the configuration and processes of the shelf, and the occurrence of a variety of geological hazards.

The semi-enclosed Black Sea received an influx of Mediterranean-sourced water during multiple openings of the Marmara Gateway during the Quaternary. **Nicholas & Chivas (2014)** provide evidence for the most recent marine transgression into the Black Sea from the stratigraphic subdivision of sediments from cores, and from the surface distribution of bivalve molluscs on the mid and outer NW shelf.

The Indian shelves are presented by **Faruque & Ramachandran (2014)** (western shelf) and **Faruque et al. (2014)** (eastern shelf). Precambrian peninsular India is tectonically stable, although there are horst- and graben-segmented sections of the west coast that control and define distinct shelf provinces. By contrast, the eastern shelf is characterized by major rivers and deltas that contribute sediment, with the interdelta regions being generally sediment-starved.

The central Myanmar (Burma) shelf (**Ramaswamy & Rao 2014**) lies in a tectonically subsiding embayment fed by some of the largest sediment-transporting rivers, including the Irrawaddy and Salween. Tidal redistribution of the suspended sediments has produced an extensive turbid zone and mud belt that characterizes the shelf.

A representative west Antarctic shelf, around Prydz Bay and Mac.Roberstson Land, is described by **O'Brien et al. (2014)**. The narrow shelf is scoured by icebergs calved from the adjacent Amery Ice Shelf. There is a thin sedimentary cover, deposited and eroded by successive glacial advances throughout the Quaternary.

Collins et al. (2014) provide an overview of the western continental margin of Australia. This is an enormous tectonically stable area with little fluvial input, spanning 20° of latitude from tropical to temperate climate and, accordingly, is largely carbonate dominated. The southern Australian margin is, by contrast, a high-energy cool-water carbonate province, although of equally large proportions (**Murray-Wallace 2014**). Siliciclastic sediments are more significant in SE Australia, where there are fluvial sediment sources.

The NE Chinese shelf (**Yang *et al.* 2014**) is fed by two of the world's largest sediment sources, the Changjiang (the Yangtze River) and the Huanghe (the Yellow River). The shelf is particularly wide, traversed by palaeovalleys, and dominated by siliciclastic sediments and eddy-current mud deposits.

Barrie *et al.* (2014) describe Canada's western continental shelf, dividing it into three geographical regions: the Salish Sea, the Pacific North Coast and the Vancouver Island Shelf. The authors reveal the contributions to each region's physiography by glaciation, tectonism, oceanography and sea-level change.

Covault & Fildani (2014) show that the tectonically active Oceanside shelf offshore southern California has served as a conveyor of sediment from land to the deep sea during millennia of significant climatic fluctuations. The authors highlight the importance of shelf physiography, coupled with climatic forcing, and timescale of observation in assessing the role of shelves as sediment capacitors or conveyors.

Dumont *et al.* (2014) identify and describe three different segments along the continental margin of Ecuador. On the basis of three-dimensional (3D) numerical modelling of a curved subduction plane, the authors demonstrate the strong relationship between the geometry of the continental boundary and the occurrence of uplift or subsidence along the continental margin.

References

ARCHER, A. A. 1973. Economics of off-shore exploration and production of solid minerals on the continental shelf. *Ocean Management*, **1**, 5–40.

ASIOLI, A., TRINCARDI, F., LOWE, J. J., ARIZTEGUI, D., LANGONE, L. & OLDFIELD, F. 2001. Sub-millennial scale climatic oscillations in the central Adriatic during the Lateglacial: palaeoceanographic implications. *Quaternary Science Reviews*, **20**, 1201–1221.

BAILEY, G. N. & FLEMMING, N. C. 2008. Archaeology of the continental shelf: marine resources, submerged landscapes and underwater archaeology. *Quaternary Science Reviews*, **27**, 2153–2165.

BARRIE, J. V. & CONWAY, K. W. 2014. Seabed characterization for the development of marine renewable energy on the Pacific margin of Canada. *Continental Shelf Research*, **83**, 45–52.

BARRIE, J. V., HETHERINGTON, R. & MACLEOD, R. 2014. Pacific margin, Canada shelf physiography: a complex history of glaciation, tectonism, oceanography and sea-level change. *In*: CHIOCCI, F. L. & CHIVAS, A. R. (eds) *Continental Shelves of the World: Their Evolution During the Last Glacio-Eustatic Cycle*. Geological Society, London, Memoirs, **41**, 303–313, http://dx.doi.org/10.1144/M41.22

BOILLOT, G. 1978. *Geology of Continental Margins*. Longman Higher Education, London.

COLLINS, L. B., JAMES, N. P. & BONE, Y. 2014. Carbonate shelf sediments of the western continental margin of Australia. *In*: CHIOCCI, F. L. & CHIVAS, A. R. (eds) *Continental Shelves of the World: Their Evolution During the Last Glacio-Eustatic Cycle*. Geological Society, London, Memoirs, **41**, 253–272, http://dx.doi.org/10.1144/M41.19

COSTANZA, R., D'ARGE, R. ET AL. 1997. The value of the world's ecosystem services and natural capital. *Nature*, **387**, 253–260.

COVAULT, J. A. & FILDANI, A. 2014. Continental shelves as sediment capacitors or conveyors: source-to-sink insights from the tectonically active Oceanside shelf, southern California, USA. *In*: CHIOCCI, F. L. & CHIVAS, A. R. (eds) *Continental Shelves of the World: Their Evolution During the Last Glacio-Eustatic Cycle*. Geological Society, London, Memoirs, **41**, 313–326, http://dx.doi.org/10.1144/M41.23

DE BATIST, M. & JACOBS, P. (eds) 1996. *Geology of Siliciclastic Shelf Seas*. Geological Society, London, Special Publications, **117**.

DUMONT, J. F., SANTANA, E., BONNARDOT, M.-A., PAZMIÑO, N., PEDOJA, K. & SCALABRINO, B. 2014. Geometry of the coastline and morphology of the convergent continental margin of Ecuador. *In*: CHIOCCI, F. L. & CHIVAS, A. R. (eds) *Continental Shelves of the World: Their Evolution During the Last Glacio-Eustatic Cycle*. Geological Society, London, Memoirs, **41**, 325–338, http://dx.doi.org/10.1144/M41.24

EARNEY, C. F. 1990. *Marine Mineral Resources*. Routledge, Ocean Management Policy Series.

EMERY, K. O. 1981. Geological limits of the continental shelf. *Ocean Development and International Law*, **10**, 1–11.

EVANS, A., FLATMAN, J. & FLEMMING, N. (eds) 2014. *Prehistoric Archaeology on the Continental Shelf: A Global Review*. Springer, Berlin.

FARUQUE, B. M. & RAMACHANDRAN, K. V. 2014. The western continental shelf of India. *In*: CHIOCCI, F. L. & CHIVAS, A. R. (eds) *Continental Shelves of the World: Their Evolution During the Last Glacio-Eustatic Cycle*. Geological Society, London, Memoirs, **41**, 211–220, http://dx.doi.org/10.1144/M41.15

FARUQUE, B. M., VAZ, G. G. & MOHAPATRA, G. P. 2014. The continental shelf of eastern India. *In*: CHIOCCI, F. L. & CHIVAS, A. R. (eds) *Continental Shelves of the World: Their Evolution During the Last Glacio-Eustatic Cycle*. Geological Society, London, Memoirs, **41**, 219–229, http://dx.doi.org/10.1144/M41.16

FERENTINOS, G., LYKOUSIS, V., PAPATHEODOROU, G. & IATROU, M. 2014. Hellenic shelf: late Quaternary tectonics, sea-level changes, sedimentation and geohazards. *In*: CHIOCCI, F. L. & CHIVAS, A. R. (eds) *Continental Shelves of the World: Their Evolution During the Last Glacio-Eustatic Cycle*. Geological Society, London, Memoirs, **41**, 185–197, http://dx.doi.org/10.1144/M41.13

FILIPPELLI, G. M. 2011. Phosphate rock formation and marine phosphorus geochemistry: the deep time perspective. *Chemosphere*, **84**, 759–766.

FRACCASCIA, S., CHIOCCI, F. L., SCROCCA, D. & FALESE, F. 2013. Very high-resolution seismic stratigraphy of Pleistocene eustatic minima markers as a tool to reconstruct the tectonic evolution of the northern Latium shelf (Tyrrhenian Sea, Italy). *Geology*, **41**, 375–378, http://dx.doi.org/10.1130/G33868.1

GAO, S. & COLLINS, M. B. 2014. Holocene sedimentary systems on continental shelves. *Marine Geology*, **352**, 268–294.

LI, M. Z., SHERWOOD, C. R. & HILL, P. R. (eds) 2012. *Sediments, Morphology and Sedimentary Processes on Continental Shelves: Advances in Technologies, Research and Applications*. International Association of Sedimentologists, Special Publications, **44**. Blackwell, Oxford.

LOBO, F. J. & RIDENTE, D. 2014. Stratigraphic architecture and spatio-temporal variability of the high-frequency (Milankovitch) depositional cycles on modern continental margins: an overview. *Marine Geology*, **352**, 215–247.

LOBO, F. J., ERCILLA, G., FERNÁNDEZ-SALAS, L. M. & GÁMEZ, D. 2014a. The Iberian Mediterranean shelves. *In*: CHIOCCI, F. L. & CHIVAS, A. R. (eds) *Continental Shelves of the World: Their Evolution During the Last Glacio-Eustatic Cycle*. Geological Society, London, Memoirs, **41**, 145–170, http://dx.doi.org/10.1144/M41.11

LOBO, F. J., LE ROY, P., MENDES, I. & SAHABI, M. 2014b. The Gulf of Cádiz continental shelves. *In*: CHIOCCI, F. L. & CHIVAS, A. R. (eds) *Continental Shelves of the World: Their Evolution During the Last Glacio-Eustatic Cycle*. Geological Society, London, Memoirs, **41**, 107–130, http://dx.doi.org/10.1144/M41.9.

MARTORELLI, E., FALESE, F. & CHIOCCI, F. L. 2014. Overview of the variability of Late Quaternary continental shelf deposits off the Italian peninsula. *In*: CHIOCCI, F. L. & CHIVAS, A. R. (eds) *Continental Shelves of the World: Their Evolution During the Last Glacio-Eustatic Cycle*. Geological Society, London, Memoirs, **41**, 169–186, http://dx.doi.org/10.1144/M41.12

MHAMMDI, N., SNOUSSI, M., MEDINA, F. & JAAÏDI, E. B. 2014. Recent sedimentation in the NW African shelf. *In*: CHIOCCI, F. L. & CHIVAS, A. R. (eds) *Continental Shelves of the World: Their Evolution During the Last Glacio-Eustatic Cycle*. Geological Society, London, Memoirs, **41**, 129–146, http://dx.doi.org/10.1144/M41.10

MILLER, K. G., BROWNING, J. V., MOUNTAIN, G. S., SHERIDAN, R. E., SUGARMAN, P. J., GLENN, S. & CHRISTENSEN, B. A. 2014. History of continental shelf and slope sedimentation on the US middle

Atlantic margin. *In*: CHIOCCI, F. L. & CHIVAS, A. R. (eds) *Continental Shelves of the World: Their Evolution During the Last Glacio-Eustatic Cycle*. Geological Society, London, Memoirs, **41**, 19–34, http://dx.doi.org/10.1144/M41.3

MURRAY-WALLACE, C. V. 2014. The continental shelves of SE Australia. *In*: CHIOCCI, F. L. & CHIVAS, A. R. (eds) *Continental Shelves of the World: Their Evolution During the Last Glacio-Eustatic Cycle*. Geological Society, London, Memoirs, **41**, 271–291, http://dx.doi.org/10.1144/M41.20

NAGAI, R. H., SOUSA, S. H. M. & MAHIQUES, M. M. 2014. The southern Brazilian shelf. *In*: CHIOCCI, F. L. & CHIVAS, A. R. (eds) *Continental Shelves of the World: Their Evolution During the Last Glacio-Eustatic Cycle*. Geological Society, London, Memoirs, **41**, 45–54, http://dx.doi.org/10.1144/M41.5

NICHOLAS, W. A. & CHIVAS, A. R. 2014. Late Quaternary sea-level change on the Black Sea shelves. *In*: CHIOCCI, F. L. & CHIVAS, A. R. (eds) *Continental Shelves of the World: Their Evolution During the Last Glacio-Eustatic Cycle*. Geological Society, London, Memoirs, **41**, 197–212, http://dx.doi.org/10.1144/M41.14

NITTROUER, C. A., AUSTIN, J. A., FIELD, M. E., KRAVITZ, J. H., SYVITSKI, J. P. M. & WIBERG, P. L. (eds) 2007. *Continental Margin Sedimentation: From Sediment Transport to Sequence Stratigraphy*. International Association of Sedimentologists, Special Publications, **37**. Blackwell, Oxford.

O'BRIEN, P. E., HARRIS, P. T., POST, A. L. & YOUNG, N. 2014. East Antarctic continental shelf: Prydz Bay and the Mac.Robertson Land shelf. *In*: CHIOCCI, F. L. & CHIVAS, A. R. (eds) *Continental Shelves of the World: Their Evolution During the Last Glacio-Eustatic Cycle*. Geological Society, London, Memoirs, **41**, 239–254, http://dx.doi.org/10.1144/M41.18

RAMASWAMY, V. & RAO, P. S. 2014. The Myanmar continental shelf. *In*: CHIOCCI, F. L. & CHIVAS, A. R. (eds) *Continental Shelves of the World: Their Evolution During the Last Glacio-Eustatic Cycle*. Geological Society, London, Memoirs, **41**, 229–240, http://dx.doi.org/10.1144/M41.17

REY, D., ÁLVAREZ-IGLESIAS, P. *ET AL*. 2014. The NW Iberian continental shelf. *In*: CHIOCCI, F. L. & CHIVAS, A. R. (eds) *Continental Shelves of the World: Their Evolution During the Last Glacio-Eustatic Cycle*. Geological Society, London, Memoirs, **41**, 89–108, http://dx.doi.org/10.1144/M41.8

RONA, P. A. 1972. *Exploration Methods for the Continental Shelf: Geology, Geophysics, Geochemistry*. National oceanic and Atmospheric Administration (NOAA) Technical Report, **ERL-238**, AOML-8 (COM-72-50920). Atlantic Oceanographic and Meteorological Laboratory, Miami, FL.

SHAW, J., TODD, B. J., LI, M. Z., MOSHER, D. C. & KOSTYLEV, V. E. 2014. Continental shelves of Atlantic Canada. *In*: CHIOCCI, F. L. & CHIVAS, A. R. (eds) *Continental Shelves of the World: Their Evolution During the Last Glacio-Eustatic Cycle*. Geological Society, London, Memoirs, **41**, 7–20, http://dx.doi.org/10.1144/M41.2

SMITH, W. H. F. & SANDWELL, D. T. 1997. Global sea floor topography from satellite altimetry and ship depth soundings. *Science*, **277**, 1956–1962, http://dx.doi.org/10.1126/science.277.5334.1956

TRUMBUL, J., LYMAN, J. & PEPPER, J. F. 1958. *An Introduction to the Geology and Mineral Resources of the Continental Shelves of the Americas*. United States Geological Survey, Bulletin, **1067**.

TSUNOGAI, S., WATANABE, S. & SATO, T. 1999. Is there a 'continental shelf pump' for the absorption of atmospheric CO_2? *Tellus*, **B51**, 701–712.

UNITED NATIONS. 2004. *Marine Mineral Resources – Scientific Advances and Economic Perspectives*. United Nations Division for Ocean Affairs and International Seabed Authority, New York.

UŚCINOWICZ, S. 2014. The Baltic Sea continental shelf. *In*: CHIOCCI, F. L. & CHIVAS, A. R. (eds) *Continental Shelves of the World: Their Evolution During the Last Glacio-Eustatic Cycle*. Geological Society, London, Memoirs, **41**, 67–89, http://dx.doi.org/10.1144/M41.7

VIOLANTE, R. A., PATERLINI, C. M. *ET AL*. 2014. The Argentine continental shelf: morphology, sediments, processes and evolution since the Last Glacial Maximum. *In*: CHIOCCI, F. L. & CHIVAS, A. R. (eds) *Continental Shelves of the World: Their Evolution During the Last Glacio-Eustatic Cycle*. Geological Society, London, Memoirs, **41**, 53–68, http://dx.doi.org/10.1144/M41.6

VITAL, H. 2014. The north and northeast Brazilian tropical shelves. *In*: CHIOCCI, F. L. & CHIVAS, A. R. (eds) *Continental Shelves of the World: Their Evolution During the Last Glacio-Eustatic Cycle*. Geological Society, London, Memoirs, **41**, 33–46, http://dx.doi.org/10.1144/M41.4

WEI, C.-L., ROWE, G. T. *ET AL*. 2010. Global patterns and predictions of seafloor biomass using random forests. *PLoS ONE*, **5**, e15323, http://dx.doi.org/10.1371/journal.pone.0015323

YANG, S., WANG, Z., DOU, Y. & SHI, X. 2014. A review of sedimentation since the Last Glacial Maximum on the continental shelf of eastern China. *In*: CHIOCCI, F. L. & CHIVAS, A. R. (eds) *Continental Shelves of the World: Their Evolution During the Last Glacio-Eustatic Cycle*. Geological Society, London, Memoirs, **41**, 291–303, http://dx.doi.org/10.1144/M41.21

Chapter 2

Continental shelves of Atlantic Canada

JOHN SHAW*, BRIAN J. TODD, MICHAEL Z. LI, DAVID C. MOSHER & VLADIMIR E. KOSTYLEV

Geological Survey of Canada (Atlantic), Bedford Institute of Oceanography, Dartmouth, Nova Scotia, Canada B2Y 4A2

Corresponding author (e-mail: johnshaw@nrcan.gc.ca)

Abstract: The wide continental shelves of Atlantic Canada are characterized by a series of banks separated by transverse troughs. These shelves have been imprinted by repeated Quaternary glaciations, so that fluvial valleys have been deepened into fjords and shelf-crossing troughs, and a suite of glacigenic sediments has been deposited. In shallow areas the seafloor is shaped by waves and currents, including the strong tidal currents of the macrotidal Bay of Fundy. Glacigenic sediments have been reworked by modern processes to yield thick muds in basins, and thinner deposits of sand and gravel on wave-dominated banks and the littoral zone. As a result of a cold climate and the Labrador Current, seasonal sea ice occurs to varying degrees across the region, and iceberg impact continues on much of the Newfoundland and Labrador shelves. For the purpose of description, we divide Atlantic continental shelves into four regions and focus on advances in understanding over the past several decades relating to: (1) processes on upper continental slopes; (2) glacial history in the last glacial cycle; (3) glacial land systems; (4) geographical changes caused by glacio-isostasy; and (5) sediment mobility on the offshore banks. We conclude with a brief overview of the biota.

The continental shelves of Atlantic Canada (Fig. 2.1, Table 2.1) range in width from 100 to 480 km. They are comparatively shallow, with a shelf break that averages −210 m but varies from as little as −75 m on parts of the Scotian Shelf to as much as −530 m off NE Newfoundland. The offshore banks that comprise the shelves are dissected by glacially overdeepened troughs and, in some areas, are separated from the coasts by shore-parallel troughs. For descriptive purpose we subdivide the shelves into four areas:

- the Scotian Shelf (including the Bay of Fundy and Gulf of St Lawrence);
- the Grand Banks of Newfoundland;
- the NE Newfoundland Shelf;
- the Labrador Shelf.

The shelves of Atlantic Canada are distinguished from shelves further south in North America by (among other factors): (1) their distinctive physiography (banks and troughs); (2) extensive modification by repeated shelf-crossing glaciations; (3) the impact of cold-water processes, primarily sea ice and icebergs; and (4) the existence of large emergent areas far offshore in early post-glacial times, concurrent with the inundation of coastal regions by high relative sea levels. These shelves were described by Keen & Williams (1990) in their benchmark publication, to which we refer readers for insights into the bedrock geology in particular. Here we focus on recent advances in understanding, particularly those resulting from advances in seafloor instrumentation, sediment-transport modelling and seafloor mapping with multibeam sonar systems, as follows: (1) processes on the upper continental slopes; (2) glacial history in the last glacial cycle; (3) new understandings of glacial land systems; (4) geographical changes caused by glacio-isostatic adjustments; (5) sediment mobility on the offshore banks and the resulting bedforms; and (6) benthic habitats.

General setting

In terms of modern climate, the continental margins of the study area range from the warm-water environment of Georges Bank in the south to the Arctic environment at the northern tip of Labrador. The major oceanographic features are the Gulf Stream and the Labrador Current. The warm-water Gulf Stream runs NE, skirting the Scotian Shelf and Grand Banks of Newfoundland. The cold Labrador Current runs south along the Labrador Shelf, branches

into the wide bays on the south coast of Newfoundland, and bifurcates east and west around Grand Bank and Flemish Cap. Cold winters result in sea ice of varying thickness and duration throughout the region; it is most extensive off NE Newfoundland and Labrador, and in the Gulf of St Lawrence, and infrequent off the Nova Scotian coast. Icebergs drifting south in the Labrador Current are common off Labrador and north and east Newfoundland, where their keels impact the seafloor, causing furrowing and pitting to water depths below 200 m (Shaw *et al.* 1999).

The eastern Canadian margin is an Atlantic-type, passive continental margin, comprising a sediment wedge located against and overlying continental crust, forming a relatively wide shelf. The margin developed along the edge of the ocean basin after the continent of Pangaea broke up. The initial North Atlantic Ocean was small and shallow; salt and other evaporates were precipitated to form thick layers at the bottom of the basin. Large sedimentary basins developed, some of which contain sedimentary rocks up to 15 km thick. The formation of petroleum deposits was influenced by the later development of complex structures and faults within these sedimentary rocks, in response to compressive stresses and the deformation of the evaporite layer.

The geology of coastal regions and the Gulf of St Lawrence generally differs from that further offshore on the banks (Fader *et al.* 1989). The Gulf of Maine and Bay of Fundy contain Palaeozoic and Triassic rocks, respectively, while Cambrian–Devonian bedrock occurs off Nova Scotia. The Gulf of St Lawrence is dominated by Upper Carboniferous–Permian rocks. Offshore Newfoundland, rocks of varying lithology range from Cambrian to Carboniferous in age, while bedrock of the inner Labrador Shelf is predominantly Precambrian in age. On the banks, by contrast, the seaward-dipping sediment prism is comprised of Cretaceous–Tertiary rocks, penetrated by salt diapers in places. Further offshore, beyond the continental slope, salt diapers reach the seabed to form seamounts. Flemish Cap is the easternmost part of the continental crust of North America (Shrivastava *et al.* 2000). Its innermost core mainly comprises Hadrynian basement rocks.

Continental shelves of Atlantic Canada

For purposes of description, the margin is divided into four regions (Fig. 2.1), each characterized by unique physiography, geological setting, Quaternary history, surficial geology and modern processes.

From: CHIOCCI, F. L. & CHIVAS, A. R. (eds) 2014. *Continental Shelves of the World: Their Evolution During the Last Glacio-Eustatic Cycle.*
Geological Society, London, Memoirs, **41**, 7–19. http://dx.doi.org/10.1144/M41.2
© The Geological Society of London 2014. Publishing disclaimer: www.geolsoc.org.uk/pub_ethics

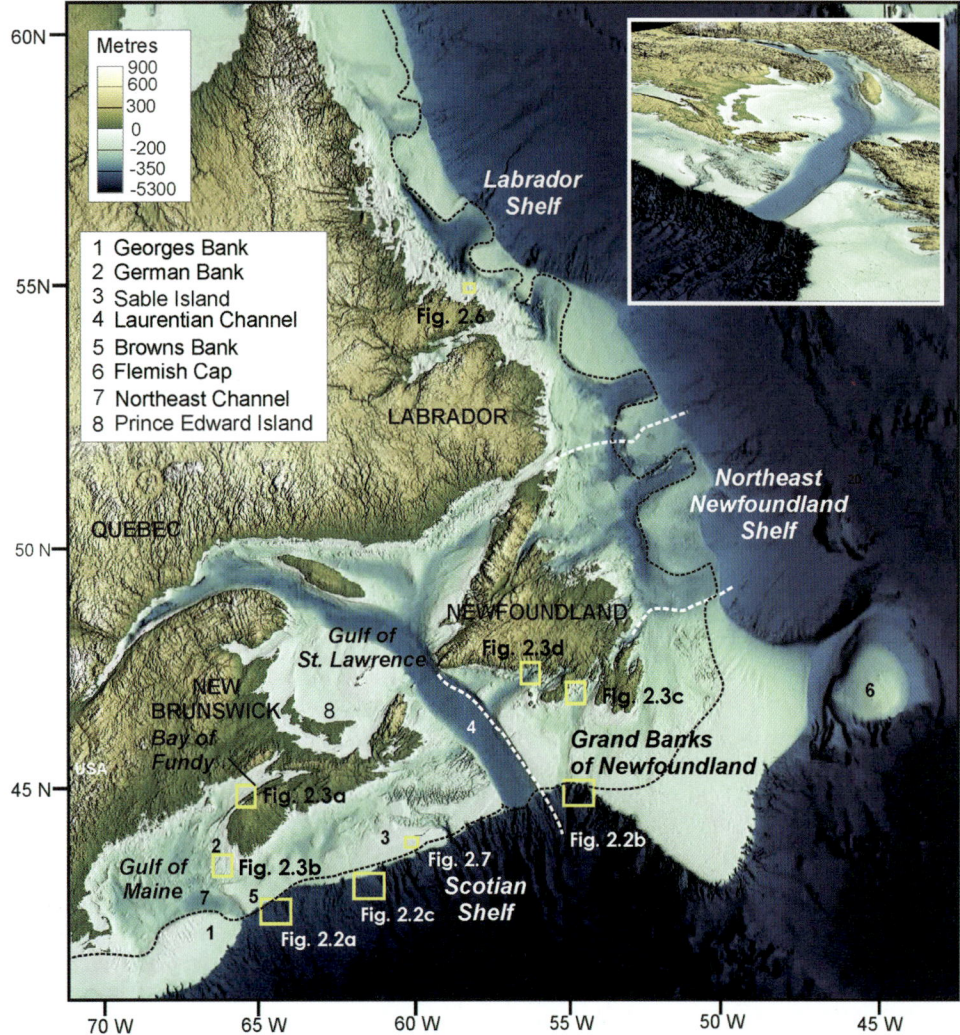

Fig. 2.1. Shaded relief map of the Atlantic Canadian margin. For descriptive purposes, we divide the margin into shelf four regions (separated by dashed lines). The inset map (upper right) shows a 3D view NW along the Laurentian Channel. The black dashed line shows the glacial ice extent at the LGM, based on Josenhans *et al.* (1986) and Shaw *et al.* (2006).

The Scotian Shelf (including the Bay of Fundy and the Gulf of St Lawrence)

This shelf comprises a series of shallow banks separated by basins and includes Sable Island, notable for its sandy shores and extensive coastal dunes. This region includes the Bay of Fundy, the 250 m-deep NE Channel, the Gulf of St Lawrence and the Laurentian Channel. The latter is 520 m deep in the Gulf of St Lawrence, and shallows to 400 m depth at the continental shelf edge. Surficial geological mapping (e.g. Fader *et al.* 1977; King & Fader

Table 2.1. *Summary of Atlantic Canadian shelves*

Length of the shelf (km)	3500
Average width (km)	230
Tidal, wave, current ranges	Tides range up to 17 m; wave heights up to 15 m; currents at the seafloor range up to 10 m s^{-1}
Dominating process	Waves and currents (Scotian Shelf), waves (Newfoundland shelves); tidal currents (Bay of Fundy, northern Labrador Shelf); sea ice (Gulf of St Lawrence, NE Newfoundland, Labrador); iceberg impact (NE Newfoundland, Labrador)
Average depth of shelf break	210 m
Sedimentation	Glacial modified by post-glacial processes
Modern/relict/palimpsest	50% modern, 50% relict
Tectonic trend over last glacial cycle	Uplifting and subsiding – glacio-isostatic/ eustatic

1986; Josenhans & Lehman 1999) shows that the region contains suites of glacigenic sediments, modified by wave processes during the Holocene transgression above a depth of about 120 m on the outer shelf. Glaciers reached the shelf edge at the Last Glacial Maximum (LGM) (Shaw *et al.* 2006).

The Grand Banks of Newfoundland

The Grand Banks of Newfoundland comprises a series of banks separated from one another by shelf-crossing troughs, and from the mainland by coast-parallel channels. Flemish Cap, separated from Grand Bank by Flemish Pass, shoals to a depth of 126 m. The deepest water on this shelf (750 m) occurs in one of the many fjords in southern Newfoundland. Compared with the Scotian Shelf, the Quaternary sediment veneer is relatively thin or absent on the banks, although thick sequences occur in some of the troughs (Fader *et al.* 1982). Iceberg scours on the Grand Banks represent two distinct populations: a modern population of scours in relatively shallow water, perhaps down to a maximum depth of approximately 220 m (Wadhams 2000); and an older population in water depths as great as 650 m (Sonnichsen & King 2005).

The NE Newfoundland Shelf

This shelf is characterized by banks and troughs that are relatively deep. Tectonic zone boundaries, deep crustal block boundaries and faults trend towards the NE, imposing a strong structural grain on the bedrock geology of the shelf. Coarse-grained glacigenic sediments predominate at the seafloor, except in basins (Shaw *et al.* 1999), and the many deep fjords in the coastal areas contain thick glaciomarine sequences overlain by post-glacial muds. The shallow inner shelf off NE Newfoundland is characterized by wave-reworked sands and gravels. The NE Newfoundland Shelf has been intensely furrowed by icebergs (Shaw *et al.* 1999), both modern and lateglacial.

The Labrador Shelf

The inner Labrador Shelf is irregular and complex, with numerous fjords in coastal areas. The north-trending Labrador Marginal Trough, >800 m deep in places, marks the boundary between the inner and outer shelves. The outer Labrador Shelf has subdued topography and comprises a series of shallow banks separated by deep east–west topographical depressions called saddles. Except for basins, most of the shelf has coarse substrates, with gravel and sand. As with the NE Newfoundland Shelf and Grand Banks, two populations of iceberg scours are found in the region. Modern icebergs drifting south in the Labrador Current turbate the seafloor (Todd *et al.* 1988; Woodward Lynas *et al.* 1991) to maximum depths of approximately 220 m (Wadhams 2000, p. 266). Unlike much of the Scotian and Newfoundland shelves, the offshore banks mostly lie below modern wave base.

Continental slopes

While the focus in this paper is on the shelves, some comments on the continental slopes are pertinent. The detailed morphology of eastern Canadian continental slopes is known through bathymetric compilations, including incorporation of multibeam and 3D seismic mapping (e.g. Mosher *et al.* 2004; Campbell *et al.* 2008*a*, *b*). While the broad geomorphological elements are dictated by the underlying tectonic framework, bedrock structure, and the Palaeogene and Neogene geological history, the modern surface is largely a result of Pleistocene glaciation. Till is known to extend to the outer shelf and uppermost slope along most of the margin (Mosher *et al.* 2004, fig. 7). It is believed that high sediment flux and intensified hyperpycnal and turbidity current flows, combined with lower sea levels during deglacial episodes,

had the most pronounced influence on modern slope morphology (Mosher *et al.* 2004; Campbell *et al.* 2008*a*, *b*).

The continental slope seafloor between SW Nova Scotia and the northern tip of Labrador is divided into three general morphological zones (Fig. 2.2) (Mosher *et al.* 2004; Deptuck *et al.* 2007; Campbell *et al.* 2008*a*, *b*), as follows:

- The SW Scotian Slope, Laurentian Channel and much of the Labrador Slope – particularly outboard of glacial troughs and channels – have a highly eroded seafloor with upper-slope ridges, gullies and buttes, and mid-slope peneplaned regions with channels merging into single broad sediment conduits. Erosion and sediment bypass is inferred from this characteristic. Outboard of the troughs and channels are trough–mouth fans (Deptuck *et al.* 2007), the Laurentian Fan being one of the largest. The general slope profile of these areas is concave in form.
- The eastern Scotian Slope and most of the Newfoundland Grand Banks slope is characterized by 500 m-deep canyons. Many canyons head below the modern shelf break; only a few incise landwards to the shelf (e.g. The Gully and Logan Canyon on the Scotian Slope). Canyons that incise deeply to the shelf tend to have a narrow thalweg in their central portions and are still moderately active, trapping sediment in their upslope portions and transferring it to ocean depths via turbidity current activity. The slope profile in these areas is approximately linear in form.
- The central portion of the Scotian slope, the St Pierre Slope of SW Newfoundland and a few broad intercanyon ridge areas are not incised and demonstrate a more complete stratigraphic succession; appearing as a relatively featureless seafloor. On closer inspection, there are many laterally extensive (tens to hundreds of kilometres long) escarpments that range from a few metres to 100 m in height in these areas. The escarpments are scars of mass-transport events and often correlate to underlying fault traces. The general slope profile in these areas is convex in form.

Mass-transport deposits are ubiquitous throughout the region, regardless of the underpinning slope morphology, and some deposits are truly massive (cf. Deptuck *et al.* 2007). From the Pleistocene onwards, mass wasting was most common during deglacial episodes owing to increased glacial rebound seismicity (Mosher *et al.* 2004; Campbell *et al.* 2008*b*). Seismicity is otherwise rare

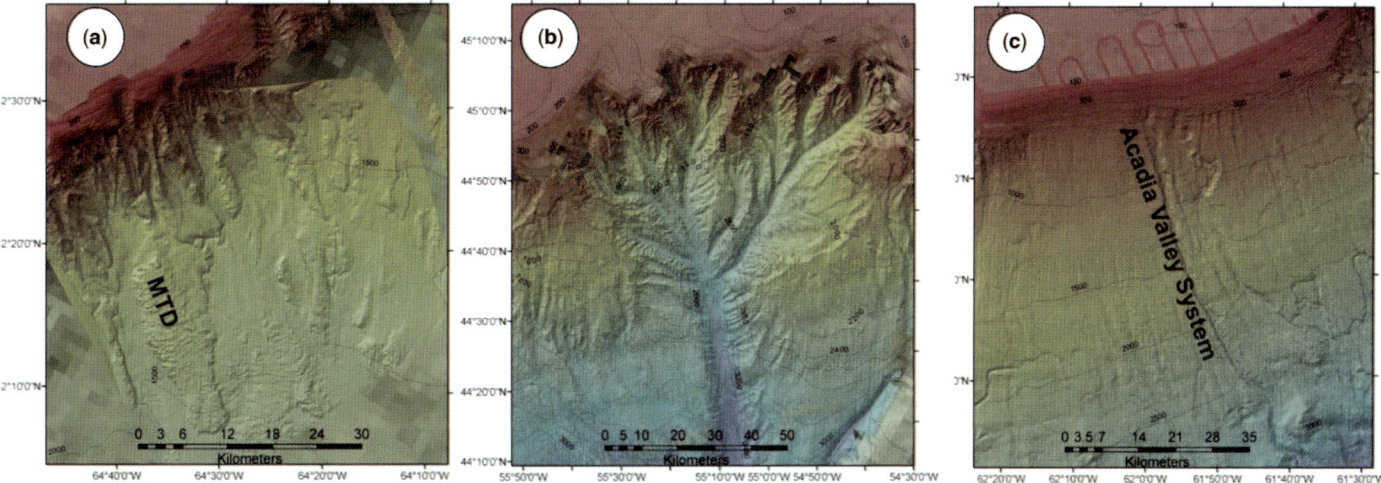

Fig. 2.2. Three seafloor renderings demonstrating the three broad geomorphological types of seafloor that exist on the continental slope, offshore eastern Canada. Hot colours are shallower water and cold colours are deep water. (**a**) The SW Scotian Slope showing Type 1 seafloor, a heavily gullied shelf break and an apparently highly eroded seafloor with a mass-transport deposit (MTD). (**b**) A Type 2 seafloor showing a canyon system from the SW Grand Banks margin, demonstrating the coalescing nature of the canyons and the broad intercanyon ridge areas. (**c**) A Type 3 geomorphology from the central Scotian Slope with a smooth transition at the shelf break and a relatively smooth seafloor, with the exception of the Acadia Valley crossing the slope. This type is rare along the margin.

on these margins (Adams & Halchuk 2004), and consequently modern landslides are not common (Mosher 2009). The 1929 Grand Banks earthquake, submarine landslide and tsunami is an exception (Mosher & Piper 2007*a*, *b*).

The Scotian margin is underpinned by Triassic and Jurassic salt. Mobility of this salt, particularly in the form of diapirs, has strongly influenced the lower-slope morphology. Large mounds (tens of kilometres in diameter) protrude on to the seafloor in many instances. These features affect subsequent sedimentation patterns by steering sediment pathways and causing local mass failures.

The modern Labrador and Newfoundland margins are strongly influenced by the Labrador and Western Boundary undercurrents. These southward-flowing currents rework, transport and redeposit sediments in mid- to lower-slope water depths. On the broad scale, where major morphological elements in the margin create a protuberance in the current pathway, large sedimentary drift deposits form, such as the Hamilton Spur, Orphan Spur and Sackville Spur. These features are hundreds of kilometres along axis and tens of kilometres across. In addition, there are contour-current-derived bedforms with 100 m-long wavelengths along the rise of the Labrador margin.

Shelf processes in the Quaternary

As a result of repeated shelf-crossing glaciations from the middle Pleistocene onwards (Piper & Normark 1989; Piper *et al.* 1994), the shelves are characterized as glaciated shelves. This is reflected in the Quaternary stratigraphy, which commonly shows five seis-mostratigraphic units (Syvitski 1991), namely: (1) bedrock; (2) ice-contact sediment (till); (3) ice-proximal and ice-distal glacio-marine sediments; (4) paraglacial sediments, deposited with waning glacial influence; and (5) post-glacial sediments, formed by the reworking of the earlier units under conditions of changing relative sea levels. Till (King 1993) commonly occurs as regional blankets, comprised of stacked layers, as shown by Josenhans & Lehman (1999) for the Gulf of St Lawrence. Till is also organized into many types of moraines including, for example, large, linear moraines (e.g. King & Fader 1986; King 1996) and fjord-mouth moraines (Shaw 2003). Glaciomarine sediments are thickest in basins (e.g. King & Fader 1986) and fjords (Shaw *et al.* 1999). Post-glacial sediments are commonly either mud in basins or sand and gravel on banks and in the littoral fringe (e.g. Fader *et al.* 1982).

Glacial history in the last glacial cycle

Dyke *et al.* (2002) and Shaw *et al.* (2006) showed that the margin of the SE Laurentide Ice Sheet lay close to the edge of the continental shelf at the LGM, in contrast to previous reconstructions (Dyke & Prest 1987) that had to rely on overly-old bulk radiocarbon dates from offshore and which depicted LGM ice margins approximately along the modern coastline. Ice streams occupied the shelf-crossing troughs, notably a major ice stream in the Laurentian Channel and secondary streams in the Bay of Fundy/Gulf of Maine, Trinity Trough, Notre Dame Trough and elsewhere (Fig. 2.1). Early retreat in the NE (Notre Dame Bay) and in the SW (Bay of Fundy/Gulf of Maine) was mainly by calving along embayments. By 18 ^{14}C ka BP a calving embayment had opened on Emerald Basin on the Scotian Shelf. Ice-margin retreat in channels marooned ice on intervening banks. A major calving episode beginning just before 14 ka BP removed ice from the Gulf of St Lawrence, triggering re-advances down secondary troughs (Josenhans & Lehman 1999). Isolation of a Newfoundland ice cap produced radial drainage via fjords and more extensive flow into the lowlands on the west coast (Shaw 2003). By 13 ka, most ice was on land, and ablating by melting rather than calving. Shelf ice

caps persisted until 11 ka on the Grand Banks and the portion of the Scotian Shelf adjacent to the Laurentian Channel.

The latest synthesis of glacial history for the Labrador Shelf (Josenhans *et al.* 1986) suggests that glacial ice crossed both the banks and troughs to reach the shelf edge but the most recent advance was confined to the troughs. With a lack of accelerator mass spectrometry (AMS) radiocarbon dating in the area, it is uncertain whether the glacial tongues in the troughs existed at LGM, as shown in Figure 2.1, or whether they represent a post-LGM re-advance following retreat from the shelf edge. Incidentally, this model is the converse of the pattern of deglaciation described by Fillon & Harmes (1982), who depicted ice lingering on banks after it had disappeared from shelf-crossing troughs. Further work is required before the glacial history of this shelf is clearly known.

Glacial land systems

Great progress was made in the 1970s and 1980s in delineating the suite of glacigenic sediments on Atlantic Canadian shelves through the use of high-resolution sub-bottom profiling systems such as the Huntec Deep Towed System (see King & Fader 1986). In the 1990s new insights into the diversity of glacial landforms and the spatial relationships between them came with the advent of multibeam sonar systems. The continental shelf is blanketed with depositional features allowing the glacial land systems approach to be employed to evaluate and interpret the terrain by linking the geomorphology and subsurface materials (Eyles 1983; Stokes & Clark 1999, 2001). The several examples described here attest to the diversity of landforms that formed under glacial ice and at ice margins.

Bay of Fundy. The Bay of Fundy glacial land system encompasses both subglacial and ice-marginal landforms (Shaw *et al.* 2008). At the mouth of the Bay of Fundy, where it joins the Gulf of Maine (Fig. 2.1), outcropping bedrock shows evidence of streamlining by an ice stream that flowed from NE to SW out of the bay. Adjacent to the bedrock are subglacial streamlined megaflutes, grading into drumlins, also orientated NE–SW, reflecting the reduction of bedform elongation with increasing distance from the Bay of Fundy ice-stream margin. Within the central Bay of Fundy, both subglacial and ice-marginal landforms demarcate former ice-sheet margins. A lobate pattern of broad and narrow recessional moraines is orientated normal to the axis of the bay reflecting the NE retreat of the Bay of Fundy ice stream (Fig. 2.3a). The southern limit of the recessional moraines marks a lateral shear margin. Drumlins occur south of the lateral shear margin; for drumlins within about 5 km of the lateral shear margin, the long-axes strikes are parallel to the shear margin. More distant from the shear margin, drumlin long axes have a west or WNW strike, less influenced by the ice-stream flow within the bay.

German Bank. The German Bank land system (Todd *et al.* 2007; Shaw *et al.* 2008) off SW Nova Scotia was created by the Bay of Fundy ice stream flowing south across the Gulf of Maine followed by steady ice retreat to the north back into the bay, punctuated by at least one major re-advance. German Bank displays both subglacial and ice-marginal landforms where glacially fluted terrain, with fields of drumlins in places, is overprinted by bank-scale regional recessional moraines hundreds of kilometres in length and local-scale De Geer moraines less than 10 km in length (Fig. 2.3b). Regional moraines with heights and widths reaching 40 and 600 m, respectively, are consistently larger than De Geer moraine dimensions. The latter are 1–8 m in height and 40–130 m in width. The distinctive geomorphology of De Geer moraines on German Bank is parallel ridges with a regular spacing of approximately 150–200 m, with some areas having ridges as closely spaced as 30–50 m.

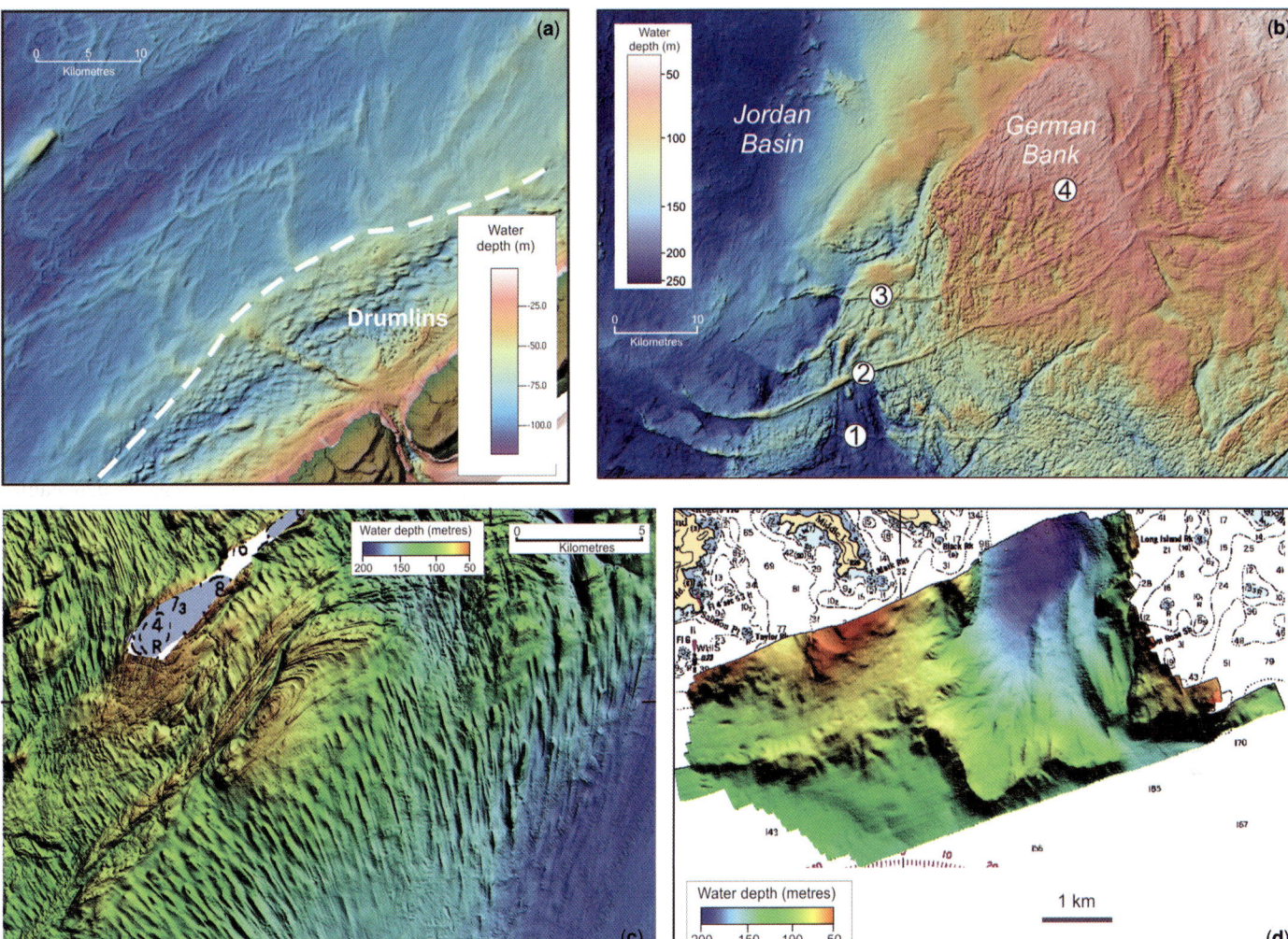

Fig. 2.3. Four examples of glacial land systems. (**a**) Recessional grounding-line wedges in the upper Bay of Fundy lie to the NW of drumlins; the two terrains are separated by a former ice shear zone (dashed line). (**b**) Regional moraines (1–4) on German Bank. Higher-resolution imagery shows that swarms of de Geer moraines are present in this area (Shaw *et al.* 2008). (**c**) Drumlins and mega-scale glacial lineations in Placentia Bay, Newfoundland, indicative of an accelerating ice flow from the north, out of the bay. (**d**) Fjord-mouth moraine off SW Newfoundland (Shaw *et al.* 2000; Shaw 2003). Such moraines are present at the boundary between basement rocks and younger rocks offshore.

Placentia Bay, Newfoundland. In Placentia Bay off southern Newfoundland, the glacial land system is composed of five elements (Brushett *et al.* 2007; Shaw *et al.* 2008): crag-and-tail forms and drumlins grading into megaflutes; swarms of De Geer moraines; ribbed (Rogen) moraines; eskers in valleys normal to the coast; and coast-parallel moraines. This land system is interpreted as evidence of the onset area for ice that formerly converged in, and streamed southward through, Placentia Bay. The increasing elongation of the drumlins and their transformation into megaflutes reflects acceleration of ice into this flow (Fig. 2.3c). The land system exhibits a marked asymmetry: drumlins and megaflutes dominate the west side of the bay, whereas the east side is populated by ribbed moraine, eskers and coast-parallel moraines. This geographical distribution is interpreted as reflecting the blockage of the flow of ice from the east by a stronger flow of ice draining into Placentia Bay from the north.

SW Newfoundland. The glacial land system offshore SW Newfoundland (Shaw *et al.* 2008) is divided into two parts. To the east, where the basement boundary is located at the coast, large submarine moraines are located at fjord mouths (Shaw *et al.* 2000, 2006; Shaw 2003). The moraines are arcuate in planform, commonly consist of several lobes and comprise up to several hundred metres of acoustically incoherent ice-contact sediment

that passes seawards into glaciomarine silt (Fig. 2.3d). To the west, the dominant glacial land system component is a 30–45 m-thick shore-parallel moraine. Flutes with relief up to 20 m, orientated normal to the moraine, extend to the north for several kilometres. This moraine is thought to have been formed by a continuous ice margin, coeval with the fjord-mouth ice margins to the east (Shaw 2003).

Post-glacial geography of the shelves

In the 1980s, a post-glacial relative sea-level lowstand of 115–120 m was well documented for the outer shelves (e.g. Fader 1989). This lowstand depth was thought applicable to very large areas. However, subsequent observations revealed continuous relative sea-level fall since deglaciation in coastal Labrador (Clark & Fitzhugh 1991), and complex relative sea-level histories around Newfoundland (Shaw & Forbes 1990) typified by shallow, early-Holocene lowstands and late-Holocene transgressions.

These observations were reconciled for three of the Atlantic margin shelves by Shaw *et al.* (2002), who combined isobase maps with a digital terrain model of Atlantic Canada to map palaeogeography from 13 ^{14}C ka BP to the present. The reconstructions revealed an archipelago on the outer shelf that persisted

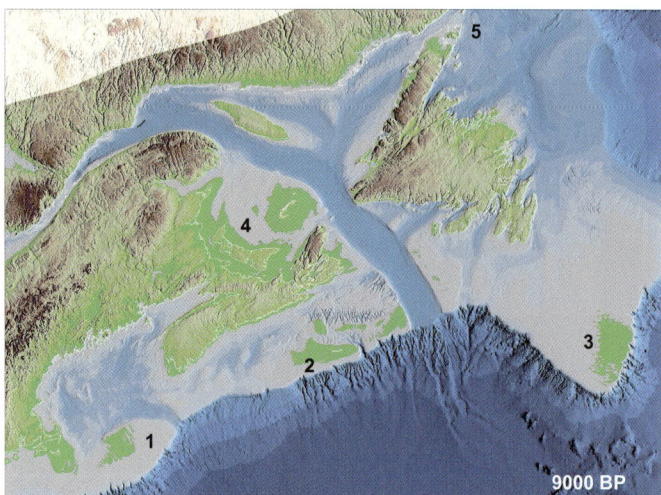

Fig. 2.4. Geography of Atlantic Canada at 9 ka BP (radiocarbon years). Emerged areas include: Georges Bank (1); Sable Island (2); and Grand Bank (3). Prince Edward Island (4) is attached to the mainland due to low relative sea levels, while parts of northern Newfoundland are submerged (5) due to high relative sea levels. The complete sequence of geographical changes since 13 ka is shown in Shaw *et al.* (2002). Outer-shelf areas are submerging at this time, and are much smaller than their earlier extent. Note that glacial ice is still present in Quebec and Labrador at this time.

from >13 ka BP until approximately 8 ka BP. The former island on Grand Bank was by far the largest subaerial region. Prince Edward Island was initially separated from the mainland, became connected after 11 ka BP and was separated again just before 6 ka BP, when Northumberland Strait formed. Figure 2.4

shows the geography at approximately 9 ka BP. Despite the vast extent of the formerly emergent areas, physical evidence of their existence has been preserved only in very particular circumstances: for example, where estuarine deposits underlie thin transgressive sands and gravels (Forbes *et al.* 1991) or in former lakes that were connected to the ocean by the late-Holocene transgression (Shaw *et al.* 2009). This analysis does not include the Labrador Shelf, which, owing to its great depth, was never subaerially exposed by Quaternary sea-level lowering and, indeed, relative sea level is still falling in coastal areas today.

Modern shelf processes

Modern hydrodynamic and sediment-transport processes on the Atlantic Shelves are controlled by tides, waves, and oceanic and storm-generated currents (Amos & Judge 1991). Tidal range on the open shelf is generally less than 2 m, but tides up to 17 m are found in the Bay of Fundy. The distribution of mean significant wave height averaged for the period 2002–5 is shown in Figure 2.5. This distribution reveals that the strongest wave effect (up to 3 m) occurs on the Grand Banks off Newfoundland. Moderate waves (up to 2 m) are found on the Scotian Shelf, the NE Newfoundland Shelf and the Labrador Shelf. In contrast, relatively low values (<1.5 m) occur in the Gulf of St Lawrence and Baffin Bay. The largest significant wave height for a normal year is between 8 and 9 m on the open shelf, and less than 7 m in the Bay of Fundy and in the Gulf of St Lawrence.

Major currents influencing shelves in the region (see Townsend *et al.* 2006, fig. 5) include the equatorward-flowing cold and relatively fresh Labrador Current of the outer Labrador Shelf, with a mean velocity in its core of 0.5 m s^{-1}. This shelf-edge current continues southwards and flows along the outer edges of Grand Banks westwards along the margin of the Grand Banks and the Scotian Shelf, with a lesser component continuing westwards along the Scotia Shelf. The NE-flowing Gulf Stream lies south of the Scotian Shelf. Vast areas of the relatively shallow Scotian and Newfoundland shelves are subject to sediment mobilization by waves and currents, in contrast to the low mobilization on

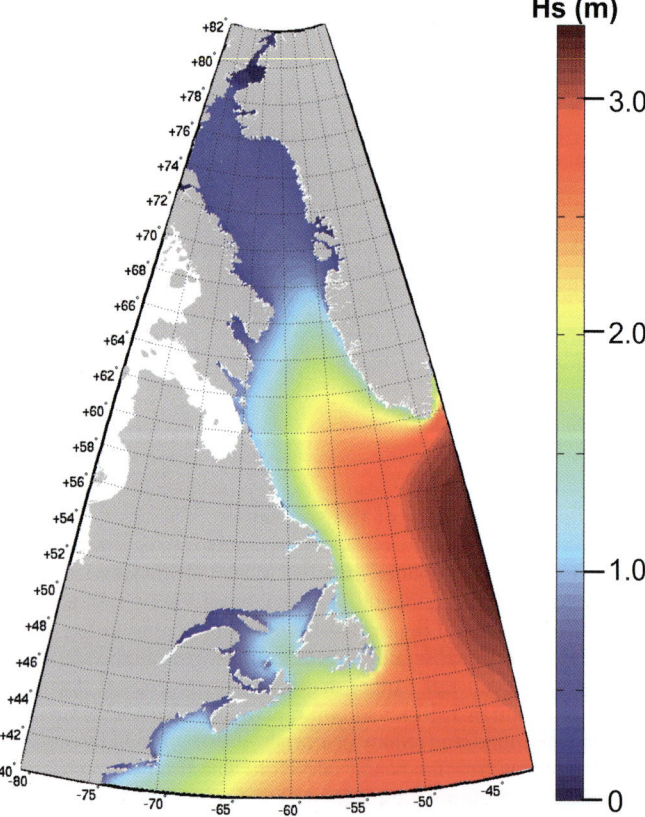

Fig. 2.5. Mean significant wave height averaged for the period of 2002–2005, showing a progressive decrease in wave effect from the open ocean, through continental shelves and to coastal zones.

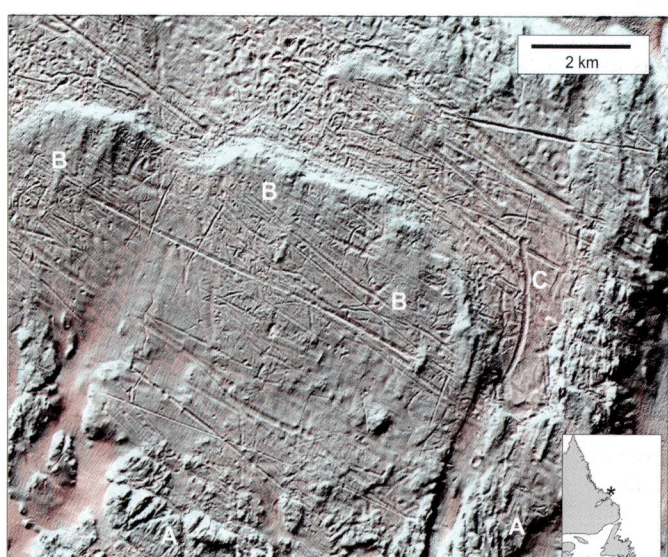

Fig. 2.6. Seafloor image off the coast of Labrador, with backscatter data superimposed on a grey shaded relief image. Water depths range from 100 (bottom left) to 300 m (upper right). Low backscatter is pink and high backscatter is greenish blue. The latter category includes bedrock outcrops (**a**) and a large area of till (**b**) formed at the former grounding line of an ice sheet. The scour at C has length of 2.2 km, a width of 120 m, depth of 4.5 m from berm crest to base, and terminates in a pit.

the relatively deep NE Newfoundland and Labrador shelves. The Bay of Fundy is dominated by tidal forces and demonstrates the highest sediment mobility.

A distinguishing feature of these continental shelves is the extent to which the seafloor has been scoured by icebergs, creating furrows and pits (Woodward-Lynas *et al.* 1991; Sonnichsen & King 2005). Two principal populations of iceberg scours occur. A relict population formed during deglaciation is observed throughout the region where it has not been buried by post-glacial sedimentation or effaced by waves. Relict iceberg impact features are widespread in the Bay of Fundy, for example (Parrott *et al.* 2009; Todd *et al.* 2011). The modern population of iceberg scours is restricted to the Labrador and Newfoundland shelves where it occurs down to depths of about 220 m (Wadhams 2000). It results from icebergs drifting south in the Labrador Current. Mapping of the inner shelf off NE Newfoundland (Shaw *et al.* 1999), for example, showed the great extent to which the seafloor bears the imprint of icebergs, although the traces of grounding are often rapidly effaced in the shallow, wave-dominated zone (0–60 m depth).

Over the last two decades multibeam sonar surveys have revealed the distribution and morphology of iceberg scours in great detail. An example of recent mapping is shown in Figure 2.6, part of the inner shelf off Labrador. With water depths of 100–300 m, this area contains both modern and relict populations of scours, both furrows and pits. Older furrows are degraded, while the most recent have a 'fresh' appearance, with well-developed berms.

Modern iceberg activity poses a hazard for hydrocarbon development on the Labrador and Newfoundland shelves. In order to determine the frequency of scour events, repetitive mapping is necessary. In the past, sidescan sonar was the primary tool but it has been replaced by multibeam sonar. Sonnichsen & King (2011) mapped a 610 km^2 area of the Grand Banks of Newfoundland with multibeam sonar, and compared the data with those from previous sidescan sonar surveys. They detected only one possible new scour event and nine possible older events; uncertainty in scour identification was ascribed to limitations in the sidescan sonar data. Today, the standard approach to determine frequency is to compare multibeam surveys only. An alternative approach is demonstrated by the report on a potential tunnel route across the Strait of Belle Isle, the water body separating northern Newfoundland from Labrador (C-CORE 2004). Radar was used to assist in estimating the frequency of scour events, and estimates of scour depths were used to calculate safe minimum depths for tunnel burial.

Sediment transport on the Atlantic shelves

From the sediment-transport perspective, the environmental conditions and seabed geology of the Atlantic shelves have been extensively documented and reviewed (e.g. Amos 1990). Synthesis and reviews of advances of sediment-transport studies on the Atlantic shelves can be found in Hodgins *et al.* (1986), Amos & Judge (1991) and Li & Heffler (2002*a*, *b*). The hydrodynamic processes related to sediment transport vary with depth, distance from the shoreline and latitude (Hodgins *et al.* 1986; Smith & Schwing 1991), and are largely storm controlled. Over the inner shelf, the major processes are currents generated by storm surges, surface gravity waves, and wind-driven and inertial currents generated by the passage of storms. On the outer shelf, impinging oceanic currents, tidal current amplified over shelf-edge banks, breaking internal waves and solitons, and storm-driven flows dominate.

Sediment transport is evident on most parts of the eastern Canadian continental shelf. It is manifested by the presence of sorted sand and gravel on the relatively shallow banks, indicative of

winnowing of fines and bedload transport of the coarser fractions, and the presence of fine-grained, ponded sediments in the basins, indicative of pelagic deposition from suspension. Bedform distribution, sediment-transport processes and pathways have all been extensively studied (e.g. Amos & King 1984; Barrie *et al.* 1984; Hoogendoorn & Dalrymple 1986; Amos & Nadeau 1988; Amos & Judge 1991; Li *et al.* 1997, 2012; Li & Amos 1999*a*, *b*; Todd *et al.* 1999; Todd 2005; Li & King 2007).

Labrador and NE Newfoundland shelves

Sediment transport on these shelves is strongly limited by relatively deep water and ice cover in the winter months. Nevertheless, the existence of ripples, subaquatic dunes and well-sorted gravel indicate intermittent transport of sediment on shelf banks in depths less than 120 m, and the net transport is believed to be to the SE under the influence of strong wind-driven and geostrophic flows. Numerical modelling predicted little transport on the NE Newfoundland Shelf even under the condition of a once in 32 years' storm (Amos & Judge 1991).

Grand Banks of Newfoundland

Tidal and geostrophic currents on Grand Banks are weak, and storms are the main cause of sediment transport in depths up to 150 m. A variety of bedforms is found on the Grand Banks, indicating that most of its seabed is active. On the north, west and SW Grand Banks, sand is winnowed and transported to the south leaving behind well-sorted gravel lag deposits. On the eastern and southern parts of the Grand Banks, the seabed is covered by medium sand with megaripples, sand waves and sand ridges. The sand ridges may be relict features but megaripples and other small superimposed bedforms are fresh and active down to a depth of 110 m, and have morphologies indicative of migration to the south. Transport of sediment was predicted over most of Grand Banks for the once in 32 years' storm. The maximum transport rate of 0.1 kg m^{-1} s^{-1} occurred in the shallow water in the central and SE areas.

Scotian Shelf: Sable Island Bank

Sable Island Bank, 255 km long and 115 km wide, is the largest bank on the Scotian Shelf. The only part of the bank above sea level is Sable Island, an arcuate sand body 40 km long by 1.6 km wide. A Holocene sand-ridge complex, the Sable Island Catena is up to 50 m thick, extends 300 km across Sable Island Bank and the adjacent Banquereau Bank (Amos & Nadeau 1988), and formed as a result of Holocene sediment transport.

The large storm-generated shoreface and offshore sand ridges (Hoogendoorn & Dalrymple 1986; Li & King 2007) are the most striking morphological features of Sable Island Bank (Fig. 2.7). They were formed by the reworking of the glacial-outwash sediments deposited during sea-level lowstands. The sand ridges generally trend NE–SW on the southern bank and NW–SE on the northern bank (Li & King 2007; Li *et al.* 2012). Spacing ranges from 0.6 to 9 km and height ranges from 0.5 to 20 m, although their size tends to decrease from west to east and from offshore to inshore. Sand ridges are active on a yearly basis in the bank-top setting. Repeat multibeam surveys over 3–4-year spans show that the mode of migration is around zero but the median rates of easterly and westerly migrations are 3–7 m a^{-1}. A suite of smaller bedforms is superimposed on the sand ridges (Fig. 2.7b): sand waves are widely superimposed on the flanks of sand ridges; 2D and 3D megaripples commonly occur in the sand-ridge troughs, on the lower western sand-ridge flanks and in the sand-wave troughs; and large-wave ripples are ubiquitously found in the troughs of all megaripples. Sidescan and multibeam

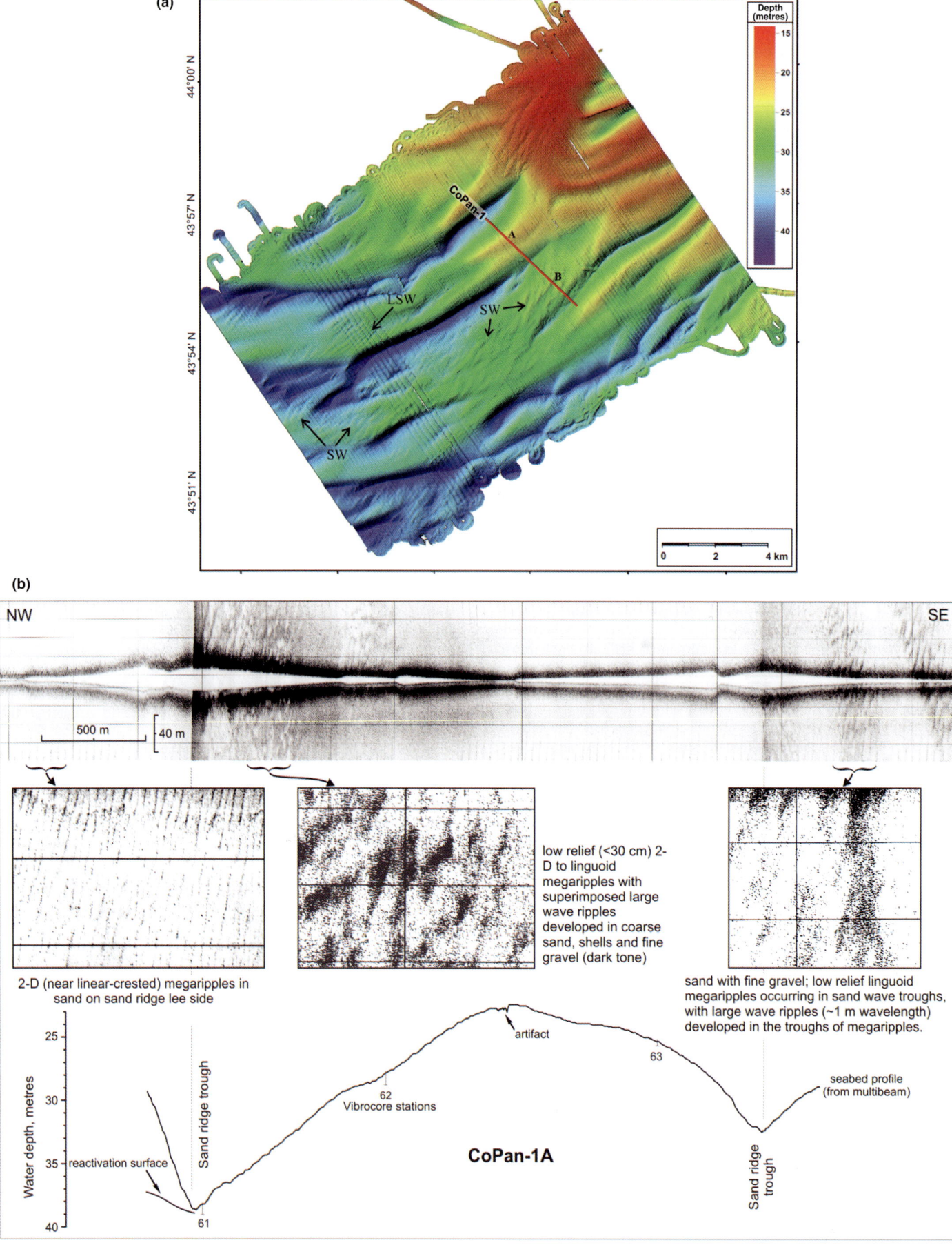

Fig. 2.7. Shaded-relief colour multibeam bathymetry (**a**) showing NE–SW-orientated sand ridges on Sable Island Bank. Linear (LSW) and regular sand waves (SW) are superimposed on the ridges. The red line marks the location of the transect sidescan image presented in (**b**), which demonstrates the superposition of sand waves, megaripples and large-wave ripples on sand ridges.

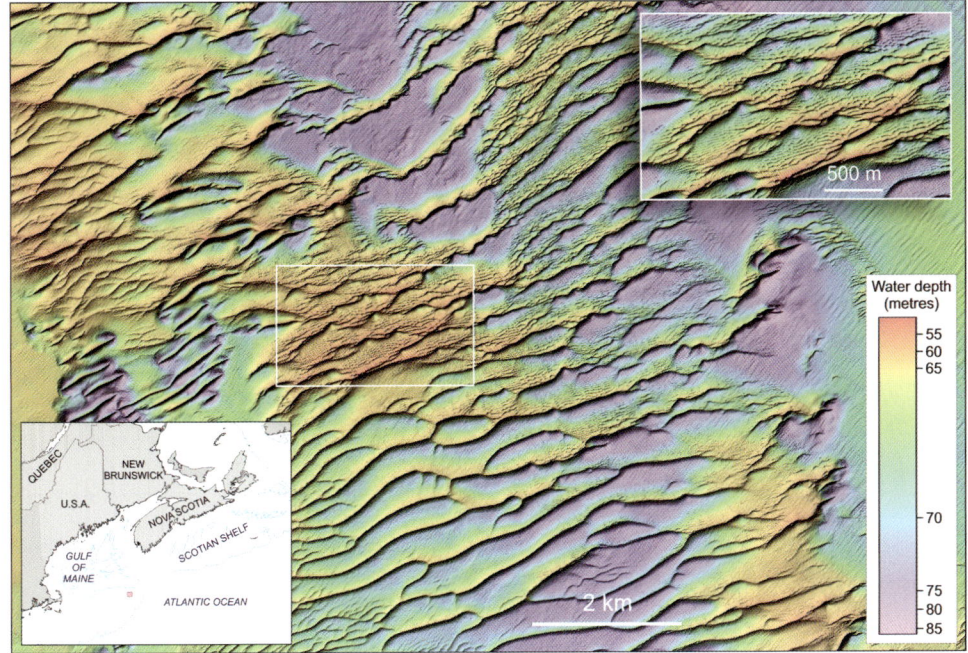

Fig. 2.8. Sand-wave field on Georges Bank, Gulf of Maine. Curvilinear, intersecting crest lines are kilometres in length. Wavelengths vary but a typical value is 300–400 m. Inset shows megaripples, with wavelengths of tens of metres, on the stoss sides of the sand waves. The area between sand waves is thin sand over gravel lag or exposed gravel lag, and is a scallop habitat (Todd & Valentine 2012). Seafloor mapping and delineation of the scallop habitat has been adopted by the commercial scallop fishery and is used by government to manage stocks (Smith *et al.* 2006).

surveys, sediment samples, and numerical model predictions of seabed disturbance were integrated to define seven bedform zones with unique bedform association, energy level and mobility characteristics (Li *et al.* 2012). Numerical model predictions demonstrate that the shear stress due to the combined effect of waves and currents can cause sediment mobility at least once a year over 93% of the bank area.

Field hydrodynamics and sediment-transport measurements obtained with deployments of instrumented seabed landers (e.g. Li *et al.* 1997; Li & Amos 1999*a, b*) demonstrate that tidal currents, waves, wind-driven currents during storms and the non-linear interaction of these processes can cause bedload transport, sediment suspension and upper-regime sheet flow. These sediment-transport processes lead to the development of current ripples, wave ripples and megaripples on the shorefaces of south and SE banks, over sand ridges and sand waves on the mid-bank, and in relatively deep water (60–70 m) on the mid to outer bank. Computed sediment-transport rates using measured wave and current parameters reach a maximum of 0.36 kg m^{-1} s^{-1}.

Amos & Nadeau 1988 and Amos & Judge 1991 show that the long-term net sediment transport is to the east and NE on Sable Island Bank. A circular and clockwise pattern of sand transport is evident on eastern Sable Island Bank. Sediment-transport direction was to the east and SE on the western and southern Sable Island Bank, and to the NE on the northern Sable Island Bank. Transport at the shelf break is restricted to fine-grained sediments that move along isobaths to the SW.

Western Scotian Shelf: Georges and Browns Banks

Most of the banks on western Scotian Shelf are dominated by a gravelly seabed but sand is present on the bank tops. Various bedforms ranging from megaripples, to sand waves and tidal sand ridges have been found. Georges Bank is approximately 280 km long and 150 km wide. A veneer of glacial sediment was emplaced on the bank at the LGM (Schnitker *et al.* 2001; Dyke *et al.* 2002). During the Holocene, sea level rose from a lowstand of 120 m below present sea level (Emery & Garrison 1967; King & Fader 1986). Georges Bank was submerged at approximately 6 ka (Shaw *et al.* 2002). Surficial sediments on the bank were reworked and redistributed by marine processes

during sea-level transgression and continue to be reworked under the modern oceanic regime (Butman 1987; Twichell *et al.* 1987; Uchupi & Austin 1987; Valentine *et al.* 1993).

At water depths of approximately 60 m or less, post-glacial/modern sediment reworking results in a hierarchy of current-generated bed forms in sand-rich areas. Sand-wave crests trend approximately SW–NE, normal to the major axis of the semi-diurnal tidal current (Butman & Beardsley 1987). Sand-wave crests display a complex anastomosing pattern in plan view (Fig. 2.8). The sand waves have wavelengths of 50–300 m and reach heights of 19 m. In cross-section they are asymmetrical with gently–sloping upcurrent, or stoss, faces, and steeply–dipping downcurrent, or lee, faces. This cross-sectional attribute, together with their crest orientation and overall distribution on the bank, makes them useful for inferring the regional direction of sediment transport. Superimposed on the sand waves and sharing the same general crest orientation are megaripples and ripples; these features have smaller wavelengths and heights, and exhibit a complex 3D pattern in plan view.

Sediment traps deployed in 60 m of water on Georges Banks in a 1989 field study (Amos & Beaver 1989) detected sand and gravel in motion under tidal currents. The movement was to the south at rates of approximately 0.5 kg m^{-1} s^{-1}. An instrumented lander deployed in 2007 over sand waves at a depth of 174 m along the southern flank of the NE Channel recorded peak near-bed mean currents of up to 40 cm s^{-1}. Bedload transport, sediment suspension and active asymmetrical current ripples were observed. Sector scanning sonar also detected the mobility of a sand-wave trough and megaripples superimposed on the sand wave. Numerical model predictions of transport of medium sand for a once in 4 years' storm under the peak ebb tidal condition show that moderate–strong sediment transport widely occurs on Georges Bank, on Browns Bank and in the NE Channel. The maximum sediment-transport rate reaches 1 kg m^{-1} s^{-1}, and sediment-transport direction is predominantly to the east and SE.

The bedform suite on Georges Bank differs from that on Browns Bank, where striking barchan dunes are found (Todd *et al.* 1999; Todd 2005). Barchan dunes on the SW bank are convex to the SE with steep lee faces to the NW, indicating a dominant NW-flowing current. Those on the northern bank are convex to the NW with steep lee faces to the SE. The orientation of the

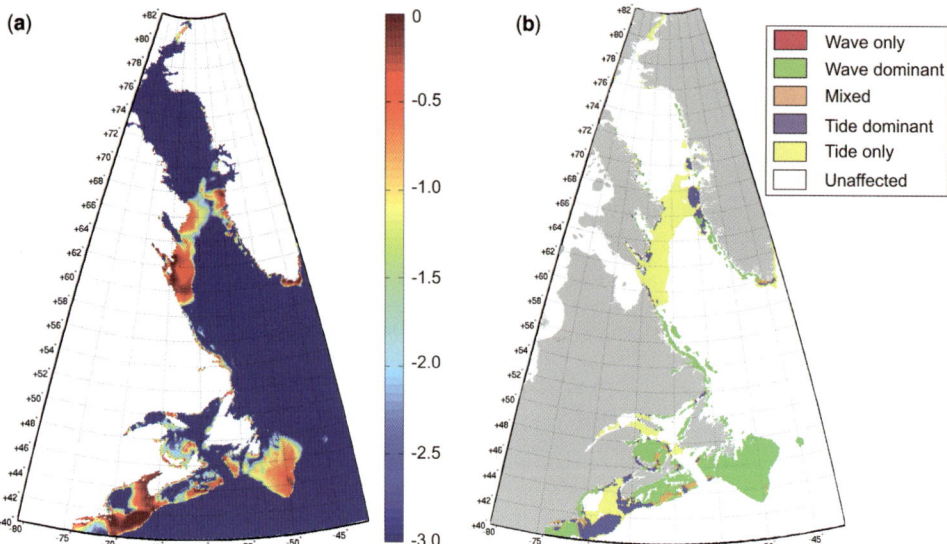

Fig. 2.9. Maps of (**a**) log frequency of medium sand mobilization and (**b**) disturbance type classification for the Atlantic shelf of Canada.

barchan dunes and obstacle marks, and the profile asymmetry of sand waves and barchan dunes, are in agreement with the anti-cyclonic pattern of mean currents on Browns Bank. This establishes the clockwise pattern of net sediment transport on the western Browns Bank.

Shelf-wide seabed disturbance and sediment mobility modelling

In a recent effort, model predictions of waves, tidal currents, and wind-driven and circulation currents over the three-year period of 2002–5 were coupled with a shelf sediment-transport model to quantify seabed shear stresses, sediment mobility, and sediment-transport pattern on the Atlantic shelf. The seabed shear stress due to the combined waves and currents is compared with the critical shear stress for bedload transport of medium sand to derive the frequency of sediment mobilization (Fig. 2.9a). The relative time percentage of sediment mobilization by various processes was used to define seabed disturbance types (Fig. 2.9b). The highest sediment mobilization frequency, reaching nearly 100% of the time, occurs in the Bay of Fundy, on Georges Bank, and on the northern Labrador Shelf near the entrance to Hudson Strait. Seabed forcing is dominated by tidal current in these areas. Sediment is also frequently mobilized on the Scotian Shelf and the Grand Banks, with the former dominated by wave-dominant and mixed disturbances, and the latter dominated by wave-dominant disturbance. Little sediment mobilization is predicted on the Labrador Shelf and on the NE Newfoundland Shelf except on the outer-shelf banks and along the coast lines. When sediment is mobilized, the seabed forcing is dominantly due to wave effects.

Benthic habitats

When classified in a natural disturbance-scope for growth framework (see Kostylev *et al.* 2005, Kostylev & Hannah 2007) the Atlantic shelves are subdivided into three latitudinal regions

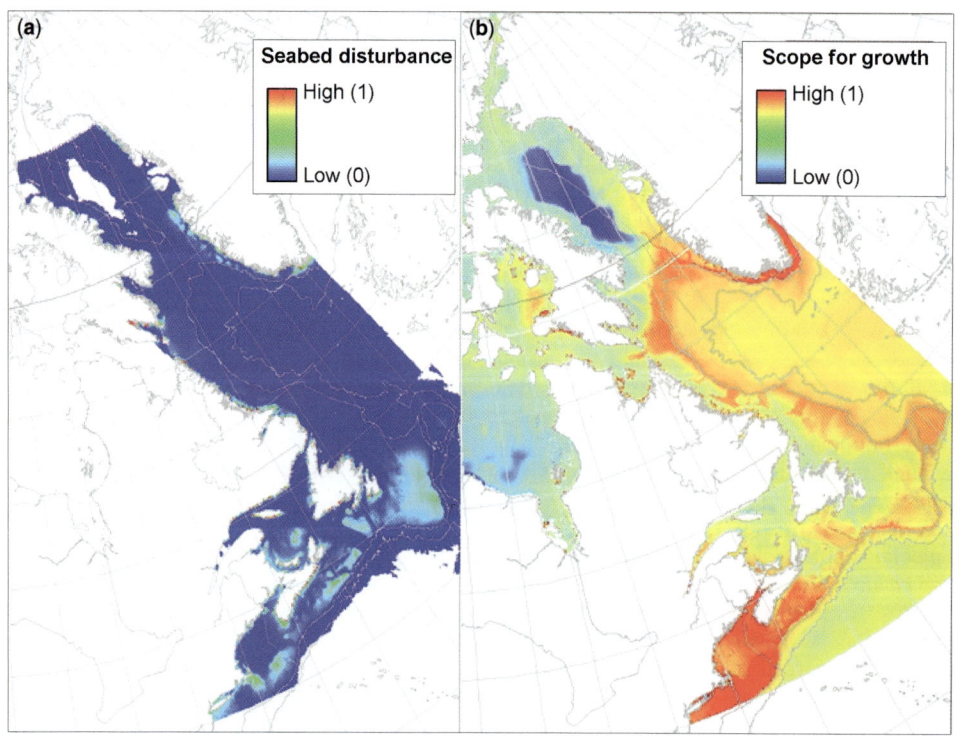

Fig. 2.10. (**a**) Seabed disturbance and (**b**) scope for growth on the continental shelves and adjacent areas of Atlantic Canada. Note the absence of disturbance on the shelves off NE Newfoundland and Labrador.

(Fig. 2.10). The southern region starting from Georges Bank and including the Gulf of Maine and the western Scotian Shelf (up to 45° latitude) has high rates of natural seabed disturbance (due to strong tidal currents and waves), high vertical mixing and, consequentially, high seabed productivity. The second region begins just north of Sable Island Bank on the Scotian Shelf and extends to the north of Grand Bank. Seabed disturbance in this region is still high mostly due to wave action but the productivity is moderate on southern Grand Bank and low in nearshore Newfoundland. From western Newfoundland, starting at latitude 48°, all the way to Hudson Strait, the shelves are relatively undisturbed by currents, and not very productive. Scope for growth is markedly higher at the shelf break and on the slope. In this region, Flemish Cap is a conspicuously stable and productive deep-water habitat, which is partially corroborated by the success of fisheries there. These findings suggest that habitats north of latitude 48° are naturally stable and can be damaged easily by anthropogenic activities. Similarly, because of the low scope for growth in this area, groundfish populations can be susceptible to overfishing. South of latitude 45°, the seabed populations have a greater likelihood to recover from fishing pressure because of the higher scope for growth, and seabed habitats are less likely to be damaged because they are adapted to naturally high levels of disturbance.

Discussion

In this paper we have deliberately focused on only a few aspects of the very large and extremely diverse continental shelves of Atlantic Canada. We have said very little about the Gulf of St Lawrence, which contains the record of rapid glacial ice retreat around 14 ka ago (Josenhans & Lehman 1999). Nor have we done justice to the fjords of Newfoundland and Labrador, and parts of southern Quebec, which rival those anywhere on the planet in terms of size and depth. The shelves have a good hydrocarbon potential, with actively producing fields on Sable Island Bank and Grand Bank. This potential has been the trigger for research into potential constraints on engineering design (Sonnichsen & King 2005), including assessment of geohazards (primarily submarine landslides) on the upper continental slope (Piper & Campbell 2005). This research topic has been increasingly important as hydrocarbon exploration migrates into deeper waters. For example, a well has recently been drilled at a water depth of 2000 m in Orphan Basin, located between Flemish Cap and the Grand Banks. However, conflict exists between hydrocarbon exploration and the rich fishing potential of the shelves. Georges Bank, as noted above, hosts rich fishing stocks and remains under an exploration moratorium until 2015.

We thank R. Parrott and G. Sonnichsen who internally reviewed this manuscript, and two external reviewers who provided comments that greatly improved the paper. This is Natural Resources Canada Earth Sciences Sector contribution number 13974.

References

ADAMS, J. & HALCHUK, S. 2004. Fourth–generation seismic hazard maps for the 2005 National Building Code of Canada. Paper 2502, presented at the 13th World Conference on Earthquake Engineering, Vancouver, Canada, August 1–6, 2004.

AMOS, C. L. 1990. Modern sedimentary processes. *In*: KEEN, M. J. & WILLIAMS, G. L. (eds) *Geology of the Continental Margin of Eastern Canada*. Geological Survey of Canada, Geology of Canada, **2**, 609–673 (also Geological Society of America, The Geology of North America, **I-1**, Chapter 11).

AMOS, C. L. & BEAVER, D. 1989. *Dawson Cruise–Georges Banks*. Bedford Institute of Oceanography Internal Cruise Report.

AMOS, C. L. & JUDGE, J. T. 1991. Sediment transport on the eastern Canadian continental shelf: Proceedings of the Canadian Continental Shelf Seabed Symposium. *Continental Shelf Research*, **11**, 1037–1068.

AMOS, C. L. & KING, E. L. 1984. Bedforms of the Canadian eastern seaboard: a comparison with global occurrences. *Marine Geology*, **57**, 167–208.

AMOS, C. L. & NADEAU, O. C. 1988. Surficial sediments of the outer banks, Scotian Shelf, Canada. *Canadian Journal of Earth Sciences*, **25**, 1923–1944.

BARRIE, J. V., LEWIS, C. F. M., FADER, G. B. J. & KING, L. H. 1984. Seabed processes on the northeastern Grand Banks of Newfoundland; Modern reworking of relict sediments. *Marine Geology*, **57**, 209–227.

BRUSHETT, D., BELL, T., BATTERSON, M. J. & SHAW, J. 2007. Ice-flow history of Placentia Bay, Newfoundland: multibeam seabed mapping. *Current Research*, Report **07-1**, 215–228,

BUTMAN, B. 1987. Physical processes causing surficial-sediment movement. *In*: BACKUS, R. H. (ed.) *Georges Bank*. Massachusetts Institute of Technology Press, Cambridge, MA, 147–162.

BUTMAN, B. & BEARDSLEY, R. C. 1987. Physical oceanography. *In*: BACKUS, R. H. (ed.) *Georges Bank*. Massachusetts Institute of Technology Press, Cambridge, MA, 88–98.

CAMPBELL, D. C., PIPER, D. J. W., MOSHER, D. C. & JENNER, K. A. 2008*a*. *Sun-illuminated Seafloor Topography, Verrill Canyon, Scotian Slope, Offshore Nova Scotia*. Geological Survey of Canada, Map Series, **2125A**.

CAMPBELL, D. C., PIPER, D. J. W., MOSHER, D. C. & JENNER, K. A. 2008*b*. *Surficial Geology and Sun-illuminated Seafloor Topography, Verrill Canyon, Scotian Slope, Offshore Nova Scotia*. Geological Survey of Canada, Map Series, **2128A**.

C-CORE 2004. *Iceberg Scour Risk in the Strait of Belle Isle*. C-CORE Report **R-04-004-011**. C-CORE, St John's, Newfoundland.

CLARK, P. U. & FITZHUGH, W. W. 1991. Postglacial relative sea level history of the Labrador coast and interpretation of the archaeological record. *In*: JOHNSON, L. L. (ed.) *Palaeoshorelines and Prehistory: An Investigation of Method*. CRC Press, London, 189–213.

DEPTUCK, M. E., MOSHER, D. C., CAMPBELL, D. C., HUGHES-CLARKE, J. E. & NOSEWORTHY, D. 2007. Along slope variations in mass failures and relationships to major Plio-Pleistocene morphological elements, SW Labrador Sea. *In*: LYKOUSIS, V., DIMITRIS, S. & LOCAT, J. (eds) *Submarine Mass Movements and their Consequences, III*. Springer, Dordrecht, 37–46.

DYKE, A. S. & PREST, V. K. 1987. Late Wisconsinan and Holocene history of the Laurentide Ice Sheet. *Géographie Physique et Quaternaire*, **41**, 237–263.

DYKE, A. S., ANDREWS, J. T., CLARK, P. U., ENGLAND, J. H., MILLER, G. H., SHAW, J. & VEILLETTE, J. J. 2002. The Laurentide and Innuitian ice sheets during the Last Glacial Maximum. *Quaternary Science Reviews*, **21**, 9–31.

EMERY, K. O. & GARRISON, L. E. 1967. Sea levels 7,000 to 20,000 years ago. *Science*, **157**, 684–687.

EYLES, N. 1983. Glacial geology: a landsystems approach. *In*: EYLES, N. (ed.) *Glacial Geology*. Pergamon, Oxford, 1–18.

FADER, G. B., KING, L. H. & MACLEAN, B. 1977. *Surficial Geology of the Eastern Gulf of Maine and Bay of Fundy*. Canada Marine Science, Paper, **19** (Geological Survey of Canada, Paper, **76-17**).

FADER, G. B., KING, L. H. & JOSENHANS, H. W. 1982. *Surficial Geology of the Laurentian Channel and the Western Grand Banks of Newfoundland*. Canada Marine Science, Paper, **21** (Geological Survey of Canada, Paper, **81–22**).

FADER, G. B., CAMERON, G. D. M. & BEST, M. A. 1989. *Geology of the Continental Margin of Eastern Canada (scale 1:5,000,000)*. Geological Survey of Canada, Map Series, **1705A**.

FADER, G. B. J. 1989. A late Pleistocene low sea-level stand of the southeast Canadian offshore. *In*: SCOTT, D. B., PIRAZOLLI, P. A. & HONIG, C. A. (eds) *Late Quaternary Sea-Level Correlation and Applications*. Kluwer Academic, Dordrecht, 71–103.

FILLON, R. H. & HARMES, R. A. 1982. Northern Labrador Shelf glacial chronology and depositional environments. *Canadian Journal of Earth Sciences*, **19**, 162–192.

FORBES, D. L., BOYD, R. & SHAW, J. 1991. Late Quaternary sedimentation and sea level changes on the inner Scotian Shelf. *Continental Shelf Research*, **11**, 1155–1179.

HODGINS, D. O., DRAPEAU, G. & KING, L. H. 1986. *Field Measurements of Sediment Transport on the Scotian Shelf. Volume 1, the Radio-isotope Experiment.* Environmental Studies Revolving Funds Report **041**.

HOOGENDOORN, E. L. & DALRYMPLE, R. W. 1986. Morphology, lateral migration, and internal structures of shoreface-connected ridges, Sable Island Bank, Nova Scotia, Canada. *Geology*, **14**, 400–403.

JOSENHANS, H. W. & LEHMAN, S. 1999. Quaternary stratigraphy and glacial history of the Gulf of St. Lawrence, Canada. *Canadian Journal of Earth Sciences*, **36**, 1327–1345.

JOSENHANS, H. W., ZEVENHUIZEN, J. & KLASSEN, R. A. 1986. The Quaternary Geology of the Labrador Shelf. *Canadian Journal of Earth Sciences*, **23**, 1190–1213.

KEEN, M. J. & WILLIAMS, G. L. (eds) 1990. *Geology of the Continental Margin of Eastern Canada.* Geological Survey of Canada, Geology of Canada, **2** (also Geological Society of America, The Geology of North America, **I-1**).

KING, L. H. 1993. Till in the marine environment. *Journal of Quaternary Science*, **8**, 347–358.

KING, L. H. 1996. Late Wisconsinan ice retreat from the Scotian Shelf. *Geological Society of America Bulletin*, **108**, 1056–1067.

KING, L. H. & FADER, G. B. J. 1986. *Wisconsinan Glaciation of the Atlantic Continental Shelf of Southeast Canada.* Geological Survey of Canada, Bulletin, **363**.

KOSTYLEV, V. E. & HANNAH, C. G. 2007. Process-driven characterization and mapping of seabed habitats. *In*: TODD, B. J. & GREENE, H. G. (eds) *Mapping the Seafloor for Habitat Characterization.* Geological Association of Canada, Special Papers, **47**, 171–184.

KOSTYLEV, V. E., TODD, B. J., LONGVA, O. & VALENTINE, P. C. 2005. Characterization of benthic habitat on northeastern Georges Bank, Canada. *In*: BARNES, P. W. & THOMAS, J. P. (eds) *Benthic Habitats and the Effects of Fishing. American Fisheries Society Symposium 41.* American Fisheries Society, Bethesda, MD, 141–152.

LI, M. Z. & AMOS, C. L. 1999*a*. Sheet flow and large wave ripples under combined waves and current: Their field observation, model prediction and effects on boundary layer dynamics. *Continental Shelf Research*, **19**, 637–663.

LI, M. Z. & AMOS, C. L. 1999*b*. Field observations of bedforms and sediment transport thresholds of fine sand under combined waves and current. *Marine Geology*, **158**, 147–160.

LI, M. Z. & HEFFLER, D. E. 2002*a*. Environmental Marine Geoscience 3. Continental shelf sediment transport studies in Canada: Theories and recent technology advances. *Geoscience Canada*, **29**, 35–48.

LI, M. Z. & HEFFLER, D. E. 2002*b*. Environmental Marine Geoscience 4. Continental shelf sediment transport studies in Canada: Principal scientific advances and future directions. *Geoscience Canada*, **29**, 53–68.

LI, M. Z. & KING, E. L. 2007. Multibeam bathymetric investigations of the morphology of sand ridges and associated bedforms and their relation to storm processes, Sable Island Bank, Scotian Shelf. *Marine Geology*, **243**, 200–228.

LI, M. Z., AMOS, C. L. & HEFFLER, D. E. 1997. Boundary layer dynamics and sediment transport under storm and non-storm conditions on the Scotian Shelf. *Marine Geology*, **141**, 157–181.

LI, M. Z., KING, E. L. & PRESCOTT, R. H. 2012. Seabed disturbance and bedform distribution and mobility on the storm-dominated Sable Island Bank, Scotian Shelf. *In*: LI, M. Z., SHERWOOD, C. & HILL, P. (eds) *Sediments, Morphology and Sedimentary Processes on Continental Shelves.* International Association of Sedimentologists, Special Publications, **44**, 197–227.

MOSHER, D. C. 2009. Submarine landslides and consequent tsunamis in Canada. *Geoscience Canada*, **36**, 179–190.

MOSHER, D. C. & PIPER, D. J. W. 2007*a*. *Multibeam Seafloor Imagery of the Laurentian Fan and the 1929 Grand Banks Landslide Area.* Geological Survey of Canada, Open File, **5638**.

MOSHER, D. C. & PIPER, D. J. W. 2007*b*. Analysis of multibeam seafloor imagery of the Laurentian Fan and the 1929 Grand Banks landslide area. *In*: LYKOUSIS, V., SAKELLARIOU, D. & LOCAT, J. (eds) *Submarine Mass Movements and Their Consequences.* Advances in Natural and Technological Hazards Research, **27**. Springer, Dordrecht, 77–88.

MOSHER, D. C., PIPER, D. J. W. P., CAMPBELL, D. C. & JENNER, K. A. 2004. Near surface geology and sediment failure geohazards of the central Scotian Slope. *American Association of Petroleum Geologists Bulletin*, **88**, 703–723.

PARROTT, D. R., TODD, B. J. *ET AL.* 2009. *Shaded Seafloor Relief, Bay of Fundy, offshore Nova Scotia/New Brunswick.* Geological Survey of Canada, Open File, **5834**.

PIPER, D. J. W. & CAMPBELL, D. C. 2005. Quaternary geology of Flemish Pass and its application to geohazard evaluation for hydrocarbon development. *In*: HISCOTT, R. N. & PULHAM, A. J. (eds) *Petroleum Resources and Reservoirs of the Grand Banks, Eastern Canadian Margin.* Geological Association of Canada, Special Papers, **43**, 29–43.

PIPER, D. J. W. & NORMARK, W. R. 1989. Late Cenozoic sea-level changes and the onset of glaciation: impact on continental slope progradation off eastern Canada. *Marine and Petroleum Geology*, **6**, 336–347.

PIPER, D. J. W., MUDIE, P. J., AKSU, A. E. & SKENE, K. 1. 1994. A 1 Ma record of sediment flux south of the Grand Banks used to infer the development of glaciation in southeastern Canada. *Quaternary Science Reviews*, **13**, 23–37.

SCHNITKER, D., BELKNAP, D. F., BACCHUS, T. S., FRIEZ, J. K., LUSARDI, B. A. & POPEK, D. M. 2001. Deglaciation of the Gulf of Maine. *In*: WEDDLE, T. K. & RETELLE, M. J. (eds) *Deglacial History and Relative Sea-level Changes, Northern New England and Adjacent Canada.* Geological Society of America Special Papers, **351**, 9–34.

SHAW, J. 2003. Submarine moraines in Newfoundland coastal waters: implications for the deglaciation of Newfoundland and adjacent areas. *Quaternary International*, **99–100**, 115–134.

SHAW, J. & FORBES, D. L. 1990. Relative sea-level change and coastal response, northeast Newfoundland. *Journal of Coastal Research*, **6**, 641–660.

SHAW, J., FORBES, D. L. & EDWARDSON, K. A. 1999. *Surficial Sediments and Placer Gold on the Inner Shelf and Coast of Northeast Newfoundland.* Geological Survey of Canada, Bulletin, **532**.

SHAW, J., GRANT, D. R., GUILBAULT, J.-P., ANDERSON, T. W. & PARROTT, D. R. 2000. Submarine and onshore end moraines in southern Newfoundland: implications for the history of late Wisconsinan ice retreat. *Boreas*, **29**, 295–314.

SHAW, J., GAREAU, P. & COURTNEY, R. C. 2002. Palaeogeography of Atlantic Canada 13–0 kyr. *Quaternary Science Reviews*, **21**, 1861–1878.

SHAW, J., PIPER, D. J. W. *ET AL.* 2006. A conceptual model of the deglaciation of Atlantic Canada. *Quaternary Science Reviews*, **25**, 2059–2081.

SHAW, J., TODD, B. J., BRUSHETT, D., PARROTT, D. R. & BELL, T. 2008. Late Wisconsinan glacial landsystems on Atlantic Canadian shelves: new evidence from multibeam and single-beam sonar data. *Boreas*, **38**, 146–159.

SHAW, J., FADER, B. J. & TAYLOR, R. B. 2009. Submerged early Holocene coastal and terrestrial landforms on the inner shelves of Atlantic Canada. *Quaternary International*, **206**, 24–34.

SHRIVASTAVA, S. P., SIBUET, J.-C., CANDE, S., ROEST, W. R. & REID, I. D. 2000. Magnetic evidence for slow seafloor spreading during the formation of the Newfoundland and Iberian margins. *Earth and Planetary Science Letters*, **182**, 1–76.

SMITH, P. C. & SCHWING, F. B. 1991. Mean circulation and variability on the eastern Canadian continental shelf. *Continental Shelf Research*, **11**, 977–1012.

SMITH, S. J., COSTELLO, G., KOSTYLEV, V. E., LUNDY, M. J. & TODD, B. J. 2006. Application of multibeam bathymetry and surficial geology to the spatial management of scallops (*Placopecten magellanicus*) in southwest Nova Scotia. *Journal of Shellfish Research*, **25**, 308.

SONNICHSEN, G. V. & KING, E. L. 2005. Grand Bank seabed and shallow subsurface geology in relation to subsea engineering and design. *In*: HISCOTT, R. N. & PULHAM, A. J. (eds) *Petroleum Resources and Reservoirs of the Grand Banks, Eastern Canadian Margin.* Geological Association of Canada, Papers, **43**, 11–27.

SONNICHSEN, G. V. & KING, T. 2011. 2004 Grand Banks iceberg scour survey. *In*: *Proceedings of the 21st International Conference on Port and Ocean Engineering under Arctic Conditions, Montreal, July 10–14, 2011.* Curran Associates, Red Hook, NY, 1–10.

STOKES, C. R. & CLARK, C. D. 1999. Geomorphological criteria for identifying Pleistocene ice streams. *Annals of Glaciology*, **28**, 67–75.

STOKES, C. R. & CLARK, C. D. 2001. Paleo-ice streams. *Quaternary Science Reviews*, **20**, 1437–1457.

SYVITSKI, J. P. M. 1991. Towards an understanding of sediment deposition on glaciated continental shelves. *Continental Shelf Research*, **11**, 897–937.

TODD, B. J. 2005. Morphology and composition of submarine barchan dunes on the Scotia Shelf, Canadian Atlantic margin. *Geomorphology*, **67**, 487–500.

TODD, B. J. & VALENTINE, P. C. 2012. Large submarine sand waves and gravel lag substrates on Georges Bank off Atlantic Canada. *In*: HARRIS, P. T. & BAKER, E. K. (eds) *Atlas of Seafloor Geomorphology as Habitat*. Elsevier, Amsterdam, 261–275.

TODD, B. J., LEWIS, C. F. M. & RYALL, P. J. C. 1988. Comparison of trends of iceberg scour marks with iceberg trajectories and evidence of paleocurrent trends on Saglek Bank, northern Labrador Shelf. *Canadian Journal of Earth Sciences*, **25**, 1374–1383.

TODD, B. J., FADER, G. B. J., COURTNEY, R. C. & PICKRILL, R. A. 1999. Quaternary geology and surficial sediment processes, Browns Bank, Scotian Shelf, based on multibeam bathymetry. *Marine Geology*, **162**, 165–214.

TODD, B. J., VALENTINE, P. C., LONGVA, O. & SHAW, J. 2007. Glacial landforms on German Bank, Scotian Shelf: evidence for Late Wisconsinan ice-sheet dynamics and implications for the formation of De Geer moraines. *Boreas*, **36**, 148–169.

TODD, B. J., SHAW, J. & PARROTT, D. R. 2011. *Shaded Seafloor Relief, Bay of Fundy, Sheet 1, Offshore Nova Scotia–New Brunswick (Scale 1:50,000)*. Geological Survey of Canada, Map Series, **2174A**.

TOWNSEND, D., THOMAS, A. C., MAYER, L. E., THOMAS, M. A. & QUINLAN, J. A. 2006. Chapter 5: Oceanography of the northwest Atlantic continental shelf (1,W). *In*: ROBINSON, A. R. & BRINK, K. H. (eds) *The Sea: The Global Coastal Ocean: Interdisciplinary Regional Studies and Syntheses*. Harvard University Press, Cambridge, MA.

TWICHELL, D. C., BUTMAN, B. & LEWIS, R. S. 1987. Shallow structure, surficial geology, and the processes currently shaping the bank. *In*: BACKUS, R. H. (ed.) *Georges Bank*. Massachusetts Institute of Technology Press, Cambridge, MA, 32–37.

UCHUPI, E. & AUSTIN, J. A., JR. 1987. Morphology. *In*: BACKUS, R. H. (ed.) *Georges Bank*. Massachusetts Institute of Technology Press, Cambridge, MA, 25–30.

VALENTINE, P. C., STROM, E. W., LOUGH, R. G. & BROWN, C. L. 1993. *Maps Showing the Sedimentary Environment of Eastern Georges Bank (Scale 1:250,000)*. United States Geological Survey, Miscellaneous Investigations Series Map, **I–2279–B**.

WADHAMS, P. 2000. *Ice in the Ocean*. Gordon & Breach, Amsterdam.

WOODWARD-LYNAS, C. M. T., JOSENHANS, H. W., BARRIE, J. V., LEWIS, C. F. M. & PARROTT, D. R. 1991. The physical process of seabed disturbance during iceberg grounding and scouring. *Continental Shelf Research*, **11**, 939–961.

Chapter 3

History of continental shelf and slope sedimentation on the US middle Atlantic margin

KENNETH G. MILLER[1]*, JAMES V. BROWNING[1], GREGORY S. MOUNTAIN[1], ROBERT E. SHERIDAN[1], PETER J. SUGARMAN[2], SCOTT GLENN[3] & BETH A. CHRISTENSEN[4]

[1]*Department of Earth and Planetary Sciences, Rutgers University, Piscataway, NJ 08854, USA*

[2]*New Jersey Geological Survey, PO Box 427, Trenton, NJ 08625, USA*

[3]*Institute of Marine and Coastal Sciences, Rutgers University, New Brunswick, NJ 08901, USA*

[4]*Environmental Studies Program, Adelphi University, Garden City, NY 11530, USA*

*Corresponding author (e-mail: kgm@rci.rutgers.edu)

Abstract: We describe sedimentation on the storm-dominated, microtidal, continental shelf and slope of the eastern US passive continental margin between the Hudson and Wilmington canyons. Sediments here recorded sea-level changes over the past 100 myr and provide a classic example of the interplay among eustasy, tectonism and sedimentation. Long-term margin evolution reflects changes in morphology from a Late Cretaceous–Eocene ramp to Oligocene and younger prograding clinothem geometries, a transition found on several other margins. Deltaic systems influenced Cretaceous and Miocene sedimentation, but, in general, the Maastrichtian–Palaeogene shelf was starved of sediment. Pre-Pleistocene sequences follow a repetitive model, with fining- and coarsening-upward successions associated with transgressions and regressions, respectively. Pleistocene–Holocene sequences are generally quite thin (<20 m per sequence) and discontinuous beneath the modern shelf, reflecting starved sedimentation under high rates of eustatic change and low rates of subsidence. However, Pleistocene sequences can attain great thickness (hundreds of metres) beneath the outermost shelf and continental slope. Holocene sedimentation on the inner shelf reflects transgression, decelerating from rates of approximately 3–4 to around 2 mm a^{-1} from 5 to 2 ka. Modern shelf sedimentation primarily reflects palimpsest sand sheets plastered and reworked into geostrophically controlled nearshore and shelf shore-oblique sand ridges, and does not provide a good analogue for pre-Pleistocene deposition.

Supplementary material: References used in the comparison of all dates for New Jersey localities in Figure 3.8 are available at http://www.geolsoc.org.uk/SUP18749.

The US middle Atlantic margin

The US middle Atlantic region (offshore New Jersey–Delaware: Figs 3.1 & 3.2) is a classic passive continental margin dominated by modern siliciclastic sedimentation, generally low sediment input during the last 10 myr and a long (180–200 myr) history of sedimentation that resulted in over 16 km of sediment buried beneath the modern shelf (Fig. 3.3) (e.g. Grow & Sheridan 1988). The margin was the subject of early studies because of its proximity to researchers using emerging technologies during the early twentieth century, and its physiography was used in defining the terms shelf, slope and rise (Heezen *et al.* 1959). Oil exploration generated great interest in the late 1970s and early 1980s, with attendant seismic profiling (e.g. Poag 1985), stratigraphic test wells (e.g. Scholle 1980), and 32 exploration wells primarily on the outer continental shelf and slope. Although oil and gas prospects were deemed uneconomical, focus on the stratigraphy of continental margins and sea-level change led to this region becoming a target of academic drilling by the Deep Sea Drilling Project (DSDP) Leg 95 (summary in Poag 1985), the Ocean Drilling Program (ODP) legs 150 (Mountain *et al.* 1994), 174A (Christie-Blick *et al.* 2003) and onshore ODP legs 150X and 174AX (summary in Browning *et al.* 2008), and the Integrated Ocean Drilling Program (IODP) Expedition 313, New Jersey Shallow Shelf (Mountain *et al.* 2010). These academic efforts have returned a wealth of information that provides unparalleled documentation of Cretaceous–Miocene sedimentation from the onshore coastal plain to the continental rise. However, Plio-Pleistocene sediments on this margin are very thin and discontinuous and, despite great efforts to core and date (e.g. Carey *et al.* 2005; Goff *et al.* 2005), are still poorly known. In this contribution, we: (1) review the physiographical and oceanographical settings and modern sedimentation on this margin; (2) discuss the geological structure and Cretaceous–Miocene sedimentation; (3) review Plio-Pleistocene sedimentation including the last glacial cycle (approximately the last 20 kyr), contrasting it with older periods; and (4) provide a brief overview of Holocene, instrumental and future sea-level changes on this margin.

Physiographical and oceanographical setting

The shelf, slope and rise physiographical provinces on the US middle Atlantic margin (offshore New Jersey and the Delmarva Peninsula: Fig. 3.1) were defined on the basis of the seafloor gradient: the shelf dips seawards with a gradient of less than 1:1000 (<0.06°), the continental slope with a gradient greater than 1:40 (>1.4°) and the continental rise with a gradient of about 1:100 (*c.* 0.6°) (Heezen *et al.* 1959; Emery 1968). The continental shelf is wide (>150 km) in this region and the water depth at the shelf–slope break averages about 135 m (Table 3.1) (Heezen *et al.* 1959). The shelf can be roughly divided into inner (0–40 m), middle (*c.* 40–100 m) and outer shelf (>100 m), marked by scarps (mid Shelf and Franklin: Fig. 3.2) that are both cut by the modern and palaeo-Hudson River valleys (Fig. 3.2) (Goff *et al.* 1999). The inner–middle shelf is flatter than the outer shelf (Fig. 3.2). The continental slope is incised by numerous submarine canyons, the largest of which indent the shelf–slope break. We focus here on the continental shelf and slope sedimentation between the Hudson and Wilmington canyons (Fig. 3.1).

Three water masses are found in the middle Atlantic shelf and slope (the following summary is largely derived from Beardsley

From: CHIOCCI, F. L. & CHIVAS, A. R. (eds) 2014. *Continental Shelves of the World: Their Evolution During the Last Glacio-Eustatic Cycle.*
Geological Society, London, Memoirs, **41**, 21–34. http://dx.doi.org/10.1144/M41.3

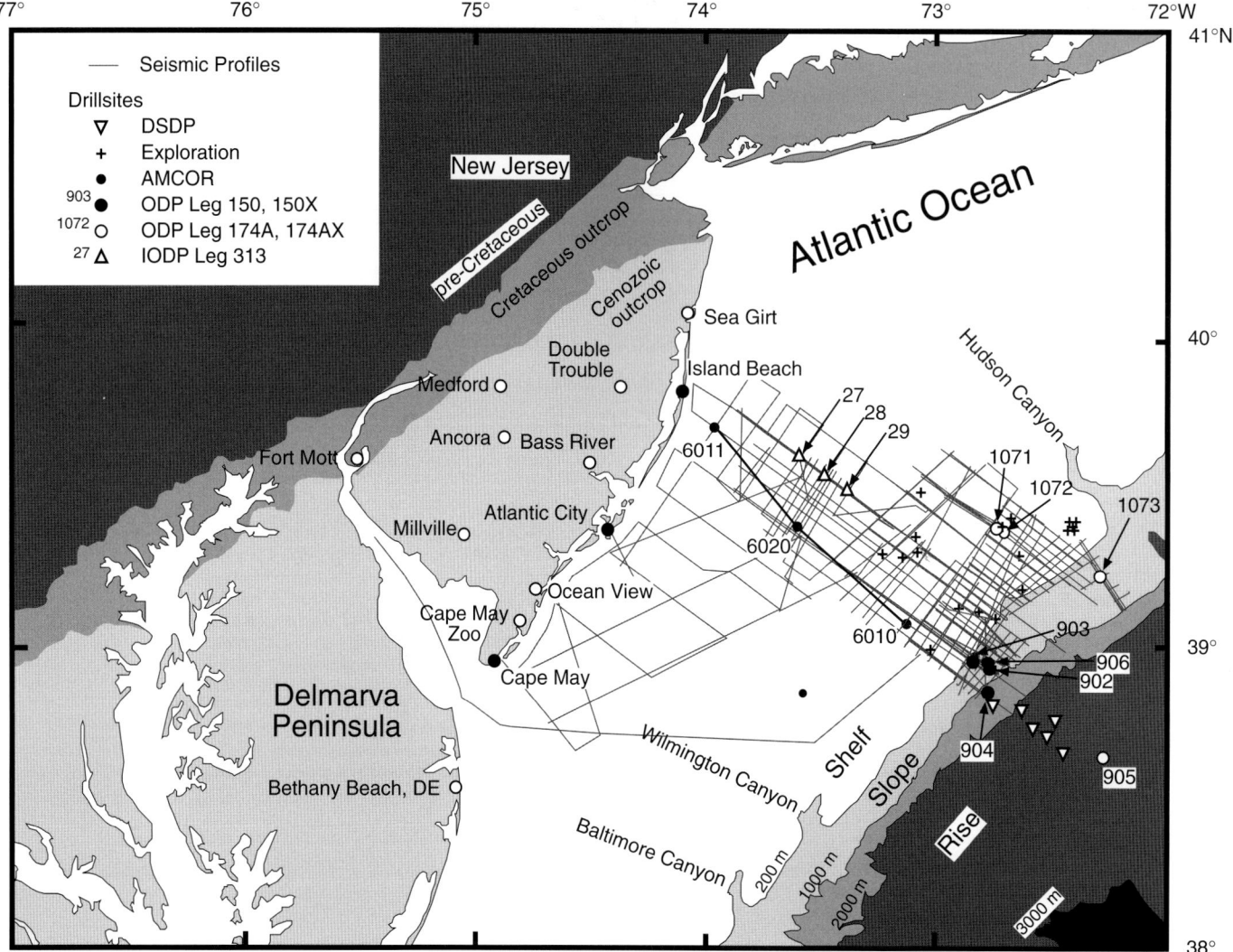

Fig. 3.1. Generalized location map showing coastal plain outcrop ages, generalized bathymetry, shelf edge, onshore and offshore coreholes, and multichannel seismic profiles collected in the 1990s that target Cenozoic strata. The black line connecting AMCOR 6011 and 6010 is shown in Figure 3.6.

& Boicourt 1981): (1) relatively fresh Shelf Water (salinity of <35‰); (2) more saline Slope Water (35–36‰); and (3) warm (>18 °C), salty (>36‰) Gulf Stream water. Slope Waters generally dominate on the continental slope, although the front between Shelf and Slope Waters is not rigidly fixed to the shelf–slope break. In addition, Gulf Stream eddies (warm core rings) advect warm, salty water to the slope and occasionally to the shelf. The source of Slope Water has been attributed to local winter cooling but it probably has a source from the Labrador Current, as indicated by the $\delta^{18}O_{seawater}$ tracer (Fairbanks 1982). A cyclonic gyre of NE-flowing Slope Water and SW-flowing Shelf Water is geostrophically balanced by regional wind forcing, although the processes controlling the position and movement of the boundary between Shelf and Slope Waters are poorly understood. The main thermocline and oxygen minimum zones in this region are shallow and seasonably stable (<400 m), with relatively high O_2 values in the minimum zone (>3 ml l^{-1}; c. 94 μm kg^{-1}: Miller & Lohmann 1982). The continental rise falls within the influence of the Western Boundary Undercurrent (WBUC), a strong SW-flowing bottom current composed primarily of North Atlantic Deep Water with an admixture of Antarctic Bottom Water (Heezen *et al.* 1966). The strongest flow of this deep geostrophic current is between water depths of 3000 and 4900 m on the continental rise, although the current has migrated up and down the rise through time,

and may have impinged on the lower slope at times (Heezen *et al.* 1966).

The physical oceanography of the US middle Atlantic continental shelf can be characterized by timescales that correspond to annual, seasonal, storm, tidal and wave components (Beardsley & Boicourt 1981). Historical analyses indicate that the long-term annual flow on the New Jersey shelf is alongshore to the SW at average velocities of less than about 20 cm s^{-1}, with values during winter storms and hurricanes exceeding 30–60 cm s^{-1} (Beardsley & Boicourt 1981; Lyne *et al.* 1990; Gong *et al.* 2010). A recent overview of annual and seasonal changes in surface water circulation on the shelf (Gong *et al.* 2010) confirms this earlier result. Mean surface flow over the New Jersey shelf for the period 2002–2007 was 2–12 cm s^{-1} along the shelf and offshore to the south (Gong *et al.* 2010). Both the Hudson Shelf Valley (Fig. 3.2) and the shelf edge act as dynamic barriers that define the continental shelf circulation. Topography, seasonal stratification and wind forcing control surface flow, which is in the approximate direction of the wind during the winter season when the water column is unstratified and more to the right of the wind during the summer season when the water column is stratified.

New views of the seasonal circulation have emerged (Gong *et al.* 2010). During the summer, SW winds drive cross-shelf offshore flows and favour upwelling. During the winter, cross-shelf

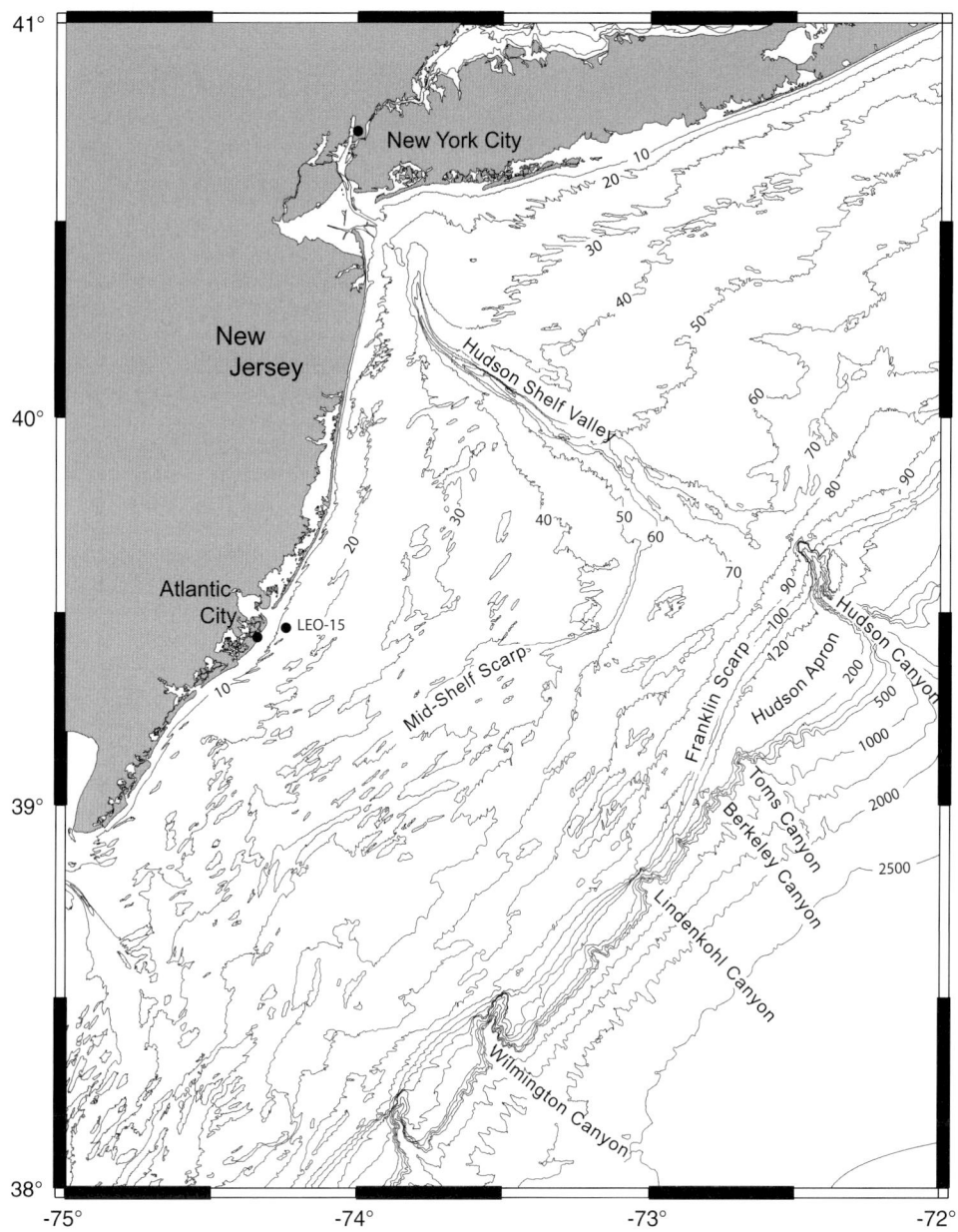

Fig. 3.2. Bathymetry of the US Atlantic margin between the Hudson and Wilmington canyons showing the 10 m contour interval.

offshore flows are driven by NW winds. Longshore NE winds associated with storm events drive energetic along-shelf flows during the autumn and spring. Thus, surface transport is cross-shelf during summer and winter, and along-shelf during the spring and autumn. Measurements indicate that the currents strong enough to entrain and transport sediment were confined to high-energy events such as storms. At higher frequencies, less energetic semi-diurnal astronomical tides with an average tidal range of 1.2 m (i.e. microtidal) drive flows that are generally orientated in the shore-normal direction. The maximum velocity of the near-surface (5 m below) tidal currents averages about 20 cm s^{-1}, and that of the near-bottom (2 m above) tidal currents averages about 10 cm s^{-1}. Although the velocities of the tidal currents are lower than other currents, most non-tidal currents flow alongshore and tidal currents often dominate the onshore/offshore signal. At very high frequencies, the energy spectrum is dominated by surface waves that are usually less than 1 m, travelling shorewards with the larger waves generated by offshore events and propagating in from offshore (Glenn *et al.* 2008).

Recent underwater measurements on the shelf by autonomous underwater vehicles ('gliders') show that the combined action and timing of surface waves, tides and storm-driven currents can explain the observed temporal variation in storm resuspension

(Glenn *et al.* 2008). Key to sediment dispersal is the role of the pycnocline, which determines how far up in the water column the sediment is resuspended and made available for transport: (1) on the inner shelf, the presence or absence of a seasonal pycnocline is a function of upwelling/downwelling; (2) on the middle shelf, a persistent seasonal stratification limits direct linkage between the upper wind-driven boundary layer and the lower combined wave and current boundary layer until mixing occurs in the full water column, usually in the autumn; and (3) on the outer shelf, the effects of surface waves are limited to only the most severe storms (Glenn *et al.* 2008).

Modern sedimentation

The modern US middle Atlantic continental shelf is generally starved because sediment input is trapped in estuaries. Surface sediments across the shelf are almost exclusively sand (i.e. >63 mm: Hollister 1973). Modern sediments fail to follow the classic graded shelf model of progressive fining offshore (Swift 1969). Instead, the complete dominance of sands on the modern shelf prompted Emery (1968) to classify modern shelf

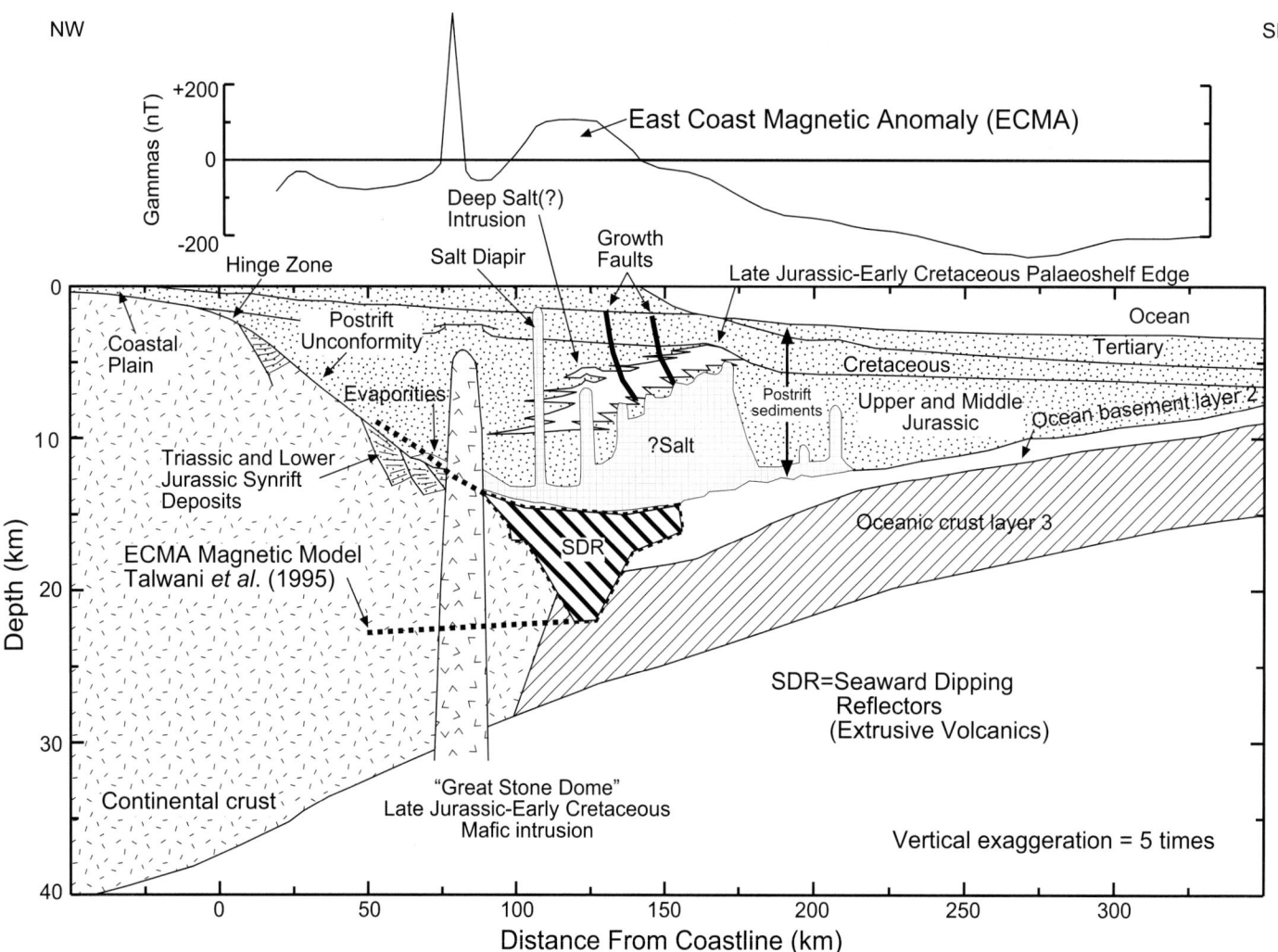

Fig. 3.3. Updated cross-section of the Baltimore Canyon Trough, modified after Grow & Sheridan (1988). The modifications are based on all publicly available seismic data, deep drilling, and most recent interpretations and magnetic modelling of the EDGE experiment (Talwani *et al.* 1995). The most significant differences from previous work are the interpretation of thick salt and clarification of the transition from ocean to continental crust.

sedimentation in this region and many others throughout the world as 'relict,' a product of the Holocene transgression and not in equilibrium with modern shelf processes. Swift *et al.* (1971) recognized that the sands were relict, in the sense that they were initially deposited in a different environment (e.g. shoreface sediments now residing on the middle shelf) but subsequently have been redeposited in a modern hydrodynamic equilibrium, and applied the term 'palimpsest'. Thus, the terms 'relict' and

Table 3.1. *Characteristics of the US middle Atlantic margin (Hudson Canyon to Wilmington Canyon)*

Length of the shelf	*c.* 200 km
Average width	150 km
Mean tide range	1.2 m
Waves	Generally <1 m
Currents	<60 cm s^{-1}
Dominating process (wave/current/tide)	Storm
Average depth of the shelf break	135 m
Siliciclastic/carbonate/authigenic/glacial sedimentation	Siliciclastic
Modern/relict/palimpsest (if possible in approximate %)	Palimpsest
Tectonic trend over the last glacial cycle (stable/uplifting/subsiding)	Stable

'palimpsest' used in this Memoir were derived first from this well-studied margin.

The sands on the modern shelf are arrayed into a series of ridges and swales that reflect multiple processes which mold shelf sedimentation today (Fig. 3.2) (e.g. Ashley *et al.* 1991; Goff *et al.* 1999, 2005). Although the source of modern quartz sand in this region is ultimately from the Appalachian Mountains via major river systems (Hudson, Delaware and Susquehanna), the maturity of the sands argues for the recycling of coastal plain sediments in the nearshore zone (Pazzaglia & Gardner 1994), with the likely source being primarily from the onshore Miocene Cohansey and Kirkwood formations and younger surficial units. These sands were reworked in nearshore/shoreface environments during the Holocene transgression across the shelf.

Today, the nearshore zone has both shore-attached and -detached sand ridges (1–12 m thickness, 2–20 km in length, 1–5 km spacing: Goff *et al.* 2005) that often run at oblique angles to the shoreline (Fig. 3.2) (Ashley *et al.* 1991). Similar ridges are found on the middle shelf off New Jersey in approximately 40 m of water (e.g. Stubblefield *et al.* 1983) and on the outer shelf (Goff *et al.* 1999). The origin of these ridges has been controversial, with early studies favouring an abandoned barrier origin (see the summary in Swift *et al.* 1973). However, numerous studies have shown that these ridges are reworked in hydrodynamic equilibrium with shelf currents (Swift *et al.* 1973; Stubblefield *et al.* 1983; Rine *et al.* 1991; Goff *et al.* 1999, 2005). Although the sands of inner and middle shelf ridges were emplaced during transgression, they are strongly modified by modern currents (Stubblefield *et al.* 1983; Goff *et al.* 2005). The role of modern currents on outer shelf ridges is still debated. Goff *et al.* (1999) suggested that the outer shelf ridges were largely erosional, whereas other studies have suggested strong modification by currents (Swift *et al.* 1973; Stubblefield *et al.* 1983; Rine *et al.* 1991). Studies in the nearshore zone at the LEO15 site (15 m present depth offshore of Tuckerton, NJ) (Fig. 3.2) show that longshore currents (particularly alongshore geostrophic storm-generated flows) are sufficiently energetic to entrain and transport sands (Styles & Glenn 2005). We suggest that reworking of shelf sand to form ridges by geostrophic currents is analogous to build-up of deep-sea drift deposits (Heezen *et al.* 1966), albeit despite differences in grain size and velocities. This explains the oblique orientation of many of the ridges and the fact that many may be erosional remnants.

Recent Chirp seismic reflection data have identified several interesting features on the outer continental shelf (Goff *et al.* 2005). In addition to sand ridges that are largely erosional remnants, NE–SW striations identified as 'sand ribbons' occur in swales and appear to be deposited by currents in water depths of 50–100 m. Elongated pits approximately 0.5–1.5 km long, 1–3 km wide and up to 10 m deep are erosional remnants. Buried Pleistocene erosional channels display both a bathymetric and backscatter expression in this region. Finally, striations interpreted as iceberg grooves have been mapped on the outer continental shelf (Goff *et al.* 1999; Goff & Austin 2009) that may have been filled by transparent and/or chaotic fill interpreted as a catastrophic flooding event (Fulthorpe & Austin 2004).

Slope sediments reflect a mixture of downslope transport and pelagic sedimentation. The modern continental shelf break at approximately 135 m in this region is related primarily to the position of sea level during the last glacial maximum that resulted in the shedding of sands directly into the deep sea, v. finer-grained muds deposited on to the slope during the Holocene. Surface sediments consist of upper slope silts that grade down to lower slope clays (Hollister 1973). Because most modern riverine sediment input is trapped in estuaries, relatively little coarse terrigenous sediment reaches the slope today, although there is evidence that more coarse material was deposited on the slope during glacials (Christensen *et al.* 1996). Thus, modern sedimentation on the slope is primarily hemipelagic and sediments are derived from muds carried as suspended material from river discharge or from resuspended shelf sediments carried off the shelf (Doyle *et al.* 1979). Slope sediments have high (>1%) but variable organic carbon values and are generally carbonate poor (<20%), with carbonate content increasing downslope as a result of increased input of pelagic carbonates and decreased input of terrigenous material (Miller & Lohmann 1982). Considerable speculation has centred on whether Holocene sedimentation on the slope is dominated by sediments transported downslope (slide, slumps, debris flows, turbidites) or pelagic/hemipelagic rain. Examination of surface samples shows a pattern of largely *in situ* benthic biofacies (Miller & Lohmann 1982). Submersible observations of the lower slope in the immediate region show that pelagic sediments drape the bottom; outcrops are restricted to occasional near-vertical walls. Visual and core evidence for large- and small-scale transport is largely limited to blocks found at the foot of the slope (2200 m in this region) and sporadic turbidity-current activity in some canyon thalwegs (McHugh *et al.* 1993; Pratson *et al.* 2007).

Geological structure and early margin history

The US middle Atlantic continental shelf and slope is a classic passive margin, and in this review we focus on the region between the Hudson and Wilmington submarine canyons that contains a thick sedimentary record of up to 16 km. Rifting and subsequent separation from NW Africa occurred during the Late Triassic–earliest Jurassic (*c.* 230–190 Ma), forming a series of rift basins that extend from the onshore today to beneath the modern shelf (Fig. 3.3: e.g. Grow & Sheridan 1988). Seafloor spreading began prior to the Bajocian (*c.* 175 Ma: Middle Jurassic), with the likely opening beginning off Georgia by around 200 Ma and progressing northwards to the US middle Atlantic margin (Withjack *et al.* 1998). This south–north 'zipper' rifting is associated with a diachronous post-rift unconformity that separates active 'rift-stage' (synrift) deposits (strongly influenced by syndepositional horst and half-graben structures) from passive margin 'drift-stage' deposits which accumulated in a progressively widening and deepening basin. Prior to the Plio-Pleistocene, post-rift history of the middle Atlantic region is dominated by simple thermal subsidence, sediment loading, lithospheric flexure, compaction and sea-level changes (Watts & Steckler 1979; Reynolds *et al.* 1991; Miller *et al.* 2005; Kominz *et al.* 2008), though mantle dynamics impacted the longer-term (more than 1 myr) record (e.g. Rowley 2013). Local normal faulting (minor, except for several large growth faults beneath the modern outer continental shelf), rare salt diapirism and a single Early Cretaceous igneous intrusion (the Great Stone Dome: Fig. 3.3) locally complicate the otherwise simple passive margin post-rift tectonic history (Fig. 3.3) (Poag 1985). Glacial Isostatic Adjustments (GIA) played a major role following the development of large northern hemisphere ice during the past 2.7 myr (Peltier 1998). Most of these are far-field effects but, during major Pleistocene glacials (Marine Isotope Chrons (MIC) 2, 6 and, perhaps, others – note that most authors use the term Marine Isotope Stage (MIS) but this is a stratigraphically incorrect usage of the term 'stage', the proper term is 'chron') (Stanford *et al.* 2001), continental ice sheets reached northern New Jersey and influenced shelf–slope sedimentation through near-field GIA effects.

Up to 16 km of post-rift sediments accumulated along the US middle Atlantic region in an offshore basin termed the 'Baltimore Canyon Trough' (BCT: Fig. 3.3). The Jurassic section is composed of shallow-water limestones and shales (typically 8–12 km) that are restricted to the offshore BCT (Fig. 3.3). In the BCT, salt of probable Jurassic age has progressively migrated as a diapiric ridge seawards over the initial ocean crust under the East Coast Magnetic Anomaly (Fig. 3.3). This salt migration is similar to what has been documented in the Gulf of Mexico and the Nova Scotian margin.

Long-term global sea-level rise plus thermal subsidence and flexural bending of the crust beneath the coastal plain led to progressive widening of the BCT during the Early Cretaceous (*c.* 120–140 Ma: Fig. 3.3) (Watts & Steckler 1979; Olsson *et al.* 1987). A fringing great barrier reef marked the edge of the shelf–slope break during the Jurassic–Early Cretaceous, and the margin experienced mixed siliciclastic and carbonate deposition (Jansa 1981; Poag 1985) (Fig. 3.3). The reef prograded seawards to a position no more than a few tens of kilometres seawards of the modern shelf edge during the Early Cretaceous and then disappeared. It is not clear what caused the demise of the barrier reef. Deltaic sediments subsequently overstepped the reef in the Early Cretaceous and, during a long interval of Late Cretaceous transgression, the shelf–slope break moved landwards once again. River input to the shelf may have overwhelmed the carbonate platform and reef, although climatic change and northward latitudinal drift may also be implicated in the demise of this great barrier reef of eastern North America. Siliciclastic input resulted in moderately thick (*c.* 2–3 km) offshore Cretaceous strata containing several major sandbodies and onshore deposits that are generally deltaically influenced (e.g. Sugarman *et al.* 1995; Kulpecz *et al.* 2008; Browning *et al.* 2008). After these Cretaceous pulses of sand input into the BCT, accumulation rates were generally low–moderate during the latest Cretaceous–Palaeogene when siliciclastic and carbonate fine-grained sediment accumulated (Poag 1985). Deposition during this latter interval occurred on a carbonate ramp with a gradient of around 1:500 (Steckler *et al.* 1999; see discussion below).

Cretaceous–Miocene sedimentation

The passive US middle Atlantic margin is a natural laboratory for unravelling sea-level history and the response of sedimentation to sea-level changes. Drilling onshore and offshore of New Jersey and Delaware has yielded a more than 100 myr record of sea-level changes (Fig. 3.4) (Miller *et al.* 2005; Browning

et al. 2008; Kominz *et al.* 2008). Fundamental to reconstructing sea level is the realization that relative sea-level falls (i.e. the combination of global sea level (eustasy) and tectonism) cause erosional unconformities and that these unconformities can be used to divide the stratigraphic record into sequences (Vail *et al.* 1977).

Early studies of the US middle Atlantic region recognized that the sea encroached the onshore coastal plain during transgressions and retreated during regressions numerous times during the Late Cretaceous–Miocene (*c.* 100–5 Ma: Owens & Sohl 1969; Olsson 1975). It is clear that these transgressions and regressions moulded the stratigraphic record buried beneath the modern shelf (Poag 1985). More recent studies placed these transgressive

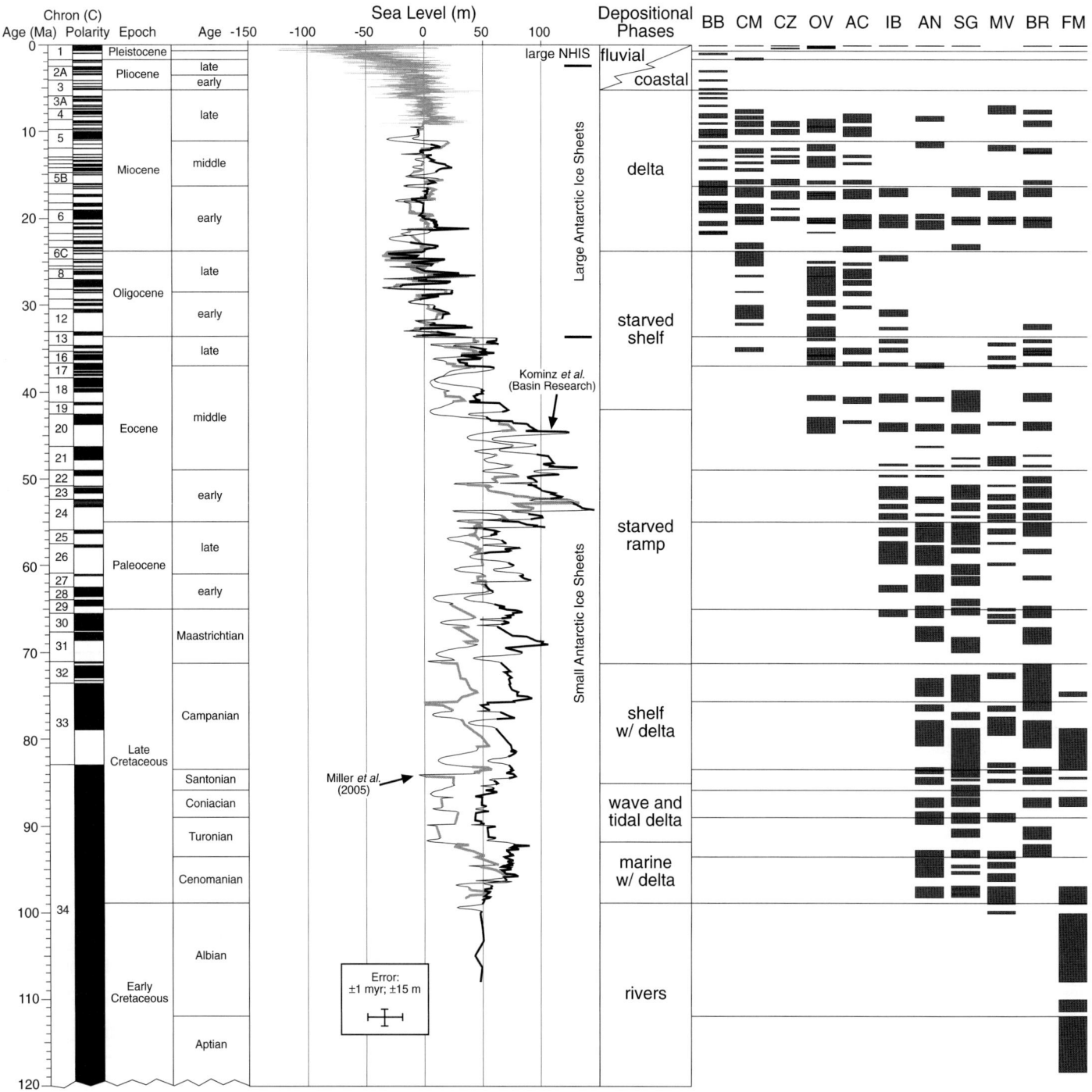

Fig. 3.4. Sea-level curves of Miller *et al.* (2005) and Kominz *et al.* (2008), depositional regimes of the onshore coreholes (Fig. 3.1), and distribution of sediments in sequences in the onshore coreholes as a function of time. BB, Bethany Beach core; CM, Cape May core; CZ, Cape May Zoo core; OV, Ocean View core; AC, Atlantic city core; IB, Island Beach core; AN, Ancora core; SG, Sea Girt core; MV, Millville core; BR, Bass River core; FM, Fort Mott core. Modified after Browning *et al.* (2008).

and regressive facies into a sequence stratigraphic framework. Continuous coring by the ODP legs 150X and 174AX onshore, and 150 and 174A offshore (see the summary in Miller *et al.* 2005; Browning *et al.* 2008), has provided one of the best-dated records of sequences (unconformity-bounded units) and a well-defined history of sea-level changes for the past 100 myr. To differentiate the effects of eustasy from other influences, a modelling technique termed 'backstripping' (Watts & Steckler 1979) was applied to this well-dated record of water-depth changes, progressively removing the effects of compaction, loading and thermal subsidence (see the summary in Miller *et al.* 2005). In the absence of regional or local tectonics, backstripping provides a global sea-level estimate, with the greatest uncertainty resulting from errors in estimates of water depth at the time of deposition. Backstripped records from 11 onshore ODP coreholes (Fig. 3.4) generally yielded similar sea-level estimates (summary in Kominz *et al.* 2008) that compare well to those from other passive margins and epicontinental seas (e.g. US Gulf Coast, NW Europe, the Russian Platform) and the oxygen isotope proxy for glacioeustasy (Miller *et al.* 2005), suggesting that global sea-level changes were a dominant process controlling million-year-scale sequences on this margin.

Although sea level appears to be a dominant control on million-year-scale sequences (Miller *et al.* 2005), facies changes within sequences reflect changes in accommodation (including effects of sea level and subsidence), sediment supply and provenance (Posamentier *et al.* 1988). In contrast to modern sedimentation, Cretaceous–Miocene nearshore and shelf sequences follow a classic pattern of a graded shelf and Walther's law, where the horizontal pattern is repeated vertically (Browning *et al.* 2008). In contrast, such patterns are not often observed in Plio-Pleistocene sequences due to low accommodation rates, low sediment supply to the shelf and high rates of global sea-level change. In Cretaceous–Miocene sections, a sequence consists of a basal unconformity overlain by transgressive sands (transgressive systems tract, TST) dominated by glauconite in the Late Cretaceous–Palaeogene and quartz in the Miocene. The TST is overlain by a coarsening-upward regressive highstand systems tract (HST). Two different facies models were derived for the region (Browning *et al.* 2008): one for riverine/deltaic influence sequences, and one for classic storm-dominated shoreface or neritic environments. Close to riverine influence, the HSTs consist of lower prodelta silts and upper delta front sands (Sugarman *et al.* 1993; Browning *et al.* 2008). Away from riverine influence, the HST consist of offshore muds, lower shoreface, shelly, heavily bioturbated, heterolithic silty fine and very fine sands, distal upper shoreface fine–medium sands with admixed silts, and upper shoreface/foreshore fine–coarse, well-sorted sands, with opaque heavy mineral laminae (see Browning *et al.* 2008). The changing water-depth patterns inferred from lithofacies changes are mirrored in the distribution of benthic foraminifera, giving confidence that the transgressive–regressive packages reflect shifting water depths on a graded shelf.

Backstripping applied to these transgressive–regressive sequences in the coastal plain wells reveal numerous Late Cretaceous–Miocene sea-level cycles: 15–17 Late Cretaceous, 6 Paleocene, 12 Eocene, 7 Oligocene and 14 Miocene sequences are shown (Fig. 3.5). The amplitudes of the eustatic estimates are best constrained (better than ± 10 m resolution) in the Oligocene by two-dimensional (2D) backstripping. They are less constrained in the Late Cretaceous–Eocene where most lowstands are missing, and are poorly constrained in the Neogene because much of each cycle is missing (Miller *et al.* 2005; Kominz *et al.* 2008). Despite this limitation, 'Icehouse' sequences of the last 33 myr show high amplitudes (up to 60 m), consistent with control by growth and decay of large (near modern-sized) Antarctic ice sheets (Miller *et al.* 2005). 'Greenhouse' Late Cretaceous–Eocene sequences show lower amplitudes (typically 25 m, although a few are 40 m); nevertheless, a 25 m global sea-level change in less than 1 myr can be explained only by the growth and decay of a significant ice sheet (i.e. about one-third of the modern ice volume). This apparent conflict with the well-documented history of warm global temperatures at this same time (e.g. Huber *et al.* 2002) has been explained by invoking small, ephemeral ice sheets in the Greenhouse world (see the summary in Miller *et al.* 2005).

The 100–1000 myr history of the US middle Atlantic margin reflects the changing influences of two large river systems during the Cretaceous and latest Oligocene–Miocene. Sediment thickness measured in onshore wells and mapped in offshore seismic profiles indicates a north source inferred to be the ancestral Hudson River draining the northern Appalachian Mountains, and a central source inferred to be the ancestral Delaware or Susquehanna rivers draining the central Appalachians (Kulpecz *et al.* 2008; Monteverde *et al.* 2008).

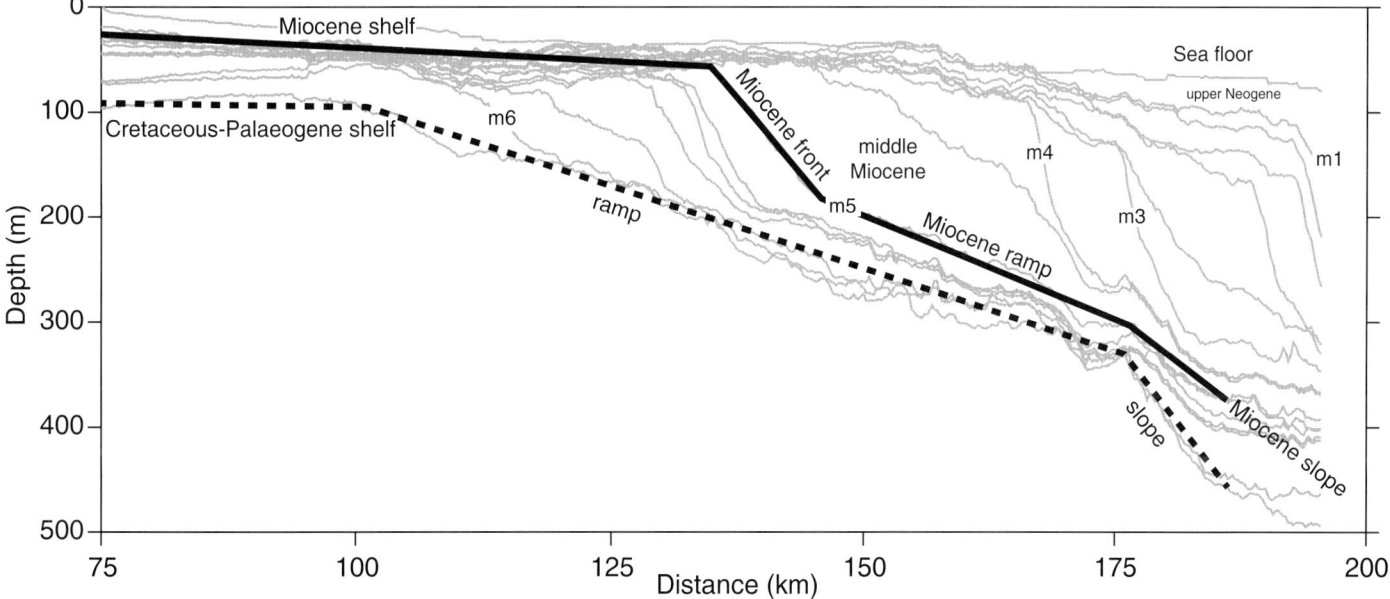

Fig. 3.5. Two-dimensional backstripped cross-sections across the New Jersey shelf. Generalized profiles for the Late Cretaceous (thick dashed line) and Miocene (thick solid line) are labelled as shelf, ramp, front (Miocene only) and slope. Modified after Steckler *et al.* (1999). Origin is the modern shoreline.

Browning *et al.* (2008) noted that the sediment facies evolved through eight depositional regimes controlled by changes in accommodation, long-term sea level and sediment supply (Fig. 3.4): (1) the Early Cretaceous consisted of anastomosing riverine environments deposited during a time of warm climates, high sediment supply and high accommodation; (2) the Cenomanian–early Turonian was dominated by marine sediments with a minor deltaic influence associated with long-term sea-level rise; (3) the late Turonian–Coniacian was dominated by non-marine fluvial wave and tidal delta systems associated with long-term sea-level fall; (4) the Santonian–Campanian consisted of marine deposition under the influence of a wave-dominated delta associated with a long-term sea-level rise and increased sediment supply; (5) Maastrichtian–middle Eocene deposition consisted primarily of starved, carbonate ramp–shelf environments associated with very high long-term sea level and low sediment supply; (6) the late Eocene–Oligocene was a starved siliciclastic shelf associated with high long-term sea level and low sediment supply; (7) early–middle Miocene sediments were deposited on a prograding shelf under a strong wave-dominated deltaic influence associated with a major increase in sediment supply and accommodation; and (8) over the past 10 myr, low accommodation and eroded coastal systems were associated with a long-term sea-level fall and low rates of sediment supply due to bypassing.

The evolution of the US middle Atlantic margin also reflects a long-term change from a carbonate ramp to a prograding siliciclastic margin. A major switch from carbonate ramp deposition to starved siliciclastic sedimentation (the 'carbonate switch') occurred progressively from the middle Eocene onshore to earliest Oligocene on the continental slope in response to global and regional cooling (Miller *et al.* 1997). Sedimentation rates increased dramatically in the late Oligocene–Miocene (Poag 1985) due to increased input from the hinterland (Poag & Sevon 1989; Pazzaglia & Gardner 1994). These thick sandbodies, arrayed as prograding units beneath nearly the entire modern shelf, have recently been continuously cored in the inner shelf region by IODP Expedition 313 (Mountain *et al.* 2010). IODP drilling recovered about 16 early–middle Miocene sequences at three sites spanning topset, inflection, foreset and bottomset deposits that provide an unprecedented coring of facies across seismically imaged sequences.

A change in margin morphology occurred during the carbonate switch, as revealed by reconstruction of past depositional surfaces using 2D backstripping (Fig. 3.5) (Steckler *et al.* 1999). Three Cretaceous–Eocene physiographical provinces can be recognized (Fig. 3.5): shelf (1:1000; 0–100 + m water depths), ramp (1:300; 100–325 m water depths) and slope (<1:100; 325–2000 m water depths). Four Miocene physiographical provinces can be recognized (Fig. 3.5): shelf (1:1000; 0–50 m water depths); front (<1:40; 50–200 m water depths); ramp (1:300; 200–350 m water depths); and slope (>1:40; 350–2000 m water depths). A flatter shelf with one sharp shelf edge developed in the Pleistocene.

The modern continental shelf break occurs in a zone of thickened oceanic crust that extends from thinned continental crust beginning approximately 10 km eastwards of the Great Stone Dome (*c.* 80 km in Fig. 3.3; *c.* 100 km in Fig. 3.5) to typical oceanic crust seawards of the dome. A change in declivity analogous to the modern shelf–slope break has existed and varied in position within this zone since rifting. At the time of separation from NW Africa (*c.* 180 Ma), the shelf–slope break was approximately 25–30 km landwards of its present location (Fig. 3.3). Two-dimensional backstripping places a break between a more steeply dipping ramp and the slope seawards of the Great Stone Dome (*c.* 175 km in Fig. 3.5; 105 km in Fig. 3.3) during the Late Cretaceous–Palaeogene. Terrestrial sediment supply decreased markedly in the Maastrichtian–Palaeogene, resulting in carbonate ramp–shelf deposition to a break in slope landward of the Great Stone Dome (Fig. 3.3). Siliciclastic input increased once again in the late Oligocene and spiked in the middle Miocene (Miller *et al.* 1997; Steckler *et al.* 1999) in what appears to be a global pattern (Bartek *et al.* 1991; Lavier *et al.* 2001) that may have been a result of global climatic cooling (Steckler *et al.* 1999). This led to two regions of distinct change in seafloor declivity for the Neogene: (1) a shelf–front break whose position was controlled by the advance and retreat of shallow-water (<100 m) clinothems; and (2) the shelf–slope break that remained close to the previous reef edge (175 km: Fig. 3.5). By the Pleistocene, the shelf–front break, controlled by clinothem location, prograded close to its present position (Fig. 3.5). In this way, the structural shelf break has existed within a 75 km-wide zone just seawards of continental crust since the time of first seafloor spreading; water depth at the shelf break has varied from around 100 m to slightly more than 300 m (Fig. 3.5), as observed on other modern margins.

Plio-Pleistocene sedimentation

The Pliocene onshore in New Jersey consists of fluvial gravels of the Pennsauken Formation that are poorly dated (Stanford *et al.* 2001). The Pliocene section in Delaware is non-marine and also poorly dated, but is fully marine in the Yorktown Formation in Virginia. This implies about 20 m of differential subsidence/uplift between Virginia and New Jersey/Delaware. The Pliocene is surprisingly poorly represented beneath the shelf and upper slope (Mountain *et al.* 2007). The reasons for this are unclear, although low sediment input and accommodation and/or mantle dynamic effects (Rowley 2013), may have been exacerbated by GIA effects of the development of large northern hemisphere ice sheets at around 2.7 Ma (Shackleton & Opdyke 1973).

Mountain *et al.* (2007) provided a detailed summary of Pleistocene sedimentation on this margin. The Pleistocene is characterized by a low-relief hinterland that provided minimal sediment input, extensive reworking on a wide shelf and little accommodation from thermal subsidence. As a result, the Pleistocene section beneath the inner–middle shelf is thin and spotty, and stacked in complex patterns owing to low sediment input and low accommodation during an interval of rapid eustatic change. For example, short-lived increases in accommodation space were fed by downcutting of channels occurring in concert with short-term sea-level rise at suborbital frequencies (e.g. meltwater pulse 1a) (Nordfjord *et al.* 2006, 2009; Christensen *et al.* 2013). Therefore, a continuous late Pleistocene record cannot be obtained at one location beneath most of the shelf. The exception is a thick accumulation beneath the outermost continental shelf and upper slope, particularly in the region of the Hudson Apron to the west of the Hudson Shelf Canyon (Mountain *et al.* 2007), as discussed below.

Two groups have worked extensively on Pleistocene sequences beneath the New Jersey shelf. The first emphasized the inner–middle shelf (Ashley *et al.* 1991; Carey *et al.* 1998, 2005; Sheridan *et al.* 2000; Wright *et al.* 2009). The second emphasized the middle–outer shelf wedge and upper slope (Duncan *et al.* 2000; Goff *et al.* 2005; Gulick *et al.* 2005; Nordfjord *et al.* 2006, 2009; Goff & Austin 2009). The latter focused on the last glacial cycle (the interval from the Last Glacial Maximum (LGM) to present) as discussed in the next section.

By using seismic stratigraphy to map upper Pleistocene sequences across the New Jersey continental shelf (Fig. 3.6), Sheridan *et al.* (2000) pieced together a history of late Pleistocene (i.e. since 130 ka) sea-level change. High-resolution seismic reflection profiles (*c.* 1–1.5 m resolution) across the New Jersey continental shelf provide a record of Pleistocene unconformity-bounded sequences (Ashley *et al.* 1991; Carey *et al.* 1998; Sheridan *et al.* 2000 and references therein). Sheridan *et al.* (2000) compiled a

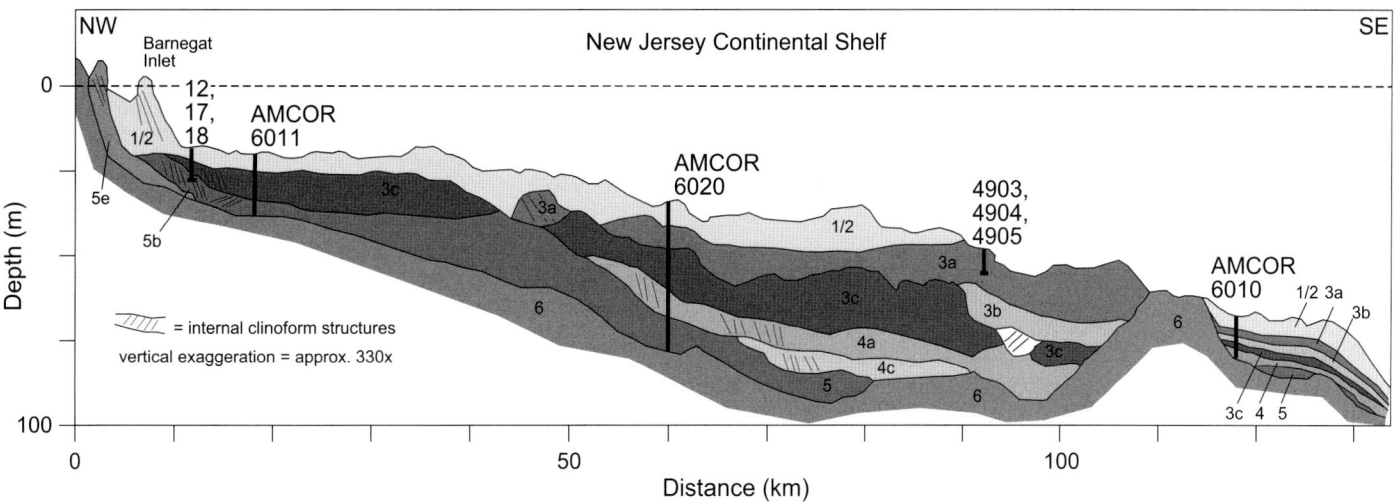

Fig. 3.6. Seismic stratigraphic cross-section of the late Pleistocene off Barnegat Inlet, New Jersey. Numbers 1/2, 3a, 3b, 3c, 4, 5, 5e, 5b and 6 refer to correlations to Marine Isotope Chrons. Shown are AMCOR sites 6011, 6020 and 6010, vibracores 12, 17 and 18 from Uptegrove (2003), and vibracores 4903, 4904 and 4905 from Knebel *et al.* (1979). Modified after Sheridan *et al.* (2000) by Wright *et al.* (2009). The profile is located in Figure 3.1.

composite seismic profile from Barnegat Inlet, New Jersey to the continental slope that showed seven seismic units (1/2, 3a, 3b, 3c, 4a, 4c and 5) above a prominent reflector assigned to MIC 6 or older (Fig. 3.6). Incision by the palaeo-Hudson River preserved thick MIC 3 and 4 deposits (up to 30 m) in the shelf valley that crossed the shelf at various places, whereas increased sediment discharge by the Hudson River is inferred for MIC 1 and 5 (Sheridan *et al.* 2000).

Although seismic profiles constrain the Pleistocene physical stratigraphy on the continental shelf, age control is difficult. Wright *et al.* (2009) reviewed age control on the Pleistocene sequence delineated previously (Carey *et al.* 1998; Sheridan *et al.* 2000) using radiocarbon dates, amino acid racemization data and superposition; they constrained the ages of large (20–80 m) sea-level falls, and correlated them with MIC 2, 3b, 4, 5b and 6 (the past 130 kyr: Fig. 3.7). They noted that, despite the proximity of New Jersey to the Laurentide ice sheet, sea-level

records for MIC 1, 2, 4, 5e and 6 are similar to those reported from New Guinea, Barbados and the Red Sea (Fig. 3.7), with some differences among records for MIC 3. The New Jersey record consistently provides the shallowest sea-level estimates for MIC 3 (*c.* 25–60 m below present: Wright *et al.* 2009), with a barrier system migrating to within about 1 km of the modern barrier (Ashley *et al.* 1991). This difference may be due to a GIA (Potter & Lambeck 2003) effect. Otherwise, the New Jersey record approximates the global record (Fig. 3.7). This can be explained by the fact that the portion of New Jersey directly influenced by the peripheral bulge of the ice sheet was north of Island Beach. A GIA correction must be applied to correct the New Jersey Pleistocene estimate for its far-field response (e.g. Peltier 1998) but this correction of 5–10 m is within the errors of the Pleistocene relative sea-level estimates.

Large volumes of sediment were deposited on the outer shelf and slope during the mid–late Pleistocene (post-750 ka),

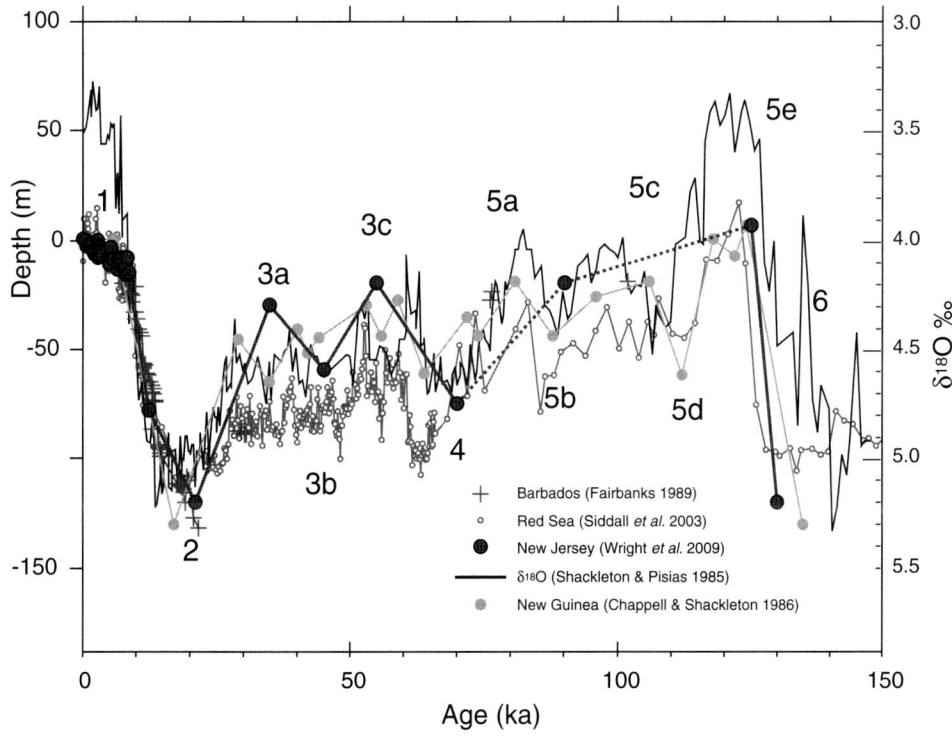

Fig. 3.7. Comparison of the sea-level record from the US middle Atlantic margin with the Huon New Guinea terraces (Chappell & Shackleton 1986), Barbados (Fairbanks 1989; Bard *et al.* 1996), the Red Sea record of Siddall *et al.* (2003) and the benthic foraminiferal $\delta^{18}O$ record from Pacific (Carnegie Ridge) core V19–30 (Shackleton & Pisias 1985: black line). After Wright *et al.* (2009).

extending the continental margin several tens of kilometres past the pre-Quaternary shelf edge (Mountain *et al.* 2007). Here we focus on erosion by submarine canyons and the rapid sedimentation that often occurs in their interfluves.

Submarine canyons off New Jersey are impressive features (e.g. Hudson Canyon is one of the largest in the world) that have a long history of study in this region (Shepard 1934; Daly 1936) and controversy on their origin. Several classes of submarine canyons cut the slope: (1) V-shaped large submarine canyons (Hudson and Wilmington: Fig. 3.2) cut deeply (*c.* 1000 m) into the slope, and much of the way across the shelf and across the continental rise; (2) smaller submarine canyons (Toms and Lindenkohl: Fig. 3.2) cut less deeply into the slope (*c.* 400 m) and breach the shelf break but cannot be traced across it; (3) U-shaped canyons are restricted to the lower slope; and (4) submarine rills and gullies are smaller features on the slope. Pratson *et al.* (2007) reviewed the processes that form these largely erosional features, caused by processes including turbidity currents and intraslope failure due to spring sapping, slides, slumps, and structural control. V-shaped canyons are cut largely by turbidity currents; U-shaped canyons are formed by other submarine processes (Mountain *et al.* 2007). Proximity to modern river systems suggests a link to sediment that most probably supplied these Pleistocene turbidity currents. Pratson *et al.* (2007) noted that submarine canyons may form during times of sea-level lowering, but also noted that intraslope processes unrelated to sea-level, including canyon piracy and intraslope failures, are potentially important in forming slope canyons.

Despite often being assumed to be Pleistocene in age, submarine canyons have a very long history. The modern Hudson River, for example, is apparently structurally controlled by the earliest Jurassic Palisades Sill, and an ancestral Hudson River has delivered large amounts of sediment to the rise near its current location since the Early Jurassic (Poag & Sevon 1989). However, the palaeo-Hudson River migrated south of its position during the Pleistocene as discussed above (Carey *et al.* 2005), and the lower slope portion of the Hudson Canyon can be dated to a seismic reflector thought to mark the approximate beginning of the growth and decay of large northern hemisphere ice sheets (*c.* 2.7 Ma: Mountain & Tucholke 1985). Buried Miocene canyons occur in the region and are likely antecedents for the Plio-Pleistocene Hudson Canyon (Miller *et al.* 1987); the evolution of a buried Miocene canyon is discussed by Mountain *et al.* (2007).

Mountain *et al.* (2007) provided an overview of Pleistocene outer shelf and slope sedimentation. Pleistocene sections at ODP sites 903 (444 m water depth near Berkeley Canyon) and 1073 (639 m water depth, Hudson Apron) recovered over 350 and 500 m, respectively, of sediment assigned to the Bruhnes Chron (last 780 kyr), with sedimentation rates greater than 60 m Ma^{-1} (Christensen *et al.* 1996; Austin *et al.* 1998; McHugh & Olson 2002). Physical properties data (e.g. magnetic susceptibility) integrated with biostratigraphy provide an astronomical chronology for these sites; this chronology shows that deposition of primarily silty clay was largely continuous, although it was punctuated by a few short hiatuses (Christensen *et al.* 1996; McHugh & Olson 2002). Mountain *et al.* (2007) mapped four Pleistocene seismic sequences on the outer shelf and upper slope on the Hudson Apron, and dated them on the slope: (1) sequence p4/yellow is early Bruhnes (MIC 19–12); (2) sequence p3/green is mid Bruhnes (MIC 12–9); (3) sequence p2/blue correlates with MIC 8; and (4) sequence p1/purple is the last interglacial to Holocene (MIC 5e-1). It appears that the sequence boundaries on the slope are generally associated with major glacials (MIC 6, 8, 12 and 20), although sedimentation was continuous across several glacial cycles, and, where there appears to be a cause–effect association, the phase relationships and the link to glacioeustasy are not simple. These sites bear testimony to the dominance of hemipelagic sedimentation between canyon thalwegs on the slope. Within canyon thalwegs and adjacent to them (particularly on the right-hand sides looking down-canyon), downslope deposition dominates.

The last deglaciation

Studies of the upper Pleistocene–Holocene on the inner continental shelf of New Jersey (Ashley *et al.* 1991; Miller *et al.* 2009) and Delaware (Ramsey & Baxter, 1996) have provided a record of sea-level change since the Last Glacial Maximum (LGM; *c.* 20–26 ka). A major erosional surface has been mapped on the inner shelf (R1 of Ashley *et al.* 1991); the erosional event occurred prior to the LGM, probably in MIC 4. During the LGM, a widespread unconformity was eroded beneath the shelf (R2 of Ashley *et al.* 1991). The relationship of the R reflector (Gulick *et al.* 2005) on the outer continental shelf is controversial and there are two interpretations: (1) an approximately 40 ka lowstand surface (Gulick *et al.* 2005; Goff & Austin 2009); and (2) correlation with MIC 6 (Sheridan *et al.* 2000) (Fig. 3.6). It is clear that this reflector is not the LGM as originally interpreted (Milliman *et al.* 1990) but is, instead, an older erosional surface.

Recent studies of the middle–outer shelf Pleistocene document that the surface stratigraphy of the last glacial cycle is more complicated than previously thought (see the summary in Mountain *et al.* 2007). The stratigraphy above the R reflector (=MIC 6 of Sheridan *et al.* 2000; =40 ka and Heinrich event 4 of Goff & Austin 2009) (Fig. 3.6) consists of the following upsection (Duncan *et al.* 2000; Nordfjord *et al.* 2006, 2009): (1) an outer shelf wedge; (2) a dendritic 'channels' reflector interpreted as fluvial channels formed during the LGM, (3) marine fill of the channels reflector during the early Holocene; and (4) a reflector formed as a transgressive ravinement surface. Detailed mapping of the shelf wedge suggest that it consists of subaqueous delta deposits, whereas the sediments above the ravinement reflector consist of lagoonal/back barrier and tidal channel deposits (Nordfjord *et al.* 2009).

The deglaciation is poorly sampled on the continental shelf. Dillon & Oldale (1978) reported dates of approximately 21 ka and a depth of 120 m for the LGM in cores from the outer shelf of New Jersey (Fig. 3.7), very similar to the estimate of 120 m from Barbados (Fairbanks 1989) that has been modelled as approximately 127 m of eustatic lowering (Peltier & Fairbanks 2006). During the MIC 2 sea-level fall, incision and reworking dominated (Goff *et al.* 2005; Christensen *et al.* 2013). Sea level rose rapidly during the last deglaciation and the timing of infilling of the channels (*c.* 16–14 ka) is consistent with the timing of meltwater pulse 1a (Christensen *et al.* 2013). Infilling occurred rapidly (0.5–1 cm a^{-1}) but in agreement with modern estuarine rates of sedimentation (Christensen *et al.* 2013). By 8.8 ka, sea level rose to a point 12 m below modern sea level, creating a thin (3–4 m) barrier system on top of an erosional transgressive ravinement surface (R3 of Ashley *et al.* 1991). The shore-attached ridges were deposited during the sea-level rise younger than approximately 9 ka, and are now reshaped and redeposited by modern currents. The present-day barrier and lagoon complex post-dates 8.8 ka (Ashley *et al.* 1991). Holocene middle–outer shelf sediments are reworked, and dates on middle–outer shelf shell material are not in stratigraphic succession, suggesting the reworking of glacial sediments (Alexander *et al.* 2003; Christensen *et al.* 2013).

The record of late Holocene sea-level rise has been complicated by the fact that the rise of approximately 9 m over the past 5 kyr is small relative to uncertainties in the method. Psuty (1986) interpreted a slowing of the rate of rise at approximately 2 ka in New Jersey. Detailed evaluation of sea-level rise in New Jersey during the Holocene was performed by evaluating the 'indicative meaning', evaluating numerous uncertainties and identifying the most reliable points (circled points in Fig. 3.8). During the Holocene, sea level rose moderately rapidly from approximately 8 to

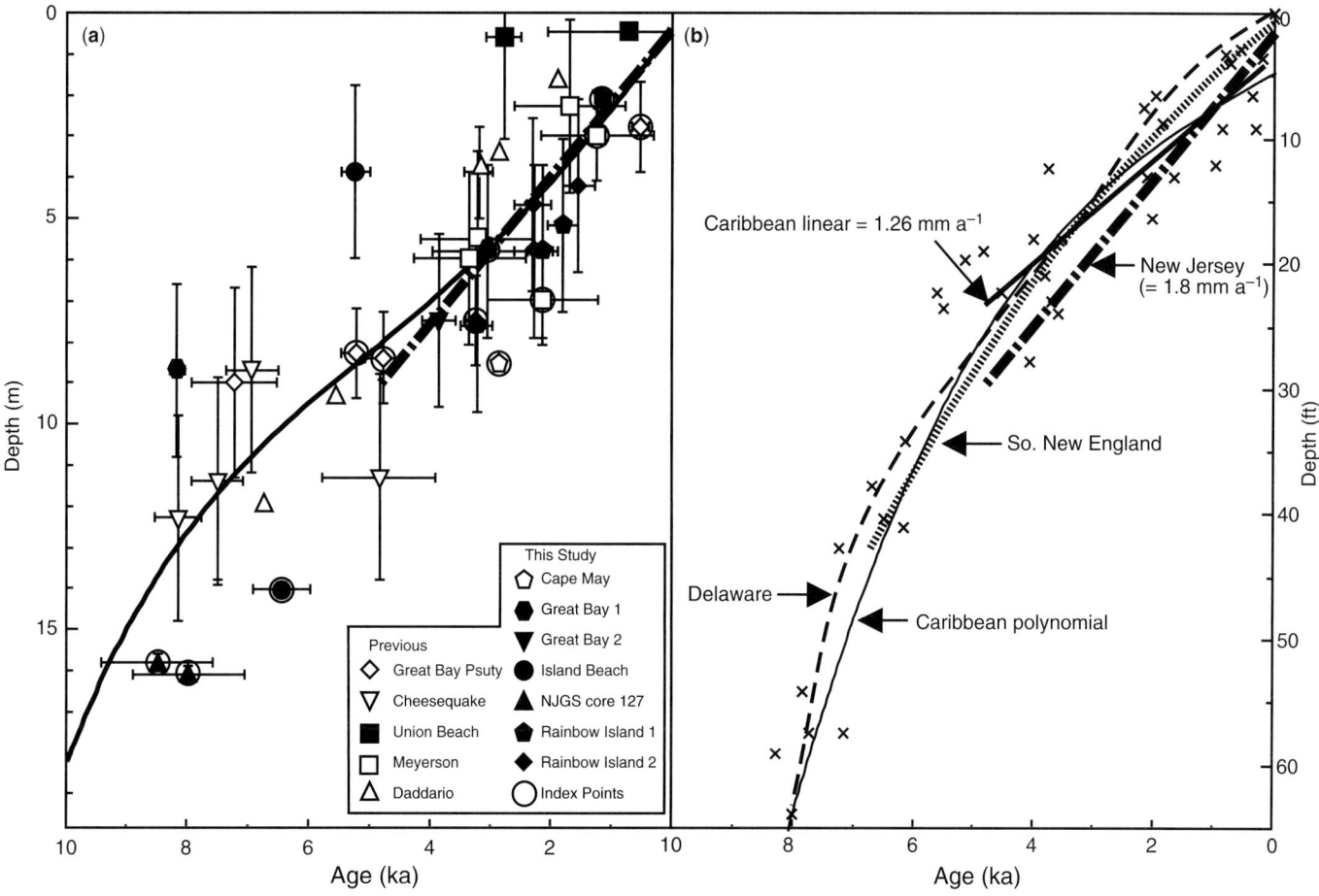

Fig. 3.8. (**a**) Comparison of all dates for New Jersey localities with error bars; data are compiled in Miller *et al.* (2009). The linear regression of 1.8 mm a^{-1} (thick dashed line) is fit to all of the data as two segments: 8–0 and 5–0 ka BP. The thick black line is a polynomial fit to the data. Note that the different regressions yield essentially the same record from 5 to 0 ka BP. Data points with no apparent depth errors are from marsh deposits with 30 cm vertical errors. (**b**) Comparison of our sea-level data and regression (red) with the sea-level record of Fairbanks (1989: crosses), which is based on the western Atlantic reef data of Lighty *et al.* (1982) for ages of less than about 6.4 ka BP. Two regressions through the reef data are shown: the first is a third-order polynomial; the other is a linear regression for all Lighty *et al.* (1982) data younger than 5.5 ka BP. The Delaware (dashed: Ramsey & Baxter 1996) and Southern New England (stippled: Donnelly *et al.* 2005) sea-level records are shown for comparison. Also shown is the record of the past 2 ka from Kemp *et al.* (2011). Modified after Miller *et al.* (2009).

5 ka (Fig. 3.8). The New Jersey record only requires a rise of about 8 m in this period (2.7 m ka^{-1} = mm a^{-1}) but the more complete record from Delaware (Ramsey & Baxter 1996) suggests upwards of 11–12 m (4 mm a^{-1}: Fig. 3.8). The rate of rise slowed around 5 ka, and has averaged 1.8 mm a^{-1} in both New Jersey and Delaware over the past 5 kyr (Miller *et al.* 2009). Much of this rise is due to GIA subsidence. Miller *et al.* (2009) assumed that the current GIA subsidence of 1 mm a^{-1} could be applied to a linear rise in sea level over the past 5 kyr, and concluded that the global rise in sea level was 0.75 ± 0.5 mm a^{-1}. However, modelling of Pacific island records requires a minimal global rise in sea level over the past 2–3 kyr (Peltier *et al.* 2002). Kemp *et al.* (2011) noted minimal eustatic change over the past 2 kyr from detailed studies in North Carolina compared to other less constrained records from throughout the world (Fig. 3.8). Miller *et al.* (2013) revisited the New Jersey sea-level record and documented a 1.4–1.6 mm a^{-1} rise from 2 ka to 1800 Common Era.

Instrument scale, present and future sea-level changes

Twentieth-century tide gauge data from the US middle Atlantic margin reveal a regional sea-level rise of approximately 3 mm a^{-1}, with Atlantic City and Sandy Hook yielding higher rates (*c.* 4 mm a^{-1}) due to compaction (Psuty 1986; Miller *et al.* 2009). This is consistent with the global increase of 1.8 mm a^{-1}

in the twentieth century derived from tide gauge data (Church & White 2006) after the GIA subsidence effects of 1.3 mm a^{-1} of subsidence are accounted for (Miller *et al.* 2013). It is clear that the rates of sea-level rise are accelerating in the twenty-first century; global sea level is rising at a rate of 3.3 mm a^{-1} in 2010 CE (Cazenave & Llovel 2010) and accelerating at a rate that intersects 80 cm of rise by 2100 CE (Rahmstorf 2007). Semi-empirical predictions of future sea-level rise (e.g. Vermeer & Rahmstorf 2009) are even higher (1.2 ± 0.6 m by 2100 CE). The US middle Atlantic margin will continue to subside due to GIA effects at 1 mm a^{-1} plus local compaction effects, bringing a minimum rise of about 1 m to this region by 2100 CE. The modern coastal environment is generally starved of sediments in this region, exacerbating the effects of a modern sea-level rise of 3 mm a^{-1} together with 1–2 mm a^{-1} of regional and local subsidence. With a 1 m rise in relative sea level by 2100, the '100 year flood' mark of 2.9 m for much of the New Jersey shoreline will be breached annually by storm surges flooding major airports and highways in the region. Beach erosion will be increasingly common and severe; following the Bruun rule, a 1 m rise in sea level would erode the shoreline by 50–100 m (Kyper & Sorensen 1985), with marsh rollback of 1 km given a 1:1000 gradient. We conclude that this region will be severely impacted by sea-level changes in the twenty-first century, affecting a coastal population of over 40 million people and an economic engine of over $50 billion annually.

We thank F. Pazzaglia, G. Ashley and an anonymous reviewer for their reviews, R. Peltier for discussions of GIA effects, and J. Goff and C. Fulthorpe for input on outer shelf sedimentation and Pleistocene stratigraphy. This overview is the product of a long history of study of this margin, and we thank R.K. Olsson, C.W. Poag and the late J. Owens for their pioneering efforts and inspiration, and NSF and ODP/IODP for funding seismic profiling and drilling.

References

ALEXANDER, C., SOMMERFIELD, C. *ET AL.* 2003. Sedimentology and age control of late Quaternary New Jersey shelf deposits. *In: Transactions of the American Geophysical Union, Fall Meeting 2003.* American Geophysical Union, Washington, DC, Abstract OS52B-0909.

ASHLEY, G. M., WELLNER, R. W., ESKER, D. & SHERIDAN, R. E. 1991. Clastic sequences developed during late Quaternary glacio-eustatic sea-level fluctuations on a passive continental margin: example from the inner continental shelf near Barnegat Inlet, New Jersey. *Geological Society of America Bulletin*, **103**, 1607–1621.

AUSTIN, J. A., JR., CHRISTIE-BLICK, N., MALONE, M., MOUNTAIN, G. S. THE LEG 174A SHIPBOARD PARTY. 1998. Miocene to Pleistocene sand-rich sequences and sea-level changes at the New Jersey outer shelf: results from Leg 174A. *JOIDES Journal*, **24**, 4–6, 19.

BARD, E., HAMELIN, B., ARNOLD, M., MONTAGGIONI, L. F., CABIOCH, G., FAURE, G. & ROUGERIE, F. 1996. Deglacial sea level record from Tahiti corals and the timing of global meltwater discharge. *Nature*, **382**, 241–244.

BARTEK, L. R., VAIL, P. R., ANDERSON, J. B., EMMET, P. A. & WU, S. 1991. Effect of Cenozoic ice sheet fluctuations in Antarctica on the stratigraphic signature of the Neogene. *Journal of Geophysical Research*, **96**, 6753–6778.

BEARDSLEY, R. C. & BOICOURT, W. C. 1981. On estuarine and continental shelf circulation in the Middle Atlantic Bight. *In:* WARREN, B. A. & WUNCH, C. (eds) *Evolution of Physical Oceanography.* MIT Press, Cambridge, MA, 198–233.

BROWNING, J. V., MILLER, K. G., SUGARMAN, P. J., KOMINZ, M. A., MCLAUGHLIN, P. P. & KULPECZ, A. A. 2008. 100 Myr record of sequences, sedimentary facies and sea-level change from Ocean Drilling Program onshore coreholes, U.S. Mid-Atlantic coastal plain. *Basin Research*, **20**, 227–248.

CAREY, J. S., SHERIDAN, R. E. & ASHLEY, G. M. 1998. Late Quaternary sequence stratigraphy of slowly subsiding passive margin; New Jersey Continental shelf. *American Association of Petroleum Geologists Bulletin*, **82**, 773–791.

CAREY, J. S., SHERIDAN, R. E., ASHLEY, G. M. & UPTEGROVE, J. 2005. Glacially-influenced late Pleistocene stratigraphy of a passive margin: New Jersey's record of the North American ice sheet. *Marine Geology*, **218**, 155–173.

CAZENAVE, A. & LLOVEL, W. 2010. Contemporary sea level rise. *Annual Review of Marine Sciences*, **2**, 145–173.

CHAPPELL, J. & SHACKLETON, N. J. 1986. Oxygen isotopes and sea level. *Nature*, **324**, 137–140.

CHRISTENSEN, B. A., HOPPIE, B. W., THUNELL, R. C., MILLER, K. G. & BURCKLE, L. 1996. Pleistocene age models. *In:* MOUNTAIN, G. S., MILLER, K. G. *ET AL.* (eds) *Proceedings of the ODP, Scientific Results*, **150**. Ocean Drilling Program, College Station, TX.

CHRISTENSEN, B. A., ALEXANDER, C., GOFF, J. A., TURNER, R. J. & AUSTIN, J. A. 2013. The Last Glacial: Insights from continuous coring on the New Jersey continental shelf. *Marine Geology*, **335**, 78–99.

CHRISTIE-BLICK, N., AUSTIN, J. A., JR & MALONE, M. J. (eds) 2003. *Proceedings of the ODP, Scientific Results*, **174A**. Ocean Drilling Program, College Station, TX.

CHURCH, J. A. & WHITE, N. J. 2006. A 20th century acceleration in global sea-level rise. *Geophysical Research Letters*, **33**, L1602.

DALY, R. A. 1936. Origin of submarine canyons. *American Journal of Science*, **31**, 401–420.

DILLON, W. P. & OLDALE, R. N. 1978. Late Quaternary sea-level curve: reinterpretation based on glacio-tectonic influence. *Geology*, **6**, 56–60.

DONNELLY, J. P., CLEARY, P., NEWBY, P. & ETTINGER, R. 2005. Coupling instrumental and geological records of sea-level: evidence from southern New England of an increase in sea-level rise in the late 19th century. *Geophysical Research Letters*, **31**, L05203.

DOYLE, L. J., PILKEY, O. H. & WOO, C. C. 1979. Sedimentation on the eastern United States continental slope. *In:* DOYLE, L. J. & PILKEY, O. H. (eds) *Geology of Continental Slopes.* SEPM Special Publications, Tulsa, **27**, 119–130.

DUNCAN, C. S., GOFF, J. A., AUSTIN, J. A. & FULTHORPE, C. S. 2000. Tracking the last sea level cycle: seafloor morphology and shallow stratigraphy of the latest Quaternary New Jersey middle continental shelf. *Marine Geology*, **170**, 395–421.

EMERY, K. O. 1968. Positions of empty pelecypod valves on the continental shelf. *Journal of Sedimentary Petrology*, **38**, 1264–1269.

FAIRBANKS, R. G. 1982. The origin of continental shelf and slope water in the New York Bight and Gulf of Maine. *Journal of Geophysical Research*, **87**, 5796–5808.

FAIRBANKS, R. G. 1989. A 17,000-year glacio-eustatic sea level record: influence of glacial melting rates on the Younger Dryas event and deep-ocean circulation. *Nature*, **342**, 637–642.

FULTHORPE, C. S. & AUSTIN, J. A., JR. 2004, Shallowly buried, enigmatic seismic stratigraphy on the New Jersey mid-outer shelf: evidence for latest Pleistocene catastrophic erosion? *Geology*, **32**, 1013–1016.

GLENN, S. M., JONES, C., TWARDOWSKI, M., BOWERS, L., KERFOOT, J., WEBB, D. & SCHOFIELD, O. 2008. Glider observations of sediment resuspension in a Middle Atlantic Bight fall transition storm. *Limnology and Oceanography*, **53**, 2180–2196.

GOFF, J. A. & AUSTIN, J. A. 2009. Seismic and bathymetric evidence for four different episodes of iceberg scouring on the New Jersey outer shelf: possible correlation to Heinrich events. *Marine Geology*, **266**, 244–254.

GOFF, J. A., SWIFT, D. J. P., DUNCAN, S. C., MAYER, L. A. & HUGHES-CLARKE, J. 1999. High-resolution swath sonar investigations of sand ridge, dune and ribbon morphology in the offshore environment of the New Jersey margin. *Marine Geology*, **161**, 307–337.

GOFF, J. A., AUSTIN, J. A., JR *ET AL.* 2005. Recent and modern marine erosion on the New Jersey outer shelf. *Marine Geology*, **216**, 275–296.

GONG, D., KOHUT, J. T. & GLENN, S. M. 2010. Seasonal climatology of wind-driven circulation on the New Jersey Shelf. *Journal of Geophysical Research*, **115**, C04006.

GROW, J. A. & SHERIDAN, R. E. 1988. U.S. Atlantic continental margin: a typical Atlantic-type or passive continental margin. *In:* SHERIDAN, R. E. & GROW, J. A. (eds) *The Atlantic Continental Margin.* The Geology of North America, 1–2. Geological Society of America, Boulder, CO, 1–7.

GULICK, S. P. S., GOFF, J. A., AUSTIN, J. A., JR, ALEXANDER, C. R., JR, NORDFJORD, S. & FULTHORPE, C. S. 2005. Basal inflection-controlled shelf-edge wedges off New Jersey track sea-level fall. *Geology*, **33**, 429–432.

HEEZEN, B. C., THARP, M. & EWING, M. 1959. *The Floors of the Oceans.* Geological Society of America, Special Papers, Boulder, **65**.

HEEZEN, B. C., HOLLISTER, C. D. & RUDDIMAN, W. F. 1966. Shaping of the continental rise by deep geostrophic contour currents. *Science*, **152**, 502–508.

HOLLISTER, C. D. 1973. *Atlantic Continental Shelf and Slope of the United States – Texture of Surface Sediments from New Jersey to Southern Florida.* United States Geological Survey, Professional Papers, **529-M**.

HUBER, B. T., NORRIS, R. D. & MACLEOD, K. G. 2002. Deep-sea paleotemperature record of extreme warmth during the Cretaceous. *Geology*, **30**, 123–126.

JANSA, L. F. 1981. Mesozoic carbonate platforms and banks of the eastern North American margin. *Marine Geology*, **44**, 97–117.

KEMP, A. C., HORTON, B. P., DONNELLY, J. P., MANN, M. E., VERMEER, M. & RAHMSTORF, S. 2011. Climate related sea-level variations over the past two millennia. *Proceedings of the National Academy of Sciences*, **108**, 11 017–11 022.

KNEBEL, H. J., WOOD, S. A. & SPIKER, E. C. 1979. Hudson River: evidence for extensive migration on the exposed continental shelf during Pleistocene time. *Geology*, **17**, 254–258.

KOMINZ, M. A., BROWNING, J. V., MILLER, K. G., SUGARMAN, P. J., MISINTSEVA, S. & SCOTESE, C. R. 2008. Late Cretaceous to Miocene sea-level estimates from the New Jersey and Delaware coastal plain coreholes: an error analysis. *Basin Research*, **20**, 211–226.

KULPECZ, A. A., MILLER, K. G., SUGARMAN, P. J. & BROWNING, J. V. 2008. Subsurface distribution of Upper Ccretaceous sequences and facies, New Jersey Coastal Plain. *Journal of Sedimentary Research*, **78**, 112–129.

KYPER, T. & SORENSEN, R. 1985. Potential impacts of selected sea level rise scenarios on the Beach and Coastal Works at Sea Bright, New Jersey. *In*: MAGOON, O. T., CONVERSE, H., MINER, D., CLARK, D. & TOBIA, L. T. (eds) *Coastal Zone '85*. American Society of Civil Engineers, New York, 2645–2655.

LAVIER, L. L., STECKLER, M. S. & BRIGAUD, F. 2001, Climatic and tectonic control on the Cenozoic evolution of the West African margin. *Marine Geology*, **178**, 63–80.

LIGHTY, R. G., MACINTYRE, I. G. & STUCKENRATH, R. 1982. *Acropora palmata* reef framework: a reliable indicator of sea level in the western Atlantic for the past 10,000 years. *Coral Reefs*, **1**, 125–130.

LYNE, V. D., BUTMAN, B. & GRANT, W. D. 1990. Sediment movement along the U.S. east coast continental shelf: I. Estimates of bottom stress using the Grant-Madsen model and near-bottom wave and current measurements. *Continental Shelf Research*, **13**, 397–428.

MCHUGH, C. M. G. & OLSON, H. C. 2002. Pleistocene chronology of continental margin sedimentation: new insights into traditional models, New Jersey. *Marine Geology*, **185**, 389–411.

MCHUGH, C. M., RYAN, W. B. F. & SCHREIBER, B. C. 1993. The role of diagenesis in exfoliation of submarine canyons. *American Association of Petroleum Geologists Bulletin*, **77**, 145–172.

MILLER, K. G. & LOHMANN, G. P. 1982. Environmental distribution of Recent benthic foraminifera on the northeast United States continental slope. *Geological Society of America Bulletin*, **93**, 200–206.

MILLER, K. G., MELILLO, A. J., MOUNTAIN, G. S., FARRE, J. A. & POAG, C. W. 1987. Middle to late Miocene canyon cutting on the New Jersey continental slope: biostratigraphic and seismic stratigraphic evidence. *Geology*, **15**, 509–512.

MILLER, K. G., BROWNING, J. V., PEKAR, S. F. & SUGARMAN, P. J. 1997. Cenozoic evolution of the New Jersey coastal plain: changes in sea level, tectonics, and sediment supply. *In*: MILLER, K. G. & SNYDER, S. W. (eds) *Proceedings of the ODP, Scientific Results*, **150X**. Ocean Drilling Program, College Station, TX, 361–373.

MILLER, K. G., KOMINZ, M. A. *ET AL.* 2005. The Phanerozoic record of global sea-level change. *Science*, **310**, 1293–1298.

MILLER, K. G., SUGARMAN, P. J. *ET AL.* 2009. Sea-level rise in New Jersey over the past 5000 years: implications to anthropogenic changes. *Global and Planetary Change*, **66**, 10–18.

MILLER, K. G., KOPP, R. E., HORTON, B. P., BROWNING, J. V. & KEMP, A. C. 2013. A geological perspective on sea-level rise and impacts along the U.S. mid-Atlantic coast. *Earth's Future*, **1**, 3–18.

MILLIMAN, J. D., JIEZAO, Z., ANCHUN, L. & EWING, J. I. 1990. Late Quaternary sedimentation on the outer and middle New Jersey continental shelf: results of local deglaciation. *Journal of Geology*, **98**, 966–976.

MONTEVERDE, D. H., MOUNTAIN, G. S. & MILLER, K. G. 2008. Early Miocene sequence development across the New Jersey margin. *Basin Research*, **20**, 249–267.

MOUNTAIN, G. S. & TUCHOLKE, B. E. 1985. Mesozoic and Cenozoic geology of the U.S. Atlantic continental slope and rise. *In*: POAG, C. W. (ed.) *Geologic Evolution of the United States Atlantic Margin*. Van Nostrand Reinhold, New York, 293–341.

MOUNTAIN, G. S., MILLER, K. G. & BLUM, P. 1994. *Proceedings of the Ocean Drilling Program, Initial reports*, **150**. Ocean Drilling Program, College Station, TX.

MOUNTAIN, G. S., BURGER, R. L. *ET AL.* 2007. The long-term stratigraphic record on continental margins. *In*: NITTROUER, C. A., AUSTIN, J. A., JR, FIELD, M. E., KRAVITZ, J. H., SYVITSKI, J. P. M. & WIBERG, P. L. (eds) *Continental-Margin Sedimentation: From Sediment Transport To Sequence Stratigraphy*. International Association of Sedimentologists, Special Publications, **37**, 381–458.

MOUNTAIN, G., PROUST, J.-N., MCINROY, D., COTTERILL, C. THE EXPEDITION 313 SCIENTISTS 2010. *Proceedings of the International Ocean Drilling Program, Expedition 313, Initial Report*. Ocean Drilling Program, College Station, TX.

NORDFJORD, S., GOFF, J. A., AUSTIN, J. A., JR. & GULICK, S. P. S. 2006. Seismic facies of incised valley-fills, New Jersey continental shelf:

implications for erosion and preservation processes acting during late Pleistocene/Holocene transgression. *Journal of Sedimentary Research*, **76**, 1284–1303.

NORDFJORD, S., GOFF, J. A., DUNCAN, L. S. & AUSTIN, J. A., JR. 2009. Shallow stratigraphy and transgressive ravinement on the New Jersey shelf: implications for sedimentary lobe deposition and latest Pleistocene-Holocene sea level history. *Marine Geology*, **266**, 232–243.

OLSSON, R. K. 1975. *Upper Cretaceous and Lower Tertiary Stratigraphy, New Jersey Coastal Plain. Second Annual Field Trip Guidebook*. Petroleum Exploration Society, New York.

OLSSON, R. K., MELILLO, A. J. & SCHREIBER, B. L. 1987. Miocene sea level events in the Maryland coastal plain and the offshore Baltimore Canyon trough. *In*: ROSS, C. & HAMAN, D. (eds) *Timing and Depositional History of Eustatic Sequences: Constraints on Seismic Stratigraphy*. Cushman Foundation for Foraminiferal Research, Special Publications, **24**, 85–97.

OWENS, J. P. & SOHL, N. F. 1969. Shelf and deltaic paleoenvironments in the Cretaceous-Tertiary formations of the New Jersey Coastal Plain. *In*: SUBITSKY, S. (ed.) *Geology of Selected Areas in New Jersey and Eastern Pennsylvania and Guidebook of Excursions*. Rutgers University Press, New Brunswick, NJ, 390–408.

PAZZAGLIA, F. J. & GARDNER, T. W. 1994. Late Cenozoic flexural deformation of the middle U.S. Atlantic passive margin. *Journal of Geophysical Research*, **99**, 12 143–12 157.

PELTIER, W. R. 1998. Postglacial variations in the level of the sea: implications for climate dynamics and solid-Earth geophysics. *Reviews of Geophysics*, **36**, 603–689.

PELTIER, W. R. & FAIRBANKS, R. G. 2006. Global glacial ice volume and last glacial maximum duration from an extended Barbados sea level record. *Quaternary Science Reviews*, **25**, 3322–3337.

PELTIER, W. R., SHENNAN, I., DRUMMOND, R. & HORTON, B. 2002. On the postglacial isostatic adjustment of the British Isles and the shallow viscoelastic structure of the Earth. *Geophysical Journal International*, **148**, 443–475.

POAG, C. W. 1985. Depositional history and stratigraphic reference section for central Baltimore Canyon trough. *In*: POAG, C. W. (ed.) *Geologic Evolution of the United States Atlantic Margin*. Van Nostrand Reinhold, New York, 217–263.

POAG, C. W. & SEVON, W. D. 1989. A record of Appalachian denudation in postrift Mesozoic and Cenozoic sedimentary deposits of the U.S. middle Atlantic margin. *Geomorphology*, **2**, 119–157.

POSAMENTIER, H. W., JERVEY, M. T. & VAIL, P. R. 1988. Eustatic controls on clastic deposition I – Conceptual framework. *In*: WILGUS, C. K., HASTINGS, B. S., KENDALL, C. G. St. C., POSAMENTIER, H. W., ROSS, C. A. & VAN WAGONER, J. C. (eds) *Sea-Level Changes: An Integrated Approach*. SEPM Special Publications, Tulsa, **42**, 109–124.

POTTER, E. & LAMBECK, K. 2003. Reconciliation of sea-level observations in the Western North Atlantic during the last glacial cycle. *Earth and Planetary Science Letters*, **217**, 171–181.

PRATSON, L. F., HUTTON, E. W. H., KETTNER, A. J., SYVITSKI, J. P. M., HILL, P. S., GEORGE, D. A. & MILLIGAN, T. G. 2007. The impact of floods and storms on the acoustic reflectivity of the inner continental shelf: a modeling assessment. *Continental Shelf Research*, **27**, 542–559.

PSUTY, N. P. 1986. Holocene sea level in New Jersey. *Physical Geography*, **7**, 156–167.

RAHMSTORF, S. 2007. A semi-empirical approach to projecting future sea-level rise. *Science*, **315**, 368–370.

RAMSEY, K. W. & BAXTER, S. J. 1996. *Radiocarbon Dates from Delaware: A Compilation*. Delaware Geological Survey, Report of Investigations, **54**.

REYNOLDS, D. J., STECKLER, M. S. & COAKLEY, B. J. 1991. The role of the sediment load in sequence stratigraphy: the influence of flexural isostasy and compaction. *Journal of Geophysical Research*, **96**, 6931–6949.

RINE, J. M., TILLMAN, R. W., CULVER, S. J. & SWIFT, D. J. P. 1991. Generation of Late Holocene ridges on the middle continental shelf of New Jersey, USA – evidence for formation in a mid-shelf setting based upon comparison with a nearshore ridge. *In*: SWIFT, D. J. P., OERTEL, G. F., TILLMAN, R. W. & THORNE, J. A. (eds) *Shelf Sand*

and Sandstone Bodies; Geometry, Facies and Sequence Stratigraphy. International Association of Sedimentologists, Special Publications, **14**, 395–426.

ROWLEY, D. B. 2013. Sea level: Earth's dominant elevation – Implications for duration and magnitudes of sea level variation. *Journal of Geology*, **121**, 445–454.

SCHOLLE, P. A. 1980. Geological studies of the COST No. B2 well, United States mid-Atlantic continental slope area. *United States Geological Circular*, **833**, 1–132.

SHACKLETON, N. J. & OPDYKE, N. D. 1973. Oxygen isotope and paleomagnetic stratigraphy of equatorial Pacific core V28–238: oxygen isotope temperatures and ice volumes on a 10^5 year and 10^6 year scale. *Quaternary Research*, **3**, 39–55.

SHACKLETON, N. J. & PISIAS, N. G. 1985. Atmospheric carbon dioxide, orbital forcing, and climate. *In*: SUNQUIST, E. T. & BROECKER, W. S. (eds) *The Carbon Cycle and Atmospheric CO_2: Natural Variations Archean to Present.* American Geophysical Union, Washington, DC, 303–317.

SHEPARD, F. P. 1934. Canyons off the New England coast. *American Journal of Science*, **27**, 24–36.

SHERIDAN, R. E., ASHLEY, G. M., MILLER, K. G., WALDNER, J. S., HALL, D. W. & UPTEGROVE, J. 2000. Onshore–offshore correlation of upper Pleistocene strata, New Jersey coastal plain to continental shelf and slope. *Sedimentary Geology*, **134**, 197–207.

SIDDALL, M., ROHLING, E. J., ALMOGI-LABIN, A., HEMLEBEN, C., MEISCHNER, D., SCHMELZER, I. & SMEED, D. A. 2003. Sea-level fluctuations during the last glacial cycle. *Nature*, **423**, 853–858.

STANFORD, S. D., ASHLEY, G. M. & BRENNER, G. J. 2001. Late Cenozoic fluvial stratigraphy of the New Jersey Piedmont: a record of glacioeustasy, planation, and incision on a low-relief passive margin. *Journal of Geology*, **109**, 265–276.

STECKLER, M. S., MOUNTAIN, G. S., MILLER, K. G. & CHRISTIE-BLICK, N. 1999. Reconstruction of Tertiary progradation and clinoform development on the New Jersey passive margin by 2-D backstripping. *Marine Geology*, **154**, 399–420.

STUBBLEFIELD, W. L., KERSEY, D. G. & McGRAIL, D. W. 1983. Development of middle continental shelf sand ridges: New Jersey. *American Association of Petroleum Geologists Bulletin*, **67**, 817–830.

STYLES, R. & GLENN, S. M. 2005. Long-term sediment mobilization at LEO-15. *Journal of Geophysical Research*, **110**, C04S90.

SUGARMAN, P. J., MILLER, K. G., OWENS, J. P. & FEIGENSON, M. D. 1993. Strontium-isotope and sequence stratigraphy of the Miocene Kirkwood Formation, southern New Jersey. *Geological Society of America Bulletin*, **105**, 423–436.

SUGARMAN, P. J., MILLER, K. G., BUKRY, D. & FEIGENSON, M. D. 1995. Uppermost Campanian–Maestrichtian strontium isotopic, biostratigraphic, and sequence stratigraphic framework of the New Jersey Coastal Plain. *Geological Society of America Bulletin*, **107**, 19–37.

SWIFT, D. J. P. 1969. Inner shelf sedimentation: process and products. *In*: STANLEY, D. J. (ed.) *The New Concepts of Continental Margin Sedimentation: Application to the Geological Record.* American Geological Institute, Washington, DC, DS-5-1–DS-5-26.

SWIFT, D. J. P., STANLEY, D. J. & CURRAY, J. C. 1971. Relict sediments, a reconsideration. *Journal of Geology*, **79**, 322–346.

SWIFT, D. J. P., DUANE, D. B. & McKINNEY, T. F. 1973. Ridge and swale topography of the middle Atlantic Bight, North America: secular response to the Holocene hydraulic regime. *Marine Geology*, **15**, 227–247.

TALWANI, M., EWING, J., SHERIDAN, R. E., HOLBROOK, W. S. & GLOVER, L., III . 1995. The EDGE experiment and the U.S. East Coast Magnetic Anomaly. *In*: BANDA, E., TALWANI, M. & TORNE, M. (eds) *Rifted Ocean–Continent Boundary, Volcanic Margin Concepts.* NATO ASI Series, **5E**, 1–26.

UPTEGROVE, J. 2003. *Late Pleistocene facies distribution, sea level changes and paleogeography of the inner continental shelf off Long Beach Island, New Jersey.* MS thesis, Rutgers University, USA.

VAIL, P. R., MITCHUM, R. M., JR & THOMPSON, S., III. 1977. Seismic stratigraphy and global changes of sea level, part 4: global cycles of relative changes of sea level. *In*: PAYTON, C. E. (ed.) *Seismic Stratigraphy – Applications to Hydrocarbon Exploration.* American Association of Petroleum Geologists, Memoirs, **26**, 83–9.

VERMEER, M. & RAHMSTORF, S. 2009. Global sea level linked to global temperature. *Proceedings of the National Academy of Science of the United States of America*, **106**, 21 527–21 532.

WATTS, A. B. & STECKLER, M. S. 1979. Subsidence and eustasy at the continental margin of eastern North America. *In*: TALWANI, M., HAY, W. W. & RYAN, W. B. F. (eds) *Deep Drilling Results in the Atlantic Ocean: Continental Margins and Paleoenvironments.* American Geophysical Union, Maurice Ewing Symposium, **3**, 218–234.

WITHJACK, M. O., SCHLISCHE, R. W. & OLSEN, P. E. 1998. Diachronous rifting, drifting, and inversion on the passive margin of central eastern North America: an analog for other passive margins. *American Association of Petroleum Geologists Bulletin*, **82**, 817–835.

WRIGHT, J. D., SHERIDAN, R. E., MILLER, K. G., UPTEGROVE, J., CRAMER, B. S. & BROWNING, J. V. 2009. Late Pleistocene sea level on the New Jersey Margin: implications to eustasy and deep-sea temperature. *Global and Planetary Change*, **66**, 93–99.

Chapter 4

The north and northeast Brazilian tropical shelves

HELENICE VITAL

Environment Monitoring, Marine Geology and Geophysics Laboratory, Department of Geology, Post-Graduation in Geodynamic and Geophysics, Federal University of Rio Grande do Norte, 59072–970 Campus Universitário, PO Box 1596, Natal, RN 59072-970, Brazil (e-mail: helenice@geologia.ufrn.br)

Abstract: The Brazilian tropical north shelf (BT N shelf) and the Brazilian tropical northeast shelf (BT NE shelf) along the Atlantic Ocean display unique conditions for tropical passive margins. Together they encompass approximately 3000 km in length, extending from Cape Orange in the north to Abrolhos Bank in the south. Both the north and NE shelves are very shallow and highly energetic systems. The first one is subject to energetic forcing from a number of different sources, including near-resonant semi-diurnal tides, large buoyancy flux from the Amazon River discharge, wind stress from the northeasterly trade winds and strong along-shelf flow associated with the North Brazil Current. The second one is subject to the full strength of the westerly-flowing South Equatorial Current, combined with high winds, moderate–high tidal range and/or waves. The BT N shelf is the largest shelf in Brazil, and is mostly covered with siliciclastic mud and sands because of the enormous water and sediment discharge from the Amazon River. In contrast, the BT NE shelf is narrow and open, and almost entirely covered by carbonate sediments due to the small amount of freshwater and sediment input.

This chapter presents a review of the current state of knowledge of the Brazilian tropical shelf encompassing its physiography and spatial distribution of sediments and bedforms in order to relate them to the sedimentological processes and physical characteristics of the environment, and to the sedimentary history of this shelf system.

The Brazilian tropical shelf, which is classified as a passive or Atlantic type, can be followed over a distance of about 3000 km from Cape Orange in northernmost Amapa to the Abrolhos Bank at the southernmost limit, extending approximately from latitudes 5°N to 22°S (Fig. 4.1). This tropical shelf varies considerably in shape and width. It is wider near the mouth of the Amazon and narrower near Salvador. Along most of its length the Brazilian tropical shelf is very narrow, with an average width of 50 km. However, it widens in its northern and southern portions, particularly in the Amazon and in the Abrolhos areas, where it extends to around 300 and 200 km, respectively, as a result of sediment supply and of volcanic intrusive activity, respectively. The shelf break is commonly at an average depth of 80 m.

The sedimentation on this shelf ranges from dominantly siliciclastic in the north to carbonate sediments in the NE. Located on the eastern margin of the South American plate, the Brazilian shelf evolved from Mesozoic times onwards, linked with the development of the Atlantic Ocean. Individual sectors along this continuous divergent margin of South America share some fundamental characteristics, including the classic rift, transitional and open marine tectonic and sedimentological patterns (Asmus 1981).

Considering the nature and orientation of the regional stress fields during rifting and the following dynamics of the divergent motion between the African and South American plates during drifting, three diverse domains are recognized along this shelf by Milani & Thomaz Filho (2000). These are, first, the dominantly extensional domains from Abrolhos to the Touros Platform, in the NE corner of Brazil, where the structural pattern of the rift phase was marked by dominantly dip-slip normal faults. Secondly, a transform segment corresponding to the margin along the Equatorial Atlantic, where right-lateral wrenching was the mechanism responsible for crustal rupture, creating a pattern of high-angle oblique faults that controlled rifting and resulted in the development of large-scale parallel-to-shore fracture zones (Gorini 1977). These fracture zones include those of Fernando de Noronha, Chain, Romanche and Saint Paul; and, thirdly, the region to the north of the mouth of the Amazon River. This constitutes a part of another extensional domain. This is the older segment

of the extensional margin of the plate, in which an early phase of rifting occurred in the Triassic. Some regions of the Equatorial Atlantic margin experienced almost continuous subsidence, resembling that of the purely extensional eastern Brazilian margin, whereas other sectors underwent strong episodes of inversion and erosion that created large lacunas in their stratigraphic record.

Research on the Brazilian tropical shelf has its roots in the 1960s and 1970s (e.g. Milliman *et al.* 1975; Summerhayes *et al.* 1975). The Federal University of Pernambuco conducted the pioneer investigations on the NE shelf mainly dealing with the character of the sediment. Brazilian Federal agencies, companies and universities supported the REMAC Project (Global Reconnaissance of the Brazilian Continental Margin) and GEOMAR (Marine Geology surveys) cruises, which included geological and geophysical examination of the Brazilian tropical shelf. Collectively, these research projects provide the initial description of the Brazilian shelf (e.g. Amaral 1979; Kowsmann & Costa 1979; Asmus 1981). Following this, new seismic and sedimentological data were incorporated by the LEPLAC project (Brazilian Continental Shelf Survey) from 1987 to 1996 and, most recently, the REMPLAC (Mineral Reconnaissance of the Brazilian Continental Margin) initiative, which began in 1997. Bilateral co-operation, such as the Brazilian German Joint Oceanographic Projects (JOPS) that were initiated in the 1990s, has also contributed to the knowledge of the tropical Brazilian shelf (e.g. Araújo 1994; Knoppers *et al.* 1999).

Identification of the complexity of oceanographic processes in coastal environments near large rivers and the recognition of the interrelationship of processes led to the interdisciplinary approach of AmasSeds (A Multidisciplinary Amazon Shelf Sediment Study) from 1989 to 1996, which was co-ordinated by academic institutions in the United States and Brazil. Results of these studies have been published for individual fields of oceanography (e.g. Beardsley *et al.* 1995; Figueiredo & Nittrouer 1995; Geyer & Kineke 1995; Nittrouer *et al.* 1996; Nittrouer & DeMaster 1996) resulting in the northern Brazilian shelf being one of the better studied maritime areas in Brazil.

However, much of the Brazilian margin has not yet been surveyed, including a great portion of the NE tropical shelf, where the rivers are short and do not contribute significant amounts of sediment. This area is characterized by extreme oligotrophic boundary currents and sedimentation of biogenic carbonates dominating great parts of the shelf. Most of the recent investigations on

From: CHIOCCI, F. L. & CHIVAS, A. R. (eds) 2014. *Continental Shelves of the World: Their Evolution During the Last Glacio-Eustatic Cycle.*
Geological Society, London, Memoirs, **41**, 35–46. http://dx.doi.org/10.1144/M41.4

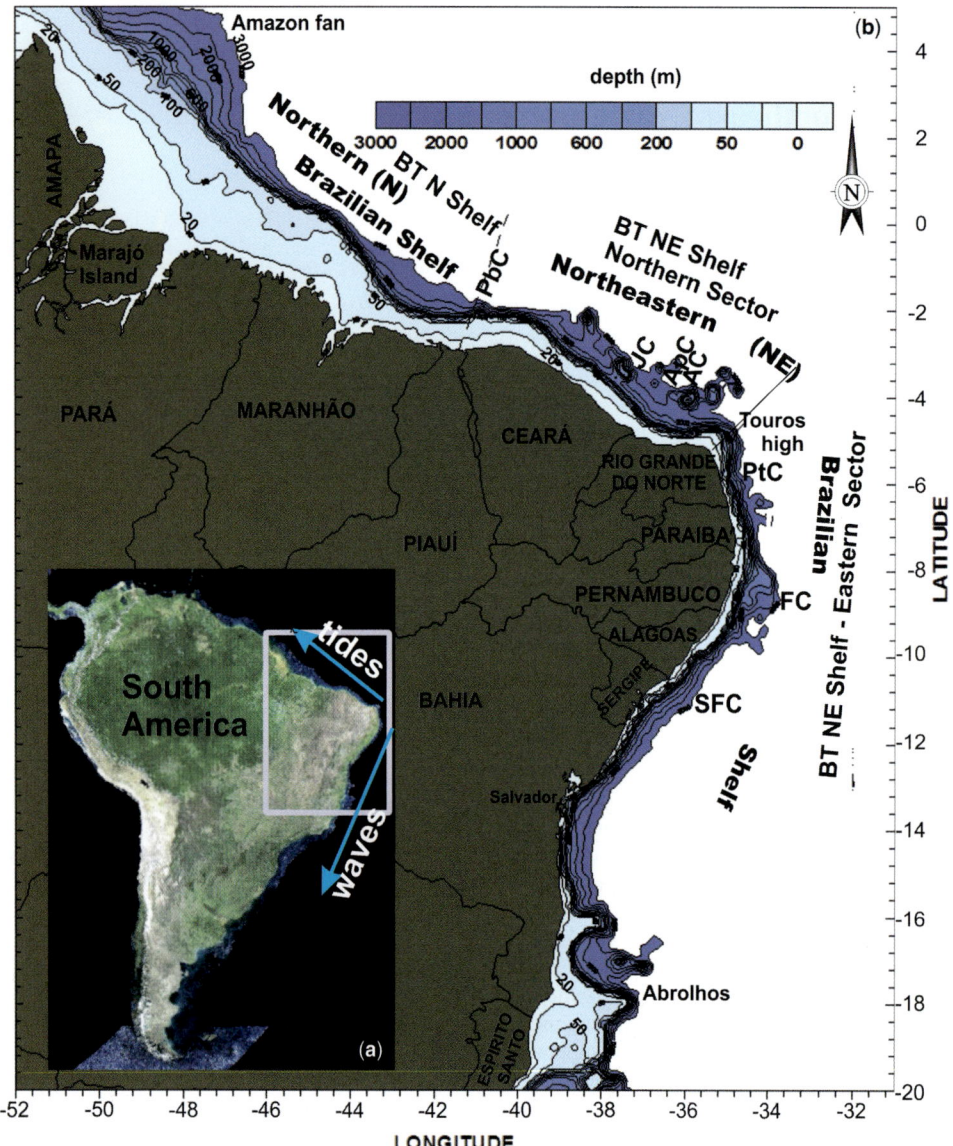

Fig 4.1. (**a**) Location map and (**b**) bathymetric map of the Brazilian tropical shelf. Canyon heads on the shelf are identified by their abbreviated names: SFC, Sao Francisco Canyon; FC, Formoso Canyon; PtC, Potengi Canyon; AC, Açu Canyon; ApC, Apodi Canyon; JC, Jaguaribe Canyon; PbC, Parnaiba Canyon; AmC, Amazon Canyon. Tidal heights increase to the north and wave heights to the south.

this shelf were based on sediments taken from the surficial seabed (e.g. Testa & Bosence 1998; Vital *et al.* 2005) or for a specific subject such as coral reefs (e.g. Leão *et al.* 2003). The existing literature contains only little information on shelf processes (e.g. Testa & Bosence 1999; Tabosa 2006) or high-resolution seismic-stratigraphy (Schwarzer *et al.* 2006; Vital *et al.* 2008).

In this chapter, a short review of the Brazilian tropical shelf is presented and, for better comprehension, the Brazilian tropical shelf is divided into two parts: the Brazilian tropical north shelf (BT N shelf), from Cape Orange to the Parnaiba Delta; and the Brazilian tropical northeast shelf (BT NE shelf) from the Parnaiba Delta to Abrolhos Bank.

Tectonic framework and geological background

As already outlined by Dominguez (2009), the general framework of the coastal zone and shelf in tropical Brazil has a strong influence on the geological heritage that goes back to the Early Proterozoic–Archaean. The north Brazilian geology is dominated by two major Palaeozoic–Mesozoic intracontinental sedimentary basins (the Amazon and Parnaiba basins), and in minor extension by the Amazon Craton (Brazilian Shield) (Fig. 4.2). The hydrography that developed in association with these intracratonic basins drains more that half of the Brazilian territory including the largest river in the world, the Amazon, and is responsible for massive influxes of sediments to the coastal zone and to the BT N shelf. In contrast, the NE Brazilian is dominated by high-grade metamorphic rocks of the Brazilian Shield. Within this shield there is one important cratonic area (the São Francisco Craton) and the Borborema Province (Fig. 4.2). Almeida *et al.* (1977, p. 382) defined this province as a 'complex mosaic-like folded region' where important tectonic, thermal and magmatic events took place during the Neoproterozoic interval assigned as the Brasiliano Cycle. Failed rifts or the rifted portion of Mesozoic sedimentary basins formed during the South America–Africa break-up also outcrop along the coastal zone (e.g. the Reconcavo, Camumu and Almada basins in the state of Bahia, and the Potiguar basin in the Rio Grande do Norte state) (Fig. 4.2). The association between the small size of the drainage basins, the low intrabasinal relief and the low precipitation values from the Alagoas to Ceará states results in the small volume of sediments from the hinterland to most of the BT NE shelf (the so-called sediment-starved coast: Dominguez 2009); from Alagoas to the extreme south of the BT NE shelf, large drainage basins with high intrabasin relief have resulted in large sediment yields for the major rivers emptying on this section, resulting in classical examples of wave-dominated deltas (Dominguez 2009).

Marginal basins associated with the Atlantic break-up border the entire Brazilian tropical shelf. These basins and the basement rocks

Fig. 4.2. Simplified geology of tropical Brazil.

have been covered by Tertiary–Quaternary sandstones, mudstones and conglomerates, notably by the Barreiras Formation (Miocene–Pliocene) which is present along almost the entire coast of the Brazilian tropical shelf. This formation has traditionally been considered to be the result of deposition in alluvial systems but in the last few years some portions of it, at least, have been interpreted as marine (Rossetti 2006). According to these recent interpretations, most of the Barreiras Formation is the result of a coastal onlap associated with mud and early Miocene high sea levels (Arai 2006).

During the Quaternary, changes in relative sea level and climate have added the younger morphological elements of the Brazilian coastal zone, including strandplains (prograded barriers) associated with wave-dominated deltas, beach rocks, transgressive retrograded barriers and transgressive dunefields along the starved sector, and mangroves to the north (Dillenburg & Hesp 2009; Dominguez *et al.* 2009; Hesp *et al.* 2009; Souza Filho *et al.* 2009; Vital 2009).

Sea-level changes

High sea levels prior to 120 ka BP have been identified and dated by Barreto *et al.* (2002) using thermoluminesce (206–220 ka) in coastal deposits in NE Brazil, which they correlated to Oxygen Isotope Stage (OIS) 7c.

Records of the high sea level of 120 ka BP are preserved as terraces of essentially sand composition (Dominguez 2009). A coral

reef beneath these terraces, in the region of Olivença (14°S), provide average values of 123.5 ± 5.7 ka BP (Bernat *et al.* 1983); in the coast of Rio Grande do Norte state this high sea level was registered by the U–Th method in beachrocks (at least 120 ka: Caldas 2002) and by luminescence dating of sands deposits (117–110 ka: Barreto *et al.* 2002), which these authors correlated to OIS 5e.

The last 7 kyr are well documented for the NE Brazilian coast (e.g. Bezerra *et al.* 2003; Martin *et al.* 2003; Caldas *et al.* 2006) and the relative sea levels for this area indicate that, in general, it has been subjected to an incipient forced regression, during the last 5.6 cal ka BP, related to the 2–5 m drop in relative sea level.

Records of higher than present sea levels are apparently absent from the northern coast of Brazil, from Ceará state to Amapá (Cohen *et al.* 2005; Souza Filho *et al.* 2006).

The Brazilian tropical north (BT N) shelf

The BT N shelf is one of the most interesting sedimentary environments of the world. It is a complex environment influenced by oceanographic processes operating in the western equatorial Atlantic Ocean, and by terrestrial processes in the Amazon, the largest river in the world (drainage basin 6.1×10^6 km^2; discharge 207.7×10^3 m^3 s^{-1}; sediment load 1.154 Mt a^{-1}: Syvitski *et al.* 2005) and the Parnaiba drainage basin. These processes affect sedimentation on the BT N shelf by controlling sediment supply and dispersal. In addition, this shelf is located at the equator; in the wet tropics, a geographical region defined by constantly high precipitation (>60 mm month^{-1}, >1500 mm a^{-1}) and temperature ($>20\,^{\circ}$C).

Holocene marine and associated sediments are found on the Amapa coastal plain. Landwards of the Holocene sediments are older coastal-plain deposits and metamorphic rocks of the Precambrian Guiana Shield, which form a hilly terrain (<500 m in elevation) and represent the source area for a number of coastal rivers. In addition to the rivers, numerous lakes (especially near Cabo Norte) are present on the coastal plain, commonly forming in abandoned channels of rivers (e.g. ox-bow lakes). The Amapa shoreline is dominated by muddy sediments. However, the southern shoreline (Cabo Norte–Cabo Cassipore) is now characterized by erosion, and has been for about the past 500 years (Allison *et al.* 1995). This is demonstrated by the destruction of jungle vegetation at the shoreline and the presence of overconsolidated sediment to about the 5 m isobaths. The only exceptions to erosion in the south are small sand bodies forming downdrift (northwards) of the coastal rivers, for which mineralogical examination indicates local sources (i.e. not the Amazon River). In contrast, the northern portion of the Amapa (Cabo Cassipore and Cabo Orange) is characterized by mud flats that have been accreting rapidly for at least 1 kyr and by Avicennia mangrove colonizing the shoreline. ^{14}C age dates indicate that the deposits presently eroding in the south were formed in similar environments about 500–900 and 2400–2900 years BP. The fluctuations are demonstrated also by chenier ridges that are observed on the coastal plain and represent periods of shoreline erosion (Allison *et al.* 1995).

However, the Para and Maranhão shorelines are characterized by a large number of small estuaries bordered by low cliffs, now in full retreat. Mangrove swamps occur at some protected places, with 7600 km^2 of continuous mangrove forests (Souza Filho 2005), helping to accentuate the irregularity of the coastline (Souza Filho *et al.* 2009). The sandy nature of the coast east of the Sao Marco Bay (Maranhao) favours the formation of sand dunes and beaches, resulting in a rather smooth coastline.

Total subsidence rates for the Amazon margin are 15–20 cm per 10 years. Local tectonics appear to be expressed on the coastal plain by abrupt diversions in the paths of local rivers. A potential driving mechanism is flexure of the Earth's crust due to the weight of the Amazon fan (Driscoll & Karner 1994). According to this model, the Amapa coastal plain and Amazon shelf are subsiding, while the Guiana Shield is being uplifted as a flexural bulge. However, timescales for the subsidence are long, isopachs are shore parallel and the process is non-reversing. Although this tectonic motion would influence accommodation space (vertical prism available to be filled with sediment), it would not explain the temporal and spatial variability of the Amazon shoreline and shelf sedimentation.

Physiography

The BT N shelf is the widest shelf in Brazil, varying between 330 km off the Amazon River and 100 km near the Parnaiba River mouth (Fig. 4.1). Much of the shelf is very shallow; the 20 m isobath extends as far as 200 km from the river mouth. The inner shelf south of the Amazon River has a gradient of $>1:1.000$, and is covered by an abundance of symmetrical and asymmetrical sand waves, 3–8 m in amplitude, with average wavelengths of between 100 and 500 m. In contrast, the inner shelf adjacent to and north of the Amazon River is flat, covered with a mud wedge, the average gradient being $<1:4.000$ (Milliman *et al.* 1975; Nittrouer *et al.* 1996).

The muds of the foreset beds are prograding over a sandy transgressive layer on the outer shelf, which is characterized by numerous channels that cut the shelf break; 3.5 kHz seismic signals identify that the underlying surface is erosional and marked

by infilled channels. This surface is interpreted as an old subaerial feature exposed during the most recent lowstand of sea level. The channels were cut by meandering rivers, much as channels are now being cut on the Amapa coastal plain (Nittrouer *et al.* 1996).

The depth of the shelf break ranges from 120 m off the Amazon to 80 m off the Parnaiba Delta. The most impressive feature of the north Brazilian tropical margin is the broad submarine Amazon fan that encompasses both slope and rise (Fig. 4.3). It was formed during the Miocene in response to the Andean orogeny, which uplifted the western edge of the South American plate and forced rivers to flow through the Amazon Basin eastwards toward the Atlantic Ocean (Damuth & Flood 1984).

The Amazon fan is a large modern deep-sea fan that extends from the north Brazilian continental shelf to water depths of about 4.700 m. The most prominent feature of the fan is the Amazon submarine canyon, which is 400 m deep and 9 km wide on the outer shelf, and runs in a NW direction (Fig. 4.3b). It extends at least from the 50 m isobath on the shelf to a depth of 1500 m on the fan (Fig. 4.3). This canyon shows a down-cut V-shaped morphology and an erosional truncation of reflectors by the steep canyon walls (Araújo 1994). Large incised valleys, not completely infilled, are associated with the Maranhão Gulf and Parnaiba River, and have a clear expression in the continental shelf bathymetry.

Oceanographic processes

As the Amazon water and sediment loads enter the ocean, they are influenced by processes operating there. An important consideration regarding their fate is salinity stratification on the adjacent shelf, which is largely controlled by tidal mixing. Stratification is well established during neap tidal conditions and poorly developed during spring tides (Geyer & Kineke 1995). Beardsley *et al.* (1995) documented the energetic character of tides on the north shelf, and the dominance of semi-diurnal tidal components (M_2, S_2, N_2). Tides are the dominant physical forcing mechanism on the shelf, with a maximum range of approximately 6 m near Cabo Norte. Throughout the tropical north shelf, tidal currents flow perpendicularly to the isobaths during ebb and flood, with velocities that range from 70 to nearly 200 cm s^{-1} between neap and spring tides. There is a persistent gradient in suspended sediment concentrations. The near-bottom sediment concentrations range from 50 to 200 g l^{-1}. These high concentrations are termed 'fluid mud' (Kineke & Sternberg 1995). The turbid plume on the Amazon shelf is swept NW by the North Brazil Current (NBC) that flows at speeds of 40–80 cm s^{-1} over the shelf. This current generates shear stresses sufficient to disperse much sediment. However, the presence of these fluid mud layers enhances tidal amplitude by decreasing bottom friction (Beardsley *et al.* 1995; Gabioux *et al.* 2005). The trade winds produce a moderate wave climate and significant wave heights. Significant wave heights as great as 3 m (13 s period) have been measured for the outer shelf during February, the period of maximum wind stress (Sternberg *et al.* 1996). The north shelf is protected from the main wave train during the year, being affected only by a secondary train (waves from the east) from September to May.

Sedimentology

The Amazon continental shelf is actively accumulating sediment, about a half a billion tons annually (Kuehl *et al.* 1986), and thereby forming strata. Most sediment are carried in suspension, and derived in part from the mountainous Andean regions (about 90%) and in part from weathering within the Amazon Basin. The Precambrian shields, as well as their sedimentary covers, contribute mainly bedload sands (Vital *et al.* 1999).

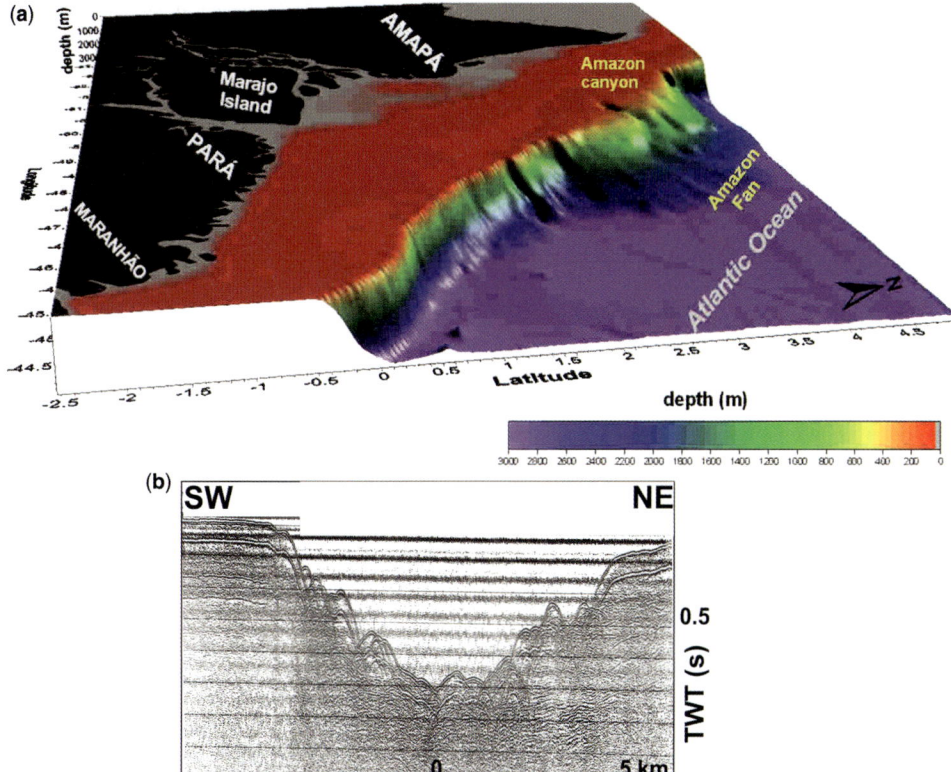

Fig. 4.3. (**a**) 3D view of the Brazilian tropical north (BT N) shelf showing the Amazon submarine canyon and fan; and (**b**) part of an airgun profile showing the Amazon Canyon.

Most of the inner and middle portion of the Amazon shelf is undergoing accumulation of mud as the subaqueous delta of the Amazon (Nittrouer *et al.* 1986; Kuehl *et al.* 1996). The accumulation terminates nearshore (water depth of 15 m) and on the outer shelf (−70 m) where shear stresses are high and sediment input is low, respectively. Between these depths, a clinoform deposit is developing, with its highest accumulation rates in water depths of 40–60 m. Modern sediment accumulation is relatively continuous along the Amazon shelf, but with partial interruption seawards of Cabo Norte and complete elimination (i.e. relict sediment exposed) across the northern boundary at −3.5″ to −4″N (Fig. 4.4). This latter region covered a larger area in the recent past, as demonstrated by the southward extension of the erosional surface as a seismic discontinuity beneath muddy deposits (Alexander *et al.* 1986; Nittrouer *et al.* 1996), but in fact modern sediment accumulation is limited to an offshore (water depth of >30 m) depocentre, which is the distal expression of the Amazon subaqueous delta foreset region, and a shoreface-based depocentre (Allison *et al.* 2000).

The deeper and older record of Amazon shelf sedimentation can be observed best in the area of the outer shelf. Here not only was the seaward progradation of clinoforms observed but also the presence of gas in the landward direction (Figueiredo & Nittrouer 1995). This gas is biogenic methane (Aller *et al.* 1996).

The muds of the foreset beds are seen to be prograding over a sandy transgressive layer (Figueiredo & Nittrouer 1995; Nittrouer *et al.* 1996). A boring obtained near the shelf break at 4″N, recovered about 8 m of sand overlying silty clay. The underlying surface is erosional and marked by infilled channels. This surface is interpreted as an old subaerial feature exposed during the most recent lowstand of sea level (Nittrouer *et al.* 1996).

The seabed is characterized by intensely reworked mud on the inner shelf with low bacterial densities, and a lack of macrofauna and meiofauna landwards of about the 15 m isobath (Kuehl *et al.* 1986, 1996).

Sediments derived from the Andes Mountains are immature, mostly deposited on the shelf north of the Amazon River, and are characterized by angular subarkosic sand, high concentrations of unstable heavy minerals and clay rich in illite. Sediments from the shelf south of the Amazon River contain subrounded–rounded orthoquartzitic sand, low concentrations of unstable heavy minerals and clay rich in kaolinite. The mature nature of these sediments reflects derivation from rivers such as the Para, which drain the low-lying rain forests of the Amazon Basin and adjacent coastal areas (Milliman 1975).

Less important are residual sediments eroded from local outcrops and sediments rich in calcium carbonate. The carbonate sediments are restricted to the banks seawards of the 70 m isobath, in the outer shelf. To the NW, carbonate banks are found near the 60 m isobath, followed by sandwaves (Fig. 4.4). These sands are of fine grain size, rich in shell fragments.

Stratigraphy

The Amapa shoreline and the Amazon subaqueous delta presently form a highstand systems tract. In a system this dynamic, several scales of erosion operate and they all tend to remove the uppermost stratigraphic units. The subaqueous delta of the Amazon River is a clinoform structure (potentially *c.* 70 m thick) with rapid accumulation in the foreset region. On various timescales, subsequent periods of erosion by energetic oceanic processes can remove the uppermost portions of the subaqueous delta (Nittrouer *et al.* 1996).

Sea-level changes expose the subaqueous delta to intense erosion, which can remove all but about 20 m of the lowermost strata (lower foreset and bottomset). A transgressive sand layer is developed above the remaining mud.

Clinoform deposits are common features in which large amounts of sediment accumulate on continental margins, and they can be created with diverse geometric characteristics. Accretion on the Amazon margin consists of a two-stage clinoform progradation – the subaqueous delta followed by the shoreline deposits.

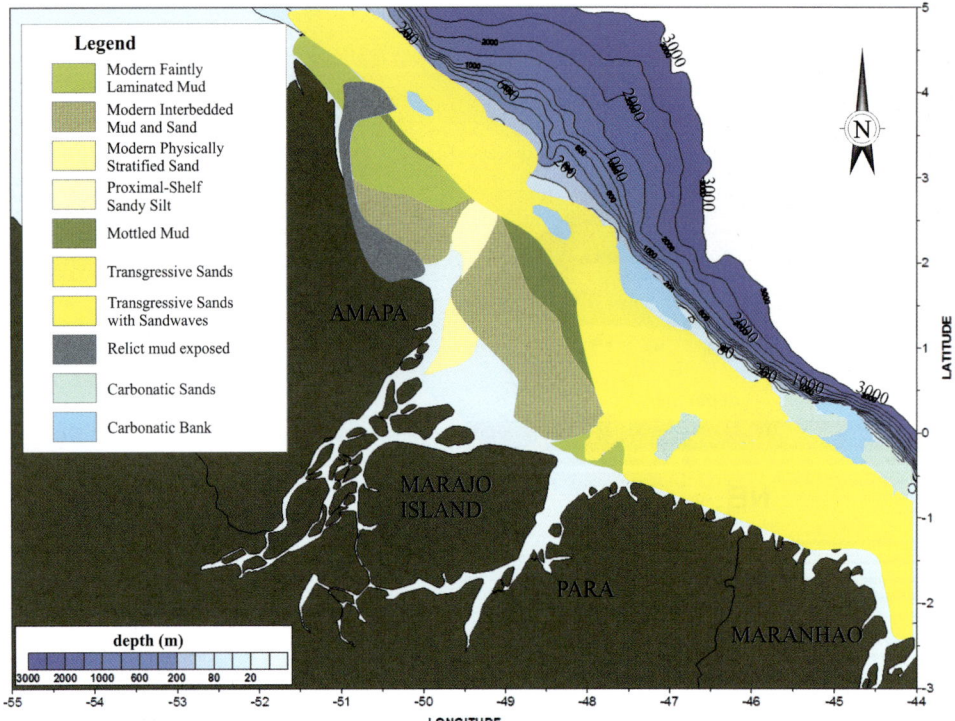

Fig. 4.4. Sediment distribution on the Brazilian tropical north (BT N) shelf (data from Projeto REMAC 1979; Kuehl *et al.* 1996; and unpublished data from JOPS-I).

The nature of sedimentation on the Amazon fan has changed during periods of low and high sea levels. The channel–levee systems of the fan developed during glacio-eustatic sea-level low-stands. During sea-level highstands, pelagic sediments dominate the sedimentation pattern. Analysis of piston cores collected near the most recently active channel–levee system revealed that all cores are capped by a surface hemipelagic layer, suggesting that the Amazon fan is inactive at the present sea-level high (Flood *et al.* 1991).

The Brazilian tropical northeast (BT NE) shelf

The BT NE shelf is situated in the NE part of Brazil along the Atlantic Ocean and comprises two different sectors: (1) the eastern sector, extending from the Abrolhos Bank to the Touros high; and (2) the northern sector, extending from the Touros high to the Parnaiba Delta (Fig. 4.1). The climate varies from tropical dry semi-arid on the northern coast to tropical humid on the eastern coast. From a morphodynamic point of view, the eastern sector is a wave-dominated coast with active sea cliffs carved into tablelands alternating with reef- or dune-barrier sections, and beach-ridge terraces, while the northern sector is a mixed-energy complex of wave-dominated and tide-dominated coast. Dunes, ebb tidal deltas, beach rocks, barrier islands and spits are present along the northern sector.

The BT NE shelf represents a modern, highly dynamic mixed carbonate–siliciclastic shelf system characterized by reduced width and shallow depths when compared with other parts of the Brazilian shelf. The shelf width averages 40 km, and its shelf break is commonly at an average depth of 60 m. The shelf passes seawards into a steep slope with a gradient up to 1:11, and reaches a maximum of 28° at the Abrolhos Bank.

This open ocean-facing shelf experiences high-energy, shore-line and shelf-margin parallel currents driven by a combination of oceanic, trade winds, and tidal and wave processes. The processes operating on the shelf also play an important role in controlling the shelf morphology and sediment distribution (Testa & Bosence 1999; Vital *et al.* 2008). The sediments of NE Brazil are unique in terms of their almost total lack of coral, and

a complete lack of ooids or other precipitated carbonate (Summer-hayes *et al.* 1975).

The BT NE shelf is located in the eastern part of the NE region of the South American Platform, encompassing the Borborema Province (Almeida *et al.* 1977) and the São Francisco Craton. The Borborema Province was submitted to regional uplift during the Phanerozoic, causing the remobilization of large amounts of continental sediments to the margin. A minimum of two more epeiro-genetic uplifts during the Cenozoic has been reported in this province. The last one was most probably associated with the intense volcanic activity on the continent, initiated about 20 Ma, and was responsible for uplift and posterior erosion of the exposed Phanerozoic covers.

Physiography

The BT NE shelf consists of a very narrow, and lower gradient, slope averaging 0.2° and 0.5° at the northern and eastern sectors, respectively, and presents an average width of 40 km along most of its extension. Most of the eastern sector is 20 km wide (Figs 4.1 & 4.5), reaching its narrowest point (5–8 km) adjacent to Salvador and Ilheus (Dominguez *et al.* 2009) (Fig. 4.1), and widening in the southern portion, particularly in the Abrolhos area, where it extends to approximately 200 km (Fig. 4.1) as a result of volcanic intrusive activity – an accretion to the shelf that was responsible for the formation of the Abrolhos Bank. The continental shelf edge runs parallel–subparallel to the coast. The shelf break around the Touros high (Fig. 4.5) and Abrolhos Bank starts at a depth of about 70–80 m, and decreases to a depth of 40–50 m westwards from the Touros high, and down to a depth of 50–60 m between the Touros high and Abrolhos Bank.

The inner, middle and outer shelf are commonly separated by their water depth as follows: inner shelf up to 15 m water depth; middle shelf between 15–20 and 25–40 m depth; and outer shelf from 25–40 m depth until the shelf break (Coutinho 1976; Vital *et al.* 2008; Gomes & Vital 2010).

A variety of bedforms ranging from tens of centimetres to the kilometre scale are present in the shallower parts of the shelf, indicating sediment-reworking processes. Because of the absence of large rivers discharging suspended sediments in

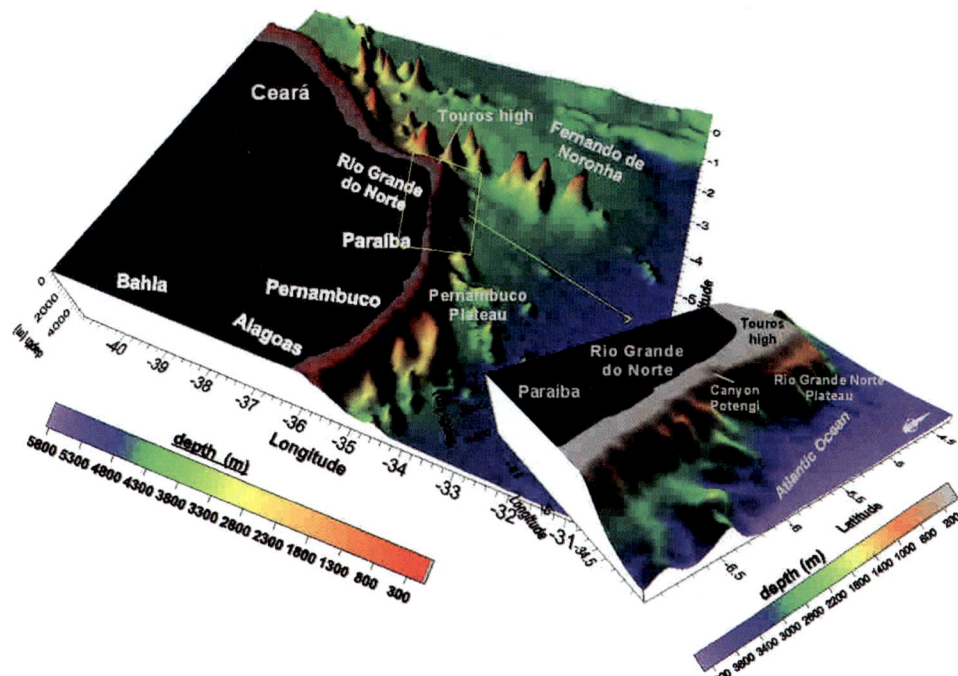

Fig. 4.5. 3D view of the Brazilian tropical northeast (BT NE) shelf showing the Touros high, where it splits into the eastern and northern sectors, the Fernando de Noronha Chain and the Pernambuco Plateau. The inset details the Rio Grande do Norte Plateau and Potengi Canyon (modified from Vital *et al.* 2010).

the shelf, distinctive seabed features on the inner shelf are well observed by Landsat imagery (e.g. Vianna *et al.* 1991; Tabosa 2006; Vital *et al.* 2005, 2008), such as: (1) very large longitudinal dunes (up to 6 m high, from 400 m to more than 900 m wide and kilometre scale in extension); (2) very large transverse dunes with crests orientated NE–SW (oblique to transversal to the coast); (3) small wave- and current-generated dunes generally superimposed on the previously described very large dunes; (4) isolated shallow-marine sand bodies (over 50 km in length); and (5) incised-valley systems (Fig. 4.6). The most remarkable features are, however, submerged beach-rock chains occurring parallel to the coast at different depths (at -10, -20–25, -30 and -40 m: Araújo *et al.* 2004; Santos *et al.* 2007; Vital *et al.* 2008, 2010).

Besides the above-described features, the most important traces of the continental terrace are the plateaux and marginal terraces off the states of Ceará, Rio Grande do Norte and Pernambuco (Fig. 4.5), as well as a series of submarine seamounts and, within this relief, the canyons of Salvador, Japaratuba, São Francisco, Formoso, Ipojuca, Potengi (Fig. 4.5), Açu, Apodi, Jaguaribe and Parnaiba.

Oceanographic processes

The BT NE shelf of Brazil is located within the trade wind belt, and experiences high-energy, coastal and shelf-parallel currents driven by combined flows due to oceanic, tidal and wave processes. The trade winds originate from the ESE (*sensu stricto* trades) in the eastern sector, and from the NE direction (return trades) on the northern sector. Since strong winds are present almost year round, water masses are well mixed without any characteristic stratification. The wave pattern is conditioned by variations in the trade winds (i.e. related to offshore high-pressure centre variations). This tropical shelf is mainly dominated by relatively high sea waves, with heights above 100 cm accounting for more than 50% (Meserve 1974).

Waves measured during the summer period on the northern sector, however, have an average height of 56 cm, with maximum heights of approximately 120 cm and minimum heights of 27 cm (Vital *et al.* 2008; Vital 2009). Waves measured during the summer period on the shelf near the Potengi River on the eastern sector have an average height of around 90 cm, with similar 120 cm maximum heights and minimum heights of

50 cm (Vital *et al.* 2008; Vital 2009). Wave climate anomalies exist, such as, for example, the occasional occurrence of swell in August–October associated with storms and hurricanes generated in the Caribbean region (Hesp *et al.* 2009).

The North Brazil Current flows relatively parallel to the coast over the narrow shelf. Current velocities reach 30–40 cm s^{-1}, overlain by tidal and wave components (Knoppers *et al.* 1999). Longshore currents in this shelf flow mainly northwards and westwards, with velocities of 20 and 105 cm s^{-1} on the eastern and northern sectors, respectively. Currents decrease their velocities to a maximum of 25 cm s^{-1} and an average of 6 cm s^{-1} close to the Touros high (Hazin *et al.* 2008)

The BT NE shelf has a mesotidal, semi-diurnal regime with maximal heights of 2.7 m and 2.0 m for spring and neap tides, respectively, in the eastern sector. Tidal measurements show that average ranges of spring and neap tides are 2.2 and 1.3 m, respectively (Vital *et al.* 2008). The relative tide range (RTR) relationship of the mean spring tide range (MSR)/wave height (Hb) for this sector is $3 < \text{RTR} < 9$, and so is classified in the wave-tide group (Vital 2009). Sea-level changes measured on the northern sector show maximum tidal heights of 3.3 and 2.5 m, respectively during spring and neap tides. The relative tide range for this sector is $4 < \text{RTR} < 15$, and so is classified to be in the mixed wave-tide group (Vital 2009).

Sedimentology

The inner shelf is characterized by a belt of siliciclastic quartz-rich sands found in the nearshore down to a water depth of 15 m, and a complex of mixed carbonate–siliciclastic medium-grained sands between depths of 15 and 25 m. High-frequency relative sea-level changes have resulted in a complex and close association of Quaternary highstand carbonates and lowstand siliciclastic facies, which are blended in the middle shelf (Fig. 4.7). The siliciclastic facies originate from river discharges, coastal erosion and reworked relict deposits of former lower sea-level stands, while the carbonates have their sources in the locally produced grains by growth and transport of calcareous organisms. Except on the inner shelf, where they are moved by waves and currents, many of these sands are relict and contain a completely reworked biogenic fraction. Some samples contain a substantial modern biogenic component and are, therefore, palimpsest (Summerhayes

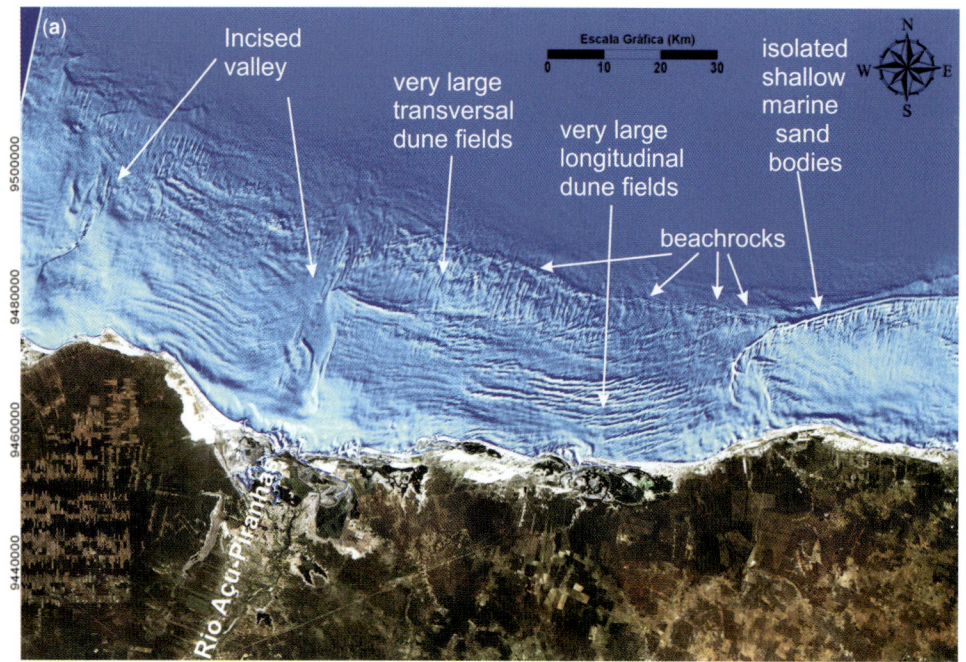

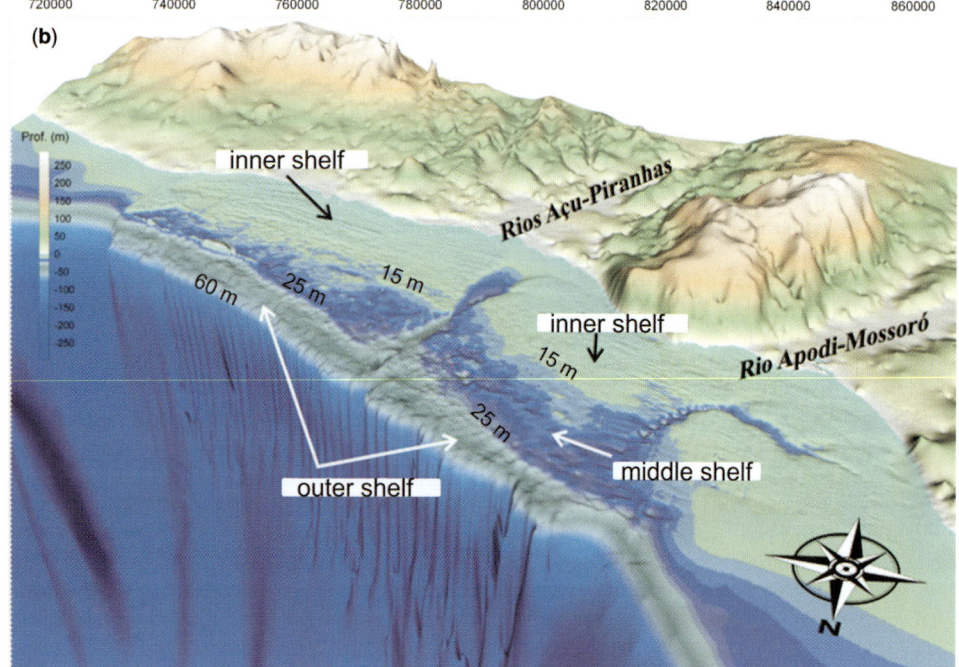

Fig. 4.6. The Brazilian tropical northeast (BT NE) shelf (northern sector) adjacent to Rio Grande do Norte State: (**a**) Landsat 7 ETM+ image showing different seabed features on this shelf. The land is in brownish colours and the shallow subaqueous shelf in blue. (**b**) Digital terrain model showing the limits of the inner, middle and outer shelf (modified from Gomes & Vital 2010).

et al. 1975; Vital *et al.* 2008). Quartz is the main component of the siliciclastic sediments (85–100%), although feldspar is also present in most samples (5–15%). The mineralogical fraction includes mica, glauconite and heavy minerals, which are present in minor amounts (Vital *et al.* 2008).

On the outer shelf, bioclastic carbonate gravel and very coarse sands dominate, and are related to the growth of branching coralline algae (Fig. 4.7). *In situ* development of rhodoliths produces a consistent coarser size fraction. Halimeda and different types of molluscs are also common but contribute variable amounts of granule- to sand-sized material to the sediments. Benthonic foraminifera, ostracods, gastropods and bivalves occur in minor amounts (Testa & Bosence 1998, 1999; Tabosa 2006). Most of the carbonate assemblage shows some degree of reworking, and consists of mixtures of recent and reworked organisms.

Fine sediments are usually found filling palaeochannels around the most important rivers and at the canyons' heads (e.g.

Japaratuba, Sao Francisco, Açu). Fine-grained sediments occupy the slope, at water depths >70 m, with increasing mud content with depth. Terrigenous muds are mainly nearshore, while planktonic foraminifera oozes dominate at depths >80 m, suggesting a relatively local low energy.

Coral knolls and patch reefs are sparse on the northern sector, and present in the inner shelf of along the eastern sector. More specifically, in its southern part, in the region of the Abrolhos Bank, the largest and richest coral reefs of Brazil are found and, indeed, of the entire SW Atlantic Ocean. The coral fauna of the Brazilian reefs has a very low diversity and a significant endemism. The major reef-building species are remnants of a most strong relict fauna dating back to the Tertiary, which was probably preserved during Pleistocene lowstands of sea level in a refugium provided by the seamountains off the Abrolhos Bank, in south Bahia. This stronger endemic fauna, more resistant and better adapted, has been withstanding the periodic turbidity of

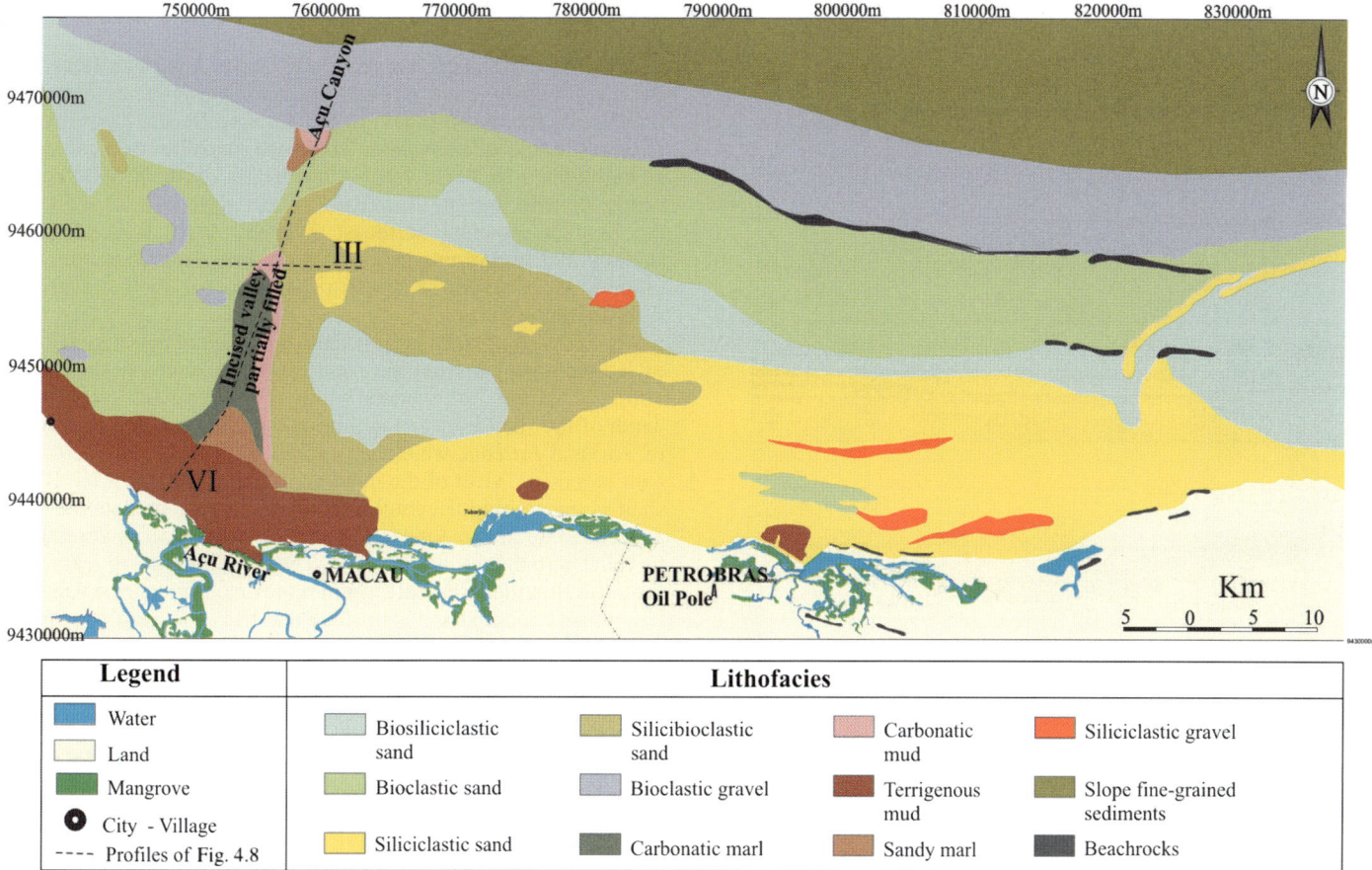

Fig. 4.7. Chart showing the distribution of surface sediments on the northern Rio Grande do Norte shelf (modified from Vital *et al.* 2005, 2008).

the Brazilian waters (Leão *et al.* 2003). The sedimentary cover of the Abrolhos Bank is biogenic, with carbonate mud derived from degradation of reefs dominant in the depression, and biodetritus with modern and fossil components at the borders.

Submerged sand banks are common along this shelf. Samples taken from diving on the northern sector, adjacent to the Rio Grande do Norte state, revealed laminated sandstone cemented by carbonate, similar to beach rocks commonly exposed along the coast. They are often overgrown by a thin corallinaceae algae encrustation, and large numbers of rhodolites and sponges. Echinoderms, molluscs, bivalves, gastropods, polychaetes and cloroficeas algae are also found but corals are almost absent (Santos *et al.* 2007; Vital *et al.* 2008, 2010).

Stratigraphy

High-resolution seismic data are scarce on the BT NE shelf. Investigations of the subsurface geology and structures of the Rio Grande do Norte coastal shelf made by Schwarzer *et al.* (2006) focused on the interpretation of seismostratigraphic units, and are used here to visualize the shelf subsurface.

Schwarzer *et al.* (2006) identified seven units (Fig. 4.8) and concluded that the BT NE shelf adjacent to Rio Grande do Norte state has experienced regressive and transgressive stages since the Pleistocene and up until recent times. The entire channel structure from the main river mouths to the shelf edge can thus be regarded as incised valley morphologies cutting into the shelf ancient deposits during sea-level lowstands. These incised valleys are currently represented by cut and fill structures (Unit IV), indicative of the erosional force of the rivers during lowstand conditions. Unit IV contains many strong reflectors, which are interpreted as pebble beds or consolidated sediments.

Hummocky structures observed in Unit V may suggest relict fluvial sedimentary structures originating from meandering rivers. Unit VII is interpreted as the surface and its underlying layers during lowstand conditions. These sediments are clearly separated from those of Unit I, which forms the present seafloor throughout the investigated area. Unit I is composed of parallel marine-sediment layers deposited at the boundary between marine sediments deposited after the Holocene transgression reached this area and terrestrial deposits formed during the Pleistocene and early Holocene.

A seismically transparent Unit III occurs immediately below horizon I, but overlies the hummocky structures (Unit V) and the incised valley deposits (Unit VI). Elsewhere, lagoonal deposits are characterized by their seismic transparency and poor internal stratification arising from their very high content of silt, clay and organic material. The location of Unit III and its seismic facies characteristics are thus consistent with early Holocene deposition in a low-energy environment sheltered by embankments or ridges.

Sigmoidal structures characterize Unit II, particularly at the transition between horizon I and the overlying subhorizontal layers of Unit I (Fig. 4.8). This configuration of seismic reflectors is indicative of sea-level rise combined with low sediment supply, which permits topset beds to aggregate simultaneously with foreset progradation. In this case, such structures would develop during sea-level rise accompanied by a prograding shoreline but the observed structures do not allow a proper distinction between these scenarios if the sigmoidal structures are arranged within topset layering. The slight offshore increase in depth of Unit II, however, may indicate the absence of topset layering but may also just stand for a period of fall in sea level, stagnation or a periodic increase in sediment supply. Such a model fits well with the interpretation by Schwarzer *et al.* (2006) of Unit III.

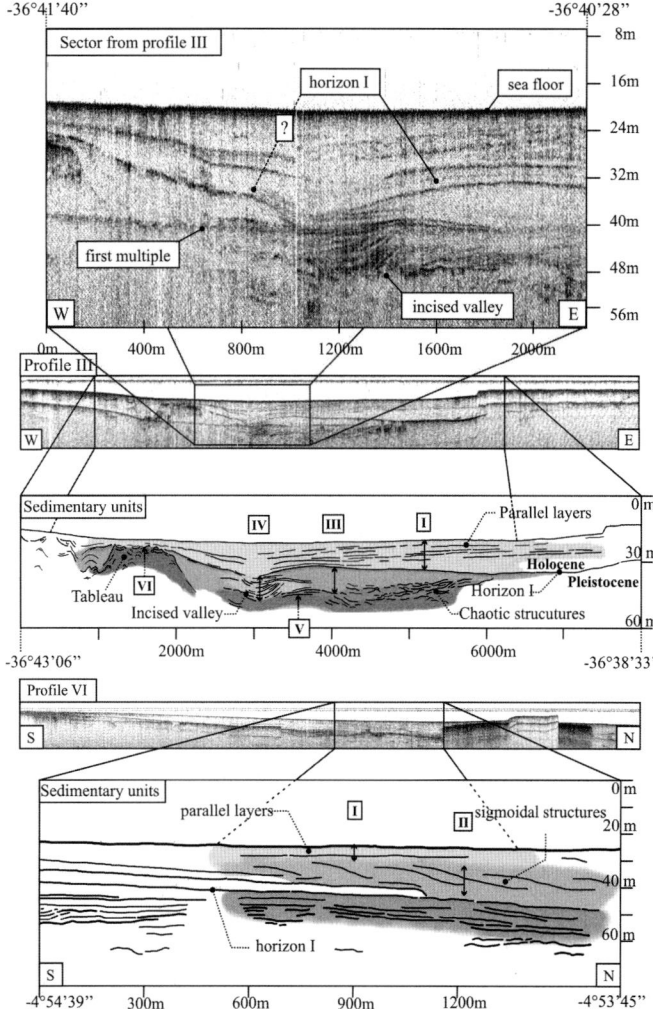

Fig. 4.8. Boomer profiles of the tropical NE shelf on the north Rio Grande do Norte State showing the Holocene stratigraphy of the area. See Figure 4.7 for the profile location (modified from Schwarzer *et al.* 2006; Vital *et al.* 2010).

The tableau-like structure (Unit VI) may suggest the presence of recent tectonic activity and is interpreted as an uplifted structure. These layers are not interrupted on the flanks and decreasing in thickness upwards. As Unit VI is buried by Unit I, the tectonic uplift must have occurred during the Pleistocene or early Holocene. A compilation of all profiles demonstrates that this tectonic activity was not restricted to profile III (Fig. 4.8), but extends over a distance of at least 10 km in a north–south direction.

Summary and concluding remarks

Located on the eastern margin of the South American Plate, the Brazilian tropical shelf evolved from Mesozoic times onwards, linked with the development of the Atlantic Ocean. This chapter reviews the recent knowledge about the Brazilian tropical shelf, as highlighted in Table 4.1. With a length of about 3000 km, it can be divided into the Brazilian tropical north (BT N) shelf and the Brazilian tropical northeast (BT NE) shelf. Both are very shallow and highly energetic systems. The main characteristics of these two sectors of the Brazilian tropical shelf are briefly summarized here.

The BT N shelf is the largest shelf as a result of massive influxes of sediment supply from the Amazon River, linked to the Andes uplift. It presents an average width of 300 km, and an average depth of the shelf break of around 100 m. Macrotides (mostly up to 6 m) and currents, which can exceed 200 cm s^{-1}, are the dominant processes. Most of the inner and middle portion of the BT N shelf is undergoing accumulation of siliciclastic mud as the subaqueous delta of the Amazon, while the outer shelf is dominated by relict carbonate sedimentation. The most impressive feature of the north Brazilian tropical margin is the broad submarine Amazon fan, which encompasses both slope and rise. The most prominent feature of the fan is the Amazon submarine canyon. The weight of the Amazon fan is resulting in the subsidence of the BT N shelf.

In contrast, as a consequence of the small drainage basins, the BT NE shelf is narrow, with an average width of 40 km, and usually starved of sediments. It reaches its narrowest point (5–8 km) adjacent to Salvador, and widens in the southern portion, in the Abrolhos area, where it extends to about 200 km as a result of volcanic intrusive activity. It presents an average depth of the shelf break of around 60 m. Waves and currents (up to 100 cm s^{-1}) are the dominant processes. The inner shelf is covered by modern and palimpsest siliciclastic–mixed sandy sediments, while the middle to outer shelf are dominated by carbonate, most modern, coarse sediments.

Distinctive seabed features, such as large dune fields, incised valleys, coral reefs and submerged chains parallel to the coast, are present on the BT NE shelf. These features are clearly observed by Landsat imagery because of the absence of large rivers discharging suspended sediments in this shelf.

Large incised valleys, not completely infilled, and canyons are associated with the main rivers along the BT NE shelf, and have a clear expression in the continental shelf bathymetry. The presence of muddy sediments on the BT NE shelf is closely associated with the incised valley filling and the heads of the canyons, indicating relatively local low energy. Moreover, submerged beach-rock chains, parallel to the coast, typical of the BT NE shelf are testimony to ancient coastlines.

The largest and richest coral reefs of Brazil and of the entire SW Atlantic Ocean are found in the region of the Abrolhos Bank, the

Table 4.1. *Summary table for the Brazilian tropical shelf*

	BT north (BT N) shelf	**BT northeast (BT NE shelf)**
Length of the shelf (km)	*c.* 1000	2000
Average width (km)	+300	40
Tide (tidal range), wave, current ranges	−8 m (macrotidal), 1–2 m wave height and 13 s period, can exceed 200 cm s^{-1}	2–3.5 m (mesotidal), 1–2 m wave height and 7.5 s period, up to 100 cm s^{-1}
Dominant process (wave/current/tides)	Tides, currents	Waves, mixed tidal and waves, currents
Average depth of the shelf break	*c.* 100 m	*c.* 60 km
Siliciclastic/carbonate/autigenic	90% siliciclastic	80–100% carbonate middle–outer shelf, siliciclastic–mixed inner shelf
Modern/relict/palimpsest	Modern inner and middle shelf, relict outer shelf	Most modern middle–outer shelf, modern and palimpsest inner shelf
Tectonic trend over the last glacial cycle	Subsiding	Uplift and subsidence

southern limit of the BT NE shelf. The coral fauna of the Brazilian reefs has a very low diversity and a significant endemism.

These remarks indicated that the Brazilian tropical shelf has experienced regressive and transgressive stages since the Pleistocene and up until recent times. Although large amounts of sediment accumulate on the Amazon margin, the Brazilian shelf was for most of the time exposed to erosion, which remove strata and, so, is marked by the lack of records. The integration of different observations on the Brazilian tropical shelf, from terrestrial and coastal environments to slope, continental rises and abyssal plains, is important for describing this particular continental shelf and needs to be encouraged.

The author wishes to express thanks to the editors (A. Chivas and F. Chiocci) for their invitation to write this chapter and for their helpful comments on the initial version, to the Brazilian National Council Research – CNPq – for their researcher fellowship (grant PQ. 303481/2009-9), to the two anonymous reviewers for their improvements to the manuscript and to P. Hesp, who kindly revised the manuscript to improve the text form. Finally, the author is indebted to the different Brazilian agencies (FINEP, CAPES, PETROBRAS, ANP), through projects POTMAR, SISPLAT, PLAT N-NE, RECIFES-Ciencias do Mar 207-10 and PRH22, and to the International Union for Quaternary Science (INQUA) for financial support to attend IGCP conferences. This is a contribution to INCT AmbTropic (CNPq/FAPESB/CAPES) and INQUA 1202.

References

ALEXANDER, C. R., JR, NITTROUER, C. A. & DeMASTER, D. J. 1986. High-resolution seismic stratigraphy and its sedimentological interpretation on the Amazon shelf. *Continental Shelf Research*, **6**, 337–357.

ALLER, R. C., BLAIR, N. E., XIA, Q. & RUDE, P. D. 1996. Remineralization rates, recycling and storage of carbon in Amazon shelf sediments. *Continental Shelf Research*, **16**, 753–786.

ALLISON, M. A., NITTROUER, C. A. & KINEKE, G. C. 1995. Seasonal sediment storage on mudflats adjacent to the Amazon River. *Marine Geology*, **125**, 303–328.

ALLISON, M. A., LEE, M. T., OGSTON, A. S. & ALLER, R. C. 2000. Origin of Amazon mudbanks along the northeastern coast of South America. *Marine Geology*, **163**, 303–328.

ALMEIDA, F. F. M., HASUI, Y., BRITO NEVES, B. B. & FUCK, R. A. 1977. Províncias Estruturais Brasileiras. *Boletim SBG, Simpósio Geologia do Nordeste*, **8**, 363–391.

AMARAL, C. A. B. (ed.) 1979. *Recursos minerais da margem continental brasileira e das áreas oceânicas adjacentes*. PETROBRAS, Rio de Janeiro, Projeto REMAC Series, **10**.

ARAI, M. 2006. A grande elevação eustática do Mioceno e sua influência na origem do Grupo Barreiras. *Revista do Instituto de Geociências – USP*, **6**, 1–6.

ARAÚJO, T. C. M. 1994. *Seismostratigraphic interpretation of the Amazon Continental Margin in view of sea level changes and subsidence of the sea floor*. PhD thesis, Christian-Albrechts-Universität zu Kiel.

ARAÚJO, T. C. M., SEOANE, J. C. S. & COUTINHO, P. N. 2004. Geomorfologia da Plataforma continental de Pernambuco. *In*: ESKINAZI-LEÇA, E., NEUMANN-LEITÃO, S. & COSTA, M. F. (eds) *Oceanografia: um cenário tropical*. Edições Bagaço, Recife, 39–57.

ASMUS, H. E. (ed.) 1981. *Estruturas e tectonismo da margem continental brasileira e suas implicações nos processos sedimentares e na avaliação do potencial de recursos minerais*. PETROBRAS, Rio de Janeiro, Projeto REMAC Series, **9**.

BARRETO, A. M. F., BEZERRA, F. H. R., SUGUIO, K., TATUMI, S. H., YEE, M., PAIVA, R. P. & MUNITA, C. S. 2002. Late Pleistocene marine terrace deposits in northeastern Brazil: sea-level change and tectonic implications. *Palaeogeography, Palaeoclimatology, Palaeoecology*, **179**, 57–69.

BEARDSLEY, R. C., CANDELA, J. *ET AL*. .1995. The Mz tide on the Amazon shelf. *Journal of Geophysical Research*, **100**, 2283–2320.

BERNAT, M., MARTIN, L., BITTENCOURT, A. C. S. P. & VILAS-BOAS, G. S. 1983. Datation Io /U du plus haut niveau marin du dernier interglaciaire sur le côte du Brésil. Utilization du 229Th comme traceur.

Comptes Rendus de l'Académie des Sciences de Paris, Série 2, **296**, 197–200.

BEZERRA, F. H. R., BARRETO, A. M. F. & SUGUIO, K. 2003. Holocene sea-level history on the Rio Grande do Norte State coast, Brazil. *Marine Geology*, **196**, 73–89.

CALDAS, L. H. O. 2002. *Late Quaternary coastal evolution of the Northern rio Grande do Norte coast, NE Brazil*. PhD thesis, Christian Albrechts University, Kiel.

CALDAS, L. H. O., STATTEGGER, K. & VITAL, H. 2006. Holocene sea-level history: evidence from coastal sediments of the northern Rio Grande do Norte coast, Ne Brazil. *Marine Geology*, **228**, 39–53.

COHEN, M. C. L., SOUZA FILHO, P. W., LARA, R. L., BEHLING, H. & ANGULO, R. 2005. A model of Holocene mangrove development and relative sea-level changes on the Bragança Peninsula (northern Brazil). *Wetlands Ecological Management*, **13**, 433–443.

COUTINHO, P. N. 1976. *Geologia Marinha da Plataforma Continental Alagoas-Sergipe. Brasil*. Livre Docência thesis. Universidade Federal de Pernambuco.

DAMUTH, J. E. & FLOOD, R. D. 1984. Morphology, sedimentation processes, and growth pattern on the Amazon deep-sea fan. *Geo-Marine Letters*, **3**, 109–117.

DILLENBURG, S. & HESP, P. (eds) 2009. *Geology and Geomorphology of Brazilian Holocene Coastal Barriers*. Springer, Heidelberg.

DOMINGUEZ, J. M. L. 2009. The coastal zone of Brazil. *In*: DILLEMBURG, S. & HESP, P. (eds) *Geology and Geomorphology of Brazilian Holocene Coastal Barriers*. Springer, Heidelberg, 17–51.

DOMINGUEZ, J. M. L., ANDRADE, A. C. S., ALMEIDA, A. B. & BITTENCOURT, A. C. S. P. 2009. The Holocene barrier strandplains of the state of Bahia. *In*: DILLEMBURG, S. & HESP, P. (eds) *Geology and Geomorphology of Brazilian Holocene Coastal Barriers*. Springer, Heidelberg, 253–288.

DRISCOLL, N. W. & KARNER, G. D. 1994. Flexural deformation due to Amazon fan loading: a feedback mechanism affecting sediment delivery to margins. *Geology*, **22**, 1015–1018.

FIGUEIREDO, A. G. & NITTROUER, C. A. 1995. New insights to high-resolution stratigraphy on the Amazon continental shelf. *Marine Geology*, **125**, 393–399.

FLOOD, R. D., MANLEY, P. L., KOWSMANN, R. O., APPI, C. A. & PIRMEZ, C. 1991. Seismic facies and late Quaternary growth of Amazon submarine fan. *In*: WEIMER, P. & LINK, M. H. (eds) *Seismic Facies and Sedimentary Processes of Submarine Fans and Turbidity Systems*. Springer, New York, 415–433.

GABIOUX, M., VINZON, S. B. & PAIVA, A. M. 2005. Tidal propagation over fluid mud layers on the Amazon shelf. *Continental Shelf Research*, **25**, 113–125.

GEYER, W. R. & KINEKE, G. C. 1995. Observations of currents and water properties in the Amazon frontal zone. *Journal of Geophysical Research*, **100**, 2321–2339.

GOMES, M. P. & VITAL, H. 2010. Revisão da Compartimentação Geomorfológica da Plataforma Continental Norte do Rio Grande do Norte–Brasil. Revista Brasileira de Geologia. *SBG*, **40**, 321–329.

GORINI, M. A. 1977. *The tectonic fabric of the equatorial and adjoining continental margins: Gulf of Guinea to northeastern Brazil*. PhD thesis, Columbia University.

HAZIN, F. H. V., WOR, C., OLIVEIRA, J. E. L., HAMILTON, S., TRAVASSOS, P. & GEBER, F. 2008. Resultados obtidos por meio do fundeio de um correntógrafo na plataforma continental do Estado do Rio Grande do Norte, Brasil. *Arquivos ciências do Mar*, **41**, 30–35.

HESP, P. A., MAIA, L. P. & CLAUDINO-SALES, V. 2009. The Holocene Barriers of Maranhão, Piauí and Ceará States, Northeastern Brazil. *In*: DILLEMBURG, S. & HESP, P. (eds) *Geology and Geomorphology of Brazilian Holocene Coastal Barriers*. Springer, Heidelberg, 325–345.

KINEKE, G. C. & STERNBERG, R. W. 1995. Distribution of fluid muds on the Amazon continental shelf. *Marine Geology*, **125**, 193–233.

KNOPPERS, B., EKAU, W. & FIGUEIREDO, A. G. 1999. The coast and shelf of east and northeast Brazil and material transport. *Geo-Marine Letters*, **19**, 171–178.

KOWSMANN, R. O. & COSTA, M. P. A. 1979. *Sedimentacao Quaternaria da margem continental Brasileira e das areas oceanicas adjacentes*. PETROBRAS, Rio de Janeiro, Projeto REMAC Series, **8**.

KUEHL, S. A., DEMASTER, D. J. & NITTROUER, C. A. 1986. Nature of sediment accumulation on the Amazon continental shelf. *Continental Shelf Research*, **6**, 209–225.

KUEHL, S. A., NITTROUER, C. A. *ET AL.* . 1996. Sediment deposition, accumulation and seabed dynamics in an energetic fine-grained coastal environment. *Continental Shelf Research*, **16**, 787–815.

LEÃO, Z. M. A. N., KIKUCHI, R. K. P. & TESTA, V. 2003. Corals and coral reefs of Brazil. *In*: CORTES, J. (ed.) *Latin American Coral Reefs*. Elsevier, Amsterdam, 9–52.

MARTIN, L., DOMINGUEZ, J. M. L. & BITTENCOURT, A. C. S. P. 2003. Fluctuating Holocene sea levels in eastern and southeastern Brazil: evidence from a multiple fossil and geometric indicators. *Journal of Coastal Research*, 101–124.

MESERVE, J. M. 1974. *US Navy Marine Climatic Atlas of The World, Vol. 1, North Atlantic Ocean*. United States Government Printing Office, Washington, DC.

MILANI, E. J. & THOMAZ FILHO, A. 2000. Sedimentary basins of South America. *In*: CORDANI, U. G., MILANI, E. J., THOMAZ FILHO, A. & CAMPOS, D. A. (eds) *Tectonic Evolution of South America: 31 International Geological Congress, August 2000, Rio de Janeiro*. In-Fólio Produção Editorial, Rio de Janeiro, 389–449.

MILLIMAN, J. D. 1975. A synthesis. Upper continental margin sedimentation off Brazil. *Contributions to Sedimentology*, **4**, 151–175.

MILLIMAN, J. D., SUMMERHAYES, C. P. & BARRETTO, H. T. 1975. Quaternary sedimentation on the Amazon continental margin: a model. *Geological Society of America Bulletin*, **86**, 610–614.

NITTROUER, C. A. & DEMASTER, D. J. 1996. The Amazon shelf setting: tropical, energetic and influenced by a large river. *Continental Shelf Research*, **16**, 553–573.

NITTROUER, C. A., KUEHL, S. A., DEMASTER, D. J. & KOWSMANN, R. 0. 1986. The deltaic nature of Amazon shelf sedimentation. *Geological Society of America Bulletin*, **97**, 444–458.

NITTROUER, C. A., KUEHL, S. A. *ET AL.* . 1996. The geological record preserved by Amazon shelf sedimentation. *Continental Shelf Research*, **16**, 817–841.

PROJETO REMAC 1979. *Reconhecimento Global da Margem Continental Brasileira – Coleção de Mapas*. Projeto REMAC Series, **11**. PETROBRAS, Rio de Janeiro.

ROSSETTI, D. F. 2006. Evolução sedimentar Miocenica nos Estados do Para e Maranhao. *Revista do Instituto de Geociências–USP, Serie Científica*, **6**, 7–18.

SANTOS, C. L. A., VITAL, H., AMARO, V. E. & KIKUCHI, R. P. 2007. Mapeamento de recifes submersos na costa do Rio Grande do Norte. *Revista Brasileira de Geofísica*, **25**, 27–36.

SCHWARZER, K., STATTEGGER, K., VITAL, H. & BECKER, M. 2006. Holocene Coastal Evolution of the Rio Açu Area (Rio Grande do Norte, Brazil). *In*: ICS '04 – International Coastal Symposium. *Journal of Coastal Research*, Special Issue, **39**, 140–144.

SOUZA FILHO, P. W. M. 2005. Costa de Manguezais de Macromaré da Amazônia: Cenários morfológicos, mapeamento e quantificação a partir de dados de sensores remotos. *Revista Brasileira de Geoísica*, **23**, 427–435.

SOUZA FILHO, P. W. M., COHEN, M. C. L., LARA, R. L., LESSA, G. C., KOCH, B. & BEHLING, H. 2006. Holocene coastal evolution and facies model of the Bragança macrotidal flat on the Amazon Mangrove Coast, Northern Brazil. *In*: ICS '04 – International Coastal Symposium. *Journal of Coastal Research*, Special Issue, **39**, 306–310.

SOUZA FILHO, P. W. M., LESSA, G. C., COHEN, M. C. L., COSTA, F. R. & LARA, R. L. 2009. The subsiding macrotidal barrier estuarine system of the eastern amazon coast, northern Brazil. *In*: DILLEMBURG, S. & HESP, P. (eds) *Geology and Geomorphology of Brazilian Holocene Coastal Barriers*. Springer, Heidelberg, 347–375.

STERNBERG, R. W., CACCHIONE, D. A., PAULSON, B., KINEKE, G. C. & DRAKE, D. E. 1996. Observations of sediment transport on the Amazon subaqueous delta. *Continental Shelf Research*, **16**, 697–715.

SUMMERHAYES, C. P., COUTINHO, P. N., FRANÇA, A. M. C. & ELLIS, J. P. 1975. Part III. Salvador to Fortaleza Northeastern Brazil. *Contributions to Sedimentology*, **4**, 44–78.

SYVITSKI, J. M. P., VÖRÖSMARTY, C. J., KETTNER, A. J. & GREEN, P. 2005. Impact of humans on the flux of terrestrial sediment to the global coastal ocean. *Science*, **308**, 376–380.

TABOSA, W. F. 2006. *Morfologia, hidrodinâmica e sedimentologia da plataforma continental brasileira adjacente a São Bento do Norte e Caiçara do Norte–RN; NE Brasil*. PhD thesis, Universidade Federal do Rio Grande do Norte.

TESTA, V. & BOSENCE, D. W. J. 1998. Carbonate–siliciclastic sedimentation on high-energy, ocean-facing, tropical ramp, NE Brazil. *In*: WRIGHT, V. P. & BURCHETTE, T. P. (eds) *Carbonate Ramps*. Geological Society, London, Special Publications, **149**, 55–71.

TESTA, V. & BOSENCE, D. W. J. 1999. Physical and biological controls on the formation of carbonate and siliciclastic bedforms on the North-East Brazilian shelf. *Sedimentology*, **46**, 279–301.

VIANNA, M. L., SOLEWICZ, R., CABRAL, A. & TESTA, V. 1991. Sandstream on the Northeast Brazilian Shelf. *Continental Shelf Research*, **2**, 509–524.

VITAL, H. 2009. The mesotidal barriers of Rio Grande do Norte. *In*: DILLEMBURG, S. & HESP, P. (eds) *Geology and Geomorphology of Brazilian Holocene Coastal Barriers*. Springer, Heidelberg, 289–324.

VITAL, H., STATTEGGER, K. & GARBE-SCHONBERG, G. 1999. Composition and trace element geochemistry of detrital clay and heavy mineral suites of the lowermost Amazon River: a provenance study. *Journal of Sedimentary Research*, **69**, 563–575.

VITAL, H., SILVEIRA, I. M. & AMARO, V. E. 2005. Carta Sedimentológica da Plataforma Continental Brasileira – Área Guamaré a Macau (NE Brasil), Utilizando Integração de Dados Geológicos e Sensoriamento Remoto. *Revista Brasileira de Geofísica, Rio de Janeiro*, **23**, 233–241.

VITAL, H., STATTEGGER, K., AMARO, V. E., SCHWARZER, K., FRAZÃO, E. P. & TABOSA, W. F. 2008. A Modern high-energy siliciclastic–carbonate platform: continental shelf adjacent to northern Rio Grande do Norte state, NE Brazil. *In*: HAMPSON, G. & DALRYMPLE, R. (eds) *Recent Advances in Models of Siliciclastic Shallow-Marine Stratigraphy*. SEPM Special Publications, Tulsa, **90**, 177–190.

VITAL, H., GOMES, M. P., TABOSA, W. F., FRAZÃO, E. P., SANTOS, C. L. A. & PLÁCIDO JUNIOR, J. S. 2010. Characterization of the Brazilian continental shelf adjacent to Rio Grande do Norte State, NE Brazil. *Brazilian Journal of Oceanography*, **58**(Special Issue 1), 43–54.

Chapter 5

The southern Brazilian shelf

RENATA H. NAGAI, SILVIA H. M. SOUSA & MICHEL M. MAHIQUES*

*Instituto Oceanográfico, Universidade de São Paulo, Praça do Oceanográfico,
191 Cidade Universitária, São Paulo, SP 05508–900, Brazil*

*Corresponding author (e-mail: mahiques@usp.br)

Abstract: The southern Brazilian shelf (extending from 34°S to 22°S) is discussed with special emphasis on its Holocene sedimentary evolution. The onset of the Rio de la Plata plume influence on the inner shelf during the Late Holocene and the reworking of seafloor sediments by the Brazil Current in the outer shelf and upper slope are key elements. High-resolution sedimentary records have revealed palaeoceanographic changes in the Brazilian shelf, highlighting the importance of this sector of the Brazilian shelf to the comprehension of the palaeoclimate of South America, especially regarding oscillations of wind regime and humidity. These records also reveal that the recent sedimentary and oceanographic contrasts north and south of São Sebastião Island (24°S) have been present in the Brazilian shelf at least since the Early Holocene.

The southern Brazilian shelf constitutes an example of a subtropical passive margin and represents one of the widest shelves in the world, extending approximately from latitude 34°S to 22°S, and covering an area of $270 \times 10^3 \, km^2$ (Mahiques *et al.* 2004). This shelf is a typical example of a hydrodynamically driven modern shelf whose present sedimentary processes are strongly dominated by oceanic dynamics and shelf circulation, as well as by an allochthonous input from fluvial sediments (Mahiques *et al.* 2004) (Fig. 5.1).

The geological evolution of the southern Brazilian shelf has its origin in the development of Atlantic-type margins, and is marked by the subsidence of the Campos and Santos basins (Cretaceous–Recent) (Meisling *et al.* 2001), by intensive Mesozoic alkaline magmatism, and by the uplift of the Serra do Mar range (Almeida & Carneiro 1998). The alignment of the Serra do Mar in relation to the present coastline is reflected in the small width of the coastal plains and in the arrangement of the isobaths. As a consequence of the Serra do Mar uplift in SE Brazil, most of the drainage systems flow inland, feeding the Paraná River and the Rio de la Plata basins; only small rivers run directly towards the sea, draining mainly Precambrian granites, gneisses and migmatites.

Bathymetric, seismic and sedimentological studies of the southern Brazilian shelf were performed during the 1960s under the auspices of the REMAC Project (Global Reconnaissance of the Brazilian Continental Margin). These studies produced a large volume of geological information, some of which still represents, for particular areas, the only available source of geological and geophysical knowledge of the Brazilian margin (see Milliman & Barreto 1975 and related papers).

More recently, sedimentological and bathymetric data were obtained in the REVIZEE project (Living Resources of the Brazilian Economic Exclusive Zone) (Figueiredo & Madureira 2004; Figueiredo & Tessler 2004; Ferreira *et al.* 2005). However, owing to the dimensions of the Brazilian shelf, there are still considerable gaps in knowledge, especially related to the evolutionary processes since the Last Glacial Cycle.

General characteristics

Zembruscki (1979) divided the area into four morphological sectors: the Cabo Frio-Cabo de São Tomé Sector; the São Paulo Bight; the Florianópolis–Mostardas Sector; and the Rio Grande Cone (Fig. 5.1). Table 5.1 summarizes the main physiographical, hydrodynamic, sedimentological and tectonic characteristics of the southern Brazilian shelf.

Geomorphological characteristics

The Cabo Frio region represents a conspicuous break in the South American coastline, and marks the limit between the São Paulo Bight and the Cabo Frio–Cabo de São Tomé Sector (Fig. 5.1). This sector is characterized by a narrow continental shelf with a predominantly north–south orientation. Carbonate sediments dominate due to the small terrigenous supply by few rivers, and the presence of warm and high salinity tropical water (TW) transported by the Brazil Current (BC) at surface levels, and which maintains the carbonate production and preservation.

The São Paulo Bight is the arc-shaped part of the southern Brazilian margin. The ocean floor shows a rather complex morphology involving channels that may or may not be related to deeper canyons present on the continental slope and which represent the main drainage systems developed during low sea-level periods (Furtado *et al.* 1996). The role of these channels in determining the present oceanographic conditions is still unknown.

The Florianópolis–Mostardas Sector corresponds to a convex portion of the coastline and the isobaths, and its characteristics are related to the Ponta Grossa Arc, which corresponds to a structural high that was subject to intense tectonic activity mainly during the Mesozoic. This warping was considered an aborted branch of a triple junction during Gondwana's fragmentation and the opening of the South Atlantic. This sector is the most regular part of the southern Brazilian shelf, with no marked channels, with a single prominent scarp and the presence of middle shelf mudbelts (Mahiques *et al.* 2010).

The Rio Grande Cone sector constitutes the southernmost and most irregular sector of the southern Brazilian shelf. It is characterized by the presence of elongated sand banks, topographical features with heights up to 30 m, and two 10 m-high scarps located along the 110 and 60 m isobaths, which have been associated with sea-level stabilization periods of the Last Glacial Maximum (LGM) and of the 11 ka BP Event (probably related to Younger Dryas), respectively (Corrêa 1996). Violante & Parker (2004) linked these features to the outflow of the Rio de la Plata

From: CHIOCCI, F. L. & CHIVAS, A. R. (eds) 2014. *Continental Shelves of the World: Their Evolution During the Last Glacio-Eustatic Cycle.*
Geological Society, London, Memoirs, **41**, 47–54. http://dx.doi.org/10.1144/M41.5

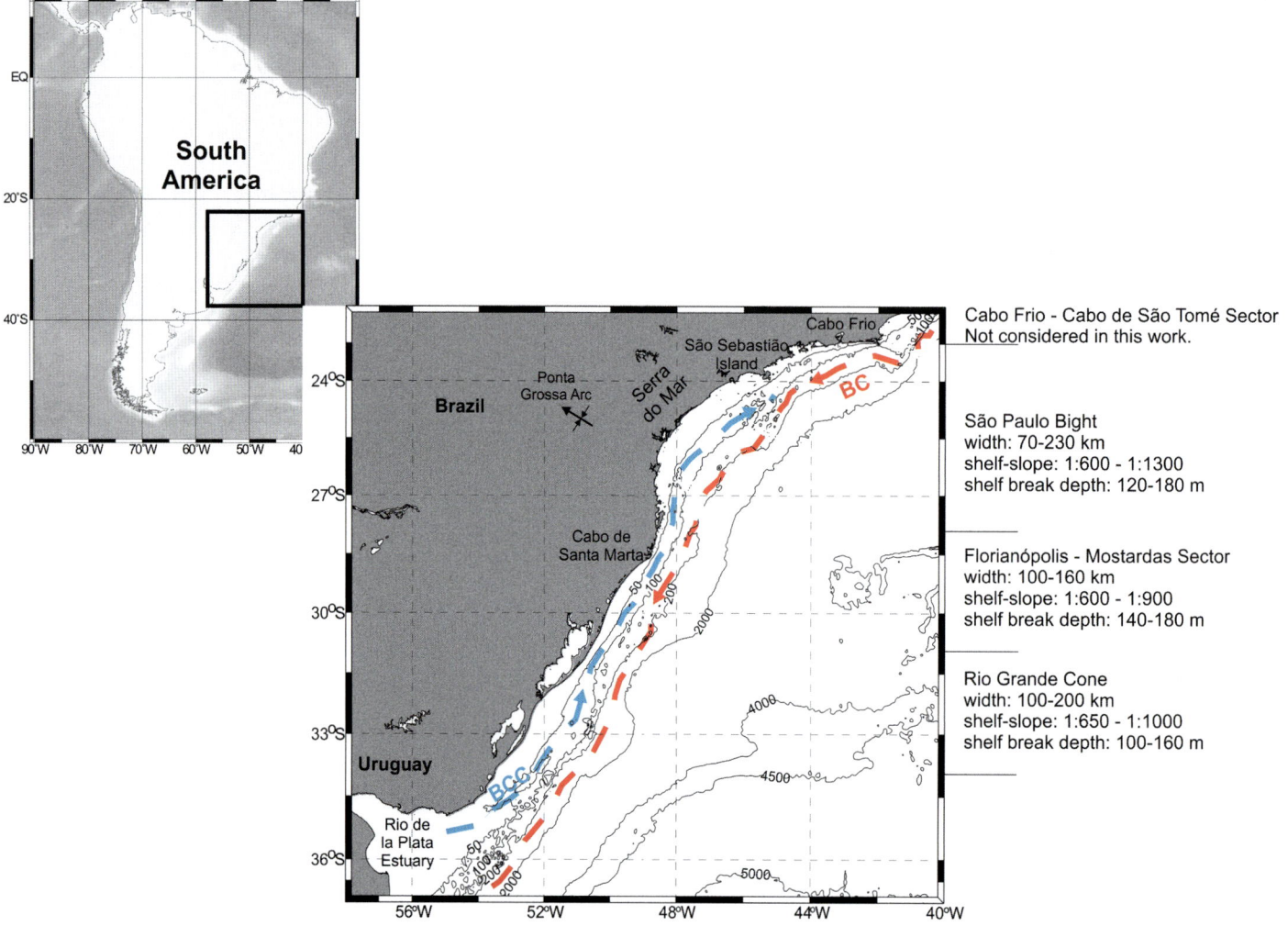

Fig. 5.1. Location map containing the description (width, shelf and slope declivity, and shelf break depth) of the sectors and main morphological features of the southern Brazilian shelf and adjacent continental areas (after Mahiques *et al.* 2010), with a schematic representation of the surface currents. The southward-flowing Brazil Current (BC): red dashed line, modified after Campos *et al.* (1999); and the northward-flowing. Brazil Coastal Current (BCC): blue dashed line, based on Souza & Robinson (2004).

during deglaciation, and recognized deposits (i.e. coastal barriers, beach ridges) in the lowermost regions of the Rio de la Plata and adjacent inner shelf during this period.

Razik *et al.* (2013) identified source and transport mode changes of the terrigenous sediments to the Rio Grande Cone due to changes in wind-driven ocean circulation during the last 14 kyr. According to these authors, during the Early and Late Holocene, the Rio Grande Cone received large amounts of silt from the Rio de la Plata due to the southerly position of the westerlies and strengthening of the South American Summer Monsoon (SASM), respectively. During the Mid-Holocene, the northern migration of the westerlies enabled the transport of sand from the Argentinian shelf to the Rio Grande Cone.

Holocene sea-level changes

Although the first references to Holocene palaeo-sea levels in Brazil were made nearly a century ago (Hartt 1870; Branner 1904), more systematic studies started in the mid-1960s. Since then, more than 100 publications with a focus on Holocene sea-level history have been published. During the 1970s and 1980s, based on thousands of radiocarbon dates, Holocene relative sea-level curves were determined for areas between 5°S and 34°S (Angulo *et al.* 2006 and references therein).

The scarcity of precise dates and the amount of unreliable palaeo-sea-level indicators characterize the sea-level curves of the southern Brazilian shelf from the Last Glacial Maximum up until 7 cal ka BP. Most of these curves are based mainly on morphosedimentary features (i.e. submerged terraces with global sea-level change curves), with the exception of the palaeo-sea-level curve presented by Corrêa (1996), which is based on more consistent indicators. According to this author, Holocene sea-level stabilization periods occurred at 9 cal ka BP (between -32 and -45 m) and 8 cal ka BP (between -20 and -25 m).

More recent data by Mahiques & Souza (1999) and Klein (2005) constitute, at this moment, the most accurate and precisely dated information on sea-level stabilization periods prior to the Holocene highstand of 5.6 ka BP. In the São Paulo state coast (23°30′S), mollusc shells collected in a palaeo-beach sediment located 6 m below the present sea level were radiocarbon dated at 7850 ± 80 cal years BP (Mahiques & Souza 1999). Additionally, four beach-rock samples located 13 ± 1 m below the present sea level showed an age of 8000 ± 50 cal years BP. These data corroborate the age of 7955 ± 170 cal years BP for a level of 1.4 ± 0.5 m below the present sea level presented by Martin *et al.* (2003).

The sea-level curves of the eastern Brazilian coastline for the Middle and Late Holocene were originally defined by various reconstructions of sea levels, both in space and over time, obtained from a database of more than 700 radiocarbon dates (Suguio

Table 5.1. *Main physiographical, hydrodynamic, sedimentological and tectonic features of the southern Brazilian shelf*

	Cabo Frio-Cabo de São Tomé Sector	São Paulo Bight	Florianópolis–Mostardas Sector	Rio Grande Cone
Length of the shelf	22–23°S[1]	23–28°S[1]	28–31°S[1]	31–35°S[1]
Average width (km)	80[1]	70–230[1]	100–170[1]	100–200[1]
Shelf-slope gradient	1:1000[1]	1:600–1:1300[1]	1:600–1:900[1]	1:650–1:1000[1]
Tidal, wave, current ranges (m s^{-1})	0.06–0.5[2]	0.25–0.70[2]	0.70–0.80[2]	0.22[2]
Dominating process (wave/current/tide)	Wind-driven circulation (inner shelf) and the BC meandering (outer shelf)[3]	Wind- and wave-driven currents (inner shelf)/BC (middle and outter shelf)[3]	BC (middle and outer shelf)[4]	Displacement of the Subtropical Shelf Water (middle shelf); BC flow (outer shelf)[5]
Average depth of the shelf break (m)	80–100[1]	120–180[1]	140–180[1]	100–160[1]
Siliciclastic/carbonate/ authigenic/glacial sedimentation (if possible rough %)	Siliciclastic sand (inner and middle shelf); mud (Cabo Frio inner shelf) and sand (Cabo Frio middle shelf); >95% carbonate sediments (outer shelf)[6]; current-reworked deposits, medium- to coarse-grained siliciclastic sands (outer shelf)[7]	Very fine siliciclastic sands and silts with variable amounts of clay and carbonate[3]	Quartzose sandy sediments (inner shelf); muddy sediments (middle and outer shelf); carbonate gravels related to relict sediments are present in a few areas[4]	Sand and siliciclastic clay with patches of bioclastic gravel (inner and middle shelves); sand facies(outer shelf)[6]
Modern/relict/palimpsest (if possible rough %)		Relict (<5%)[3]		Palimpsest[8]
Tectonic trend over the last glacial cycle (stable/uplifting/ subsiding)	Stable	Stable	Stable	Stable

[1]Zembruski (1979); [2]Castro *et al.* (2006); [3]Mahiques *et al.* (2002); [4]Figueira *et al.* (2006); [5]Mahiques *et al.* (2008); [6]Kowsmann & Costa (1979); [7]Viana *et al.* (1998); [8]Martins *et al.* (2003)

et al. 1985; Angulo & Suguio 1995). Based on these data, palaeo-sea-level trends have been determined for several sectors of the Brazilian coast.

A Mid-Holocene sea-level highstand, followed by a decrease towards the present time, is evident in these studies, although the Holocene relative sea-level history still has many controversial aspects. The main debates concerning the palaeo-sea level for the last 7 kyr include the uncertainity in the elevation of the maximum highstand and the presence or absence of high-frequency sea-level oscillations (Martin *et al.* 2003; Angulo *et al.* 2006).

One model proposes two or three periods of sea level lower than the present after 5.1 ka BP (Suguio *et al.* 1985; Angulo & Suguio 1995), and a second model, based entirely on vermetide tubes, assumes that the present sea level constitutes the lowest level of approximately the last 7 kyr (Angulo & Lessa 1997; Angulo

et al. 1999). According to these latter authors, the great majority of the indicators used to infer the secondary sea-level oscillations in previous studies derive from shell middens, which can be unreliable palaeo-sea-level indicators. The latter sea-level model has since been corroborated by Ybert *et al.* (2003).

Martin *et al.* (2003) produced the first reservoir-corrected, astronomically calibrated sea-level change curve for a sector of the Brazilian coast. The results obtained by these authors suggest the occurrence of three main events of submergence of the coast (7.8–5.6, 3.7–3.5 and 2.3–2.1 cal ka BP) interspersed among periods of emergence. Controversially, Angulo *et al.* (2006) suggested that a progressive decline in the sea level occurred since the Mid-Holocene maximum (Fig. 5.2), based on a large dataset.

The main controversy of these sea-level curves lies in the magnitude of the sea-level oscillations during the Holocene. However,

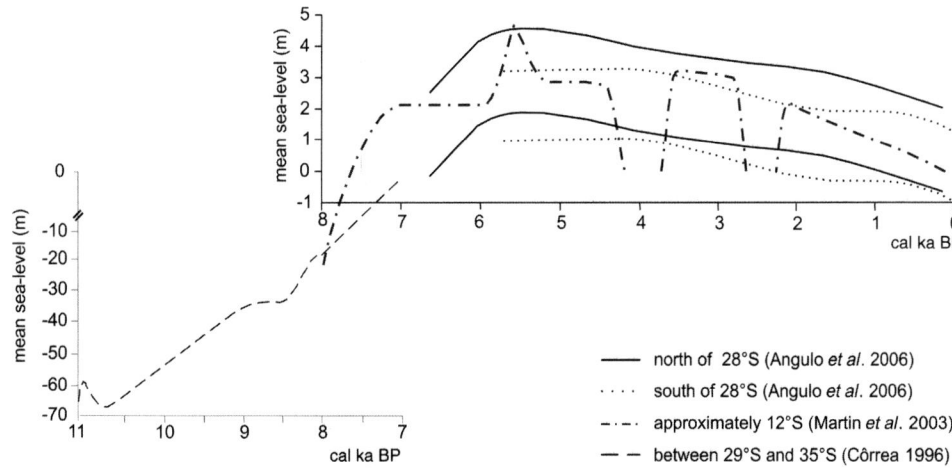

— north of 28°S (Angulo *et al.* 2006)
···· south of 28°S (Angulo *et al.* 2006)
—·—· approximately 12°S (Martin *et al.* 2003)
— — between 29°S and 35°S (Côrrea 1996)

Fig. 5.2. Curves of relative sea-level variations for the Brazilian coast plotted on the same age and elevation scales, modified after Angulo *et al.* (2006), Martin *et al.* (2003) and Corrêa (1996).

independent of the model adopted, the range of these oscillations during the Mid- and Late Holocene would be of approximately 3 m (Fig. 5.2). Although changes at the coastline of the study area might reflect sea-level oscillations, such as by the development of some lagoonal systems, the depositional processes on the shelf would not have been affected by Mid- and Late-Holocene sea-level oscillations.

Modern oceanographic and climatic conditions

The southern Brazilian shelf is characterized by the confluence of two main current systems (Fig. 5.1). The outer shelf is dominated by the southward flow of the Brazil Current (BC) (Mahiques *et al.* 2004) and, in the inner shelf, the northward flow of the Brazilian Coastal Current (BCC) is dominant (Souza & Robinson 2004) (Fig. 5.1). The BCC transports sediments from the Rio de la Plata (Mahiques *et al.* 2008) in the low-temperature and low-salinity plume (Möller *et al.* 2008), and this wind-dependent northward displacement of cold and less saline waters also controls seasonal variation in the primary productivity of the area (Ciotti *et al.* 1995).

The inner and middle shelf dynamics in the study area are determined by the displacement of three water masses with strong seasonal variation (Castro *et al.* 1987). Between November and March, the South Atlantic Central Water (SACW, $T = 14.0$ °C, $S = 35.5‰$) moves close to the seafloor towards the coast, leading to the oceanward displacement of the less dense Coastal Water (CW, $T = 22.0$ °C, $S < 35.0‰$) and keeping the Tropical Water (TW, $T = 25.0$ °C, $S = 37.1‰$) relatively distant from the coastline. This period also corresponds to the rainy season in SE Brazil and, thus, to increase of terrigenous flux to the coastal waters. The displacement of the CW is the most important factor in the transport of terrigenous organic matter towards the deeper areas of the shelf. From March to November, the retreat of the SACW leads to a greater influence of the TW on shelf processes (Mahiques *et al.* 1999).

The study area is located near the boundary between tropical and subtropical zones, and is in the continental climate region classified as Tropical Atlantic. Changes in the wind and rainfall regimes in the region have been attributed to several factors: variations of the anticyclone associated with the South Atlantic Subtropical High (SAS), which is responsible for the dominance of NE winds both in spring and summer (Bastos & Ferreira 2000); latitudinal shifts of the Inter Tropical Convergence Zone (ITCZ); and the penetration of cold fronts, which is associated with the northward displacement of the Polar Anticyclone. The cold front systems are active throughout the year, and have a strong influence on temperature and rainfall regimes (Nobre & Shukla 1996).

The rainfall regime in the southern Brazilian region follows a seasonal pattern. According to Vera *et al.* (2002), regional precipitation during winter and early spring (May–September) is mainly due to the extratropical circulation regime as a result of migratory cyclones along the subtropical Atlantic coast. Conversely, the warm season precipitation (September–April) is associated with the activity of the South American Summer Monsoon (SASM). During this time, the SASM is associated with the South Atlantic Convergence Zone (SACZ), which is responsible for the intensity and location of the summer precipitation. The SACZ exhibits significant variations in terms of intensity and geographical extensions in different timescales (Paegle & Mo 2002). The monsoon starts to decay when the convection changes northward to the equator, following the decrease of solar heating in the South American subtropical zone.

Although the SASM is weaker and has a shorter life span than the Asian monsoon system, the SASM is still responsible for more than 80% of the annual mean precipitation in its activity centre (23°S) and for 50% of the total summer precipitation in the coast south of 25°S (Cruz *et al.* 2006).

Proxy records have revealed that changes in the wind and rainfall regimes also occurred during the Holocene in tropical and subtropical South America (Behling 1998; Ledru *et al.* 1998; Gilli *et al.* 2005). These changes have been attributed to many factors, such as variations in the intensity of El Niño-Southern Oscillation (ENSO) (Martin *et al.* 1993), latitudinal shifts of the ITCZ (Jaeschke *et al.* 2007) and latitudinal changes in the Meridional Overturning Circulation (Cruz *et al.* 2009).

Modern sedimentation

Hydrodynamic processes, oceanic water mass dynamics and shelf circulation are the main controlling factors for the primary productivity and sediment redistribution in the upper continental margin off SE Brazil, leading to the establishment of differences in sedimentation rates and sedimentary facies (Mahiques *et al.* 2004). Modern sedimentation rates vary from 2 to 68 cm ka^{-1}, and are controlled by shelf and upper-slope morphology, the BC meander dynamics and the CW offshore motion. The highest sedimentation rates are found in a low-energy (ria type) coastal system, as well as in the upwelling zones of Santa Catarina and Cabo Frio. On the southern sector of the shelf, sedimentation is strongly influenced by the terrigenous input from the Rio de la Plata outflow, which is transported via the BCC, leading to the development of inner- and middle-shelf mudbelts. The lowest rates occur on the outer shelf and shelf break, confirming a strong dependency of the coupled BC–Intermediate Western Boundary Current (BC–IWBC) system on the sedimentary processes (Mahiques *et al.* 2011). Sediment distribution exhibits latitudinal changes and bathymetric control. The area off São Sebastião Island (24°S–45°30′W) marks a boundary between two main sedimentary zones that are characterized by differences in the organic and inorganic fractions of their sediments (Fig. 5.3).

South of São Sebastião Island, the depositional processes are related to the shelf penetration of the shelf water associated with the Rio de la Plata runoff. This water plume, carried northwards along the southern Brazilian shelf by the BCC (Souza & Robinson 2004), has been identified as far north as 24°S (Campos *et al.* 1999). However, Nd and Pb isotope data (Mahiques *et al.* 2008) indicate that the influence of the Rio de la Plata estuary is limited to the south of 28°S, as the sector between latitudes 28°S and 24°S is a transitional zone. Furthermore, the pelagic processes associated with the northward propagation of southerly origin cold waters and the meandering of the BC, as well as the shelf morphology, favour the establishment of high productivity zones, with correspondingly high sedimentation rates south of São Sebastião Island (Mahiques *et al.* 2004, 2005).

North of 24°S, the relative heterogeneity of sediments is related to more complex hydrodynamics, especially on the middle and inner shelves, where seafloor morphology, the presence of islands and the shape of the coastline lead to the establishment of a complex sedimentary mosaic. Also, the sedimentary processes exhibit a conspicuous bathymetric control, such that, on the inner shelf, Holocene sedimentation fills the irregular relief developed during low sea-level stages.

On the outer shelf, the seafloor characteristics result from the more intense action of the meandering nature of the Brazil Current with the exposure of extensive relict surfaces (Mahiques *et al.* 2002). This characteristic is clearly observed along the whole shelf as far south as 28°S, where seismic and echo-sounding profiles together with sediment samples reveal the existence of a rough surface and ages as old as 2060 years BP in non-hermatypic corals sampled in core tops.

North of São Sebastião Island, low sedimentation rates in the outer shelf are associated with the main flow of the BC, which acts as a 'floor-polisher' on the seafloor (Mahiques *et al.* 2002).

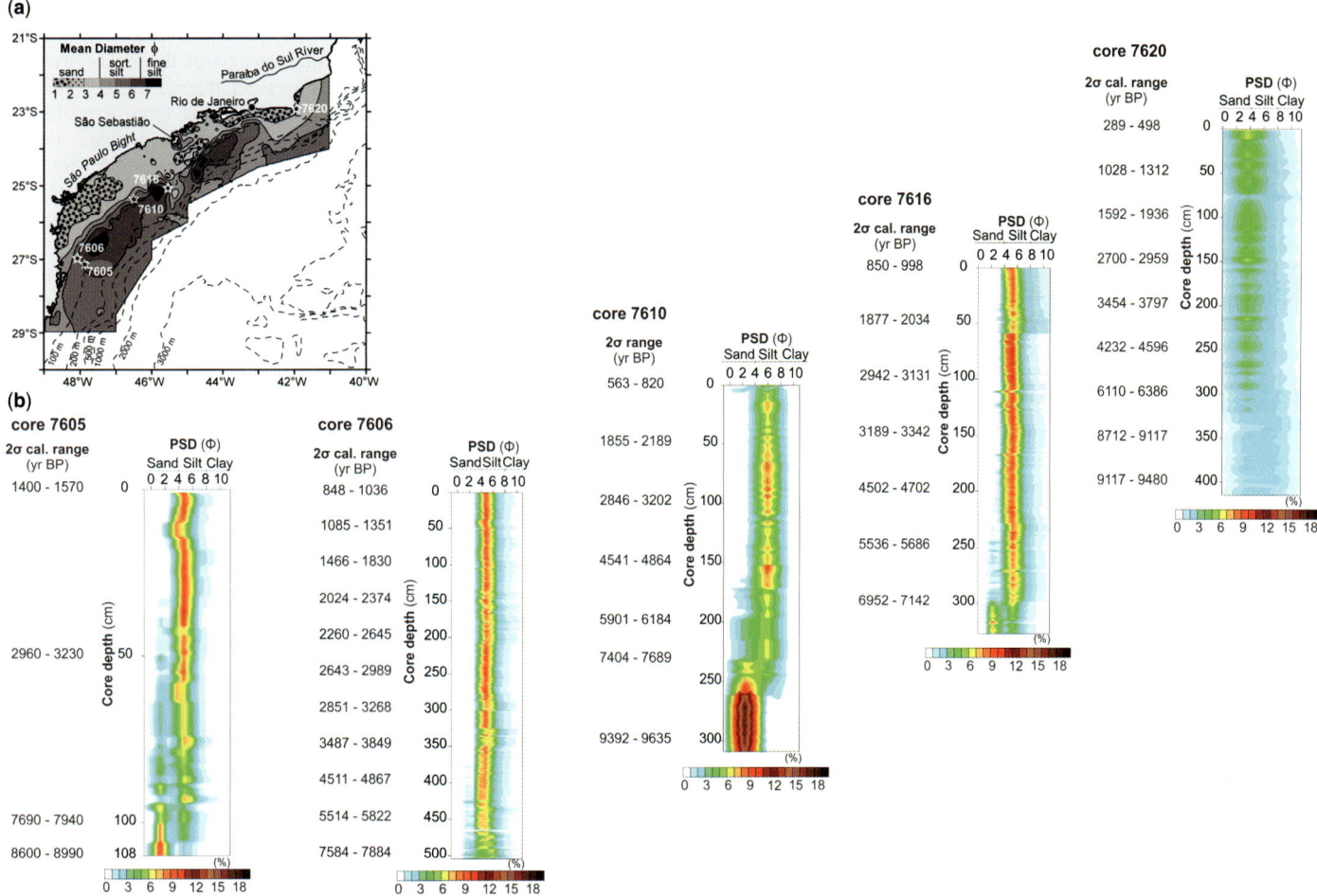

Fig. 5.3. (**a**) Location map with core sites and the distribution of the mean particle size diameter (φ) for the surface sediments of the study area, from Gyllencreutz *et al.* (2010); coring locations are indicated with white stars. (**b**) Particle-size results for cores 7605, 7606, 7610, 7616 and 7620 represented by particle-size distribution (PSD), where the frequency percentage in each size class is indicated by coloured/shaded contours.

Higher rates are apparently correlated with areas of upwelling processes related to the BC's meandering nature, as observed in the Cabo Frio region (23°S–42°W).

The present oceanographic and climatic conditions offer the southern Brazilian shelf a privileged status for the study of short-term (seasonal and decadal) changes in the wind-driven currents and freshwater discharge regimes of the SW Atlantic Ocean.

Holocene palaeo-oceanographic evolution

During the Late Quaternary, the morphotectonic control of sedimentation was overprinted by the transgressive–regressive events related to sea-level changes, mainly the desiccation and submersion of the shelf during the Last Glacial Cycle. During the LGM, the coastline in the study area corresponded roughly to the 120 m isobath. Thus, the outer-shelf and upper-slope topography must have been the main factors that then controlled the dynamics of the water masses. In this period, the NE–SW general orientation of the São Paulo Embayment, together with the absence of a flat, wider shelf and eventually a stronger trade-wind regime (Abrantes 2000; Flores *et al.* 2000), may have been responsible for changes in the sedimentary pattern through displacement of the Brazil Current's meandering dynamics in the study area.

The hydrodynamic control, together with the relative tectonic stability and the absence of post-glacial rebound, makes the SSE Brazilian continental margin a favourable site for

investigations of Late Quaternary climatic changes of the SW Atlantic (Mahiques *et al.* 2011). Yet, little is known about the SW Atlantic continental margin Quaternary history and evolution.

For this purpose, five piston cores (7605, 7606, 7610, 7616 and 7620) that cover a time span of approximately 9 kyr were retrieved from the inner–middle southern Brazilian shelf on board the RV *Prof. W. Besnard* in 2005. Cores 7605 (27°6.24′S and 047°48.24′W, water depth of 93 m) and 7606 (26°59.28′S and 048°4.56′W, water depth of 60 m) were collected from the southernmost part of the study area, where the Rio de la Plata sediments have a greater influence. Cores 7610 (25°30.48′S and 046°38.1′W, water depth of 89 m) and 7616 (25°5.88′S and 045°38.6′W, water depth of 100 m) were also collected south of São Sebastião Island, with the latter retrieved close to the convergence region of the main currents in the area, corresponding to the northernmost influence of the BCC current, according to Gyllencreutz *et al.* (2010). Finally, core 7620 (22°56.52′S and 041°58.8′W, water depth of 43 m) was collected north of this island, in the upwelling region of Cabo Frio (Fig. 5.3).

These high-resolution Holocene sediment profiles obtained on the Brazilian shelf reveal a considerable variability in grain size (Fig. 5.3), sediment composition (calcium carbonate, organic carbon and elemental contents, i.e. Al, Ba, Fe, Ti) and benthic foraminifer assemblages, indicating that the sediment input, productivity and current strength and/or variability were affected in approximately the last 9 kyr by sea-level oscillations and changes in the wind and Rio de la Plata discharge (Mahiques *et al.* 2009; Gyllencreutz *et al.* 2010).

Early Holocene (9–7 cal ka BP)

The Early Holocene is represented in the base of cores 7605, 7610 and 7620. In the southernmost cores (7605, 7610), a pronounced bimodal distribution with a sand mode (2φ) is observed, this mode experiences a conspicuous decrease beginning at approximately 7 cal ka BP (Fig. 5.3). If we assume that the sea level in the Early Holocene was 50–40 m below the present mean sea level (Fig. 5.2), we conclude that the occurrence of sandy sediments is probably related to changes in sediment availability, and, hence, provenance and/or current variability.

In the northernmost core (7620), collected from the inner shelf, the deposition of muddy sediments with high organic matter content and benthic foraminifer species typical of shallow coastal environments reflects low sea-level conditions and a less effective action of the BC over the shelf prior to 7 cal ka BP (Nagai *et al.* 2009). Hydrodynamic changes in the Brazilian continental margin, derived from sea-level fluctuations, have been explored previously (Mahiques *et al.* 2007; Nagai *et al.* 2010). According to these authors, low sea-level conditions promote offshore displacement of the BC, resulting in a less intense action of this current on the shelf.

Mid-Holocene (7–5 cal ka BP)

From approximately 7 to 5 cal ka BP, a decrease in sandy sediment is observed in all the cores except in core 7620 (Fig. 5.3). The disappearance of the sand modes, however, took place at different time intervals in each core: in the northernmost core (7616), the sand mode disappeared at approximately 5.6 cal ka BP; in core 7610, at approximately 5 cal ka BP; in the southernmost cores, the sand mode disappears at approximately 4.5 cal ka BP (7606); and, in core 7605, the sand mode continues into the Late Holocene, disappearing at approximately 1.8 cal ka BP. The second (and more important) grain-size mode is represented by finer sediments centred at approximately 5φ (medium–coarse silts) (Fig. 5.3).

During the Mid-Holocene, a southward migration of the ITCZ from 6 cal ka BP is proposed (Haug *et al.* 2001; Wanner *et al.* 2008), and which would promote a rainfall regime increase in the continental area of SE South America and, consequently, an increase in the Rio de la Plata discharge. Conversely, the Mid-Holocene highstand would have weakened the hydraulic gradient, significantly reducing the fluvial freshwater current strength (Gyllencreutz *et al.* 2010).

Taking into account the relative sea-level oscillations and environmental variations, Gyllencreutz *et al.* (2010) proposed that the Mid-Holocene sand found in cores south of 25°S was derived from the Argentinian shelf. According to these authors, Mid-Holocene coarser sediment deposition would be the result of the combination of higher sea-level conditions, inhibiting the Rio de la Plata outflow hydraulic gradient and SW winds that enhanced the northward sediment transport along the SE South American coast. It is important to highlight that this hypothesis, however, was based solely on grain-size data, and that sand content and sortable silt size lacked correlation. Indeed, latitudinal (south–north) differences in the sand-mode representativeness in grain-size populations suggest a southern source for these coarser sediments. However, grain-size data alone cannot assign the origin of the Mid-Holocene sands, since it is also possible that the Mid-Holocene sands represent reworked southern Brazil shelf sediments made available by the Early–Middle Holocene sea-level rise (Fig. 5.2).

Meanwhile, north of São Sebastião Island (core 7620), the increase in sandy sediment (centred at around 3φ) indicates increasing current speed or velocity variability during this time interval, which is corroborated by foraminiferal data (Nagai *et al.* 2009; Gyllencreutz *et al.* 2010).

Late Holocene (5 cal ka BP–present)

In cores collected in the southernmost part of the study area, an increase in muddy sediment is observed from 5 up to 3 cal ka BP, with a significant increase in mud content from 3 cal ka BP (Fig. 5.3). During this time, the mean sea-level position would not have affected the fluvial freshwater current strength.

The progressive southward migration of the ITCZ would probably be responsible for the frequency and amplitude increase in ENSO events since 5 cal ka BP (Haug *et al.* 2001), and changes in the wind regime and periods of high humidity would be expected, promoting intense discharge events from the Paraná River (the major contributor to the Rio de la Plata). Owing to the fact that the northward displacement of the Rio de la Plata plume depends both on precipitation over its drainage basin and favourable SSW winds (Möller *et al.* 2008; Piola *et al.* 2008). Mahiques *et al.* (2009) attributed their results to climatic oscillations, namely increased moisture conditions and intensification of southerly winds in the Late Holocene. Moreover, an increase in precipitation over SE South America has also been reported as a main factor influencing the increase in terrigenous supply for the Uruguayan slope from the Mid- to Late Holocene (Chiessi *et al.* 2010).

Taking into account that during El Niño events longshore NNE winds counteract the northward plume penetration (Piola *et al.* 2005), Gyllencreutz *et al.* (2010) suggested that the maximum northward extensions of the Rio de la Plata plume water are associated with low discharge years (La Niña events), as fine sediments deposited in the Rio de la Plata area during El Niño events could be resuspended by the stronger wind-driven currents during subsequent La Niña events, reaching latitudes as far north as 25°S (Fig. 5.3).

In the northern part of the area, the sedimentation history is different, with a predominance of sandy sediment since 3 cal ka BP and an upwelling enhancement in the Cabo Frio area (Nagai *et al.* 2009). This process is highly correlated to NE winds (Rodrigues & Lorenzzetti 2001), which are more intense during El Niño events. A prevailing NE wind direction and southward flow of the BC would explain the high proportion of sand in core 7620, which is apparently also related to the Paraíba do Sul River outflow (Gyllencreutz *et al.* 2010).

Final remarks

Although there is a lack of knowledge of the sedimentary processes and still much controversy about the sea-level change curve, even for the Middle and Late Holocene, this study has shown that the southern Brazilian shelf is key to the study of the palaeoenvironmental changes of eastern South America, especially regarding oscillations in the wind regime and humidity.

The dichotomy in the modern sedimentary and oceanographic processes observed south and north of São Sebastião Island has been present in the Brazilian shelf at least since the Early Holocene.

Finally, owing to its relative tectonic stability and non-glaciated character, the southern Brazilian shelf may also be considered to be a favourable area for studies on sea-level changes during the Last Glacial Cycle. This is not yet possible, as the number of available long cores that reach Pleistocene strata is very small, as is also the number of high-resolution shallow seismic profiles.

This work is a contribution by the Brazilian representatives to IGCP-464 and IGCP-526 projects. Thanks are due to Dr A. Chivas (University of Wollongong, Australia) and Dr F.L. Chiocci (Università di Roma 'La Sapienza', Italy) for the opportunities given during the development of the project. We would also like to thank Dr R. Violante for his review, comments and suggestions. The authors are also indebted to the Fundação de Amparo à Pesquisa do Estado de São Paulo (FAPESP, grant numbers 2003/10740-0 and 2010/04617-5, and scholarship 2009/01594-6) and the Conselho de Desenvolvimento Científico e Tecnológico (CNPq, grant number 301106/2010-0) for the financial support provided for studies on the continental shelf.

References

ABRANTES, F. 2000. 200 000 yr diatom records from Atlantic upwelling sites reveal maximum productivity during LGM and a shift in phytoplankton community structure at 185 000 yr. *Earth and Planetary Science Letters*, **176**, 7–16.

ALMEIDA, F. F. M. & CARNEIRO, C. D. R. 1998. Origem e evolução da Serra do Mar. *Revista Brasileira de Geociências*, **28**, 135–150.

ANGULO, R. J. & LESSA, G. 1997. The Brazilian sea-level curves: a critical review with emphasis on the curves from Paranaguá and Cananéia regions. *Marine Geology*, **140**, 141–166.

ANGULO, R. J. & SUGUIO, K. 1995. Re-evaluation of the maxima of the Holocene sea-level curve for the State of Paraná, Brazil. *Palaeogeography, Palaeoclimatology, Palaeoecology*, **112**, 385–393.

ANGULO, R. J., GIANNINI, P. C. F., SUGUIO, K. & PESSENDA, L. C. R. 1999. The relative sea-level changes in the last 5500 years southern Brazil (Laguna-Imbituba region, Santa Catarina State) based on vermetid ^{14}C ages. *Marine Geology*, **159**, 327–339.

ANGULO, R. J., LESSA, G. C. & SOUZA, M. C. 2006. A critical review of mid to late-Holocene sea-level fluctuations on the eastern Brazilian coastline. *Quaternary Science Reviews*, **25**, 486–506.

BASTOS, C. C. & FERREIRA, N. J. 2000. Análise climatológica da alta subtropical do Atlântico Sul. [Climatology of the South Atlantic Subtropical High.] Paper presented at the Congresso Brasileiro de Meteorologia, 11, Rio de Janeiro, 2000 Anais, Sociedade Brasileira de Meteorologia (SBMET), Rio de Janeiro.

BEHLING, H. 1998. Late Quaternary vegetational and climatic changes in Brazil. *Review of Palaeobotany and Palynology*, **99**, 143–156.

BRANNER, J. C. 1904. The stone reefs of Brazil, their geological and geographical relations. *Bulletin of the Museum of Comparative Zoology*, **44**, 207–275.

CAMPOS, E. J. D., LENTINI, C. D., MILLER, J. L. & PIOLA, A. R. 1999. Interannual variability of the sea surface temperature in the South Brazil Bight. *Geophysical Research Letters*, **26**, 2061–2064.

CASTRO, B. M., MIRANDA, L. B. & MIYAO, S. Y. 1987. Condições hidrográficas na plataforma continental de Ubatuba: variações sazonais e de média escala. *Boletim do Instituto Oceanográfico*, **35**, 135–151.

CASTRO, B. M., LORENZZETTI, J. A., SILVEIRA, I. C. A. & MIRANDA, L. B. 2006. Estrutura termohalina e circulação na região entre o Cabo de São Tomé (RJ) e o Chuí (RS). [The thermohaline structure and oceanic circulation between Cabo de São Tomé (RJ) and Chuí (RS).]. *In*: MADUREIRA, C. L. D. B. (ed.) *O ambiente oceanográfico da plataforma continental e do talude na região sudeste-sul do Brasil.* [The oceanographic environment of the continental shelf and slope in south-southeastern Brazil.] L.S. Editora, Universidade de São Paulo, São Paulo, 11–120.

CHIESSI, C. M., MULITZA, S., PÄTZOLD, J. & WEFER, G. 2010. *How Different Proxies Record Precipitation Variability over Southeastern South America.* Institute of Physics, Conference Series: Earth and Environmental Science, **9**.

CIOTTI, A. M., ODEBRECHT, C., FILMAN, G. & MOLLER, O. JR. 1995. Freshwater outflow and Subtropical Convergence influence on phytoplankton biomass on the southern Brazilian continental shelf. *Continental Shelf Research*, **15**, 1737–1756.

CORRÊA, I. C. S. 1996. Les variations du niveau de la mer durant les derniers 17.500 ans BP: l'exemple de la plate-forme continentale de Rio Grande do Sul–Brésil. *Marine Geology*, **130**, 163–178.

CRUZ, F. W., BURNS, S. J., KARMANN, I., SHARP, W. D. & VUILLE, M. 2006. Reconstruction of regional atmospheric circulation features during the late Pleistocene in subtropical Brazil from oxygen isotope composition of speleothems. *Earth and Planetary Science Letters*, **248**, 494–506.

CRUZ, F. W., VUILLE, M. ET AL. 2009. Orbitally driven east–west antiphasing of South American precipitation. *Nature Geoscience*, **2**, 210–214.

FERREIRA, C. S., MADUREIRA, L. S. P., KLIPEL, S., WEIGERT, S., HABIAGA, R. G. P. & DUVOISIN, A. C. 2005. *Mapas do relevo marinho das regiões 25 sudeste, sul e central do Brasil: acústica e altimetria por satélite.* [Maps of the marine morphology of the Brazilian southeast, south and central regions: acoustics and satellite altimetry.] Série Documentos REVIZEE: Score Sul. Instituto Oceanográfico, São Paulo.

FIGUEIRA, R. C. L., TESSLER, M. G., MAHIQUES, M. M. & CUNHA, I. I. L. 2006. Distribution of ^{137}Cs, ^{238}Pu and $^{239+240}$Pu in sediments of the southeastern Brazilian shelf–SW Atlantic margin. *Science of the Total Environment*, **357**, 146–159.

FIGUEIREDO, A. G. & MADUREIRA, L. S. P. 2004. *Topografia, composição, refletividade do substrato marinho e identificação de províncias sedimentares na Região Sudeste-Sul do Brasil.* [Marine topography, substrate composition, reflectivity and sedimentary provinces identification of south-southeastern Brazil.] Série Documentos REVIZEE: Score Sul. Instituto Oceanográfico, São Paulo.

FIGUEIREDO, A. G. & TESSLER, M. G. 2004. *Topografia e composição do substrato marinho da Região Sudeste-Sul do Brasil.* [Marine topography and substrate composition of south-southeastern Brazil.] Série Documentos REVIZEE: Score Sul. Instituto Oceanográfico, São Paulo.

FLORES, J. A., BARCENA, M. A. & SIERRO, F. J. 2000. Ocean-surface and wind dynamics in the Atlantic Ocean off Northwest Africa during the last 140 000 years. *Palaeogeography, Palaeoclimatology, Palaeoecology*, **161**, 459–478.

FURTADO, V. V., BONETTI FILHO, J. & CONTI, L. A. 1996. Paleo river valley morphology and sea-level changes. *Anais da Academia Brasileira de Ciências*, **68**, 163–169.

GILLI, A., ARIZTEGUI, D., ANSELMETTI, F. S., MCKENZIE, J. A., MARKGRAF, V. & HAJDAS, I. 2005. Mid-Holocene strengthening of the Southern Westerlies in South America – Sedimentological evidences from Lago Cardiel, Argentina (49°S). *Global and Planetary Change*, **49**, 75–93.

GYLLENCREUTZ, R., MAHIQUES, M. M., ALVES, D. V. P. & WAINER, I. K. C. 2010. Mid- to late-Holocene paleoceanographic changes on the southeastern Brazilian shelf based on grain size records. *The Holocene*, **20**, 1–13.

HARTT, C. F. 1870. *Geology and Physical Geography of Brazil.* Boston, Fields, Osgood & Co., Boston, MA.

HAUG, G. H., HUGHEN, K. A., SIGMAN, D. M., PETERSON, L. C. & RÖHL, U. 2001. Southward migration of the Intertropical Convergence Zone through the Holocene. *Science*, **293**, 1304–1308.

JAESCHKE, A., RÜHLEMANN, C., ARZ, H., HEIL, G. & LOHMANN, G. 2007. Coupling of millennial-scale changes in sea surface temperature and precipitation off northeastern Brazil with high-latitude climate shifts during the last glacial period. *Paleoceanography*, **22**, PA4206, http://dx.doi.org/10.1029/2006PA001391

KLEIN, D. A. 2005. *Registros de variações ambientais no Canal de São Sebastião (Estado de São Paulo), durante o Último Ciclo Glacial.* [Records of environmental changes in the São Sebastião Channel (São Paulo State), during the Last Glacial Cycle.] PhD thesis, University of São Paulo.

KOWSMANN, R. O. & COSTA, M. P. A. 1979. *Sedimentação Quaternária da Margem Continental Brasileira e das Areas Adjacentes.* [Quaternary sedimentation of the Brazilian continental margin and adjacent areas.] PETROBRAS, Rio de Janeiro, Projeto REMAC Series, **8**.

LEDRU, M. P., SALGADO-LABOURIAU, M. L. & LORSCHEITTER, M. L. 1998. Vegetation dynamics in southern and central Brazil during the last 10 000 yr BP. *Review of Palaeobotany and Palynology*, **99**, 131–142.

MAHIQUES, M. M. & SOUZA, L. A. P. 1999. Shallow seismic reflectors and upper Quaternary sea-level changes in the Ubatuba region, São Paulo State, Southeastern Brazil. *Revista Brasileira de Oceanografia*, **47**, 1–10.

MAHIQUES, M. M., MISHIMA, Y. & RODRIGUES, M. 1999. Characteristics of the sedimentary organic matter on the inner and middle continental shelf between Guanabara Bay and São Francisco do Sul, eastern Brazilian margin. *Continental Shelf Research*, **19**, 775–798.

MAHIQUES, M. M., SILVEIRA, I. C. A., SOUSA, S. H. M. & RODRIGUES, M. 2002. Post-LGM sedimentation on the outer shelf-upper slope of the northermost part of São Paulo Bight, southeastern Brazil. *Marine Geology*, **181**, 387–400.

MAHIQUES, M. M., TESSLER, M. G. *ET AL.* 2004. Hydrodynamically-driven patterns of recent sedimentation in the shelf and upper slope off southeast Brazil. *Continental Shelf Research*, **24**, 1685–1697.

MAHIQUES, M. M., BÍCEGO, M. C., SILVEIRA, I. C. A., SOUSA, S. H. M., LOURENÇO, R. A. & FUKUMOTO, M. M. 2005. Modern sedimentation in the Cabo Frio upwelling system, Southeastern Brazilian shelf. *Annais da Acadêmia Brasileira de Ciências*, **77**, 535–548.

MAHIQUES, M. M., FUKUMOTO, M. M., SILVEIRA, I. C. A., FIGUEIRA, R. C. L., BICEGO, M. C., LOURENCO, R. A. & SOUSA, S. H. M. 2007. Sedimentary changes on the Southeastern Brazilian upper slope during the last 35,000 years. *Anais da Academia Brasileira de Ciências*, **79**, 171–181.

MAHIQUES, M. M., TASSINARI, C. C. G. *ET AL.* 2008. Nd and Pb isotope signatures on the Southeastern South American upper margin: implications for sediment transport and source rocks. *Marine Geology*, **250**, 51–63.

MAHIQUES, M. M., WAINER, I. K. C. *ET AL.* 2009. A high-resolution Holocene record on the Southern Brazilian shelf: paleoenvironmental implications. *Quaternary International*, **206**, 52–61.

MAHIQUES, M. M., SOUSA, S. H. M. *ET AL.* 2010. The Southern Brazilian Shelf: general characteristics, Quaternary evolution and sediment distribution. *Brazilian Journal of Oceanography*, **58**, 25–34.

MAHIQUES, M. M., SOUSA, S. H. M. *ET AL.* 2011. Radiocarbon geochronology of the sediments of the Sao Paulo Bight (southern Brazilian upper margin). *Anais da Academia Brasileira de Ciências*, **83**, 817–834.

MARTIN, L., FOURNIER, M., MOURGUIART, P., SIFEDDINE, A., TURCQ, B., ABSY, M. L. & FLEXOR, J.-M. 1993. Southern Oscillation signal in South American palaeoclimatic data of the last 7000 years. *Quaternary Research*, **39**, 338–346.

MARTIN, L., DOMINGUEZ, J. M. L. & BITTENCOURT, A. C. S. P. 2003. Fluctuating Holocene sea-levels in eastern and southeastern Brazil: evidence from multiple fossil and geometric indicators. *Journal of Coastal Research*, **19**, 101–124.

MARTINS, L. R., MARTINS, I. R. & URIEN, C. M. 2003. Aspectos sedimentares da plataforma continental na área de influência do Rio de La Plata. *Gravel*, **1**, 68–80.

MEISLING, K. E., COBBOLD, P. R. & MOUNT, V. S. 2001. Segmentation of an obliquely rifted margin, Campos and Santos Basins, Southeastern Brazil. *American Association of Petroleum Geologists Bulletin*, **85**, 1903–1924.

MILLIMAN, J. D. & BARRETTO, H. T. 1975. Background. Upper continental margin sedimentation off Brazil. *Contributions to Sedimentology*, **4**, 1–10.

MÖLLER, O. O., JR., PIOLA, A. R., FREITAS, A. C. & CAMPOS, E. J. D. 2008. The effects of river discharge and seasonal winds on the shelf off southeastern South America. *Continental Shelf Research*, **28**, 1607–1624.

NAGAI, R. H., SOUSA, S. H. M., BURONE, L. & MAHIQUES, M. M. 2009. Paleoproductivity changes during the Holocene in the inner shelf of Cabo Frio, southeastern Brazilian continental margin: benthic foraminifera and sedimentological proxies. *Quaternary International*, **206**, 62–71.

NAGAI, R. H., SOUSA, S. H. D. E., LOURENCO, R. A., BICEGO, M. C. & MAHIQUES, M. M. 2010. Paleoproductivity changes during the Late Quaternary in the Southeastern Brazilian upper continental margin of the Southwestern Atlantic. *Brazilian Journal of Oceanography*, **58**, 31–41.

NOBRE, P. & SHUKLA, J. 1996. Variations of sea surface temperature, wind stress, and rainfall over the tropical Atlantic and South America. *Journal of Climate*, **9**, 2464–2479.

PAEGLE, J. N. & MO, K. C. 2002. Linkages between summer rainfall variability over South America and sea surface temperature anomalies. *Journal of Climate*, **15**, 1389–1407.

PIOLA, A. R., MOLLER, O. O. & PALMA, E. 2005. O impacto do Prata sobre o oceano Atlântico. *Ciência Hoje*, **36**, 30–37.

PIOLA, A. R., ROMERO, S. I. & ZAJACZKOVSKI, U. 2008. Space-time variability of the plata plume inferred from ocean color. *Continental Shelf Research*, **28**, 1556–1567.

RAZIK, S., CHIESSI, C. M., ROMERO, O. E. & DOBENECK, T. 2013. Interaction of the South American Monsoon System and the Southern Westerly Wind Belt during the last 14 kyr. *Palaeogeography, Palaeoclimatology, Palaeoecology*, **374**, 28–40.

RODRIGUES, R. R. & LORENZZETTI, J. A. 2001. A numerical study of the effects of bottom topography and coastline geometry on the Southeast Brazilian coastal upwelling. *Continental Shelf Research*, **21**, 371–394.

SOUZA, R. B. & ROBINSON, I. S. 2004. Lagrangian and satellite observations of the Brazilian Coastal Current. *Continental Shelf Research*, **24**, 241–262.

SUGUIO, K., MARTIN, L., BITTENCOURT, A. C. S. P., DOMINGUEZ, J. M. L., FLEXOR, J. M. & DE AZEVEDO, A. E. G. 1985. Flutuações do nível relativo do mar durante o quaternário superior ao longo do litoral brasileiro e suas implicações na sedimentação costeira. *Revista Brasileira de Geociências*, **15**, 273–286.

VERA, C. S., VIGLIAROLO, P. K. & BERBERY, E. H. 2002. Cold season synoptic scale waves over subtropical South America. *Monthly Weather Review*, **130**, 684–699.

VIOLANTE, R. A. & PARKER, G. , 2004. The post-last glacial maximum transgression in the La Plata River and adjacent inner continental shelf, Argentina. *Quaternary International*, **114**, 167–181.

VIANA, A. R., FAUGÈRES, J.-C. & STOW, D. A. V. 1998. Bottom-current-controlled sand deposits: a review of modern shallow- to deep-water environments. *Sedimentary Geology*, **115**, 53–80.

WANNER, H., BEER, J. *ET AL.* 2008. Mid- to Late Holocene climate change: an overview. *Quaternary Science Reviews*, **27**, 1791–1828.

YBERT, J. P., BISSA, W. M., CATHARINO, E. L. M. & KUTNER, M. 2003. Environmental and sea-level variations on the southeastern Brazilian coast during the Late Holocene with comments on prehistoric human occupation. *Palaeogeography, Palaeoclimatology, Palaeoecology*, **189**, 11–24.

ZEMBRUSCKI, S. G. 1979. Geomorfologia da margem continental sul brasileira e das bacias oceânicas adjacentes. *In*: CHAVES, H. A. F. (ed.) *Geomorfologia da margem continental brasileira e das bacias oceânicas adjacentes*, PETROBRAS, Rio de Janeiro, Projeto REMAC Series, **7**, 129–177.

Chapter 6

The Argentine continental shelf: morphology, sediments, processes and evolution since the Last Glacial Maximum

R. A. VIOLANTE[1]*, C. M. PATERLINI[1], S. I. MARCOLINI[1], I. P. COSTA[1], J. L. CAVALLOTTO[1], C. LAPRIDA[2,3], W. DRAGANI[4], N. GARCÍA CHAPORI[2], S. WATANABE[5], V. TOTAH[5], E. I. ROVERE[6] & M. L. OSTERRIETH[7]

[1]*Division of Marine Geology and Geophysics, Argentine Hydrographic Survey, Avenida Montes de Oca 2124, Buenos Aires C1270AVB, Argentina*

[2]*Department of Geology, University of Buenos Aires, IDEAN-SACMA, Ciudad Universitaria, Buenos Aires C1428EHA, Argentina*

[3]*CONICET, National Council of Science, Buenos Aires, Argentina*

[4]*Division of Coastal Oceanography, Argentine Hydrographic Survey, Avenida Montes de Oca 2124, Buenos Aires C1270AVB, Buenos Aires, Argentina*

[5]*Argentine Museum of Natural Sciences 'B. Rivadavia', CONICET (National Council of Science), Avenida Angel Gallardo 470, Buenos Aires C1405DJR, Argentina*

[6]*Division of Regional Geology, Argentine Geological Survey, Parque Tecnológico Miguelete, Avenida General Paz 5445, Edificio 25, San Martín, Buenos Aires B1650WAB, Argentina*

[7]*Institute of Coastal Geology and Quaternary, University of Mar del Plata, Funes 3350, Mar del Plata B7602AYL, Argentina*

Corresponding author (e-mail: violante@hidro.gov.ar)

Abstract: The Argentine continental shelf is one of the largest and smoothest siliciclastic shelves in the world. Although it is largely emplaced in a passive continental margin, the southernmost regions are related to transcurrent and active margins respectively associated with the Malvinas Plateau and Scotia Arc. Sea-level fluctuations, sediment dynamics and climatic/oceanographic processes were the most important conditioning factors in the modelling of the shelf, with a minor influence from isostatic and tectonic factors that are more relevant in the southernmost regions. The shelf is shaped by diverse geomorphic features, among which the most significant are four sets of terraces genetically associated to sea-level stillstands during the post-glacial transgression; the final one occurred at around 11 ka and is associated with the Younger Dryas event. The Last Glacial Maximum (LGM) sedimentary sequence is composed of, on average, 5–15 m-thick terrigenous, siliciclastic, relict–palimpsest sands mainly sourced from the Andean region, with minor amounts of bioclast and gravels, resulting from the reworking of pre-transgressive coastal environments.

The Argentine continental shelf (ACS) is one of the largest and smoothest siliciclastic shelves in the world. Its morphological and evolutionary characteristics are primarily the consequence of its geotectonic emplacement, associated with a passive margin on most of its surface. Processes involved in its final stages of evolution during glacial and post-glacial times comprise climatic and oceanographic changes, sea-level fluctuations, ocean dynamics and sedimentary processes. As a result of its large latitudinal extension and the proximity of its southernmost regions to the glaciated areas of Patagonia and Antarctica, most of the modelling processes that occurred during post-LGM times changed in magnitude from north to south, resulting in varied and complex morpho-sedimentary features.

Bathymetric and high-resolution seismic surveys, as well as coring and surface-sediment samplings, have been systematically carried out in the northern (Pampean) region of the ACS in the last 30 years by the Argentine Hydrographic Survey (see Parker & Violante 1982; Parker et al. 1982, 1996, 1997, 1999, 2008; Violante et al. 1992, 2007, 2012; Violante 2004, 2005; Violante & Parker 2004; Violante & Rovere 2005; Violante & Cavallotto 2011) in order to study Quaternary morpho-sedimentary, stratigraphic, dynamic and evolutionary aspects of the shelf. Those surveys complemented previous studies performed by the same institution (see Urien 1970; Urien & Ewing 1974; Urien et al. 1979). Other regions of the ACS, such as those adjacent to Patagonia, have not been studied in detail to date, at least not for the

uppermost (Quaternary) sequences, although valuable regional information is available (see Urien 1970; Ewing & Lonardi 1971; Urien & Ewing 1974; Zambrano & Urien 1974; Urien et al. 1979; Ramos & Turic 1996).

Despite the presently uneven knowledge, with relatively well-known areas and others where studies are still lacking, an updated revision of the geological aspects of the ACS is necessary to obtain a framework for future studies. This contribution attempts to synthesize such a framework.

Geomorphological and geotectonic setting of the ACS

The ACS is part of the Argentine continental margin (ACM), which covers an area of around 2×10^6 km^2. Basic characteristics of the shelf are summarized in Table 6.1. The margin developed on different geotectonic settings, from the stable pericratonic areas of the Brazilian shield in the vicinity of the de la Plata River to the tectonically and isostatically active southern Patagonian regions. It comprises major physiographical features, such as the continental shelf, slope and rise, as well as the Malvinas Plateau and Scotia Arc (Fig. 6.1).

The ACM is located on the South American plate in a region of cortical extension associated with the break-up of Gondwana and constitutes, on most of its surfaces, a typical Atlantic passive continental margin. Only in the southern regions of the margin is it

From: CHIOCCI, F. L. & CHIVAS, A. R. (eds) 2014. *Continental Shelves of the World: Their Evolution During the Last Glacio-Eustatic Cycle.* Geological Society, London, Memoirs, **41**, 55–68. http://dx.doi.org/10.1144/M41.6

Table 6.1. *Summary of basic Argentine continental shelf characteristics*

Length of the shelf (km)	2400
Average width (km)	400
Tidal, wave, current ranges	Tides: between 1.5 and 12 m. Waves: 1–4 m. In both cases increasing to the south
Dominating process (wave/current/tide)	Waves dominate in northern sector (Pampas) Tides dominate in southern sector (Patagonia)
Average depth of the shelf break (m)	115
Siliciclastic/carbonate/authigenic/glacial sedimentation	100% siliciclastic
Modern/relict/palimpsest	98% modern sediments, *c.* 50/50% relict/palimpsest
Tectonic trend over the last glacial cycle (stable/uplifting/subsiding)	Stable, slight uplifting increasing to the south

transcurrent (around Malvinas Islands) and active (in the Southern Scotia ridge) (Ramos & Turic 1996). Despite these differences, the shelf does not change significantly according to its location on one type of margin or another, although distinct sectors can be recognized in relation to different deep tectonic and structural aspects (Table 6.2). Cavallotto *et al.* (2011) described the complex climatic, oceanographic and tectonic processes involved in the evolution of the ACM.

The base of the Quaternary deposits in the ACS was defined as a seismic discontinuity at a water depth of 140 m (seismic horizon 'b': Ewing & Lonardi 1971) that extends north of 43°S. Consequently, the substratum of the post-LGM deposits is represented there by almost complete Plio-Pleistocene marine sequences. However, south of 43°S, the substratum is represented mainly by late Tertiary pre-glacial continental and marine sequences with scattered and incomplete patches of Quaternary deposits. The discontinuous and reduced Quaternary deposits in the southern region can be attributed to a lack of space for deposition due to

the post-glacial isostatic rebound, which was more significant there, closer to the glaciated areas of Patagonia.

The Argentine continental shelf

Morphology

The ACS covers an area of $9.6 \times 10^5 \text{ km}^2$, trending north–south for 2400 km between the de la Plata River (35°30′S) and Cape Horn (57°S) (Fig. 6.1). The adjacent coastline is around 5300 km long if major irregularities are considered. Shelf width varies between 170 and 850 km. The inner (shorewards) shelf edge is represented by a 10–20 m-high shoreface. The outer (offshore) shelf edge (shelf break) follows a NE–SW direction between the de la Plata River and 44°S, from where it gradually changes to a north–south direction down to 50°S, then acquires an easterly direction towards Malvinas Island and, after surrounding them,

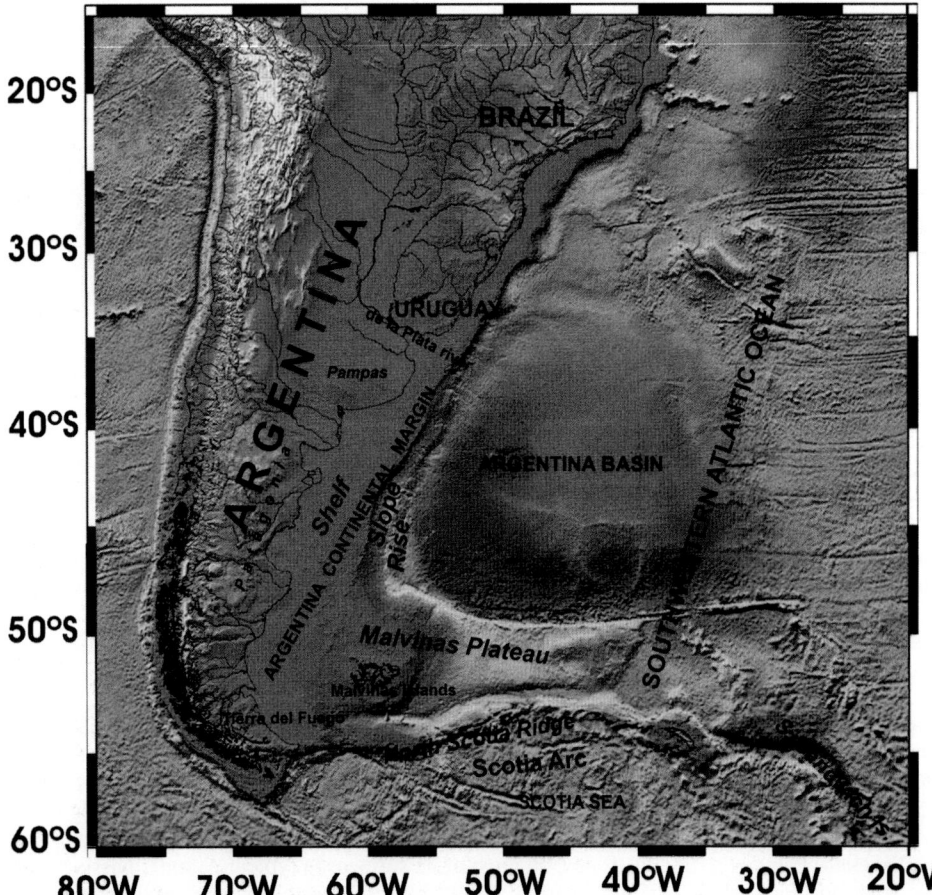

Fig. 6.1. Location map of the Argentine continental shelf.

Table 6.2. *Margin sectors on the basis of structure and stratigraphy*

Structure	North of 43°S		Typical lower-plate passive margin. An old basement and thick continental crust. Evolution controlled by cortical discontinuities and transverse extensional systems with little basaltic magmatism
	Between 43°S and 49°S		Same type of margin as above, although the crust is younger and thinner. Pre-rift associations and longitudinal rifts characterized by acid volcanism
	South of 49°S		Transcurrent and convergent margin as a result of interaction between the South American and Scotian plates
Tectonic processes	North of 55°S		Shelf affected by downwarping conditioned by isostatic equilibrium and sediment overloading. Thick sedimentary deposits as a result of increasing tilting of the continent towards the east occurred after the Andes cordillera rise in the Miocene
	South of 55°S		Complex tectonic processes conditioned a thin Cenozoic sedimentary cover that experienced intense marine erosion
Substratum	Below the shelf		Lies over continental crust
	Below the slope and rise. Malvinas Plateau and Scotia arc		Lies over oceanic crust
Stratigraphy (based on offshore oil drillings and reflection seismic data)	Pre-Cretaceous basement	North of 39°S	Precambrian metamorphic and intrusive rocks from the Brazilian shield, Silurian–Triassic sediments and Cretaceous basalts; seismic velocities of 5.5 km s^{-1}
		Between 39°S and 43°S	Igneous basement covered by metamorphosed Palaeozoic continental–marine sediments; seismic velocities between 5 and 5.5 km s^{-1}
		South of 44°S	Above a Precambrian metamorphic basement there are upper Palaeozoic–lower Mesozoic metamorphic and acid intrusive rocks with seismic velocities of up to 6 km s^{-1} segment, followed by Silurian–Jurassic pyroclastic and acid-mesosilicic extrusive rocks interbedded with continental sediments with seismic velocities of between 4.2 and 5.1 km s^{-1}
	Post-Cretaceous sedimentary filling of the basins		Post-Cretaceous sequence thickness of 6–8 km, composed of continental–marine shales, filites, lutites, limestones, sandstones and conglomerates

Compiled from Zambrano & Urien (1974), Ramos (1996, 1999), Turic *et al.* (1996) and Urien & Zambrano (1996).

goes back again to the west and reaches regions near the continent at 51°30′S. The depth of the shelf break is variable between 70 and 190 m (Parker *et al.* 1996; Violante & Cavallotto 2011), showing a broad north–south deepening (Fig. 6.2). Surface gradient is relatively smooth, with slopes ranging between 1:500 and 1:10 000. In the areas adjacent to the Pampean region, it has a convex profile (steeper gradient towards the outer shelf), whereas in the areas adjacent to Patagonia it is concave (steeper gradient towards the continent). North of 38°S, the slope gradient changes from 1:2000 above 90 m water depth to 1:500 below. Between 38°S and 48°S the shelf surface is more uniform and subhorizontal, with gradients of around 1:10 000. South of 51°S, the gradient is around 1:3000/1:4000. Relative relief does not exceed 20 m. Table 6.1 details most of the basic shelf characteristics.

The shelf is shaped in several terraced surfaces separated by high-gradient steps (Groeber 1948; Parker *et al.* 1997; Violante 2005; Ponce *et al.* 2011). Four terraces (named I, II, III and IV) are recognized, showing a predominant NNE–SSW direction subparallel to the coastline (Fig. 6.3, Table 6.3). Post-LGM transgressive deposits averaging 10 m thick constitute the subhorizontal surface of the terraces, particularly those closer to the coast and in the northern parts of the ACS, whereas the steps separating them are usually devoid of transgressive sediments; on most of these steps the relicts of Plio-Pleistocene marine transgressions and interbedded continental sediments crop out.

Exogenous conditioning factors involved in the morpho-sedimentary configuration

The morpho-sedimentary configuration of the ACS depends mainly on two major factors: (a) the heritage of the above-mentioned

regional geotectonic framework (i.e. from the endogenous conditioning factors primarily associated with the margin structure and evolution); and (b) the external factors of climate, oceanography and associated processes (sea-level fluctuations and sediment dynamics), which become more significant (with respect to the endogenous) during the late Cenozoic.

Climate since LGM times. The climate in southern South America during glacial and post-glacial times was conditioned by global oceanographic and climatic factors, although it was also influenced by regional and local factors, such as: (a) atmospheric conditions imposed by the interaction between the South Pacific and South Atlantic anticyclonic centres that affected regional wind patterns; (b) the proximity to both the southern Andes and the Antarctic ice masses; and (c) the highly variable relationship between emerged and drowned lands throughout the glacial–interglacial cycle, which at Patagonian latitudes represented, respectively, a duplication and a reduction to half of the continental area with consequent 'continentalization' or 'oceanization' of climates as a result of the changing sea moderating effect. After the extreme cooling during the LGM, significant climatic changes and environmental instability characterized late glacial times with several glacial re-advances and recessions, including the Younger Dryas (Rabassa *et al.* 2011 and references therein). The major peak of the Hypsithermal occurred at 6 ka, with sharp climatic changes afterwards that evolved towards present climatic conditions.

Oceanography. The main sources of the ACS water masses are Subantarctic water flowing from the northern Drake Passage between the coast and the Malvinas Islands (Hart 1946), and the Malvinas Current in the outer continental slope (Bianchi *et al.* 2005). In the vicinity of the northernmost part of the shelf, the

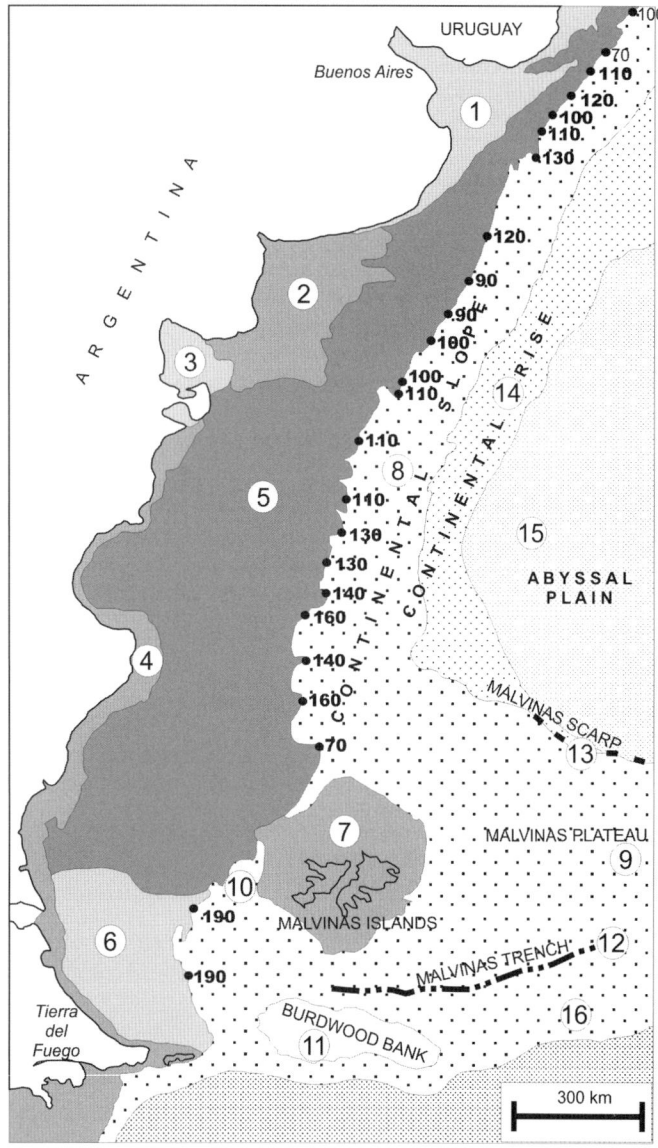

Fig. 6.2. Major physiographical features of the Argentine continental shelf (modified after Parker *et al.* 1996). Shelf is in grey shades. Numbers in bold at the shelf–slope boundary indicate the depth of the shelf break. 1, Rioplatense Terrace; 2, deltaic front of the Colorado and Negro rivers; 3, northern Patagonian gulfs; 4, Patagonian inner shelf; 5, Patagonian outer shelf; 6, Tierra del Fuego shelf; 7, Malvinas shelf; 8, continental slope; 9, Malvinas Plateau; 10, Malvinas depression; 11, Burdwood Bank; 12, Malvinas Trench; 13, Malvinas escarpment; 14, continental rise; 15, abyssal plain; 16, Scotia Arc.

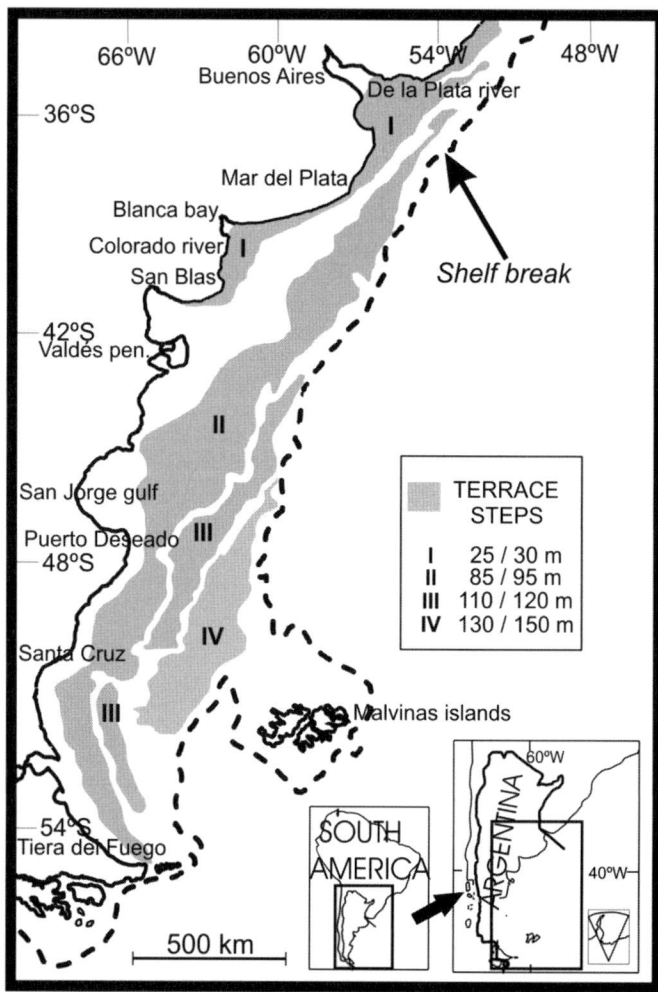

Fig. 6.3. Submarine terraces in the continental shelf (modified after Parker *et al.* 1997).

Confluence Zone between the Malvinas (flowing to the north) and Brazil (flowing to the south) currents occurs. The main freshwater source comes from the de la Plata River (around $25\,000\ \mathrm{m}^3\ \mathrm{s}^{-1}$: Simionato *et al.* 2007; Campos *et al.* 2008*a*), with much less influence ($<2000\ \mathrm{m}^3\ \mathrm{s}^{-1}$) from Patagonian rivers (Gaiero *et al.* 2003).

The mean shelf water circulation has a predominant NNE direction with a velocity of up to $0.30\ \mathrm{m}\ \mathrm{s}^{-1}$, slightly decreasing with depth. Forbes & Garrafo (1988) estimated an averaged depth intensity ranging from 0.01 to $0.07\ \mathrm{m}\ \mathrm{s}^{-1}$ in winter and from 0.02 to $0.04\ \mathrm{m}\ \mathrm{s}^{-1}$ in summer. Tidal amplitude varies between 1.5 m in the eastern Buenos Aires province and 12.3 m in southern Patagonia (Servicio de Hidrografía Naval 2011), with tidal waves propagating northwards. Persistent and strong wind blowing from the south and SE, coinciding with large or even moderately high tides, can induce surges that produce significant coastal erosion and offshore sand transport.

Wind waves in the Buenos Aires province coast show a wave height, period and direction of 0.87 m, 9.2 s and ESE to SE, respectively, indicating a predominant northward flow and sand transport maintained along that region. In the Tierra del Fuego inner continental shelf, the most frequent wave trains propagate from the sector comprising the southwesterly and northwesterly directions, coinciding with predominant winds. However, waves with the greatest heights are frequently propagated from the sector comprising the north and northeasterly directions, associated with the most severe storms. Owing to the fact that westerlies are dominant between 40°S and 55°S, these characteristics could be extended to the whole Patagonian shelf. Dragani *et al.* (2010) modelled an increase in wind wave heights between 32°S and 40°S. In general, for the entire continental shelf, wave heights vary between 1 and 4 m, increasing to the south.

According to the aforementioned aspects, the ACS is classified as wave-dominated in the northern region (adjacent to the Pampas) and tide-dominated in the south (Patagonian region).

Palaeoceanographic changes can by synthesized from the variability of oceanic temperatures, and water-mass displacement between the continents and the oceans during glacial–interglacial periods. According to CLIMAP Project Members (1981), sea-surface temperatures (SSTs) in glacial times were in the region of between 2 and 4 °C lower than present, undoubtedly affecting seawater evaporation and oceanic circulation. Berger & Wefer (1996) pointed out that, during the last glaciation, the North Atlantic Deep Water mass weakened in the SW Atlantic, at the same time that the Antarctic Bottom Water layer thickened. Foraminifera-based studies (Laprida *et al.* 2011; Groeneveld &

Table 6.3. *Terraces on the shelf*

Terrace	Level	Depth (m)	Location
I	A	30	Rioplatense Terrace associated with the de la Plata River deltaic body Bahía Blanca–San Blas area related to the deltaic bodies of the Colorado and Negro rivers Also in reduced sectors adjacent to the Patagonian coasts, as in Puerto Deseado
	B	50	Middle part of the outer step of Rioplatense Terrace
	C	70	Base of the step offshore the de la Plata River mouth, as well as offshore Blanca Bay where it is blurred by the deltaic deposits of the Colorado and Negro rivers Equivalent levels in some places in Patagonia (Valdés Peninsula, San Jorge Gulf and offshore Tierra del Fuego)
II	D	80	Very extensive, very low gradient and smooth, reaching its best expression between Mar del Plata and the Valdés peninsula Covered by dark, fine and very fine silty–clayey sands
	E	90	Subhorizontal, smooth ramp-like feature extended south of the Valdés Peninsula that has its most extensive development south of the Santa Cruz River Covered by dark fine and very fine silty–clayey sands
	F	100	Very extensive and subhorizontal, covered by clean fine and medium sands with gravel concentrations offshore southern Patagonia Dissected by erosive channels and scours partially buried by Holocene deposits, representing an ancient fluvial network
III		110–120	Covered by bioclastic sands and gravels of probable glacio-fluvial origin Evidence of a relict fluvial network Small terraces without regional significance occur at the same levels in the upper edge of the slope offshore the de la Plata River
IV		130–150	Covered by bioclastic sands and gravels of probable glacio-fluvial origin Small terraces without regional significance occur at the same levels in the upper edge of the slope offshore the de la Plata River

Chiessi 2011; García Chapori 2013) document the increasing in intensity of the Malvinas Current during glacial times with the consequent northward displacement of the Confluence Zone, as well as offshore displacement due to reducing water depth. Most probably, mean wind–wave conditions on the whole ACS, and the magnitude and frequency of storm surges in the Buenos Aires coast, could have been a little different in glacial times due to stronger low-level winds resulting from a larger mean latitudinal atmospheric temperature gradient.

Sea-level fluctuations. The extension and geotectonic setting of the ACS introduced complex variables that influenced sea-level fluctuations. Rostami *et al.* (2000) considered that regional differences are evident in the fact that the predictions of sea-level fluctuations and models of deglaciation coincide for Northern Patagonia but not for Southern Patagonia. Several curves exist in different coastal regions of Argentina for the last part of the post-glacial transgressive event (Urien 1970; Farinati 1984; Peltier 1988; Isla 1989; Pirazzoli 1991; Aguirre & Whatley 1995; Gómez & Perillo 1995; Cavallotto *et al.* 2004 recently calibrated by Gyllencreutz *et al.* 2010). These curves match each other only in the general tendency of sea-level changes, not in the details. This is considered to be the consequence of 'local' factors characterizing each region. The only curve comprising the entire transgressive cycle since the LGM was published by Guilderson *et al.* (2000) (Fig. 6.4), which is considered by these authors as 'eustatic' after applying models for isostatic-tectonic compensation. This curve does not match the global eustatic sea-level curve established by Fleming *et al.* (1998) in what can be preliminary related to tectonic and hydro-isostatic influence. Neotectonism has been mentioned as being responsible for the recent reactivation of old faults in marine regions (Zambrano & Urien 1974). Codignotto *et al.* (1992) inferred a relative uplift along the Argentinean coast of 0.12–1.63 m ka^{-1} (higher in the interbasins than in the basins), with a general trend of 0.7 m ka^{-1} in the last 9.5 kyr decreasing from south to north. However, Rostami *et al.* (2000) and Schellmann & Radtke (2000) considered more uniform isostatic readjustment without significant differences between basins and interbasins. A model-based study performed by Rostami *et al.* (2000) obtained a relatively uniform regional

uplift of 0.9 m ka^{-1} for the last 300 kyr, with progressively higher elevations to the south.

Based on this evidence, the sea-level curve by Guilderson *et al.* (2000) must be considered as 'relative'. According to this curve, the LGM sea-level lowstand was approximately −105 m at around 18 ka (Fig. 6.4). The following relative sea-level rise occurred rapidly at the earlier stages of the transgression, at a rate of 11–12 mm a^{-1} (Cavallotto *et al.* 2004; Violante & Parker 2004; Schnack *et al.* 2005). No sea-level fluctuation is depicted around 11 ka in coincidence with the Younger Dryas event, although it could be the consequence of the scarcity of ^{14}C datings in that part of the curve. At 8.6 ka, the sea level was at −18 m below present, rising at a rate of 9.4 mm a^{-1}, and then decelerating before reaching its highest position (+6 m without subtracting tides and waves influence) at 6 ka (Cavallotto *et al.* 2004). The calibration of the Cavallotto *et al.* (2004) curve performed by Gyllencreutz *et al.* (2010) points to a sea-level maximum of 6.5 m above present at 7 cal ka BP, a conclusion that does not match the regional evidence, which indicates that at that time the sea level was still below its present position. Ponce *et al.* (2011) used digital models to depict palaeogeographical maps that show different positions of the coastline at different times during the post-glacial transgression, on the basis of the global sea-level curve proposed by Fleming *et al.* (1998). The authors state that at approximately 15.3 ka, the great emerged shelf plain was reduced to half of its original size.

The middle–late Holocene regressive event is not definitively well constrained. The differences in the tendencies of sea-level fall in different regions are the result of local tectonic–isostatic–subsidence characteristics. In general, it is considered that a constant but stepped decreasing in sea level occurred. Rapid drops in sea level at around 5–3 ka have been recorded by lithological and foraminiferal information by Laprida *et al.* (2007), and a possible drop below present sea-level position was mentioned at 2.6 ka offshore Blanca Bay (Gómez *et al.* 2006). Most of the authors that worked on relatively stable areas of Argentina concur with the age of the transgressive maximum occurring between 6 and 5 ka, although heights reached by the sea are not in agreement (they vary between 6 and 2.5 m). However, in Patagonian regions, the maximum Holocene beaches are documented

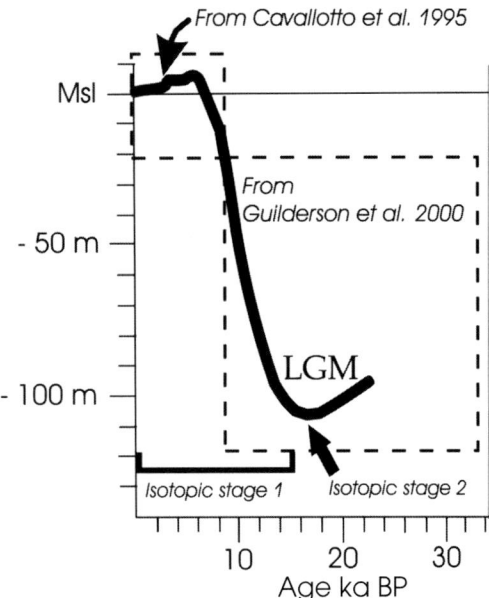

Fig. 6.4. Relative sea-level fluctuation curve for the Argentine continental shelf (from Cavallotto *et al.* 1995; Guilderson *et al.* (2000).

at >10 m (Codignotto *et al.* 1992; Rostami *et al.* 2000 and references therein), mostly dependent on isostatic influence.

Sedimentology and sediment dynamics

Sediment distribution on the shelf (Fig. 6.5) is the result of complex processes in which some of the 'external' forcing factors, particularly the way in which sea-level fluctuations affected the shelf surface as well as the oceanographic conditions, played a substantial role. Furthermore, in order to better understand the distribution patterns, the characteristics of the sediments themselves as well as their previous 'history' throughout the whole sedimentary cycle, including the source areas and sediment dynamics, must be taken into account.

Sediment provenance and source areas. The ACS sediments originated in two main source areas: the Andean region and the Brazilian Shield (Teruggi 1954; Etchichuri & Remiro 1963; Depetris & Griffin 1968; Berkowsky 1986; Campos *et al.* 2008*b*). The sedimentary and volcanic products originated in the Andean region were transported mainly to the east, conditioned by climate (dominant westerly winds) and morphology, and were partially trapped and/or reworked by fluvial and aeolian processes in the Pampean and Patagonian regions, and finally reached the coast and sea. The volcaniclastic composition of the shelf sediments, as well as the evidence of volcanic ash levels preserved in submarine cores, document the importance of these processes (Violante & Rovere 2005). However, the cratonic (igneous–metamorphic) regions of the Brazilian Shield provided sediments that were transported almost exclusively by streams through the de la Plata fluvial basin to the sea. The predominance of the volcanic Andean-sourced sediments with respect to the cratonic Brazilian-sourced sediments allowed Potter (1994) to classify the coastal regions of Argentina as 'Andean' in terms of the provenance of sand. In concordance with this, Mahiques *et al.* (2008) used neodymium and lead isotopes to interpret sediment transport and source rocks in the Argentine, Uruguayan and Brazilian shelves, concluding that the isotopic signature of most of the ACS sediments are typical of Andean rocks, whereas north of the de la Plata River it is concordant with the basaltic province from NE Argentina and southern Brazil. Campos *et al.* (2008*b*) discussed the types and spatial distribution of clay minerals according to the

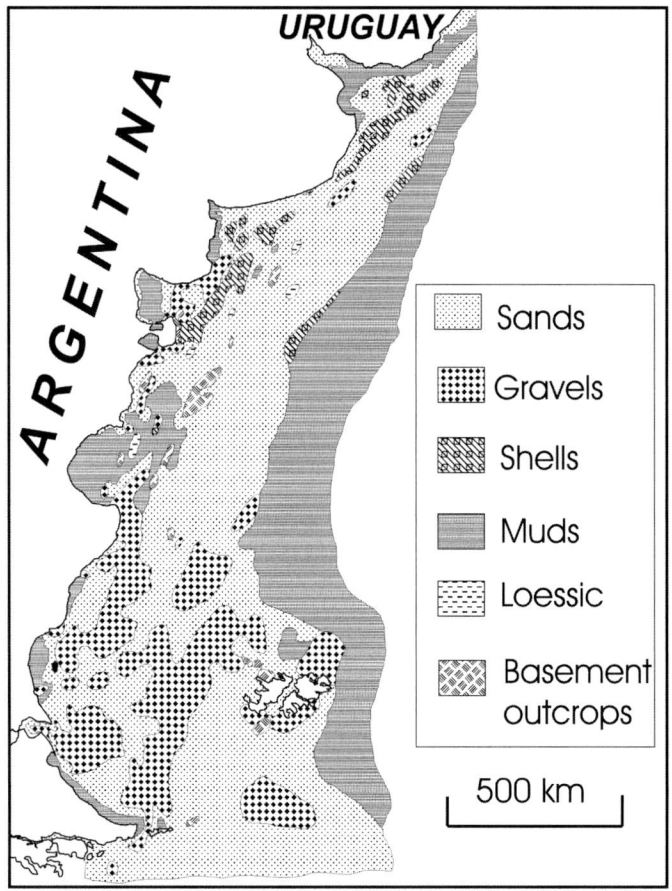

Fig. 6.5. Sediment distribution in the continental shelf and upper slope (modified after Parker *et al.* 1996).

source areas and the transport through the de la Plata River. Bozzano *et al.* (2011) revealed, after studying rock fragments contained in contourite deposits of the upper slope offshore Buenos Aires province, that both the basement of the Brazilian craton and the volcanic rocks from northern Patagonia supplied sediments that, after crossing the shelf, were transported towards the deeper-marine regions.

Sediment dynamics. Terrigenous sediments introduced into the coastal system by coastal erosion, as well as by fluvial and aeolian transport, were then delivered offshore, transported to the north as a result of the dominant northward littoral currents and, finally, deposited on the shelf where they became relict or palimpsest depending on how they had been reworked by marine processes. These processes occurred during the entire post-glacial cycle under the conditions imposed by sea-level fluctuations, climate, coastal water circulation and sediment supply. In the case of the post-LGM transgression, it provoked the sweep of the underlying pre-transgressive substratum through the ravinement process that resulted from the erosive coastal retreat and partial sediment transfer offshore (Urien & Ewing 1974; Parker & Violante 1982; Violante & Parker 2004; Parker *et al.* 2008). Isla & Cortizo (2005) estimated that 243.8 Mt a^{-1} of sediments are eroded from the Patagonian cliffs and introduced into the sea. Sediment supply by fluvial input is relatively low, as the low-flow Patagonian and Pampean rivers carry small amounts of sediments, whereas the larger rivers usually have estuarine environments that retain most of the sedimentary load. However, the de la Plata River seems to discharge large amounts of sediment to the shelf, mainly silts, ranging between 57 and 130 Mt a^{-1} (Depetris & Griffin 1968; Giberto *et al.* 2004; Campos *et al.* 2008*a*). Streams were more significant in pre-Holocene times as demonstrated by the existence of oversized fluvial valleys with respect to the present fluvial dynamic, as well as by the large amount of gravels on the southern shelf surface that cannot be transported by present streams. Kokot (2004) estimated that the Santa Cruz River in Patagonia presently has a discharge equal to one-tenth of the discharge in the Pleistocene. Aeolian activity has a greater significance as a provider of sediment to the shelf than streams do. The total amount of terrigenous sediments transferred to the sea bypassing the Patagonian coasts was estimated to be 70 Mt a^{-1} (Pierce & Siegel 1979; Gaiero *et al.* 2003), from which 56% (39 Mt a^{-1}) corresponds to coastal erosion, 41% (29 Mt a^{-1}) to atmospheric processes (dust transport) and 3% (2 Mt a^{-1}) to fluvial activity. Based on these data, the ACS can be classified as passive and 'autochthonous' in terms of the sedimentary regime (although sediments are allochthonous) following the concepts described by Swift (1968). Violante (2004) has also previously stated this classification. Although the mechanisms of sediment transfer from the shelf edge and upper slope to the head of the submarine canyons are not yet well known, the general consensus is that most of the submarine canyons are disconnected from the shelf. Pierce & Siegel (1979) estimated a shelf sediment export to the slope of 17 Mt a^{-1}.

Sediment composition and facies. Surface sedimentary facies are represented in decreasing order of abundance by sands (65% of the shelf surface), shells (12.5%), gravels (12.5%), muds (8%), and consolidated sediments and rocks (2%) (Parker *et al.* 1997) (Fig. 6.5). Table 6.4 shows details of these sediment types. Sands dominate over the entire shelf, whereas shells are more abundant in the northern (offshore Pampas) regions, and gravels in the southern (Patagonian) areas. Shelf sands are composed of two main mineralogical associations according to the source areas already described: the volcanic–pyroclastic association that dominates south of the de la Plata River (and so distributed on most of the ACM); and the igneous–metamorphic association that dominates only to the north of the de la Plata River.

Benthic foraminifer assemblages preserved in sediments collected in cores are good palaeoenvironmental/palaeoceanographic indicators, as suggested for different shelf settings (Boltovskoy 1954*a*, *b*, 1973; Giussani & Watanabe 1980; Ferrero 2005; Gómez *et al.* 2006; Laprida *et al.* 2007; Bernasconi & Cusminsky 2009; García Chapori 2013). North of 41°S at water depths >100 m, faunas in the lowermost levels of the cores (>4–5 m depth in the cores) suggest pre-LGM inner-shelf environments; at approximately 3–4 m, they reveal littoral–upper sublittoral environments related to the LGM lowstand, and in the uppermost levels (near the shelf surface) they indicate inner-shelf environments related to the early Holocene transgression influenced by the Malvinas Current. At water depths of between 50 and 100 m, faunal assemblages indicate a littoral and inner-shelf environment associated with the beginning of the post-glacial transgression; at about water depths of 70–100 m, isolated species typical of the Malvinas Current suggest short-term variations in its western bottom boundary during the early Holocene; currently, this boundary lies between 80 and 100 m north of 42°30′S, and between 110 and 115 m from 43°S–47°30′S. At water depths shallower than 50 m, faunas are exclusively Holocene, indicating coastal settings related to successive positions of the coastline during the sea-level rise prior to 6 ka; typical inner-shelf deposits were recognized at water depths of 12 m close to the south of the de la Plata River mouth. Salt marsh deposits from approximately 6.35 ka finally evolved to tidal flats at about 2.3 ka, and therefore concluded in high-energy coastal environments related to an increasing sea level. In the northern Patagonian shelf (41°S–46°30′S, water depth *c.* 50–100 m), foraminiferal assemblages indicate late Pleistocene lowstand deposits below about 4 m depth in the cores, and early Holocene inner-shelf facies above 4 m. Isolated Malvinas Current specimens were found in cores at approximately 75–100 m water depth. In Nuevo Gulf, faunas

suggest a transition from normal marine conditions at 8–7.7 ka to marginal marine conditions in the late Holocene, accompanied by a change in the circulation dynamics. In the southern Patagonian shelf (45°S–47°S, water depth *c.* 40–143 m) Holocene sediments dominated by inner-shelf species are reported, with no outstanding vertical variations in faunal composition.

The post-LGM sedimentary sequence

Processes involved in the evolution of the shelf during post-LGM times produced a complex sedimentary sequence (SS), which was defined as a 'depositional sequence' (or 'seismic-stratigraphic unit') based on high-resolution seismic surveys undertaken in the northern region of the shelf (Violante *et al.* 1992; Parker *et al.* 1999, 2008; Violante & Parker 2004). The SS is bounded at its base by the transgressive surface, whereas its top is represented by the present topographical surface. The thickness of the SS averages 5–10 m, although it is thicker (10–15 m) in the shelf adjacent to the Pampas, and thinner (<5 m) in the Patagonian shelf. In the regions where the SS was studied in detail, it has a distinctive and homogeneous seismic-reflection pattern mainly represented by a chaotic and non-transparent character, which indicates a high sand content. In places where muddy content is high, as in coastal estuarine environments, transparent, either parallel or reflections-free patterns, are common. Piston cores and bottom grab samples recovered sediments that allowed the sedimentary facies that compose the SS to be depicted, which is in general represented by terrigenous, siliciclastic, relict–palimpsest deposits presenting varied facies associated with the different environments developed during the regional evolution.

Three systems tracts can be recognized in the SS: lowstand, transgressive and highstand.

Lowstand. Some seaward-prograding seismic-stratigraphic units made up of soft muddy sediments found in different positions in the upper slope could correspond to these deposits, although this needs to be confirmed as the regions beyond the shelf break are still under study. A recent finding (Violante *et al.* 2014) in a core obtained at the shelf break at 100 m water depth offshore the southern Pampean region (around 40°S) provided a 2.75 m-long sequence composed of fine sands of possible nearshore-beach origin in the base, followed upwards by sandy clay sediments containing mixohaline–freshwater microfaunas with vegetal remains in a heterolithic structure, indicating marginal–inshore lacustrine environments, then a shelly deposit representing a beach ridge and, finally, on top of the sequence, the present shelf sands. The marginal–inshore environments have been dated at 15 ^{14}C ka, thus indicating that the sequence represents the first record of a coastal environment in the Argentinian outer shelf associated with the first stages of the post-LGM transgression.

Transgressive. This constitutes the upper layer that covers the entire continental shelf. Although the deposits are mainly sandy, ridge-like features made up of coarse sands, gravels and shells are also common, as well as depressed sectors filled with muds. The surface levels are represented in the nearshore areas of eastern Pampas by shoal-retreat massifs containing linear-shoal complexes associated with sediment reworking during the late Holocene regressive event (Urien & Ewing 1974; Parker *et al.* 1982), as well as by estuarine facies in the vicinity of the de la Plata River mouth (Violante *et al.* 1992; Cavallotto *et al.* 2004; Violante & Parker 2004; Cavallotto 2008).

Highstand. The highest sea-level positions reached approximately +4/6 m at 6 ka. Deposits are found along most of the Argentine coasts at altitudes always above present sea level, ranging from 3 m in eastern Buenos Aires to 10 m in Tierra del Fuego, these differences being associated with different tectonic and isostatic

Table 6.4. *Sedimentary facies characteristics*

Sediment main type	Areal distribution	Description	Texture	Colour	Regional distribution	Morphologies
Sand	65% of the shelf surface.	Fine–medium sand. Subordinated coarse and very fine sand. Bioclastics	Loose deposits. Texturally mature. Moderate–well sorted	Yellowish, brownish and greyish	North and central Patagonia: fine–medium fractions. Pampean and Tierra del Fuego littorals: coarser fractions. Semi-enclosed areas (de la Plata River, Blanca Bay, and San Matías and San Jorge gulfs): very fine, sometimes silty-sands	San Antonio cape (NE Buenos Aires province, eastern Pampas): submerged dune systems and linear shoals related to tidal current action. Between Blanca Bay and the San Matías Gulf: giant submerged dunes. Buenos Aires province (Pampean coasts): relicts of barriers and beach-ridges. Offshore de la Plata and Colorado river mouths: submerged deltaic systems
Shells	12.5%	Entire shells and fragments up to several cm in diameter. Constituted by diverse species of pelecipods, brachiopods and arthropods (barnacles), fish bones and rest made up of echinoids	Loose, slightly consolidated or cemented coquinas. Sometimes as a subordinated fraction of sands	White, yellowish, pale grey	North of 43°S constitutes ridges that indicate the position of ancient coastlines	Ridge-like morphologies
Gravels	12.5%	Generally rounded gravels up to several cm in diameter	Loose gravelly deposits	Depends on source rocks	Offshore the Patagonian rivers mouth: very large gravel concentrations as the source is the glacifluvial deposits that extend over most of Patagonia	None evident
Muds	8%	Silts and/or clays or different kinds of combinations. High content of organic matter	Predominantly cohesive	Dark brown and green	Mainly located in semi-enclosed coastal regions (estuaries, bays and gulfs)	None evident
Consoli dated sediments and rocks	2%	Outcrops of different pre-transgressive substrata	Depends on the rock and sediment types	Depends on the rock and sediment types	Pre-transgressive substratum	Depends on the pre-transgressive morphology

behaviours. Deposits are mainly represented by beach ridges, tidal flats, estuaries, coastal lagoons and beach-dune complexes. A compilation of the available literature on the subject was carried out by Cavallotto (2008).

Evolution

The evolution of the ACS during post-LGM times resulted from the interaction of diverse factors, such as relative sea-level changes, climate, oceanographic processes, sediment dynamics and isostatic/tectonic components. The evidence of the evolution is recorded not only in the post-glacial morpho-sedimentary features and sedimentary sequences but also in the characteristics of the transgressive surface above which the last transgression took place.

The transgressive surface

During glacial times the ACS was an extensive subaerial plain with particular morphological, hydrographical, pedological and climatic characteristics. The post-glacial transgression, although it substantially modified this, did not completely eliminate some pre-transgressive features, which remained preserved in many places. The transgressive surface was the consequence of the ravinement process that occurred as a result of coastal erosional retreat during post-LGM times and, hence, it is time-transgressive (Parker *et al.* 1999, 2008; Violante & Parker 2004). Its characteristics resulted from: (a) the relief and lithological constitution of the pre-transgressive substratum; (b) the subaerial processes (fluvial, aeolian and lacustrine action, as well as soil formation, desiccation and oxidation) during pre-transgressive times; (c) the rate and variability of the relative sea-level rise; (d) the balance between erosive–depositional processes at each stage of the transgression, which depended on the littoral dynamics, the accommodation of the transgressive sediments to the rising base level and the sediment supply. The resulting sequences preserved in the shelf substratum can, therefore, change from exclusively marine sequences to mixed marine–coastal–fluvial–aeolian–palaeosol sequences, depending on its location (Fig. 6.6). The identification of pre-and post-transgressive sequences is clear in seismic records and cores when different environments are recorded but, when similar environments are superposed, sedimentological and geotechnical characteristics can serve to aid differentiation (Table 6.5). Major features preserved on the transgressive surface are incised valleys, evidence of glacial activity, palaeosols and relicts of ancient coastlines.

Incised valleys. Relicts of a palaeo-fluvial network are evidenced on the shelf surface. As it has not been completely obliterated during the transgression, the palaeodrainage pattern can be recognized (Parker *et al.* 1996; Violante *et al.* 2007). Although, in a regional sense, the drainage pattern is dendritic, north of 42°S the ancient fluvial valleys are subparallel and regionally orientated to the SE. South of 42°S, the pattern is more chaotic with numerous distributaries; main valleys are orientated to the south and then change to the SE towards their lowermost sections. Between 42°S and 46°S, some valleys seem to be connected with submarine canyons. South of 46°S, the valleys tend to converge towards the Malvinas depression located south of the Malvinas Islands. In relation to relict fluvial networks in coastal semi-enclosed areas, they show particular patterns more similar to the irregular branching of tributary valleys associated with submerged deltas (like the Colorado–Negro deltaic system) or a semi-radial, centripetal drainage in semi-enclosed basins (San Matías, Nuevo and San Jorge gulfs). Ponce *et al.* (2011) stated that the larger extension of emerged regions during glacial times favoured a better distribution and integration of the drainage network.

Glacial features. The southern extreme of the Andes was covered by ice during glacial times. Owing to the narrowness of the continent there, ice masses and/or glacio-fluvial deposits extended on to the coastal and nearshore regions of Tierra del Fuego, where relicts of glacial features are present in the shelf (Isla & Schnack 1995; Mouzo 2005). However, gravels that extended on to the shelf surface south of 46°S are associated with ancient fluvial and glacio-fluvial deposits.

Palaeosols and related aspects. In the vicinity of the de la Plata River outlet, Osterrieth *et al.* (2005) and Violante *et al.* (2007) described probable palaeosols based on the finding of silicophytoliths in sediment cores at water depths exceeding 80 m. Two palaeosols were found initially, one developed on continental late Pleistocene sediments and the other one on pre-transgressive coastal Holocene deposits. Cione *et al.* (2005) described mammal remains at water depths of 45 m in continental late Pleistocene–early Holocene sediments outcropping in the inner shelf.

Ancient coastlines. Longitudinal and parallel-to-the-coastline deposits mainly composed of coarse sand, gravel and shells were recorded in the outer border of the terraces, particularly on terrace I (TI), which have been interpreted as relicts of ancient coastlines (Urien & Ewing 1974; Urien *et al.* 1979; Parker *et al.* 1996, 1997). As mentioned above, a record of an ancient coastline was found in a core containing coastal (beaches to inshore lacustrine) sediments sandwiched between nearshore and shelf deposits at a water depth of 100 m (Violante *et al.* 2014).

The post-glacial morpho-sedimentary features

Main features of the ACS are terraces stepped at different depths, increasing offshore, and with different regional extensions. Its

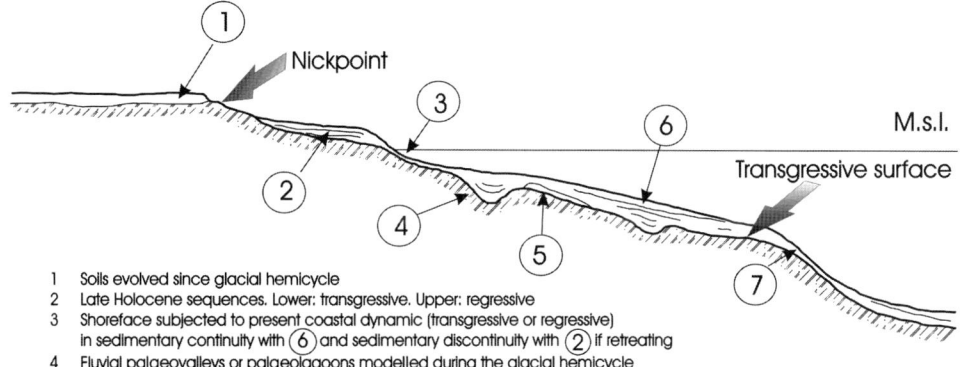

1 Soils evolved since glacial hemicycle
2 Late Holocene sequences. Lower: transgressive. Upper: regressive
3 Shoreface subjected to present coastal dynamic (transgressive or regressive)
   in sedimentary continuity with ⑥ and sedimentary discontinuity with ② if retreating
4 Fluvial palaeovalleys or palaeolagoons modelled during the glacial hemicycle
5 Soils evolved between glacial and deglacial hemicycles
6 Post-LGM transgressive deposits
7 Step at the outer edge of inner shelf modelled in Pre-transgressive sediments

Fig. 6.6. Schematic cross-section showing the relative position of the different parts of the post-LGM sequence above the transgressive surface (from Violante & Parker 2004).

Table 6.5. *Sedimentological and geotechnical differences between pre- and post-LGM sequences*

	Lithology/environment	Shearing resistance (Parker *et al.* 1976)
Pre-LGM sequences	Continental (aeolian–lacustrine) semi-consolidated, reddy, browny and yellowish, silty to loessic sediments. Exceptionally littoral sands	2.40–4.60 g cm^{-2} increasing downcore
Post-LGM sequences	Littoral–shallow marine, not consolidated, yellowish, browny, greyey and greenish bioclastic fine sands to sandy muds.	Normally under 0.55 g cm^{-2} (exceptionally reaching 1.65 g cm^{-2})

origin and shaping was associated (Groeber 1948; Parker *et al.* 1997; Violante 2005; Perillo & Kostadinoff 2005; Ponce *et al.* 2011) with interruptions in the rate of sea-level rise during the post-glacial transgression. These interruptions were driven by climatic changes, with the consequent establishment of a stationary shoreline at that position for a relatively long time, so allowing the modelling of the outer edge of the terraces to occur by erosional coastal retreat.

The age of the terraces was estimated by Violante (2005) after comparing the depth of the base of the terraces and the local sea-level curve (Guilderson *et al.* 2000). TI coincides with the age of the Younger Dryas event, so enabling the possibility to be considered that this cold period could have induced a stillstand in sea level at a depth around 70–80 m, with consequent rapid coastal erosion and the modelling of the terrace front. After sea level began, once more, to rise, the diminishing coastal erosive processes were not able to substantially modify the terrace morphology. When sea level reached a position of around 20 m below present, in close agreement with a new decrease in the rate of relative sea-level rise observed at around 8.6 ka, the surface of TI was levelled by persistent wave action and covered by sediments that resulted from the reworking of the substratum. Based on the same comparisons, TII, TIII and TIV could have been modelled during sea-level stabilizations at around 12–13, 14–15 and 16–18 ka, respectively, the last one probably indicating the LGM. The modelling of TII at around 12–13 ka could possibly be related to some sea-level stabilization associated with the Antarctic Cold Reversal Event mentioned by McCullock *et al.* (2000).

Ponce *et al.* (2011) made similar comparisons based on the eustatic curve by Fleming *et al.* (1998) and the considerations by Hodgson *et al.* (2009). As a result, the ages of these terraces are different from those established by Violante (2005), the differences increasing as we go back in time. The extreme case is that when using the curve by Fleming *et al.* (1998), Ponce *et al.* (2011) considered that TIV could have an age of 1 Ma. These differences indicate that this statement needs more research. The modelling of terraces II, III and IV occurred during the first stages of the post-LGM transgression that corresponded with the unstable climatic oscillations which preceded the Younger Dryas. In addition, those terraces, unlike TI, could have been partially influenced by isostatic uplifting readjustment as they extend adjacent to Patagonia.

Apart from the sea-level fluctuations that modelled the terraces, the complex processes involved in the morpho-sedimentary evolution of the shelf resulted in the shaping of diverse features that have been defined as 'geomorphological provinces' (Parker *et al.* 1997) (Fig. 6.2):

- Rioplatense terrace – this is located in the northernmost part of the shelf corresponding to the northern sector of TI. It has been modelled during the post-LGM transgression in several erosional and agradational features represented by pre-transgressive relict reliefs and transgressive sand bodies.
- Deltaic front of the Colorado and Negro rivers – this is a feature formed by successive deltas evolved during the Pleistocene–Holocene, and shows a typical lobate shape and a gravelly–sandy–bioclastic composition. It is part of the southern sector of TI.
- North Patagonian gulfs – these are semi-enclosed coastal basins, with the peculiarity that the maximum depths of the gulfs are deeper than the shelf break. The gulfs are separated

Table 6.6. *Geomorphology of the shelf*

Sub-environments		
Inner shelf	From the coastline to around 30 m water depth	Sediments adjusted to the present nearshore hydrodynamic conditions (palimpsests). Active morphosedimentary features like shoal retreat massifs, linear shoals and giant subaqueous dunes are influenced by nearshore sedimentation, coastal currents, waves and tidal action. This sub-environment is more precisely defined in the Rioplatense terrace, above a water depth of 30 m
Middle shelf	From around 30–90 m depth	Represented by sedimentologically more stable areas with low sediment mobilization (relict sediments); most of the Patagonian shelf is of this type
Outer shelf	From around 90 m depth to the shelf break.	Close to the shelf edge where sediment dynamic is associated with the shelf–slope transition zone; boundary currents and upper-slope to submarine canyon-head processes (such as turbiditic and debris flows) occur there
Regions according to balance between continental and marine processes		
North	Offshore the eastern Pampas	Fluvial activity dominates over marine processes owing to the presence of the de la Plata fluvial–estuarine environment, active since the Pliocene
Central	Offshore the southern Pampas and most of the Patagonian regions	Predominance of wave action in the north and tidal action in the south Fluvial influence is of minor and local importance as the sediment provider to the shelf The coastal retreat occurred as a result of the balance between post-glacial sea-level rise and isostatic uplifting. A ravinement surface was produced with the consequent formation of a relict sandy mantle
South	Southern tip of Patagonia and Tierra del Fuego	The very narrow continent was almost completely covered by ice during LGM times, when glaciers reached positions close to the sea. Glacial and glacio-fluvial deposits were covered by the sea during the post-glacial transgression

from the open sea by sills, the tops of which are at water depths of 50–70 m. The origin of the gulfs was attributed to aeolian activity in continental depressions, later flooded by marine waters during the Quaternary transgressions and, consequently, lacustrine facies could have probably occupied the depressions during sea-level lowstands.

- Patagonian inner shelf – this sector, located between the Nuevo Gulf and the Santa Cruz River, shows a strong relief marked by coastal lobate morphologies probably related to small deltaic environments associated with the Patagonian rivers.
- Patagonian outer shelf – this is the largest province of the ACS that comprises terraces II, III and IV. On most of its surface, a partially buried drainage system develops, which constitutes incised valleys excavated by subaerial processes during pre-transgressive times.
- Tierra del Fuego shelf – this has a predominant gravelly composition and the presence of moraine-like topographies, resulting from glacial processes, occurred around LGM times.
- Malvinas islands shelf – this is a flat feature whose upper part constitutes the islands, and represents a morphological and geological extension of the Patagonian outer shelf.

Terraces in the ACS show different regional distributions, with deeper (older) terraces disappearing in the northern part of the shelf and very little development of TI in the south (Fig. 6.3). It is considered that the present conditions of Patagonian coasts (cliffed reliefs, high-energy) have also characterized previous stages of the late-glacial evolution, in such a way that deeper terraces have been shaped in the same way and therefore reached a more significant development. However, northern (Pampean) coastal regions, today characterized by low-energy coasts with lowlands, coastal plains and estuaries, have evolved with the same features in the past, and therefore no relicts of significant terraces remain recorded in the shelf, except for TI. This terrace represents a particular case as it probably responded to more complex and energetic interactive processes, among which the most significant can be attributed to the deep erosive processes driven by a stationary sea level during the Younger Dryas, and the influence of large estuarine and deltaic environments (the de la Plata and Colorado rivers) during the last stages of the regional evolution.

It can be synthesized that the modelling of the ACS, depending on the interaction between global, regional and local factors, such as sea-level fluctuations, climate, subaerial and subaqueous processes, isostasy and tectonism, originated in different geomorphological regions. Three distinct sub-environments are recognized: the inner, the middle and the outer shelf. However, regional differences due to the balance between continental and marine processes allow three geographical regions to be considered: North, Central and South. Table 6.6 gives the main characteristics of all three regions.

Conclusions

The ACS is one of the most extensive shelves in the world, showing diverse geological characteristics as a result of its emplacement in different geotectonic and oceanographic settings. Owing to its enormous size, knowledge of the ACS is uneven and still incomplete and, to learn more about it, is a short- to mid-term challenge for present and future generations of marine scientists.

The main conclusions arising from our present knowledge of the ACS can be summarized in the following points.

- The shelf has developed mainly on a passive continental margin, although it is also influenced, particularly in the southern regions, by other (active) types of margins. Consequently, northern regions of the shelf (adjacent to the Pampas

and northern Patagonia) show different morpho-sedimentary features to those in the southern regions (southern Patagonia).
- Four terraces with different extensions and characteristics, as well as several geomorphological provinces of complex origin, constitute the main reliefs of the shelf.
- Morpho-sedimentary features and sediment facies document the increasing influence of hydro-isostasy to the south.
- Sedimentary cover of the shelf is terrigenous and siliciclastic, constituted by sediments mainly sourced in Andean Patagonian regions. Sediments are considered as relict and palimpsest as a result of reworking by marine processes by means of the progressive sweeping of the shelf surface during the post-LGM transgression.
- Although the post-LGM sea-level rise was relatively uniform, some fluctuations are evident through the disposition of morpho-sedimentary features, the most important ones probably being associated with the Younger Dryas event.

Many regional geological aspects related to the evolution of the ACS during post-LGM times still remain incompletely solved, particularly: (a) the origin and evolution of terraces and related steps, incised valleys and north Patagonian gulfs; (b) the real significance of cold events (i.e. Younger Dryas, Antarctic Cold Reversal and neoglacial periods) and the evidence for this; (c) why the shelf border changes greatly in depth in different regions; (d) better interpretation of features resulting from relative sea-level fluctuations, particularly the lowstand deposits present in the shelf edge and upper slope; and (e) the processes that regulated the last subaerial exposure of the shelf.

To solve these problems a significant effort must be made in order to organize multidisciplinary projects aimed at carrying out research activities on the entire shelf based on geophysical–geological surveys focused on Late Quaternary sequences, in a continuous and systematic way, and with the objective of progressively covering successive areas until the entire shelf has been surveyed. It is necessary to do this under 'sea–land correlation' and 'source-to-sink' integrated perspectives. A programme initiated in the 1980s by the Group of Marine Geology and Geophysics of the Argentine Hydrographic Survey, and still continuing today, was the starting point for such a research line.

The summary that this contribution represents must serve as a framework for further studies and to encourage future –and essential – research in the region.

This contribution is part of the Framework Research Project MARGIN 'Geological and Geophysical Reconnaissance of the Argentine Continental Margin', carried out by the Group of Marine Geology and Geophysics from the Argentine Hydrographic Survey. Authors are indebted to this Institution and to the National Agency of Scientific Research for supporting and financing the studies on the continental shelf. Special gratitude is given to the late Dr G. Parker, who led the Group for 30 years, settled the basis for the development of the project and encouraged the Argentine marine geological community to continue working on the matter. Thanks are due to the leaders of IGCP 464 and IGCP 526 for supporting and encouraging participation in the projects and their active interaction, and UNESCO and INQUA for financial support to attend the project's meetings. We also want to thank an anonymous reviewer and Dr F. Chiocci (Volume editor), for constructive comments and suggestions that enabled us to improve the manuscript.

References

AGUIRRE, M. L. & WHATLEY, R. C. 1995. Late Quaternary marginal marine deposits and palaeoenvironments from northeastern Buenos Aires Province, Argentina: a review. *Quaternary Science Review*, **14**, 223–254.

BERGER, W. H. & WEFER, G. 1996. Expeditions into the past: Paleoceanographic studies in the South Atlantic. *In*: WEFER, G., BERGER, W. H., SIEDLER, G. & WEBB, D. J. (eds) *The South Atlantic: Present and Past Circulation*. Springer, Berlin, 363–410.

BERKOWSKY, F. 1986. Arenas del Río de la Plata: una excepción a la relación entre composición de areniscas y la tectónica de placas. *In*: *Actas de Resúmenes*, Primera Reunión Argentina de Sedimentología, La Plata. Asociación Argentina de Sedimentología, La Plata, Buenos Aires, 263–266.

BERNASCONI, E. & CUSMINSKY, G. 2009. Estudio paleoecológico de Foraminíferos de testigos del Holoceno de Golfo Nuevo (Patagonia, Argentina). *Geobiología*, **42**, 435–450.

BIANCHI, A., BIANUCHI, L., PIOLA, A., RUIZ PINO, D., SCHLOSS, I., POISSON, A. & BALESTRINI, C. 2005. Vertical stratification and air-sea CO_2 fluxes in the patagonian shelf. *Journal of Geophysical Research*, **110**, C07003, http://dx.doi.org/10.1029/2004JC002488

BOLTOVSKOY, E. 1954a. Foraminíferos de la Bahía de San Blas. *Revista del Museo Argentino de Ciencias Naturales 'Bernardino Rivadavia' Ciencias Geologia*, **3**, 247–300.

BOLTOVSKOY, E. 1954b. Foraminíferos del Golfo San Jorge. *Revista del Museo Argentino de Ciencias Naturales 'Bernardino Rivadavia' Ciencias Geologia*, **3**, 85–246.

BOLTOVSKOY, E. 1973. Estudio de testigos submarinos del Atlántico Sudoccidental. *Revista del Museo Argentino de Ciencias Naturales 'Bernardino Rivadavia' Ciencias Geologia*, **7**, 215–240.

BOZZANO, G., VIOLANTE, R. A. & CERREDO, M. E. 2011. Middle slope contourite deposits and associated sedimentary facies off NE Argentina. *Geo-Marine Letters*, **31**, 495–507, http://dx.doi.org/10.1007/s00367-011-0239-x

CAMPOS, E. J. D., MULKHERJEE, S., PIOLA, A. & DE CARVALHO, F. M. S. 2008a. A note on the mineralogical analysis of the sediments associated with the Plata River and Patos Lagoon outflows. *Continental Shelf Research*, **28**, 1687–1691.

CAMPOS, E. J. D., PIOLA, A. R., MATANO, R. P. & MILLER, J. L. 2008b. PLATA: a synoptic characterization of the Southwest Atlantic shelf under influence of the Plata River and Patos Lagoon outflows. *Continental Shelf Research*, **28**, 1551–1555.

CAVALLOTTO, J. L. 2008. Geología y geomorfología de los ambientes costeros y marinos. *In*: BOLTOVSKOY, D. (ed.) *Atlas de Sensibilidad Ambiental del Mar y la costa Patagónica. Project ARG 02/018*. Conservación de la Diversidad Biológica y Prevención de la Contaminación Marina en Patagonia, **GEF 28385**. Secretaría de Ambiente y Desarrollo Sustentable de la Nación, Buenos Aires.

CAVALLOTTO, J. L., PARKER, G. & VIOLANTE, R. A. 1995. Relative sea-level changes in the Rio de la Plata during the Holocene. *International Geological Correlation Programme Project 375, Late Quaternary coastal records of rapid change: application to present and future conditions*. Second Annual Meeting, Abstract, Antofagasta, Chile, 19–20.

CAVALLOTTO, J. L., VIOLANTE, R. A. & PARKER, G. 2004. Sea level fluctuations during the last 8600 yrs in the Río de la Plata (Argentina). *Quaternary International*, **114**, 155–165.

CAVALLOTTO, J. L., VIOLANTE, R. A. & HERNÁNDEZ MOLINA, F. J. 2011. Geological aspects and evolution of the Patagonian continental margin. *In*: *Palaeogeography and Palaeoclimatology of Patagonia: Implications for Biodiversity. Biological Journal of the Linnean Society*, **103**(Special Issue), 346–362.

CIONE, A. L., TONNI, E. P. & DONDAS, A. 2005. A mastodon (Mammalia, Gomphotheriidae) from the Argentinian Continental Shelf. *Neues Jahrbuch für Geologie und Paläontologie, Mitteilungen*, **10**, 614–630.

CLIMAP PROJECT MEMBERS 1981. *Seasonal Reconstruction of the Earth's Surface at the Last Glacial Maximum*. Geological Society of America, Map and Chart Series, **MC-36**, 1–18.

CODIGNOTTO, J. O., KOKOT, R. R. & MARCOMINI, S. C. 1992. Neotectonism and sea-level changes in the coastal zone of Argentina. *Journal of Coastal Research*, **8**, 125–133.

DEPETRIS, P. J. & GRIFFIN, J. J. 1968. Suspended load in the Rio de la Plata drainage basin. *Sedimentology*, **11**, 53–60.

DRAGANI, W. C., MARTÍN, P., CAMPOS, M. I. & SIMIONATO, C. 2010. Are wind wave heights increasing in South-eastern South American continental shelf between 32°S and 40°S. *Continental Shelf Research*, **30**, 481–490, http://dx.doi.org/10.1016/j.csr.2010.01.002

ETCHICHURI, M. C. & REMIRO, J. R. 1963. *La corriente de Malvinas y los sedimentos pampeano-patagónicos*. Comunicaciones del Museo Argentino de Ciencias Naturales "Bernardino Rivadavia", Ciencias Geológicas, **1**.

EWING, M. & LONARDI, A. G. 1971. Sediment transport and distribution in the Argentine Basin. 5. *In*:AHREMS, L., PRESS, F., RUNKORN, S. K. & UREY, H. C. (eds) *Sedimentary Structure of the Argentine Margin, Basin, and Related Provinces*. Physics and Chemistry of the Earth, **8**. Pergamon Press, Oxford, 125–251.

FARINATI, E. A. 1984. Dataciones radiocarbónicas en depósitos holocenos de los alrededores de Bahía Blanca, Provincia de Buenos Aires, Argentina. *In*: *Simposio Internacional sobre cambios del nivel del mar y evolución costera en el Cuaternario tardío, Mar del Plata, Resúmenes*. IUGS-UNESCO, Mar del Plata, Argentina, 27–31.

FERRERO, L. 2005. Foraminíferos y ostrácodos cuaternarios de dos testigos de la plataforma continental argentina al sudeste de Mar del Plata. *Ameghiniana*, **42**, 29R.

FLEMING, K., JOHNSTON, P., ZWARTZ, D., YOKOYAMA, Y., LAMBECK, K. & CHAPPELL, J. 1998. Refining the eustatic sea-level curve since the Last Glacial Maximum using far- and intermediate-field sites. *Earth and Planetary Science Letters*, **163**, 327–342.

FORBES, C. & GARRAFO, Z. 1988. A note on the mean seasonal transport on the Argentinian Shelf. *Journal of Geophysical Research*, **93**, 2311–2319.

GAIERO, D. M., PROBST, J. L., DEPETRIS, P. J., BIDART, S. M. & LELEYTER, L. 2003. Iron and other transition metals in Patagonian riverborn and windborne materials: geochemical control and transport to the South Atlantic Ocean. *Geochimica et Cosmochimica*, **67**, 3603–3623.

GARCÍA CHAPORI, N. 2013. *Reconstrucción paleoceanográfica del talud bonaerense a partir de testigos cuaternarios*. PhD thesis, Departamento de Ciencias Geológicas, Facultad de Ciencias Exactas y Naturales, Universidad de Buenos Aires, Argentina.

GIBERTO, D. A., BREMEC, C. S., ACHA, E. M. & MIANZÁN, H. W. 2004. Large-scale spatial patterns of benthic assemblages in the SW Atlantic: the Río de la Plata estuary and adjacent shelf waters. *Estuarine, Coastal and Shelf Science*, **61**, 1–13.

GIUSSANI, G. & WATANABE, S. 1980. Foraminíferos bentónicos como indicadores de la corriente de Malvinas. *Revista Española de Micropaleontología*, **12**, 169–177.

GÓMEZ, E. A. & PERILLO, G. M. E. 1995. Submarine outcrops underneath shoreface-connected sand ridges, outer Bahía Blanca Estuary, Argentina. *Quaternary of South America and Antarctic Peninsula*, **9**, 23–37.

GÓMEZ, E. A., MARTÍNEZ, D. E., BOREL, C. M., GUERSTEIN, G. R. & CUSMINSKY, G. C. 2006. Negative sea level oscillation in Bahía Blanca Estuary related to a global climatic change around 2650 yr BP. *In*: *ICS '04 – International Coastal Symposium. Journal of Coastal Research*, Special Issue **39**, 181–185.

GROEBER, P. 1948. Las plataformas submarinas y su edad. *Revista Ciencia e Investigación*, **6**, 224–231.

GROENEVELD, J. & CHIESSI, C. M. 2011. Mg/Ca of Globorotalia inflata as a recorder of permanent thermocline temperatures in the South Atlantic. *Paleoceanography*, **26**, PA2203, http://dx.doi.org/10.1029/2010PA001940

GUILDERSON, T. P., BURKLE, L., HEMMING, S. & PELTIER, W. R. 2000. Late Pleistocene sea level variations derived from the argentine shelf. *Geochemistry, Geophysics, Geosystems*, **1**, 1055, http://dx.doi.org/10.1029/2000GC000098

GYLLENCREUTZ, R., MAHIQUES, M. M., ALVES, D. V. P. & WAINER, I. K. C. 2010. Mid- to late-Holocene paleoceanographic changes on the southeastern Brazilian shelf based on grain size records. *The Holocene*, **20**, 863–875, http://dx.doi.org/10.1177/09596836 10365936

HART, T. 1946. Report on trawling survey of the Patagonian Continental Shelf. *Discovery Reports*, **23**, 223–248.

HODGSON, D. A., VERLEYEN, E., VYVERMAN, W., SABBE, K., LENG, M. J., PICKERING, M. D. & KEELY, B. J. 2009. A geological constraint on relative sea-level in marine isotope stage 3 in the Larsemann Hills, Lamber Glacier region, East Antarctica (31366–33 228 cal yr. B.P.). *Quaternary Science Reviews*, **28**, 2689–2696.

ISLA, F. I. 1989. Holocene sea-level fluctuation in the southern hemisphere. *Quaternary Science Reviews*, **8**, 359–368.

ISLA, F. I. & CORTIZO, L. C. 2005. Patagonian cliff erosion as sediment input to the continental shelf. *XVI Congreso Geológico Argentino, La Plata, Actas*, **4**, 773–778.

ISLA, F. I. & SCHNACK, E. 1995. Submerged moraines offshore northern Tierra del Fuego, Argentina. *Quaternary of South America and Antarctic Peninsula*, **9**, 205–222.

KOKOT, R. R. 2004. Erosión en la costa por cambio climático. *Asociación Geológica Argentina*, **59**, 715–726.

LAPRIDA, C., GARCÍA CHAPORI, N., VIOLANTE, R. A. & COMPAGNUCCI, R. H. 2007. Late Holocene evolution and paleoenvironments of northeastern Argentine shoreface–offshore transition: new evidences based on benthic foraminifera. *Marine Geology*, **240**, 43–56.

LAPRIDA, C., GARCÍA CHAPORI, N., CHIESSI, C. M., VIOLANTE, R. A., WATANABE, S. & TOTAH, V. 2011. Middle Pleistocene sea surface temperature in the Brazil–Malvinas Confluence Zone: Paleoceanographic implications based on planktonic foraminifera. *Micropaleontology*, **57**, 183–195.

MAHIQUES, M. M., TASSINARI, C. C. G. *ET AL.* 2008. Nd and Pb isotope signatures on the Southeastern South America upper margin: implicances for sediment transport and source rocks. *Marine Geology*, **250**, 51–63, http://dx.doi.org/10.1016/j.margeo.2007.11.007

MCCULLOCK, R. D., BENTLEY, M. J., PURVES, R. S., HULTON, N. R. J., SUGDEN, D. E. & CLAPPERTON, C. 2000. Climatic inferences from glacial and palaeoecological evidence at the last glacial termination, Southern South America. *Journal of Quaternary Science*, **15**, 409–417.

MOUZO, F. 2005. Límites de las glaciaciones Plio-Pleistocenas en la plataforma continental al noreste de la Tierra del Fuego. *XVI Congreso Geológico Argentino, Actas*, **3**, 787–792.

OSTERRIETH, M., VIOLANTE, R. A. & BORRELLI, N. 2005. Silicophytoliths in sediments from submarine cores in the northern region of the Argentine Continental Shelf. The Phytolitharien. *Bulletin of the Society for Phytolith Research*, **17**, 18–19.

PARKER, G. & VIOLANTE, R. A. 1982. Geología del frente de costa y plataforma interior entre Pinamar y Mar de Ajó, Prov. de Buenos Aires. *Acta Oceanográphica Argentina*, **3**, 57–91.

PARKER, G., PERILLO, G. M. E., RIVES, G. E. & MARTÍNEZ, H. C. 1976. Geología costera de superficie y subsuelo. VI: Propiedades ingenieriles de los suelos. Proyecto COPUAP: Complejo Portuario de Ultramar en Aguas Profundas. Unpublished Technical Report **6**, Argentina Hydrographic Survey.

PARKER, G., LANFREDI, N. & SWIFT, D. J. P. 1982. Seafloor response to flow in a southern hemisphere sand ridge field: Argentine inner shelf. *Sedimentary Geology*, **33**, 195–216.

PARKER, G., VIOLANTE, R. A. & PATERLINI, C. M. 1996. Fisiografía de la Plataforma Continental. *In*: RAMOS, V. & TURIC, M. (eds) *Geología y Recursos Naturales de la Plataforma Continental Argentina. Relatorio XIII Congreso Geológico Argentino*. Asociación Geológica Argentina, Buenos Aires, 1–16.

PARKER, G., PATERLINI, C. M. & VIOLANTE, R. A. 1997. El fondo marino. *In*: BOSCHI, E. (ed.) *El Mar argentino y sus Recursos Marinos*. INIDEP, Mar del Plata, **1**, 65–87.

PARKER, G., PATERLINI, C. M., VIOLANTE, R. A., COSTA, I. P., MARCOLINI, S. I. & CAVALLOTTO, J. L. 1999. *Descripción Geológica de la Terraza Rioplatense (Plataforma Interior del Noreste Bonaerense)*. Boletín, Servicio Geológico y Minero Argentino, **273**.

PARKER, G., VIOLANTE, R. A., PATERLINI, C. M., MARCOLINI, S., COSTA, I. P. & CAVALLOTTO, J. L. 2008. Las secuencias sismoestratigráficas del Plioceno-Cuaternario en la Plataforma Submarina adyacente al litoral del este bonaerense. *Latin American Journal of Sedimentology and Basin Analysis*, **15**, 105–124.

PELTIER, W. R. 1988. Global sea level and Earth rotation. *Science*, **240**, 895–901.

PERILLO, G. M. E. & KOSTADINOFF, J. 2005. Margen Continental de la Provincia de Buenos Aires. *In*: DE BARRIO, R. E., ETCHEVERRY, R. O., CABALLÉ, M. F. & LLAMBÍAS, E. (eds) *Geología y Recursos Minerales de la Provincia de Buenos Aires. Relatorio XVI Congreso Geológico Argentino*. La Plata, Argentina, 277–292.

PIERCE, J. W. & SIEGEL, F. R. 1979. Suspended particulate matter on the Southern Argentina Shelf. *Marine Geology*, **29**, 73–91.

PIRAZZOLI, P. A. 1991. *World Atlas of Holocene Sea-Level Changes*. Elsevier Oceanographic Series, **58**. Elsevier, Amsterdam.

PONCE, J. F., RABASSA, R., CORONATO, A. & BORROMEI, A. M. 2011. Paleogeographic evolution of the Atlantic Coast of Pampa and Patagonia since the Last Glacial Maximum to the Middle Holocene. *Biological Journal of the Linnean Society*, **103**, 363–379.

POTTER, P. E. 1994. Modern sands of South America: composition, provenance and global significance. *Geologische Rundschau*, **83**, 212–232.

RABASSA, J. L., CORONATO, A. & MARTÍNEZ, O. 2011. Late Cenozoic glaciations in Patagonia and Tierra del Fuego: an updated review. *Biological Journal of the Linnean Society*, **103**, 316–335.

RAMOS, V. A. 1996. Evolución tectónica de la plataforma continental. *In*: RAMOS, V. A. & TURIC, M. A. (eds) *Geología y Recursos Naturales de la Plataforma Continental Argentina*. XIII Congreso Geológico Argentino y III Congreso de Exploración de Hidrocarburos, Buenos Aires, 1996. Asociación Geológica Argentina-Instituto Argentino del Petróleo, Relatorio, **21**, 405–422.

RAMOS, V. A. 1999. Rasgos Estructurales del Territorio Argentino. Evolución Tectónica de la Argentina. In: CAMINOS, R. (ed.) *Geología Argentina*. Instituto de Geología y Recursos Minerales, SEGEMAR, Anales, **29**, 715–784.

RAMOS, V. A. & TURIC, M. (eds) 1996. *Geología y Recursos Naturales de la Plataforma Continental Argentina*. XIII Congreso Geológico Argentino, Buenos Aires.

ROSTAMI, K., PELTIER, W. R. & MANZINI, A. 2000. Quaternary marine terraces, sea-level changes and uplift history of Patagonia, Argentina: comparisons with predictions of the ICE-4G (VM2) model of the global process of glacial isostatic adjustment. *Quaternary Science Review*, **19**, 1495–1525.

SCHELLMANN, G. & RADTKE, U. 2000. ESR dating of stratigraphically well-constrained marine terraces along the Patagonian Atlantic coast (Argentina). *Quaternary International*, **68–71**, 261–273.

SCHNACK, E. J., ISLA, F. I., DE FRANCESCO, F. D. & FUCKS, E. E. 2005. Estratigrafía del Cuaternario marino tardío en la Provincia de Buenos Aires. *In*: DE BARRIO, R. E., ETCHEVERRY, R. O., CABALLÉ, M. F. & LLAMBÍAS, E. (eds) *Geología y Recursos Minerales de la Provincia de Buenos Aires. Relatorio XVI Congreso Geológico Argentino*. La Plata, Argentina, 159–182.

SERVICIO DE HIDROGRAFÍA NAVAL 2011. *Tablas de Marea*. Servicio de Hidrografía Naval, Buenos Aires, Publication, **H-610**.

SIMIONATO, C. G., MECCIA, V., GUERRERO, R., DRAGANI, W. C. & NUÑEZ, M. 2007. Te Rio de la Plata estuary response to wind variability in synoptic to intraseasonal scales: II currents vertical structure and its implications on the salt wedge structure. *Journal of Geophysical Research, Oceans*, **112**, C07005, http://dx.doi.org/10.1029/2006JC003815

SWIFT, D. J. P. 1968. Coastal erosion and transgressive stratigraphy. *Journal of Geology*, **76**, 444–456.

TERUGGI, M. E. 1954. El material volcánico-piroclástico en las sedimentación cuaternaria argentina. *Revista Asociación Geológica Argentina*, **9**, 184–191.

TURIC, M. A., NEVISTIC, A. V. & REBAY, G. 1996. Geología y Recursos Naturales de la Plataforma Continental. *In*: RAMOS, V. A. & TURIC, M. A. (eds) *Geología y Recursos Naturales de la Plataforma Continental Argentina*. XIII Congreso Geológico Argentino y III Congreso de Exploración de Hidrocarburos, Buenos Aires. Asociación Geológica Argentina-Instituto Argentino del Petróleo. Relatorio, **22**, 405–423.

URIEN, C. M. 1970. Les rivages et plateau continental du Sud du Brésil, de l'Uruguay et de l'Argentine. *Quaternaria*, **12**, 57–69.

URIEN, C. M. & EWING, M. 1974. Recent sediments and environments of Southern Brazil, Uruguay, Buenos Aires and Río Negro Continental Shelf. *In*: BURK, C. & DRAKE, CH. (eds) *The Geology of Continental Margins*. Springer, Heidelberg, 157–177.

URIEN, C. M. & ZAMBRANO, J. J. 1996. Estructura de la Plataforma Continental. *In*: RAMOS, V. A. & TURIC, M. A. (eds) *Geología y Recursos Naturales de la Plataforma Continental Argentina*. XIII Congreso Geológico Argentino y III Congreso de Exploración de Hidrocarburos, Buenos Aires, 1996. Asociación Geológica Argentina-Instituto Argentino del Petróleo, Relatorio, **1**, 29–66.

URIEN, C. M., MARTINS, L. R. & MARTINS, I. R. 1979. Modelos deposicionales de la Plataforma Continental de Rio Grande do Sul, Uruguay y Buenos Aires. *7° Congreso Geológico Argentino, Neuquén, Actas*, **2**, 639–658.

VIOLANTE, R. A. 2004. Coastal-marine processes and sediment supply during the post-LGM transgression in the northern part of the Argentine Continental Shelf. *In*: *Abstracts of the 4th Annual Conference Project IGCP 464 'Continental Shelves During the Last Glacial Cycle', Roma-Ponza, Italy*. UNESCO-IGCP, Paris, 58–60.

VIOLANTE, R. A. 2005. Submerged terraces in the continental shelf of Argentina and its significance as paleo-sea level indicators: the example of the Rioplatense Terrace. *In*: *Abstracts of the 5th Annual Conference Project IGCP 464 'Continental Shelves During the Last Glacial Cycle', St. Petersburg, Russia*. UNESCO-IGCP, Paris, 97–99.

VIOLANTE, R. A. & CAVALLOTTO, J. L. 2011. The record of the LGM sea-level position at the Argentina Continental Shelf: an evidence of the complex processes involved in the post-glacial relative sea-level rise. *In*: *Abstracts of the 5th IGCP 526 Conference 'Continental Shelves: Risks, Resources and Record of the Past', Victoria (BC), Canada*. UNESCO-IGCP, Paris, 128–129.

VIOLANTE, R. A. & PARKER, G. 2004. The post-Last Glacial Maximum transgression in the de la Plata river and adjacent Inner Continental Shelf, Argentina. *Quaternary International*, **114**, 167–181.

VIOLANTE, R. A. & ROVERE, E. I. 2005. Los sedimentos de la Plataforma Submarina y su relación con el volcanismo andino neógeno. *16° Congreso Geológico Argentino, Actas*, **3**, 239–246.

VIOLANTE, R. A., PARKER, G., CAVALLOTTO, J. L. & MARCOLINI, S. 1992. La Secuencia Deposicional del Holoceno en el 'Río' de la Plata y Plataforma del noreste bonaerense. *4ª Reunión Argentina de Sedimentología, Actas*, **1**, 275–282.

VIOLANTE, R. A., OSTERRIETH, M. & BORRELLI, N. 2007. Evidences of subaerial exposure of the Argentine Continental Shelf during the Last Glacial Maximum. *In*: CHEN, M.-T., LIU, Z. & CATTO, N. (eds) *Quaternary of East Asia and Western Pacific: a Regional to Global Perspective: Part I*. Quaternary International, **167–168**, 434.

VIOLANTE, R. A., LAPRIDA, C. *ET AL.* 2014. Registro paleoambiental del Estadio Isotópico 2 en la plataforma continental exterior del sureste bonaerense: un nuevo aporte a la evolución regional. *Revista Asociación Geológica Argentina*, **71**, MS 2883.

ZAMBRANO, J. J. & URIEN, C. M. 1974. Pre-Cretaceous basins in the Argentine continental shelf. *In*: BURK, C. A. & DRAKE, C. L. (eds) *The Geology of Continental Margins*. Springer, Berlin, 463–470.

Chapter 7

The Baltic Sea continental shelf

SZYMON UŚCINOWICZ

*Branch of Marine Geology, Polish Geological Institute – National Research Institute, Koscierska 5,
80–328 Gdańsk, Poland (e-mail: szymon.uscinowicz@pgi.gov.pl)*

Abstract: The Baltic Sea is a semi-enclosed intracontinental sea of approximately 418 500 km^2, surrounded by Scandinavia and the lowlands of Central and Eastern Europe. Its catchment area is about four times larger than the area of the sea itself. The Baltic Sea has an average depth of around 50 m, and is connected to the Atlantic Ocean via the shallow and narrow Danish Straits. A characteristic feature of the Baltic Sea is thermohaline stratification, in particular the occurrence of a permanent halocline.

The northern part of the Baltic Sea is situated within the Precambrian Baltic Shield, whereas the southern part lies on the East European Platform. A small SW part lies on the Palaeozoic West European Platform. During the Pleistocene period, Scandinavian ice sheets advanced on the area of the present Baltic Sea several times, resulting in damage to the bedrock and a deepening of the Baltic Basin. The present-day Baltic Sea had its beginnings in the retreat of the last Scandinavian ice sheet, which melted approximately 15.5–14.5 ka ago in the area of the southern part of today's Baltic Sea.

The Quaternary cover of the seabed in the area of the whole Baltic Sea is formed from Pleistocene sediments of glacial, glacio-fluvial and limnoglacial origin, as well as from sediments of Holocene marine accumulation. In the southern part of the Baltic, during the Holocene, the sea constantly encroached upon the land, destroying Pleistocene sediments. In the northern part, as a result of uplift of the land and regression of the sea, Pleistocene glacial and glaciofluvial deposits, as well as early Holocene Baltic sediments, are being eroded. In the Baltic Sea, sandy and gravelly sediments are present above the pycnocline, while muds have been deposited below the pycnocline.

The Baltic Sea is an intracontinental sea, surrounded by land in all directions: in the NW by Scandinavia, and in the south and east by the lowlands of Central and Eastern Europe (Fig. 7.1). Its catchment area is 1 748 300 km^2, which is more than four times the area of the Baltic Sea itself.

Both the Baltic Sea and its catchment area extend in a north–south direction. The Baltic Sea spreads in this direction over an area of about 1300 km, between latitudes 53°95′N and 65°46′N. The meridional extent of the catchment area, from the Carpathians in the south to the northern part of the Scandinavian Mountains beyond the Polar Circle, is approximately 2300 km (Fig. 7.1). The latitudinal extent of the sea and its catchment area is smaller; it spreads maximally between longitudes 10° and 30°E, while its width, in general, is considerably smaller. In the southern part, between Jutland and the Lithuanian coast, the latitudinal extent of the sea is approximately 750 km, while in the central part, between the Swedish and Estonian coasts, it is approximately 280 km. In the narrowest parts, in the area of the Åland Islands, the width of the sea is approximately 180 km, while in the area of the Kvarken Archipelago, between the towns of Vaasa and Umeå, it is only around 100 km. The latitudinal extent of the catchment area is only slightly larger, between 8° and 38°E.

In the NW, the Baltic Sea catchment area is limited by the Scandinavian Mountains, and, in the south, by the Sudetes and the western part of the Carpathians. From the west and the SE, the east and the NE, the catchment area is limited mainly by moraine hills formed during the Pleistocene glaciations. Owing to the geological structure, around 76% of the catchment area is comprised of lowlands with heights of up to 200 m above sea level (asl). Uplands cover around 17%, while highlands (500–2655 m asl) cover 7% of the catchment area. The area of the Baltic Sea, together with the Danish Straits and the Kattegat, is approximately 418 500 km^2. It is a shallow sea, connected to the Atlantic Ocean by the narrow, shallow Danish Straits.

Owing to the hydrographical conditions and the shape of the coastline, a number of regions can be distinguished within the area of the Baltic Sea: Bothnian Bay, the Bothnian Sea (including the Archipelago Sea), the Gulf of Finland, the Gulf of Riga, the Baltic Proper, the Belt Sea (including Western Baltic and the Sound) and the Kattegat (Fig. 7.1). Similarly, a number of basins and depths divided by seabed elevations and escarpments can be distinguished in the Baltic Sea. Starting from the west, in the southern part of the Baltic Proper, these are the Arkona Basin (max. depth 53 m), the Bornholm Basin (max. depth 105 m) and the Gdańsk Basin (max. depth 107 m). The Bornholm Basin is connected to the Gdańsk Basin by the Słupsk Furrow, which has a maximum depth of about 95 m (Mojski *et al.* 1995). The northern part of the Baltic Proper has the West Gotland Basin (which in its northern part includes the 459 m Landsort Deep, the deepest area of the Baltic Sea), the East Gotland Basin (max. depth 245 m) and the North-Central Basin (max. depth *c.* 200 m). To the north of the North-Central Basin lies Åland Deep, reaching depths of up to 301 m. The deepest place in the Bothnian Sea (294 m) is Härnösand Deep (Uvödjupet) in the NW. In Bothnian Bay, nowhere is deeper than 147 m, while the maximum depth of the Gulf of Riga is 67 m and of the Gulf of Finland is 123 m (Winterhalter *et al.* 1981). Despite these deep areas within the basins, the average depth of the Baltic Sea is about 50 m.

Climate and general oceanography of the Baltic Sea

The geographical situation of the Baltic Sea means that its climate is strongly diversified. The western and southern parts are under the strong influence of the Atlantic Ocean; therefore, marine and temperate climates prevail. The eastern and, in particular, northern parts of the Baltic Sea, beyond 60°N, have attributes of intermediate, temperate and cold climates, with features of a polar continental climate. Polar-sea and polar-continental air masses prevail over the Baltic Sea; less often, tropical or arctic air masses occur that cause frequent and rapid changes in weather conditions (Majewski & Lauer 1994).

A characteristic feature of the Baltic Sea waters is their thermohaline stratification, particularly the occurrence of a constant halocline dividing the low-salinity surficial layers with their seasonally changing thermics from the deep waters of higher salinity and stable temperature (Fig. 7.2).

In winter, in Bothnian Bay, the Bothnian Sea and the Gulfs of Finland and Riga, the water temperature at the surface of the sea drops to 0 °C and even to the freezing point of brackish water

From: CHIOCCI, F. L. & CHIVAS, A. R. (eds) 2014. *Continental Shelves of the World: Their Evolution During the Last Glacio-Eustatic Cycle.*
Geological Society, London, Memoirs, **41**, 69–89. http://dx.doi.org/10.1144/M41.7

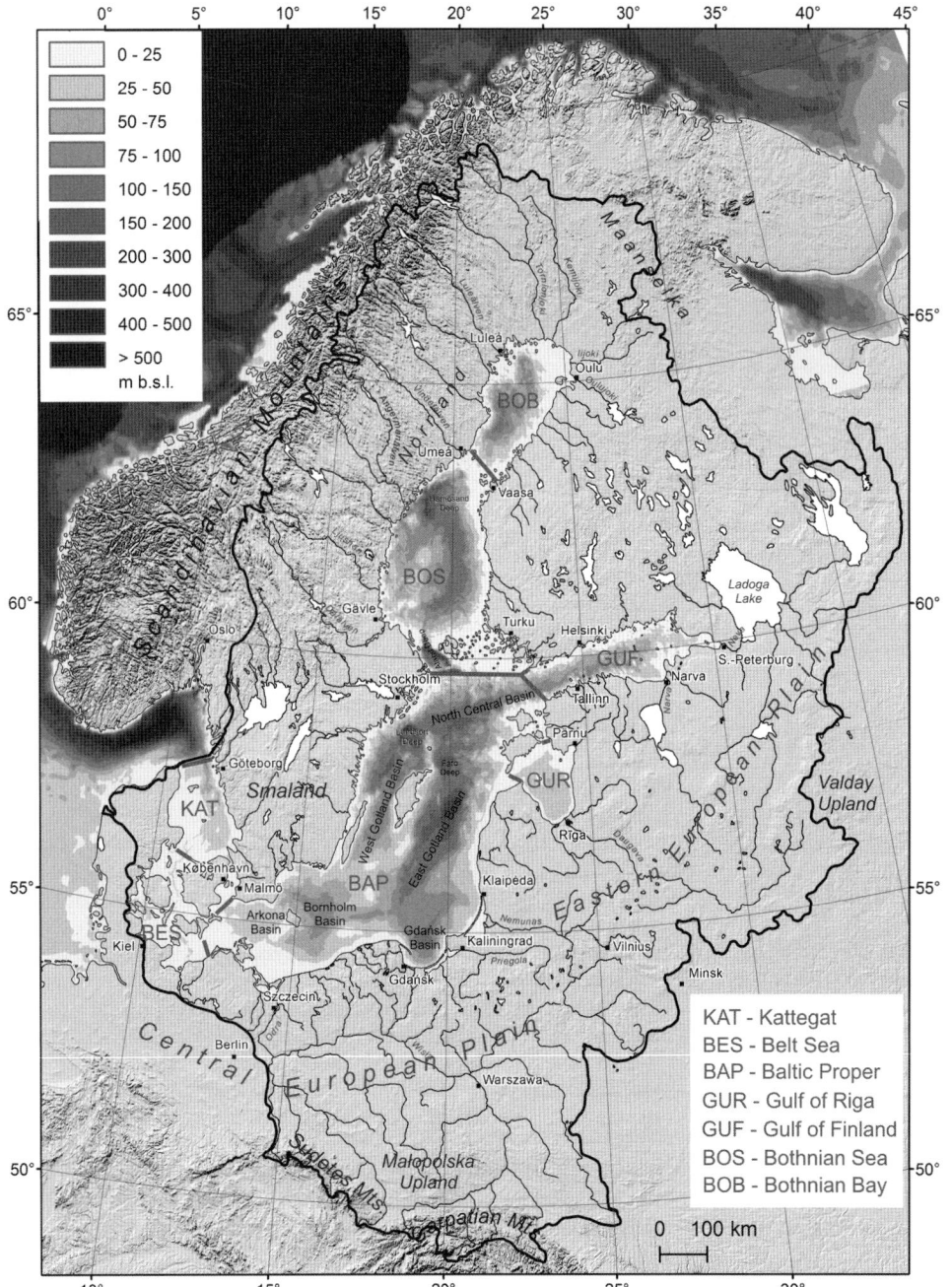

Fig. 7.1. Baltic Sea catchment area and the division of the Baltic Sea into major hydrographical regions and basins. Land area: shaded terrain model after USGS GTOPO30. Bathymetry of the Baltic Sea after iowtopo2 (Seifert *et al.* 2001).

KAT - Kattegat
BES - Belt Sea
BAP - Baltic Proper
GUR - Gulf of Riga
GUF - Gulf of Finland
BOS - Bothnian Sea
BOB - Bothnian Bay

(*c.* −0.5 °C). In the Baltic Proper, the winter surficial water temperature very rarely drops below 1–2 °C. In summer, the temperature at the surface of Bothnian Bay is approximately 12–13 °C, while in the Baltic Proper it is around 16–17 °C. Water-temperature variability within the year concerns not only the surface layer but also the depths of the sea, causing thermal stratification of the waters. In summer, water is heated up to a depth of 20–30 m, below which the colder water of the previous winter can be found. In autumn and spring, temperature compensation occurs from the surface to the bottom. In winter, the lowest temperature occurs near the surface, increasing with depth to approximately 4–6 °C on the seabed (Łomniewski 1975).

Haline stratification of Baltic Sea waters results from the positive water balance and limited exchange of waters with the ocean. On average, annually, 428 km³ of river water flows into the Baltic Sea, precipitation delivers 237 km³, while evaporation removes 184 km³ (Lass & Matthäus 2008). The Baltic Sea is connected with the Atlantic Ocean via the Danish Straits, the Little Belt, Great Belt, Øresund and the Kattegat. Sills limiting the exchange of waters are situated at depths of 18 m in the Great Belt and 8 m in Øresund. Surficial brackish waters flow out through the Danish Straits, while salty waters rich in oxygen flow in near the bottom. Inflow of salty waters is constant but limited. Episodically, under specific meteorological situations, strong inflows of oceanic waters occur, increasing the salinity and oxygen supply of bottom waters.

The salinity of surface waters of the Baltic Sea decreases from the Kattegat to Bothnian Bay. In the Danish Straits surface water, salinity is between 25 and 10‰, dropping to 6‰ in the northern part of the Baltic Proper, 6–5‰ in the Gulf of Riga and 5–4‰ in the Gulf of Finland. Surface salinity decreases northwards in the Bothnian Sea from 6 to 5‰, and, in Bothnian Bay, from 4 to 2–1‰ in the most northern part of the bay. The average salinity of bottom waters in the Danish Straits is approximately 30‰. In the Baltic Proper, salinity drops from around 16‰ in the SW part (Bornholm Basin) to approximately 10‰ in the northern part. Salinity of bottom waters in the Gulf of Finland decreases eastwards from 9 to 5‰. In the Bothnian Sea,

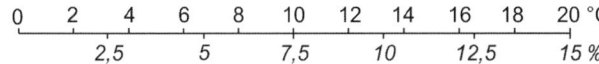

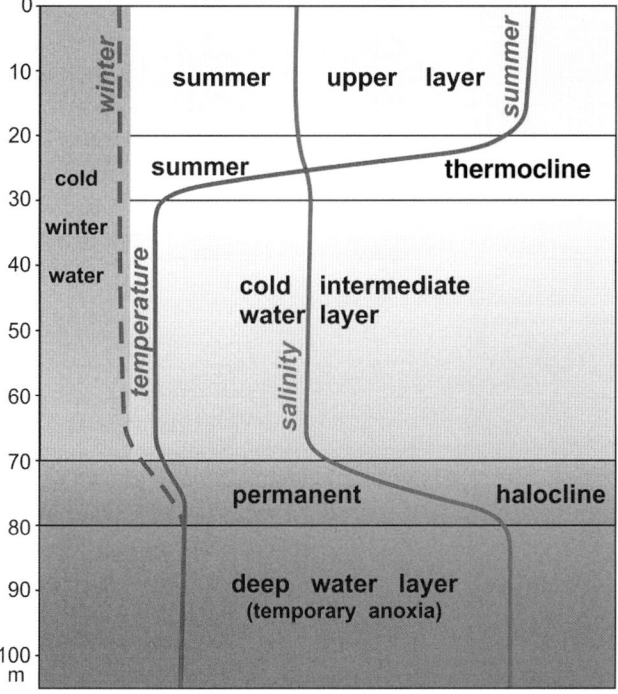

Fig. 7.2. Thermohaline stratification of the Baltic Sea (an example from the Gdańsk Basin, generalized).

salinity of deep waters is approximately 6–5‰, while in the Bothnian Bay it is around 4‰ (Kullenberg 1981).

In the periods between inflows of oceanic waters, the bottom waters of the deep sedimentary basins undergo periods of stagnation when the oxygen is utilized in the decomposition of organic matter. This results in an oxygen deficiency and often also in the occurrence of hydrogen sulphide. This phenomenon has major significance for the chemical and biological conditions of the bottom waters, as well as the geochemistry of sediments and the sedimentary environment. Benthic organisms disappear and conditions for formation of laminated sediments occur.

The Baltic Sea is a tideless sea; therefore, the most important climatic element that influences the dynamics of the water masses is the wind. It causes the formation of wind waves, drift currents and the swelling of water levels, as well as the mixing of upper layers of seawater and heat exchange. In the coastal zone, wind induces coastal currents and upwelling. In the Baltic Sea area, wind is characterized by high variability, both in terms of speed and direction. Over the Baltic Proper, winds from the SW and west dominate all year, constituting 35–40% of winds, depending on the season. In addition, storm winds (>15 m s^{-1}) most often come from the SW and west. Relatively often over the Baltic Proper, winds also occur from the NW and south in 7–15% and 7–14%, respectively, of all cases. In the north, over the Bothnian Sea and Bothnian Bay, winds from the south and SW are most frequent (30–55%). Similarly frequent (20–30%) are winds from the north and NE (Defant 1972).

Average yearly speeds of winds over the Baltic Sea are within the range of 6–8 m s^{-1}. Average top speeds of up to 9 m s^{-1} occur in the autumn and winter in the northern part of the Baltic Proper. Winds occurring over the Baltic Sea, particularly strong storm winds with speeds exceeding 15–17 m s^{-1}, are related to the movement of low-pressure centres, particularly frequent at the turn of autumn and winter. Their paths lead over the Baltic Sea predominantly from the North Sea or the Norwegian Sea. From 1970 to 2003, in the southern part of the Baltic Proper,

each year between five and 37 storms occurred (average annual count of 19). From 1948 to 2003, in the NE part of the Baltic Proper, near the coast of the island of Saaremaa, between two and 36 storms were noted annually (average of 20).

A high variability of pressure systems and winds means that the Baltic Sea has no stable system of currents. Currents and waves created during storms are especially important for the processes of coastal and seabed erosion, as well as for the transport of sediments. Then, near-bottom currents can have speeds exceeding 1 m s^{-1} (Majewski & Lauer 1994; Lass & Matthäus 2008). During strong storms, with wind speeds up to 25–30 m s^{-1}, one can observe waves at an average height of 3–3.5 m and period of 7–8 s on the Baltic Sea. The largest waves observed in different parts of the Baltic Sea were 5–10 m high and 90–160 m long, with periods of 10–13 s (Majewski & Lauer 1994; Schmager et al. 2005).

Overlapping of wind surge and low atmospheric pressure, together with storm waves, leads to the occurrence of maximum levels of water, causing substantial – even catastrophic – damage to the shore. Maximum levels of the sea observed in the SW part of the Baltic Proper exceeded the zero level by 2.83 m in Warnemünde and 2.11 m in Kołobrzeg.

Apart from the movements of water masses generated by wind, the seabed is also affected by internal waves occurring in the layers of the pycnocline. A high diversification of vibration periods is observed, from minutes up to several days. The maximum amplitude of internal waves is related to the deep-sea pycnocline within the upper part of the salty layer. The second maximum occurs in the summer thermocline. Speeds of internal wave-related currents can reach 0.8 m s^{-1}, influencing the graining and thickness of sediments (Tareew 1965).

Tectonic framework and pre-Quaternary geology

The Baltic Sea is an intracontinental sea and, therefore, covers an area geologically similar to the land that surrounds it. The northern part of the Baltic Sea is situated within the Precambrian Baltic Shield, whereas the southern part lies on the East European Platform. A small SW part of the Baltic Sea lies on the Palaeozoic West European Platform, separated from the East European Platform by the Teisseyre–Tornquist Fault Zone (Fig. 7.3). In the part situated on the Baltic Shield, Precambrian crystalline rocks create outcrops in the coastal area and on bottom elevations. The Archaean, represented by metamorphic and volcanic rocks (gneiss, enderbite and migmatite) dated to 3–2.5 Ga, occurs on the Baltic Sea bed only in the extreme northern coastal part of Bothnian Bay, directly under a thin layer of Quaternary sediments or on the bottom surface (Lundqvist & Bygghammar 1994) (Fig. 7.4).

Proterozoic rocks have significantly wider range. Outcropping directly on the seabed or under Quaternary rocks, they lie along the coasts of Bothnian Bay, the Bothnian Sea, and along the northern coasts of the Gulf of Finland in the area of the Archipelago Sea and along the eastern coasts of Sweden, as well as along the eastern shores of the Kattegat. The northern part is dominated by granitoids and rarer dolerites, as well as metamorphosed sedimentary and volcanic rocks. Gneisses, less metamorphosed sedimentary and volcanic rocks occur mostly in the Kattegat. At the bottom of the Baltic Sea, in the area of the Åland Islands and in the NE part of the Gulf of Finland, intrusions of rapakivi granites that had formed in the Mesoproterozoic (c. 1.5–1.65 Ga) can also be found.

Proterozoic sedimentary rocks are represented by Mesoproterozoic, 1.5–1.2 Ga Jotnian sandstones and late Neoproterozoic (Ediacaran) clayey deposits. Jotnian sandstones remain on magmatic rocks in depressions of Bothnian Bay and the Bothnian Sea, and locally also on the coasts (Lundqvist & Bygghammar

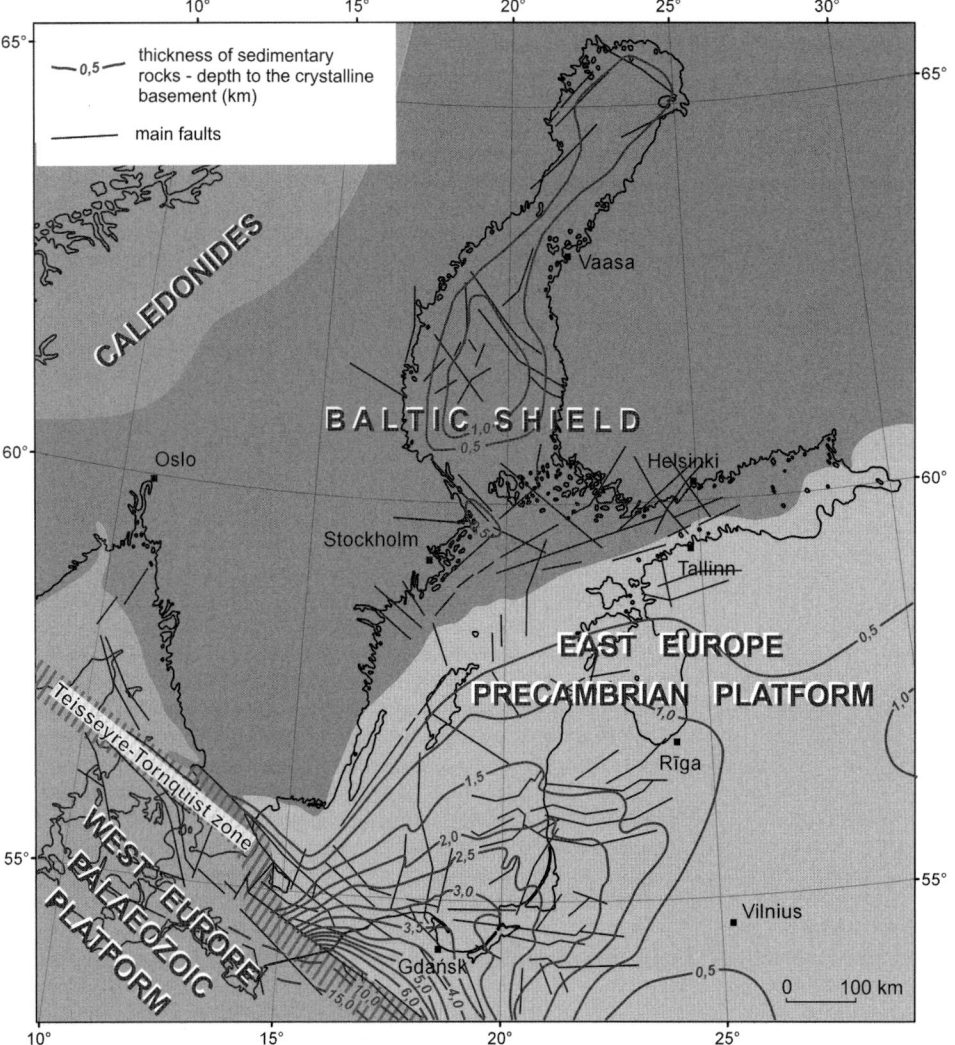

Fig. 7.3. Tectonic features of the Baltic Sea area.

1994). Ediacaran clayey deposits, termed 'blue clay', are known in the eastern part of the Gulf of Finland and in the vicinity of St Petersburg (Russian Federation).

In the northern part of Bothnian Bay and the western part of the Bothnian Sea, early Palaeozoic sedimentary rocks lie on Jotnian sandstones: Cambrian sandstones in the bay; and Cambrian sandstones and Ordovician limestones in the Bothnian Sea (Fig. 7.4).

The Precambrian ceiling lowers towards the south and SE, forming the crystalline foundations of the East European Platform (Figs 7.3 & 7.4). On the northern coast of Estonia, near Tallinn, the Precambrian foundation is present at a depth of approximately 200–400 m, while in the southern parts of Gotland and Saaremaa it lies at 800 m. Along the eastern coast, the crystalline foundation lowers to a depth of around 1000 m in the southern part of the Gulf of Riga, to about 2200 m on the Lithuanian coast in the area of Klaipeda and to 3500–3000 m in the area of the Gulf of Gdańsk. On the southern coasts of the Baltic Sea, on the NE side of the Teisseyre–Tornquist zone, the ceiling of Precambrian rocks lies at a depth of approximately 6.5–6.0 km (Fig. 7.3) (Pokorski & Modliński 2007).

In the area of the East European Platform, the thickness of the Phanerozoic sedimentary cover increases towards the SE and the south along with a lowering of the ceiling of the crystalline foundation. The platform cover is formed by two sedimentary complexes: an Ediacarian–early Palaeozoic and Devonian–Carboniferous complex lies concordantly on the crystalline basement; Permian–Mesozoic and Palaeogene–Neogene deposits rest discordantly on the older complex.

On the Baltic Proper seabed, sedimentary rocks occur under the Quaternary, from Cambrian in the north and NW to Neogene in the south. In the northern part of the Baltic Proper, Palaeozoic sediments lie directly under the Quaternary cover, while in the eastern and southern parts they are covered with Mesozoic and Cenozoic sediments (Fig. 7.4).

At the western side of the Teisseyre–Tornquist Fault Zone, in the area of the West European Palaeozoic Platform, the ceiling of Precambrian rocks lies at a depth of about 10–15 km (Fig. 7.3). Folded sediments from the early Palaeozoic (Cambrian, Ordovician and Silurian) lie above this, covered by sediments of late Palaeozoic and Mesozoic age – locally also of Palaeogene age. These sediments form two platform complexes: Carboniferous–Devonian and Permo-Mesozoic, which, at the turn of the Cretaceous and Palaeogene periods, were divided into smaller units. Several local anticlines are present here, with Jura and even Trias in the core, as well as synclines with axes orientated NW–SE, divided by faults into smaller tectonic blocks. In the Baltic Sea area west of the Teisseyre–Tornquist zone, at the bottom of the sea, directly under the Quaternary sediments, only Mesozoic and Palaeogene sediments are present.

Cambrian sandstones appear along the northern coasts of Estonia and further in the Baltic Sea bed towards the SW along the scarp (erosion escarpment), the so-called 'Cambrian glint'. Further on, they can be found along the coasts of Sweden southwards to Scania (Fig. 7.4). In the central and southern part of the Baltic Proper, deposits of gas and oil are related to Cambrian sandstones.

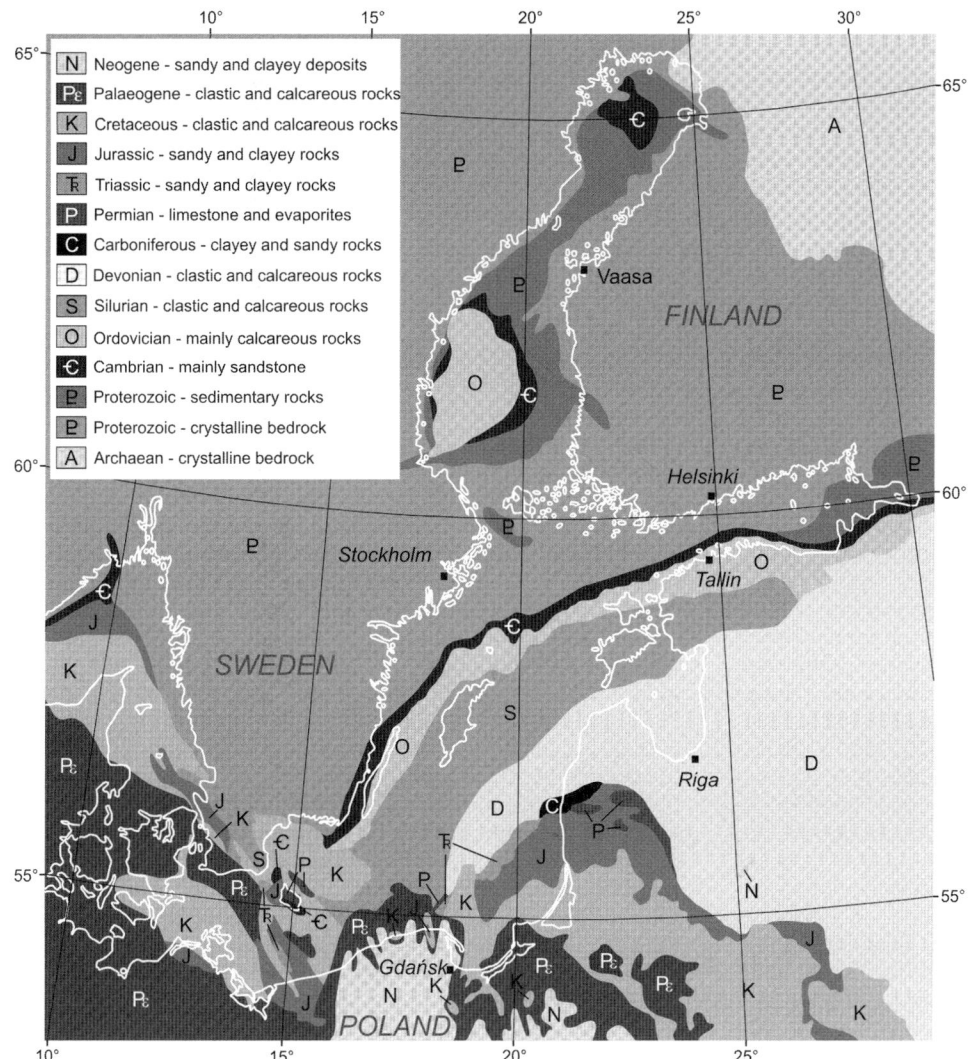

Fig. 7.4. Pre-Quaternary geology of the Baltic Sea area (based on Norling 1994; Sigmond 2002, simplified).

Ordovician red and grey limestones are present on the bottom of the sea also along the 'Ordovician glint' scarp, stretching from the southern side parallel to the Cambrian outcrop. Deposits of oil and gas are also related to the Ordovician sediments.

Silurian sedimentary rocks appear under Quaternary sediments in significant areas of the Baltic Proper (Fig. 7.4). The erosive range of those sediments stretches from Hiiumaa, turning towards the SW, northwards to Gotland, creating the Silurian scarp (termed the 'Silurian glint') and forming, among others, the western coasts of Gotland. The 'Silurian glint' is built of reef limestones. Southwards, the Silurian sediments become clayey facies, deposited in deeper sea.

Devonian sedimentary rocks are present only in the SE part of the Baltic Proper in the form of a strip leading from the southern coasts of Estonia, through the Gulf of Riga, and further westwards to the Latvian and Lithuanian coasts. The Early Devonian is represented mainly by terrigenic sandstones formed in continental environments; Middle Devonian sediments are formed by sandstones covered by dolomites; the Late Devonian is represented mainly by dolomites, and layers of gypsum and anhydrite.

Carboniferous fine-grained sandstones with thin inserts of clays remaining on Devonian deposits occur within a very limited area. Under Quaternary rocks they appear only near the southern coasts of Latvia and stretch eastwards in the form of a narrow strip.

Permian sedimentary rocks are represented only by Permian limestone carbonate–evaporite facies. In the southern part of the Baltic Proper they are common; however, directly under Quaternary rocks, they are present in very small areas, eastwards from the

Lithuanian coast and near the Polish coast in the region north of Gdańsk (Fig. 7.4).

The Triassic is present only in the southern part of the Baltic Proper. In the area of the East European Platform, it occurs directly under Quaternary sediments in a narrow strip from the southern coasts of Latvia to the northern parts of the Gdańsk Basin, while in the area of the West European Platform it appears locally in the axes of anticlines (Fig. 7.4). In the area of the East European Platform, continental sediments prevail: red sandstones and mudstones, as well as limestones, deposited in river environments, lakes and lagoons. In the area of the West European Platform, Trias sediments are also mainly continental sediments; however, brackish–marine sediments are also present locally (Winterhalter *et al.* 1981).

The Jurassic, like the Triassic, is present only in the southern part of the Baltic Proper. In the area of the East European Platform, directly under the Quaternary layer, Jurassic sediments appear in a strip stretching from southern Lithuanian coasts towards the SW, towards the NE part of Gdańsk Basin. In the area of the West European Platform, similarly to Triassic sediments, they appear in the axes of anticlines. On the East European Platform, neritic clastic sediments (sandstones, mudstones) of the Middle and Late Jurassic period are present. On the Palaeozoic Platform a more complete profile of the Jurassic period is present. The Early Jurassic consists mostly of clastic inland sediments: fluvial, deltaic and lacustrine with interbedded clayey marine or brackish sediments. Marine sediments of the Middle Jurassic period are alternately clayey and sandy sediments. Late Jurassic sediments are clastic formations at the beginning, with an increasing contribution of marl and clayey rocks up-profile. Jurassic sediments occur under Quaternary sediments, on the wings and axes of anticlines.

The Cretaceous period is present over large areas in the southern part of the Precambrian platform, from the Lithuanian coast to Scania. Marine sediments of the Late Cretaceous period are mainly fine clastic with larger contributions of carbonate–silica sediments (opokas and gaizes) (Kramarska 1999). Directly under the Quaternary layers, Cretaceous sediments are present in the form of a strip from the southern part of the Curonian Lagoon to the central and southern part of the Gulf of Gdańsk, as well as in the NW part of Gdańsk Basin. Within the Palaeozoic Platform, Cretaceous sediments occur under Cenozoic sediments over quite a large area. The Early Cretaceous period is developed in clastic facies of mixed origin: marine and inland. The Late Cretaceous period is represented mainly by carbonate sediments with sandy layers in its lower part and a thicker layer with flints in the upper part. Directly under the Quaternary layers, Cretaceous sediments occur on significant fragments of the seabed in the SW part of the Baltic Proper, to Rügen and Zealand. Furthermore, Cretaceous sediments appear in the Kattegat (Winterhalter et al. 1981; Kramarska 1999).

The Palaeogene in the area of the Precambrian platform occurs directly under Quaternary sediments in the strip from Sambia westwards along the Polish coast. Paleocene marine sediments remained in the bottom of this part of the Baltic Sea only along the coast of the Sambian Peninsula and in the southern part of the Bornholm Basin. They are present in carbonate–clayey facies and in calcareous sands with glauconite. In the area of the Palaeozoic platform, the Paleocene is present locally in small areas of the southern part of the Bornhom Basin, in the northern part of the Arkona Basin and in the SW part of Scania, as well as in the area of the Belt Sea (Danish Straits) including the southern part of the Kattegat. Eocene marine sediments in the Baltic seabed occur mainly within the Precambrian East European platform. In the region north of the Polish coast, they are sandy mudstones with phosphoritic concretions, locally clayey sediments and quartz-glauconite. Eocene sediments are also present near western and eastern slopes of the Gulf of Gdańsk. In this area deltaic sediments containing large amounts of amber occur. In the area of the West European Palaeozoic Platform, Eocene sediments directly appear in the Bay of Meklemburg and locally in the Danish Straits. Oligocene sediments in the area of the Precambrian East European Platform are most probably present locally, in the coastal area of the southern part of the Baltic Sea. In the NW part of the Gulf of Gdańsk they occur as brackish clayey and sandy sediments with coal dust. West of the Teisseyre–Tornquist zone, marine clayey and sandy Oligocene sediments occur locally near the coasts of the Bay of Meklemburg and the Jutland Peninsula.

The Neogene occurs only in the area of the Precambrian East European Platform and is represented only by Miocene sediments. Sandy silts with inserts of brown coal occur locally in the coastal zone of the Sambian Peninsula and from the NW coasts of the Gulf of Gdańsk to the central part of the Polish coast (Kramarska 1999).

The origin of the Baltic Sea

The Baltic Sea is a very young sea. In its present shape, it has existed for only a few thousand years. The processes that caused the initial formation of a depression in the Earth's crust, followed by filling with waters of the present Baltic Sea, started far earlier. From the Devonian, sea transgressions did not reach further north than the central part of the Baltic Proper. Mesozoic and Cenozoic seas were limited to the southern and SW parts of the present Baltic Sea (Winterhalter et al. 1981; Norling 1994).

During the Eocene in the Baltic area, the climate was warm and humid, and the range of the sea was limited to the southern parts of the present Baltic Proper. The major part of Scandinavia at that time was land, where coniferous trees, producing large amounts of resin, grew in forests. Sediments from Scandinavia, along with fragments of resin, were transported southwards by rivers and deposited on the coasts of the Eocene sea. This is indicated by Eocene deltaic sediments containing resin transformed into amber, currently present on the coasts of the Gulf of Gdańsk (Jaworowski 1987; Piwocki & Olkowicz-Paprocka 1987; Blazhchishin 1999).

During the Miocene the level of the ocean dropped so much that the whole area of today's Baltic Sea became land. Cooling of the climate at the end of the Miocene and in the Pliocene fostered development of the processes of weathering of rocks and strong erosion. Weathered, loose rocky fragments were easily removed by rivers flowing from higher areas of Scandinavia through the area of the present Baltic Sea towards the basin of the North Sea (e.g. Overeem et al. 2001; Kuhlman et al. 2004). In that time, the original shapes of the Baltic Basin were formed.

Around 900 ka ago, the climate became so cold that the first continental glaciers appeared in Scandinavia to spread over the area of the present Baltic Sea and the lowlands of Central and Eastern Europe. In the last several hundred thousand years, glacials alternated cyclically with short periods of interglacials when the ice sheets disappeared. Each time, continental glaciers propagated southwards and eastwards through the area of the present Baltic Sea. Each subsequent advance of the ice sheet caused damage to the bedrock and deepening of the Baltic basin. During the interglacials, the Baltic Basin was occupied by seas similar to the present Baltic Sea, although information on possible marine interglacial sediments is scarce. On the coasts of the Baltic Sea, sediments remained locally of the Holsteinian and Eemian seas, which existed about 420–360 and 130–115 ka ago, respectively (Winterhalter et al. 1981; Makowska 1986; Eronen 1988; Lundqvist & Robertson 1994).

The last cold period (the Vistulian Glaciation) lasted from 115 to 11.5 ka ago. During this time a continental glacier appeared three times in Scandinavia and entered into the basin of the Baltic Sea. However, the chronology of ice-sheet advances, as well as the number and limits of the main glacial phases during the Vistulian period, is still controversial. Different authors have opted for one, two or three major ice-sheet advances (stadials) when an ice sheet crossed the present southern coastline of the Baltic Sea and reached the Lower Vistula region (Marks 2010). During the interstadials, the ice margin retreated to northernmost Scandinavia. Remains of terrestrial deposits of interstadials are known from Norrland and Skåne (Lagerlund 1987; Lundqvist & Robertson 1994). Riverine and limnic deposits, with radiocarbon ages from 46.5 to 21.48 ka BP, occur in the region of Odra Bank (SW Baltic Proper) (Kramarska 1998). Interstadial lacustrine and terrestrial organic sediments with ^{14}C dates between 36 and 41 ka BP have been documented in the shallows of Kriegers Flak in the western Baltic (Andrén et al. 2011). Within the depression of the Baltic Basin area, in its deepest parts, large lakes developed at that time (Lundqvist & Robertson 1994). There is also evidence for the occurrence of marine–brackish sediments, dated to Marine Isotope Stage (MIS) 3, in Hanö Bay and Kriegers Flak (Andrén et al. 2011).

According to radiocarbon dates of organic silts from Odra Bank (SW Baltic Proper) and the middle part of the Polish coast (Kramarska 1998, Rotnicki & Borówka 1995), the last ice sheet crossed the area of the Baltic Sea approximately 26 cal ka ago, destroying a large part of the sediments formed both during earlier glaciations and during the interglacials and interstadials. The beginnings of the present-day Baltic Sea are related to the retreat of the last Scandinavian ice sheet. Ice-sheet decay started in the European lowlands around 22–20 ka ago (Rinterknecht et al. 2006; Marks 2010). In the area of the most southerly part of the present Baltic Sea, the ice sheet melted approximately 15.5–14.5 ka ago (Fig. 7.5).

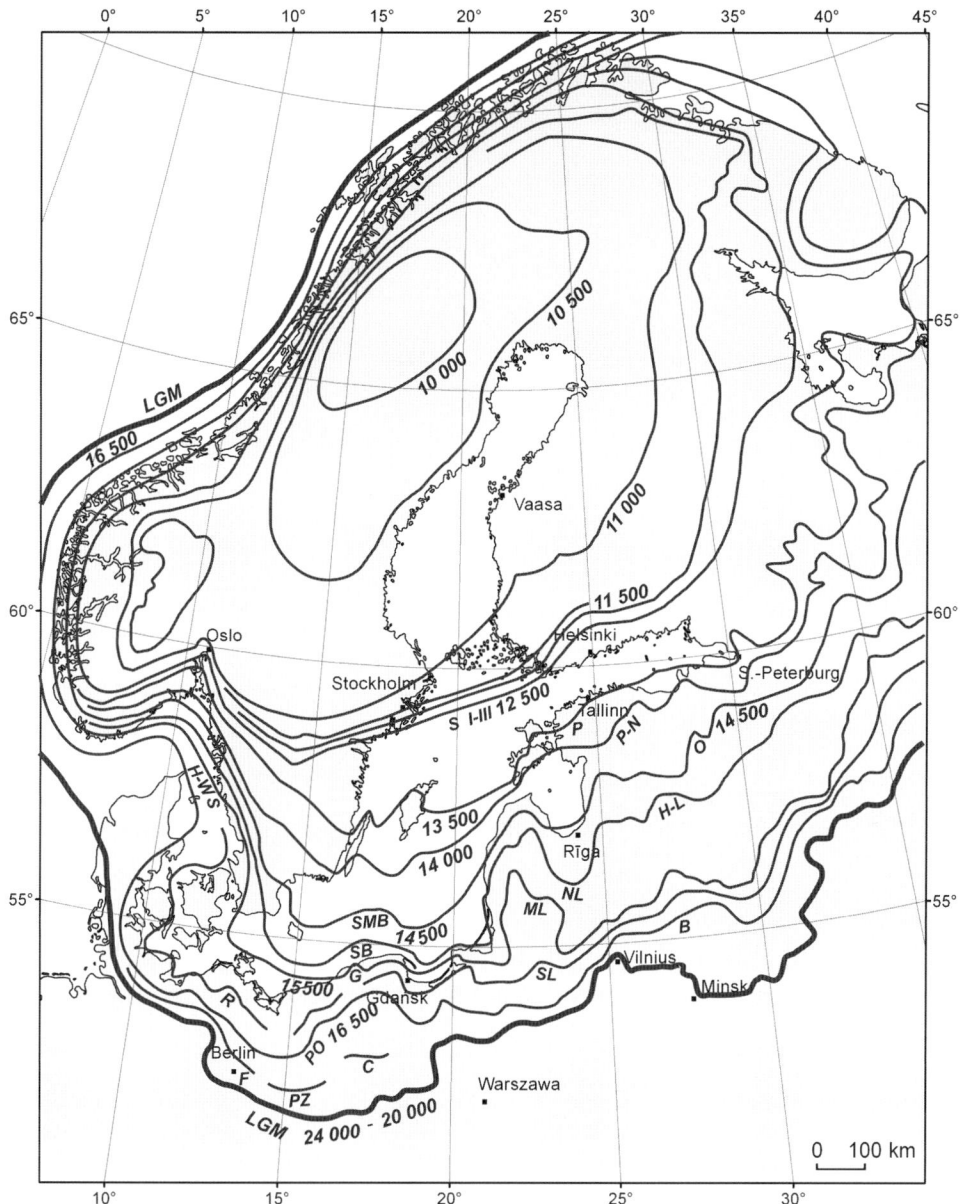

Fig. 7.5. Ice-sheet retreat from the Baltic Sea basin area (compiled after Lundqvist 1994; Uścinowicz 1999; Berglund 2004; Rinterknecht *et al.* 2006; Gyllencreutz *et al.* 2007; Johnsen *et al.* 2009; Marks 2010). LGM, Last Glacial Maximum; F, Frankfurt Phase; PZ, Poznań Phase; C, Chodzież Phase; PO, Pomeranian Phase; B, Baltija Phase; R, Rosenthal Phase; SL, South Lithuanian Phase; H-WS, Halland–West Skåne Phase; G, Gardno Phase; ML, Middle Lithuanian Phase; SB, Słupsk Bank Phase; NL, North Lithuanian Phase; H-L, Haanja–Luga Phase; SMB, Southern Middle Bank Phase; Otepää Phase; P-N, Pandivere–Neva Phase; P, Palivere Phase; S I–III, Salpausselkä Phases.

Melt waters gathered in front of the ice sheet as the glacier retreated, creating increasingly large ice-dammed lakes. Sediments from ice-dammed lakes are known from many places on the bottom of the southern part of the Baltic Sea. Whereas areal deglaciation dominated during ice-sheet retreat in the area north of the Pomeranian Phase moraines, in the area of the southern part of the present-day Baltic Sea this became subaqueus deglaciation. In areas where proglacial lakes were shallower, the ice sheet rested on the bottom; in deeper areas, the edge of the glacier formed an ice cliff. Rare relicts of forms related to ice-sheet retreat, such as end moraines, glaciofluvial deltas, eskers or de Greer moraines, can be found in the area of the southern part of the Baltic Sea (Fig. 7.6). In the area of the southern part of the Baltic Sea, those forms marking stages of deglaciation created in the marginal zones were, in major part, eroded during the Holocene transgression. Forms created in deeper parts of the Baltic Sea that were not encompassed by the transgression are covered with a thicker layer of sediments and, mostly, are not distinguished in the seabed relief.

Around 14.5 ka ago, local marginal lakes that existed at that time in the Bornholm and Gdańsk basins merged. A newly formed water reservoir, the Baltic Ice Lake, stretched from the Lithuanian coasts in the east to Denmark in the west. Its area, although limited by the edge of the glacier in the north, enlarged

as the ice sheet melted (Fig. 7.7). Results from palynological and diatomaceous analyses of sediment cores from the southern Baltic Sea show that in the Allerød and Younger Dryas in the late Pleistocene it was a freshwater, cold reservoir (Zachowicz *et al.* 2008).

At the end of the Allerød, the front of the glacier was in line with the Salpausselkä and central Sweden moraines (Fennoscandian ice-marginal zone). Rapid cooling of the climate in the Younger Dryas caused the ice sheet to re-advance on the central Swedish lowlands. Oscillations of the glacier front in this area, opening and closing connections with the ocean, caused rapid changes in the water level in the Baltic Ice Lake. The end of the Younger Dryas saw fast ice-sheet recession. At the beginning of the Holocene, within about 2–2.5 ka, the ice sheet disappeared from the Bothnian Sea, Bothnian Bay and, finally, from northern Sweden (Fig. 7.5).

Ice-sheet retreat caused uplift of the Earth's crust. Glacioisostatic rebound on the southern coasts of the Baltic Sea started approximately 17 ka ago and finished around 10–9 ka ago (Uścinowicz 2003). In the centre of the glaciation, the uplift started about 15.5 ka ago and has continued until today (e.g. Mörner 1979, 1980). Total uplift in the southern part of the Baltic Proper reached around 100–150 m; however, in the centre of the glaciation, it reached as much as 800 m (Figs 7.8 & 7.9).

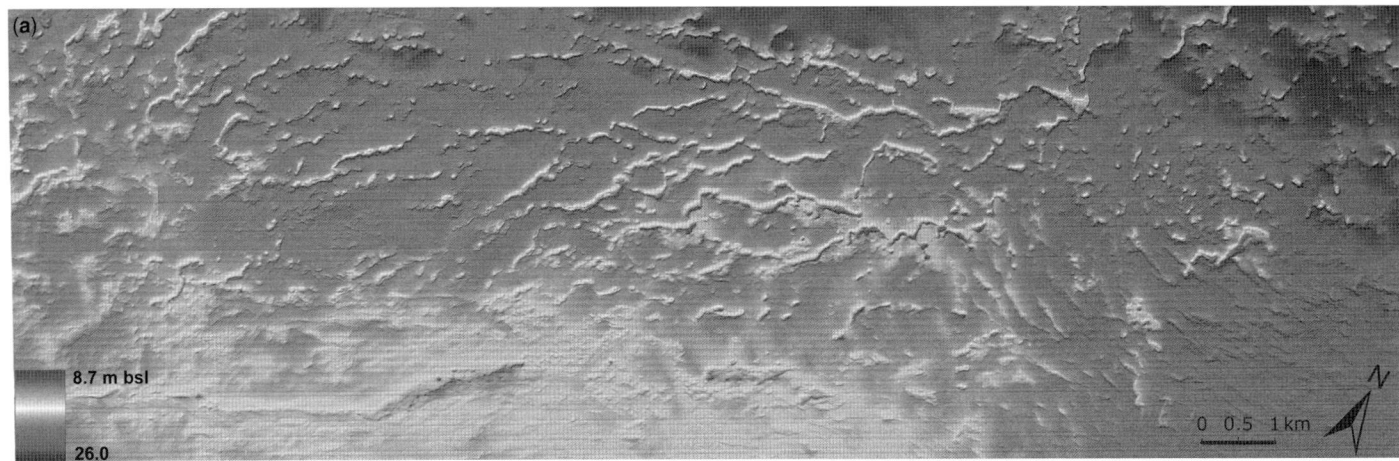

Fig. 7.6. Relicts of the de Geer moraines: (**a**) drowned on the Słupsk Bank (southern part of the Baltic Proper), digital elevation model of the seabed, from EEA grant 'Mapping Polish Waters to Safeguard Biodiversity'; (**b**) emerging from under the sea surface on the Kverken Archipelago (northern Baltic Sea). Photograph by S. Uścinowicz.

The varied size and duration of the glacioisostatic rebound of the southern and the northern parts of the Baltic Sea influenced both the sedimentation processes and the evolution of the coasts. In the northern regions of the Baltic Sea, north of the Stockholm–Turku line (i.e. in the areas of the Bothnian Sea and Bothnian Bay), in the whole period after deglaciation the uplift processes more than balanced the eustatic rise in water levels. Therefore, in those regions, permanent regression occurred

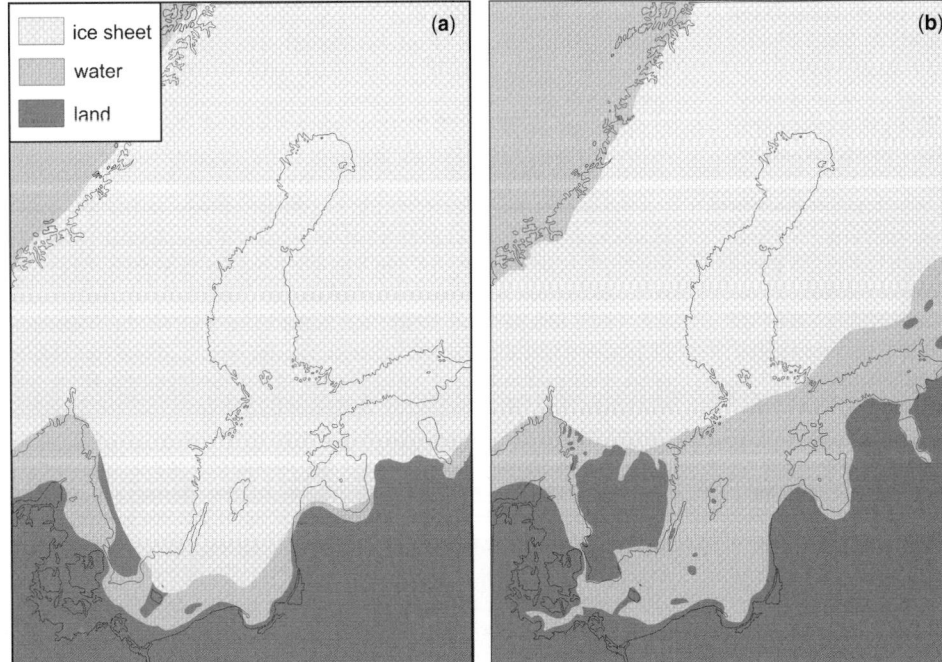

Fig. 7.7. Development of the Baltic Ice Lake – the Baltic Sea: (**a**) around 14.5 ka ago, during the Southern–Middle Bank Phase (the Oteppä Phase); (**b**) between 12.5 and 11.5 ka ago during the Salpausselkä phases.

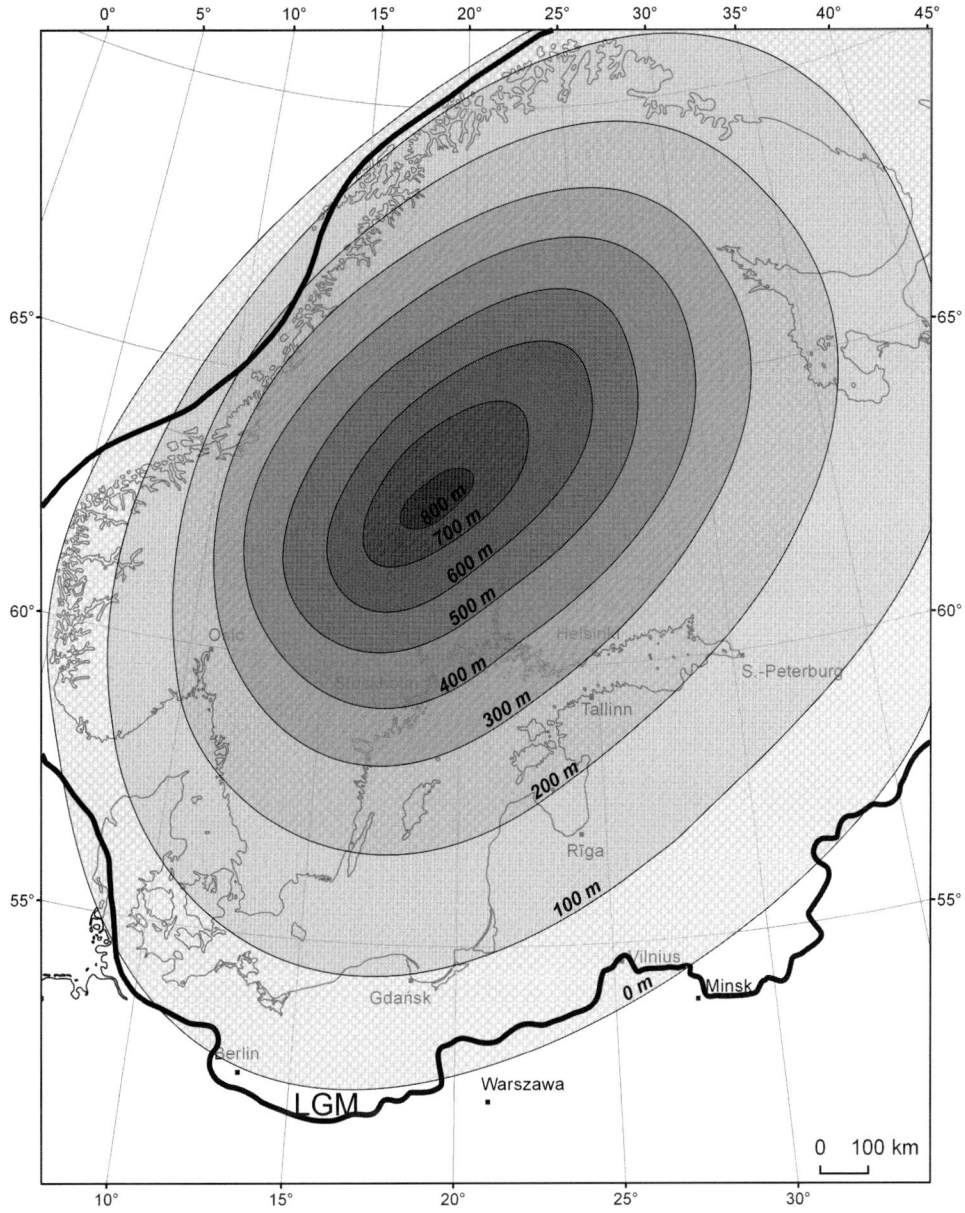

Fig. 7.8. Total post-glacial uplift and maximum extent of the last glaciation (thick black line marks the LGM). After Erikson & Henkel (1994); southern part changed by the author.

(e.g. Berglund 2004; Lindén *et al.* 2006) (Fig. 7.10) and still exists today. Areas of the seabed originally located under water gradually emerged and this emergence still continues. As a result of uplift, glacial and glaciofluvial relief that had formed under the glacier and at its front, as well as those sediments deposited in the early phases of Baltic Sea development, were first exposed to the erosive impact of waves and next emerged over the sea surface (Fig. 7.11). The highest coastlines located in Sweden, on the SW coasts of Bothnian Bay, have been elevated to a height of about 280 m above present sea-level since the time of deglaciation (Lundqvist 1994).

Near the end of the Pleistocene, the southern part of the Baltic Proper area, adjacent to the coasts of southern Denmark, Germany, Poland and Lithuania, was affected by rapid variations in water level. Two drainage events caused by ice-sheet front oscillation occurred in the Baltic Ice Lake. During the Holocene, this area was permanently exposed to the processes of marine transgression (Fig. 7.12).

The sea entered land areas, eroding relief forms, glacial and glaciofluvial deposits and also sediments of land environments formed after ice-sheet retreat (e.g. Uścinowicz 2003, 2006). In certain places on the bottom of this part of the Baltic Sea, relicts of the lacustrine sediments, peats and even tree trunks rooted *in situ* occur under marine sands (Fig. 7.13). At the bottom of the

Gulf of Gdańsk, large areas of the Vistula delta from the period of the late Pleistocene and the first half of the Holocene are also located.

A more complex history of sea-level change during the Holocene needs unravelling in the central part of the Baltic Proper coasts (i.e. on the coasts of southern Sweden and Latvia, Estonia and southern Finland). In the periods when fast eustatic rise in sea level outpaced the rate of uplift of the Earth's crust, transgressions occurred. Along with a slowing in the eustatic rise of sea level, uplift processes began to prevail and regression occurred. The transgressions and regressions started and finished at different times in different regions. This was due to different relationships between the rate of eustatic water-level rise and the rate of land uplift (e.g. Miettinen 2004; Rosentau *et al.* 2011). The zero-line (i.e. the areas where eustasy and land uplift were in balance) shifted north as the ice sheet vanished.

Holocene history of the Baltic Sea

Around 11.7–11.5 ka ago the ice sheet finally melted in the lowlands of central Sweden. The Baltic Ice Lake could connect with the ocean for the second time. At that time, the level of

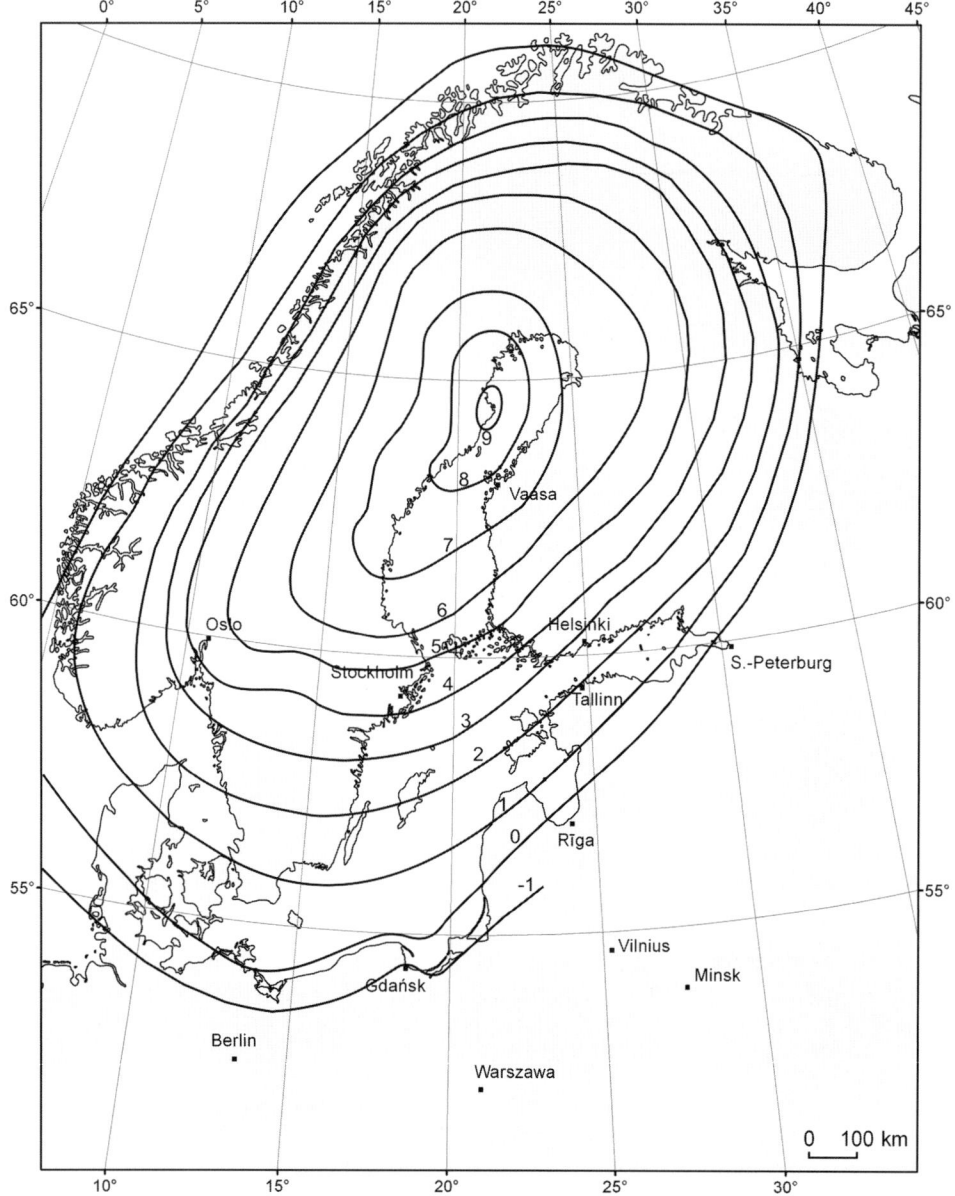

Fig. 7.9. Recent uplift of Fennoscandia in mm a^{-1}. After Ekman (1996); southern part changed by the author.

water in the ocean was approximately 25–27 m lower than in the Baltic Ice Lake, which resulted in waters from the lake flowing into the ocean. The Baltic Ice Lake surface lowered by about

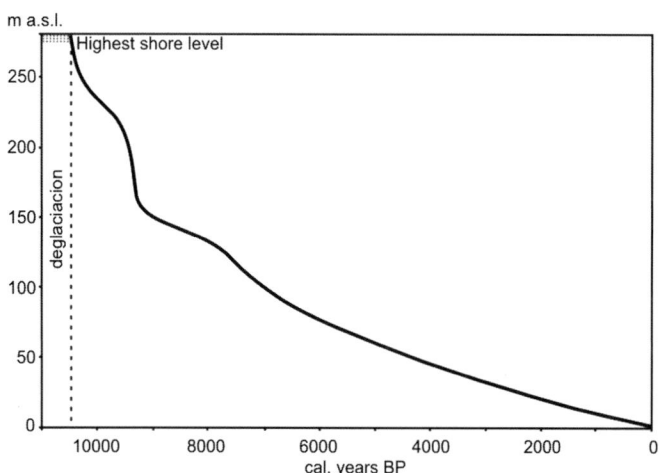

Fig. 7.10. Relative sea-level curve for the SW coast of Bothnian Bay (Scandinavian glacioisostatic uplift centre). After Berglund (2004).

25 m within a few decades (Björck 1995). The area of the water body reduced significantly and its southern coasts moved northwards. At the turn of the Pleistocene and Holocene, successive rises in the level of waters in the ocean resulted in salty oceanic waters starting to flow into the Baltic Sea, forming the water body traditionally called the Yoldia Sea (Fig. 7.14). The name of that phase of Baltic Sea development comes from the sea clam *Yoldia arctica*. Fossils of *Y. arctica* are known from Swedish uplifted coasts of the northern Baltic Proper and are not found in sediments of the Baltic seafloor. Most probably, the extension of brackish water was limited to the areas in the vicinity of the strait connecting the Baltic Basin with the ocean. The rest of the Baltic remained freshwater or only very slightly brackish at that time. In the northern parts of the Yoldia Sea, the proximity of the glacier caused Arctic conditions to occur.

The rate of uplift of Scandinavia was faster at that time than the rise in the level of the ocean, causing the Yoldia Sea to become isolated from the ocean at about 10.8–10.7 ka. Without an inflow of salty oceanic waters, the sea quickly became a fully freshwater lake, termed the Ancylus Lake (Fig. 7.15). The name of that phase of Baltic Sea development comes from a freshwater snail *Ancylus fluviatilis*. In the northern areas of the Yoldia Sea and Ancylus Lake, glacioisostatic uplift of Scandinavia caused the

Fig. 7.11. Erratic boulder, near which flintstone artefacts, charcoal and animal bones were found, and radiocarbon-dated to 3.8 ka BP. The boulder, currently located 46 m asl, was on an island at that time and was used by Stone Age people as a shelter during fishing and seal-hunting trips. SE coast of Bothnian Bay, area of Vaasa in Finland. Photograph by S. Uścinowicz.

relative level of waters to lower. In the southern part, transgression occurred throughout.

The first, although weak, signs of saline water entering the Baltic Basin via the Danish Straits were ^{14}C dated to about 9.8 ka ago (Andrén *et al.* 2011). However, the end of the freshwater phase (Ancylus Lake) in the southern part of the Baltic Sea is palynologically dated to the end of the Boreal period (Zachowicz *et al.* 2008). Marine influences manifested first in the western and southern parts of the Baltic Proper, and gradually propagated northwards. The process of transformation of a freshwater lake into a brackish-water sea lasted for around 1000 years. The Litorina Sea (Fig. 7.16), named after the slug *Litorina littorea*, was saltier and slightly warmer than the present-day Baltic Sea. This is confirmed both by *L. littorea*'s range reaching further north than today and by the results of diatomaceous analyses. Mesohalobe and euhalobe diatoms have dominated sediments of the southern part of the Baltic Proper since the early Atlantic period (Zachowicz *et al.* 2008). Within this period, glacioisostatic movements on the southern coasts stopped (Uścinowicz 2003), so that the rate of sea-level rise here was similar to the rate of the rise of ocean waters, causing the sea to quickly enter those areas that, thus far, had been land (Figs 7.12 & 7.16).

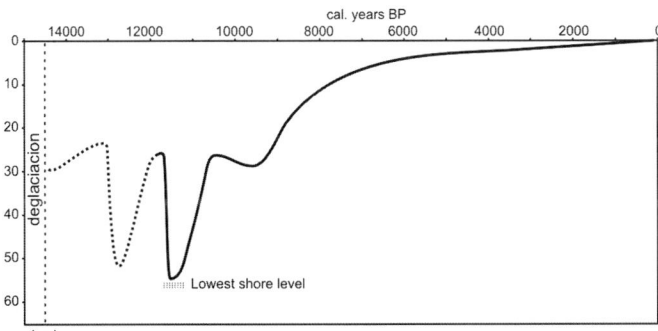

Fig. 7.12. Relative sea-level curve for the southern coast of the Baltic Proper (Polish coast). After Uścinowicz (2003, 2006).

Around 5.5–5.0 ka ago, a slight uplift of the Earth's crust in the area of the Danish Straits hindered free exchange of waters with the North Sea. This meant that the Baltic Sea became a less salty sea, the Post-Litorina Sea. The gradual reduction in salinity, shown by the increase in significance of mesohalobe species, is dated palynologically to the Sub-Boreal period (Zachowicz *et al.* 2008). The Baltic Sea coastline continued to change in a similar manner as earlier – the sea retreated from areas in the north and entered the land in the southern parts.

Quaternary deposits

Very often, Pleistocene sediments over large areas of the bed of the Baltic Sea, both in deep-water basins and on their shallow peripheries, are represented by a single layer of till from the last glaciation. Glaciofluvial deltas and eskers occur locally on the Baltic Sea bed, both in its southern and northern parts, but their preserved condition is different. In deep-water basins they are often masked by a younger sedimentary cover, while in shallow areas they are eroded in a varied way and transformed in the marine environment. Thin layers of glacial deposits (<10 m) occur in areas of the present deep-water basins where exaration processes dominated in the Pleistocene period. The thinnest deposit, often less than 5 m, is associated with those areas of the northern and southern parts of the Baltic Sea where marine erosion also dominated in the Holocene, causing a lack of Holocene sediments.

Glacial and glaciofluvial sediments of older glaciations remained mainly in deep tunnel valleys or in tectonic troughs. Those depressions, formed in the pre-Quaternary bedrock, are also the places where thick Pleistocene deposits occur. For example, in the northern part of the Baltic Proper, in the Landsort Depth, which is a typical tectonic trough deepened by exaration, about 150 m of Quaternary sediments remain (Winterhalter *et al.* 1981). The greatest thickness of Pleistocene deposits, up to 200–300 m, is recognized in tunnel valleys near the southern coasts of the Baltic Sea.

The Pleistocene structure on the slopes of basins is significantly more complex, particularly in the shallow area of the southern and

Fig. 7.13. Pine trunk (radiocarbon age 9.58–9.43 ka BP) rooted in sediments of an ice-marginal lake. Currently it lies 23 m below sea level, around 15 km north of the central part of the Polish coast (photograph by W. Ossowski, courtesy of Polish Central Maritime Museum, taken during the project 'Managing Cultural Heritage Underwater (MACHU)').

eastern part of the Baltic Proper. Except for till horizons, in this area accumulations of glaciofluvial and glaciolacustrine sediments are present; locally, interglacial marine sediments as well as sediments of different terrestrial environments also occur (e.g. peat, limnic gyttja).

Post-glacial (late Pleistocene and Holocene) sediment cover and the associated stratigraphic units formed in the Baltic Sea basin are generally the same over the whole area of the Baltic Sea, from the Arkona Basin in the SW to Bothnian Bay in the north. However, the distribution and layout are different in the southern and northern parts of the Baltic Sea. In the southern part, in areas where the bedrock is built of sedimentary rocks and the morphology of the seabed is relatively soft, the sedimentary cover – both Pleistocene and Holocene – is relatively homogeneous over large expanses (Fig. 7.17). In the northern part of the Baltic Sea, in areas where the bedrock is composed of magmatic and metamorphic rocks, the morphology of the seabed is uneven with narrow and relatively deep depressions at the bottom with steep slopes. Here, the

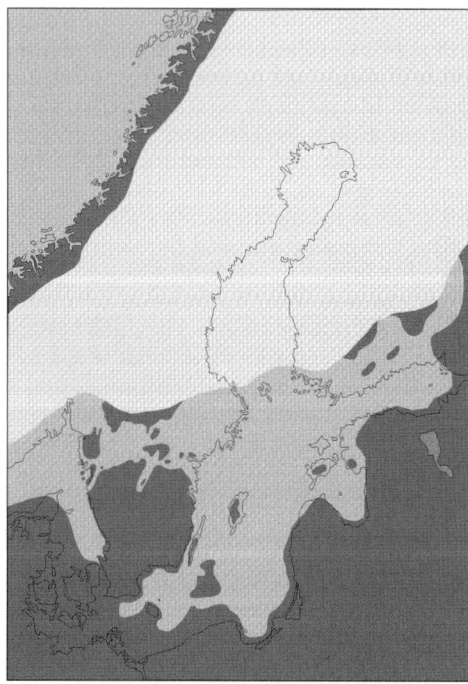

Fig. 7.14. The Baltic Sea around 11.3 ka ago – the Yoldia Sea. After Eronen (1988); Björck (1995); southern part changed by the author.

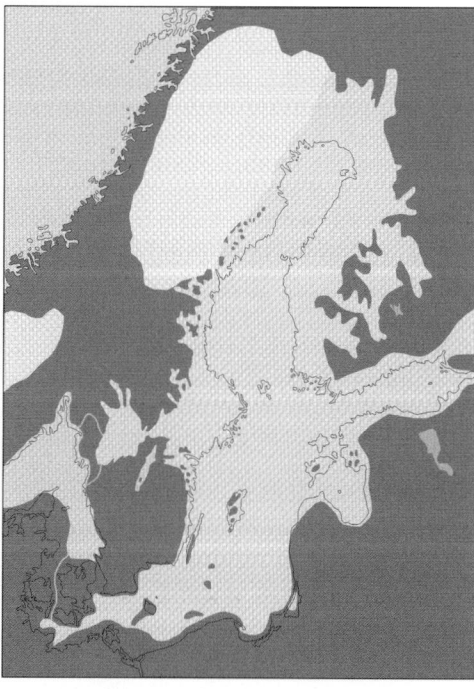

Fig. 7.15. The Baltic Sea around 10.4 ka ago – the Ancylus Lake. After Eronen (1988); Björck (1995); Lemke (1998); southern part changed by the author.

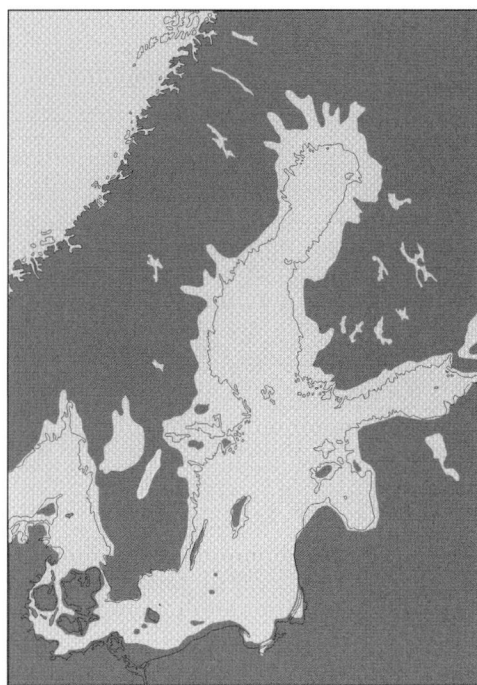

Fig. 7.16. The Baltic Sea around 8 ka ago – the early Litorina Sea. After Eronen 1988; southern part changed by the author.

sedimentary cover has a mosaic nature. In small areas, the layout and thickness often change and, on elevations of the bottom, bedrocks often appear (Fig. 7.18).

Late Pleistocene and Holocene (post-glacial) silty–clayey sediments present on the bottom of sedimentary basins form three major lithostratigraphic units (Fig. 7.17): brown Baltic clays of limnoglacial origin (Baltic Ice Lake); grey Baltic clays, also mainly of limnoglacial origin (Yoldia Sea and Ancylus Sea); and olive-grey Baltic muds, of marine origin (Litorina and Post-Litorina Sea). The lower parts of the stratigraphic sequence (the late Pleistocene and early Holocene sediments) are diachronic, whereas the upper part (middle and late Holocene sediments) is synchronic. The units distinguished are: glacial clays and silts; transition clays; and post-glacial mud, respectively, as specified by Winterhalter *et al.* (1981) and Winterhalter (1992). The units distinguished on the grounds of physical characteristics by Harff *et al.* (2001) are: A1, brown Baltic clays; A2–A6, grey Baltic clays; and units B1–B6, olive-grey Baltic muds, respectively.

Brown Baltic clays are varved clayey–silty sediments transforming upwards into microlaminated clays and then into homogeneous clays. The term 'clay' has been used with a broad meaning here. However, mostly the clays occurring here (i.e. sediments dominated by fractions <0.004 mm) are varved sediments consisting of alternating clayey and silty laminas, but clays with significant admixtures of silt and sand, and silts with admixtures of clay and/or sand, are also present. In these sediments, sandy inserts and individual gravel grains occur. The total thickness of varve, microlaminated and homogeneous sediments lying concordantly on the sculptured surface of the till is the greatest in the southern parts of the Baltic Sea, reaching around 10 m in central parts of deep-water basins. Distinctive features of those sediments are the low content of organic matter, usually below 1.5%, and the large amount of carbonates (up to 20%). In the topmost parts, homogeneous clays can be limeless. Locally, in brown calcareous Baltic clays, black laminas or spots of iron sulphide occur, and the colour changes upwards from brown to light brown and brown-grey. In the southern Baltic Sea, brown Baltic clays began forming in Bølling and their lithological features indicate deposition in the neighbourhood of the melting ice sheet. At first, this was close to where varved sediments were deposited and, as the deglaciation proceeded, microlaminated and homogeneous clays were deposited further from the glacier front. These sediments were formed in the Baltic Ice Lake phase (Winterhalter *et al.* 1981; Winterhalter 1992; Zachowicz *et al.* 2008). In the northern part of the Baltic Sea, in the Bothnian Sea and Bothnian Bay, sediments corresponding to this unit had already been deposited in the early Holocene because the ice sheet had persisted there earlier.

Brown Baltic clays lack or contain only insignificant amounts of pollen and diatoms, which indicate both that poor plant communities were present on areas released from the ice sheet, and that waters were cold and limited in nutritive components.

Grey Baltic clays are deposited concordantly on lower-lying brown clays. In this regard, maintaining sedimentary continuity, the lower boundary of the layer is weakly indicated. As the name suggests, they are most often sediments dominated by clayey fractions. However, here sediments containing clayey and silty fractions are also present, sometimes also sandy ones in different proportions. Such sediments remain on the periphery of sedimentary basins and were deposited in a slightly shallower environment than clays. The distinctive features of the layer are a lack or low content of calcium carbonate, a low (up to 2%) content of organic matter, and the presence of black laminas and irregular spots of iron sulphide. The colour of the sediments varies from brownish-grey in the lower parts of the layer to grey

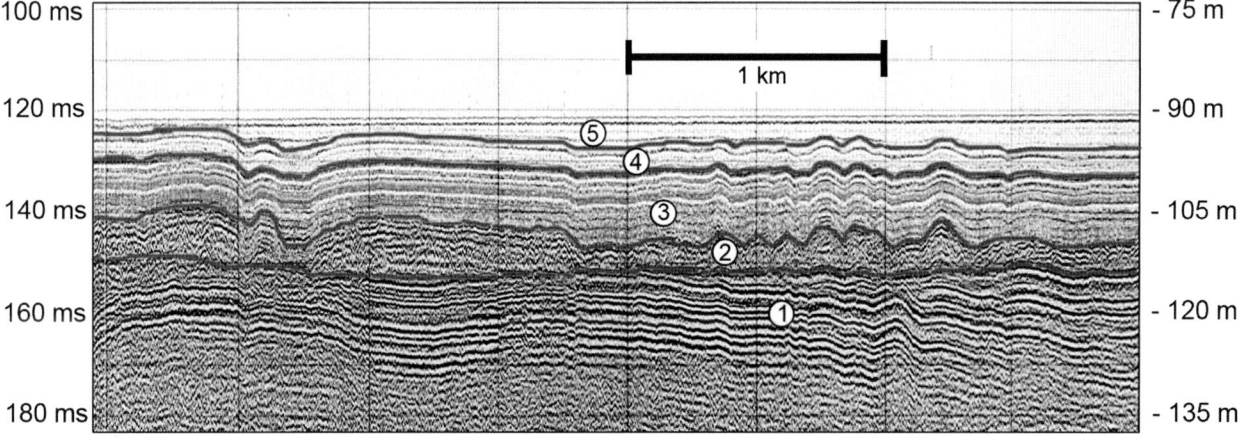

Fig. 7.17. Seismo-acoustic profile (boomer) from the southern part of the Baltic Proper (from the central part of the Bornholm Basin) with the lithostratigraphic units highlighted. 1, pre-Quaternary bedrock (Silurian); 2, till (Pleistocene); 3, brown Baltic clay (late Pleistocene – Baltic Ice Lake); 4, grey Baltic clay (early Holocene – Yoldia Sea and Ancylus Lake); 5, Baltic mud (middle and late Holocene – Litorina and Post-Litorina Seas).

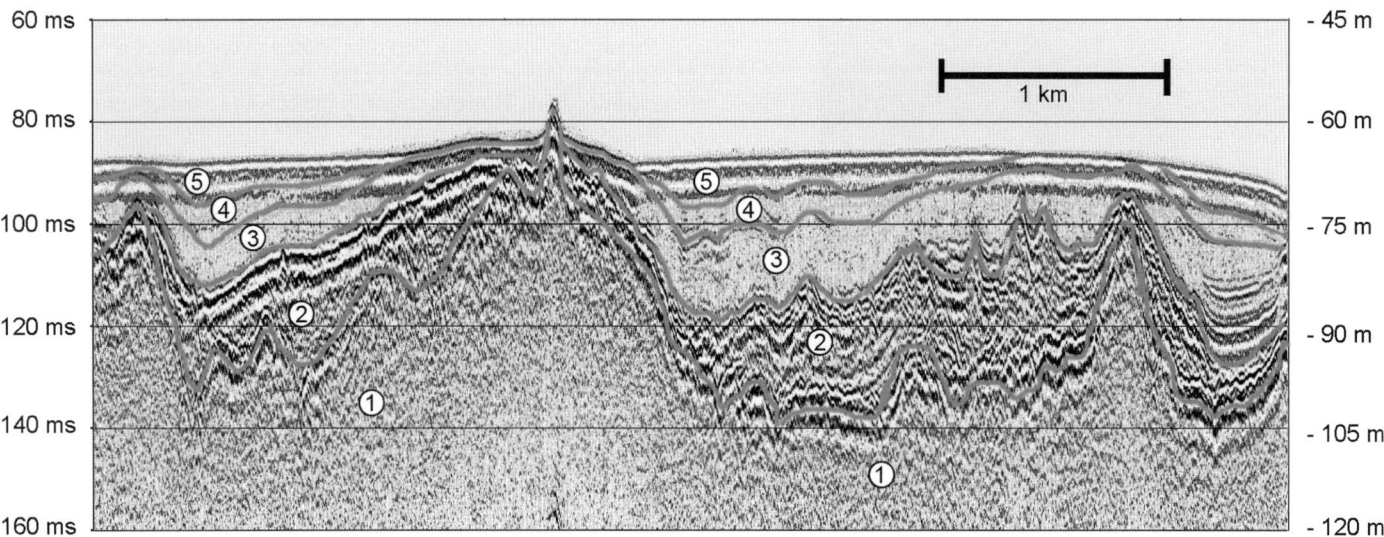

Fig. 7.18. Seismo-acoustic profile (sparker) from the eastern part of the Gulf of Finland with lithostratigraphic units highlighted. Key as in Figure 7.17, except for: 1, pre-Quaternary crystalline bedrock (Proterozoic). After Petrov (2010), courtesy of V. Zhamoida and the team of authors from VSEGEI.

and light grey, often with a bluish hue in the topmost parts. Besides the colour change, the content of organic matter also increases slightly upwards. By palynological analysis, the time of formation of the grey Baltic clays is in the early Holocene – the Preboreal and early Boreal periods. Diatomaceous spectra indicate that these sediments were generally formed in a freshwater environment (limnoglacial), and only in the early Preboreal period was there a short episode of insignificant water salinity, spatially limited to the NW and central parts of the Baltic Proper (the Gotland Basin). Similarly to the brown Baltic clays, the lower part of the grey Baltic clays is diachronic over the whole Baltic Sea (Winterhalter *et al.* 1981; Winterhalter 1992); in the area of the Baltic Proper, it is synchronic. Grey clays, with a total thickness of up to 5–8 m, were formed in the phases of the Yoldia Sea and the Ancylus Lake.

Olive-grey Baltic muds lie discordantly on grey Baltic clays. The term 'mud' has been used with a broad meaning here, and encompasses a range of granulometric types specified on the grounds of the proportion of silt and clay fractions. The layer of olive-grey Baltic muds is formed by two lower-grade lithological units: laminated mud of a grey colour with an olive hue; and muddy–clayey dark grey sediments, locally also with a characteristic olive hue. The sediments of the latter unit are usually homogeneous, with traces of bioturbation; however, in places they are also laminated. Both layers are enriched with organic matter, up to 10–15%, and they usually do not contain calcium carbonates. According to palynological and radiocarbon dating, the lower layer – laminated grey mud with an olive hue – was formed in the late Boreal period and early Atlantic period. This layer was formed in a brackish-water environment, indicated by an increase in the content of brackish (mezohalobe) diatoms. The beginning of deposition of the upper layer of olive-grey mud is attributed to the end of the early Atlantic period; this layer is definitely dominated by euhalobe and mesohalobe diatoms, indicating that this sedimentary layer was deposited in marine conditions. The thickness of olive-grey muds reaches 4–5 m and locally exceeds 6 m. The boundary between grey Baltic clays and olive-grey Baltic mud is often of an erosive nature. On the periphery of sedimentary basins, on elevations inside the basins and on dividing sills, the beginning of the marine phase, an increase in salinity and the formation of haline stratification were marked by erosion, in places emphasized by a thin layer of sand. Sediments forming the layer of olive-grey muds were deposited in marine phases of the Baltic Sea development: that is, in the Litorina Sea and the Post-Litorina Sea.

Sandy facies, the same age as brown Baltic clays and genetically related to them, occur probably only on the bottom of the southern part of the Baltic Sea. In the northern part, in areas of glacioisostatic uplift, they are currently on the land. It is similar in the case of shallow-water sediments, corresponding in terms of age to the grey Baltic clays. In sandy facies they occur, among other places, on the southern slopes of the Bornholm Basin. Starting from Latvian coasts northwards, shallow-water sediments are currently present on the land, and, in Estonia, Finland and Sweden, they most often occur in coarse-grained (gravel cobble, boulders) facies.

Beside sandy facies associated with brown and grey Baltic clays, in the southern part of the Baltic Proper and in the Belt Sea (i.e. in the areas of middle Holocene transgression) relicts of the sediments of land environments – wetlands, lacustrine and deltaic (peats, gyttjas, silts and sands) – have remained locally at the bottom of the Baltic Sea.

Sandy and gravelly facies of the same age as olive-grey Baltic muds – that is, marine sediments formed in the middle and late Holocene (Litorina and Post-Litorina seas) – occur on the slopes of sedimentary basins and on nearshores. In major areas of the shallow-water zone of the Baltic Proper, the thickness of marine sands and gravels is usually less than 2 m and, over large areas, it does not reach even 1 m. The thickness of marine sands exceeds 3 m only locally. Thicker sandy lithosomes, occurring on the southern and eastern Baltic Sea coasts, are related to barriers that had developed during the Atlantic (Litorina) transgression.

Recent surficial sediments and sedimentary processes

The surface of the Baltic Sea catchment area is covered mainly by Pleistocene glacial and glaciofluvial deposits. They are transported to the Baltic Sea by rivers or arrive there directly as a result of erosion of coasts and also, partly, the seabed. The processes of erosion, redeposition and deposition that transform glacial, glaciofluvial and fluvial deposits into marine sediments began at the beginning of the Atlantic period. In the southern parts of the Baltic, the sea constantly enters the land, destroying Pleistocene sediments in the coastal area. However, for around 5 kyr, during which time the rate of sea-level rise in the southern part of the Baltic Sea decreased, sedimentation conditions have been reasonably stable. In the northern part, as a result of the constant land uplift and sea regression, Pleistocene glacial and glaciofluvial deposits, as well as Holocene sediments of the earlier development phases of the Baltic Sea that were previously deposited at the bottom out of the reach of waves, are being eroded.

In the Baltic Sea, in general, sandy and gravelly sediments are present above the pycnocline. Hydrodynamic processes in this layer of water make permanent deposition of muddy–clayey sediments impossible. The content of fractions finer than 0.063 mm is usually less than 1%. Fine-grained sediments, mud and clays are usually deposited below the pycnocline.

Sandy and sandy–gravelly sediments cover major areas in the southern and SE part of the Baltic Proper, north of the German and Polish coasts, and west of the Russian, Lithuanian and Latvian coasts. Significant areas of the bottom are also covered by sand in shallower, coastal parts of the Belt Sea. In these parts of the Baltic Sea, sands and gravels lying on the surface of the bottom are the result of long-term and repeated redeposition of glacial and fluvioglacial sediments, the original characteristics of which have been blurred and transformed into the features typical for sediments of a shallow epicontinental sea. Pratje (1948) and Kolp (1966) drew attention to the zonal presence of sediments on the surface of the seabed in the southern part of the Baltic Proper.

In the coastal zone, up to a depth of 10 m, fine-grained sands dominate. Coarse-grained sediments occur locally, on abraded fragments of the coastal zone, particularly at the foot of cliffs. Sediments on the coastal zone are often redeposited by surf waves. During strong storms, washing of Pleistocene deposits underlying the marine sediments within the underwater coastal slope can occur. Fine-grained sands of the coastal zone are characterized by the frequent occurrence of negative skewness of grain-size distribution, with good and very good sorting. Typical forms of the coastal zone are sandbars and rip-current canals and cones.

Outside the coastal zone, at depths from around 10 to 25–30 m, medium- and coarse-grained sands, as well as sandy–gravelly sediments, are present most frequently. Fine-grained sands occur locally at this depth. Sandy and sandy–gravelly sediments at this depth are moderately sorted. In certain places, boulders and cobbles – lag deposits of washed-out Pleistocene sediments – also occur. On the surface of sands, at depths from 10 m to around 30 m, ripplemarks and also sand waves occur. Sand-wave crests are about 50–1000 m long (Fig. 7.19). The spacing between the crests is 25–200 m, while the height of the crests is 0.5–2 m. The profile of the sand waves is slightly asymmetrical.

On coarse-grained and gravelly sands, ripples occur with a spacing of 0.5–1.5 m, a height of between 0.08 and 0.3 m, and a length of up to tens of metres. On the surface of gravel and sandy–gravelly sediments, larger seabed forms also occur: sand patches and structures similar to sand ribbons. The size of sand patches varies from several metres to several hundred metres. Their shapes are irregular, sometimes oval or elongated. Forms similar to sand ribbons are up to 500 m long, 40–50 m wide and of varied spacing (Uścinowicz 1995).

Sediments present in this area are within the range of the impact of regular storm waves. Information on the thickness of the dynamic layer mobilized during storms is given by the vertical distribution of caesium-137 (^{137}Cs) in sediment cores (Fig. 7.20). Caesium-137 is an artificial element found in the environment after atomic weapons testing in the 1950s and the first half of the 1960s, and as a result of the failure of the Chernobyl nuclear power plant in 1986. In the Gdańsk Basin, at a depth of 10–30 m, ^{137}Cs is present in the layer of sands between 0.2 and 1.3 m, indicating the thickness of the layer exposed to transport processes during storms (Fig. 7.20). These depths are the same as the thickness of the dynamic layer of sands specified on the basis of the height of sand waves.

Below the average base of storm waves (i.e. below 25–30 m), fine-grained sands dominate. They are characterized by good and very good sorting, and almost symmetrical or positive skewness of grain-size distributions. The thickness of fine-grained sands usually does not exceed 2 m, indicating that they are probably transported only during extremely strong storms, with deposition taking place only periodically. In this area on the surface of the seabed, Kolp (1966) found small ripple marks with crest heights of up to 1 cm. It is apparent, however, that biogenic structures are more typical and more popular in the fine-grained sands: creeping traces left by *Mesidotea entomon* crustaceans, domichnia of oligochaeta and dwelling traces of *Macoma baltica* clams (Uścinowicz 1995).

The northern part of the Baltic Sea does not have the widespread sand covers that are typical for the southern and eastern part of the Baltic Proper. In the Gulf of Finland, the Bothnian Sea and Bothnian Bay, sands occur commonly in a patchy manner. In coastal areas, thin and discontinuous gravel and stony residual covers or outcrops of glacial sediments occur. Sandy and sandy–gravelly sediments are distributed to a significantly smaller extent. These are very often glaciofluvial sediments (e.g. sands and gravels of glaciofluvial deltas or eskers), currently washed and transformed

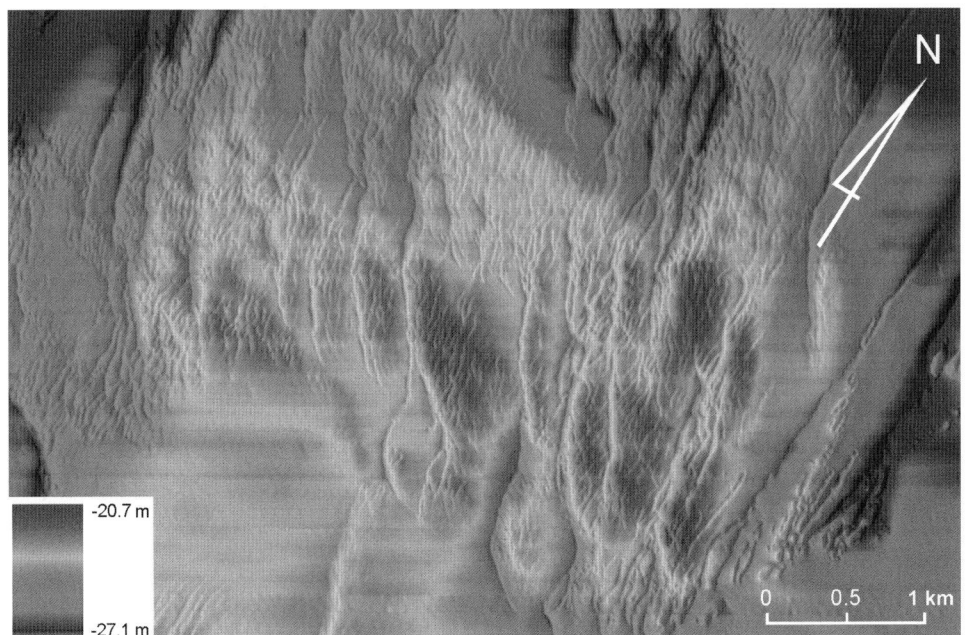

Fig. 7.19. Digital elevation model of the seabed (southern part of the Baltic Proper, *c.* 10 km north of the Polish middle coast, water depth 20–27 m). Two generations of sand waves are visible: larger – with spacing between the crests of about 200–300 m and a crest height of 1–2 m; and smaller – with spacing between the crests of about 40–70 m and a crest height of 0.2–0.3 m (from the 'Managing Cultural Heritage Underwater (MACHU)' EU Culture 2000 programme, courtesy of the Polish Central Maritime Museum).

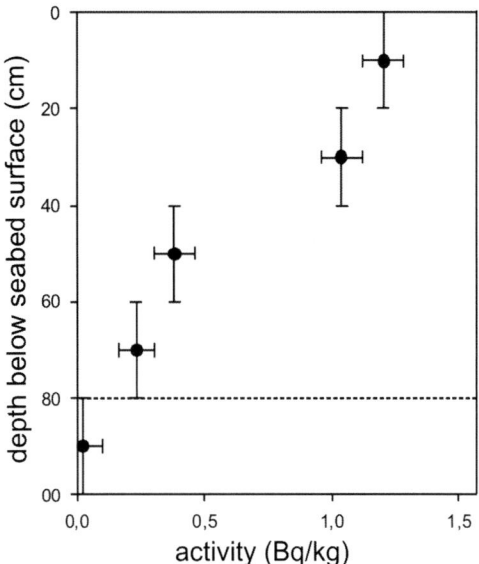

Fig. 7.20. [137]Cs activity in sediment core (medium sand, water depth 15 m, southern part of the Baltic Proper, 5 km north from the eastern Polish coast).

into marine sediments (Winterhalter 1972; Winterhalter *et al.* 1981; Cato & Kjellin 1994).

Recent muddy–clayey sediments, in which finer fractions from 0.063 mm occur in amounts larger than 75%, cover large areas of the sea bottom in all sedimentary basins of the Baltic Sea. In the Baltic Proper, these areas are the Arkona, the Bornhom, the Gdańsk, the East and West Gotland, and the North-Central basins. These sediments also cover the deepest parts of the Bothnian Sea and Bothnian Bay, as well as the Gulf of Finland and Riga. Smaller areas are covered by muddy–clayey sediments in the SW Baltic Sea, in the Bays of Meklemburg and Kiel, as well as in the Kattegat. The coastline shape and seabed relief are of particular significance in the western and northern parts of the Baltic Sea basin, where numerous islands limit the free circulation of waters and reduce the wave base so that fine-grained sediments accumulate even in shallow basins often above the pycnocline in a patchy manner.

The current rate of accumulation of muddy–clayey sediments varies from 0.5 to 2 mm a^{-1} (Winterhalter *et al.* 1981; Pempkowiak 1992; Walkusz *et al.* 1992; Cato & Kjellin 1994; Szczepańska & Uścinowicz 1994; Hille *et al.* 2006; Mattila *et al.* 2006). In central parts of deep-water basins the rate of accumulation is higher than on the peripheries, and regional differences also occur.

Differentiated sedimentary internal structures occur depending on oxygen conditions. Bioturbative structures related to the dwelling of benthonic organisms (mainly *M. baltica* and *M. entomon*) occur in muddy–clayey sediments that remain in contact with water masses containing oxygen. In deeper areas, located below the pycnocline, laminated sediments reflecting the annual rhythm of sedimentation have formed in the oxygen-deficient conditions. These sediments cover the largest areas in the West and East Gotland basins, also occurring in the Gulf of Finland, the Gdańsk Basin and the Bornholm Basin. Laminated sediments cover increasingly larger areas on the bottom. Particularly rapid expansion has been noted since the end of the 1940s, along with deterioration of oxygen conditions in demersal waters (Jonsson 1992; Kotilainen *et al.* 2007).

In the transition zone between muddy–clayey sediments and sandy sediments, on the peripheries of deep-water sedimentary basins, sandy–muddy sediments or mixtites (i.e. sandy–gravelly–muddy sediments) occur. These sediments most often occur in the Baltic Proper, in the areas of the sea bottom where the pycnocline is in contact with the seabed, and they are characterized by high variability in, and differentiation of, graining. They consist of gravelly (64–2 mm), sandy (2–0.062 mm), silty (0.063–0.004 mm) and clayey (<0.004 mm) fractions mixed in different and variable proportions. They are poorly and very poorly sorted. The thickness of sandy–muddy sediments and mixtites is often less than 0.2 m, sometimes less than 0.1 m, and Pleistocene glacial deposits or clayey sediments of early phases of Baltic Sea development (i.e. brown and grey Baltic clays) occur below. Lithological features and sonar images of the seabed from the area of sandy–muddy sediments and mixtites indicate that there are demersal currents here with significant speeds. An important role is also played by internal waves created within the pycnocline. Sandy–muddy sediments and mixtites lying on glacial clays cover significantly larger areas of the sea bottom in the Archipelago Sea and the Bothnian Sea, as well as in Bothnian Bay, where often older sediments also appear on the surface of the bottom.

Ferromanganese concretions are important constituents of sediments in some regions of the Baltic Sea, where three main types occur: spheroidal, discoidal and crusts (Glasby *et al.* 1997). The content of Mn varies from 8 to 30% and of Fe from 10 to 23%, depending on the type of concretion and region of occurrence. Concretions from Bothnian Bay and the eastern Gulf of Finland are the most abundant, where the abundance reaches 40 kg m^{-2} in places. Within the boundaries of concretion-rich fields in the Russian sector of the Gulf of Finland, covering an area of approximately 300 km^2, the average abundance of spheroidal concretions is about 20–30 kg m^{-2}, sometimes reaching 50 kg m^{-2} (Zhamoida *et al.* 2007). These concretions are mostly spheroidal, up to 30 mm in diameter, display an Mn/Fe ratio of 0.68–0.88 and are associated with muddy, organic-rich sediments adjacent to the depressions. The abundance of concretions in the Bothnian Sea, central Gulf of Finland and in the Gulf of Riga is common to abundant (15–18 kg m^{-2}). In the Bothnian Sea, flat crusts are widely distributed, whereas, in the Gulf of Riga, spheroidal and discoidal concretions occur with an Mn/Fe ratio of around 0.43. Concretions from the Baltic Proper are found mainly around the margins of the deep basins. They are mainly discoidal, 20–150 mm in diameter, with an Mn/Fe ratio of about 0.5–0.6. Their abundance is sporadic to common, only locally reaching 10–16 kg m^{-2}. In the Belt Sea, concretions are known from the Kiel, Lübeck and Mecklemburg bays, and occur as spheroidal concretions with diameters of 10–30 mm and discoidal ones with diameters of 20–100 mm. The concretions have Mn/Fe ratios varying from 2.6 to 2.9 (Glasby *et al.* 1997). Discoidal concretions and crusts occur in all regions, mainly in the transitional zone between sands and muds, and are associated with sandy–muddy sediments, mixtites and lag deposits underlain by tills or by brown clay of Baltic Ice Lake.

The mineral composition of recent sediments of the Baltic Sea is closely connected with the material constituting their direct source (i.e. with Pleistocene glacial and glaciofluvial sediments). The mineral composition of Pleistocene deposits indicates regional differences depending on the sedimentary bedrock lithology.

The petrographical and mineral composition of the Baltic Sea sediments changes according to the size of the clastic components. Fragments of crystalline and sedimentary rocks dominate among boulders, cobbles and gravels. As the fragments decrease in size, there are fewer sedimentary rocks – first carbonate, then sandstones – and the percentage of crystalline rocks increases. Among crystalline rocks, the most common are granites, diorites and gneisses. In relation to the source of the Pleistocene sediments, the percentage of components resistant to destruction (i.e. crystalline rocks) increases in marine sediments. As the grain size decreases, the percentage of quartz, which dominates in the sandy fractions, clearly increases. Feldspars, fragments of crystalline and sedimentary rocks, and heavy minerals occur as admixtures. In fine-grained sands, the quartz content often exceeds 90%. The processes of grain selection during sediment transport result in fine-grained sands in the shallow-water areas containing heavy minerals (mainly garnets, amphiboles, epidotes, tourmalines and also zircon, among others) that locally constitute up to several per cent.

The mineral composition of muddy–clayey sediments also reflects the geology of the areas from which these components are transported to the sea. Fine-grained sediments mainly consist of minerals of terrigenic origin: quartz, feldspars, illites, chlorites, and, in smaller quantities, kaolinite and carbonates. These minerals occur in variable proportions in all basins of the Baltic Sea (e.g. Blazhchishin 1982; Gingele & Leipe 1997; Uścinowicz *et al.* 2003). Calcium carbonate, both of terrigenic and biogenic origin, is quickly dissolved in Baltic waters. Significant additions of carbonates occur only in the region of outcrops of Silurian and Ordovician carbonate rocks in the northern part of the Baltic Proper.

Quartz in contemporary muddy–clayey sediments is present mainly in amounts of less than 40%. The smallest amounts of quartz occur in clays lying in the deepest parts of the sedimentary basins, far from primary sources. The lowest quartz content was noted in muddy clays of the East Gotland Basin. In fine-grained sediments, silica of biogenic origin also occurs (diatomaceous opal) but it does not constitute a significant amount. Feldspars, mainly represented by plagioclases and potassium feldspars, are present in varied amounts, ranging from 5% in the Landsort Depth to 25% in the Kattegat.

Illite is a mineral dominating clayey minerals, and only rarely and locally occurs in smaller amounts than other clayey minerals. Chlorite usually occurs in smaller amounts than illite. However, locally in the gulfs of Riga and Finland, chlorite was present in larger amounts than illite, while, in the East Gotland Basin, it was the second-most common mineral after kaolinite. Kaolinite occurs regionally, and has been identified in the sediments of the southern part of the Gulfs of Finland and Riga, the West and East Gotland basins, and in the NE part of Gdańsk Basin (Uścinowicz *et al.* 2003). The presence of kaolinite in these regions is most probably related to Palaeozoic rocks that contain the mineral, which are present in Estonia, Latvia and in the north of Lithuania. Other areas where kaolinite occurs in present-day muddy sediments are Kiel Bay and Kattegat in the western part of the Baltic Sea.

Illite–montmorillonite has been found only in the northern part (Bothnian Bay, and the Bothnian and Archipelago seas) and in the southern part of the Baltic Sea (the SW part of the Gdańsk Basin, and in the Bornholm and Arkona basins). Other terrigenic minerals, such as calcite, manganic calcite, dolomite, montmorillonite, nontronite, vermiculite, beidellite and mixed package minerals (illite–montmorillonite, illite–chlorite), occur locally and often only in trace amounts (Stoch *et al.* 1980; Blazhchishin 1982; Śliwiński & Uścinowicz 1983).

In muddy–clayey sediments, authigenic minerals also occur: kutnahorite, rhodochrosite, witherite, vivianite, pyrite, marcasite, goethite, barite and, locally, also gypsum. Authigenic minerals occur in far smaller amounts than terrigenic minerals. The variability in the presence of authigenic minerals, both regionally and in vertical profiles of sediments, is considerably higher than in the case of terrigenic minerals. It results from the variability of physiochemical conditions in demersal and interstitial waters in different parts of the Baltic Sea, both now and in the past, as well as from the complexity of the processes of early diagenesis (Alvi & Winterhalter 2001). Minerals, such as pyrite (also hydrotroilite), kutnahorite and vivianite, are formed in oxygen-deficient conditions, whereas goethite is formed in an oxygenated environment.

Chemistry of recent surficial sediments

Recent surface sediments of the Baltic Sea show similarities to the chemical composition of Pleistocene source material. The ratios between the different elements are modified by granulometric differentiation and autogenic processes. The content of silicon, aluminium and iron depends strongly on sediment grain size. Si clearly dominates in sands, whereas Al and Fe form significant admixtures in fine-grained sediments. The concentration of titanium and manganese does not show any distinct correlation with grain size (Table 7.1). The elevated content of TiO_2 (more than *c.* 0.2–0.3%) is associated with concentrations of heavy minerals, including rutile, ilmenite and leucoxene, occurring in fine-grained sands from shallow-water zones. The variability in manganese content within a large range is related to the fact that elevated amounts of manganese occur not only in a scattered form in fine-grained sediments but also as coatings on mineral grains in places where Mn–Fe concretions occur. Calcium carbonate, both terrigenous grains and biogenic detritus, is rapidly mechanically destroyed and undergoes chemical dissolution because Baltic Sea waters are unsaturated with $CaCO_3$. A slightly elevated $CaCO_3$ content ($>1\%$) is found in the northern part of the Baltic Proper only, in the regions where Silurian and Ordovician limestones crop out.

For several hundred years, concomitant with the development of civilization, the usual natural circulation of chemical elements has been affected by processes related to human economic activity. The Baltic Sea is prone to degradation, which, among other processes, results from stratification and a limited exchange of waters, as well as from the small area of the sea in relation to its catchment area. In the twentieth century, the natural circulation of chemical elements has been most affected by eutrophication and contamination of the sea by different substances. The most important groups of substances flowing into the Baltic Sea are nutrients, heavy metals and persistent organic pollutants.

The result of eutrophication of waters is excessive accumulation of organic matter in sediments. Total organic carbon (TOC) in surficial sediments of the Baltic Sea covers a broad range from near-zero values up to around 16% (e.g. Carman & Rahm 1996; Emelyanov 2001; Leipe *et al.* 2011). A low TOC content, usually $<1\%$, occurs in sandy sediments, mainly in shallow coastal regions, whereas the highest ones occur in muddy–clayey sediments in deep basins. Another effect of eutrophication is that of a higher content of phosphorus and nitrogen in surficial sediments. The nutrient content in sediments is strictly related to

Table 7.1. *Main constituents of the Baltic Sea sediments (values most frequently found)*

	SiO_2 (%)	Al_2O_3 (%)	TiO_2 (%)	Fe_2O_3 (%)	MnO (%)	$CaCO_3$ (%)
Sand (<20% particles <0.063 mm)	80–95	0.5–5	<0.1–1.1	0.05–1.5	<0.01–1.0	<1–5
Sandy mud and mud (>20% particles <0.063 mm)	35–75	7–18	0.2–0.9	2.5–12	0.1–16	<1–7

Compiled from many sources.

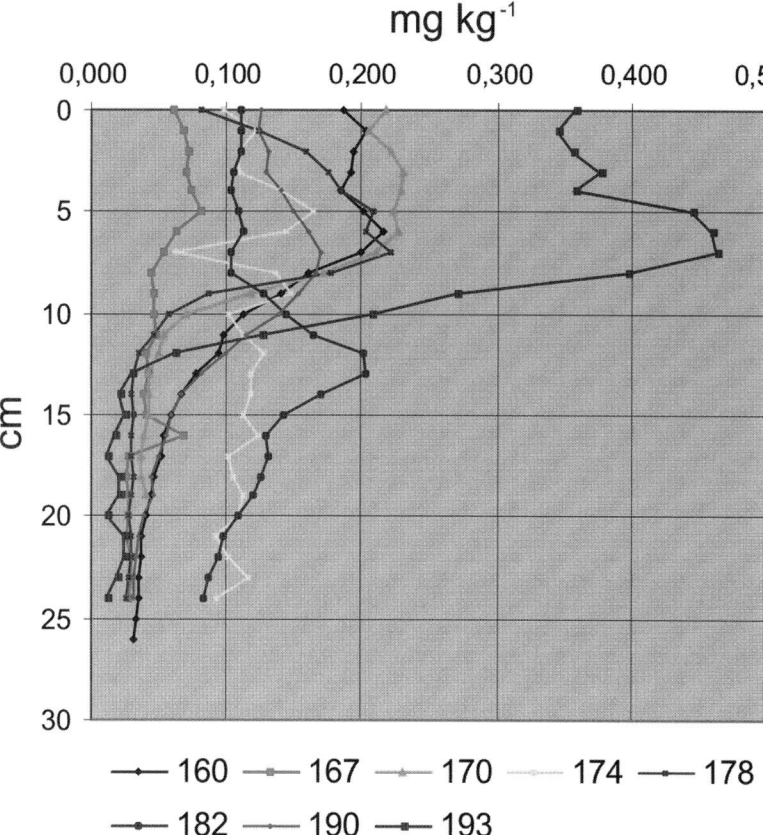

Fig. 7.21. Mercury content variability in the vertical profile of muds in different regions of the Baltic Sea. Sampling stations after Perttilä (2003): 160, SW Balic–Lübeck Bay; 167, Bornholm Basin; 170, NE Gdańsk Basin; 174, Riga Bay; 178, W Gotland Basin; 182, Gulf of Finland; 190, Bothnian Sea; 193, Bothnian Bay (courtesy of Mirja Leivuori and Matti Perttila).

their lithology: phosphorus and nitrogen in fine-grained surficial sediments varies from 48 to 523 μmol g^{-1} dry wt and from 107 to 1157 μmol g^{-1} dry wt, respectively (Carman 2003). The nutrient content in sandy sediments is usually much smaller.

The content of trace elements and other substances in muddy sediments from major sedimentary basins is presented on the basis of the results of the 'Baltic Sea sediments baseline study' realized in the early 1990s under the auspices of ICES and the Helsinki Commission (HELCOM) (Perttilä 2003).

Variability in the content of trace elements and persistent organic pollutants in the profiles of muddy–clayey cores indicates that at depths of 20–25 cm below the bottom surface the content of these substances is stable and their concentrations can be considered natural, which means that they were deposited in the pre-Industrial era and typical for the geochemical background. From around 15 cm below the bottom surface, a clear increase in the content of trace elements and persistent organic pollutants occurs. The maximum content, depending on the region of the sea and the analysed component, occurs within the range of depths from 2 to 8 cm below the bottom surface. Closer to the bottom surface, concentrations of the components decrease but they do not return to the geochemical background value.

A good example is shown by the curves illustrating changes in mercury content (Fig. 7.21). The lowest values occur in lower parts of the core, which is followed by a more or less distinct increase in the content from the 15–10 cm level upwards. The maximum content of mercury (Hg) mostly occurs at the level of 7–5 cm, with the values oscillating from 0.08 mg kg^{-1} in the Bornholm Basin to 0.46 mg kg^{-1} in Bothnian Bay. In the surficial layer, a slight decrease occurs. In lower segments of the cores (20–25 cm), Hg concentration does not exceed 0.05 mg kg^{-1}. In the gulfs of Riga and Finland, there are exceptional cores in which the Hg content at this level ranges from 0.08 to 0.11 mg kg^{-1} (Fig. 7.21). It can be assumed that the geochemical background values for Hg are usually less than 0.05 mg kg^{-1}, maximally reaching 0.1 mg kg^{-1}.

Mercury in the surficial layer of muddy–clayey sediments occurs in the largest amounts (0.46 mg kg^{-1}) in sediments from Bothnian Bay and the NE part of the Gulf of Finland (0.25–0.22 mg kg^{-1}) (Fig. 7.22) (Leivuori 1998; Albrecht et al. 2003). The higher level of mercury in the Gulf of Finland is undoubtedly the result of the activity of the paper industry in Finland, whereas the genesis of the higher concentrations of this component in Bothnian Bay has not been completely explained.

The geochemical background values for certain other trace elements in muddy–clayey sediments of the Baltic Sea (in

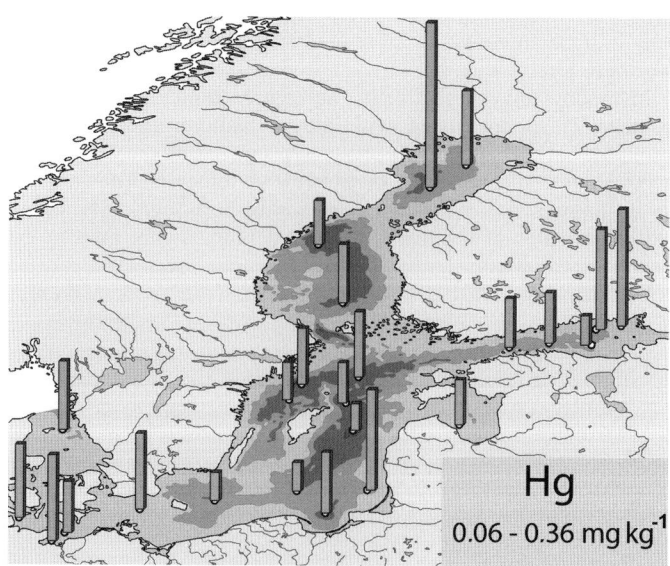

Fig. 7.22. Mercury content in surficial (0–1 cm) muddy sediments of the Baltic Sea (results from 1993; courtesy of Mirja Leivuori and Matti Perttila).

mg kg^{-1}) are: As <5–20, Cd < 0.05–1, Cr < 75–100, Cu < 25–50, Pb < 30–60 and Zn < 100–200. The maximum content of individual trace elements (except chromium) are present at the level of 0–15 cm below the bottom surface: As, 383 mg kg^{-1} in Bothnian Bay; Cd, 4.38 mg kg^{-1} in the northern part of the Baltic Proper; Cu, 279 mg kg^{-1} in the North-Central Basin; Hg, 0.46 mg kg^{-1} in Bothnian Bay; Pb, 199 mg kg^{-1} in Lübeck Bay (SW part of the Baltic Sea); Zn, 880 mg kg^{-1} in the West Gotland Basin.

Similar regularities also occur in the case of persistent organic pollutants: polychlorinated biphenyls (PCB), chloro-organic pesticides or policyclic aromatic hydrocarbons (PAH), among others. For example, the largest PCB amounts have been recognized in sediments dated to the 1970s. A decrease in PCB content has been recognized in present-day laminated muddy sediments (Jonsson *et al.* 2000; Olsson *et al.* 2004). The highest PCB content, reaching 92.84 ng g^{-1}, has been noted in the northern part of the Baltic Proper, in the layer of sediments lying at depths of 5–7 cm below the bottom surface, whereas in the surficial layer (0–2 cm) the content reached 14.08 ng g^{-1}.

Despite the surficial sediments being repeatedly enriched by harmful substances, absolute concentrations usually do not exceed the norms assumed for sediments and soils that are considered contaminated. Furthermore, sediments deposited 30–40 years ago contain more harmful substances than sediments deposited currently, clearly indicating the decreasing inflow of contaminants to the Baltic Sea.

In conclusion, sediments deposited in the last decade of the twentieth century and in the current century contain far less harmful substances, such as heavy metals or persistent organic pollutants (among others WWA, PCB), than sediments deposited within the period from the nineteenth century to the 1980s.

References

ALBRECHT, H., PERTTILÄ, M. & LEIVUORI, M. 2003. Distribution of trace metals in the Baltic Sea sediments. *In*: PERTTILÄ, M. (ed.) *Contaminants in the Baltic Sea Sediments*. Results of the 1993 ICES/HELCOM Sediment Baseline Study. MERI Report Series of the Finnish Institute of Marine Research, **50**, 40–44.

ALVI, K. & WINTERHALTER, B. 2001. Authigenic mineralization in the temporary anoxic Gotland Deep, the Baltic Sea. *Baltica*, **14**, 74–83.

ANDRÉN, T., BJÖRCK, S., ANDRÉN, E., CONLEY, D., ZILLÉN, L. & ANJAR, J. 2011. The development of the Baltic sea basin during the last 130 ka. *In*: HARFF, J., BJÖRCK, S. & HOTH, P. (eds) *The Baltic Sea Basin*. Springer, Heidelberg, 75–97.

BERGLUND, M. 2004. Holocene shore displacement and chronology in Ångermanland, eastern Sweden, the Scandinavian glacio-isostatic uplift centre. *Boreas*, **33**, 48–60.

BJÖRCK, S. 1995. A review of the history of the Baltic Sea, 13.0–8.0 ka BP. *Quaternary International*, **27**, 19–40.

BLAZHCHISHIN, A. I. 1982. Skład mineralny osadów dennych Morza Bałtyckiego. *In*: GUDELIS, V. K. & EMELIANOV, E. M. (eds) *Geologia Morza Bałtyckiego*. Wydawnictwa Geologiczne, Warszawa, 222–256.

BLAZHCHISHIN, A. I. 1999. Eocene palaeogeography and sedimentation in the Baltic deep and adjoning areas. *In*: KOSMOWSKA-CERANOWICZ, B. & PANER, H. (eds) *Investigations into Ambers*. Archaeological Museum in Gdańsk, Museum of the Earth, Polish Academy of Sciences, Gdańsk, 19–26.

CARMAN, R. 2003. Carbon and nutrients. *In*: PERTTILÄ, M. (ed.) *Contaminants in the Baltic Sea Sediments*. MERI Report Series of the Finnish Institute of Marine Research, **50**, 40–44.

CARMAN, R. & RAHM, L. 1996. Early diagenesis and chemical characteristics of interstitial water and sediments in the deep deposition bottoms of the Baltic proper. *Journal of Sea Research*, **37**, 25–47.

CATO, I. & KJELLIN, B. 1994. Quaternary deposits on the sea floor. *In*: FREDÉN, C. (ed.) *National Atlas of Sweden – Geology*. Geological Survey of Sweden, Uppsala, 150–153.

DEFANT, F. 1972. Klima und wetter der Ostsee. *Kieler Meeresforschungen*, **28**, 1–30.

EKMAN, M. 1996. A consistent map of the postglacial uplift of Fenoscandia. *Terra Nova*, **8**, 158–165.

EMELYANOV, E. M. 2001. Biogenic components and elements in sediments of the Central Baltic and their redistribution. *Marine Geology*, **172**, 23–41.

ERIKSON, L. & HENKEL, H. 1994. Geophysics. *In*: FREDÉN, C. (ed.) *National Atlas of Sweden – Geology*. Geological Survey of Sweden, Uppsala, 76–101.

ERONEN, M. 1988. A scrutiny of the Late Quaternary history of the Baltic Sea. *In*: WINTERHALTER, B. (ed.) *The Baltic Sea*. Geological Survey of Finland, Special Papers, **6**, 11–18.

GINGELE, F. & LEIPE, T. 1997. Clay mineral assemblages in the western Baltic Sea: recent distribution and relation to sedimentary units. *Marine Geology*, **140**, 97–115.

GLASBY, G. P., EMELYANOV, E. M., ZHAMOIDA, V. A., BATURIN, G. N., LEIPE, T., BAHLO, R. & BONACKER, P. 1997. Environments formation of ferromanganese concretions in the Baltic Sea: a critical review. *In*: NICHOLSON, K., HEIN, J. R., BÜHN, B. & DASGUPTA, S. (eds) *Manganese Mineralization: Geochemistry and Mineralogy of Terrestrial and Marine Deposits*. Geological Society, London, Special Publications, **119**, 213–237.

GYLLENCREUTZ, R., MANGERUD, J., SVENDSEN, J.-I. & LOTNE, Ø. 2007. DATED – a GIS-based reconstruction and dating database of the Eurasian deglaciation. *In*: JOHANSSON, P. & SARALA, P. (eds) *Applied Quaternary Research in the Central Part of Glaciated Terrain*. Proceedings of the INQUA Peribaltic Group Field Symposium 2006, Oulanka Biological Research Station, Finland, September 11–15. Geological Survey of Finland, Special Papers, **46**, 113–120.

HARFF, J., BOHLING, G. *ET AL.* 2001. Physico-chemical stratigraphy of Gotland Basin Holocene sediments, the Baltic Sea. *Baltica*, **14**, 58–66.

HILLE, S., LEIPE, T. & SEIFERT, T. 2006. Spatial variability of recent sedimentation rates in the eastern Gotland Basin (Baltic Sea). *Oceanologia*, **48**, 1–21.

JAWOROWSKI, K. 1987. Origin of amber-bearing Palaeogene sediments in the area of Chłapowo. *Biuletyn Instytutu Geologicznego*, **356**, 89–102 (in Polish with English summary).

JOHNSEN, T. F., ALEXANDERSON, H., FABEL, D. & FREEMAN, S. P. H. T. 2009. New [10]Be cosmogenic ages from the Vimmerby moraine confirm the timing of Scandinavian Ice Sheet Deglaciation in southern Sweden. *Geografiska Annaler*, **91A**, 113–120.

JONSSON, P. 1992. Laminated sediments in the Baltic Proper. *In*: PERTTILÄ, M. (ed.) *Review of Contaminants in Baltic Sediments*. Cooperative Research Report, **180**. International Council for the Exploration of the Sea (ICES), Copenhagen, 33–36.

JONSSON, P., ECKHÉLL, J. & LARSSON, P. 2000. PCB and DDT in laminated sediments from offshore and archipelago areas of the NW Baltic Sea. *Ambio*, **29**, 268–276.

KOLP, O. 1966. *Rezente Fazies der westlichen und südlichen Ostsee*. Petermans Geographische Mittelungen, **110**. Quartalsheft, Leipzig, 1–18.

KOTILAINEN, A., VALLIUS, H. & RYABCHUK, D. 2007. Seafloor anoxia and modern laminated sediments in coastal basins of the eastern Gulf of Finland, Baltic Sea. *In*: VALLIUS, H. (ed.) *Holocene Sedimentary Environment and Sediment Geochemistry of the Eastern Gulf of Finland, Baltic Sea*. Geological Survey of Finland, Special Papers, **45**, 49–62.

KRAMARSKA, R. 1998. Origin and development of the Odra Bank in the light of the geological structure and radiocarbon dating. *Geological Quarterly*, **42**, 277–288.

KRAMARSKA, R. (ed.) 1999. *Geological Map of the Baltic Sea Bottom without Quaternary Deposits, 1:500 000*. Państwowy Instytut Geologiczny, Warszawa.

KUHLMAN, G., DE BOER, P. L., PEDERSEN, R. B. & WONG, T. E. 2004. Provenance of Pliocene sediments and paleoenvironmental changes in the southern North Sea region using Samarium-Neodymium

(Sm/Nd) provenance ages and clay mineralogy. *Sedimentary Geology*, **171**, 205–226.

KULLENBERG, G. 1981. Physical oceanography. *In*: VOIPIO, A. (ed.) *The Baltic Sea*. Elsevier Oceanography Series, **30**. Elsevier, Amsterdam, 135–181.

LAGERLUND, E. 1987. An alternative Weichselian glaciation model, with reference to the glacial history of Skåne. *Boreas*, **16**, 433–459.

LASS, H.-U. & MATTHÄUS, W. 2008. General Oceanography of the Baltic Sea. *In*: FEISTEL, R., NAUSCH, G. & WASMUND, N. (eds) *State and Evolution of the Baltic Sea, 1952–2005*. Wiley, Hoboken, NJ, 5–43.

LEIPE, T., TAUBER, F. *ET AL*. 2011. Particulate organic carbon (POC) in surface sediments of the Baltic Sea. *Geo-Marine Letters*, **31**, 175–188.

LEIVUORI, M. 1998. Heavy metal contamination in surface sediments in the Gulf of Finland and comparison with the Gulf of Bothnia. *Chemosphere*, **36**, 43–59.

LEMKE, W. 1998. *Sedimentation und paläogeographische Entwicklung im westlichen Ostseeraum (Mecklenburger Bucht bis Arkonabecken) vom Ende der Weichselvereisung bis zur Litorinatransgression*. Meereswissenschaftliche Berichte **31**. Institut Für Ostseeforschung Warnemünde.

LINDÉN, M., MÖLLER, P., BJÖRCK, S. & SANDGREN, P. 2006. Holocene shore displacement and deglaciation chronology in Norrbotten, Sweden. *Boreas*, **35**, 1–22.

LUNDQVIST, J. 1994. The deglaciation. *In*: FREDÉN, C. (ed.) *National Atlas of Sweden – Geology*. Geological Survey of Sweden, Uppsala, 124–135.

LUNDQVIST, J. & ROBERTSON, A.-M. 1994. Glacials and interglacials. *In*: FREDÉN, C. (ed.) *National Atlas of Sweden – Geology*. Geological Survey of Sweden, Uppsala, 120–124.

LUNDQVIST, T. & BYGGHAMMAR, B. 1994. The Swedish Precambrian. *In*: FREDÉN, C. (ed.) *National Atlas of Sweden – Geology*. Geological Survey of Sweden, Uppsala, 16–21.

ŁOMNIEWSKI, K. 1975. Stosunki termohaliczne w Morzu Bałyckim. *In*: ŁOMNIEWSKI, K., MAŃKOWSKI, W. & ZALEWSKI, J. (eds) *Morze Bałtyckie*. PWN, Warszawa, 156–187.

MAJEWSKI, A. & LAUER, Z. (eds) 1994. *Atlas Morza Bałtyckiego*. Instytut Meteorologii i Gospodarki Wodnej, Warszawa.

MAKOWSKA, A. 1986. Pleistocene seas in Poland; sediments, age and palaeogeography [in Polish with English Summarry]. *Prace Instytutu Geologicznego*, **120**, 1–74.

MARKS, L. 2010. Timing of the Late Vistulian (Weichselian) glacial phases in Poland. *In*: LÜTHGENS, C., LEE, J., BÖSE, M. & ROSE, J. (eds) *Quaternary Glaciation History of Northern Europe. Quaternary Science Reviews*, **44**, 81–88.

MATTILA, J., KANKAANPÄÄ, H. & ILUS, E. 2006. Estimation of recent sediment accumulation rates in the Baltic Sea using artificial radionuclides ^{137}Cs and ^{239,240}Pu as time markers. *Boreal Environment Research*, **11**, 95–107.

MIETTINEN, A. 2004. Holocene sea-level changes and glacio-isostasy in the Gulf of Finland, Baltic Sea. *Quaternary International*, **120**, 91–104.

MOJSKI, J. E., DADLEZ, R., SŁOWAŃSKA, B., UŚCINOWICZ, SZ. & ZACHOWICZ, J. (eds) 1995. *Geological Atlas of the Southern Baltic*. Państwowy Instytut Geologiczny, Warszawa.

MÖRNER, N. A. 1979. The Fennoscandian uplift and Late Cenozoic geodynamics: geological evidence. *GeoJournal*, **3**, 287–318.

MÖRNER, N. A. 1980. Late Quaternary sea-level changes in northwestern Europe: a synthesis. *Geologiska Förhandlingar*, **100**, 381–400.

NORLING, E. 1994. Bedrock of the Swedish continental shelf. *In*: FREDÉN, C. (ed.) *National Atlas of Sweden – Geology*. Geological Survey of Sweden, Uppsala, 38–43.

OLSSON, B., BRADLEY, B., GILEK, M., REIMER, O., SHEPARD, J. L. & TADENGREN, M. 2004. Physiological and proteomic responses in *Mytilus edulis* exposed to PCBs and PAHs extracted from Baltic Sea sediments. *Hydrobiologia*, **514**, 15–27.

OVEREEM, I., WELTJE, G. J., BISHOP-KAY, C. & KROONENBERG, S. B. 2001. The Late Cenozoic Eridanos delta system in the Southern North Sea Basin: a climate signal in sediment supply. *Basin Research*, **13**, 293–312.

PEMPKOWIAK, J. 1992. Enrichment factor of heavy metals in Southern Baltic surface sediments dated with Pb-210, Cs-137 and Cs-134. *Environment International*, **17**, 421–428.

PERTTILÄ, M. (ed.) 2003. *Contaminants in the Baltic Sea Sediments. Results of the 1993 ICES/HELCOM Sediment Baseline Study*. MERI Report Series of the Finnish Institute of Marine Research, **50**.

PETROV, O. V. (ed.) 2010. *Atlas of Geological and Environmental Geological Maps of the Russian area of the Baltic Sea*. VSEGEI, St Petersburg.

PIWOCKI, M. & OLKOWICZ-PAPROCKA, I. 1987. *Lithostratigraphy of the Palaeogene: methods and outlooks of amber prospecting in northern Poland*. Biuletyn Instytutu Geologicznego, **356**, 7–28 (in Polish with English summary).

POKORSKI, J. & MODLIŃSKI, Z. (eds) 2007. *Geological Map of the Western and Central Part of the Baltic Depression without Permian and Younger Formations, 1:750 000*. Państwowy Instytut Geologiczny, Warszawa.

PRATJE, O. 1948. *Die Bodenbedeckung der südlichen und mittleren Ostsee und ihre Bedeutung für die Ausdeutung fossiler Sedimente*. Deutsche Hydrographische Zeitschrift, **1** (2/3).

RINTERKNECHT, V. R., CLARK, P. U. *ET AL*. 2006. The last deglaciation of the southeastern sector of the Scandinavian Ice Sheet. *Science*, **311**, 1449–1452.

ROSENTAU, A., VESLI, S. *ET AL*. 2011. Palaeogeographic model for the SW Estonian coastal zone of the Baltic Sea. *In*: HARFF, J., BJÖRCK, S. & HOTH, P. (eds) *The Baltic Sea Basin*. Springer, Heidelberg, 165–188.

ROTNICKI, K. & BORÓWKA, K. 1995. The last cold period in the Gardno-Łeba Coastal Plain. *Journal of Coastal Research*, Special Issue **22**, 225–229.

SCHMAGER, G., FRÖHLE, P., SCHRADER, D., WEISSE, R. & MÜLLER-NAVARRA, S. 2005. Sea state, tides. *In*: FEISTEL, R., NAUSCH, G. & WASMUND, N. (eds) *State and Evolution of the Baltic Sea, 1952–2005*. Wiley, Hoboken, NJ, 143–198.

SEIFERT, T., TAUBER, F. & KAYSER, B. 2001. A high resolution spherical grid topography of the Baltic Sea. 2nd edn. Poster #147 presented at the Baltic Sea Science Congress, Stockholm, 25–29 November 2001.

SIGMOND, E. M. O. 2002. *Geological Map, Land and Sea Areas of Northern Europe, Scale 1:4 million*. Geological Survey of Norway, Uppsala.

ŚLIWIŃSKI, Z. & UŚCINOWICZ, SZ. 1983. Lithology of surface sediments southern part of the Bornholm Basin. *Kwartalnik Geologiczny*, **27**, 631–644 (in Polish with English summary).

STOCH, L., GÖRLICH, K. & PIECZKA, F. B. 1980. Lithology and mineral composition of sea floor deposits in the Gdańsk Basin. (in Polish with English summary). *Kwartalnik Geologiczny*, **24**, 395–413.

SZCZEPAŃSKA, T. & UŚCINOWICZ, SZ. 1994. *Geochemical Atlas of the Southern Baltic, 1:500 000*. Państwowy Instytut Geologiczny, Warszawa.

TAREEW, B. A. 1965. Wnutriennye baroklinnye wolny pri obtiekanii nierownostiey dna i ich wlijanie na procesy osadkoobrazowanja w okeane. *Okieanologia*, **1**.

UŚCINOWICZ, SZ. 1995. Recent sedimentary processes. *In*: MOJSKI, J. E., DADLEZ, R., SŁOWAŃSKA, B., UŚCINOWICZ, SZ. & ZACHOWICZ, J. (eds) *Geological Atlas of the Southern Baltic*. Państwowy Instytut Geologiczny, Warszawa, 51–55.

UŚCINOWICZ, SZ. 1999. Southern Baltic area during the last deglaciation. *Geological Quarterly*, **43**, 137–148.

UŚCINOWICZ, SZ. 2003. *The Southern Baltic Relative Sea Level Changes, Glacio-isostatic Rebound and Shoreline Displacement*. Polish Geological Institute, Special Papers, **10**.

UŚCINOWICZ, SZ. 2006. A relative sea-level curve for Polish Southern Baltic Sea. *Quaternary International*, **145–146**, 86–105.

UŚCINOWICZ, SZ., NARKIEWICZ, W. & SOKOŁOWSKI, K. 2003. Mineralogical composition and granulometry. *In*: PERTTILÄ, M. (ed.) *Contaminants in the Baltic Sea Sediments. Results of the 1993 ICES/HELCOM Sediment Baseline Study*. MERI Report Series of the Finnish Institute of Marine Research, **50**, 21–24.

WALKUSZ, J., ROMAN, S. & PEMPKOWIAK, J. 1992. Contamination of the southern Baltic surface sediments with heavy metals. *Biuletyn Morskiego Instytutu Rybackiego*, **1**, (125), 33–37.

WINTERHALTER, B. 1972. On the geology of the Bothnian Sea: an epeiric sea that has undergone Pleistocene glaciation. *Geological Survey of Finland Bulletin*, **258**, 66.

WINTERHALTER, B. 1992. Late-Quaternary stratigraphy of Baltic Sea basins – a review. *Bulletin of the Geological Society of Finland*, **64**, (Part 2), 189–194.

WINTERHALTER, B., FLODÉN, T., IGNATIUS, H., AXBERG, S. & NIEMISTÖ, L. 1981. Geology of the Baltic Sea. *In*: VOIPIO, A. (ed.) *The Baltic Sea*. Elsevier Oceanography Series, **30**. Elsevier, Amsterdam, 1–121.

ZACHOWICZ, J., MIOTK-SZPIGANOWICZ, G., KRAMARSKA, R., UŚCINOWICZ, SZ. & PRZEZDZIECKI, P. 2008. A critical review and reinterpretation of bio-, litho- and seismostratigraphic data of the southern Baltic deposits. *In*: UŚCINOWICZ, SZ. & ZACHOWICZ, J. (eds) *Proceedings of the Workshop 'Relative Sea Level Changes – From Subsiding to Uplifting Coasts', June 19–20, 2005, Gdańsk, Poland*. Polish Geological Institute, Special Papers, **23**, 117–138.

ZHAMOIDA, V., GRIGORIEV, A., GRUZDOV, K. & RYABCHUK, D. 2007. The influence of ferromanganese concretions-forming processes in the eastern Gulf of Finland on the marine environment. *In*: VALLIUS, H. (ed.) *Holocene Sedimentary Environment and Sediment Geochemistry of the Eastern Gulf of Finland*. Geological Survey of Finland, Special Papers, **45**, 21–32.

Chapter 8

The NW Iberian continental shelf

D. REY[1]*, P. ÁLVAREZ-IGLESIAS[1], M. F. ARAÚJO[2], A. M. BERNABEU[1], M. COMAS[3],
M. DECASTRO[4], M. DRUET[5], E. FERREIRA DA SILVA[6], A. FERRÍN[1], M. GESTEIRA[4],
V. MARTINS[6,7,8], K. J. MOHAMED[1], B. RUBIO[1] & F. VILAS[1]

[1]GEOMA, Dpt. Geociencias Marinas, Universidad de Vigo, 36310 Vigo, Spain

[2]Centro de Ciências e Tecnologias Nucleares, Instituto Superior Técnico, Universidade de Lisboa,
Estrada Nacional 10 (km 139,7), 2695-066 Bobadela, Lisboa, Portugal

[3]Instituto Andaluz de Ciencias de la Tierra (CSIC – Universidad de Granada), Campus Fuentenueva,
18002 Granada, Spain

[4]EPhysLab (Environmental Physics Laboratory), Facultad de Ciencias, Universidad de Vigo, 32004 Ourense, Spain

[5]Instituto Español de Oceanografía, Calle del Corazón de María, 8, 28020 Madrid, Spain

[6]GeoBioTec, Dpto. Geociências, Universidade de Aveiro, Campus de Santiago, 3810-193 Aveiro, Portugal

[7]PPG da Faculdade de Geologia, Universidade do Estado do Rio de Janeiro e UERJ, Brazil

[8]CESAM, Dpto. Geociências, Universidade de Aveiro, Campus de Santiago, 3810-193 Aveiro, Portugal

*Corresponding author (e-mail: danirey@uvigo.es)

Abstract: The continental shelf of NW Iberia is of particular interest due to the complex North Atlantic rifting tectonics and modern oceanographic processes that led to its current geomorphological configuration and sediment dynamics. The shelf forms a narrow slow-dipping north-bearing geomorphological structure with a well-defined shelf break located at water depths of 160–180 m. It is a continental margin with sedimentation rates of about 1.5 ± 2.0 mm a^{-1} (the highest values in recent bottom sediments of the ría) subjected to a highly energetic seasonal regime of waves and tides, seasonal upwelling and coast-parallel currents, significantly modulated by the North Atlantic Oscillation. Sediment provenance is dominated by inputs of continental sediments via runoff and riverine discharges and, to a lesser extent, by an upwelling-enhanced bioclastic input. Present-day sedimentation in the area occurs closely associated with two geomorphologically distinct sedimentary environments: the rías inner shelf sedimentary systems and the open continental shelf, and their subsequent evolution since the last glacial.

The build-up and evolution of the NW Iberian continental shelf up to its present configuration have been chiefly driven by three major geological events of very different timescale: the opening of the Atlantic Ocean and the Gulf of Biscay (some 259 and 126 Ma ago, respectively), and the approximately 120 m of sea-level rise that has taken place since the end of the last glaciation about 20 ka ago and that peaked out about 6 ka ago, during the Holocene Climatic Optimum. These three global-scale events are alone responsible for the main submerged and subareal geomorphological features of the NW Iberian continental margin such as the shelf width, the occurrence of rías, and the position, extension and height of islands and submarine banks. Oceanic currents, wave climate, local weather and their millennial (oceans) and decadal (atmosphere) cyclic variations, the influence of local tectonics and changes in the sediment local production are the lesser order variables that have modulated this evolution over the last 100 kyr.

The objective of this paper is to give a concise vision of the most distinct features and processes on the NW Iberian continental shelf in an area extending between Cape Finisterre (43°N) and a latitude close to 41°N (Duero River) that forms a morphosedimentary unit sharing common oceanographic and climatic conditions. The paper will focus on the Holocene sediment-forming system, and the Pleistocene–Holocene evolution and climate variability. For a better understanding of the key geological and oceanographic constraining issues, such as the nature of the pre-Cenozoic substrate and subsequent Cenozoic shelf construction, or the processes related to the present-day water masses, ocean currents and coastal processes will be approached first. The figures and table are included not only to illustrate but also to complete the information given in the main text.

General geotectonic setting

The continental margin west of Galicia is the northernmost segment of the three that comprise the western Iberia Margin (Fig. 8.1). Several bathymetric features stand out in the Galician Margin, the most striking one being the seamounts area, with the Galicia Bank among them. The Galicia Interior Basin separates the seamounts region from the continental shelf, which in this segment is relatively narrow (30–50 km wide) with a well-defined shelf break located at water depths of 160–180 m. One of the most characteristic present-day features in the area is the occurrence of the Rías Baixas, which are structurally controlled incised valleys (Fig. 8.1), bounded by steep hills (horsts), only inundated in the later stages of the last sea-level rise.

Geodynamic evolution

During the Triassic and Early Jurassic (Lias), a major extensive tectonic event occurred between the African and North American plates (Figs 8.2 & 8.3). This tectonic episode affected current Europe and eastern Canada by the generation of extensional basins that were filled with clastic and evaporite materials, such

From: CHIOCCI, F. L. & CHIVAS, A. R. (eds) 2014. *Continental Shelves of the World: Their Evolution During the Last Glacio-Eustatic Cycle.*
Geological Society, London, Memoirs, **41**, 91–108. http://dx.doi.org/10.1144/M41.8

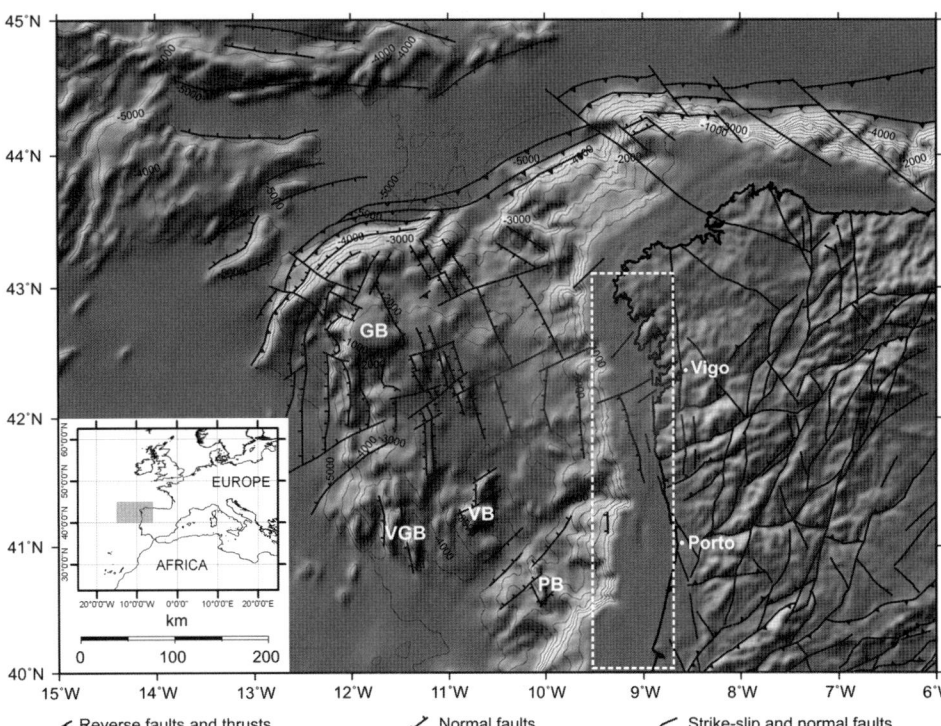

Fig. 8.1. Location of the study area (dashed square), including a structural scheme for the NW margin of Iberia. Shaded relief map from GEBCO Digital Atlas data. Modified from Groupe Galice (1979), Boillot *et al.* (1979, 1988), Grimaud *et al.* (1982), Boillot & Malod (1988), Thommeret *et al.* (1988), Murillas *et al.* (1990), Álvarez-Marrón *et al.* (1997), Ramírez *et al.* (2006) and Vázquez *et al.* (2008). GB, Galicia Bank; VGB, Vasco da Gama Bank; VB, Vigo Bank; PB, Porto Bank.

as the Lusitanian Basin in Portugal (Wilson *et al.* 1989; Murillas *et al.* 1990). During the Middle Jurassic, while the Central Atlantic was opening, the area between Iberia and North America remained tectonically quiet, and carbonates started filling up the graben basins (Boillot *et al.* 1987; Sawyer *et al.* 1994; Whitmarsh *et al.* 1998). The opening of the North Atlantic between Galicia and its conjugate margin, the Flemish Cap, began in Valanginian times and lasted until the Aptian. Continental drift is assumed to have started in this area in the Aptian (Boillot & Winterer 1988), although other further studies describe a much shorter rifting duration between the Berriasian and Valanginian (Wilson *et al.* 1996, 2001).

From the Early Cretaceous, the Iberian plate evolved intermittently coupled to the African and Eurasian plates, with successive jumps in the plate boundary between the two. While the opening of the North Atlantic Ocean in the Galicia Margin was taking place during the Aptian (Figs 8.3 & 8.4), the opening of the Gulf of Biscay began, where the plate boundary between Eurasia and Africa was placed until chron C19 (Montadert *et al.* 1979*a*; Srivastava *et al.* 1988, 1990). This plate boundary became unstable and jumped towards the structure that, today, is formed by the King's Trough–Azores–Biscay Rise–North Iberia Trough (Le Pichon & Sibuet 1971; Le Pichon *et al.* 1977; Searle & Whitmarsh 1978; Grimaud *et al.* 1982; Schouten *et al.* 1984; Klitgord & Schouten 1986; Srivastava & Tapscott 1986). This plate boundary coexisted with that of the Azores–Gibraltar Transform, further south, during the time that the Iberian plate was moving independently of the African and Eurasian plates (Roest & Srivastava 1991). Since the Late Oligocene (chron C6c), Iberia has moved jointly with the Eurasian plate, separated from Africa by the Azores–Gibraltar fracture zone (Srivastava *et al.* 1990; Roest & Srivastava 1991).

From the Late Cretaceous and during Cenozoic times (Figs 8.2 & 8.3), significant deformation occurred due to the convergence between the African and Eurasian plates, responsible for the elevation of the Pyrenees and the Cantabrian Mountains, the initiation of subduction on the north Iberia Margin, and the compression in the Mediterranean (Ferrer *et al.* 2008). During the Miocene, as a result of this compressive regime, ancient structures (both Hercynian and Mesozoic) were inverted in the

NW corner of Iberia, which also affected their present-day offshore extension. From the Late Miocene to present, the Iberian Peninsula has been under a strike-slip stress regime, with maximum compression directed NW–SE (Galindo-Zaldívar *et al.* 1993; De Vicente *et al.* 1996, 2008; Herraiz *et al.* 2000; De Vicente 2004).

Structural scheme

According to the geodynamic evolution of Iberia since Mesozoic times, the main tectonic structures observed in the west Galicia Margin are (Fig. 8.1): north–south (N340–N020°) normal faults, parallel or subparallel to the shoreline; NE–SW (N055–N070°) and NW–SE (N115–N135°) strike-slip faults (e.g. Groupe Galice 1979; Thommeret *et al.* 1988); and NE–SW to ENE–WSW thrust and reverse faults (Vázquez *et al.* 2008).

The principal structural trends differ to the north and south of the 42°N latitude (Groupe Galice 1979; Sibuet *et al.* 1987): faults trending N040–N150° are typical to the north; whereas faults to the south occur mainly in the N060° and north directions. Onshore, basement structures in the west Galicia Margin have similar structural guidelines. Tardihercynian NNE–SSW and ENE–WSW strike-slip faults behaved as normal faults during the Mesozoic rift, especially in the Portuguese margin (Pinheiro *et al.* 1996), and sometimes with a reverse component during the Cenozoic compression (Boillot *et al.* 1979; Alvarado 1983; Masson *et al.* 1994). These faults appeared to structurally define the geometry of the Mesozoic rift.

Cenozoic shelf construction and evolution of the central section

Rifting between Europe, Africa and North America preceding the Jurassic–Early Cretaceous opening of the North Atlantic formed the topographical grain of the western Iberian Margin. Metamorphic and plutonic rocks along the west coast occur in the 'Galicia Interior Basin' and the 'Western Banks' (Fig. 8.1). The Galician shelf is characterized by the segmentation of small tectonic units, mostly by dextral strike-slip faults (WSW–ENE),

AGE Ma	Magnet. anom.	Period	Epoch	Age	Iberia dynamics	Plate boundaries	West Galicia margin
23.03	6, 7	CENOZOIC / PALAEOGENE (NEÓG.)	Miocene	Aquitanian	Iberia moves with Eurasia.	Iberia-Eurasia plate boundary in KT-ABR-NIT, coeval with the Az-G fracture zone boundary, further south	From Late Cretaceous and during Cenozoic times, the Galicia margin is under the influence of a N-S to NW-SE compressive stress field.
	10		Oligocene	Chattian	Iberia is moving independent (Roest & Srivastava 1991) Pyrenean and Mediterranean compression.		
33.9±0.1	13			Rupelian			
	19			Priabonian		Strike-slip movement in the Bay of Biscay leading to an unstable R-R-T triple junction, so in C19 plate boundary jumps to the south.	
	20		Eocene	Bartonian	Iberia moves with Africa (Kristoffersen 1978; Montadert et al. 1979).		
	21			Lutetian			
55.8±0.2	24			Ypresian			
	25		Paleocene	Thanetian			
65.5±0.3	29			Danian			
	30	MESOZOIC / CRETACEOUS	Late (Senonian)	Maastrichtian			
	32			Campanian		Bay of Biscay opening (Bay of Biscay Ridge, R-R-R triple junction with Mid Atlantic Ridge), which is the main plate boundary between Africa and Europe in the next Ma (Montadert et al. 1979; Srivastava et al. 1988).	
	33			Santonian			
	34			Coniacian			
				Turonian	Until M0 anomaly, Iberian and African movements are similar. Iberia is moving, intermittently, coupled to Africa or Eurasia.		
99.6±0.9				Cenomanian			
112±1.0	M0		Early (Neocomiam)	Albian			The west Galicia rift axis jumps to te west. Opening of the Atlantic Ocean in the Galicia margin segment of the west Iberia margin.
125±1.0	M1			Aptian			
130±1.5	M5			Barremian			Opening of the Atlantic Ocean in the south Iberia Abyssal Plain segment. Initiation of Galicia Interior Basin.
	M10			Hauterivian			
	M12			Valanginian			
140.2±3.0	M14, M16			Berriasian			
145.5±4.0	M18	JURASSIC	Late	Tithonian			Opening of the Atlantic Ocean in the Tagus Abyssal Plain segment of West Iberia.
150.8±4.0	M22, M25			Kimmeridgian			
	M29			Oxfordian			

Fig. 8.2. Synthesis of the geodynamic evolution of Iberia and its plate boundaries from Late Jurassic to Miocene times, after Kristoffersen (1978), Montadert *et al.* (1979*b*), Srivastava *et al.* (1988) and Roest & Srivastava (1991).

roughly perpendicular to the coastline, as well as north–south and NE–SW faults (García-Gil *et al.* 1999).

A seismic stratigraphic analysis of the Cenozoic sediment infill comes from Ferrin (2005), who correlated and mapped four second-order composite sequences (S1, S2, S3 and S4: Fig. 8.4) within the sedimentary succession of the SW Galician shelf (Fig. 8.5). According to the author, deposition in the shelf has occurred since the Late Cretaceous, during relative sea-level global supercycles UZA4, TA1, TA2, TA3, TB2 and TB3 (Haq *et al.* 1987). Major discontinuities separate these composite sequences (A, B, C and D) developed during major eustatic and tectonic events that have occurred since the opening of the Atlantic.

S1, S2 and S3 show an overall progradation of the shelf towards the basin between the Late Cretaceous and the Middle Eocene. This period was dominated by global sea-level highstands, including the Late Cretaceous highstand peak (Vail *et al.* 1977), where sea level was above the shelf in the North Atlantic. S1, S2 and S3 are intensely deformed (both faulted and folded); what Vail *et al.* (1977) discussed may be the result of the superposition of the Alpine and the Betic orogenies between the Middle Eocene and the Middle Miocene. Surface D, a major unconformity separating S3 from S4, represents a major depositional hiatus related to exhumation along the NW portion of the Iberian plate (Figs 8.4 & 8.5).

S4, the youngest of the second-order composite sequences, was deposited between the Middle Miocene and present, during relative sea-level global supercycles TB2 and TB3. S4 is composed of five third-order composite sequences (sequences (Sqs) 4.1–4.5) interpreted to have been deposited during third-order relative sea-level cycles TB3.1–TB3.10 (Haq *et al.* 1987).

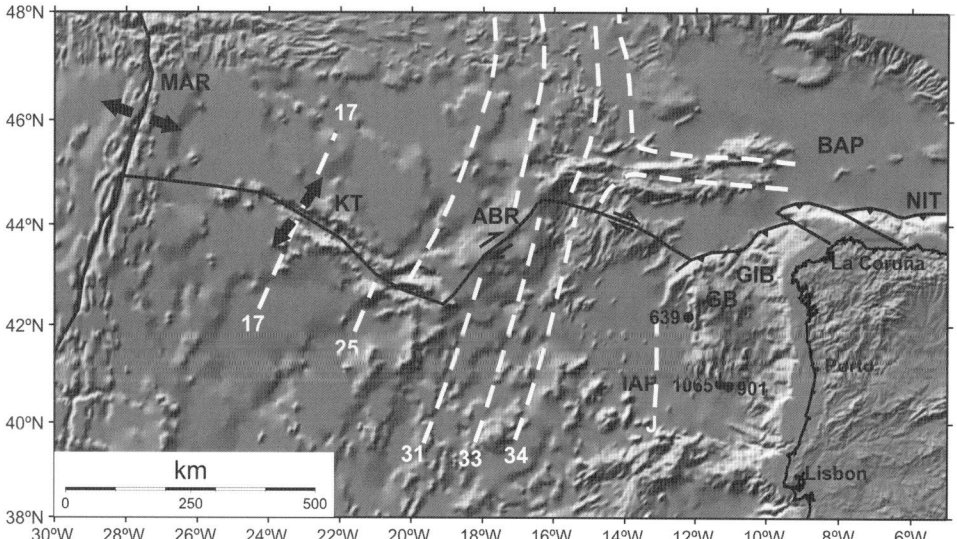

Fig. 8.3. Shaded relief map (GEBCO Digital Atlas) of the western Iberian continental margin. Main fracture zones and plate boundaries during the Eocene are shown (modified from Grimaud *et al.* 1982). MAR, Mid-Atlantic Ridge; KT, King's Trough; BBR, Azores–Biscay Rise; IAP, Iberia Abyssal Plain; BAP, Biscay Abyssal Plain; GIB, Galicia Interior Basin; GB, Galicia Bank; NIT, North Iberia Trough. White dashed lines show magnetic anomalies on the oceanic crust (Srivastava *et al.* 1990). Numbered black dots are Ocean Drilling Program sites from where much of the sedimentary sequence information comes.

Unconformities E, F, G and H were developed during global sea-level lowstands affecting the Atlantic from the Upper Miocene to the present.

Sq 4.1–Sq 4.5 show three different stacking patterns: (1) Sqs 4.1–4.3 are mainly retrogradational to aggradational (Fig. 8.4), with preservation of the transgressive systems tract (TST) and early highstand systems tract (HST) deposits accumulated during third-order global relative sea-level cycles TB 3.1–3.4. Sequence 4.3 probably represents the oldest depositional sequence recorded within the rías, overlying the Palaeozoic basement; (2) Sq 4.4 is mainly aggradational in the inner shelf, and progradational to strongly progradational in the middle–outer shelf, with lateral preservation of early–late HST deposits, respectively. This sequence accumulated during third-order global relative sea-level cycles TB 3.5–3.7; and (3) the uppermost depositional sequence in the succession, Sq 4.5, represents the entire Quaternary depositional record, characterized by low sediment input and the shelf's sediment starvation (Ferrin 2005).

Water masses and coastal processes around the Galician coast

Water masses

The water masses surrounding the Galician coast, located at the NW corner of the Iberian Peninsula (IP), were first described by Fraga (1987). According to this author, three different levels can be distinguished depending on depth (Table 8.1):

- Deep waters – they are divided into two different layers. The lower layer is the North Atlantic Deep Water (NADW), which can be observed below a depth of 3500 m. This water mass is characterized by a temperature (T) of of 2.2 °C and a salinity (S) of 34.91%. The upper layer is the Labrador Sea Water (LSW), which is located between NADW and the Mediterranean Water. This water constitutes a layer of minimum salinity and maximum oxygen concentration. Its main properties in the formation zone are $T = 3.4$ °C and $S = 34.89\%$.
- Intermediate waters – the Mediterranean Water (MW) characterizes them. The MW, which attains the Galician coast with low dilution, is formed in the Mediterranean Sea and splits into two veins after leaving the Gibraltar Strait. The lower vein flows northwards along the coast of the Iberian Peninsula, at a depth of 1200 m (Diaz del Río *et al.* 1998). At the equilibrium depth, its temperature ranges from 9.5 to 12.5 °C and its salinity ranges from 35.8 to 37.5%. The upper core flows

at a depth of 800 m, with temperature ranging from 10.5 to 13.5 °C and salinity ranging from 35.8 to 36.8%. The distribution along the Galician coast is rather irregular both in space and in time.

- Surface waters – Different layers can be considered, the subsurface water and the near-surface water. Eastern North Atlantic Central Water (ENACW) (Pérez *et al.* 1995) constitutes the main subsurface water near the Galician shelf. Its lower limit is located below 300–400 m, depending on the branch, and its upper limit depends on position, season and meteorological forcing. ENACW is formed in two main areas: a subpolar branch (10.5–12.5 °C and 35.55–35.70%), which is formed to the south of the North Atlantic Current and spreads south or SE (Pollard *et al.* 1996); and a subtropical branch (temperature and salinity values greater than 12.5 °C and 35.75‰, respectively) formed at the northern margin of the Azores Current and which moves NE towards the Iberian coast (Pingree 1997). The properties of near-surface water depend on local meteorological conditions, especially in the first 20–30 m. In the lower part (30–50 m) there is a seasonal thermocline that disappears only in winter, when the solar irradiance attains a minimum. This water is also locally affected by the presence of river plumes.

Coastal processes

Different coastal processes characterize the physical properties of the Galician coast. We will briefly summarize the most important ones; namely, coastal upwelling, poleward current and river plumes.

Coastal upwelling. Coastal upwelling along the western coast of the Iberian Peninsula is a frequent phenomenon during the spring–summer months and was extensively studied both inside the estuaries (Wooster *et al.* 1976; Fraga 1981; Blanton *et al.* 1984; Tenore *et al.* 1984; Álvarez-Salgado *et al.* 1993; Pérez *et al.* 1995; Rosón *et al.* 1995; Gómez-Gesteira *et al.* 2006; Martins *et al.* 2011) and at the adjacent shelf (Torres *et al.* 2003; Torres & Barton 2006, 2007; Miguez *et al.* 2005; Álvarez *et al.* 2007, 2008). Nevertheless, upwelling occurrence can also be observed during autumn and winter (Álvarez *et al.* 2003, 2009, 2012; deCastro *et al.* 2006b, 2008).

Long-term studies allowed the analysis of the summer upwelling variability and the influence of atmospheric modes on coastal upwelling along the Galician coast (Herrera *et al.* 2008; Álvarez *et al.* 2011; deCastro *et al.* 2011). Although there is an

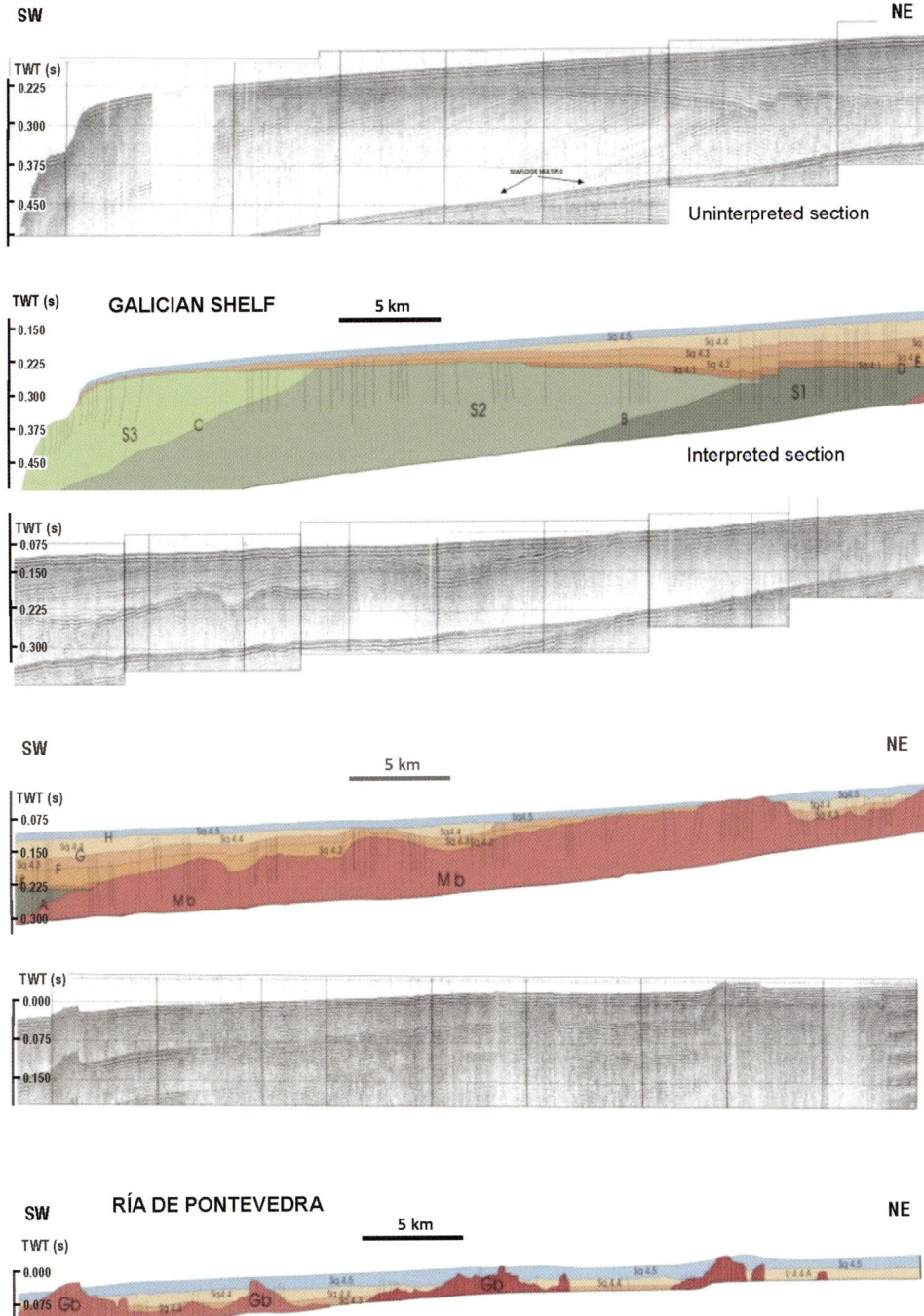

Fig. 8.4. Interpreted RSP-1 seismic profile collected in July 1997 by the Instituto Español de Oceanografía cruise aboard the R/V *Francisco de Paula Navarro*. Seismic penetration, approximately 0.3 s below seafloor (TWT) is sufficient to image the Upper Cretaceous–Recent time sedimentary section. Core information from the Pontevedra Marino 1 exploration well (PMB-1), drilled by ENIEPSA in 1984 (3538 m measured total depth), was also used in the study. See the location of the line in Figure 8.6 (after Ferrin 2005).

almost general consensus about the existence of a negative trend in upwelling intensity along the western Iberian Peninsula coast, the precise weakening rate is highly variable depending on the particular location examined, the database and the length of the series (Lemos & Pires 2004; Pérez *et al.* 2010).

The biochemical and biological implications of summer upwelling events on primary production inside the estuaries have been analysed along the Galician coast over the last decade or so (Castro *et al.* 2000; Barbosa *et al.* 2001; Gago *et al.* 2003; Varela *et al.* 2005; Prego *et al.* 2007; Álvarez-Salgado *et al.* 2008, 2011; Lønborg *et al.* 2010). It has an important effect on the production of biogenic carbonate and total organic carbon (TOC) exportation to the sediment (Vilas *et al.* 2005).

Poleward current. The Iberian Poleward Current (IPC) has been characterized over the last two decades (Pingree & LeCann 1989; Frouin *et al.* 1990; Haynes & Barton 1990; Peliz *et al.*

2005; Torres & Barton 2006). The IPC is a narrow (25–40 km), slope-trapped tongue-like structure that flows northerly along a distance in excess of 1500 km off the coasts of the Iberian Peninsula and SE France. It is a salty surface current (about 200 m deep), geostrophically trapped by the bathymetric discontinuity at the shelf break–upper slope zone and is capable of transporting very fine-grained detrital material in suspension. Satellite observations indicate that the thermal signature of the current is between 1.0 and 1.5 °C warmer than the surrounding ones. Meridional density gradients, the slope and wind-forcing interactions are the mechanisms for generation of the IPC (Gil & Gomis 2008).

It is well documented that the IPC normally arrives in the Cantabrian Sea at the beginning of every winter (e.g. Ambar *et al.* 1986; Garcia-Soto *et al.* 2002). The January warm-water extension of the IPC along the Cantabrian Sea has been referred to as 'Navidad' (Christmas) because it begins to become to be evident

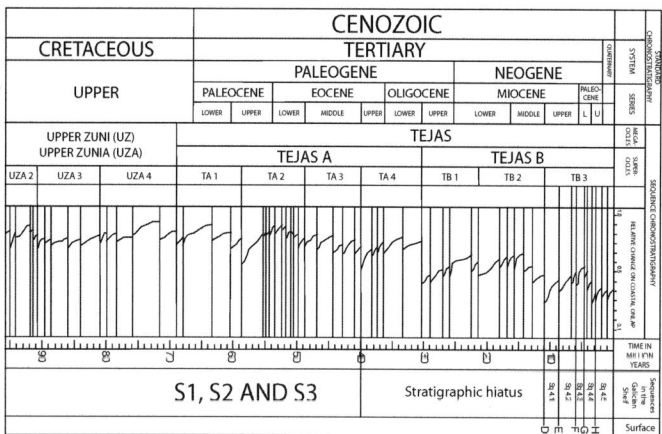

Fig. 8.5. Correlation between the interpreted depositional sequences on the Galician shelf and the global sequence chronostratigraphic chart (Haq *et al.* 1987).

near Christmas and New Year. Owing to the inability of the poleward flow to follow the abrupt changes of topography, the IPC exhibits a turbulent nature and produces the appearance of some unstable structures and separating eddies in the Bay of Biscay region (e.g. Pingree & LeCann 1993; Garcia-Soto *et al.* 2002). The variability of the IPC along the NW coast of the Iberian Peninsula and its intrusion in the Cantabrian Sea in terms of atmospheric forcing was analysed by deCastro *et al.* (2011).

River plumes. The discharge of freshwater from river outflows is particularly important in near-coastal regions. At mid latitudes and near western coasts of the northern hemisphere, such as in the study area, fresh river water usually spreads northwards and alongshore due to the influence of prevailing SW winds in winter coinciding with periods of high river runoff. Thus, freshwater can flood the major coastal estuaries located north of the river estuary for prolonged periods, reversing the normal estuarine density and salinity gradients. The two main rivers flooding the study area are the Duero and Miño.

The Duero River is the most important river that flows into the Portuguese NW coast. This river has a drainage area of 97 682 km^2 and an annual average discharge of 710 m^3 s^{-1} (Loureiro *et al.* 1986). The monthly discharge presents high variations according to the precipitation levels, varying between 50 (dry season) and 3000 m^3 s^{-1} (wet season), in this case often provoking floods.

The Miño River is the most important river that flows into the Galician west coast. This river has a catchment area of 17 081 km^2 and an annual average discharge of 430.8 m^3 s^{-1}. The monthly average discharge oscillates between 100 m^3 s^{-1} in August and 1000 m^3 s^{-1} in February, following a pattern similar to the rainfall.

Table 8.1. *Summary of water masses*

Water masses	Depth (m)	Potential temperature (°C)	Salinity (%)
ENACWst	<300	13.13–18.50	35.80–36.75
ENACWsp	<400	10.00–12.20	35.40–35.66
MW	400–1500	9.5–13.5	35.8–37.5
LSW	1500–3000	3.4	34.89
NADW	<3500	2.2	34.91

Adapted from different sources: Ambar & Howe (1979*a*, *b*), Ambar (1983), Emery & Meincke (1986), van Acken & Becker (1996), Pollard *et al.* (1996) and OSPAR (2000). The existence of different sublayers has been omitted for MW. Similarly for ENACW, only the two main branches have been considered.

The hydrographical behaviour of the Galician Rías Baixas was also studied to determine the influence exerted by the Miño River outflow (Álvarez *et al.* 2006; deCastro *et al.* 2006*a*).

Wave climate

The wind regime in this area also conditions the wave climate. The Galician Atlantic Margin is generally exposed to energetic swell waves from the NW to SW (Fig. 8.1). The most dominant and prevailing waves come from the fourth quadrant, mainly in the sector NW to WNW, with a probability of occurrence of 45%; the swell waves from the west have a probability of occurrence of around 30%, and the SW waves occur less frequently (<8%) but are normally associated with stormy conditions. The mean significant wave height (H_{sm}) and peak period (T_p) are 2 m and 10 s, respectively, increasing up to values of $H_{s12} = 8$ m and $T_p = 16$ s for a typical annual storm condition, described by the significant wave height exceeded for 12 h in a year and its associated peak period. The Galician continental shelf, therefore, is under a high wave energy regime.

Based on wave linear theory, the wave base level varies from 78 m for mean conditions to 200 m for extreme events, approximately coinciding with the shelf break. The inner shelf, including the coastal embayments called rías, is constantly affected by swell waves, exerting an important influence on the different sedimentary processes. Previous studies (Rey *et al.* 2005; Vilas *et al.* 2005) have revealed the swell waves as the forcing process that controls both the sediment surficial distribution and the subsequent diagenetic evolution of the sediments in the rías.

During extreme events, the whole shelf is under the wave's influence. Based on direct measured data, Huthnance *et al.* (2002) established that the contribution of waves on the significant bottom stress is 5–10 times greater than that of mean currents and, consequently, also 5–10 times greater on the moving sediments. Each wave cycle is relevant to the threshold for sediment movement and the rate of grain entrainment to the water column. The presence of waves greatly increases the sediment transport on the Galician continental shelf, even at the shelf-break at 200 m.

The North Atlantic Oscillation

The climate and oceanography variability of the region is controlled by the North Atlantic oscillation (NAO), particularly in winter. The NAO is a climatic fluctuation caused by recurrent changes in the intensity and relative position of the Icelandic low (IL) and the Azores high (AH) pressure systems (Kutzbach 1970; Wallace & Gutzler 1981; Hurrell 1995; Hurrell *et al.* 2001, 2010). This results in an oscillation with east–west motion that controls the strength and direction of storm paths, and westerly winds in the North Atlantic, producing two different phases that – once initiated – prevail for some years (generally 5–7 years), modulated by the local meteorological conditions. During the negative phase, winters in NW Iberia are wet and warm with intense storms increasing riverine discharge and sediment resuspension on the shelf (Trigo *et al.* 2004; Lebreiro *et al.* 2006; Otero *et al.* 2010). During the positive phase, winters are cold and dry with weaker winter storms.

The Holocene sediment-forming system

The northern Iberian Margin is a continental margin subjected to a highly energetic seasonal regime of waves and tides. Sediment provenance is dominated by inputs of continental sediments via runoff and riverine discharges and, to a lesser extent, by an

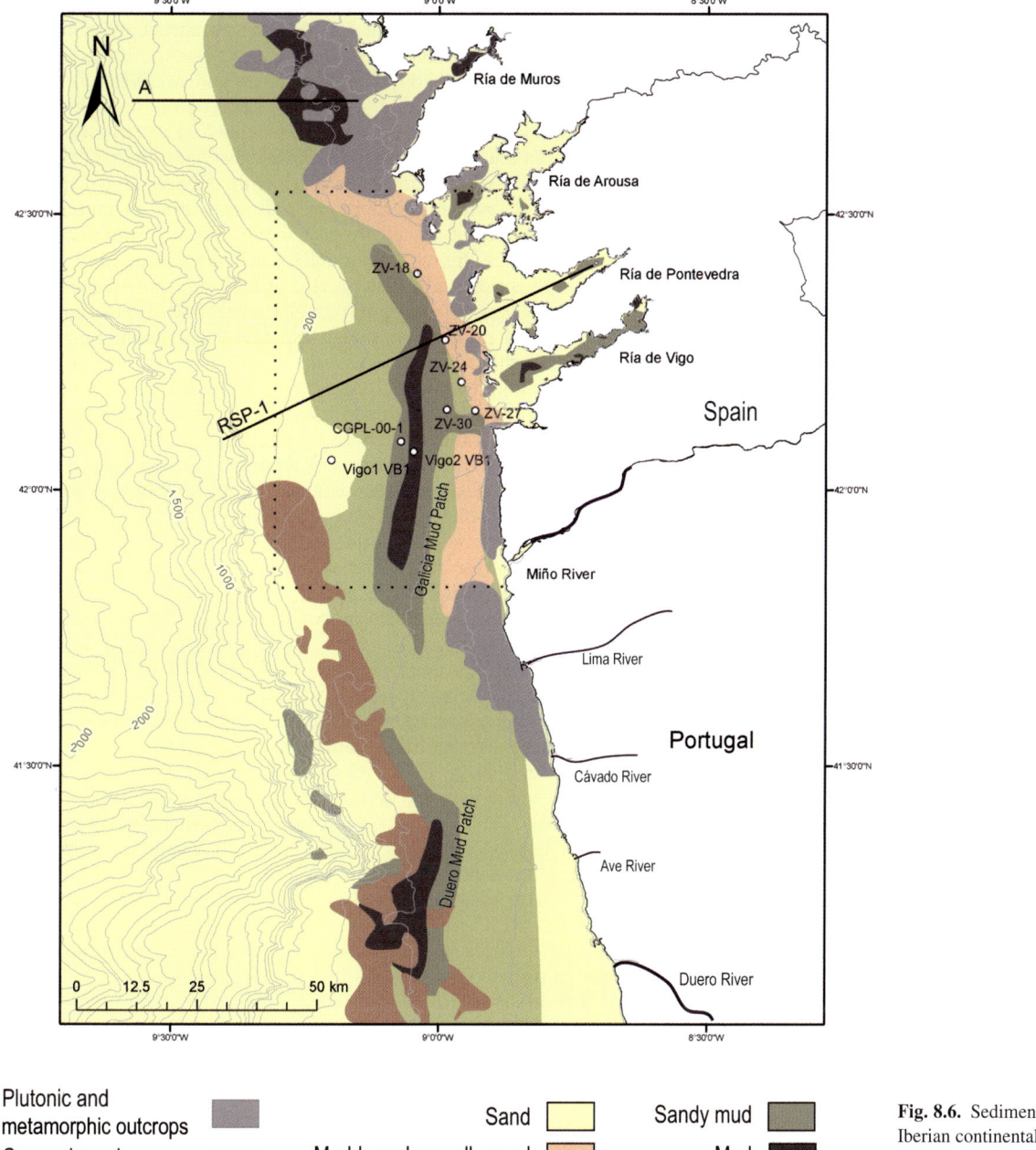

Fig. 8.6. Sediment distribution map of the NW Iberian continental shelf, based on Dias *et al.* (2002*a*), Vilas *et al.* (2005) and Mohamed *et al.* (2010). The location of key cores and seismic lines mentioned in the text are also plotted.

upwelling-enhanced bioclastic input. The present-day sedimentation in the area occurs closely associated with two geomorphological distinct sedimentary environments: the rías and the open continental shelf. Their structurally controlled NE–SW and NNW–SSE orientations have produced a contrasting sediment-facies distribution (Fig. 8.6), generally forced by common coastal processes and strongly conditioned by sediment provenance and early diagenetic transformations.

The Rías Baixas and the innermost shelf

The Rías Baixas are a characteristic geomorphological coastal feature of the NW Iberian Margin consisting of four deep and narrow V-shaped Tertiary river valleys that were drowned during the last sea-level transgression. Their orientation, tectonically controlled, is roughly perpendicular to the main coastline (Figs 8.1 & 8.6). All but the northernmost Ría de Muros are

shielded from the high-energy wave conditions common in the open shelf by a set of islands located at their mouth. Fluvial discharge in the rías is small. Therefore, estuarine conditions can be found only in the innermost parts of the rías. The balance between the predominance of wave processes v. fluvial influence determines the sedimentary characteristics of the ría sedimentary environments.

Sediment composition inside the rías is predominantly siliciclastic as a result of the granitic and metamorphic rocks of their catchment areas. Coarse calcareous bioclastic sediments favoured by upwelling fertilization of the rías can make a significant proportion of sediments in external areas. Exchange with the open sea is very limited.

The distribution of sediments in the rías is largely controlled by depth and distance from the open shelf (Vilas *et al.* 2005), reflecting the relative importance of fluvial, tidal and wave processes in each sector. Fluvial influence, albeit moderate, leads to the development of estuarine conditions in the innermost parts of the rías. Siliciclastic gravels and sands transported by the rivers are rapidly deposited near the river mouths in the estuarine zone, where they remain due to the low-energy characteristic of the innermost ría sector (Vilas *et al.* 2005). Tidal influence is also relatively important in this sector, as opposed to the predominance of waves in the more external sectors. As a result, intertidal mud flats crossed by sandy and gravelly tidal channels are also common sedimentary environments of the inner ría (Vilas *et al.* 2005; Pérez-Arlucea *et al.* 2007).

The frequent exposure to intense wave action in the external sector of the rías and the coastal margins results in the predominance of sands and gravels (Fig. 8.6). These coarse sediments contain up to 60% of CaCO₃ due to the higher marine influence in this sector further from the river mouths, particularly in the most external areas.

Wave energy is lowest in the deep central axis and the inner zones, which favours the deposition of muds. As a result, these sediments are composed of more than 60–80% mud (Rey *et al.* 2000). Mud accumulation is also promoted by the agglutinating effect of organic matter (OM). As a result, OM content is higher in muds and increases towards the inner ría to values in excess of 10% (Vilas *et al.* 2005).

Siliciclastic riverine gravels and sands are rapidly deposited close to the river mouths, while muds are trapped in intertidal flats or in the central axis. Therefore sediment export from the Rías Baixas to the adjacent continental shelf is negligible, essentially acting as sediment traps (Rey 1993). However, some export may occur under downwelling conditions, where approximately 40% of the water piled up in the shelf enters the rías and returns to the shelf through the bottom ongoing flux (Álvarez-Salgado *et al.* 1993). Sediments exported by this process may correspond to the fine-grained sediments enriched in organic matter off the southern Ría de Vigo described by Mohamed *et al.* (2010), in agreement with the proposed supply from the Ría de Vigo to explain the high chlorite content in the Galician Mud Belt (GMB) (Oliveira *et al.* 2002*a*).

In the rías, grain-size distributions reflect the presence of two main populations: one dominated by silt and clay, derived mainly from terrestrial sources; and the other by fine sand to coarse silt, which is derived mainly from continental shelf and ría mouth sources (Rubio *et al.* 2001). Mineralogical analysis shows an abundance of terrestrial intensive-weathering products near the ría head and a dominance of shelf-derived sediment towards the mouth. Muscovite/illite is the predominant mineral in inner ría sediments (Rubio *et al.* 2001), whilst quartz content and carbonates (aragonite and calcite) increase towards the open sea, both related to the enhancement of sand and bioclast content, respectively (Vilas *et al.* 2005, 2008). Clay minerals (kaolinite, mixed-layer and 14 Å minerals) and gibbsite are more abundant towards the head of the ría and decrease towards the entrance. Rubio *et al.* (2001) suggested that this material, derived mainly from terrestrial sources, is diluted towards the open sea.

The open continental shelf

Beyond the rías and the inner shelf, the surficial sediments of the NW Iberian continental shelf are predominantly terrigenous and biogenic sands (Figs 8.6 & 8.7). Terrigenous sediment supply by the Miño and Duero rivers, the two major rivers in the region, is estimated at 2.25 Mt a⁻¹ (Mota Oliveira *et al.* 1982), with a 79% contribution from the Duero alone (Dias *et al.* 2002*b*). These sediments are then redistributed by waves and currents acting on the shelf, resulting in differential transport and segregation according to their size. Sands and gravels are deposited close to the major river mouths in the area, the Duero and Miño, and transported southwards along the coast by the dominant southwards littoral drift. However, fine-grained sediments are transported

northwards and offshore by the combination of SW storm waves, downwelling and riverine low-density surface nepheloid layers (Oliveira *et al.* 2002*b*).

During autumn and winter, SW storms and swells induce strong bottom shear velocities (>3.5 cm s⁻¹), enough to resuspend fine-grained sediments and create bottom nepheloid layers several tens of metres thick (Vitorino *et al.* 2002). For more than 70% of the time, these storms are associated with downwelling (Vitorino *et al.* 2002), which transports the sediments offshore and into the poleward current forced by SW winds. Fine-grained sediments transported in this way eventually reach shelf areas less affected by wave action and are deposited at two mid-shelf deposits, the Duero (D) and Galician (G) Mud Belts (MB). This mechanism of transport in suspension also explains the detachment of these mud deposits from their sources.

Both mud belts are located on the middle shelf and their position is controlled topographically (Fig. 8.6). The DMB is constrained on the west by rocky outcrops less than 5 m high on the west, which prevents offshore export of resuspended sediments in all but the most extreme storms (Fig. 8.8). The GMB occurs at the change of slope between the steeper inner shelf and the more gradual middle and outer shelf. The lack of a topographical barrier to the west of the GMB makes export of resuspended fines more likely, which may explain the coarser grain sizes found in the GMB compared to the DMB (Vitorino *et al.* 2002).

The outermost shelf is composed of relict sands. Sediment supply to this sector is very small, slightly higher in the Galician sector. The relict nature of these sands is confirmed by the occurrence of significant amounts of glauconite, which have been dated at 5–6 Ma (Odin & Lamboy 1988). Maxima of mature glauconite are found along the 100 m isobath (Odin & Lamboy 1988; Fernández-Bastero *et al.* 2000), in the middle shelf. This is consistent with the landward migration of the outer shelf relict sands forced by sea-level rise (Rey 1993; Lantzsch *et al.* 2010; Vilas *et al.* 2010) and is consistent with pre-Holocene deposits with similar characteristics found by Mohamed *et al.* (2010) along a north–south transect along the approximately 100 m isobath (Fig. 8.7a).

Export of sediments offshore towards the slope is small and takes place essentially by entrapment in canyons like the Porto Canyon (Oliveira *et al.* 2002*b*) or through intense resuspension and transport by downwelling in the most intense SW storms.

Provenance and transport pathways

Oliveira *et al.* (2002*a*) showed that distributions of clay minerals in the sedimentary cover of the continental shelf are related to their continental sources, and also that they reflect the influence of winter storms and longshore currents. These deposits directly influenced by riverine discharges have a higher kaolinite content (20%), whereas those that are not directly influenced show a higher illite content (80%). Chlorite is a minor constituent of the clay fractions in the shelf sediments, with a content of about 7%. This clay mineral is more abundant in the inner and middle shelf sediments related to the proximity of the river mouths. These data support the idea that no significant quantities of terrigenous particles are being discharged from the Galician rías, which have acted as sediment traps since the last transgression, as proposed by Rey (1993), Vilas *et al.* (2005) and others (Fig. 8.7b).

The general elemental composition of the fine-grained fractions of the seabed sediments is consistent with the importance of fluvial inputs (Araújo *et al.* 2002, 2007). The main source of the fine-grained detrital particles to the shelf region is the River Duero discharge, despite contributions from the Miño River (Araújo *et al.* 2002; Oliveira *et al.* 2002*a*). These particles settle on the middle shelf, below the 60 m isobath. During storm events, particles are resuspended and advected northwards towards the Galician shelf

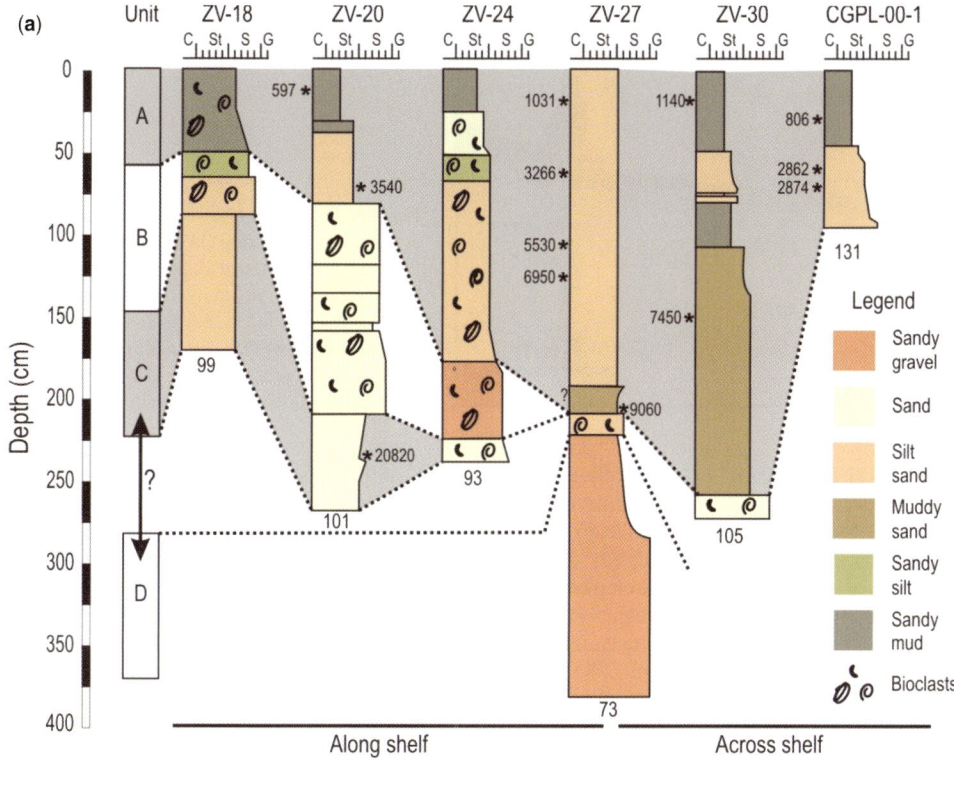

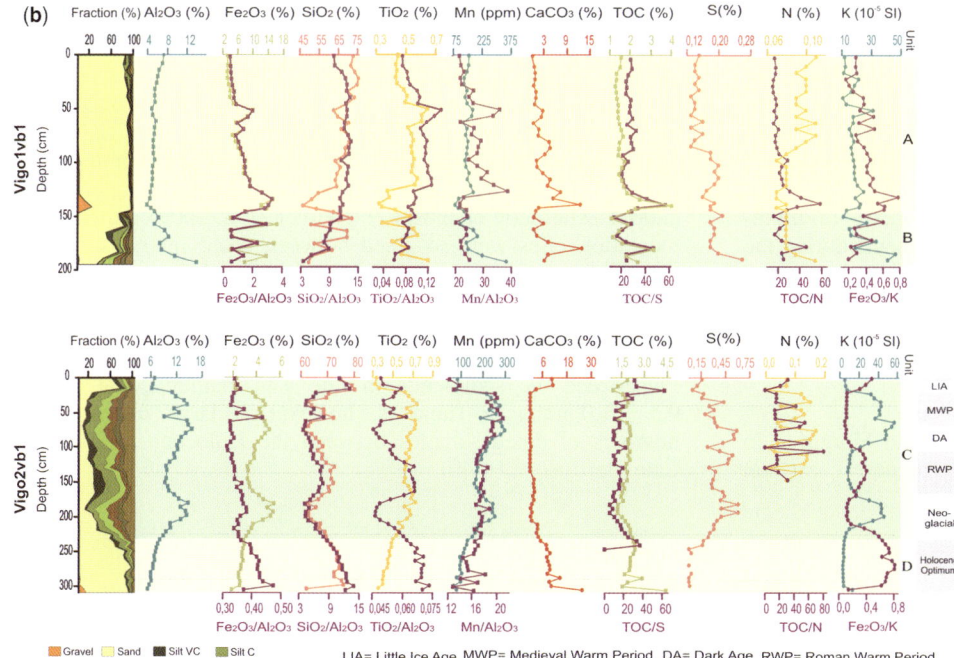

Fig. 8.7. (**a**) Core correlation illustrating west–east (rías to outer shelf) and north–south facies variability (location in Fig. 8.6). Asterisks mark dated levels and indicate the average calendar age. The numbers at the bottom of each log indicates the water depth in metres at each core location (after Mohamed *et al.* 2010). (**b**) Composition of two representative cores from sandy and muddy (Galician mud patch) areas (locations in Fig. 8.6). Units A and D show the typical high-energy signature during cold and relatively low sea-level periods consisting of higher carbonate bioclast content. Subsequent high sea-level stabilization is marked by decreasing grain size and by a relative increase in ferrimagnetic minerals of riverine provenance during warmer Holocene periods (RWP and MWP) and more stable hydrodynamics (after Ares *et al.* 2008).

or into deeper domains. Thus, the distribution of the clays indicates that there is a net transport of fine-grained sediments both north-wards and off-shelf.

Published geochemical data (Araújo *et al.* 2002; Ares *et al.* 2008; Bernárdez *et al.* 2008*a*, *b*) confirm this interpretation with a typical dominance of detrital or terrigenous elements (e.g. Fe, Al, Ti and lithogenic Si) mostly restricted to confined Holocene fine-grained depocentres, which are located in the middle-shelf position (Lantzsch *et al.* 2009*a*, *b*). Rare earth elements distribution patterns are highly comparable to the European shales, and exhibit a negative Eu-anomaly that is typically associated with K-rich granites, which dominate the continental rock lithology (Araújo *et al.* 2007). Geochemical parameters for organic matter in surface sediments, comprising TOC and TN content,

and stable isotope ratios ($\delta^{13}C$, $\delta^{15}N$) indicate its primarily marine origin, except in coastal regions receiving significant amounts of terrigeneous sediments (i.e. Duero: Alt-Epping *et al.* 2007).

Proxies for the lithogenic input show an increase at around 2–1.8 cal. ka BP linked to warmer conditions and high precipitation patterns during the Roman Warm Period, and also to soil erosion due to forest degradation and other anthropogenic activities (Martins *et al.* 2006*a*; Bernárdez *et al.* 2008*a*, *b*; Mohamed *et al.* 2010). On the outer continental shelf, a high Fe content is characteristic and related to the already mentioned glaucony-containing sandy facies that mostly formed by the reworking of older deposits (relict glauconite) during the last-glacial sea-level rise, as well as by authigenic mineral formation (Lantzsch *et al.* 2010). High-resolution measurements of TOC, TN and $\delta^{13}C$ in the muddy

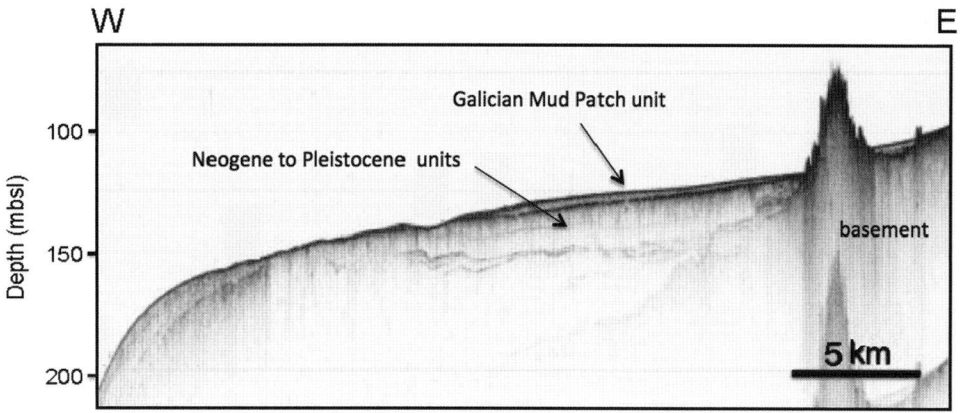

Fig. 8.8. High-resolution seismic boomer profiles obtained during GALIOMAR I Cruise P342 (after Hanebuth *et al.* 2007) showing the basic sedimentary structure of the recent shelf deposits (located in Fig. 8.6 as line A). Notice that the uppermost Galician mud patch Unit (dark grey) occurrs in the area associated with basement highs. This is not always the case.

sedimentary record, covering the Holocene period, also confirm the previous zonation, with large amounts of land-derived organic matter characterized by high C/N ratios and low δ^{13}C in the muddy facies (Ares *et al.* 2008; Burdloff *et al.* 2008) (i.e. Fig. 8.7b). Besides, variations in the contribution of terrestrial organic matter to the marine environment could be associated with palaeoclimatic events: for example, the highest continental supplies were found both in the Younger Dryas event and in the Upper Holocene (Burdloff *et al.* 2008).

Furthermore, Schmidt *et al.* (2010) distinguished three main regions with distinct OM provenance by analysing several biomarkers: (1) the inner-shelf region, dominated by fresh marine OM; (2) the middle-shelf mudbelt, characterized by high inputs of terrestrial OM and high total OM content; and (3) the outer shelf, with lower concentrations of relatively degraded OM with increased proportions of refractory terrestrial components. According to Burdloff *et al.* (2008) diagenetic effects on the sedimentary OM were probably of minor influence compared with the major change in OM supplies that has accompanied the Holocene period related to climate changes: for example, the rapid diminution in the terrigenous OM content of the sediments during the Little Ice Age.

Geochronology, sedimentation rates, environments and depocentres

A detailed revision of studies on the NW Iberian Margin – in particular the Galician rías, the northern Portuguese coast and

their adjacent continental shelf – has given a total of 235 absolute or relative datings (Fig. 8.9). The techniques commonly used are based on ^{14}C (71% of the datings) and ^{210}Pb dating (20%), whereas dating by ^{137}Cs, luminescence or by identification of pollen markers is very scarce (e.g. Jouanneau *et al.* 2002; Álvarez-Iglesias *et al.* 2007, 2009; Muñoz Sobrino *et al.* 2007; Pérez-Arlucea *et al.* 2007; González-Villanueva *et al.* 2009; Lantzsch *et al.* 2009*a*, *b*). A lack of a chronological framework constructed from absolute and/or relative dating techniques in many of the studies has become apparent, although reliable age–depth models are needed to study the different processes acting on the continental shelf. Most of the radionuclide datings were obtained from sediments of the continental shelf (45%), followed by coastal lagoons (27%) and the Galician rías (21%), especially in the Rías Baixas (e.g. Jouanneau *et al.* 2002; Lorenzo *et al.* 2007; Pena *et al.* 2007; Álvarez-Iglesias & Rubio 2009*a*; Costas *et al.* 2009; Lantzsch *et al.* 2009*a*, 2009*b*; Arribas *et al.* 2010; Mohamed *et al.* 2011). All ^{14}C dates were recalibrated with Calib 6.0.0 (Stuiver & Reimer 1993), selecting the appropriate dataset–Marine09 or IntCal09 (Reimer *et al.* 2009). Linear sedimentation rates were obtained between dated intervals and for the different sedimentary environments (Fig. 8.9).

The averaged sedimentation rate calculated from the 240 available datings is 1.5 ± 2.0 mm a^{-1}. The average sedimentation rate during the Holocene was 1.7 ± 2.0 mm a^{-1}, whereas for the Late Pleistocene the sedimentation rate was maintained at around 0.5 ± 1.0 mm a^{-1}. There is, however, a large difference in numbers of dated samples between the Holocene (>200) and Late Pleistocene (<40).

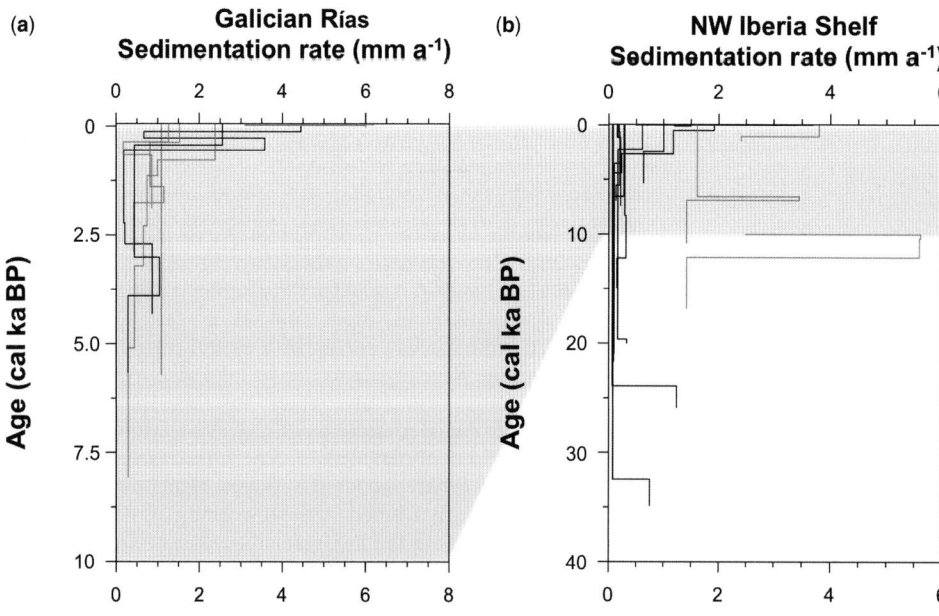

Fig. 8.9. Average sedimentation rates for selected cores and areas: (**a**) for the last 8 kyr for Galician rías (grey line: Álvarez-Iglesias *et al.* 2007; Pena *et al.* 2007; Pérez-Arlucea *et al.* 2007) and coastal lagoons (black line: Vilas *et al.* 1991; Bao *et al.* 2007; González-Villanueva *et al.* 2009); (**b**) main estuaries (grey line: Drago *et al.* 2006) and continental shelf (black line: Jouanneau *et al.* 2002; Lantzsch *et al.* 2009*a*, *b*; Mohamed *et al.* 2010).

Average sedimentation rates in the Galician rías varied between 0.2 and 10 mm a^{-1} during the last 8 kyr (Fig. 8.9). The highest sedimentation rates were detected for the last century and in areas close to river mouths (Cabanas *et al.* 1979; Álvarez-Iglesias *et al.* 2007; Lorenzo *et al.* 2007; González-Villanueva *et al.* 2009) and, currently, below mussel rafts. The antropogenically forced increase in sedimentation rates can be tracked back to the last two millennia, coincident with deforestation related to human activities (Allen *et al.* 1996; Muñoz Sobrino *et al.* 2007) and, in general, sedimentation rates increased towards recent times. It is noticeable that peaks in sedimentation rates are coincident with the intervals of increased fluvial activity in Spain described by Thorndycraft & Benito (2006). Ages of peat levels detected in lagoons (e.g. Vilas *et al.* 1991; Bao *et al.* 2007; González-Villanueva *et al.* 2009; Arribas *et al.* 2010) are, in general, also coincident with these intervals.

Studies on the northern Portuguese coast have been centred on the mouths of the rivers Miño and Duero (Drago *et al.* 2006; Álvarez-Iglesias *et al.* 2009), where average sedimentation rates varied between 1.4 and 5.6 mm a^{-1} during the last 17 kyr (Fig. 8.9). Maxima sedimentation rates in these rivers were also detected for periods of increased fluvial activity (Thorndycraft & Benito 2006).

Rivers Miño and Duero are considered to be the main sediment suppliers to the middle continental shelf mud belts formed during the last 5.3 kyr. These patchy elongated bodies that have concentrated the attention of most research effort (Jouanneau *et al.* 2002; González-Álvarez *et al.* 2005; Lantzsch *et al.* 2009a, b; Mohamed *et al.* 2010) show a high content of coarse silt, whilst the remainder of the surficial shelf sediments mostly consist of sands and gravels (Dias *et al.* 2002a). The mud patches overlie coarser sediments, mostly sands (Martins *et al.* 2006a; Lantzsch *et al.* 2009a, b). Average sedimentation rates in the NW Iberian continental shelf have varied between 0.1 and 5.8 mm a^{-1} for the last 35 kyr (Fig. 8.9), with the highest values observed at the mud belts, which are still receiving terrigenous inputs (Araújo *et al.* 2002; Jouanneau *et al.* 2002; Oliveira *et al.* 2002a). Exceptionally high sedimentation rates (>10 mm a^{-1}) were detected related to the deposition of relatively thick sandy layers by storm events (González-Álvarez *et al.* 2005). Storms and currents caused a sorting of the sediments that can be observed in the sedimentary record (Dias *et al.* 2002a; Lantzsch *et al.* 2009b; Mohamed *et al.* 2010). These processes, together with bioturbation, also generated a surficial mixed layer of a few centimetres (Jouanneau *et al.* 2002). In general, the highest average sedimentation rates are observed over the last 5.4 kyr, and especially in the most recent deposits. Nevertheless, relative sedimentation rates peaked at certain deep levels coincident with cold and arid periods and/or with Heinrich events, usually described in the literature (Thouveny *et al.* 2000; Naughton *et al.* 2007b).

The variability observed for sedimentation rates within and between the different areas of deposition (rías, estuaries and shelf) could be related to sea-level variation. According to the relative sea-level curve established by Dias *et al.* (2000) for the Portuguese coast, the sea level was close to the shelf break during the Last Glacial Maximum (LGM). At this time, the rías emerged (García-García *et al.* 2005; Durán *et al.* 2007), river courses extended far to the shelf and their sediment load was almost directly discharged on to the continental slope (Dias *et al.* 2000). Afterwards, sea level rose (c. 16 ka BP) and stabilized (for about 3 kyr). Since 13 ka BP, the rise in sea level has been rapid, causing estuaries to become sediment traps (sedimentation rates peaked during this period in the Duero estuary record) and decreasing the export of sediments to the shelf (Dias *et al.* 2000). The sea level dropped to −60 m at the Younger Dryas event, causing erosion of the sedimentary infill of estuaries by rivers, and export of sediments to the shelf. The shallower areas of the rías previously submerged experienced subaerial exposure and channel incision (Durán *et al.* 2007). Later, the sea

level again rose quickly (between 10 and 8 ka BP), and thick deposits accumulated in river mouths during this time (Dias *et al.* 2000). The sea level was around −20 m at 8 ka BP and, at this time, most of the rías were flooded to some extent. Rías acted as sediment traps during the last transgression and thick sedimentary sequences developed (García-Gil *et al.* 1999; Durán *et al.* 2007; Vilas *et al.* 2010). Stratigraphic studies of the rías showed a progressive enlargement of their sedimentary basins, shorter pulses of sea-level rise and stillstand, and a migration of the sediment depocentres (García-Gil *et al.* 1999; Durán *et al.* 2007). The rates of sea-level rise slowed down thereafter, and the rise tended to occur in short pulses (Durán *et al.* 2007). The present sea level was reached after 6 ka BP (Dias *et al.* 2000), and the strongly attenuated sea-level rise led to the development of coastal features such as barriers, spits and lagoons (Dias *et al.* 2002a; Bao *et al.* 2007; González-Villanueva *et al.* 2009).

Environmental quality of seabed sediments

An anthropogenic influence in the sediments of the margin is, in general, reflected by trace metal anomalous concentration. However, in spite of the heavy metal contamination detected in the downstream region of some of the rivers draining the NW coastal area, including the Ave and the Duero rivers (the major source of fine-grained shelf sediments), no significant metal enrichments were detected in shelf sediments (Araújo *et al.* 1998, 2002; Corredeira *et al.* 2005, 2009; Mil-Homens *et al.* 2006). This is much more clearly demonstrated in the Galician rías (Prego & Cobelo-Garcia 2003), where the contamination level is considerably higher, and the geographical distribution pattern of the typical anthropogenic metals (Cu, Zn, Cr, Pb) results from the local discharge of pollutants and the adsorption capabilities of the sediments. Both factors depend on the textural and compositional variability of the different sediment types present in the study area. The highest concentration of trace metals occurs in the muddiest and organic-rich sediments of the rías (e.g. Rubio *et al.* 2000; Corredeira *et al.* 2005, 2009; Álvarez-Iglesias *et al.* 2006; Álvarez-Iglesias & Rubio 2009b; Rubio *et al.* 2010) but they are confined to only the uppermost layers of the cores, representing about half of the last century. Examples of this anthropogenic point source come from the work on the inner Ría de Vigo carried out by Álvarez-Iglesias *et al.* (2003, 2006). Sediments in this area are fine grained and enriched in organic matter; hence, they have a higher potential to adsorb contaminants than the coarser-grained, less organic-rich sediments of the middle and outer ría. In general terms, a significant fraction of these elements mainly concentrates in the residual fraction, especially the most harmful metals and semi-metals, such as Cu, Pb, As and Hg (Rubio *et al.* 2001; Álvarez-Iglesias *et al.* 2003; León *et al.* 2004; Álvarez-Iglesias & Rubio 2008). However, there are relatively high trace metal proportions in the oxidizable fractions (sulphides and organic matter) of the sediments from the inner rías. This means that the fate and bioavailability of these pollutants will depend on the intensity and speed of bacterial-mediated redoxomorphic post-depositional processes (Álvarez-Iglesias & Rubio 2008, 2009b; Rubio *et al.* 2010).

Pleistocene–Holocene evolution and climate cycles

High-resolution pollen and marine proxy analyses (δ^{18}O of *Globigerina bulloides*, ice-rafted detritus (IRD) and *Neogloboquadrina pachyderma*) from the Galician Margin composite cores studied by Sánchez Goñi *et al.* (2000) show a synchronicity of the vegetation response to the North Atlantic climatic variability during Heinrich events 1 and 2 (H1, H2), LGM, Bølling–Allerød (B–A), and Younger Dryas (YD) events. These events are not only characterized by low sea surface temperature (SST) but also by a sharp increase in aridity in the Iberian Peninsula, indicated by increases in steppe vegetation pollen or enhanced inputs of Saharan dust in the studied marine sediments (Sánchez Goñi *et al.* 2000; Boessenkool *et al.* 2001; Bout-Roumazeilles *et al.* 2007; Fletcher & Sánchez Goñi 2008). According to Naughton *et al.* (2007*a*), the pollen count record shows that the beginnings of both the H2 and H1 cold events are associated with *Pinus* forest reduction in northern Iberia. It also shows the presence of two vegetation phases within the H1 and H2 events, associated with an initial cold and wet episode followed by a cool and, particularly, dry episode during H1. Relatively warm and humid conditions (indicated by the expansion of *Quercus*) prevailed in western Iberia during the Bølling–Allerød interstadial. Such conditions were interrupted by a sudden decrease in both temperature and humidity at the transition of the Allerød–Younger Dryas. The cooling of the surface ocean occurred within 400 years and coincided with a change in the pollen record (expansion of steppe vegetation and a decline in *Quercus*).

The climatic amelioration marked the beginning of the Holocene – the current warm interglaciation of the last 10 kyr. In Iberia, vegetation responded to these climatic conditions in a similar way to that of the Bølling–Allerød event (Naughton *et al.* 2007*a*). The Holocene was also characterized by a millennial-scale climatic variability, as has been revealed by studies based on marine records from the North Atlantic Ocean (Bond *et al.* 1997, 1999, 2001; Bianchi & McCave 1999; Chapman & Shackleton 2000). This variability was of low amplitude, estimated at about 2 °C in seawater surface temperatures (Bond *et al.* 1997).

During the Late Pleistocene–Holocene, climatic warming and the melting of the ice sheets (Ruddiman & McIntyre 1981; Pirazzoli 1996) influenced the rise in sea level in the last eustatic hemicycle (18–6 ka BP), which influenced the sedimentary dynamics on continental shelves and the establishment of a dynamic current system on the shelf (Abrantes 1988; Caralp 1992).

According to several works on the Miño and Duero estuaries (Drago 2005; Guerreiro *et al.* 2005; Naughton *et al.* 2007*a*; Álvarez-Iglesias *et al.* 2009), two distinct phases of geomorphological change can be considered to have existed in the North Iberian

estuaries during the Holocene. The first, between *c.* 10.7 and 6.5 cal ka BP, in the transgressive phase, was characterized by a warm and humid climate, and by estuarine and/or marine sedimentary facies. The second phase extends after the transgressive maximum until the present. During this phase, a gravel barrier developed in the Duero Estuary as a consequence of both the attenuation in sea-level rise and the high fluvial hydrodynamism, leading to changes in coastal morphology. At the same time, estuarine sedimentary facies are detected in the Miño Estuary. Afterwards, in both estuaries, a sand barrier stage was described that developed until the present (Drago 2005).

Stabilization of the sea level during the last 6.5 kyr has led to the formation of a sedimentary wedge on the shelf over the deposits of the previous transgressive sedimentary environments and the establishment of the present coastal environment (Swift *et al.* 1991). The deposition of fine-grained sediments controlled by the shelf morphology and seasonal circulation patterns gave rise to the formation of the mud depocentres previously mentioned (Dias *et al.* 2002*a*, *b*). In recent years, these mud deposits became of major interest due to their role in material budgeting (Jouanneau *et al.* 2002) and as high-resolution climate archives (González-Álvarez *et al.* 2005; Martins *et al.* 2005, 2006*a*, *b*, 2007; Bartels-Jónsdóttir *et al.* 2006; Hanebuth *et al.* 2007; Bernárdez *et al.* 2008*b*). The Holocene climatic changes had a significant influence on the hydrodynamic regimes of the shelf circulation controlling sediment grain-size and composition fluctuations (González-Álvarez *et al.* 2005; Bartels-Jónsdóttir *et al.* 2006; Martins *et al.* 2006*a*, 2007; Bernárdez *et al.* 2008*a*). Similar influences are also detected in cores collected in the Galician rías (Diz *et al.* 2002; Desprat *et al.* 2003; Álvarez *et al.* 2005) and in Portuguese estuaries (Naughton *et al.* 2007*b*).

Solar radiative budget and oceanic circulation seem to be the main mechanisms forcing the Holocene climatic cyclicity in NW Iberia (Desprat *et al.* 2003) or oscillations in the atmospheric pressure system, such as the North Atlantic Oscillation (NAO) index (e.g. González-Álvarez *et al.* 2005; Bartels-Jónsdóttir *et al.* 2006; Martins *et al.* 2006*b*, 2007; Bernárdez *et al.* 2008*a*). These works demonstrate the importance of the NAO oscillations on the NW Iberian region in influencing the intensity and wind

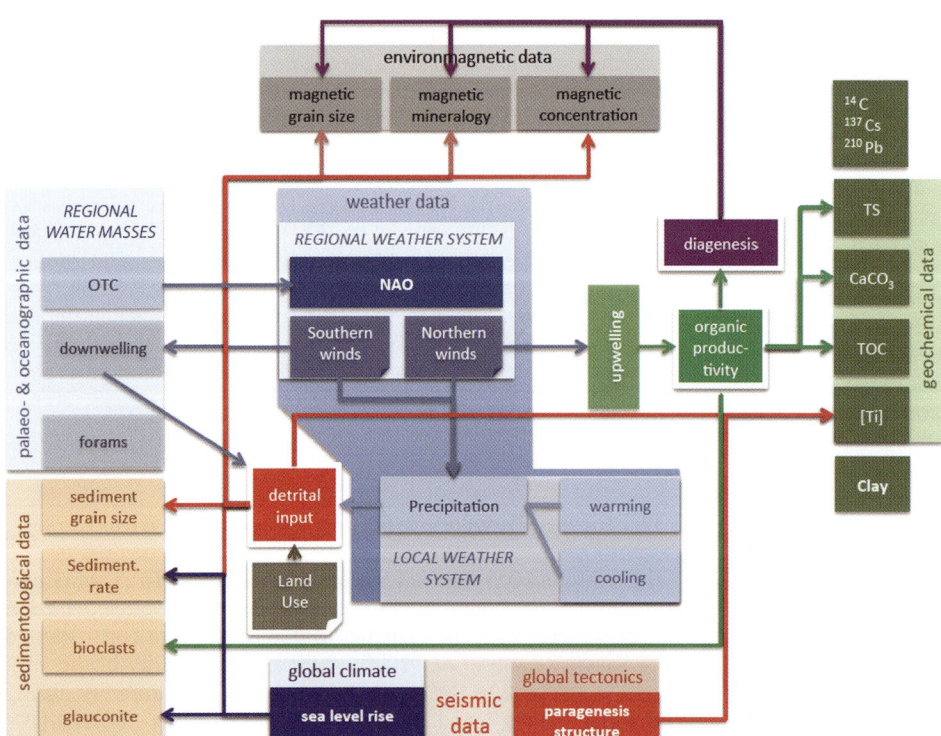

Fig. 8.10. Flow chart showing the main data sources and forcings involved in the NW Iberian shelf sediment-forming system. OTC, overturning circulation; NAO, North Atlantic Oscillation; TS, total sulphur; TOC, total organic carbon; [Ti], Ti concentration.

direction triggered by the position of the high-/low-pressure systems and the magnitude of floods, as well as the upwelling and sediment-transport patterns.

Final summary and conclusions

Integrated analyses of hydrographical, sedimentological, palaeoceanographic, environmagnetic and geochemical data from the NW Iberian continental shelf allow a good approximation of the continental shelf dynamics in NW Iberia and its climatic reconstruction since the Last Glacial Maximum (LGM), summarized in Figure 8.10.

The NW Iberian continental shelf extends between Cape Finisterre (43°N) and a latitude close to 41°N (Duero River), forming a 200 km-long and 30–50 km-wide morphosedimentary unit that shares common oceanographic and climatic conditions. The regional weather system is modulated by the North Atlantic Oscillation (NAO) that controls the local weather system and, subsequently, the on-land humidity and coastal processes affecting the regional water masses, such as currents, coastal upwelling and river plumes. Sedimentation rates (averaging 1.5 ± 2.0 mm a^{-1}, ranging between 1.4 and 5.6 mm a^{-1}) and provenances are defined by inputs of continental sediments via runoff and riverine discharges and, to a lesser extent, by the upwelling-enhanced bioclastic input. Since the LGM, the relative intensity of these processes depends on the decadal-modulated NAO alternation of increased detrital input, alternating with intervals of lower detrital input and intensified upwelling in the area. It is notorious that the sedimentation rates have undergone an anthropogenically forced increase of 3 mm a^{-1} over the last 500 years.

Two Holocene units are easily identified in the area, showing a sediment succession from fluvial and subaerial settings to high- and, finally, low-energy marine deposits subsequent to the post-LGM sea-level rise.

The distribution of surficial sediments is closely associated with four distinct sedimentary environments: the open continental shelf; the middle-shelf mud patches; the inner-shelf–Galician rías system; and the Duero and Miño estuaries. The open shelf sediments comprise mixed siliciclastic carbonatic sands and muds that distribute parallel to the coast, with grain size decreasing towards the open sea. The fine-grained sediments in the open shelf may contain variable amounts of glauconite. Three fine-grained patches of variable length and width occur in the middle shelf below the 60 m isobath, disrupting this sediment distribution. The source of this fine-grained material is mostly from the Duero River discharge, notwithstanding contributions from the Miño River.

Finally, the Galician rías are LGM-incised valleys in which typical estuarine sediments are strongly affected by inner-shelf processes. In sedimentological terms, the confinement of the rías in the north and the strong geomorphological control over the fill and facies distribution in this part of the inner shelf divide the shelf into two segments. In the southern segment, the Duero and Miño rivers are the main detrital sediment suppliers, increasing the sedimentation rates at their mouths and in the middle shelf. In the northern segment, the inner shelf is deprived of local fluvial provenances entrapped in the Galician rías as the last sea-level rise progressively enlarged their sedimentary basins.

Our understanding of the NW Iberian continental shelf mainly comes from the long-term funding of the Portugese and Spanish national research programmes, implemented by their university systems; particularly the universities of Aveiro, Lisboa, Porto, Granada and Vigo, and public research institutes such as the ITN, CSIC and IEO. There are also significant contributions from the multinational efforts such as the ODP (especially legs 149 and 173), UE-funded OMEX, HOLSMEER and, most recently, Germany's GALIOMAR cruises led by the University of Bremen.

This work was supported by the Spanish Ministry of Science and Technology through projects CTM2007-61227/MAR, GCL2010-16688 and IPT-310000-2010-17, by IUGS-UNESCO through project IGCP-526, and by the Xunta de Galicia through projects 09MMA012312PR and 10MMA312022PR. P. Álvarez-Iglesias is grateful for the support from the Xunta de Galicia (Spain) through a post-doctoral contract and short-stay fellowships under the Ángeles Alvariño program. KJM acknowledges the receipt of a Marie Curie fellowship.

References

ABRANTES, F. 1988. Diatom assemblages as upwelling indicators in surface sediments off Portugal. *Marine Geology*, **85**, 15–39.

ALLEN, J. R. M., HUNTLEY, B. & WATTS, W. A. 1996. The vegetation and climate of northwest Iberia over the last 14 000 yr. *Journal of Quaternary Science*, **11**, 125–147.

ALT-EPPING, U., MIL-HOMENS, M., HEBBELN, D., ABRANTES, F. & SCHNEIDER, R. R. 2007. Provenance of organic matter and nutrient conditions on a river- and upwelling influenced shelf: a case study from the Portuguese Margin. *Marine Geology*, **243**, 169–179.

ALVARADO, M. M. 1983. Evolución de la Placa Ibérica. *In*: RÍOS, J. M. (ed.) *Geología de España*. Instituto Geológico y Minero de España, Libro Jubilar, 21–55.

ÁLVAREZ, I., DECASTRO, M., PREGO, R. & GÓMEZ-GESTEIRA, M. 2003. Hydrographic characterization of a winter-upwelling event in the Ria of Pontevedra (NW Spain). *Estuarine Coastal Shelf Science*, **56**, 869–876, http://dx.doi.org/10.1016/S0272-7714(02)00309-8

ÁLVAREZ, I., DECASTRO, M., GÓMEZ-GESTEIRA, M. & PREGO, R. 2006. Hydrographic behavior of the Galician Rias Baixas (NW Spain) under the spring intrusion of the Miño River. *Journal of Marine Systems*, **60**, 144–152, http://dx.doi.org/10.1016/j.jmarsys.2005.12.005

ÁLVAREZ, I., GÓMEZ-GESTEIRA, M., DECASTRO, M. & NOVOA, E. M. 2007. Ekman transport along the Galician Coast (NW, Spain) calculated from QuikSCAT winds. *Journal of Marine Systems*, **72**, 101–115, http://dx.doi.org/10.1016/j.jmarsys.2007.01.013

ÁLVAREZ, I., GÓMEZ-GESTEIRA, M., DECASTRO, M. & DIAS, J. M. 2008. Spatiotemporal evolution of upwelling regime along the western coast of the Iberian Peninsula. *Journal of Geophysical Research*, **113**, C07020, http://dx.doi.org/10.1029/2008JC004744

ÁLVAREZ, I., OSPINA-ÁLVAREZ, N. ET AL. 2009. A winter upwelling event in the Northern Galician Rias: frequency and oceanographic implications. *Revista: Estuarine Coastal and Shelf Science*, **82**, 573–582, http://dx.doi.org/10.1016/j.ecss.2009.02.023

ÁLVAREZ, I., GÓMEZ-GESTEIRA, M., DECASTRO, M., LORENZO, M. N., CRESPO, A. J. C. & DIAS, J. M. 2011. Comparative analysis of upwelling influence between the western and northern coast of the Iberian Peninsula. *Continental Shelf Research*, **31**, 388–399, http://dx.doi.org/10.1016/j.csr.2010.07.009

ÁLVAREZ, I., PREGO, R., DECASTRO, M. & VARELA, M. 2012. Galicia upwelling revisited: out of season event in Rias (1967–2009). *Ciencias Marinas*, **38**, 143–159.

ÁLVAREZ, M. C., FLORES, J. A., SIERRO, F. J., DIZ, P., FRANCÉS, G., PELEJERO, C. & GRIMALTT, I. 2005. Millennial surface wáter dynamics in the Ría de Vigo during the last 3000 years as recealed by coccoliths and molecular biomarkers. *Palaeogeography, Palaeoclimatology, Palaeoecology*, **218**, 1–13.

ÁLVAREZ-IGLESIAS, P. & RUBIO, B. 2008. The degree of trace metal pyritization in subtidal sediments of a mariculture area: application to the assessment of toxic risk. *Marine Pollution Bulletin*, **56**, 973–983.

ÁLVAREZ-IGLESIAS, P. & RUBIO, B. 2009a. Geochemistry of marine sediments from inner Ría de Vigo (NW of Spain). *Journal of Radioanalytical and Nuclear Chemistry*, **281**, 247–251.

ÁLVAREZ-IGLESIAS, P. & RUBIO, B. 2009b. Redox status and heavy metal risk in intertidal sediments in NW Spain as inferred from the degrees of pyritization of iron and trace elements. *Marine Pollution Bulletin*, **58**, 542–551.

ÁLVAREZ-IGLESIAS, P., RUBIO, B. & VILAS, F. 2003. Pollution in intertidal sediments of San Simón Bay (Inner Ría de Vigo, NW of Spain): total heavy metal concentrations and speciation. *Marine Pollution Bulletin*, **46**, 491–503.

ÁLVAREZ-IGLESIAS, P., RUBIO, B. & PÉREZ-ARLUCEA, M. 2006. Reliability of subtidal sediments as 'geochemical recorders' of pollution input: San Simón Bay (Ría de Vigo, NW Spain). *Estuarine Coastal and Shelf Science*, **70**, 507–521.

ÁLVAREZ-IGLESIAS, P., QUINTANA, B., RUBIO, B. & PÉREZ-ARLUCEA, M. 2007. Sedimentation rates and trace metal input history in intertidal sediments derived from ^{210}Pb and ^{137}Cs chronology. *Journal of Environmental Radioactivity*, **98**, 229–250.

ÁLVAREZ-IGLESIAS, P., ARAÚJO, M. F., GOUVEIA, A. & DRAGO, T. 2009. Geochemical analysis of Minho Estuary sedimentary record and its contribution to palaeoenvironmental evolution. *Journal of Radioanalytical and Nuclear Chemistry*, **281**, 237–240.

ÁLVAREZ-MARRÓN, J., RUBIO, E. & TORNE, M. 1997. Subduction-related structures in the North Iberian Margin. *Journal of Geophysical Research*, **102**, 22 497–22 511.

ÁLVAREZ-SALGADO, X. A., ROSON, G., PEREZ, F. F. & PAZOS, Y. 1993. Hydrographic variability off the Rias Baixas (NW, Spain) during the upwelling season. *Journal of Geophysical Research*, **98**, 14 447–14 455, http://dx.doi.org/10.1029/93JC00458

ÁLVAREZ-SALGADO, X. A., LABARTA, U. *ET AL.* 2008. Renewal time and the impact of harmful algal blooms on the extensive mussel raft culture of the Iberian coastal upwelling system (SW Europe). *Harmful Algae*, **7**, 849–855, http://dx.doi.org/10.1016/j.hal.2008.04.007

ÁLVAREZ-SALGADO, X. A., FIGUEIRAS, F. G., FERNANDEZ-REIRIZ, M. J., LABARTA, U., PETEIRO, L. & PIEDRACOBA, S. 2011. Control of lipophilic shellfish poisoning outbreaks by seasonal upwelling and continental runoff. *Harmful Algae*, **10**, 121–129, http://dx.doi.org/10.1016/j.hal.2010.08.003

AMBAR, I. 1983. A shallow core of Mediterranean water off western Portugal. *Deep-Sea Research*, **30**, 677–680.

AMBAR, I. & HOWE, M. R. 1979*a*. Observations of the Mediterranean outflow I: mixing in the Mediterranean outflow. *Deep-Sea Research*, **26A**, 535–554.

AMBAR, I. & HOWE, M. R. 1979*b*. Observations of the Mediterranean outflow II: the deep circulation in the vicinity of the Gulf of Cadiz. *Deep-Sea Research*, **26A**, 555–568.

AMBAR, I. J., FIUZA, A. F. G., BOYD, T. & FROUIN, R. 1986. Observations of a warm oceanic current flowing northwards along the coasts of Portugal and Spain during Nov–Dec 1983. *Eos, Transactions of the American Geophysical Union*, **67**, 1054.

ARAÚJO, M. F., VALÉRIO, P. & JOUANNEAU, J. M. 1998. Heavy metal assessment in sediments of the Ave river basin (Portugal) by EDXRF. *X-Ray Spectrometry*, **27**, 305–312.

ARAÚJO, M. F., JOUANNEAU, J.-M. *ET AL.* 2002. Geochemical tracers of northern Portuguese estuarine sediments on the shelf. *Progress in Oceanography*, **52**, 277–297.

ARAÚJO, M. F., CORREDEIRA, C. & GOUVEIA, A. 2007. Distribution of the rare earth elements in sediments of the Northwestern Iberian Continental Shelf. *Journal of Radioanalytical and Nuclear Chemistry*, **271**, 255–260.

ARES, A., REY, D., RUBIO, B., MOHAMED, K., BERNABEU, A. & VILAS, F. 2008. Reconstrucción paleoclimática de la plataforma continental gallega basada en datos geoquímicos y magnéticos. *Geogaceta*, **44**, 87–90.

ARRIBAS, J., ALONSO, A., PAGÉS, J. L. & GONZÁLEZ-ACEBRÓN, L. 2010. Holocene transgression recorded by sand composition in the mesotidal Galician coastline (NW Spain). *The Holocene*, **20**, 375–393, http://dx.doi.org/10.1177/0959683609353429

BAO, R., ALONSO, A., DELGADO, C. & PAGÉS, J. L. 2007. Identification of the main driving mechanisms in the evolution of a small coastal wetland (Traba, Galicia, NW Spain). *Palaeogeography, Palaeoclimatology, Palaeoecology*, **247**, 296–312.

BARBOSA, A. B., GALVÃO, H. M., MENDES, P. A., ÁLVAREZ-SALGADO, X. A., FIGUEIRAS, F. G. & JOINT, I. 2001. Short-term variability of heterotrophic bacterioplankton during upwelling off the NW Iberian margin. *Progress in Oceanography*, **51**, 339–359, http://dx.doi.org/10.1016/S0079-6611(01)00074-X

BARTELS-JÓNSDÓTTIR, H. B., KNUDSEN, K. L., ABRANTES, F., LEBREIRO, S. & SIRÍKSSON, J. 2006. Climate variability during the last 2000 years in the Tagus Prodelta, western Iberian Margin: benthic foraminifera and stable isotopes. *Marine Micropaleontology*, **59**, 83–103.

BERNÁRDEZ, P., GONZÁLEZ-ÁLVAREZ, R., FRANCÉS, G., PREGO, R., BÁRCENA, M. A. & ROMERO, O. E. 2008*a*. Late Holocene history of the rainfall in the NW Iberian peninsula – evidence from a marine record. *Journal of Marine Systems*, **72**, 366–382.

BERNÁRDEZ, P., GONZÁLEZ-ÁLVAREZ, R., FRANCÉS, G., PREGO, R., BÁRCENA, M. A. & ROMERO, O. E. 2008*b*. Palaeoproductivity changes and upwelling variability in the Galicia Mud Patch during the last 5000 years: geochemical and microfloral evidence. *Holocene*, **18**, 1207–1218.

BIANCHI, G. G. & McCAVE, I. N. 1999. Holocene periodicity in North Atlantic climate and deep-ocean flow south of Iceland. *Nature*, **397**, 515–517.

BLANTON, J. O., ATKINSON, L. P., CASTILLEJO, F. & MONTERO, A. L. 1984. Coastal upwelling of the Rias Bajas, Galicia, northwest Spain, I; hydrographic studies. *Rapports et Procès-verbeaux des Rèunions Conseil International pour l'Exploration de la merMer*, **183**, 179–190.

BOILLOT, G. & MALOD, J. 1988. The North and North-West Spanish Continental Margin: a review. *Revista de la Sociedad Geológica de España*, **1**, 295–316.

BOILLOT, G. & WINTERER, E. L. 1988. Drilling on the Galicia Margin: retrospect and prospect. *In*: BOILLOT, G., WINTERER, E. L. & MEYER, A. W. (eds) *Proceedings of the ODP, Scientific Results (Part B)*, **103**. Ocean Drilling Program, College Station, TX, 809–828.

BOILLOT, G., AUXIETRE, J. L., DUNAND, J. P., DUPEUBLE, P. A. & MAUFFRET, A. 1979. The northwestern Iberian Margin: a Cretaceous passive margin deformed during Eocene. *In*: TALWANI, M., HAY, W. W. & RYAN, W. B. F. (eds) *Deep Drilling Results in the Atlantic Ocean: Continental Margins and Paleoenvironment*. American Geophysical Union, Maurice Ewing Series, **3**, 138–153.

BOILLOT, G., WINTERER, E. L. & MEYER, A. W. (eds) 1987. *Proceedings of the ODP, Initial Reports (Part A)*, **103**. Ocean Drilling Program, College Station, TX.

BOILLOT, G., GIRADEAU, J. & KORNPROBST, J. 1988. Rifting of the Galicia Margin: crustal thinning and emplacement of mantle rocks on the seafloor. *In*: BOILLOT, G., WINTERER, E. L. & MEYER, A. W. (eds) *Proceedings of the ODP, Scientific Results*, **103**. Ocean Drilling Program, College Station, TX, 741–756.

BOESSENKOOL, K. P., BRINKHUIS, H., SCHÖNFELD, J. & TARGARONA, J. 2001. North Atlantic sea-surface temperature changes and the climate of western Iberia during the last deglaciation: a marine palynological approach. *Global and Planetary Change*, **30**, 33–39.

BOND, G., SHOWERS, W. *ET AL.* 1997. A pervasive millennial-scale cycle in North Atlantic Holocene and glacial climates. *Science*, **278**, 1257–1266.

BOND, G. C., SHOWERS, W. *ET AL.* 1999. The North Atlantic's 1–2 kyr climate rhythm: relation to Heinrich Events, Dansgaard/Oeschger Cycles and the Little Ice Age. *In*: CLARK, P. U., WEBB, R. S. & KEIGWIN, L. D. (eds) *Mechanisms of Global Climate Change at Millennial Time Scales*. American Geophysical Union, Geophysical Monograph Series, **112**, 35–58.

BOND, G., KROMER, B. *ET AL.* 2001. Persistent solar influence on North Atlantic climate during the Holocene. *Science*, **294**, 2130–2136.

BOUT-ROUMAZEILLES, V., COMBOURIEU NEBOUT, N., PEYRON, O., CORTIJO, E., LANDAIS, A. & MASSON-DELMOTTE, V. 2007. Connection between South Mediterranean climate and North African atmospheric circulation during the last 50 000 yr BP North Atlantic cold events. *Quaternary Science Reviews*, **26**, 3197–3215.

BURDLOFF, D., ARAÚJO, M. F., JOUANNEAU, J.-M., MENDES, U., MONGE SOARES, A. M. & DIAS, J. M. D. 2008. Sources of organic carbon in the Portuguese continental shelf sediments during the Holocene period. *Applied Geochemistry*, **23**, 2857–2870.

CABANAS, J. M., GONZÁLEZ, J. J. & MARIÑO, J. 1979. Estudio del mejillón y de su epifauna en los cultivos flotantes de la ría de Arousa. II. Observaciones previas sobre la retención de partículas y la biodeposición de una batea. *Boletín del Instituto Español de Oceanografía (V)*, **268**, 45–80.

CARALP, M.-H. 1992. Late glacial to recent deep-sea benthic foraminifera from the Northeastern Atlantic (Cadiz Gulf) and Western

Mediterranean (Alboran Sea): paleooceanographic results. *Marine Micropaleontology*, **13**, 265–289.

CASTRO, C. G., PEREZ, F. F., ÁLVAREZ-SALGADO, X. A. & FRAGA, F. 2000. Coupling between the thermohaline, chemical and biological fields during two contrasting upwelling events off the NW Iberian Peninsula. *Continental Shelf Research*, **20**, 189–210, http://dx.doi.org/10.1016/S0278-4343(99)00071-0

CHAPMAN, M. R. & SHACKLETON, N. J. 2000. Evidence of 550-year and 1000-year cyclicity in North Atlantic circulation patterns during the Holocene. *The Holocene*, **10**, 287–291.

CORREDEIRA, C., ARAÚJO, M. F., GOUVEIA, A. & JOUANNEAU, J.-M. 2005. Geochemical characterization of sediment cores from the continental shelf off the western rias are (NW Iberian Peninsula). *Ciencias Marinas*, **31**, 319–325.

CORREDEIRA, C., ARAÚJO, M. F., GOUVEIA, A. & JOUANNEAU, J.-M. 2009. Sediments of Galician Continental Shelf: elemental composition and accumulation rates. *Journal of Radioanalytical and Nuclear Chemistry*, **281**, 265–268.

COSTAS, S., MUÑOZ SOBRINO, C., ALEJO, U. & PÉREZ-ARLUCEA, M. 2009. Holocene evolution of a rock-bounded barrier-lagoon system, Cíes Islands, northwest Iberia. *Earth Surface Processes and Landforms*, **34**, 1575–1586.

DE VICENTE, G. 2004. Estructura alpina del Antepaís Ibérico. *In*: VERA, J. A. (ed.) *Geología de España*. Sociedad Geológica de España e Instituto Geológico y Minero de España, Madrid, 587–634.

DE VICENTE, G., GINER, J., MUÑOZ-MARTÍN, A., GONZÁLEZ-CASADO, J. M. & LINDO, R. 1996. Determination of present-day stress tensor and neotectonic interval in the Spanish Central System and Madrid Basin, central Spain. *Tectonophysics*, **22**, 405–424.

DE VICENTE, G., CLOETINGH, S. ET AL. 2008. Inversion of moment tensor focal mechanisms for active stresses around the microcontinent Iberia: tectonic implications. *Tectonics*, **27**, TC1009, http://dx.doi.org/10.1029/2006TC002093

DECASTRO, M., ÁLVAREZ, I., VARELA, M., PREGO, R. & GÓMEZ-GESTEIRA, M. 2006a. Miño River dams discharge on neighbor *Galician Rias Baixas* (NW Iberian Peninsula): hydrological, chemical and biological changes in water column. *Estuarine Coastal and Shelf Science*, **70**, 52–62.

DECASTRO, M., DALE, A. W., GÓMEZ-GESTEIRA, M., PREGO, R. & ÁLVAREZ, I. 2006b. Hydrographic and atmospheric analysis of an autumnal upwelling event in the Ria of Vigo (NW Iberian Peninsula). *Estuarine Coastal and Shelf Science*, **68**, 529–537, http://dx.doi.org/10.1016/j.ecss.2006.03.004

DECASTRO, M., GÓMEZ-GESTEIRA, M., ÁLVAREZ, I., CABANAS, J. M. & PREGO, R. 2008. Characterization of fall-winter upwelling recurrence along the Galician western coast (NW Spain) from 2000 to 2005: dependence on atmospheric forcing. *Journal of Marine Systems*, **72**, 145–158, http://dx.doi.org/10.1016/j.jmarsys.2007.04.005

DECASTRO, M., GÓMEZ-GESTEIRA, M., ÁLVAREZ, I. & CRESPO, A. J. 2011. Atmospheric modes influence on Iberian Poleward Current variability. *Continental Shelf Research*, **48**, 333–341, http://dx.doi.org/10.1016/j.csr.2010.03.004

DESPRAT, S., SÁNCHEZ GOÑI, M. F. & LOUTRE, M. 2003. Revealing climatic variability of the last three millennia in northwestern Iberia using pollen influx data. *Earth and Planetary Science Letters*, **213**, 63–78.

DIAS, J. M. A., BOSKI, T., RODRIGUES, A. & MAGALHÃES, F. 2000. Coast line evolution in Portugal since the Last Glacial Maximum until present: a synthesis. *Marine Geology*, **170**, 177–186.

DIAS, J. M. A., GONZALEZ, R., GARCIA, C. & DIAZ-DEL-RIO, V. 2002a. Sediment distribution patterns on the Galicia-Minho continental shelf. *Progress in Oceanography*, **52**, 215–231.

DIAS, J. M. A., JOUANNEAU, J. M. ET AL. 2002b. Present day sedimentary processes on the northern Iberian shelf. *Progress in Oceanography*, **52**, 249–259.

DIAZ DEL RÍO, G., GONZALEZ, N. & MARCOTE, D. 1998. The intermediate Mediterranean water inflow along the northern slope of the Iberian Peninsula. *Oceanologica Acta*, **21**, 157–163.

DIZ, P., FRANCÉS, G., PELEJERO, C., GRIMALT, J. O. & VILAS, F. 2002. The last 3000 years in the Ría de Vigo (NW Iberian Margin): climatic and hydrographic signals. *The Holocene*, **12**, 459–468.

DRAGO, T. 2005. Late Quaternary environmental changes of northern Portuguese estuaries. *In*: FREITAS, M. C. & DRAGO, T. (eds) *Iberian Coastal Holocene Paleoenvironmental Evolution – COASTAL HOPE 2005 Proceedings, Lisbon, Portugal*. Faculdade de Ciências, Universidade de Lisboa, 46–51.

DRAGO, T., FREITAS, C. ET AL. 2006. Paleoenvironmental evolution of estuarine systems during the last 14000 years – the case of Douro estuary (NW Portugal). *Journal of Coastal Research*, **SI39**, 186–192.

DURÁN, R., GARCÍA-GIL, S., DIEZ, R. & VILAS, F. 2007. Stratigraphic framework of gas accumulations in the Ría de Pontevedra (NW Spain). *Geo-Marine Letters*, **27**, 77–88.

EMERY, W. J. & MEINCKE, J. 1986. Global water masses: summary and review. *Oceanologica Acta*, **9**, 383–391.

FERNÁNDEZ-BASTERO, S., VELO, A., GARCÍA, T., GAGO-DUPORT, L., SANTOS, A., GARCÍA-GIL, S. & VILAS, F. 2000. The glaucony from the Galician continental shelf (N.W. Spain): Geochemical markers of the maturity degree. *Journal of Iberian Geology*, **26**, 233–247.

FERRER, O., ROCA, E., BENJUMEA, B., MUÑOZ, J. A., ELLOUZ, N. MARCONI TEAM 2008. The deep seismic reflection MARCONI-3 profile: role of extensional Mesozoic structure during the Pyrenean contractional deformation at the Eastern part of the Bay of Biscay. *Marine and Petroleum Geology*, **25**, 714–730.

FERRIN, A. 2005. *Cenozoic seismic stratigraphy of the SW Galician continental shelf. Comparative study with the Canterbury shelf (SE New Zealand) during the Quaternary*. PhD thesis, Universidad de Vigo, Spain.

FLETCHER, W. J. & SÁNCHEZ GOÑI, M. F. 2008. Orbital- and sub-orbital scale climate impacts on vegetation of the western Mediterranean basin over the last 48 000 yr. *Quaternary Research*, **70**, 451–464.

FRAGA, F. 1981. Upwelling off the Galician Coast, Nortwest Spain. *In*: RICHARDS, F. A. (ed.) *Coastal Upwelling*. American Geophysical Union, Washington, DC, 176–182.

FRAGA, F. 1987. *Oceanografía de la plataforma gallega*. Academia Galega de Ciencias, Santiago de Compostela.

FROUIN, R., FIUZA, A. F. G., AMBAR, I. & BOYD, T. J. 1990. Observations of a poleward surface current off the coasts of Portugal and Spain during winter. *Journal of Geophysical Research*, **95**, 679–691.

GAGO, J., ÁLVAREZ-SALGADO, X. A., GILCOTO, M. & PEREZ, F. F. 2003. Assessing the contrasting fate of dissolved and suspended organic carbon in a coastal upwelling system ('Ria de Vigo', NW Iberian Peninsula). *Estuarine, Coastal and Shelf Science*, **56**, 271–279, http://dx.doi.org/10.1016/S0272-7714(02)00186-5

GALINDO-ZALDÍVAR, J., GONZÁLEZ-LODEIRO, F. & JABALOY, A. 1993. Stress and paleostress in the Betic- Rif cordilleras (Miocene to present). *Tectonophysics*, **227**, 105–126.

GARCÍA-GARCÍA, A., GARCÍA-GIL, S. & VILAS, F. 2005. Quaternary evolution of the Ría de Vigo, Spain. *Marine Geology*, **220**, 153–179.

GARCÍA-GIL, S., VILAS, F., MUÑOZ, A., ACOSTA, J. & UCHUPI, E. 1999. Quaternary Sedimentation in the Ría de Pontevedra (Galicia) Northwestern Spain. *Journal of Coastal Research*, **15**, 1083–1090.

GARCIA-SOTO, C., PINGREE, R. D. & VALDÉS, L. 2002. Navidad development in the southern Bay of Biscay: climate change and swoddy structure from remote sensing and in situ measurements. *Journal of Geophysical Research*, **107**, C8, http://dx.doi.org/10.1029/2001JC001012

GIL, J. & GOMIS, D. 2008. The secondary ageostrophic circulation in the Iberian Poleward Current along the Cantabrian Sea (Bay of Biscay). *Journal of Marine Systems*, **74**, 60–73, http://dx.doi.org/10.1016/j.jmarsys.2007.11.005

GÓMEZ-GESTEIRA, M., MOREIRA, C., ÁLVAREZ, I. & DECASTRO, M. 2006. Ekman Transport along the Galician Coast (NW, Spain) calculated from forecasted winds. *Journal of Geophysical Research*, **111**, C10005, http://dx.doi.org/10.1029/2005JC003331

GONZÁLEZ-ÁLVAREZ, R., BERNÁRDEZ, P., PENA, L. D., FRANCÉS, G., PREGO, R., DIZ, P. & VILAS, F. 2005. Paleoclimatic evolution of the Galician continental shelf (NW of Spain) during the last 3000 years: from a storm regime to present conditions. *Journal of Marine Systems*, **54**, 245–260.

GONZÁLEZ-VILLANUEVA, R., PÉREZ-ARLUCEA, M., ALEJO, I. & GOBLE, R. 2009. Climatic-related factors controlling the sedimentary architecture of a Barrier–Lagoon complex in the context of the Holocene transgression. *In: Proceedings of the 10th International Coastal Symposium ICS 2009, Volume I.* Journal of Coastal Research, Special Issue, **56**, 627–631.

GRIMAUD, S., BOILLOT, G., COLLETE, B., MAUFFRET, A., MILES, P. R. & ROBERTS, D. B. 1982. Western extension of the Iberian–European plate boundary during the early Cenozoic (Pyrenean) convergence: a new model. *Marine Geology*, **45**, 63–77.

GROUPE GALICE. 1979. The continental margin of Galicia and Portugal, acoustic stratigraphy, dredge stratigraphy and structural evolution. *In:* RYAN, W. B. F. & SIBUET, J. C. (eds) *Proceedings of the Deep Sea Drilling Project*, **47**. US Government Printing Office, Washington, DC, 633–662.

GUERREIRO, C., CACHÃO, M. & DRAGO, T. 2005. Calcareous nannoplankton as a tracer of the marine influence on the NW coast of Portugal over the last 14000 years. *Journal of Nannoplankton Research*, **27**, 159–172.

HANEBUTH, T. J. J., BENDER, V. B. *ET AL.* 2007. *Report and First Results of the Poseidon Cruise P342 Galiomar, Vigo-Lisboa, Portugal, August 19th–September 06th, 2006. Distribution Pattern, Residence Times and Export of Sediments on the Pleistocene/Holocene Galician Shelf (NW Iberian Peninsula).* Berichte aus dem Fachbereich Geowissenschaften, Universität Bremen, **255**.

HAQ, B. U., HARDENBOL, J. & VAIL, P. R. 1987. Chronology of fluctuating sea levels since the Triassic (250 million years ago to present). *Science*, **235**, 1156–1167.

HAYNES, R. & BARTON, E. D. 1990. A poleward flow along the Atlantic coast of the Iberian Peninsula. *Journal of Geophysical Research*, **95**, 11 425–11 441.

HERRAIZ, M., DE VICENTE, G. *ET AL.* 2000. The recent (upper Miocene to Quaternary) and present tectonic stress distributions in the Iberian Peninsula. *Tectonics*, **19**, 764–786.

HERRERA, J. L., ROSÓN, G., VARELA, R. A. & PIEDRACOBA, S. 2008. Variability of the western Galician upwelling system (NW Spain) during an intensively sampled annual cycle. An EOF analysis approach. *Journal of Marine Systems*, **72**, 200–217.

HURRELL, J. W. 1995. Decadal trends in the North Atlantic Oscillation: regional temperatures and precipitation. *Science*, **269**, 676–679.

HURRELL, J. W., KUSHNIR, Y. & VISBECK, M. 2001. The North Atlantic Oscillation. *Science*, **291**, 603–605.

HURRELL, J. W., KUSHNIR, Y., OTTERSEN, G. & VISBECK, M. 2010. *The North Atlantic Oscillation: Climate Significance and Environmental Impact.* American Geophysical Union, Geophysical Monograph Series, **134**.

HUTHNANCE, J. M., HUMPHERY, J. D., KNIGHT, P. J., CHATWIN, P. G., THOMSEN, L. & WHITE, M. 2002. Near-bed turbulence measurements, stress estimates and sediment mobility at the continental shelf edge. *Progress in Oceanography*, **52**, 171–194.

JOUANNEAU, J. M., WEBER, O. *ET AL.* 2002. Recent sedimentation and sedimentary budgets on the western Iberian shelf. *Progress in Oceanography*, **52**, 261–275.

KLITGORD, K. D. & SCHOUTEN, H. 1986. Plate kinematics of the central Atlantic. *In:* VOGT, P. R. & TUCHOLKE, B. E. (eds) *The Geology of North America (Vol. M): The Western North Atlantic Region.* Geological Society of America, Boulder, CO, 351–378.

KRISTOFFERSEN, Y. 1978. Sea-floor spreading and the early opening of the North Atlantic. *Earth and Planetary Science Letters*, **38**, 273–290.

KUTZBACH, J. E. 1970. Large-scale features of monthly mean Northern Hemisphere anomaly maps of sea-level pressure. *Monthly Weather Review*, **98**, 708–716.

LANTZSCH, H., HANEBUTH, T. J. J. & BENDER, V. B. 2009a. Holocene evolution of mud depocentres on a high-energy, low-accumulation shelf (NW Iberia). *Quaternary Research*, **72**, 325–336.

LANTZSCH, H., HANEBUTH, T. J. J., BENDER, V. B. & KRASTEL, S. 2009b. Sedimentary architecture of a low-accumulation shelf since the Late Pleistocene (NW Iberia). *Marine Geology*, **259**, 47–58.

LANTZSCH, H., HANEBUTH, T. J. JM. & HENRICH, R. 2010. Sediment recycling and adjustment of deposition during deglacial drowning of a low-accumulation shelf (NW Iberia). *Continental Shelf Research*, **30**, 1665–1679.

LE PICHON, X. & SIBUET, J. C. 1971. Western extension of boundary between European and Iberian plates during the Pyrenean orogeny. *Earth and Planetary Science Letters*, **12**, 83–88.

LE PICHON, X., SIBUET, J. C. & FRANCHETEAU, J. 1977. The fit of the continents around the North Atlantic Ocean. *Tectonophysics*, **38**, 169–209.

LEBREIRO, S. M., FRANCÉS, G. *ET AL.* 2006. Climate change and coastal hydrographic response along the Atlantic Iberian margin (Tagus Prodelta and Muros Ría) during the last two millennia. *Holocene*, **16**, 1003–1015.

LEMOS, R. T. & PIRES, H. O. 2004. The upwelling regime off the west Portuguese coast, 1941–2000. *International Journal of Climatology*, **24**, 511–524, http://dx.doi.org/10.1002/joc.1009

LEÓN, I., MÉNDEZ, G. & RUBIO, B. 2004. Geochemical phases of Fe and degree of pyritization in sediments from Ría de Pontevedra (NW Spain): implications of mussel raft culture. *Ciencias Marinas*, **30**, 585–602.

LØNBORG, C., ÁLVAREZ-SALGADO, X. A., DAVIDSON, K., MARTINEZ-GARCIA, S. & TEIRA, E. 2010. Assessing the microbial bioavailability and degradation rate constants of dissolved organic matter by fluorescence spectroscopy in the coastal upwelling system of the Ría de Vigo. *Marine Chemistry*, **119**, 121–129, http://dx.doi.org/10.1016/j.marchem.2010.02.001

LORENZO, F., ALONSO, A., PELLICER, M. J., PAGÉS, J. L. & PÉREZ-ARLUCEA, M. 2007. Historical analysis of heavy metal pollution in three estuaries on the north coast of Galicia (NW Spain). *Environmental Geology*, **52**, 789–802.

LOUREIRO, J. J., MACHADO, M. L. *ET AL.* 1986. *Monografias hidrológicas dos principais cursos de água de Portugal continental.* Direcção Geral dos Serviços Hidráulicos, Lisbon, Portugal.

MARTINS, V., ROCHA, F., GOMES, C., GOMES, V., JOUANNEAU, J., WEBER, O. & DIAS, J. 2005. Geochemical, textural, mineralogical and micropaleontological data used for climatic reconstruction during the Holocene in the Galician sector of the Iberian continental margin. *Ciencias Marinas*, **31**, 239–308.

MARTINS, V., JOUANNEAU, J.-M., WEBER, O. & ROCHA, F. 2006a. Tracing the late Holocene evolution of the NW Iberian upwelling system. *Marine Micropaleontology*, **59**, 35–55.

MARTINS, V. A., PATINHA, C., SILVA, E. F., JOUANNEAU, J.-M., WEBER, O. & ROCHA, F. 2006b. Holocene record of productivity in the NW Iberian continental shelf. *Journal of Geochemical Exploration*, **88**, 408–411.

MARTINS, V., DUBERT, J. *ET AL.* 2007. A multiproxy approach of the Holocene evolution of shelf-slope circulation on the NW Iberian Continental Shelf. *Marine Geology*, **239**, 1–18.

MARTINS, V., ARAÚJO, M. F. *ET AL.* 2011. Upwelling dominated oceanographic periods in the Ria de Vigo during the late Holocene. *In: Proceedings of the 11th International Coastal Symposium.* Journal of Coastal Research, Special Issue, **64**, 1998–2001.

MASSON, D. G., CARTWRIGHT, J. A., PINHEIRO, L. M., WHITMARSH, R. B., BESLIER, M. O. & ROESER, H. 1994. Compressional deformation at the continent–ocean transition in the NE Atlantic. *Journal of the Geological Society, London*, **15**, 607–613.

MIGUEZ, B. M., VARELA, R. A., ROSÓN, G., SOUTO, C., CABANAS, J. M. & FARIÑA-BUSTO, L. 2005. Physical and biogeochemical fluxes in shelf waters of the NW Iberian upwelling system. Hydrography and dynamics. *Journal of Marine Systems*, **54**, 127–138.

MIL-HOMENS, M., STEVENS, R. L., ABRANTES, F. & CATO, I. 2006. Heavy metal assessment for surface sediments from three areas of the Portuguese continental shelf. *Continental Shelf Research*, **26**, 1184–1205.

MOHAMED, K. J., REY, D., RUBIO, B., VILAS, F. & FREDERICHS, T. 2010. Interplay between detrital and diagenetic processes since the last glacial maximum on the northwest Iberian continental shelf. *Quaternary Research*, **73**, 507–520.

MOHAMED, K. J., REY, D., RUBIO, B., DEKKERS, M. J., ROBERTS, A. P. & VILAS, F. 2011. Onshore-offshore gradient in reductive early diagenesis in coastal marine sediments of the Ria de Vigo, Northwest Iberian Peninsula. *Continental Shelf Research*, **31**, 433–447.

MONTADERT, L., DE CHARPAL, O., ROBERTS, D. G., GENNOC, P. & SIBUET, J.-C. 1979a. Northeast Atlantic passive margins: rifting and subsidence processes. *In:* TALWANI, M., HAY, W. W. & RYAN, W. B. F.

(eds) *Deep Drilling Results in the Atlantic Ocean: Continental Margins and Paleoenvironment*. American Geophysical Union, Maurice Ewing Series, **3**, 154–186.

MONTADERT, L., ROBERTS, D. G., DE CHARPAL, O. & GUENNOC, P. 1979b. Rifting and subsidence of the Northern continental margin of the Bay of Biscay. *In*: MONTADERT, L., ROBERTS, D. G. & AUFFRET, G. A. (eds) *Initial Reports of the Deep Sea Drilling Project*, **48**. US Government Printing Office, Washington, DC, 1025–1060.

MOTA OLIVEIRA, I. B., VALLE, A. J. S. F. & MIRANDA, F. C. C. 1982. Littoral problems in the Portuguese west coast. *In*: *Proceedings of the 18th Coastal Engineering Conference, Volume 3, ASCE, Cape Town, South Africa, November 14–19 1982*. ASCE, Reston, VA, 1950–1969.

MUÑOZ SOBRINO, C., GARCÍA GIL, S., DIEZ, J. B. & IGLESIAS, J. 2007. Palynological characterization of gassy sediments in the inner part of Ría de Vigo (NW Spain). *Geo-Marine Letters*, **27**, 289–302.

MURILLAS, J., MOUGENOT, D., BOILLOT, G., COMAS, M. C., BANDA, E. & MAUFFRET, A. 1990. Structure and evolution of the Galicia Interior Basin (Atlantic western Iberian continental margin). *Tectonophysics*, **184**, 297–319.

NAUGHTON, F., SANCHEZ GOÑI, M. F. *ET AL.* 2007a. Present-day and past (last 25000 years) marine pollen signal off western Iberia. *Marine Micropaleontology*, **62**, 91–114.

NAUGHTON, F., SANCHEZ GOÑI, M. F., DRAGO, T., FREITAS, M. C. & OLIVEIRA, A. 2007b. Holocene changes in the Douro estuary (Northwestern Iberia). *Journal of Coastal Research*, **23**, 711–720.

ODIN, G. S. & LAMBOY, M. 1988. Glaucony from the margin off northwestern Spain. *In*: ODIN, G. S. (ed.) *Green Marine Clays*. Developments in Sedimentology, **45**. Elsevier, Amsterdam, 249–275.

OLIVEIRA, A., ROCHA, F., RODRIGUES, A., JOUANNEAU, J., DIAS, A., WEBER, O. & GOMES, C. 2002a. Clay minerals from the sedimentary cover from the Northwest Iberian shelf. *Progress in Oceanography*, **52**, 233–247.

OLIVEIRA, A., VITORINO, J., RODRIGUES, A., JOUANNEAU, J. M., DIAS, J. A. & WEBER, O. 2002b. Nepheloid layer dynamics in the northern Portuguese shelf. *Progress in Oceanography*, **52**, 195–213.

OSPAR COMMISSION. 2000. *Quality Status Report 2000. Region IV – Bay of Biscay and Iberian Coast*. OSPAR Commission, London.

OTERO, P., RUIZ-VILLARREAL, M., PELIZ, A. & CABANAS, J. M. 2010. Climatology and reconstruction of runoff time series in northwest Iberia: influence in the shelf buoyancy budget off Ría de Vigo. *Scientia Marina*, **74**, 247–266.

PELIZ, A., DUBERT, J., SANTOS, A. M. P., OLIVEIRA, P. B. & LE CANN, B. 2005. Winter upper ocean circulation in the Western Iberian Basin – Fronts, Eddies and Poleward Flows: an overview. *Deep Sea Research Part I: Oceanographic Research Papers*, **2**, 621–646.

PENA, L. D., FRANCÉS, G., DIZ, P., NOMBELA, M. A. & ALEJO, I. 2007. Climatic fluctuations during the Holocene in NW Iberia: high and low latitude linkages. *Climate of the Past Discussions*, **3**, 1283–1309.

PÉREZ, F. F., RIOS, A. F., KING, B. A. & POLLARD, R. T. 1995. Decadal changes of the θ-S relationship of the Eastern North Atlantic Central Water. *Deep Sea Research Part I: Oceanographic Research Papers*, **42**, 1849–1864, http://dx.doi.org/10.1016/0967-0637(95)00091-7

PÉREZ, F. F., PADÍN, X. A. *ET AL.* 2010. Plankton response to weakening of the Iberian coastal upwelling. *Global Change Biology*, **16**, 1258–1267, http://dx.doi.org/10.1111/j.1365-2486.2009.02125.x

PÉREZ-ARLUCEA, M., ÁLVAREZ-IGLESIAS, P. & RUBIO, B. 2007. Holocene evolution of estuarine and tidal-flat sediments in San Simón Bay (Galicia, NW Spain). *In*: *International Coastal Symposium 2007, Gold Coast, Queensland, Australia. Journal of Coastal Research, Special Issue*, **50**, 163–167.

PINGREE, R. D. 1997. The eastern subtropical gyre (North Atlantic): Flow rings recirculations structure and subduction. *Journal of the Marine Biological Association of the UK*, **78**, 351–76.

PINGREE, R. D. & LECANN, B. 1989. Celtic and Armorican slope and shelf residual currents. *Progress in Oceanography*, **23**, 303–338.

PINGREE, R. D. & LECANN, B. 1993. A shallow Meddy (a Smeddy) from the secondary Mediterranean salinity maximum. *Journal of Geophysical Research*, **98**, 20,169–20,185.

PINHEIRO, L. M., WILSON, R. C. L., PENA DOS REIS, R., WHITMARSH, R. B. & RIBEIRO, A. 1996. The western Iberia margin: a geophysical and geological overview. *In*: WHITMARSH, R. B., SAWYER, D. S. & KLAUS, A. (eds) *Proceedings of the ODP, Scientific Results*, **149**. Ocean Drilling Program, College Station, TX, 3–23.

PIRAZZOLI, P. A. 1996. *Sea-level Changes: The Last 20,000 Years*. Wiley, Chichester.

POLLARD, R. T., GRIFFITHS, M. J., CUNNINGHAM, S. A., READ, J. F., PÉREZ, F. F. & RÍOS, A. F. 1996. Vivaldi 1991 – a study of the formation, circulation and ventilation of Eastern North Atlantic Central Water. *Progress in Oceanography*, **37**, 167–192.

PREGO, R. & COBELO-GARCIA, A. 2003. Twentieth century overview of heavy metals in the Galician Rias (NW Iberian Peninsula). *Environmental Pollution*, **121**, 425–452.

PREGO, R., GUZMÁN-ZUÑIGA, D., VARELA, M., DECASTRO, M. & GÓMEZ-GESTEIRA, M. 2007. Consequences of winter upwelling events on biogeochemical and phytoplankton patterns in a western Galician ria (NW Iberian peninsula). *Estuarine Coastal and Shelf Science*, **73**, 409–422.

RAMÍREZ, M., LUCINI, M., PLAZA, J., CARREÑO, A., MARTÍNEZ, J. M. & DE VICENTE, G. 2006. *Determinación de fallas de Primer Orden mediante el análisis integrado de datos geológicos, Colección Documentos I + D, 15.2006*. Consejo de Seguridad Nuclear (CSN), Madrid.

REIMER, P. J., BAILLIE, M. G. L. *ET AL.* 2009. IntCal09 and Marine09 radiocarbon age calibration curves, 0–50,000 years cal BP. *Radiocarbon*, **51**, 1111–1150.

REY, J. 1993. *Relación morfosedimentaria entre la plataforma continental de Galicia y las Rias Bajas y su evolución durante el Cuaternario*. Instituto Español de Oceanografía, Publicaciones Especiales, **17**.

REY, D., LÓPEZ-RODRÍGUEZ, N., RUBIO, B., VILAS, F., MOHAMED, K., PAZOS, O. & BÓGALO, M. F. 2000. Magnetic properties of estuarine-like sediments. The study case of the Galician Rias. *Journal of Iberian Geology*, **26**, 239–257.

REY, D., MOHAMED, K. J., RUBIO, B., BERNABEU, A. & VILAS, F. 2005. Early diagenesis of magnetic minerals in marine transitional environments: geochemical signatures of hydrodynamic forcing. *Marine Geology*, **215**, 215–236.

ROEST, W. R. & SRIVASTAVA, S. P. 1991. Kinematics of the plate boundaries between Eurasia, Iberia and Africa in the North Atlantic from the late Cretaceous to present. *Geology*, **19**, 613–616.

ROSÓN, G., PEREZ, F. F., ÁLVAREZ-SALGADO, X. A. & FIGUEIRAS, F. G. 1995. Variation of both thermohaline and chemical properties in an estuarine upwelling ecosystem: Ria de Arousa; I. time evolution. *Estuarine, Coastal and Shelf Science*, **41**, 195–213, http://dx.doi.org/10.1006/ecss.1995.0061

RUBIO, B., ÁLVAREZ-IGLESIAS, P. & VILAS, F. 2010. Diagenesis and anthropogenesis of metals in the recent Holocene sedimentary record of the Ría de Vigo (NW Spain). *Marine Pollution Bulletin*, **60**, 1122–1129.

RUBIO, B., NOMBELA, M. A. & VILAS, F. 2000. Geochemistry of major and trace elements in sediments of the Ria de Vigo (NW Spain): an assessment of metal pollution. *Marine Pollution Bulletin*, **40**, 968–980.

RUBIO, B., PYE, K., RAE, J. E. & REY, D. 2001. Sedimentological characteristics, heavy metal distribution and magnetic properties in subtidal sediments, Ría de Pontevedra, NW Spain. *Sedimentology*, **48**, 1277–1296.

RUDDIMAN, W. F. & MCINTYRE, A. 1981. The mode and mechanism of the last deglaciation: Oceanic evidence. *Quaternary Research*, **16**, 125–134.

SÁNCHEZ GOÑI, M. F., TURON, J. L., EYNAUD, F. & GENDREAU, S. 2000. European climatic response to millennial-scale changes in the atmosphere–ocean system during the last glacial period. *Quaternary Research*, **54**, 394–403.

SAWYER, D. S., WHITMARSH, R. B., KLAUS, A. SHIPBOARD SCIENTIFIC PARTY. 1994. *Proceedings of the ODP, Initial Reports*, **149**. Ocean Drilling Program, College Station, TX.

SCHMIDT, F., HINRICHS, K.-U. & ELVERT, M. 2010. Sources, transport, and partitioning of organic matter at a highly dynamic continental margin. *Marine Chemistry*, **118**, 37–55.

SCHOUTEN, H., SRIVASTAVA, S. P. & KLITGORD, K. 1984. Iberian plate kinematics: jumping plate boundaries, an alternative to ball-bearing tectonics. *Eos, Transactions of the American Geophysical Union*, **65**, 190.

SEARLE, R. C. & WHITMARSH, R. B. 1978. The structure of King's Trough, northeast Atlantic, from bathymetric, seismic and gravity studies. *Geophysical Journal of the Royal Astronomical Society*, **53**, 259–287.

SIBUET, J. C., MAZÉ, J. P., AMORTILLA, P. & LE PICHON, X. 1987. Physiography and structure of the western Iberian continental margin off Galicia, from sea beam and seismic data. *In*: BOILLOT, G., WINTERER, E. L. & MEYER, A. W. (eds) *Proceedings of the ODP, Initial Reports (Part A)*, **103**. Ocean Drilling Program, College Station, TX, 77–97.

SRIVASTAVA, S. P. & TAPSCOTT, C. R. 1986. Plate kinematics of the North Atlantic. *In*: VOGT, P. R. & TUCHOLKE, B. E. (eds) *The Geology of North America (Vol. M): The Western North Atlantic Region*. Geological Society of America, Boulder, CO, 379–404.

SRIVASTAVA, S. P., VERHOEF, J. & MACNAB, R. 1988. Results from a detailed aeromagnetic survey across the northeast Newfoundland margin. Part I and II. *Marine and Petroleum Geology*, **5**, 306–337.

SRIVASTAVA, S. P., ROEST, W. R., KOVACS, L. C., LEVESQUE, S., VERHOEF, J. & MACNAB, R. 1990. Motion of Iberia since the Late Jurassic: results from detailed aeromagnetic measurements in the Newfoundland basin. *Tectonophysics*, **184**, 229–260.

STUIVER, M. & REIMER, P. J. 1993. Extended ^{14}C data base and revised CALIB 3.0 ^{14}C age calibration program. *Radiocarbon*, **35**, 215–230.

SWIFT, S. A., HOSKINS, H. & STEPHEN, R. A. 1991. Seismic stratigraphy in a transverse ridge, Atlantis II Fracture Zone. *In*: Proceedings of the ODP, Scientific Results, 118. Ocean Drilling Program, College Station, TX, 219–226.

TENORE, K. R., CAL, R. M., HANSON, R. B., LOPEZ-JAMAR, E., SANTIAGO, G. & TIETJEN, J. M. 1984. Coastal upwelling off the Rias Bajas, Galicia, Northwest Spain. II Benthic studies. *Rapports et Proces-Verbaux des Reunions, Conseil International pour l'Exploration de la Mer*, **183**, 91–100.

THOMMERET, M., BOILLOT, G. & SIBUET, J. C. 1988. Structural map of the Galicia margin. *In*: BOILLOT, G. & WINTERER, E. L. (eds) *Proceedings of the ODP, Scientific Results*, **103**. Ocean Drilling Program, College Station, TX, 31–36.

THORNDYCRAFT, V. R. & BENITO, G. 2006. The Holocene fluvial chronology of Spain: evidence from a newly compiled radiocarbon database. *Quaternary Science Reviews*, **25**, 223–234.

THOUVENY, N., MORENO, E., DELANGHE, D., CANDON, L., LANCELOT, Y. & SHACKLETON, N. J. 2000. Rock magnetic detection of distal ice-rafted debries: clue for the identification of Heinrich layers on the Portugues margin. *Earth and Planetary Science Letters*, **180**, 61–75.

TORRES, R. & BARTON, E. D. 2006. Onset and development of the Iberian poleward flow along the Galician coast. *Continental Shelf Research*, **26**, 1134–1153, http://dx.doi.org/10.1016/j.csr.2006.03.009

TORRES, R. & BARTON, E. D. 2007. Onset of the Iberian upwelling along the Galician coast. *Continental Shelf Research*, **27**, 1759–1778, http://dx.doi.org/10.1016/j.csr.2007.02.005

TORRES, R., BARTON, E. D., MILLER, P. & FANJUL, E. 2003. Spatial patterns of wind and sea surface temperature in the Galician upwelling region. *Journal of Geophysical Research*, **108**, 3130, http://dx.doi.org/10.1029/2002JC001361

TRIGO, R. M., POZO-VÁZQUEZ, D., OSBORN, T. J., CASTRO-DÍEZ, Y., GÁMIZ-FORTIS, S. & ESTEBAN-PARRA, M. J. 2004. North Atlantic oscillation influence on precipitation, river flow and water resources in the Iberian Peninsula. *International Journal of Climatology*, **24**, 925–944.

VAIL, P. R., MITCHUM, R. M. & THOMPSON, S. 1977. Seismic stratigraphy and global change of sea level, part 4. *In*: PAYTON, C. E. (ed.) *Seismic Stratigraphy; Aplications to Hydrocarbon Exploration*. American Association of Petroleum Geologists, Memoirs, **26**, 83–97.

VAN ACKEN, H. M. & BECKER, G. 1996. Hydrography and through-flow in the north-eastern North Atlantic Ocean: the NANSEN project. *Progress in Oceanography*, **38**, 297–326.

VARELA, M., PREGO, R., PAZOS, Y. & MORONO, A. 2005. Influence of upwelling and river runoff interaction on phytoplankton assemblages in a Middle Galician Ria and Comparison with northern and southern rias (NW Iberian Peninsula). *Estuarine, Coastal and Shelf Science*, **64**, 721–737, http://dx.doi.org/10.1016/j.ecss.2005.03.023

VÁZQUEZ, J. T., MEDIALDEA, T. *ET AL.* 2008. Cenozoic deformational structures on the Galicia Bank Region (NW Iberian continental margin). *Marine Geology*, **249**, 128–149.

VILAS, F., SOPEÑA, A., REY, L., RAMOS, A., NOMBELA, M. A. & ARCHE, A. 1991. The Corrubedo beach-lagoon complex, Galicia, Spain: dynamics, sediments and recent evolution of a mesotidal coastal embayment. *Marine Geology*, **97**, 391–404.

VILAS, F., BERNABEU, A. M. & MÉNDEZ, G. 2005. Sediment distribution pattern in the Rias Baixas (NW Spain): main facies and hydrodynamic dependence. *Journal of Marine Systems*, **54**, 261–276.

VILAS, F., REY, D., RUBIO, B., BERNABEU, A. M., MÉNDEZ, G., DURÁN, R. & MOHAMED, K. 2008. Los fondos de la Ría de Vigo: Composición, distribución y origen del sedimento. *In*: GONZÁLEZ-GARCÉS, A., VILAS, F. & ÁLVAREZ, X. A. (eds) *Una aproximación integral al ecosistema marino de la Ría de Vigo*. Instituto de Estudios Vigueses, Vigo, 17–50.

VILAS, F., BERNABEU, A. M., RUBIO, B. & REY, D. 2010. Rías, estuarios y llanuras intermareales. *In*: ARCHE, A. (ed.) *Sedimentología. Del proceso físico a la cuenca sedimentaria*. CSIC, Madrid, 619–673.

VITORINO, J., OLIVEIRA, A., JOUANNEAU, J. M. & DRAGO, T. 2002. Winter dynamics on the northern Portuguese shelf. Part 2: bottom boundary layers and sediment dispersal. *Progress in Oceanography*, **52**, 155–170.

WALLACE, J. M. & GUTZLER, D. S. 1981. Teleconnections in the geopotential height field during Northern Hemisphere winter. *Monthly Weather Review*, **109**, 784–812.

WHITMARSH, R. B., BESLIER, M. O. & WALLACE, P. J. 1998. Proceedings of the ODP, Initial Reports, 173. Ocean Drilling Program, College Station, TX.

WILSON, R. C. L., HISCOTT, R. N., WILLIS, M. G. & GRADSTEIN, F. M. 1989. The Lusitanian Basin of west-central Portugal: Mesozoic and Tertiary tectonic, stratigraphic and subsidence history. *In*: TANKARD, A. J. & BALKWILL, H. R. (eds) *Extensional Tectonics and Stratigraphy of the North Atlantic Margins*. American Association of Petroleum Geologists, Memoirs, **46**, 341–362.

WILSON, R. C. L., SAWYER, D. S., WHITMARSH, R. B., ZERONG, J. & CARBONELL, J. 1996. Seismic stratigraphy and tectonic history of the Iberia abyssal plain. *In*: WHITMARSH, R. B., SAWYER, D. S. & KLAUS, A. (eds) *Proceedings of the ODP, Scientific Results*, **149**. Ocean Drilling Program, College Station, TX, 617–633.

WILSON, R. C. L., MANATSCHAL, G. & WISE, S. 2001. Rifting along non-volcanic passive margins: stratigraphic and seismic evidence from the Mesozoic successions of the Alps and western Iberia. *In*: WILSON, R. C. L., WHITMARSH, R. B., TAYLOR, B. & FROITZHEIM, N. (eds) *Non-Volcanic Rifting of Continental Margins: A Comparison of Evidence from Land and Sea*. Geological Society, London, Special Publications, **187**, 429–452.

WOOSTER, W. S., BAKUN, A. & McCLAIN, D. R. 1976. The seasonal upwelling cycle along the eastern boundary of the north Atlantic. *Journal of Marine Research*, **34**, 131–141.

Chapter 9

The Gulf of Cádiz continental shelves

F. J. LOBO[1]*, P. LE ROY[2], I. MENDES[3] & M. SAHABI[4]

[1]*Instituto Andaluz de Ciencias de la Tierra, CSIC-Universidad de Granada,
Avenida de las Palmeras n° 4, 18100 Armilla, Granada, Spain*

[2]*Université Européenne de Bretagne, Université de Brest-CNRS, UMR 6538 Domaines Océaniques,
Institut Universitaire Européen de la Mer, Place Copernic, 29280 Plouzané, France*

[3]*CIMA, Universidade do Algarve, Edifício 7, Campus de Gambelas, 8005-139 Faro, Portugal*

[4]*Laboratoire Géosciences Marines, Département des Sciences de la Terre, Faculté des Sciences, 24000 El Jadida, Morocco*

*Corresponding author (e-mail: pacolobo@iact.ugr-csic.es)

Abstract: The continental shelves fringing the Gulf of Cádiz are shallow-water environments of high scientific interest as they record the complex interplay between competing sedimentary, tectonic and oceanographic processes. These shelves receive the contributions of fluvial sources of varying significance, evolving laterally to shelf segments dominated by a vigorous oceanographic regime. In the long term, the Gulf of Cádiz margins show lateral physiographical compartmentalization related to the structural imprint.

Recent approaches on the Iberian Shelf have focused on the study of shelf sedimentary processes and long-term evolutionary patterns through the application of high-resolution seismic-sequence stratigraphy concepts. The Late Quaternary record exhibits extensive regressive-lowstand shelf-margin deltas mostly comprising nested sequences disposed in a hierarchical pattern. This complex architecture has been attributed to the leading control of orbitally driven, climatic–eustatic changes. As a result, the Iberian Shelf constitutes an excellent location to study the interaction between sedimentary processes, climatic and sea-level changes. Recently collected data in the Moroccan Shelf allow a high-resolution seismic-sequence stratigraphy scheme to be established. Significant latitudinal changes on depositional sequence architecture are driven by differential accommodation led by the reactivation of deformational fronts during the Pleistocene. As a consequence, this southern shelf constitutes a good example of tectonic control on sequence stratigraphy.

The shelves of the Gulf of Cádiz constitute one of the most studied shelf environments of the entire eastern Atlantic façade. Although pioneering studies dealing with the structural setting and geodynamic evolution of the northern margin were completed during the 1970s and 1980s mainly by French researchers (e.g. Malod & Mougenot 1979; Malod 1982), it is during the last 20 years that a boom in geological studies dealing with the Gulf of Cádiz shelves has taken place. In this way, a *Marine Geology* Special Issue devoted to the Spanish sector of the Iberian Shelf (Table 9.1) highlighted the importance of this margin and, by extension, of its continental shelf, located in a key area near the Atlantic–Mediterranean boundary (Maldonado & Nelson 1999*b*). As a consequence of these efforts, we now have a better knowledge of the shelf tectonic and stratigraphic record, and of the main sedimentation processes active on this shelf (Oliveira *et al.* 1992; Maldonado *et al.* 2003; Martín-Serrano *et al.* 2005). Concerning the Moroccan Shelf (Table 9.1), initial work was devoted mainly to describing the lithospheric structure of the margin and the mechanism of the opening of the central Atlantic Ocean (e.g. Von Rad *et al.* 1979; Hinz *et al.* 1982; Le Roy *et al.* 1997; Contrucci *et al.* 2004; Sahabi *et al.* 2004). Most recent studies have focused on the active deformation caused by the relative movements of Africa and Eurasia (Gutscher *et al.* 2009; Maad *et al.* 2010). These different studies allow the structural segmentation of the Moroccan Shelf to be highlighted, and a link to be established between the onshore and offshore observations conducted along the coastal area of the Rif Cordillera.

The continental shelves of the Gulf of Cádiz deserve attention for a number of reasons. First, they are located in a seismically active tectonic setting, close to the Eurasia–Africa plate boundary and influenced by the westward displacement of the Betic–Rif Arc. These tectonic processes have influenced the geological evolution of this margin, with important spatial and temporal variations in subsidence. On the shelf, the complex tectonic setting interacting with laterally changing sediment supplies has

generated distinct shelf architectures at different timescales. In the northern Iberian Margin the essentially starved shelves in the proximity of the Cape São Vicente and the Strait of Gibraltar are separated by the prograding shelves off the Guadiana and Guadalquivir rivers. A similar geometry is observed along the Morocan Margin where the starved Mesetan and Western Rifan shelves bound the prograding Rharb Margin off the Sebou River. In addition, Quaternary shelf sedimentation dynamics have also been influenced by an active oceanographic regime, controlled by a vigorous exchange of water masses through the Strait of Gibraltar, and by the influence of rapid, high-frequency sea-level changes, the seismic stratigraphy record of which is particularly outstanding and well documented on the Iberian Shelf. The existence of sediment sources of different sizes, including two of the most major rivers in the Iberian Peninsula, and the main rivers of the Moroccan Atlantic Coast, has generated very diverse coastal–shelf depositional systems in response to the Holocene stabilization of sea level and the establishment of active shallow-water dynamics. Finally, several human interventions in the river drainage basins and the coastal zone have modified these shallow-water processes during the last centuries.

General setting

The Gulf of Cádiz is a crescent-shaped basin located in the NE Atlantic Ocean (Fig. 9.1). Its northern margin (Iberian Shelf) extends off the coasts of the SW Iberian Peninsula, from Cape São Vicente to the west to the Strait of Gibraltar to the east (Fig. 9.2). The following sectors are distinguished in the Iberian Shelf: (a) the southern Algarve Shelf is the westernmost sector between Cape São Vicente and the Guadiana River, which marks the Portuguese–Spanish border at the coastline. This shelf is the seaward prolongation of the Mesozoic Algarve Basin.

From: CHIOCCI, F. L. & CHIVAS, A. R. (eds) 2014. *Continental Shelves of the World: Their Evolution During the Last Glacio-Eustatic Cycle.*
Geological Society, London, Memoirs, **41**, 109–130. http://dx.doi.org/10.1144/M41.9

Table 9.1. *Summary table of the main descriptive parameters of the Iberian and Moroccan shelves in the Gulf of Cádiz*

	Iberian Shelf	Moroccan Shelf
Shelf length (km)	350	370
Shelf width	From a few km to 45 km	From a few km to 45 km
Tidal conditions	Mesotidal	Mesotidal
Wave conditions	Dominance (50%) of southwesterly and westerly waves	Dominance of westerly waves
Currents (m s^{-1})	Up to 1	Up to 1
Dominating process	Tides and currents	Tides and currents
Shelf break depth (m)	110–150	150
Sedimentation type	Siliciclastic	Siliciclastic
Modern/relict/palimpsest	Inner–middle shelf modern sediments, outer shelf relict and palimpsest	Inner–middle shelf modern sediments, outer shelf relict and palimpset
Tectonic trend over the last glacial cycle	Local tectonic activity in the Algarve Shelf and off the Guadalquivir River	Local tectonic activity along the northern and southern borders of the Rharb Shelf

(b) The Huelva Shelf extends off the coasts between the Guadiana and Guadalquivir rivers, constituting the seaward prolongation of the foreland Guadalquivir Basin. (c) The western Cádiz Shelf is the SE shelf sector between the Guadalquivir River and the Strait of Gibraltar. This shelf is the transition to the prolongation to the west of the Betic Cordillera. For the sake of simplicity, these three shelf sectors will be referred in the text as the Algarve, Huelva and Cádiz shelves, respectively (Fig. 9.2).

The southern margin of the Gulf of Cádiz extends off the NW Moroccan coasts from Cape Spartel, marking the limit of the Strait of Gibraltar to the north, to El Jadida city to the south (Fig. 9.2). Three structural domains are distinguished from north to south along a 500 km-long coastline: (a) the Western Rifan

Shelf occurs adjacent to the Rif Belt, and is characterized by a pronounced vergence of thrust nappes towards the south and WSW resting on the External Zones of the Betic–Rif Cordillera. (b) The Rharb Shelf is the western termination of the Rharb Basin (Chalouan *et al.* 2008). This shelf extends from Rabat to Moulay Bouselham cities. This domain is divided into a foredeep basin, the Rharb Basin *sensu stricto* parallel to the southern edge of the Rif and a shallower foreland basin constituting the southern hinge of the western Meseta domain. (c) The Mesetan Shelf extends between the cities of El Jadida and Rabat. This shelf lies on the western Meseta structural domain, considered as essentially stable since the central Atlantic rifting during Late Triassic–Early Liassic times (Fig. 9.2).

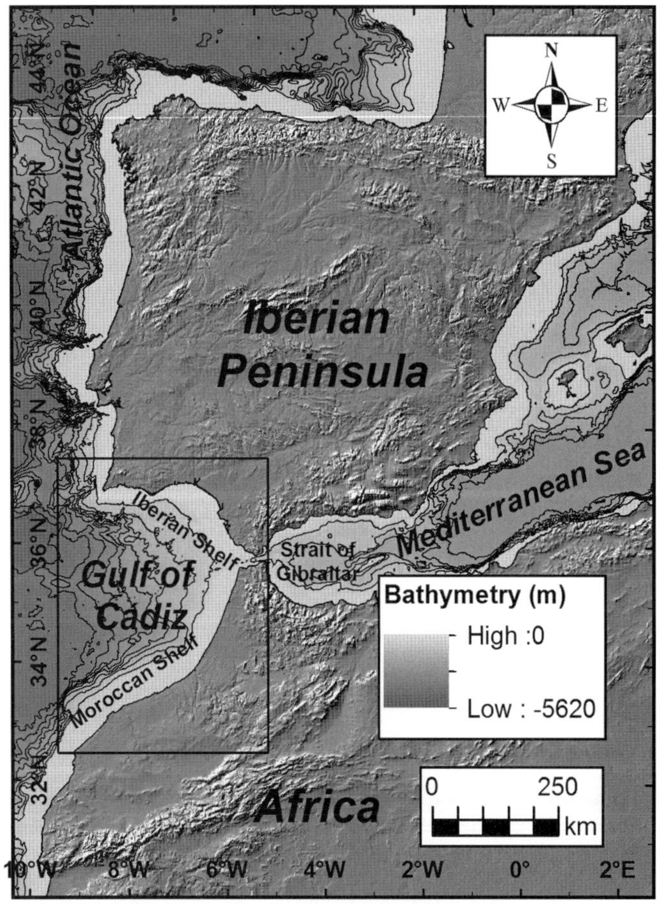

Oceanography

The Iberian Shelf

Wave climate can be classified as being of medium to low energy. Dominant waves approach from the west and SW (about 71% of occurrences); SE waves are less frequent (about 23% of occurrences) but more energetic (Costa *et al.* 2001). The average offshore significant wave height is 0.9 m (with maximum values of 4.9 m and minimum values of 0.14 m), with an average peak period of 8 s (Costa 1994). Under such wave-climate conditions, very coarse sands can be remobilized down to a depth of 30 m during storms, whereas finer sediments (coarse silt) can reach to water depths of up to 50 m (Mendes *et al.* 2004). The coast is mesotidal with a tidal range of between 0.2 and 3.8 m (e.g. González *et al.* 2001). On the shelf, bidirectional tidal currents show an increasing intensity towards the Strait of Gibraltar, with measured velocity values of around 1 m s^{-1} (Besio & Losada 2008).

The surface circulation is governed by a branch of the larger-scale Portuguese–Canary Eastern Boundary Current, which leads to a general anticyclonic circulation due to the SE movement of the Surface Atlantic Water (SAW), which is the upper part of the North Atlantic Surface Water (NACW) modified by air–sea interactions (Criado-Aldeanueva *et al.* 2006*a*; García-Lafuente *et al.* 2006). Towards the Strait of Gibraltar, the current follows two paths (Fig. 9.2). One branch enters the Mediterranean Sea through the strait (i.e. Atlantic Inflow) to balance evaporation and buoyancy losses within this sea and countering the outflow of Mediterranean Water (Criado-Aldeanueva *et al.* 2006*a, b*). The other branch turns southwards, forming a meander that joins the Canary Current (Criado-Aldeanueva *et al.* 2006*b*). In addition, Warm Shelf Waters (WSW) may occur on the shelf, when the SAW is influenced by continental freshwater inputs, generating a warmer and fresher water mass (Criado-Aldeanueva *et al.* 2006*b*). A nearly persistent upwelling, considered a prolongation

Fig. 9.1. Geographical location of the Gulf of Cádiz within the general context of the NE Atlantic Ocean.

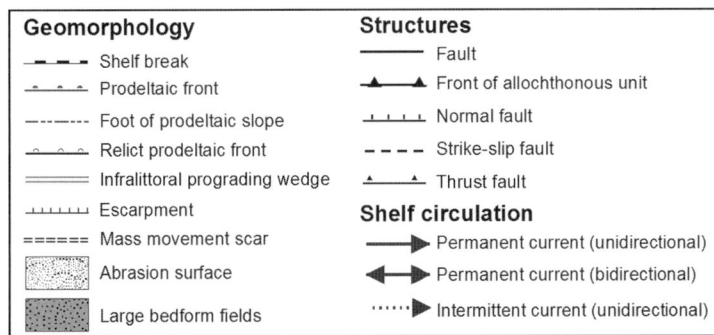

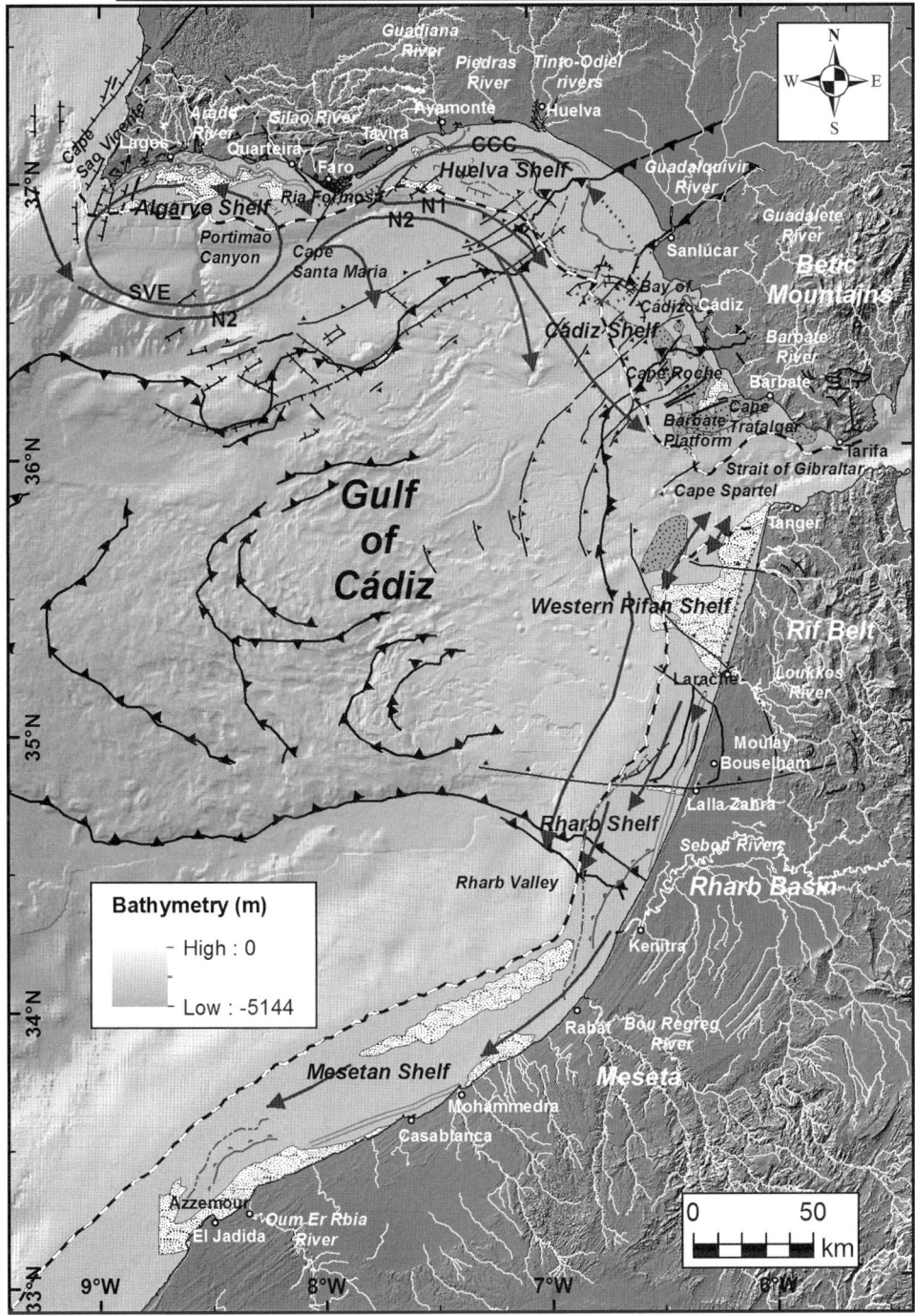

Fig. 9.2. Shelf geomorphology, main tectonic structures and current patterns in the Gulf of Cádiz. The bathymetric information derives from the GEBCO database and from the 250 m resolution SWIN digital map published by Zitellini *et al.* (2009) and available at http://www.sciencedirect.com/science/article/pii/S0012821X0800753X. Geomorphological information from the Iberian Shelf modified from Martín-Serrano (2005). Tectonic features are synthesized from several sources: Gràcia *et al.* (2003), Medialdea *et al.* (2009) and Maad *et al.* (2010). Shallow circulation information modified from García-Lafuente *et al.* (2006) and Criado-Aldeanueva *et al.* (2006*a*, *b*). N1, shelf-break (Huelva) front; N2, branch of the Portuguese–Canary Eastern Boundary Current; SVE, cyclonic eddy off Cape São Vicente; CCC, coastal counter-current.

of the western coastal upwelling system, occurs around Cape São Vicente (Fiúza *et al.* 1982; Fiúza 1983; Vargas *et al.* 2003; Criado-Aldeanueva *et al.* 2006*b*; García-Lafuente & Ruiz 2007).

Landwards of the general circulation, two cyclonic cells have been detected (García-Lafuente *et al.* 2006). A cyclonic circulation occurs to the east of Cape Santa Maria, linked to the westward movement of a counter-current composed of WSW over the inner shelf, and whose generation is associated with coastal processes. An additional cyclonic circulation is identified between capes São Vicente and Santa Maria, in relation to open-sea processes (García-Lafuente *et al.* 2006; García-Lafuente & Ruiz 2007) (Fig. 9.2).

Meteorological conditions play a role in controlling shelf circulation patterns in the Gulf of Cádiz. Under westerlies' dominance, coastal upwelling of NACW is favoured on the Algarve Shelf (Vargas *et al.* 2003; Criado-Aldeanueva *et al.* 2006*a*; García-Lafuente & Ruiz 2007). The upwelled waters tend to move eastwards following the main current until 7.5°W, where they form a filament (Criado-Aldeanueva *et al.* 2006*b*). Under easterlies' dominance, however, the upwelling is confined to the west of Cape Santa Maria, and the WSW composing the coastal counter-current moves from east to west, displacing the upwelled waters offshore and linking the two cyclonic cells. The WSW often turns poleward around Cape São Vicente (e.g. Relvas & Barton 2002, 2005; Criado-Aldeanueva *et al.* 2006*a*; García-Lafuente *et al.* 2006; García-Lafuente & Ruiz 2007; Relvas *et al.* 2007; Teles-Machado *et al.* 2007).

Internal waves are another significant oceanographic phenomena documented in the Gulf of Cádiz. These waves may reach speeds of 0.56 m s^{-1} (Global Ocean Associates 2004), and they have been proposed to affect the distribution of outer-shelf sediments (Nelson *et al.* 1999).

The Moroccan Shelf

In the Moroccan Shelf, prevailing waves propagate towards the SE. During storm events, waves may reach heights of 7–9 m and transport medium-sized sands up to 100 m water depths (Jaaidi 1993). As in the rest of the Gulf of Cádiz, the tidal regime is mesotidal and induces weak currents able to mobilize silts and fine sands with flood currents preferentially directed to the north and ebb currents to the south (Jaaidi 1993).

At a larger scale, the surface circulation of the Moroccan Shelf derives from the general anticyclonic circulation in the Gulf of Cádiz. The southern branch is directed towards the SSW along the Moroccan coast, with average velocities of about 30 cm s^{-1} (Criado-Aldeanueva *et al.* 2006*b*) (Fig. 9.2).

Upwelling phenomena have been described in the Rifan and Rharb shelves, with seasonal fluctuations related to the NE trade winds. Upwelled waters emanate from the NACW at water depths of 100–600 m (Treguer & Le Corre 1979); however, their relative contribution to the shelf circulation remains poorly documented. Furthermore, the comparison of upper-slope (700–1000 m water depths) microfossil assemblages of the SW Iberian and NW Atlantic Morrocan margins shows major differences over the last 30 kyr. This contrast is especially marked during climatic deteriorations linked to Heinrich events, suggesting the existence of a strong hydrological frontal system in the vicinity of the Gulf of Cádiz (Penaud *et al.* 2010).

Climate

In the Iberian Margin, the climate is Mediterranean with Atlantic influence. Average temperatures are of 17–19 °C, and maximum summer temperatures range between 35 and 40 °C. Average annual rainfall is below 600 mm, with two maxima in November–December and spring (Cendrero *et al.* 2004). Episodic floods play a major role in the supply of sediment from the river basin to the continental shelf (Morales 1997; Portela 2006). The Iberian Margin is also influenced by the North Atlantic Oscillation (NAO), which is the leading pattern of weather and climate variability over the northern hemisphere (e.g. Hurrell 1995). A negative phase of the NAO in the Iberia Peninsula results in more rainfall, and subsequent flooding in the river basin during winter months, also confirmed for the Guadiana River basin (Dias *et al.* 2004; Trigo *et al.* 2004). The climate is also Mediterranean along the NW Atlantic Morroco, although the climate of the western Rifan Domain is subhumid, with an average annual rainfall of 900 mm at the location of Tanger (Sauvage 1963).

Sediment sources

Fluvial supply is moderate on the southern Algarve Margin. In the western Algarve Coast between Cape São Vicente Cape and Quarteira, Mesozoic cliffs are progressively substituted to the east by non-consolidated Miocene formations (Vanney & Mougenot 1981), only interrupted by small fluvio-estuarine systems (e.g. Arade Estuary) (Moita 1986). The erosion of these non-consolidate cliffs acts as a source of coarse sediments to the shelf and the Ria Formosa barrier islands (Dias & Neal 1992; Rosa *et al.* 2013). The Ria Formosa barrier–lagoon system is composed of five barrier islands and two peninsulas extending about 50 km between Quarteira and Tavira. Their origin was linked to spits over a platform subsequently transformed into islands as the platform was transgressed (Pilkey *et al.* 1989). The eastern Algarve coast between Tavira and the Guadiana River mouth is composed of low-lying sandy beaches with some small estuaries (e.g. the Gilão-Almargem Estuary).

The Huelva Margin receives the most significant fluvial supplies to the northern Gulf of Cádiz, from the Guadiana, Piedras, Tinto-Odiel and Guadalquivir rivers. The fluvial discharge of these rivers is very irregular, with significant seasonal and interannual variability (e.g. Borrego *et al.* 1995). The Guadiana River is the second river in terms of discharge into the Gulf of Cádiz. Peak discharges occur in winter months (values higher than 3000 m^3 s^{-1}), and very low discharges in summer months (10 m^3 s^{-1}). Mean annual discharges also show considerable variability (80–140 m^3 s^{-1}) (Palanques *et al.* 1995; Morales 1997). Estimates of annual values of sediment discharge are of 57.90 × 10^4 m^3 a^{-1} for the average suspended load and 43.96 × 10^4 m^3 a^{-1} for bedload between 1946 and 1990 (Morales 1997). For the winter of 2000–1, an exceptionally rainy year compared to the previous 10 years, the estimated sand on to the inner shelf was about 7.5 × 10^5–9.5 × 10^5 m^3 (González *et al.* 2007), about twice the average estimated by Morales (1997). Towards the river mouth, the Guadiana River has formed a narrow, basement-controlled estuary that widens in the most seaward portion, where large marsh systems and tidal channels are protected by several coastal spits (Morales *et al.* 1994; Morales 1997; González *et al.* 2001). The coastal stretch between the Guadiana and Guadalquivir rivers is fed by the Piedras and Tinto-Odiel rivers, with low mean annual discharges (between 1 and 10 m^3 s^{-1}) (Borrego *et al.* 1995; Palanques *et al.* 1995). The Guadalquivir River, however, is a fluvial system with the highest mean water discharge (164 m^3 s^{-1}) to the Gulf of Cádiz (Palanques *et al.* 1995). This river supplies sediments from its Neogene basin, but it also drains the Palaeozoic materials of the Iberian Massif (Sierra Morena Mountains) and the the Internal Zones of the Betic Mountains to the north and south of the Guadalquivir Basin, respectively (Gutiérrez-Mas *et al.* 1994, 1996; López-Galindo *et al.* 1999).

To the south of the Guadalquivir River, the Cádiz coast exhibits a main NNW–SSE trend with small east–west coastal stretches. The main coastal forms are extensive beaches interrupted by small cliffs, whose erosion provides most of the sediment supply (López-Galindo *et al.* 1999). The main rivers are the Guadalete River, which debouches into the Bay of Cádiz, and the Barbate River. Both rivers drain the Guadalquivir Depression and the western units of the Betic Cordillera, most notably the Gibraltar Flysch (Gutiérrez-Mas *et al.* 1996; López-Galindo *et al.* 1999). To the south of Cape Trafalgar, the coast is dominated by abrupt cliffs carved into the Betic Mountains.

The second regional sediment source is the littoral drift, which produces an east- and SE-directed sandy sediment transport from the southern Portuguese coast towards the eastern portion of the Gulf of Cádiz of around 180 × 10^3 m^3 a^{-1} (Morales 1997; González *et al.* 2001).

The Morrocan coastline is surrounded by cliffs exposed to the north of Larache and to the south of El Jadida, and carved, respectively, into Neogene and Late Cretaceous formations. Elsewhere, the coastline exhibits alternating massive sand dunes and small bluffs cut into Quaternary deposits and including marine and aeolian units (Saiidi 1979; Aberkan *et al.* 1993; Plaziat *et al.* 2008). Numerous river mouths spread out along these segments and locally develop estuarine environments (so-called oueds) at the front of larger rivers (the Loukkos, Sebou, Bou Regreg, Oum Er Rbia). These estuarine environments are protected by south-trending spits composed of massive sands.

Fluvial supplies show significant differences along the NW Atlantic Moroccan coast. The 600 km-long Oum Er Rbia River located at the southern tip of the Mesetan Shelf is the only permanent Moroccan river. Its 32 000 km^2 drainage basin extends from the middle Atlasic Chain to the Atlantic Ocean. To the north, the Bou Regreg River (9700 km^2 drainage basin and average annual sediment discharge of 1.2×10^6 m^3 a^{-1}) also provides sediment to the Mesetan Shelf. The Sebou River is fed by a 40 000 km^2 drainage basin extending across the Rif Cordillera, the Meseta and the Middle Atlas, and which debouches in the Rharb Shelf. This river supplies the highest water and sediment discharges along the entire Moroccan Shelf. Its water discharge ranges from 3 to 10 000 m^3 s^{-1} and the mechanical denudation of its drainage basin provides a sediment yield of 2000 t km^{-2} a^{-1}, with a suspended load of about 10.6×10^6 m^3 a^{-1} reaching the estuary (Snoussi *et al.* 1990; Jaaidi 1993). The main sediment contributor to the Western Rifan Shelf is the Loukkos River (with a 3750 km^2 drainage basin).

The dominant littoral drift generated by prevailing SE waves is directed southwards. Nevertheless, local seasonal conditions are able to reverse the orientation of sediment transport along the coastline, as shown by studies based on heavy mineral distribution between the cities of Casablanca and Rabat (Niftah *et al.* 2005).

Tectonic framework and geological background

The Gulf of Cádiz is located close to the eastern end of the Azores–Gibraltar Fracture Zone, which is part of the Eurasia–Africa plate boundary and regarded as a diffuse deformation zone (Sartori *et al.* 1994). This seismically active area has been under convergence since the Miocene, and it is the source of large earthquakes and tsunamis (e.g. the 1755 Lisbon earthquake) (Gràcia *et al.* 2003). During the Triassic, passive margins were created in southern Iberia and Africa as a consequence of the break-up of Pangaea and subsequent rifting (Baldy *et al.* 1977; Maldonado *et al.* 1999). The Mesozoic evolution was related to the establishment of the Azores–Gibraltar Fracture Zone as a major transcurrent boundary and very active rifting in the Gulf of Cádiz (Fig. 9.3) (Maldonado & Nelson 1999a).

The relative motions of Eurasia and Africa along the Azores–Gibraltar Fracture Zone have been complex since the Tertiary, involving strike-slip right-lateral motion (Buforn *et al.* 1988) with a component of slow oblique convergence between both plates leading to compressional deformation. An active margin was established from the Late Eocene to the Early Miocene (Fig. 9.3), as Iberia became a plate independent of Africa and the Gulf of Cádiz was part of the extensive area of deformation located along a transcurrent fault system between both plates (Maldonado *et al.* 1999). In the Algarve Basin, the convergence between Iberia and Africa caused a tectonic inversion of Mesozoic extensional structures (Baldy *et al.* 1977); in addition, salt tectonic activity occurred in the Algarve Margin owing to the existence of a thick evaporitic unit that enabled the existence of synchronous extensional and compressional structures (Lopes *et al.* 2006). The compressive regime during the Late Tortonian and Messinian caused uplifting and closure of the Betic and Rif straits. However, the most significant event was the emplacement off the Strait of Gibraltar of a giant Allochthonous Unit (or olistostrome in

early studies) during the Late Miocene as a consequence of the north–south to NNW–SSE convergence between Iberia and Africa, and the displacement to the west of the Betic–Rifean Arc (Fig. 9.2) (Roberts 1970; Torelli *et al.* 1997; Medialdea *et al.* 2004). Some alternative interpretations consider the Allochthonous Unit as an accretionary wedge emplaced as a consequence of eastward subduction beneath Gibraltar (Gutscher *et al.* 2002). The giant Allochthonous Unit induced the formation of the Guadalquivir Basin to the NW through a WNW–ESE-orientated thrust fault with a strike-slip component (Medialdea *et al.* 2004).

The stress field changed to north–south at the end of the Messinian, and pull-apart basins were created under a transtensional regime, reconnecting the Atlantic and the Mediterranean (Maldonado & Nelson 1999a). During the Plio-Quaternary, faults and folds underwent several reactivations under a dominant transpressive regime (Rodero *et al.* 1999; Zitellini *et al.* 2009; Carvalho *et al.* 2012). The dominant tectonic regime at present is characterized by moderate tectonic subsidence and local transpressive tectonics. Abundant geophysical evidence indicates the recent (up to post-glacial times) activity of faults and folds (Lobo *et al.* 2003b), in relation to the reactivation of deeper structures, such as Plio-Quaternary listric faults in the Cádiz Shelf, and basement faults in the Algarve Shelf. In most of the cases, the recent tectonic activity is linked to halokinetic movements (Vázquez *et al.* 2010).

In the Moroccan Shelf, Quaternary tectonic events mainly occurred along the Western Rifan Shelf. The deformation did not involve the reactivation of the blind front of the Prerifan Nappe but mainly affected the segment located between the Lalla Zahra Ridge (south of Moulay Bouselham city) and the outcropping Neogene nappes just to the north of Larache (Fig. 9.2). The deformation includes an arc-shaped succession of extensional faults and pressure ridges with a westward open concavity, interpreted as the upslope proximal part of a gravity spreading lobe. The gravitational mechanisms were induced by the reactivation of the large frontal Prerifan structures and the offshore continuation of the Lalla Zahra coastal structure attributed to a major Quaternary transpressive boundary. Such style of deformation is compatible with recent GPS measurements showing a current displacement of the central Rifan block with respect to the stable African block toward the SSW at an average speed of 3.4 mm a^{-1} (Tahayt *et al.* 2008).

Uplift/subsidence of the margin

In the Iberian Shelf, the offshore extension of the Guadalquivir Foreland Basin (Huelva Shelf) is the main subsiding sector (Riaza & Martínez del Olmo 1996). There, depositional sequences stack vertically and attain significant thickness. Laterally, the stacking pattern is less aggradational, linked to tectonic tilting on the Algarve Shelf and to intense sequence deformation (e.g. diapir uplift) and shallow basement occurrence in the SE (Lobo *et al.* 1999; Nelson *et al.* 1999).

The subsidence regime has undergone significant temporal variations, mostly controlled by the emplacement of the Allochthonous Unit during the Late Miocene (Fig. 9.3). As a consequence, sediment thickness depicts a very irregular distribution, with main post-Allochthonous Unit depocentres reaching more than 4000 m in thickness (Maldonado & Nelson 1999a). Subsidence rates were very high during the Early Pliocene owing to the development of extensional collapses in the Allochthonous Unit nappes (Medialdea *et al.* 2004). During the Late Pliocene, subsidence rates decreased substantially (Maldonado & Nelson 1999a). On the Algarve Margin, spatial and temporal changes of subsidence were mainly due to margin flexure (Lopes *et al.* 2006a).

Along the Moroccan Margin, the Rharb Shelf – containing more than 5000 m of post-Palaeozoic deposits – is the main subsiding sector. This shelf underwent high subsidence rates in the Tortonian

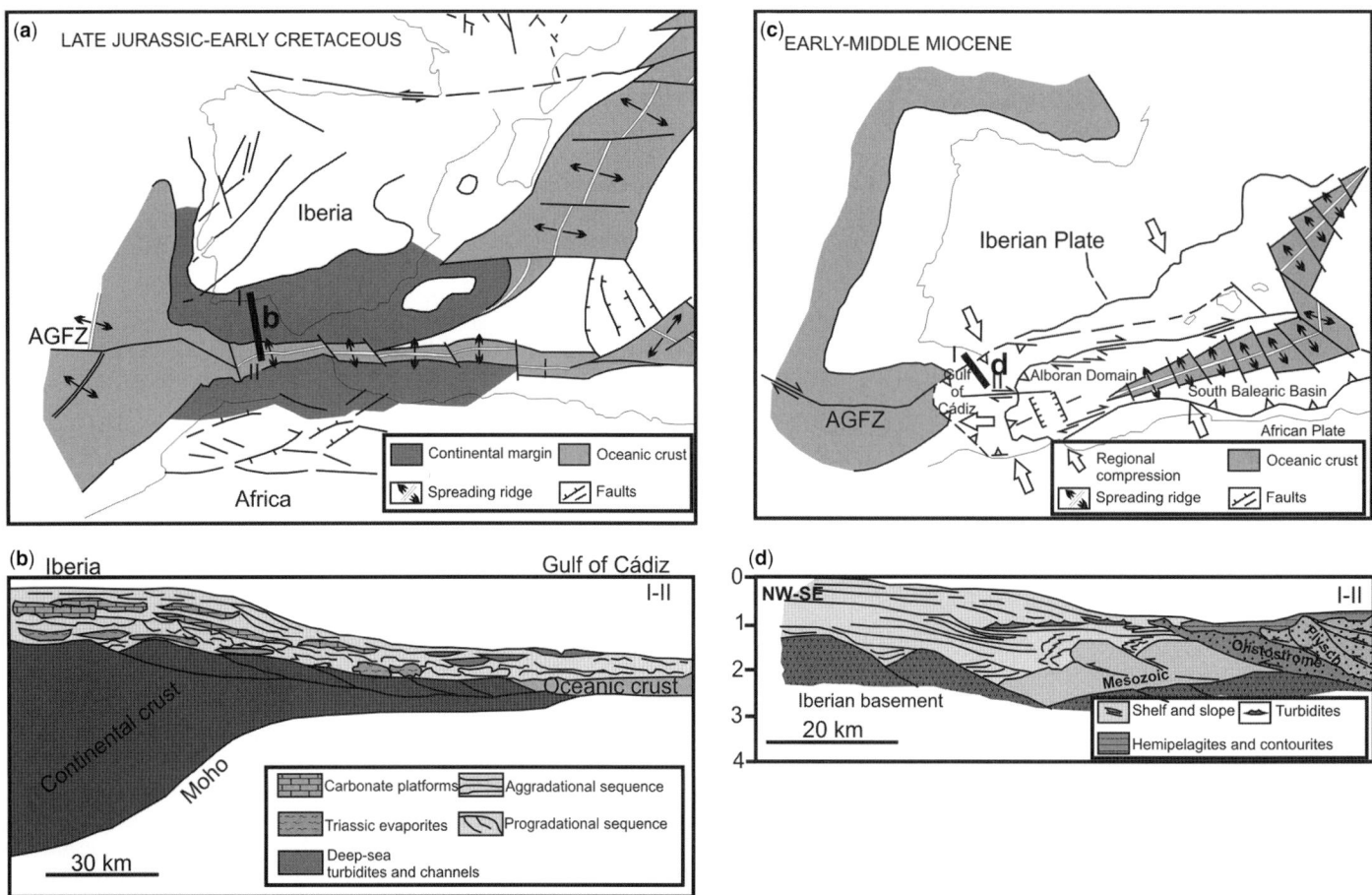

Fig. 9.3. Long-term geological evolution of the Gulf of Cádiz: (**a**) geological setting during the Late Jurassic and Early Cretaceous; (**b**) interpretative cross-section during the Late Jurassic and Early Cretaceous; (**c**) geological setting during the Early–Middle Miocene; and (**d**) interpretative cross-section during the Early–Middle Miocene. Modified from Maldonado & Nelson (1999*a*).

due to the stacking of pre-Rif nappes. As in the Iberian Margin, the subsidence regime underwent spatial variations as the result of coeval compression and extensional collapses of the nappes during the Late Miocene. During the Pliocene and Pleistocene, the highest subsidence occurred in the southern part of the Rharb Shelf. There, the present-day coastline acted as a hinge zone that separated uplifted alluvial plain deposits from subsiding shelf-margin siliciclastic wedges (Flinch & Vail 1999). In contrast, the reactivation of large frontal Prerifan structures in the northern Rharb Shelf induced transpressional, compressional and gravitational processes during the Plio-Pleistocene (Flinch 1993; Maad *et al.* 2010).

Uplifting was dominant both in the north and in the south of the Rharb Shelf. In the north, the Western Rifan Shelf underwent a general uplift during the Pleistocene, as inferred from differential elevations of ancient beaches and terraces along the Atlantic coast (Cadet *et al.* 1977). Vertical displacements of 25–100 m occurred between Larache and Tanger during the last 780 kyr. The low preservation of marine deposits off El Jadida and the abundant fluvial terraces hanging along the banks of the Oum Er Rbia River also suggest a widespread uplift of the Mesetan Shelf and the adjacent coastal plain during the Quaternary (Ouadia 1998).

Shelf description: physiography and sediments

Shelf physiography

The Iberian Shelf shows marked changes in width according to the significance of depositional forms and the structural setting

(Nelson *et al.* 1999; Lobo 2000; Maldonado *et al.* 2003). The Algarve Shelf is narrow but has laterally variable width (Fig. 9.2). Its minimum width (5–7 km) occurs off Cape Santa Maria and it widens both to the west (25 km wide at around 8°W) and to the east (24 km wide off the Guadiana River) (Moita 1986). The Algarve Shelf is steep, with an average value of 0.4° and maximum values of up to 1.5° (Roque 1998). The shelf break is located at water depths of 110–150 m (Vanney & Mougenot 1981). The western part of the Algarve Shelf between Cape São Vicente and Quarteira is covered by erosive and tectonic features, such as Mesozoic rocky outcrops and abrasion platforms (Fig. 9.2). Depositional morphologies are more frequent landwards, where a steep slope outlines a shallow-water wedge laterally connected to a prodeltaic feature off the Arade River. Prograding wedges related to palaeo-valleys and ancient coastal deposits have been described seawards (Vanney & Mougenot 1981). The central Algarve Shelf between Faro and Tavira shows a well-developed Infralittoral Prograding Wedge (IPW), that is narrow in the dip direction (a few kilometres) but extensive in the along-shelf direction (>10 km long) (Hernández-Molina *et al.* 2000*a*). The eastern Algarve Shelf is marked by the distal termination of a poorly developed inner-shelf prodelta connected to the Guadiana River; seawards, several wave-cut terraces occur at mid depths (50–84 m) (Roque 1998). The rest of the eastern Algarve Shelf exhibits a relatively smooth surface generated by muddy belts (Fig. 9.2).

In the Huelva Shelf, shelf width increases to the east, attaining more than 30 km off the Guadalquivir River. In the Cádiz Shelf, maximum shelf width (41 km) occurs off Cape Roche. Average shelf slopes decrease to the east (<0.3° off the Guadiana River and <0.2° off the Guadalquivir River) (Lobo 2000). The outer shelf is bounded by the shelf break at a water depth of about

130 ± 20 m (Nelson *et al.* 1999; Maldonado *et al.* 2003). The shelf break is particularly deep (around 150 m water depth) off the Guadiana River and off Cape Roche (Lobo 2000; Maldonado *et al.* 2003).

The shelf shows three physiographical domains (inner, middle and outer) between the Guadiana River mouth and Cape Roche (Nelson *et al.* 1999). The inner shelf is partially covered by elongated prodeltaic lobes off the main rivers (Fig. 9.2), such as the Guadiana, Tinto-Odiel and Guadalquivir (Fernández-Salas *et al.* 1999). Inner-shelf interprodeltaic areas show a more irregular morphology owing to the common occurrence of erosional features, such as erosional surfaces and Pliocene–Quaternary rocky outcrops (Gutiérrez-Mas *et al.* 1996; Fernández-Salas *et al.* 1999). Other depositional morphologies such as infilled depressions and bedform fields show a patchy distribution (Nelson *et al.* 1999; Lobo *et al.* 2000). The middle shelf is covered by several muddy depocentres, extending from the shelf to the east of the Guadiana River to the Bay of Cádiz (Nelson *et al.* 1999). The middle shelf off the Guadiana River, however, exhibits a more irregular morphology and a step-like profile owing to the occurrence of several backstepping sediment wedges laterally related to marine terraces (Fernández-Salas *et al.* 1999; Lobo *et al.* 2001). The distal toesets of the muddy shelf depocentres generate low seafloor gradients on the outer shelf (Nelson *et al.* 1999; Rodero *et al.* 1999).

To the SE of Cape Roche, the average shelf width is 35 km and the shelf break occurs at a water depth of 120–140 m (Fig. 9.2). The distinction between the inner and middle shelf is not clear between capes Roche and Trafalgar due to the existence of a relatively flat, shallow platform up to 50 m water depth, named the Barbate High or Platform (Lobo *et al.* 2000, 2010). This shallow platform is covered by several large-scale sandy sediment bodies with superimposed submarine dunes (Nelson *et al.* 1999; Lobo *et al.* 2000, 2010). To the south of Cape Trafalgar, shelf width decreases to 15 km and the shelf break occurs at 100 m water depth. The shelf becomes narrower toward the Strait of Gibraltar (Maldonado *et al.* 2003), where it shows an abrupt physiography generated by outcropping Flysch units (Esteras *et al.* 2000; Luján *et al.* 2011).

In the southern shelf of the Gulf of Cádiz, the Mesetan Shelf is 40 km wide, gently dipping (0.1–0.6°) up to the 150 m-deep shelf break (El Foughali & Griboulard 1985; Ruellan 1985). A 30–100 m-deep sedimentary wedge marks the morphology of the shelf off the Oum Er Rbia River (Fig. 9.2). Cretaceous formations outcrop extensively between the coastline and 130–150 m water depths, generating a highly rugged seafloor. A wide abrasion surface is also exposed in the middle shelf between the latitudes of the cities of Casablanca and Kenitra, and close to Mohammedia city where a coastal dune complex is submerged.

The offshore prolongation of the Rharb Basin is characterized by a 20–35 km-wide shelf that is influenced by a major wedge-shaped progradational prodelta off the Sebou River. The prodelta extends across the shelf and thins toward the shelf break where it is connected to a major canyon named the Rharb Valley. To the north, an IPW covers the inner shelf up to 30 m water depth between the cities of Moulay Bouselham and Larache (Fig. 9.2).

The bathymetry of the Western Rifan Shelf reflects the structure of the Rif Belt to the north of Lalla Zahra (Fig. 9.2). The Western Rifan Shelf is 25–30 km wide, with a 150 m-deep shelf break. The offshore prolongation of the huge tectonosedimentary complex generates a highly irregular seafloor. The NW termination of the Western Rifan Shelf also exhibits giant submarine dunes stacked at the entrance of the Strait of Gibraltar between water depths of 120 and 220 m (Trentesaux *et al.* 2009).

Shelf sediments

In the Iberian Shelf, surficial sediments depict two distinct distribution areas (Fig. 9.4). From the Algarve Shelf to the Bay of Cádiz, shelf sediments occur as relatively continuous bands from the inner shelf to the shelf break. The inner shelf is covered by a continuous belt of sandy sediments, with local occurrence of gravels and rocky outcrops (Instituto Hidrográfico 1985; Rey & Medialdea 1989; Gutiérrez-Mas *et al.* 1996; Fernández-Salas *et al.* 1999; Nelson *et al.* 1999; Maldonado *et al.* 2003; González *et al.* 2004; Rosa *et al.* 2013) (Fig. 9.4). This general trend is interrupted in the proximity of the most important river mouths (Guadiana, Guadalquivir) where muddy patches occur (e.g. López-Galindo *et al.* 1999; González *et al.* 2004). The mid–outer shelf is covered mostly by muds, the main distribution area of which occurs on the middle shelf to the east of the Guadiana River, forming an elongated, laterally continuous muddy layer (Nelson *et al.* 1999). However, sandy sediments may also occur in distal settings, such as the middle shelf between Lagos and Cape Santa Maria (Moita 1986), and the middle shelf off the Guadiana River (Fernández-Salas *et al.* 1999; González *et al.* 2004) (Fig. 9.4). Further seawards, different types of sediments such as clayey sands, sandy and silty clays occur on the outer shelf. Large patches of sands and gravels occur around the shelf break, and they become more widespread to the SE (López-Galindo *et al.* 1999; Nelson *et al.* 1999; González *et al.* 2004).

SE of the Bay of Cádiz, and particularly over the Barbate Platform, most of the shelf sediment cover is composed of reworked relict sands with a high content of ultrastable heavy minerals and bioclastic remains (Gutiérrez-Mas *et al.* 1994; López-Galindo *et al.* 1999). Towards the Strait of Gibraltar, the gravel content increases and Gibraltar Flysch rocky outcrops become more common (Rey & Medialdea 1989; López-Galindo *et al.* 1999; Nelson *et al.* 1999) (Fig. 9.4).

Along the Moroccan Shelf, the sediment distribution is quite similar to the pattern observed in the northern Iberian Shelf (Fig. 9.4). The inner shelf is covered by a continuous belt of fine quartz sands with abundant sandstones and quartzite lithoclasts (Cirac *et al.* 1979). These deposits are locally interrupted by outcropping lithified formations and by muddy sands around the Loukkos River Estuary. The mid–outer shelf is covered mostly by muds, particularly across the Rharb Shelf where they derive from the Sebou River (Fig. 9.4). The mud belt is surrounded by sandy muds formed by a progressive mixing of glauconitic fine and bioclastic sands exported from the inner and outer shelves. Offshore the Mesetan Domain, the outer shelf–upper slope (120–150 m water depths) is also covered by coarse-grained sediments, such as worn-out rubified bioclasts.

Pre-Quaternary shelf sedimentation

Pre-orogenic sedimentation styles underwent a gradual transition from carbonate to terrigenous in the Gulf of Cádiz. The initial development of passive margins during the Mesozoic favoured the development of carbonate platforms, which become mixed (calcareous–terrigenous) during the Late Cretaceous–Early Tertiary (Riaza & Martínez del Olmo 1996). Terrigenous margins developed during the Oligocene and Early Miocene over transcurrent margins. The post-orogenic sedimentation (after the emplacement of the allochthonous body) was initiated by a very significant progradation associated with the closure of the Atlantic–Mediterranean gateways (Maldonado & Nelson 1999a).

In the Iberian Shelf, two major sequences were generated during the Pliocene. The Early Pliocene sequence shows an aggradational stacking pattern, and is capped by a discontinuity that initiated the generation of a shelf–slope profile. The Late Pliocene sequence shows a more distinctive progradational pattern, and is eroded by a Late Pliocene (*c.* 2.4 Ma) regional discontinuity. These major sequences are composed mainly of both Regressive and Lowstand Systems Tracts (RST and LST) (Hernández-Molina *et al.* 2002).

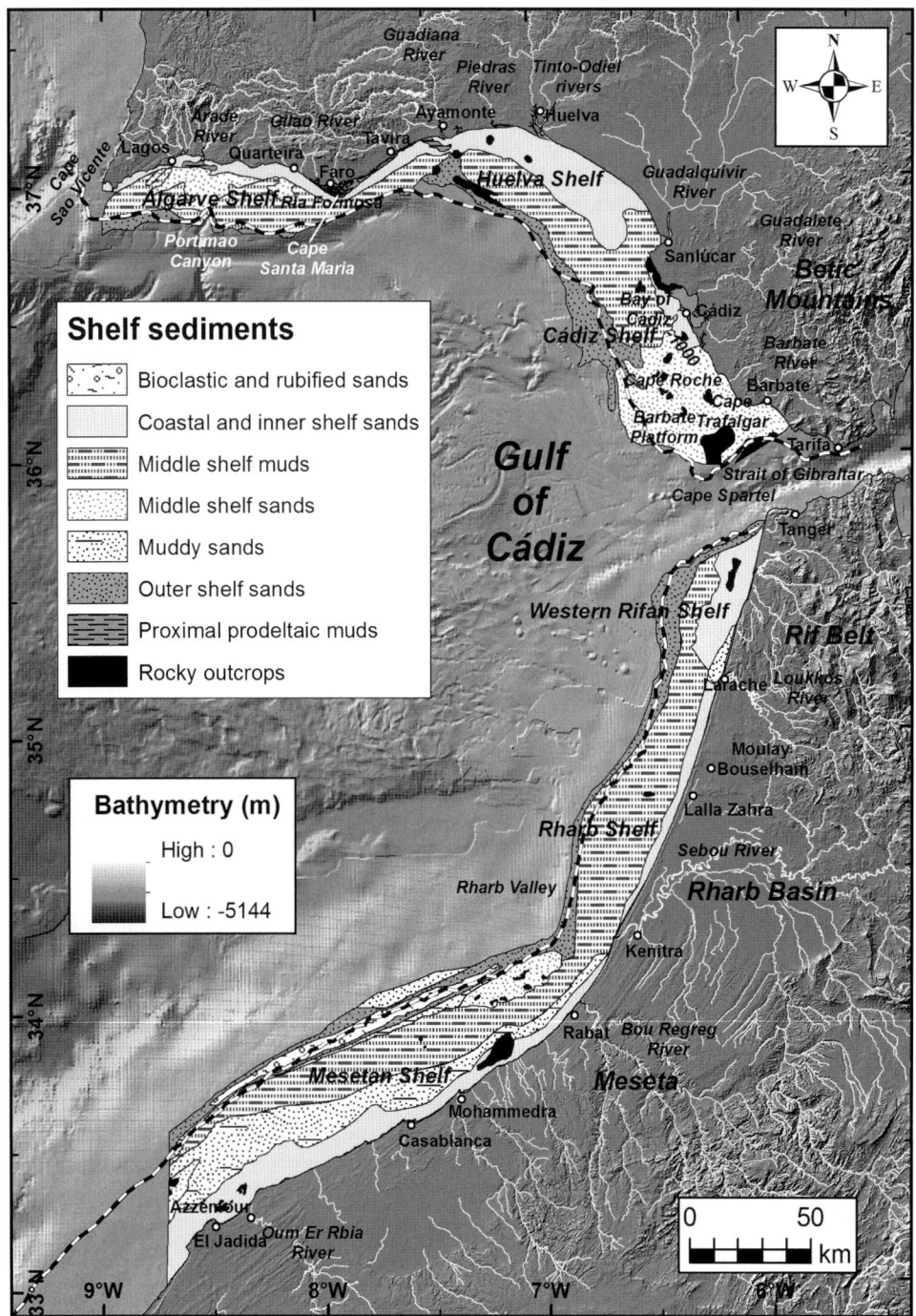

Fig. 9.4. Surface sediment distribution over the shelves around the Gulf of Cádiz. Data from the Iberian shelf was compiled from Instituto Hidrográfico (1985), Rey & Medialdea (1989), López-Galindo *et al.* (1999), Nelson *et al.* (1999) and González *et al.* (2004). Data from the Moroccan Shelf was compiled from Cirac *et al.* (1979) and Jaaidi (1993).

In the Moroccan Shelf, pre-orogenic Mesozoic sedimentary deposits have been described along the offshore foreland Rharb Basin, where they are overlain by continuous high-amplitude reflectors and crop out onshore along the edge of the Meseta Domain. They are arranged in two Infra-Nappe Sequences (INS1 and INS2) (Fig. 9.5). Sequence INS1 is composed of Triassic shales and sandstones filling basement-tilted blocks related to the Triassic rifting. Sequence INS2 is made of Late Cretaceous transgressive calcareous sediments. Towards the axis of the basin, the two sequences are buried by the Prerifan Nappe Unit (PNU) that is characterized both by imbricate thrusts along its southern front and extensional faults affecting the upper part of the northward-thickening wedge. Along the slowly subsiding Mesetan Shelf to the south, the Late Cretaceous carbonate platform (INS2) outcrops at the seabed between the coastline and the shelf break, generating NE–SW-trending cuestas that delimit a major erosional surface from 60 to 120 m water depth (Le Roy *et al.* 2004) (Fig. 9.6).

The post-orogenic pre-Quaternary deposits overlying the Prerifan nappes system are recognized in the Rharb Shelf where they are constituted by a thick (locally >2500 m) Supra-Nappe Sequence (SNS1) showing aggrading–prograding configurations directed to the west. Biostratigraphic analyses have established that the age of Sequence SNS1 ranges from Late Tortonian–Messinian to Gelasian (Fig. 9.5). Sequence SNS1 is sealed by an erosional unconformity (SNR1) (Fig. 9.7a) that initiates the transition to a shelf–slope profile. This unconformity is correlated with the Moghrebian Pliocene surface (*c.* 4 Ma) described onshore (Benson & Rakic-El-Bield 1996).

Quaternary shelf sedimentation

The Quaternary sediment record in the Gulf of Cádiz shows a composite character, as sequences with different periodicities are

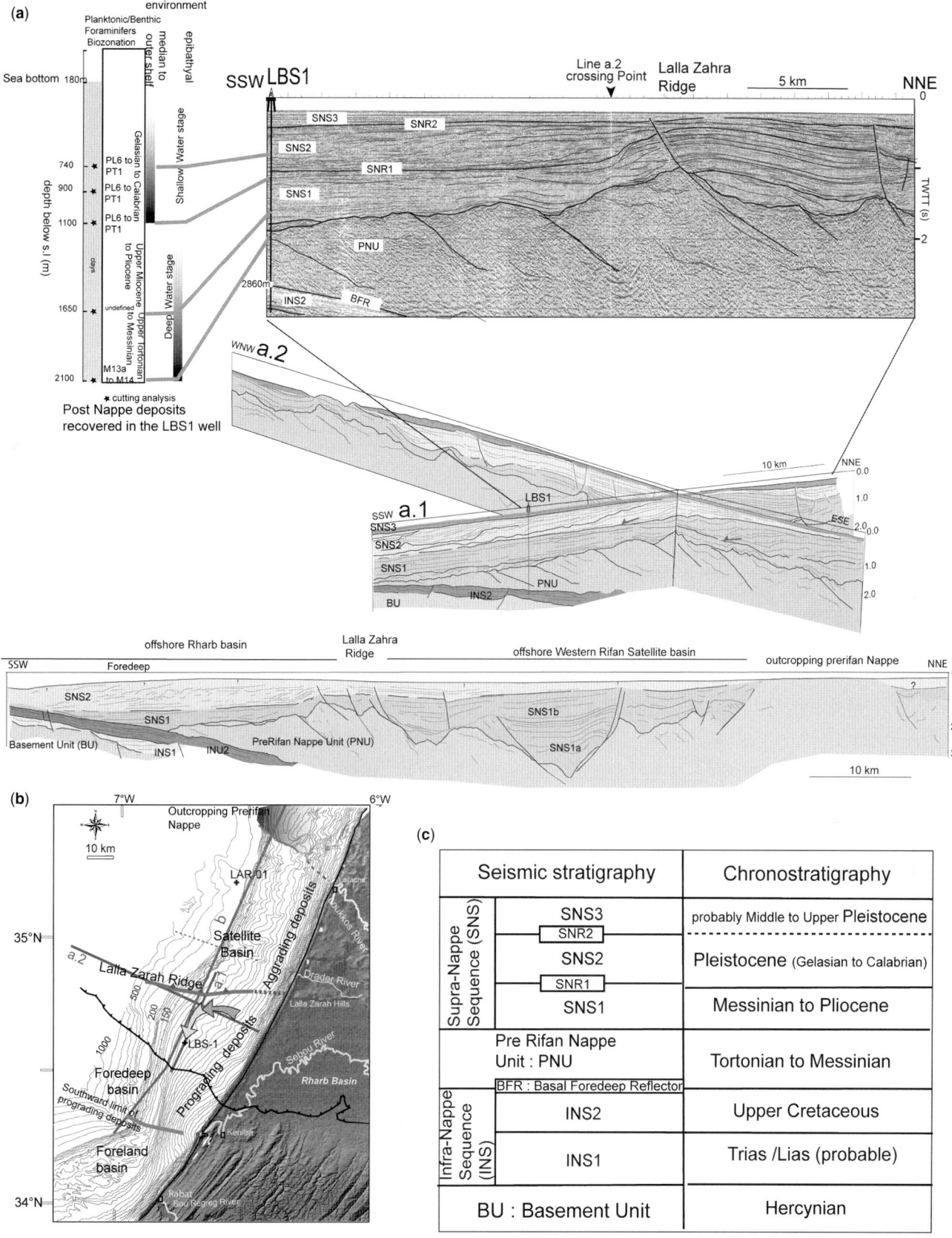

Fig. 9.5. Original line drawings with location of representative industrial seismic lines of the Rharb Shelf by courtesy of the Office National des Hydrocarbures et des Mines (ONHYM) and general seismic stratigraphy table of the NW Atlantic continental shelf. (**a**) Cross-sections displaying the southward shift of prograding Supra-Nappe Sequences (SNS1–SNS3) in response to the uplift of the Lalla Zahra Ridge. The chronostratigraphic attribution is based on the correlation with the industrial well LBS1. The correlation was achieved through an interval velocity analysis made on the multi-channel industrial seismic line (by courtesy of ONHYM) and a micropalaeontological analysis of the cuttings from the LBS-1 well conducted by the ERADATA Company. The lithology of the deposits recovered in the well is mainly clayey. In the location map, arrows display directions of progradation of seismic sequences SNS1 (dark colour) and SNS2 (light colour). (**b**) Strike line along the middle Rharb Shelf showing the flexured Infra-Nappe Mesetan Sequences (INS1 and INS2) to the SSW. (**c**) Correlation between the seismic stratigraphy interpretation of industrial and academic lines and the chronostratigraphic attribution modified from Flinch (1993). The vertical scales of the seismic profiles are given in two-way travel time (TWTT).

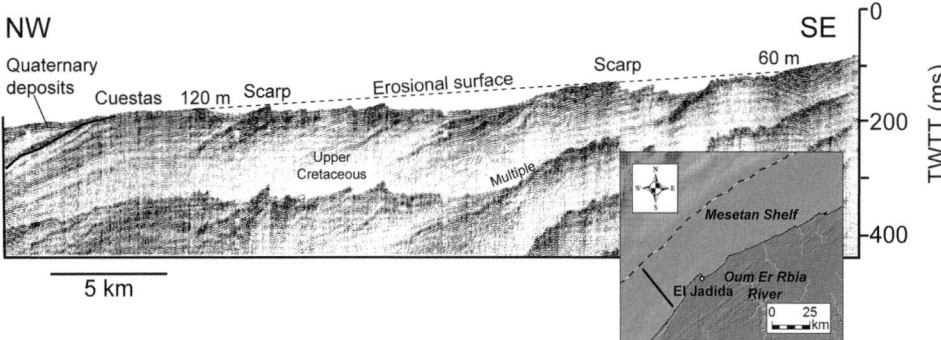

Fig. 9.6. Industrial seismic line showing cuesta morphology constructed on the Late Cretaceous carbonate platform in the Mesetan Shelf. The vertical scale of the seismic profile is given in two-way travel time (TWTT). Modified from Le Roy *et al.* (2004).

stacked in a hierarchical fashion (Figs 9.7b–9.9). On the Iberian Shelf, two major sequences correlated with 800 kyr cycles and separated by a prominent discontinuity at about 1–0.9 Ma, proposed as the Middle Pleistocene Revolution (MPR) discontinuity, were described (Fig. 9.8). Each major sequence is internally composed by higher-frequency sequences, formed in response to different cyclicities: 400, 200 and 100 kyr. These last 100 kyr sequences started to develop after the MPR and are composed mainly of regressive shelf-margin deltas (Hernández-Molina *et al.* 2002).

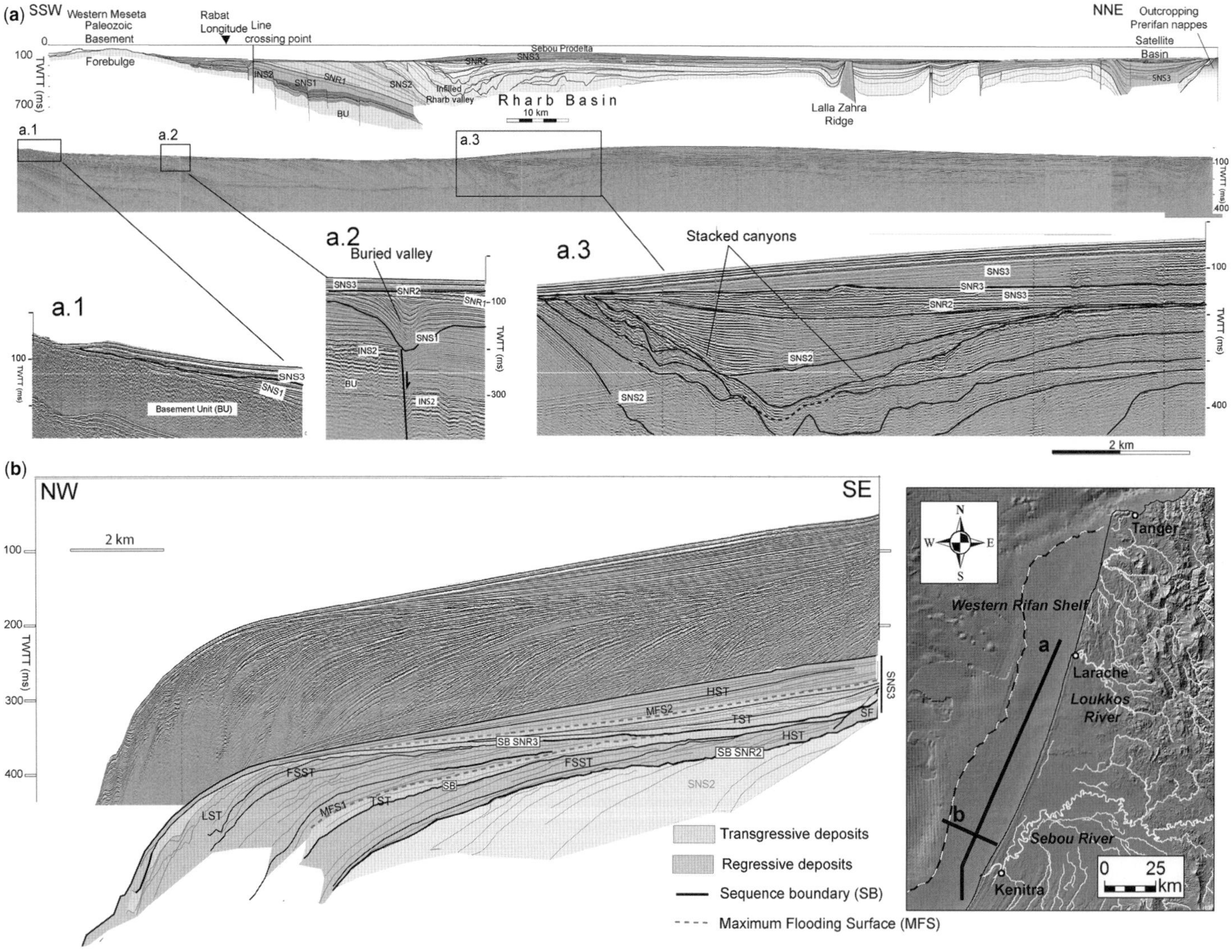

Fig. 9.7. Geometry of Pleistocene deposits in the Moroccan Shelf. (**a**) Shelf-parallel seismic line and corresponding line drawing. Significant shelf stratigraphic features are enlarged: (a.1) uplifted Pleistocene deposits onlapping the Palaeozoic basement outcropping at the basin forebulge; (a.2) buried valley located along a fault plane affecting Neogene deposits and infra-nappe units; (a.3) buried stacked canyons and the Sebou River prodelta. Seismic stratigraphy nomenclature indicated in (a) is summarized in Figure 9.5c. (**b**) Sparker seismic line located in the middle Rharb Shelf. The corresponding line drawing shows 14 seismic units dominated by shelf-margin wedges generated during falling, lowstand and highstand sea-level stages. SB, sequence boundary; FSST, Falling-Stage Systems Tract; LST, Lowstand Systems Tract; TST, Transgressive Systems Tract; HST, Highstand Systems Tract; MFS, Maximum Flooding Surface; SNR2, SNR3, sequence boundaries associated with major sea-level lowstands. The vertical scales of the seismic profiles are given in two-way travel time (TWTT).

On the Moroccan Shelf, the styles of Quaternary shelf sedimentation exhibit important along-strike variability. In the south, the Quaternary record is restricted to inner-shelf prodeltaic and coastal-wedge deposits, and to complex shelf-margin wedges. The Quaternary sedimentation is more remarkable across the Rharb Shelf where the Pleistocene is recovered at 1009 m below sea level in the LBS1 well (Maad *et al.* 2010). Quaternary deposits exhibit numerous sequences with different periodicities and are organized into two megasequences (SNS2 and SNS3) separated by a regional unconformity (SNR2). Quaternary sequences show a prograding pattern towards the SW due to the differential distribution of accommodation (Fig. 9.7b), driven mainly by the reactivation of the large frontal Prerifan structures to the north of the Rharb Basin. There, a major east–west-trending folded corridor occurs off Lalla Zahra Ridge. A coeval uplift of the southern edge of the Rharb Shelf fringing the Mesetan Domain was also recorded during the Quaternary. This southern edge is interpreted as the forebulge of the active foreland Rharb Basin, which is flanked to the north by a narrow corridor (the Rharb Valley) progressively infilled with sediments from the Sebou River.

Late Quaternary

The Late Quaternary record in the Iberian estuarine systems shows stacked deposits generated during several highstand intervals, such as Marine Isotopic Stages (MISs) 5 and 3 (Dabrio *et al.* 2000; Boski *et al.* 2002; Lobo *et al.* 2003a). On the shelf, the Late Quaternary sediment record has been studied by a wide high-resolution seismic database. The generation of stacked depositional sequences in the Cádiz Shelf was considered to be a response to the 100 kyr periodicity, which was dominant after the MPR. Most of the sequences were developed through regressive periods separated by hiatuses that reflect non-deposition and erosion during sea-level rises (Rodero *et al.* 1999). However, the most exhaustive account of the Late Quaternary shelf stratigraphic architecture was provided for the Huelva Shelf, where two fourth-order depositional sequences generated in response to the two most recent 100 kyr glacio-eustatic cycles have been nicely imaged (Somoza *et al.* 1997; Hernández-Molina *et al.* 2000b) (Fig. 9.9). These two recent fourth-order sequences are composed internally of higher-frequency (fifth-order) sequences linked to sea-level cycles with lower periodicity (about 20 kyr). Therefore, the Late Quaternary stratigraphic architecture in the Gulf of Cádiz Shelf is controlled by the imprint of Milankovitch-scale cycles. Both types of sequences (fourth and fifth order) show an asymmetrical character due to the higher preservation of Forced Regressive (FRWST) and LST, which led the main margin progradation (Somoza *et al.* 1997; Hernández-Molina *et al.* 2000b) (Fig. 9.9). For the last 80 kyr, shelf architecture even records the influence of climatic and sea-level cycles of sub-Milankovitch scale (Hernández-Molina *et al.* 2000b).

The last fourth-order sequence is composed of several fifth-order sequences, strongly dominated by shelf-margin wedges that exhibit low-angle oblique seismic configurations (Lobo *et al.* 1999). These shelf-margin wedges were interpreted as the distal portions of coastal–prodeltaic deposits generated during falling to lowstand sea levels (Lobo *et al.* 1999), and, thus, composing both the FRWST and the LST (Hernández-Molina *et al.* 2000b). Later work revealed that most of the wedges are composed mainly of FRWSTs generated during stepped sea-level falls (Lobo *et al.* 2005a) (Fig. 9.10).

Transgressive (TST) and Highstand Systems Tracts (HST) were considered to be under-represented on the shelf. However, in restricted shelf sectors of the Huelva Shelf, transgressive–highstand deposition (sandy barrier islands and coastal lagoons) was favoured (Lobo *et al.* 1999). In addition, over the middle–outer shelf, a backstepping sediment wedge was related to relative highstand conditions during the last glacial cycle, possibly during MIS 3 preceding the last depositional sequence (Lobo *et al.* 2002).

During the Last Glacial Maximum (LGM), present-day and recent estuaries and lagoons were fluvial valleys incising into previously deposited coarse-grained deposits, generating an erosional surface interpreted as a sequence boundary (Dabrio *et al.* 2000; Hilbich *et al.* 2008). During the LGM, sediments did not accumulate significantly, because of increased sediment bypass toward the middle–outer shelf (Lobo *et al.* 2003a). There, a laterally continuous, anomalous thick progradational–aggradational shelf-margin wedge recorded a long-lived sea-level lowstand after a prolonged fall (Lobo *et al.* 2005a) (Fig. 9.10).

On the Morrocan Shelf, the Late Quaternary stratigraphic architecture is documented in the Rharb Shelf, where seismic stratigraphic analysis reveals the dominance of shelf-margin wedges generated during falling, lowstand and highstand sea-level conditions (Fig. 9.7b). The stacking pattern shows three distinct seismic packages showing different sedimentation trends. The lower and upper packages correspond to a succession of regressive deposits representing a set of HST, FRWST and LST. They are separated by an intermediate package composed of a sheet-like unit interpreted as a TST. The chronostratigraphic attribution of the Late Quaternary sequences is based on limited age dating of a major erosional unconformity (SNR3) extending through the whole shelf and capping the upper regressive package. This surface has been dated at 15 000 ± 250 years BP) and corresponds to the LGM erosional surface (Fig. 9.7b). The interbedded TST extends toward the continent in the northern part of the Rharb Shelf suggesting a major flooding event. It could correspond either to the penultimate glacial termination (MISs 6–5) or, alternatively, to a sea-level rise during MIS 3, as recognized in the northern Iberian Shelf.

Post-glacial transgression

During the post-glacial transgression, estuaries along the Iberian Margin were progressively flooded and sediments accumulated in the deeper areas. The initial marine influences in estuaries date back to 10 ka BP and enabled deposition of clayey sediments (Dabrio *et al.* 2000; Boski *et al.* 2002). Subsequently, a more or less continuous sedimentation was established in most estuaries and persisted after the sea-level stabilized in a highstand position (Lario *et al.* 2002).

In the shelf, the post-glacial transgression is represented by a TST developed in the 14–6.5 ka BP time interval (Hernández-Molina *et al.* 1994; Gutiérrez-Mas *et al.* 1996). Initial interpretations mostly based on sedimentological analyses considered the post-glacial TST as basically composed of a sandy bed overlying a ravinement surface and outcropping over extensive outer shelf areas (Nelson *et al.* 1999). The extensive sandy layer that covers most of the shelf off Cape Trafalgar was also considered to be transgressive in origin (Gutiérrez-Mas *et al.* 1996; López-Galindo *et al.* 1999). Later interpretations based on seismic profiling interpretation registered up to four seismic units on the Huelva Shelf regarded as post-glacial transgressive deposits (Lobo *et al.* 2001). The initial shelf flooding was recorded by a low-angle progradational, laterally continuous outer-shelf deposit (Lobo *et al.* 2001), linked to the activity of overflow channels during large-scale flood events (González *et al.* 2004). The ensuing shelf flooding enabled the formation of two distinct depositional environments on the Huelva Shelf. The sector off the Guadiana River, with irregular seafloor topography and sandy composition, shows three lenticular-shaped deposits interpreted to have been deposited in a high-energy environment (Lobo *et al.* 2001) (Fig. 9.11). The internal configuration and sediment facies of these parasequences suggest different depositional mechanisms, with the influence both of flood events in the river basins and

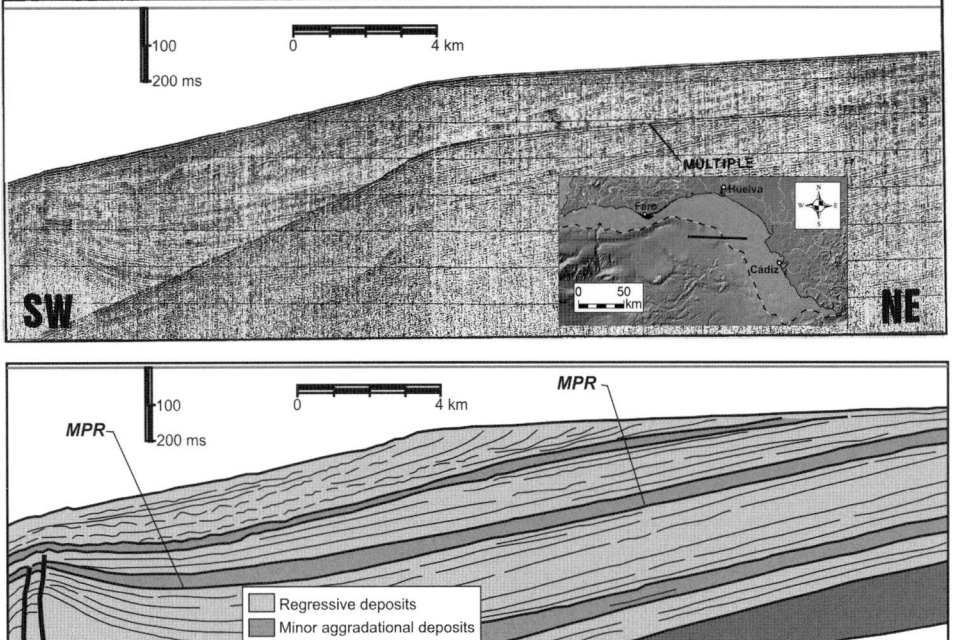

Fig. 9.8. Representative seismic profile and interpretation showing the Quaternary stratigraphic architecture of the Iberian Shelf off the Guadalquivir River. The vertical scale of seismic profile is given in two-way travel time (TWTT). Modified from Hernández-Molina *et al.* (2002).

storm-induced seaward-sediment transport. In contrast, the sector between the Piedras and Guadalquivir rivers was a low-gradient shelf where extensive barrier–lagoon systems developed (Lobo *et al.* 2001).

The generation of these post-glacial transgressive deposits was linked with the stepped and fluctuating pattern of sea level, as periods of increased sea-level rise were punctuated by periods of reduced rises or even by short-term sea-level stillstands. This sea-level pattern was linked to climatic fluctuations that also triggered significant modifications of the sediment export to the shelf. The Younger Dryas event (11–10 ^{14}C ka BP) probably caused the most dramatic consequences (Hernández-Molina *et al.* 1994; Lobo *et al.* 2001), as the supplies of terrestrial sediments were particularly high. This event was terminated by increased marine sediment supply during the subsequent sea-level rise (Burdloff *et al.* 2008). However, possibly other post-glacial

climatic events also left their sediment signature on the shelf (Lobo *et al.* 2001).

On the Moroccan Shelf, the post-glacial TST is located above the Lowstand Unconformity SNR3 and shows a backstepping-stacking pattern towards the middle–inner shelf. The initial shelf flooding is characterized by low-angle reflections onlapping previous shoreface deposits. They are overlain by low-amplitude aggrading wavy reflectors attributed to low-energy deposits. Sedimentological and age data provided by coring collected across the distal part of the TST suggest that the low-energy deposits could be formed during the early Holocene, which was marked by an increase in pluviosity and the development of a mid-shelf muddy belt (Jaaidi 1993). Furthermore, several terraces located between water depths of 90 and 60 m along the southern Mesetan Shelf could be correlated with sea-level stillstands during the last sea-level rise. The end of the post-glacial

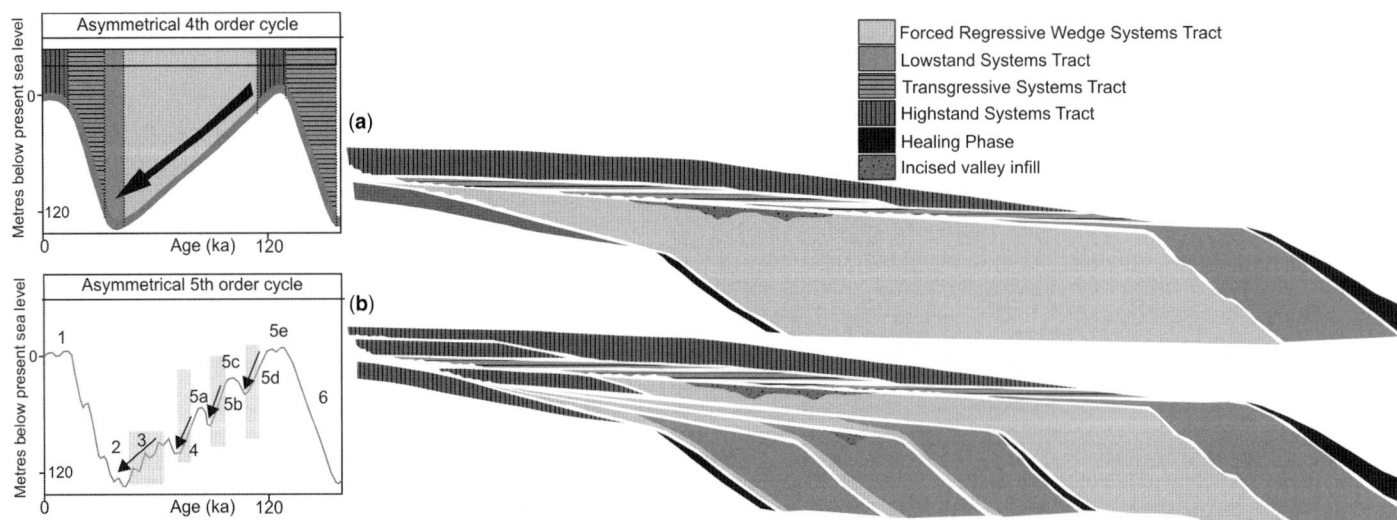

Fig. 9.9. Sequence stratigraphy interpretation of Late Quaternary depositional sequences on the Iberian Shelf of the Gulf of Cádiz. (**a**) Sequence stratigraphy of the last fourth-order depositional sequence. (**b**) Sequence stratigraphy of higher-frequency fifth-order depositional sequences composing the lower-frequency sequence. Modified from Hernández-Molina *et al.* (2000*b*).

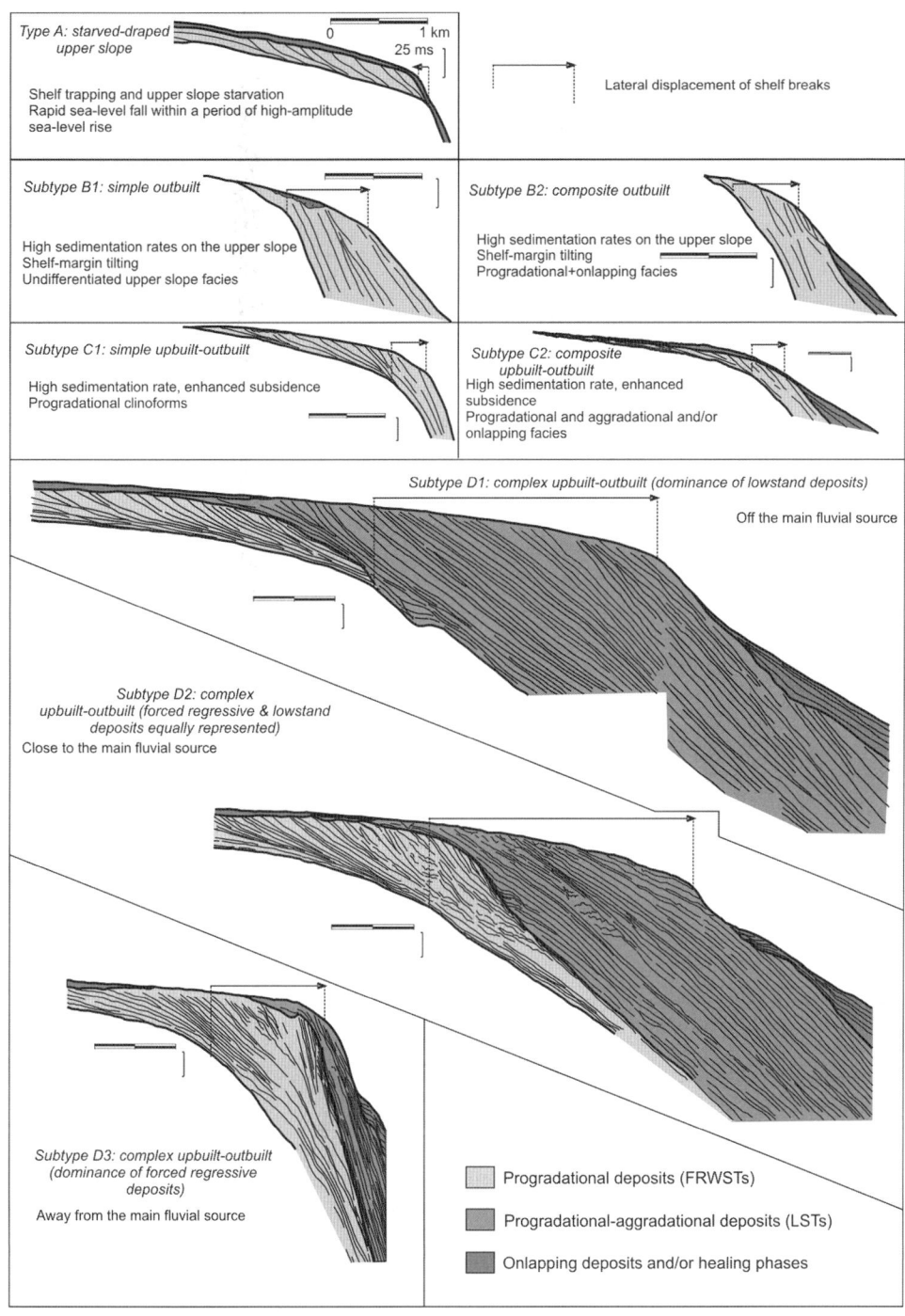

Fig. 9.10. Different types of stratigraphic architectures of regressive deposits in the Iberian Shelf of the Gulf of Cádiz. Modified from Lobo *et al.* (2005a).

transgression between 8 and 6.5 ka BP is recorded in coastal areas by widespread erosion of lithified aeolian dune systems generated during the previous lowstand. The sea level reached +2 m during the middle Holocene at 6.5–4.5 ka BP, invading topographical lows and generating an extensive lagoonal environment (Carruesco 1989).

Holocene highstand

In coastal embayments of the Iberian Margin, the Holocene sea-level stabilization favoured the closure of lagoonal bodies by sand spits in the vicinity of the main estuarine systems and the accretion of tidal flats (Dabrio *et al.* 2000). In the deeper part of estuarine channels, sandy progradational sediment bodies such as estuarine bars and tidal deltas were developed (Boski *et al.*

2002, 2008; Lobo *et al.* 2003a). In coastal spits, several climatically induced progradational events have been described during the last sea-level Holocene stabilization (Zazo *et al.* 1994, 2008; Lario *et al.* 1995; Goy *et al.* 1996; Rodríguez-Ramírez *et al.* 1996). The progradational phases occurred in response to increased aridity and low relative sea levels (Zazo *et al.* 2008), in relation to abrupt cooling events (i.e. Bond events) in the North Atlantic (Fletcher *et al.* 2007). High-energy events, such as tsunamis, have also been recorded in several coastal embayments of the Gulf of Cádiz, such as the Bay of Cádiz, the Guadalquivir River marshes and the Huelva Estuary (Luque *et al.* 2001; Morales *et al.* 2008; Gutiérrez-Mas *et al.* 2009a, b).

In the shelf, the Holocene sedimentation is represented by a HST composed of different depositional systems (Fig. 9.12) (Gutiérrez-Mas *et al.* 1996; Nelson *et al.* 1999). On the Algarve Shelf, the genesis of the Faro–Tavira IPW was linked to enhanced littoral drift and seaward-sediment transport by downwelling currents,

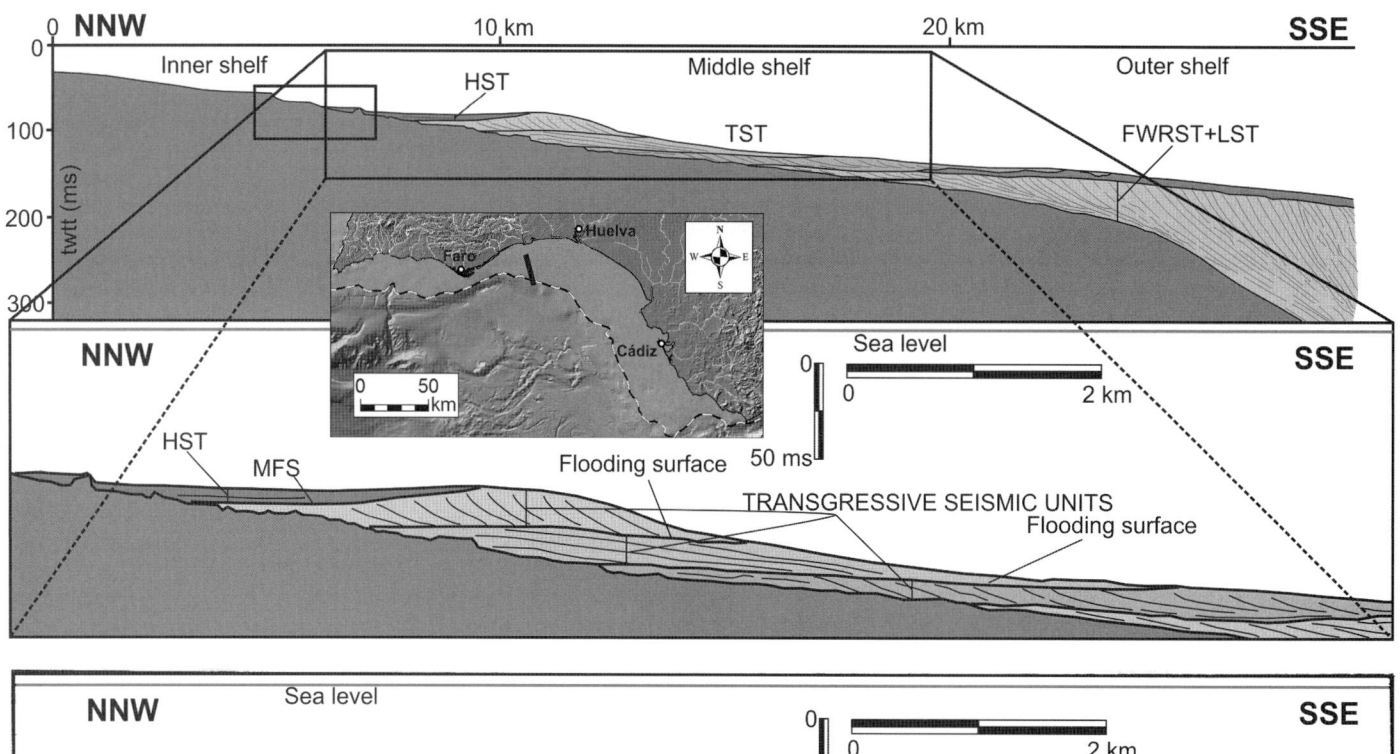

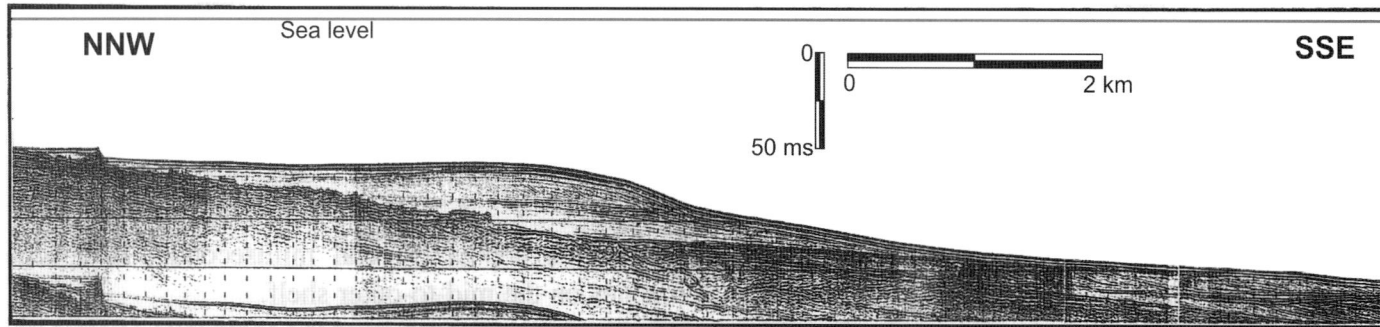

Fig. 9.11. Post-glacial transgressive architecture in the shelf off the Guadiana River, highlighting the typical backstepping parasequence pattern. The vertical scales of the seismic profiles are given in two-way travel time (TWTT). Modified from Lobo *et al.* (2001).

generated after energetic storm activity (Hernández-Molina *et al.* 2000*a*) (Fig. 9.12a). The seaward export of fine-grained sediments led to the construction of outer muddy belts on the Algarve Shelf (Lobo *et al.* 2004*b*). The lateral elongation of most of these distal muddy accumulations is dictated by the dominant large-scale anticyclonic circulation, which causes eastward-directed sediment dispersal (Lobo *et al.* 2004*b*).

A significant highstand depositional system is connected to the Guadiana River, where a process of depositional system partitioning has been reported (Fig. 9.12b, c). There, a proximal prodelta evolves distally to a distal muddy belt (Lobo *et al.* 2004*b*). The proximal prodelta is composed of a mixture of fine and coarse sediments, although the upper part is dominated by muds (from 1.5 to 0.2 cal ka BP). This recent change is associated with increased human pressure in the drainage basin, linked with changes in the climate, which led to distributary infilling and enhanced seaward-sediment transport from the present-day estuary (Mendes *et al.* 2010). The distal muddy belt is composed mostly of fine sediments, showing a mean fining-upward grain-size trend (from 11.5 cal ka BP to Recent times). In the prodelta and distal muddy belt, Holocene sedimentary processes were controlled mainly by natural (sea-level changes and climate variations) and human-induced processes (e.g. deforestation, agriculture). The establishment of the distal muddy belt occurred during approximately the last 1.5–1 cal ka BP, and only after the prodelta formation took place (Mendes *et al.* 2012*b*).

The main highstand depocentre in the Gulf of Cádiz shelf is the fluvially dominated Guadalquivir River prodelta (Fig. 9.12d) that covers the entire shelf and shows the highest Holocene

sedimentation rates (Nelson *et al.* 1999) due to high fluvial supply and moderate reworking (Lobo *et al.* 2004*b*). Acoustic masking was detected within the upper layers of this prodeltaic deposit, at shallow waters between the river mouth and the Bay of Cádiz. Its genesis was initially related to shelly sand layers eroded from a nearby inner-shelf outcrop (Acosta 1984). However, subsequent interpretations related the acoustic masking to gas-bearing sediments (Fernández-Salas *et al.* 2003).

The main prodeltaic accumulation extends both to the SE over the middle shelf and to the NW over the inner shelf, in response to the complex oceanographic patterns established over the shelf. The prolongation to the SE of the prodeltaic deposit connected to the Guadalquivir River is controlled by the displacement of the Surface Atlantic Water (SAW) over the shelf (Gutiérrez-Mas *et al.* 1994, 1996). In the Huelva Shelf, the inner-shelf depocentre connected to the Guadalquivir River reflects the influence of the SAW during easterlies' dominance (Lobo *et al.* 2004*b*).

The internal architecture of many of these shelf depositional systems also reveals a fluctuating environmental pattern during the Holocene highstand (Fig. 9.13). Depositional systems such as the Guadalquivir and Guadiana river prodeltas and the Faro–Tavira IPW show two cycles of progradation–aggradation internally composed of minor-scale motifs (Lobo *et al.* 2005*b*). The existence of this hierarchical, stratigraphic pattern in shallow-water deposits was related to different forcing factors. The major-scale architecture was related to the effect of high-frequency small-amplitude sea-level changes that also involved changes in sediment supply (Fernández-Salas *et al.* 2003; Lobo *et al.* 2004*b*). In contrast, the smaller-scale architectures were in

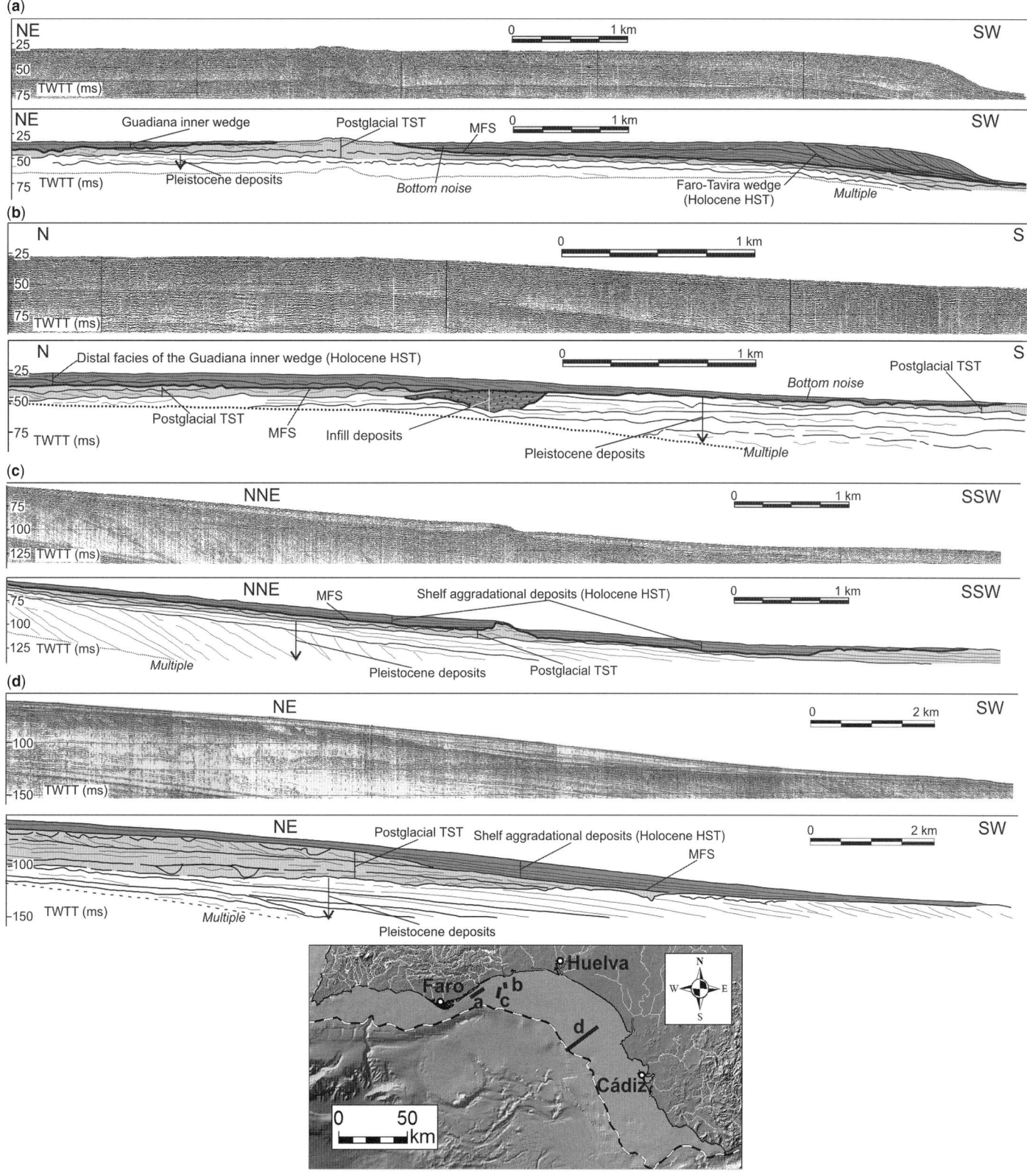

Fig. 9.12. Diverse types of depositional systems compose the Holocene HST in the Iberian Shelf of the Gulf of Cádiz. (**a**) The Faro–Tavira IPW is a shallow sediment wedge located on the Algarve Shelf, in an area with minimal fluvial supplies. (**b**) The proximal prodelta system off the Guadiana River. (**c**) Distal muddy belts occur on the shelf off and around the Guadiana River. (**d**) The prodeltaic deposit connected to the Guadalquivir River covers most of the shelf with an aggradational configuration. The vertical scales of the seismic profiles are given in two-way travel time (TWTT). Modified from Lobo *et al.* (2004*b*).

agreement with the patterns of coastal progradation observed in coastal spits, indicating the influence of changes in the direction and/or the magnitude of storms, as well as minor changes of sediment supply triggered by climatic changes (Lobo *et al.* 2005*b*).

The muddy belt between the Guadiana and Guadalquivir rivers has been constructed through the export of fine sediments toward the middle shelf due to enhanced flooding events in the river basin (Rosa *et al.* 2011; Mendes *et al.* 2012*b*). The proximal and distal boundaries of the mid-shelf muddy belts are controlled by

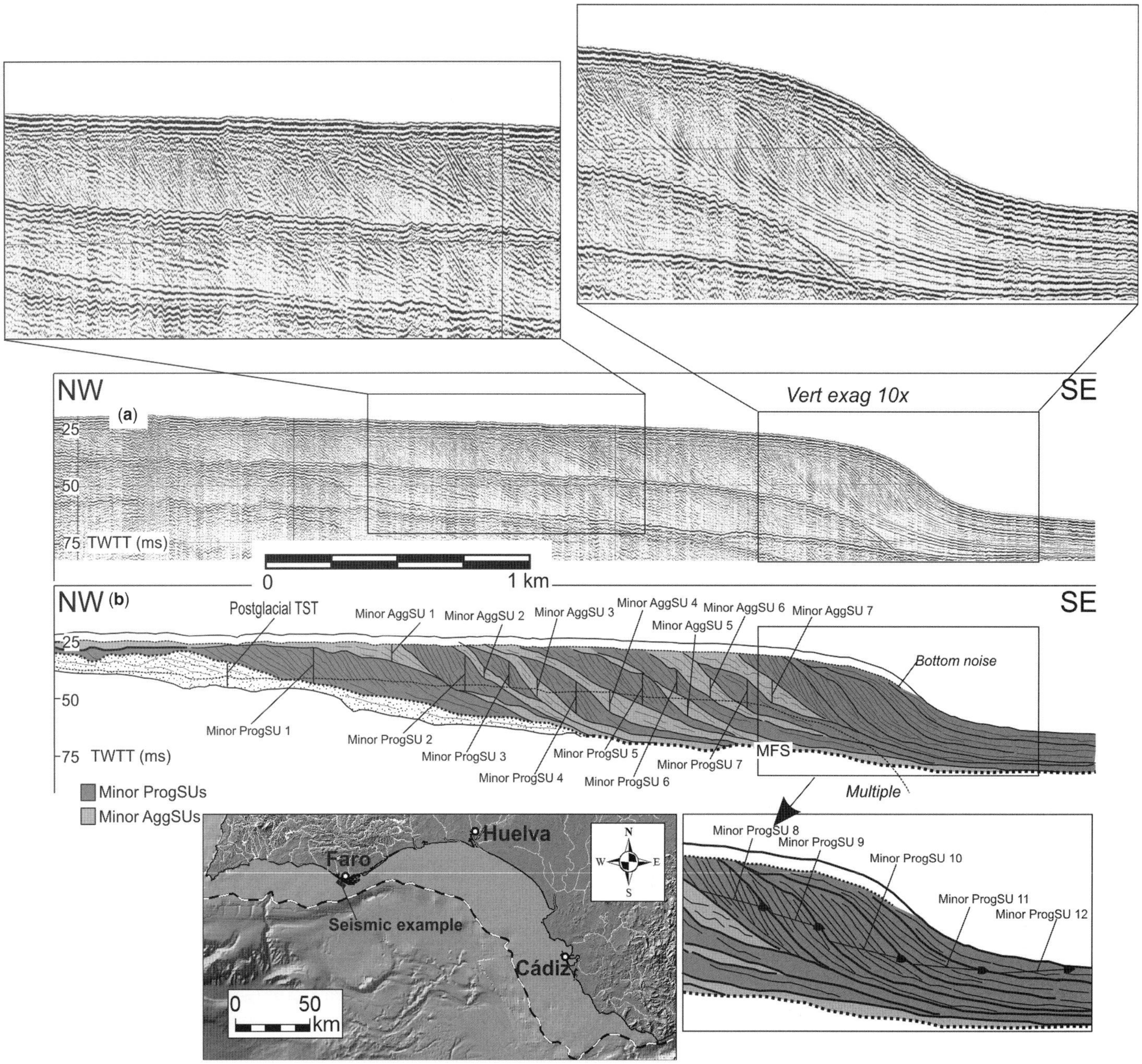

Fig. 9.13. (**a**) High-resolution seismic section and (**b**) interpretation of the western part of the Faro–Tavira IPW, highlighting the complex internal architecture of the shallow-water sediment wedge. The deposit is mainly composed of minor progradational seismic units (ProgSUs) with some alternations of minor aggradational seismic units (AggSUs). The vertical scales of the seismic profiles are given in two-way travel time (TWTT). Modified from Lobo *et al.* (2005*b*).

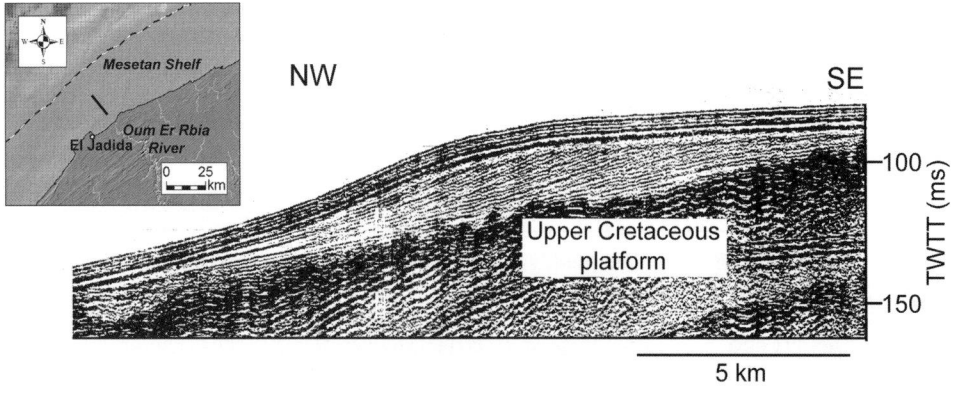

Fig. 9.14. Extract of Sparker seismic line illustrating the Oum Er Rbia prodeltaic deposits prograding over the Late Cretaceous carbonate platform on the Mesetan Shelf. The vertical scale of the seismic profile is given in two-way travel time (TWTT).

wave action, complex shelf-edge currents and internal waves (Nelson *et al.* 1999). On the Cádiz Shelf, the progressive decrease in thickness of the mid-shelf muddy belts to the SE reflects the decreasing river supply, and also the strengthening of the SAW towards the Strait of Gibraltar (Nelson *et al.* 1999). The main fluvial input occurs off the Bay of Cádiz, where a prodeltaic deposit has formed by supply from the Guadalete River and by resuspension of fine-grained sediment by ebb-tidal currents and energetic storm events (López-Galindo *et al.* 1999).

To the SE of the Bay of Cádiz, the fluvially derived muds are replaced by sandy facies locally moulded into bedforms that are indicative of an energetic shelf dominated by oceanographic agents, such as the lower energy SAW and more intense storm events (Nelson *et al.* 1999). In addition, the occurrence of eastward-directed dunes over the inner shelf was attributed to the influence of shoaling waves (Nelson *et al.* 1999). Over the Barbate Platform, the morphological arrangement of sand banks and ridges with superimposed dunes indicate long-term sediment transport patterns toward the SW and south. These seafloor features have been linked to the combined activity of two main current flows with different directions: like the SAW, which sweeps the shelf towards the SE; and the episodical influence of current reversal linked to the tidal cycle in the Strait of Gibraltar (Lobo *et al.* 2000, 2010). Towards the strait, sandy features and bedforms are substituted by bedrock outcrops owing to the increasing speeds of bottom currents (Nelson *et al.* 1999).

On the Moroccan Shelf, the Holocene HST is constituted by spatially confined prodeltaic lobes that laterally disappear as many parts of the shelf are covered by relict, erosional features (Fig. 9.14). The most significant lobes are connected to the Sebou and Oum Er Rbia rivers. The muddy prodeltaic deposits off the Sebou River spread across the shelf and are deflected towards the north in response to contemporary coastal drift. The Sebou River prodeltaic facies are affected by acoustic masking and evolve distally into muddy belts.

The prodeltaic Oum Er Rbia River prodeltaic wedge occurs at 30–100 m water depths, where it is constrained by a NNE–SSW-trending syncline fold affecting the Cretaceous basement. This sediment wedge is composed of intercalated muds and silty sands, which develop sigmoid–oblique configurations dipping to the NW (Le Roy *et al.* 2004) (Fig. 9.14).

These prodeltas were built up from 6 to 2 ka BP during the sea-level stabilization. An increase in fluvial sediment supply was led by humid climatic conditions, as documented by thick alluvial deposits recorded in the coastal plain (Weisrock *et al.* 1985). Sediment supply was drastically reduced after this period when arid climatic conditions set in across Morocco. More recent processes include the influence of high-energy events, such as tsunami waves (i.e. the AD 1755 Lisbon tsunami), which are recorded along the coastline by a large number of giant boulders lying upon lithified dune systems (Mhammdi *et al.* 2008).

Modern shelf sedimentation

In coastal environments of the Iberian Margin, the active sedimentary processes are controlled by the geomorphology and by the interaction between hydrodynamic agents. Modern processes have been particularly documented in the Guadiana River Estuary. In the main estuarine channel, tidal currents are the main sediment-transport agents, particularly during low river discharge periods (Lobo *et al.* 2004*a*). Downstream, the distal part of the main estuarine channel is a bypassing channel where the sediment transport is directed seawards due to the influence of asymmetrical tidal currents (Lobo *et al.* 2004*a*; Morales *et al.* 2006). Further seawards, the distal estuarine system is a wave-dominated environment, where the sediments are redistributed by prevailing waves toward the east. As a consequence, a series

of coastal spits have developed, mainly in the eastern margin of the estuary (Morales 1997). The Guadiana River Estuary also holds drastic hydrographic changes, such as a freshwater deficit and salinity increases, as a consequence of both natural (prolonged droughts) and anthropogenic influences (river regulation by dams). These processes lead to modifications in the distribution and biomass abundance of nursery sites (Chícharo *et al.* 2006) and affect the amount of sediment supplied to the continental shelf (González *et al.* 2001; Dias *et al.* 2004).

Modern shelf sediment dynamics are governed by the sediment export towards the shelf and the subsequent formation of sediment plumes. The most significant plumes are generated by the Guadal-quivir and Guadiana rivers and, to some extent, from the Tinto–Odiel rivers (Palanques *et al.* 1986–1987, 1995). River plumes tend to be transported east and SE due to the influence of the SAW (Periáñez 2009). Recent sedimentary processes off the Guadiana River have been studied with greater detail. The formation of river plumes is enhanced in the coastal domain after peaks of fluvial discharge (Cravo *et al.* 2006). There, sands tend to be deposited on the shoreface and inner shelf, where a main eastward-directed littoral drift has been observed (González *et al.* 2007). However, in some cases, the coastal sediment plume related to the Guadiana River moves westwards, indicating more complex oceanographic conditions (Cravo *et al.* 2006). In the long term, fine sediments tend to be resuspended and deposited on the middle shelf (González *et al.* 2007; Rosa *et al.* 2011; Mendes *et al.* 2012*b*).

The dynamics of biological communities are more complex as they are controlled by different processes, such as the nature of the substratum, fluvial inputs and the occurrence of upwelling. Maximum abundances of phytoplankton, demersal and benthic foraminiferal species tend to occur in shallow waters under the influence of the Guadalquivir River discharges (Abrantes 1990; Catalán *et al.* 2006; Prieto *et al.* 2009; Mendes *et al.* 2012*a*). The spatial distribution of benthic foraminiferal assemblages on the continental shelf between the Guadiana and Guadalquivir rivers is closely associated with river discharge, bathymetry, sea-bottom sedimentary environments and primary productivity (Mendes *et al.* 2004, 2012*a*). Increased biological diversity is also influenced by the onset of wind-induced upwelling in the Algarve Shelf, as reflected by the high abundances of diatoms (Abrantes & Sancetta 1985).

Along the Morrocan Shelf, the knowledge of modern sedimentation processes is restricted mainly to the coastal environments. Estuarine systems, such as the Loukkos and Sebou estuaries, undergo drastic seasonal hydrographic changes, leading to a freshwater deficit and salinity increases (Snoussi 1984; Cheggour *et al.* 2004). Other coastal systems, such as the Bou Regreg and Oum Er Rbia estuaries, have been strongly influenced by dam construction in recent times. As a consequence, the estuarine hydrology is determined mainly by tidal conditions owing to significant decreases in freshwater inputs (Cheggour *et al.* 2004).

Final considerations

Future geological studies conducted in the Gulf of Cádiz shelves should try to fill the current gaps in information and knowledge. At present, complete multibeam bathymetric data are lacking and only specific sectors have been surveyed. These data would be very helpful in mapping the spatial distribution of post-glacial processes and products, and to enhance our knowledge of recent shelf evolution. In addition, the age control of shelf deposits is limited mostly to the Holocene. We propose that the shelves of the Gulf of Cádiz represent a very interesting target for drilling. They provide an excellent sediment record where the concepts of sequence stratigraphy can be tested in order to decipher the complex relationships between climate change, rapid sea-level change and shelf sedimentation patterns.

We would like to dedicate this summary to the intensive work of many people (scientists, technicians and crew members) from different institutions of Spain (University of Cádiz, Instituto Español de Oceanografía, Instituto Geológico y Minero de España and Consejo Superior de Investigaciones Científicas), Portugal (University of Algarve and Instituto Hidrográfico), Morocco (Universities of El Jadida and Rabat, Institute of Sciences of Rabat and Office National des Hydrocarbures et des Mines) and France (INSU, universities of Brest and Bordeaux, and IFREMER) in the waters of the Gulf of Cádiz during the last 20 years. Research on the Iberian Shelf was executed under the umbrella of different research projects (PB-91-0622-C03, PB-94-1090-C03, CTM2005-04960 and CGL2011-30302-C02-02). Research on the Moroccan Shelf has been carried out within the Action Intégrée project MA/08/192 associated with the partenariat Hubert Curien Volubilis, the French ANR project Isis ('Programme Catastrophes Telluriques et Tsunami') and the European project Nearest (Integrated observations from NEAR shore sourcES of Tsunamis). Juan Tomás Vázquez (IEO) provided a synthetic figure displaying the tectonic setting of the Gulf of Cádiz. The Portuguese Instituto da Água provided GIS data of the Algarve region. I. Mendes thanks the Portuguese Science Foundation for grant SFRH/BPD/72869/2010. Improvements to the original version were provided by F.J. Hernández-Molina (University of Vigo) and by an anonymous reviewer.

References

ABERKAN, M., AKIL, M. & OUADIA, M. 1993. Plioquaternary atlantic littoral formations (Larache to Safi): sedimentology, paleopedology, neotectonics and modern sedimentary dynamics. *In: Field-trip Guidebook of the 14th IAS Regional Meeting on Sedimentology*, Volume 2. Publication Universitaires du Maghreb, Marrakesh, 277–288.

ABRANTES, F. 1990. The influence of the Guadalquivir river on modern surface sediments diatom assemblages: Gulf of Cadiz. *Comunicaciones del Servicio Geológico de Portugal*, **76**, 23–31.

ABRANTES, F. & SANCETTA, C. 1985. Diatom assemblages in surface sediments reflect coastal upwelling off Southern Portugal. *Oceanologica Acta*, **8**, 7–12.

ACOSTA, J. 1984. Occurrence of acoustic masking in sediments in two areas of the continental shelf of Spain: Ria de Muros (NW) and Gulf of Cadiz (SW). *Marine Geology*, **58**, 427–434.

BALDY, P., BOILLOT, G., DUPEUBLE, P.-A., MALOD, J. A., MOITA, I. & MOUGENOT, D. 1977. Carte géologique du plateau continental sud-portugais et sud-espagnol (Golfe de Cadix). *Bulletin de la Societé Geologique de France*, **XIX**, 703–724.

BENSON, R. H. & RAKIC-EL-BIED, K. 1996. The Bou Regreg Section, Morocco: proposed Global Boundary Stratotype Section and Point of the Pliocene. *Notes et Mémoires du Service Géologique du Maroc*, **383**, 51–150.

BESIO, G. & LOSADA, M. A. 2008. Sediment transport patterns at Trafalgar offshore windfarm. *Ocean Engineering*, **35**, 653–665.

BORREGO, J., MORALES, J. A. & PENDÓN, J. G. 1995. Holocene estuarine facies along the mesotidal coast of Huelva, south-western Spain. *In:* FLEMMING, B. W. & BARTHOLOMÄ, A. (eds) *Tidal Signatures in Modern and Ancient Sediments*. International Association of Sedimentologists, Special Publications, **24**, 151–170.

BOSKI, T., MOURA, D., VEIGA-PIRES, C., CAMACHO, S., DUARTE, D., SCOTT, D. B. & FERNANDES, S. G. 2002. Postglacial sea-level rise and sedimentary response in the Guadiana Estuary, Portugal/Spain border. *Sedimentary Geology*, **150**, 103–122.

BOSKI, T., CAMACHO, S. *ET AL*. 2008. Chronology of the sedimentary processes during the postglacial sea level rise in two estuaries of the Algarve coast, Southern Portugal. *Estuarine, Coastal and Shelf Science*, **77**, 230–244.

BUFORN, E., UDÍAS, A. & COLOMBÁS, M. A. 1988. Seismicity, source mechanisms and tectonics of the Azores-Gibraltar plate boundary. *Tectonophysics*, **152**, 89–118.

BURDLOFF, D., ARAÚJO, M. F., JOUANNEAU, J. M., MENDES, I., MONGE SOARES, A. M. & DIAS, J. M. A. 2008. Sources of organic carbon in the Portuguese continental shelf sediments during the Holocene period. *Applied Geochemistry*, **23**, 2857–2870.

CADET, J. P., FOURNIGUET, J., GIGOUT, M., GUILLEMIN, M. & PIERRE, G. 1977. L'histoire tectonique récente (Tortonien à Quaternaire) de l'Arc de Gibraltar et des bordures de la mer d'Alboran.

III – néotectonique des littoraux. *Bulletin de la Société Géologique de France*, **19**, 600–605.

CARRUESCO, C. 1989. *Genèse et évolution de trois lagunes du littoral atlantique depuis l'Holocène: Oualidia–Moualy Bou Salham (Maroc) et Arcachon (France)*. PhD thesis, University of Bordeux I.

CARVALHO, J., MATIAS, H., RABEH, T., MENEZES, P. T. L., BARBOSA, V. C. F., DIAS, R. & CARRILHO, F. 2012. Connecting onshore structures in the Algarve with the southern Portuguese continental margin: the Carcavai fault zone. *Tectonophysics*, **570–571**, 151–162.

CATALÁN, I. A., JIMÉNEZ, M. T., ALCONCHEL, J. I., PRIETO, L. & MUÑOZ, J. L. 2006. Spatial and temporal changes of coastal demersal assemblages in the Gulf of Cadiz (SW Spain) in relation to environmental conditions. *In:* RUIZ, J. & GARCÍA-LAFUENTE, J. (eds) *The Gulf of Cadiz Oceanography: A Multidisciplinary View. Deep Sea Research Part II: Topical Studies in Oceanography*, **53**, 1402–1419.

CENDRERO, A., SÁNCHEZ-ARCILLA, A. & ZAZO, C. 2004. Impactos sobre las zonas costeras. *Impactos del Cambio Climático en España*, 469–528.

CHALOUAN, A., MICHARD, A., EL KADIRI, K., NEGRO, F., FRIZON DE LAMOTTE, D. & SADDIQI, O. 2008. The Rif Belt. *In:* MICHARD, A., CHALOUAN, A., SADDIQI, O. & FRIZON DE LAMOTTE, D. (eds) *Continental Evolution: the Geology of Morocco. Structure, Stratigraphy, and Tectonic of the Africa-Atlantic-Mediterranean Triple Junction*. Springer, Berlin, 203–302.

CHEGGOUR, M., CHAFIK, A., FISHER, N. & BENBRAHIM, S. 2004. Metal concentrations in sediments and clams in four Moroccan estuaries. *Marine Environmental Research*, **59**, 119–137.

CHÍCHARO, M. A., CHÍCHARO, L. & MORAIS, P. 2006. Inter-annual differences in ichthyofauna structure of the Guadiana estuary and adjacent coastal area (SE Portugal/SW Spain): Before and after Alqueva dam construction. *Estuarine, Coastal and Shelf Science*, **70**, 39–51.

CIRAC, P., FAUGERES, J. C. & GAYET, J. 1979. Résultats préliminaires d'une reconnaissance sédimentaire du plateau atlantique marocain. *Bulletin de l'Institut Géologique du Bassin d'Aquitaine*, **25**, 69–81.

CONTRUCCI, I., KLINGELHOEFER, F. *ET AL*. 2004. The crustal structure of the NW-Moroccan continental margin from wide angle and reflection seismic data. *Geophysical Journal International*, **159**, 117–128.

COSTA, C. 1994. *Wind-wave Climatology of the Portuguese Coast*. Final Report of Sub-Project A. NATO PO-WAVES Report **6/94-A**.

COSTA, M., SILVA, R. & VITORINO, J. 2001. Contribuição para o estudo do clima de agitação marítima na costa portuguesa. *In: Proceedings das 2as Jornadas Portuguesas de Engenharia Costeira e Portuária*. International Navigation Association PIANC, Sines, Portugal (CD-ROM).

CRAVO, A., MADUREIRA, M., FELÍCIA, H., RITA, F. & BEBIANNO, M. J. 2006. Impact of outflow from the Guadiana River on the distribution of suspended particulate matter and nutrients in the adjacent coastal zone. *Estuarine, Coastal and Shelf Science*, **70**, 63–75.

CRIADO-ALDEANUEVA, F., GARCÍA-LAFUENTE, J., VARGAS, J. M., DEL RIO, J., SANCHEZ, A., DELGADO, J. & SANCHEZ, J. C. 2006a. Wind induced variability of hydrographic features and water masses distribution in the Gulf of Cadiz (SW Iberia) from in situ data. *Journal of Marine Systems*, **63**, 130–140.

CRIADO-ALDEANUEVA, F., GARCÍA-LAFUENTE, J., VARGAS, J. M., DEL RIO, J., VAZQUEZ, A., REUL, A. & SANCHEZ, A. 2006b. Distribution and circulation of water masses in the Gulf of Cadiz from in situ observations. *In:* RUIZ, J. & GARCÍA-LAFUENTE, J. (eds) *The Gulf of Cadiz Oceanography: A Multidisciplinary View. Deep Sea Research Part II: Topical Studies in Oceanography*, **53**, 1144–1160.

DABRIO, C. J., ZAZO, C. *ET AL*. 2000. Depositional history of estuarine infill during the last postglacial transgression (Gulf of Cadiz, Southern Spain). *Marine Geology*, **162**, 381–404.

DIAS, J. M. A. & NEAL, J. 1992. Sea cliff retreat in Southern Portugal: profiles, processes and problems. *Journal of Coastal Research*, **8**, 641–654.

DIAS, J. M. A., GONZALEZ, R. & FERREIRA, Ó. 2004. Natural versus Anthropic Causes in Variations of Sand Export from River Basins: an example from the Guadiana River Mouth (Southwestern Iberia). *Polish Geological Institute Special Papers*, **11**, 95–102.

EL FOUGHALI, A. & GRIBOULARD, R. 1985. Les grands traits structuraux et lithologiques de la marge atlantique marocaine de Tanger au Cap

Cantin. *Bulletin de l'Institut Géologique du Bassin d'Aquitaine*, **38**, 179–211.

ESTERAS, M., IZQUIERDO, J., SANDOVAL, N. G. & BAHMAD, A. 2000. Evolución morfológica y estratigráfica Plio-Cuaternaria del Umbral de Camarinal (Estrecho de Gibraltar) basada en sondeos marinos. *Revista de la Sociedad Geológica de España*, **13**, 539–550.

FERNÁNDEZ-SALAS, L. M., REY, J. ET AL. 1999. Morphology and characterisation of the relict facies on the internal continental shelf in the Gulf of Cadiz, between Ayamonte and Huelva (southern Iberian Peninsula). *Boletín del Instituto español de Oceanografía*, **15**, 123–132.

FERNÁNDEZ-SALAS, L. M., LOBO, F. J., HERNÁNDEZ-MOLINA, F. J., SOMOZA, L., RODERO, J., DÍAZ DEL RÍO, V. & MALDONADO, A. 2003. High-resolution architecture of late Holocene highstand prodeltaic deposits from southern Spain: the imprint of high-frequency climatic and relative sea-level changes. *Continental Shelf Research*, **23**, 1037–1054.

FIÚZA, A. 1983. Upwelling patterns off Portugal. *In*: SUESS, E. & THIEDE, J. (eds) *Coastal Upwelling*. Plenum, New York.

FIÚZA, A. F. G., MACEDO, M. E. & GUERREIRO, M. R. 1982. Climatological space and time variation of the Portuguese coastal upwelling. *Oceanologica Acta*, **5**, 31–40.

FLETCHER, W. J., BOSKI, T. & MOURA, D. 2007. Palynological evidence for environmental and climatic change in the lower Guadiana valley, Portugal, during the last 13 000 years. *The Holocene*, **17**, 481–494.

FLINCH, J. F. 1993. *Tectonic evolution of the Gibraltar Arc*. PhD thesis, Rice University.

FLINCH, J. F. & VAIL, P. R. 1999. Plio-Pleistocene sequence and tectonics of the Gibraltar Arc. *In*: DE GRACIANSKY, P.-C., HARDENBOL, J., JACQUIN, T. & VAIL, P. R. (eds) *Mesozoic and Cenozoic Sequence Stratigraphy of European Basins*. SEPM Special Publications, Tulsa, **60**, 201–208.

GARCÍA-LAFUENTE, J. & RUIZ, J. 2007. The Gulf of Cádiz pelagic ecosystem: a review. *Progress in Oceanography*, **74**, 228–251.

GARCÍA-LAFUENTE, J., DELGADO, J., CRIADO-ALDEANUEVA, F., BRUNO, M., DEL RIO, J. & MIGUEL VARGAS, J. 2006. Water mass circulation on the continental shelf of the Gulf of Cadiz. *In*: RUIZ, J. & GARCÍA-LAFUENTE, J. (eds) *The Gulf of Cadiz Oceanography: A Multidisciplinary View. Deep Sea Research Part II: Topical Studies in Oceanography*, **53**, 1182–1197.

GLOBAL OCEAN ASSOCIATES. 2004. Iberian Peninsula – Atlantic Coast. *In*: *An Atlas of Oceanic Internal Solitary Waves*. Office of Naval Research, Code – 322 PO. Global Ocean Associates, Alexandria, VA, 87–97.

GONZÁLEZ, R., DIAS, J. M. A. & FERREIRA, O. 2001. Recent rapid evolution of the Guadiana Estuary (Southern Portugal/Spain). *In*: *ICS 2000 – International Coastal Symposium. Journal of Coastal Research*, Special Issue **34**, 516–527.

GONZÁLEZ, R., DIAS, J. M. A., LOBO, F. & MENDES, I. 2004. Sedimentological and paleoenvironmental characterisation of transgressive sediments on the Guadiana Shelf (Northern Gulf of Cadiz, SW Iberia). *Quaternary International*, **120**, 133–144.

GONZÁLEZ, R., ARAÚJO, M. F. ET AL. 2007. Sediment and pollutant transport in the Northern Gulf of Cadiz: a multi-proxy approach. *Journal of Marine Systems*, **68**, 1–23.

GOY, J. L., ZAZO, C., DABRIO, C. J., LARIO, J., BORJA, F., SIERRO, F. J. & FLORES, J. A. 1996. Global and regional factors controlling changes of coastlines in southern Iberia (Spain) during the Holocene. *Quaternary Science Reviews*, **15**, 773–780.

GRÀCIA, E., DAÑOBEITIA, J., VERGÉS, J., BARTOLOMÉ, R. & CÓRDOBA, D. 2003. Crustal architecture and tectonic evolution of the Gulf of Cadiz (SW Iberian margin) at the convergence of the Eurasian and African plates. *Tectonics*, **22**, 1033.

GUTIÉRREZ-MAS, J. M., DOMÍNGUEZ-BELLA, S. & LÓPEZ AGUAYO, F. 1994. Present-day sedimentation patterns of the Gulf of Cadiz northern shelf from heavy mineral analysis. *Geo-Marine Letters*, **14**, 52–58.

GUTIÉRREZ-MAS, J. M., HERNÁNDEZ-MOLINA, F. J. & LÓPEZ AGUAYO, F. 1996. Holocene sedimentary dynamics on the Iberian continental shelf of the Gulf of Cadiz (SW Spain). *Continental Shelf Research*, **16**, 1635–1653.

GUTIÉRREZ-MAS, J. M., JUAN, C. & MORALES, J. A. 2009*a*. Evidence of high-energy events in shelly layers interbedded in coastal Holocene sands in Cadiz Bay (south-west Spain). *Earth Surface Processes and Landforms*, **34**, 810–823.

GUTIÉRREZ-MAS, J. M., LÓPEZ-ARROYO, J. & MORALES, J. A. 2009*b*. Recent marine lithofacies in Cadiz Bay (SW Spain): sequences, processes and control factors. *Sedimentary Geology*, **218**, 31–47.

GUTSCHER, M.-A., MALOD, J., REHAULT, J.-P., CONTRUCCI, I., KLINGELHOEFER, F., MENDES-VICTOR, L. & SPAKMAN, W. 2002. Evidence for active subduction beneath Gibraltar. *Geology*, **30**, 1071–1074.

GUTSCHER, M.-A., DOMINGUEZ, S. ET AL. 2009. Tectonic shortening and gravitational spreading in the Gulf of Cadiz accretionary wedge: observations from multi-beam bathymetry and seismic profiling. *Marine and Petroleum Geology*, **26**, 647–659.

HERNÁNDEZ-MOLINA, F. J., SOMOZA, L., REY, J. & POMAR, L. 1994. Late Pleistocene–Holocene sediments on the Spanish continental shelves: model for very high resolution sequence stratigraphy. *Marine Geology*, **120**, 129–174.

HERNÁNDEZ-MOLINA, F. J., FERNÁNDEZ-SALAS, L. M., LOBO, F., SOMOZA, L., DÍAZ-DEL-RÍO, V. & ALVEIRINHO DIAS, J. M. 2000*a*. The infralittoral prograding wedge: a new large-scale progradational sedimentary body in shallow marine environments. *Geo-Marine Letters*, **20**, 109–117.

HERNÁNDEZ-MOLINA, F. J., SOMOZA, L. & LOBO, F. 2000*b*. Seismic stratigraphy of the Gulf of Cádiz continental shelf: a model for Late Quaternary very high-resolution sequence stratigraphy and response to sea-level fall. *In*: HUNT, D. & GAWTHORPE, R. L. G. (eds) *Sedimentary Responses to Forced Regressions*. Geological Society, London, Special Publications, **172**, 329–362.

HERNÁNDEZ-MOLINA, F. J., SOMOZA, L., VÁZQUEZ, J. T., LOBO, F., FERNÁNDEZ-PUGA, M. C., LLAVE, E. & DÍAZ DEL RÍO, V. 2002. Quaternary stratigraphic stacking patterns on the continental shelves of the southern Iberian Peninsula: their relationship with global climate and palaeoceanographic changes. *Quaternary International*, **92**, 5–23.

HILBICH, C., MÜGLER, I., DAUT, G., FRENZEL, P., VAN DER BORG, K. & MÄUSBACHER, R. 2008. Reconstruction of the depositional history of the former coastal Lagoon of Vilamoura (Algarve, Portugal): a sedimentological, microfaunal and geophysical approach. *Journal of Coastal Research*, **24**, 83–91.

HINZ, K., WINTERER, E. L. ET AL. 1982. Preliminary results from DSDP Leg79 seaward of the Mazagan Plateau of Morocco. *In*: VON RAD, U., HINZ, K., SARNTHEIN, M. & SEIBOLD, E. (eds) *Geology of the Northwest African Continental Margin*. Springer, Berlin, 23–33.

HURRELL, J. W. 1995. Decadal trends in the North Atlantic Osicilation: regional temperatures and precipitation. *Science*, **269**, 676–679.

INSTITUTO HIDROGRÁFICO. 1985. *Carta dos Sedimentos da Plataforma Continental, Folha SED 7 e 8 do Cabo de S. Vicente ao Rio Guadiana. Escala 1:150.000*. Instituto Hidrográfico Português, Lisboa.

JAAIDI, E. B. 1993. *La couverture sedimentaire post-glaciaire de la plate-forme continentale atlantique ouest-rifaine (Maroc Nord-Occidental): exemple d'une séquence transgressive*. PhD thesis, University Mohammed V.

LARIO, J., ZAZO, C., DABRIO, C. J., SOMOZA, L., GOY, J. L., BARDAJI, T. & SILVA, P. G. 1995. Record of Recent Holocene sediment input on spit bars and deltas of south Spain. *Journal of Coastal Research*, **17**, 241–245.

LARIO, J., ZAZO, C. ET AL. 2002. Changes in sedimentation trends in SW Iberia Holocene estuaries (Spain). *Quaternary International*, **93–94**, 171–176.

LE ROY, P., PIQUÉ, A., LE GALL, B., AÏT BRAHIM, L., MORABET, A. M. & DEMANATI, A. 1997. Les bassins côtiers triasico-liasiques du Maroc occidental et la diachronie du rifting intra-continental de l'Atlantique Central. *Bulletin de la Société Géologique de France*, **168**, 637–648.

LE ROY, P., SAHABI, M., LAHSINI, S., MEHDI, K. & ZOURARAH, B. 2004. Seismic stratigraphy and Cenozoic evolution of the mesetan moroccan atlantic continental shelf. *Journal of African Earth Sciences*, **39**, 385–392.

LOBO, F. J. 2000. *Estratigrafía de alta resolución y cambios del nivel del mar durante el Cuaternario del margen continental del Golfo de Cádiz (S de España) y del Roussillon (S de Francia): Estudio comparativo*. PhD thesis, University of Cádiz.

LOBO, F. J., HERNÁNDEZ-MOLINA, F. J., SOMOZA, L. & DÍAZ DEL RÍO, V. 1999. Palaeoenvironments, relative sea-level changes and tectonic influence on the Quaternary seismic units of the Huelva continental shelf (Gulf of Cadiz, southwestern Iberian Peninsula). *Boletín del Instituto español de Oceanografía*, **15**, 161–180.

LOBO, F. J., HERNÁNDEZ-MOLINA, F. J., SOMOZA, L., RODERO, J., MALDONADO, A. & BARNOLAS, A. 2000. Patterns of bottom current flow deduced from dune asymmetries over the Gulf of Cadiz shelf (southwest Spain). *Marine Geology*, **164**, 91–117.

LOBO, F. J., HERNÁNDEZ-MOLINA, F. J., SOMOZA, L. & DÍAZ DEL RÍO, V. 2001. The sedimentary record of the post-glacial transgression on the Gulf of Cadiz continental shelf (Southwest Spain). *Marine Geology*, **178**, 171–195.

LOBO, F. J., HERNÁNDEZ-MOLINA, F. J., SOMOZA, L., DÍAZ DEL RÍO, V. & DIAS, J. M. A. 2002. Stratigraphic evidence of an upper Pleistocene TST to HST complex on the Gulf of Cádiz continental shelf (southwest Iberian Peninsula). *Geo-Marine Letters*, **22**, 95–107.

LOBO, F. J., DIAS, J. M. A., GONZÁLEZ, R., HERNÁNDEZ-MOLINA, F. J., MORALES, J. A. & DÍAZ DEL RÍO, V. 2003a. High-resolution seismic stratigraphy of a narrow, bedrock-controlled estuary: the Guadiana estuarine system, SW Iberia. *Journal of Sedimentary Research*, **73**, 973–986.

LOBO, F. J., DIAS, J. M. A., VÁZQUEZ, J. T., DÍAZ DEL RÍO, V., GONZÁLEZ, R. & FERNÁNDEZ-PUGA, M. C. 2003b. New data about neotectonic activity in the Eastern Algarve shelf, Gulf of Cadiz, SW Iberian Peninsula. *Thalassas*, **19**, 63–64.

LOBO, F. J., PLAZA, F., GONZÁLEZ, R., DIAS, J. M. A., KAPSIMALIS, V., MENDES, I. & DÍAZ DEL RÍO, V. 2004a. Estimations of bedload sediment transport in the Guadiana Estuary (SW Iberian Peninsula) during low river discharge periods. *In*: Sediment Transport in European Estuarine Environments: Proceedings of the STRAEE Workshop (Winter 2004). *Journal of Coastal Research*, Special Issue **41**, 12–26.

LOBO, F. J., SÁNCHEZ, R. *ET AL.* 2004b. Contrasting styles of the Holocene highstand sedimentation and sediment dispersal systems in the northern shelf of the Gulf of Cadiz. *Continental Shelf Research*, **24**, 461–482.

LOBO, F. J., DIAS, J. M. A., HERNÁNDEZ-MOLINA, F. J., GONZÁLEZ, R., FERNÁNDEZ-SALAS, L. M. & DÍAZ DEL RÍO, V. 2005a. Late Quaternary shelf-margin wedges and upper slope progradation in the Gulf of Cadiz margin (SW Iberian Peninsula). *In*: HODGSON, D. M. & FLINT, S. S. (eds) *Submarine Slope Systems: Processes and Products*. Geological Society, London, Special Publications, **244**, 7–25.

LOBO, F. J., FERNÁNDEZ-SALAS, L. M., HERNÁNDEZ-MOLINA, F. J., GONZÁLEZ, R., DIAS, J. M. A., DÍAZ DEL RÍO, V. & SOMOZA, L. 2005b. Holocene highstand deposits in the Gulf of Cadiz, SW Iberian Peninsula: a high-resolution record of hierarchical environmental changes. *Marine Geology*, **219**, 109–131.

LOBO, F. J., MALDONADO, A. & NOORMETS, R. 2010. Large-scale sediment bodies and superimposed bedforms on the continental shelf close to the Strait of Gibraltar: interplay of complex oceanographic conditions and physiographic constraints. *Earth Surface Processes and Landforms*, **35**, 663–679.

LOPES, F. C., CUNHA, P. P. & LE GALL, B. 2006. Cenozoic seismic stratigraphy and tectonic evolution of the Algarve margin (offshore Portugal, southwestern Iberian Peninsula). *Marine Geology*, **231**, 1–36.

LÓPEZ-GALINDO, A., RODERO, J. & MALDONADO, A. 1999. Surface facies and sediment dispersal patterns: southeastern Gulf of Cadiz, Spanish continental margin. *Marine Geology*, **155**, 83–98.

LUJÁN, M., CRESPO-BLANC, A. & COMAS, M. 2011. Morphology and structure of the Camarinal Sill from high-resolution bathymetry: evidence of fault zones in the Gibraltar Strait. *Geo-Marine Letters*, **31**, 163–174.

LUQUE, L., LARIO, J., ZAZO, C., GOY, J. L., DABRIO, C. J. & SILVA, P. G. 2001. Tsunami deposits as paleoseismic indicators: examples from the Spanish coast. *Acta Geológica Hispánica*, **36**, 197–211.

MAAD, N., LE ROY, P., SAHABI, M., GUTSCHER, M. A., BABONNEAU, N., RABINEAU, M. & VAN VLIET LANOË, B. 2010. The seismic stratigraphy of the NW Moroccan Atlantic continental shelf and Quaternary deformations at the offshore termination of the Southern Rif front. *Comptes Rendus Geoscience*, **342**, 731–740.

MALDONADO, A. & NELSON, C. H. 1999a. Interaction of tectonic and depositional processes that control the evolution of the Iberian Gulf of Cadiz margin. *In*: MALDONADO, A. & NELSON, C. H. (eds) Marine Geology of the Gulf of Cadiz. *Marine Geology*, Special Issue **155**, 217–242.

MALDONADO, A. & NELSON, C. H. (eds) 1999b. *Marine Geology of the Gulf of Cadiz*. Marine Geology, Special Issue **155**.

MALDONADO, A., SOMOZA, L. & PALLARÉS, L. 1999. The Betic orogen and the Iberian-African boundary in the Gulf of Cadiz: geological evolution (central North Atlantic). *Marine Geology*, **155**, 9–43.

MALDONADO, A., RODERO, J. *ET AL.* 2003. *Mapa Geológico de la Plataforma Continental Española y Zonas Adyacentes. Escala 1:200.000. Hojas n°86-86S-87S (Cádiz)*. Instituto Geológico y Minero de España, Madrid.

MALOD, J. A. 1982. *Comparaison de l'évolution des marges continentales au nord et au sud de la Péninsule Ibérique*. PhD thesis, University Pierre et Marie Curie.

MALOD, J. A. & MOUGENOT, D. 1979. L'histoire géologique néogène du Golfe de Cadix. *Bulletin de la Societé Geologique de France*, **XXI**, 603–611.

MARTÍN-SERRANO, A. (ed.) 2005. *Mapa Geomorfológico de España y del margen continental. Escala 1:1.000.000*. Instituto Geológico y Minero de España, Madrid.

MEDIALDEA, T., VEGAS, R. *ET AL.* 2004. Structure and evolution of the 'Olistostrome' complex of the Gibraltar Arc in the Gulf of Cádiz (eastern Central Atlantic): evidence from two long seismic cross sections. *Marine Geology*, **209**, 173–198.

MEDIALDEA, T., SOMOZA, L. *ET AL.* 2009. Tectonics and mud volcano development in the Gulf of Cádiz. *Marine Geology*, **261**, 48–63.

MENDES, I., GONZALEZ, R., DIAS, J. M. A., LOBO, P. & MARTINS, V. 2004. Factors influencing recent benthic foraminifera distribution on the Guadiana shelf (Southwestern Iberia). *Marine Micropaleontology*, **51**, 171–192.

MENDES, I., ROSA, F., DIAS, J. A., SCHÖNFELD, J., FERREIRA, Ó. & PINHEIRO, J. 2010. Inner shelf paleoenvironmental evolution as a function of land–ocean interactions in the vicinity of the Guadiana River, SW Iberia. *Quaternary International*, **221**, 58–67.

MENDES, I., DIAS, J. A., SCHÖNFELD, J. & FERREIRA, Ó. 2012a. Distribution of living benthic foraminífera on the northern Gulf of Cadiz continental shelf. *Journal of Foraminiferal Research*, **42**, 18–38.

MENDES, I., DIAS, J. A., SCHÖNFELD, J., FERREIRA, Ó., ROSA, F., GONZALEZ, R. & LOBO, F. J. 2012b. Natural and human-induced Holocene paleoenvironmental changes on the Guadiana shelf (northern Gulf of Cadiz). *The Holocene*, **22**, 1011–1024.

MHAMMDI, N., MEDINA, F., KELLETAT, D., AHMAMOU, M. & ALOUSSI, L. 2008. Large boulders along the Rabat coast (Morocco); possible emplacement by the November, 1st, 1755 A.D. Tsunami. *Science of Tsunami Hazards*, **27**, 17–30.

MOITA, I. 1986. *Plataforma Continental. Carta dos Sedimentos Superficiais. Notícia explicativa da folha sed 7 e 8. Cabo de S. Vicente ao rio Guadiana*. Instituto Hidrográfico, Lisboa.

MORALES, J. A. 1997. Evolution and facies architecture of the mesotidal Guadiana River delta (S.W. Spain-Portugal). *Marine Geology*, **138**, 127–148.

MORALES, J. A., PENDÓN, J. G. & BORREGO, J. 1994. Origen y evolución de flechas litorales recientes en la desembocadura del estuario mesomareal del río Guadiana (Huelva, SO de España). *Revista de la Sociedad Geológica de España*, **7**, 155–167.

MORALES, J. A., DELGADO, I. & GUTIÉRREZ-MAS, J. M. 2006. Sedimentary characterization of bed types along the Guadiana estuary (SW Europe) before the construction of the Alqueva dam. *Estuarine, Coastal and Shelf Science*, **70**, 117–131.

MORALES, J. A., BORREGO, J., SAN MIGUEL, E. G., LÓPEZ-GONZÁLEZ, N. & CARRO, B. 2008. Sedimentary record of recent tsunamis in the Huelva Estuary (southwestern Spain). *Quaternary Science Reviews*, **27**, 734–746.

NELSON, C. H., BARAZA, J., MALDONADO, A., RODERO, J., ESCUTIA, C. & BARBER, J. H. 1999. Influence of the Atlantic inflow and Mediterranean outflow currents on Late Quaternary sedimentary facies of the Gulf of Cadiz continental margin. *Marine Geology*, **155**, 99–129.

NIFTAH, S., DEBÉNATH, A. & MISKOVSKY, J.-C. 2005. Origine du remplissage sédimentaire des grottes de Témara (Maroc) d'après l'étude des minéraux lourds et l'étude exoscopique des grains de quartz. *Quaternaire*, **16**, 73–83.

OLIVEIRA, J. T., PEREIRA, E., RAMALHO, M., ANTUNES, M. T. & MONTEIRO, J. H. (coords) 1992. *Carta Geológica de Portugal a escala 1:500000*. Serviços Geológicos de Portugal, Lisboa.

OUADIA, M. 1998. *Les formations plio-quaternaires dans le domaine Mésétien occidental du Maroc entre Casablanca et Safi: géomorpholohie, sédimentologie, paléoenvironnements quaternaires et évolution actuelle*. PhD thesis, University Mohammed V-Agdal, Rabat.

PALANQUES, A., PLANA, F. & MALDONADO, A. 1986–1987. Estudio de la materia en suspensión en el Golfo de Cádiz. *Acta Geológica Hispánica*, **21–22**, 491–497.

PALANQUES, A., DÍAZ, J. I. & FARRÁN, M. 1995. Contamination of heavy metals in the suspended and surface sediment of the Gulf of Cadiz (Spain): the role of sources, currents, pathways and sinks. *Oceanologica Acta*, **18**, 469–477.

PENAUD, A., EYNAUD, F. ET AL. 2010. Contrasting paleoceanographic conditions off Morocco during Heinrich events (1 and 2) and the Last Glacial Maximum. *Quaternary Science Reviews*, **29**, 1923–1939.

PERIÁÑEZ, R. 2009. Environmental modelling in the Gulf of Cadiz: Heavy metal distributions in water and sediments. *Science of the Total Environment*, **407**, 3392–3406.

PILKEY, O. H., JR, NEAL, W. J., MONTEIRO, J. H. & DIAS, J. M. A. 1989. Algarve Barrier Islands: a noncoastal-plain system in Portugal. *Journal of Coastal Research*, **5**, 239–261.

PLAZIAT, J. C., ABERKAN, M., AHMAMOU, M. & CHOUKRI, A. 2008. The quaternary deposits of Morocco. In: MICHARD, A., CHALOUAN, A., SADDIQI, O. & FRIZON DE LAMOTTE, D. (eds) *Continental Evolution: The Geology of Morocco. Structure, Stratigraphy, and Tectonic of the Africa–Atlantic–Mediterranean Triple Junction*. Springer, Berlin, 359–376.

PORTELA, L. I. 2006. Sediment delivery from the Guadiana Estuary to the Coastal Ocean. *Journal of Coastal Research*, **SI 39**, 1819–1823.

PRIETO, L., NAVARRO, G., RODRÍGUEZ-GÁLVEZ, S., HUERTAS, I. E., NARANJO, J. M. & RUIZ, J. 2009. Oceanographic and meteorological forcing of the pelagic ecosystem on the Gulf of Cadiz shelf (SW Iberian Peninsula). *Continental Shelf Research*, **29**, 2122–2137.

RELVAS, P. & BARTON, E. D. 2002. Mesoscale patterns in the Cape São Vicente (Iberian Peninsula) upwelling region. *Journal of Geophysical Research*, **107**, 3164.

RELVAS, P. & BARTON, E. D. 2005. A separated jet and coastal counterflow during upwelling relaxation off Cape São Vicente (Iberian Peninsula). *Continental Shelf Research*, **25**, 29–49.

RELVAS, P., BARTON, E. D., DUBERT, J., OLIVEIRA, P., PELIZ, Á., DA SILVA, J. C. B. & SANTOS, A. M. P. 2007. Physical oceanography of the western Iberia ecosystem: Latest views and challenges. *Progress in Oceanography*, **74**, 149–173.

REY, J. & MEDIALDEA, T. 1989. Morfología y sedimentos recientes del margen continental de Andalucía Occidental. In: DÍAZ DEL OLMO, F. & RODRÍGUEZ VIDAL, J. (eds) *El Cuaternario en Andalucía Occidental*. AEQUA, Monografías, **1**, 133–144.

RIAZA, C. & MARTÍNEZ DEL OLMO, W. 1996. Depositional model of the Guadalquivir–Gulf of Cadiz Tertiary Basin. In: FRIEN, P. F. & DABRIO, C. J. (eds) *Tertiary Basins of Spain: The Stratigraphic Record of Crustal Kinematics*. Cambridge University Press, Cambridge, 330–338.

ROBERTS, D. G. 1970. The Rif–Betic orogen in the Gullf of Cadiz. *Marine Geology*, **9**, M31–M37.

RODERO, J., PALLARES, L. & MALDONADO, A. 1999. Late Quaternary seismic facies of the Gulf of Cadiz Spanish margin: depositional processes influenced by sea-level change and tectonic controls. *Marine Geology*, **155**, 131–156.

RODRÍGUEZ-RAMÍREZ, A., RODRÍGUEZ VIDAL, J. ET AL. 1996. Recent coastal evolution of the Doñana National Park (SW Spain). *Quaternary Science Reviews*, **15**, 803–809.

ROQUE, C. 1998. *Análise morfosedimentar da sequência deposicional do Quaternário Superior da plataforma continental Algarvia entre Faro e a foz do Rio Guadiana*. Master thesis, University of Lisbon.

ROSA, F., DIAS, J. A., MENDES, I. & FERREIRA, Ó. 2011. Mid to late Holocene constraints for continental shelf mud deposition in association with river input: the Guadiana Mud Patch (SW Iberia). *Geo-Marine Letters*, **31**, 109–121.

ROSA, F., RUFINO, M. M., FERREIRA, Ó., MATIAS, A., BRITO, A. C. & GASPAR, M. B. 2013. The influence of coastal processes on inner shelf sediment distribution: the Eastern Algarve Shelf (Southern Portugal). *Geologica Acta*, **11**, 59–73.

RUELLAN, E. 1985. *Géologie des marges continentales passives. Evolution de la marge atlantique du Maroc (Mazagan); étude par submersible, Seabeam et sismique-réflexion. Comparaison avec la marge N. W africaine et la marge homologue E. américaine*. PhD thesis, University of Bretagne Occidentale.

SAIIDI, E. K. 1979. *Etude géologique du Quaternaire de la Méséta côtière marocaine. Terrasses fluviatiles et autres types d'épendages*. PhD thesis, University Mohammed V.

SAHABI, M., ASLANIAN, D. & OLIVET, J. L. 2004. A new starting point for the history of the central Atlantic. *Comptes Rendus Geoscience*, **336**, 1041–1052.

SARTORI, R., TORELLI, L., ZITELLINI, N., PEIS, D. & LODOLO, E. 1994. Eastern segment of the Azores-Gibraltar line (central-eastern Atlantic): an oceanic plate boundary with diffuse compressional deformation. *Geology*, **22**, 555–558.

SAUVAGE, C. 1963. Etages bioclimatiques. In: *Atlas du Maroc, section II, carte 6b. Notice explicative par C. Sauvage*. Comité National de Géographie du Maroc, Rabat.

SNOUSSI, M. 1984. Comportement du Pb, Zn, Ni et Cu dans les sédiments de l'estuaire du Loukkos et du proche plateau continental (côte atlantique marocaine). *Bulletin de l'Institut Géologique du Bassin d'Aquitaine*, **35**, 23–30.

SNOUSSI, M., JOUANNEAU, J. M. & LATOUCHE, C. 1990. Flux de matieres issues de bassins versants de zones semi-arides (Bassins du Sebou et du Souss, Maroc). Importance dans le bilan global des apports d'origine continentale parvenant à l'océan Mondial. *Journal of African Earth Sciences*, **11**, 43–54.

SOMOZA, L., HERNÁNDEZ-MOLINA, F. J., DE ANDRÉS, J. R. & REY, J. 1997. Continental shelf architecture and sea-level cycles: Late Quaternary high-resolution stratigraphy of the Gulf of Cádiz, Spain. *Geo-Marine Letters*, **17**, 133–139.

TAHAYT, A., MOURABIT, T. ET AL. 2008. Mouvements actuels des blocs tectoniques dans l'arc Bético-Rifain à partir des mesures GPS entre 1999 et 2005. *Comptes Rendus Geoscience*, **340**, 400–413.

TELES-MACHADO, A., PELIZ, A., DUBERT, J. & SÁNCHEZ, R. F. 2007. On the onset of the Gulf of Cadiz Coastal Countercurrent. *Geophysical Research Letters*, **34**.

TREGUER, P. & LE CORRE, P. 1979. The ratios of nitrate, phosphate and silicate during uptake and regeneration phases of the Moroccan upwelling regime. *Deep-Sea Research*, **26**, 163–184.

TRENTESAUX, A., LE ROY, P. ET AL. 2009. Giant dunes at the Atlantic Entrance of the Mediterranean Sea setting and preservation processes. In: *12ème Congrès Français de Sédimentologie*, Rennes, 25–31 Octobre 2009. Association des Sédimentologues Français, Publication, **66**.

TRIGO, R. M., POZO-VÁZQUEZ, D., OSBORN, T. J., CASTRO-DÍEZ, Y., GÁMIZ-FORTIS, S. & ESTEBAN-PARRA, M. J. 2004. North Atlantic Oscillation influence on precipitation, river flow and water resources in the Iberian Peninsula. *International Journal of Climatology*, **24**, 925–944.

TORELLI, L., SARTORI, R. & ZITELLINI, N. 1997. The giant chaotic body in the Atlantic Ocean off Gibraltar: new results from a deep seismic reflection survey. *Marine and Petroleum Geology*, **14**, 125–138.

VANNEY, J.-R. & MOUGENOT, D. 1981. *La plate-forme continentale du Portugal et les provinces adjacentes: Analyse geomorphologique*. Memórias dos Serviços Geológicos de Portugal, **28**.

VARGAS, J. M., GARCIA-LAFUENTE, J., DELGADO, J. & CRIADO, F. 2003. Seasonal and wind-induced variability of Sea Surface Temperature patterns in the Gulf of Cadiz. *Journal of Marine Systems*, **38**, 205–219.

VÁZQUEZ, J. T., FERNÁNDEZ-PUGA, M. C. ET AL. 2010. Fracturación normal durante el Cuaternario Superior en la plataforma continental septentrional del Golfo de Cádiz (SO de Iberia). Paper presented at the Resúmenes de la 1ª Reunión Ibérica sobre Tectónica Activa y Paleosismología, Sigüenza, Spain.

VON RAD, U., RYAN, W. B. F. *ET AL.* 1979. *Initial Reports of the Deep Sea Drilling Project, Part 1*, **47**. US Government Printing Office, Washington, DC.

WEISROCK, A., DELIBRIAS, G., ROGNON, P. & COUDÉ-GAUSSEN, G. 1985. Variations climatiques et morphogenèse au Maroc atlantique (30–33°N) à la limite Pléistocène-Holocène. *Bulletin de la Société de Géologique de France*, **I**, 565–569.

ZAZO, C., GOY, J. L. *ET AL.* 1994. Holocene sequence of sea-level fluctuations in relation to climatic trends in the atlantic-mediterranean linkage coast. *Journal of Coastal Research*, **10**, 933–945.

ZAZO, C., DABRIO, C. J. *ET AL.* 2008. The coastal archives of the last 15 ka in the Atlantic–Mediterranean Spanish linkage area: sea level and climate changes. *Quaternary International*, **181**, 72–87.

ZITELLINI, N., GRÀCIA, E. *ET AL.* 2009. The quest for the Africa–Eurasia plate boundary west of the Strait of Gibraltar. *Earth and Planetary Science Letters*, **280**, 13–50.

Chapter 10

Recent sedimentation in the NW African shelf

NADIA MHAMMDI[1]*, MARIA SNOUSSI[2], FIDA MEDINA[3,4] & EL BACHIR JAAÏDI[2]

[1]*Géophysique de subsurface et Environnement, Laboratoire de Physique du Globe, Institut Scientifique,
Université Mohammed V-Agdal, BP 703 Agdal, Rabat, Morocco*

[2]*Département des Sciences de la Terre, Faculté des Sciences, Université Mohammed V-Agdal,
BP 1014 R.P., Rabat, Morocco*

[3]*Laboratoire de Géologie et Télédétection (URAC 46), Institut Scientifique, Université
Mohammed V-Agdal, BP 703 Agdal, Rabat, Morocco*

[4]*Present address: Rue Oued Draa, No. 28, Agdal, Rabat, Morocco*

*Corresponding author (e-mail: mhammdi@israbat.ac.ma)

Abstract: A review of the data gathered over the last five decades on the NW African continental shelf within numerous research programmes shows that the sedimentary processes along the shelf are driven by long-term factors such as Quaternary glacial–interglacial periods and shelf morphology, and by short-term factors such as fluvial and aeolian sediment supply, local climate (temperature, rainfall and wind) and hydrodynamic conditions (tides, swell, longshore current, the Canary Current and upwelling). Based on the sedimentary characteristics, the margin has been subdivided into four segments: northern Morocco (30–36°N) and the Grande Côte of Senegal (15–16°30′N) show long mid-shelf mud belts controlled by fluvial input and low hydrodynamic energy; southern Morocco–Mauritania (16°30′N–30°N) and the Petite Côte south of Cape Vert, Gambia and Casamance (12–15°N) are dominated by biogenic and aeolian sands or no sediment input. In this chapter, we present a synopsis of the state of knowledge on the NW African shelf sediments and processes based on various published and unpublished documents.

Over the past 50 years, there has been an increase in scientific interest, research effort and information gathering on the continental margin off NW Africa, as part of the general geomarine investigation programmes carried out on a global scale after World War II (Bond & Kominz 1988). These studies have focused on the physical and chemical oceanography (Furnestin 1959; Barton 1998), on the geological and deep structures (e.g. Von Rad *et al.* 1982; Hafid *et al.* 2006), on the sedimentary characteristics and environmental significance of deposits (Summerhayes *et al.* 1976; Jaaïdi & Cirac 1987) and, more recently, on the hydrocarbon potential (e.g. Davison 2005). Consequently, data and information gathered from these numerous publications have greatly improved the understanding of the geological origin and evolution of the margin.

However, since the outstanding paper of the Woods Hole Oceanographic Institution researchers (Summerhayes *et al.* 1976), and with the exception of the work by Jaaïdi (1993) on the whole Moroccan platform, most of the published findings on the shelf concern relatively small areas or restricted issues. This is either as a result of limited data coverage or because the previously observed geological processes were active only within restricted areas, so these data have been combined and integrated rarely.

The main objective of our contribution to this Memoir is to present a synopsis of the published literature on the NW African continental shelf (abridged to NWAS hereafter) extending from the Strait of Gibraltar in the north to southern Senegal in the south. Figure 10.1 shows the location and bathymetry of the studied area from the Gibraltar Strait at 36°N to Cape Vert at 15°N, using the bathymetry taken from the gridded chart of the General Bathymetric Chart of the Oceans (GEBCO).

In the following sections, we first expose the general setting of the margin (climate, and oceanographic and hydrological setting), then we review the sedimentation regime and sources, and the tectonic and geological setting. In the last sections, we assess the Pre-Quaternary and Quaternary sedimentary evolution, describe the shelf sediments distribution and briefly discuss the sedimentary processes.

General setting

Climate

The modern climate of NW Africa depends on the latitudinal position of the considered segment. At its northern part, the climate is under the influence of the mid-latitude westerlies, while the tropical southern part is under the control of the seasonal position of the Intertropical Convergence Zone (ITCZ), to which is tightly related the wind field that brings drought or humidity (Nicholson 2000).

In the northern segment (36–30°N), the climate is typically temperate with four distinct seasons, including a really cold season due to the arrival of polar air masses. Thus, snow is common above 1500 m and even persistent in summer, as on Mount Toubkal. In the western Rif chain, the climate is sub-humid with a mean annual rainfall of up to 900 mm in Tangier (Sauvage 1963; Stour & Agoumi 2008). Southwards, in Casablanca, the annual average temperature is 18 °C; the mean summer temperature ranges between 28 and 30 °C, and the annual rainfall decreases to 450 mm.

From Agadir to northern Senegal (30–20°N), the climate becomes Saharan. According to the maps of Nicholson (2000), the mean annual rainfall in this region is only 25–50 mm from 1925 to 1990. However, the effectively recorded annual rainfall at Nouakchott from 1940 to 2005 was from 0 (years 1977, 1984) to 240 mm (1944, 1956) but the mean sharply decreased from 150 mm in 1940 to 75 mm in 2005. Maximum drought was recorded in the 1980s (Ould Sidi Cheikh *et al.* 2007).

South of 20–24°N, summer monsoonal rains play a major role in subtropical African climate. During boreal summer (June–August), the ITCZ and associated Tropical Rain Belt migrate northwards (19°N) and bring moisture-laden air over the Senegal River basin. For instance, in Senegal, during summer (27 °C and 254 mm precipitation during August on average), strong Earth-surface turbulence is associated with the monsoonal front system. During winter (December–February), the ITCZ migrates

From: CHIOCCI, F. L. & CHIVAS, A. R. (eds) 2014. *Continental Shelves of the World: Their Evolution During the Last Glacio-Eustatic Cycle.*
Geological Society, London, Memoirs, **41**, 131–146. http://dx.doi.org/10.1144/M41.10

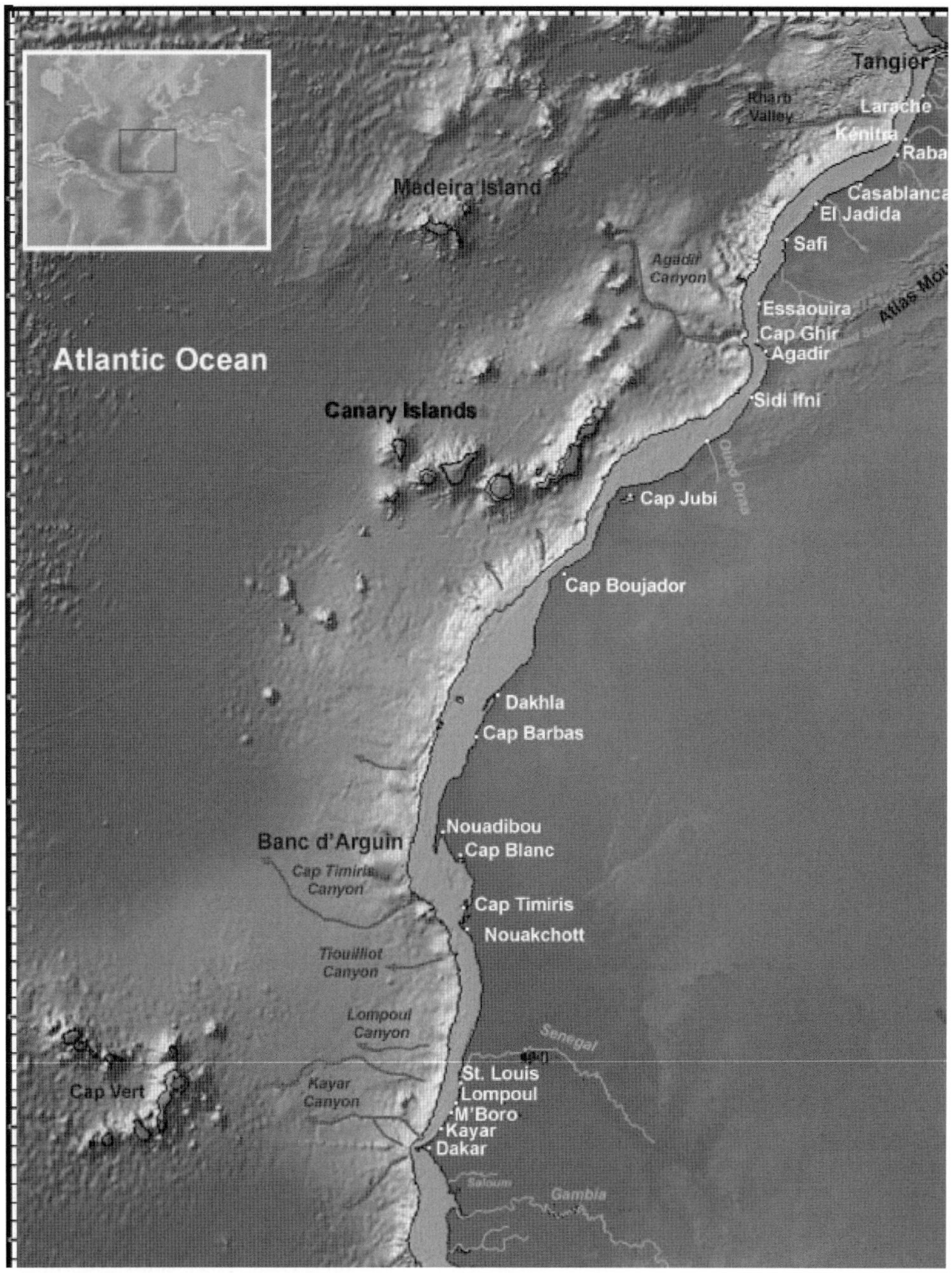

Fig. 10.1. Bathymetry of the NW African continental margin, taken from the gridded chart of the General Bathymetric Chart of the Oceans (GEBCO).

toward the south (5°N), which induces cool and dry conditions over the Senegal River basin (Leroux 2001). During this season (22 °C and no precipitation in January on average), the NE Trade Winds are the dominant atmospheric feature.

In the southernmost part of the margin, the climate becomes much wetter, and the interannual rainfall variability is controlled by the monsoon. For example, at Saloum, the annual rainfall was 1200–1400 mm prior to the 1970s but a decrease down to 500 mm has been recorded since then (Diara 1999).

Oceanographic and hydrological setting

Over the last few decades, the use of databases from terrestrial and satellite sources coupled with modelling has greatly improved our knowledge of ocean circulation along the NW African margin, as well as the fishery and mineral resources related to upwelling. For instance, oceanic circulation has been modelled using the ROMS software (Regional Ocean Modelling System: Moujane *et al.* 2011*b*), wind forcing using data from NOGAPS (Navy Operational Global Atmospheric Prediction System) (Batteen *et al.* 2007), ALADIN and QuickScat (Moujane *et al.* 2011*a*, *b*), and

sea surface temperatures and salinity from AVHRR (Advanced Very High Resolution Radiometer) (Van Camp *et al.* 1991; Atillah *et al.* 2005; Pelegrí *et al.* 2005) and COADS (Comprehensive Ocean–Atmosphere Data Set) (Batteen *et al.* 2007). Integrated models were also used, such as POM (Princeton Ocean Model: Batteen *et al.* 2007). Field control of surface current velocities was mainly carried out by tracking drifting buoys (Sena Martins *et al.* 2002; Barton *et al.* 2004) or by the use of current meters.

Tides. At the northern part of the NWAS, tides are mesotidal (amplitude of 1–3.4 m) and semi-diurnal (24 h 50 min). The flood tides induce northerly currents and, oppositely, southerly currents during ebb tides. Observed velocities at 1 m depth are 20–30 cm s^{-1} during spring tides (Jaaïdi 1993). The tide amplitude along the Senegal coast is 1.1–1.6 m during spring tide and 0.5–0.9 m during neap tide (Diara 1999). The mean velocity is 10–15 cm s^{-1} (Domain 1977). Further south, the tide amplitude is 0.9–1.6 m during spring tide and 0.4–0.9 m during neap tide, and surface velocities up to 110 cm s^{-1} were recorded at the mouth of the Casamance River (Diop 1990).

Swell. Swell characteristics in several Moroccan harbours were regularly collected during the period 1928–52 by the Service de Physique du Globe et Météorologie of the Institut Scientifique Chérifien (Simonet & Tanguy 1956). At Rabat, the main swell directions during storms were SW, NW and west, with an amplitude of 0.5–1.5 m. Extreme amplitudes of 7 m were observed mainly in winter (January–February). Northwards, on the Larache coast, Tejera de Leon & Duplantier (1981) indicated that the main swell direction (amplitude of 0.5 m) is NW, but swell with maximum energy had a SW direction.

In Mauritania off Nouakchott, the major swell directions are NW, WNW and west, with periods of 8–11 s and significant heights of 1.64–2.26 m (Ould El Moustapha *et al.* 2007).

Off the Senegal coast, two swell systems occur: (1) a major high-energy NW swell from October to May, with a direction of propagation that fluctuates from 10°N to 30°N, disturbing the water column down to a depth of 30 m (wave length of 300 m, period varying from 8 to 12 s, height of 1 m and a velocity of 21 m s^{-1}: Pinson-Mouillot 1980); (2) a weaker south to SW swell from June to September, with a period of 5–8 s, an amplitude of 0.8 m and significant heights of 0.75 and 2.25 m; this subordinate swell may interfere with the dominant one, enhancing or diminishing the littoral drift (Masse 1968; Pinson-Mouillot 1980).

The Canary Current. The NW African continental shelf is mainly under the influence of the southward-flowing Canary Current (CC) and, to a lesser extent, that of a northward-flowing counter-current located immediately to the west (Mittelstaedt 1991; Pelegrí *et al.* 2005). The CC represents the eastern branch of the North Atlantic Subtropical Gyre, and is one of the main Eastern Boundary Currents (EBC) in the world (Allain 1970; Hagen 2001; Carr 2002). Other currents appear close to the Equator (15–10°N), such as the SW-flowing North Equatorial Current (NEC), the eastward-flowing North Equatorial Counter Current (NECC) and the poleward-flowing currents (Mittelstaedt 1991).

The CC displaces water masses down to a depth of 500 m in the open sea over distances of 1000 km. Reported velocities are 25–75 cm s^{-1} (US Naval Oceanographic Office 1965 *in* Summerhayes *et al.* 1976) and 10–30 cm s^{-1} (Zhou *et al.* 2000) but may reach 75 cm s^{-1} near the Canary Islands (Fedoseev 1970). Modelling by researchers at the Hamburg Hydrographic Institute has given mean seasonal values of <10 cm s^{-1} (Mittelstaedt 1991). More detailed descriptions of the general circulation off NW Africa are given by Mittelstaedt (1983, 1991), Barton (1998) and Batteen *et al.* (2007).

At the northern boundary of the CC, the surface circulation of the Moroccan Shelf derives from the general anticyclonic circulation in the Gulf of Cádiz. Its southern branch is directed towards the SSW along the Moroccan coast with average velocities of about 30 cm s^{-1} before joining the CC (Criado-Aldeanueva *et al.* 2006), also fed by the Azores Current flowing eastwards at about 35°N (Knoll *et al.* 2002).

South of the Canary Islands, there is a small-scale formation of cyclonic and anticyclonic eddies, which result from the lee-side hydrographic and atmospheric conditions (Arístegui *et al.* 1994; Barton 1998).

Offshore Cape Blanc the interaction of these currents results in a cyclonic eddy circulation and a deflection of the CC from the coast (Bein & Fütterer 1977); however, only a minor part of the CC flow continues as far south as Cape Verde, then detaches from the African coast under the influence of the North Equatorial Counter Current merging into the Guinea Current (Wooster *et al.* 1976; Barton 1998; Batteen *et al.* 2000) (Fig. 10.2).

Wind forcing and coastal upwelling. It is well known that the atmospheric circulation pattern off NW Africa depends on the relative position and strength of the Icelandic Low relative to the Azores High, and the displacement of the ITCZ. Accordingly, the most relevant wind systems along the NW African coast are the NE Trade Winds and the Saharan Air Layer (SAL). The main surface wind directions at selected meteorological stations from 12°N to 36°N and their seasonality are illustrated by Kirk & Speth (1985), and the southern segment from Mauritania to Guinea-Bissau by Faye (2010). Recent satellite observations based on radar provide more continuous – although less precise – spatial–temporal measurements (Moujane *et al.* 2011*a*).

Along the NW African coast, the NE Trade Winds act on the CC with stresses of 0.9–1.9 dyn cm^{-1} (Mittelstaedt 1991) and force the upwelling of cold water masses in response to Ekman transport over the continental shelf area (Hagen 2001). The upwelling zone is restricted to a coastal strip of about 50–70 km (Mittelstaedt 1991; Hagen 2001; Makaoui *et al.* 2005), with the highest upwelling intensity being located at capes, such as Cape Ghir, Cape Jubi, Cape Blanc and Cape Vert (Makaoui *et al.* 2012).

The seasonal and spatial coverage dependence of the upwelling system has been extensively described by several authors (Wooster *et al.* 1976; Mittelstaedt 1991; Van Camp *et al.* 1991; Nykjaer & Van Camp 1994; Abrantes *et al.* 2002; Freudenthal *et al.* 2002; El Moussaoui *et al.* 2003; Helmke 2004; Makaoui *et al.* 2005). In general, seasonal migration of the strong upwelling areas and of dust transport trajectories appears related to the displacement of the Azores High, and the seasonal variability of the trade-wind belt and the ITCZ. Consequently, upwelling occurs during summer and autumn north of 25°N, is persistent throughout the year with a maximum intensity during the spring and autumn between 20°N and 25°N, and only during winter south of 20°N (Mittelstaedt 1991).

A major feature of the upwelling zone is the formation of filament structures generated by the interaction of the current system and the coastal cape morphology (Hagen *et al.* 1996; Stevens & Johnson 2003; Helmke 2004), which transport the upwelled water several hundreds of kilometres offshore (Van Camp *et al.* 1991; Nykjaer & Van Camp 1994; Hernandez-Guerra & Nykjaer 1997; Hagen 2001; Davenport *et al.* 2002; Helmke 2004). Kostianoy & Zatsepin (1996) identified up to 60 filaments

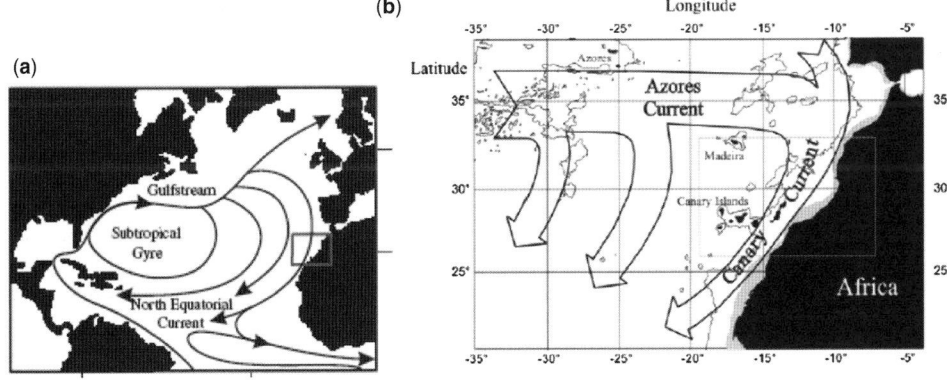

Fig. 10.2. The Atlantic currents of NW Africa: (a) Atlantic Subtropical Gyre circulation pattern; and (b) circulation pattern along the eastern branch of the Atlantic Subtropical Gyre.

over 1000 km of coastline, the most conspicuous being the 'giant' filament off Cape Blanc. Smaller filaments were also found north of the latter, in the area of Cape Ghir (Hagen *et al.* 1996), attaining 15 °C above a depth of 100 m (Head *et al.* 1996; Helmke 2004), and Cape Juby (Arístegui *et al.* 1994; Barton 1998).

SST and vertical profiles. Sea surface temperatures (SST) measured directly along the NWAS do not generally exceed 20 °C. In the northern segment (36–25°N), maximum surface temperatures may reach 19 °C in winter, 20 °C in spring, 23 °C during summer and 22 °C in autumn. A clear latitudinal gradient can be observed. In the southern segment (25–12°N), maximum surface temperatures are 25 °C in winter, 26 °C in spring, 28 °C during summer and 25 °C in autumn. A latitudinal gradient is also observed.

However, during upwelling activity, the temperature decreases to 16 °C along the coasts. This SST seasonality is illustrated by Mittelstaedt (1991, fig. 4), who showed that, with respect to the open sea, the maximum temperature differences (−3 to −6 °C) between 10°N and 25°N from December to May migrate polewards to the area between 20°N and 36°N from June to November. Similar results were found from using data taken from satellites (Marcello *et al.* 2011). However, the differences between temperatures measured by ships and those taken from satellites (NOAA/AVHRR) may be up to several degrees (Atillah *et al.* 2005).

Recent detailed vertical profiles down to a depth of 200 m along the shelf and slope off Agadir in 2008–9 (Makaoui *et al.* 2012) showed that, during upwelling periods, the 'normal' deep horizontal to oblique isotherms become parallel to the shelf surface, lowering the temperatures by 4–5 °C towards the shoreline.

Sedimentation regime and sources

Along the northern segment of the NWAS, a number of seasonal rivers transport sediment derived from the Rif and Atlas mountains hinterland to the continental shelf (Snoussi 1984; Snoussi *et al.* 1990; Wynn *et al.* 2000). Offshore Senegal, primary fluvial sediment is also delivered from the Senegal, Gambia and Casamance rivers (Michel 1973; Gac & Kane 1986; Orange & Gac 1990;

Diara 1999), while the Saloum River has little or no solid load (Diara 1999). However, between 21°N and 30°N, the fluvial contributions are almost absent from the shelf sediments, and transport is related to the mid-tropospheric Harmattan winds originating from the central Sahara that lead to the deposition of significant amounts of Saharan dust in the northern Atlantic (Sarnthein *et al.* 1982; Mittelstaedt 1983; Matthewson *et al.* 1995; Swap *et al.* 1996; Torres-Padrón *et al.* 2002; Holz *et al.* 2004).

Fluvial input

Along the northern segment, the total sediment discharge of NW African rivers, including the Mediterranean ones, has been estimated to be 110×10^6 t a^{-1} (Milliman & Meade 1983; Hillier 1995). The estimated individual contributions of each river are shown in Table 10.1. Comparison of the proposed values shows that the amounts of transported solid matter are very different depending on the authors due to variable input parameters and calculation methods. However, all of the authors agree that the Sebou River (10×10^6–33×10^6 t a^{-1}) provides the largest amount of sediments, followed by the Oum Er Rbii River with 6.6×10^6–10×10^6 t a^{-1}.

For the Senegal River, the alluvial valley covers an area of approximately 0.343×10^6 (Nizou *et al.* 2010) to 0.369×10^6 km^2 (Ludwig & Probst 1998). The average water discharge over the year at the last downstream point prior to the estuary (Dagana) is around 641 m^3 s^{-1}; however, this discharge is highly seasonal due to the monsoonal conditions. The annual sediment load delivered to the shelf is 1.9×10^6–2×10^6 t a^{-1} (Gac & Kane 1986; Kattan *et al.* 1987; Milliman & Syvitski 1992) and the sediment yield is 8 (Milliman & Syvitski 1992) to 65 t km^{-2} a^{-1} (Ludwig & Probst 1998). Nizou *et al.* (2010) stated that, before the construction of the Diama dam in 1985, the low-flow regime and a very low riverbed allowed ocean waters to penetrate up to Bogué several hundreds of kilometres inland. This is also observed in the Saloum estuary (Diara 1999).

More southwards, the Gambia River (watershed of 0.063×10^6 km^2) provides a sedimentary yield of only 5 t km^{-2} a^{-1} (Ludwig & Probst 1998). The solid charge attains 0.66×10^6 t a^{-1} at 530 km from the river mouth (Lerique 1975) but only a small part of this arrives at the ocean.

Table 10.1. *Summary of the sedimentary characteristics of NW African rivers*

	Watershed ($\times 10^6$ km^2)	Load ($\times 10^6$ t)	Yield (t km^{-2} a^{-1})
Loukkos	0.0037 (BB10)	2.0 (PAS92)	750–1500 (PAS92)
Sebou	0.04 (MS92)	33.8 (PAS92)	930 (MS92)
	0.026 (EG81)	10 (SN90)	
		26 (SN90)	
Oum Er-Rabii	0.034 (EG81)	6.6 (PAS92)	–
	0.046 (BB10)	10 (SN90)	
Tensift	0.018 (EG81)	0.9 (PAS92)	–
	0.024 (BB10)	7.4 (SN90)	
		6 (SN90)	
Souss	0.16 (MS92)	1 (MS92, SN90)	260 (MS92)
	0.059 (BB10)	1.6 (MS92)	
		2.2 (PAS92)	
Draa	0.088 (BB10)	13.9 (PAS92)	–
		5 (SN90)	
Saguia El Hamra	0.319 (BB10)	2 (SN90)	–
Senegal	0.343 (NZ12)	1.9–2 (GK86, KAT87, MS92)	8 (MS92)
	0.369 (LP98)		65 (LP98)
Gambia	0.063 (LR75)	0.66 (LR75)	5 (LP98)

MS92, Milliman & Syvitski (1992); NZ12, Nizou *et al.* (2010); LR75, Lerique (1975); GK86, Gac & Kane (1986); LP98, Ludwig & Probst (1998); KAT87, Kattan *et al.* (1987); PAS92, Probst & Amiotte Suchet (1992); EG81, El Gharbaoui (1981); BB10, Bouaicha & Benabdelfadel (2010); SN90, Snoussi *et al.* (1990).

Aeolian transport

D'Almeida (1989) estimated the amount of Saharan dust to be $0.6 \times 10^{15} - 0.7 \times 10^{15}$ g a^{-1}, of which about one-third is deposited in the North Atlantic Ocean (Duce *et al.* 1991). Off Mauritania and Senegal, the NE Trade Winds deliver dust material during the cool and dry boreal winter, and the monsoonal front system during summer (e.g. Sarnthein *et al.* 1982; Stuut *et al.* 2005), which lifts the dust to a height of 3000–7000 m. Dust is transported westwards by the Sahara Air Layer (see Stuut *et al.* 2005 for a review). Prominent dust source regions that are frequently suggested in the literature are the Bodélé depression in Chad (Washington *et al.* 2006), and a low relief area extending over eastern Mauritania, western Mali and southern Algeria (McTainsh & Walker 1982).

Structural framework

As revealed by numerous authors, the sediments of the NW African margin derive from numerous structural domains, the greatest diversity of which is observed in Morocco (see Michard *et al.* 2008 for a recent synopsis).

The northern Moroccan hinterland comprises, from north to south: (i) the Alpine Rif Mesozoic–Cenozoic marly and sandy units from Tangier to Rabat; (ii) the sedimentary and crystalline Palaeozoic, and calcareous-marly and phosphate Cretaceous–Tertiary formations of the western Meseta, together with the Middle and High Atlas Mesozoic–Quaternary formations from Rabat to Essaouira; and (iii) the Mesozoic formations in front of the High Atlas in Agadir.

South of Sidi Ifni, where the fluvial input is small with respect to the northern segment, the southern Moroccan and Mauritanian–Senegambian coasts comprise the crystalline Precambrian and sedimentary Palaeozoic formations of the Dorsale Reguibat and Mauritanide chain (Bellion 1991*a*), with their Mesozoic–Cenozoic cover of the Tarfaya–Laayoune and Senegal–Mauritanian basin (Bellion 1991*b*).

A conspicuous feature is the presence of sand dunes in the hinterland belonging to the Sahara desert. As stated by several authors from the analysis of dust recurrences in shelf sediment, desertification of Africa during the Quaternary resulted from aridity and destruction of the vegetation (e.g. Hanebuth & Henrich 2009). This led to an increasing uptake and transport of available soils by wind (Nizou *et al.* 2011).

Tectonic uplift

After the Pliocene regression, the analysis of the geometrical relationship between the Quaternary deposits along the NW African coastal areas shows that their position is globally related to sea-level changes, with a minor role of tectonics (see the section on 'Late Quaternary evolution of the shelf' later). However, Mary (1982) suggested that the +2 m level (Mellahian in Morocco, and Nouakchottian in Mauritania and Senegal) that can be observed might have been uplift caused as an isostatic response to the melt of polar ice.

Despite this general tectonic quiescence, the effects of vertical tectonics are obvious in the marine continuation of the Rif and High Atlas Alpine chains. For instance, in the Agadir area, the oldest Quaternary deposits (the 'Moghrebian stage': *c.* 2 Ma) located above anticlinal folds are lifted up to 300 m at Cape Ghir, and 200 m at the Kasbah of Agadir (Weisrock 1981; Meghraoui *et al.* 1999; Boudad *et al.* 2010). This implies a mean uplift rate of 0.1–0.2 m ka^{-1} at the Kasbah during the Quaternary (Meghraoui *et al.* 1999). Seawards, the mean (relative to undeformed strata) uplift rate of the offshore Kasba anticline since the Pliocene is only 0.04 m ka^{-1}, while that since the Miocene is 0.128 m ka^{-1} (Mridekh *et al.* 2009).

Uplift of Quaternary fossil beaches was also reported from the Canary Islands (e.g. Zazo *et al.* 2002 for Lanzarote and Fuerteventura; Kröchert *et al.* 2008 for Tenerife). However, comparison of the actual beach altitudes with sea-level changes suggests either stability or uplift/subsidence related to topographical changes induced by deep magma chamber emplacement, as in the recent eruption at Hierro (Instituto Geografico Nacional, Madrid).

Description of the margin

Topography and morphology

The shape and orientation of the NW African margin was inherited from the fault system related to the break-up of Pangaea and the Triassic rifting (e.g. Sahabi *et al.* 2004). This initial shape was modified by the subsequent development of sedimentary wedges from Mesozoic to present times and the disruption of volcanic islands (Canary Islands north of 25°N) since the Miocene (Carracedo *et al.* 2001). The general trend is NE–SW from Tangier to Cape Blanc, and then turns to north–south until Cape Vert. However, several parts, such as the Essaouira–Agadir (north–south) and Tarfaya (ENE–WSW) segments, show different orientations with respect to the mean trend.

Taking the shelf break at an average water depth of 100 m, the width of the continental shelf off NW Africa, typically of the order of 50–150 km, appears as relatively narrow in comparison with other continental shelves. The mean width is 30–40 km but isolated sectors may attain 100 km off Mauritania, or, oppositely, only 10 km off Cape Vert Peninsula (Fig. 10.1).

The bottom topography of the shelf appears to be flat, except off sea cliffs of basement rocks and in areas where beach rocks are present. Irregularities also result from pre-existing structures, neotectonics, halokinesis and prevailing conditions.

The continental shelf and slope of the region are dissected by submarine valleys or canyons such as the Gharb Valley and the Agadir canyon system off Morocco, the Tiouillit Canyon in front of Cape Timiris (Mauritania), and the Kayar Canyon off Senegal.

Morocco. The Atlantic continental margin of Morocco (Fig. 10.1), from 36°N to 20°N, is characterized by a relatively shallow (<150 m) continental shelf that extends offshore from about 25 km at Cape Ghir to 75 km north of Cape Juby (Summerhayes *et al.* 1976). The main geomorphological features have been described by numerous authors (Erimesco 1965; McMaster & Lachance 1968; Robb 1971; Summerhayes *et al.* 1971; Tooms *et al.* 1971; Dillon 1974; Seibold & Hinz 1974; Cirac *et al.* 1979; Von Rad *et al.* 1979; Griboulard 1980, 1983; Jaaïdi 1981, 1993; El Foughali 1982; Hinz *et al.* 1982; Seibold 1982; Von Rad & Wissman 1982; Ruellan 1985; El Foughali & Griboulard 1985; Mathieu 1986).

In general, the surface of the continental shelf is smooth, with locally up to 10 m-high vertical reliefs (McMaster & Lachance 1968; Robb 1971; Summerhayes *et al.* 1971; Newton *et al.* 1973; Dillon 1974; Cirac *et al.* 1993; Jaaïdi 1993). Most of these topographical irregularities seem to be caused by low ridges that may represent basement outcrops, beach rocks and lithified dunes (Summerhayes *et al.* 1976) or palaeo-positions of the Holocene shoreline.

Based upon slope gradient and the presence of canyons, the continental shelf can be subdivided into four major zones:

1. To the north, the zone off Tangier–Rabat is characterized by a narrow continental shelf (20–35 km) with a 150 m water-depth shelf break, a relatively gentle slope of 3.25–6.66% around the 'Rharb Valley' (El Foughali 1982). This zone is marked by a major wedge-shaped progradational body corresponding to the Sebou prodelta. This structure extends across the whole shelf and thins toward the shelf break where it

seems to be connected to a major canyon, the Rharb Valley, located off the Sebou River mouth, which feeds the Seine abyssal plain. The NW termination of the Western Rif shelf also exhibits giant submarine dunes stacked at the entrance of the Strait of Gibraltar at water depths between 120 and 220 m (Trentesaux *et al.* 2008). By using side-scan sonar, a recent study by Geawhari *et al.* (2012) reported the presence of sandwaves in front of the Cape Spartel.

2. Between Rabat and Essaouira, the continental shelf is 40 km wide, gently dipping (0.1–0.6°) seawards to the 150 m-deep shelf break (El Foughali & Griboulard 1985; Ruellan 1985). A wide 30–100 m-deep sedimentary wedge characterizes the morphology of the shelf off the Oum Er Rbii estuary mouth. Cretaceous formations outcrop extensively between the coastline and 130–150 m water depths, generating locally a high seabed roughness. A wide abrasion surface is also exposed in the middle shelf between the latitudes of Rabat and Casablanca, and close to Mohammedia where a coastal dune complex is submerged.

3. The offshore segment between Essaouira and Sidi Ifni is dominated by the Agadir Canyon system that originates on the shelf as an individual tributary feeding into a large submarine canyon; the trajectory of the canyon presents sharp changes in its direction, mainly controlled by volcanic seamounts and salt diapirs (Uchupi *et al.* 1976).

4. In the southernmost zone, most of the shelf is shallower than 100 m and the shelf width at Oued Draa is 75 km.

Mauritania–Gambia. This part of the NW African continental margin extends from 17°N to 12°N and includes, from north to south: a part of the Mauritanian continental shelf (17°N–16°03′N); the Senegambian continental shelf (16°03′N–12°20′N); and a part of the Guinea-Bissau continental shelf (12°20′N–12°N), the width of which varies from approximately 9 km at the Almadies Pointe to around 126 km at Guinea-Bissau.

The Mauritanian continental shelf has a surface of approximately 39 000 km², of which 9000 km² belong to the Bay of Lévrier–Banc d'Arguin.

Between the Cape Blanc peninsula and Cape Timiris, the shelf widens significantly to a maximum of 150 km, forming the very shallow Golfe d'Arguin bight (<10 m), which is thought to have been a former Tertiary delta, as suggested by its lobe (McMaster & Lachance 1969). The continental slope off the Banc d'Arguin is incised by numerous canyons (e.g. the Cape Timiris Canyon), but only a few dissect the shelf (e.g. Diester-Haass 1975; Bein & Fütterer 1977).

From Cape Timiris to Kayar (also written as Cayar) Canyon, the shelf is generally narrow (30–40 km: Meagher *et al.* 1977; Ruffman *et al.* 1977) and shows relatively steep slope angles of 2.5–3° (Antobreh & Krastel 2006). North of the Senegal River mouth, the shelf is narrow (Pinson-Mouillot 1980) and the shelf break is situated, on average, at a water depth of 100 m. At the Senegal River mouth the shelf is around 40 km wide and its slope is only 0.2%. This segment hosts the Peul Canyon off Lompoul. From M'Boro to Kayar, the shelf is narrow (*c.* 20 km), shows a slope ranging from 1.8 to 0.3% and hosts two canyons: the Kayar Canyon off Kayar and the Djoloff Canyon off M'Boro, which spectacularly cut back into the outer shelf.

The Senegambian continental shelf (0–200 m; surface area 27 600 km²) lies between latitudes 17°N and 12°S in the eastern Central Atlantic (Fig. 10.3). The Senegal continental shelf is located between 12°08′N and 16°41′N, and 11°21′W and 17°32′W. Its coast marks the westernmost part of the African continent.

The Gambian continental shelf is about 480 km long and 50 km wide at its widest end facing the Atlantic Ocean, and tapers towards the east to a width of about 30 km. It has an 80 km-long coastline and a continental shelf area of about 4000 km², rich in marine fish resources (GEF 2002).

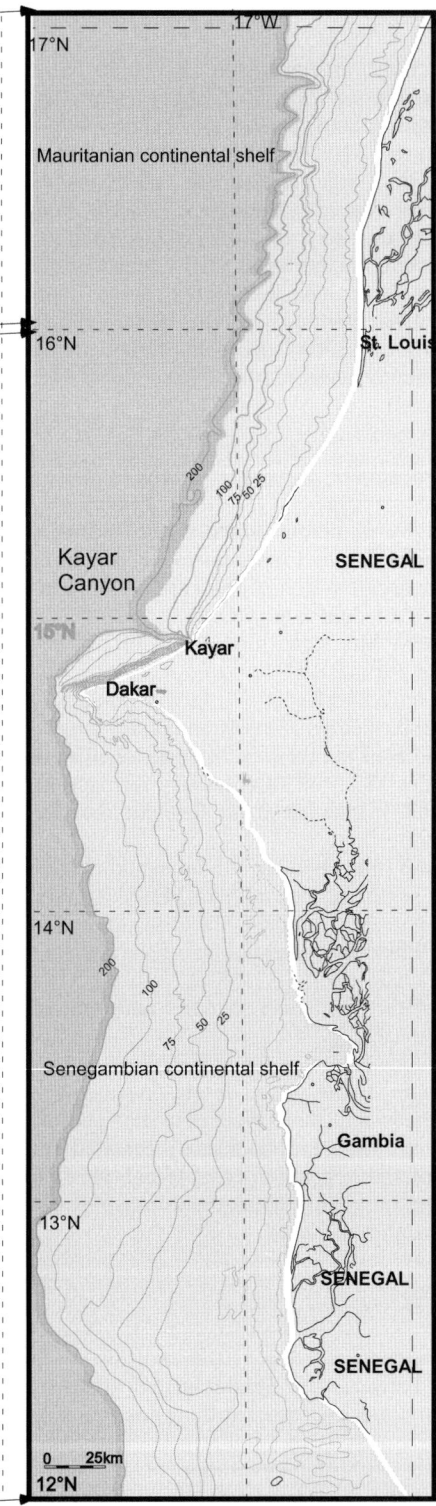

Fig. 10.3. Bathymetry of the Mauritanian and Senegambian continental shelf (simplified after Domain 1976, 1985).

Pre-Quaternary sedimentary units

Long-term Pre-Quaternary shelf sedimentation from Jurassic to Pliocene times along the NWAS is the result of the interaction of several parameters such as thermotectonic subsidence, sediment input and load, global eustatic changes, Atlantic circulation changes, and climatic variations related to the latitudinal migration of the African plate. These aspects were largely studied in the setting of the Central Atlantic evolution (Von Rad *et al.* 1982; Ruellan 1985; Medina 1995) but, in recent years, there has been a renewed interest driven by hydrocarbon exploration (Davison 2005; Hafid *et al.* 2008; Tari *et al.* 2012).

After a relatively monotonous development, only disturbed by slow halokinesis off Morocco (Tari *et al.* 2012), Mauritania and southern Senegal (Bellion 1991*b*), the NW African margin, especially its northern part, underwent some 'exceptional' events such as: (i) the Oligocene lowstand and subsequent submarine erosion, which removed part of the margin sediments (Flamant-Lieffrig 1979); (ii) the disruption of the Canary volcanic islands from the Miocene to the present (Carracedo *et al.* 2001, p. 182) and of several other volcanic bodies, such as the Leona Dome south of Saint-Louis and the Kayar Seamount NW of Dakar (Bellion 1991*b*); (iii) the deformations related to the collision of the African and Eurasian plates, which had some influence off the new mountain chains, namely the Rif and the Atlas; and (iv) the latest Cenozoic regression at the northern end, which led to the development of the modern fluvial system (El Gharbaoui 1981).

Late Quaternary evolution of the shelf

The nature of Late Quaternary climate changes in NW Africa has been the subject of a large number of investigations in recent years (Lioubimtseva 2004). The latest Pleistocene sea-level history is still poorly documented off NW Africa, but it seems that the sea-level lowstand was around −110 m during the Last Glacial Maximum (LGM), followed by a rapid deglacial rise at varying rates (Giresse 1987; Pegler 1999; Hanebuth & Lantzsch 2008; Hanebuth & Henrisch 2009). The following phases of glaciation and temperature rise, accompanied by regression and transgression of the sea, are the main factors that have determined the evolution of the NW African coastal plain and continental shelf.

Evidence of sea-level changes has been reported by several authors, as summarized in the following paragraphs (Fig. 10.4).

During the last maximum glacial cycle, at around 18 ka BP, there is a relative agreement as to the position of the sea level at about 110–115 m below the present sea level (Martin 1972; Delibrias *et al.* 1973; Einsele *et al.* 1977; Giresse & Barusseau 1986; Jaaïdi 1993; Bard *et al.* 1996). This low level, which lasted probably some 3000 years, was followed by an active eustatic period,

from 15 to 7 ka BP. Between 6 ka BP and the present, sea level has been relatively stable with fluctuations of only a few metres above or below the present sea level, depending on geographical location.

In general, the West African platform presents a record of postglacial shorelines at about −80, −50, −35 and −20 m, which, however, were not sufficiently dated. The shoreline accumulations are contemporaneous with phases of a relative sea-level stabilization.

Indices of these phases have been identified on the continental shelf of Morocco as a succession of breaks in the slope, at the following depths: 130–150 115–125, 75–80, 45–55, 25–35 and 5–10 m (Griboulard 1983; Jaaïdi 1981, 1993; El Foughali & Griboulard 1985). Along the coast south of Rabat, Gigout (1959) described, at Temara, a decimetre-scale sandstone with a similar lithology and palaeontological content to those of the present-day beach sands. This level located at +2 m is known as the 'Mellahian'. Similar deposits were later observed in other places, such as in the Loukkos estuary at Larache (Carmona Gonzalez 2005) and Agadir (Boudad *et al.* 2010). Radiocarbon dating of marine shells provided ages ranging from 2.7 to 5.97 ka (Gigout 1959; Stearns & Thurber 1965; Choubert *et al.* 1967; Carmona Gonzalez 2005; Choukri *et al.* 2007; Carmona & Ruiz 2009; Weisrock 2009; Boudad *et al.* 2010).

Along the coast of Mauritania, this highstand is reflected by a 5.5 cal ka BP palaeo-shoreline that is often located 10 km or more landwards of the modern shoreline (Hébrard 1973; Einsele *et al.* 1974; Elouard *et al.* 1977; Barusseau *et al.* 1995). This highstand lasted from 6 to 4 cal ka BP (the 'Nouakchottian'), and swung back towards modern levels about 3 ka ago (the 'Tafolian').

Along the Senegal coast, Faure *et al.* (1980) proposed a sealevel reconstruction that was estimated by ^{14}C-dated fossil assemblages. This reconstruction is assumed to document a sea-level highstand at around 5.5 cal ka BP and a subsequent regression with two minor peaks above modern sea level at around 2.85 and 1.7 cal ka BP.

Although these observations are in accordance with many worldwide low-latitude sea-level reconstructions (e.g. Clark *et al.* 1978; Pirazzoli & Montaggioni 1988; Haworth *et al.* 2002; Lambeck 2002), other studies support that the so-called 'nouakchottian transgression' (e.g. Barusseau *et al.* 1995) was a marine incursion linked to climatic and morphosedimentary phenomena (Ausseil-Badie *et al.* 1991; Barusseau *et al.* 2007, 2009; Vernet 2007). Marine ingression and salt-water intrusion allowing deposition of shells into the lower Senegal River valley could also have appeared owing to a reduction in freshwater input related to dry climate (Monteillet *et al.* 1981; Nizou *et al.* 2010). Drought conditions leading to reduced freshwater discharge by the Senegal River could explain the peaks at 50 cm above modern sea level observed by Faure *et al.* (1980) at around 2.85 and 1.7 cal ka BP (Bouimetarhan *et al.* 2009; Nizou *et al.* 2010)

Modern shelf sedimentation

Facies

Along the Moroccan (including the former Spanish Sahara) coasts, Summerhayes *et al.* (1976) mapped five lithological types based on granulometry: coarse and shelly sand; fine sand; muddy sand/sandy mud; mud; and, finally, algal mats. Later, Jaaïdi (1993) mapped four lithological types along the shelf from Tangier to Agadir: terrigenous sands; bioclatic sands; muds; and bedrock. Southwards, along the Mauritanian and Senegambian coast, Domain (1977) distinguished bedrocks and unconsolidated sediments. The latter were subdivided based on BRGM's norms into four facies depending on the ratio of lutites (size < 63 μm): sands (<5% lutites); muddy sands (5–25% lutites); sandy muds (25–90% lutites); and, finally, muds (>90% lutites). South of

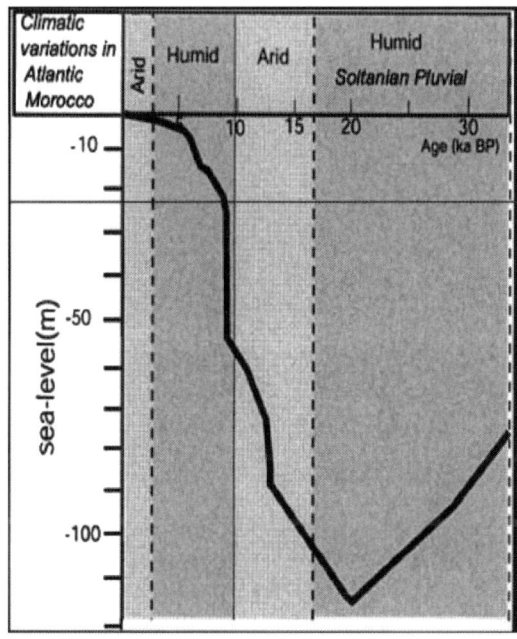

Fig. 10.4. Climatic variations through North Atlantic Morocco during the last 30 kyr (Weisrock *et al.* 1985) and the corresponding global eustatic curve (Bard *et al.* 1996).

the Cape Vert peninsula, Barusseau (1984) described seven types, which, however, can be reduced to four because of the repetition of three granulometric classes within distinct domains: 'sablons' (40–105 μm); terrigenous sands; bioclastic sands; and fine sands.

We describe hereafter the facies according to Jaaïdi (1993) because of the availability of detailed quantitative data. Complementary data by Summerhayes *et al.* (1976) are also included.

Terrigenous sands of the inner shelf are characterized by the prevalence of the inorganic fraction (quartz and mica) with a fraction of probably reworked foraminifera. Grain-size analysis shows that the sands may be fine (63 < mode < 100 μm) or medium (100 < mode < 200 μm). The median is between 125 and 250 μm (fine to medium sand), and the transport mode of grains was saltation. Phosphate is locally abundant in the form of sand-size detrital grains. In the southern part of the NWAS, aeolian deposits blown offshore are an important component of the fine-grained, carbonate-free sediments. The mean of the quartz grain size (36 μm) matches that of wind-carried dust (Hanebuth & Lantzsch 2008).

Bioclastic sands have a carbonate ratio >50% and locally >70%; they may be coarse (median > 500 μm) or medium size (200 < median < 500 μm) sands. The carbonated phase consists primarily of broken, used and rubefied bioclasts of bivalves and gastropods and, to a lesser extent, of echinoids, bryozoans, brachiopods and sponge spicules; foraminifers (*Miniacina miniacea*) representing up to 65% of the carbonate content were also observed by Summerhayes *et al.* (1976) on the outer shelf (depths of 95–110 m); the mineral fraction consists of quartz and micas, as well as authigenic glauconite grains, common on the outer shelf and upper slope, and phosphatic aggregates. Granulometric analysis revealed transport by saltation or gradual suspension.

Mud deposits of the mid-shelf have a content higher than 70% of lutites and 10–30% of carbonates. Three granulometric 'facies' characterize these mud deposits according to the method of Rivière (1977): a 'parabolic facies', where silts dominate, and clays (size <2 μm) represent only 10%; a 'logarithmic facies' where clays represent 10–45%; and a 'hyperbolic facies', with clays representing 55–80%. The analyses indicate a transport per graduated suspension of raised energy, transport by uniform

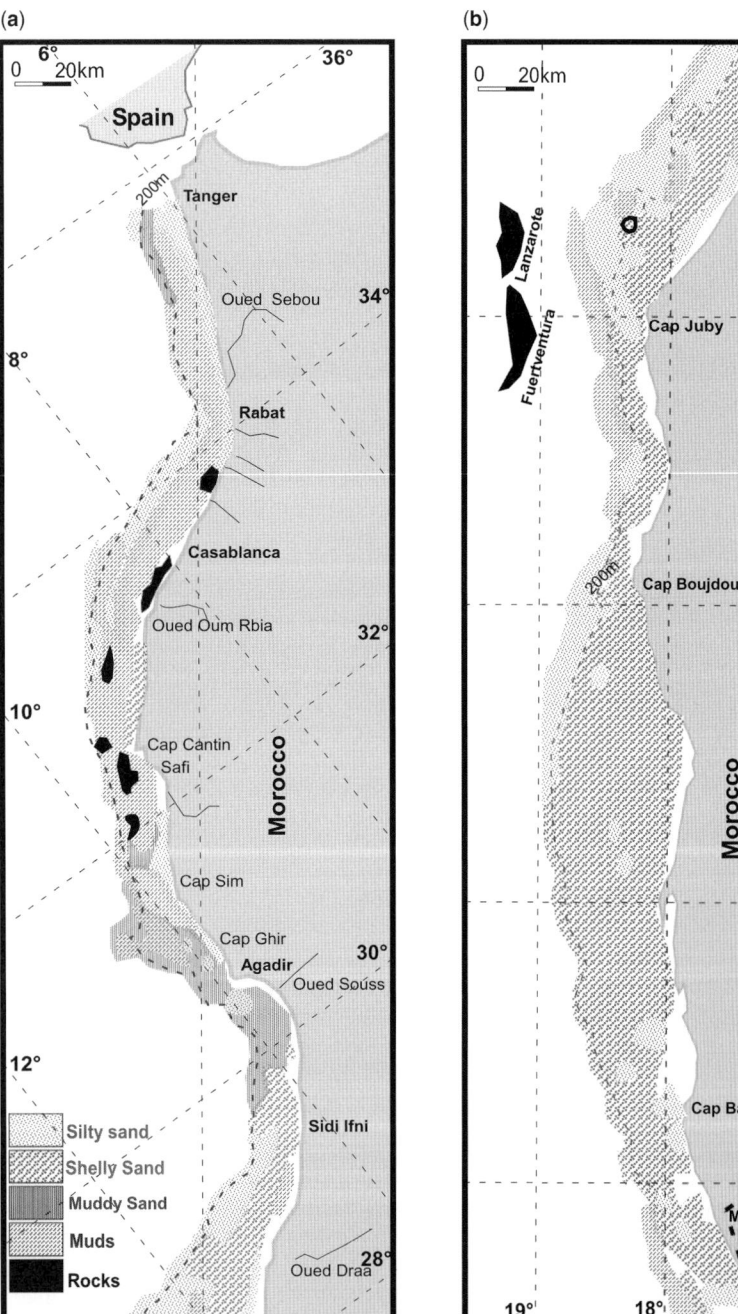

Fig. 10.5. Modern sediment distribution on (**a**) the North Moroccan continental shelf from Tangier to Sidi Ifni; and (**b**) the South Moroccan continental shelf from Sidi Ifni to Cape Barbas (after Summerhayes *et al.* 1976; Jaaïdi & Cirac 1987).

suspension of energy lower than that prevailing and transport in pelagic suspension. Main clay minerals are illite and smectite (montmorillonite), with small amounts of kaolinite and chlorite. As will be shown in the next subsection, the amounts of each mineral strongly depend on the geographical situation (latitude and position in the shelf and with respect to river mouths).

The bio-constructions described by Summerhayes *et al.* (1976) correspond to ahermatypic corals of the species *Dendrophyllia ramea.*

Distribution

On the base of the grain size of the unconsolidated deposits, the NWAS can be subdivided into four segments: (1) a northern muddy segment from Tangier to Agadir (36–30°N); (2) a northern sandy segment from Agadir to north of Saint-Louis (16°30′N); (3) a southern muddy segment from Saint-Louis to Dakar (15°N); and, finally, (4) a southern sandy segment from Dakar to the mouth of the Casamance River (12°30′N).

Northern Morocco mud province. Between 36°N and 30°N (Fig. 10.5a), the sediments mainly consist of mud, fluvial terrigenous sands and biogenic sands (Summerhayes *et al.* 1976; Jaaïdi 1981, 1993). The sand fraction is carbonate-rich and is considered to be mainly relict from the Pleistocene, when sea level was lower. This relict sand is mixed with, and locally buried by, Holocene detrital silts; these are concentrated on the middle shelf and the slope (Summerhayes *et al.* 1971).

From Tangier at 36°N to El Jadida (former Mazagan) at 33°N, the dominant mud deposits occupy, in general, a central position on the continental shelf, at water depths of between 70 and 110 m, and are surrounded by terrigenous sands of the inner shelf and bioclastic sands of the outer shelf. Such mud deposits present a maximum extension off the Gharb Plain drained by the Sebou River, where they can exceed 30 m in thickness; however, their areal extent decreases towards the south and the north. The volume of sediments is evaluated at $28 \times 10^9 \text{ m}^3$ corresponding to a weight of $46 \times 10^9 \text{ t}$ (Jaaïdi 1993). The bio-constructions correspond to ahermatypic corals (*Dendrophyllia ramea*). Such sediments are fixed to the rock exposures of the outer continental shelf, offshore El Jadida and Kénitra.

Between El Jadida and north of Cape Sim (32°N), the sedimentary cover consists mainly of shelly sand in the outer shelf and terrigenous sand in the inner shelf.

Further south, between 32°N and the mouth of the Oued Souss at 30°N, the central mud belt appears again, although it is narrower than the northern one.

Southern Morocco: Mauritania sand province. From the mouth of the Oued Souss to the latitude north of Saint-Louis (16°30′N), the most common surface sediment type is biogenic sand (McMaster & Lachance 1968; Summerhayes *et al.* 1976). This facies has been recovered between Ifni and Cape Blanc (21°N) (Fig. 10.5b). Between 30°N and 21°N, sediments are of biological origin and comprise a large organic fraction. They are distributed within 10–100 m water depth. Thicknesses of the unconsolidated sediments on the shelf are usually <10 m, and often <2 m (Summerhayes *et al.* 1971).

Terrigenous sands are distributed over the internal part of the continental shelf, and constitute a littoral prism of variable extension supplied with the products of erosion of the beaches and the destruction of Plio-Quaternary rocks.

Off the Mauritanian coastline, the sedimentary cover differs north and south of Timiris Cape (Fig. 10.6). To the north, it is characterized by a large area of coarse sands and high percentages of $CaCO_3$. In the south, very fine sands prevail with a low content of $CaCO_3$. In general, the content in lutites is higher north of Cape Timiris (Domain 1980) where two mud wedges,

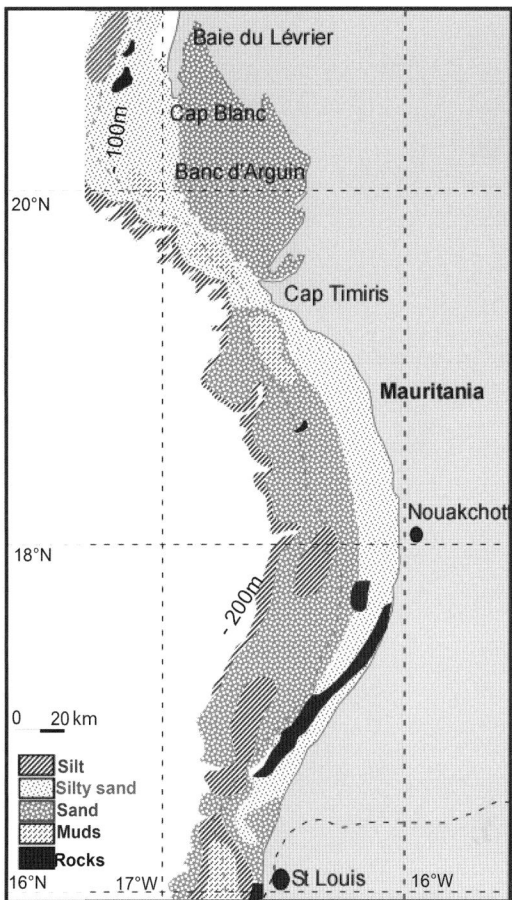

Fig. 10.6. Modern sediment distribution on the Mauritanian continental shelf (simplified after Domain 1976, 1985; Augris *et al.* 2000).

the Arguin and Timiris mud wedges of aeolian origin, have been observed in the outer shelf (Hanebuth & Lantzsch 2008). The distribution of the various types of sediments appears as follows (Domain 1986): (i) sands (mainly quartzose) from 0 to 35–40 m water depth; (ii) muddy sand containing from 5 to 25% of lutites, from 40 to 100–150 m water depth; and (iii) sandy mud in water depths greater than 150 m, the sediments contain between 25 and 75% of lutites. The mean of the quartz grain size (36 μm) matches that of wind-carried dust (Hanebuth & Lantzsch 2008).

Grande Côte mud province. North of Dakar, the modern shelf sediments of the Senegal margin consist mainly of mud, muddy sand and sand, locally containing up to 70% of biogenic carbonate (Domain 1977; Pinson-Mouillot 1980). Scattered rocky outcrops are present close to the coastline. The Senegal River sediment output results in an accumulation of a conspicuous mud belt in front of Saint-Louis, extending from 15°30′N to 16°30′N (Domain 1977; Nizou *et al.* 2010; see Fig. 10.6), which can be subdivided into a northern inactive and a southern active segment (Nizou *et al.* 2010). Coring also showed that thickness of this mud belt is 17 m at the river mouth (Nizou *et al.* 2010). The southern limit of this mud belt is truncated by the Kayar Canyon, while the northern part grades into the northern sand province.

Petite Côte–Casamance sand province. From Cape Vert to the mouth of the Casamance River, the wide shelf is dominated mainly by bioclastic sands and rocky submarine outcrops (Fig. 10.7). Fine sediment belts were mapped off the Gambia River in front of Banjul, and in the middle–outer shelf area between 13°N and 14°N (Domain 1977). This deposit is characterized by more than 75% of fine-grained sediments (Seibold &

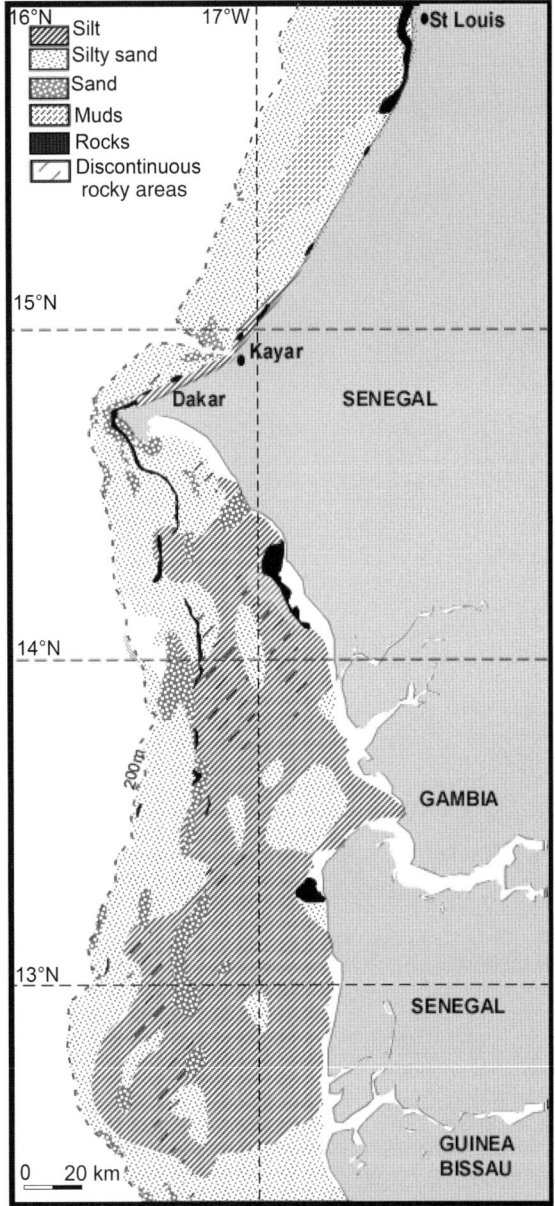

Fig. 10.7. Modern sediment distribution on the Senegal and Gambia continental shelf (simplified after Domain 1977, Augris *et al.* 2000).

0–20 m (Barusseau 1984). Mud also exists as a prodeltaic body off the Saloum River.

Relict sediments enriched in green minerals (verdine) are found at a water depth of about 100 m (Pinson-Mouillot 1980; Odin & Masse 1988). Similar sediments have been reported from several locations along the African coast (e.g. Delibrias *et al.* 1973; Giresse *et al.* 1976).

Sedimentary processes and controlling factors

As long-term processes such as sea-level rise have been described in a previous section, in this last section we briefly review the short-term sedimentary processes and controlling factors on modern sediments, in particular the transport and dispersal of sediments by currents and the role of morphology.

Transport by currents

At the inner shelf, scattered data on the hydrological conditions close to the shoreline show that erosion, transport and deposition of sediments provided by rivers and/or wind are related to the hydrodynamic parameters of, in decreasing order of importance, longshore currents, storm-related waves, tides and upwelling.

Longshore currents related to swell are a powerful process that can displace a large volume of sediments. In Table 10.2, we expose examples of the volumes of reworked sediments that have been calculated by various authors along the NWAS. It is certain that a large part of the sediments, especially that carried by longshore currents, is accumulated on nearby beach segments (Idrissi *et al.* 2004) but a fraction may be transported seawards by submarine gullies.

Spring tides with velocities of 20–30 m s^{-1} at 1 m depth were recorded during the campaign Geomar II in April 1984 along the northern Moroccan shelf (Jaaïdi 1993). This velocity can also produce reworking of fine to medium sands and maintain turbidity on muddy bottoms.

Data on *in situ* measurements of bottom current directions and velocity on the middle and outer shelf are scarce. At the northern part of the margin, Cirac *et al.* (1989) measured velocities of 9.6–31.4 cm s^{-1} at mid-depths in relatively calm weather. The directions are NE and SW. The lowest values (10–20 cm s^{-1}) were recorded on the inner shelf, whereas the highest values (20–30 cm s^{-1}) were measured on the middle and external shelf.

North of Cape Vert, measured surface currents (at a depth of 9 m) have mean monthly velocities of 2–10 cm s^{-1} in December and January. Maximum velocities are in the range 17–32 cm s^{-1}.

In order to find the possible role of middle and outer shelf and bottom currents in the transport of sediments, we added the reported main current directions to the maps of sediments.

From the physical aspect, the role of upwelling in the dispersal of sediments appears to be minor, as the vertical velocities of

Fütterer 1982). Carbonated sands are the major component off the Petite Côte south of Cape Vert peninsula, where they constitute 75% of the content of the collected sediments, whereas mud (pelites) is only present in the coast south of Dakar at depths of

Table 10.2. *Sedimentary volumes transported by littoral drift/longshore currents*

Region	Volume of sediment (m^3 a^{-1})	Reference
North of Casablanca	0.21×10^6	Idrissi *et al.* (2004)
Cape Timiris–Cape Vert	0.45×10^6–1.1×10^6	Barusseau (1985)
		Shi-Leng & Teh-Fu (1987)
		Ould El Moustapha (2000)
Senegal River mouth	0.365×10^6	Sall (1982)
	0.470×10^6	Barusseau (1985)
North of Mbour	72×10^3	Sall (1982)
South of Mbour	258×10^3	Sall (1982)
La Petite Côte Saloum	10^4–25×10^3	Barusseau (1980, 1985)
	10^4–150×10^3	Diara (1999)

upwelling waters are very small, ranging from 10^{-2} to 10^{-1} cm s^{-1} (Mittelstaedt 1983). However, at very shallow shelf depths, upwelling may indirectly induce strong bottom currents as, for instance, at the Banc d'Arguin. On this shallow shelf, the surface waters undergo a high degree of evaporation and, because of their increased density, sink beneath less dense upwelling waters. These bottom waters can follow submarine canyons, and may reach depths of 400 m with velocities of up to 50 cm s^{-1} (Mittelstaedt 1983).

Role of shelf morphology

The role of shelf morphological features as sedimentary pathways (gullies, valleys) or traps (closed depressions) has been empha-sized by several authors. In particular, mud belts at mid-shelf depths are thought to have developed against morphological scarps. Remarkable examples of such morphological traps were described from the Galicia mud belt off NW Iberia (e.g. Lantzsch et al. 2009).

Along the NWAS, shallow seismic profiles across the mud belts observed off northern Morocco and the Banc d'Arguin off Maur-itania clearly show that the Holocene sediments and capping present-day deposits are trapped within depressions surrounded by structural/topographical highs. The morphological highs may correspond to basement rocks, pre-Holocene consolidated dunes and beach sandstones, and coastline features (cliffs, notches), while the depressions may represent lowstand coastal lagoons (Cirac et al. 1979; Jaaïdi 1993).

These mud belts and patches are detached from the fluvial source, and can therefore be assigned to type 2b mud depocentres according to the recent classification of Lantzsch et al. (2009). However, the north Moroccan mud belt is very long and, therefore, may also be assigned to type 2a of Lantzsch et al. (2009).

In contrast, the deposits of the mud belt developed off the Senegal River mouth show a general upward-convex shape, although cross-sections show small depocentres (Nizou et al. 2010). This mud belt appears as still attached to the river, as during its Holocene history (Nizou et al. 2010), so it can be con-sidered as type 1a of Lantzsch et al. (2009), where the role of mor-phology is minor with respect to hydrodynamics.

In conclusion, it appears that the location of each type of sedi-ment on the shelf depends of the input by rivers, and the hydrodyn-amics and the morphology of the shelf bottom.

Conclusions

In this chapter, we have tried to give a synoptic view on the NWAS based on recent published articles and unpublished reports.

The wide variety of deposits and sedimentary processes recog-nized in this area has greatly contributed to our understanding of shelf sedimentation. However, there are still major gaps in our knowledge of features along the margin, as demonstrated by the recent discovery of a large meandering submarine channel system offshore the Sahara Desert. Our understanding of principal sedimentary processes, with special reference to their control mechanisms, is also limited; in particular those acting during the Holocene.

Sediment transport offshore NW Africa is found to be very sen-sitive to climate spatial–temporal changes. Fluvial discharge to the Atlantic Ocean is restricted currently to the Moroccan conti-nental margin, north of 29–30°N and, to a lesser extent, between the Senegal River and Cape Vert. Between these two segments, the terrigenous component of the marine sediments is character-ized by aeolian dust fallout. The recent studies suggest that during the early Holocene and the preceding Late Glacial period, the climatic regime in the Sahara Desert was very different to

that of the present day, with numerous fluvial systems supplying clastic sediment to the outer continental shelf.

Mid-shelf facies are found where the shelf is narrow and the sediment supply is ample. Where the shelf is wide, fine sediments are deposited in estuaries and on the inner shelf. Also, a calcareous sediment type is found where the shelf is tectonically stable and the sediment supply inadequate. Finally, carbonate values of less than 50% occur where a modern shelf facies is being deposited (McMaster & Lachance 1968).

Finally, there are few datasets available on hydrological par-ameters such as wave and flow dynamics, sediment transport path-ways and bedforms in the North African shelf, and there is an effort to produce more in order to understand the mechanisms.

Here we have reviewed mainly the physical (hydrological and morphological) aspects but others, such as the chemical processes (CO_2, salinity, authigenic minerals, carbonate and clay chemistry, and phosphate transport) that certainly play a role in, for example, the type of sediments and the lithification, were beyond this review. The biological effects of benthic fauna are also numerous and diverse; these are neither well mapped nor well understood.

This contribution was initiated from the activities of the Research Team 'Marges Méso-Cénozïques Marocaines (MMCM)' at the Laboratoire de Géologie et Télé-détection of the Institut Scientifique, Rabat, which was active from 2007 to 2009. The authors are grateful to Dr F. L. Chiocci (Universitá di Roma 'La Sapienza', Italy) and Dr A. Chivas (University of Wollongong, Australia) for the opportunity to contribute to this Memoir as a result of the IGCP-464 project.

The authors are also indebted to Prof. M. Collins of Ocean and Earth Sciences, University of Southampton, National Oceanography Center for his very construc-tive comments and discussions on an earlier version of the manuscript. We are grateful to Prof. G. Ercilla and an anonymous reviewer for helpful comments on the manuscript.

References

ABRANTES, F., MEGGERS, H. ET AL. 2002. Fluxes of micro-organisms along a productivity gradient in the Canary Islands region (291N): implications for paleoreconstructions. Deep Sea Research Part II: Topical Studies in Oceanography, **49**, 3599–3629.

ALLAIN, C. 1970. Les conditions hydrologiques sur la bordure Atlantique de l'Afrique du nord-ouest. Rapports et Procès verbaux Conseil International de l'Exploration Scientifique de la mer Méditeranée, **159**, 25–29.

ANTOBREH, A. A. & KRASTEL, S. 2006. Morphology, seismic character-istics and development of Cap Timiris Canyon, offshore Mauritania: a newly discovered canyon preserved-off a major arid climatic region. Marine and Petroleum Geology, **23**, 37–59.

ARÍSTEGUI, J., SANGRÁ, P., HERNÁNDEZ-LEON, S., CANTÓN, M., HERNÁNDEZ-GUERRA, A. & KERLING, J. L. 1994. Island-induced eddies in the Canary Islands. Deep Sea Research Part I: Oceano-graphic Research Papers, **41**, 1509–1525.

ATILLAH, A., ORBI, A., HILMI, K. & MANGIN, A. 2005. Produits opération-nels d'océanographie spatiale pour le suivi et l'analyse du phénomène d'upwelling marocain. Géo-Observateur, **14**, 49–62.

AUGRIS, F., DOMAIN, V., PASQUALINI, A., CRUSSON, M. & N'DIAYE, 2000. Carte des formations superficielles du plateau continental: partie méridionale de la Mauritanie, Sénégal et Gambie. IFREMER, Brest.

AUSSEIL-BADIE, J., BARUSSEAU, J. P., DESCAMPS, C., DIOP, E. H. S., GIRESSE, P. & PAZDUR, M. 1991. Holocene deltaic sequence in the Saloum estuary, Senegal. Quaternary Research, **36**, 178–194.

BARD, E., HAMELIN, B., ARNOLD, M., MONTAGGIONI, L., CABIOCH, G., FAURE, G. & ROUGERIE, F. 1996. Deglacial sea-level record from Tahiti corals and the timing of global melt water discharge. Nature, **382**, 241–244.

BARTON, E. D. 1998. Eastern boundary of the North Atlantic: Northwest Africa and Iberia. In: ROBINSON, A. R. & BRINK, K. H. (eds) The Sea, Volume 11. Wiley, New York, 633–657.

BARTON, E. D., ARÍSTEGUI, J., TETT, P. & NAVARRO-PÉREZ, E. 2004. Variability in the Canary Islands area of filament-eddy exchanges. In: BARTON, E. D. & ARÍSTEGUI, J. (eds) The Canary Islands

Coastal Transition Zone – Upwelling, Eddies and Filaments. Progress in Oceanography, **62**, 71–94.

BARUSSEAU, J.-P. 1980. Essai d'évaluation des transports littoraux sableux sous l'action des houles entre Saint Louis et Joal (Sénégal). *Bulletin Asequa, Dakar,* **58–59**, 31–39.

BARUSSEAU, J.-P. 1984. *Analyse sédimentologique des fonds marins de la 'Petite Côte' (Sénégal).* Institut Sénégalais des Recherches Agricoles, Centre de Recherches Océanographiques de Dakar-Tiaroye, Document Scientifique, **94**.

BARUSSEAU, J.-P. 1985. *Évolution de la ligne de rivage en république islamique de Mauritanie.* Unesco, Division des Sciences de la Mer, unpublished report, Contract sc 217.614.4.

BARUSSEAU, J.-P., Bâ, M., DESCAMPS, C., DIOP, S., GIRESSE, P. & SAOS, J.-L. 1995. Coastal evolution in Senegal and Mauritania at 103, 102 and 10-year scales: natural and human records. *Quaternary International,* **29**, 61–73.

BARUSSEAU, J.-P., VERNET, R., SALIÈGE, J. F. & DESCAMPS, C. 2007. Late Holocene sedimentary forcing and human settlements in the Jerf el Oustani–Ras el Sass region (Banc d'Arguin, Mauritania). *Geomorphologie-Relief Processus Environnement,* **1**, 7–18.

BARUSSEAU, J.-P., ROGER, J., NOEL, B. J., SERRANO, O. & DUVAIL, C. 2009. *Notice explicative de la carte géologique du Sénégal à 1/200 000, feuilles de Saint-Louis–Dagana, Podor–Saldé, Matam Semmé.* Ministère des Mines, de l'Industrie et des PME, Direction des Mines et de la Géologie, Dakar.

BATTEEN, M. L., MARTINEZ, J. R., BRYAN, D. W. & BUCH, E. J. 2000. A modeling study of the coastal eastern boundary current system off Iberia and Morocco. *Journal of Geophysical Research,* **105**, 14,173–14,195.

BATTEEN, M. L., KENNEDY, C. L., JR & MILLER, H. A. 2007. A process-oriented numerical study of currents, eddies and meanders in the Leeuwin Current system. *In*: HUVENNE, V. & DAVIES, J. (eds) *Submarine Canyons: Complex Deep-Sea Environments Unravelled by Multidisciplinary Research. Deep Sea Research Part II: Topical Studies in Oceanography,* **54**, 859–883, http://dx.doi.org/10.1016/j.dsr2.2006.09.006

BEIN, A. & FÜTTERER, D. K. 1977. Texture and composition of continental shelf to rise sediments off the northwest coast of Africa: an indication for downslope transportation. *Meteor Forschungsergebnisse,* **27**, 46–74.

BELLION, Y. 1991*a*. Les grands ensembles lithostratigraphiques et structuraux de la Mauritanie. *In*: CARUBA, R. & DARS, R. (eds) *Géologie de la Mauritanie.* Institut Supérieur Scientifique, Nouakchott, 18–22.

BELLION, Y. 1991*b*. Le bassin marginal de Mauritanie; bassin sédimentaire côtier. *In*: CARUBA, R. & DARS, R. (eds) *Géologie de la Mauritanie.* Institut Supérieur Scientifique, Nouakchott, 96–123.

BOND, G. C. & KOMINZ, M. A. 1988. Evolution of thought on passive continental margins from the origin of geosynclinal theory (~1860) to the present. *Geological Society of America Bulletin,* **100**, 1909–1933.

BOUAICHA, R. & BENABDELFADEL, A. 2010. Variabilité et gestion des eaux de surface au Maroc. *Sécheresse,* **21**, 1–5.

BOUDAD, L., WEISROCK, A., AIT HSSAINE, A., CHAKIR, L., OCCHIETTI, S., OUAMMOU, A. & BOUAJAJA, M. 2010. *La Quaternaire du Maroc sem-aride dans la région d'Agadir.* AFEQ-AMEQ, Fieldtrip Guides, http://www.afeq.cnrs-bellevue.fr/excursion/Guide%20AFEQ-2010.pdf.

BOUIMETARHAN, I., DUPONT, L., SCHEFUSS, E., MOLLENHAUER, G., MULITZA, S. & ZONNEVELD, K. 2009. Palynological evidence for climatic and oceanic variability off NW Africa during the late Holocene. *Quaternary Research,* **72**, 188–197.

CARMONA, P. & RUIZ, J. M. 2009. Geomorphological evolution of the river Loukkos estuary around the Phoenician city of Lixus on the Atlantic littoral of Morocco. *Geoarchaeology,* **24**, 821–845.

CARMONA GONZALEZ, C. 2005. Cambios geomorfologicos y maleogeografia del litoral de Lixus (Larache, Marruecos). *In*: HABIBI, M. & ARANEGUI, C. (eds) *Estudio Malacológico.* Lixus-2 ladera sur, Saguntum, **extra 6**, Knochenarbeit, Valencia, 5–11.

CARR, M.-E. 2002. Estimation of potential productivity in eastern boundary currents using remote sensing. *In*: DONEY, S., SARMIENTO, J. & FALKOWSKI, P. (eds) *The US JGOFS Synthesis and Modeling Project: Phase 1. Deep Sea Research Part A. Oceanographic Research Papers,* **49**, 59–80.

CARRACEDO, J.-C., RODRÍGUEZ BADIOLA, E., GUILLOU, H., DE LA NUEZ, J. & PEREZ TORRADO, F. J. 2001. Geology and volcanology of La Palma and El Hierro (Canary Islands). *Estudios Geologicos,* **57**, 175–273.

CHOUBERT, G., FAURE-MURET, A. & MAARLEVELD, G. C. 1967. Nouvelles dates isotopiques du Quaternaire marocain et leur signification. *Comptes Rendus de l'Académie des Sciences, Paris,* **264**, 434–437.

CHOUKRI, A., HAKAM, O.-K., REYSS, J.-L. & PLAZIAT, J.-C. 2007. Radiochemical dates obtained by alpha spectrometry on fossil mollusc shells from the 5e Atlantic shoreline of the High Atlas, Morocco. *Applied Radiations Isotopes,* **65**, 883–890.

CIRAC, P., FAUGÈRES, J. C. & GAYET, J. 1979. Résultats préliminaires d'une reconnaissance sédimentaire du plateau atlantique marocain. *Bulletin de l'Institut de Géologie du Bassin d'Aquitaine, Bordeaux,* **25**, 69–81.

CIRAC, P., DE RESSEGUIER, A. & WEBER, O. 1989. Situation courantologique et hydrologique sur le plateau continental nord-marocain au cours de la mission GEOMAR II. *Bulletin de l'Institut de Géologie du Bassin d'Aquitaine, Bordeaux,* **46**, 81–95.

CIRAC, P., FRAPPA, M. & JAAÏDI, E. B. 1993. Evolution morpho-structurale récente de la plate-forme continentale ouest-africaine (Maroc nord-atlantique). *Oceanologica Acta,* **16**, 1–9.

CLARK, J. A., FARRELL, W. E. & PELTIER, W. R. 1978. Global changes in postglacial sea level: a numerical calculation. *Quaternary Research,* **9**, 265–278.

CRIADO-ALDEANUEVA, F., GARCIA-LAFUENTE, J., VARGAS, J. M., DEL RIO, J., VAZQUEZ, A., REUL, A. & SANCHEZ, A. 2006. Distribution and circulation of water masses in the Gulf of Cadiz from in situ observations. *In*: RUIZ, J. & GARCÍA-LAFUENTE, J. (eds) *The Gulf of Cadiz Oceanography: A Multidisciplinary View. Deep Sea Research Part II: Topical Studies in Oceanography,* **53**, 1144–1160.

D'ALMEIDA, G. A. 1989. Desert aerosol: characteristics and effects on climate. *In*: LEINEN, M. & SARNTHEIN, M. (eds) *Palaeoclimatology and Palaeometeorology: Modern and Past Patterns of Global Atmospheric Transport.* Kluwer, Dordrecht, 311–338.

DAVENPORT, R., NEUER, S., HELMKE, P., PEREZ-MARRERO, J. & LLINAS, O. 2002. Primary productivity in the northern Canary Islands region as inferred from SeaWiFS imagery. *In*: PARILLA, G. (ed.) *CANIGO I. Deep Sea Research Part II: Topical Studies in Oceanography,* **49**, 3481–3496.

DAVISON, I. 2005. Central Atlantic margin basins of North West Africa: geology and hydrocarbon potential (Morocco to Guinea). *Journal of African Earth Sciences,* **43**, 254–274.

DELIBRIAS, G., GIRESSE, P. & KOUYOUMONTZAKIS, G. 1973. Géochronologie des divers stades de la transgression Holocène au large du Congo. *Comptes Rendus de l'Academie des Sciences, Paris,* **276**, 1389–1391.

DIARA, M. 1999. *Formation et évolution fini-holocènes et dynamique actuelle du delta Saloum – Gambie (Sénégal – Afrique de l'Ouest).* Doctorat thesis, Université de Perpignan, France.

DIESTER-HAASS, L. 1975. Late Quaternary sedimentation and climate in the East Atlantic between Senegal and Cap Verde Islands. *Meteor Forschungsergebnisse, Reihe C,* **20**, 1–32.

DILLON, W. P. 1974. Structure and development of the Southerm Moroccan Continental Shelf. *Marine Geology,* **16**, 121–143.

DIOP, S. 1990. *La côte ouest-africaine du Saloum (Sénégal) à la Mellacorée (Rép. de Guinée).* ORSTOM, Paris.

DOMAIN, F. 1977. *Carte sédimentologique du plateau sénégambien. Extension 1% une partie du plateau continental de la Mauritanie et de la Guinée-Bissau.* ORSTOM, Paris, Notice Explicative, **68**.

DOMAIN, F. 1980. *Contribution à la connaissance de l'écologie des poissons démersaux du plateau continental sénégalo-mauritanien. Les ressources démersales dans le contexte général du Golfe de Guinée.* Thése Doctorat Etat, Université Pierre et Marie Curie, Paris VI et Museum National Histoire Naturelle.

DOMAIN, F. 1985. *Carte sédimentologique du plateau continental mauritanien. (entre le Cap Blanc et 17°N) au 1/200.000 (feuilles Nouadhibou et Nouakchott).* Institut Français de Recherche Scientifiquepour le Développement en Coopération/Centre National de Recherches Océanographiques et des Pêches de Nouadhibou, Notice Explicative, **105**.

DOMAIN, F. 1986. Les fonds de pêche et les ressources. *In*: JOSSE, E. & GARCIA, S. (eds) *Description et évaluation des ressources halieutiques de la ZEE mauritanienne: rapport du groupe de travail CNROP/FAO/ORSTOM sur la Description et Evaluation des Ressources Halieutiques de la ZEE Mauritanienne, 1985/09/16–27, Nouadhibou*. FAO, Rome.

DUCE, R. A., LISS, P. S., MERRILL, J. T., ATLAS, E. L. & BUAT-MENARD, P. 1991. The atmospheric input of trace species to the world ocean. *Global Biochemical Cycles*, **5**, 193–259.

EINSELE, G., HERM, D. & SCHWARZ, H. U. 1974. Holocene eustatic (?) sea level fluctuations at the coasts of Mauritania. *'Meteor' Forschung ergebnisse, C*, **18**, 43–62.

EINSELE, G., ELOUARD, P., HERM, D., KÖGLER, F. C. & SCHWARZ, H. U. 1977. Source and biofacies of late Quaternary sediments in relation to sea level on the shelf off Mauritania. *'Meteor' Forschung ergebnisse, C*, **26**, 1–43.

EL FOUGHALI, A. 1982. *Analyse morpho-structurale du plateau continental atlantique marocain de Tanger à Cap Cantin*. Maroc: Thèse 3ème cycle, Université de Bordeaux l.

EL FOUGHALI, A. & GRIBOULARD, R. 1985. Les grands traits structuraux et lithologiques de la marge atlantique marocaine de Tanger à Cap Cantin. *Bulletin de l'Institut de Géologie du Bassin d'Aquitaine, Bordeaux*, **38**, 179–211.

EL GHARBAOUI, A. 1981. *La terre et l'Homme dans la Péninsule tingitane, étude sur l'homme et le milieu naturel dans le Rif occidental*. Travaux de l'Institut Scientifique, Rabat, Série Géologie & Géographie Physique, **15**.

EL MOUSSAOUI, A., DJENIDI, S. & KOSTIANOY, A. 2003. Physical processes study in the Tansition Zone of the Northwest African Upwelling: climatological data analysis. *Geophysical Research Abstracts*, **5**, 11465.

ELOUARD, P., FAURE, H. & HÉBRARD, L. 1977. Variation du niveau de la mer au cours des 15 000 dernières années autour de la presqu'île du Cap Vert, Dakar, Sénégal. *Bulletin d l'Association Sénégalaise pour l'Etude du Quaternaire Africain*, **50**, 29–49.

ERIMESCO, P. 1965. La mer et l'atmosphère des côtes marocaines. *Bulletin de l'Institut des Pêches du Maroc, Casablanca*, **13**, 3–12.

FAURE, H., FONTES, J. C., HEBRARD, L., MONTEILLET, J. & PIRAZZOLI, P. A. 1980. Geoidal change and shore-level tilt along Holocene estuaries: Senegal river area, West Africa. *Science*, **210**, 421–423.

FAYE, I. B. N. 2010. *Dynamique du trait de côte sur les littoraux sableux de la Mauritanie à la Guinée-Bissau (Afrique de l'Ouest): approches régionale et locale par photointerprétation, traitement d'images et analyse de cartes anciennes*. Thesis, Université de Bretagne occidentale, Brest.

FEDOSEEV, A. 1970. Geostrophic circulation of surface waters on the shelf of Northwest Africa. *Rapports et Procès-Verbaux des Réunions, Conseil International pour l'Exploration de la Mer*, **159**, 30–37.

FLAMENT-LIEFFRIG, D. 1979. *La marge continentale africaine du Sud des Iles Canaries au détroit de Gibraltar. Géologie des bassins El Aiun-Tarfaya-Essaouira*. Thèse de 3ème cycle, Université Paris VI.

FREUDENTHAL, T., MEGGERS, H., HENDERIKS, J., KUHLMANN, H., MORENO, A. & WEFER, W. 2002. Upwelling intensity and filament activity off Morocco during the last 250 000 years. *In*: PARILLA, G. (ed.) *CANIGO I. Deep Sea Research Part II: Topical Studies in Oceanography*, **49**, 3655–3674.

FURNESTIN, J. 1959. Hydrologie du Maroc atlantique. *Revue et Travaux de l'Institut des Sciences et Techniques des Pêches Maritimes, Paris*, **23**, 5–77.

GAC, J. Y. & KANE, A. 1986. Le fleuve Sénégal: I. Bilan hydrologique et flux continentaux de matières particulaires à l'embouchure. *Sciences Géologiques Bulletin*, **39**, 99–130.

GEAWHARI, M. A., MHAMMDI, N. & AMMAR, A. 2012. Etude morphologique par Sonar à balayage latéral des sédiments meubles de la plateforme interne atlantique au Sud de Tanger (Maroc). *Bulletin de l'Institut Scientifique, Rabat, Section Sciences de la Terre*, **34**, 57–67.

GEF. 2002. *Development and Protection of the Coastal and Marine Environment in Subsaharan Africa. The Gambia National Report, Phase 1: Integrated Environment Problem Analysis*. Global Environmental Facility, Washington, DC.

GIGOUT, M. 1959. Ages par radiocarbone, de deux formations des environs de Rabat (Maroc). *Comptes Rendus de l'Académie des Sciences, Paris*, **249**, 2802–2803.

GIRESSE, P. 1987. Quaternary sea-level changes on the Atlantic coast of Africa. *In*: TOOLEY, M. J. & SHENNON, I. (eds) *Sea-level Changes*. Institute of British Geographers, Special Publications, **20**, 249–271.

GIRESSE, P. & BARUSSEAU, J. P. 1986. La succession des lignes de rivage quaternaires du continent africain. *In*: FAURE, H., FAURE, L. & DIOP, E. S. (eds) *Changements globaux en Afrique, passé – présent – futur. International Symposium INQUE/ASEQUA, Dakar*. Travaux et Documents, **197**. ORSTOM, Paris, 165–168.

GIRESSE, P., KOUYOUMONTZAKIS, G. & MOGUEDET, G. 1976. Le Quatemaire supérieur du plateau continental congolais. Exemple de l'évolution paléocéanographique d'une plateforme depuis environ 50 000 ans. *Paleoecology of Africa*, **10/11**, 193–217.

GRIBOULARD, R. 1980. *Relation entre morphologie tectonique et lithologie dans le domaine côtier et sous-marin de la Meseta septentrionale marocaine*. Thèse 3ème cycle. Univ- Bordeaux 1.

GRIBOULARD, R. 1983. Analyse morpho-structurale de la Meseta côtière septentrionale et du proche plateau continental (Maroc). *Bulletin de I'Institut de Géologie du Bassin d'Aquitaine, Bordeaux*, **33**, 25–37.

HAFID, M., ZIZI, M., BALLY, A. W. & AIT SALEM, A. 2006. Structural styles of the western onshore and offshore termination of the High Atlas, Morocco. *Comptes Rendus Geoscience*, **338**, 50–64.

HAFID, M., TARI, G. ET AL. 2008. Atlantic basins. *In*: MICHARD, A., SADDIQI, O., CHALOUAN, A. & FRIZON DE LAMOTTE, D. (eds) *Continental Evolution: the Geology of Morocco. Structure, Stratigraphy, and Tectonics of the Africa–Atlantic–Mediterranean Triple Junction*. Lecture Notes in Geosciences, **116**, Springer, Berlin, 303–329.

HAGEN, E. 2001. Northwest African upwelling scenario. *Oceanologica Acta*, **24**, 113–128.

HAGEN, E., ZULICKE, C. & FEISTEL, R. 1996. Near-surface structures in the cape Ghir filament off Morocco. *Oceanologica Acta*, **19**, 577–598.

HANEBUTH, T. J. J. & HENRICH, R. 2009. Recurrent decadal-scale dust events over Holocene western Africa and their control on canyon turbidite activity (Mauritania). *Quaternary Science Reviews*, **28**, 261–270.

HANEBUTH, T. J. J. & LANTZSCH, H. 2008. A Late Quaternary sedimentary shelf system under hyperarid conditions: unravelling climatic, oceanographic and sea-level controls (Golfe d'Arguin, Mauritania, NW Africa). *Marine Geology*, **256**, 77–89.

HAWORTH, R. J., BAKER, R. G. V. & FLOOD, P. G. 2002. Predicted and observed Holocene sealevels on the Australian coast: what do they indicate about hydro-isostatic models in far-field sites? *Journal of Quaternary Science*, **17**, 581–591.

HEAD, E. J. H., HARRISON, W. G., IRWIN, B. I., HORNE, E. P. W. & LI, W. K. W. 1996. Plankton dynamics and carbon flux in an area of upwelling off the coast of Morocco. *Deep Sea Research Part I: Oceanographic Research Papers*, **43**, 1713–1738.

HÉBRARD, L. 1973. *Contribution à l'étude géologique du Quaternaire du littoral mauritanien entre Nouakchott et Nouadhibou (18°–21° lat. N), Volume 1*. Laboratoire de géologie de la faculté des sciences de l'université de Dakar, Lyon.

HELMKE, P. 2004. *Remote sensing of the Northwest African upwelling and its production dynamics*. Dissertation zur Erlangung des Doktorgrades der Naturwissenschaften am Fachbereich 5, Geowissenschaften der Universität Bremen.

HERNANDEZ-GUERRA, A. & NYKJAER, L. 1997. Sea surface temperature variability off Northwest Africa: 1981–1989. *International Journal of Remote Sensing*, **18**, 2539–2558.

HILLIER, S. 1995. Erosion, sedimentation, and sedimentary origin of clays. *In*: VELDE, B. (ed.) *Origin and Mineralogy of Clays – Clays and the Environment*. Springer, Berlin, 162–219.

HINZ, k., DOSTMANN, H. & FRITSCH, J. 1982. The continental margin of Morocco. Seismic sequences structural elements and geological development. *In*: VON RAD, U., HINZ, k., SARNTHEIN, M. & SEIBOLD, E. (eds) *Geology of the North West African Continental Margin*. Springer, Berlin, 34–60.

HOLZ, C., STUUT, J.-B. W. & HENRICH, R. 2004. Terrigenous sedimentation processes along the continental margin off NW-Africa:

implications from grain-size analysis of seabed sediments. *Sedimentology*, **51**, 1145–1154.

IDRISSI, M., AIT LAAMEL, M., HOURIMECHE, A. & CHAGDALI, M. 2004. Impact of the swell on the current morphological and sedimentary evolution of the coastal zone of Casablanca–Mohammedia (Morocco). *Journal of African Earth Sciences*, **39**, 541–548.

JAAÏDI, E. B. 1981. *Les environnements sédimentaires actuels et Pléistocènes du plateau continental atlantique marocain entre Larache et Agadir.* Thèse de 3ème Cycle, Océanologie, Université de Bordeaux I.

JAAÏDI, E. B. 1993. *La couverture sédimentaire post glaciaire de la plate forme continentale atlantique Ouest-Rifaine (Maroc Nord-Occidental): Exemple d'une séquence transgressive).* Thèse d'Etat, Océanologie, Université Bordeaux I.

JAAÏDI, E. B. & CIRAC, P. 1987. La couverture sédimentaire meuble du plateau continental atlantique marocain entre Larache et Agadir. *Bulletin de l'Institut de Géologie du Bassin d'Aquitaine, Bordeaux*, **42**, 33–51.

KATTAN, Z., GAC, J.-Y. & PROBST, J. L. 1987. Suspended sediment load and mechanical erosion in the Senegalbasin; estimation of surface runoff concentration and relative contributions of channel and slope erosion. *Hydrology*, **92**, 59–76.

KIRK, A. & SPETH, P. 1985. Wind conditions along the coast of Northwest Africa and Portugal during 1972–79. *Tropical Ocean–Atmosphere Newsletter*, **30**, 15–16.

KNOLL, M., HERNANDEZ-GUERRA, A., LENZ, B., LOPEZ LAATZEN, F., MACHIN, F., MULLER, T. J. & SIEDLER, G. 2002. The eastern boundary current system between the Canary Islands and the African coast. *In*: PARILLA, G. (ed.) *CANIGO I. Deep Sea Research Part II: Topical Studies in Oceanography*, **49**, 3427–3440.

KOSTIANOY, A. G. & ZATSEPIN, A. G. 1996. The West African coastal upwelling filaments and cross-frontal water exchange conditioned by them. *Journal of Marine Systems*, **7**, 349–359.

KRÖCHERT, J., MAURER, H. & BUCHNER, E. 2008. Fossil beaches as evidence for significant uplift of Tenerife, Canary Islands. *Journal of African Earth Sciences*, **51**, 220–234.

LAMBECK, K. 2002. Sea level change from mid Holocene to recent time: an Australian example with global implications. *In*: MITROVICA, J. X. & VERMEERSEN, B. L. A. (eds) *Ice Sheets, Sea Level and the Dynamic Earth*. American Geophysical Union, Geodynamics Series, **29**, 33–50.

LANTZSCH, H., HANEBUTH, T. J. J. & BENDER, V. B. 2009. Holocene evolution of mud depocentres on a high-energy, low-accumulation shelf (NW Iberia). *Quaternary Research*, **72**, 325–336.

LERIQUE, J. 1975. *Les transports solides en suspension de la Gambie à Kédougou et Goulombo. Résultats de la campagne 1974*. PNUD, Projet Régional, **60**. ORSTOM, Dakar.

LEROUX, M. 2001. *The Meteorology and Climate of Tropical Africa*. Springer, Berlin.

LIOUBIMTSEVA, E. 2004. Climate change in arid environments: revisiting the past to understand the future. *Progress in Physical Geography*, **28**, 502–530.

LUDWIG, W. & PROBST, J. L. 1998. River sediment discharge to the oceans: present-day controls and global budgets. *American Journal of Science*, **298**, 265–295.

MAKAOUI, A., ORBI, A., HILMI, K., ZIZAH, S., LARISSI, J. & TALBI, M. 2005. L'upwelling de la côte atlantique du Maroc entre 1994 et 1998. *Comptes Rendus Geoscience*, **337**, 1518–1524.

MAKAOUI, A., ORBI, A., ARESTIGUI, J., BEN AZZOUZ, A., LAARISSI, J., AGOUZOUK, A. & HILMI, K. 2012. Hydrological seasonality of cape Ghir filament in Morocco. *Natural Science*, **4**, 5–13.

MARCELLO, J., HERNANDEZ-GUERRA, A., EUGENIO, F. & FONTE, A. 2011. Seasonal and temporal study of the northwest African upwelling system. *International Journal of Remote Sensing*, **32**, 1843–1859.

MARTIN, L. 1972. Variations du niveau de la mer et du climat en Côte d'Ivoire depuis 25 000 ans. *Cahiers de l'ORSTOM, Série Géologie*, **IV**, **2**, 93–103.

MARY, G. 1982. Rôle probable de l'isostasie dans les modalités de la transgression holocène sur la côte atlantique de l'Europe et de l'Afrique. *Bulletin de l'Association française pour l'étude du quaternaire*, **19**, 39–45.

MASSE, J. P. 1968. *Contribution à l'étude des sédiments actuels du plateau continental de la région de Dakar. Essais d'analyse de la sédimentation biogène*. Rapport du Laboratoire de Géologie de la Faculté des Sciences de l'Université de Dakar, **23**.

MATHIEU, R. 1986. *Sédiments et Foraminifères actuels de la marge atlantique du Maroc*. Thèse d'Etat. Université Pierre et Marie Curie, Paris.

MATTHEWSON, A. P., SHIMMIELD, G. B., KROON, D. & FALLICK, A. E. 1995. A 300 kyr high-resolution aridity record of the North African continent. *Paleoceanography*, **10**, 677–692.

MCMASTER, R. L. & LACHANCE, T. P. 1968. Seismic reflectivity studies on the North Western African continental shelf. *American Association of Petroleum Geologists Bulletin*, **52**, 2387–2395.

MCMASTER, R. L. & LACHANCE, T. P. 1969. Northwestern African continental shelf sediments. *Marine Geology*, **7**, 57–67.

MCTAINSH, G. H. & WALKER, P. H. 1982. Nature and distribution of Harmattan dust. *Zeitschrift für Geomorphologie NF*, **26**, 417–435.

MEAGHER, L. J., RUFFMAN, A. S., STEWART, J. Mc G., ZUKAUSKAS, W. & VAN DER LINGEN, W. J. M. 1977. *CSS Baffin Offshore Survey: Senegal and The Gambia. Volume 2: A Contribution to the Geophysics and Geology of the Continental Shelf and Margin of Senegal and The Gambia, West Africa*. Departments of Fisheries and the Environment and of Energy, Mines and Resources, Ottawa.

MEDINA, F. 1995. Syn- and post-rift evolution of the El Jadida-Agadir basin (Morocco): constraints for the rifting models of the Central Atlantic. *Canadian Journal of Earth Sciences*, **32**, 1273–1291.

MEGHRAOUI, M., OUTTANI, F., CHOUKRI, A. & FRIZON DE LAMOTTE, D. 1999. Coastal tectonics across the South Atlas thrust front and the Agadir active zone, Morocco. *In*: STEWART, I. S. & VITA-FINZI, C. (eds) *Coastal Tectonics*. Geological Society, London, Special Publications, **146**, 239–253.

MICHARD, A., SADDIQI, O., CHALOUAN, A. & FRIZON DE LAMOTTE, D. (eds) 2008. *Continental Evolution: The Geology of Morocco. Structure, Stratigraphy, and Tectonics of the Africa–Atlantic–Mediterranean triple Junction*. Lecture Notes in Geosciences, **116**, Springer, Berlin.

MICHEL, P. 1973. *Géologie des régions traversées par les fleuves Sénégal et Gambie, geomorphological map*. ORSTOM, Paris.

MILLIMAN, J. D. & MEADE, R. H. 1983. World-wide delivery of river sediment to the oceans. *Journal of Geology*, **91**, 1–21.

MILLIMAN, J. D. & SYVITSKI, J. P. M. 1992. Geomorphic/tectonic control of sediment discharge to the ocean: the importance of small mountainous rivers. *Journal of Geology*, **100**, 525–544.

MITTELSTAEDT, E. 1983. The upwelling area off northwest Africa: a description of phenomena related to coastal upwelling. *Progress in Oceanography*, **12**, 307–331.

MITTELSTAEDT, E. 1991. The ocean boundary along the northwest African coast: circulation and oceanographic properties at the sea surface. *Progress in Oceanography*, **26**, 307–355.

MONTEILLET, J., FAURE, H., PIRAZZOLI, P. A. & RAVISE, A. 1981. Invasion saline du Ferlo (Senegal) à l'Holocène supérieur (1900 BP). *Paleoecology of Africa*, **13**, 205–215.

MOUJANE, A., BENTAMY, A., CHAGDALI, M. & MORDANE, S. 2011*a*. Analysis of high spatial and temporal surface winds from Aladin model and from remotely sensed data over the Canarian upwelling region. *Revue Télédétection*, **10**, 11–22.

MOUJANE, A., CHAGDALI, M., BLANKE, B. & MORDANE, S. 2011*b*. Impact des vents sur l'upwelling au sud du Maroc; apport du modèle ROMS forcé par les données ALADIN et QuikSCAT. *Bulletin de l'Institut Scientifique, Section Sciences de la Terre*, **33**, 53–64.

MRIDEKH, A., MEDINA, F., MHAMMDI, N., SAMAKA, F. & BOUATMANI, R. 2009. Structure of the Kasbah fold zone (Agadir bay, Morocco). Implications on the chronology of the recent tectonics of the western High Atlas and on the seismic hazard of the Agadir area. *Estudios Geológicos*, **65**2, 121–132.

NEWTON, R. S., SEIBOLD, E. & WERNER, F. 1973. Facies distribution patterns on the Spanish Sahara continental shelf, mapped with side scan sonar. *Meteor Forschungsergebnisse, Reihe C*, **15**, 55–57.

NICHOLSON, S. E. 2000. The nature of rainfall variability over Africa on time scales of decades to millennia. *Global and Planetary Change*, **26**, 137–158.

Nizou, J., Hanebuth, T. J. J., Heslop, D., Schwenk, T., Palamenghi, L., Stuut, J.-B. & Henrich, R. 2010. The Senegal River mud belt: a high-resolution archive of paleoclimatic change and coastal evolution. *Marine Geology*, **278**, 150–164.

Nizou, J., Hanebuth, T. J. J. & Vogt, C. 2011. Deciphering signals of late Holocene fluvial and aeolian supply from a shelf sediment depocentre off Senegal (north-west Africa). *Journal of Quaternary Science*, **26**, 353–456.

Nykjaer, L. & Van Camp, L. 1994. Seasonal and interannual variability of coastal upwelling along northwest Africa and Portugal from 1981 to 1991. *Journal of Geophysical Research*, **99**, 14 197–14 207.

Odin, G. S. & Masse, J. P. 1988. The verdine facies of the Senegalese continental shelf. *In*: Odin, G. S. (ed.) *Green Marine Facies*. Developments in Sedimentology, **45**. Elsevier, Amsterdam, 83–104.

Orange, D. & Gac, J. Y. 1990. Bilan géochimique des apports atmosphériques en domaines sahélien et soudano-guinéen d'Afrique de l'Ouest (bassins supérieurs du Sénégal et de la Gambie). *Géodynamique*, **5**, 51–65.

Ould El Moustapha, A. 2000. *Influence d'un ouvrage portuaire sur l'équilibre d'un littoral soumis à un fort transit sédimentaire. L'exemple du port de Nouakchott (Mauritanie)*. Thèse de doctorat d'université, Université de Caen/Basse Normandie.

Ould El Moustapha, A., Levoy, F., Monfort, O. & Koutitonsky, V. G. 2007. A numerical forecast of shoreline evolution after harbour construction in Nouakchott, Mauritania. *Journal of Coastal Research*, **23**, 1409–1417.

Ould Sidi Cheikh, M. A., Ozer, P. & Ozer, A. 2007. Risques d'inondations dans la ville de Nouakchott (Mauritanie). *Géo-Eco-Trop*, **31**, 19–42.

Pegler, E. A. 1999. Mid- to late Quaternary environment and stratigraphy of the southern Sierra Leone shelf, West Africa. *Journal of the Geological Society, London*, **156**, 977–990.

Pelegrí, J. L., Arístegui, J. et al. 2005. Coupling between the open ocean and the coastal upwelling region off northwest Africa: water recirculation and offshore pumping of organic matter. *Journal of Marine Systems*, **54**, 3–37.

Pinson-Mouillot, J. 1980. *Les environnements sédimentaires actuels et quaternaires du plateau continental sénégalais (Nord de la presqu'île du Cap-Vert)*. PhD thesis, University of Bordeaux, France.

Pirazzoli, P. A. & Montaggioni, L. F. 1988. Holocene sea-level changes in French Polynesia. *Palaeogeography, Palaeoclimatology, Palaeoecology*, **68**, 153–175.

Probst, J. L. & Amiotte Suchet, P. 1992. Fluvial suspended sediment transport and mechanical erosion in the Maghreb. *Hydrological Sciences – Journal des Sciences Hydrologiques*, **37**, 621–637.

Rivière, A. 1977. *Méthodes granulo-métriques, techniques et interprétation*. Masson, Paris.

Robb, J. M. 1971. Structure of continental margin between Cap Ghir and Cap Sim, Morocco. *American Association of Petroleum Geologists, Bulletin*, **55**, 643–650.

Ruellan, E. 1985. *Géologie des marges continentales passives: évolution de la marge atlantique du Maroc (Mazagan): étude par submersible seabeam et sismique réflexion. Comparaison avec la marge NW africaine et la marge homologue E américaine*. Thèse de Doctorat, Université de Bretagne Occidentale, Brest.

Ruffman, A., Meagher, L. J. & Stewart, J. M. G. 1977. Bathymétrie du talus et du plateau continental du Sénégal et de la Gambie (Afrique de l'Ouest). *In*: *Le Baffin, Levé au large du Sénégal et de la Gambie, Vol. 1*. Ministère des Pêches et de l'Environnement du Canada, Gatineau, 23–38.

Sahabi, M., Aslanian, D. & Olivet, J.-L. 2004. Un nouveau point de départ pour l'histoire de l'Atlantique central. *Comptes Rendus Géosciences*, **336**, 1041–1052.

Sall, M. 1982. *Dynamique et morphogenèse actuelles au Sénégal occidental*. Thèse, Université de Strasbourg.

Sarnthein, M., Thiede, J. et al. 1982. Atmospheric and oceanic circulation patterns off northwest Africa during the past 25 million years. *In*: Von Rad, U., Hinz, K., Sarnthein, M. & Seibold, E. (eds) *Geology of the Northwest African Continental Margin*. Springer, Berlin, 545–604.

Sauvage, C. 1963. Le Quotient pluviométrique d'Emberger, son utilisation et la représentation géographique de ses variations au Maroc. *Annales du Service de Physique du Globe et de Météorologie, Institut Scientifique Chérifien*, **20**, 11–23.

Seibold, E. 1982. The N.W. African continental margin, an introduction. *In*: Von Rad, U., Hinz, K., Sarnthein, M. & Seibold, E. (eds) *Geology of the North West African Continental Margin*. Springer, Berlin, 3–23.

Seibold, E. & Fütterer, D. 1982. Sediment dynamics on the Northwest African continental margin. *In*: Scrutton, R. & Talwani, M. (eds) *The Ocean Floor: Bruce Heezen Commemorative Volume*. Wiley, Chichester, 147–163.

Seibold, E. & Hinz, K. 1974. Continental slope construction and destruction. W Africa. *In*: Burke, C. A. & Drake, C. L. (eds) *Geology of the Continental Margin*. Springer, Berlin, 179–196.

Sena Martins, C., Hamann, M. & Fiúza, A. F. G. 2002. Surface circulation in the eastern North Atlantic, from drifters and altimetry. *Journal of Geophysical Research*, **107**, 3217, http://dx.doi.org/10.1029/2000JC000345

Shi-Leng, X. & Teh-Fu, L. 1987. Long-term variation of longshore sediment transport. *Coastal Engineering*, **11**, 131–140.

Simonet, R. & Tanguy, R. 1956. Etude statistique de la houle dans les différents ports marocains pour la période 1928–1952. *Annales du Service de Physique du Globe et de Météorologie, Institut Scientifique Chérifien*, **16**, 109–130.

Snoussi, M. 1984. Comportement du Pb. Zn. Ni et Cu dans les sédiments de l'estuaire du Loukkos et du proche plateau continental (côte atlantique marocaine). *Bulletin de l'Institut de Géologie du Bassin d'Aquitaine, Bordeaux*, **35**, 23–30.

Snoussi, M., Jouanneau, J. M. & Latouche, C. 1990. Flux de matières issues de bassins versants de zones semi-arides (Bassins du Sebou et du Souss, Maroc). Importance dans le bilan global des apports d'origine continentale parvenant à l'Océan Mondial. *Journal of African Earth Sciences*, **11**, 43–54.

Stearns, C. E. & Thurber, D. L. 1965. Th230/U234 dates of late Pleistocene marine fossils from the Mediterranean and Moroccan littorals. *Progress in Oceanography*, **4**, 293–305.

Stevens, I. & Johnson, J. 2003. A numerical modelling study of upwelling filaments off the NW African coast. *Oceanologica Acta*, **26**, 549–564.

Stour, L. & Agoumi, A. 2008. Sécheresse climatique au Maroc durant les dernières décennies. *Hydroécologie Appliquée*, **16**, 215–232.

Stuut, J.-B., Zabel, M., Ratmeyer, V., Helmke, P., Schefuss, E., Lavik, G. & Schneider, R. 2005. Provenance of present-day eolian dust collected off NW Africa. *Journal of Geophysical Research*, **110**, D04202, http://dx.doi.org/10.1029/2004JD005161

Summerhayes, C. P., Nutter, A. H. & Tooms, J. S. 1971. Geological structure and development of the continental margin of North Western Africa. *Marine Geology*, **11**, 1–25.

Summerhayes, C. P., Milliman, J. D., Briggs, S. R., Bee, A. G. & Hogan, C. 1976. Northwest African shelf sediments: influence of climate and sedimentary processes. *Journal of Geology*, **84**, 277–300.

Swap, R., Garstang, M. et al. 1996. The long-range transport of southern African aerosols to the tropical South Atlantic. *Journal of Geophysical Research*, **101**, 23 777–23 791.

Tari, G., Brown, D., Jabour, H., Hafid, M., Louden, K. & Zizi, M. 2012. The conjugate margins of Morocco and Nova Scotia. *In*: Roberts, D. J. & Bally, A. W. (eds) *Regional Geology and Tectonics: Phanerozoic Passive Margins, Cratonic Basins and Global Tectonic Maps*. Elsevier, Amsterdam, 285–318.

Tejera de Leon, J. & Duplantier, F. 1981. Origine et signification sédimentologique des dépôts meubles de l'estuaire du Loukkos et du prisme littoral adjacent (littoral nord atlantique marocain). *Bulletin de l'Institut de Géologie du Bassin d'Aquitaine*, **29**, 133–159.

Tooms, J. S., Summerhayes, C. P. & McMaster, , 1971. Marine geological studies of the NW African margin: Rabat Dakar. *In*: Delaney, F. M. (ed.) *The Geology of the East Atlantic Continental Margin*. Institute of Geological Sciences, Report, **70-16**, 9–25.

Torres-Padrón, M. E., Gelado-Caballero, M. D., Collado-Sánchez, C., Siruela-Matos, V., Cardona-Castellano, P. & Hernández-Brito, J. J. 2002. Variability of dust inputs to the

CANIGO zone. *In*: PARILLA, G. (ed.) *CANIGO I. Deep Sea Research Part II: Topical Studies in Oceanography*, **49**, 3455–3464.

TRENTESAUX, A., MALENGROS, D. *ET AL.* 2008. Uncommonly-deep giant dunes offshore the Moroccan Atlantic Coast Marine and river dune dynamics. *In*: PARSONS, D., GARLAN, T. & BEST, J. (eds) *Proceedings of Marine and River Dune Dynamics III, International Workshop*, University of Leeds, UK, 1–3 April 2008, 297–300.

UCHUPI, E. K., EMERY, K. O. *ET AL.* 1976. Continental margin off western Africa from Senegal to Portugal. *American Association of Petroleum Geologists Bulletin*, **60**, 809–878.

VAN CAMP, L., NYKJAER, L., MITTELSTAEDT, E. & SCHLITTENHARDT, P. 1991. Upwelling and boundary circulation off Northwest Africa as depicted by infrared and visible satellite observations. *Progress in Oceanography*, **26**, 357–402.

VERNET, R. 2007. *Le golfe d'Arguin de la préhistoire à l'histoire – littoral et plaines intérieures*. Collection PNBA, **3**, Parc national du Banc D'Arguin, Nouakchott.

VON RAD, U. & WISSMANN, G. 1982. Cretaceous–Cenozoic history of the West Sahara Continental Margin (NW Africa): development, destruction and gravitational sedimentation. *In*: VON RAD, U., HINZ, K., SARNTHEIN, M. & SEIBOLD, E. (eds) *Geology of the Northwest African Continental Margin*. Springer-Verlag, Berlin, 106–132.

VON RAD, U., RYAN, W. B. F. *ET AL.* 1979. *Initial Reports of the Deep Sea Drilling Project*, **47**. US Government Printing Office, Washington, DC.

VON RAD, U., HINZ, K., SARNTHEIN, M. & SEIBOLD, E. (eds) 1982. *Geology of the North West African Continental Margin*. Springer, Berlin.

WASHINGTON, R., TODD, M. C. *ET AL.* 2006. Links between topography, wind, deflation, lakes and dust: The case of the Bodélé Depression, Chad. *Geophysical Research Letters*, **33**, L09401, http://dx.doi.org/10.1029/2006GL025827

WEISROCK, A. L. E. 1981. Neotectonic and coastal morphology in the Atlantic Atlas (Morocco). *Zeitschrift für Geomorphologie NF*, **40**, 175–182.

WEISROCK, A. 2009. Revision du Quaternaire marin de l'Atlas atlantique. RQM6 Fès, unpublished.

WEISROCK, A., DELIBRIAS, G., ROGNON, P. & COUDE-GAUSSEN, G. 1985. Variations climatiques et morphogénèse au Maroc atlantique (30–35°N) à la limite Pleistocène–Holocène. *Bulletin de la Société Géologique de France*, **8**, 565–569.

WOOSTER, W. S., BAKUM, A. & MCLAIN, D. R. 1976. The seasonal upwelling cycle along the eastern boundary of the North Atlantic. *Journal of Marine Research*, **34**, 131–140.

WYNN, R. B., MASSON, D. G., STOW, D. A. V. & WEAVER, P. P. E. 2000. The Northwest African slope apron: a modern analogue for deep-water systems with complex seafloor topography. *Marine and Petroleum Geology*, **17**, 253–265.

ZAZO, C., GOY, J. L. *ET AL.* 2002. Raised marine sequences of Lanzarote and Fuerteventura revisited – a reappraisal of relative sea-level changes and vertical movements in the eastern Canary Islands during the Quaternary. *Quaternary Science Reviews*, **21**, 2019–2046.

ZHOU, M., PADUAN, J. D. & NIILER, P. P. 2000. Surface currents in the Canary Basin from drifter observations. *Journal of Geophysical Research*, **105**, 21 893–21 911.

Chapter 11

The Iberian Mediterranean shelves

F. J. LOBO[1]*, G. ERCILLA[2], L. M. FERNÁNDEZ-SALAS[3] & D. GÁMEZ[4]

[1]*Instituto Andaluz de Ciencias de la Tierra (CSIC-Universidad de Granada), Avenida de las Palmeras no 4, 18100 Armilla, Granada, Spain*

[2]*Institut de Ciències del Mar (CSIC), Grup de Marges Continentals, Passeig Marítim de la Barceloneta no 37–49, 08003 Barcelona, Spain*

[3]*Instituto Español de Oceanografía, Centro Oceanográfico de Cádiz, Puerto Pesquero, Muelle de Levante s/n, 11006 Cádiz, Spain*

[4]*ExxonMobil Development Company, 12450 Greenspoint Drive, Houston, TX 77060, USA*

Corresponding author (e-mail: pacolobo@iact.ugr-csic.es)

Abstract: The Iberian Mediterranean shelves are divided into three different geographical segments (the Northeastern Shelf, the Southeastern Shelf and the Northern Alboran Sea Shelf), the understanding of which has evolved over the years. The best known sector is the Northeastern Shelf, comprising the narrow, abrupt and prograding Catalonia Shelf and the wider, prograding Ebro Shelf–Gulf of Valencia, where pioneering Spanish marine geology studies have been conducted since the 1970s. The knowledge of the Quaternary stratigraphic architecture of the Northeastern Shelf is very detailed, and provides an outstanding example of regressive–transgressive cycles leading to shelf build-up with various margin configurations. The Southeastern Shelf exhibits a change of margin configuration from intermediate to abrupt in response to declining fluvial influence. The knowledge of this shelf is limited in comparison with the rest of the Iberian Mediterranean shelves. Abundant studies have also been performed on the Northern Alboran Sea Shelf, which, in contrast to the Northeastern Shelf, does not have a major fluvial source but numerous short, mountain rivers draining from the Betic Cordillera. For this shelf, a high-resolution sequence stratigraphy model has been proposed for the most recent Late Quaternary depositional sequence.

The Iberian shelves facing the Mediterranean Sea extend from the Gulf of Lions in the NW Mediterranean to the Strait of Gibraltar at the Atlantic–Mediterranean boundary. This shelf segment receives the fluvial supply of the Ebro River, which is one of the main sediment sources of the Mediterranean Basin. In spite of this, the dominant depositional systems consist of relatively small, steep and lobate prodeltas laterally connected to elongated progradational prisms off coasts with negligible fluvial inputs. This prevailing depositional pattern results from the main sedimentary fingerprint of short, mountain rivers, which tend to have a torrential character, influenced by a Mediterranean climate, increasingly arid towards SE Iberia, and by an abrupt hinterland in many places.

For the purposes of this summary, the Iberian Mediterranean shelves have been divided into three main sectors (Table 11.1). The Northeastern Shelf comprises the poorly developed Catalonia Shelf receiving supply from small rivers draining the Catalan Coastal Ranges and the Pyrenees, the wide, extensive Ebro Shelf, supplied by the main fluvial source in the Iberian Peninsula, and the Gulf of Valencia, which also shows a prograding configuration, as it is mainly influenced by the southward advection of sediments from the Ebro River. The Southeastern Shelf displays two arc-shaped segments, the morphological configuration of which is influenced by the Betic Mountains and a decrease in fluvial supply to the south. The Northern Alboran Sea Shelf is narrow, with numerous small mountain rivers and streams draining from the Betic Mountains.

The above-mentioned division of the Iberian Mediterranean shelves is rooted in the historical evolution of shelf environment knowledge in Spain. The majority of the pioneering studies of marine geology in Spain were conducted on the Ebro Shelf by various research groups, mostly based in Barcelona. As a consequence, the Northeastern Shelf has been the most thoroughly studied in Spain over the last 40 years, and a *Marine Geology* Special Issue devoted to the Ebro Margin was published more than 20 years ago (Nelson & Maldonado 1990*b*). The Northern

Alboran Sea Shelf has also been thoroughly studied, especially during the last 20 years, over which time numerous studies using similar methodologies to those already applied on the Northeastern Shelf have been conducted, and which include a *Geo-Marine Letters* Special Issue focusing on the geology and geophysics of the Alboran Sea, published in 1992 (Maldonado 1992). In contrast to these two relatively well-known shelf sectors, the Southeastern Shelf remains comparatively less documented. Although, during the 1980s, a large number of sedimentological and seismic studies were conducted on the Southeastern Shelf, most were of limited relevance as they were published in Spanish journals and books and followed pre-sequence stratigraphy approaches. In addition, the geological interest in this shelf declined in subsequent years, when most of the geological information was summarized in several cartographical works executed within a programme devoted to the systematic mapping of the Spanish shelves.

The Northeastern Shelf

General setting

The Northeastern Shelf is located in the eastern Iberian Margin and extends from the southernmost Pyrenean Mountains to Cape La Nao (Fig. 11.1).

Oceanography. The coastal environment is affected by a low-energy wave climate, with mean wave heights of 1 m and a mean period of about 4 s (Gracia *et al.* 1989). The coast and adjacent shelf are mainly exposed to swell waves coming from the NE to the SW (Cendrero *et al.* 2005; Nuez *et al.* 2008). The most important easterly storms have a typical duration of few days and are often associated with cyclonic activity in the western Mediterranean Sea (Ojeda & Guillén 2006).

The prevailing currents on the shelf are controlled by the cyclonic eddy of the western Mediterranean derived from the Northern

From: CHIOCCI, F. L. & CHIVAS, A. R. (eds) 2014. *Continental Shelves of the World: Their Evolution During the Last Glacio-Eustatic Cycle.*
Geological Society, London, Memoirs, **41**, 147–170. http://dx.doi.org/10.1144/M41.11

Table 11.1. *Summary table of the main descriptive parameters of the three shelf sectors (the Northeastern Shelf, Southeastern Shelf and Northern Alboran Sea Shelf) comprising the Iberian Mediterranean Shelves*

	Northeastern Shelf	Southeastern Shelf	Northern Alboran Sea Shelf
Shelf length	550 km	345 km in the Peninsular sector and 860 km around the Balearic Islands	350 km
Shelf width	From a few to 80 km	From a few to >40 km	From a few km to >20 km
Tidal conditions	Microtidal	Microtidal	Microtidal
Wave conditions	Swell waves from NE to SW (dominance of waves from the ENE)	Waves from the SE in summer, and from the NW in winter	Equal dominance (40%) of waves from the east and west
Currents (cm s^{-1})	Up to 30–40	–	Up to 1
Dominating process	Waves and currents (strong seasonality)	Waves and currents	Waves and currents
Shelf break depth (m)	150–200	140–170	100–125
Sedimentation type	Mostly siliciclastic	Mostly siliciclastic, carbonates in the shelf of the Balearic Islands and in the southern part of the Southern Arc	Mostly siliciclastic, carbonates in the easternmost shelf
Modern/relict/palimpsest	Inner–middle shelf modern sediments, middle–outer shelf relict and palimpsest	Inner shelf modern sediments, outer shelf relict and palimpsest	Inner shelf modern sediments, outer shelf relict and palimpsest
Tectonic trend over the last glacial cycle	Mostly subsidence (sedimentary and tectonic). Local uplifting	In the Northern Arc, mostly subsidence; in the Southern Arc, differential subsidence	Laterally variable (uplifting v. subsidence)
Maximum thickness of Quaternary deposits (km)	>3	>0.4	>0.5

Current (Millot 1999), which flows southwards with velocities of about 30–40 cm s^{-1} (Hopkins 1985; Font *et al.* 1988; Castellón *et al.* 1990; La Violette *et al.* 1990; Flexas *et al.* 2002) (Fig. 11.1a). The Northern Current is fed by continental freshwater, generating a permanent density front (Catalan Front) (López-García *et al.* 1994). The main circulation is occasionally modified by small gyres formed by the effect of local winds and enhanced by the topography of the shelf and/or canyon heads (Arnau *et al.* 2004).

Climate. The regional climate is characterized by a marked seasonal regime with wet temperate circulation during winter and dry subtropical circulation in the summer (Martín-Vide & López-Bustins 2006). As a consequence, the Catalan Sea is also strongly influenced by seasonality. Sea surface temperatures range from less than 13 °C in winter to more than 27 °C in summer (Picco 1991; Millot 1999). This seasonality leads to the formation of a thermocline between early spring and late autumn, whereas in winter the water column is non-stratified (Salat *et al.* 2002).

Sedimentation regime. The Northeastern Shelf is fed mainly by rivers and streams (Fig. 11.2). The river basins are characterized by contrasting areal extensions, ranging from 85 835 km^2 for the Ebro River to a few thousands of km^2 for rivers such as the Fluviá, Ter, Besós and Llobregat, and a few hundreds of km^2 for, for example, the Muga and Tordera rivers. The Ebro River is one of the most important rivers draining into the Western Mediterranean Sea (Palanques *et al.* 1990; Liquete *et al.* 2007, 2009). The contribution of this river generates a mud-accumulating sedimentary regime on the shelf off the river mouth and to the south in the Gulf of Valencia (Maldonado *et al.* 1983). The suspended sediment load of the Ebro River was drastically reduced in the twentieth century due to dam construction, from around 17×10^6 to about 1.5×10^5 t a^{-1} after widespread damming (Palanques *et al.* 1990).

The Northeastern Shelf is also fed by numerous short (2–3 km-long), ephemeral streams that drain areas of a few hundreds of km^2, providing input to coastal sediment bodies during short-lived flood events while, at other times, remaining dry. The total suspended sediment matter supplied by those systems is highly variable, ranging between 0.4 and 2000 mg l^{-1} (Liquete *et al.* 2009), according to seasonal and irregular flow regimes.

The predominant littoral drift is towards the SSW, parallel to the coastline (Flos 1985; Font *et al.* 1988; Medialdea *et al.* 1989). This drift is intercepted by numerous marinas and breakwaters. As a result, sand accumulates updrift of the marinas, resulting in a negative balance downdrift (Durán *et al.* 2008; Nuez *et al.* 2008).

Tectonic framework and geological background

The Northeastern passive margin is located on the western branch of the Valencia Trough, between the Iberian Peninsula and the Balearic Promontory. This margin is bounded by several convergent structural elements, from north to south: (a) the east–west-trending Eastern Pyrenees; (b) the Catalan Coastal Ranges, which trend NE parallel to the coastline; and (c) the NW-trending eastern end of the Iberian Range (Fig. 11.1b).

The Valencia Trough is a NE-trending tectonic depression with steep narrow margins (Fig. 11.1b) that opened during the Late Oligocene–Early Miocene, related to the propagation of the Western European rift system and the oceanic accretion led by the SE drift of the Corsica–Sardinia block (Rehault *et al.* 1984; Olivet 1996). Opening was facilitated by the formation of normal faults (Fig. 11.1b) due to the inversion of Palaeogene thrusts and strike-slip faults, generating horst-and-graben structures (Maillard *et al.* 1992; Maillard & Mauffret 1999; Roca *et al.* 1999). Sedimentation was restricted to the graben, and the sediment-transport direction changed from longitudinal (NE–SW) to transverse (NW–SE) (Guimerà 1984; Roca *et al.* 1999). The post-rift stage of the Valencia Trough was characterized by attenuated tectonic activity including extensional reactivation (Díaz del Río *et al.* 1986; Bartrina *et al.* 1992; Maillard & Mauffret 1999; Roca *et al.* 1999).

Uplift/subsidence of the margin. The continental margin has been subjected to irregular subsidence due to both sedimentary loading and tectonics throughout the Quaternary. This subsidence regime has enabled the preservation of Quaternary progradational geometries (Fig. 11.3). In the vicinity of the Cape Creus Canyon, the Plio-Quaternary sediment record is up to 1 km thick, revealing substantial subsidence (García *et al.* 2011). On the Ebro Shelf, increased accommodation creation during the Pleistocene was caused by compaction of underlying strata and enhanced margin aggradation (Kertznus & Kneller 2009). In the Gulf of Valencia, a high subsidence regime allowed the formation of thick prograding deposits due to the activity of epeirogenic flexural movements and tectonic structures (Maldonado & Zamarreño 1983; Rey & Díaz del Río 1983; Díaz del Río *et al.* 1986). In spite of the

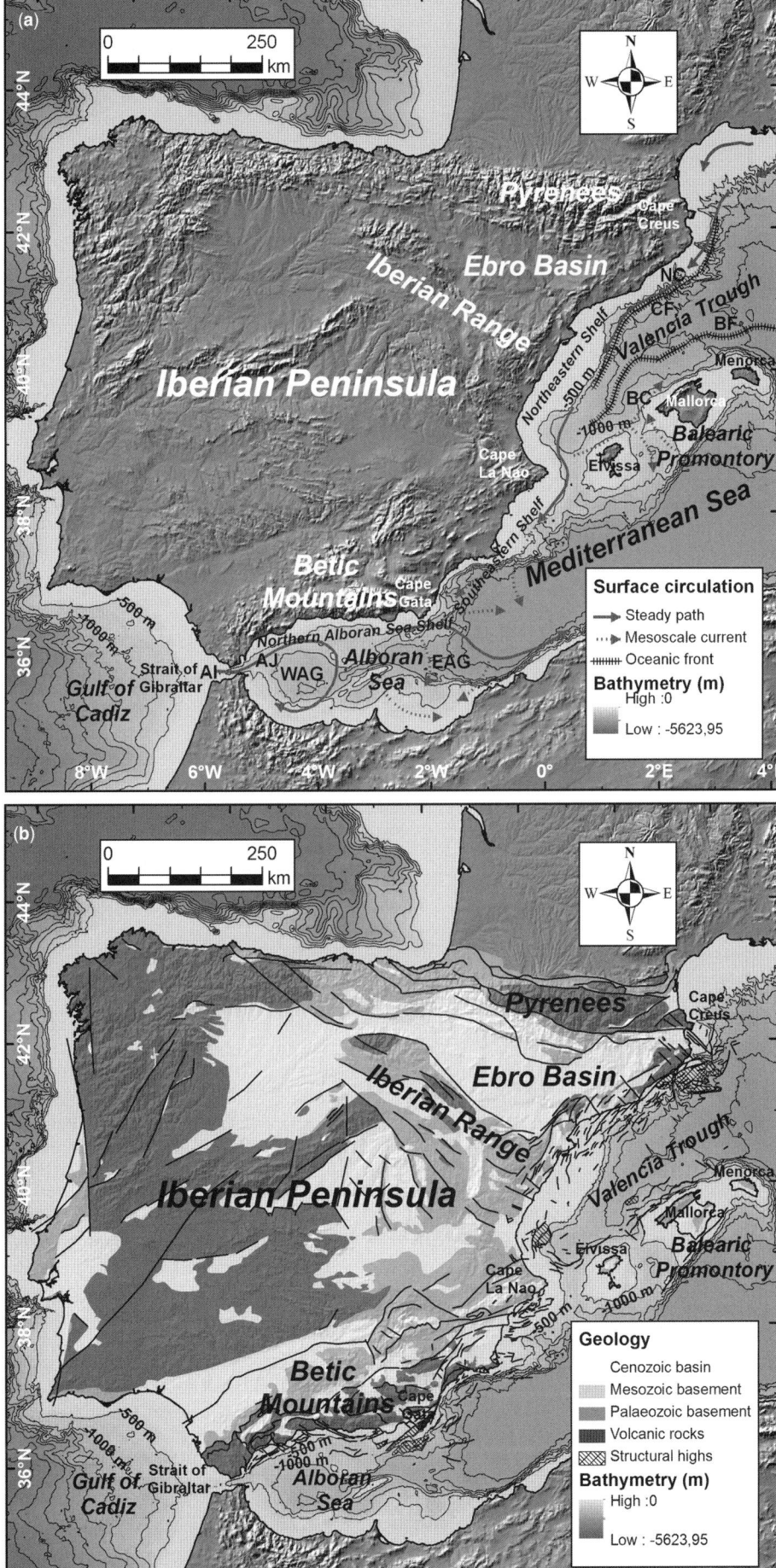

Fig. 11.1. Geographical location of the Iberian Mediterranean shelves in the eastern part of the Iberian Peninsula and the western Mediterranean Sea. Three main sectors are considered according to level of knowledge, and specific physiographical and sediment-supply conditions: the Northeastern Shelf; the Southeastern Shelf, including the shelf around the Balearic Islands; and the Northern Alboran Sea Shelf. 500 m bathymetric contours are displayed. (**a**) Main surface circulation patterns along the Iberian Mediterranean shelves compiled from López-García *et al.* (1994) and Millot (1999). NC, Northern Current; BC, Balearic Current; CF, Catalan Front; BF, Balearic Front; AI, Atlantic Inflow; AJ, Atlantic Jet, WAG, Western Alboran Gyre; EAG, Eastern Alboran Gyre. (**b**) Main faults along the Iberian Mediterranean margins compiled from Catafau *et al.* (1990, 1994), Comas *et al.* (1992) and Maillard & Mauffret (1999). The onland geology of the Iberian Peninsula (Andeweg 2002) is also displayed.

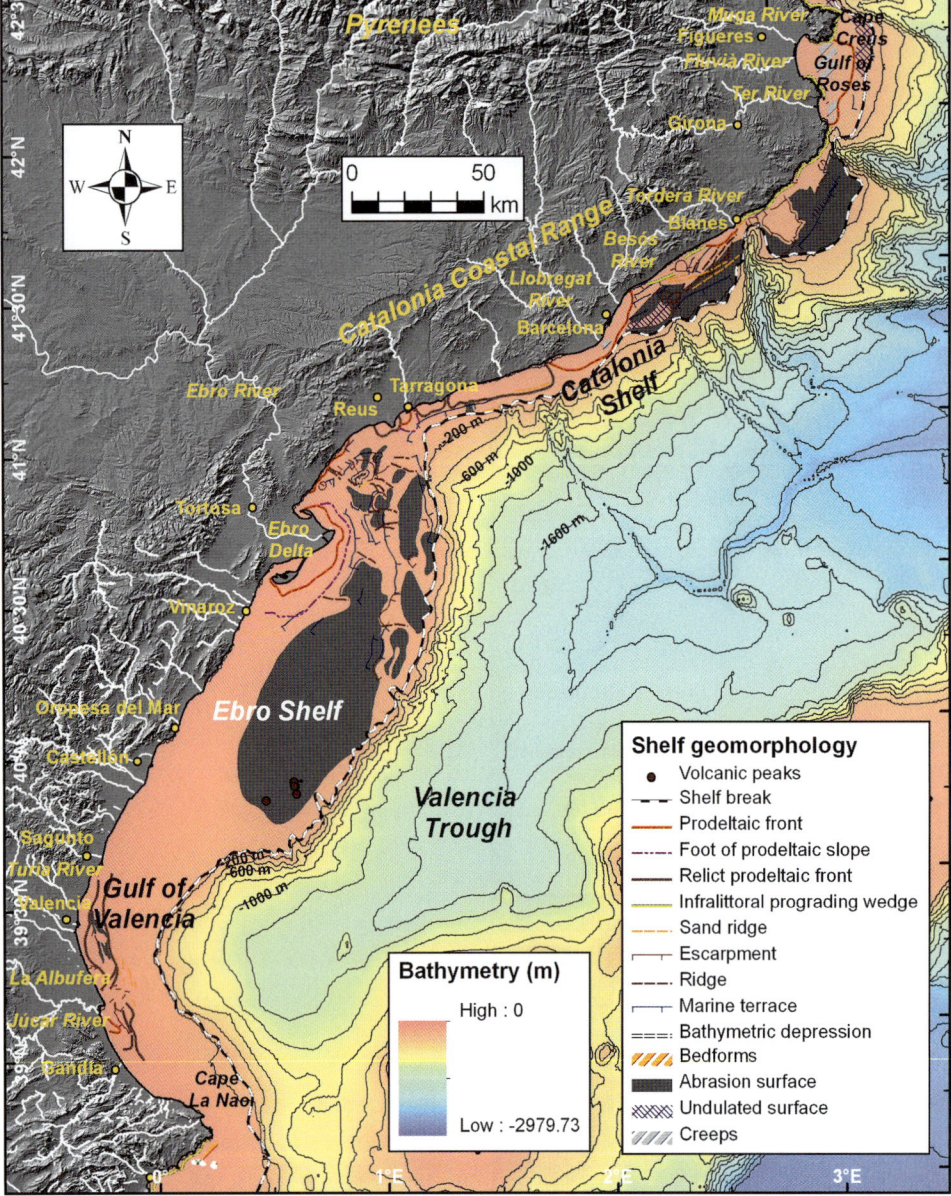

Fig. 11.2. Main morphosedimentary features of the Northeastern Shelf. 200 m bathymetric contours are displayed. Modified after Martín-Serrano (2005).

dominance of subsidence processes, local uplifting can also be seen, as, for example, off the Creus Peninsula (Pyrenean Axial Zone), favouring the rejuvenation of the coastal relief and erosional processes on the shelf (e.g. Farrán & Maldonado 1990; Ercilla *et al.* 1994*b*; Medialdea *et al.* 1994; Gámez *et al.* 2009).

Shelf description: physiography and sediments

The Northeastern Shelf comprises the Catalonia and Ebro shelves, and the Gulf of Valencia (Fig. 11.2). The name 'Catalonia Shelf' refers to the sector of the Spanish Mediterranean Shelf adjacent to the three Catalan provinces facing the sea, which are from north to south: Girona, Barcelona and Tarragona. The last province is affected by the Ebro River and its related delta. A large amount of marine geology literature refers to this last sector as the Ebro Shelf, and therefore the name 'Ebro Shelf' will be used hereafter in this paper.

The geomorphology of the Northeastern Shelf is related to the physiographical configuration of the onshore areas. The shelf exhibits a SW trend, except to the north where it trends north–south (Fig. 11.2). The shelf extends down to 140–200 m water depth, and its width varies greatly from 4 to 80 km off the Ebro

River; this variability results from the complex structural configuration of the margin and/or the uneven long-term sedimentation and preservation. The distribution of seafloor gradients reveals steep seafloors generated by infralittoral wedges (up to 5°) and prodeltaic deposits (up to 3°) in proximal settings. The shelf dips gently (<0.3°) seawards up to the outer shelf, which exhibits lower gradients (0.1–0.2°) (Liquete *et al.* 2007; Urgeles *et al.* 2007; Ercilla *et al.* 2010). The shelf break is mostly parallel to the coastline and displays a sinuous pattern due to the presence of canyon heads (indenting up to 15 km on the shelf) but the shelf break also builds out off large point sources (e.g. the Ebro River) (Fig. 11.2).

Sedimentation in the Northeastern Shelf is mainly siliciclastic with the sediment deriving from rivers and streams, although carbonate particles also arrive at the shelf from the coastal erosion of nearby outcrops of Cretaceous calcareous massifs. The surficial sediment cover comprises nine types of sediments (sands, clayey sands, silty sands, sandy silts, clayey silts, sandy clays, silty clays, sands and muds) as well as rocks (Fig. 11.4). The cluster of sands includes coastal sands composed of mixed amounts of clastics and carbonates along the present-day infralittoral environment and on the outer shelf. The clusters of silts and clays are associated with prodeltaic bodies that exhibit seaward changes

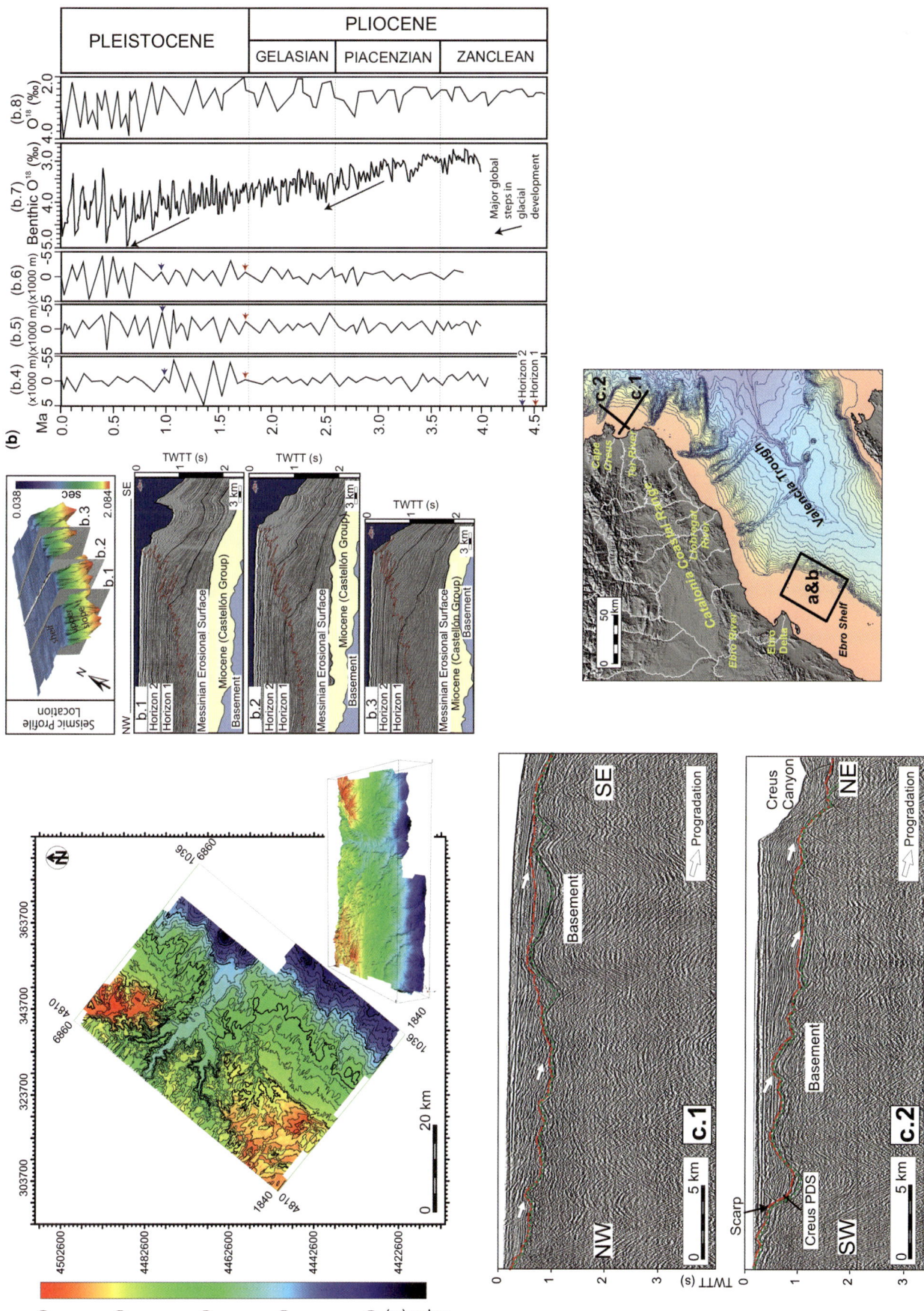

Fig. 11.3. Summary of stratigraphic observations regarding the Messinian and the Plio-Quaternary of the Northeastern Shelf. (**a**) Morphology of the Messinian Erosional Surface in a shelf sector located off the Ebro Delta. Modified after Urgeles *et al.* (2011*a*). (**b**) Plio-Quaternary stratigraphy of the shelf off Ebro Delta: (b.1–b.3) seismic profiles showing the Plio-Quaternary stratigraphic architecture; (b.4–b.6) progradational pulses measured in three across-margin depositional profiles; (b.7) oxygen isotope record from Ocean Drilling Program Site 846; (b.8) composite smoothed record based on DSDP/ODP sites. Modified after Kertznus & Kneller (2009). (**c**) Examples of seismic profiles collected off Cape Creus highlighting the Messinian Erosional Surface (in red) and the Plio-Quaternary sedimentary record of the continental margin. Modified after García *et al.* (2011). PDS, palaeo-drainage system; TWTT, two-way travel time.

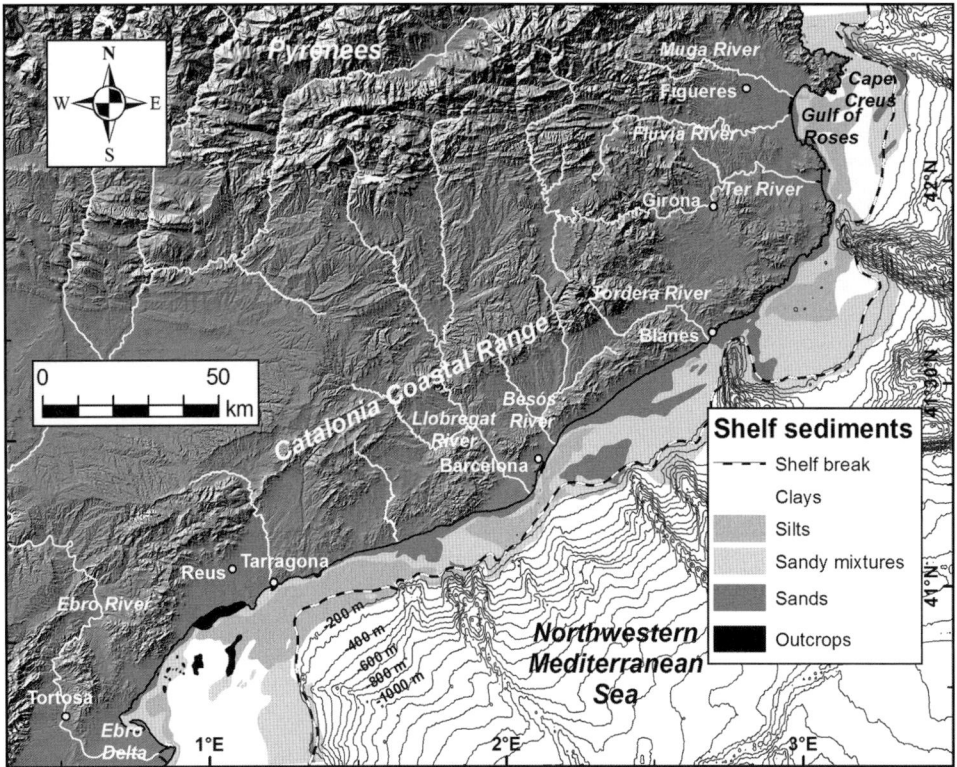

Fig. 11.4. Surficial sediment distribution of the Northeastern Shelf to the north of the Ebro Delta. 100 m bathymetric contours are displayed.

from silts to clays, and to the surrounding inner- and middle-shelf domains (Fig. 11.4). These surficial sediments rest over a basal layer of gravels and sands exposed in confined inner- and mid-shelf settings (Maldonado *et al.* 1983). This surficial distribution is the result of a significant change from a transgressive lineal sediment source to a modern highstand point sediment source laterally redistributed by the southward-directed geostrophic flow (Giró & Maldonado 1983; Maldonado & Zamarreño 1983; Medialdea *et al.* 1986, 1989, 1994).

Pre-Quaternary shelf sedimentation

Knowledge of the long-term sedimentation patterns in the Northeastern Shelf derives from the correlation of diverse types of seismic data with well-log information (Kertznus & Kneller 2009; Urgeles *et al.* 2011*a*) (Fig. 11.3). The initial infilling of Early Miocene horst-and-graben structures was from the draining of the Ebro River into the Mediterranean Sea, which generated a prograding megasequence (Castellón Group) during the Middle–Late Miocene (Urgeles *et al.* 2011*a*). The Messinian Salinity Crisis caused a significant sea-level drop and widespread margin erosion, forming the Messinian Erosional Surface (MES) (Fig. 11.3a). The physiography of the MES was dictated by structural features off Cape Creus, whereas, to the south, wide flat platforms were developed. The main valleys were also controlled by underlying structures, although some drainage networks also evolved as the result of subaerial erosion confined by basement highs (García *et al.* 2011; Urgeles *et al.* 2011*a*). As a result, depths of incision reached up to 400 m (Frey-Martínez *et al.* 2004). The main fluvial system was the Messinian Ebro River, which developed a dendritic drainage pattern within an area similar to that of the present one (Frey-Martínez *et al.* 2004; García *et al.* 2011; Urgeles *et al.* 2011*a*) (Fig. 11.3a).

Early Pliocene clinoforms (5.38–3.5 Ma) forming canyon-head deltas were essentially confined to Messinian valleys in the Ebro Margin. Once the valleys were infilled, numerous transgressive–regressive pulsations led to widespread progradation of shallow-marine deposits and generated a nearly straight shelf edge (Figs. 11.3b, c). During this phase (3.5–1.8 Ma) short-term sediment changes were led mainly by 41 ka climatic and sea-level fluctuations (Field & Gardner 1990; Nelson & Maldonado 1990*a*; Kertznus & Kneller 2009). The top of the Pliocene is marked by a regional erosional surface (Alonso *et al.* 1990; Ercilla *et al.* 1994*b*).

Quaternary shelf sedimentation

In contrast to the Pliocene, the Quaternary witnessed dramatic changes on the Northeastern Shelf, with higher rates of margin aggradation and a progressive increase in relief of margin clinoforms (maximum thickness of Quaternary deposits >3 km on the Ebro Shelf) (Fig. 11.3b). These changes were triggered by the increased amplitude of climatic fluctuations that favoured sediment loading and compaction. The rate of margin progradation decreased as a consequence of repeated shelf-margin collapse (Kertznus & Kneller 2009).

On the shelf, Quaternary deposits extend from the inner and middle shelf, where they onlap previous Pliocene, Miocene, Palaeogene and Palaeozoic deposits, to the slope (Medialdea *et al.* 1986, 1989, 1994). Their stratigraphic architecture is characterized by the stacking of numerous seismic units (from 9–10 at Girona to 13 at Ebro) pinching out or thinning in a landward direction and displaying a longitudinal (north–south), subtabular geometry (Medialdea *et al.* 1986, 1989; Farrán & Maldonado 1990; Ercilla *et al.* 1994*b*; Chiocci *et al.* 1997). They have been mapped over most of the Northeastern Shelf (Fig. 11.5); locally some units only occur at the inner–middle shelf (e.g. Ebro Shelf) (Fig. 11.5f) and others are limited to the shelf break (e.g. northernmost Girona Shelf) (Fig. 11.5a–c). The regional units are internally defined by low-angle oblique clinoforms on the inner–middle shelf that become progressively more inclined seawards; this configuration is better observed off large rivers (e.g. the Fluviá–Muga system, the Ter, Llobregat and Ebro) (Fig. 11.5c, d, g). These deposits are affected by erosive features, such as palaeoriver-fill deposits on the inner shelf (e.g. the Llobregat River on the Barcelona Shelf) and palaeocanyon-fill deposits

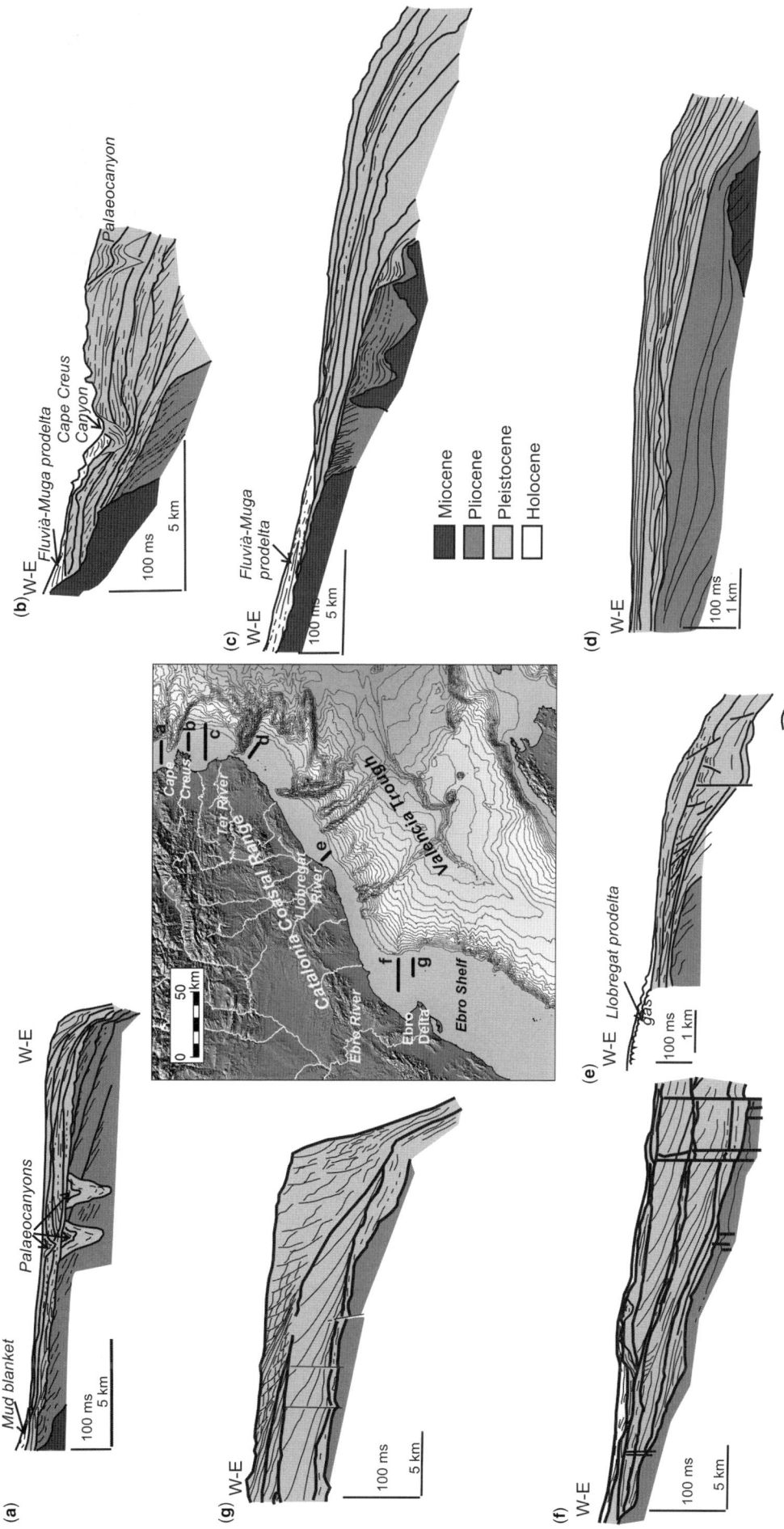

Fig. 11.5. Line drawings showing the stratigraphic architecture of the Northeastern Shelf. Note that the Quaternary deposits onlap Miocene and Pliocene deposits, and extend from the inner and middle shelf merging down-dip with the continental slope. Their stratigraphic architecture is mainly characterized by the vertical stacking of numerous units bounded by unconformities.

at the outer shelf (e.g. the La Escala and Cape Creus canyons on the Girona Shelf) (Fig. 11.5a, b). Likewise, deformational features, such as slides and creeps, affect high-angle clinoforms over the (palaeo)-shelf break (e.g. off the Fluviá–Muga, Tordera and Ebro river mouths) (Fig. 11.5e, g).

Units of limited distribution that display subtabular–wedge aggradational and subtabular–lensoindal progradational geometries occur intercalated between progradational units (Fig. 11.5). These units have been identified in proximal locations off the Ebro and Llobregat deltas (Farrán & Maldonado 1990; Somoza *et al.* 1998; Liquete *et al.* 2008; Gámez *et al.* 2009).

The Quaternary chronostratigraphic framework of specific sectors of the Northeastern Shelf was obtained through a correlation between wide seismic databases, sediment cores and boreholes that provided limited radiocarbon datings (Liquete *et al.* 2008; Gámez *et al.* 2009). The geometry and seismic facies of the progradational units extending over most of the Northeastern Shelf are indicative of the regressive shelf-margin deltas generated by the fourth-order (*c.* 100 ka) asymmetrical sea-level changes that have prevailed during the Late Quaternary (Medialdea *et al.* 1986, 1989, 1994; Checa *et al.* 1988; Farrán & Maldonado 1990; Díaz & Ercilla 1993; Ercilla *et al.* 1994*b*; Liquete *et al.* 2008; Gámez *et al.* 2009). Some of these regressive deltas (e.g. the Ebro and Llobregat) display a more complex stratal pattern, indicating higher-frequency cycles. The shelf-margin deltas led to the outbuilding of the shelf. The shelf-break progradation (4–11 km) (Fig. 11.5) was non-uniform due to a combination of factors such as total subsidence of the margin (i.e. thermal/tectonic subsidence plus sedimentary loading), uneven sediment supply from rivers, streams and coastal erosion of the Catalonia Coastal Range (seacliffs, open beaches and deltas), and the non-uniform southwards sediment dispersion by along-shelf flows influenced by the irregular shelf-margin physiography (Medialdea *et al.* 1986, 1989, 1994; Checa *et al.* 1988; Font *et al.* 1988; Farrán & Maldonado 1990; Masó *et al.* 1990; Ercilla *et al.* 1994*b*; Ercilla & Alonso 1996; Somoza *et al.* 1998; Rubio *et al.* 2005; Liquete *et al.* 2008; Gámez *et al.* 2009).

The external geometry and seismic facies of intercalated aggradational and progradational units with restricted distribution in proximal locations (Fig. 11.5) are, respectively, indicative of transgressive, patchy coastal deposits and low-energy prodeltaic deposits deposited during short-lived sea-level rises and stillstands (Farrán & Maldonado 1990; Gámez *et al.* 2009). In the Gulf of Valencia, several coastal sandy barriers documented in the inner shelf are associated with successive sea-level stillstands (Albarracín *et al.* 2013; Alcántara-Carrió *et al.* 2103). The combination of sediment supply and moderate–high subsidence would have played a main role in their deposition and/or preservation on the shelf (Farrán & Maldonado 1990; Chiocci *et al.* 1997; Somoza *et al.* 1998; Gámez *et al.* 2009; Albarracín *et al.* 2013). These deposits are substituted in the rest of the Northeastern Shelf by erosional and non-depositional surfaces separating the shelf-margin deltas and/or thin depositional bodies (<1 m), which are under the resolution of geophysical methods (Medialdea *et al.* 1986, 1989, 1994; Checa *et al.* 1988; Díaz & Maldonado 1990; Díaz & Ercilla 1993; Ercilla *et al.* 1995; Chiocci *et al.* 1997; Somoza *et al.* 1998; Liquete *et al.* 2008).

Last depositional sequence. The last depositional sequence developed during the last fifth-order (*c.* 20 ka) sea-level cycle comprises lowstand (LST), transgressive (TST) and highstand (HST) systems tracts separated by unconformities. The LST corresponds to the youngest shelf-margin delta showing similar seismic facies and geometry to previous Late Quaternary marginal deposits.

The TST and related depositional and erosive features occur from the inner to outer shelf as an extensive sand sheet with isolated sandbodies (also called shoals, ridges or bars: e.g. in Barcelona, Tarragona, Valencia), dunes (e.g. Girona), bioconstructions (e.g. Girona), terraces (e.g. Barcelona and Girona) and onlapping infralittoral prograding wedges (Medialdea *et al.* 1986, 1989, 1994; Checa *et al.* 1988; Díaz & Maldonado 1990; Díaz & Ercilla 1993; Ercilla *et al.* 1995; Chiocci *et al.* 1997; Somoza *et al.* 1998; Liquete *et al.* 2007) (Fig. 11.6). They have been mapped, seismically and/or sedimentologically, on the present-day shelf in places not covered by the highstand deposition. These sediment bodies were formed as overstepped barrier beaches or as submarine sand ridges on a distal-retreating shoreface during the Post-glacial Transgression (Fig. 11.6a) (Maldonado *et al.* 1983; Young *et al.* 1983; Liquete *et al.* 2007). The Post-glacial Transgression was interrupted by brief periods of coastal progradation that enabled the formation of coastal deposits and partial burial of the transgressive sand sheet. Consequently, their formation and preservation have been controlled mainly by the balance between sediment supply and wave energy during sea-level stillstands, and their interaction with seafloor morphology and the hydrodynamic regime (both offshore downwelling bottom currents and southward shelf flow) (Maldonado *et al.* 1983).

Equivalent transgressive deposits have been identified, restricted to the coastal domain. The transgressive and overlying coastal highstand deposits form a prograding coastal lithosome wedge (i.e. an infralittoral prograding wedge, IPW) in the present-day infralittoral environment (Medialdea *et al.* 1986, 1989, 1994; Sardá *et al.* 2000; Liquete *et al.* 2007; Serra *et al.* 2007; Gámez *et al.* 2009; Ercilla *et al.* 2010) (Fig. 11.6b). The IPW has been mapped along the coast as a sandy belt lying subparallel to the coastline, which extends down to 20–30 m water depth or is restricted to small bays within cliffed coast (e.g. on the Girona Coast) (Medialdea *et al.* 1986, 1989, 1994). Their genesis is linked to storm processes producing rip and undertow currents with seaward sediment transport and deposition down to the storm-base slope, a limit outlined by the base of the prograding foresets (Sardá *et al.* 2000; Serra *et al.* 2007; Gámez *et al.* 2009; Ercilla *et al.* 2010). IPW stratigraphy is characterized by divisions hierarchized into lower and upper sequences, each formed by two–three units (Ercilla *et al.* 2010) (Fig. 11.6b). This stratal architecture suggests that progradation occurred during different Holocene sea-level positions, representing the coastal response to transgressive (lower sequence) and highstand (upper sequence) conditions, further modulated by millennial- and centennial-scale sea-level oscillations (1–2 m). These minor sea-level changes interacted with local factors such as shelf topography (structural and sedimentary), (palaeo)-coastline orientation v. hydrodynamic conditions and river/stream regimes. As a consequence, a laterally variable seaward sediment supply and dispersion generated non-correlatable units along the coast (Somoza *et al.* 1998; Ercilla *et al.* 2010).

Highstand sedimentation also occurred in front of the main rivers, whose supply favoured seaward progradation down to water depths of 60 (Ebro) to 120 m (Fluviá–Muga complex, Ter) forming prodeltaic bodies and mud belts and/or blankets (Fig. 11.2). The following prodeltas occur on the Northeastern Shelf, from north to south: Fluviá–Muga, Ter, Tordera, Besós–Llobregat complex and Ebro (Medialdea *et al.* 1986, 1989, 1994; Checa *et al.* 1988; Díaz *et al.* 1990, 1996; Grimalt & Albaigés 1990; Díaz & Ercilla 1993; Ercilla *et al.* 1994*b*, 1995; Chiocci *et al.* 1997; Liquete *et al.* 2007; Serra *et al.* 2007; Gámez *et al.* 2009). Their areal extension ranges from the large Ebro River prodelta of 2300 km^2 to >200 km^2 of the Fluviá–Muga and Llobregat–Besós river prodeltas, and <200 km^2 of the Ter and Tordera river prodeltas. The prodeltas began to develop over the transgressive sand sheet after the shoreline transgressed the inner–middle shelf. The main prodeltaic depocentres (20–60 m thick) formed off river mouths (Fig. 11.6c), and the seaward transport (SE, south and SW) of suspended load formed extensive mud blankets on the shelf (Ercilla *et al.* 1994*b*; Díaz *et al.* 1996).

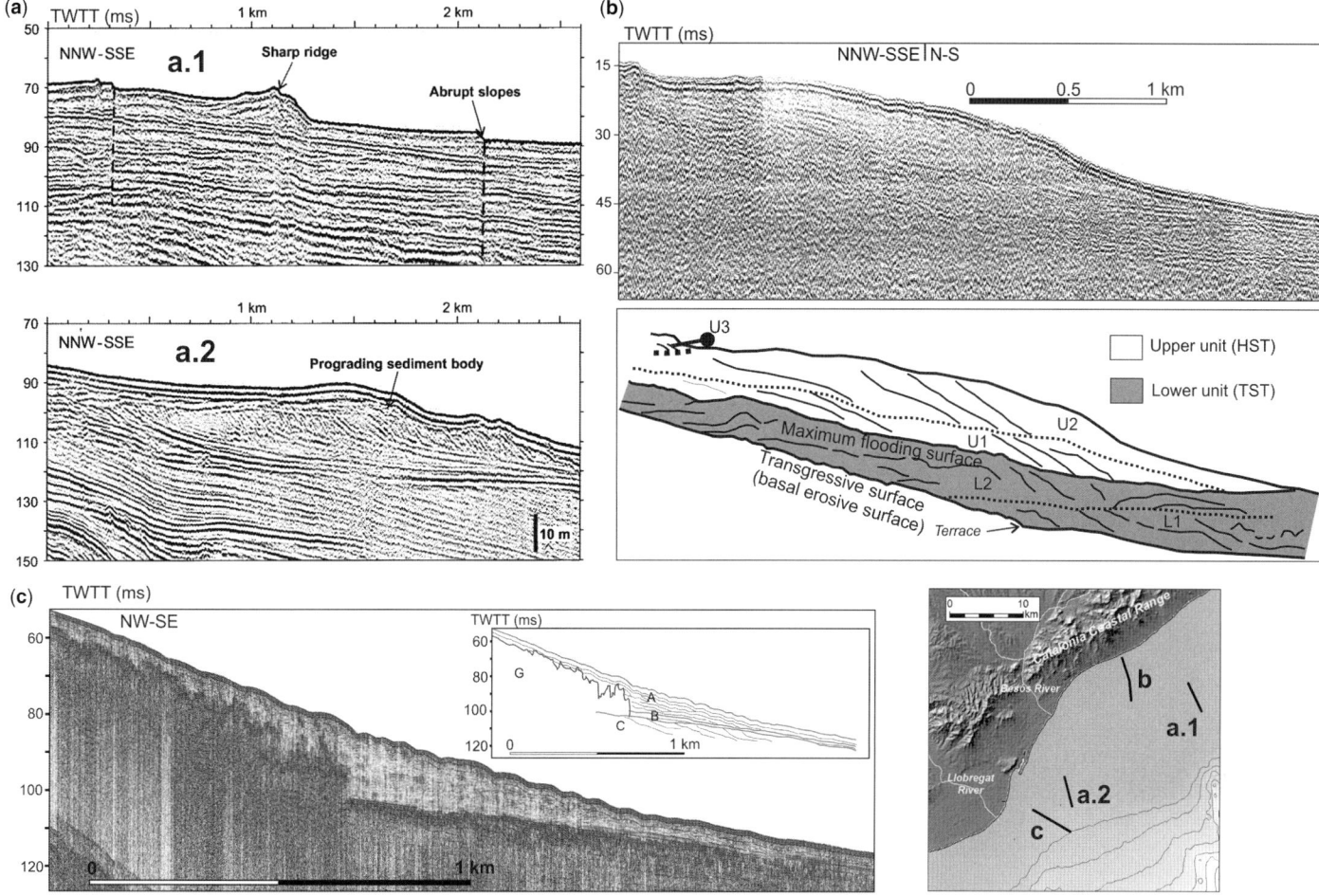

Fig. 11.6. Summary of post-glacial deposits in the Northeastern Shelf. (**a**) Transgressive deposits developed at or near ancient coastlines, such as sediment ridges (a.1) and prograding sediment bodies (a.2). Modified after Liquete *et al.* (2007). (**b**) Seismic profile and interpretation of the internal structure of an Infralittoral Prograding Wedge composed of both transgressive and highstand deposits. Modified after Ercilla *et al.* (2010). (**c**) Seismic profile illustrating the prodeltaic deposit off the Llobregat River. Modified after Urgeles *et al.* (2007). TWTT, two-way travel time.

The stratal architecture of these prodeltas mostly exhibits sigmoid–oblique clinoforms with SE-, south- and SW-directed progradations. In some prodeltas (e.g. Ebro and Llobregat) some internal discontinuities can be identified. Gas-bearing sediments frequently occur in proximal–middle prodeltaic locations (Fig. 11.6c). The long-term sedimentary build-up of these prodeltas has been controlled by autocyclic processes, such as channel avulsion in alluvial plains and delta-lobe switching, as a result of the interaction between climatic events, neotectonics and human activities (Checa *et al.* 1988; Díaz *et al.* 1990, 1996; Díaz & Ercilla 1993; Ercilla *et al.* 1995; Somoza *et al.* 1998; Gámez *et al.* 2009). River mouth changes would also have been influenced by small-amplitude (1–2 m), millennial- to centennial-scale sea-level oscillations that probably controlled the longitudinal (north–south) seismic facies variations and the internal discontinuities observed in some prodeltas (e.g. Fluviá–Muga, Llobregat–Besós, Ebro). In addition, the general south-directed shelf flow (Font *et al.* 1988; La Violette *et al.* 1990; Millot 1999) may have promoted the interfingering of neighbouring prodeltas, favouring the formation of single, laterally continuous and relatively thick muddy depocentres (e.g. Fluviá–Muga with Ter, Besós with Llobregat).

Modern shelf sedimentation

The distribution of the morphosedimentary bodies and surficial sediments of the present-day Northeastern Shelf reflect not only the two aforementioned sedimentary scenarios, transgressive and highstand, but also the modern scenario. This last scenario is dominated by processes that rework and redistribute highstand sediments (Fig. 11.7). Thus, the seafloor of IPWs may be affected by erosive furrows, slides, bedform fields, and anthropogenic dredging trenches and pits (Fig. 11.7a). The genesis of many of these features has been attributed to rip and undertow current scouring, hyperpycnal flows, storm wave currents and/or flash stream floods (Serra *et al.* 2007; Ercilla *et al.* 2010).

The modern scenario is also characterized by suspended sediment dynamics over the prodeltaic areas. The direct correlation between these processes and the resulting seismic and sedimentary facies (Grimalt & Albaigés 1990; Díaz *et al.* 1990, 1996; Díaz & Ercilla 1993; Ercilla *et al.* 1995) allows an environmental classification of modern prodeltas: (a) continuous stratified facies characterize a silty proximal prodelta environment; (b) discontinuous stratified facies describe a clayey silty middle environment; and (c) transparent facies indicate a silty clay distal environment. The fronts of some prodeltas on the Catalonia Shelf (Fluviá–Muga, Ter and Llobregat) are affected by sediment undulation fields (Fig. 11.7b), interpreted either as multiple rotational slides (e.g. on the Fluviá–Muga and Ter river prodeltas) favoured by the interplay between steep gradients (0.6°), interstitial gas and storm wave action (Díaz & Ercilla 1993; Ercilla *et al.* 1995) or as sediment waves formed by sediment resuspension driven by hyperpycnal flows (e.g. on the Llobregat River prodelta: Urgeles *et al.* 2007; 2011*b*).

Hydrodynamic processes, bottom morphology and biological productivity may affect the plumes of suspended particulate matter formed by river supply, storm waves and/or downwelling current

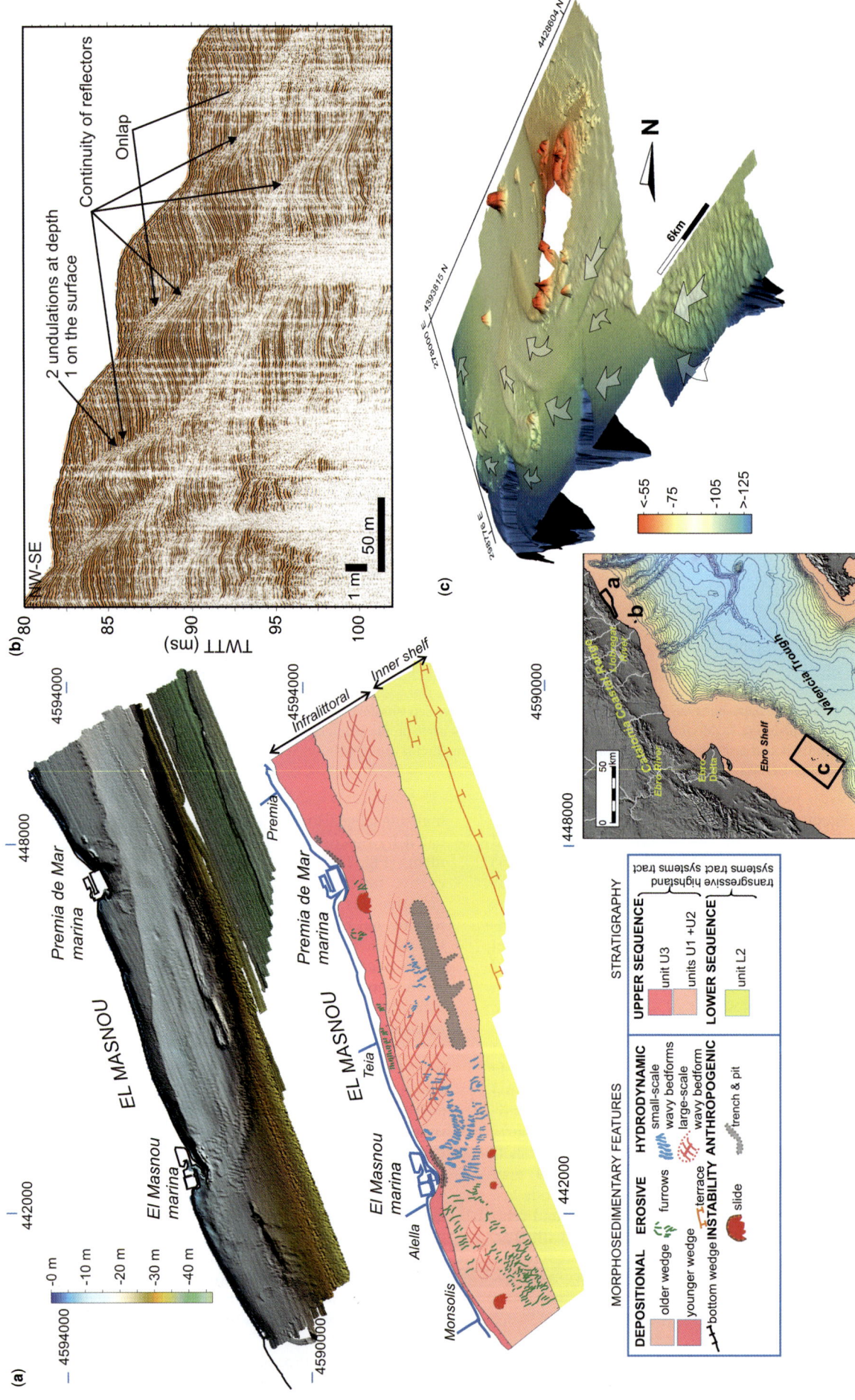

Fig. 11.7. Examples of recent sedimentary processes in the Northeastern Shelf. (**a**) Surface morphology and interpretative sketch of recent processes moulding an Infralittoral Prograding Wedge. Modified after Ercilla *et al.* (2010). (**b**) Seafloor undulations over the Llobregat Prodelta. Modified after Urgeles *et al.* (2011*b*). (**c**) Three-dimensional bathymetric block showing a set of submarine dunes on the southern Ebro Shelf that documents an active hydrodynamic regime. Modified after Lo Iacono *et al.* (2010). TWTT, two-way travel time.

resuspension, or sediment instability processes (Palanques & Drake 1990; Díaz *et al.* 1996; Puig & Palanques 1998; Puig *et al.* 1999). The suspended particulate matter entering prodeltaic areas may also be transported beyond the limits of the prodeltas to relatively low-energy depositional environments, forming isolated muddy patches of a few to several tens of km^2 (e.g. Ebro, Girona, Valencia), or mixing with transgressive sands to form palimpsest clayey sands and sandy clays (Palanques *et al.* 1990; Ercilla *et al.* 1994b, 1995; Díaz *et al.* 1996; Puig & Palanques 1998; Puig *et al.* 1999). The southwards decrease in accumulation rates suggests that the Ebro River is at present the most significant sediment source, as suspended fine sediments move southwards along the shelf with the prevailing flow (Maldonado *et al.* 1983). However, the supply of particulate matter from resuspension processes seems to have increased during the last decades (as far back as 2000 years ago for the Ebro Delta) due to the paucity of river supply caused by deforestation, dams, water management for urban, agriculture and industrial use (e.g. the Ebro, Llobregat and Besós rivers), and the increase in building for residential and industrial purposes in the hinterland coastal domain (Nelson 1990; Palanques *et al.* 1990; Thorndycraft *et al.* 2005; Liquete *et al.* 2007, 2009; Ercilla *et al.* 2010).

Several areas of the shelf are dominated by non-deposition, bypass or very low accumulation rates. In the Gulf of Valencia, the inner shelf is affected by lateral redistribution processes induced by storm-flow activity that favours the seaward transport of coarse sediments and the generation of bedform fields (Giró & Maldonado 1983; Young *et al.* 1983). Other middle and outer shelf areas are carpeted by transgressive sands indicating episodic activity of hydrodynamic agents. For example, large and very large subaqueous dunes cover relict sandbodies on the outer Ebro Shelf (Fig. 11.7c). These bedforms are considered to be active under the influence of energetic hydrodynamic events, linked to the action of the Northern Current (Lo Iacono *et al.* 2009).

The Southeastern Shelf

General setting

The Southeastern Shelf extends between capes La Nao and Gata in the Iberian Peninsula, and it also includes the shelf of the Balearic Promontory, which is attached to the Iberian Margin (Fig. 11.1).

Oceanography. The shallow margin is considered to be affected by part of the Northern Current that continues southwards through the Ibiza Channel (Fig. 11.1a), although with decreasing intensity and increasing mesoscale variability. The current eventually enters the Alboran Sea, where it tends to be deflected towards the Algerian Basin due to the influence of Atlantic Water in the Alboran Sea (Millot 1999). The shelf around the Balearic Islands is affected by the Balearic Front (Fig. 11.1a), fed by the entrance of Modified Atlantic Water in the Valencia Trough through the Ibiza Channel (López-Jurado *et al.* 1995). The intrusion of Atlantic Water generates the Balearic Current, the geostrophic flow associated with the Balearic Front (López-García *et al.* 1994).

Climate. High temperatures, scarce rainfall (<700 mm a^{-1}) and the wind regime produce an extreme arid climate. Summer temperatures are high (an average of 20–30 °C), although winter temperatures are mild in the nearshore region (January average temperature being 10 °C).

Sedimentation regime. The Southeastern Shelf is fed by rivers (the Turia, Júcar, Segura and Almanzora) and streams, both of which exhibit seasonal and irregular flow regimes, conditioned by the marked aridity of the onshore regions (Fig. 11.8). A southward-directed littoral drift has formed several spit bars enclosing relatively large coastal lagoons, such as the Mar Menor Lagoon (Fig. 11.8). On the shelf, the main sedimentation processes are linked to the action of major storms that remobilize the sediment and transport it offshore. Biogenic production is significant on the shelf around the Balearic Islands and in the vicinity of Cape Gata (Maldonado & Zamarreño 1983).

Tectonic framework and geological background

On its southern side, the Valencia Trough is flanked by the Betic–Balearic Domain, constituted by the easternmost sector of the External Zones of the Betic Cordillera (Fig. 11.1b). This Alpine Orogen extends towards the NE into the Balearic Promontory, which delineates the SE edge of the Valencia Trough (Dañobeitia *et al.* 1990). These compressive domains evolved synchronously with and subsequent to the main extensive phase of the Valencia Trough, although their exact relationships are controversial (e.g. Sàbat *et al.* 1997; Maillard & Mauffret 1999). The basement of the shelf is formed of horst-and-graben systems delimited by NE- to east-trending faults (Fig. 11.1b) (Catafau *et al.* 1994). There is evidence for Quaternary activity of faults and folds in the shelf off Torrevieja (Perea *et al.* 2012).

Uplift/subsidence of the margin. Tectonic features related to the Betic Cordillera have modified the uplift/subsidence shelf regime and limited the thickness of the sediment cover (Fig. 11.9). The acoustic basement outcrops, or is very shallow, off the main coastal promontories such as the La Nao and Palos capes (Rey & Díaz del Río 1983). However, thick Neogene–Quaternary depocentres occur in fault-controlled horsts and graben affected by differential subsidence south of Cape La Nao (Medialdea *et al.* 1982; Rey & Díaz del Río 1983; Catafau *et al.* 1994).

Shelf description: physiography and sediments

The Southeastern Shelf is built up over the Betic Margin, which is an intermediate to abrupt margin comprising two arc-shaped segments (northern and southern) bounded by Cape Palos (Fig. 11.8). The Balearic Islands occurring above the Balearic Promontory are also included within the Betic Margin.

The Northern Arc extends between capes La Nao and Palos (Fig. 11.8). In general, shelf width decreases to the south; the maximum extension (40 km) occurs off Altea, decreasing to 23 km off Santa Pola. Accordingly, seafloor slopes increase to the south from 0.1° to 0.2°. The shelf break occurs at water depths of 100–130 m (Rey & Díaz del Río 1983; Catafau *et al.* 1994).

The Southern Arc is an abrupt margin with a narrow shelf located between capes Palos and Gata (Fig. 11.8). The shelf is 13 km wide in the vicinity of Cape Palos but it becomes narrower (<6 km) to the south. Seafloor slopes are less than 1° in the vicinity of Cape Palos but to the south the shelf is steeper than 2°. The shelf break is 100–110 m deep (Medialdea *et al.* 1982; Rey & Díaz del Río 1983; Catafau *et al.* 1990).

The surface distribution of geomorphological types has been provided by several regional studies (Medialdea *et al.* 1982; Rey & Díaz del Río 1983; Catafau *et al.* 1990, 1994; Rey & Fumanal 1996; Díaz del Río & Fernández-Salas 2005) (Fig. 11.8). Small lobate prodeltas occur off the main rivers (e.g. Segura and Almanzora) and laterally evolve into elongate IPWs, with local development of seagrass meadows along cliffed coasts. Other depositional features such as littoral bars, ridges and bedforms are frequent on inner–middle shelf areas off coastal lagoons. A widespread abrasion surface is located south of Alicante, indicating the scarcity of fluvial supply. On the outer shelf, several shelf-margin deltas have been identified, although most of the outer shelf is covered by erosional and/or tectonically controlled

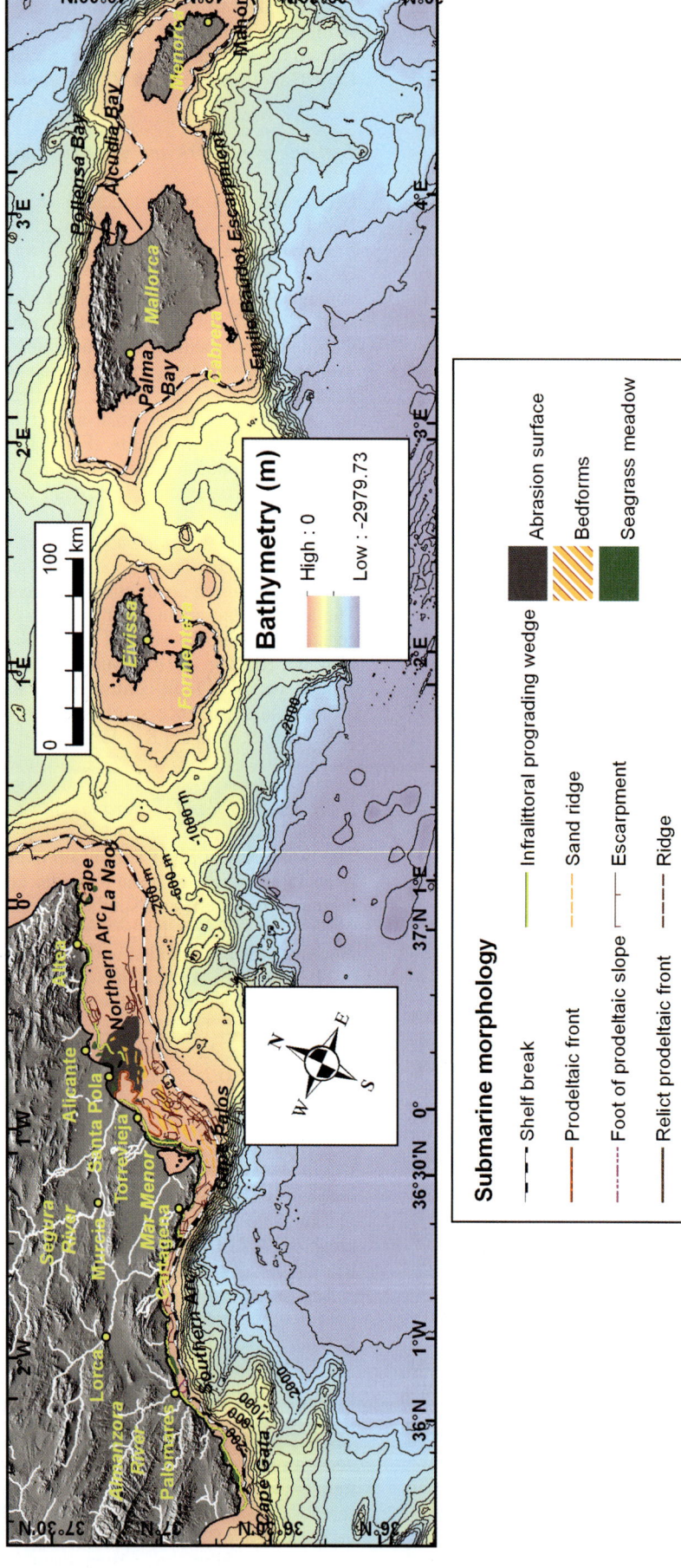

Fig. 11.8. Main morphosedimentary features of the Southeastern Shelf. The picture is rotated to include the shelves around the Balearic Promontory. 200 m bathymetric contours are displayed. Modified after Martín-Serrano (2005).

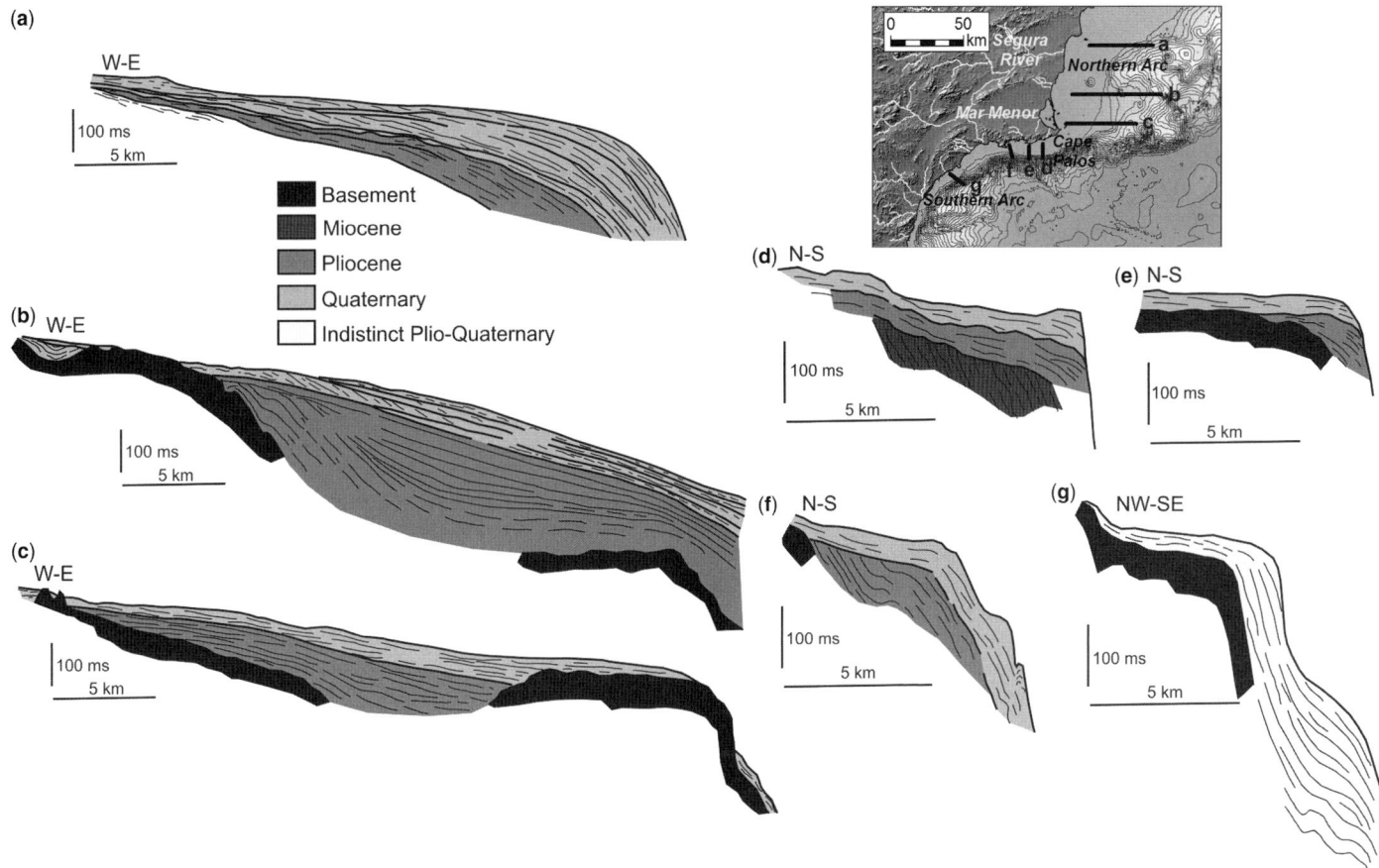

Fig. 11.9. Examples of interpretative seismic sections documenting the long-term stratigraphic architecture (mostly Plio-Quaternary) in the Southeastern Shelf. (a–c) The Plio-Quaternary record shows considerable development in the Northern Arc, where numerous offlapping sequences can be recognized. (d–g) In the Southern Arc, Plio-Quaternary shelf development is more limited. Modified after Catafau *et al.* (1990).

features (escarpments and submarine terraces, remnants of previous coastlines, abrasion and undulating surfaces, wavy surfaces, depressions, channel-like morphologies and hard bottoms).

The Balearic Shelf comprises the larger Mallorca–Menorca Shelf to the east and the smaller Eivissa–Formentera Shelf to the west (Fig. 11.8). The islands of Mallorca and Menorca have a common shelf including the smaller Cabrera Island, with a total surface of 6418 km². Eivissa and Formentera have a common shelf with a total surface area of 2709 km². The septentrional shelf of the Balearic Islands is 10–20 km wide, with the shelf break at water depths ranging between 139 and 160 m (Acosta *et al.* 2003). This shelf is mostly covered by biogenic sands, with a mid-shelf continuous belt of littoral bars. In contrast, the septentrional Eivissa Shelf is controlled by structural lineaments (Acosta *et al.* 2003; Díaz del Río & Fernández-Salas 2005). The central shelf of the Balearic Islands exhibits several bays (Palma, Pollensa and Alcudia) covered by extensive fields of *Posidonia*, as well as diverse depositional features (bedforms, bars and channel infillings) and erosional–tectonic elements (escarpments, terraces and abrasion surfaces) (Palomino *et al.* 2009). The meridional shelf of the Balearic Islands has an abrupt morphology, with a minimum width of 3 km, and the shelf break at water depths of between 120 and 150 m delineated by the Emile Baudot Escarpment along the island of Mallorca. Terrace-like reefal build-ups are frequent on this shelf (Díaz del Río & Fernández-Salas 2005).

Surficial sediments are siliciclastic in the Northern Arc, where the inner shelf is covered by coarse-grained sediments with locally high carbonate content, and scattered seagrass meadows. On the mid–outer shelf, terrigenous muds prevail to the north of Alicante, whereas, to the south, most of the shelf is covered by palimpsest coarse-grained sediments (Catafau *et al.* 1994).

Widespread terrigenous sediments also occur in the northern half of the Southern Arc. Siliciclastic sediments occur mainly around the most important river mouths (e.g. Almanzora) and streams, evolving seawards to finer-grained sediments. In contrast, biogenic coarse-grained carbonates prevail in the southern-most shelf next to Cape Gata (Maldonado & Zamarreño 1983; Zamarreño *et al.* 1983). Shelf sediments around the Balearic Islands are composed primarily of biogenic sands and gravels, with variable and low fine-component content (Alonso *et al.* 1988; Fornós & Ahr 1997).

Pre-Quaternary shelf sedimentation

The correlation of exploration wells with low-resolution multi-channel seismic data enabled the identification of the Neogene–Quaternary sedimentary cover on the Southeastern Shelf (Rey & Díaz del Río 1983; Catafau *et al.* 1994) (Fig. 11.9). The influence of the Betic Mountains on the shelf causes the occurrence of intermediate and abrupt margin sectors, with drastic thickness variations in the Neogene–Quaternary sediment cover. In the Northern Arc (intermediate margin), the basement outcrops, or is very shallow, in several places (Fig. 11.9a–c). The sedimentary thickness and distribution of the Miocene series are very variable, as major depocentres are controlled by graben location. Pliocene sediments show a more widespread distribution, covering most of the shelf and outcropping above basement highs (Fig. 11.9a–c). Miocene and Pliocene sediments are affected by reactivated faults and folding in many places (Catafau *et al.* 1994). In the Southern Arc (abrupt margin), the sedimentary cover is even thinner (Fig. 11.9d–g), although small sedimentary basins have been identified (Serra *et al.* 1979).

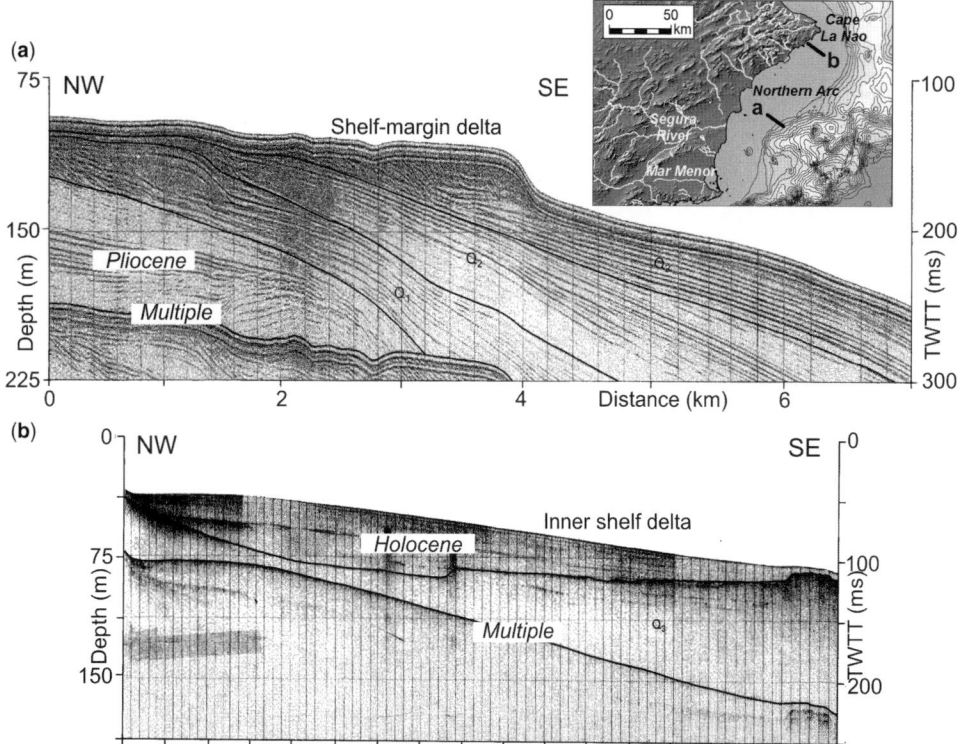

Fig. 11.10. Examples of Late Quaternary deposits identified in the Southeastern Shelf: (**a**) shelf-margin delta in the middle part of the Northern Arc; (**b**) highstand delta in the vicinity of Cape La Nao. Modified after Catafau *et al.* (1994). TWTT, two-way travel time.

Quaternary shelf sedimentation

The knowledge of Quaternary sediment patterns in the Southeastern Shelf is based mostly on old, pre-sequence stratigraphy-based observations. Under such schemes, the structural framework was considered the primary factor controlling the margin physiography and the deposit geometry owing to its long-term influence on the balance between subsidence and sediment supply (Serra *et al.* 1979; Maldonado & Zamarreño 1983). In contrast, Quaternary climatic–eustatic oscillations were considered to be somehow a subsidiary controlling factor (Serra *et al.* 1979). At short timescales, it was believed that the sedimentary facies were controlled by the oceanographic regime and the water-depth variations (Maldonado & Zamarreño 1983). A very recent study has described several wedge-shaped units separated by erosional surfaces linked to major sea-level lowstands (Perea *et al.* 2012).

A general trend of laterally constant, seawardly increasing thicknesses of the Quaternary sedimentary cover is observed, although the thickness is also controlled by underlying horsts and graben, as described before (Fig. 11.9). Thus, the Quaternary sedimentary cover is very thin (tens of metres) over basement highs, such as those located off and to the south of Alicante. Quaternary deposits may reach several hundreds of metres in thickness, with maximum values of >400 m in the main depositional areas of the Northern Arc (Fig. 11.9a–c). Quaternary deposits are composed of several offlapping wedges, erosionally truncated at the top, whose generation was related to different regressive phases (Catafau *et al.* 1990, 1994). In the Southern Arc, Quaternary deposits reach a moderate development due to the narrowness of the shelf, and can be locally eroded at the shelf break by submarine canyons (Fig. 11.9d–g).

Last depositional sequence. The study of recent sedimentary systems was extracted mainly from a correlation of shoreline markers observed in high-resolution seismic data with sea-level curves (Catafau *et al.* 1990, 1994). The Last Glacial Maximum is recorded by shelf-margin deltas occurring at the present-day shelf break (Fig. 11.10a). During the Post-glacial Transgression, a process of quasi-continuous retreat of a barrier–lagoon system was recorded

by sediment ridges and bedforms on several areas of the shelf, with the most extensive field located off the Mar Menor Lagoon. In addition, several wedge-shaped deposits identified at water depths of 20–60 m have been related with cold climatic events, such as the Younger Dryas and the 8.2 ka cooling event (Tent-Manclús *et al.* 2009; de la Vara *et al.* 2011). In the Northern Arc, post-transgressive sedimentary processes included the formation of relatively small, coarse-grained prodeltaic deposits (Fig. 11.10b) related to the input of the most important rivers, laterally connected through elongate IPWs (Giró & Maldonado 1983). Muddy sediments were transported seawards and advected southwards, and their deposition generated a continuous muddy belt from the Gulf of Valencia to the shelf off Alicante (Maldonado & Zamarreño 1983). In the Southern Arc, the most recent accumulations are restricted to small prodeltaic deposits off the main fluvial inputs, such as the Segura and Almanzora rivers (Rey & Díaz del Río 1983; Catafau *et al.* 1990, 1994).

Modern shelf sedimentation

The knowledge of recent–modern sedimentation processes on the Southeastern Shelf is very limited. Suspended fine sediments move southwards along the shelf with the prevailing flow (Maldonado *et al.* 1983). Fine-sediment transport towards the south, coupled with the abrupt shelf configuration, leads to fine-sediment capture by submarine canyons, and its export towards deeper domains in the Southern Arc. In addition, scarce terrigenous supplies and high biogenic production favour the development of extensive seagrass meadows in the southernmost part of the Southeastern Shelf (Zamarreño *et al.* 1983).

The Northern Alboran Sea Shelf

General setting

The Alboran Sea is a 150 km-wide, 350 km-long basin located between Spain and Morocco, surrounded by the Gibraltar Arc

(Fig. 11.1). The shelf of the northern margin of the Alboran Sea represents the southern part of the Iberian Mediterranean shelves. This narrow shelf extends between Cape Gata in the east and the Strait of Gibraltar in the west (Fig. 11.2). The nearby Betic Mountains dictate an abrupt margin physiography with input from short, mountain rivers.

Oceanography. Dominant low-energy waves approach both from the east and the west, although the easterly waves are slightly more energetic. Significant wave heights are lower than 2 m (Lobo *et al.* 2006). Tidal currents are weak as on the other Iberian Mediterranean shelves, where tidal oscillations are in the range of few centimetres.

Surface current patterns are controlled by the entrance of Atlantic waters, designated either as the Atlantic Inflow (AI) or as Surface Atlantic Water (SAW), into the Mediterranean Sea through the Strait of Gibraltar with an estimated speed of 1 m s^{-1} (Fig. 11.1a). The AI is mixed with variable quantities of water masses of Mediterranean origin generating the Atlantic Jet (AJ) that feeds two anticyclonic gyres, the quasi-permanent Western Alboran Gyre (WAG) and the more elusive Eastern Alboran Gyre (EAG) (Tintoré *et al.* 1988; Perkins *et al.* 1990) (Fig. 11.1a). The AJ originates a strong thermohaline front between the cold and dense Mediterranean waters to the left and the Atlantic waters to the right of the jet (Vargas-Yáñez & Sabatés 2007). In addition, two cyclonic gyres are frequently found over the shelf.

This mesoscale pattern composed of the AJ and the two anticyclonic gyres has a high temporal variability due to changes in the AI through the Strait of Gibraltar, and the atmospheric pressure patterns between the Atlantic Ocean and the Mediterranean Sea (Vargas-Yáñez *et al.* 2002). In particular, the AJ fluctuates in a north–south direction owing to changes in the WAG and also due to changes in wind dominance (Sarhan *et al.* 2000). Under the dominance of westerlies, the AJ is constrained close to the coast, favouring the development of coastal upwelling; whereas, when easterlies dominate, the entrance of Atlantic Water is diminished and the AJ moves southwards, while on the shelf an enhanced cyclonic circulation is established (Macías *et al.* 2008).

Climate. The presence of the Betic Cordillera near the coast induces a marked climatic zonation. The coastal domain has a mild Mediterranean climate (average annual temperature of 18–19 °C), although it becomes increasingly desertified towards the east, with more extreme temperatures and scarce rainfall (less than 300 mm a^{-1}). In contrast, climatic conditions in the mountains are more severe, with lower temperatures (average annual value of 13 °C) and higher rainfall (Liquete *et al.* 2005).

Sedimentation regime. Over most of the Northern Alboran Sea Shelf, deltaic coastlines alternate with coastal sectors where sediment dynamics is governed by littoral drift. However, biogenic sedimentation and deposition of cool-water carbonates occur on the easternmost shelf around Gata Cape (Zamarreño *et al.* 1983). Whatever the coastal environment, fluctuations of the North Atlantic Oscillation are suggested to cause either increased fluvial discharges or enhanced AI activity (Goy *et al.* 2003; Liquete *et al.* 2005).

In deltaic coastlines, the majority of sediment is supplied via relatively short, mountain streams. The longest rivers with the most extensive drainage basins are, from west to east, the Guadiaro, Guadalhorce, Gudalfeo, Adra and Andarax rivers (Fig. 11.11). In general, mean discharges decrease from west to east but, in contrast, mean sediment loads and yields tend to increase eastwards, indicating a progressively more torrential character (Liquete *et al.* 2005). This supply pattern is conditioned by: (a) the abrupt coastal physiography, as the Betic Mountains are located only a short distance from the coast and, in several places, show recent tectonic activity leading to steepening of physiographical profiles and river incision (Carvajal & Sanz de

Galdeano 2008); and (b) the Mediterranean climatic conditions, with increasing aridity towards the east. Owing to their torrential character, most of the rivers are very effective in transporting sediment from the drainage basins towards the shelf and, ultimately, into deeper water (Liquete *et al.* 2005). During the Holocene, the greatest sediment supply occurred after the Holocene Climatic Optimum due to a combination of natural factors (e.g. increasing aridity) and anthropic activity (burning) (Gil-Romera *et al.* 2010).

Littoral drift is highly variable, although eastward-directed drift is more common (Lario *et al.* 1999; Goy *et al.* 2003) owing to the interaction between prevailing waves and coastline configuration. In coastal sectors dominated by along-shore processes, beach ridges and spit bars were constructed during the Holocene Highstand. The most significant examples are found in Carchuna and Campo de Dalías (Lario *et al.* 1999; Goy *et al.* 2003). The growth of these beach ridge systems apparently occurred through successive millennial-scale progradational phases in relation to high sea levels, enhanced AI and increased sediment input (Goy *et al.* 2003).

Tectonic framework and geological background

The Alboran Sea is located along the Eurasia–Africa plate boundary. The formation of the Alboran Sea within the Betic–Rif Orogen has been explained by several hypotheses: extensional collapse of thickened crust, delamination processes or westward retreating subduction (Mauffret *et al.* 2007). The basin underwent significant extension and crustal thinning through a number of rifting stages that created several graben and half-graben during the Oligocene–Late Miocene. The shelf domain was structured by NE-trending normal faults in the vicinity of the Strait of Gibraltar changing to roughly east–west-trending faults to the east of Málaga (Comas *et al.* 1992) (Fig. 11.1b).

The tectonic regime changed during the Tortonian, coincident with the termination of the rifting process. The NW–SE convergence between Eurasia and Africa favoured strike-slip motion along pre-existing normal faults and folds. The faults were arranged in a conjugate system of NE-trending left-lateral and NW-trending right-lateral strike-slip faults (Ballesteros *et al.* 2008) (Fig. 11.1b). The convergence changed to roughly north–south during the Messinian and facilitated the closure of the Strait of Gibraltar. At the end of the Messinian, the stress regime allowed extension in an east–west direction. The stress regime continued during the Early Pliocene, with a dominant transtensive regime in the western half of the basin (Campillo *et al.* 1992) and extensional tectonics to the NE of the basin (Estrada *et al.* 1997).

The stress regime changed during the late Pliocene to a NNW–SSE to north–south compression. During the Quaternary, a transpressive regime prevailed and previous structures were affected by strike-slip movements (Campillo *et al.* 1992; Estrada *et al.* 1997). A major fault off Adra affected Holocene sediments and is considered to be active nowadays (Gràcia *et al.* 2012).

Uplift/subsidence of the margin. The dominant Quaternary compressive regime has produced lateral changes of the subsidence/uplifting regime, modifying shelf-growth patterns. For example, the westernmost shelf close to the Strait of Gibraltar has undergone tectonic tilting, favouring outer shelf uplift and erosion of deposits (Ercilla *et al.* 1992). Other sectors, such as the shelves off the Fuengirola and Guadalfeo rivers, have been subsiding, favouring sediment accumulation and strata preservation (Ercilla *et al.* 1992; Lobo *et al.* 2008) (Fig. 11.12).

Shelf description: physiography and sediments

The shelf is narrow (only a few kilometres wide), with a minimum width of 2 km off Cape Sacratif, where several submarine canyons

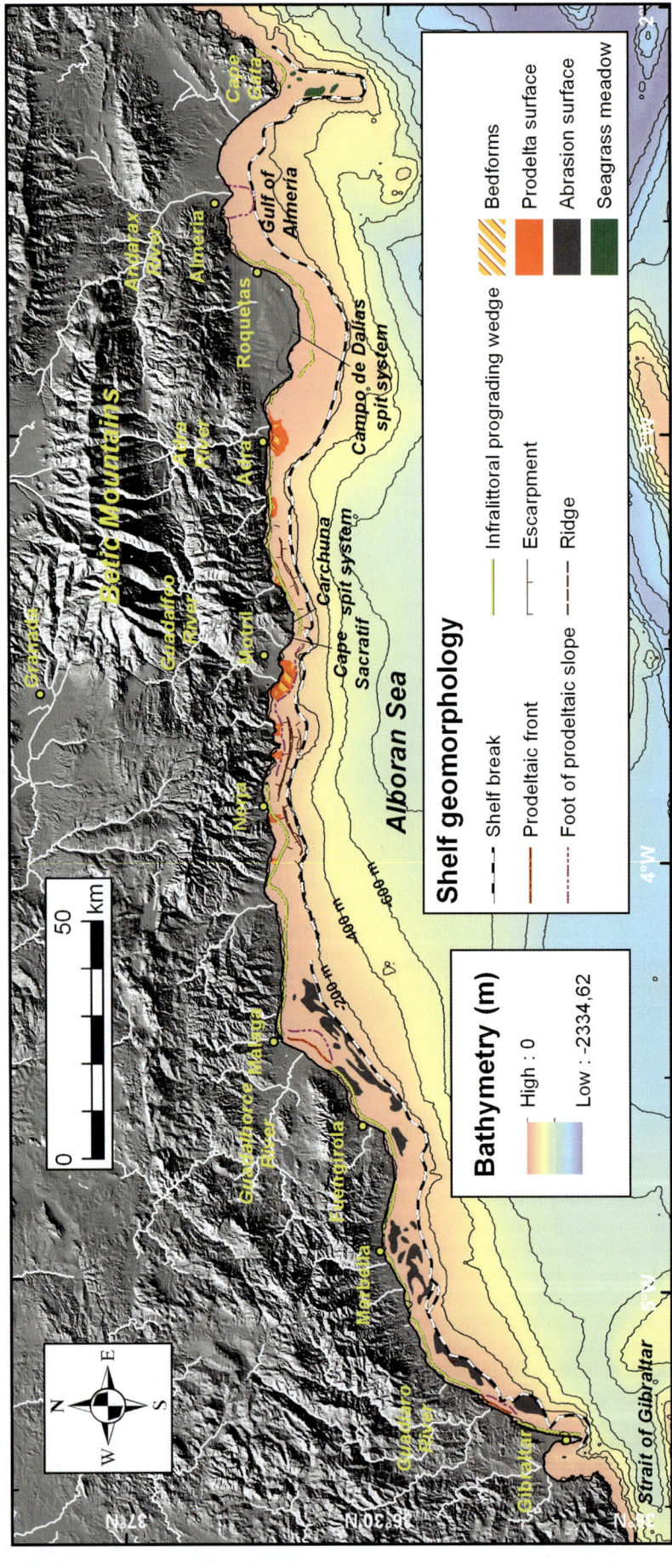

Fig. 11.11. Main morphosedimentary features of the Northern Alboran Sea Shelf. 200 m bathymetric contours are displayed. Modified from Martín-Serrano (2005), with additional information from Lobo *et al.* (2006), Bárcenas *et al.* (2009) and Jabaloy-Sánchez *et al.* (2010).

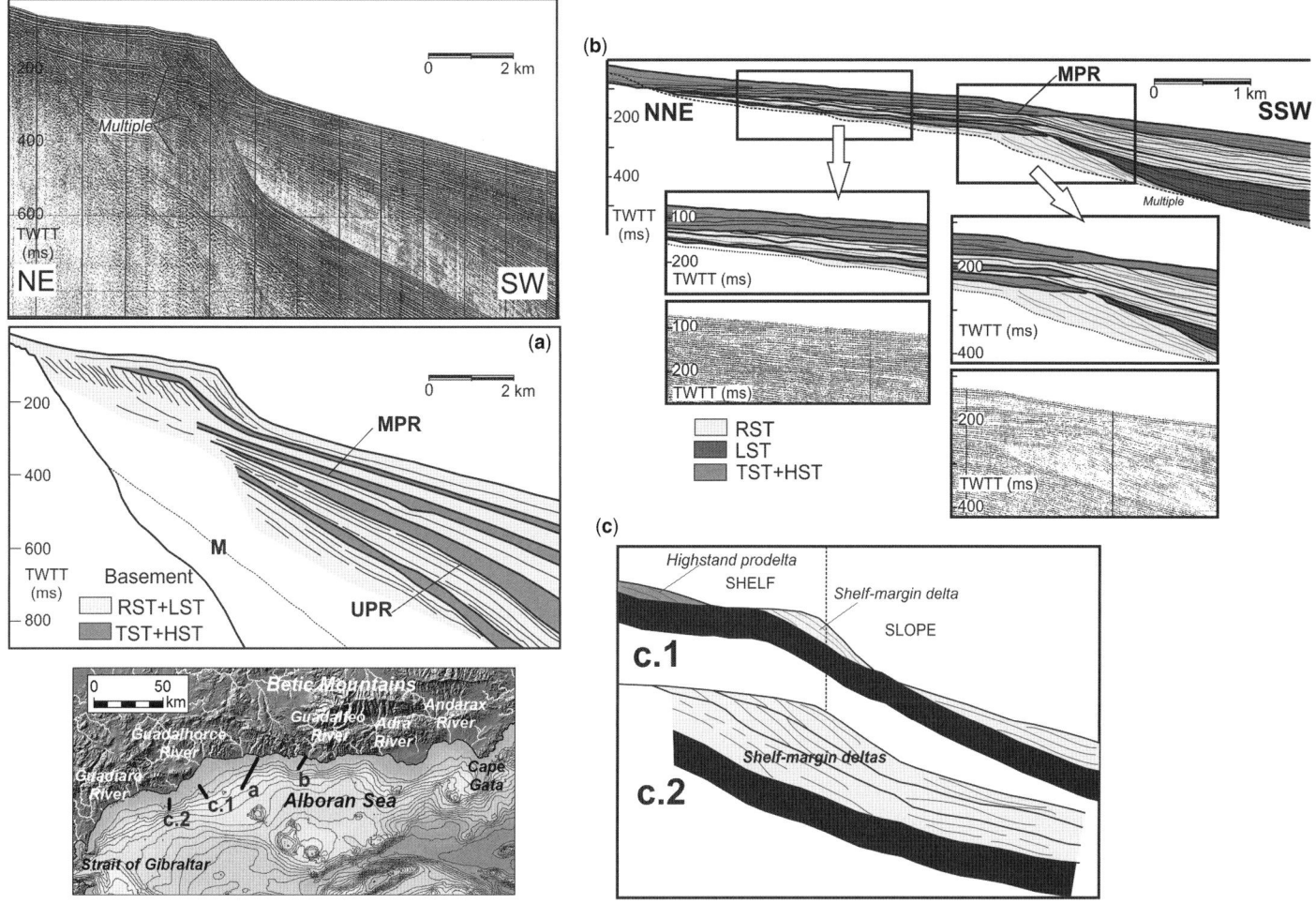

Fig. 11.12. Plio-Quaternary stratigraphic architecture and stacking patterns of the Northern Alboran Sea Shelf. (**a**) Seismic profile and interpretation of the general architecture of the northern margin of the Alboran Sea, with dominance of progradational stacking pattern. Modified after Hernández-Molina *et al.* (2002). (**b**) Plio-Quaternary shelf stratigraphic architecture off Guadalfeo River highlighting the aggradational stacking pattern after the Middle Pleistocene Revolution (MPR). Modified after Lobo *et al.* (2008). (**c**) Shelf stratigraphic sections displaying the lateral variability of stacking patterns resulting from the dominance of either uplift (c.1) or subsidence (c.2) in the eastern part of the Northern Alboran Sea Shelf. Modified after Ercilla & Alonso (1996). M, Messinian Unconformity; UPR, Upper Pliocene Revolution; MPR, Middle Pleistocene Revolution; RST, Regressive Systems Tract; LST, Lowstand Systems Tract; TST, Transgressive Systems Tract; HST, Highstand Systems Tract. TWTT, two-way travel time.

incise the shelf (Carter *et al.* 1972). The shelf is also narrow and abrupt towards the Strait of Gibraltar (Vázquez 2005). In contrast, the shelf may reach a width of more than 20 km off the Málaga and Almería provinces due to sediment supplied by major rivers (Fig. 11.11). In addition, a Neogene volcanic seamount has generated a 28 km-wide north-trending platform off Cape Gata (Medialdea *et al.* 1982; Muñoz *et al.* 2008).

The most significant inner shelf morphologies include prodeltaic bodies in front of the main fluvial inputs and infralittoral prograding wedges (IPWs) lateral from the main fluvial entries (Fig. 11.11). Wide segments of the outer shelf, particularly to the west, are covered either by sand ridges or by erosional and/or tectonic morphologies, such as abrasion surfaces, submarine terraces and escarpments (Hernández-Molina *et al.* 1994, 1996; Vázquez 2005; Lobo *et al.* 2006).

The shelf break is located at water depths of 100–125 m, although off Cape Gata it lies in deeper water (>175 m) (Fig. 11.11). In most places the shelf break is smooth due to the influence of shelf-margin deltas; however, on the western shelf, it is abrupt due to faulting (Muñoz *et al.* 2008).

Sedimentation is mainly siliciclastic in the western and central parts of the Northern Alboran Sea Shelf, whereas it is mixed (terrigenous–carbonate) on the shelf off the Almería province. Siliciclastic sediments are derived mainly from suspended load

deposition, developing muddy prodeltaic facies with coarsening fining-upward sequences and variable amounts of sands (Ercilla *et al.* 1994*a*). Recent studies using backscatter data reveal the widespread distribution of sandy substratums with variable amounts of gravel and mud (Bárcenas *et al.* 2011). The sediments are transported laterally either by littoral drift or under the influence of the AI. Relict sediments composed of reworked sands and gravels consist of a thin carpet mainly occurring on the outer shelf, and which are interpreted to have been deposited during the Postglacial Transgression (Ercilla *et al.* 1994*a*; Hernández-Molina *et al.* 1994). Spillover facies, mainly composed of sands, tend to occur in the proximity of the shelf break; these facies are related to reworking and resuspension by storm currents and by upper-slope gravitational processes (Ercilla *et al.* 1994*a*).

Carbonate sediments are widespread on most of the shelf off the Almería province, being an example of carbonate development in temperate waters. The most common sediment types are relict and palimpsest facies that tend to cover the inner shelf. The occurrence of these carbonate facies is attributed to the lack of terrigenous input due to the sub-arid climatic conditions and the common occurrence of morphological highs. On this shelf, siliciclastic sediments are restricted to the Gulf of Almería, where proximal gravels and sands evolve seawards to muds (Zamarreño *et al.* 1983).

Pre-Quaternary shelf sedimentation

The closure of the connection between northern Africa and southern Iberia in the Messinian led to the Messinian Salinity Crisis, an event that had a dramatic influence on the shaping of the Mediterranean margins between 5.96 and 5.33 Ma (Duggen *et al.* 2003). During the Messinian, a large fall in water level caused extreme erosion and incision on the margins leading to the development of a pronounced unconformity and large canyon systems, as well as the deposition of an evaporitic sequence in the deeper basin.

In the northern margin of the Alboran Sea, the chronostratigraphic framework of the Plio-Quaternary sedimentary succession overlying the Messinian Unconformity was obtained by correlating a dense grid of seismic data with DSDP Site 121 and several well logs obtained on the shelf and slope (Alonso & Maldonado 1992; Hernández-Molina *et al.* 2002). Two main constructional episodes occurred during the Pliocene. Following the major Messinian lowstand, the re-opening of the Strait of Gibraltar during the Early Pliocene, together with the global Pliocene sea-level rise, resulted in a highstand situation during which sandy sediments were trapped along the coast and a major sequence was generated on the shelf. This interval was interrupted by a significant sea-level fall at about 4.2 Ma, when a well-marked shelf–slope profile was established in the margin. This initial profile influenced subsequent physiographical profiles during successive sea-level lowstands (Hernández-Molina *et al.* 2002). The Late Pliocene was dominated by falling sea-level conditions (Alonso & Maldonado 1992) that led to the formation of another major sequence boundary at about 2.4 Ma (Hernández-Molina *et al.* 2002) (Fig. 11.12a). The regressive deposits formed shelf-margin deltas, whose seaward progradation was controlled by the existence of laterally variable uplifting/subsidence (Lobo *et al.* 2008).

Quaternary shelf sedimentation

In the northern margin of the Alboran Sea, a strong dominance of shelf-progradational packages has been documented for the Quaternary, interrupted only by isolated carbonate platforms (Alonso & Maldonado 1992). A chronostratigraphic scenario was obtained by correlating the stacking pattern of depositional sequences with global sea-level cycles and Quaternary isotopic stages (Hernández-Molina *et al.* 2002). At the largest scale, two major sequences are separated by a major unconformity related to the Middle Pleistocene Revolution (MPR) at approximately 900 ka BP (Hernández-Molina *et al.* 2002) (Fig. 11.12a).

The complex architecture of the post-MPR shelf sedimentary record reflects the onset of 100 kyr, high-amplitude asymmetrical sea-level cycles (Hernández-Molina *et al.* 2002). The 100 kyr shelf sequences are composed mainly of shelf-margin wedges, generated during relative sea-level falls and lowstands, and separated by polygenetic erosional surfaces related to transgresssive intervals. As a consequence, most Late Quaternary sequences are composed mainly of Regressive (RST) to Lowstand Systems Tracts (LST) (Ercilla *et al.* 1994a) (Fig. 11.12a, b).

The stacking pattern of post-MPR sequences is laterally variable according to the uplift/subsidence regime of the continental margin. Forestepping patterns are found in several sectors, such as the shelf off Málaga (Fig. 11.12a) (Hernández-Molina *et al.* 2002). In areas subjected to strong tectonic tilting, shelf sequences are poorly preserved due to the lack of accommodation space and/or erosion (Ercilla *et al.* 1994a) (Fig. 11.12c.1). However, in areas with high subsidence rates such as shelf sectors off the Guadalfeo and Fuengirola rivers, Late Quaternary progradational deposits exhibit aggradational to backstepping patterns and are better preserved (Ercilla & Alonso 1996; Lobo *et al.* 2008)

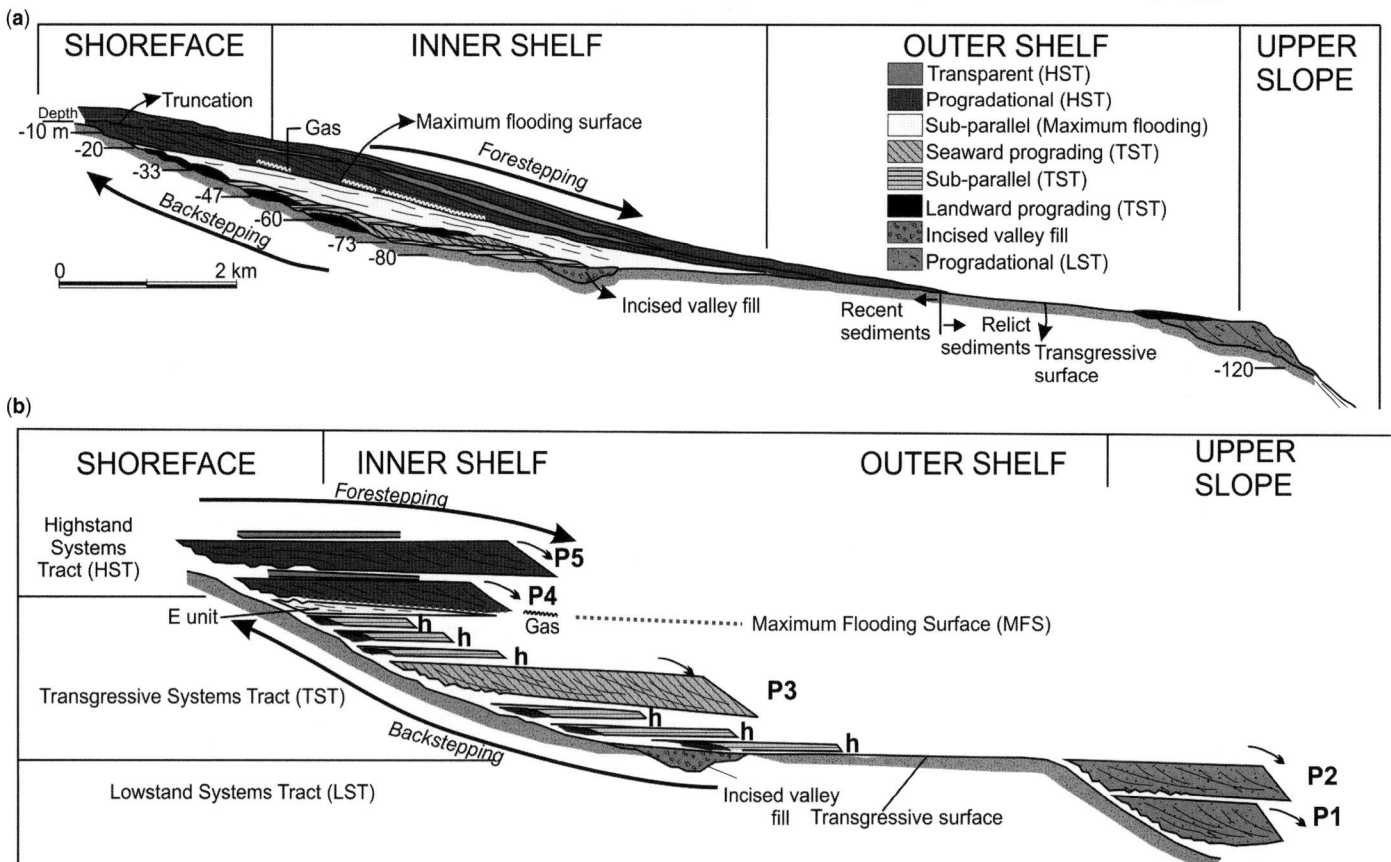

Fig. 11.13. Synthetic sequence stratigraphic model of the last Late Quaternary depositional sequence in the Northern Alboran Sea Shelf (developed from the shelf off Málaga). (**a**) Shelf deposits attributed to the different systems tracts. (**b**) High-resolution architecture of P and h progradational bodies, indicating periods of major and minor progradation within each segment of the glacio-eustatic curve (lowstand, transgression and highstand). Modified after Hernández-Molina *et al.* (1994).

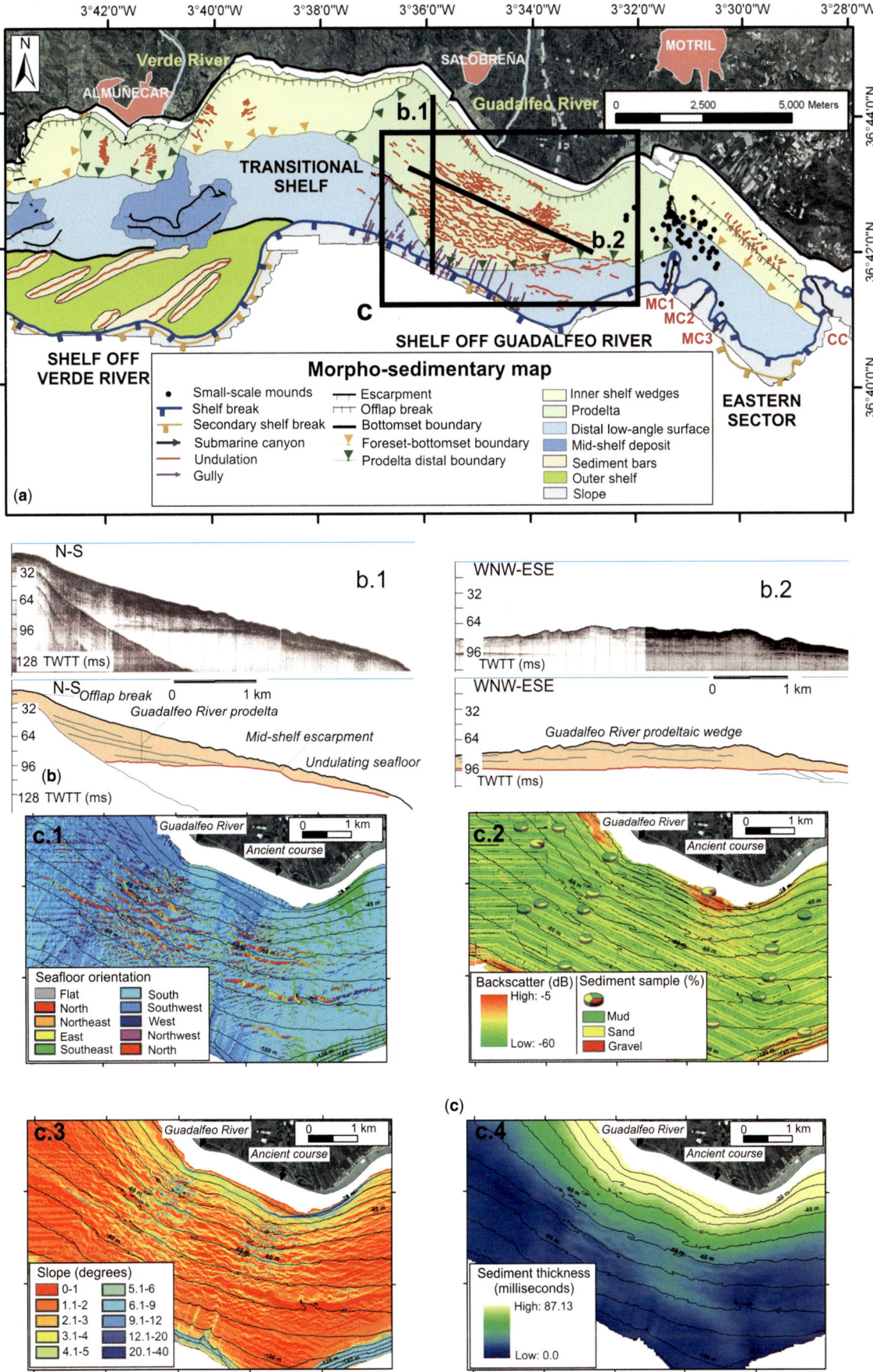

Fig. 11.14. Recent sedimentary processes in a sector of the Northern Alboran Sea Shelf, off the Guadalfeo River, as depicted by multibeam bathymetry and high-resolution seismic profiles. (**a**) Seafloor geomorphological interpretation. Modified after Lobo *et al.* (2006). (**b**) Examples of seismic sections collected in the Guadalfeo River prodeltaic body. Modified after Lobo *et al.* (2006). (**c**) Multibeam bathymetry and derived information from a field of prodeltaic undulations found over the Guadalfeo River prodeltaic feature: (c.1) seafloor orientation; (c.2) backscatter and superimposed sediment samples; (c.3) seafloor slope; (c.4) sediment thickness of the recent prodeltaic wedge. Modified after Fernández-Salas *et al.* (2007). TWTT, two-way travel time.

(Fig. 11.12b, c.2). These marginal wedges usually have a deltaic origin, as they are connected to fluvial palaeo-channels (Ercilla *et al.* 1994*b*). However, shoreface–nearshore ridge systems may occur laterally as a result of declining fluvial input (Lobo *et al.* 2008).

Last depositional sequence. The sequence stratigraphy of the last depositional sequence has been derived from high-resolution seismic analysis coupled with surficial and core sediment data that provided information about sedimentary facies. In addition, the correlation of coastline markers such as submarine terraces improved the correlation with the post-glacial sea-level curve (Ercilla *et al.* 1994*a*; Hernández-Molina *et al.* 1994; Fernández-Salas *et al.* 2003).

A high-frequency sequence composed of LST, TST and HST was developed during the last 20 kyr (Ercilla *et al.* 1994*a*; Hernández-Molina *et al.* 1994, 1996) (Fig. 11.13). The systems tracts show a complex internal architecture, recording the superposition of higher-frequency cycles (i.e. P and h cycles as depicted in Fig. 11.13, although a level of higher stratigraphic complexity (c cycles) was also documented) (Hernández-Molina *et al.* 1994, 1996). These cycles involved asymmetrical relative sea-level changes with slow sea-level falls and rapid rises (Hernández-Molina *et al.* 1996).

The rapid post-glacial sea-level rise (17–5 ka BP) generated different types of deposits. The most complete record shows several backstepping deposits laterally associated with submarine terraces, and interpreted as beach deposits generated during sea-level stillstands (Fig. 11.13). In places of low sediment supply, the transgressive record is represented by a sand sheet exposed over the outer shelf (Ercilla *et al.* 1994*a*). After the Post-glacial Transgression, two basic depositional styles were developed during the Holocene sea-level stabilization: prodeltaic wedges and IPWs (HST in Fig. 11.13).

Prodeltaic wedges occur off major entry points, such as the Guadiaro, Guadalhorce, Guadalfeo, Adra and Andarax river outflows, where they are progradationally configured and display terrigenous muddy compositions (Ercilla *et al.* 1994*a*; Hernández-Molina *et al.* 1994). Their internal architecture shows a thin TST (green units in Fig. 11.13) and a HST thickening landwards (grey units in Fig. 11.13). The bulk of the deposits was generated during the Holocene Highstand (last 6–7 kyr), when two progradational–aggradational motifs (P4 and P5 grey units) (Fig. 11.13) are thought to reflect the influence of low-amplitude, high-frequency sea-level oscillations that triggered significant sediment supply variations (Hernández-Molina *et al.* 1994, 1996; Fernández-Salas *et al.* 2003). The sediment supply fluctuations were also governed by climatic changes, such as increased rainfall (Senciales & Malvárez 2003) and marked seasonal climatic conditions, prone to the generation of episodic torrential discharges, and also enhanced by the abrupt onland physiography. As a consequence, prodeltaic wedges exhibit very steep clinoform geometry, with the offlap break very close to the coastline and coarse-sediment composition. The generation of these prodeltaic wedges has been related to rapid deposition after high-energy sediment flows and limited lateral redistribution (Lobo *et al.* 2006) (Fig. 11.14a, b).

In places of minor direct fluvial input, such as coastal stretches off the Málaga, Granada and Almería provinces, the Holocene highstand sedimentary record is thought to be composed of sigmoid–oblique progradational wedges whose offlap breaks tend to parallel the adjacent coastline. These coastal deposits were interpreted as IPWs, and their origin linked to offshore sediment transport below storm wave-base level led by downwelling currents (Hernández-Molina *et al.* 2000). More recent work has revealed a strong parallelism between the internal architecture of these shallow-water sediment wedges and the stratigraphic architecture of adjacent coastal plains, as in the case of Carchuna or Campo de Dalías, suggesting a genetic relationship (Fernández-Salas *et al.* 2009). Here, the pattern of progradation

of IPWs is oblique to parallel in relation to the main shoreline in coastal segments with enhanced littoral drift, suggesting that further work should be conducted in order to explain the generation and development of these enigmatic shallow-water sediment wedges.

Modern shelf sedimentation

Recent processes reshaping the original morphology of the deltaic wedges include storm activity, high-density sediment flows and sediment deformation. In some deltaic settings, coastline orientation controls the sites of deposition/erosion associated with strong storm events, whose effects seem to be reflected preferentially on the shoreface (Backstrom *et al.* 2008). Fine sediments are transported seawards by high-energy events, either by storms or seasonal floods, creating a net granulometric contrast between proximal and distal shelf settings (Maldonado & Zamarreño 1983).

In recent years, the interpretation of multibeam mosaics has provided a new wealth of information concerning recent sedimentation processes on the Northern Alboran Sea Shelf. The occurrence of crenulated seafloor is a common phenomenon in many prodeltaic environments (Bárcenas *et al.* 2009) (Fig. 11.14c). Seafloor gradients, river-mouth position and sediment-supply patterns seem to exert a major influence on the genesis and development of those shallow-water undulations (Bárcenas *et al.* 2009). In the case of the undulations found off the Guadalfeo River, their location off the present and ancient river mouth suggests that they are generated by strong sediment flows normal to bathymetric contours coupled with slow sediment deformation (Fernández-Salas *et al.* 2007). In and around prodeltaic environments, such as the Guadalfeo River prodelta, incision by canyon heads or gullies of the prodelta and lateral deposits indicate recent sediment bypass on the shelf and sediment export to the slope (Lobo *et al.* 2006) (Fig. 11.14a).

Human intervention in the river basins has also modified the recent dynamics and growth patterns of prodeltaic systems. For example, deforestation has increased the amount of available sediment and enhanced delta growth; in contrast, river regulation operations such as dam construction and river deviation have led to a decrease in sediment supply, widespread coastal erosion and shoreline retreat (Senciales & Malvárez 2003; Jabaloy-Sánchez *et al.* 2010). Human influence on deltaic environments is particularly noticeable on the Adra Delta, where two major prodeltaic lobes were recognized (Jabaloy-Sánchez *et al.* 2010). The western lobe was generated as a consequence of continuous sediment supply during the natural growth of the prodelta, with negligible anthropic influence, which lasted until the final part of the nineteenth century. A major channel deviation promoted the generation of an eastern prodeltaic protuberance, whose development was subsequently influenced by enhanced lateral redistribution processes (Jabaloy-Sánchez *et al.* 2010).

We thank the Secretaría General del Mar and the Instituto Español de Oceanografía for providing access to the data collected in the ESPACE project. The Institut Cartogràfic Valencià provided onland cartographic information. The following people provided very useful information and/or data to complete the paper: S. Contreras, F. Estrada, M. Farrán, M. García, V. Kertznus, B. Kneller, C. Lo Iacono, J. M. López García, L. Pomar and R. Urgeles. C. Ellison sharpened the English style of the manuscript. Constructive reviews by M. Rabineau (CNRS, Brest) and S. García-Gil (University of Vigo) significantly improved the original manuscript. Additional remarks were also provided by the Guest Editor, F. L. Chiocci.

References

ACOSTA, J., CANALS, M. *ET AL.* 2003. The Balearic Promontory geomorphology (western Mediterranean): morphostructure and active processes. *Geomorphology*, **49**, 177–204.

ALBARRACÍN, S., ALCÁNTARA-CARRIÓ, J., BARRANCO, A., SÁNCHEZ GARCÍA, M., FONTÁN BOUZAS, Á. & REY SALGADO, J. 2013. Seismic evidence for the preservation of several stacked Pleistocene coastal barrier/lagoon systems on the Gulf of Valencia continental shelf (western Mediterranean). *Geo-Marine Letters*, **33**, 217–223.

ALCÁNTARA-CARRIÓ, J., ALBARRACÍN, S., MONTOYA MONTES, I., FLOR-BLANCO, G., FONTÁN BOUZAS, Á. & REY SALGADO, J. 2013. An indurated Pleistocene coastal barrier on the inner shelf of the Gulf of Valencia (western Mediterranean): evidence for a prolonged relative sea-level stillstand. *Geo-Marine Letters*, **33**, 209–216.

ALONSO, B. & MALDONADO, A. 1992. Plio-Quaternary margin growth patterns in a complex tectonic setting: Northeastern Alboran Sea. *Geo-Marine Letters*, **12**, 137–143.

ALONSO, B., GUILLÉN, J. ET AL. 1988. Los sedimentos de la plataforma balear. *Acta Geológica Hispánica*, **23**, 185–196.

ALONSO, B., FIELD, M. E., GARDNER, J. V. & MALDONADO, A. 1990. Sedimentary evolution of the Pliocene and Pleistocene Ebro margin, northeastern Spain. *Marine Geology*, **95**, 313–331.

ANDEWEG, B. 2002. *Cenozoic Tectonic Evolution of the Iberian Peninsula: Effects and Causes of Changing Stress Fields*. PhD thesis, University of Amsterdam.

ARNAU, P., LIQUETE, C. & CANALS, M. 2004. River mouth plume events and their dispersal in the northwestern Mediterranean Sea. *Oceanography*, **15**, 22–30.

BACKSTROM, J. T., JACKSON, D. W. T., COOPER, J. A. G. & MALVÁREZ, G. C. 2008. Storm-driven shoreface morphodynamics on a low-wave energy delta: the role of nearshore topography and shoreline orientation. *Journal of Coastal Research*, **24**, 1379–1387.

BALLESTEROS, M., RIVERA, J., MUÑOZ, A., MUÑOZ-MARTÍN, A., ACOSTA, J., CARBÓ, A. & UCHUPI, E. 2008. Alboran Basin, southern Spain – Part II: neogene tectonic implications for the orogenic float model. *Marine and Petroleum Geology*, **25**, 75–101.

BÁRCENAS, P., FERNÁNDEZ-SALAS, L. M., MACÍAS, J., LOBO, F. J. & DÍAZ DEL RÍO, V. 2009. Estudio morfométrico comparativo entre las ondulaciones de los prodeltas de los ríos de Andalucía Oriental. *Revista de la Sociedad Geológica de España*, **22**, 43–56.

BÁRCENAS, P., LOBO, F. J., MACÍAS, J., FERNÁNDEZ-SALAS, L. M. & DÍAZ DEL RÍO, V. 2011. Spatial variability of surficial sediments on the northern shelf of the Alboran Sea: the effects of hydrodynamic forcing and supply of sediment by rivers. *Journal of Iberian Geology*, **37**, 195–214.

BARTRINA, M. T., CABRERA, L. L., JURADO, M. J., GUIMERÀ, J. & ROCA, E. 1992. Cenozoic evolution of the central Catalan margin (Valencia Trough, western Mediterranean). Geology and Geophysics of the Valencia Trough, Western Mediterranean. *Tectonophysics*, **203**, 219–248.

CAMPILLO, A. C., MALDONADO, A. & MAUFFRET, A. 1992. Stratigraphic and tectonic evolution of the western Alboran Sea: Late miocene to recent. *Geo-Marine Letters*, **12**, 165–172.

CARTER, T. G., FLANAGAN, J. P. ET AL. 1972. A new bathymetric chart and physiography of the Mediterranean Sea. *In*: STANLEY, D. J., KELLING, G. & WEILER, Y. (eds) *The Mediterranean Sea: A Natural Sedimentation Laboratory*. Hutchinson & Ross, Dowden, 1–23.

CARVAJAL, R. & SANZ DE GALDEANO, C. 2008. Aplicación de índices geomorfológicos al estudio de la cuenca del Río Adra (Almería). *Revista de Cuaternario y Geomorfología*, **22**, 17–31.

CASTELLÓN, A., FINT, J. & GARCIA LADONA, E. 1990. The Liguro-Provençal-Catalan current (NW Mediterranean) observed by Doppler profiling in the Balearic Sea. *Science on the March*, **45**, 269–276.

CATAFAU, E., DÍAZ, J. I., MEDIALDEA, T., SAN GIL, C., VÁZQUEZ, J. T. & WANDOSSELL, J. 1990. *Mapa geológico de la plataforma continental española y zonas adyacentes. Escala 1:200.000. Hojas 79–79E (Murcia)*. Instituto Geológico y Minero de España, Madrid.

CATAFAU, E., GAYTÁN, M., PEREDA, I., VÁZQUEZ, J. T. & WANDOSSELL, J. 1994. *Mapa geológico de la plataforma continental española y zonas adyacentes. Escala 1:200.000. Hojas 72–73 (Elche-Alicante)*. Instituto Geológico y Minero de España, Madrid.

CENDRERO, A., SÁNCHEZ-ARCILLA, A. & ZAZO, C. 2005. Impactos sobre la costa. Impactos sobre las zonas costeras. *In*: MORENO, J. M. (ed.) *Evaluación preliminar de los impactos en España por efecto del cambio climático*. Ministerio de Medio Ambiente, Madrid, 469–524.

CHECA, A., DÍAZ, J. I., FARRÁN, M. & MALDONADO, A. 1988. Sistemas deltaicos holocenos de los ríos Llobregat, Besós y Foix: modelos evolutivos transgresivos. *Acta Geológica Hispánica*, **23**, 241–255.

CHIOCCI, F. L., ERCILLA, G. & TORRES, J. 1997. Stratal architecture of Western Mediterranean Margins as the result of the stacking of Quaternary lowstand deposits below 'glacio-eustatic fluctuation base-level'. *Sedimentary Geology*, **112**, 195–217.

COMAS, M. C., GARCÍA-DUEÑAS, V. & JURADO, M. J. 1992. Neogene tectonic evolution of the Alboran Sea from MCS data. *Geo-Marine Letters*, **12**, 157–164.

DAÑOBEITIA, J. J., ALONSO, B. & MALDONADO, A. 1990. Geological framework of the Ebro continental margin and surrounding areas. *Marine Geology*, **95**, 265–288.

DE LA VARA, A., TENT-MANCLÚS, J. E., ESTÉVEZ, A., SORIA, J. M. & REY, J. 2011. El prisma sedimentario submarino ligado al Younger Dryas en la plataforma continental de Benidorm (Alicante, SE de España). *Geogaceta*, **50**, 137–140.

DÍAZ, J. I. & ERCILLA, G. 1993. Holocene depositional history of the Fluvia-Muga prodelta, northwestern Mediterranean Sea. *Marine Geology*, **111**, 83–92.

DÍAZ, J. I. & MALDONADO, A. 1990. Transgressive sand bodies on the Maresme continental-shelf western Mediterranean-sea. *Marine Geology*, **91**, 53–72.

DÍAZ, J. I., NELSON, C. H., BARBER, J. H. JR. & GIRÓ, S. 1990. Late Pleistocene and Holocene sedimentary facies on the Ebro continental shelf. *Marine Geology*, **95**, 333–352.

DÍAZ, J. I., PALANQUES, A., NELSON, C. H. & GUILLÉN, J. 1996. Morpho-structure and sedimentology of the Holocene Ebro prodelta mud belt (northwestern Mediterranean Sea). *Continental Shelf Research*, **16**, 435–456.

DÍAZ DEL RÍO, V. & FERNÁNDEZ-SALAS, L. M. 2005. El margen continental del Levante español y las Islas Baleares. *In*: MARTÍN-SERRANO, A. (ed.) *Mapa Geomorfológico de España y del Margen Continental*. Instituto Geológico y Minero de España, Madrid, 177–188.

DÍAZ DEL RÍO, V., REY, J. & VEGAS, R. 1986. The Gulf of Valencia continental shelf: Extensional tectonics in Neogene and Quaternary sediments. *Marine Geology*, **73**, 169–179.

DUGGEN, S., HOERNIE, K., VAN DEN BOGAARD, P., RUPKE, L. & PHIPPS MORGAN, J. 2003. Deep roots of the Messinian salinity crisis. *Nature*, **422**, 602–605.

DURÁN, R., NUEZ, M., ALONSO, B., ERCILLA, G., ESTRADA, F., CASAS, D. & FARRÁN, M. 2008. Assessment of sand trapped by coastal structures towards better management. El Masnou (Maresme, Catalunya). *Geotemas*, **10**, 511–514.

ERCILLA, G. & ALONSO, B. 1996. Quaternary siliciclastic sequence stratigraphy of western Mediterranean passive and tectonically active margins: the role of global v. local controlling factors. *In*: DE BATIST, M. & JACOBS, P. (eds) *Geology of Siliciclastic Shelf Seas*. Geological Society, London, Special Publications, **117**, 125–137.

ERCILLA, G., ALONSO, B. & BARAZA, J. 1992. Sedimentary evolution of the Northwestern Alboran Sea during the Quaternary. *Geo-Marine Letters*, **12**, 144–149.

ERCILLA, G., ALONSO, B. & BARAZA, J. 1994a. Post-Calabrian sequence stratigraphy of the northwestern Alboran Sea (southwestern Mediterranean). *Marine Geology*, **120**, 249–265.

ERCILLA, G., FARRÁN, M., ALONSO, B. & DÍAZ, J. I. 1994b. Pleistocene progradational growth pattern of the northern Catalonia continental shelf (northwestern Mediterranean). *Geo-Marine Letters*, **14**, 264–271.

ERCILLA, G., DÍAZ, J. I., ALONSO, B. & FARRÁN, M. 1995. Late Pleistocene-Holocene sedimentary evolution of the northern Catalonia continental shelf (north-western Mediterranean Sea). *Continental Shelf Research*, **15**, 1435–1451.

ERCILLA, G., ESTRADA, F., CASAS, D., DURÁN, R., NUEZ, M., ALONSO, B. & FARRÁN, M. 2010. The Masnou Infralittoral sedimentary environment (Barcelona province, NW Mediterranean Sea): morphology and late Holocene seismic stratigraphy. *Scientia Marina*, **74**, 180–195.

ESTRADA, F., ERCILLA, G. & ALONSO, B. 1997. Pliocene-Quaternary tectonic-sedimentary evolution of the NE Alboran Sea (SW Mediterranean Sea). *Tectonophysics*, **282**, 423–442.

FARRÁN, M. & MALDONADO, A. 1990. The Ebro continental shelf: Quaternary seismic stratigraphy and growth patterns. *Marine Geology*, **95**, 289–312.

FERNÁNDEZ-SALAS, L. M., LOBO, F. J., HERNÁNDEZ-MOLINA, F. J., SOMOZA, L., RODERO, J., DÍAZ DEL RÍO, V. & MALDONADO, A.

2003. High-resolution architecture of late Holocene highstand prodeltaic deposits from southern Spain: the imprint of high-frequency climatic and relative sea-level changes. *Continental Shelf Research*, **23**, 1037–1054.

FERNÁNDEZ-SALAS, L. M., LOBO, F. J., SANZ, J. L., DÍAZ DEL RÍO, V., GARCÍA, M. C. & MORENO, I. 2007. Morphometric analysis and genetic implications of pro-deltaic sea-floor undulations in the northern Alboran Sea margin, western Mediterranean Basin. *Marine Geology*, **243**, 31–56.

FERNÁNDEZ-SALAS, L. M., LOBO, F. J., DABRIO, C. J. *ET AL.* 2009. Land-sea correlation between Late Holocene coastal and infralittoral deposits in the SE Iberian Peninsula (Western Mediterranean). *Geomorphology*, **104**, 4–11.

FIELD, M. E. & GARDNER, J. V. 1990. Pliocene–Pleistocene growth of the Rio Ebro margin, northeast Spain: a prograding-slope model. *Geological Society of America Bulletin*, **102**, 721–733.

FLEXAS, M. M., DURRIEU DE MADRON, X., GARCÍA, M. A., CANALS, M. & ARNAU, P. 2002. Flow variability in the Gulf of Lions during the Mater HFF Experiment (March–May 1997). *Journal of Marine Systems*, **33/34**, 197–214.

FLOS, J. 1985. The driving machine. *In*: MARGALEF, R. (ed.) *Key Environments. Western Mediterranean*. Pergamon Press, Oxford, 60–90.

FONT, J., SALAT, J. & TINTORÉ, J. 1988. Permanent features of the circulation in the Catalan Sea. *Oceanologica Acta*, **S-9**, 51–57.

FORNÓS, J. J. & AHR, W. M. 1997. Temperate carbonates on a modern low-energy, isolated ramp: the Balearic platform, Spain. *Journal of Sedimentary Research*, **67**, 364–373.

FREY-MARTÍNEZ, J., CARTWRIGHT, J. A., BURGESS, P. M. & BRAVO, J. V. 2004. 3D seismic interpretation of the Messinian Unconformity in the Valencia Basin, Spain. *In*: DAVIES, R. J., CARTWRIGHT, J. A., STEWARD, S. A., LAPPIN, M. & UNDERHILL, J. R. (eds) *3D Seismic Technology: Application to the Exploration of Sedimentary Basins*. Geological Scoiety, London, Memoirs, **29**, 91–100.

GÁMEZ, D., SIMÓ, J. A., LOBO, F. J., BARNOLAS, A., CARRERA, J. & VÁZQUEZ-SUÑÉ, E. 2009. Onshore–offshore correlation of the Llobregat deltaic system, Spain: Development of deltaic geometries under different relative sea-level and growth fault influences. *Sedimentary Geology*, **217**, 65–84.

GARCÍA, M., MAILLARD, A., ASLANIAN, D., RABINEAU, M., ALONSO, B., GORINI, C. & ESTRADA, F. 2011. The Catalan Margin during the Messinian Salinity Crisis: physiography, morphology and sedimentary record. *Marine Geology*, **284**, 158–174.

GIL-ROMERA, G., CARRIÓN, J. S., PAUSAS, J. G., SEVILLA-CALLEJO, M., LAMB, H. F., FERNÁNDEZ, S. & BURJACHS, F. 2010. Holocene fire activity and vegetation response in Southeastern Iberia. *Quaternary Science Reviews*, **29**, 1082–1092.

GIRÓ, S. & MALDONADO, A. 1983. Definición de facies y procesos sedimentarios en la plataforma continental de Valencia (Mediterráneo occidental). *In*: CASTELLVÍ, J. (ed.) *Estudio Oceanográfico de la Plataforma Continental*. Comité Conjunto Hispano–Norteamericano para la Cooperación Cultural y Educativa, Cádiz, 75–96.

GOY, J. L., ZAZO, C. & DABRIO, C. J. 2003. A beach-ridge progradation complex reflecting periodical sea-level and climate variability during the Holocene (Gulf of Almeria, Western Mediterranean). *Geomorphology*, **50**, 251–268.

GRÀCIA, E., BARTOLOME, R. *ET AL.* 2012. Acoustic and seismic imaging of the Adra Fault (NE Alboran Sea): in search of the source of the 1910 Adra earthquake. *Natural Hazards and Earth System Sciences*, **12**, 3255–3267.

GRACIA, V., COLLADO, F., GARCÍA, M., MONSO, J. L. 1989. Análisis y porpuesta de soluciones para estabilizar el delta del Ebro: Clima de oleaje. *In*: DIRECCIO GENEREAL DE PORTS I COSTES (ed.) *Technical Report LT-2/5*. Generalitat de Catalunya, Barcelona, Spain.

GRIMALT, J. O. & ALBAIGÉS, J. 1990. Characterization of the depositional environments of the Ebro Delta (western Mediterranean) by the study of sedimentary lipid markers. *Marine Geology*, **95**, 207–224.

GUIMERÀ, J. 1984. Paleogene evolution of deformation in the northeastern Iberian Peninsula. *Geological Magazine*, **121**, 413–420.

HERNÁNDEZ-MOLINA, F. J., SOMOZA, L., REY, J. & POMAR, L. 1994. Late Pleistocene-Holocene sediments on the Spanish continental shelves: model for very high resolution sequence stratigraphy. *Marine Geology*, **120**, 129–174.

HERNÁNDEZ-MOLINA, F. J., SOMOZA, L. & REY, J. 1996. Late Pleistocene–Holocene high-resolution sequence analysis on the Alboran Sea continental shelf. *In*: DE BATIST, M. & JACOBS, P. (eds) *Geology of Siliciclastic Shelf Seas*. Geological Society, London, Special Publications, **117**, 139–154.

HERNÁNDEZ-MOLINA, F. J., FERNÁNDEZ-SALAS, L. M., LOBO, F., SOMOZA, L., DÍAZ-DEL-RÍO, V. & ALVEIRINHO DIAS, J. M. 2000. The infralittoral prograding wedge: a new large-scale progradational sedimentary body in shallow marine environments. *Geo-Marine Letters*, **20**, 109–117.

HERNÁNDEZ-MOLINA, F. J., SOMOZA, L., VÁZQUEZ, J. T., LOBO, F., FERNÁNDEZ-PUGA, M. C., LLAVE, E. & DÍAZ DEL RÍO, V. 2002. Quaternary stratigraphic stacking patterns on the continental shelves of the southern Iberian Peninsula: their relationship with global climate and palaeoceanographic changes. *Quaternary International*, **92**, 5–23.

HOPKINS, T. S. 1985. Physics of the Sea. *In*: MARGALEF, R. (ed.) *Key Environments. Western Mediterranean*. Pergamon Press, Oxford, 100–125.

JABALOY-SÁNCHEZ, A., LOBO, F. J., AZOR, A., BÁRCENAS, P., FERNÁNDEZ-SALAS, L. M., DÍAZ DEL RÍO, V. & PÉREZ-PEÑA, J. V. 2010. Human-driven coastline changes in the Adra River deltaic system, southeast Spain. *Geomorphology*, **119**, 9–22.

KERTZNUS, V. & KNELLER, B. 2009. Clinoform quantification for assessing the effects of external forcing on continental margin development. *Basin Research*, **21**, 738–758.

LA VIOLETTE, P. E., TINTORÉ, J. & FONT, J. 1990. The surface circulation of the Balearic Sea. *Journal of Geophysical Research*, **95**, 1559–1568.

LARIO, J., ZAZO, C. & GOY, J. L. 1999. Fases de progradación y evolución morfosedimentaria de la flecha litoral de Calahonda (Granada) durante el Holoceno. *Estudios Geológicos*, **55**, 247–250.

LIQUETE, C., ARNAU, P., CANALS, M. & COLAS, S. 2005. Mediterranean river systems of Andalusia, southern Spain, and associated deltas: a source to sink approach. *Marine Geology*, **222–223**, 471–495.

LIQUETE, C., CANALS, M. *ET AL.* 2007. Long-term development and current status of the Barcelona continental shelf: a source-to-sink approach. *Continental Shelf Research*, **27**, 1779–1800.

LIQUETE, C., CANALS, M., DE MOL, B., DE BATIST, M. & TRINCARDI, F. 2008. Quaternary stratal architecture of the Barcelona prodeltaic continental shelf (NW Mediterranean). *Marine Geology*, **250**, 234–250.

LIQUETE, C., CANALS, M., LUDWIG, W. & ARNAU, P. 2009 Sediment discharge of the rivers of Catalonia, NE Spain, and the influence of human impacts. *Journal of Hydrology*, **366**, 76–88.

LOBO, F. J., FERNÁNDEZ-SALAS, L. M., MORENO, I., SANZ, J. L. & MALDONADO, A. 2006. The sea-floor morphology of a Mediterranean shelf fed by small rivers, northern Alboran Sea margin. *Continental Shelf Research*, **26**, 2607–2628.

LOBO, F., MALDONADO, A., HERNÁNDEZ-MOLINA, F., FERNÁNDEZ-SALAS, L., ERCILLA, G. & ALONSO, B. 2008. Growth patterns of a proximal terrigenous margin offshore the Guadalfeo River, northern Alboran Sea (SW Mediterranean Sea): glacio-eustatic control and disturbing tectonic factors. *Marine Geophysical Researches*, **29**, 195–216.

LO IACONO, C., GUILLÉN, J. *ET AL.* 2010. Large-scale bedforms along a tideless outer shelf setting in the western Mediterranean. *Continental Shelf Research*, **30**, 1802–1813.

LÓPEZ-GARCÍA, M. J., MILLOT, C., FONT, J. & GARCÍA-LADONA, E. 1994. Surface circulation variability in the Balearic Basin. *Journal of Geophysical Research*, **99**, 3285–3296.

LÓPEZ-JURADO, J. L., GARCÍA LAFUENTE, J. & CANO LUCAYA, N. 1995. Hydrographic conditions of the Ibiza Channel during November 1990, March 1991 and July 1992. *Oceanologica Acta*, **18**, 235–243.

MACÍAS, D., BRUNO, M., ECHEVARRÍA, F., VÁZQUEZ, A. & GARCÍA, C. M. 2008. Meteorologically-induced mesoscale variability of the Northwestern Alboran Sea (southern Spain) and related biological patterns. *Estuarine, Coastal and Shelf Science*, **78**, 250–266.

MAILLARD, A. & MAUFFRET, A. 1999. Crustal structure and riftogenesis of the Valencia Trough (north-western Mediterranean Sea). *Basin Research*, **11**, 357–379.

MAILLARD, A., MAUFFRET, A., WATTS, A. B., TORNÉ, M., PASCAL, G., BUHL, P. & PINET, B. 1992. Tertiary sedimentary history and structure of the Valencia Trough (western Mediterranean). *Tectonophysics*, **203**, 57–75.

MALDONADO, A. (ed.) 1992. Alboran Sea. *Geo-Marine Letters*, **12**, 61–186.

MALDONADO, A. & ZAMARREÑO, I. 1983. Modelos sedimentarios en las plataformas continentales del Mediterráneo español: factores de control, facies y procesos que rigen su desarrollo. *In*: CASTELLVÍ, J. (ed.) *Estudio Oceanográfico de la Plataforma Continental*. Comité Conjunto Hispano–Norteamericano para la Cooperación Cultural y Educativa, Cádiz, 15–52.

MALDONADO, A., SWIFT, D. J. P. *ET AL*. 1983. Sedimentation on the Valencia continental shelf: preliminary results. *Continental Shelf Research*, **2**, 195–211.

MASÓ, M., LA VIOLETTE, P. E. & TINTORÉ, J. 1990. Coastal flow modification by submarine canyons along the NE Spanish coast. *Scientia Marina*, **54**, 343–348.

MARTÍN-SERRANO, A. (ed.) 2005. *Mapa Geomorfológico de España y del margen continental. Escala 1:1.000.000*. Instituto Geológico y Minero de España, Madrid.

MARTÍN-VIDE, J. & LÓPEZ-BUSTINS, J. 2006. The Western Mediterranean oscillation and rainfall in the Iberian Peninsula. *International Journal of Climatology*, **26**, 1455–1475.

MAUFFRET, A., AMMAR, A., GORINI, C. & JABOUR, H. 2007. The Alboran Sea (Western Mediterranean) revisited with a view from the Moroccan Margin. *Terra Nova*, **19**, 195–203.

MEDIALDEA, J., BAENA, J. *ET AL*. 1982. *Mapa Geológico de la Plataforma Continental Española y Zonas Adyacentes. Escala 1:200.000. Hoja no 84–85/84S-85S (Almería-Garrucha/Chella-Los Genoveses)*. Instituto Geológico y Minero de España, Madrid.

MEDIALDEA, J., MALDONADO, A. *ET AL*. 1986. *Mapa geológico de la plataforma continental española y zonas adyacentes. E 1:200000. Hojas 41–42 (Tarragona)*. Instituto Geológico y Minero de España, Madrid.

MEDIALDEA, J., MALDONADO, A. *ET AL*. 1989. *Mapa geológico de la plataforma continental española y zonas adyacentes. E 1:200000. Hojas 35–42E (Barcelona)*. Instituto Geológico y Minero de España, Madrid.

MEDIALDEA, J., MALDONADO, A. *ET AL*. 1994. *Mapa geológico de la plataforma continental española y zonas adyacentes. E 1:200000. Hojas 25–25E (Figueres)*. Instituto Geológico y Minero de España, Madrid.

MILLOT, C. 1999. Circulation in the Western Mediterranean Sea. *Journal of Marine Systems*, **20**, 423–442.

MUÑOZ, A., BALLESTEROS, M., MONTOYA, I., RIVERA, J., ACOSTA, J. & UCHUPI, E. 2008. Alboran Basin, southern Spain – Part I: geomorphology. *Marine and Petroleum Geology*, **25**, 59–73.

NELSON, C. H. 1990. Estimated post-Messinian sediment supply and sedimentation rates on the Ebro continental margin, Spain. *Marine Geology*, **95**, 395–418.

NELSON, C. H. & MALDONADO, A. 1990*a*. Factors controlling late Cenozoic continental margin growth from the Ebro Delta to the western Mediterranean deep sea. *Marine Geology*, **95**, 419–440.

NELSON, C. H. & MALDONADO, A. (eds) 1990*b*. *The Ebro Continental Margin, Northwestern Mediterranean Sea. Marine Geology*, **95**.

NUEZ, M., DURÁN, R., ALONSO, B., ERCILLA, G., ESTRADA, F., CASAS, D. & FARRÁN, M. 2008. A morphological evolution model of the sandy coastline Premià – El Masnou (NE Spain). *Geotemas*, **10**, 643–646.

OJEDA, E. & GUILLÉN, J. 2006. Monitoring beach nourishment based on detailed observations with video measurements. *Journal of Coastal Research*, **48**, 100–106.

OLIVET, J.-L. 1996. Kinematics of the Iberian plate. *Bulletin des Centres de Recherches Elf Exploration Production*, **20**, 191–195.

PALANQUES, A. & DRAKE, D. E. 1990. Distribution and dispersal of suspended particulate matter on the Ebro continental shelf, northwestern Mediterranean Sea. *Marine Geology*, **95**, 193–206.

PALANQUES, A., PLANA, F. & MALDONADO, A. 1990. Recent influence of man on the Ebro margin sedimentation system, northwestern Mediterranean Sea. *Marine Geology*, **95**, 247–263.

PALOMINO, D., VÁZQUEZ, J. T., DÍAZ DEL RÍO, V. & FERNÁNDEZ-SALAS, L. M. 2009. Estudio de los procesos sedimentarios recientes de la

Bahía de Palma a partir del análisis de la morfología y la respuesta acústica (Islas Baleares, Mediterráneo Occidental). *Revista de la Sociedad Geológica de España*, **22**, 79–93.

PEREA, H., GRÀCIA, E. *ET AL*. 2012. Quaternary active tectonic structures in the offshore Bajo Segura basin (SE Iberian Peninsula – Mediterranean Sea). *Natural Hazards and Earth System Sciences*, **12**, 3151–3168.

PERKINS, H., KINDER, T. & LA VIOLETTE, P. E. 1990. The Atlantic inflow in the Western Alboran Sea. *Journal of Physical Oceanography*, **20**, 242–263.

PICCO, P. 1991. Climatological Atlas of the Western Mediterranean Sea. *Journal of Marine Systems*, **20**, 423–442.

PUIG, P. & PALANQUES, A. 1998. Nepheloid structure and hydrographic control on the Barcelona continental margin, northwestern Mediterranean. *Marine Geology*, **149**, 39–54.

PUIG, P., PALANQUES, A., SÁNCHEZ-CABEZA, J. A. & MASQUÉ, P. 1999. Heavy metals in particulate matter and sediments in the southern Barcelona sedimentation system (North-western Mediterranean). *Marine Chemistry*, **63**, 311–329.

REHAULT, J. P., BOILLOT, G. & MAUFFRET, A. 1984. The western Mediterranean basin geological evolution. *Marine Geology*, **55**, 447–477.

REY, J. & DÍAZ DEL RÍO, V. 1983. Aspectos geológicos sobre la estructura poco profunda de la plataforma continental del levante español. *In*: CASTELLVÍ, J. (ed.) *Estudio Oceanográfico de la Plataforma Continental*. Comité Conjunto Hispano–Norteamericano para la Cooperación Cultural y Educativa, Cádiz, 53–74.

REY, J. & FUMANAL, M. P. 1996. The Valencian coast (western Mediterranean): neotectonics and geomorphology. *Quaternary Science Reviews*, **15**, 789–802.

ROCA, E., SANS, M., CABRERA, L. & MARZO, M. 1999. Oligocene to Middle Miocene evolution of the central Catalan margin (northwestern Mediterranean). *Tectonophysics*, **315**, 209–233.

RUBIO, A., ARNAU, P. A. *ET AL*. 2005. A field study of the behaviour of an anticyclonic eddy on the Catalan continental shelf (NW Mediterranean. *Progress in Oceanography*, **66**, 142–156.

SÀBAT, F., ROCA, E. *ET AL*. 1997. Role of extension and compression in the evolution of the eastern margin of Iberia: the ESCI-València trough seismic profile. *Revista de la Sociedad Geológica de España*, **8**, 431–448.

SALAT, J., GARCÍA, M. A. *ET AL*. 2002. Seasonal changes of water mass structure and shelf–slope exchanges at the Ebro shelf (NW Mediterranean). *Continental Shelf Research*, **22**, 327–348.

SARDÁ, R., PINEDO, S., GREMARE, A. & TABOADA, S. 2000. Changes in the dynamics of shallow sandy-bottom assemblages due to sand extraction in the Catalan Western Mediterranean Sea. *ICES Journal of Marine Sciences*, **57**, 1446–1453.

SARHAN, T., GARCÍA LAFUENTE, J., VARGAS, M., VARGAS, J. M. & PLAZA, F. 2000. Upwelling mechanisms in the northwestern Alboran Sea. *Journal of Marine Systems*, **23**, 317–331.

SENCIALES, J. M. & MALVÁREZ, G. 2003. La desembocadura del río Vélez (provincia de Málaga, España). Evolución reciente de un delta de comportamiento mediterráneo. *Cuaternario y Geomorfología*, **17**, 47–61.

SERRA, J., MALDONADO, A. & RIBA, O. 1979. Caracterización del margen continental de Cataluña y Baleares. *Acta Geológica Hispánica*, **14**, 494–504.

SERRA, J., VALOIS, X. & PARRA, D. 2007. Estrucutura del prodelta de la Tordera (costa del Maresme, NO Mediterráneo) a partir del analisis sísmico de alta resolución. *Geogaceta*, **41**, 211–213.

SOMOZA, L., BARNOLAS, A., ARASA, A., MAESTRO, A., REES, J. & HERNÁNDEZ-MOLINA, F. J. 1998. Architectural stacking patterns of the Ebro delta controlled by Holocene high-frequency eustatic fluctuations, delta-lobe switching and subsidence processes. *Sedimentary Geology*, **117**, 11–32.

TENT-MANCLÚS, J. E., ESTÉVEZ, A. *ET AL*. 2009. Registro del evento 8.2 ka en la plataforma continental de Alicante (SE, España). *Geogaceta*, **47**, 97–100.

THORNDYCRAFT, V. R., BENITO, G., RICO, M., SOPEÑA, A., SANCHEZ-MOYA, Y. & CASAS, A. 2005. A long-term flood discharge record derived from slackwater flood deposits of the Llobregat River, NE Spain. *Journal of Hydrology*, **313**, 16–31.

TINTORÉ, J., LAVIOLETTE, P. E., BLADE, I. & CRUZADO, A. 1988. A Study of an Intense Density Front in the Eastern Alboran Sea: the Almeria-Oran Front. *Journal of Physical Oceanography*, **18**, 1384–1397.

URGELES, R., DE MOL, B. *ET AL.* 2007. Sediment undulations on the Llobregat prodelta: Signs of early slope instability or bottom current activity? *Journal of Geophysical Research*, **112**, B05102.

URGELES, R., CAMERLENGHI, A. *ET AL.* 2011*a*. New constraints on the Messinian sealevel drawdown from 3D seismic data of the Ebro Margin, western Mediterranean. *Basin Research*, **23**, 123–145.

URGELES, R., CATTANEO, A. *ET AL.* 2011*b*. A review of undulated sediment features on Mediterranean prodeltas: distinguishing sediment transport structures from sediment deformation. *Marine Geophysical Researches*, **32**, 49–69.

VARGAS-YÁÑEZ, M. & SABATÉS, A. 2007. Mesoscale high-frequency variability in the Alboran Sea and its influence on fish larvae distributions. *Journal of Marine Systems*, **68**, 421–438.

VARGAS-YÁÑEZ, M., PLAZA, F., GARCÍA-LAFUENTE, J., SARHAN, T., VARGAS, J. M. & VÉLEZ-BELCHI, P. 2002. About the seasonal variability of the Alboran Sea circulation. *Journal of Marine Systems*, **35**, 229–248.

VÁZQUEZ, J. T. 2005. El margen continental del Mar de Alborán. *In*: MARTÍN-SERRANO, A. (ed.) *Mapa Geomorfológico de España y del margen continental*. Instituto Geológico y Minero de España, Madrid, 191–198.

YOUNG, R. A., SWIFT, D. J. P., NITTROUER, C. A., DEMASTER, D. & BERGENBACK, B. 1983. Event-dominated sediment transport on the valencian continental shelf, Spain, and its effect on sediment accumulation and Holocene stratigraphy. *In*: CASTELLVÍ, J. (ed.) *Estudio Oceanográfico de la Plataforma Continental*. Comité Conjunto Hispano–Norteamericano para la Cooperación Cultural y Educativa, Cádiz, 1–14.

ZAMARREÑO, I., VÁZQUEZ, A. & MALDONADO, A. 1983. Sedimentación en la plataforma de Almería: Un ejemplo de sedimentación mixta silícico-carbonatada en clima templado. *In*: CASTELLVÍ, J. (ed.) *Estudio Oceanográfico de la Plataforma Continental*. Comité Conjunto Hispano–Norteamericano para la Cooperación Cultural y Educativa, Cádiz, 97–119.

Chapter 12

Overview of the variability of Late Quaternary continental shelf deposits of the Italian peninsula

ELEONORA MARTORELLI[1]*, FRANCESCO FALESE[2] & FRANCESCO L. CHIOCCI[2]

[1]*CNR-IGAG, Piazzale Aldo Moro, 00185 Rome, Italy,*

[2]*Dipartimento di Scienze della Terra, Sapienza Università di Roma, Piazzale Aldo Moro 5, 00185 Rome, Italy*

Corresponding author (e-mail: eleonora.martorelli@uniroma1.it)

Abstract: This paper documents the Late Quaternary (120 ka BP–present) stratigraphic architecture of Italian continental shelves through the interpretation of single-channel, very high-resolution seismic reflection profiles and from data derived from published studies, by using the high-resolution sequence stratigraphy framework. The result of this analysis provides a detailed reconstruction of the variability of shelf stratigraphy in relation to differences in physiography, sediment supply, structural framework and local factors. We distinguish four stratigraphic types: (1) wide shelves with high sediment supply (e.g. the central Adriatic and the central-northern Tyrrhenian shelves); (2) wide but sediment-starved shelves (e.g. the southern Latium shelf); (3) narrow shelves with high sediment supply (e.g. shelves located in the Ionian and southern Tyrrhenian Sea); and (4) narrow and sediment-starved shelves (shelves located around islands and archipelagos). Besides physiography, sediment supply and vertical movements, local factors and specific processes play a significant role in determining Late Quaternary stratal architecture and sedimentary facies. For example, the attributes of antecedent topography commonly seem to govern the formation and preservation of transgressive deposits, whereas the formation and preservation of falling stage and lowstand systems deposits can be hindered by the presence of canyons indenting the shelf edge and/or slope instability.

Late Quaternary shelf deposits formed during the last glacio-eustatic cycle (i.e. since Marine Isotope Stage 5, MIS 5) have been documented in detail both for their stratigraphic architecture, and for the factors controlling their formation and preservation (e.g. Posamentier *et al.* 1992; Ercilla & Alonso 1996; Chiocci *et al.* 1997; Chiocci 2000; Trincardi & Correggiari 2000; Ridente & Trincardi 2002; Lobo *et al.* 2004; Labuane *et al.* 2005). However, at present, only a few studies dealing with their variability along large spans of continental margins are available. In Italy, Tortora *et al.* (2001) first undertook such an analysis, although focusing on depositional and erosional processes that occurred during the Late Quaternary sea-level fluctuation rather than attempting a comparison between different margins.

The continental shelf of the Italian peninsula represents a very favourable setting to study the variability of shelf deposits as its width, gradient and sediment supply vary considerably over short lateral distances. This variability is due both to the young age of the continental margins and to the differences in the geodynamic setting: for example, the foredeep area of the Adriatic Sea; the young passive margin of the NE Tyrrhenian Sea; and the tectonically active margin of the Ionian and SE Tyrrhenian Sea.

This work provides an overview of Late Quaternary deposits from the eastern Tyrrhenian, western Adriatic, Ionian and Ligurian shelves (Fig. 12.1). Their stratigraphic variability will be related to differences in sediment supply, and shelf physiography and tectonics. These factors represent local conditions that vary along the margin; by contrast, eustasy and sea-level fluctuations due to climatic changes can be considered as constant boundary conditions.

Italian Late Quaternary shelf deposits were examined through the interpretation of single-channel, very high-resolution seismic profiles (more than 50 000 km of Chirp, Sub-bottom profiler, 1 kJ Sparker and Uniboom seismic profiles) and from data taken from published studies, in the framework of sequence stratigraphy concepts, modified to take into account the specificity of high-order eustatic cycles: that is, sea-level oscillations characterized by high-magnitude (about 120 m), high-frequency (about 100 kyr) and asymmetry (about 90% of the time for sea-level fall and lowstand: see, e.g. Hernandez-Molina *et al.* 1994; Chiocci *et al.* 1997).

In general, sequence stratigraphy is a very useful tool for depositional systems reconstruction, with relevant prediction capability of the sedimentary facies, as well as for the definition of a chrono-stratigraphic scheme. The sea-level variation during the last glacio-eustatic cycle is relatively well known (Chappell & Shackleton 1986; Fairbanks 1989; Bard *et al.* 1996, 2010; Helland-Hansen & Martinsen 1996; Siddall *et al.* 2003, 2006; Deschamps *et al.* 2012). The last glacio-eustatic cycle can be subdivided into four phases: (1) falling stage (*c.* 120–20 ka BP); (2) lowstand (*c.* 20–18 ka BP); (3) sea-level rise (*c.* 18–6 ka BP); and (4) highstand (*c.* 6 ka BP–present). Both the lowering and rising of sea level were punctuated by higher-frequency (fifth-order) cycles (Waelbroeck *et al.* 2002; Bard *et al.* 2010).

The whole suite of deposits formed during the last glacio-eustatic cycle make up a fourth-order depositional sequence (Late Quaternary depositional sequence, LQDS) including four systems tracts: the Falling Stage Systems Tract (FSST: Posamentier *et al.* 1992; Helland-Hansen & Gjelberg 1994; Hunt *et al.* 1995; Plint & Nummedal 2000); the Lowstand Systems Tract (LST: Posamentier & Vail 1988); the Transgressive Systems Tract (TST: Posamentier & Vail 1988); and the Highstand Systems Tract (HST: Posamentier & Vail 1988). The deposits of the TST and of the HST can be referred to post-glacial or post-Last Glacial Maximum (LGM) deposits.

The recognition of forced regressive shelf deposits opens a controversy regarding the position of the sequence boundary. As the surface that separates FSST and LST is commonly undetectable, these systems tracts are frequently described as an undifferentiated stratigraphic unit. In this paper we follow the indication of Posamentier & Vail (1988), placing the sequence boundary at the base of forced regressive shelf deposits.

The complete succession of Late Quaternary systems tracts rarely occurs in the same site. Indeed, the LQDS is usually incomplete and/or systems tracts are detached. In particular, the TST is commonly absent or represented by a thin lag of coarse, bioclastic sediment; FSST is preserved mainly below the shelf break; and the LST is commonly absent or represented by discontinuous shelf-margin wedges. By contrast, the HST is usually well developed, forming a muddy wedge that covers at least the inner-mid part of the shelf, depending on shelf width and sediment supply.

From: CHIOCCI, F. L. & CHIVAS, A. R. (eds) 2014. *Continental Shelves of the World: Their Evolution During the Last Glacio-Eustatic Cycle.*
Geological Society, London, Memoirs, **41**, 171–186. http://dx.doi.org/10.1144/M41.12

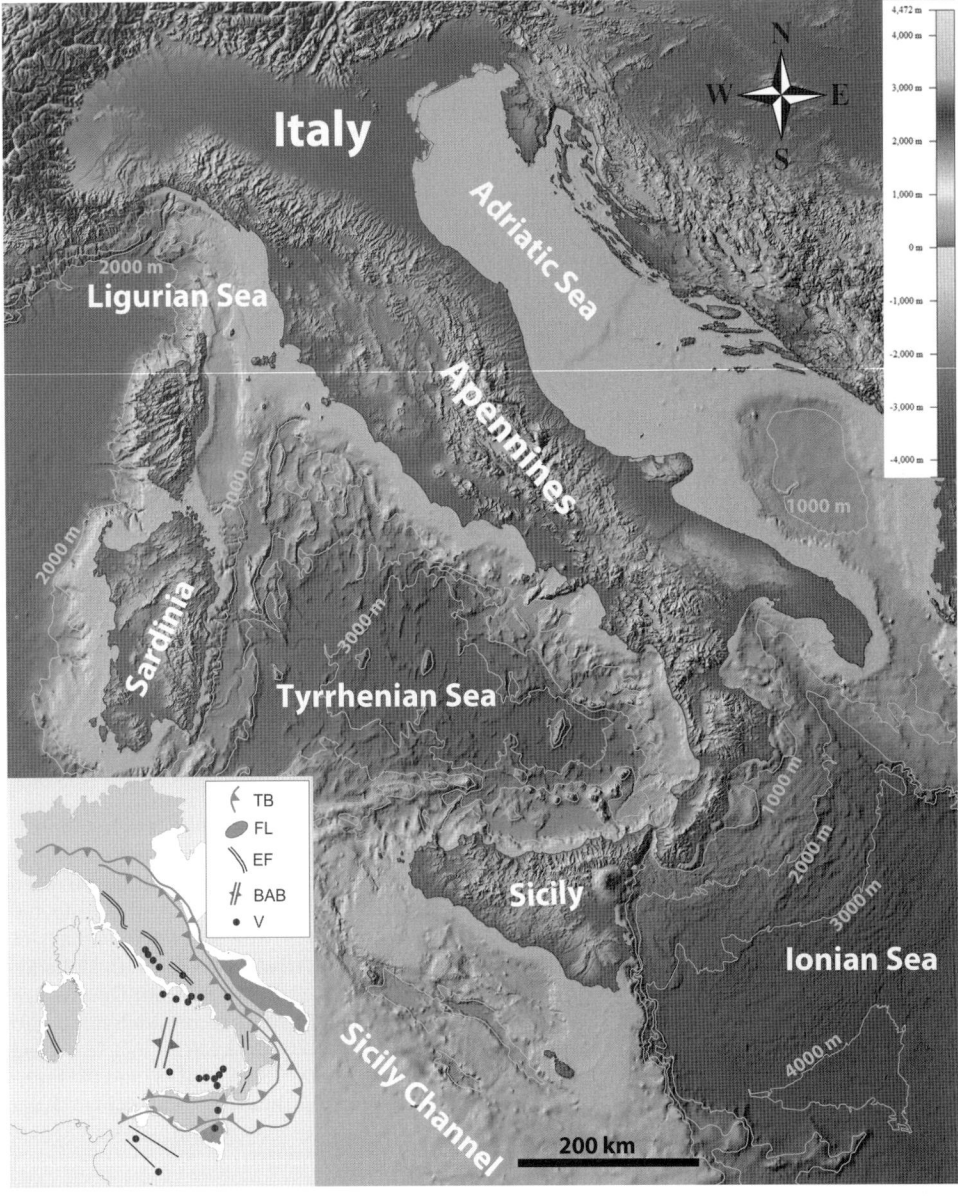

Fig. 12.1. Shaded relief map and bathymetry of the study area (bathymetric data from GEBCO Digital Atlas). Inset map sketches the main tectonic elements (TB, thrust belts; FL, forelands; EF, extensional faults; BAB, back-arc basin) and Quaternary volcanoes (V).

Following Belknap & Kraft (1981), Demarest & Kraft (1987) and Meijer *et al.* (2007), the main factors controlling the formation–preservation of deposits during Quaternary glacial–interglacial cycles can be summarized as: (1) sediment supply; (2) magnitude, rate and duration of sea-level fluctuation; (3) vertical movements; (4) pre-existing topography; and (5) post-depositional erosional processes related to sea-level fluctuation (e.g. shoreline erosion during forced regression and formation of the ravinement surface during transgression: e.g. Swift 1968; Nummedal & Swift 1987). During a transgression–regression, the topography of the translation surface represents a key aspect for the formation and preservation of the deposit (e.g. Tortora *et al.* 2001; Cattaneo & Steel 2003 and reference therein). In more detail, different mechanisms of shoreline retreat have been proposed within the transgressive dynamics (e.g. shoreface retreat, in-place drowning) in order to explain differences in the preservation potential of coastal lithosomes (Sanders & Kumar 1975; Rampino & Sanders 1980; Cowell *et al.* 1995; Snedden *et al.* 1999).

Overview of the Italian continental shelves

The Italian shelves, located between 36° and 45°N, belong to the temperate climate zone (Fig. 12.1, Table 12.1). Depending on

the geological setting, some Italian shelves are fed by major rivers (North Adriatic and central-northern Tyrrhenian Sea) or by multiple streams (southern Tyrrhenian Sea, Ligurian and Ionian Sea). Sedimentary deposits are usually siliciclastic, with some cases of temperate carbonate sediments (e.g. rhodalgal and bryomol assemblages) related to underfed shelves (usually islands and sectors of southern Italy). As for other Mediterranean margins, Italian continental shelves are wave-dominated because of the microtidal regime and the medium energy of marine currents (typically $<10-40$ cm s^{-1}: Istituto Idrografico della Marina 1982). As Italian shelves were built mainly during the Plio-Pleistocene, they have an immature physiography that mostly reflects the tectonic evolution and growth history of the margin (Tortora *et al.* 2001). Compared with continental shelves developed on older passive margins (e.g. the Atlantic margin), they are extremely narrow and steep: the mean width of the Tyrrhenian shelf is 13.6 km and the mean slope is 0.65° (Selli 1970). The Adriatic shelf differs substantially from other Italian shelves in terms of physiography and stratigraphy, as it developed on a shallow, semi-enclosed epicontinental sea that corresponds to a foredeep basin between the Apennines and the Dinarides (Argnani *et al.* 1993).

The complex tectonic setting caused a wide variability of vertical displacements along the margins during the Quaternary (Table 12.1). MIS 5.5 markers (Tyrrhenian highstand: last

Table 12.1. *Main characteristics of the continental shelves analysed in this study*

Shelf	Shelf sector	Width (km)	Extent (km)	Gradient (°)	Shelf break depth (m)	Tectonic trend*	Sediment supply	Sediment type
Western Adriatic†	Northern	300	150–200	0.02	120–150	Strong subsidence	Starved	Siliciclastic and locally bioclastic
	Central	30–50	150	0.2	120–140	Slight uplift	High	Siliciclastic
	Southern	15–90	300	0.2	100–160	Stable-slight uplift	Starved	Siliciclastic and bioclastic
Eastern Tyrrhenian‡	Northern (Tuscany)	30–50	240	0.2–0.3	120–150	Stable	Medium-starved	Siliciclastic and locally bioclastic
	Central (Latium–Campania)	10–30	450	0.2–1	120–160	Stable-localised uplift and subsidence	High-starved	Siliciclastic and locally bioclastic
	Southern (Calabria–Basilicata)	<10	200	>1	90–160	Strong uplift	High	Siliciclastic
Ligurian§		<10	240	>1	50–130	Stable-slight uplift	High	Siliciclastic
Ionian‖	Western	<10	310	>1	50–130	Strong uplift	High	Siliciclastic
	Eastern	10–20	120	0.2–0.8	60–120	Stable-slight uplift	Starved	Bioclastic

*Tectonic trend along the coast over the last glacial cycle derived by Ferranti *et al.* (2006).

Selected references on Late Quaternary shelf deposits are shown below:

†Trincardi *et al.* (1994); Correggiari *et al.* (1996, 2001, 2005); Cattaneo *et al.* (1997); Cattaneo & Trincardi (1999); Trincardi & Correggiari (2000); Ridente & Trincardi (2002); Cattaneo *et al.* (2003); Ridente & Trincardi (2005); Storms *et al.* (2008).

‡Marani *et al.* (1986); Chiocci *et al.* (1989, 1997, 2004); Tortora (1989, 1996); Trincardi & Field (1991); Bellotti *et al.* (1994, 1995); Chiocci & La Monica (1996); Chiocci & Orlando (1996); Milia & Torrente (2000); Budillon *et al.* (2004); De Pippo & Pennetta (2004); Roveri & Correggiari (2004); Aiello *et al.* (2005); Falese *et al.* (2008); Martorelli *et al.* (2008, 2010, 2011); Ridente *et al.* (2012); Di Bella *et al.* (2014).

§Corradi *et al.* (1984, 2004); Bozzano *et al.* (2006); Fierro *et al.* (2010).

‖Senatore *et al.* (1980); Pennetta (1985); Pennetta *et al.* (1986); Pescatore & Senatore (1986); Toscano & Sorgente (2002); Senatore (2004); Tessarolo *et al.* (2008).

125 kyr) indicate that the southern Tyrrhenian and Ionian coastal regions are characterized by strong uplift (up to 1.5 mm a⁻¹ in southern Calabria); most of eastern Tyrrhenian sector, from Tuscany to northern Calabria, is essentially stable or slowly subsiding, whereas the northern Adriatic is affected by subsidence up to 1 mm a⁻¹ (Ferranti *et al.* 2006).

Pre-Late Quaternary shelf architecture is made up of Plio-Pleistocene siliciclastic deposits. The most recent part of the shelf is composed of Upper Pleistocene units, forming fourth-order depositional sequences. On shelves affected by strong uplift, these sequences are mainly composed of regressive lowstand–falling stage slope units (Chiocci & Orlando 1995) as transgressive and highstand shelf deposits formed during pre-Late Quaternary cycles have been eroded during sea-level falls and lowstands (Chiocci *et al.* 1997). On the northern Latium and Tuscan shelves, Pleistocene shelf stratigraphy is somewhat different as moderate regional subsidence favoured the preservation of erosional unconformities (Chiocci 2000) and pre-Late Quaternary highstand deposits (Ridente *et al.* 2012). This trend continues towards the Ligurian shelf, which is characterized by shelf aggradation and extensive preservation of highstand deposits (Fanucci *et al.* 1974; Corradi *et al.* 1984). On the Adriatic shelf, Upper Pleistocene depositional sequences are well developed between the western flank of the Meso-Adriatic depression and the area south of the Gargano Promontory (Trincardi & Correggiari 2000; Ridente & Trincardi 2002). Each sequence includes muddy forced regression units and deposits formed during short-lived highstands, recording both the 100 kyr glacio-eustatic cyclicity and higher-frequency cycles (Ridente & Trincardi 2005; Ridente *et al.* 2008, 2009).

Eastern Tyrrhenian continental shelves

Eastern Tyrrhenian shelves range from narrow (a few kilometres) and steep (up to 1–2°) examples located to the south (e.g.

Calabrian shelf), to wide (some tens of kilometres) and gentle (0.2–0.5°) segments located to the north (Tuscany – Latium and Campanian shelves: Table 12.1). The widening of the eastern Tyrrhenian shelves occurred mainly by progradation of slope deposits during the Plio-Pleistocene (Chiocci *et al.* 1997; Chiocci 2000). As a whole, they are relatively narrow and steep because of their young age and recent tectonic deformation. Moreover, as discussed in the previous section, the variability in physiography and stratigraphic architecture is due to differences in vertical displacements from north (subsidence) to south (rapid uplift) during the Quaternary (Fanucci *et al.* 1974; Savelli & Wezel 1980; Chiocci *et al.* 1997).

In particular, the Calabrian continental shelf (C in Fig. 12.2; Table 12.1) is very narrow (less than 8 km), steep (>1°), and affected by tectonic deformation and strong seismicity. It is well supplied both by small rivers (the Amato and Savuto rivers) and by a network of very small ephemeral streams ('fiumare'). Canyons have developed on the continental slope (e.g. Angitola, Gioia-Mesima), reaching the inner–middle continental shelf and even the coastline. Above the Pleistocene prograding units, Late Quaternary transgressive and highstand deposits are present. They have a significant thickness (up to 60 m) and form a wedge-shaped lithosome that covers the whole shelf. In particular, transgressive deposits are very thick and cover wide shelf sectors (Chiocci *et al.* 1989; Martorelli *et al.* 2010). Locally, lowstand deposits form well-defined submerged depositional terraces (SDT: Chiocci & Orlando 1996).

The Campanian shelf (Ca in Fig. 12.2; Table 12.1) is gently sloping (<1°) and up to 20 km wide; the sectors offshore the southern flank of Penisola Sorrentina and south of Capo Palinuro are, instead, narrow and steep as they are structurally controlled. The shelf break is at a water depth of 140–150 m. It is cut by submarine canyons (in the Gulf of Naples, and Dohrn and Magnaghi canyons) that acted as sediment transfer features during major eruptive phases of the Ischia and Procida islands (Aiello *et al.* 2005).

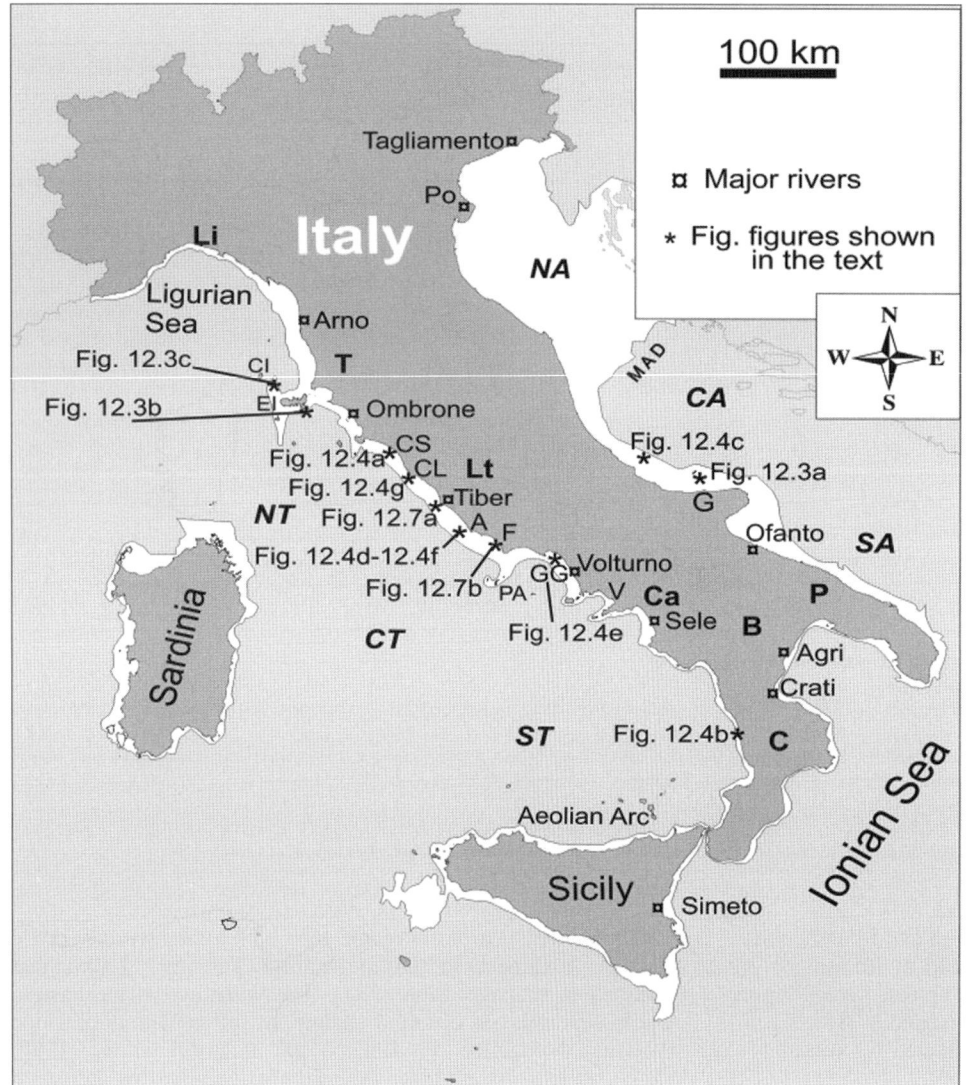

Fig. 12.2. Location of seismic profiles shown in the text. The white area delineates the continental shelf. Main rivers are also indicated. Li, Liguria; T, Tuscany; Lt, Latium; Ca, Campania; C, Calabria; B, Basilicata; P, Puglia; CI, Capraia Island; EI, Elba Island; CS, Chiarone Stream; CL, Capo Linaro; A, Anzio; F, Fogliano; PA, Pontine Archipelago; GG, Gaeta Gulf; Is, Ischia Island; V, Vesuvio; G, Gargano. MAD, Meso-Adriatic Depression. NT, North Tyrrhenian Sea; CT, central Tyrrhenian Sea; ST, South Tyrrhenian Sea; NA, North Adriatic Sea; CA, central Adriatic Sea; SA, South Adriatic Sea.

The Campanian shelf is strongly conditioned by the presence of a major active volcanic belt, with frequent high-explosive volcanic activity during post-LGM and historical time (i.e. the Vesuvius, Campi Flegrei, Ischia and Procida volcanic complexes). Major explosive events from the Phlegrean fields occurred during the last sea-level fall and subsequent transgression, producing huge pyroclastic deposits such as the Neapolitan Yellow Tuff (covering at least 1000 km², *c.* 15 ka BP) and the Campanian Ignimbrite (*c.* 39–41 ka BP). On the Volturno River continental shelf, both the transgressive and highstand systems tracts represent the bulk of Late Quaternary deposits (Martorelli *et al.* 2008). These deposits are also well represented off Pozzuoli Bay (e.g. Milia & Torrente 2000). FSST and LST deposits were identified off the Gulf of Naples (Milia & Torrente 2000), the Sele River–Punta Licosa sector (mid-shelf and shelf-margin downlapping deposits: Trincardi & Field 1991), and off the Policastro Gulf and Penisola Sorrentina (SDTs: Budillon *et al.* 2004; De Pippo & Pennetta 2004).

The Latium continental shelf (Lt in Fig. 12.2; Table 12.1) is relatively wide (15–30 km) and gentle (0.2–0.4°), and has a shelf break at a water depth of 120–160 m. It is supplied by the major watercourse of the Tyrrhenian Sea: that is, the Tiber River (catchment area of *c.* 17 000 km²).

Highstand deposits represent the bulk of Late Quaternary sedimentation, reaching a maximum thickness (>60 m) offshore the Tiber mouth, where a wide Holocene delta is present (Bellotti *et al.* 1994). Other highstand deposits are related to coastal prisms located in the inner sector of the northern Latium shelf (Tortora 1989). Transgressive deposits are rather sparse and volumetrically subordinate (Chiocci & La Monica 1996). Lowstand deposits are present locally at the shelf break where they form SDTs (Chiocci & Orlando 1996). Some evidence of the Late Quaternary FSST is related to the outer shelf tabular bodies described by Marani *et al.* (1986).

The Latium shelf includes a volcanic archipelago (the Pontine Archipelago) where temperate carbonate sedimentation occurs as it is far from the main sediment supply of the Italian peninsula (e.g. Martorelli *et al.* 2011). In this area, the bulk of the Late Quaternary deposits is composed of FSST and LST deposits shaped in well-preserved SDTs (Chiocci & Orlando 1996).

The Tuscan shelf (T in Fig. 12.2; Table 12.1) has the more mature physiography of the eastern Tyrrhenian margin. It is fairly wide (up to 30–40 km), has a gentle slope (0.2–0.3°) and is fed by two major rivers (the Arno and Ombrone rivers). Holocene highstand deposits represent the bulk of Late Quaternary sedimentation, with a muddy wedge mostly confined to the inner–mid shelf. Maximum sediment thickness (>50 m) occurs offshore the Ombrone River subaqueous delta (Di Bella *et al.* 2014). Transgressive deposits are rather sparse. They reach a substantial thickness (*c.* 10 m) only offshore Massa (Falese *et al.* 2008). Lowstand deposits form well-defined SDT offshore the Capraia, Giglio and Elba islands (Roveri & Correggiari 2004; Falese *et al.* 2008). Falling stage deposits are represented by outer shelf progradational deposits (Ridente *et al.* 2012).

Moreover, some areas of the Tuscan shelf (e.g. the Elba Ridge, Tuscan Archipelago and sectors of the outer shelf) are characterized by temperate carbonate sedimentation.

Ligurian continental shelf

In contrast to the adjacent Tuscan shelf, the Ligurian shelf (Li in Fig. 12.2; Table 12.1) is very narrow (less than 10 km), steep and largely dissected by submarine canyons (e.g. Levante, Taggia, Bisagno). The Ligurian shelf is part of a tectonically active margin characterized by strong seismicity (Larroque *et al.* 2011). It is supplied by very small rivers (e.g. the Roja and Centa) and several ephemeral streams that periodically carry a large amount of sediment (Bozzano *et al.* 2006). Post-LGM deposition (usually highstand mud) is discontinuous, with common sediment-starved areas. Post-LGM deposits are moderately thick offshore main streams or in embayed areas (Corradi *et al.* 1984), while transgressive deposits are locally present with well-preserved parasequences (Corradi *et al.* 2004). LST deposits locally occur as shelf-margin prograding deposits (Fierro *et al.* 2010).

Ionian continental shelf

The Ionian continental shelf (Fig. 12.2, Table 12.1) is very narrow (less than 10 km), steep ($1-3°$) and abundantly cut by submarine canyons. It is supplied by rather small rivers (e.g. the Crati and Agri) and several streams. The most significant morphological feature of this area is the Taranto Valley, which marks a change in the physiography and stratigraphy of the margin. In fact, it separates the allochthonous thrust sheets of the southern Apennine–Calabrian Arc to the west from the foreland sector of the southern Apennine (Apulian Foreland) to the east (Rossi & Sartori 1981; Critelli & Le Pera 1998). Therefore it represents the submerged continuation of the Bradanic foredeep, funnelling sediments eroded from the thrust belt to the deep sea. On the western sector (Calabrian side), the continental shelf width is usually <10 km and the average gradient is 1° (Pescatore & Senatore 1986). The shelf is fed by rivers (e.g. the Crati and Agri) and torrential streams, and is affected by strong uplift (up to 1 mm a^{-1}), tectonic deformation and seismicity. Accordingly, the whole shelf is characterized by very high sediment input, by deformed Plio-Pleistocene units, and is strongly dissected by submarine gullies and canyons (e.g. the Caulonia and Marina di Gioiosa canyons: Tessarolo *et al.* 2008). Since, in this area, tectonic uplift and sediment supply are prominent controlling factors, shelf morphology and stratigraphy are fairly similar to those of the Calabro-Tyrrhenian margin. By contrast, the eastern sector (Apulian Foreland) is characterized by a wider (up to about 20 km) and gentler shelf, with the shelf break located at an average water depth of 110 m. This sector is affected by normal faults and graben filled with Plio-Quaternary sediments (Pescatore & Senatore 1986; Senatore 1987). Overall, this sector is underfed, thus Late Quaternary deposits are very thin, whereas rock outcrops and bioclastic sediments (coralline algal banks and bioclastic sands) are widespread (Pennetta 1985; Toscano & Sorgente 2002). Moreover, terraced morphologies formed during the last transgression are present (Senatore *et al.* 1980; Pennetta *et al.* 1986), as well as the lowstand SDT (Senatore 2004).

Western Adriatic continental shelf

The Adriatic is an epicontinental sea characterized by very shallow bathymetry (<100 m) in the northern sector, while a deeper seafloor (maximum water depth of 1200 m) occurs in the southern sector (Fig. 12.2, Table 12.1). In the central sector, the Meso-Adriatic depression (MAD) is a small remnant basin of the Adriatic Sea. The Adriatic Basin is characterized by a broad (about 300 km) and very gently sloping (0.02°) continental shelf in the northern sector, and by a slightly steeper (about 0.2°) shelf on the central-southern sector (Cattaneo & Steel 2003). The northern-central sector is fed by the lateral advection of fine sediment supplied by multiple sources (mainly the Po River and Apennine rivers). During the LGM, the northern Adriatic was emergent and dissected by an enlarged Po River system that reached the MAD, where an extensive lowstand prodelta developed (Trincardi & Correggiari 2000).

The Late Quaternary depositional sequence consists of falling-stage, lowstand, transgressive and highstand deposits, with variable distribution on the shelf. North of the modern Po delta, the HST encompasses undersupplied barrier–lagoon systems and a starved shelf. However, south of the Po delta, a dominant muddy deposition characterizes the highstand interval, forming an extensive muddy wedge of deltaic and shallow-marine deposits. This muddy wedge develops continuously, subparallel to the coast and confined to the inner shelf by the geostrophic flow (e.g. Trincardi *et al.* 1994; Cattaneo *et al.* 2003). East and SE of Gargano Promontory, the HST includes a prograding deposit located downdrift of the fluvial sediment sources of the Apennines that is interpreted as a 'subaqueous delta' (Cattaneo *et al.* 2003). The TST is composed of various deposits (e.g. marine muddy units, drowned barrier–lagoon systems and marine bedforms) characterized by highly variable thickness and extent; across outer shelf areas, transgressive deposits typically condense to a thin lag layer of shelly and muddy sediment (Trincardi *et al.* 1994). In the north, the TST is typically a few metres thick and consists mainly of drowned barrier–lagoon systems. In the central Adriatic, the TST is thicker and composed of backstepping wedges of muddy marine deposits (Cattaneo *et al.* 1997). LST deposits are well developed on the shelf margin NW of the MAD, forming a more than 200 m-thick progradational wedge, made up of delta and prodelta deposits of the Po River (Trincardi & Correggiari 2000), whereas FSST deposits are widespread from the western flank of the MAD to the area south of Gargano Promontory (Ridente & Trincardi 2002, 2005).

Description of Italian Late Quaternary shelf deposits

Falling stage and lowstand deposits (c. 120–18 ka BP)

The last sea-level fall produced a ubiquitous and profound environmental modification of the whole Italian continental shelf, with formation of a major subaerial unconformity, extending across the entire shelf. During the sea-level fall, the northern Adriatic experienced an eightfold widening of the subaerial area (e.g. Cattaneo & Trincardi 1999); as a result, the largest lowstand progradational wedge of the Italian shelves formed (i.e. the lowstand Po system: Trincardi & Correggiari 2000).

Falling stage deposits occur widely on the central sector of the western Adriatic shelf as prograding deposits separated from the post-LGM and LST shelf deposits by a shelf-wide unconformity (Fig. 12.3a). They are mainly composed of muddy sediment of outer shelf marine facies, probably representing the offshore counterpart of shoreface deposits; their depocentre distribution is subparallel to the shelf break, reaching a maximum of more than about 45 m (Ridente & Trincardi 2002, 2005). In the Tyrrhenian Sea, FSST deposits occur offshore of Elba Island (Chiocci *et al.* 1997) where they form a thick prograding wedge (Fig. 12.3b), on the Tuscan shelf as outer shelf progradational deposits (Ridente *et al.* 2012), on the southern Latium shelf (outer shelf tabular bodies of Marani *et al.* 1986) and on the Campanian shelf as prograding deposits located on the mid-outer shelf (Trincardi & Field 1991).

Typical lowstand deposits are represented by coastal prograding wedges, deltaic wedges and SDTs (i.e. wedge-shaped prograding

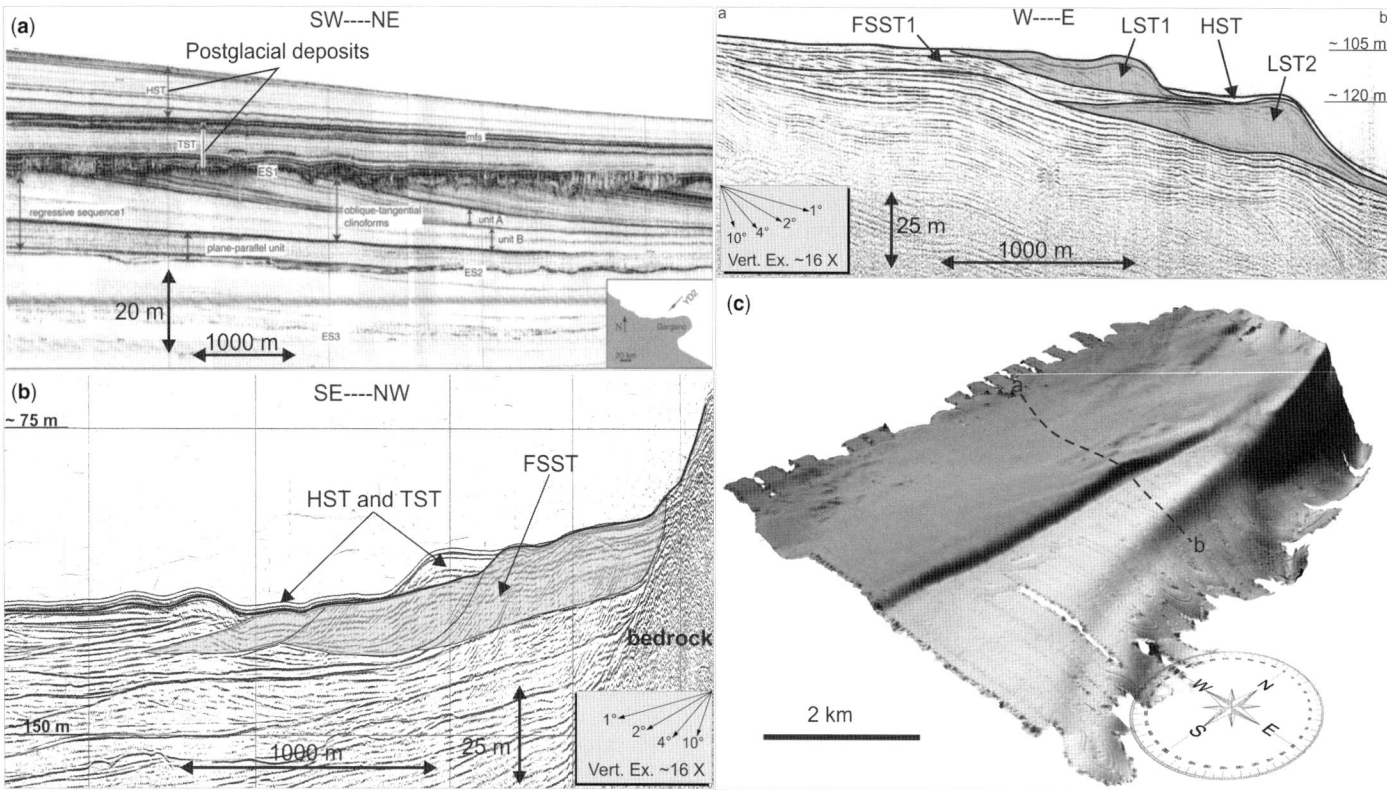

Fig. 12.3. Examples of falling stage and lowstand deposits. (**a**) Falling stage deposits identified in the central Adriatic shelf (modified from Ridente & Trincardi 2002). FSST deposits are bounded by ES1 unconformity; above the ES1 unconformity, the TST and HST deposits of the last eustatic cycle are present. ES2 and ES3 are older unconformities bounding Upper Pleistocene fourth-order depositional sequences. (**b**) Falling stage deposits located on the Tuscan shelf, off Elba Island (modified after Chiocci *et al.* 1991). (**c**) Submerged depositional terraces offshore Capraia Island (Tuscan shelf) formed during the last sea-level fall and lowstand (FSST1 and LST1); (upper panel) 1 kJ seismic profiles; (lower panel) 3D perspective view of multibeam bathymetry. See Figure 12.2 for the location.

deposits (Fig. 12.3c) formed below the wave-base level: Chiocci & Orlando 1996); in some cases, SDTs also include forced regressive units located landwards of the shelf break. Usually these deposits form vast sandy lithosomes (several kilometres long and up to several tens of metres thick), commonly characterized by a predominant bioclastic fraction (Chiocci *et al.* 2004). They occur mainly on insular shelves (e.g. the Pontine Archipelago, Aeolian Islands and some areas of the Tuscan Archipelago) and on tectonically controlled areas (e.g. the Calabrian and Campanian shelves).

During the sea-level fall and lowstand, incised valleys formed locally, cutting into underlying Pleistocene units (Fig. 12.4a). These palaeovalleys were filled mainly during the following sea-level rise. In the Tyrrhenian Sea, fluvial palaeovalleys are usually located from the present coastline down to a water depth of 60–70 m and are mostly found in front of the mouth of even very small rivers or streams. In the Adriatic Sea, the palaeovalleys are mainly related to the Po system, and are mostly found at water depths of between 60 and 110 m (Trincardi *et al.* 1994).

Transgressive deposits (18–6 ka BP)

The last sea-level rise was very rapid, with an average rate of approximately 10 mm a^{-1}, and exceeding 40 mm a^{-1} during Meltwater Pulse 1A (Bard *et al.* 2010; Deschamps *et al.* 2012). Over most of the Italian shelf, the high rates of sea-level rise limited the formation of transgressive deposits, determining a non-depositional transgression. In such cases the subaerial unconformity, the transgressive surface, the ravinement surface and the maximum flooding surface merge in a composite and diachronous surface (Chiocci & Milli 1995).

The transgressive deposits are usually thin and areally confined, showing highly variable stratigraphic patterns and patchy distribution. As indicated by the sequence stratigraphy model, transgressive deposits are usually organized in backstepping parasequences, bounded at their top by the maximum flooding surface and at their base by the transgressive surface or (more commonly) by the erosional unconformity. However, well-preserved backstepping parasequences (Fig. 12.4b, c and column 1; Fig. 12.5) are relatively uncommon; they have been identified in the Ligurian shelf (Corradi *et al.* 2004), the central-southern Tyrrhenian shelf (Martorelli *et al.* 2008, 2010) and the central Adriatic shelf (Cattaneo *et al.* 1997; Cattaneo & Trincardi 1999). In other cases, Italian transgressive deposits are represented by different types, such as sand ridges, incised-valleys fills, large-scale bedforms and transgressive lags. Transgressive sand ridges are typically 3–10 m thick, 1–4 km long and 300–600 m wide. They were locally identified in the Adriatic (column 4; Fig. 12.5) (Trincardi *et al.* 1994), Tuscan and Latium shelves (Fig. 12.4d, e). These deposits are somewhat similar to the US Atlantic sand ridges (e.g. Swift & Field 1981; McBride & Moslow 1991; Trowbridge 1995) and, analogous to them, they possibly originated by the transgressive reworking of previously formed nearshore deposits. Incised-valleys fills are mainly composed of transgressive deposits. Generally, they form elongated bodies crossing the shelf, bounded by the LGM unconformity at their base and overlain by highstand marine deposits (Fig. 12.4a). They are relatively uncommon deposits, mainly developed on medium–low gradient shelves (e.g. NW Adriatic (column 3; Fig. 12.5) and Latium–Tuscan shelves). Large-scale bedforms are sandy dunes, typically a few metres high with a wavelength of several hundreds of metres. They were identified in the NW Adriatic shelf (Correggiari *et al.* 1996) and the Latium shelf (Fig. 12.4f), and are thought to form by the reworking of drowned coastal deposits. Transgressive lags occur in sediment-starved areas (e.g. most of the North Adriatic and North Tyrrhenian shelf), either outcropping at the seafloor or

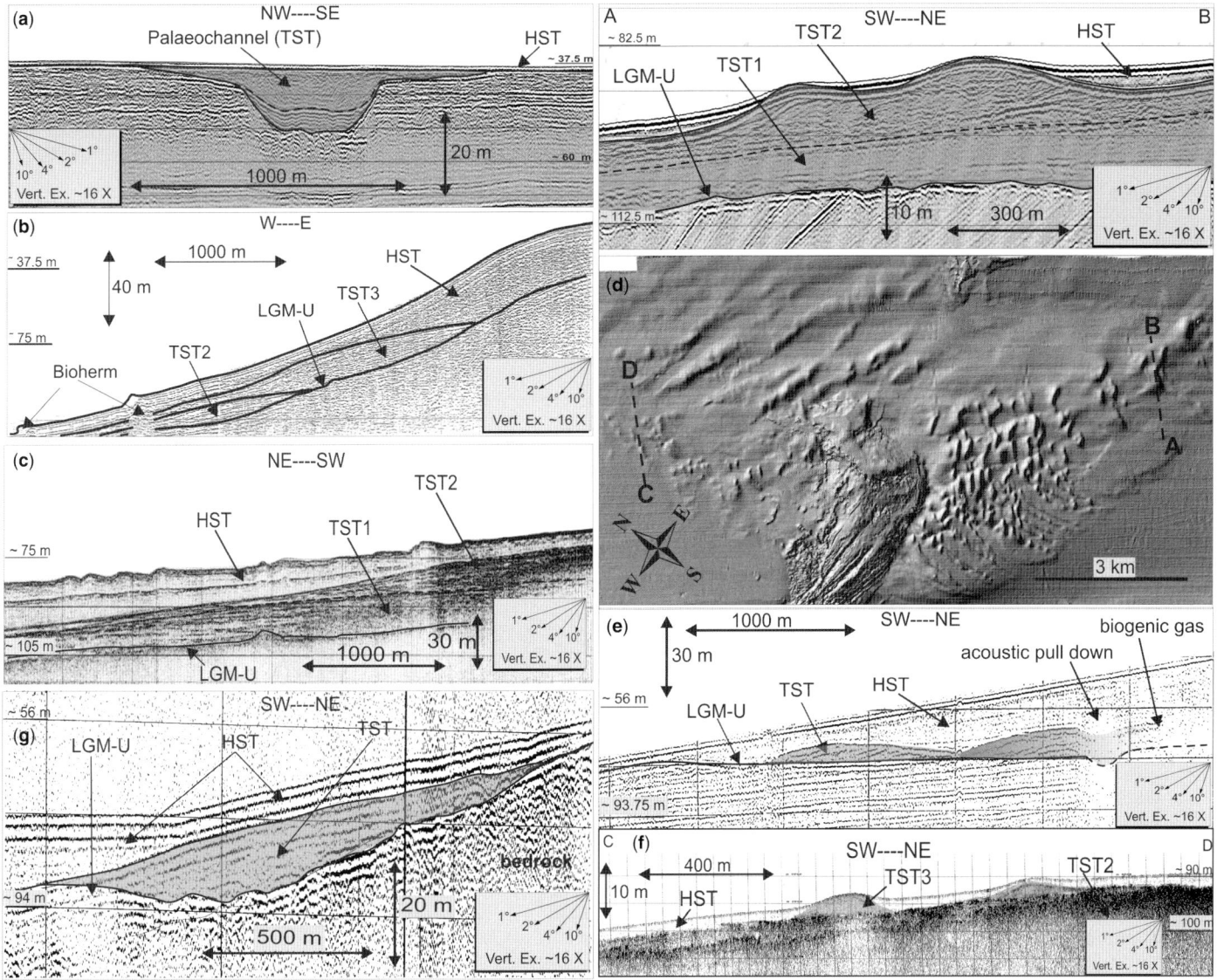

Fig. 12.4. Examples of transgressive deposits. (**a**) Fluvial palaeochannel formed during the last lowstand and infilled by transgressive deposits identified off the Chiarone stream. Palaeochannel-fill deposits are locally present on the Latium continental shelf and the North Adriatic shelf. (**b**) & (**c**) Post-LGM deposits from the Calabro-Tyrrhenian and the central Adriatic shelves with well-developed backstepping parasequences. (**d**) Transgressive sand ridges identified in the Latium shelf (Anzio); (upper panel) 1 kJ Sparker profile; (lower panel) shaded relief of multibeam bathymetry. (**e**) Bubble-pulser profile across transgressive sand ridges in the Gaeta Gulf (Latium shelf). (**f**) Transgressive submarine dunes mainly composed of fine sands and usually superimposed on sandy deposits (Anzio, Latium shelf). (**g**) Headland-attached transgressive deposits identified in the Latium shelf (Capo Linaro). See Figure 12.2 for the location.

buried by highstand deposits. They represent the most common and widespread expression of transgressive deposits occurring on the Italian shelves.

Further types of transgressive deposit are represented by drowned barrier–lagoon systems (North Adriatic, column 2; Fig. 12.5) (e.g. Storms *et al.* 2008) and nearshore bodies located where abrupt slope changes of the basal surface occur (Fig. 12.4g). This last type is commonly related to morphological highs of the basal surface and could be defined as headland-attached transgressive deposits; Roy *et al.* (1994) interpreted similar deposits as drowned sand barriers.

Highstand deposits (since about 6 ka BP)

The highstand phase started approximately 6 ka BP when the rate of sea-level rise decreased and, thus, sea level gradually approached the present-day position. On the middle and outer continental shelves, muddy sedimentation occurs, whereas sandy

deposits are confined into the nearshore area or on delta-front sectors where large rivers flow into the sea; muddy sediments usually form wedge-like bodies thinning offshore.

HST deposits encompass mainly muddy wedges, submerged deltaic bodies, coastal prisms and barrier–lagoon systems.

The largest Holocene delta system occurs in the Adriatic Sea, related to the main Italian watercourse (the Po River). Indeed, large-volume highstand deltaic deposits (up to 40–50 m thick) are related to areas fed by major rivers (Fig. 12.6). Other significant highstand deposits occur as quasi-continuous wedge-shaped depocentres, up to 35 m thick and tens to hundreds of kilometres long, on the Adriatic shelf along the Apennine coast (e.g. Trincardi *et al.* 1994; Cattaneo *et al.* 2003) and on the Calabro-Tyrrhenian shelf off multiple small rivers or creeks with high sediment discharge (Martorelli *et al.* 2010).

Major Italian submarine deltas (e.g. the Po, Tiber, Arno, Volturno and Ombrone delta) are characterized by a similar morphostratigraphic architecture encompassing the main features of wave-dominated deltas (Fig. 12.7a): that is, the almost flat

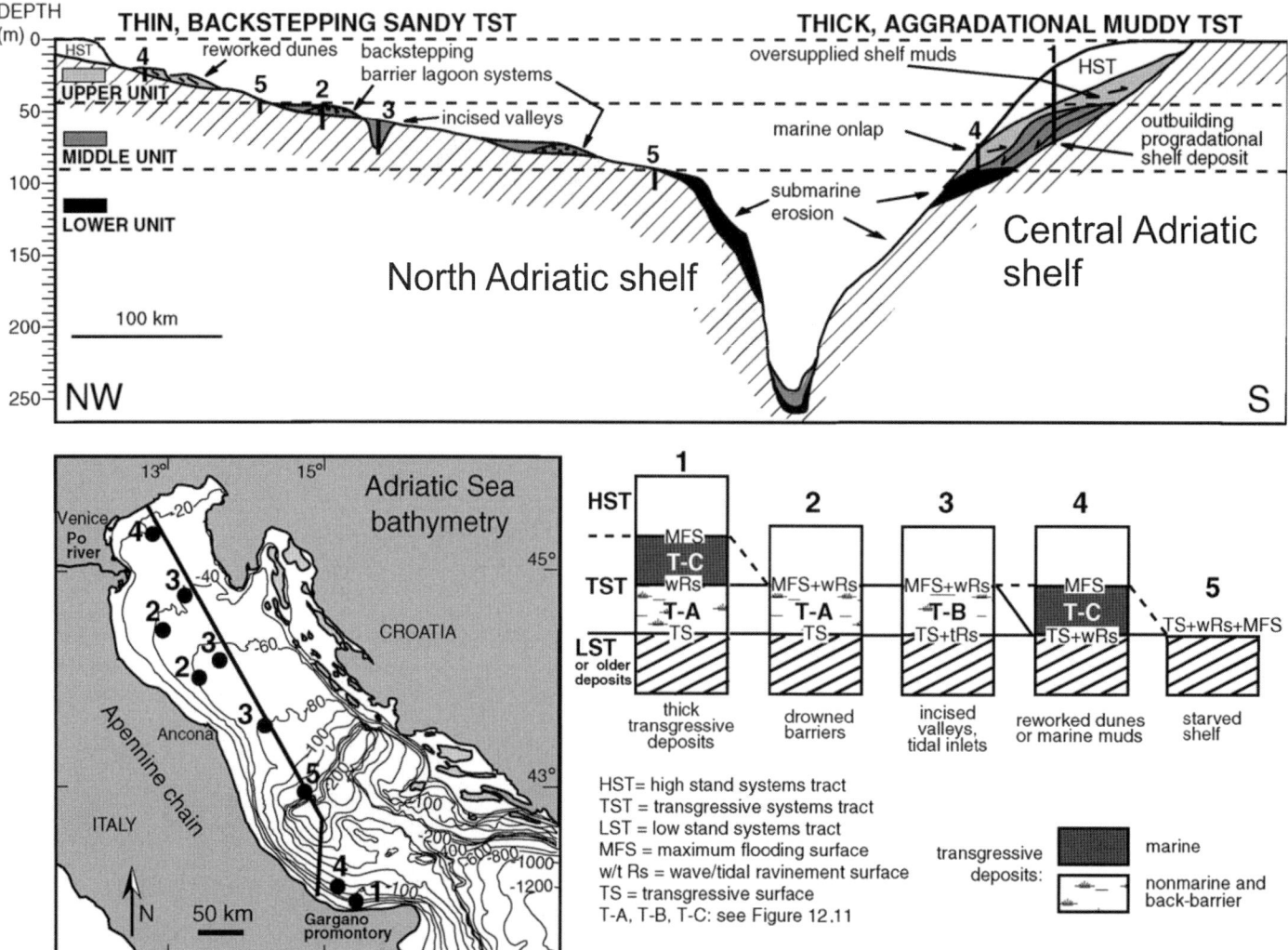

Fig. 12.5. Transgressive deposits identified in the Adriatic shelf (modified from Cattaneo & Steel 2003). The cross-section shows the distribution of transgressive deposits in the northern and central sectors of the shelf. The stratigraphic columns show their variability along the shelf.

delta-front area dominated by sandy sediments redistributed by wave processes, and the gentle sloping prodelta dominated by deposition of muddy sediments carried in buoyant plumes (e.g. Bellotti *et al.* 1994). The Po delta shows a more complex structure that includes an extensive prodelta composed of multiple lobes extending to the south and coalescing into an undifferentiated distal prodelta system (Correggiari *et al.* 2005). Gas-charged sediments are commonly present on the delta front and inner prodelta areas (Fig. 12.7a). The occurrence of gas, as well as the low shear strength of unconsolidated muddy sediments, can favour the development of soft-sediment deformation, such as synsedimentary creep (Chiocci *et al.* 1996).

Analogous to worldwide observations by Stanley & Warne (1994), the initiation of the Italian delta systems took place during the Holocene (Amorosi & Milli 2001). Major delta systems of the Tyrrhenian Sea passed through common evolutionary phases. In fact, after their growth as lagoonal delta bodies during transgression, they prograded in historical periods across the inner shelf (i.e. the onset of the modern wave-dominated delta, encompassing beach ridge formation); the progradation ended in the twentieth century as beach erosion took place (Bellotti *et al.* 1995). The Holocene Po delta experienced a similar evolution but, after land reclamation works endorsed by the Venice Republic (AD 1600–1604, *c.* 0.4 ka BP), it developed as a multi-lobe, supply-dominated delta (Amorosi & Milli 2001).

Highstand coastal prisms refer to non-deltaic deposits forming wedge-shaped bodies pinching out seawards at water depth of about 30–40 m (Fig. 12.7b). They are characterized by sandy sediments, with a progressive reduction in grain size towards the sea, where they intermingle with shelf mud. Similar deposits have been reported from the Spanish shelf by Hernandez-Molina *et al.* (2000). Thick coastal prisms have been identified in the Tyrrhenian Sea off the Pontine Archipelago and Latium shelf.

Finally, barrier–lagoon systems occur in the northern Adriatic, and in localized areas dispersed along the Italian peninsula and the main islands. They develop along low-gradient subsiding areas with transgressive shorelines, and are mainly composed of sandy barrier and lagoonal muds. Where present, lagoons trap the sediment, leaving the facing shelf starved and characterized by thin highstand deposits, as observed in the North Adriatic shelf, offshore the Venice Lagoon (Correggiari *et al.* 1996).

Discussion on the variability of Italian Late Quaternary shelf deposits

Knowledge of the Italian Late Quaternary shelf deposits obtained by analysis of seismostratigraphic data collected over the last 20 years and from published studies enabled us to assess their character and variability. The observed variability is summarized

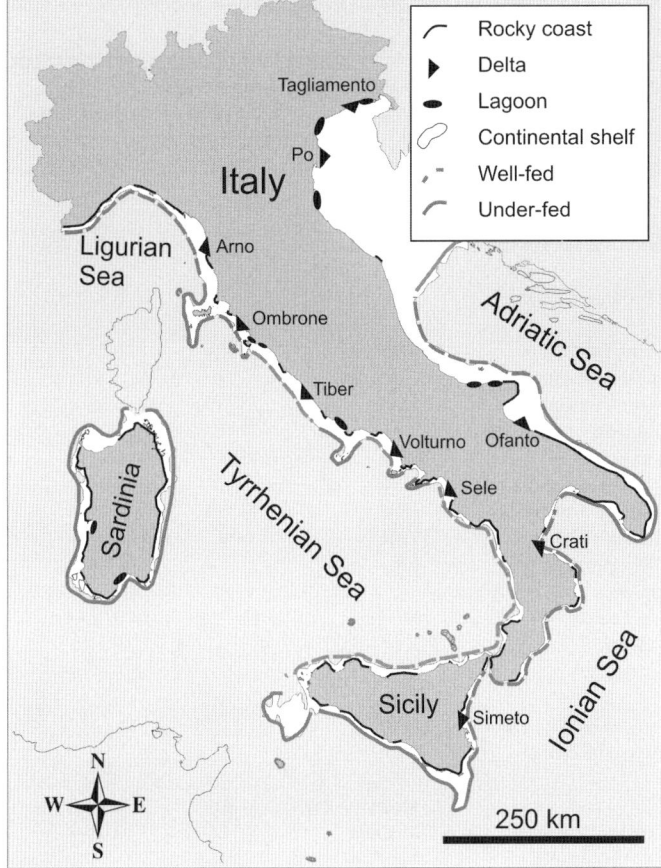

Fig. 12.6. Distribution of the Italian continental shelves (in white) with an indication of the sediment supply.

in Table 12.2. Hereafter it is considered within the sequence stratigraphic framework and synthesized in representative architectures.

FSST and LST deposits

Falling Stage Systems Tract (FSST) shelf deposits are mainly represented by progradational units located on the outer shelf.

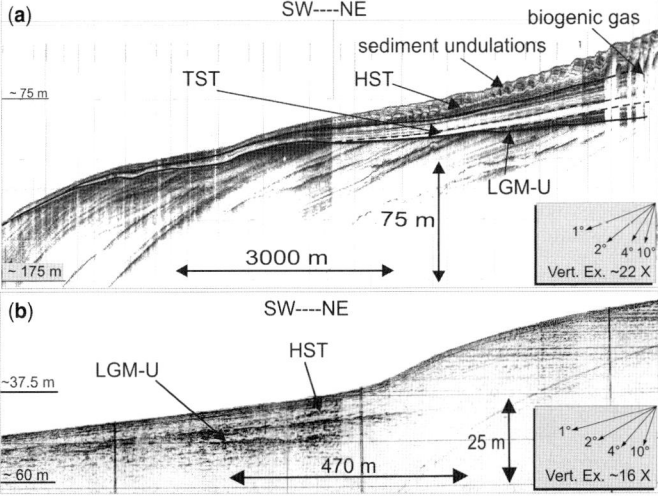

Fig. 12.7. Examples of highstand deposits: (**a**) submarine portion of the Tiber delta; and (**b**) highstand coastal prism identified off Fogliano (Latium shelf). See Figure 12.2 for the location.

On low-gradient, well-fed shelves, such as the central Adriatic and Tuscan shelves, the falling stage deposits are mostly muddy deposits composed of distal marine facies, whereas sandy deposits may occur on high-gradient shelves, such as the shelves surrounding islands (e.g. the Tuscan, Pontine and Aeolian archipelagos).

Lowstand Systems Tract (LST) deposits mainly develop offshore the shelf break where they constitute the last set of progradational units on the continental slope. In Italy, the most relevant example of lowstand deposits is represented by the deltaic deposits of the palaeo-Po system (Trincardi & Correggiari 2000). However, on high-gradient shelves, widespread sandy deposits related to lowstand SDTs are present. These deposits are mostly constituted by inner shelf marine facies and can reach a significant extent (up to 25 km) and thickness (up to 45 m: Chiocci et al. 2004).

Channel-fill lowstand deposits occur only locally, on medium–low gradient shelves with moderate–high sediment supply (e.g. the Latium, Tuscan and NW Adriatic shelves). In the Tyrrhenian Sea, fluvial palaeochannels are recognized mainly off small rivers or streams. However, offshore the major rivers (e.g. the Tiber, Volturno and Ombrone rivers), the limited penetration of the seismic signal as a result of the thickness of overlying deposits and the presence of biogenic gas inhibits their identification.

In general, the formation and preservation of lowstand shelf-margin deposits are mainly controlled by the sediment supply and depth reached by the lowstand shoreline relative to the shelf break (Trincardi & Field 1991).

TST deposits

Transgressive Systems Tract (TST) deposits show the maximum variability within the Late Quaternary depositional sequence. In fact, the record of the last transgression varies from thin (a few decimetres) bioclastic lag deposits to thick (tens of metres) parasequences. As a whole, their patchy occurrence and variability is in agreement with the variable stratigraphic expression observed by other authors (Tortora et al. 2001; Cattaneo & Steel 2003; Nordfjord et al. 2009).

The best examples of thick backstepping sandy parasequences come from the narrow Calabro-Tyrrhenian shelf, where they developed under high sediment supply conditions and on a high-gradient shelf (Chiocci et al. 1989; Martorelli et al. 2010). Similarly, the central Adriatic shelf is characterized by thick muddy parasequences in response to its higher gradient than the northern counterpart (Cattaneo & Trincardi 1999). By contrast, low-gradient areas (e.g. <0.2°), such as most of the central-northern Tyrrhenian shelves and the northern Adriatic, are typically characterized by thin deposits or transgressive lags. The low-gradient areas are also characterized by sand ridges and dunes derived from drowned transgressive deposits reworked by marine processes.

In general, the relevance of local controlling factors, such as the morphology of the basal surface (e.g. the presence of topographical lows, and abrupt changes in slope gradient and headlands), emerges as a key factor for the formation and preservation of TST deposits, especially on morphologically complex areas such as the southern Tyrrhenian shelf.

Regarding the stratigraphic response to the last sea-level rise, the North Adriatic and South Tyrrhenian shelves therefore can be considered as two end members, as the high-gradient/high-sediment supply of the South Tyrrhenian shelves supported the formation and preservation of thick deposits, whereas very-low-gradient/low-sediment supply of the North Adriatic shelf enhanced transgressive reworking. Moreover, in this last setting, transgressive reworking was maximized as a result of the progressive intensification of the oceanographic regime, in response to the widening of the basin and increased wind fetch (Correggiari et al. 1996).

Table 12.2. *Main characteristics of Late Quaternary shelf deposits analysed in this study, expressed in terms of their systems tracts, related to sediment supply, shelf width and gradient*

Systems tract	Deposit/facies	Distribution	Area	Sediment supply	Shelf width (km)	Shelf gradient (°)
FSST	Muddy wedges and muddy progradational deposits – *distal marine facies*	Limited	Central Adriatic; central Tyrrhenian	High	30–70	0.2
	Sandy wedges – *marine facies*	Limited	Central-northern Tyrrhenian (Elba Island)	Medium–high	15–40	0.2–1
	Inner part of sandy SDTs* – *inner shelf facies*	Limited	Tyrrhenian (Pontine–Capraia)	Low–starved	2–15	0.2–1.5
LST	Outer part of sandy SDT – *inner shelf facies*	Limited	Central-southern Tyrrhenian; Tyrrhenian islands; Ionian	Medium–starved	2–20	0.2–1.5
	Palaeochannel-fill – *fluvial facies*	Limited	Central-northern Tyrrhenian; northern Adriatic	Medium–low	20–300	0.02–0.3
	Deltaic wedges	Limited	Adriatic (Po river system)	High	300	0.02
TST	Sandy parasequences	Limited	Southern Tyrrhenian; Liguria	High	<10	>1
	Muddy–sandy parasequences	Limited	Central Adriatic; central Tyrrhenian	High	20–40	0.2
	Barrier–lagoon	Limited	Northern Adriatic	Low	300	0.02
	Transgressive lag	Wide	Central-northern Adriatic; Tyrrhenian	Low	–	–
	Palaeochannel-fill; sand ridges; reworked dunes; headland-attached transgressive deposits*	Limited	Central Tyrrhenian; northern Adriatic	Medium	15–300	0.02–0.4
HST	Shelf mud	Wide	Everywhere in non-starved areas	High	–	–
	Deltaic bodies	Offshore main rivers	Adriatic; Tyrrhenian; Ionian	High	–	–
	Coastal prisms (non-deltaic)	Limited	Tyrrhenian	Medium–low	2–20	0.2
	Barrier–lagoon (non-deltaic)	Limited	Northern Adriatic; Tyrrhenian	Medium	20–300	0.02–0.2
	Starved	Limited	Tyrrhenian islands; Apulia; northern Adriatic	Very low	–	–

HST deposits

The Highstand Systems Tract (HST) deposits developed since the sea level gradually approached the present-day position and are mostly composed of a variable thickness (from a few metres up to 40–50 m) of muddy sediment supplied by rivers. HST deposits show a lower variability than transgressive and forced regressive deposits as they are unaffected by the reworking processes caused by shoreline migration.

The thickness of HST deposits may reflect either the dominance of a point-sourced or linear-sourced fluvial input. This last case is well represented by the HST deposits of the southern Tyrrhenian and western Ionian shelves; these shelves are very narrow (<10 km), steep (*c.* 1°) and are fed by a set of 'fiumare' (ephemeral but powerful streams) flowing through high-relief terrain. In such settings, HST deposits with a thickness comparable to those of the major subaqueous deltaic systems can form (Martorelli *et al.* 2010).

Along deltaic coasts, the submarine portion of deltaic depositional systems represents the main HST deposit. The thickness of the prodelta muddy wedges is mainly controlled by sediment supply; however, redistribution processes operated by geostrophic currents are able to disperse fine sediment downstream of the river mouth over a large area. This situation is well demonstrated both by the Latium shelf, where Tiber mud is dispersed several tens of kilometres towards the NW (Bellotti *et al.* 1994), and the NW Adriatic shelf, where muddy sediments from the Po and Apennine rivers are redistributed for several tens of kilometres to the SE (Correggiari *et al.* 2001; Cattaneo *et al.* 2003).

Along non-deltaic coasts, inner shelf deposits vary in response to the migration trend of the shoreline. Sandy deposits of the modern coastal prism are characteristic of prograding areas (e.g. sectors of the Latium and Pontine coasts), whereas

barrier–lagoon systems are typical of transgressive shorelines characterized by low gradient and sediment supply (e.g. the northern Adriatic, and some sectors of the Latium and Tuscan shelves).

Far from river mouths, the mud blanket thins to a layer a few centimetres thick. In areas where it is absent, the shelf is floored by relict sediments (e.g. transgressive or lowstand–falling stage deposits). This situation is typical of the middle–outer sector of wide shelves. Other starved shelves are those facing coastal lagoons or marshland (i.e. almost the entire northern Adriatic) or rimming islands without significant watercourses. These shelves are precious environments as they usually host extensive and luxuriant meadows of *Posidonia oceanica*, and coastal detritic-*Coralligène* assemblages (Pérès & Picard 1964), colonizing both

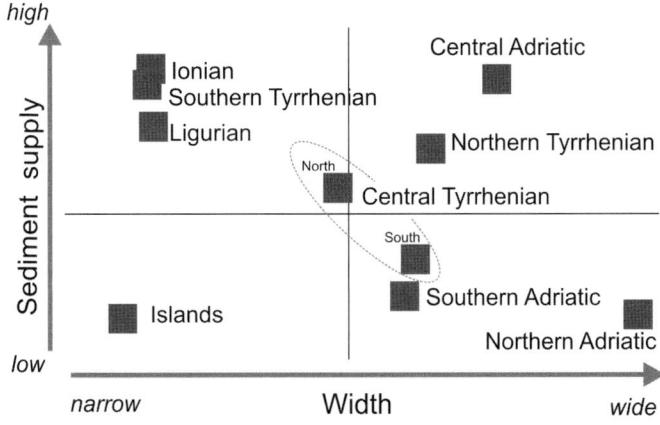

Fig. 12.8. Classification of the shelves analysed in this study based on sediment supply and shelf width.

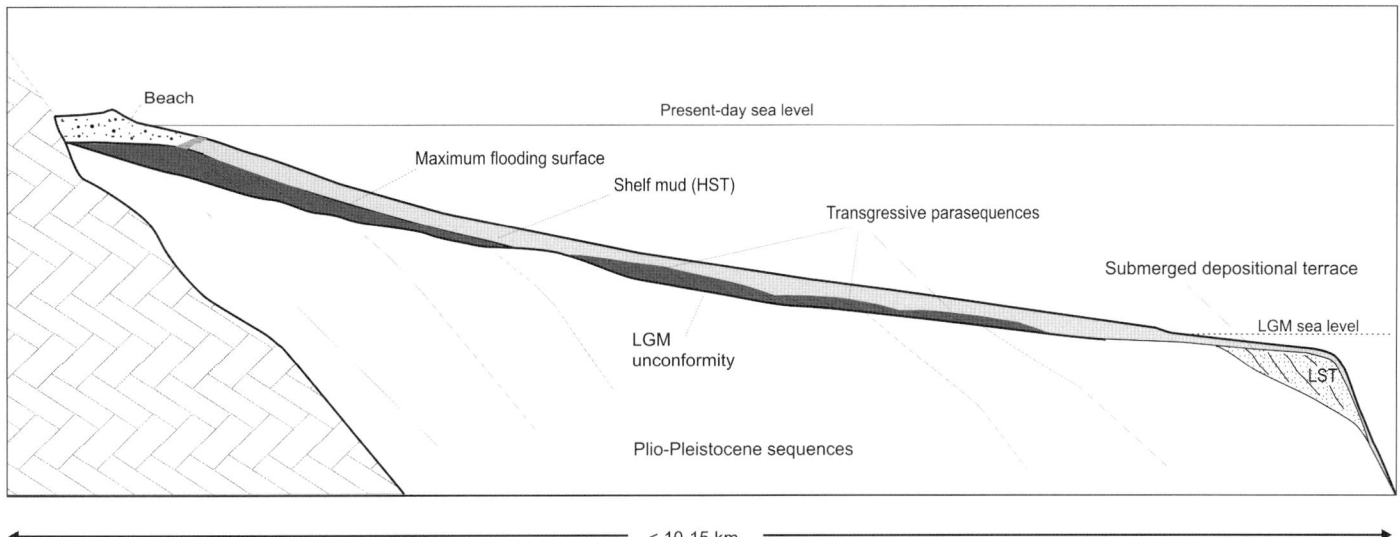

Fig. 12.9. Sketch of typical shelf deposits formed during the last glacio-eustatic cycle on narrow shelves with a high sediment supply.

rocky substrates and detrital sediments; these biocenoses are conversely underdeveloped off the mouths of main rivers as they experience turbid waters.

Types of stratigraphic architecture of Italian shelves

Based on the sediment supply and shelf physiography variability (Fig. 12.8), four types of shelves are identified:

- narrow shelves with a high sediment supply from linear sources;
- narrow shelves with a low sediment supply and intrabasinal sedimentation (usually islands);
- wide shelves with a low sediment supply related to semi-arid/karstic hinterlands or the presence of coastal lagoons and marshlands;
- wide shelves with a high sediment supply from large rivers.

Narrow shelves with a high sediment supply (Fig. 12.9)

The southern Tyrrhenian, western Ionian and sectors of the Ligurian shelf are characterized by very steep slopes ($>1°$), a limited width (a few kilometres) and high sediment supply provided by ephemeral but powerful watercourses, draining coastal ranges with a high relief close to the shoreline. Therefore, the shelf is usually covered by highstand mud that reaches the shelf break, blanketing the whole shelf with a significant thickness (tens of metres in the inner–middle sector). The coastal sector is characterized by coarse-grained (sand and gravel) beaches between rocky cliffs. Transgressive deposits are usually well developed, reaching the maximum thickness for Italian shelves (c. 35 m), owing to the lack of a coastal plain able to retain sediment during the sea-level rise. As a result, sandy parasequences with a retrogradational stacking pattern formed, generally as thick as the HST parasequences. FSST mainly develop on the continental slope. On the shelf margin, lowstand deposits are commonly present, forming well-developed SDTs characterized by a notable sediment thickness (c. 25–30 m).

Narrow shelves with a low sediment supply (Fig 12.10)

Most of the Italian islands are rimmed by narrow (a few kilometres wide), steep ($>0.5–1°$) and morphologically complex shelves. The lack of fluvial input, as well as the high hydrodynamic

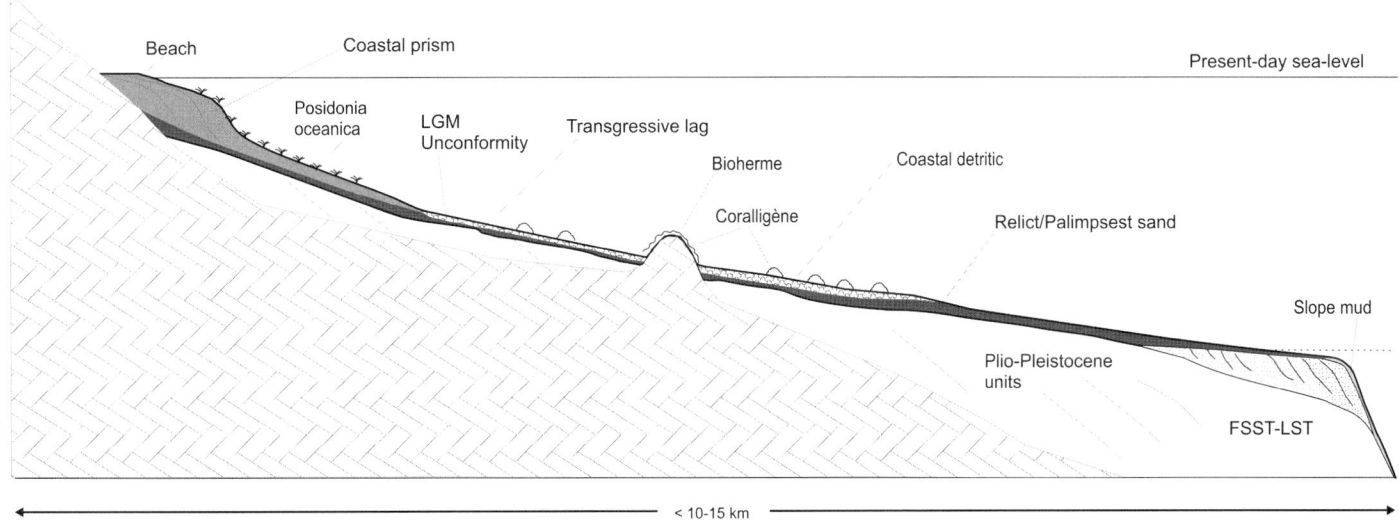

Fig. 12.10. Sketch of typical shelf deposits formed during the last glacio-eustatic cycle on narrow shelves with a low sediment supply (sediment-starved areas).

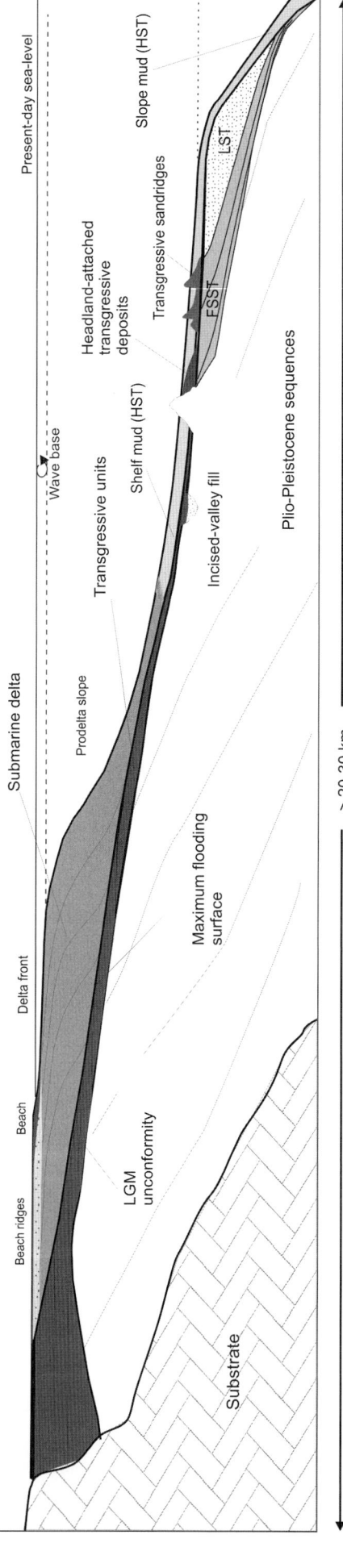

Fig.12.11. Sketch of typical shelf deposits formed during the last glacio-eustatic cycle on wide shelves with a high sediment supply.

regime, allows these shelves to be almost mud-free. Along the coast, sandy or pebble pocket beaches are present along an otherwise retreating cliffed coast. On the inner shelf, small coastal prisms are locally present, they are mainly composed of terrigenous sediments derived from cliff erosion and then swept offshore by storm currents. On the middle and outer shelf, HST and TST deposits of the last eustatic cycle are very thin or absent.

Shelf sediments are mostly composed of bioclastic sands produced by benthic communities related to *Posidonia oceanica* in the inner shelf, and to coastal detritic and *Coralligène* on the middle–outer shelf. The lack of muddy sedimentation supports the development of *Coralligène* that widely colonizes the hard rocky substratum, producing bioherms that control the morphology of the seafloor. Bioclastic sediments are thin and clustered on the seafloor. This small amount of highstand sediment is mixed with relict sediment characterized by reworked grains. In such settings, SDTs commonly represent the bulk of the LQDS.

Wide shelves with a low sediment supply

Wide shelves that are not directly fed by rivers show a situation similar to the previous one, with enhanced bioclastic sedimentation and relict sediment cropping out on the seafloor. However, in this case the morphological control produced by the exposed rocky substrate is usually absent. Sediment-starved shelves may occur offshore of coastal lagoons or marshlands, where fine sediment is trapped onshore (e.g. southern Latium, the Adriatic north of the modern Po delta) and offshore semi-arid regions or regions with karst hinterland (e.g. Apulia). In addition, areas where deltaic sediments are confined to the inner–middle shelf by geostrophic circulation (e.g. the Adriatic south of the modern Po delta) can develop a sediment-starved condition.

Most of the North Adriatic shelf can be considered an extreme case of a sediment-starved system during the post-LGM period. In fact, post-LGM shelf stratigraphy encompasses thin or absent highstand deposits, drowned barrier–lagoon systems and transgressive lags. Conversely, the very different setting characteristic of the glacial interval supported the formation of a huge prograding deposit built by the palaeo-Po river system, as well as the incision of several fluvial valleys on the exposed shelf. The different stratigraphic architecture of the North Adriatic Sea derives from the peculiar characteristics of a semi-enclosed epicontinental sea that enhanced the role of specific factors, such as extreme changes in basin extent, sediment supply and oceanographic regime.

Wide shelves with high sediment supply (Fig. 12.11)

The central Adriatic and central-northern Tyrrhenian shelves are several tens of kilometres wide, and have gentle gradients (<0.4°). They are fed by large rivers, producing delta systems (e.g. Volturno, Tiber, Ombrone), or by several medium-sized rivers (e.g. Apennines rivers for the central Adriatic), and sandy beaches are usually present. This setting causes a mud-covered shelf even if, on wider shelves, mud distribution terminates before the shelf break, causing relict sediment to rest on the seafloor.

TST deposits are rare, and occur locally as a result of sediment availability and favourable conditions related to the morphology of the transgressed surface. Transgressive sand ridges and headland-attached deposits typically occur in areas far from the delta systems, whereas in offshore major deltas they are replaced by condensed units with an acoustically transparent seismic facies. In fact, during the sea-level rise, watercourse bedload was trapped in coastal lagoons developing within fluvial valleys; only finer sediments were able to escape the lagoon, forming condensed units on the shelf.

The LST systems tract is locally present with shelf-margin prograding wedges. FSST deposits are mainly represented by muddy progradational units composed of distal marine facies located on the outer shelf.

Conclusions

Italy is a geologically young and complex region where very different geological–geodynamic settings are present over short lateral distances. Such complexity is mirrored by the variability of morphological and stratigraphic architecture of the continental shelves, and can be used to unravel the role of the different factors controlling their evolution. Late Quaternary sedimentary deposits show great variability in lithology, thickness, external form and internal stratigraphy, as well as volumetric and spatial relationships between systems tracts. Within the Late Quaternary depositional sequence, the transgressive systems tract shows the maximum variability. The 'end members' that have been defined encompass: (1) wide and well-fed shelves typical of nearly stable margins, such as the central Adriatic and the central-northern Tyrrhenian seas; (2) wide but sediment-starved shelves that match margins with a semi-arid/karst hinterland or with coastal lagoons and marshlands, such as the Apulia and spots of margin elsewhere; (3) narrow and well-fed shelves corresponding to very young and geologically active margins with coastal ranges close to the shoreline, such as most of the southern Italy; and (4) narrow and sediment-starved shelves usually corresponding to shelves rimming volcanic islands.

This scheme is very general and it is worth noting that, in detail, local factors (such as the distance from river mouths, oceanographic conditions, attributes of the antecedent topography, and specific palaeomorphological and palaeogeographical settings) may play a central role in determining the Late Quaternary stratal architecture and facies. The influence of local factors seems to be greater for the transgressive systems tract than for the highstand or lowstand systems tracts. The falling stage and lowstand systems tracts are rather more dependent on vertical movement (subsidence/uplift) of the continental margin, which that favours or hinders their preservation on the shelf. The northern Adriatic shelf represents a very specific type, which differs significantly from other Italian–Mediterranean shelves as it developed on a semi-enclosed epicontinental sea that developed under the control of specific factors, such as extreme changes in basin extent, sediment supply and oceanographic regime.

The proposed models of the architecture of the Italian shelves may provide a useful scheme for the stratigraphic interpretation of Late Quaternary shelf deposits formed under a microtidal, wave-dominated and temperate climate regime.

We acknowledge the editor, A. Chivas, for helpful comments on the manuscript. The critical reviews by three reviewers, G. Ercilla, P. Lobo and V. Lykousis, were greatly appreciated and significantly improved the manuscript. We wish to also thank D. Ridente for fruitful suggestions. Part of this work has been developed within the framework of the Flagship Project RITMARE (SP4-WP1-Action 3), coordinated by the Italian National Research Council and funded by the Italian Ministry of Education, University and Research.

References

AIELLO, G., ANGELINO, A., D'ARGENIO, B., MARSELLA, E., PELOSI, N., RUGGIERI, S. & SINISCALCHI, A. 2005. Buried volcanic structures in the Gulf of Naples (Southern Tyrrhenian sea, Italy) resulting from high resolution magnetic survey and seismic profiling. *Annals of Geophysics*, **48**, 1–15.

AMOROSI, A. & MILLI, S. 2001. Late Quaternary depositional architecture of Po and Tevere river deltas (Italy) and worldwide comparison with coeval deltaic successions. *Sedimentary Geology*, **144**, 357–375.

ARGNANI, A., FAVALI, P. *ET AL.* 1993. *Foreland deformational pattern in the southern Adriatic Sea. Annals of Geophysics*, **36**, 229–247.

BARD, E., HAMELIN, B., ARNOLD, M., MONTAGGIONI, L., CABIOCH, G. & FAURE, G. 1996. Deglacial sea-level record from Tahiti corals and the timing of global meltwater discharge. *Nature*, **382**, 241–244.

BARD, E., HAMELIN, B., DELANGHE, R. & SABATIER, D. 2010. Deglacial Meltwater Pulse 1B and Younger Dryas levels revisited with boreholes at Tahiti. *Science*, **327**, 1235–1237.

BELKNAP, D. F. & KRAFT, J. C. 1981. Preservation Potential of Transgressive Coastal Lithosomes on the U. S. Atlantic Shelf. *Marine Geology*, **42**, 429–442.

BELLOTTI, P., CHIOCCI, F. L., MILLI, S., TORTORA, P. & VALERI, P. 1994. Sequence Stratigraphy and depositional setting of the Tiber Delta: integration of high-resolution seismics, well-log and archaeological data. *Journal of Sedimentary Research*, **B64**, 416–432.

BELLOTTI, P., MILLI, S., TORTORA, P. & VALERI, P. 1995. Physical stratigraphy and sedimentology of the Late Pleistocene–Holocene Tiber Delta depositional sequence. *Sedimentology*, **42**, 617–634.

BOZZANO, A., CORRADI, N., FANUCCI, F. & IVALDI, R. 2006. Late Quaternary deposits from the Ligurian continental shelf (NW Mediterranean): a response to problems of coastal erosion. *Chemistry and Ecology*, **22**(Suppl. 1), S349–S359.

BUDILLON, F., CRISTOFALO, G. & TONIELLI, R. 2004. Terrazzi deposizionali sommersi in Penisola Sorrentina (Campania). *In*: CHIOCCI, F. L., D'ANGELO, S. & ROMAGNOLI, C. (eds) *Atlante dei Terrazzi Deposizionali Sommersi lungo le coste italiane*. Memorie Descrittive della Carta Geologica d'Italia, **LVIII**, 49–56.

CATTANEO, A. & STEEL, R. J. 2003. Transgressive deposits: a review of their variability. *Earth-Science Reviews*, **62**, 187–228.

CATTANEO, A. & TRINCARDI, A. 1999. The late-Quaternary transgressive record in the Adriatic Epicontinental Sea: basin widening and facies partitioning. *In*: BERGMAN, K. & SNEDDEN, J. (eds) *Isolated Shallow Marine Sand Bodies: Sequence Stratigraphic Analysis and Sedimentologic Interpretation*. International Association of Sedimentologists, Special Publications, **64**, 127–146.

CATTANEO, A., TRINCARDI, F. & ASIOLI, A. 1997. Shelf sediment dispersal in the late-Quaternary trasgressive record around the Tremiti High. *Giornale di Geologia*, **59**, 217–244.

CATTANEO, A., CORREGGIARI, A., LANGONE, L. & TRINCARDI, F. 2003. The late-Holocene Gargano subaqueous delta, Adriatic shelf: sediment pathways and supply fluctuations. *Marine Geology*, **193**, 61–91.

CHAPPELL, J. & SHACKLETON, N. J. 1986. *Oxygen isotopes and sea level*. Nature, **324**, 137–140.

CHIOCCI, F. L. 2000. Depositional response to Quaternary fourth-order sea-level fluctuations on the Latium margin (Tyrrhenian Sea, Italy). *In*: HUNT, D. & GAWTHORPE, R. L. (eds) *Sedimentary Responses to Forced Regressions*. Geological Society, London, Special Publications, **172**, 271–290.

CHIOCCI, F. L. & LA MONICA, G. B. 1996. Analisi sismostratigrafica della piattaforma continentale. *In*: *Il Mare del Lazio*. Università 'La Sapienza' di Roma, Regione Lazio, 41–61.

CHIOCCI, F. L. & MILLI, S. 1995. Construction of a chronostratigraphic diagram for a high-frequency sequence: the 20 ky B.P. to present Tiber depositional sequence. *Il Quaternario, Italian Journal of Quaternary Sciences*, **8**, 339–348.

CHIOCCI, F. L. & ORLANDO, L. 1995. Effects of high-frequency Pleistocene sea-level changes on a highly deforming continental margin: Calabrian shelf (southern Tyrrhenian Sea, Italy). *Bollettino di Geofisica Teorica e Applicata*, **37**, 39–58.

CHIOCCI, F. L. & ORLANDO, L. 1996. *Lowstand terraces on Tyrrhenian Sea steep continental slopes*. Marine Geology, **134**, 127–143.

CHIOCCI, F. L., D'ANGELO, S., ORLANDO, L. & PANTALEONE, A. 1989. Evolution of the Holocene shelf sedimentation defined by high-resolution seismic stratigraphy and sequence analysis (Calabro-Tyrrhenian continental shelf). *Memorie della Società Geologica Italiana*, **48**, 359–380.

CHIOCCI, F. L., ORLANDO, L. & TORTORA, P. 1991. Small-scale seismic stratigraphy and paleogeographical evolution of the continental shelf facing the SE Elba island (northern Tyrrhenian Sea, Italy). *Journal of Sedimentary Petrology*, **61**, 506–526.

CHIOCCI, F. L., ESU, F., TOMMASI, P. & CHIAPPA, V. 1996. Stability analysis of the submarine slope of the Tiber River delta. *In*: SENNESET, K. (ed.) *Landslides – Glissements de terrain*. Balkema, Rotterdam, 521–526.

CHIOCCI, F. L., ERCILLA, G. & TORRES, J. 1997. Stratal architecture of Western Mediterranean Margins as the result of the stacking of Quaternary lowstand deposits below glacio-eustatic fluctuation base-level. *Sedimentary Geology*, **112**, 195–217.

CHIOCCI, F. L., D'ANGELO, S. & ROMAGNOLI, C. (eds) 2004. *Atlante dei Terrazzi Deposizionali Sommersi lungo le coste italiane*. Memorie Descrittive della Carta Geologica d'Italia, **LVIII**.

CORRADI, N., FANUCCI, F., FIERRO, G., FIRPO, M., PICAZZO, M. & MIRABILE, L. 1984. *La piattaforma continentale ligure: caratteri, struttura ed evoluzione*. Rapporto Tecnico Finale del Progetto Finalizzato: 'Oceanografia e Fondi Marini'. CNR, Rome.

CORRADI, N., IVALDI, R., BALDUZZI, I. & BOZZANO, A. 2004. La ricerca delle sabbie sulla piattaforma continentale ligure: campagna di geologia marina per la localizzazione di depositi sedimentari idonei al ripascimento dei litorali. *In*: *La ricerca di sabbie nel Mar Ligure*. Regione Liguria, Genova, 29–59.

CORREGGIARI, A., FIELD, M. E. & TRINCARDI, F. 1996. Late Quaternary transgressive large dunes on the sediment-starved Adriatic shelf. *In*: DE BATIST, M. & JACOBS, P. (eds) *Geology of Siliciclastic Shelf Seas*. Geological Society, London, Special Publications, **117**, 155–169.

CORREGGIARI, A., TRINCARDI, F., LANGONE, L. & ROVERI, M. 2001. Styles of failure in late Holocene highstand prodelta wedges on the Adriatic shelf. *Journal of Sedimentary Research*, **71**, 218–236.

CORREGGIARI, A., CATTANEO, A. & TRINCARDI, F. 2005. The modern Po Delta system: Lobe switching and asymmetric prodelta growth. *Marine Geology*, **222–223**, 49–74.

COWELL, P. J., ROY, P. S. & JONES, R. A. 1995. Simulation of large-scale coastal change using a morphological behavior model. *Marine Geology*, **126**, 45–61.

CRITELLI, S. & LE PERA, E. 1998. Post-Oligocene sediment dispersal systems and unroofing history of the Calabrian microplate, Italy. *International Geology Review*, **40**, 609–637.

DE PIPPO, T. & PENNETTA, M. 2004. Terrazzi deposizionali sommersi nel Golfo di Policastro (Campania). *In*: CHIOCCI, F. L., D'ANGELO, S. & ROMAGNOLI, C. (eds) *Atlante dei Terrazzi Deposizionali Sommersi lungo le coste italiane*. Memorie Descrittive della Carta Geologica d'Italia, **LVIII**, 57–62.

DEMAREST, J. M. & KRAFT, J. C. 1987. Stratigraphic record of Quaternary sea levels: implications for more ancient strata. *In*: NUMMEDAL, D., PILKEY, O. H. & HOWARD, J. D. (eds) *Sea Level Fluctuation and Coastal Evolution*. SEPM Special Publications, Tulsa, **41**, 223–240.

DESCHAMPS, P., DURAND, N. *ET AL.* 2012. Ice sheet collapse and sea-level rise at the Bølling warming, 14,600 yr ago. *Nature*, **483**, 559–564.

DI BELLA, L., FREZZA, V. *ET AL.* 2014. Foraminiferal record and high-resolution seismic stratigraphy of the late Holocene succession of the submerged Ombrone River delta (Northern Tyrrhenian Sea, Italy). *In*: PASCUCCI, V., COLTORTI, M. & PIERUCCINI, P. (eds) *SEQS 2012 Sardinia: at the Edge of the Sea. Quaternary International*, **328–329**, 287–300.

ERCILLA, G. & ALONSO, B. 1996. Modern siliciclastic shelves: architecture, sea level, tectonics and sediment supply. *In*: DE BATIST, M. & JACOBS, P. (eds) *Geology of Siliciclastic Shelf Seas*. Geological Society, London, Special Publications, **117**, 125–137.

FAIRBANKS, R. G. 1989. 17 000-year glacio-eustatic sea level record: influence of glacial melting rates on the Younger Dryas event and deep-ocean circulation. *Nature*, **342**, 637–642.

FALESE, F. G., MARTORELLI, E. & MINORENTI, V. 2008. Mitigation of coastal erosion by the use of marine sands: examples from the Tyrrhenian Sea continental shelves. *Rendiconti della Società Geologica Italiana*, **3**, 363–364.

FANUCCI, F., FIERRO, G. & REHAULT, J. P. 1974. Evoluzione quaternaria della piattaforma continentale ligure. *Memorie della Società Geologica Italiana*, **13**, 233–240.

FERRANTI, L., ANTONIOLI, F. *ET AL.* 2006. Markers of the last interglacial sea-level high stand along the coast of Italy: Tectonic implications. *Quaternary International*, **145–146**, 30–54.

FIERRO, G., CORRADI, N. ET AL. 2010. *La géologie sous-marine de la Mer Ligure: une synthèse.* Bulletin Société Géographique de Liège, **54**, 31–40.

HELLAND-HANSEN, W. & GJELBERG, J. G. 1994. Conceptual basis and variability in sequence stratigraphy: a different perspective. *Sedimentary Geology*, **92**, 31–52.

HELLAND-HANSEN, W. & MARTINSEN, O. J. 1996. Shoreline trajectories and sequence: description of variable depositional-dip scenarios. *Journal of Sedimentary Research*, **66**, 670–688.

HERNANDEZ-MOLINA, F. J., SOMOZA, L., REY, J. & POMAR, L. 1994. Late Pleistocene–Holocene sediments on the Spanish continental shelves: model for very high resolution sequence stratigraphy. *Marine Geology*, **120**, 129–174.

HERNANDEZ-MOLINA, F. J., FERNANDEZ-SALAS, L. M., LOBO, F., SOMOZA, L., DIAZ-DEL-RIO, V. & ALVEIRINHO DIAS, J. M. 2000. The infralittoral prograding wedge: a new large-scale progradational sedimentary body in shallow marine environments. *Geo-Marine Letters*, **20**, 109–117.

HUNT, D., GAWTHORPE, R. & DOCHERTY, M. (eds) 1995. *Proceedings of the Conference on Sedimentary Responses to Forced Regression: Recognition, Interpretation and Reservoir Potential.* Geological Society, London.

ISTITUTO IDROGRAFICO DELLA MARINA. 1982. *Atlante delle correnti superficiali dei mari italiani.* Istituto Idrografico della Marina, Genova.

LABAUNE, C., TESSON, M. & GENSOUS, B. 2005. Integration of high and very high-resolution seismic reflection profiles to study Upper Quaternary deposits of a coastal area in the western Gulf of Lions, SW France. *Marine Geophysical Researches*, **26**, 109–122.

LARROQUE, C., DE LÉPINAY, B. M. & MIGEON, S. 2011. Morphotectonic and fault–earthquake relationships along the northern Ligurian margin (western Mediterranean) based on high resolution, multibeam bathymetry and multichannel seismic-reflection profiles. *Marine Geophysical Research*, **32**, 163–179.

LOBO, F. J., TESSON, M. & GENSOUS, B. 2004. Stratal architectures of late Quaternary regressive–transgressive cycles in the Roussillon Shelf (SW Gulf of Lions, France). *Marine and Petroleum Geology*, **21**, 1181–1203.

MARANI, M., TAVIANI, M., TRINCARDI, F., ARGNANI, A., BORSETTI, A. M. & ZITELLINI, N. 1986. Pleistocene progradation and post-glacial events of the NE Tyrrhenian continental shelf between the Tiber river delta and Capo Circeo. *Memorie della Società Geologica Italiana*, **36**, 67–89.

MARTORELLI, E., PIERALISI, F. & CHIOCCI, F. L. 2008. Effects of changes in volcanic activity on post LGM seismic stratigraphy: example from the northern Campanian Shelf (Tyrrhenian Sea, Italy). Paper presented at the 33rd International Geological Congress, Symposium: EME-09 Risks, Resources, and Record of the Past on the Continental Shelf, Oslo, Norway, 6–14 August 2008.

MARTORELLI, E., CHIOCCI, F. L. & ORLANDO, L. 2010. Imaging continental shelf shallow stratigraphy by using different high-resolution seismic sources: an example from the Calabro-Tyrrhenian margin (Mediterranean Sea). *Brazilian Journal of Oceanography*, **58** (Special Issue, IGCP 526), 55–66.

MARTORELLI, E., D'ANGELO, S., FIORENTINO, A. & CHIOCCI, F. L. 2011. Non-tropical carbonate shelf sedimentation. The Archipelago Pontino (Central Italy) case history. *In*: HARRIS, P. T. & BAKER, E. K. (eds) *Seafloor Geomorphology as Benthic Habitat: GeoHab Atlas of Seafloor Geomorphic Features and Benthic Habitats.* Elsevier Insights, Amsterdam, 449–455.

MCBRIDE, R. A. & MOSLOW, T. F. 1991. Origin, evolution and distribution of shoreface sand ridges, Atlantic inner shelf, USA. *Marine Geology*, **97**, 57–85.

MEIJER, X. D., POSTMA, G., BURROUGH, P. A. & DE BOER, P. L. 2007. Modelling the preservation of sedimentary deposits on passive continental margins during glacial–interglacial cycles. *In*: DE BOER, P. L., POSTMA, G., VAN DER ZWAN, C. J., BURGESS, P. M. & KUKLA, P. (eds) *Analogue and Numerical Forward Modelling of Sedimentary Systems; from Understanding to Prediction.* International Association of Sedimentologists, Special Publications, **39**, 223–238.

MILIA, A. & TORRENTE, M. M. 2000. Fold uplift and syn-kinematic stratal architectures in a region of active transtensional tectonics and volcanism, Eastern Tyrrhenian Sea. *Geological Society of America Bulletin*, **112**, 1531–1542.

NORDFJORD, S., GOFF, J. A., AUSTIN, J. A. JR. & DUNCAN, L. S. 2009. Shallow stratigraphy and complex transgressive ravinement on the New Jersey middle and outer continental shelf. *Marine Geology*, **266**, 232–243.

NUMMEDAL, D. & SWIFT, D. J. P. 1987. Transgressive stratigraphy at sequence-bounding unconformities: some principles derived from Holocene and Cretaceous examples. *In*: NUMMEDAL, D., PILKEY, O. H. & HOWARD, S. D. (eds) *Sea Level Fluctuation and Coastal Evolution.* SEPM Special Publications, Tulsa, **41**, 241–260.

PENNETTA, M. 1985. *Caratteri Granulometrici dei sedimenti del Golfo di Taranto (Alto Ionio).* Estratto da Annali dell'Istituto Universitario Navale di Napoli, **54**.

PENNETTA, M., PESCATORE, T. & SENATORE, M. R. 1986. I tipi di piattaforma continentale del Golfo di Taranto (Alto Jonio, Italia). *In*: Convegno ENEA *'Evoluzione dei litorali', Policoro, 16–17 October 1986.* ENEA, Rome, 195–214.

PÉRÈS, J. M. & PICARD, J. 1964. Nouveau Manuel de Bionomie Benthique de la Mer Mediterranee. *Recueil des Travaux de la Station Marine d'Endoume, France*, **31**, 5–137.

PESCATORE, T. & SENATORE, M. R. 1986. A comparison between a present-day (Taranto Gulf) and a Miocene (Irpinian Basin) foredeep of the Southern Apennines (Italy). *In*: ALLEN, P. A. & HOMEWOOD, P. (eds) *Foreland Basins.* International Association of Sedimentologists, Special Publications, **8**, 169–182.

PLINT, A. G. & NUMMEDAL, D. 2000. The falling stage systems tract: recognition and importance in sequence stratigraphic analysis. *In*: HUNT, D. & GAWTHORPE, R. (eds) *Sedimentary Responses to Forced Regressions.* Geological Society, London, Special Publications, **172**, 1–17.

POSAMENTIER, H. W. & VAIL, P. R. 1988. Eustatic controls on clastic deposition: II. Sequences and systems tract model. *In*: WILGUS, C. K., HASTINGS, B. S., KENDALL, C. G. ST. C., POSAMENTIER, H. W., ROSS, C. A. & VAN WAGONER, J. C. (eds) *Sea-Level Changes – An Integrated Approach.* SEPM Special Publications, Tulsa, **42**, 125–154.

POSAMENTIER, H. W., ALLEN, G. P., JAMES, D. P. & TESSON, M. 1992. Forced regressions in a sequence stratigraphic framework: concepts, examples, and exploration significance. *American Association of Petroleum Geologists Bulletin*, **76**, 1687–1709.

RAMPINO, M. R. & SANDERS, J. E. 1980. Holocene Transgression in South-Central Long Island, New York. *Journal of Sedimentary Petrology*, **50**, 1063–1079.

RIDENTE, D. & TRINCARDI, F. 2002. Eustatic and tectonic control on deposition and lateral variability of Quaternary regressive sequences in the Adriatic basin. *Marine Geology*, **184**, 273–293.

RIDENTE, D. & TRINCARDI, F. 2005. Pleistocene 'muddy' forced-regression deposits on the Adriatic shelf: a comparison with prodelta deposits of the late Holocene highstand mud wedge. *Marine Geology*, **222–223**, 213–233.

RIDENTE, D., TRINCARDI, F., PIVA, A., ASIOLI, A. & CATTANEO, A. 2008. Sedimentary response to climate and sea level changes during the past 400 ka from borehole PRAD1.2 (Adriatic margin). *Geochemistry, Geophysics, Geosystems*, **9**, Q09R04.

RIDENTE, D., TRINCARDI, F., PIVA, A. & ASIOLI, A. 2009. The combined effect of sea level and supply during Milankovitch cyclicity: evidence from shallow-marine $\delta^{18}O$ records and sequence architecture (Adriatic margin). *Geology*, **37**, 1003–1006.

RIDENTE, D., PETRUNGARO, R., FALESE, F. & CHIOCCI, F. 2012. Middle–Upper Pleistocene record of 100-ka depositional cycles on the Southern Tuscany continental margin (Tyrrhenian Sea, Italy). Sequence architecture and margin growth pattern. *Marine Geology*, **326–328**, 1–13.

ROSSI, S. & SARTORI, R. 1981. A seismic reflection study of the external Calabrian Arc in the N Ionian Sea (eastern Mediterranean). *Marine Geophysical Research*, **4**, 403–426.

ROVERI, M. & CORREGGIARI, A. 2004. Submerged depositional terraces in the Tuscan Archipelago (Eastern margin of the Corsica Basin). *Memorie Descrittive della Carta Geologica*, **58**, 11–15.

ROY, P. S., COWELL, P. J., FERLAND, M. A. & THOM, B. G. 1994. Wave-dominated coasts. *In*: CARTER, R. W. G. & WOODROFFE,

C. D. (eds) *Coastal Evolution*. Cambridge University Press, Cambridge, 121–186.

SANDERS, J. E. & KUMAR, N. 1975. Evidence of shoreface retreat and in-place 'drowning' during Holocene submergence of barriers, shelf off Fire Island, New York. *Geological Society of America Bulletin*, **86**, 65–76.

SAVELLI, D. & WEZEL, F. C. 1980. Morphological map of the Tyrrhenian Sea. *In: C.N.R., P.F. Oceanografia e Fondi Marini (Theme ≪Bacini Sedimentari≫) (Scale 1:1 250 000)*. Litografia Artistica Cartografica, Firenze.

SELLI, R. 1970. Cenni morfologici generali sul Mar Tirreno. *Giornale di Geologia*, **37**, 4–29.

SENATORE, M. R. 1987. Caratteri sedimentari e tettonici di un bacino di avanfossa. Il Golfo di Taranto. *Memorie della Società Geologica Italiana*, **38**, 177–204.

SENATORE, M. R. 2004. Submerged depositional terraces along the Ionian margin of Puglia. *Memorie Descrittive della Carta Geologica*, **58**, 141–146.

SENATORE, M. R., MIRABILE, L., PESCATORE, T. & TRAMUTOLI, M. 1980. La piattaforma continentale del settore nord-orientale del Golfo di Taranto (Piattaforma Pugliese). *Geologia Applicata e Idrogeologia*, **15**, 33–50.

SIDDALL, M., ROHLING, E. J., ALMOGI-LABIN, A., HEMLEBEN, CH., MEISCHNER, D., SCHMELZER, I. & SMEED, D. A. 2003. Sealevel fluctuations during the last glacial cycle. *Nature*, **423**, 853–858.

SIDDALL, M., CHAPPELL, J. & POTTER, E.-K. 2006. Eustatic sea level during past interglacials. *In*: SIROCKO, F., CLAUSSEN, M., SANCHEZ GONI, M. F. & LITT, T. (eds) *The Climate of Past Interglacials*. Elsevier, Amsterdam, 75–92.

SNEDDEN, J. W., KREISA, R. D., TILLMAN, R. W., CULVER, S. J. & SCHWELLER, W. J. 1999. An expanded model for modern shelf sand ridge genesis and evolution on the New Jersey Atlantic shelf. *In*: BERGMAN, K. & SNEDDEN, J. (eds) *Isolated Shallow Marine Sand Bodies: Sequence Stratigraphic Analysis and Sedimentologic Interpretation*. International Association of Sedimentologists, Special Publications, **64**, 147–163.

STANLEY, D. J. & WARNE, A. G. 1994. Worldwide initiation of Holocene marine deltas by deceleration of sea-level rise. *Science*, **265**, 228–231.

STORMS, J. E. A., WELTJE, G. J., TERRA, G. J., CATTANEO, A. & TRINCARDI, F. 2008. Coastal dynamics under conditions of rapid sea-level rise: Late Pleistocene to Early Holocene evolution of barrier–lagoon systems on the northern Adriatic shelf (Italy). *Quaternary Science Reviews*, **27**, 1107–1123.

SWIFT, D. J. P. 1968. Coastal erosion and transgressive stratigraphy. *Journal of Geology*, **76**, 444–456.

SWIFT, D. J. P. & FIELD, M. E. 1981. Evolution of a classic sand ridge field: Maryland sector, North American inner shelf. *Sedimentology*, **28**, 461–482.

TESSAROLO, C., MALINVERNO, E., AGATE, M., DI GRIGOLI, G. & CORSELLI, C. 2008. Preliminary data concerning the morphology of a Calabrian Ionian margin area: Caulonia and Marina di Gioiosa canyons. *Chemistry and Ecology*, **24**, (S1), 225–242.

TORTORA, P. 1989. La sedimentazione olocenica nella piattaforma continentale interna tra il promontorio di M. Argentario e la foce del Fiume Mignone (Tirreno centrale). *Giornale di Geologia*, **51**, 93–117.

TORTORA, P. 1996. Depositional and erosional coastal processes during the last postglacial sea-level rise: an example from the central Tyrrhenian continental shelf (Italy). *Journal of Sedimentary Research*, **66**, 391–405.

TORTORA, P., BELLOTTI, P. & VALERI, P. 2001. Late-Pleistocene and Holocene deposition along the coasts and continental shelves of the Italian peninsula. *In*: VAI, G. B. & MARTINI, I. P. (eds) *Anatomy of an Orogen: the Apennines and Adjacent Mediterranean Basins*. Kluwer Academic, London, 455–478.

TOSCANO, F. & SORGENTE, B. 2002. Rhodalgal–Bryomol temperate carbonates from the Apulian shelf (Southeastern Italy), relict and modern deposits on a current dominated shelf. *Facies*, **46**, 103–118.

TRINCARDI, F. & CORREGGIARI, A. 2000. Quaternary forced-regression deposits in the Adriatic basin and the record of composite sea level cycles. *In*: HUNT, D. & GAWTHORPE, R. (eds) *Depositional Response to Forced Regression*. Geological Society, London, Special Publications, **172**, 245–269.

TRINCARDI, F. & FIELD, M. E. 1991. Geometry, lateral variation, and preservation of downlapping regressive shelf deposits: Eastern Tyrrhenian Sea Margin, Italy. *Journal of Sedimentary Research*, **61** 775–790.

TRINCARDI, F., CORREGGIARI, A. & ROVERI, M. 1994. Late Quaternary transgressive erosion and deposition in a modern epicontinental shelf: the Adriatic semienclosed basin. *Geo-Marine Letters*, **14**, 41–51.

TROWBRIDGE, J. H. 1995. A mechanism for the formation and maintenance of shore-oblique sand ridges on storm-dominated shelves. *Journal of Geophysical Research*, **100**, 16071–16086.

WAELBROECK, C., LABEYRIE, L. *ET AL*. 2002. Sea-level and deep water temperature changes derived from benthic foraminifera isotopic records. *Quaternary Science Reviews*, **21**, 295–305.

Chapter 13

Hellenic shelf: late Quaternary tectonics, sea-level changes, sedimentation and geohazards

G. FERENTINOS[1,2]*, V. LYKOUSIS[1,2], G. PAPATHEODOROU[1,2] & M. IATROU[1,2]

[1]Laboratory of Marine Geology and Physical Oceanography, Department of Geology, University of Patras,
University Campus, 26504 Rio Patras, Greece

[2]Institute of Oceanography, Hellenic Centre for Marine Research (HCMR), PO Box 712,
46.7 km Athens–Sounio Avenue, 19 013 Anavyssos, Greece

*Corresponding author (e-mail: gferen@upatras.gr)

Abstract: The Hellenic shelf lies within and around the Aegean microplate, which is one of the world's most seismically active areas, and has experienced extreme tectonism throughout Quaternary time. This activity, together with eustatic sea-level changes and water-circulation patterns over the same time period, controls the overall configuration of the Hellenic shelf, the rates of uplift and subsidence, and determines the sediment supply and depot centre, as well as the sediment-transfer processes. The above-mentioned geological processes are the causative factors for the frequent occurrence of a variety of geological hazards, such as active faults, submarine gravitational mass movements, tsunami and active gas seeping from the seabed.

The Hellenic shelf lies within and around the Aegean microplate, which is one of the most rapidly deforming areas in the world (Fig. 13.1) (Jackson 1994). It comprises the Aegean, an epicontinental sea (Stanley & Perissoratis 1977), and the western margin of the Hellenic peninsula in the Ionian Sea (Fig. 13.1). The Hellenic shelf can therefore be considered as a natural laboratory for the study of continental tectonics such as fault geometry and kinematics, basin formation and filling processes, sediment-transport processes, and quantitative modelling of erosion and deposition in relation to tectonic uplift and subsidence.

Furthermore, the Hellenic shelf, due to strong and frequently occurring earthquakes, can be considered an ideal area for the study of a variety of geohazards, such as active faults, submarine landslides, tsunami and active gas seeping from the seabed, and their causative mechanisms.

Tectonic framework

The Quaternary active deformation of the Hellenic peninsula, and the surrounding Ionian and Aegean seas' shelf, is the result of: (i) the westward movement of the Anatolian microplate between the North Anatolian Fault (NAF) and the East Anatolian Fault (EAF) (Fig. 13.1a), at a rate of 20 mm a^{-1}, in response to the northward collision of the Arabian plate into the Eurasian plate; and (ii) the southwestward movement of the Aegean plate at 30–32 mm a^{-1}, and its collision with the Apulian microplate to the west and the African plate to the south and SW (Fig. 13.1a; Taymaz et al. 1991, 2007; McClusky et al. 2000). The effect of the movement of these plates during the Quaternary is reflected in the spatial and temporal variation of the observed fault pattern and stress field across the above-mentioned area (Doutsos & Kokkalas 2001; Kokkalas et al. 2006).

The Hellenic shelf

The Hellenic shelf along the western margin of the Hellenic peninsula in the Ionian Sea lies over the Hellenic Compressional Belt in the outer arc domain (Fig. 13.1). The Hellenic shelf in the Aegean, an epicontinental sea (Stanley & Perissoratis 1977), overlies the continental crust of the Alpine Hellenides tectonic fabric of folds, thrust faults and sutures formed during the Alpine orogeny. The complex deformation of the crust in the Hellenic shelf, which lies within and around the Aegean microplate, during the Quaternary may be the result of: (1) slab retreat along the subduction zone and consequent back-arc extension; (2) collapse of an overthickened crust; (3) the westward escape of Anatolia along its plate boundaries; and (4) differential rates of convergence between the NE-directed subduction of the African plate in relation to the Anatolian plate (Taymaz et al. 2007).

The Hellenic shelf in the Ionian Sea

The Hellenic shelf in the Ionian Sea, based on the major morphotectonic features that shape the Hellenic Compressional Belt, can be divided into three segments: (i) the Illyrian folding belt (IFb); (ii) the Lefkada–Kefallinia escarpment (LKe); and (iii) the Hellenic subduction zone (HSz) (Fig. 13.1).

The Hellenic shelf in the Ionian Sea extends up to the 120 m isobath, has a width ranging from 3 to 50 km and dips seawards at an average gradient of 0.6°.

The shelf along the IFb runs along the coastal zone of southern Albania and continues south along the margin of western Greece as far as the northern end of the Lefkada Island (Fig. 13.1). It overlies the Apulian, pre-Apulian and Ionian zones of the External Hellenides (Monopolis & Bruneton 1982, figs 1 & 3), which is characterized by NE–SW-trending active shortening with folding, reverse faulting and thrusting (Monopolis & Bruneton 1982; Hatzfeld et al. 1995).

The flatness of the seafloor and the absence of any compressional features that affect the sedimentary cover suggest that the underlain reverse faults and thrusts do not affect the seafloor (Ferentinos & Papatheodorou 1994). The slope is narrow with a gradient of between 1.6° and 5.7°, is covered by a thin veneer of unconsolidated sediments (<20 m) and is probably the morphological expression of the thrust front (Monopolis & Bruneton 1982, fig. 4a).

The Hellenic shelf along the Lefkada–Kefallinia escarpment segment develops between the island (Lefkada, Kefallinia and Zakynthos) chain and the Greek mainland (Fig. 13.1). The shelf

From: CHIOCCI, F. L. & CHIVAS, A. R. (eds) 2014. Continental Shelves of the World: Their Evolution During the Last Glacio-Eustatic Cycle.
Geological Society, London, Memoirs, **41**, 187–197. http://dx.doi.org/10.1144/M41.13

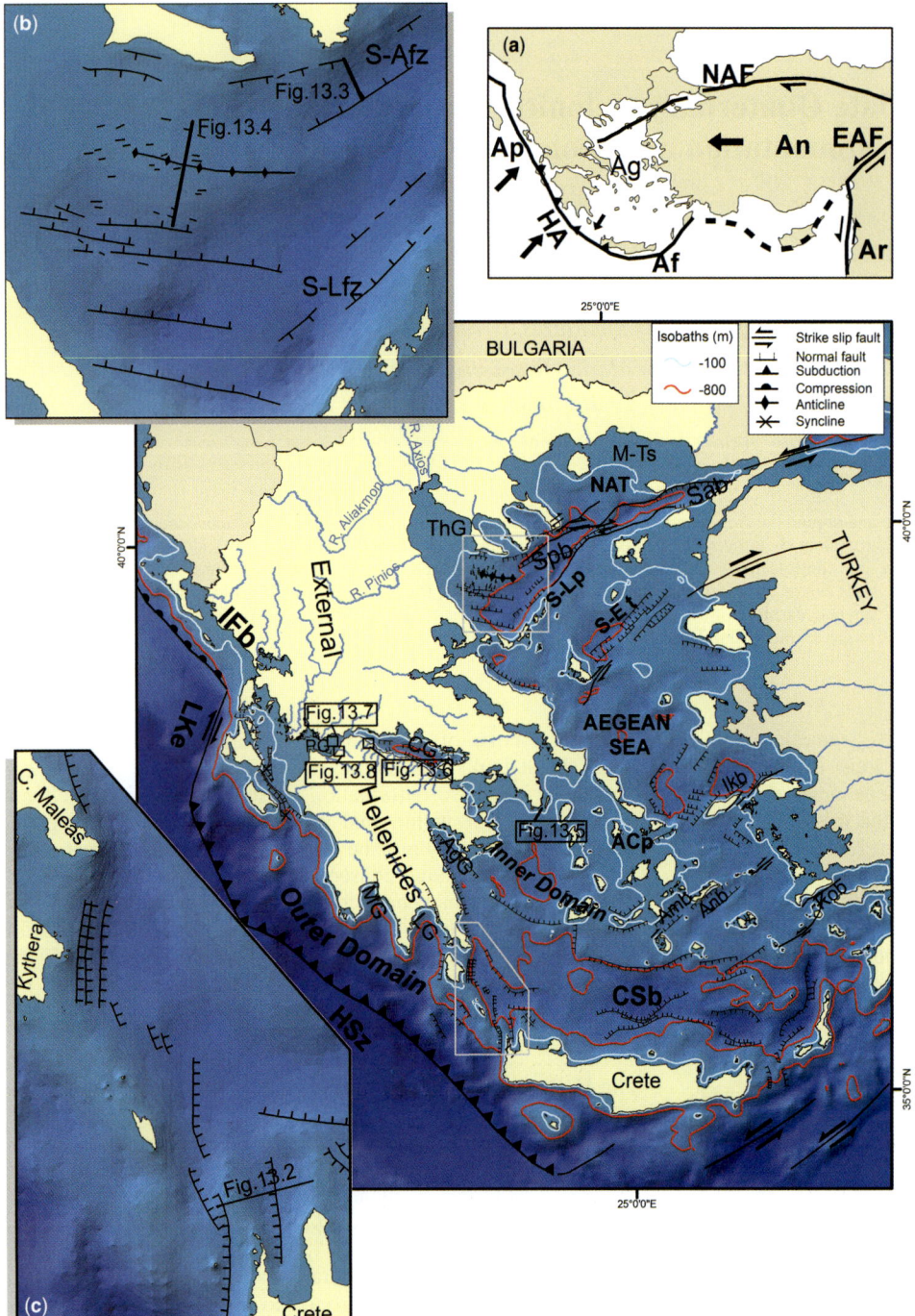

Fig. 13.1. Regional map of the Hellenic Arc shelf showing the bathymetric and active tectonic features (compiled from Bartole *et al.* 1983; Brooks & Ferentinos 1984; Mascle & Martin 1990; Ferentinos 1992; Lykousis *et al.* 1995; Koukouvelas & Aydin 2002; Piper & Perissoratis 2003; Kokkalas *et al.* 2006; Kokkalas & Aydin 2012). HSz, Hellenic subduction zone; LKe, Lefkada–Kefalonia escarpment; IFb, Iliyrian folding belt; NAT, North Aegean Trough; CSb, Cretan Sea forearc basin; ACp, Attico–Cyclades platform; S-Lp, Sporades–Limnos platform; S-Et, Skyros–Edremit trough; Sab, Saros Basin; Spb, Sporades Basin; ThG, Thermaikos Gulf, M-Ts, Macedonian–Thracian shelf, MG, Messiniakos Gulf; LG, Lakonikos Gulf; AgG, Argolikos Gulf; PG, Patras Gulf; CG, Corinth Gulf; Ikb, Ikaria Basin; Amb, Amorgos Basin; Anb, Anafi Basin; Kob, Kos Basin. (**a**) Schematic map showing the geodynamic framework in the eastern Mediterranean (modified from McClusky *et al.* 2000). NAF, North Anatolia Fault; EAF, East Anatolia Fault; An, Anatolian microplate; Ar, Arabian plate; Ag, Aegean microplate; Af, African plate; Ap, Apulian microplate; HT, Hellenic trench. (**b**) Map showing the bathymetry and fault lines in the Sporades Basin, based on two marine geophysical surveys carried out in 1978 (RSS *Shackleton* cruise) and 1982 (*Discovery* cruise) by the Department of Oceanography, University of Swansea, Wales. S-Afz, Sithonia–Athos fault zone; S-Lfz, Sporades–Limnos fault zone. (**c**) Map showing the bathymetry and fault lines in the Maleas–Kythera–Crete Ridge, based on a marine survey carried out in 1988 by the Laboratory of Marine Geology and Physical Oceanography, University of Patras, Greece. Grb, Gramvousa Basin; Kb, Kissamos Basin.

has a width ranging from 25 to 50 km. It overlies the Apulian, pre-Apulian and Ionian zones of the External Hellenides (Monopolis & Bruneton 1982).

The shelf morphology is dominated by two narrow, NW–SE- to NNW–SSE-trending depressions elongated parallel to the External Hellenides, each about 60 and 50 km long; the Zakynthos depression, reaching water depths of 520 m, and the Kefallinia depression, reaching water depths of 340 m, separated by a NNW–SSE-trending ridge (Brooks & Ferentinos 1984, figs 6 & 7; Ferentinos & Papatheodorou 1998). This morphological configuration of the shelf reflects the complex deformation of the underlying Pre-Apulian and Ionian zones, which is controlled by reverse faulting, thrusting and associated diapirism due to the active NE–SW compression (Monopolis & Bruneton 1982, figs 3 & 4; Brooks & Ferentinos 1984, figs 6–8).

The slope develops to the west of the islands and is associated with the Lefkada–Kefallinia escarpment, which is about 2 km

in height and is characterized by steep slopes ranging between 9.4° and 18.4°. It is the surface expression of the Lefkada–Kefallinia transform fault (Kokkinou *et al.* 2006, figs 6 and 7).

The Hellenic shelf along the Hellenic subduction zone in the Ionian Sea runs from Zakynthos Island to the SW tip of the Peloponnesus and, from there, continues to western Crete (Fig. 13.1). The shelf is narrow (2.5–5 km), flat and extends up to the 120 m isobath (Papanikolaou *et al.* 1988; Poulos *et al.* 2002). The shelf configuration is controlled by active faults trending NW to NNW and NNE. The NW to NNW faults are responsible for the formation of NW–SE- and NNW–SSE- trending onshore and off-shore elongated basins associated with the Messiniakos, Lakonikos and Argolikos gulfs in southern Peloponnesus (Kokkalas *et al.* 2006, fig. 6), and the formation of two elongated offshore basins in western Crete: the Gramvousa and Kissamos basins (Fig. 13.1c). A sparker profile, which runs across the two basins (Fig. 13.2), shows that the faults are closely spaced with steep

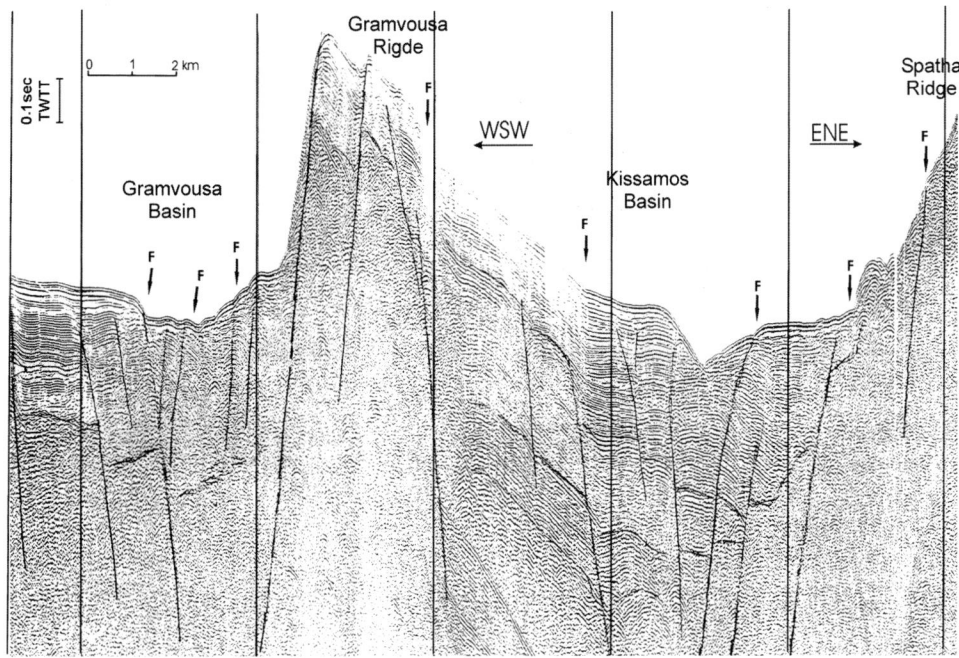

Fig. 13.2. Sparker profile across the Gramvousa and Kissamos basins in western Crete (for the profile location see Fig. 13.1c). The overall configuration pattern that the faults exhibit on the sparker profile indicates oblique normal faulting. The profile was collected during the RRS *Shackleton* 1972 and 1974 cruises in the Cretan Sea.

fault planes that seem to converge with depth and displace individual tilting blocks. This fault pattern indicates oblique normal faulting. Taking into consideration the similarities they exhibit with the nearby faults on land (Kokkalas *et al.* 2006, fig. 6), it is concluded that these two offshore faults are oblique normal faults. The NNW- and NNE-trending faults border the Maleas–Kythera–Crete ridge (Fig. 13.1c) and are responsible for its formation.

The study of 3.5 kHz, high-resolution seismic profiles across the Ionian Sea shelf shows that the sediments deposited on the shelf overlie an erosion terrace cut by the rising sea during the post-Last Glacial Transgression (18 ka BP) (Ferentinos & Papatheodorou 1994; Poulos *et al.* 2002, fig. 3; Ferentinos *et al.* 2012, fig. 6). These deposits correspond to the late Holocene Highstand Systems Tract (HST), and are characterized by a typical sigmoid progradational configuration consisting of: (i) discontinuous and discordant reflectors along the present-day river mouths, representing topsets; (ii) stratified inclined reflectors, representing foresets; and (iii) stratified horizontal reflectors, representing bottomsets (Poulos *et al.* 2002, fig. 3). The HST sedimentary sequence may or may not have prograded across the shelf edge draping the shelf and slope.

The Hellenic shelf in the Aegean Sea

The Hellenic shelf in the south Aegean Sea occupies the inner arc domain of the Hellenic Arc, which consists of three morphotectonic units: the Cretan outer arc; the Cretan Sea forearc basin (CSb); and the Attico–Cyclades Volcanic Arc/Metamorphic Core Complex continental platform (ACp) (Fig. 13.1) The south Aegean is dominated by a complex tectonic regime (Fig. 13.1) (Mascle & Martin 1990; Kokkalas & Doutsos 2001; Piper & Perissoratis 2003; Kokkalas *et al.* 2006; Kokkalas & Aydin 2012) characterized by: (i) an older set of mid–late Miocene east–west-trending normal faults as a result of pull-apart between stable Europe and the south-moving Aegean microplate; (ii) a younger set of Pliocene NE–SW-trending strike-slip faults that corresponds to the time when the North Anatolia Fault propagated to the NE Aegean; and (iii) a set of mid-Quaternary–present NNW–SSE- to NNE–SSW-trending normal faults in SW Peloponnesus and the SW Aegean Sea, and ENE–WSW-trending strike-slip faults in the SE Aegean Sea, which created and/or

reactivated secondary normal faults trending NNE–SSW, NNW–SSE and east–west (Mascle & Martin 1990, fig. 1; Piper & Perissoratis 2003, fig. 20; Kokkalas & Aydin 2012, fig. 13). The above-mentioned tectonic regime controls the formation of the Cretan Sea, which is an arc-shaped basin consisting of a fault-controlled chain of smaller basins trending NW–SE and WNW–ESE in the west, west–east in the central part of the arc and NE–SW in the east, and ranging in depth from 1400 to 2658 m (Bartole *et al.* 1983, fig. 4), and of the Ikaria, Amorgos, Anafi and Kos basins in the eastern part of the Attico–Cyclades continental platform (ACp). The Cretan Sea basin is bordered to the south by the Cretan shelf, which extends to the 120 m isobath, and to the north by the Attico–Cyclades continental platform shelf, which also extends to the 120 m isobath. Both shelves are generally flat and are covered by a thin sequence ($<c$. 20 m) of layered unconsolidated sediments, which overlie the post-Last Glacial Transgression erosion surface (18 ka BP) (Ferentinos & Papatheodorou 1993, 1995). This sequence corresponds to the late Holocene HST. In the Attico–Cyclades shelf, the presence of between two and four successive oblique progradational delta sequences at the western (Kapsimalis *et al.* 2009, figs 2–4), southern (Piper & Perissoratis 2003, fig. 11) and eastern (Lykousis *et al.* 1995, fig. 2) margin suggest that the shelf continuously subsides. These successive progradational delta sequences were deposited during low sea-level stands during the middle and late Quaternary glacial stages corresponding to the Oxygen Isotopic Stages (IOSs) 2.2 (18 ka BP), 6.2 (146 ka BP), 8.2 (250 ka BP) and 10 (340 ka BP). Similar stacked delta deposits have been found in Argolikos Gulf (Piper & Perissoratis 2003, fig. 4) and in Izmir Bay in eastern Aegean (Aksu *et al.* 1987, figs 6 & 7), suggesting that the shelf over the Cyclades subsides. The estimated subsidence rates range from 0.35 to 0.90 m ka^{-1} and vary in space and time (Lykousis *et al.* 1995, 2007; Lykousis 2009, table 1).

The Hellenic shelf in the northern Aegean Sea includes the Macedonian–Thracian shelf (M-Ts) in the north and the Sporades–Limnos continental platform (S-Lp) in the south, separated by the North Aegean Trough (NAT) and the Skyros–Edremit Trough (S-Et) (Fig. 13.1). The two troughs are considered morphological expressions of the westward extension of the northern and southern branches of the North Anatolian transform fault. The North Aegean Trough (NAT) is bounded to the north by the M-Ts and to the south by the S-Lp. It is about 346 km long and 30 km wide, and is divided into two basins: the Saros Basin

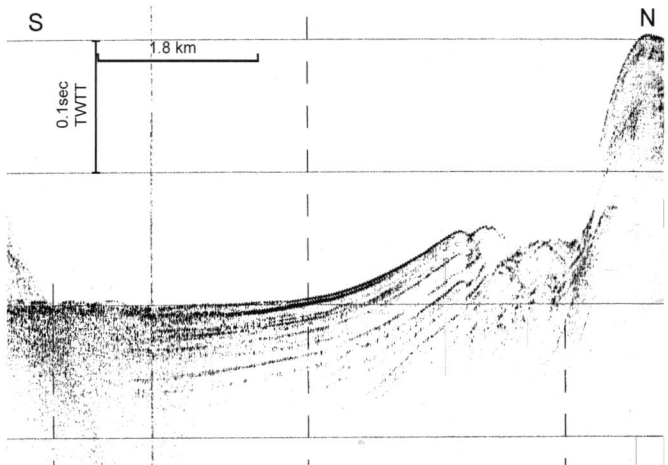

Fig. 13.3. High-resolution 3.5 kHz seismic profile across the Sithonia–Athos fault zone (for the profile location see Fig. 13.1b) displaying Holocene–Upper Quaternary layered sediments, which are affected by thrusts faulting, crumbled against the Sithonia–Athos fault zone, indicative of active transpressional forces.

(Sab) to the east, trending ENE–WSW; and the Sporades Basin (Spb) to the west, trending NE–SW (Stanley & Perissoratis 1977; Brooks & Ferentinos 1980). The Saros Basin is bounded to the north and south by the Ganos right-lateral strike-slip fault zone (Koukouvelas & Aydin 2002). The Sporades Basin at the eastern end is bounded to the south by the Sporades–Limnos fault zone (S-Lfz) and to the north by the Sithonia–Athos fault zone (S-Afz), which are characterized by normal and right-lateral strike-slip components (Roussos & Lyssimachou 1991, figs 2, 5–7; Koukouvelas & Aydin 2002, figs 7 & 8). The faults are active as they offset Pleistocene and Holocene strata (Fig. 13.3). The Sporades Basin at the western end changes direction from NE–SW to WNW–ESE due to the activity of listric faulting (Fig. 13.1c). The basin to the SSW is flanked by a series of active synthetic listric faults dipping to the NNE (Ferentinos 1992, fig. 2; Laigle *et al.* 2000, fig. 2), whilst to the NNE it is flanked by an active listric fault that dips to the SSE (Ferentinos 1992, fig. 2). Between the NNE- and SSW-dipping listric faults an anticlinal structure trending WWS–ESE is formed due to the space reduction caused by the opposite moving hanging wall of the listric faults. Airgun seismic profiles across the anticlinal structure show evidence of a close association between folding and faulting (Brooks & Ferentinos 1980, fig. 7). These faults offset the seafloor and face upslope, are closely spaced and form individual tilted

blocks. The wedged-shape layers in each block indicate growth faulting (Fig. 13.4). The Macedonian–Thracian shelf has a width ranging from 22 to 34 km. Since the late Quaternary, the shelf has undergone a tectonic subsidence. At the western end, the Thermaikos shelf has been subsiding to the SSW for the last 450 ka BP under the influence of the listric faults flanking the basin to the SSW. The Thermaikos shelf subsides at an average rate of 0.93 m ka^{-1}, as is indicated by the presence of a sequence of deltaic units with a cumulative thickness of 600 m. The older unit (IOS 12) now underlies a water depth of 700 m, whilst the younger (IOS 2.2) is under a water depth of only 120 m (Lykousis 1991*a*, figs 4 & 8). At the eastern end, the shelf edge subsides at an average rate of between 0.3 and 0.5 cm ka^{-1}, as is indicated by the displacement of a delta progradation sequence that corresponds to the IOS 6 (146 ka BP) lowstand down to a deeper water depth due to faulting (Piper & Perissoratis 1991, fig. 9). The Sporades–Limnos continental platform consists of Oligocene–Miocene volcanic rocks and Eo-Oligocene sedimentary rocks, and is fault-bounded to the north and south by ENE–WSW-trending faults (Mascle & Martin 1990).

Late Quaternary sedimentation

Over the last five decades a large number of papers have been published relating to: (i) the character of surficial sediments (Lykousis *et al.* 1981; Anagnostou *et al.* 1993, 1998; Volakis & Anagnostou 1993; Giresse *et al.* 2003; Poulos 2009); (ii) transport processes and deposition (Ferentinos *et al.* 1985*a, b*, 1988; Lykousis & Chronis 1989; Lykousis 2001; Hasiotis *et al.* 2005; Lykousis *et al.* 2009); and (iii) stratigraphy and palaeoclimatology/palaeoceanography (Aksu *et al.* 1995*a, b*; Geraga *et al.* 2000, 2005, 2008; Roussakis *et al.* 2004).

Surficial sediments

The unconsolidated surficial sediments covering the seafloor of the Hellenic shelf, in terms of texture and composition, are mainly terrigenous muds (M) and sandy muds (sM) (Poulos 2009, figs 6, 8 & 9). In the shelf areas, where the riverine input is minimal, the surficial sediments are mainly muddy sands (mS) and sands (S) of biogenic origin (Anagnostou *et al.* 1993, 1998; Volakis & Anagnostou 1993; Georgiadis *et al.* 2009).

The terrigenous sediments found on the shelf are considered to be late Holocene prodelta sediment sequences draping the shelf (Lykousis & Chronis 1989; Piper & Perissoratis 1991; Poulos *et al.* 2002; Lykousis *et al.* 2005). This sedimentation has been

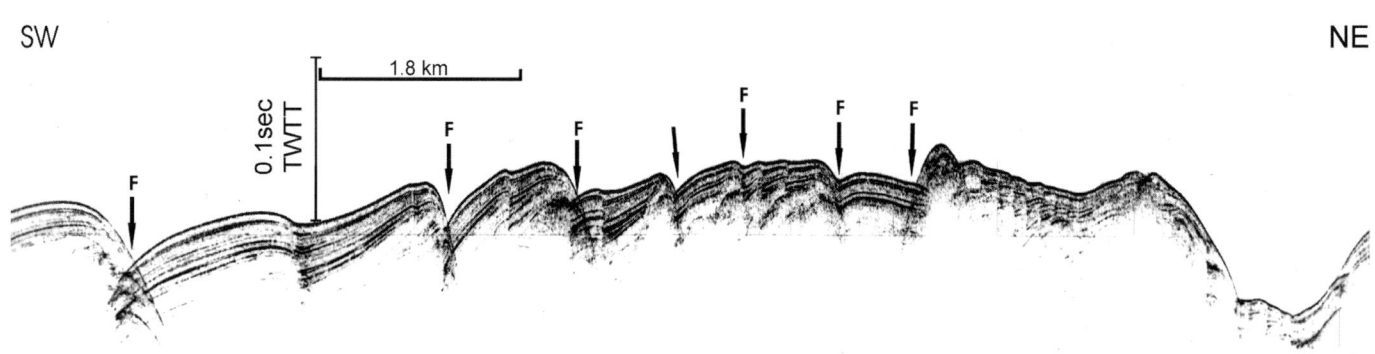

Fig. 13.4. 3.5 kHz seismic profile across an anticlinal structure formed between two antithetic listric faults (for the profile location see Fig. 13.1b), showing the formation of a graben at the crest of the anticline and upslope-facing antithetic growth faults in the flank. The approximate 30 m-thick sedimentary sequence that drapes the slope corresponds to hemipelagic sedimentation. The upper 3 m of the sequence is deposited during the late Holocene highstand, whilst the rest is deposited during the IOS 2.2 lowstand.

active since 6 ka BP when the sea attained its present level (Lykousis & Chronis 1989; Piper & Perissoratis 1991; Lykousis *et al.* 2005). The increased amount of sand observed in the mid and outer shelf represents either lowstand deltaic sediments or reworked shallow coastal sands deposited during the early stages of the post-glacial transgression 18–6 ka BP (Piper & Perissoratis 1991; Lykousis *et al.* 2005).

The biogenic sediments that cover the Attico–Cyclades and Sporades–Limnos continental platform consist mainly of coraline algal debris (*Lithothannium* and *Phytomatoliton* sp.), coral debris (*Derdrophylia* sp.) and mollusc shell debris (Anagnostou *et al.* 1993, 1998; Volakis & Anagnostou 1993; Georgiadis *et al.* 2009). In the Attico–Cyclades continental platform shelf, in water depths of between 40 and 120 m, the floor is covered by patches of coraligenous formations that have been growing since the beginning of the post-glacial transgression (Georgiadis *et al.* 2009, figs 2–6).

Further offshore, over the tectonically controlled deep basins that dissect the Hellenic shelf, hemipelagic sedimentation has prevailed over the last 6 kyr. In the northern Aegean, in the settling particulate matter, terrigenous material is dominant, whilst, in the southern Aegean, biogenic material is more pronounced (Poulos 2009). This hemipelagic sedimentation, in almost every basin, is often interrupted by gravity flows (Giresse *et al.* 2003; Roussakis *et al.* 2004; Geraga *et al.* 2000, 2005).

Sediment-transfer processes

Sediment-transfer processes from land to shelf, then to slope and, subsequently, to the deep basin seafloor that operate in the Hellenic shelf can be synoptically demonstrated by the following three key examples, dealing with three different shelf environments. The first demonstrates the transfer processes that operate in a wide and low-gradient (less than 1°) shelf with high river sediment input (Thermaikos Gulf), the second demonstrates the transfer processes that operate in a very narrow low-gradient shelf with episodic, but high, river discharge (Corinth Gulf) and the third is an island complex shelf with restricted sediment supply (Attico–Cyclades continental platform:

- The Thermaikos Gulf shelf at the western end of the Macedonian–Thracian shelf is about 12 km wide and is flat, with a gradient less than 1°. At the Gulf head, three major rivers discharge an annual mean average of 25×10^6 t a^{-1} of sediments. A typical prodelta wedge showing sigmoid progradational configuration has developed in the Thermaikos Gulf shelf during the last 6 kyr when the sea attained its present level (Lykousis *et al.* 2005, figs 2 & 5). The coarser material is deposited at the mouth of the rivers as topsets, the finer material at the delta slope as foresets, and the finest is blanketing the shelf and slope seafloor as bottomsets (Lykousis & Chronis 1989, fig. 3; Lykousis *et al.* 2005). The thin sediment blanket that covers the shelf overlies coarse sediments deposited during the post-Last Glacial Transgression from 18 to 6 ka. These deposits overlie river deposits accumulated between 22 and 18 ka when the sea level was standing 120 m below present during the Last Glacial Maximum. At the shelf edge, at a water depth of 200 m, seafloor deltaic sediments crop out that correspond to IOS 2.2, which is when the sea level was standing at 125 m below the present (Lykousis & Chronis 1989, fig. 6). The approximate 30 m-thick layered sedimentary sequence that drapes the slope consists of an upper layered unit that corresponds to the late Holocene HST, which conformably overlies a lower unit that corresponds to the IOS 2.2 lowstand (Lykousis *et al.* 2005, fig. 9). The bulk of the sediment deposited over the outer Thermaikos Gulf shelf and slope over the last 6 kyr is composed of terrestrial sediments supplied by the rivers and the surrounding land masses. These sediments

have been dispersed over the shelf and slope via nepheloid layers, and deposited as hemipelagic muddy deposits, as can be deduced from the distribution of silts, clays and clay minerals (Lykousis & Chronis 1989; Lykousis *et al.* 2005). Nepheloid layers have been observed to form at the present time at the surface, and at bottom water depths all the way from the river mouths down to the shelf edge and then to the slope (Durrieu de Mardon *et al.* 1992, figs 2, 5 & 9; Karageorgis & Anagnostou 2001, fig. 2). At the same time, dense water has been observed to form at the head of the Gulf in winter, by cooling during extremely cold northerly winds (Estournel *et al.* 2005). This dense water sweeps the seafloor across the shelf and slope with speeds of between 5–15 and 30 cm s^{-1} in the shelf and slope, respectively, transporting sediments to the basin.

- The southern margin of the Corinth Gulf is the surface expression of step-like active fault segments (Stefatos *et al.* 2002). The margin is characterized by a narrow (less than 2 km) low-gradient shelf and a steep slope with a gradient ranging from 10° to 40°. The slope is segmented and related to fault escarpment. The study of high-resolution seismic profiles shows that the shelf and slope are dissected by canyons, which are either directly connected to ephemeral rivers on land or begin at the shelf edge. The former directly transports the river sediment load to the basin floor as density flows and turbidity currents, whilst the latter delivers sediments from the shelf to the basin floor through retrogressive slumping at the shelf edge, and subsequent disintegration of the slumped masses to debris/mud flows and turbidity currents (Ferentinos

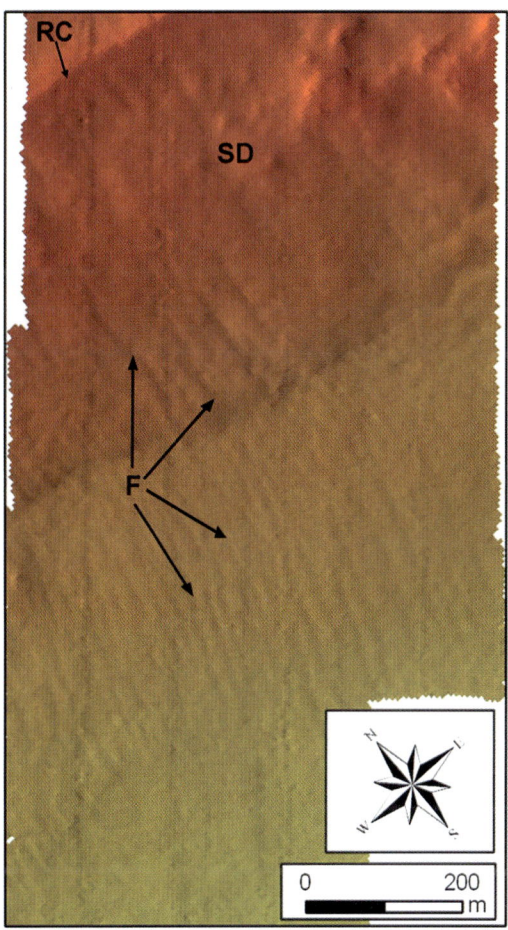

Fig. 13.5. Multibeam seafloor images (for the location see Fig. 13.1) showing the formation of a sand drift deposit at the lee side of a ridge due to the overflow of bottom currents over the ridge crest. The lineations on the surface of the sand deposits correspond to erosional furrows behind obstacles indicating the bottom flow direction. RC, ridge crest; SD, sand drift deposits; F, furrows.

et al. 1988). These processes control the sediment transport
from land to shelf and deep basin seafloor during low and
high sea-level stands, resulting in an average sedimentation
rate of between 100 and 190 cm ka^{-1} (Lykousis *et al.* 2007;
Bell *et al.* 2009). The mass movements are triggered by earthquakes, heavy rain and stormy weather, and occur at least once
every 2 years. Individual events can occur in two or more
localities simultaneously along the shelf and slope, and the
sediment masses can travel between 2 and 6 km on the basin
floor (Ferentinos *et al.* 1988).

- The Attica–Cyclades continental platform shelf constitutes a
  shallow-water barrier, with an average depth of 100 m, which
  separates the northern Aegean from the southern Aegean. The
  outflow of the low-salinity Black Sea Water mass on the
  surface, the underlying Levantine Intermediate Water mass
  and the occasional cold, dense water masses forming during
  the dry, cold winter periods lead to the generation of an intensive current that flows over the continental platform shelf
  through the island straits to the southern Aegean Sea. The
  current crossing through the straits sweeps all the sediments
  from the seafloor, leaving only outcrops of barren rock (Lykousis 2001, figs 2, 7 & 10). At the Straits between the islands, the
  current can reach speeds of up to 0.8 m s^{-1} (Ferentinos &
  Papatheodorou 1993), transporting sediment from the northern
  Aegean Sea upslope over the plateau and then downslope to the

southern Aegean Sea (Ferentinos & Papatheodorou 1993, fig 5).
Away from the straits over the plateau, the currents sculpt the
seafloor sediments, forming a variety of transverse and longitudinal bedforms, such as obstacle marks, sand ribbons,
burchan-shaped dunes, sand waves, megaripples and drift
deposits behind obstacles (Fig. 13.5) (Lykousis 2001, figs 2, 7
& 10; Ferentinos & Papatheodorou 1993; Ferentinos *et al.*
2010). This current transports the sediments to the shelf edge
and cascades them into the deep basin (Lykousis 2001;
Canals *et al.* 2010).

Sedimentation rates

The average sediment thickness deposited on the erosion terrace
cut across the Hellenic shelf by the rising sea during the Last
Glacial Transgression, at about 18 ka BP, ranges between 10 and
30 m, giving an average sedimentation rate on the shelf of
between 55 and 165 cm ka^{-1}, respectively. These sedimentation
rates are much higher than the sedimentation rates observed in
the Hellenic shelf basins, where sedimentation rates for the Holocene and for the last 18 ka are between 4 and 22 cm ka^{-1} (Geraga
et al. 2000, 2005, 2008; Giresse *et al.* 2003; Piper & Perissoratis
2003, table 1; Roussakis *et al.* 2004). Roussakis *et al.* (2004),
in the North Aegean Trough, observed a decrease in the

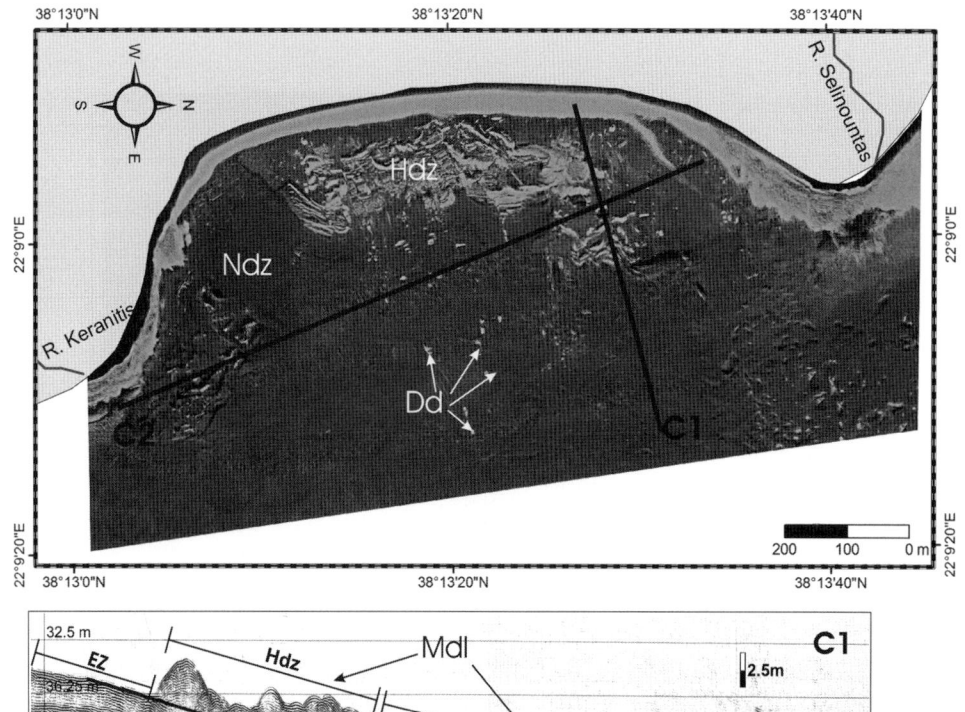

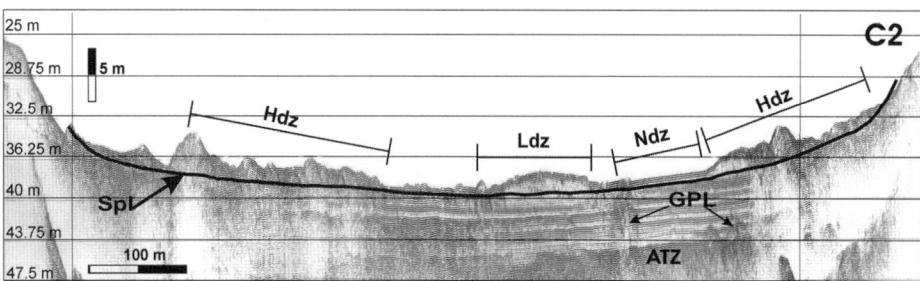

Fig. 13.6. Side scan sonar mosaic and 3.5 kHz
seismic profiles in the (C1) dip direction and in
the (C2) strike direction along the Selinountas
and Keranitis rivers delta front in the Corinth
Gulf (for the location see Fig. 13.1) showing the
areal extent of a complex translational slide less
than 5 m thick with lateral deformational
complexity. The slide covers a seafloor area of
about 600 000 m^2 and occurred in 1995 by a
6.2R earthquake. Ez, evacuation zone; Spl, slide
plane; Mdl, main depositional lobe; Hdz, high
deformational zone; Ldz, low deformational
zone; Ndz, non-deformational zone; PR,
pressure ridges; Dd, distal deposits; ATZ,
acoustic turbidity zone, indicating the presence
of bubble phase gas; GPL, gas plumes.

sedimentation rates from $30 \, \mathrm{cm \, ka^{-1}}$ at the onset of the Last Glacial Transgression to $15.4 \, \mathrm{cm \, ka^{-1}}$ during the period of the sapropel formation and $14 \, \mathrm{cm \, ka^{-1}}$ in the post-sapropel period.

The high sedimentation rates prevailing on the erosion terraces during the post-Last Glacial Transgression in the shelf compared to the low sedimentation rates on the basin floors suggest that most of the terrigenous land-derived material remains on the shelf. The high sedimentation rates of about $30 \, \mathrm{cm \, ka^{-1}}$ observed at the onset of the Last Glacial Transgression can be attributed to the direct discharge of river sediment on to the basins rims during the Last Glacial Lowstand. Furthermore, the relatively increased rates of deposition of $15-25 \, \mathrm{cm \, ka^{-1}}$ observed in some basins are attributed to additional sediment input by gravitational processes. The sedimentation rates in the land-locked and tectonically active Corinth and Patras gulfs are, in general, higher than in the Ionian and Aegean shelves. The sediment thickness deposited on the erosion terrace is about 50 m in the last 18 kyr, and about 20 m in the last 12 kyr in the Patras and Corinth gulfs, giving average sedimentation rates of 2.7 and $1.7 \, \mathrm{m \, ka^{-1}}$, respectively (Ferentinos et al. 1985a, b; Chronis et al. 1991).

However, in the Corinth Gulf basin, the sedimentation rate is between 5 and 20 times higher than in the basins of the Aegean shelf. The sedimentation rate in the Corinth Basin in the last 12 kyr is about $100 \, \mathrm{cm \, ka^{-1}}$, whilst for the period between 12 and 250 ka BP it is about $50 \, \mathrm{cm \, ka^{-1}}$ (Bell et al. 2009).

Geohazards

Gravitative mass movements

Gravitational mass movements have been observed all across the shelf in the Aegean and Ionian seas (Lykousis & Chronis 1989; Lykousis 1991b; Ferentinos 1992; Papatheodorou & Ferentinos 1993, 1997; Hasiotis et al. 2002; Lykousis et al. 2009) affecting the late Pleistocene and Holocene deposits.

Three main types of slope failure were identified in the acoustically stratified late Quaternary layers: (i) sliding of masses with no or only slight internal deformation on a basal planar or concave shear surface; (ii) sliding of masses with disintegration of sediment fabric into flows; and (iii) slow downslope creep. Many of the disintegrating slides as they move downslope display a continuum of transformation from slides to debris flows to turbidity currents.

Slides are associated with: (i) active delta fronts and delta slopes with gradients ranging from $2°$ to $5°$ (Fig. 13.6); (ii) active delta fronts and delta slopes with gradients ranging from $6°$ to $20°$; (iii) inactive delta fronts with slope gradients from $2°$ to $5°$; (iv) well-stratified gentle slopes with gradients from $1°$ to $5°$; and (v) fault escarpments with slope gradients from $10°$ to $40°$ (Ferentinos & Papatheodorou 1993, fig. 9).

The slides, the thickness of which is between 5 and 25 m, take place over slip planes that are: (i) bedding planes parallel to the

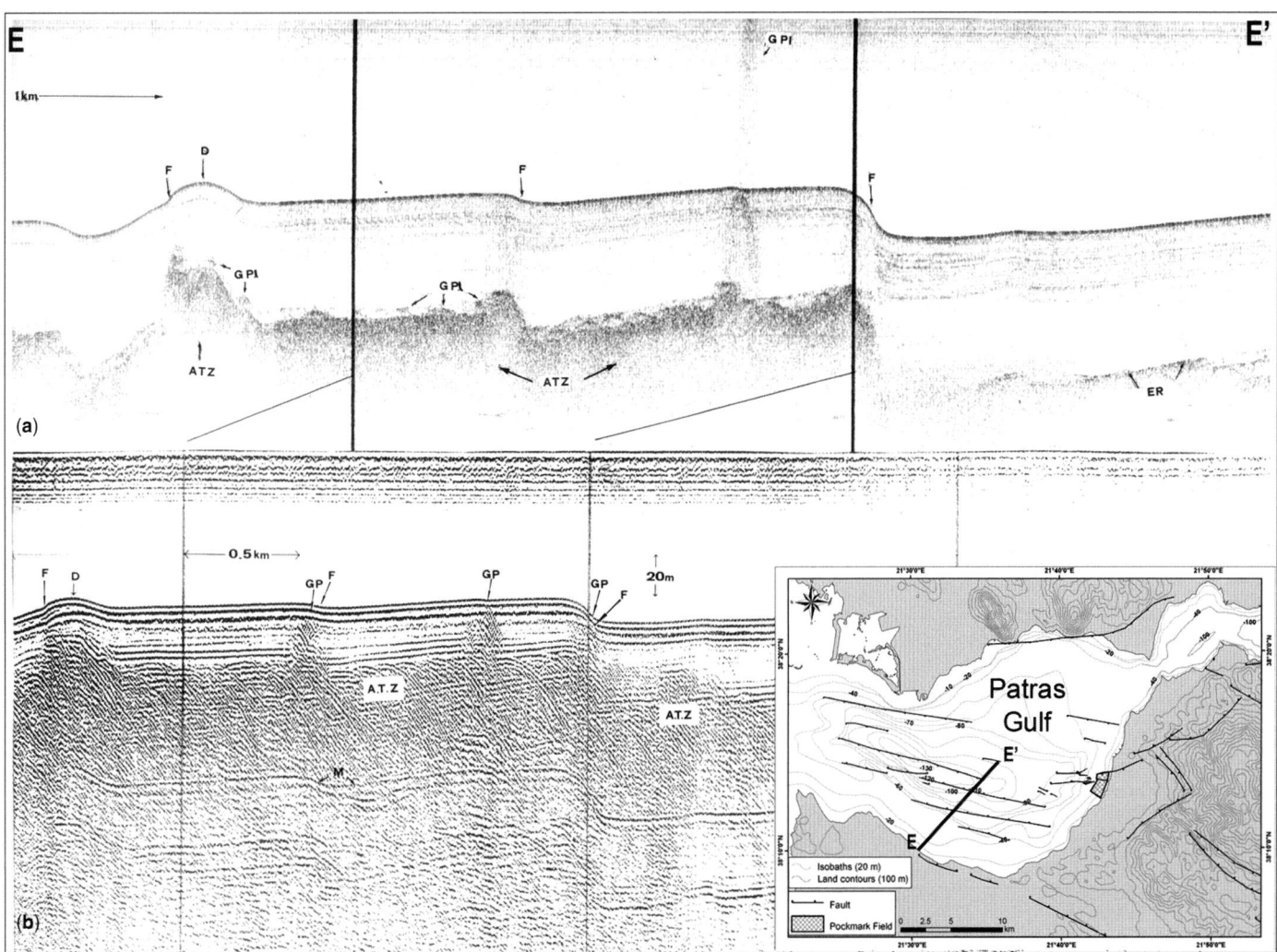

Fig. 13.7. (a) 3.5 kHz and (b) airgun seismic profiles across active faults in the Patras Gulf (for the locations of the profiles see the inset) showing: (i) gas accumulation in the Pleistocene(?)–Holocene interface (ATZ, acoustic turbid zone); (ii) active faults (F) displacing the Pleistocene(?)–Holocene interface and the seafloor; and (iii) gas plumes (GP/GPl) rising along the fault planes to the Holocene cover, forming domes (D), and to the water column.

slope with a dip from 1° to 5°; (ii) unconformities parallel to the slope with a dip from 2° to 7°; (iii) fault planes with a dip of between 10° and 40°; and (iv) deltaic foresets or bottomsets with a dip of between 1° and 10°.

Most of the dated mass movements occurred during and after the Last Glacial Trangression (Lykousis & Chronis 1989). Some landslides occurred 2–3 ka BP (Stefatos *et al.* 2006) and others during the last 300 years (Galanopoulos *et al.* 1966; Papatheodorou & Ferentinos 1997; Hasiotis *et al.* 2002; Lykousis *et al.* 2005, 2007). The latter group of landslides occurred in the Corinth Gulf and were caused by known earthquake events (Papatheodorou & Ferentinos 1997; Hasiotis *et al.* 2002; Lykousis *et al.* 2005, 2007).

The causative mechanisms responsible for triggering the landslides in the Hellenic shelf are mainly cyclic loading stresses, induced by earthquakes, and an increase in the excess pore pressure, which induces sediment liquefaction. Other factors that can also contribute to sliding initiation are slope steepening through tectonic control, and gas (methane) in the sediment pores (Lykousis 1991*b*; Ferentinos 1992; Papatheodorou & Ferentinos 1993).

Based on the existing data, the areas where most mass movements are observed are the Corinth Gulf, the shelf adjacent to the Lefkada–Kefallinia transform fault and the shelves bordering the North Aegean Trough. These areas coincide with the areas of maximum expected peak ground accelerations in Greece.

Active gas seeping

Gas in the sediment pores has been observed all across the Hellenic shelf (Papatheodorou *et al.* 1993; Hasiotis *et al.* 1996; Soter 1999; Chistodoulou *et al.* 2003).

The gas-charged sediments are found in Pleistocene and present-day land-locked gulfs, Pleistocene and present-day deltaic environments, open-sea shelf environments, and lakes. The gas is usually associated with the: (i) Holocene–Pleistocene(?) boundary, which seems to be an accumulative horizon (Fig. 13.7); (ii) dipping Pleistocene or Pliocene bedding planes through which the gas migrates up-dip; and (iii) fault planes through which the gas migrates (Fig. 13.7). The gas is also associated with morphological features either on the seabed, or under the seabed, such as surface and buried pockmarks (Fig. 13.8), intrasedimentary and seabed dome-shaped mounds, gas-pockets, and cemented sediments on the seafloor.

The gas found in the Pleistocene and present-day land-locked gulfs and deltaic environments is of biogenic origin, while that

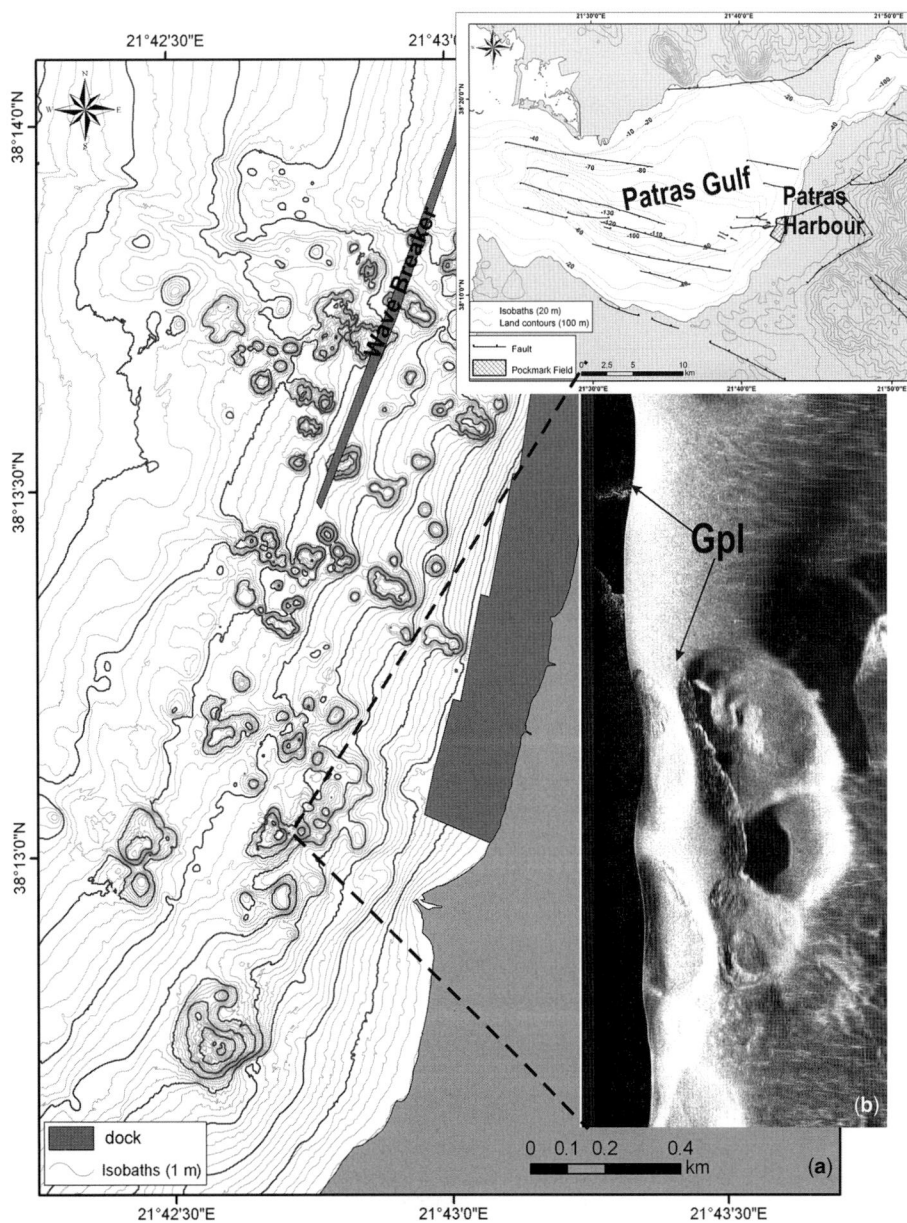

Fig. 13.8. (a) Detailed bathymetric map showing a large pockmark field off Patras Harbour in the Patras Gulf (for the location see the inset). (b) The side scan sonar record shows gas plumes (GPl) rising from the pockmarks to the water column about 24 h after a nearby 6.2R earthquake in 2008.

found in the pre-Quaternary open-sea shelf environments, which are associated with faults and salt doming, is probably of thermogenic origin.

Detailed studies in two pockmark fields in the Patras and Corinth gulfs, respectively, and in a gas-seeping field over a salt dome in Katakolon (Ionian Sea shelf) have shown that: (i) the Corinth Gulf pockmark field was formed by groundwater seepage, whilst that of the Patras Gulf is by biogenic methane seepage (Christodoulou et al. 2003, 2010); (ii) the methane seeping from the seafloor in the Katakolo gas-seeping field is of thermogenic origin (Etiope et al. 2006; Christodoulou 2010); (iii) the methane concentration in the water near the seabed inside the pockmarks ranges between 0.4 and 1.1 μmol l^{-1} over a background concentration of between 0.002 and 0.2 μmol l^{-1} (Christodoulou et al. 2003; Christodoulou 2010); (iv) the estimated contribution of methane to the atmosphere from the pockmark field is between 0.09 and 0.35 t a^{-1}; (v) the methane concentration in the water near the seabed in the Katakolo field is over 4.5 μmol l^{-1}, whilst the estimated contribution of methane to the atmosphere is between 1262 and 1500 t a^{-1} (Christodoulou 2010). The pockmark field in Patras Gulf, which was activated twice by earthquakes in 1993 and 2008 (Fig. 13.8), had an impact on the stability of the breakwater and also on the stability of the harbour at Patras (Chistodoulou 2010). Similarly, the gas seepage in Katakolo is considered hazardous to human activities and health owing to the presence of methane and hydrosulphide at levels that are potentially inflammable and toxic (Etiope et al. 2006).

Tsunami

The study of tsunami catalogues reveals a high frequency of tsunami occurrence over the Hellenic shelf. A total of 160 probable and definite tsunami events have been recorded since AD 1628. The two most recent destructive tsunami in the Hellenic shelf occurred in 1956 at Amorgos Island and in 1963 in the western Corinth Gulf (Papadopoulos & Chalkis 1984). The causal mechanisms of the former tsunami were the co-seismic displacement of the seafloor and the triggering of a submarine landslide (Perissoratis & Papadopoulos 1999), whilst the causal mechanism of the latter tsunami was a coastal submarine landslide (Stefatos et al. 2006). A tsunami risk-assessment study in the Corinth Gulf, a tsunamigenic zone (Papadopoulos & Chalkis 1984), based on a wealth of offshore geological data, has shown that: (i) a maximum probable 6.7 (M_w) offshore earthquake would produce a rupture in the seafloor with a maximum displacement of 1.08 m and a potential wave of the same height; and (ii) a submarine mass failure would have the potential of generating a tsunami with maximum wave height of 4.04 m over the source (Stefatos et al. 2006).

Conclusions

The Hellenic shelf is located within one of the world's most seismically active areas and has experienced extreme tectonism throughout Quaternary time. These tectonic forces are responsible for: (i) the overall configuration of the Hellenic shelf, as well as for the formation and orientation of the basins within it; (ii) the spatial and temporal variability in the uplift and subsidence rates observed in the Hellenic shelf; (iii) the sediment yields and sedimentary rates; and (iv) the frequent occurrence of a variety of offshore geological hazards.

Sea-level changes that occurred throughout the last 400 kyr have also played an important factor in the configuration of the Hellenic shelf with the formation of: (i) transgressive and regressive depositional sequences; and (ii) the transport and deposition of sediment over the shelf.

References

AKSU, A. E., PIPER, D. & KONUK, T. 1987. Quaternary growth patterns of Buyuk Menderes and Kuyuk Menderes deltas, western Turkey. *Sedimentary Geology*, **52**, 227–250.

AKSU, A. E., YASAR, D. & MUDIE, P. J. 1995a. Origin of late Glacial–Holocene hemipelagic sediments in the Aegean Sea: clay mineralogy and carbonate cementation. *Marine Geology*, **123**, 33–59.

AKSU, A. E., YASAR, D., MUDIE, P. J. & GILLESPIE, H. 1995b. Late Glacial–Holocene paleoclimatic and paleoceanographic evolution of the Aegean Sea: micropaleontological and stable isotopic evidence. *Marine Paleontology*, **25**, 1–28.

ANAGNOSTOU, CH., TSORMBATZOGLOU, G. & SIOULAS, A. 1993. The recent sedimentation in the central part of Cyclades. Paper presented at the Proceedings of the 4th National Symposium on Oceanography and Fisheries, Rhodes, 26–29 April.

ANAGNOSTOU, CH., RICHTER, D., RIEDEL, D. & TRAPP, T. 1998. Recent sediments in the south Cyclades marine area, Aegean Sea. *Bulletin of the Geological Society of Greece*, **XXXII**, 193–203.

BARTOLE, R., CATANI, G., LENARDON, G. & VINCI, A. 1983. Tectonics and sedimentation of the Southern Aegean Sea. *Bollettino di Oceanologia Teorica ed Applicata*, **1**, 319–340.

BELL, R. E., MCNEILL, L. C., BULL, J. M., HENSTOCK, T. J., COLLIER, R. E. L. & LEEDER, M. R. 2009. Fault architecture, basin structure and evolution of the Gulf of Corinth Rift, central Greece. *Basin Research*, **21**, 824–855.

BROOKS, M. & FERENTINOS, G. 1980. Structure and evolution of the Sporadhes basin of the North Aegean trough, northern Aegean Sea. *Tectonophysics*, **68**, 15–30.

BROOKS, M. & FERENTINOS, G. 1984. Tectonics and sedimentation in the Gulf of Corinth and the Zakynthos and Kefallinia Channels, Western Greece. *Tectonophysics*, **101**, 25–54.

CANALS, M., DANOVARO, R. ET AL. 2010. Cascades in Mediterranean Submarine Grand Canyons. *Oceanography*, **22**, 26–43.

CHRISTODOULOU, D. 2010. *Geophysical, Geotechnical, Sedimentological and Remote Sensing Monitoring of Active Pockmarks Fields in High Seismicity Region, Western Greece*. PhD thesis, Department of Geology, University of Patras.

CHRISTODOULOU, D., PAPATHEODOROU, G., FERENTINOS, G. & MASSON, M. 2003. Active seepage in two contrasting pockmark fields in the Patras and Corinth gulfs, Greece. *Geo-Marine Letters*, **23**, 194–199.

CHRONIS, G., PIPER, D. J. W. & ANAGNOSTOU, CH. 1991. Late Quaternary evolution of the Gulf of Patras, Greece: Tectonism, deltaic sedimentation and sea-level change. *Marine Geology*, **97**, 191–209.

DOUTSOS, TH. & KOKKALAS, S. 2001. Stress and deformation patterns in the Aegean region. *Journal of Structural Geology*, **23**, 455–472.

DURRIEU DE MARDON, X., NYFELLER, F., BALOPOULOS, S. & CHRONIS, G. 1992. Circulation and suspended matter in the Sporades Basin (northwestern Aegean). *Journal of Marine Systems*, **3**, 237–248.

ESTOURNEL, C., ZERVAKIS, V., MARSALEIX, A., PAPADOPOULOS, A., AUCLAIR, F., PERIVOLIOTIS, L. & TRAGOU, E. 2005. Dense water formation and cascading in the Thermaikos Gulf (North Aegean) from observation and modelling. *Continental Shelf Research*, **25**, 2366–2386.

ETIOPE, G., PAPATHEODOROU, G., CHRISTODOULOU, D., FERENTINOS, G., SOKOS, E. & FAVALI, P. 2006. Methane and hydrogen sulfide seepage in the NW Peloponnesus petroliferous basin (Greece): origin and geohazard. *American Association of Petroleum Geologists Bulletin*, **90**, 701–713.

FERENTINOS, G. 1992. Recent gravitative mass movements in a highly tectonically active arc system: the Hellenic Arc. *Marine Geology*, **104**, 93–107.

FERENTINOS, G. & PAPATHEODOROU, G. 1993. *150 kv Submarine Power Link Evia–Andros–Tinos, Greece*. Land and Marine Survey. Interim Report. Laboratory of Marine Geology & Physical Oceanography, Department of Geology, University of Patras.

FERENTINOS, G. & PAPATHEODOROU, G. 1994. *Submarine Telecommunication Link Preveza-Corfu, Greece*. Land and Marine Survey. Interim Report. Laboratory of Marine Geology & Physical Oceanography, Department of Geology, University of Patras.

FERENTINOS, G. & PAPATHEODOROU, G. 1995. *Submarine Telecommunication Link Lagonisi (Athens)–Milos Isl.–Chania (Crete), Greece.* Land and Marine Survey. Interim Report. Laboratory of Marine Geology & Physical Oceanography, Department of Geology, University of Patras.

FERENTINOS, G. & PAPATHEODOROU, G. 1998. *150 kv Submarine Power Link Zakynthos–Kyllini, Greece.* Land and Marine Survey. Interim Report. Laboratory of Marine Geology & Physical Oceanography, Department of Geology, University of Patras.

FERENTINOS, G., BROOKS, M. & DOUTSOS, T. 1985a. Quaternary tectonics in the Gulf of Patras, Western Greece. *Journal of Structural Geology*, **7**, 713–717.

FERENTINOS, G., COLLINS, M., PATTIARATCHI, C. B. & TAYLOR, P. G. 1985b. Mechanisms of sediment transport and dispersion in a tectonically active submarine valley/canyon system: Zakynthos Straits, NW Hellenic Trench. *Marine Geology*, **65**, 243–269.

FERENTINOS, G., PAPATHEODOROU, G. & COLLINS, M. 1988. Sediment transport processes on an active submarine fault escarpment: Gulf of Corinth, Greece. *Marine Geology*, **83**, 43–61.

FERENTINOS, G., PAPATHEODOROU, G. ET AL. 2010. *Ag. Georgios Isl. Windfarm. Submarine Power Cable Line Route Survey Report.* Interim Report. Laboratory of Marine Geology & Physical Oceanography, Department of Geology, University of Patras.

FERENTINOS, G., GKIONI, M., GERAGA, M. & PAPATHEODOROU, G. 2012. Early seafaring activity in the southern Ionian Islands, Mediterranean Sea. *Journal of Archaeological Science*, **39**, 2167–2176, http://dx.doi.org/10.1016/j.jas.2012.01.032

GALANOPOULOS, A., DELIMBASIS, N. D. & COMNINAKIS, P. E. 1966. A tsunami generated by a slide without a seismic shock. *Geological Chronicles of Greece*, **16**, 93–110 (in Greek).

GERAGA, M., TSAILA, M. S., IOAKIM, CH., PAPATHEODOROU, G. & FERENTINOS, G. 2000. Evolution of palaeoevironmental changes during the last 18.000 years in the Myrtoon Basin SW Aegean Sea. *Palaegeography, Palaeoclimatology, Palaeoecology*, **156**, 1–17.

GERAGA, M., TSAILA-MONOPOLI, S., IOAKIM, CH., PAPATHEODOROU, G. & FERENTINOS, G. 2005. Short-term climate changes in the southern Aegean Sea over the last 48 000 years. *Palaeogeography, Palaeoclimatology, Palaeoecology*, **220**, 311–332.

GERAGA, M., MYLONA, G., TSAILA-MONOPOLI, S., PAPATHEODOROU, G. & FERENTINOS, G. 2008. Northeastern Ionian Sea: Palaeoceanographic variability over the last 22 ka. *Journal of Marine Systems*, **74**, 623–638.

GEORGIADIS, M., PAPATHEODOROU, G., TZANATOS, E., GERAGA, M., RAMFOS, A., KOUTSIKOPOULOS, C. & FERENTINOS, G. 2009. Coralligène formations in the eastern Mediterranean Sea: morphology, distribution, mapping and relation to fisheries in the southern Aegean Sea (Greece) based on high-resolution acoustics. *Journal of Experimental Marine Biology and Ecology*, **368**, 44–58.

GIRESSE, P., BUSCAIL, R. & CHARRIERE, B. 2003. Late Holocene multisource material input into the Aegean Sea: depositional and post-depositional processes. *Oceanologica Acta*, **26**, 657–672.

HASIOTIS, T., PAPATHEODOROU, G., KASTANOS, N. & FERENTINOS, G. 1996. A pockmark field in the Patras (Greece) and its activations during the 14/7/93 seismic event. *Marine Geology*, **130**, 333–344.

HASIOTIS, T., PAPATHEODOROU, G., BOUCKOVALAS, G., CORBAU, C. & FERENTINOS, G. 2002. Earthquake-induced coastal sediment instabilities in the western Gulf of Corinth, Greece. *Marine Geology*, **186**, 319–335.

HASIOTIS, T., PAPATHEODOROU, G. & FERENTINOS, G. 2005. A high resolution approach in the recent sedimentation processes at the head of Zakynthos Canyon, western Greece. *Marine Geology*, **214**, 49–73.

HATZFELD, D., KASSARAS, I., PANAGIOTOPOULOS, D. G., AMORESE, D., MAKROPOULOS, K., KARAKAISIS, G. F. & COUTANT, O. 1995. Microseismicity and strain pattern in North-western Greece. *Tectonics*, **14**, 773–785.

JACKSON, J. A. 1994. Active tectonics of the Aegean region. *Annual Review Earth and Planetary Science*, **22**, 239–271.

KAPSIMALIS, V., PAVLOPOULOS, K., PANAGIOTOPOULOS, I., DRAKOPOULOS, P., VANDARAKIS, P., SAKELARIOU, D. & ANAGNOSTOU, CH. 2009. Geoarchaological challenges in the Cyclades continental shelf (Aegean Sea). *Zeitschrift für Geomorphologie*, **53**, (Suppl. 1), 169–190.

KARAGEORGIS, A. & ANAGNOSTOU, CH. 2001. Particulate matter spatial-temporal distribution and associated surface sediment properties in Thermaikos Gulf and Sporades Basin, NW. Aegean Sea. *Continental Shelf Research*, **21**, 2141–2153.

KOKKALAS, S. & AYDIN, A. 2012. Is there a link between faulting and magmatism in the south-central Aegean Sea? *Geological Magazine*, **150**, 193–224, http://dx.doi.org/10.1017/S0016756812000453

KOKKALAS, S. & DOUTSOS, TH. 2001. Strain-dependent stress field and plate motions in the south-east Aegean region. *Journal of Geodynamics*, **32**, 311–332.

KOKKALAS, S., XYPOLIAS, P., KOUKOUVELAS, I. & DOUTSOS, T. 2006. Post-collisional contractional and extensional deformation in the Aegean region. *In*: DILEK, Y. & PAVLIDES, S. (eds) *Post Collisional Tectonics & Magmatism in the Mediterranean Region and Asia.* Geological Society of America Special Papers, **409**, 97–123.

KOKKINOU, E., PAPADIMITRIOU, E., KARAKOSTAS, V., KAMPERIS, E. & VALLIANATOS, F. 2006. A review of the Kefalonia Transform Zone (offshore Western Greece) with special emphasis to its prolongation towards the Ionian Abyssal Plain. *Marine Geophysical Researches*, **27**, 241–252.

KOUKOUVELAS, I. & AYDIN, A. 2002. Fault structure and related basins of the north Aegean Sea and its surroundings. *Tectonics*, **21**, 1–7.

LAIGLE, M., HIRN, A., SACHPAZI, M. & ROUSSOS, N. 2000. North Aegean crustal deformation: an active fault imaged to 10 km depth by reflection seismic data. *Geology*, **28**, 71–74.

LYKOUSIS, V. 1991a. *Sea-Level Changes and Sedimentary Evolution during Quaternary in the Northwest Aegean Continental Margin, Greece. In:* MACDONALD, D. I. M. (ed.) *Sedimentation, Tectonics and Eustasy: Sea-Level Changes at Active Margins.* International Association of Sedimentologists, Special Publications, **12**, 123–131.

LYKOUSIS, V. 1991b. Submarine slope instabilities in the Hellenic Arc Region, Northeastern Mediterranean sea. *Marine Geotechnology*, **10**, 83–96.

LYKOUSIS, V. 2001. Subaqueous bedforms on the Cyclades Plateau (NE Mediterranean)-evidence of Cretan deep water formation? *Continental Shelf Research*, **21**, 495–507.

LYKOUSIS, V. 2009. Sea-level changes and shelf break prograding sequences durin the last 400 ka in the Aegean margins: subsidence rates and Palaeogeographic implications. *Continental Shelf Research*, **29**, 2037–2044.

LYKOUSIS, V. & CHRONIS, G. 1989. Mechanisms of sediment transport and deposition: sediment sequences and accumulation during Holocene on the Thermaikos plateau, the continental slope and basin (Sporades basin), northwestern Aegean Sea, Greece. *Marine Geology*, **87**, 15–26.

LYKOUSIS, V., COLLINS, M. B. & FERENTINOS, G. 1981. Modern sedimentation in the NW Aegean Sea. *Marine Geology*, **43**, 111–130.

LYKOUSIS, V., ANAGNOSTOU, C., PAVLAKIS, P., ROUSSAKIS, G. & ALEXANDRI, M. 1995. Quaternary sedimentary history and neotectonic evolution of the eastern part of Central Aegean Sea, Greece. *Marine Geology*, **128**, 59–71.

LYKOUSIS, V., KARAGEORGIS, A. & CHRONIS, G. 2005. Delta progradation and sediment fluxes since the last glacial in the Thermaikos Gulf and the Sporades Basin, NW Aegean Sea, Greece. *Marine Geology*, **222–223**, 381–397.

LYKOUSIS, V., SAKELLARIOU, D., MORETTI, I. & KABERI, H. 2007. Late Quaternary basin evolution of the Gulf of Corinth: sequence stratigraphy, sedimentation, fault- slip and subsidence rates. *Tectonophysics*, **440**, 29–51.

LYKOUSIS, V., SAKELLARIOU, D. & ROUSSAKIS, G. 2009. Slope failures and stability analysis of shallow water prodeltas in the active margins of Western Greece, northeastern Mediterranean Sea. *International Journal of Earth Sciences*, **98**, 807–822.

MASCLE, J. & MARTIN, L. 1990. Shallow structure and recent evolution of the Aegean Sea: a synthesis based on continuous reflection profiles. *Marine Geology*, **94**, 271–299.

MCCLUSKY, S., BALASSANIAN, S. ET AL. 2000. Global positioning system constraints on plate kinematics and dynamics in the eastern Mediterranean and Caucasus. *Journal of Geophysical Research*, **105**, 5695–5719.

MONOPOLIS, D. & BRUNETON, A. 1982. Ionian Sea (Western Greece): its structural outline deduced from drilling and geophysical data. *Tectonophysics*, **83**, 227–242.

PAPADOPOULOS, G. A. & CHALKIS, B. J. 1984. Tsunamis observed in Greece and the surrounding area from Antiquity up to the present times. *Marine Geology*, **56**, 309–317.

PAPANIKOLAOU, D., LYKOUSIS, V., CHRONIS, G. & PAVLAKIS, P. 1988. A comparative study of neotectonic basins across the Hellenic arc: the Messiniakos, Argolikos, Saronikos and Southern Evoikos Gulfs. *Basin Research*, **1**, 167–176.

PAPATHEODOROU, G. & FERENTINOS, G. 1993. Sedimentation processes and basin filling depositional architecture in an active graben: Strava graben, Gulf of Corinth, Greece. *Basin Research*, **5**, 235–253.

PAPATHEODOROU, G. & FERENTINOS, G. 1997. Submarine and coastal sediment failure triggered by the 1995, Ms = 6.1R Aegion earthquake, Gulf of Gorinth, Greece. *Marine Geology*, **137**, 287–304.

PAPATHEODOROU, G., HASIOTIS, T. & FERENTINOS, G. 1993. Gas-charged sediments in the Aegean and Ionian Seas, Greece. *Marine Geology*, **112**, 171–184.

PERISSORATIS, C. & PAPADOPOULOS, G. 1999. Sediment instability and slumping in the southern Aegean Sea and the case history of the 1956 tsunami. *Marine Geology*, **161**, 287–305.

PIPER, D. J. W. & PERISSORATIS, C. 1991. The Quaternary sedimentation of the North Aegean continental margin, Greece. *American Association of Petroleum Geologists Bulletin*, **75**, 46–61.

PIPER, D. J. W. & PERISSORATIS, C. 2003. Quaternary Neotectonics of the South Aegean arc. *Marine Geology*, **198**, 259–288.

POULOS, S. 2009. Origin and distribution of the terrigenous component of the unconsolidated surface sediment of the Aegean seafloor: a synthesis. *Continental Shelf Research*, **29**, 2045–2060.

POULOS, S., VOULGARIS, G., KAPSIMALIS, V., COLLINS, M. & LYKOUSIS, V. 2002. Sediment fluxes and evolution of a riverine-supplied tectonically active coastal system: Kyparissiakos Gulf, Ionian Sea (eastern Mediterranean). *In*: JONES, S. J. & FROSTICK, L. E. (eds) *Sediment Flux to Basins: Causes, Controls and Consequences*. Geological Society, London, Special Publications, **191**, 247–266.

ROUSSAKIS, G., KARAGEORGIS, A. P., CONISPOLIATIS, N. & LYKOUSIS, V. 2004. Last glacial-Holocene sediment sequences in N. Aegean basins: structural, accumulation rates and clay mineral distribution. *Geo-Marine Letters*, **24**, 97–111.

ROUSSOS, N. & LYSSIMACHOU, T. 1991. Structure of the Central North Aegean trough: an active strike slip deformation zone. *Basin Research*, **3**, 39–48.

SOTER, S. 1999. Macroscopic seismic anomalies and submarine pockmarks in the Corinth–Patras rift, Greece. *Tectonophysics*, **308**, 275–290.

STANLEY, D. & PERISSORATIS, K. 1977. Aegean Sea ridge-barrier and basin sedimentation patterns. *Marine Geology*, **24**, 97–107.

STEFATOS, A., PAPATHEODOROU, G., FERENTINOS, G., LEEDER, M. & COLLIER, R. 2002. Seismic reflection imaging of active offshore faults in the Gulf of Corinth: their seismotectonic significance. *Basin Research*, **14**, 487–502.

STEFATOS, A., CHARALAMBAKIS, M., PAPATHEODOROU, G. & FERENTINOS, G. 2006. Tsunamigenic sources in an active European half-graben (Gulf of Corinth, Central Greece). *Marine Geology*, **232**, 35–47.

TAYMAZ, T., JACKSON, J. & MCKENZIE, D. 1991. Active tectonics of the north and central Aegean Sea. *Geophysical Journal International*, **106**, 433–490.

TAYMAZ, T., YILMAZ, Y. & DILEK, Y. 2007. The geodynamics of the Aegean and Anatolia: introduction. *In*: TAYMAZ, T., YILMAZ, Y. & DILEK, Y. (eds) *The Geodynamics of the Aegean and Anatolia*. Geological Society, London, Special Publications, **291**, 1–16.

VOLAKIS, S. & ANAGNOSTOU, CH. 1993. Modern sedimentation at the North part of the Cyclades Plateau. *In*: PAPATHANASSIOU, E. & CHAROU, E. (eds) *Proceedings of the Fourth National Symposium on Oceanography and Fisheries, Rhodes Island, 26–29 April 1993*. National Centre for Marine Research, Athens, 101–104.

Chapter 14

Late Quaternary sea-level change on the Black Sea shelves

W. A. NICHOLAS[1,2]* & ALLAN R. CHIVAS[3]

[1]*GeoQuEST Research Centre, School of Earth & Environmental Sciences, University of Wollongong, Wollongong, NSW 2522, Australia*

[2]*Present address: Geoscience Australia, GPO Box 378, Canberra, ACT 2601, Australia*

[3]*GeoQuEST Research Centre, School of Earth & Environmental Sciences, University of Wollongong, NSW 2522, Australia*

*Corresponding author (e-mail: wan734@uowmail.edu.au)

Abstract: The continental shelf of the microtidal Black Sea, with its 4125 km-long coastline, has an average width of *c.* 40 km and may be subdivided into seven regions, the largest of these being the NW Shelf and the SW Shelf. Palimpsest shelf environments and terrestrially derived siliciclastic sedimentation are dominant. The Marmara Gateway, bisecting the southwestern shelf, controls the influx of Mediterranean-sourced water into the semi-enclosed Black Sea, and has been episodically open during the Quaternary. The best evidence for the most recent marine transgression of the pre-existing and isolated lake is found on the mid- and outer NW Shelf. Radiocarbon dating of marine and non-marine bivalve molluscs and peat indicates the most recent re-connection of the Black Sea with the oceanic reservoir occurred between 8200 and 8600 ^{14}C years BP. Water levels during the transgression in the enclosed lake rose from a level 107 m or more below present values, to above the Bosphorus Sill (-35 m) in about 400 ^{14}C years. However, the transition from low-salinity (0–5‰) lacustrine conditions to near-modern values (18‰) took $\geq$1000 years, indicated by the progression from freshwater to brackish molluscs and gastropods, dominated by *Dreissena*, to Mediterranean-derived estuarine taxa (e.g. *Mytilus*), suited to nutrient-rich conditions.

The Black Sea is a semi-enclosed and restricted environment, connected to the Mediterranean Sea and, therefore, to the global ocean, only by the narrow Bosphorus and Dardanelles straits and the Sea of Marmara. As a result, interaction of the Black Sea with the oceanic marine reservoir is limited. The Black Sea Basin is situated within Paratethys, a continental-scale region to the north of the Mediterranean–Tethys tectonic system, notable because it consisted of episodically and internally interconnected basins situated between the European Alps and the Aral Sea, some of which no longer act as major sedimentary sinks. The sedimentary record of the Black Sea documents the development of Ponto-Caspian fauna during the Quaternary, and the subsequent decline of the Paratethys region (Black Sea, Dacian, Caspian, Aral and Pannonian basins) as sedimentary basins. Since its formation during the Cretaceous in a back-arc setting, the Black Sea region has repeatedly become isolated from neighbouring depressions; principally the Caspian and Mediterranean basins. The marine gateway to the Black Sea, the Dardanelles–Marmara–Bosphorus seaway, controls the movement of eustatically derived marine water into this, the Pontic Basin. During the Quaternary, this and other peri-Tethyan gateways acted as physical and biogeographical restrictions across which fauna could or could not cross. This has resulted in the development of an endemic Ponto-Caspian fauna restricted at present to lagoons and deltas on the north and NW shelves of the Black Sea, and, primarily, to the neighbouring Caspian Sea. Faunal and floral changes in the Black Sea have been used as relative chronometers and environmental markers. Didacnid and Dreissenid molluscs, in particular, are associated with brackish water, and indicative of separation of the Pontic and adjacent basins from a marine source, and comparatively cool and dry climatic conditions. In contrast, the presence of Mediterranean-derived fauna within the Black Sea Basin indicates a marine environment, with warm and relatively humid climatic conditions.

Unravelling the Quaternary history of the Black and neighbouring seas is difficult because of the tectonic and geographical setting. The task of correlating at fine scales of resolution across adjacent coasts and regions of the Black Sea is made difficult because of differences in language and style of reporting, uncertainties in and the lack of high-resolution neotectonic data, and also the lack of numerical chronological control. The latter is perhaps the most important, and is reflected in continuing efforts to comprehend the history of the Last Glacial Cycle in this region. Therefore, this study aimed at reviewing recent chronological evidence for the youngest transgression of the Black Sea, and at presenting critical new accelerator mass spectrometry (AMS) radiocarbon ages from bivalve molluscs and peat samples that indicate a rapid transition from the restricted and isolated lacustrine environment to marine conditions during the early Holocene.

Tectonic framework and geological setting

The Black Sea (Fig. 14.1) is bordered by six countries – Bulgaria, Romania, Ukraine, Russia, Georgia and Turkey – and two principal mountain chains – the Crimea–Caucasus and the Balkan–Pontides, both formed during the Alpine Orogeny. This region is tectonically active and, although extensional in nature, it is surrounded by compressive orogenic belts: the Pontides of northern Turkey; the Caucasus of Georgia and Russia; the Crimea in Ukraine; and the Balkanides of Bulgaria (Robinson *et al.* 1996). Neotectonism presently occurs in response to compression because of the northward movement of the Arabian plate, and in response to westward escape of the Anatolian block along the North and East Anatolian Faults (Rangin *et al.* 2002; Shillington *et al.* 2008).

The present basin developed from two pre-existing Cretaceous basins originated within a peri-Tethyan back-arc setting, driven by the northward subduction of the Arabian plate (Nikishin *et al.* 2003). The Bosphorus Strait, Marmara Sea and Dardanelles Strait, together, constitute the Marmara Gateway. Connection of Paratethys to the Tethys Ocean has, since the Miocene, been through the Marmara Gateway (Çağatay *et al.* 2006). This physical and biogeographical restriction has been episodically utilized in the migration of marine organisms into the variously isolated and commonly freshwater to brackish-water Paratethyan environments. The Marmara Gateway may have been the sole structural

From: CHIOCCI, F. L. & CHIVAS, A. R. (eds) 2014. *Continental Shelves of the World: Their Evolution During the Last Glacio-Eustatic Cycle.* Geological Society, London, Memoirs, **41**, 199–212. http://dx.doi.org/10.1144/M41.14

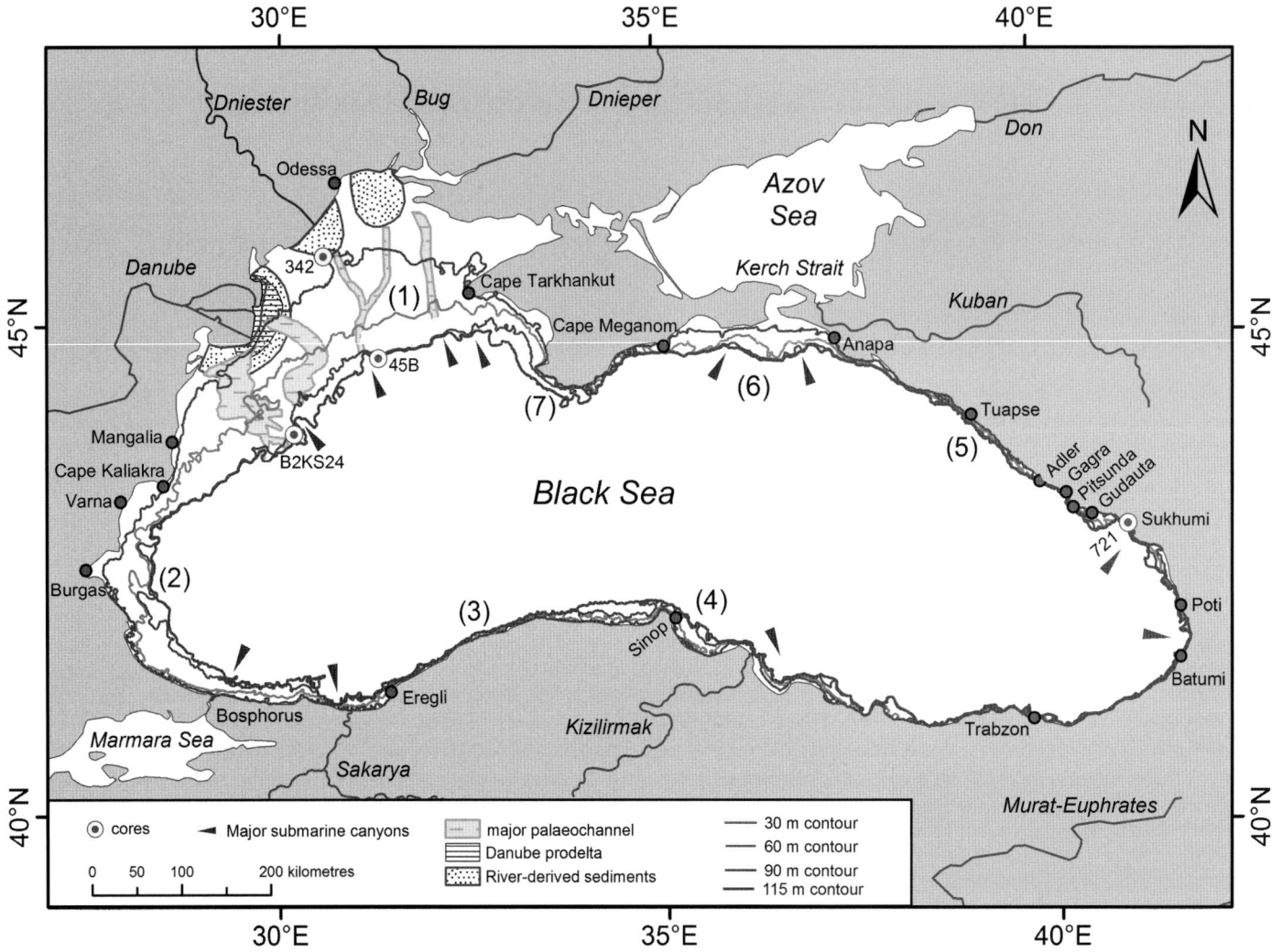

Fig. 14.1. Selected principal features of the shelves of the Black Sea adapted from Panin (2005) and Lericolais *et al.* (2009). Relict features from previous lowstands include palaeo-channels of the Danube, Dniester and Dnieper. The largest section, the NW Shelf, is similar to other Black Sea shelves by being predominantly sediment-starved. Numerous canyon systems are present on the slopes, with sediment ultimately being deposited in the basin at depths of 1000–2000 m. Selected contours illustrated at 30, 60, 90 and 115 m water depth.

control responsible for regulating freshwater and glacial meltwater outflow from the Black Sea Basin in the Mediterranean region during the Quaternary (Yaltirak *et al.* 2002). Based on the $\delta^{18}O$ values of speleothems from Sofular cave, northern Turkey, this gateway has been open at least 12 times over the past 670 kyr (Badertscher *et al.* 2011).

Oceanography and climate

The oceanography of the Black Sea is constrained by the restricting bathymetry, a cyclonic wave-climate regime and limited water exchange through the Bosphorus. These conditions result in a positive water balance, created by an excess of fluvial discharge (*c.* 350 km³ a⁻¹) and precipitation (*c.* 300 km³ a⁻¹) over evaporation (*c.* 350 km³ a⁻¹), and maintained by an excess outflow (*c.* 1 km³ per day) of the low-salinity (17–19‰) surface layer over inflow of dense (36–38‰) saline water through the Bosphorus Strait (Unluata *et al.* 1990; Özsoy & Ünlüata 1997; Oguz *et al.* 2004). This is a microtidal basin, more influenced by wave-storm climate than by its mean semi-diurnal tides of 7–12 cm. Storm surges commonly elevate sea level by 1–2 m above datum (Giosan *et al.* 2005). Mean spring tides in Odessa may raise the water level up to 17 cm (Kosarev *et al.* 2008). Maximum tidal range on the Danube delta coast is 12 cm (Vespremeanu-Stroe

et al. 2007), while elsewhere (e.g. Tuapse and Varna) tidal ranges are commonly 0–5 cm. Mean yearly sea level oscillates by 7–8 cm owing to changes in atmospheric pressure predominantly controlled by eastward-tracking cyclonic conditions driven by North Atlantic Oscillation (NAO) mechanisms (Stanev & Peneva 2002).

The modern Black Sea Basin is an example of the classical euxinic basin (Degens & Ross 1974; Ogawa *et al.* 2009), having restricted circulation and anaerobic conditions in 99% of the water column. Warm marine water (*c.* 15 °C, salinity = 36‰) flows into the Black Sea underneath the outgoing and less saline, cooler and lower-density surface water through the Bosphorus Strait. The water column within the Black Sea can be differentiated into three principal units: an upper water column at 0–300 m; an intermediate water mass between 300 and 1000 m; and basal waters below 1000 m depth (Korotaev *et al.* 2006). Ventilation through density stratification occurs only in the upper 60 m (Oguz *et al.* 2002). Ventilation of deep-water layers is extremely limited, and residence time in the deep water is approximately 300 years (Murray *et al.* 2007). A permanent halocline at 100–150 m water depth differentiates the oxygenated surface waters from the anoxic–sulphide basinal waters. This differentiation leads to oxygenated shelves, suboxic water at or just below shelf break and anoxic water from the upper slope below the 200 m isobath to the abyssal plain. Shelf systems in the Black Sea are thus dominated by brackish, Mediterranean-derived estuarine taxa, suited to nutrient-rich marine conditions. These ecosystems are limited to the upper approximately 150 m of the water column. Remnant Caspian fauna inhabit lagoons, rivers and lakes in the NW and north.

Nutrient-rich conditions are the result of elevated primary production, strong density stratification and weak ventilation with high O_2 demand resulting from an enhanced primary productivity, particularly during peak flood discharge during spring and early summer (Sur *et al.* 1996). An estuarine–brackish water mass (distal to estuaries and deltas) with a seasonally variable salinity of 17–18‰ occupies the upper 150 m. Water below 100 m is sulphide-rich. Salinity in close proximity to deltas can be as low as 1–5‰ (Stanev *et al.* 2002) and, at about 40 m water depth on the NW Shelf, salinity is in the region of 6–9‰. Surface waters above the 40 m isobath in the Danube delta prodelta region have a salinity of approximately 6‰. Underneath, a higher-salinity intermediate water body (300–1000 m depth) with temperatures of 6.5–8 °C overlies the slightly warmer and more saline basal waters (8–8.5 °C).

Climate

Climate in the Black Sea varies between temperate in the north and NW, and subtropical in the south and SE. A dry temperate climate is present on the western seaboard. Mean annual air temperature varies from 10 °C in the NW to 15 °C in the SE. During winter, mean minimum daily air temperatures on northern coasts may be as low as −25 °C (Kosarev *et al.* 2008), although reaching 8–10 °C during the day. Winter air temperatures on the southern Crimean coast rarely go below 0 °C because the mountain chain prevents cold air from the north penetrating into the coastal zone. Temperatures also rarely drop below 0 °C on the Anatolian and Caucasian coasts during winter. Occasionally, summer temperatures can reach a maximum of 35–40 °C but mean daily temperatures of up to 25–30 °C occur during summer for all coastal regions of the Black Sea. Mean daily values of 35–40 °C are never reached (Kosarev *et al.* 2008).

Precipitation in the Black Sea is controlled primarily by cyclonic activity (Kosarev *et al.* 2008), both from northern Europe and from the Mediterranean. Cyclonic activity in the Mediterranean strongly influences precipitation on the SW Shelf (Kwiecien *et al.* 2009) where more rainfall is received in coastal mountains than on the arid Anatolian Plateau. In these southern Black Sea coastal regions, precipitation is driven to a lesser extent by convection. Rainfall is greatest in the SE and eastern coastal regions where the Caucasian Mountains exist close to the coastal zone. Higher humidity occurs during the summer in the eastern coastal regions but varies little between hotter and cooler seasons. Rainfall is lowest on the NW Shelf, with a mean annual rainfall of 345 mm on the dry temperate Danube delta (Vespremeanu-Stroe & Preoteasa 2007). Approximately one-third of the area of Europe drains into this semi-enclosed sea, predominantly via rivers that discharge on to the NW Shelf. Shelf currents are dominated by the wind-driven cyclonic Rim Current, which has several anticyclonic gyres inshore of the influence of the anticlockwise-directed flow. These include the Caucasus, Sevastopol, Bosphorus and Sakarya gyres. Inshore of these gyres on the NW Shelf, wind-generated longshore currents transport littoral sediment southwards.

Present shelf environments

The continental shelf of the Black Sea may be conveniently subdivided into seven regions (Fig. 14.1). These regions include the two larger shelf areas: the NW Shelf, which includes the Romanian, Ukrainian and northern part of the Bulgarian Shelf; and the SW

Table 14.1. *Characteristics of the continental shelves of the Black Sea*

Length of shelf (km)	4125
Average width (km)	41 (average of 35 measurements on Fig. 14.1)
Tidal, wave, current ranges Dominating processes	Microtidal to non-existent tides Inshore: variable wave-driven longshore drift, south-directed on the NW and Caucasian shelves Offshore: basin-wide anticlockwise-directed offshore suspension currents
Average depth of shelf break	Estimated 115 m but variable between 100 and 150 m
Siliciclastic/carbonate/ authigenic/glacial sedimentation	Shelves presently dominated by terrestrially derived siliciclastics; offshore seasonal deposition of coccolithophore-derived carbonate occurs
Modern/relict/palimpsest	30% modern, 70% relict–palimpsest (variable)
Tectonic trend over the last glacial cycle	Variably stable. Anatolian and Caucasian coasts are likely to have experienced uplift, whereas central Georgia is likely to have undergone minor subsidence since the LGM. By comparison, the NW and SW shelves are likely to have been relatively stable over this time period. The exception to this is the extent to which the NW Shelf has been depressed by loading since the most recent transgression.

Shelf, which includes the southern Bulgarian Shelf, the Bosphorus Channel region and the Sakarya delta to the east. By comparison, the east and west Anatolian shelves on the southern coast, the Caucasian Shelf in the east and SE, the Kerch–Taman, and the Crimean shelves are significantly smaller, and commonly very narrow. Overall, the coastal system of the Black Sea is dominated by erosional processes (Table 14.1). The majority of the wide NW Shelf is currently sediment-starved. The best evidence for the most recent transgression of the Black Sea by Mediterranean-sourced water is found on this shelf.

The NW Shelf

The NW Shelf occupies approximately 94% of the total (135 000 km^3) shelf area of the Black Sea, and has a water volume of 6500 km^3 (Panin & Jipa 2002). The NW Shelf receives the major proportion of fluvial discharge into the Black Sea via the Danube, Dnieper and Dniester rivers. This segment of shelf extends from Cape Tarkhankut on the Crimean coast to Cape Kaliakra in Bulgaria. In contrast with other shelves in this basin, the NW Shelf appears to have remained comparatively stable throughout the Quaternary and, therefore, sea-level changes documented on this broad submarine surface suggest a eustatic rather than a tectonic origin (Robinson *et al.* 1996).

The influence of the 2857 km-long Danube River (Stanica *et al.* 2007) is predominant in terms of sediment discharge and water volume, sourced from a drainage basin approximately twice the area of the Black Sea. Palaeo-deltas of the Danube, Dnieper and Dniester and their related and sediment-filled NE–SW-trending palaeo-channels are located on the outer shelf and upper continental slope (Popescu *et al.* 2004; Naudts *et al.* 2008; Lericolais *et al.* 2009). The shelf break at the location of the palaeo-Dnieper delta occurs at a water depth of 105 m. The Ukrainian rivers, Dniester, Dnieper and Southern Bug, deposit sediment in coastal lakes and lagoons prior to discharging in a SW direction on to the shelf. Onshore, mid- and late-Pleistocene fluvial terraces of the

Dnieper and adjacent rivers provide a ready source of sandy sediment that is subsequently redeposited within the lagoons (limans) and around coastal bars on the Ukrainian coast (Konikov *et al.* 2007).

The Romanian coastal zone is 240 km long (Caraivan *et al.* 2003). The northern portion of this shelf includes the Danube delta front (*c.* 1300 km^2) and pro-delta ($\geq$6000 km^2). The delta front extends to water depths of 15–45 m, while that of the pro-delta extends to depths of 50–60 m. Coarse-grained palimpsest sediments form the most shelfward extent of the delta (Giosan *et al.* 2005). Shoreline retreat occurs between the Sulina mouth and the Saint George (Sfantu Gheorghe) distributary (Giosan *et al.* 2005). The dominant longshore current direction, driven by wave climate in front of the Danube delta littoral zone, is southwards. Within the Danube littoral zones, superficial sediments are mainly fine to very fine well-sorted sands, sourced from the Danube and redistributed onshore by waves and currents (Stanica *et al.* 2007). Sediments in the delta front consist overwhelmingly of terrestrially derived black muds. Bivalve molluscs (e.g. *Cardium edule*) and similar shells that inhabit the littoral zone are reworked into these fine-grained sediments. However, the high sediment discharge rate adjacent to the distributaries prevents colonization of the delta front, and the muds tend to be high in H$_2$S (Oaie *et al.* 2005).

South of the Danube delta, the Romanian Shelf consists of an inner (0–30 m water depth), mid (30–60 m) and outer shelf. The inner shelf is approximately 5 km wide in the south, opposite Mangalia, and 15 km wide in its northern section. Sedimentation on the inner shelf is driven by fair-weather and storm-wave-generated processes, occasionally producing sand sheets (Caraivan *et al.* 2003). However, the littoral zone in this region is sediment starved and, apart from the northern side of Midia Harbour, the shoreline is retreating. Shorelines south of Constanta receive no sandy sediment from the Danube, and beaches tend to be dominated in places by *Mytilus* shell debris (concentration deposits). In contrast, the inner and mid Romanian Shelf above 60 m water depth is fed by suspension-load from the Danube River (Panin & Jipa 2002). These silts and clays are transported southwards by shelf currents, reaching the slope and abyssal plain after travelling across the Bulgarian Shelf, occasionally reaching the Bosphorus region (Panin & Jipa 2002; Giosan *et al.* 2005).

The Southwestern Shelf

The Southwestern Shelf includes the Bulgarian Shelf south of Cape Kaliakra, and the Turkish Shelf NW (Thracian) and SE (Anatolian) of the Bosphorus Channel, extending to and including the Sakarya delta. The Anatolian Shelf is incised by the 80–90 m-deep and 200–500 m-wide Bosphorus Channel – an extension of the Bosphorus Strait (Di Iorio & Yüce 1999; Aksu *et al.* 2002; Ongan *et al.* 2009). The channel divides the shelf region into a 25–35 km-wide western segment of approximately 10 000 km^2 and a 15–20 km-wide eastern sector. The shelf break to the NW and SE of the Bosphorus Channel occurs at 110–120 m water depth (Aksu *et al.* 2002; Ongan *et al.* 2009) with slopes of 5–9° leading to the Euxine abyssal plain at −2200 m. On the Bulgarian, and other shelves, methane seeps are commonly inhabited by molluscan communities, including *Mytilus* sp.

The Sakarya fluvial system may have been an alternative or additional connection between the Sea of Marmara and the Black Sea (Pfannenstiel 1944; Yanko-Hombach *et al.* 2007; Gürbüz & Leroy 2009). The shelf at the Sakarya delta extends east and west by approximately 25 km in each direction. The shelf at this location is generally less than 10 km wide and the shelf break is at 100 m water depth. The submarine canyon associated with this delta (Algan *et al.* 2002) is located 50 km from the coast, dissects the submarine Sakarya delta and adjoins an adjacent canyon at −800 m.

The east and west Anatolian shelves

The Pontide mountain chain extends from eastern Turkey to the Sakarya River in the west. The Anatolian coasts are therefore dominated by steep rocky cliffs and, as elsewhere, these shelves are predominantly erosional in nature. The maximum width of these shelves is approximately 30 km but may be as narrow as 1–4 km in width (Yilmaz 2007; Ignatov 2008). The western Anatolian Shelf extends from the western margin of the Sakarya delta near Ereğli, eastwards to Sinop. The eastern Anatolian Shelf stretches between the Sinop Peninsula and the Georgian border. These shelves are dissected by numerous submarine canyons (Ignatov 2008). In eastern Anatolia, marine terraces have been noted at Sinop, Gerze and Trabzon at up to 250 m above present sea level, although the precise ages of these sedimentary units are uncertain.

The Caucasian Shelf

The Caucasian Shelf, consisting of Georgian and Russian coastal zones, occurs in close proximity to the Caucasus Mountains. This particularly narrow shelf extends from Anapa in the north to the Ch'oroki River, west of Batumi near the Turkish–Georgian border. The coastal sections from Anapa to Sukhumi commonly consist of cliffs up to 200 m high and narrow beaches. The average shelf width is 12 km but extends to 30 km wide between Cape Pitsunda and the Gumista River, and in the area between Cape Iskuriya and Cape Anakliya (Ignatov 2008). This shelf is narrowest off Leslidze and Sukhumi, with a width of 4 km, and many submarine canyons occur within a few kilometres of the coast. From NE to SE, the source regions are increasingly elevated (from a few tens of metres to over 3000 m in height), drainage basins are enlarged, precipitation increases and the distance to the foothills of the mountains increases. Thus, the highest sediment discharge occurs on the Georgian Shelf. Most rivers are short, with small drainage basins (as little as 14 km^2), and much of the sediment is transported directly into canyon systems, bypassing the littoral zone. Much of this shelf is sediment-starved as a result, and Caucasian beaches are commonly composed of gravel and pebbles rather than fine sand. The Georgian region around Potihas subsided by approximately 400 m (1 m per 7 kyr) during the Quaternary (Tsagareli 1974). In comparison, subsidence at Sukhumi and Batumi is near zero. However, for most of the Black Sea region, the extent of neotectonic deformation is poorly constrained.

The Kerch–Taman Shelf

The 60 km-wide gently sloping Kerch–Taman Shelf is located between Cape Meganom (Crimea) and Cape Utrish, near Anapa (Russian Caucasia). The outer shelf has both sediment-starved and/or erosional sections, as well as accumulative bars (Ignatov 2008). The major rivers, Don and Kuban, discharge into the Sea of Azov, and fine sediment from these fluvial sources is transported through the shallow Kerch Strait and across the Kerch–Taman Shelf.

The southern Crimean Shelf

The Crimean Shelf is between 5 and 15 km wide in the east, and up to 40 km wide in the southernmost section, and extends from Cape Meganom in the east to Cape Tarkhankut in the west. The Crimean Shelf is undergoing uplift, the exact value of which is uncertain. Similar to other mountain-backed coasts of the Black Sea, beaches on the Crimea are predominantly composed of gravel because finer material is transported down the submarine canyons. Consequently, turbidites are common on the outer shelf and slopes.

AMS ^{14}C chronology of post-glacial shelf sedimentation: the most recent transgression

Three principal hypotheses for the deposition of recent sediments on the shelves of the Black Sea have been proposed. The concept of an instantaneous flooding of the southern Crimean shelves and, therefore, of the Black Sea (Shcherbakov 1991) predates the catastrophic flood hypotheses (Ryan *et al.* 1997) that claimed an abrupt transgression of the pre-existing and isolated basin by Mediterranean-sourced marine water into a lake with water level below that of the present-day Bosphorus Sill. In contrast, a persistent outflow (Aksu *et al.* 2002) of lacustrine water from the pre-exisiting lacustrine phase of the Pontic Basin into the Sea of Marmara during the early Holocene has also been proposed. This latter hypothesis requires the water level in the lake to be above that of the Sea of Marmara at the time of reconnection. A third possibility is that water levels fluctuated over this time, permitting periodic migration of Mediterranean fauna into the restricted basin (Chepalyga 1984; Yanko-Hombach 2007; Yanko-Hombach *et al.* 2007). Each of these hypotheses has been modified subsequently (Major *et al.* 2002, 2006; Ryan *et al.* 2003; Ryan 2007; Hiscott *et al.* 2007, 2010; Lericolais *et al.* 2009, 2010; Marret *et al.* 2009, 2010; Nicholas *et al.* 2011; Yanko-Hombach *et al.* 2014), yet the controversy remains (Buynevich *et al.* 2011). Each of these hypotheses has been dependent on radiocarbon dating for numerical chronology.

The traditional biostratigraphic chronology for the Black Sea was based on the presence of either Mediterranean or Ponto-Caspian taxa (e.g. Nevesskaya 1974; Nevesskaja 2007). In material sufficiently young, these were commonly dated using conventional (β-counting) radiocarbon methods on bulk samples (i.e. multiple mollusc specimens). However, conventional radiocarbon dating on bulk mollusc samples, composed of multiple individuals of fossil molluscs, can result in time-averaged ages that may be significantly older than the depositional event in question and, therefore, not all will be credible (Brückner *et al.* 2010;

Soulet *et al.* 2011). In contrast, AMS dating of single molluscs permits the dating of individual components of a sample, thereby reducing the effect of time-averaging (Nicholas *et al.* 2011). Prior to the 'Flood' hypothesis (Ryan *et al.* 1997), conventional radiocarbon ages were used as chronological evidence for the 'instantaneous flooding' of Crimean shelves (i.e. the Black Sea) by Shcherbakov (1991). However, the majority of previously derived ^{14}C measurements using conventional (β-counting) radiocarbon dating on multiple mollusc specimens, many of which were composed of mixed and potentially reworked species, do not support the flood hypothesis (e.g. Panin *et al.* 1983; Balabanov 2007). The use of conventional and AMS-based radiocarbon ages on molluscs has meant that two disparate radiocarbon-based chronologies have been in use in the Black Sea (Nicholas *et al.* 2011). Until recently, the general lack of unambiguous ages on stratigraphic units in the Black Sea has clouded other issues, including the application of radiocarbon reservoir ages (Jones & Gagnon 1994; Kwiecien *et al.* 2008; Ménot & Bard 2012), neotectonics (Brückner *et al.* 2010) and issues of human dispersal (Turney & Brown 2007; Dolukhanov *et al.* 2009).

Recent sediments in the Black Sea may be differentiated into three principal stratigraphic units (Ross & Degens 1974; Major *et al.* 2002) whose characteristics vary depending on whether deposited on shelves or in deeper-water settings. The stratigraphic subdivision of Holocene sediments (Ross & Degens 1974) is reflected in the distinctive spatial distribution (Fig. 14.2) of shell and/or coccolith-rich strata on the NW and western shelves (Lericolais *et al.* 2009, 2010; Ongan *et al.* 2009; Oaie & Melinte-Dobrinescu 2012). This is partially a response to recent sediment starvation and erosive processes on the outer shelf, as well as to physical and faunal changes in these shelf environments since the last glacial lowstand. Three principal genera of bivalve molluscs, *Modiolus*, *Mytilus* and *Dreissena*, occur in these commonly mud-rich, horizons. *Modiolus*-rich mud commonly occurs at the top of these sedimentary layers at water depths of between 50 and 125 m, and these deposits do not normally exceed 30 cm in thickness. *Mytilus*-rich mud is generally found between the shelf break and 40–50 m depth, although this material may also occur in shelf regions less than 30 m deep. *Modiolus* overlies *Mytilus*, and these are underlain by *Dreissena*-rich sediments. The *Dreissena*-rich layer occurs as the uppermost layer at the

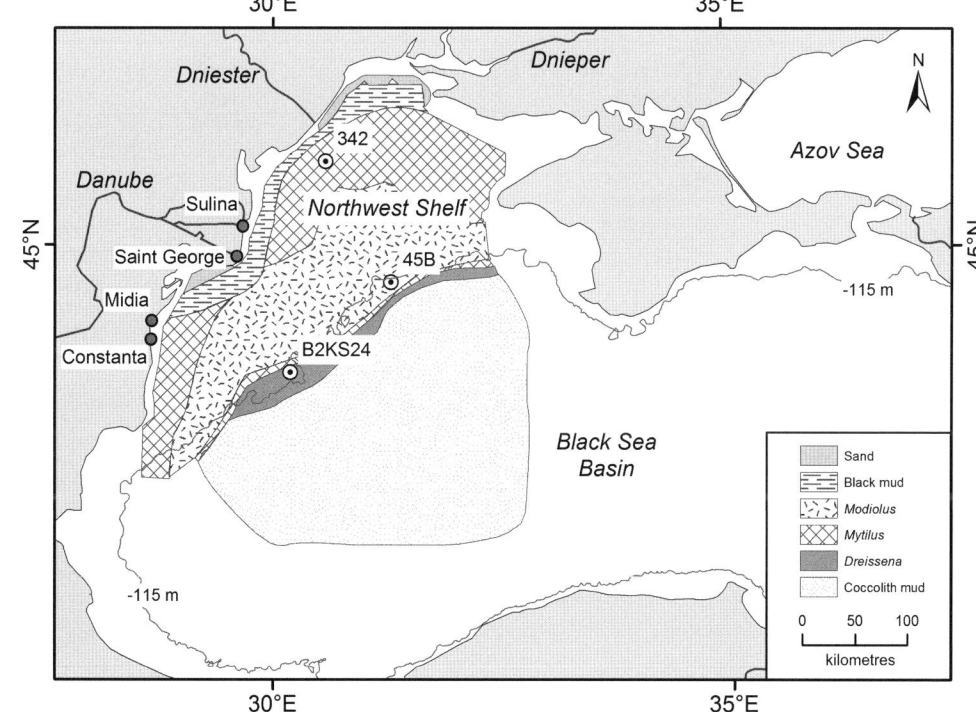

Fig. 14.2. Spatial distribution of bivalve molluscs on the present-day NW Shelf (redrawn from Lericolais *et al.* 2010). The locations from which cores 45B and B2KS24 were recovered are located near the outer margin of the *Modiolus* zone. Black mud in the Danube prodelta zone in Figure 14.1 results from fluvial discharge, with low to zero (in places) shell content.

top of the sediment-starved continental slope, and is commonly composed of bleached and partially to almost completely dissolved *Dreissena* valves in a matrix of carbonate mud and fine sand. The *Dreissena*-rich strata are concentration deposits and seem unlikely to represent shells that have been transported downslope, as suggested by Yanko-Hombach *et al.* (2014). The subaqueous downslope transport of shells would result in widely disseminated *Dreissena* fragments, unlike the coquina in core 45B and more similar, perhaps, to the widespread distribution of *Mytilus* and *Modiolus* fragments in a mud matrix on the NW Shelf. This stratum contrasts strongly with the two overlying units with unbleached and undissolved *Modioulus* and *Mytilus* – although both are commonly strongly fragmented. The sedimentary succession from *Dreissena* to *Mytilus* to *Modiolus* in cores indicates a change from low-salinity conditions to a marine Black Sea, the timing of which is the focus of debate. To address these questions, and taking a whole of basin view, Nicholas *et al.* (2011) undertook AMS ^{14}C and amino acid racemization (AAR) dating of selected peat and bivalve mollusc shells to understand the extent to which reworked shells contribute to the differences in views on sea-level change in the Black Sea. Because the focus of much of this debate has been on the radiocarbon chronology, the methods employed by Nicholas *et al.* (2011) to derive a radiocarbon chronology are presented below.

Methods

AMS determinations were undertaken at the laboratories of the Australian Nuclear Science and Technology Organisation (ANSTO) (Hua *et al.* 2001; Fink *et al.* 2004). Fifteen ^{14}C AMS ages were obtained, four on peat samples and 11 on bivalve mollusc shells (Table 14.2). Detrital peat samples from core 342, consisting of plant matter, were picked using 0.5 mm-width stainless steel tweezers after soaking the sediment sample, composed primarily of silt-grade terrestrially derived siliciclastics, briefly in Milli-Q water (18.2 MΩ), followed by ultrasonic disaggregation and rinsing, also in Milli-Q water over a 63 μm sieve. Dried and granular peat samples (interpreted as indicating desiccation) from core 721 were free of sediment prior to picking with fine tweezers for AMS ^{14}C dating. Plant material was subjected to a

1 mol HCl rinse to remove carbonate, and these samples were used only after inspection with a binocular microscope to rule out the presence of microfossils prior to submission at ANSTO. Single pieces of identifiable bivalve mollusc shell were used from cores 342 and 45B for AMS dating. Each shell sample from core 721 consisted of multiple individuals of microscopic (juvenile) bivalve molluscs. Radiocarbon ages on peat were calibrated using the INTCAL09 calibration curve (Reimer *et al.* 2009).

Results and discussion

Two cores on the outer Romanian and Ukrainian shelves (Figs 14.1 & 14.3), core 45B (107 m water depth: Nicholas *et al.* 2011) and core B2KS24 (96 m water depth: Giunta *et al.* 2007; Lericolais *et al.* 2007a, b, 2009), demonstrate the stratigraphic relationship alluded to above (Ross & Degens 1974; Major *et al.* 2002). In core 45B (Fig. 14.3), a 20 cm-thick *Dreissena rostiformis* coquina consisting primarily of bleached and partially to almost completely dissolved and disarticulated *Dreissena* valves ($n = 3$, mean age 8562 ^{14}C years BP; individually 8170 $\pm$ 60, 8695 $\pm$ 40 and 8820 $\pm$ 70 ^{14}C years BP) is overlain by terrestrially derived mud rich in unbleached fragments of the bivalve mollusc *Mytilus* ($n = 1$, 6530 $\pm$ 45 ^{14}C years BP), which is in turn overlain by *Modiolous*-rich muds. The presence of the bivalve mollusc *Dreissena rostiformis* is significant as it has a Ponto-Caspian affinity and is indicative of low-salinity brackish water (semi-freshwater: Chepalyga 1984; upper and lethal limit of tolerance 6–8‰: Orlova *et al.* 2005). These indicate restricted lacustrine conditions in the pre-existing lake phase of the Black Sea. The upper two units (*Mytilus* and *Modiolus*, both species of Mediterranean-derived bivalve molluscs, commonly represented by broken valves in mud) indicate more saline and marine conditions and, therefore, a connection of the Black Sea with the Aegean through the Dardanelle and Bosphorus Straits. Significantly, the boundary between the respective time-equivalent lacustrine and marine units in core 45B and core B2KS24 (Giunta *et al.* 2007; Lericolais *et al.* 2007a, b, 2009; Nicholas *et al.* 2011) represents a time gap of approximately 2 kyr.

Core B2KS24, a correlative of core 45B, is located approximately 130 km distant on the southern margin to the Danube

Table 14.2. *AMS ^{14}C ages on bivalve molluscs and peat*

Core	Water depth (m)	Depth in core (m)	Depth of sample below sea level (m)	Material	^{14}C age (years)	1σ	δ^{13}C*†	1σ	PMC‡	1σ	Calibrated age (years)	1σ	ANSTO code
721	14.9	2	16.9	Micromolluscs	1790	90	0	0	80.06	0.81			OZL305
721	14.9	8	22.9	Micromolluscs	2530	80	0	0	73.02	0.72			OZL305
721	14.9	15	29.9	Micromolluscs	3190	90	0.1	0	67.24	0.75			OZL306
721	14.9	22	36.9	Micromolluscs	8210	120	0	0	36	0.52			OZL307
721	14.9	26.2	41.1	Peat	8530	70	−25	0	34.58	0.47	9510	40	OZK766
721	14.9	26.5	41.4	Peat	9370	70	−29.1	0.1	31.15	0.24	10 600	90	OZK767
342	30	0.85	30.85	*Mytilus* (A) single valve	5765	35	0.1	0.1	48.78	0.2			OZM332
342	30	0.85	30.85	*Mytilus* (B) single valve	4365	30	−1.2	0.1	58.09	0.21			OZM333
342	30	0.85	30.85	*Dreissena polymorpha* single valve	9620	60	−8.7	0.2	30.19	0.2			OZL579
342	30	1.15	31.15	Peat	9020	70	−24.9	0.3	32.55	0.28	10 200	50	OZL577
342	30	1.65	31.65	Peat	8920	60	−28.1	0	32.94	0.23	10 000	70	OZL581
45B	107	0.31	107.31	*Mytilus* (C) single valve	6530	45	−0.7	0.1	44.36	0.24			OZL583
45B	107	0.44	107.44	*Dreissena rostiformis* (A) single valve	8820	70	0.4	0.1	33.37	0.25			OZL580
45B	107	0.44	107.44	*Dreissena rostiformis* (B) single valve	8695	50	0.4	0.1	33.88	0.2			OZL578
45B	107	0.44	107.44	*Dreissena rostiformis* (C) single valve	8170	60	−0.2	0.2	36.19	0.26			OZM331

*δ^{13}C values relate solely to the graphite derived from the sample used for the radiocarbon measurement. It may be the case that the value obtained from the graphite is not the same as that of the original bulk material.
†δ^{13}C values of 0.0 are assumed. Measured values are not available.
‡PMC is per cent modern carbon.

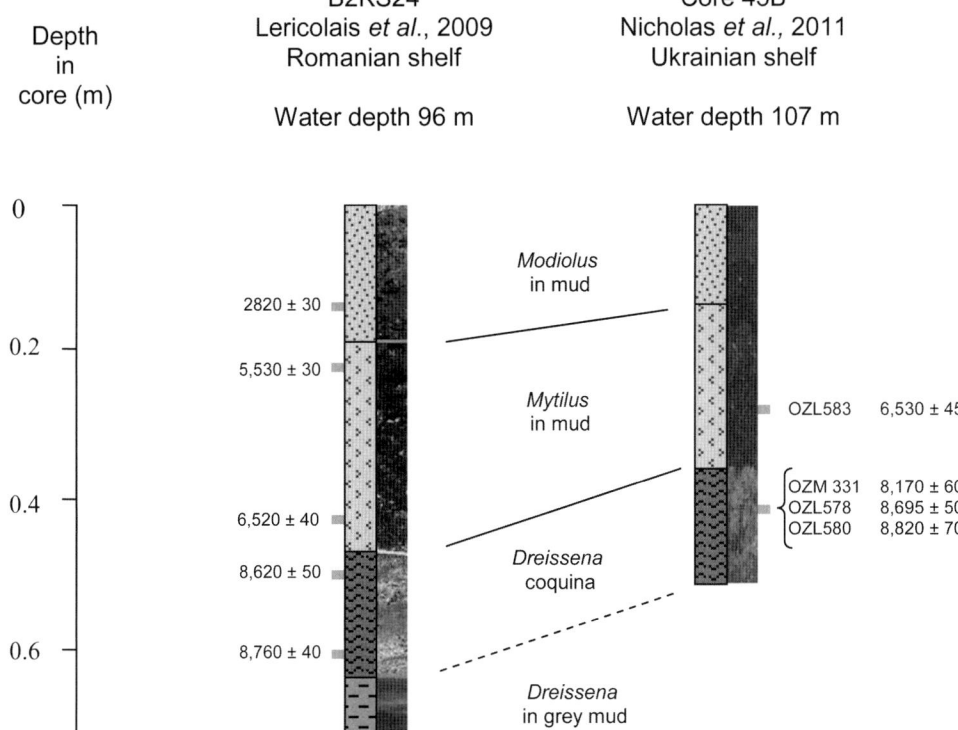

Depth
in
core (m)

B2KS24
Lericolais *et al.*, 2009
Romanian shelf

Water depth 96 m

Core 45B
Nicholas *et al.*, 2011
Ukrainian shelf

Water depth 107 m

2820 ± 30

5,530 ± 30

6,520 ± 40

8,620 ± 50

8,760 ± 40

11,060 ± 40

Modiolus
in mud

Mytilus
in mud

Dreissena
coquina

Dreissena
in grey mud

OZL583   6,530 ± 45

OZM 331   8,170 ± 60
OZL578   8,695 ± 50
OZL580   8,820 ± 70

Fig. 14.3. Comparison and correlation between cores from the Ukrainian (core 45B, this study) and Romanian shelves (core B2KS24; Lericolais *et al.* 2007*a*, *b*, 2009). The narrow range of AMS radiocarbon ages among the different *Dreissena* shells in these and similar deposits (Fig. 14.6) indicates that these coquina are time-averaged over hundreds of years and probably formed over short time intervals.

canyon (Giunta *et al.* 2007; Lericolais *et al.* 2009). The evident similarity of these cores (Fig. 14.3) in age and composition indicate formation in very similar environments at approximately the same time. The *Dreissena* coquina from the NW Shelf and slope have been described as shoreline beach deposits (Ballard *et al.* 2000; Ryan *et al.* 2003; Lericolais *et al.* 2007*a*, *b*, 2009, 2010). There is disagreement on the mode of emplacement of the *Dreissena* coquina, with some workers favouring a depth of emplacement of 30 m or more below sea level for these concentration deposits (Yanko-Hombach *et al.* 2014). Because of the high density of *Dreissena rostiformis* bivalve shells, the absence of rounding, moderate fragmentation and low articulation of valves in these concentrated shell accumulations, we favour the argument that they were emplaced at or close to the water level, and were deposited by storm waves and currents (cf. Bressan & Palma 2010). However, the depths of these end-Neoeuxinian submerged coastal deposits vary (Yanko-Hombach 2007; Yanko-Hombach *et al.* 2007). These concentration deposits have been recovered from water depths of 120 m on the NW Shelf (Ryan *et al.* 1997, 2003), 112 and 95 m on the outer Romanian Shelf (Giunta *et al.* 2007; Lericolais *et al.* 2007*a*, *b*, 2009, 2010), and 100 m at the Sakarya delta (Algan *et al.* 2007). Taken together, the bleached condition of the shells and the age–depth relationships among these deposits points to a rapidly transgressing water level of between 107 and 49.5 m below present sea level, taking approximately 400 ^{14}C years to raise the water level to that of the Bosphorus Sill (estimated at 8.2 ^{14}C ka BP). The proliferation of *Dreissena rostiformis* in coastal environments, dated to between 8.6 ^{14}C ka BP (mean value, $n = 3$, 107 m water depth: this study) and 8.25 ^{14}C ka BP (mean value, $n = 2$, 49.5 m water depth: Major *et al.* 2006), and occurring predominantly as *Dreissena rostiformis* coquina in cores, may have been a response by this species to rising salinity because it is more tolerant of higher-salinity levels than, for example, the freshwater species, *Dreissena polymorpha*. Hence, the large number of *Dreissena* coquina that mark the beginning of the transition from non-marine to marine conditions (the Initial

Marine Inflow (IMI) and Disappearance of Lacustrine Species (DLS): Soulet *et al.* 2011; Yanko-Hombach *et al.* 2014). These have been alternatively described as forming the base of the Shallow Unit in B2SK24 (equivalent to units I and II: Jones & Gagnon 1994) above the lacustrine Unit III (Giunta *et al.* 2007), or as Neoeuxinian (Görür *et al.* 2001) forming the uppermost layer in Unit III (e.g. Nevesskaya 1974).

Coastal peat deposits

While there are no fixed sea-level indicators, such as coral, to be found in the Black Sea, the coastal margins of the Black Sea are noted for the large spatial extent of peat deposits. In core 342, recovered from 30 m on the inner Ukrainian Shelf (Figs 14.1 & 14.4; Table 14.2), lagoonal peat, dated by AMS methods to 9020 ± 70 and 8920 ± 60 ^{14}C years BP, is overlain by a variably mixed shell coquina, consisting of mixed marine and non-marine shells, the youngest individual fragment (*Mytilus*) dated to 4365 ± 30 ^{14}C years in age. This age is slightly younger than the age obtained on the same species in core 45B. These mixed coquina from core 342 have been reworked and deposited under marine conditions when sea level in the Black Sea was nearer present values. Yet, they are underlain by non-marine deposits indicative of freshwater (i.e. alluvial) to lagoonal (coastal) environments (Yanko-Hombach *et al.* 2014). Thus, based on the presence of peat of similar age on opposite coasts of the lake (the Ukraine Shelf and Georgian coast), some of which is below the level of the Bosphorus Sill (core 721), and the presence of *Dreissena rostiformis* coquina, it is suggested that generally freshwater (alluvial) environments existed on the innermost NW Shelf of the Black Sea when very-low-salinity conditions existed within the lake. The latter indicated by the absence of peat and presence of low-salinity tolerant *Dreissena rostiformis* on the outer NW Shelf, and with *Dreissena polymorpha* on the inner NW Shelf. Therefore, the lake must have been at a lower level than the former level of the Bosphorus Sill prior to its most recent transgression.

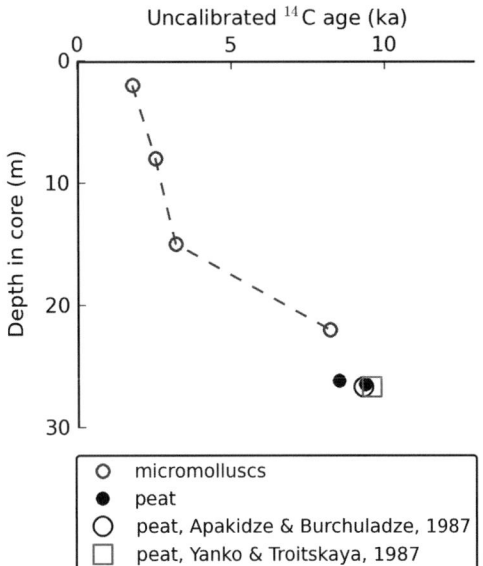

Fig. 14.4. Radiocarbon age v. depth for peat and micromollusc samples from core 721, Sukhumi Bay, Georgia, SE Black Sea. Here we note the similarity of three radiocarbon ages on peat samples at similar depths: one from this study (9370 ± 70 ^{14}C years BP, 26.5 m depth in core) using AMS, and two previously published ages (9310 ± 80 and 9580 ^{14}C years BP, both at 26.7 m depth in core: Apakidze & Burchuladze 1987; Yanko & Troitskaya 1987) obtained using conventional scintillation-counting methods (Balabanov 2007; Yanko-Hombach 2007; Yanko-Hombach *et al.* 2007).

Radiocarbon ages, of which there are a large number, obtained using conventional β- or scintillation-counting methods from coastal peat deposits have previously been utilized to determine the extent and timing of sea-level and environmental changes in the Black Sea (Görür *et al.* 2001; Balabanov 2007; Filipova-Marinova 2007; Yanko-Hombach 2007; Yanko-Hombach *et al.* 2007). Because the credibility of radiocarbon ages obtained on bulk samples has been viewed as problematic (Giosan 2007), the previously published radiocarbon ages on peat were reassessed in the light of new ages from core 721 (Fig. 14.4) (14.9 m water depth: Table 14.2) (Nicholas *et al.* 2011). These new results indicate the broad equivalence of radiocarbon ages on peat in the Black Sea derived from AMS and scintillation-counting methods, as would be expected in similarly time-averaged sediments. Together, these radiocarbon-dated peat deposits indicate a similarly rising water level in the early Holocene (Fig. 14.5) for the SW, NW, western and eastern coasts (i.e. basin-wide; Bulgaria, Ukraine, Russia, Georgia, and Turkey). The age–depth relationship of these peat deposits suggest a water level rising from below that of the Bosphorus Sill (35 m below present sea level).

Combining these data, while being aware of the limitations of knowledge on marine reservoir ages in the Black Sea, enables a sea-level curve for the Holocene Black Sea to be derived (Fig. 14.6). This curve indicates a lowered water level in the pre-existing palaeo-lake following the Younger Dryas. Previously, the work of Hiscott *et al.* (2007) and Marret *et al.* (2009) had suggested the presence of marine shells on the shelf at approximately 78 m depth on the SW Shelf during this time. However, these identifications have been changed to non-marine species; *Dreissena*, *Didacna* and *Theodoxus* (Hiscott *et al.* 2010; Marret *et al.* 2010). These (revised) species are indicative of freshwater–low-salinity brackish water (maximum 8‰: Orlova *et al.* 2005), and thus, this negates the possibility of a major marine influence prior to the transgression. These more recent descriptions also counter outflow hypotheses because the species–age–depth location of these samples on Figure 14.6 are consistent with a prompt transgression of marine water into a

lake that was below the level of the Bosphorus Sill. The post-Younger Dryas lowstand depicted here is interpreted from AMS ^{14}C ages on *Dreissena* coquina, and on the evidence indicating the presence of coastal deposits on the Ukrainian and Romanian outer shelves (Dolukhanov *et al.* 2009; Lericolais *et al.* 2009). In addition, peat was forming on the shelves of the Black Sea during this time (Görür *et al.* 2001; Balabanov 2007; Filipova-Marinova 2007; Yanko-Hombach *et al.* 2014). The presence of *Pisidium* (pea clam: Lericolais *et al.* 2010) and Ponto-Caspian fauna (*Dreissena* and *Didacna*: Hiscott *et al.* 2010; Marret *et al.* 2010) at levels substantially below that of the Bosphorus Sill (i.e. below 35 m water depth) provide additional support for a lowered water level during a period of regional climatic aridity (Shumilovskikh *et al.* 2012; Connor *et al.* 2013). This radiocarbon-based evidence indicates that euryhaline Mediterranean molluscs colonized the Black Sea after a gradual change in salinity following the prompt transgression. The evidence provided by the presence of bivalve molluscs *Monodacna caspia* and *Dreissena* (*D. rostiformis*, maximum salinity tolerance of 6–8‰: Orlova *et al.* 2005) in the Black Sea between 8.6 and 7.13 ^{14}C ka BP (the age of the youngest *Monodacna* on the shelf at 49 m below present sea level: Major *et al.* 2006) suggest a period (≥1.5 kyr) in which salinity rose and conditions changed from that suitable for Ponto-Caspian fauna to that of an environment capable of supporting a Mediterranean-type fauna. Furthermore, the age–depth location of the non-marine bivalve mollusc *Monodacna* overlaps to some extent with that of marine fauna, suggesting a gradual transition. This period of overlap is contiguous with the earliest deposition of sapropel in the Holocene Black Sea, further supporting the gradual salinization hypothesis.

Notably, the sea-level curves for the *Dreissena* coquina and for the coastal peat deposits meet naturally (Fig. 14.6). A near-zero ^{14}C reservoir age is suggested for these *Dreissena* shells (cf. Kwiecien *et al.* 2008). This indicates that the environment which *Dreissena* inhabited was probably ventilated, and suggests possible post-mortem deposition in retreating delta-front or similar coastal environments during the marine ingress. This sea-level curve is similar to that proposed by proponents of the Flood hypothesis (Ryan *et al.* 1997, 2003; Major *et al.* 2002, 2006; Lericolais *et al.* 2007a, 2009) and, more recently, to the work of Yanko-Hombach *et al.* (2014). These data do not support previous arguments for repeated marine invasions nor rapid oscillations in water level of the Holocene Black Sea (Chepalyga 1984; Martin *et al.* 2007; Yanko-Hombach 2007; Yanko-Hombach *et al.* 2007). While Yanko-Hombach *et al.* (2014) do not agree with the Flood hypothesis, their acceptance of an initial marine inflow rather than an outflow from the Black Sea in the early Holocene means that the only major obstacle to understanding the truth of the Black Sea Flood hypothesis is the depth of water in the pre-existing isolated lacustrine phase. Unfortunately, specific water-level indicators have been difficult to identify for that time. However, from a basin-wide perspective, when taken together, the *Dreissena* coquina, evidence for drowned coastal dunes (Lericolais *et al.* 2007a, b) and a distinct absence of other subfossil shells from the outer NW Shelf at depths below the peat deposits at the time of the marine ingress (Fig. 14.6) strongly suggests an exposed outer shelf immediately prior to the transgression.

The volume of marine water (≤1% of the total volume of the Black Sea) likely to have been involved in the prompt marine transgression by Mediterranean-sourced water into an originally lacustrine environment at a level below that of the present-day eroded Bosphorus Sill (Gökasan *et al.* 2005) is unlikely to have resulted in immediate salinization of the pre-existing freshwater–semi-freshwater lake (Chepalyga 1984). This is because the volume of salt in the transgressive marine water body (equivalent to c. 1% of the total volume of the basin at 35‰ salinity) would have been insufficient to suddenly raise salinity from 0–5‰ to the present-day value of 18–22‰ for the bulk water column. It would

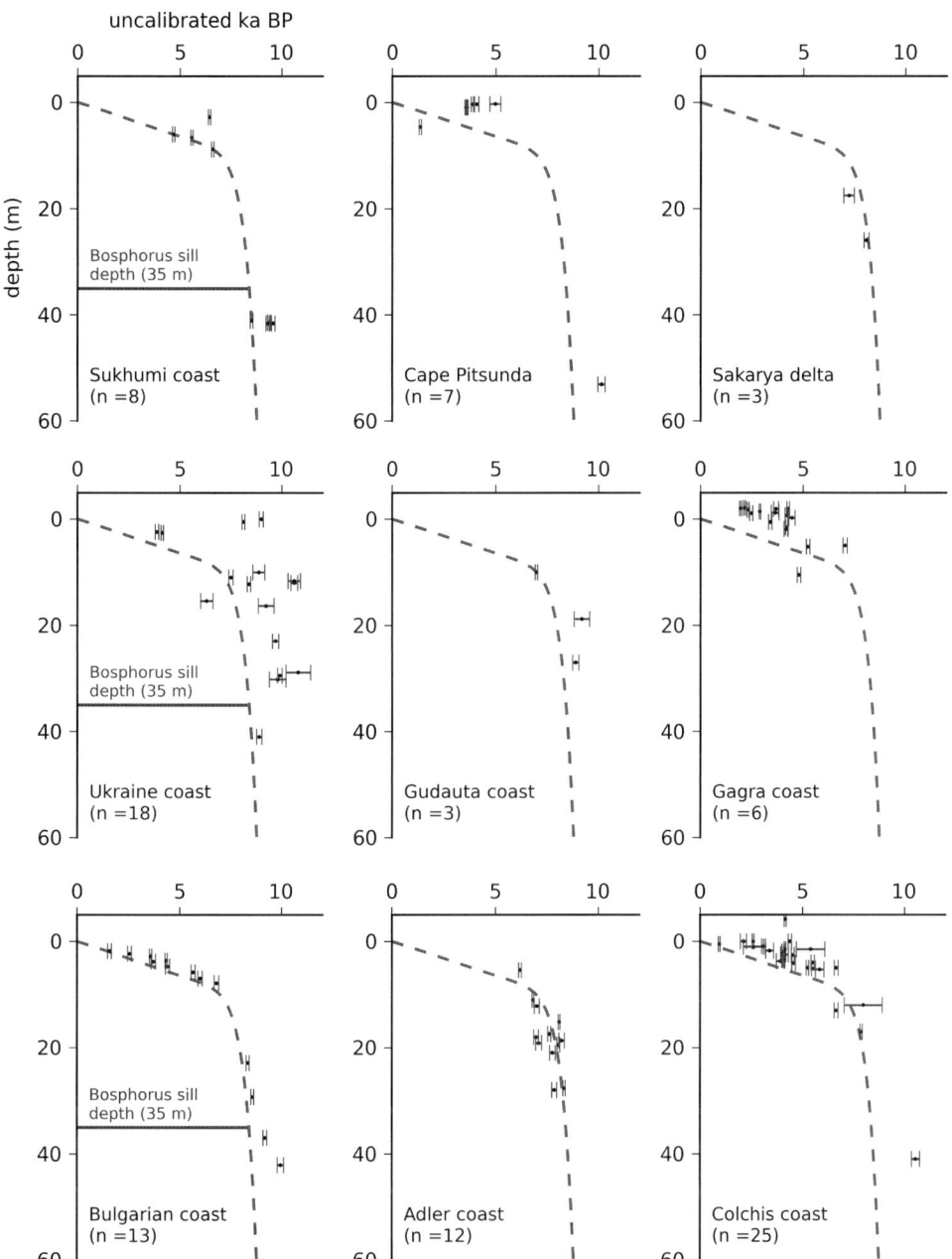

Fig. 14.5. Comparison of radiocarbon ages on peat recovered from present, basin-wide, coastal regions (Görür *et al.* 2001; Balabanov 2007; Filipova-Marinova 2007). Values in brackets indicate the number of samples from each location. These data indicate that the sea-level curve for Sukhumi Bay can be used as an approximation, at the basin-wide scale, for the Black Sea. Differences are likely to reflect variations, including the degree of tectonic movement over the Holocene and compaction.

have required approximately 5 times the volume of transgressing water (at 35‰) to raise salinity in the bulk water column from 5 to 20‰. This strongly suggests a gradual salinization following the transgression. This gradual salinization model is supported by the distribution of AMS radiocarbon ages on bivalve molluscs (Fig. 14.6), the foraminiferal record in core 721 (Yanko-Hombach 2007; Yanko-Hombach *et al.* 2007) and recent work on dinoflagellate cysts from the SW Black Sea (Bradley *et al.* 2012). The concept of gradual salinization is further supported by the time necessary for salinity to rise sufficiently, following the marine ingress, to force bottom-water oxygen depletion and to assist in the formation of sapropels (Major *et al.* 2006), and this has been estimated at approximately 900 years (Soulet *et al.* 2011; Lericolais *et al.* 2013). These conclusions are unlike that of Mertens *et al.* (2012), who suggested a drop in salinity upon reconnection of the Black Sea with marine water (Mertens *et al.* 2012), the opposite of that expected to happen to a near-freshwater body upon it becoming connected to a marine source.

The phase of significant salinization, and the development of the pycnocline and stratified water column following the mixing

phase during the transgression, seems to have occurred prior to, and during, the period in which the first sapropels formed (Fig. 14.7) (base of Unit 2 of Ross & Degens 1974; Major *et al.* 2002). The earliest sapropel development occurred at 7450–7550 AMS ^{14}C years BP (Ross & Degens 1974; Coolen *et al.* 2009), although an earlier estimate also exists (7920 ± 65 AMS ^{14}C years BP: Strechie-Sliwinski 2007). Therefore, the conclusion drawn (Fig. 14.7) is that prompt transgression, occurring over approximately 400 years, of the pre-existing lake was followed by a gradual salinization that took ≥1 kyr, sufficient for conditions to change and enable mass colonization by Mediterranean-derived fauna that normally inhabit more saline- and nutrient-rich coastal environments (Nicholas *et al.* 2011).

Euryhaline Mediterranean fauna predominantly appear directly at the same time or just after the first sapropel deposits (*c.* 7.5–8 AMS ^{14}C ka BP; Fig. 14.7), and earlier than the 7.16 ^{14}C ka BP for the Unit 2–Unit 3 boundary estimated by Giunta *et al.* (2007). *Dreissena* younger than 7.5 AMS ^{14}C ka BP have not been recovered from the mid and outer shelves of the Black Sea. This indicates that this genus and its supporting habitat have not

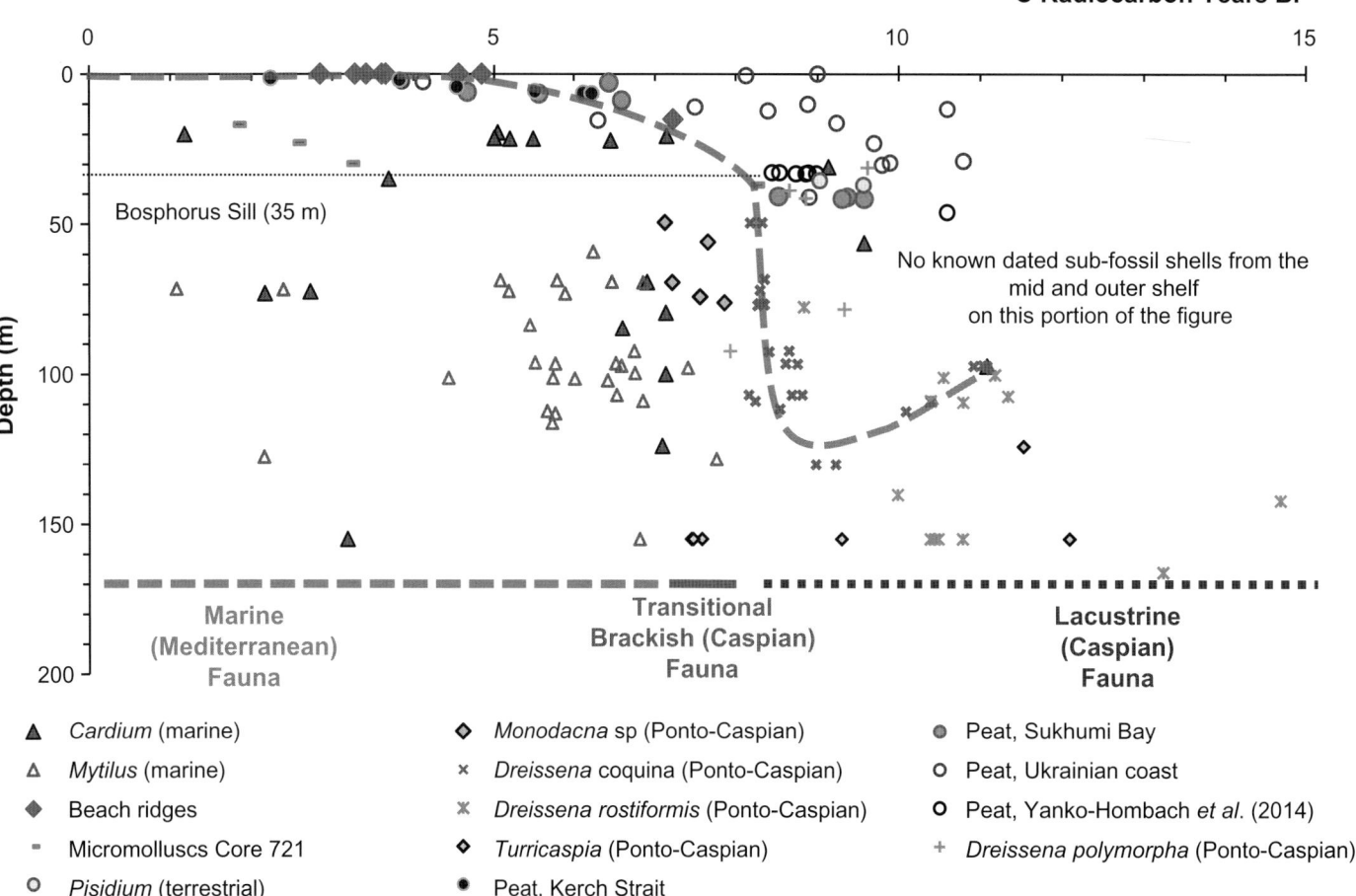

Fig. 14.6. Composite sea-level approximation, with radiocarbon age–depth relationships among marine, brackish and freshwater molluscs, peat samples and beach ridges from the Black Sea (Ryan *et al.* 1997; Ballard *et al.* 2000; Görür *et al.* 2001; Aksu *et al.* 2002; Major *et al.* 2006; Algan *et al.* 2007; Hiscott *et al.* 2007, 2010; Ivanova *et al.* 2007; Giosan *et al.* 2009; Lericolais *et al.* 2009, 2010; Marret *et al.* 2009, 2010; Brückner *et al.* 2010; Erginal *et al.* 2013). The transition from the lake to the Black Sea is based on AMS ¹⁴C ages on clearly identified bivalve molluscs and gastropods, and ¹⁴C ages from AMS and β-counting ¹⁴C dating methods on peat. The post-Younger Dryas lowstand depicted here is interpreted from AMS ¹⁴C ages on *Dreissena coquina*, and on the evidence indicating the presence of coastal deposits on the Ukrainian and Romanian outer shelves (Dolukhanov *et al.* 2009; Lericolais *et al.* 2009; Larchenkov & Kadurin 2011; Yanko-Hombach *et al.* 2014).

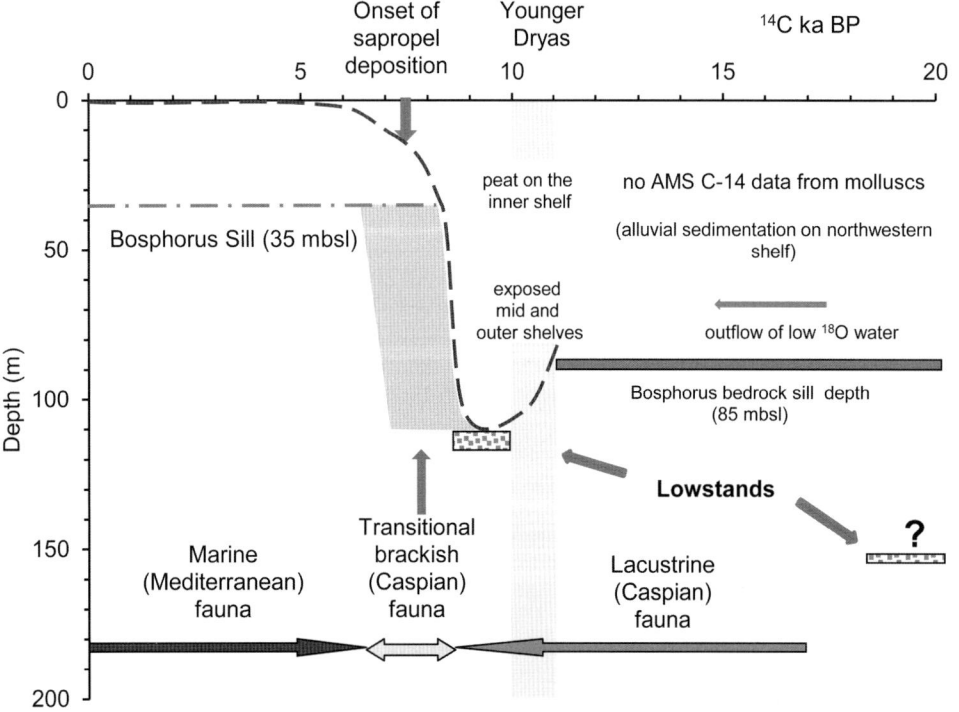

Fig. 14.7. Summary figure of the principal phases of the lacustrine–marine Black Sea during the period of time from the Younger Dryas to the late Holocene. We have not found any data during this work indicating AMS ¹⁴C ages older than approximately 12 ¹⁴C ka BP on shell material from the shelves of the Black Sea, although ages from scintillation counting on bulk samples of shells do exist (Balabanov 2007). The depth of the LGM lake level is estimated to be below the *Dreissena*-rich coquina because there is no evidence – based on the dating of individual shells by AMS methods – that a LGM shoreline exists above the *Dreissena* coquina. (mbsl, metres below sea level.)

existed on these shelf environments since that time. Furthermore, this indicates that the demise of the Dreissenids on the shelves of the Black Sea is a direct consequence of salinization, perhaps higher nutrient levels and perhaps also of warming of the previously isolated lake by water sourced from the Mediterranean–Marmara corridor. Modern conditions must have existed since at least the middle Holocene (Giosan *et al.* 2006; this study), and significantly earlier than the estimated 3330 ^{14}C years BP suggested by Giunta *et al.* (2007).

Notably, there are no AMS ^{14}C radiocarbon ages on shell material from the shelves of the Black Sea older than 12 ^{14}C ka BP despite there being conventional (β-counting) ages on mollusc samples (e.g. Caraivan *et al.* 2003). This is problematic because such data have been obtained using β-counting methods on bulk samples commonly consisting of different genera from, for example, the Danube delta (Noakes & Herz 1983). The reason for this difference in ages probably originates in the incorporation of shells of various ages into individual conventionally dated (i.e. bulk) radiocarbon samples. It has been suggested that erosion and subaerial exposure of the shelves occurred during the Younger Dryas drawdown period (Lericolais *et al.* 2009), except perhaps in water-filled depressions. As a result, little if any dateable shell material has remained. However, this may be an oversimplification because sediments from the Bosphorus Strait deposited on bedrock indicate outflow at approximately 16.6 ^{14}C ka BP (Algan *et al* 2001) and therefore an open seaway. Given that a prompt ingress of Mediterranean water would have occurred only if the Bosphorus was 'closed', this closure must have happened after outflow but prior to the transgression.

One reason for the absence of AMS ^{14}C ages on molluscs older than 12 ^{14}C ka BP from the shelves may be that post-Last Glacial Maximum (LGM) water level in the lake did not rise either for long enough or to a sufficient height on the shelves to allow the establishment of typical lacustrine fauna (e.g. Caspian Dreissenids) until water levels in the Black Sea rose upon reconnection with, and transgression by, Mediterranean-sourced water. In this case, if there were outflow from the post-LGM lake, the Bosphorus Strait was open and perhaps did not provide as significant a restriction in the Marmara Gateway as at present. It is also possible that the level of the lake never reached higher than 70–90 m below present sea level during the post-LGM outflow phase as suggested by facies-based and geomorphic studies of the NW Shelf (Larchenkov & Kadurin 2011). If so, this requires an open connection to the Sea of Marmara – that is, the shallow Bosphorus Sill (35 m water depth) may not have existed prior to 16.6 ^{14}C ka BP. This hypothesis requires the Bosphorus to have 'closed' or infilled during or after the post-LGM outflow and prior to the Younger Dryas lowstand by sedimentation and/or tectonics. Closure would have led to the development of geomorphological conditions ripe for a prompt transgression once sea level in the Sea of Marmara was of sufficient height: that is, driving the prompt transgression.

Thus, the summary sea-level curve for the post-glacial Black Sea (Fig. 14.7) indicates an outflow by glacially derived meltwater prior to drawdown initiated during the Younger Dryas. This lowered water level existed until the transgression at 8.6–8.2 ^{14}C ka BP. Water level in the Black Sea at 5 ^{14}C ka BP was close to present values (Giosan *et al.* 2006; Brückner *et al.* 2010; Erginal *et al.* 2013; this study) and, at the basin-wide scale, has remained so since then.

Conclusions

- The Marmara Gateway has, to a large extent, controlled changes in eustatic water level (transgressions) and biogeographical change in the Black Sea during the Quaternary.
- There is now broad agreement that AMS radiocarbon ages on bivalve molluscs indicate that the most recent transgression

by Mediterranean-derived marine water occurred between 8.6 and 8.2 ^{14}C ka BP (approximately 9.6–9.2 cal ka BP).
- This transgression began when the isolated lake had a water depth of approximately 107 m or more below present sea level, and is recorded by the age–depth distribution of AMS ^{14}C dated and subaerially exposed shells, particularly *Dreissena rostiformis*, recovered from coquina in cores obtained from the mid and outer Ukrainian and Romanian shelves.
- The prompt transgression of the Black Sea was accompanied by a gradual salinization, the latter taking approximately 1 kyr or more to occur, and indicated by the gradual transition from non-marine to marine bivalve mollusc species.
- The initial colonization of the former lake by Mediterranean-derived taxa, particularly *Mytilus* sp. and *Cardium* sp., began as the earliest Holocene sapropel deposits formed.
- Sea level in the Black Sea has been close to modern levels and relatively stable since approximately 5 ^{14}C ka BP.

We are grateful to Prof. V. Yanko-Hombach, and Dr S. Kadurin for their help during this research by supplying samples from cores 721, 45B and 342. AMS radiocarbon ages were funded by AINSE grant 05/220 and through a collaborative Ccash project with D. Fink (ANSTO). We gratefully acknowledge the contributions of two anonymous reviewers, who provided valuable constructive comments on the text.

References

AKSU, A. E., HISCOTT, R. N., YAŞAR, D., IŞLER, F. I. & MARSH, S. 2002. Seismic stratigraphy of Late Quaternary deposits from the southwestern Black Sea shelf: evidence for non-catastrophic variations in sea-level during the last 10 000 years. *Marine Geology*, **190**, 61–94.

ALGAN, O., ÇAĞATAY, N., TCHEPALYGA, A., ONGAN, D., EASTOE, C. & GÖKASAN, E. 2001. Stratigraphy of the sediment infill in Bosphorus Strait: water exchange between the Black and Mediterranean Seas during the last glacial Holocene. *Geo-Marine Letters*, **20**, 209–218.

ALGAN, O., GÖKASAN, E. *ET AL.* 2002. A high-resolution seismic study in Sakarya Delta and submarine canyon, southern Black Sea shelf. *Continental Shelf Research*, **22**, 1511–1527.

ALGAN, O., ERGIN, M., KESKIN, Ş., GÖKASAN, E., ALPAR, B., ONGAN, D. & KIRICI-ELMAS, E. 2007. Sea-level changes during the Late Pleistocene-Holocene on the southern shelves of the Black Sea. *In*: YANKO-HOMBACH, V., GILBERT, A., PANIN, N. & DOLUKHANOV, P. M. (eds) *The Black Sea Flood Question: Changes in Coastline, Climate and Human Settlement*. Springer, Dordrecht, 603–631.

APAKIDZE, A. M. & BURCHULADZE, A. A. 1987. *Radiouglerodnoe datirovanie arkheologicheskikh I paleobotanicheskikh abraztsov Gruzii* [Radiocarbon dating of archaeological and paleobotanical samples in Georgia]. Metsnierreba, Tbilisi, Georgia [in Russian].

BADERTSCHER, S., FLEITMANN, D. *ET AL.* 2011. Pleistocene water intrusions from the Mediterranean and Caspian seas into the Black Sea. *Nature Geoscience*, **4**, 236–239.

BALABANOV, I. P. 2007. Holocene sea-level changes of the Black Sea. *In*: YANKO-HOMBACH, V., GILBERT, A., PANIN, N. & DOLUKHANOV, P. M. (eds) *The Black Sea Flood Question: Changes in Coastline, Climate and Human Settlement*. Springer, Dordrecht, 711–730.

BALLARD, R. D., COLEMAN, D. F. & ROSENBERG, G. D. 2000. Further evidence of abrupt Holocene drowning of the Black Sea shelf. *Marine Geology*, **170**, 253–261.

BRADLEY, L. R., MARRET, F., MUDIE, P. J., AKSU, A. E. & HISCOTT, R. N. 2012. Constraining Holocene sea-surface conditions in the southwestern Black Sea using dinoflagellate cysts. *Journal of Quaternary Science*, **27**, 835–843.

BRESSAN, G. S. & PALMA, R. M. 2010. Taphonomic analysis of fossil concentrations from La Manga Formation (Oxfordian), Neuquén Basin, Mendoza Province, Argentina. *Journal of Iberian Geology*, **36**, 55–71.

BRÜCKNER, H., KELTERBAUM, D., MARUNCHAK, O., POROTOV, A. & VÖTT, A. 2010. The Holocene sea level story since 7500 BP – Lessons from the eastern Mediterranean, the Black and the Azov Seas. *Quaternary International*, **225**, 160–179.

BUYNEVICH, I. V., YANKO-HOMBACH, V., GILBERT, A. & MARTIN, R. E. (eds) 2011. *Geology and Geoarchaeology of the Black Sea Region: Beyond the Flood Hypothesis.* Geological Society of America, Special Papers, **473**.

ÇAĞATAY, M. N., GÖRÜR, N. ET AL. 2006. Paratethyan-Mediterranean connectivity in the Sea of Marmara region (NW Turkey) during the Messinian. *Sedimentary Geology*, **188–189**, 171–187.

CARAIVAN, G., POPESCU, D., PADURARU, G. & PANTELIMON, C. 2003. Black Sea level rising and coastal erosion. *Ovidius University Annals Series: Civil Engineering*, **1**, 99–104.

CHEPALYGA, A. I. 1984. Inland Sea Basins. *In*: BARNOSKY-CATHY, W. (ed.) *Late Quaternary Environments of the Soviet Union.* University of Minnesota Press, Minneapolis, MN, 229–247.

CONNOR, S. E., ROSS, S. A., SOBOTKOVA, A., HERRIES, A. I. R., MOONEY, S. D., LONGFORD, C. & ILIEV, I. 2013. Environmental conditions in the SE Balkans since the Last Glacial Maximum and their influence on the spread of agriculture into Europe. *Quaternary Science Reviews*, **68**, 200–215.

COOLEN, M. J. L., SAENZ, J. P., GIOSAN, L., TROWBRIDGE, N. Y., DIMITROV, P., DIMITROV, D. & EGLINTON, T. 2009. DNA and lipid molecular stratigraphic records of haptophyte succession in the Black Sea during the Holocene. *Earth and Planetary Science Letters*, **284**, 610–621.

DEGENS, E. T. & ROSS, D. A. (eds) 1974. *The Black Sea – Geology, Chemistry, and Biology.* American Association of Petroleum Geologists, Memoirs, **20**.

DI IORIO, D. & YÜCE, H. 1999. Observations of Mediterranean flow into the Black Sea. *Journal of Geophysical Research*, **104**, 3091–3108.

DOLUKHANOV, P. M., KADURIN, S. V. & LARCHENKOV, E. P. 2009. Dynamics of the coastal North Black Sea area in Late Pleistocene and Holocene and early human dispersal. *Quaternary International*, **197**, 27–34.

ERGINAL, A. E., EKINCI, Y. L., DEMIRCI, A., BOZCU, M., OZTURK, M. Z., AVCIOGLU, M. & OZTURA, E. 2013. First record of beachrock on Black Sea coast of Turkey: implications for Late Holocene sea-level fluctuations. *Sedimentary Geology*, **294**, 294–302.

FILIPOVA-MARINOVA, M. 2007. Archaeological and paleontological evidence of climate dyanamics, sea-level change, and coastline migration in the Bulgarian sector of the circum-Pontic region. *In*: YANKO-HOMBACH, V., GILBERT, A. S., PANIN, N. & DOLUKHANOV, P. M. (eds) *The Black Sea Flood Question: Changes in Coastline, Climate, and Human Settlement.* Springer, Dordrecht, 453–481.

FINK, D., HOTCHKIS, Q. ET AL. 2004. The ANTARES AMS Facility at ANSTO. *Nuclear Instruments and Methods in Physics Research B*, **223–224**, 109–115.

GIOSAN, L. 2007. Book Review: The Black Sea Flood Question: Changes in Coastline, Climate and Human Settlement. *In*: YANKO-HOMBACH, V., GILBERT, A. S., PANIN, N. & DOLUKHANOV, P. M. (eds) *Quaternary Science Reviews*, **26**, 1897–1900.

GIOSAN, L., DONNELLY, J. P., VESPREMEANU, E., BHATTACHARYA, J. P. & OLARIU, C. 2005. River delta morphodynamics: examples from the Danube Delta. *In*: GIOSAN, L. & BHATTACHARYA, J. P. (eds) *River Deltas – Concepts, Models and Examples.* SEPM Special Publications, Tulsa, **83**, 393–411.

GIOSAN, L., DONNELLY, J. P. ET AL. 2006. Young Danube delta documents stable Black Sea level since the middle Holocene: morphodynamic, paleogeographic, and archaeological implications. *Geology*, **34**, 757–760.

GIOSAN, L., FILIP, F. & CONSTATINESCU, S. 2009. Was the Black Sea catastrophically flooded in the early Holocene? *Quaternary Science Reviews*, **28**, 1–6.

GIUNTA, S., MORIGI, C., NEGRI, A., GUICHARD, F. & LERICOLAIS, G. 2007. Holocene biostratigraphy and paleoenvironmental changes in the Black Sea based on calcareous nannoplankton. *Marine Micropaleontology*, **63**, 91–110.

GÖKASAN, E., TUR, H. ET AL. 2005. Evidence and implications of massive erosion along the Strait of Istanbul (Bosphorus). *Geo-Marine Letters*, **25**, 324–342.

GÖRÜR, N., ÇAĞATAY, M. N. ET AL. 2001. Is the abrupt drowning of the Black Sea shelf at 7150 yr a myth? *Marine Geology*, **176**, 65–73.

GÜRBÜZ, A. & LEROY, S. A. G. 2009. Science versus myth: was there a connection between the Marmara Sea and Lake Sapanca? *Journal of Quaternary Science*, **25**, 103–114.

HISCOTT, R. N., AKSU, A. E. ET AL. 2007. A gradual drowning of the southwestern Black Sea shelf: evidence for a progressive rather than abrupt Holocene reconnection with the eastern Mediterranean Sea through the Marmara Sea gateway. *Quaternary International*, **167–168**, 19–34.

HISCOTT, R. N., AKSU, A. E. ET AL. 2010. Corrigendum to 'A gradual drowning of the southwestern Black Sea shelf: Evidence for a progressive rather than abrupt Holocene reconnection with the eastern Mediterranean Sea through the Marmara Sea Gateway' [*Quaternary International*, **167–168** (2007) 19–34]. *Quaternary International*, **226**, 160.

HUA, Q., JACOBSEN, G. E., ZOPPI, U., LAWSON, E. M., WILLIAMS, A. A., SMITH, A. M. & MCGANN, M. J. 2001. Progress in radiocarbon target preparation at the ANTARES AMS centre. *Radiocarbon*, **43**, 275–282.

IGNATOV, E. I. 2008. Coastal and bottom topography. *In*: KOSTIANOY, G. & KOSAREV, A. N. (eds) *The Black Sea Environment.* The Handbook of Environmental Chemistry, **5**, Part Q. Springer, Berlin, 47–62.

IVANOVA, E. V., MURDMAA, I. O. ET AL. 2007. Holocene sea-level oscillations and environmental changes on the Eastern Black Sea shelf. *Palaeogeography, Palaeoclimatology, Palaeoecology*, **246**, 228–259.

JONES, G. A. & GAGNON, A. R. 1994. Radiocarbon chronology of Black Sea sediments. *Deep Sea Research Part I: Oceanographic Research Papers*, **41**, 531–557.

KONIKOV, E., LIKHODEDOVA, O. & PEDAN, G. 2007. Paleogeographic reconstructions of sea-level change and coastline migration on the northwestern Black Sea shelf over the past 18 kyr. *Quaternary International*, **167–168**, 49–60.

KOROTAEV, G., OGUZ, T. & RISER, S. 2006. Deep Sea Research Part II: Topical Studies in Oceanography. *In*: MURRAY, J. W. (ed.) *Black Sea Oceanography. Deep Sea Research Part II: Topical Studies in Oceanography*, **53**, 1901–1910.

KOSAREV, A. N., ARKHIPKIN, V. S. & SURKOVA, G. V. 2008. Hydrometeorological conditions. *In*: KOSTIANOY, G. & KOSAREV, A. N. (eds) *The Black Sea Environment.* The Handbook of Environmental Chemistry, **5**, Part Q. Springer, Berlin, 135–158.

KWIECIEN, O., ARZ, H. W., LAMY, F., WULF, S., BAHR, A., RÖHL, U. & HAUG, G. H. 2008. Estimated reservoir ages of the Black Sea since the Last Glacial. *Radiocarbon*, **50**, 99–118.

KWIECIEN, O., ARZ, H. W., LAMY, F., PLESSEN, B., BAHR, A. & HAUG, G. H. 2009. North Atlantic control on precipitation pattern in the eastern Mediterranean/Black Sea region during the last glacial. *Quaternary Research*, **71**, 375–384.

LARCHENKOV, E. & KADURIN, S. 2011. Palaeogeography of the Pontic Lowland and northwestern Black Sea shelf for the past 25 k.y. *In*: BUYNEVICH, I. V., YANKO-HOMBACH, V., GILBERT, A. S. & MARTIN, R. E. (eds) *Geology and Geoarchaeology of the Black Sea Region: Beyond the Flood Hypothesis.* Geological Society of America Special Papers, **473**, 71–87.

LERICOLAIS, G., POPESCU, I., GUICHARD, F. & POPESCU, S. M. 2007a. A Black Sea lowstand at 8500 yr B.P. indicated by a relict coastal dune system at a depth of 90 m below sea level. *In*: HARFF, J., HAY, W. W. & TETZLAFF, D. M. (eds) *Coastline Changes: Interrelation of Climate and Geological Processes.* Geological Society of America Special Papers, **426**, 171–188.

LERICOLAIS, G., POPESCU, I., GUICHARD, F., POPESCU, S. M. & MANOLA-KAKIS, L. 2007b. Water-level fluctuations in the Black Sea since the Last Glacial Maximum. *In*: YANKO-HOMBACH, V., GILBERT, A., PANIN, N. & DOLUKHANOV, P. M. (eds) *The Black Sea Flood Question: Changes in Coastline, Climate and Human Settlement.* Springer, Dordrecht, 437–452.

LERICOLAIS, G., BULOIS, C., GILLET, H. & GUICHARD, F. 2009. High frequency sea level fluctuations recorded in the Black Sea since the LGM. *Global and Planetary Change*, **66**, 65–75.

LERICOLAIS, G., GUICHARD, F., MORIGI, C., MINEREAU, A., POPESCU, I. & RADAN, S. 2010. A post Younger Dryas Black Sea regression identified from sequence stratigraphy correlated to core analysis and dating. *Quaternary International*, **225**, 199–209.

LERICOLAIS, G., BOURGET, J., POPESCU, I., JERMANNAUD, P., MULDER, T., JORRY, S. & PANIN, N. 2013. Late Quaternary deep-sea sedimentation in the western Black Sea: new insights from recent coring and seismic data in the deep basin. *Global and Planetary Change*, **103**, 232–247.

MAJOR, C., RYAN, W., LERICOLAIS, G. & HAJDAS, I. 2002. Constraints on Black Sea outflow to the Sea of Marmara during the last glacial-interglacial transition. *Marine Geology*, **190**, 19–34.

MAJOR, C. O., GOLDSTEIN, S. L., RYAN, W. B. F., LERICOLAIS, G., PIOTROWSKI, A. M. & HAJDAS, I. 2006. The co-evolution of Black Sea level and composition through the last deglaciation and its paleoclimatic significance. *Quaternary Science Reviews*, **25**, 2031–2047.

MARRET, F., MUDIE, P., AKSU, A. & HISCOTT, R. N. 2009. A Holocene dinocyst record of a two-step transformation of the Neoeuxinian brackish water lake into the Black Sea. *In*: YANKO-HOMBACH, V. & SMYNTYNA, O. V. (eds) *Quaternary History of the Black Sea and Adjacent Regions: Selected Papers, IGCP 521-INQUA0501, Odessa and Gelendzhik–Kerch Plenary Meetings. Quaternary International*, **197**, 72–86.

MARRET, F., MUDIE, P., AKSU, A. & HISCOTT, R. N. 2010. Corrigendum to 'A Holocene dinocyst record of a two-step transformation of the Neoeuxinian brackish water lake into the Black Sea' [*Quaternary International*, **197** (2009) 72–86]. *Quaternary International*, **226**, 161.

MARTIN, R. E., LEORRI, E. & MCLAUGHLIN, P. P. 2007. Holocene sea level and climate change in the Black Sea: multiple marine incursions related to freshwater discharge events. *Quaternary International*, **167–168**, 61–72.

MÉNOT, G. & BARD, E. 2012. A precise search for drastic temperature shifts of the past 40 000 years in southeastern Europe. *Paleoceanography*, **27**, PA2210, http://dx.doi.org/10.1029/2012PA002291

MERTENS, K. N., BRADLEY, L. R. *ET AL.* 2012. Quantitative estimation of Holocene surface salinity variation in the Black Sea using dinoflagellate cyst process length. *Quaternary Science Reviews*, **39**, 45–59.

MURRAY, J. W., STEWART, K., KASSAKIAN, S., KRYNYTZKY, M. & DIJULIO, D. 2007. Oxic, suboxic and anoxic conditions in the Black Sea. *In*: YANKO-HOMBACH, V., GILBERT, A., PANIN, N. & DOLUKHANOV, P. M. (eds) *The Black Sea Flood Question: Changes in Coastline, Climate and Human Settlement*. Springer, Dordrecht, 1–21.

NAUDTS, L., GREINERT, J., ARTEMOV, Y., BEAUBIEN, S. E., BOROWSKI, C. & DE BATIST, M. 2008. Anomalous sea-floor backscatter patterns in methane venting areas, Dniepr paleo-delta, NW Black Sea. *Marine Geology*, **251**, 253–267.

NEVESSKAYA, L. A. 1974. Molluscan shells in deep-water sediments of Black Sea. *In*: DEGENS, E. T. & ROSS, D. A. (eds) *The Black Sea – Geology, Chemistry and Biology*. American Association of Petroleum Geologists, Memoirs, **20**, 349–352.

NEVESSKAJA, L. A. 2007. History of the genus *Didacna* (Bivalvia: Cardiidae). *Palaeontological Journal*, **41**, 861–949.

NICHOLAS, W. A., CHIVAS, A. R., MURRAY-WALLACE, C. V. & FINK, D. 2011. Prompt transgression and gradual salinization of the Black Sea during the early Holocene constrained by amino acid racemization and radiocarbon dating. *Quaternary Science Reviews*, **30**, 3769–3790.

NIKISHIN, A. M., KORATAEV, M. V., ERSHOV, A. V. & BRUNET, M-F. 2003. The Black Sea basin: tectonic history and Neogene-Quaternary rapid subsidence modelling. *Sedimentary Geology*, **156**, 149–168.

NOAKES, J. E. & HERZ, N. 1983. University of Georgia radiocarbon dates VII. *Radiocarbon*, **25**, 919–929.

OAIE, G. & MELINTE-DOBRINESCU, M. C. 2012. Holocene litho- and biostratigraphy of the NW Black Sea (Romanian shelf). *Quaternary International*, **261**, 146–155.

OAIE, G., SECRIERU, D. & SHIMKUS, K. 2005. Black Sea basin: sediment types and distribution, sediment processes. *GeoEcoMarina*, **9–10**, 21–30.

OGAWA, Y., TAKAHASHI, K., YANANAKA, T. & ONODERA, J. 2009. Significance of euxinic condition in the middle Eocene paleo-Arctic basin: a geochemical study on the IODP Arctic Coring Expedition 302 sediments. *Earth and Planetary Science Letters*, **285**, 190–197.

OGUZ, T., DESHPANDE, A. G. & MALANOTTE-RIZZOLI, P. 2002. The role of mesoscale processes controlling biological variability in the Black Sea coastal waters: inferences from SeaWIFS-derived surface chlorophyll field. *Continental Shelf Research*, **22**, 1477–1492.

OGUZ, T., TUGRUL, S., KIDEYS, A. E., EDIGER, V. & KUBILAY, N. 2004. Physical and biogeochemical characteristics of the Black Sea. *In*: ROBINSON, A. R. & BRINK, K. H. (eds) *The Sea, Volume 14B: The Global Coastal Ocean: Interdisciplinary Regional Studies and Synthesis*. Harvard University Press, Cambridge, MA, 1331–1369.

ONGAN, D., ALGAN, O., KAPAN-YEŞILYURT, S., NAZIK, A., ERGIN, M. & EASTOE, C. 2009. Benthic faunal assemblages of the Holocene sediments from the southwest Black Sea shelf. *Turkish Journal of Earth Sciences*, **18**, 1–59.

ORLOVA, M. I., THERRIAULT, T. W., ANTONOV, P. I. & SHCHERBINA, G. KH. 2005. Invasion ecology of quagga mussels (*Dreissena rostiformis bugensis*): a review of evolutionary and phylogenetic impacts. *Aquatic Ecology*, **39**, 401–418.

ÖZSOY, E. & ÜNLÜATA, Ü. 1997. Oceanography of the Black Sea: a review of some recent results. *Earth Science Reviews*, **42**, 231–272.

PANIN, N. 2005. The Black Sea Coastal Zone – an overview. *GeoEcoMarina*, **11**, 21–40.

PANIN, N. & JIPA, D. 2002. Danube river sediment input and its interaction with the north-western Black Sea. *Estuarine, Coastal and Shelf Science*, **54**, 551–562.

PANIN, N., PANIN, S., HERZ, N. & NOAKES, J. E. 1983. Radiocarbon dating of Danube Delta deposits. *Quaternary Research*, **19**, 249–255.

PFANNENSTIEL, M. 1944. Diluviale Geologie des Mittelmeergebietes, die Diluvialen Entwicklundstadien und die Urgeschichte von Dardanellen, Marmara Meer und Bosphorus. *Geologische Rundschau*, **43**, 334–434.

POPESCU, I., LERICOLAIS, G., PANIN, N., NORMAND, A., DINU, C. & LE DREZEN, E. 2004. The Danube submarine canyon (Black Sea): morphology and sedimentary processes. *Marine Geology*, **206**, 249–265.

RANGIN, C., BADER, A. G., PASCAL, G., ECEVITOĞLU, B. & GÖRÜR, N. 2002. Deep structure of the Mid Black Sea High (offshore Turkey) imaged by multi-channel seismic survey (BLACKSIS) cruise. *Marine Geology*, **182**, 265–278.

REIMER, P. J., BAILLIE, M. G. L. *ET AL.* 2009. Intcal09 and Marine09 radiocarbon age calibration curves, 0–50 000 Years Cal BP. *Radiocarbon*, **51**, 1111–1150.

ROBINSON, A. G., RUDAT, J. H., BANKS, C. J. & WILES, R. L. F. 1996. Petroleum geology of the Black Sea. *Marine and Petroleum Geology*, **13**, 195–223.

ROSS, D. A. & DEGENS, E. T. 1974. Recent sediments of Black Sea. *In*: DEGENS, E. T. & ROSS, D. A. (eds) *The Black Sea – Geology, Chemistry and Biology*. American Association of Petroleum Geologists, Memoirs, **20**, 183–199.

RYAN, W. B. F. 2007. Status of the Black Sea flood hypotheses. *In*: YANKO-HOMBACH, V., GILBERT, A., PANIN, N. & DOLUKHANOV, P. M. (eds) *The Black Sea Flood Question: Changes in Coastline, Climate and Human Settlement*. Springer, Dordrecht, 63–88.

RYAN, W. B. F., PITMAN, W. C., III. *ET AL.* 1997. An abrupt drowning of the Black Sea shelf. *Marine Geology*, **138**, 119–126.

RYAN, W. B. F., MAJOR, C. O., LERICOLAIS, G. & GOLDSTEIN, S. L. 2003. Catastrophic flooding of the Black Sea. *Annual Review of Earth and Planetary Sciences*, **31**, 525–554.

SHCHERBAKOV, F. A. 1991. Lithodynamic conditions of sedimentation on the Black Sea shelf of Crimea. *Oceanology*, **31**, 353–356.

SHILLINGTON, D. J., WHITE, N., MINSHULL, T. A., EDWARDS, G. R. H., JONES, S. M., EDWARDS, R. A. & SCOTT, C. L. 2008. Cenozoic evolution of the eastern Black Sea: a test of depth-dependent stretching models. *Earth and Planetary Science Letters*, **265**, 360–378.

SHUMILOVSKIKH, L., TARASOV, P. *ET AL.* 2012. Vegetation and environmental dynamics in the southern Black Sea region since 18 kyr BP derived from the marine core 22-GC3. *Palaeogeography, Palaeoclimatology, Palaeoecology*, **337–338**, 177–193.

SOULET, G., MÉNOT, G., LERICOLAIS, G. & BARD, E. 2011. A revised calendar age for the last reconnection of the Black Sea to the global ocean. *Quaternary Science Reviews*, **30**, 1019–1026.

STANEV, E. V. & PENEVA, E. L. 2002. Regional sea level response to global climatic change: Black Sea examples. *Global and Planetary Change*, **32**, 33–47.

STANEV, E. V., BECKERS, J. M., LANCELOT, C., STANEVA, J. V., LE TRAON, P. Y., PENEVA, E. L. & GREGOIRE, M. 2002. Coastal-open ocean exchange in the Black Sea: observations and modelling. *Estuarine, Coastal and Shelf Science*, **54**, 601–620.

STANICA, A., DAN, S. & UNGREANU, V. G. 2007. Coastal changes at the Sulina mouth of the Danube River as a result of human activities. *Marine Pollution Bulletin*, **55**, 555–563.

STRECHIE-SLWINSKI, C. 2007. Changements environmentaux récents dans la zone de Nord-Ouest de la Mer Noire. *GeoEcoMarina*, **13** (Special Publications, **1**).

SUR, H. İ., ÖZSOY, E., ILYIN, Y. P. & ÜNLÜATA, Ü. 1996. Coastal/deep ocean interactions in the Black Sea and their ecological/environmental impacts. *Journal of Marine Systems*, **7**, 293–320.

TSAGARELI, A. L. 1974. Geology of Western Caucasus. *In*: DEGENS, E. T. & ROSS, D. A. (eds) *The Black Sea – Geology, Chemistry and Biology*. American Association of Petroleum Geologists, Memoirs, **20**, 77–89.

TURNEY, C. S. M. & BROWN, H. 2007. Catastrophic early Holocene sea level rise, human migration and the Neolithic transition in Europe. *Quaternary Science Reviews*, **26**, 2036–2041.

ÜNLÜATA, Ü., OǦUZ, T., LATIF, M. A. & ÖZSOY, E. 1990. On the physical oceanography of the Turkish Straits. *In*: PRATT, L. J. (ed.) *The Physical Oceanography of Sea Straits*. NATO/ASI Series, **318**. Kluwer, Dordrecht, 25–60.

VESPREMEANU-STROE, A. & PREOTEASA, L. 2007. Beach–dune interactions on the dry-temperate Danube delta coast. *Geomorphology*, **86**, 267–282.

VESPREMEANU-STROE, A., CONSTANTINESCU, S., TĂTUI, F. & GIOSAN, L. 2007. Multi-decadal evolution and North Atlantic Oscillation influences on the dynamics of the Danube Delta shoreline. *Journal of Coastal Research*, **50**, 157–162.

YALTIRAK, C., SAKINÇ, M., AKSU, A. E., HISCOTT, R. N., GALLEB, B. & ULGEN, U. B. 2002. Late Pleistocene uplift history along the southwestern Marmara Sea determined from raised coastal deposits and global sea-level variations. *Marine Geology*, **190**, 283–305.

YANKO, V. V. & TROITSKAYA, T. S. 1987. *Pozdnechetvertichnye Foraminifery Chernogo Moria* [Late Quaternary Foraminifera of the Black Sea]. Nauka, Moscow [in Russian].

YANKO-HOMBACH, V. V. 2007. Controversy over Noah's Flood in the Black Sea: Geological and foraminiferal evidence from the shelf. *In*: YANKO-HOMBACH, V., GILBERT, A. S., PANIN, N. & DOLUKHANOV, P. M. (eds) *The Black Sea Flood Question: Changes in Coastline, Climate and Human settlement*. Springer, Dordrecht, 149–203.

YANKO-HOMBACH, V., GILBERT, A. S. & DOLUKHANOV, P. 2007. Controversy over the great flood hypotheses in the Black Sea in light of geological, paleontological, and archaeological evidence. *Quaternary International*, **167–168**, 91–113.

YANKO-HOMBACH, V., MUDIE, P. J., KADURIN, S. & LARCHENKOV, E. 2014. Holocene marine transgression in the Black Sea: new evidence from the northwestern Black Sea shelf. *Quaternary International*, http://dx.doi.org/10.1016/j.quaint.2013.07.027

YILMAZ, Y. 2007. Morphotectonic evolution of the Southern Black Sea Region and the Bosphorus Channel. *In*: YANKO-HOMBACH, V., GILBERT, A., PANIN, N. & DOLUKHANOV, P. (eds) *The Black Sea Flood Question: Changes in the Coastline, Climate and Human Settlement*. Springer, Dordrecht, 537–569.

Chapter 15

The continental shelf of western India

B. M. FARUQUE[1]* & K. V. RAMACHANDRAN[2]

[1]*Geological Survey of India, Marine Wing, Bhu-Vijnan Bhavan, Kolkata, India*

[2]*Geological Survey of India, Marine Wing, Mangalore, India*

Corresponding author (e-mail: bmfaruque@gmail.com)

Abstract: The west coast of India, extending from Sir Creek in the north to Kanyakumari in the south, is a trailing passive margin that bears the imprint of generally shore-parallel and less common orthogonal structural elements. The geomorphology of the marginal part of western India, mostly controlled by the three principal structural trends of NNW–SSE Dharwar, NE–SW Aravalli orogenies and west–east Narmada graben, displays a straight coastline. The coastal ocean current operates from April to October and the net sediment transport is from north to south with the longshore current. The continental shelf width varies from 345 km off Daman in the north to 120 km off Goa and tapers to 60 km off Kochi in the south. The western continental shelf of India has an area of about 310 000 km^2, and is divided into an inner shelf with modern clayey silt and silty clay sediments with high organic matter and low carbonate content, and an outer shelf having relict carbonate sediments, coarse sands with low organic matter and high carbonates. The mid-shelf is rather uneven topographically, and the outer shelf is commonly interrupted by shore-parallel ridges and reefs with a relief of 2–18 m. Phosphatic limestone micronodules occur in the outer shelf.

The continental shelf basins of India have evolved due to divergent plate tectonics, active since the early Mesozoic. The western coastline of India stretches from Sir Creek to Kanyakumari. The northern part of the shelf features longitudinal extension faults in parallel sets giving rise to a series of narrow horst–graben structures. Such extensional faulting is responsible for a wider shelf in the northern segment. The orientation of faulting is controlled by three major orogenic trends in the western Indian region: the NE–SW Aravalli, the east–west Narmada lineament and the NNW–SSE Dharwar trend (Biswas 1989). Most epeirogenic movements, at the western continental margin, take place along these three directions. The western continental shelf margin is characterized by a wide continental shelf extending NNW–SSE, with a remarkably straight shelf edge breaking at 140–180 m depth. The shelf is 345 km wide off Daman, but gradually narrows to 100 km in the north off Kutchh, and tapers in the south, to 120 km off Goa and 60 km off Quilon in Kerala (Fig. 15.1). The western continental shelf is segmented into several shelf sub-basins by transverse basement arches of fault-bounded highs. These are from north to south: (1) Kutchh–Surat; (2) the Ratnagiri shelf; (3) Konkan; (4) the Kerala Basin; and are separated by: (i) the Saurashtra Arch; (ii) the Bombay High; (iii) the Vengurla Arch; and (iv) the Tellicherry Arch, respectively.

The western continental shelf of India has an area of about 310 000 km^2, and is divided into an inner shelf and an outer shelf. It comprises three sedimentary facies: beach sand; inner shelf silt and clay; and outer shelf carbonate sand. The first two units are in conformity with the expected modern sedimentation pattern. However, the third zone is anomalous and a relict formation deposited during the low sea level of the Last Glacial Maximum. The ooids have been radiometrically dated to 9 ka BP (Nair *et al.* 1979). The sediments in the shelf may be categorized as: modern clayey silt and silty clay sediments with high organic matter and low carbonate content in the inner shelf; and relict carbonate sediments of the outer shelf, which are coarse sands with low organic matter and high carbonate content of late Pleistocene–early Holocene age, 9–11 ka BP (Pandarinath *et al.* 2001).

Oceanography in the Indian continental margin is dominated by three seasons, namely: the SW monsoon from June to September; the NE monsoon from October to January; and fair-weather summer from February to May. During the SW monsoon, currents in the Arabian Sea are clockwise and these reverse for 2–4 months during the NE monsoon. The longshore current is southerly during

January–May and October. Although sediment transport is northerly in Gujarat and balanced off Maharashtra, net transport is southerly on the west coast. The average salinity values of the adjacent Arabian Sea are 34–37 ppt. The current speed in the open sea is 1.4 m s^{-1}. Upwelling nutrient-rich cool water occurs along the Karnataka and Kerala coastlines. Accordingly, the western coastal waters are highly productive based on the distribution of phytoplankton, as well as fishery production. The tidal range is extremely high in the north, with 8.5 m in the Gulf of Khambat to 6 m in the Gulf of Kutchh, and 3 m at Okha and Mumbai, decreasing to 2 m in Goa, 1.75 m at Karwar and 1.0 m at Kochi. Thirty-five per cent of the west coast's total length is affected by coastal erosion (Sanil Kumar *et al.* 2006).

Climate

The area falls within the tropical zone and the climate in the north around Kutchh is semi-arid, whereas south of the Narmada lineament the entire coastline from Mumbai to Kanyakumari receives very high rainfall of more than 1600 mm annually. The summer maximum temperature fluctuates between 30 and 45 °C in most places. This region experiences a mild winter of 23–30 °C, except in Kutchh where it is 10 °C.

Drainage

The Narmada and Tapti are two major rivers draining into the Arabian Sea (Fig. 15.1). Several smaller rivers like the Sabarmati, Shetrunji, Mandovi, Netravati and Periyar, along with innumerable seasonal streams, run for short distances during the rainy season along the western coast. The Gujarat and Maharashtra shelves receive mechanically and chemically weathered detritus from the Deccan volcanics, the southern part off Konkan and Kerala receives weathered material from argillite and arenaceous gneissic rocks besides laterite.

Neotectonism

The shelf off Kutchh is a tectonically active zone. Spatial analysis of the raised marine terraces along the Maharashtra coast show vertical differential movement and displacements in the continental

From: Chiocci, F. L. & Chivas, A. R. (eds) 2014. *Continental Shelves of the World: Their Evolution During the Last Glacio-Eustatic Cycle*.
Geological Society, London, Memoirs, **41**, 213–220. http://dx.doi.org/10.1144/M41.15

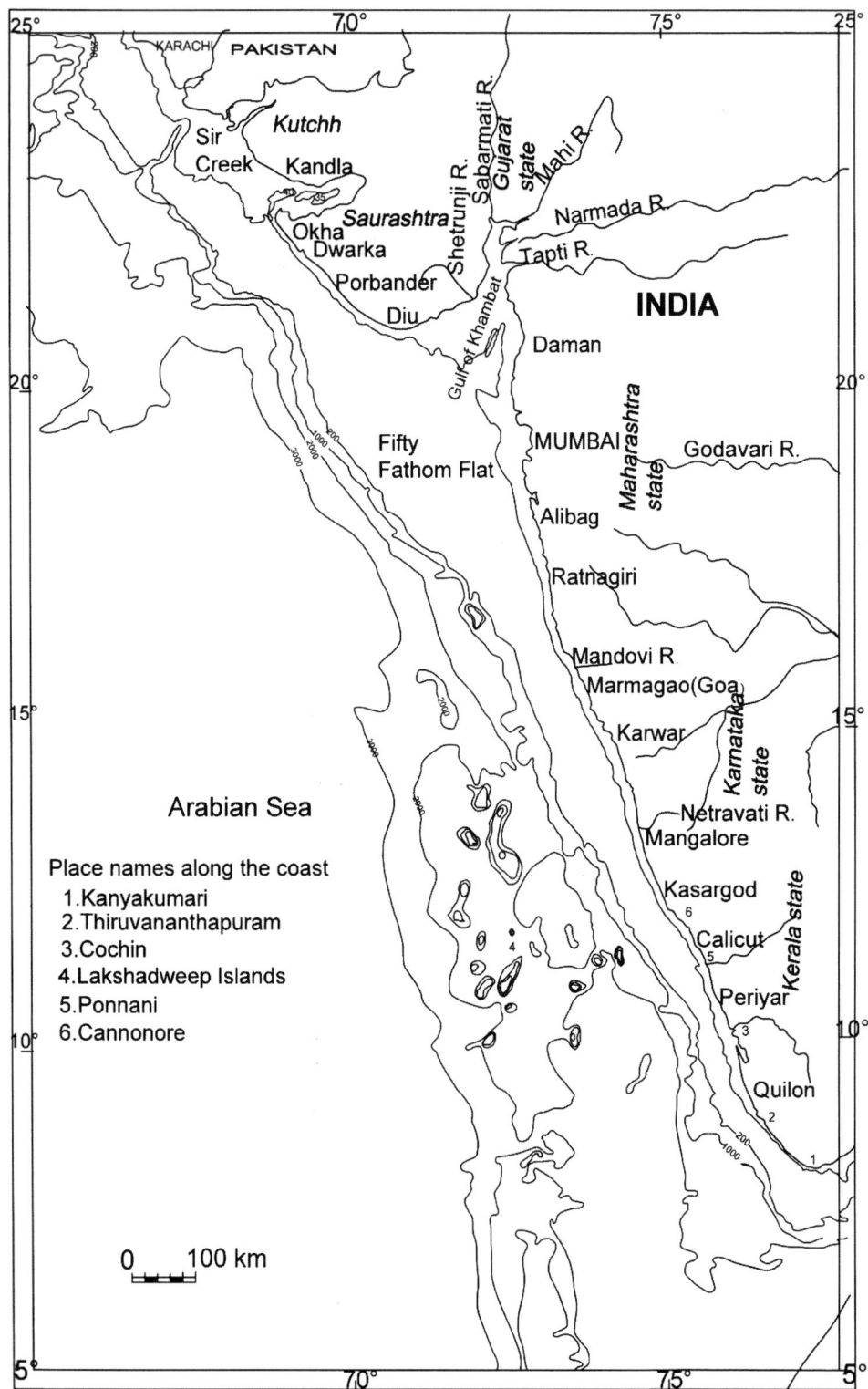

Fig. 15.1. Map showing the bathymetry of the shelf of the western coast of India.

margins concurrent to changing sea levels. Low–moderate seismicity was reported in the Palghat Gap region by Rajendran & Rajendran (1996). The structural constriction of the shelf in the Ponnani offshore may be a result of deep cut valleys (roughly 30 km long and 8 km wide) in the shelf formed along the structural continuity of Palghat Gap governed by east–west-striking faults. The tide gauge records of Mumbai Harbour reveal subsidence of 1.28 mm a^{-1} relative to sea level, which is also attributed to neotectonism (Manjunatha & Shankar 1992). Black phosphorite and creamy limestone were recovered from a water depth of 473 m from the upper continental slope off Kutchh. It is inferred that phosphatization took place in shallow-water conditions of

the shelf environment, which subsided to this depth due to tectonic epeirogenic movement.

The variation in the characteristics of the western shelf is described under three segments: (1) Kutchh–Khambat; (2) Daman–Karwar; and (3) Karwar–Kanyakumari (see Table 15.1).

Kutchh–Khambat

The Kutchh shelf is a gently sloping platform with two regional highs. The basin architecture displays a characteristic marginal

Table 15.1. *Summary table of the shelves along the west coast of India*

Western continental shelf of India	Kutchh shelf			Daman–Karwar	Karwar–Kanyakumari
	Kutchh shelf	**Gulf of Kutchh**	**Gulf of Khambat**		
Length (km)	662	65 (within Kutchh shelf)	135 (within Daman shelf)	652	860
Average width (km)	120	120	350	150	70
Tidal, wave and current ranges	Tide 3 m	Tide 6 m, current 1.5–4.5 knots	Tide 8.5 m, current 2.5 knots	Tide 1.75 m	Tide 1 m
Dominating process (wave, current, tide)	Tide	Tide and current	Tide and current	Wave	Wave
Sediment type: Siliciclastic/ carbonate/ authigenic/ palimpsest (%)	20/70/5/5	65/35/0/0	90/10/0/0	65/30/0/5	65/25/5/5
Modern/relict/palimpsest	90% modern, 7% relict, 3% palimpsest	75% modern, 25% relict	80% modern, 15% relict	85% modern, 10% relict, 5% palimpsest	80% modern, 15% relict, 5% palimpsest
Tectonic trend over LGC	Stable	Stable, partial subsidence	Graben, low-magnitude subsidence	Stable, very-low-magnitude subsidence	Stable

rift-basin system developing a horst–graben feature. The Narmada, Kutchh and Khambat basins are three prominent geological features in the western periphery. These basins were formed by rifting along the three Precambrian orogenic trends at different stages of evolution of the Indian sub-continent. The width of the continental shelf surrounding the Gujarat Peninsula varies from 140 km off Kori creek to 100 km off Dwarka. The shelf follows almost parallel to the shoreline up to Porbandar. Further south, the shoreline turns eastwards along the Narmada–Surat depression, whereas the shelf margin extends straight in a SSE direction, widening to 200 km off Diu, and reaches as much as 345 km off Daman. The shelf slopes gently to the west from 1:394 to 1:3050. Around Kutchh, vertical displacement of the shelf has been mild, increasing slightly southwards along the Daman shelf where faults and folds are significant. The shelf break occurs at a water depth of around 180 m.

The Cenozoic sequence is marked by carbonates and shales of Paleocene age, and carbonate formations of Eocene, Oligocene, Miocene and Pliocene age. The Saurashtra Basin displays shelf to basin carbonate facies from the Paleocene to Miocene, with a homoclinal platform dipping towards the west. The basin formation started in Late Cretaceous–Paleocene time, with rifting synchronous with the eruption of Deccan basalt.

There is an abundance of reefal terraces in the carbonate-rich zone in the outer shelf–shelf edge zone. The terraces, 2–4 km wide, are defined by a ridge on one side representing a palaeo-dune or barrier spit of the lowered sea level. The Khambat shelf is an intracratonic rift graben between the Saurashtra uplift and the Aravalli Range, extending in a north–south alignment. The Khambat Basin developed during the late Mesozoic and was influenced by Deccan volcanism. Evolution of the basin is due to deep-seated faults along the Dharwarian north–south trend, which cut across the east–west trend of Satpura and the NE–SW trend of Aravalli. In the offshore region, the graben loses its identity due to intense tensional faulting.

The textural and mineralogical characteristics of subsurface sediments, along with ground-penetrating radar (GPR) data collected along the western margin of India, were used to understand the depositional environments and the provenance of sediments. Particle size parameters such as mean, standard deviation, skewness and kurtosis show fluctuations due to fluvio-marine interaction, and it is well understood that the bottom sediments are enriched with muddy sand, which might have been deposited by means of tidal invasion. The abundance of very coarse silt to very fine sand indicates the prevalence of a low-energy

environment (Trivedi *et al.* 2012). Modern sands of the shore zone occur within a few kilometres of the coast and extend to a water depth of about 10 m. These sands are predominantly quartz with a small proportion of heavy minerals. Beyond a water depth of 10–20 m there is silty clay representing terrigenous clastic material, texturally silt with low carbonate content, although with a significant level of organic matter. Carbonate material is present on the outer shelf in the form of precipitated carbonate cement with lithification of sand grains. The Gulf of Kutchh has large areas covered by limestone and dead coral reefs. The continental shelf is covered by grey calcareous sandy and clayey sediments with skeletal debris and ooids. Grey and white ooids occur at 55 and 110 m isobaths, respectively. The terrigenous sand content is negligible (1–2%) and rarely occurs beyond the 100 m isobath. In the SW Gujarat coastal tract, a semi-arid condition prevails with a paucity of fluvial sediment input to the shelf.

A significant occurrence of lime mud in the subsurface levels of the outer continental shelf, shelf edge and continental slope sediments off Gulf of Kutchh was first brought to light during cruise SM 75 of the Geological Survey of India (GSI) research vessel *Samudra Manthan* (Vaz *et al.* 1993). The CaO content ranges from 43 to 53%. SiO$_2$ varies from 1.70 to 4.63% (Fig. 15.2). Sediment cover over the lime mud varies from 7 to >150 cm. Lime mud thickness ranges from 1.5 to 3 m and, in the majority of the cores, the base of the lime mud horizon was not intersected. In the southern sector of the shelf edge, the sediment cover is very thin to negligible, resulting in cropping out of the lime mud. The lime mud is soft, amorphous and creamy white, and occurs below an average 40 cm-thick dark brown to greenish grey clay cover. The contact between the lime mud and overlying clay is sharp, indicating a hiatus in sedimentation. This lime mud is traceable as a continuous blanket to the outer shelf off Ratnagiri, spread over a large platform. The upper slope region off Kutchh has authigenic formation of phosphorite within a calcium carbonate matrix (Vaz *et al.* 1996). The steep slope angles at the shelf edge made the sediment lose equilibrium and have triggered gravity flows characterized by the upward-fining sequence within the lime. Oolitic limestones are younger than the ooid grains by about 1.5 kyr. The age of the lime muds recovered in cores ranges between 16.1 and 11.9 ka BP. After 11.9 ka BP, the lime mud sediment deposition gradually changed to terrigenous-dominated sediments (Rao *et al.* 2012). *Halimeda* bioherms occur as 2–14 m-high linear ridges and massive structures in the outer shelf. *Halimeda* bioherms grew on the carbonate platforms until 8.3 ka BP.

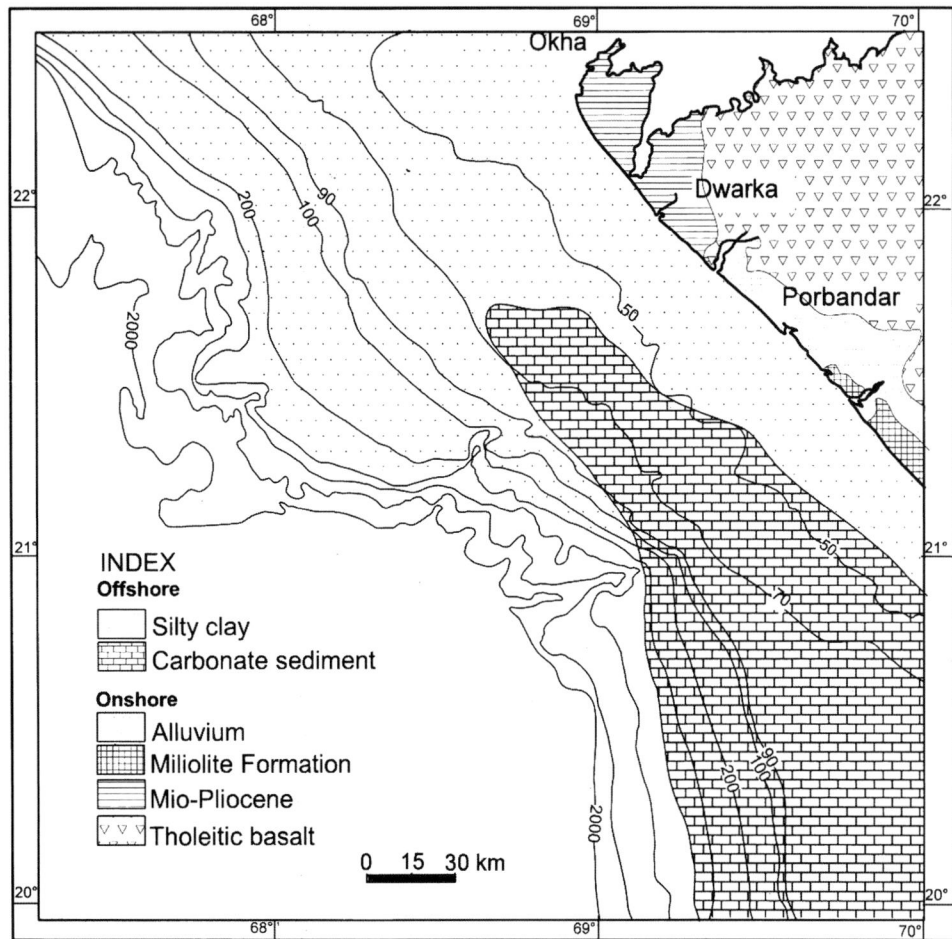

Fig. 15.2. Physiography and sediment distribution in the shelf off the Gujarat coast.

Daman–Karwar

The geomorphology of the Maharashtra coast is characterized by a coastal plain of variable altitudes and widths, backed by the hilly Western Ghats on the east and Arabian Sea on the west. A greater part of the Maharashtra coast is covered by basaltic flows of Deccan volcanics, and in places overlain by Cenozoic formations comprising laterite, submarine fossiliferous sediments and littoral deposits. Beneath the laterites are carbonaceous clays, peat and sporadically occurring grey/bluish grey clays and fossil-bearing sandstones of Neogene age showing diverse environments from terrestrial to aquatic (Kumaran *et al.* 2004). The Maharashtra coast is a submerging type, lying south of Daman, is rocky in places with promontories, cliffs and embayments.

The continental shelf is 345 km wide off Daman (Fig. 15.1) and in the southern part it reduces to 90 km off Ratnagiri. The linearity of the western continental margin is regarded as being due to a Pliocene fault system. The shelf break occurs between 130 and 180 m. Topography of the inner shelf in many places is moderately smooth, with a gentle gradient of 1:695–1:325. A drowned carbonate platform is the largest topographical feature on the north-central continental shelf of western India. The platform forms the outer shelf with relict sandy carbonate sediments between water depths of 56 and 121 m, extending from the shelf off Ratnagiri in the central part to the area off Porbandar in the north. Gravity cores collected from this area contain abundant lime muds in the lower section, with siliciclastic sediments dominant in the upper section. Radiocarbon age determination has placed the lime muds within the age range of 16.1–11.9 ka BP (Rao *et al.* 2012) and to 11.3 ka BP (Vaz *et al.* 1993).

The near-oval-shaped extremely even Fifty Fathom Flat (FFF) is a prominent topographical feature located roughly between 90 and 260 km from the coast, it covers an area of more than

28 000 km^2, and almost the entire area is occupied by carbonate oolite and sand formed between the Eocene and Holocene. The carbonate in the FFF seems to be receiving sediments from the Narmada and Tapti rivers, while the silt and clay finer particles were retained in the inner shelf (Mukhopadhyay *et al.* 2010). The continental shelf edge along the north Maharashtra region merges gently with the slope, whereas in the south the shelf break is incised by V-shaped channels seawards of the break. The depth of the incision increases further south from 15 to 50 m. Between a depth of 95 and 115 m, a series of terraces has developed due to the still stands of the Holocene transgression. The terraces are categorized into: (a) wave-cut terraces; (b) coral/algal reef-induced terraces; and (c) palaeo-beach barrier terraces, and their evolution has been ascribed to reef growth, progradation and wave activity during lowstands of sea level. Buried features such as palaeo-channels and wave-cut terraces have been interpreted from the seismic records between Daman and Goa. These are buried under 10–15 m-thick Holocene sediments and, perhaps, represent relict extension that might have traversed the width of the exposed shelf during the lowered sea level of the late Pleistocene (Rao & Wagle 1997). A series of linear horsts and graben occurs east of the Bombay High, extending south into the Ratnagiri shelf.

The terrigenous clay zone occurs along the inner shelf up to the 60 m isobath off Daman, Mazgaon and Dabhol. This comprises a sand size fraction up to a maximum of 30%, with a $CaCO_3$ content of less than 20%. The skeletal carbonate components are molluscs and foraminifers, and non-skeletal carbonate is not present. The clay content decreases gradually towards the outer shelf. South of Ratnagiri, a major part of the shelf, except the inner shelf and the shelf edge, is occupied by skeletal sand facies. Northwards, it continues in two bands: the one on the seaward margin of the inner shelf is narrow and discontinuous due to the influence of terrigenous clay facies; the other band extends northwards along the shelf edge. The skeletal matter comprises coralline debris, and fragments of calcareous algae, foraminifers, molluscs, echinoderms and bryozoa. The non-skeletal component gradually disappears towards the south where the lithology is predominantly skeletal sands. Ooids occur along two separate geomorphic zones, namely: adjacent to the shelf break between the 90 and 115 m isobaths; and also within water depths of 75–85 m. A major part of the outer continental shelf is dominated by peloids with minor proportions of ooids. The Daman–Goa shelf is nearly horizontal with a very low gradient and smooth surface, and is categorized as tropical open shelf. The distribution of foraminifers in the surface sediments of the Arabian Sea shows a dominance of specific taxonomic groups in relation to the geomorphic domains. The dominance of *Amphistegina* spp. within open ocean plankton dominant area indicates an environment conducive to reefal growth. In the middle shelf, *Stainforthia pontoni*, *Bulimina denudate*, *Cibicides lobatulus*, *Florilus asterizans*, *F. scapha* and *Cassidulina limbata* are diagnostic, and first appear around the 50 m isobath. The outer shelf is characterized by *Bolivina amygdalaeformis*, *Hosglundina elegans*, *Uvigerina cushmani*, *Baggina philippinensis* and *Cassidulina alcide* between the 100 and 150 m isobaths. Sediments in the inner shelf off Ratnagiri contain sand–silt–clay and silty clays, while the outer shelf is dominated by silty sand and sand.

The sediments off Mumbai display a progressive coarsening from sand–silt–clay to silty sand on the slope to sand in the outer shelf. Limestone from FFF contains Quaternary *Halimeda* bioherms, and aragonite constitutes a substantial part of the sediments. Numerous reefs are found in the inner shelf near FFF. The highest carbonate content is found on the NW part of the shelf and the maximum ($>75\%$) in the region occurs in the outer shelf off Mumbai. The sediment is coarse grained and the carbonate content is mostly in the coarse fraction ($>65 \mu m$) of the sediments. The carbonate assemblage consists of skeletal and non-skeletal components. Non-skeletal carbonates are represented by ooids and peloids. Skeletal components are composed of bivalves, gastropods, pteropods, foraminifers and calcareous algae. Reef growth appears to be prolific on the shelf between Ratnagiri and Marmugao due to the favourable environment caused by the absence of, or low, river discharge. The entire shelf is segmented by basement-controlled NW–SE and north–south faults giving rise to many horst–graben features.

Phosphatized coral, algal nodules, encrustations on carbonate grains and pelletal limestones occur between the 60 and 100 m isobaths on the western continental shelf of India. Phosphate occurs as brownish-grey infillings in boring or septal cavities of corals, trapped within algal laminations and as matrix in pelletal limestones. Phosphate occurs in various forms such as dumb-bell, rods and oval-shaped apatite microparticles or coalesced aggregates or microbial filaments. Relict phosphatized limestones with 2–10% P_2O_5 have been reported from almost the entire length of the western continental shelf of India (Rao & Lamboy 1996). The CaO content of the phosphatized limestone varies from 7 to 62%. These are products of biochemical processes mediated by fungi and bacteria genetically related to upwelling waters (Rao & Nair 1988). Thick Neogene and Palaeogene carbonates with minor shales underlie a late Miocene–Holocene finer clastic sequence with a homoclinal westerly dip.

Karwar–Kanyakumari

In this sector the continental margin is defined by NNW–SSE-orientated en echelon faults and fractures, and ESE–WNW-trending shear zones, resulting in the development of 40–60 m-deep graben off the Karnataka coast. Other significant topographical expressions in the shelf and margins include a series of coast-parallel coral–algal ridges, submarine terraces at water depths of 35–122 m and palaeo-strandlines in the outer shelves, from Goa to Kanyakumari (Ramachandran & Faruque 2005; Faruque & Ramachandran 2008). The shelf width in the Karnataka coast is 130 km near Karwar and 80 km off Mangalore (Fig. 15.1). Exposures of relict bedrock as small islands in the inner shelf waters off Karwar and Malpe alter locally the geometry of the sea bottom. Palaeo-channels, sunken structures and rock outcrops are common in the inner shelf and nearshore regions. The inner shelf is relatively smooth. The mid shelf in the Karnataka sector (water depths of 50–75 m) is flat and even. Typical features of the mid shelf include a series of submarine terraces, carved on and locally blanketed by late Pleistocene biotic and terrigenous accumulations, and coast-parallel colonial bioherm accumulations.

There are at least four independent NNW–SSE-trending reefal structures present in the middle and outer shelf regions off Karwar at depths of between 45 and 75 m, 86 and 89 m, and 96 and 102 m. The outer shelf beyond a water depth of 75 m is one of the most prominent physiographical sections of the shelf, comprising entirely carbonate bioherms, oolites and relict sands. Radiocarbon ages have substantiated that the submerged terraces off Karwar at water depths of 75–92 m correspond to the stillstand periods of global lowered sea level during the early Holocene. Beyond the shelf edge at a depth of 130 m, the gradients become steeper from 1:700 to 1:400.

The shelf width is 90 km off Calicut in the north, 70 km in Ponnani, 65 km in Kochi and 60 km off Quilon in the south (Faruque & Ramachandran 2008). The constriction of the shelf width offshore Ponnani is thought to have resulted from deep-cut valleys (roughly 30 km long and 8 km wide) in the shelf formed along the structural continuity of Palghat Gap governed by east–west-striking faults. The inner shelf off Kerala is more or less even and smooth with a silty clay sediment cover. The seaward part beyond a water depth of 25–30 m is covered by sediments of various mixtures of sand, silt, clay and shell carbonates. The marine sediments between 20 and 80 m water depth in the shelf carry appreciable amounts of silica sands up to a subsurface level of 2–3 m, off Kannur, Chavakkad, Beypore-Ponnani and Kochi-Kollam of Kerala (Sukumaran *et al.* 2010).

The outer shelf at the present water depths of 60–135 m is characterized by submarine terraces (Wagle *et al.* 1994; Faruque & Ramachandran 2008) located at depths of 55–60, 65–70, 75–80, 85–90 and 95–100 m (Ramachandran & Faruque 2005). Detailed studies on the shelf off the Kerala and Karnataka regions have indicated as many as nine submarine terraces in the shallow waters (Ramachandran *et al.* 2008). The prominent terraces occur between water depths of 80 and 122 m. The sea level was lowered to as much as 122 m below the present sea level during the peak of the Last Glaciation, which caused the development of the carbonate ridges at 120–122 m below the present mean sea level in the entire central part of the shelf (Fig. 15.3). Echograms, shallow seismic records and sonar images (Fig. 15.4) of the seafloor reveal the presence of a carbonate reef at a water depth of between 115 and 127 m off Quilon, Mangalore and Karwar. The carbonate reefs are located close to the shelf break. A series of coral algal ridges, algal-knolls and reefal structures of varying shapes and dimensions, present almost at the shelf break, make a conspicuous geomorphic feature of the continental shelf. The depths of occurrence of some of the ridges within the Quilon shelf, along with their elevations from the seafloor in parentheses, are 127.3 (17.5 m), 102.8 (14.9 m) 91.6 (12.4 m) and 85.5 m (1.98 m) (Faruque & Ramachandran 2008).

Radiocarbon ages of shells from the reefs in the eastern continental shelf of India have placed the reefs at 13.82 ka BP in the north Andhra Pradesh shelf, and 18.39 ka BP for the coral reef off Karaikal, in the eastern continental shelf of India. These

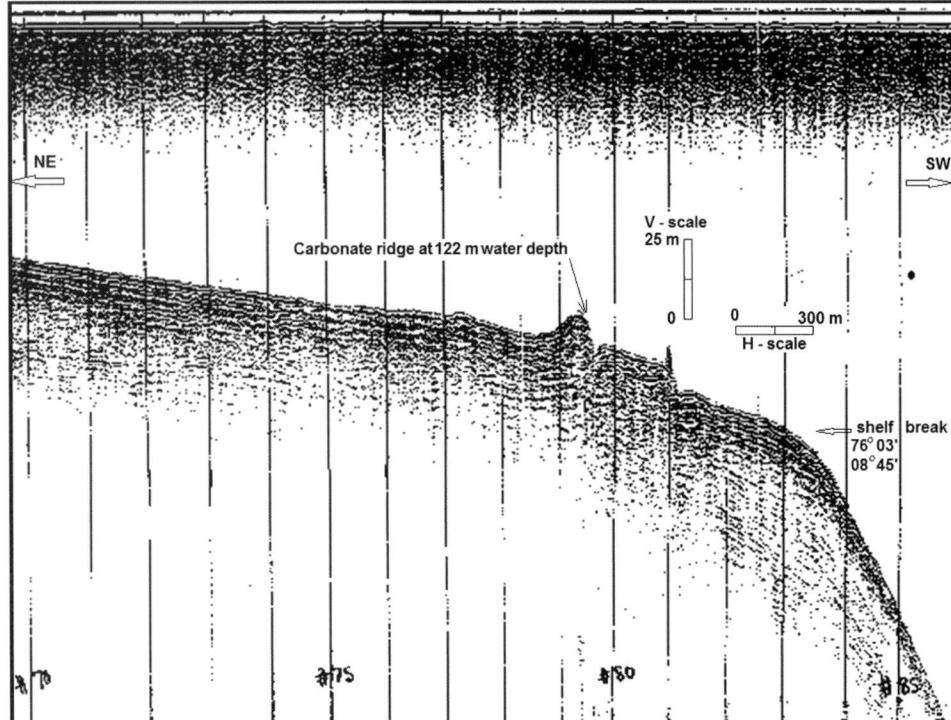

Fig. 15.3. Seismic record of the carbonate reef structure (76°05′E: 08°47′N) at the palaeo-shoreline position of the Last Glacial Maximum off Quilon.

reefal structures were formed during the low sea-level stands of the Last Glacial Maximum (LGM) (Faruque & Ramachandran 2008). The LGM reef has been located at a water depth of 122 m off Gopalpur and Visakhapatnam in the eastern continental shelf of India (Faruque 2012). The depth of formation of the beach ridges and carbonate reef along the LGM sea level has been recorded from most locations at a water depth of between 120 and 125 m. In the areas without any sign of neotectonic process, these reefs have been consistently recorded at water depths of 122 m for long distances off the east and west coasts of India. However, in the Kerala continental shelf, the reefal ridges were recorded at depths of 115 m (southern transect) and 127 m (northern transect) off Quilon. The difference in the level of the occurrences of reefs, off Quilon, adds a tectonic element to the stable position of the shelf. The reefal structure in the shelf edge off Quilon has a maximum relief of 17.5 m from the seafloor

(Figs 15.3 & 15.4). Vora *et al.* (1996) have reported the presence of such shelf edge reefs at depths of between 85 and 136 m in the shelf off Ratnagiri and Marmugao, and again from Mangalore to Quilon. According to Vora *et al.* (1996), the presence of coral reefs at around 120 m in the south (Kerala) and at a water depth of 136 m in the north near Marmugao in Goa is due to subsidence of the shelf around Marmugao after the formation of the reefal limestone.

Holocene features

There are reports of pockmarks, seepages and plumes on the inner and middle shelf and upper continental slopes of western India at water depths of between 20 and 260 m (Karisiddaiah & Veerayya 2002). Seismic reflection profiles reveal 5–35 m-thick, weakly

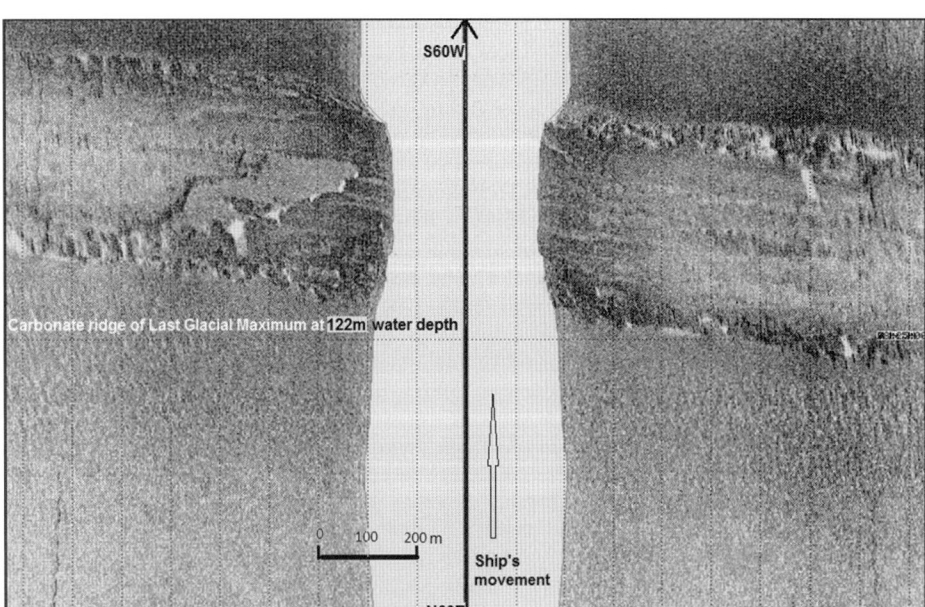

Fig. 15.4. Sonar image of the carbonate reefal structure (76°05′E: 08°47′N) trending N30°W at a water depth of 122 m off Quilon, shown in Figure 15.3.

stratified acoustic layers of clays in the inner shelf. Characteristic gas seepages are noticed off Coondapur at 20–25 m depths and clustered pockmarks at water depths of 25 m off Kasaragod. These isolated pockmarks are circular in shape, enclosed by depressions. Active pockmarks of biogenic gas seepage are also reported in the inner shelf off Quilon, Kerala (Faruque & Rama-chandran 2008), which explains the extent of peaty material in dark grey clayey sediments as due to marshy horizons below a clayey to silty layer (up to water depths of 50 m). Individual pockmarks off Quilon have a relief of 2–3 m, with a basal diameter of about 10 m occurring as raised domal-shaped features with or without a small crater-like pit on the top.

The inner, middle and outer shelf of the SW continental margin of India is covered by Holocene sediment over an older carbonate platform. The peat and carbonized wood in the sediment sequence of the shelf suggests shallow-marine, intertidal conditions and the drowning of mangrove forests along the palaeo-coastline during Holocene transgressions. The presence of iron-stained quartz grains, reworked microforaminifers and beachrock frag-ments in the outer shelf sediments indicate subaerial exposure of a palaeo-beach. The inner shelf is mostly Holocene terrigenous sediments of silty clay and gravelly mud. The inner shelf is charac-terized by clays mixed with reworked palaeo-beach sediments, detrital sand and silt from the headlands comprising coastal sedi-ments. The outer shelf is a sediment-starved Pleistocene plain. The paucity of sedimentation in the outer shelf is due to meagre sediment input from land and also due to the reefal structures on the shelf edge, obstructing the movement of outer shelf modern sediments. The reefs that occur along the shelf edge are very hard, with coral–algal facies in association with ooids, reworked palaeo-strandline deposits, bioclasts, coral debris and oolites. Authigenic and relict biogenic carbonates comprising *Halimeda* lithofacies, rhodalgal–coral facies occur in northern, central and SW shelves. Glauconite, a K- and Fe-rich mica-structure clay mineral, occurs as flakes and pellets, and also as infilling in fora-miniferal tests and sponge spicules, in the outer shelves and slopes off the Ratnagiri–Cochin coastal tract (Rao & Wagle 1997).

The distribution pattern of terrigenous sediments in the SW con-tinental shelf of India is more or less patchy coast-parallel and len-ticular, with a concentration of sediments more on the shore side. Late Holocene sediment comprising various proportions of sand, silt and clays is deposited over Quaternary sediments in the inner and outer shelves, and tapers and pinches out seawards. Places where the silty and clayey sediments are reworked by wave actions and the ocean currents concentrate terrigenous minerals (such as ilmenite, garnet, sillimanite, rutile, magnetite, kyanite, zircon, pyriboles, biotite and epidote) in the shelf, with minor addition of gravels of laterite, beachrock and skeletal carbonate. The main sources for the terrigenous sediments in the shelf are from the hinterland Precambrian rocks and Cenozoic formations.

During the SW monsoon season, patches of calm turbid water with a significant load of suspended sediment appear close to the shores of Kerala, identified as mud banks. At least 20 such reported seasonal features exist along the mouths of rivers and lagoons between Cannanore and Quilon. Mudbanks are known to occur on the SW coast off Purakkad in Kerala.

Conclusions

The continental shelf of western India owes its configuration to the rifting of the Indian subcontinent from Madagascar. The continen-tal margin defines an almost straight line from Sir Creek in the north to the Kerala coast, and is considered to be the result of an offshore fault. The northern part of the shelf was influenced by Deccan volcanic activity, the north–south Dharwar trends, the east–west Satpura lineament and the NE–SW structural trend of the Aravalli. In the north, the continental shelf is 345 km wide off the Mumbai coast and narrows down to a 60 km shelf off

southern Kerala. The inner shelf is more or less smooth and the outer shelf beyond the 90 m isobath displays an uneven surface due to a NNW–SSE-trending series of ridges with 2–18 m of relief from the seafloor. The shelf break occurs at 180 m in the north to 140 m in the south off Quilon.

The sediments in the shelf can be broadly described as belonging to three different facies: (a) beach and nearshore terrigenous sand of modern age, occupying the seafloor up to 10 m isobaths; (b) recent terrigenous silt and clay spread over the seafloor up to water depths of 60 m; and (c) relict carbonate-rich medium- to coarse-grained sediment with shell fragments, remains of micro-fauna and coral debris. The NW part of the western continental shelf is characterized by lime mud-dominated sediments, a few tens of centimetres below the seafloor, at water depths of between 56 and 121 m, overlain by younger siliciclastic-dominated terrige-nous sediment. The abrupt transition from lime mud to siliciclastic terrigenous sediment in the shelf is attributed to climate change and the rapid rise of sea level during the early Holocene (Rao *et al.* 2012). Sedimentological studies of the lime mud layers indi-cate their source from the carbonate shoal of the outer continental shelf and localization through transport by currents.

The phosphatic limestones are products of biochemical pro-cesses mediated by fungi and bacteria genetically related to upwel-ling waters. It is inferred that phosphatization took place in shallow-water conditions of the shelf environment, which subsided to this depth due to tectonic epeirogenic movements.

The outer shelf is characterized by a series of reefal structures and carbonate beach ridges, although discontinuous at places, between depth of 76 and 125 m, from Daman to Quilon. The evol-ution of these ridges is ascribed to reef growth, progradation and wave activity during the lowstands of sea level, and reflects epi-sodes of stillstands of late Pleistocene–mid-Holocene age.

The Holocene mudbanks have a socio-economic implication as these areas are known for high biological productivity (fisheries) and they prevent the otherwise rampant sea erosion during the SW monsoon (Tatavarti *et al.* 1999).

Offshore oil wells are located in the shelf off Kutchh and north Maharashtra, which includes the Bombay High. The continental margin off southern Maharashtra, Goa and northern Kerala hold good prospects for the occurrence of gas hydrate. The offshore occurrence of heavy mineral sands containing 4–7% heavy min-erals off Quilon–Varkkala is of economic significance. The marine sand sediments occurring off the Kerala coast at a water depth of between 20 and 80 m carry appreciable amounts of silica sand, which is being explored by the Geological Survey of India (GSI) as a construction material as it is a rare commodity onshore for the state of Kerala due to environmental consider-ations. The western continental shelf of India shows a number of characteristic physiographical and sedimentary features, and is classified as a stable shelf.

The first author (B. M. Faruque) expresses his gratitude to P. C. Shrivastava, former Deputy Director General of the GSI, for his guidance and motivation during the cruise and laboratory studies on the Arabian Sea. The authors are indebted to A. R. Chivas, whose review of the manuscript has helped improve the quality of the paper. The author (B. M. Faruque) is grateful to F. L. Chiocci for setting the guidelines for this work. Finally, the authors are grateful to UNESCO-IGCP for supporting this project IGCP 588.

References

BISWAS, S. K. 1989. Hydrocarbon exploration in western offshore basins of India. *In*: DHOUNDIAL, D. P. (ed.) *Recent Geoscientific Studies in the Arabian Sea off India*. Geological Survey of India, Special Publi-cations, **24**, 185–194.

FARUQUE, B. M. 2012. Records of deglaciation and Little Ice Ages in the sediments off Kerala coast. *In*: *Seminar Proceedings on Environment and Climate Change*. Utkal University, Bhubaneswar, India, 7–15.

FARUQUE, B. M. & RAMACHANDRAN, K. V. 2008. Signatures of glacio-eustasy due to Last Glacial Cycle in the shelf sediments off Kasargod–Mangalore coast. *SGAT Bulletin*, **9**, 9–13.

KARISIDDAIAH, S. M. & VEERAYYA, M. 2002. Occurrence of pockmarks and gas seepages along the central western continental margin of India. *Current Science*, **82**, 52–57.

KUMARAN, K. P. N., SHINDIKAR, M. & LIMAYE, R. B. 2004. Mangrove associated lignite beds of Malvan, Konkan: evidences for higher sea-level during the Late Tertiary (Neogene) along west coast of India. *Current Science*, **86**, 335–340.

MANJUNATHA, B. R. & SHANKAR, R. 1992. A note on the factors controlling the sedimentation rate along the western continental shelf of India. *Marine Geology*, **104**, 219–224.

MUKHOPADHYAY, R., FERNANDES, W. A., NAIK, Y. S. & KARISIDDAIAH, S. M. 2010. An insight into the 'Fifty Fathom Flat' off India's west coast. *Geomorphology*, **118**, 465–470.

NAIR, R. R., HASHIMI, N. H. & GUPTA, M. V. S. 1979. Holocene limestones of part of western continental shelf of India. *Journal of the Geological Society of India*, **20**, 17–20.

PANDARINATH, K., SHANKAR, R. & YADAVA, M. G. 2001. Late Quaternary changes in sea level and sedimentation rate along the SW coast of India: evidence from radiocarbon dates. *Current Science*, **81**, 594–600.

RAJENDRAN, C. P. & RAJENDRAN, K. 1996. Low moderate seismicity in the vicinity of Palghat Gap South India, and it implications. *Current Science*, **70**, 304–307.

RAMACHANDRAN, K. V. & FARUQUE, B. M. 2005. Signatures of sea level changes during the Last Glacial Cycle along the south western continental shelf of India. *In: Abstract Volume of the International Conference on Mineral Deposits of the shelf and IGCP 464 Continental shelves during the Last Glacial Cycle*. VNIIOkeangeologia, Russian Academy of Sciences, Saint Petersburg, Russia.

RAMACHANDRAN, K. V., FARUQUE, B. M., BANERJEE, K., ZAHEER, B., GANGADHARAN, A. V. & JAYAPRAKASH, C. 2008. Sea level stands along western continental shelf off Cape Comorin – Quilon area during the Last Glacial Cycle. *Proceedings Volume IGCP 464 Workshop on IGCP 464 Continental Shelves During the Last Glacial Cycle*. Geological Survey of India, Kolkata, India, Special Publication, **96**.

RAO, V. P. & LAMBOY, M. 1996. Genesis of apatite in the phosphatised limestones of the western continental shelf of India. *Marine Geology*, **136**, 41–53.

RAO, V. P. & NAIR, R. R. 1988. Microbial origin of the phosphorites of the western continental shelf of India. *Marine Geology*, **84**, 105–110.

RAO, V. P. & WAGLE, B. G. 1997. Geomorphology and surficial geology of the western continental slope of India: a review. *Current Science*, **73**, 330–350.

RAO, V. P., KUMAR, A. A., NAQVI, S. W. A., CHIVAS, A. R., SEKAR, B. & KESSARKAR, P. M. 2012. Lime muds and their genesis off Northwestern India during the Late Quaternary. *Journal of Earth System Science*, **121**, 769–779.

SANIL KUMAR, V., PATHAK, K. C., PEDNEKAR, P., RAJU, N. S. N. & GOWTHAMAN, R. 2006. Coastal processes along Indian coastline. *Current Science*, **4**, 530–536.

SUKUMARAN, P. V., UNNIKRISHNAN, E. *ET AL.* 2010. Marine sand resources in the southwest continental shelf of India, *Indian Journal Geo-Marine Sciences*, **39**, 572–578.

TATAVARTI, R., NARAYANA, A. C., KUMAR, P. M. & CHAND, S. 1999. Mudbank regime off the Kerala coast during monsoon and non-monsoon seasons. *Proceedings of the Indian Academy of Sciences (Earth and Planetary Sciences)*, **108**, 57–68.

TRIVEDI, D., RAICY, M. C. *ET AL.* 2012. Sediment characteristics of tidal deposits at Mandvi, Gulf of Kuchchh, Gujarat, India: geophysical, textural and mineralogical attributes. *International Journal of Geosciences*, **3**, 515–524.

VAZ, G. G., BISWAS, N. R. *ET AL.* 1993. Lime mud in continental shelf edge and slope off Kuchchh. *Indian Journal of Marine Sciences*, **22**, 209–215.

VAZ, G. G., FARUQUE, B. M., BISWAS, N. R., VIJAYKUMAR, P., MISRA, U. S., SANKAR, J. & KRISHNA RAO, J. V. 1996. Phosphorite from the upper continental slope off Kuchchh, northwest coast of India. *Indian Journal of Marine Sciences*, **25**, 20–24.

VORA, K. H., WAGLE, B. G., VEERAYYA, M., ALMEIDA, F. & KARISIDDAIAH, S. M. 1996. 1300 km long late Pleistocene-Holocene shelf edge barrier reef system along the western continental shelf of India: occurrence and significance. *Marine Geology*, **134**, 145–162.

WAGLE, B. G., VORA, K. H., KARISIDDAIAH, S. M., VEERAYYA, M. & ALMEIDA, F. 1994. Holocene submarine terraces on the western continental shelf of India; implications for sea level changes. *Marine Geology*, **117**, 207–225.

Chapter 16

The continental shelf of eastern India

B. M. FARUQUE*, G. G. VAZ[†] & G. P. MOHAPATRA

Geological Survey of India, Marine Wing, Bhu-Vijnan Bhavan, Salt Lake, Kolkata, India

*Corresponding author (e-mail: bmfaruque@gmail.com)

Abstract: The eastern continental margin of India is aligned NE–SW due to the trend of the Eastern Ghat orographic belt in the northern part and almost north–south in the southern part. It is of passive type, the origin of which can be traced back to its separation from Antarctica around 127 Ma ago. The 2493 km-long shoreline of eastern India is fringed with a continental shelf with a variable width of 35 km off Tamil Nadu to 60 km off north Andhra Pradesh and 120 km around Digha. The shelf has a gentle slope in the northern sector and is moderately steep in the south. Net longshore sediment transport along the east coast is from south to north. The east continental shelf of India is characterized by four major deltas; the Ganges, Mahanadi, Krishna–Godavari and Cauvery. The interdeltaic shelves are generally sediment starved. The inner shelf is silty to clayey silt and sandy, whereas the outer shelf has carbonate sands with coral debris and shell fragments. The outer shelf is characterized by carbonate sands, lime mud and ooids in the north Andhra Pradesh sector off the Kakinada–Kalingapatnam sector. The outer shelf off Chennai–Mahabalipuram, Nagapattinam–Point Calimere is characterized by a thin veneer of authigenic sediment represented by verdine, glaucony and phosphatic sediment. In Tamil Nadu, the region between 10°N and 12°30′N is characterized by two megalineaments and associated tectonics, suggesting that the area is tectonically active.

The ^{14}C dating of relict corals, from water depths of 120 m, off Karaikal indicates an age of 18 390 ± 210 years BP, establishing the low sea-level position of the Last Glacial Maximum. The large volume of sediment input from the deltaic system of two major drainages, Krishna and Godavari, has a significant influence on the morphology and sedimentation of the continental shelf. The influence of glacio-eustasy is noticeable in outer shelf sediments along almost the entire length of the shelf.

Geological investigations of the seabed around the Bay of Bengal have attracted geoscientists for five decades, but studies were restricted to sporadic and selected sectors. Studies carried out by the Geological Survey of India, Department of Ocean Development, Oil and Natural Gas Corporation, Oil India Ltd and the National Institute of Oceanography have gradually contributed information on surface and subsurface sediments, bathymetry, magnetics and shallow seismics, which, when combined, afford a synoptic view.

The eastern continental margin of India is aligned NE–SW due to the trend of the Eastern Ghat orographic belt in the northern part and almost north–south in the southern part. The margin is of passive type, the origin of which can be traced back to its separation from Antarctica around 127 Ma ago. Although sedimentation commenced in the Early Cretaceous, the evolution of the continental shelf and slope started in the Palaeogene (Bharali *et al.* 1992). The coastal zone of the study area is a wave-dominated regime, the sedimentation of which is variable as the major rivers of India – the Ganges-Brahmaputra, Mahanadi, Godavari, Krishna, Penner and Cauvery drain into the Bay of Bengal. The morphology of the continental margin off the coastal sedimentary basins, such as Mahanadi, Krishna-Godavari, Palar, Penner and Cauvery, is markedly different from the rest of the area (Fig 16.1).

Coastal regions of the east coast of India are very vulnerable to severe flooding due to storm surges associated with intense tropical cyclones originating in the Bay of Bengal. Wind is the main generating mechanism of these storm surges, and a rise in sea level would inundate low-lying areas (Jain *et al.* 2010). The sea surface temperature of the North Indian Ocean, which includes the Bay of Bengal, ranges from 21 to 24 °C in February and from 28 to 30 °C during July. The dominant semi-diurnal tidal range increases northwards from 0.75 m in Tutticorin along the Tamil Nadu coast to 1.4 m at Visakhapatnam, 1.75 m at Paradip and 3–5 m in the Sundarban delta in the extreme north. Except for the Sundarban delta and its neighbouring areas, which are meso- to macrotidal, the entire east coast comes under the microtidal category. In general, the salinity increases from north to south and is highest (34 ppt) from February to April. The high runoff entering the Bay of Bengal dominates the ocean water condition, and surface currents are driven by the monsoon winds corresponding to the three seasons: June–September, SW monsoon; October–January, NE monsoon; and February–May, summer. Longshore transport along the east coast is southerly from November to February and northerly from April to September, and variable during March and October. The net sediment transport varies from sector to sector; off the Tamil Nadu coast, net transport is equal in both directions. The net annual transport off Visakhapatnam, Gopalpur and Puri is northerly. Net longshore transport for the entire east coast is also northerly.

The Gangetic deltaic plain is characterized by a number of ENE–WSW-trending basement faults, which were reactivated during Cenozoic and Quaternary periods as indicated from gravity, magnetic and seismic refraction studies. Subsidence of the Mahanadi Basin continued throughout the Cenozoic, Pleistocene and Holocene. Variable thicknesses of Quaternary sediment in resulting horsts and graben explain the continuity of the same tectonic movements through time (Vaidyanadhan & Ghosh 1993).

- Orissa-Andhra Pradesh: the maximum depth to the base of the graben off Paradip is of the order of 6 km where the flanks on either side varied from 1.5 km shorewards to 3 km seawards, while another trend off Chilika Lake forms the northward extension of the 85°E ridge lineation from the deep-sea Bengal Fan. Earthquakes off Visakhapatnam at water depths of 40 m result from the reactivation of the offshore fragment of the converging point of two lineaments NW–SE and east–west traversing across Visakhapatnam, abutting the east coast at 17°45′ and extending into the Bay of Bengal. The Kandivalasa River fault reactivation appears to be the cause of tremors reported off Vizianagaram (Subrahmanya 2003).

- Tamil Nadu: the emerging nature of coasts near Mahabalipuram and Porto Novo, and subsidence of the coast between Puducherry and Cuddalore substantiate the phenomenon of epeirogenic movement. The observed canyons in the Bay of Bengal off Cuddalore, Puducherry and Palar support the fault block structure of the Tamil Nadu coast. Narasimhan (1990) has shown that Palar had three palaeochannels, all of which were to the north of the present channel. The reason for the

[†]Deceased October 2009

From: CHIOCCI, F. L. & CHIVAS, A. R. (eds) 2014. *Continental Shelves of the World: Their Evolution During the Last Glacio-Eustatic Cycle.*
Geological Society, London, Memoirs, **41**, 221–229. http://dx.doi.org/10.1144/M41.16

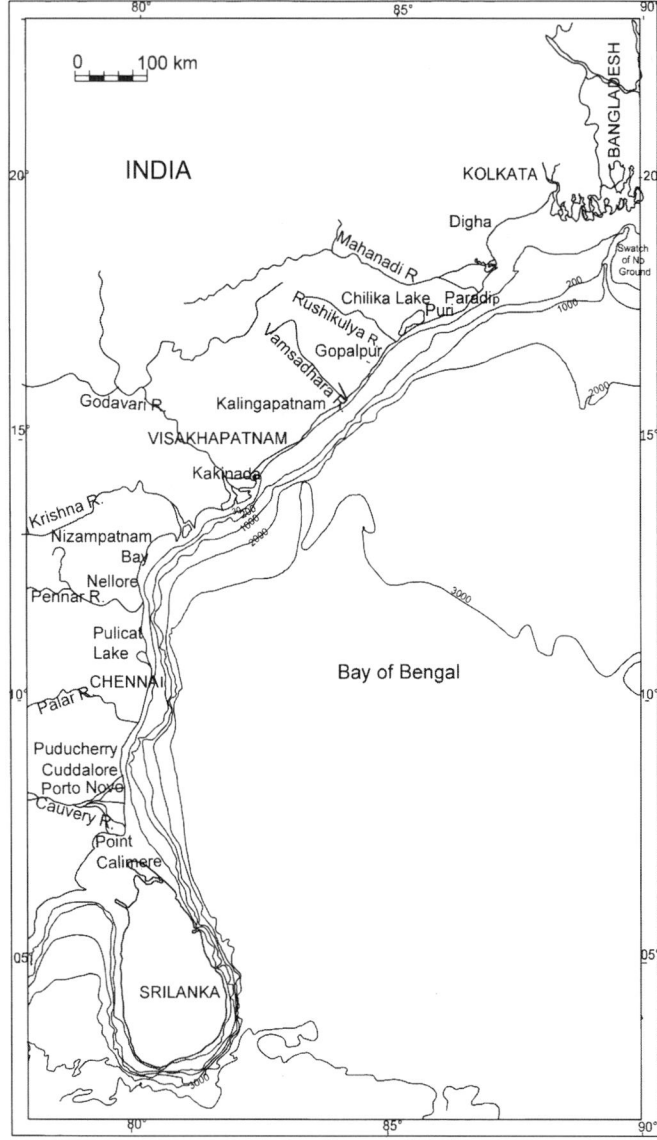

Fig. 16.1. The eastern continental shelf of India with major rivers entering the Bay of Bengal.

shift of the drainage from north to south has been ascribed to a gradual uplift of the northern parts of the region. Venkataraman & Rangaraju (1968) presented evidence to show that Mahabalipuram and the city of Pumpuhar in Tamil Nadu have undergone submergence, whereas some areas along the Ramnad and Tirunelveli coasts have emerged recently.

The vast stretch of the east coast (2493 km) and the wide spectrum of coastal ecosystems make it necessary to deal with the continental shelf characters by sectors based on regional physiography as follows: (i) Ganges–Mahanadi–Rushikulya, with abundant deltaic deposits; (ii) Sonapurpeta–Kakinada: a sediment-starved shelf devoid of major rivers; (iii) Godavari–Krishna: deltaic; (iv) Nizampatnam Bay–Pulicat Lake: a sheltered bay enriched by deltas of the Krishna and Penner rivers; and (v) Pulicat Lake–Point Calimere: with coastal lagoons and a delta with moderate sedimentation (see Table 16.1).

Ganges–Mahanadi–Rushikulya

The Ganges system discharges into the Bay of Bengal along a delta front with a width of 380 km. Delta tidal currents are, perhaps, the strongest hydrodynamic influence on the subaerial delta front and the subaqueous part of the Ganges delta (Allison 1998). Sediment partitioning across the river–ocean interface has led to the formation of a subaqueous mud clinoform on the continental shelf adjacent to the river mouths. The topset bed within water depths of 30 m dips gently at 1:1600. The surface sediments are sandy silt, tidally laminated, in the area shorewards of the 15 m isobath. Progradation of a coarse-grained subaerial delta-front clinoform over the subaqueous mud clinoform is characteristic of the Ganges not seen elsewhere. The Swatch-of-No-Ground submarine canyon incises the shelf within 40 km of the shoreline. Evidence of growth faults and slumps near the head of the Swatch-of-No-Ground submarine canyon, which incises the shelf to about the 20 m isobath, suggests that the canyon intercepts deltaic sediment transported alongshore to the west, funnelling a fraction of this material to the deep sea (Kuehl *et al.* 1997). Near the canyon head, strata become more irregular and are commonly chaotic, with discontinuous and truncated beds. Seismic records indicate that an acoustically transparent surface sediment layer, of about 1 m in thickness, covers most of the seafloor. Core samples from this area demonstrate sequences of sandy mud overlain by muddy fossiliferous sand rich in brown iron oxide-stained ooids (0.5–1.5 mm in diameter, biogenic nuclei). The microfacies characteristics point to formation in a subaqueous shallow-water, high-energy and nearshore environment.

The compound delta of Mahanadi–Brahmani–Baitarani has an arcuate front developed in a strong current-dominated environment, which has given rise to spits and barrier islands along the shoreline. The coastal belt is fringed by a continental shelf that varies from about 50 km wide off Sonapurpeta to about 60 km off Chilika Lake. Further east, the shelf gradually widens to 65 km. Shelf width is as much as 100 km off the Subarnarekha River and about 120 km off Digha. The gradient of the shelf is generally 1:250–1:280 in the western sector but moderates to 1:600–700 in the eastern sector off the West Bengal coast. The shelf edge is encountered at water depths of 150–180 m. The continental shelf off Rushikulya–Chilika Lake is nourished by input from the Rushikulya River. Off the Chilika mouth, the thickness of the sediment over the sub-bottom reflector increases from about 5–10 m in the inner shelf to over 40 m in the outer shelf (Faruque 2006). The Quaternary sediments in the shelf up to −100 m below seafloor consist generally of three lithological units, namely: (i) chaotic facies representing the surface of an early–middle Pleistocene sedimentary sequence; (ii) another unit composed of sandy to clayey layers, about 30 m thick in the mid shelf region; and (iii) a sequence of alternating dark and transparent sedimentary layers, overlying the chaotic facies, with a thickness varying from 30 to 80 m (Murthy 1989).

The inner continental shelf down to a depth of 30–50 m is moderately smooth, with a seaward gradient of around 1:115, except off these deltaic regions. The outer shelf is generally flat and dotted with pinnacle-shaped irregularities, small ridges and shallow depressions (Fig. 16.2). These morphological features are prominent off Gopalpur and towards the east these are subdued due to the masking effect of the Mahanadi deltaic sediments. Sandy sediments occur on a regional scale as a more or less continuous body in the nearshore area. In the shelf off Rushikulya-Chilika, the sand extends to the 35 m isobath, whereas to the north of Chilika it is restricted to the 10 m isobath. The continental shelf off Chilka Lake is 55–60 km wide (Fig. 16.1). However, off Puri, the continental shelf narrows to almost 25 km and the shelf breaks at a depth of 50 m (Fig. 16.2). To the immediate east of Puri, the shelf attains its usual width of 60 km and widens further east. The textural gradation from coarse sand to finer sediment occurs from the beach to the 15 m isobath within the nearshore sand facies. Sand occurring beyond the 15 m isobath is relatively coarser, commonly grading to gravel size. In the shelf off Gopalpur at 33 m and 56 m depth,

Table 16.1. *Summary table of the shelves along the east coast of India*

Eastern continental shelf of India	Ganges	Mahanadi	Sonapurpeta– Kakinada	Krishna– Godavari	Nizampatnam– Pulicat	Pulicat–Point Calimere
Length (km)	446	280	410	311	231	387
Average width (km)	120	65	50	20	45	35
Tidal, wave, current ranges	Tide 3–5 m	Tide 2 m	Tide 1.5 m	Tide 1.3 m	Tide 1 m	Tide 0.9 m
Dominating process (wave, current, tide)	Tide and episodic storm	Wave and episodic storm	Wave and episodic storm	Wave	Wave	Wave
Sediment type: Siliciclastic/ carbonate/authigenic/ palimpsest (%)	95/5/0/0	95/5/0/0	70/25/5/0	95/5/0/0	65/25/5/5	65/20/10/5
Modern/relict/palimpsest	100% modern	100% modern	80% modern, 15% relict, 5% palimpsest	100% modern	80% modern, 15% relict, 5% palimpsest	85% modern, 10% relict, 5% palimpsest
Tectonic trend over LGC	Stable, mostly subsidence	Stable, partial subsidence	Stable	Stable, partial subsidence	Stable	Stable, uplifting (north), subsidence (south)

medium sand from topographical highs shows high a concentration of relict heavy minerals. The erosional unconformity with younger sediments can be attributed to the lowering of the sea level to −122 m and the consequent exposure of the continental shelf to subaerial conditions. Short cores recovered from this area show a plastic brownish clay/sandy clay horizon within 3 m from the top, being directly overlain by soft clay with carbonaceous matter in places and, in turn, by grey and brown sand (Mohapatra *et al.* 1992). This stratigraphic sequence is marked by a thick pile of deltaic sediments off Mahanadi and in the north due to input from the Ganges. Illite, smectite and kaolinite are the major clay minerals found in the surface sediments of the shelf off Gopalpur and the Mahanadi delta (Faruque *et al.* 2002).

Sonapurpeta–Kakinada

The shelf between Kakinada and Sonapurpeta is sediment-starved due to the absence of any major river. The major source of sediment is from the Eastern Ghats and the coastal plain is occupied with red semi-consolidated sediment in places.

The general trend of the continental shelf is NE–SW parallel to the shoreline and with the major trend of the adjacent Eastern Ghat belt. The width of the shelf varies from a minimum of 8 km off Pentakota (gradient 1:520) to 57 km off Pudimadaka–Sonapurpeta (gradient 1:480–1:105). The shelf is intersected by the valley system off the Vamsadhara–Nagavalli River, dissecting the outer shelf up to the 70 m isobath. The gradient of the shelf is near horizontal (1:520–1:410) up to a water depth of 100 m, beyond which the gradient increases to 1:480–1:105. The shelf break occurs at 150 m off Santapalle and 280 m off Pentakota. The uppermost sediment cover laid down recently over an erosional surface is not thick enough to cover these irregularities and, hence, abuts against ridges and domes. Towards the Godavari delta, however, the thickness of the sediment gradually increases and obscures the older erosional surface. The dissected topography of the NE part of the continental slope, as shown by the presence of gullies and canyons, is the effect of erosion and mass movement due to turbidity currents.

The surf zone extends up to a depth of 2–3 m, except in the area of rocky cliffs, platforms and bays. The shore face from the surf zone extends up to an average depth of 12 m and the gradient of

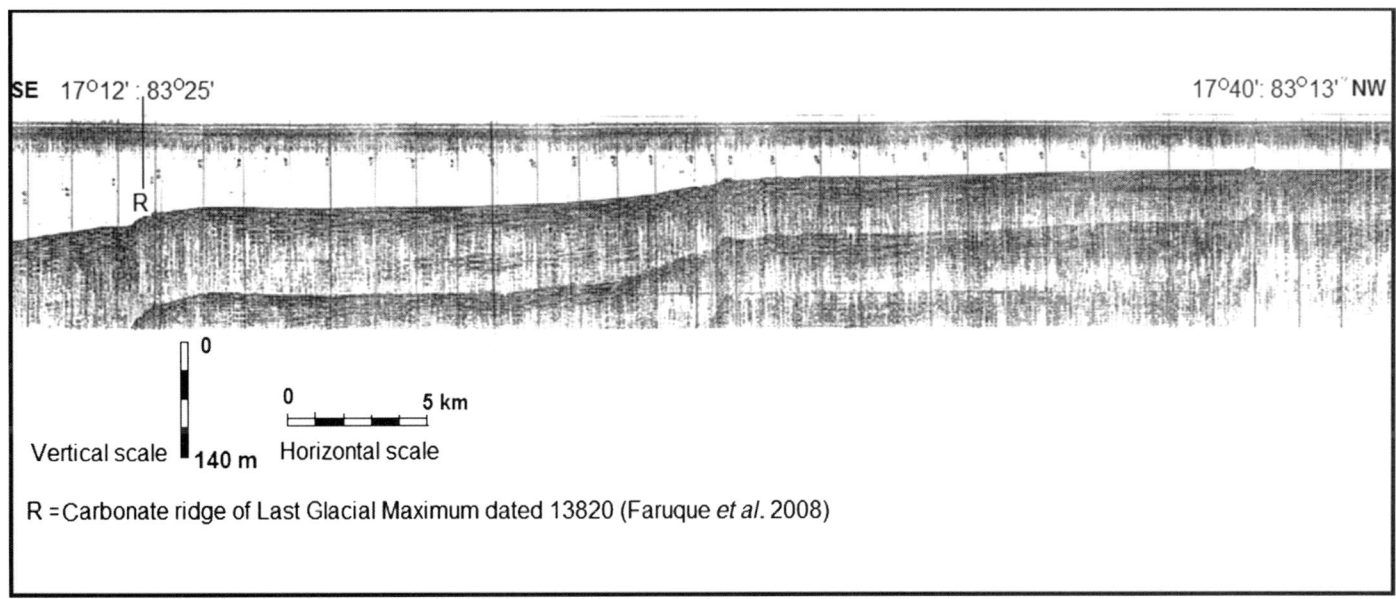

Fig. 16.2. A shallow seismic profile in a NW–SE direction orthogonal to the shoreline, across the width of the shelf, off Gopalpur (seismic data from GSI cruises).

this zone is of the order of 1:50. Rocky promontories extending offshore are common, particularly between Sonapurpeta and Pudimadaka. The physiography is, in many parts, smooth with the shoreface having a steeper gradient, which merges with the inner shelf within variable depths of 5–15 m, with a distinct break in gradient. The proximal part up to the 50 m isobath exhibits a 1:250 gradient and is almost smooth except for some isolated ridges off Gopalpur and a subsurface ridge that continues from Gopalpur to Bavanapadu at the 30 and 50 m isobaths. The ridge around 63 m continues from Gopalpur in the north to Kalingapatnam above the seafloor, and is buried off Pudimadaka in the south. The surface sediment north of Kalingapatnam is predominantly sandy, whereas it is clayey sand to silty clay further south. The central part of the inner shelf is flatter with a gradient of 1:650. In the sector off Sonapurpeta–Kakinada, the proximal part of the inner shelf is floored by medium to coarse brown sand up to an average depth of −35 m, in places spreading up to −50 m off Bavanapadu. This sand body is persistant at −5 m below the seafloor and contains placer minerals such as ilmenite, garnet, sillimanite, monazite and rutile. Further offshore of this sandy zone, the sediment, in general, is clayey-silty sand with variable proportions of silt and clay that lie within a depth of 35–62 m. The muddy sediment has greenish rounded granules and oval faecal pellets mixed with terrigenous and biogenic sediment. The sand content diminishes south of Kalingapatnam where mud predominates over sand. The sediment broadly exhibits fining offshore from clay-silt to silty-clay. The clayey-silty sand is dominated by well-rounded quartz grains with a yellowish brown surficial tint. The biogenic component is dominated by molluscs, followed by foraminifers, bryozoans, ostracods and coral fragments. The continental shelf has a distinct break in gradient at around the 100 m isobath that divides the shelf into an inner portion with a 1:300 gradient and an outer portion with a 1:100 gradient. Lime mud occurs in the outer shelf between the 100 and 225 m isobath. The outer shelf zone exhibits a smooth profile with parallel sub-bottom reflectors and occasional growth faults in the seismic records.

The shelf off Pentakota–Santapalle, between the 80 and the 100 m isobaths, is covered by carbonate sand such as ooids, peloids, and skeletal debris, and the deeper parts beyond 100 m by silty clay and clayey sediments. The subsurface lime mud mostly comprises silt-sized material (38–95%). The clay-sized fraction varies from 2 to 14%. The capping that overlies the lime mud has sand varying from 1.6% at the top to 54% in the bottom. The lime mud is greyish white to creamy white, fine-grained, compact material associated with skeletal and non-skeletal fragments, and terrigenous detrital minerals. The maximum thickness of lime mud encountered is 2.89 m, and is expected to continue further in the subsurface. Lime mud chiefly comprises aragonite (85–90%), with Mg-calcite, calcite and halite. Calcium carbonate content varies from 73.7 to 92.9%, strontium from 0.5 to 0.6% and lithium from 50 to 475 ppm. The 200 km-long zone of lime mud is not present in the vicinity of the Godavari delta in the south, nor near the Vamsadhara River in the north (Murty *et al.* 2006). The lack of terrigenous influx, the slow pace of transgression and the possibly higher upwelling resulted in carbonate deposition under open shelf conditions (Ginsburg & James 1974). Aragonite needle mud occurs on the offshore side of the carbonate sand. That the lime and ooids were formed in shallow water is supported by faunal evidence, particularly foraminifers.

In the segment off Gopalpur–Kalingapatnam, a linear reef-like structure is confined to the 62–65 m isobaths. Recovery of seaferns, coral fragments, large shells of lamellibranchs and other skeletal matter, together with the morphological expression, support a coral reef build-up in this zone. In the northern part of the slope valley, the surface and subsurface sediments are generally muddy, dominated by terrigenous mud in the NE sector, while the southern part comprises lime mud (Murty *et al.* 2006).

The shelf break does not occur at uniform depth, and occurs between depths of 180 and 230 m. The most prominent of several ridges occurs at the depth of 122 m, with a relief of 3–17 m, and is composed of coralline debris, molluscs, shell fragments, ostracods accompanied by ooids, and is traceable for several tens of kilometres (Faruque & Lahiri 2002). The outer shelf carbonate sand occurs between the 90 and 120 m isobaths in the sector off Gopalpur–Kalingapatnam as linear patches around the 100 m isobath. Skeletal components are represented by, for example, molluscs, foraminifers, bryozoans and ostracods. The benthic foraminifers are commonly represented by *Amphistegina*, *Quinqueloculina*, *Elphidium* and *Ammonia*, followed by, for example, *Lenticulina*, *Nummulites*, *Cancris*, *Cibicides*, *Eponides*, *Bolivina*, *Calcarina* and *Triloculina*. Planktic foraminifers are represented mainly by *Globorotalia*, *Globigerinoides* and *Pulleniatina*. Brown-coloured ooids are coated with iron oxide, suggesting not only a shallow-water origin for the ooids but also subaerial exposure (Faruque *et al.* 2008). As the river discharge during the lowstands was directly channelled into the continental slopes and deeper parts, the bypassed shallower parts near the strandline were impoverished of terrigenous influx. These conditions, perhaps, enabled carbonate precipitation (Milliman *et al.* 1975).

Carbonate sediments on the east coast at a water depth of 115 m (Vaz 1996) have ^{14}C dates of 14 510 ± 190 years BP, while at 100 and 85 m the ages are 12 500 ± 170 and 10 790 ± 170 years BP, respectively. The lime mud, on the outer shelf, is inferred to have been formed contemporaneously with the carbonate sand deposition. The −122 m coralline ridge was formed during the Last Glacial Maximum and the radiocarbon age corresponds to 13 820 years BP (Faruque 2006) (Fig. 16.2). The continental slope off Kalingapatnam is characterized by the formation of three major canyons (Fig. 16.1), besides several smaller ones. The canyons from north to south have been named Krishna with a 340 m-deep valley, Mahadevan at 570 m and Andhra at 450 m (La Fond 1965).

Godavari–Krishna deltas

The prodeltaic sediments of the Godavari and Krishna rivers cover the entire continental shelf off their mouths, whereas some parts of the shelf distal from the deltas expose relict sediments due to a lower influx of recent sediments. The width of the shelf is between 10 and 15 km (Fig. 16.3). The shelf break is poorly defined where it smoothly merges with the continental slope. The inner continental shelf down to water depths of 30–50 m is even, with a seaward inclination of around 1:115, except off the Godavari and Krishna deltas where the shelf is almost flat. The outer shelf is generally flat, broken by low-relief ridges and moderate depressions. The outer shelf off deltas is relatively steeper, with seaward gradients of 1:115–1:70.

The large sediment inputs from deltaic systems off the two major rivers, Krishna and Godavari, have a significant influence on the morphology and sedimentation of the continental shelf. Sandy sediments are scarce off the deltas and, if present, they are mostly confined to a depth of 5 m, in the form of bars, spits, etc. Beyond the 5 m isobath, sands merge into sandy silts–clayey sands, clayey silts and silty clays indicating a progressive seaward fining. Prodeltaic mud from Godavari–Krishna prevails in the shelf, south of Kakinada–Kottapatnam. The calcium carbonate content in these sediments is less than 10%. Delta front lobes, crescentic bars and spits are common at the river mouths. Prominent among these is the 16 km-long sand spit off Kakinada enclosing Kakinada Bay.

The sand is characteristically terrigenous in nature and mainly composed of quartz with lenses of heavy mineral concentration. The biogenic and authigenic fraction includes skeletal grains of

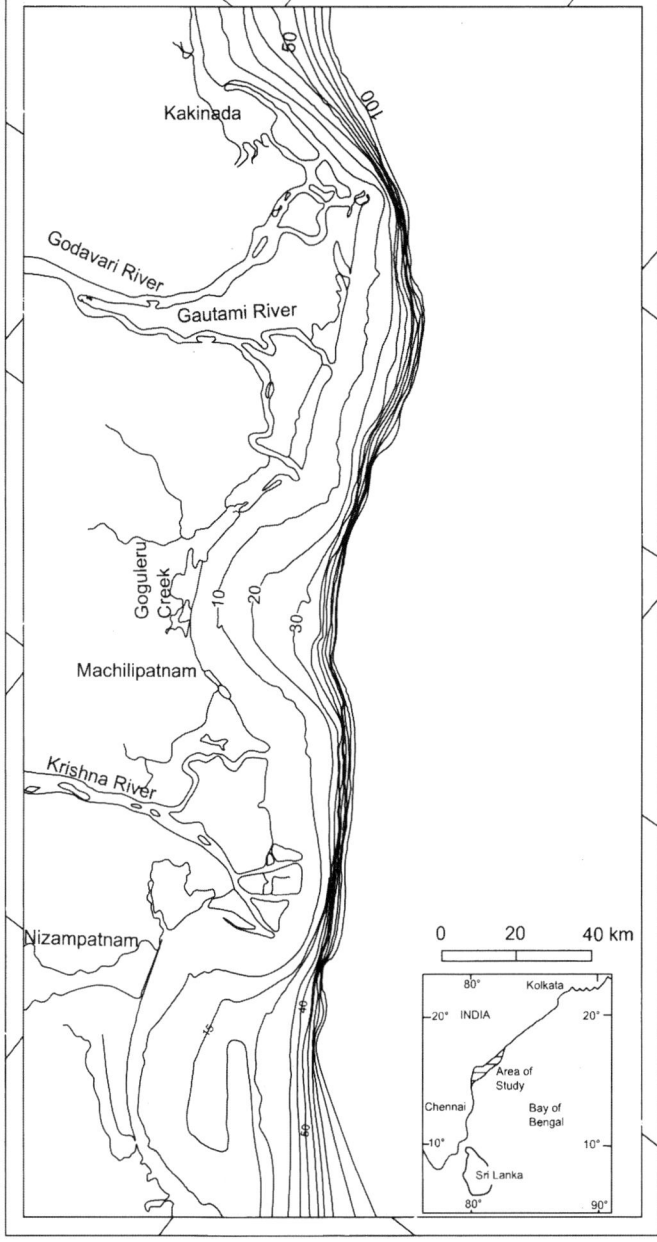

Fig. 16.3. The narrow width of the continental shelf off Krishna–Godavari delta in the central part of the east coast of India (bathymetric data from GSI cruises).

Based on grain-size parameters, the coarse brown sand occurring beyond the 15 m isobath is considered to be of relict nature. Short cores recovered in the area show a plastic brownish clay–sandy clay horizon within 3 m from the top, being directly overlain by clay with carbonaceous matter, and, in turn, by a grey sand and, finally, brown sand. The bottommost plastic clay–sandy clay, which resembles the brown 'latosol' of the adjoining coastal belt, is possibly of pre-Holocene age, formed due to subaerial exposure. The grey sand in the cores represents transgressional deposit spread over lagoonal or back swamp sediments.

The inner shelf area beyond the reach of direct input from major rivers has a thin veneer of clayey sediments. The basal reflector where exposed on the outer shelf is represented by shallow-water assemblages such as ooids and foraminifers (shallow-water species) representing a depositional surface of an early Holocene epoch. The basal surface is characterized by a number of mounds, ridges and pinnacles, some of which are interpreted to be carbonate build-ups by Rao *et al.* (1980) and Murthy (1989).

The outer shelf carbonate sedimentation might have taken place during the late Pleistocene–early Holocene epochs. The carbonate-rich facies continued up to -80 m from the shelf edge, and carbonate and clastic mixed facies from -80 to -60 m. The calcium carbonate content of the sediment is found to be proportional to the sand-size fraction, indicating that the silt- and clay-sized grains are of a terrigenous nature. It is inferred that after stabilization of sea level at around 6 ka BP, delta progradation started on the east coast.

Nizampatnam Bay–Pulicat Lake

The Nizampatnam Bay adjoining the Krishna delta is a crescentic sheltered part of the western Bay of Bengal, occupying an area of 1825 km[2]. It is a shallow-marine environment just south of the Krishna delta on the east coast of India. Even though the sediment types display a wide range, the sediments in the bay are predominantly composed of sandy clay and sand. A lenticular sandy ridge located at about 25 km from the present coastline in the Nizampatnam Bay is interpreted as a palaeoshoreline evolved during a stillstand position of the Holocene transgression. [14]C dating of a mollusc shell from the beach ridge has an age of 8200 ± 120 years BP, suggesting a Holocene strandline at a water depth of 17 m. The sediment is basically mud with varying proportions of sand (5%), silt (50%) and clay (45%). The submerged beach ridges are composed mainly of coarse sand, negatively skewed and better sorted, with sand 91%, silt 7% and clay 2%. The deposition of fine-grained sediments from the Krishna River covered the sand in the northern part, leaving the southern bay sand exposed up to the present coastline (Rao *et al.* 1990). The continental shelf break occurs at depths of 180–220 m, and is commonly marked by minor domal-shaped irregularities, small terraces and less pronounced rises with a relief of around 5–10 m.

With a long stretch of sandy beaches and microtidal energy, the coast can be classified as wave dominated. The nearshore sand prism with textural fining offshore is impacted by a littoral current directed towards the NE during the SW monsoon, and to the SW during the NE monsoon. The intensity of littoral transport is higher during the SW monsoon, as a result of which most of the spits, shoals and bars have grown in a NE direction. Beyond the deltaic influence, south of Nizampatnam Bay up to Pulicat Lake, the width of the nearshore varies from 10 to 15 km. Shoals occur in this zone that are either subparallel or at an angle with the sinuous coastline. A distinct morphological break marked by abrupt change in gradient characterizes the boundary between the shoreface and inner shelf. The inner shelf gradient varies from 1:100 to 1:425, whereas the steeper shoreface has a gradient

molluscs or their fragments and calcareous concretions of possible algal origin. Recent study of the coastline off Godavari delta with the aid of satellite imagery indicates that, during the last quarter century, the shoreline has been retreating landwards in the absence of fluvial input, possibly due to neotectonic activity. The present river mouth bars are the products of the reworking of the older sand bars rather than being accretionary (Mohapatra *et al.* 2002). Coastal erosion in the area north of Kakinada could be due to non-replenishment of sediment combined with neotectonic epeirogenic movement along the lineament. The islands of the Krishna delta front are intertidal and submerged to a large extent during spring tide.

The sedimentation pattern on the shelf appears to be influenced largely by a complex set of process variables, both of the present and the past, resulting from the differences in fluvial inputs, ocean currents, waves and littoral processes. These are subjected to redistribution processes resulting in depositional forms such as bars, spits and delta front lobes.

of 1:50–1:90. In the southern sector, off the Kottapatnam coast, the proximal part is characterized by linear bay mouth bars in the Nizampatnam Bay, and sandy shoals off Pulicat and Ennore, besides minor undulations within depths of 54–62 m. South of Nizampatnam Bay, the inner shelf sand body is confined within the 30 m depth contour. The significant terraces (T) and ridges (R) that could be the imprints of stillstands of a sea-level oscillation occur at 122 (R1), 99 (T1), 84 (T2), 76 (R2), 62 (R3), 54 (T3) and 48 m (T4) depths.

The distal part of the mid shelf within water depths of 62 and 120 m has a number of ridges, pinnacles and terraces with terrigenous sand/mud up to the 80 m contour followed by carbonate sand from depths of 80 to 120 m. The carbonate sand comprises ooids and peloids within depths of 90–120 m . Besides the non-skeletal grains, ridges of hard substratum with corals and coralline material are encountered within depths of 80–120 m. The seaward side of the 122 m ridge/terrace has a platform that has a slope of 1:170–1:45 before the commencement of the continental slope.

The sea-level rise following sedimentation along the beach ridge was much faster and transgressed the entire inner shelf within a span of 2200 years (Mohapatra *et al.* 2002). Molluscan shells and wood from the subsurface samples in the adjacent northern part of the area at water depths of 19.5 and 12.8 m are radiocarbon dated at 7940 ± 130 and 7470 ± 80 years BP, respectively (Mohapatra & Murty 1999). An irregular hard substratum under the present seafloor is interpreted as a late Pleistocene erosional surface upon which carbonate reefs were formed during the last transgression (Murthy 1989). Some of the linear ridges at water depths of 62 and 75 m, and terraces at depths of 54 and 48 m, can be explained due to intermittent still stands, at these levels, during the rise of sea level. A beach ridge at a depth of 28 m prior to the Last Glacial Maximum (LGM) indicates the possibility of some remnant morphological features of an earlier regression. From the available dates (Bruckner 1988; Rao *et al.* 1990; Vaz & Banerjee 1997), it can be inferred that the sea level stabilized around 6.5 ka BP on the east coast of India. This was followed by the deltaic build-up, particularly of the Godavari and Krishna rivers (Sambasiva Rao *et al.* 1978; Nageswara Rao & Sadakata 1993). A rapid rise of sea level from -50 m seems to have flooded a considerable part of the shelf as indicated by cessation of coral reef growth above the 50 m level, decoupling the rivers from their slope valley heads which terminate around the 50 m isobath. It is also marked by a change in the sedimentation pattern from carbonate to terrigenous clastics. This change is due to an influx of freshwater into the bay, which restricted carbonate growth.

Pulicat Lake–Point Calimere

The width of the continental shelf varies from 20 km off Pulicat Lake to about 50 km off Chennai, and further south the shelf narrows to 25 km off Puducherry–Porto Novo. The shelf is about 35 km off Karaikal and widens to as much as 80 km off Point Calimere. The nearshore zone up to a water depth of 20 m is between 5 and 8 km wide. The inner shelf has an average gradient of 1:400. In the area off Pulicat Lake–Mahabalipuram, the inner shelf sediments are dominated by silty clays and clayey–silty sands, and off the Puducherry coast by coarse sands, except for sporadic patches of silty clays localized near the river mouths. The shelf off Pulicat–Mahabalipuram and the seabed off Coleroon–Point Calimere are dominated by silty clay and clayey silt, whereas the shelf sediment off Puducherry–Cuddalore is very coarse grained and commonly gravelly. The outer shelf sediments comprise <5% of terrigenous components. Echoprofiles indicate that the continental shelf has an average gradient of 1:115 and gradually widens from north to south. Shore-parallel

minor-scale terraces were recorded at water depths of 65, 80, 100 and 130 m off Mahabalipuram. The outer shelf beyond 120 m depth has a steep gradient at 1:80 and the seafloor is irregular at 185 m due to exposure of a hard substratum. Shore-parallel patches of skeletal carbonate-rich sediments consisting of shallow-water skeletal components and ooids indicative of palaeostrandline positions occur between depths of 60 and 120 m. The outer shelf beyond a depth of 130 m off Chennai–Mahabalipuram and Nagapattinam–Point Calimere is covered by clayey-sand.

The coastline is inflected landwards off Palar–Karaikal, where the slope widens to 90 km. Submarine valleys off the Palar–Porto Novo coast have been named from north to south as Palar, Puducherry and Cuddalore. During seabed mapping by RV *Samudra Manthan* of the Geological Survey of India (GSI) (cruises SM 72 and SM 110), two more submarine valleys – Coleroon and Cauvery – were mapped. The Coleroon Valley originates at a depth of 100 m, trending east–west for long distances. The Cauvery submarine valley starts at a depth of 300 m south of Coleroon.

A study carried out by Thakur & Kumar (2007) reveals that narrowing of the shelf and the steep gradient in the vicinity of Nagapattinam, Cuddalore and Chennai provide appropriate conditions for enhancement of tidal waves, and thus explains the maximum damage at this site during the 26 December 2004 tsunami.

Offshore Chennai–Palar River, magnetic and gravity signals indicate the offshore existence of granulite, which corresponds to the area between the Palar and Cauvery rivers (Fig. 16.4). The shelf edge invariably exposes the hard rock charnockite continuing from the land offshore. Hard, rugged karstic surfaces have been recorded off Chennai and Nagapattinam (Vaz 1995). Fossils such as *Crasatella* sp. and the foraminifer *Globigerina linaperta* (Vaz *et al.* 1999) from the outer shelf off Chennai–Mahabalipuram and Nagapattinam–Point Calimere attest to Late Cretaceous–lower Miocene ages. Phosphatic algal nodules and rock fragments were reported from the outer shelf in water depths of 100–225 m off Chennai. The sediments comprise relict clayey sands, ooids, carbonate skeletal matter and phosphatic nodules. Phosphatic rock fragments are dark brown and grey in colour, irregular and 1–5 cm in length. They are composed of phosphatized foraminiferal tests filled with amorphous phosphate, black peloids and quartz grains. A zone of weak upwelling established off Chennai–Karaikal implies a zone of low biological activity. Relict sediments in the outer continental shelf without overlying recent sediments indicate a low rate of sedimentation. The NE–SW-oriented Ariyalur–Puducherry tectonic depression comprising upper Cretaceous–Eocene sediments may extend NE into the offshore domain. The assimilation of phosphates from seawater by bacteria usually takes place in shallow-water conditions during low rates of sedimentation. The occurrence of phosphatic algal nodules, with carbonate-rich relict sediments, indicates their formation in a shallow-water environment (Vaz 1995).

The outer shelf off Chennai–Mahabalipuram, Nagapattinam–Point Calimere is characterized by a thin veneer of authigenic sediment represented by verdine, glaucony and phosphatic sediment. The presence of verdine at 180 m on the outer shelf can be related to its formation during lowered sea level of the Last Glacial Cycle, as the verdinization process generally takes place within a water depth of 60 m (Vaz 2000). The verdine and glaucony zones in the outer continental shelf and subsurface levels of the upper continental slope, respectively, clearly demarcate two zones. The morphology of the grains suggests that foraminiferal tests and faecal pellets have acted as substrates for the formation of verdine and glaucony grains. The P_2O_5 content of the verdine and glaucony grains varies from 2 to 3%. Contemporaneous formations of verdine–phosphorite and glaucony–phosphorite have been reported in marine sediments. Phosphatization and

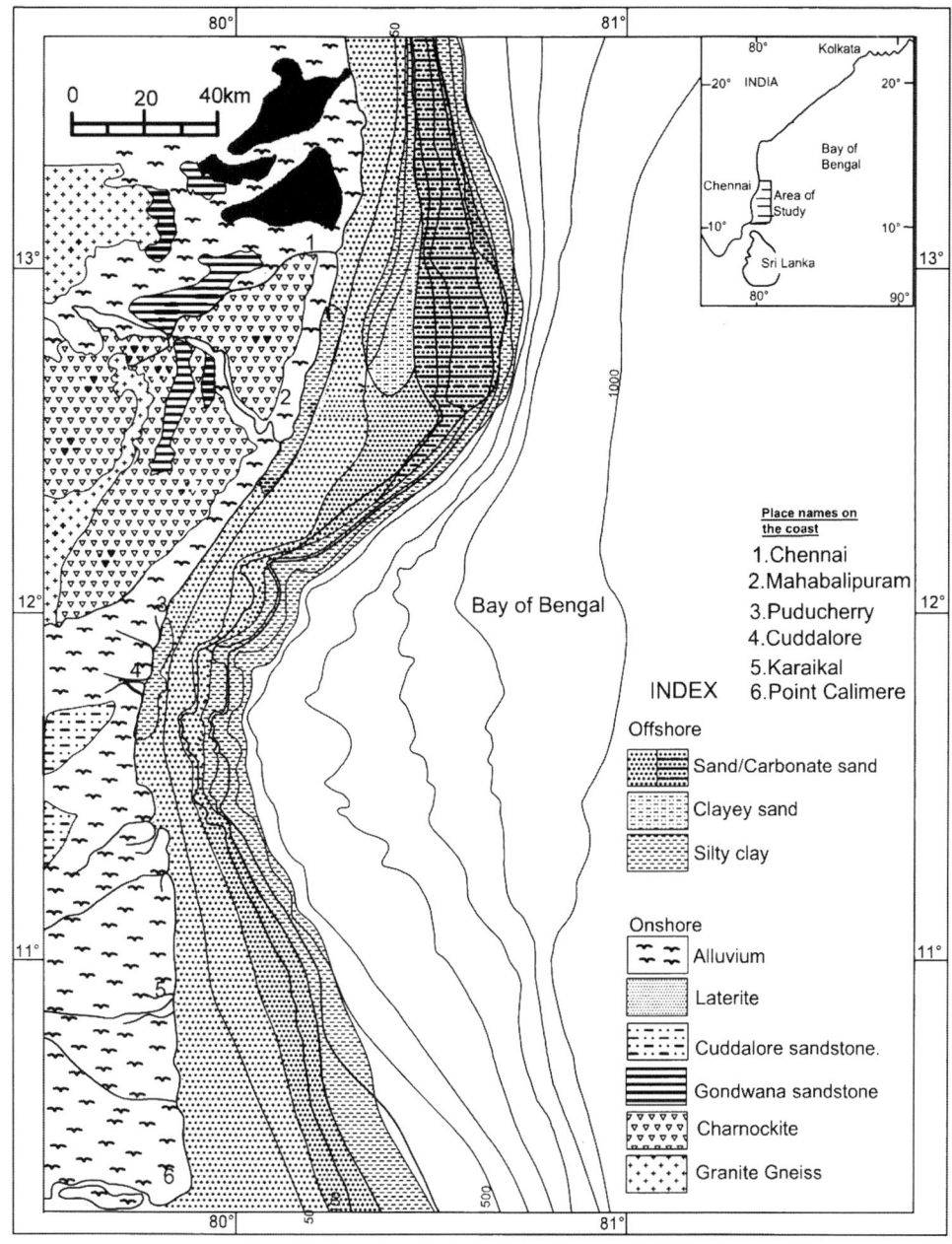

Fig. 16.4. The continental shelf off the Chennai–Puducherry coast in the southern part of Indian peninsula and its sediments (bathymetric and sediment data from GSI cruises).

glauconitization need warm and cold climates, respectively. Glauconitization usually takes place during a marine regression. It is likely that these two processes occurred over the same substrate during regressive and transgressive phases of the Last Glacial Cycle. The grains from the upper slope constitute glauconitic smectite and glauconitic mica of the glaucony facies. The ^{14}C dating of relict corals, from a depth of 120 m, off Karaikal indicated an age of 18 390 ± 210 years BP (Vaz 2000) establishing the low sea-level position of the Last Glacial Maximum. Another ^{14}C date of 14 510 ± 190 years BP of a relict coral reef from a depth of 115 m off Mahabalipuram signifies a stillstand at the same level (Vaz 1996). The abundance of alkali feldspar on the outer shelf sediments has been interpreted to show the prevalence of an arid climate during the late Pleistocene epoch within peninsular India.

The gravity data reveal a basement morphology composed of alternating horsts and graben that form basement ridges and depressions. These tectonic elements trend in a NE–SW direction. The various basement features are: Aryalur Puducherry depression extending offshore; Porto Novo High, a large domal feature lying offshore; followed by the Thanjavur Depression.

Geophysical studies by Subrahmanyam *et al.* (1995) indicate that the Cauvery offshore basin is a fault-controlled basin. Gravity and magnetic data indicate offshore extension of two major Precambrian lineaments, namely: the Moyar Bhavani Attur (MBA) lineament in the northern part; and the Palghat–Cauvery lineament in the southern part of the Cauvery Basin (Murthy *et al.* 2010). The region between 10°N and 12°30′N of the south Indian continent is characterized by two megalineaments, and associated neotectonics suggest that the area is tectonically active (Murthy *et al.* 2012).

Conclusions

The continental shelf of eastern India owes its geometric shape to the separation of the Indian subcontinent from the other Gondwana continents. The continental shelf fringes the Indian east coast with minor variations in width and gradient. The influence of Ganges sediment is seen in the Upper Bengal Fan region, dominated by illite and chlorite among the clay components, while the surface sediment in the inner shelf is illite and kaolinite, that of the outer shelf is carbonate off Kalingapatnam–Visakhapatnam. Smectite is present in the shelf sediments around the Krishna and Godavari deltas. In the south, it is illite and kaolinite again, with clastic silica, carbonate and phosphatic minerals in places. Clastic silicates in the inner shelf and carbonate sediments in the outer shelf are typical of the entire shelf of eastern India, barring the areas off major deltas.

The shelf off the Sundarban has been scoured in its centre by the discharge from the Ganges, known as the Swatch-of-No-Ground. The width of the shelf varies from 120 to 125 km off Sundarban. Progressively away from the Swatch-of-No-Ground, it narrows to 70 km off Mahanadi, 60 km width off Chilika Lake, and much of the length of the shelf. Further south, it attains a width of 50 km (Faruque 2012) with very narrow shelves of 15 and 25 km off the Godavari–Krishna rivers and Cuddalore coast, respectively. The oolite zone extends from Gopalpur in the north to Chennai in the south at water depths of >90 m to as much as 180 m. The east coast is dominantly progradational due to deltaic and littoral processes. The physical oceanographic conditions in the Bay of Bengal are, to a great extent, influenced by the change in the seasonal cycle of monsoons. The sedimentation pattern on the shelf is influenced largely by a complex set of processes, resulting from the variation in fluvial inputs, ocean currents, waves, littoral processes and the transgressive phase of the Last Glacial Cycle.

The first author (B. M. Faruque) expresses his gratitude to P. C. Shrivastava, former Deputy Director General of the GSI for guidance, encouragement and technical support during the studies in parts of eastern continental shelf of India, and to A. R. Chivas, Project Leader of IGCP 464 for his critical review of the manuscript and for suggestions to improve the quality of this paper. B. M. Faruque records his indebtedness to F. L. Chiocci, Project Leader IGCP 464, for suggestions made on the structure of this write-up. The authors would like to record their thanks to UNESCO-IGCP for extending all support to this project IGCP 464.

References

ALLISON, M. A. 1998. Geologic framework and environmental status of the Ganges-Brahmaputra Delta. *Journal of Coastal Research*, **14**, 826–836.

BHARALI, B., SRIVASTAVA, S. K. & RAVI CHANDRAN, V. 1992. Seismo-stratigraphic analysis of Cretaceous–Tertiary sequence of the Mahanadi offshore basin. *In*: *Recent Geoscientific Studies in the Bay of Bengal and the Andaman Sea*. Geological Survey of India, Special Publications, **29**, 247–254.

BRUCKNER, H. 1988. Indications for formerly higher sea levels along the east coast of India and the Andaman Islands. *Hamburger Geographische Studian*, **44**, 47–72.

FARUQUE, B. M. 2006. Continental shelf of Orissa. *In*: MAHALIK, N. K. (ed.) *Geology and Mineral Resources of Orissa*. Society of Geoscientist and Allied Technologist, Bhubaneswar, Odisha, 112–126.

FARUQUE, B. M. 2012. Geology of the continental shelf. *In*: MAHALIK, N. K. (ed.) *Coastal Tract of Odisha*. Geomin Consultants, Bhubaneswar, Odisha, 53–64.

FARUQUE, B. M. & LAHIRI, A. 2002. Post Wisconsin glaciation sea level stands off parts of east coast, Andhra Pradesh. *In*: *2002 In-House Workshop on Quaternary Geology of Coromandel Coast and Drainage Basins of Andhra Pradesh*. Geological Survey of India, Hyderabad, 7–8.

FARUQUE, B. M., CHOUDHURI, K. & MUKHERJEE, S. 2002. Temporal and spatial variation of clay mineral assemblage. in the Bay of Bengal sediments. *In*: *Proceedings Volume of Four Decades of Marine Geosciences in India – A Retrospect*. Geological Survey of India, Special Publications, **74**, 36–42.

FARUQUE, B. M., RAO, B. R., LAHIRI, A., SATYANARAYANA, B., BRAHMAM, C. V. & BANERJEE, K. 2008. Geomorphic and sedimentary imprints of low sea level stands in the shelf off Kalingapatnam–Rattikonda, Andhra Pradesh. *In*: *Proceedings Volume of the Workshop on IGCP 464*. Geological Survey of India, Kolkata, 2–10.

GINSBURG, R. N. & JAMES, N. P. 1974. Holocene carbonate sediment of continental shelves. *In*: BURK, C. A. & DRAKE, C. L. (eds) *Geology of Continental Margin*. Springer, Berlin, 137–155.

JAIN, I, RAO, A. D., JITENDRA, V. & DUBE, S. K. 2010. Computation of expected total water levels along the east coast of India. *Journal of Coastal Research*, **26**, 681–687.

KUEHL, S. A., LEVY, B. M., MOORE, W. S. & ALLISON, M. A. 1997. Subaqueous delta of the Ganges-Brahmaputra river system. *Marine Geology*, **144**, 81–96.

LA FOND, E. C. 1965. Indian Ocean. *In*: *McGraw Hill Yearbook of Science and Technology*. McGraw-Hill, New York, 213–216.

MILLIMAN, J. D., SUMMERHAYES, C. P. & BARRETTO, H. J. 1975. Quaternary sedimentation on the Amazon continental margin – a model. *Geological Society of America Bulletin*, **86**, 610–614.

MOHAPATRA, G. P. & MURTY, P. S. N. 1999. Morphology of the continental shelf off North Andhra Pradesh and its bearing on Holocene transgression. *Indian Journal of Geomorphology*, **4**, 35–44.

MOHAPATRA, G. P., RAO, B. R. & BISWAS, N. R. 1992. Morphology and surface sediments of the eastern continental shelf of peninsular India. *In*: *Recent Geoscientific Studies in the Bay of Bengal and the Andaman Sea*. Geological Survey of India, Special Publications, **29**, 229–243.

MOHAPATRA, G. P., VAZ, G. G. & RAO, R. B. S. 2002. Morphology and surface sediments off southern part of eastern continental margin of India in the Bay of Bengal sediments. *In*: *Proceedings Volume of Four Decades of Marine Geosciences in India – A Retrospect*. Geological Survey of India, Special Publications, **74**, 24–31.

MURTHY, K. S. R. 1989. Seismic stratigraphy of Ongole Paradip continental shelf, east coast of India. *Indian Journal of Earth Sciences*, **16**, 47–58.

MURTHY, K. S. R., SUBRAHMANYAM, V., SUBRAHMANYAM, A. S., MURTY, G. P. S. & SARMA, K. V. L. N. S. 2010. Land ocean tectonics and the associated seismic hazard over the eastern continental margin of India. *Natural Hazards*, **55**, 167–175.

MURTHY, K. S. R., SUBRAHMANYAM, A. S. & SUBRAHMANYAM, V. 2012. *Tectonics of the Eastern Continental Margin of India*. The Energy and Resources Institute (TERI), New Delhi.

MURTY, P. S. N., RAMAMURTY, M. & MOHAPATRA, G. P. 2006. Origin and significance of subsurface lime mud in the outer shelf off Visakhapatnam, central east coast of India. *Journal of the Geological Society of India*, **68**, 623–629.

NAGESWARA RAO, K. & SADAKATA, N. 1993. Holocene evolution of deltas on the east coast of India. *In*: Coastal Engineering Considerations in Coastal Zone Management. Proceedings of Coastal Zone 93, the Eighth Symposium on Coastal and Ocean Management, New Orleans, LA, July 19–23, 1993. Coastlines of the World Series. American Society of Civil Engineers, New York, 1–15.

NARASIMHAN, T. N. 1990. Palaeo channels of Palar River west of Madras city: possible implications for vertical movement. *Journal of the Geological Society of India*, **36**, 471–474.

RAO, K. M., RAO, N. V. N. D. P. & RAO, T. C. S. 1990. Holocene sea levels on the Visakhapatnam shelf, east coast of India. *In*: *Abstracts Volume. Seminar on Continental Margins of India*, Andhra University, Visakhapatnam, 19–20.

RAO, T. C. S., MACHADO, X. T. & MURTHY, K. S. R. 1980. Topographic features over the continental shelf off Visakhapatnam. *Mahasagar*, **13**, 23–28.

SAMBASIVA RAO, M., NAGESWARA RAO, K. & VAIDYANATHAN, R. 1978. Morphology and evolution of Mahanadi and Brahmani–Baitarani deltas. *In*: *Proceedings of Symposium on Morphology and Evolution of Landforms*. Department of Geology, Delhi University, Delhi, 241–248.

SUBRAHMANYA, K. R. 2003. Neotectonics and coastal dynamics of the Indian peninsula. *In*: *Proceedings of the 4th South Asia Geological Congress (GEOSAS-IV)*. Geological Survey of India, New Delhi, 163–174.

Subrahmanyam, A. S., Lakshminarayana, S., Chandrasekhar, D. V., Murthy, K. S. R. & Rao, T. C. S. 1995. Offshore structural trends from magnetic data over Cauvery basin, east coast of India. *Geological Society of India Journal*, **46**, 269–273.

Thakur, N. K. & Kumar, A. P. 2007. Role of bathymetry in tsunami devastation along the east coast of India. *Current Science*, **92**, 432–434.

Vaidyanadhan, R. & Ghosh, R. N. 1993. Quaternary of the east coast of India. *Current Science*, **64**, 804–816.

Vaz, G. G. 1995. Phosphatic nodules in the outer continental shelf off Madras, Bay of Bengal. *Indian Journal of Marine Sciences*, **24**, 8–12.

Vaz, G. G. 1996. Relict coral reef and evidence of pre-Holocene sea level stand off Mahabalipuram, Bay of Bengal. *Current Science*, **71**, 240–241.

Vaz, G. G. 2000. Age of relict coral reef off Karaikal, Bay of Bengal: evidence of Last Glacial Maxima. *Current Science*, **79**, 228–230.

Vaz, G. G. & Banerjee, P. K. 1997. Middle and Late Holocene sea level changes in and around Pulicat Lagoon, Bay of Bengal, India. *Marine Geology*, **138**, 261–271.

Vaz, G. G., Vijay Kumar, P. & Rao, B. L. 1999. Phosphorite from the continental margin off Madras, Bay of Bengal. *Marine Georesources and Geotechnology*, **17**, 33–48.

Venkataraman, S. & Rangaraju, M. 1968. Oscillatory movement in Cretaceous–Tertiary basin of South India. *In: Cretaceous–Tertiary Formations of South India*. Geological Society of India, Memoirs, **2**, 201–207.

Chapter 17

The Myanmar continental shelf

V. RAMASWAMY* & P. S. RAO

National Institute of Oceanography, Dona Paula, Goa 403004, India

*Corresponding author (e-mail: rams@nio.org)

Abstract: The Myanmar (Burma) coastline is about 2280 km long, with the continental shelf covering an area of approximately 230 000 km^2. The Myanmar coastline may be divided into the northern Rakhine (Arakan) coast, the central Ayeyarwady coast and the southern Tanintharyi (Tenasserim) coast. The Rakhine coast lies adjacent to the Bay of Bengal, while the Ayeyarwady and Tanintharyi coast faces the Andaman Sea. The continental shelf of Myanmar receives sediments from some of the largest rivers in the world like the Ayeyarwady (Irrawaddy), the Salween (Thanlwin) and the Ganges–Brahmaputra rivers. The Myanmar shelf is characterized by a seasonal reversal of monsoon winds and coastal currents, periodic tropical cyclones and storm surges, meso- to macrotidal conditions, and neotectonic activity. The most prominent bathymetric feature on the Ayeyarwady continental shelf is the 120 km-wide Martaban Depression, at the centre of which is located the Martaban Canyon. Most of the suspended sediment discharge of the Ayeyarwady is transported eastwards by coastal and tidal currents, and trapped in the Gulf of Martaban, resulting in the formation of an extensive mud belt covering an area of more than 45 000 km^2. Because of these features, the Myanmar shelf ranks amongst the most globally important continental shelves in the world but remains inadequately studied owing to its inaccessibility.

The 2280 km-long Myanmar continental shelf (Table 17.1) lies partly within the Bay of Bengal and partly in the Andaman Sea, covering an area of 230 000 km^2. Some of the world's largest rivers like the Ayeyarwady (Irrawaddy), Salween and Ganges–Brahmaputra discharge their sediment on to this shelf. The Ayeyarwady and Salween together contribute more than 500 MT of sediment (Robinson *et al.* 2007), which presently accumulates mostly on the continental shelf. These rivers have contributed significantly to the deep Andaman Basin in the past, especially during low sea-level stands (Rodolfo 1969*a*). The continental shelf of Myanmar is strongly influenced by monsoonal processes with strong wind and wave regimes, and strong coastal currents, especially during the summer monsoon between mid-May and October (Rodolfo 1969*a*; Ramaswamy *et al.* 2004). Macrotidal conditions prevailing in the northern Andaman Sea play a major role in redistributing sediments delivered by rivers (Ramaswamy *et al.* 2004; Rao *et al.* 2005). This shelf is characterized by earthquakes and neotectonic activity along the Andaman–Nicobar island chain and on the Ayeyarwady shelf. The continental shelf sediments are also periodically buffeted by severe tropical cyclones and storm surges, the frequency of which has increased in recent years (Fritz *et al.* 2009).

Myanmar's coastline can be divided into three coastal regions (Fig. 17.1): the northern Rakhine coastal region in the Bay of Bengal; the central Ayeyarwady coastal region, which includes the Ayeyarwady delta and the Gulf of Martaban; and the southern Tanintharyi coastal region in the Andaman Sea. Most information on the Myanmar shelf was collected during the International Indian Ocean Expedition (IIOE) carried out in the 1960s and a more recent expedition of the Indian research vessel ORV *Sagar Kanya* in 2002. The IIOE results from the Andaman Sea have been summarized by Rodolfo (1969*a, b*), while the ORV *Sagar Kanya* results are reported in some more recent publications (Ramaswamy *et al.* 2004, 2008; Rao *et al.* 2005; Ramaswamy & Rao 2006). The above studies mostly relate to the Ayeyarwady continental shelf, and very little information is available on the Rakhine and Tanintharyi shelf. Limited information on coral reefs in the Andaman Sea has been reviewed by Brown (2007). In this paper we summarize available information on the Myanmar continental shelf with emphasis on recent sediment distribution and sedimentary processes on the Ayeyarwady continental shelf.

Tectonic framework and geological background

The geological and tectonic development of the Myanmar shelf and Andaman Sea is highly complex (Fig. 17.1 inset). The NE-moving Indian plate is colliding with the nearly stationary Eurasian or SE Asian plate, and subduction of the Indian plate occurs all along the Sunda arc (Rodolfo 1969*b*; Raju *et al.* 2004; Curray 2005). The oblique plate convergence has led to strike-slip faulting parallel to the trench axis, the formation of a sliver plate, back-arc extension and the formation of the Andaman Basin. Initiation of the spreading in the Andaman Sea has occurred with an opening rate of 3.72 cm a^{-1}. This spreading centre is connected to the Sagaing Fault System in the eastern Burma highlands in the north and to the Semangko Fault System in the south. A complex system of north–south-trending dextral strike-slip faults runs through the Gulf of Martaban and the Ayeyarwady shelf; the most prominent of these is the Sagaing Fault System that extends southwards and joins the Central Andaman Rift (Curray *et al.* 1979; Raju *et al.* 2004). The seismicity in the region shows a general concentration of events along the Andaman–Nicobar ridge and along the spreading centre in the Andaman trough, as well as along the Sagaing Fault (http://earthquake.usgs.gov/).

Weather and climate

Seasonally reversing monsoon winds play a major role in controlling rainfall patterns in Myanmar and the northern Indian Ocean. Physiographically, Myanmar can be divided into the Eastern Mountain System, the Western Mountain Belt and the Central Lowland Belt. The Eastern Mountain System includes the long, narrow, Tanintharyi coast bordering the Andaman Sea and the hilly Shan Plateau to the north (Fig. 17.1 inset). The Western Mountain Belt includes a group of low mountains called the Rakhine Mountains (Arakan Yoma), which extend from Myanmar–India border to the Bay of Bengal. This belt also includes a narrow plain of rich agricultural lands bordering the Bay. The Central Belt consists chiefly of the Ayeyarwady and Sittang river valleys.

The SW monsoon is active between mid-May and October, while the NE monsoon is active between December and February.

From: CHIOCCI, F. L. & CHIVAS, A. R. (eds) 2014. *Continental Shelves of the World: Their Evolution During the Last Glacio-Eustatic Cycle.* Geological Society, London, Memoirs, **41**, 231–240. http://dx.doi.org/10.1144/M41.17

Table 17.1. *Summary table of the Myanmar continental shelf*

Length of the shelf	2280 km
Average width	152 km
Tidal, wave and current range	Maximum tidal range 2.4–7 m, waves 0.1–3.7 m, currents up to 3 m s^{-1}
Dominating process (wave/ current/tide)	Tide throughout the year. Wave during the SW monsoon period (June–September)
Average depth of the shelf break	110 m
Siliciclastic/carbonate/authigenic/ glacial sedimentation	90% siliciclastic, 10% carbonate
Modern/relict/palimpsest	50% modern, 50% relict
Tectonic trend over the last glacial cycle	Subsiding

The dry season is from mid-October to mid-May, while the main wet spell is from mid-May to the end of September. During January and February, dry NE winds (15–20 km h^{-1}) blow out of a high-pressure cell situated over central Asia. Rainfall during this period is less than 30 mm. During the summer monsoon, winds blow steadily at rates more than twice that of the NE monsoon (averaging 30 km h^{-1}) with occasional gale force winds along the coast (Bay of Bengal Pilot 1978), especially during tropical cyclones. During the rest of the year, mostly calm conditions prevail.

The coastal regions and the Rakhine and Tanintharyi ranges receive more than 5000 mm of precipitation annually, while the delta regions receive about 2500 mm (Bender 1983). The central region, being a rain-shadow region of the Rakhine Mountains, receives an annual rainfall of only 500–1000 mm. The Shan Plateau, because of its elevation, usually receives an annual rainfall of around 2000 mm.

River discharge

Myanmar has many major and minor rivers, flowing into either the Andaman Sea or the Bay of Bengal. Apart from small rivers like the Mayu and Kaladan draining the Rakhine Mountains, the Rakhine coast is also influenced by fluvial discharge from the Ganges–Brahmaputra river system. The Ayeyarwady, Salween (Thanlwin), Sittang, Ye, Dawei (Tavoy) and Tanintharyi (Great Tenasserim) rivers discharge into the Andaman Sea (Fig. 17.1). Freshwater and sediment discharge in all the rivers is highly seasonal, with more than 80% of the annual discharge occurring during the SW monsoon. Compared to the Ayeyarwady, Salween and Sittang rivers, the smaller rivers on the Tanintharyi coast are a relatively minor source of sediments.

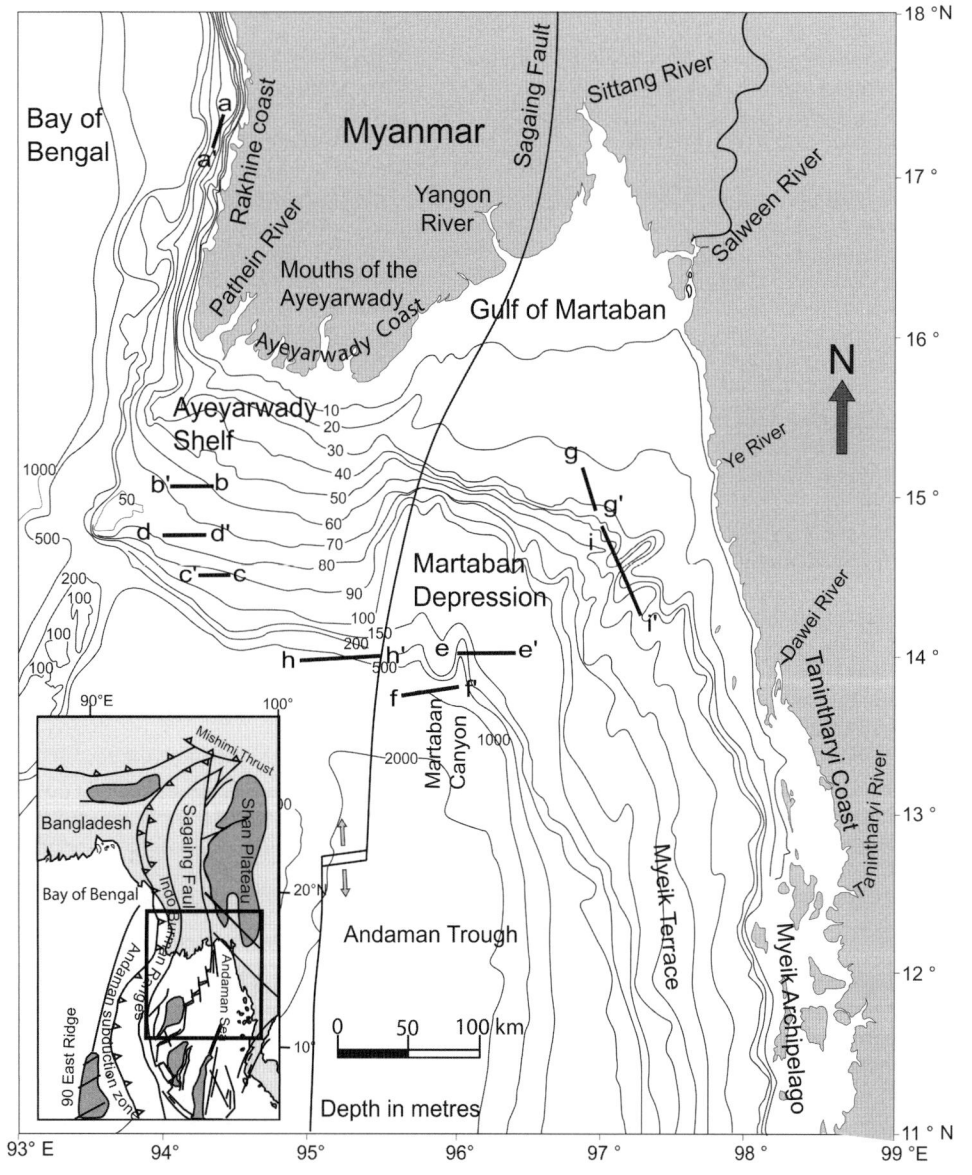

Fig. 17.1. Simplified bathymetry of the Myanmar continental shelf. The lines **a–i** indicate the location of the echograms given in Figure 17.2. Inset shows the tectonic features of the region.

The Ayeyarwady River is about 2170 km long and has a drainage basin of 0.413×10^6 km^2. The catchment area of this river is underlain by Neogene sediments that are prone to erosion, especially during flash floods. The river flows through a short flood plain and enters the huge Ayeyarwady delta, it then breaks into a vast network of streams before emptying into the Andaman Sea through multiple mouths. Measurement of water discharge and sediment load at the head of the delta from 1869 to 1879 carried out by Gordon (1885) indicates that over 265 MT of sediment is brought to the delta annually, and about 82% of this load is discharged from June to October. Robinson et al. (2007) re-examined the data of Gordon (1885) and revised the sediment flux to 364 ± 60 MT a^{-1}. Data collected between 1969 and 1996 provide a more recent estimate of the sediment discharge of the Ayeyarwady to be about 379 ± 47 MT (Furuichi 2009).

The Salween River is about 2415 km long and has a drainage basin of 0.272×10^6 km^2. The Salween River, although as long as the Irrawaddy, has a much smaller drainage area. It flows mainly through deep gorges in resistant granites, metamorphic rocks and Palaeozoic sedimentary rocks. The annual sediment load of the Salween has been estimated to be between 148 and 206 MT (Gaillardet et al. 1999; Bird et al. 2008).

The Sittang River flows into the Gulf of Martaban in the Andaman Sea. The river is about 418 km long, and, for a comparatively short river, has a large valley and delta with a significant sediment discharge and a large tidal bore at its mouth. The Ayeyarwady–Salween river system is, at present, the third largest in the world in terms of suspended sediment discharge (Robinson et al. 2007; Bird et al. 2008). No major dams or barrages have been constructed on these rivers. The Ayeyarwady delta, however, has been extensively modified. The forest and mangroves have been replaced with rice fields and there is an extensive system of canals (Fritz et al. 2009). The combined Ayeyarwady–Salween river system contributes more than 500 MT of sediment and 5.7–8.8 MT of organic carbon annually, suggesting that presently it may be the second largest point source of organic carbon to the global ocean after the Amazon (Bird et al. 2008).

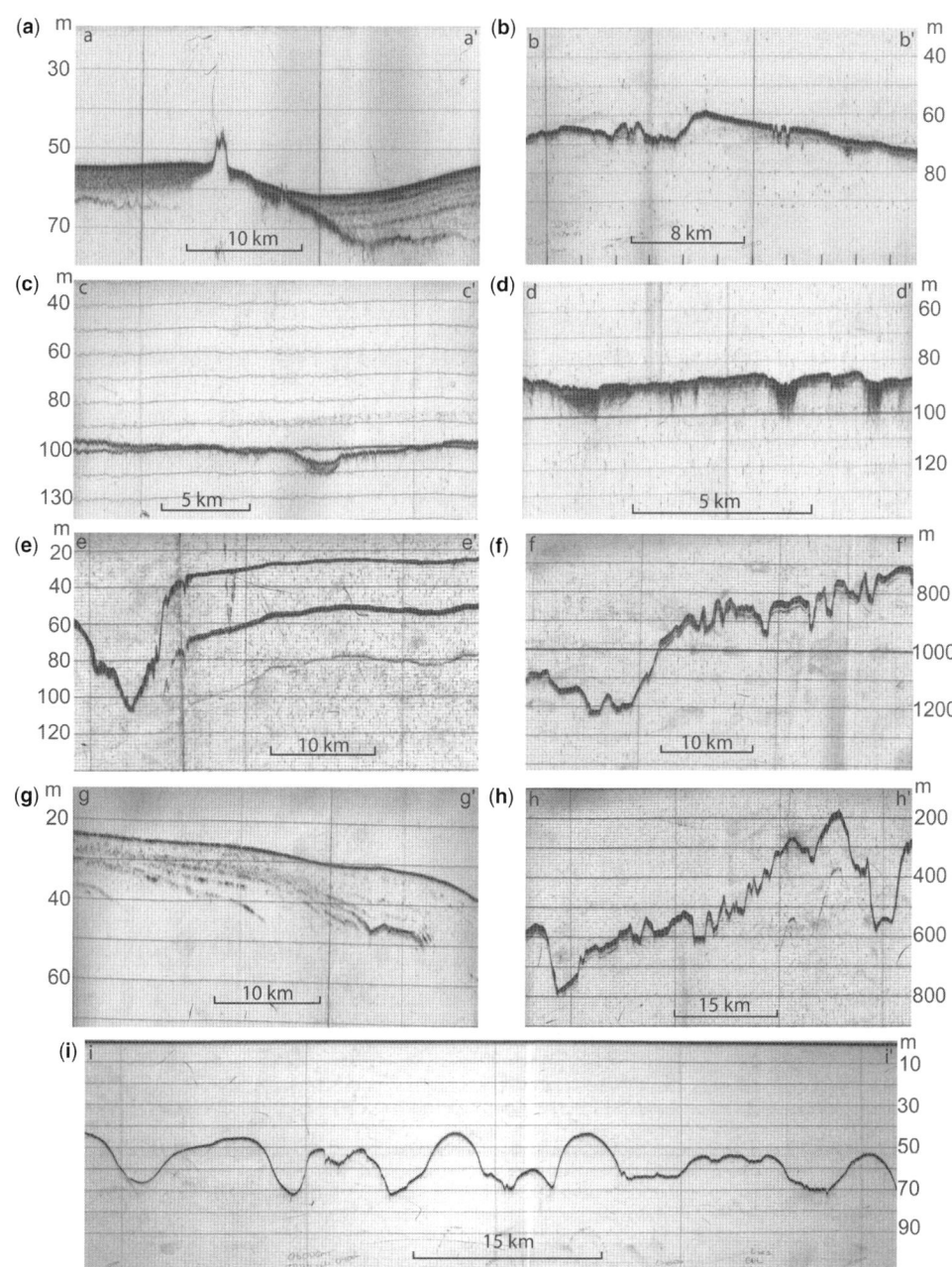

Fig. 17.2. Echograms of the Myanmar continental shelf. (**a**) A pinnacle on the Rakhine continental shelf partially covered with modern sediment. (**b**) Rough topography on the Ayeyarwady outer continental shelf. (**c**, **d**) Buried channels on the Ayeyarwady continental shelf. (**e**) Profile across the Martaban Canyon. (**f**) Slump scars, erosion channels, V-shaped notches and other mass-wasting features on the Martaban Depression. (**g**) Thick deposits of modern sediments on the eastern part of the Ayeyarwady inner shelf. (**h**) Rugged topography and erosion features on the Ayeyarwady continental slope. (**i**) Sand ridges on the eastern part of the Ayeyarwady shelf.

Coastal oceanography

The entire northern Indian Ocean including the Andaman Sea is strongly influenced by the seasonally reversing Asian monsoon (Wyrtki 1973). Circulation in the Andaman Sea is cyclonic during the SW monsoon (May–September) and anti-cyclonic during the NE monsoon (December–February). A comprehensive review of the oceanography of the Bay of Bengal and Andaman Sea has been provided by Varkey et al. (1996), and a simulation model driven by monthly mean winds has been provided by Potemra et al. (1991). In response to the monsoons, the oceanic flow in the Andaman Sea changes direction twice a year with a cyclonic flow in spring and summer, and anti-cyclonic flow for the rest of the year (Potemra et al. 1991). Model layer thicknesses reveal coastal Kelvin waves propagating along the coast, and travelling the entire perimeter of the Andaman Sea and the Bay of Bengal. This wave also excites westward-propagating Rossby waves into the interior of the bay.

The Andaman Sea is also noted for large-amplitude long-interval waves (solitons) with associated surface waves (Osborne & Burch 1980; Osborne 1990; Hyder et al. 2005). These internal waves, which are apparently due to non-linear internal tides, have been observed as surface rips in satellite imagery of the Andaman Sea (Alpers et al. 1997). Hyder et al. (2005) have shown that the solitons are generated during spring tides when tidal ranges are more than 1.5 m with a higher probability of occurrence with an increase in tidal range. Because of the presence of large solitons and macrotides, mixing of waters is expected to be comparatively more intense on the Ayeyarwady shelf.

During the dry season prior to the SW monsoon, salinity values along the Ayeyarwady coast are around 31‰ (or ppt) due to freshwater influx from the Ganges–Brahmaputra and Ayeyarwady–Salween river systems. During the summer monsoon period, salinities can drop to less than 26‰ as a result of increased river discharge and precipitation. Below the surface layer, a strong halocline develops as salinities grade to a maximum of about 35‰ at a depth of 1500 m. Southward flow of low-salinity waters from the Ganges–Brahmaputra and extensive precipitation lowers salinities along the Rakhine coast to less than 18‰, which increases to 34‰ in the dry season due the northward flow of highly saline waters. In the central Andaman Basin, salinity values during the SW monsoon can be less than 25‰ (Tomczak & Godfrey 1994). Off the Tanintharyi coast, salinities are also low, with values below 32‰ during the pre-monsoon period due to southward flow of the freshwater discharge of the Ayeyarwady–Sittang–Salween rivers. Surface mixed layer depths are less than 5 m along the Ayeyarwady coast, deepening to 57 m south of 12°N. High sea surface temperatures (SSTs) that energize tropical storms in the Bay of Bengal are maintained, in part, by a freshwater cap produced from river runoff and open-ocean rainfall. This freshwater cap creates thin surface mixed layers that isolate the surface waters from colder waters of the thermocline.

The Rakhine coast and mouths of the Ayeyarwady River are mesotidal (2–4 m), while the Gulf of Martaban and Tanintharyi coast are macrotidal (>4 m) (Indian Tide Tables 2002; United Kingdom Hydrographic Office 2004). Maximum tidal ranges in the Gulf of Martaban are between 4 and 7 m (Table 17.1). The tides in this area are mostly semi-diurnal with M2 and S2 being the major components. During spring tide, tidal currents up to 2.5–3 m s^{-1} have been observed in the Gulf of Martaban (Bay of Bengal Pilot 1978). Strong tidal currents are also observed in the Tanintharyi coast and between the islands of the Myeik Archipelago.

The Rakhine Coast and continental shelf

The northern portion of the Rakhine coast, between 20°N and 18.5°N latitude, is highly crenulated and has numerous islands,

while south of 18.5°N the coastline is much simpler and rocky. The Rakhine coast receives about 5000 mm of rain annually, which gives rise to numerous short, rapid streams. Of these, the Kaladan and Mayu rivers form broad river deltas. The rivers in the central and south Rakhine coast have short estuaries, and flow directly into the Bay of Bengal. The south Rakhine coast is also volcanic, the last eruption being reported in 1952 (Gordon & Thwin 1998).

The Rakhine shelf is virtually unexplored, except for some information between 18°N and 16°N collected during the ORV *Sagar Kanya* cruise of 2002. The Rakhine continental shelf is about 85 km wide in the north, which narrows towards the south to less than 40 km in places. The shelf topography is complex due to the presence of numerous islands and submerged shoals. The shelf break occurs at a water depth of about 100 m in the north and shallows to about 70 m in the south. The shelf break towards the south is intercepted by shoals and topographical prominences (Fig. 17.2a). Some of the topographical highs have flat tops and V-shaped valleys adjacent to the flanks. Beyond the shelf break, the depth falls steeply to more than 2000 m. The upper part of the shelf is shallow and deltaic (Gordon & Thwin 1998). The central part is more or less rocky and devoid of sediment. The southern part consists of silty-clays on the inner shelf, while the outer shelf is shelly or coralline. On the southern part of the Rakhine shelf, sand content is <18%, while silt and clay are between 32 and 59%, and 30 and 66%, respectively (Rao et al. 2005).

The Ayeyarwady coast and morphology of the continental shelf

The Ayeyarwady coast is aligned east–west in the northern Andaman Sea, and includes the delta and river mouths of the Ayeyarwady River and also the entire Gulf of Martaban. The coast is made up of either large mudflats or mangrove swamps. Recent satellite images show that most of the mangroves of the Ayeyarwady delta have been removed for charcoal and rice paddies, while pristine, dense mangrove jungles are restricted to a few protected areas. The Ayeyarwady delta is characterized by enormous sediment discharge and the data from recent hydrographic surveys show that there are four newly formed islands (Gordon & Thwin 1998). The central and eastern portion of the Ayeyarwady delta has a very low gradient, and is prone to flooding, as happened when tropical cyclone *Nargis* hit the Ayeyarwady coast in May 2008 causing catastrophic destruction and human fatalities exceeding 138 000 (www.cmdat.be; Fritz et al. 2009; McPhaden et al. 2009). Flooding on the Ayeyarwady delta also occurs during exceptional rain events. Bank overflow and flooding have been controlled by construction of numerous embankments that aid freshwater and sediment discharge to the sea (Woodroffe 2000). The Ayeyarwady delta coast has thin strips of sandy beaches with numerous shoals and fringing mangrove swamps. Changes since the surveys of 1860 indicate that the shoreline is advancing at a rate of about 2.5 km per 100 years (Rodolfo 1969a). No appreciable change was observed at the western part of the delta. The Salween has a very small delta and almost the entire river sediment discharge is deposited on the continental shelf. Since a substantial portion of the suspended sediment discharge of both the rivers is accumulating in the Gulf of Martaban, the 40 m-depth contour is advancing at an estimated rate of 55 km every 100 years (Gordon & Thwin 1998).

The Ayeyarwady continental shelf width is about 170 km off the Ayeyarwady river mouths and more than 250 km from the head of the Gulf of Martaban (Fig. 17.1). The depth of the Gulf of Martaban is less than 30 m and the gradient within the Gulf up to the 30 m contour is less than 0.01°. A bathymetric depression about 120 km wide and trending north–south is present on the

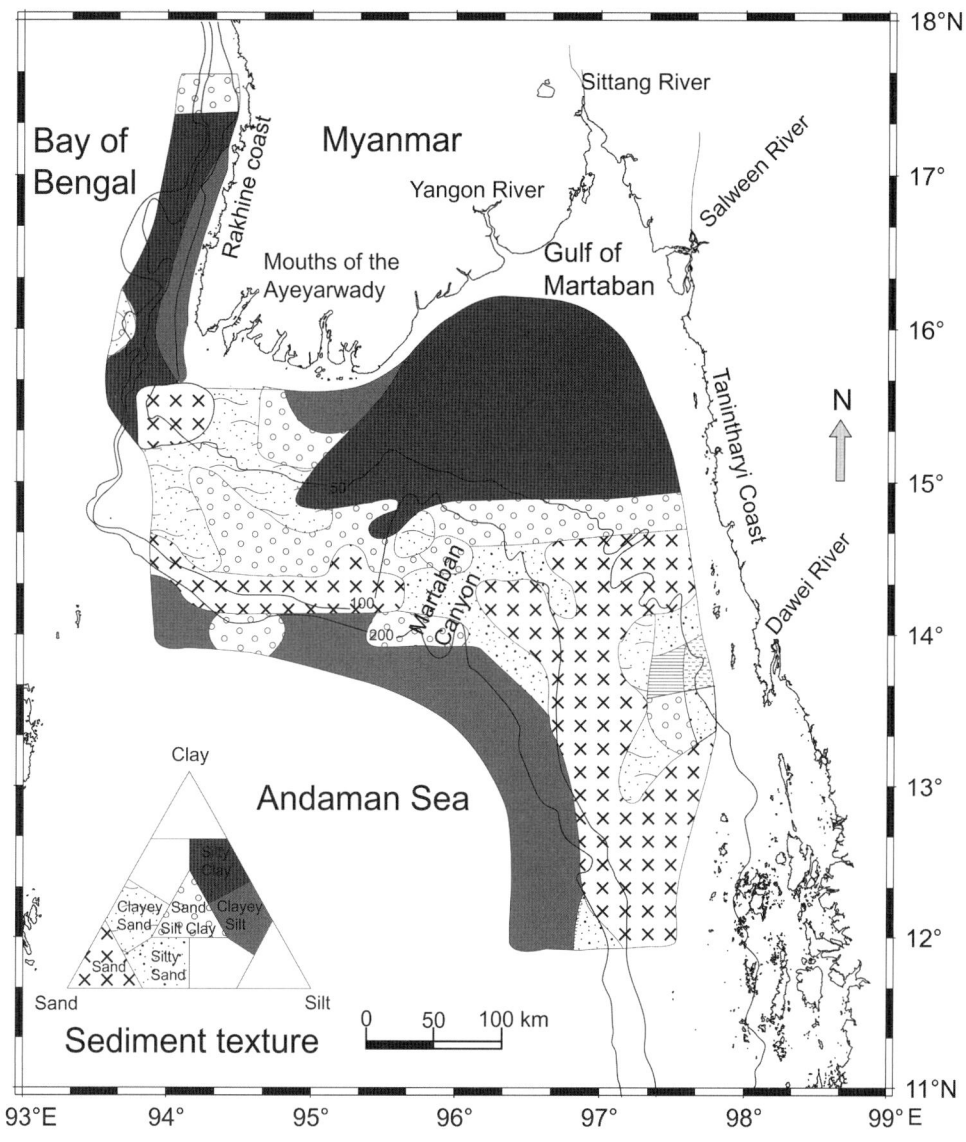

Fig. 17.3. Textural variations of the surficial sediments on the Myanmar continental shelf.

outer shelf between the Sagaing Fault System and the Malay continental margin (Ramaswamy *et al.* 2004; Rao *et al.* 2005). Numerous faults, slump scars, erosion channels, V-shaped notches and other mass-wasting features are common within this bathymetric depression (Fig. 17.2f). The Martaban Canyon lies within this bathymetric low and appears to be controlled by the north–south-trending fault systems (Fig. 17.1). A network of irregular channels runs southwards in the area of low bathymetry and joins at the head of the Martaban Canyon. The Martaban Canyon closely follows the Sagaing Fault and joins the Central Andaman Trough in the deep Andaman Sea. The canyon is well developed on the outer shelf and upper slope region. The width of the canyon is more than 5 km and has a V-shaped cross-section at a depth of 450 m (Fig. 17.2e). Further south, the main channel of the canyon branches into two channels (Fig. 17.2f, h), that merge into a broad U-shaped valley before joining the Central Andaman Trough. The continental shelf off the Ayeyarwady river mouths has a gentle slope of 0.04° from coast to shelf break. The seafloor in the Gulf of Martaban and the adjacent inner shelf up to a depth of 30 m is generally smooth, whereas the outer shelf has a rough surface with relief of 2–20 m, and has topographical features such as pinnacles, highs and valleys, buried channels, and scarps (Rao *et al.* 2005). The pinnacles, most probably shelf-parallel ridges with coarse-grained bioclastic sediments, are commonly 1–2 m in height, reaching a maximum of 6 m in places. The troughs are commonly 2 m deep, occasionally reaching up to 8 m deep. A number of buried channels,

5–10 m deep and up to 250 m wide, are observed over the western part of Ayeyarwady shelf. Bathymetric data indicate that the shelf break is at the 110 m isobath. Beyond the shelf break the seafloor depth increases rapidly to approximately 2000 m, except in the Martaban Depression. Echograms and sub-bottom profiler records in the Gulf of Martaban reveal a minimum of 18 m-thick strata of modern muds (Fig. 17.2g). At the outer boundary of the Gulf of Martaban (15°N latitude), brown muds overlie coarse sands indicating that modern deltaic sediments are prograding on to the shelf.

Well-defined submerged sand ridges aligned in an ENE–WSW direction are observed on the middle and outer shelf between water depths of 50 and 70 m in the eastern part of the Ayeyarwady shelf. These ridges are 20–25 m high and 5–10 km wide, and several tens of kilometres long. According to the definitions of Flemming (1978) and Ashley (1990), they may be classified into very large subaqueous bedform/dune type. Some appear to be moribund and are characterized by rounded to flat crests and smooth surfaces with a more or less symmetrical shape, while others display superimposed sand waves and megaripple-like features (Fig. 17.2i).

The Tanintharyi coast and continental shelf

The 1200 km-long Tanintharyi coast extends from the Gulf of Martaban to the Thailand border. The northern portion of the coast is aligned north–south, while the southern portion is highly crenulated and includes the Myeik (Mergui) Archipelago which comprises more than 800 islands covering an area of about 34 340 km². Survey of the Tanintharyi coast is difficult because of the remoteness and the treacherous waters due to the presence of numerous island, shoals and strong tidal currents. Data have been collected during the ORV *Sagar Kanya* cruise north of 12°N only (Rao *et al.* 2005). South of 12°N limited data are available only for the outer shelf and terrace (Rodolfo 1969*a*).

South of the Salween River, the coastline is fringed with coral reefs (Chhibber 1934; Brown 2007). Sheltered coasts of the peninsula and of the larger islands are extensive mudflats, which merge into swamps and marshes. Exposed coasts are wave cut, steep and rocky, and the outer islands generally have sandy beaches. Coral reefs surround the outer islands of the Myeik Archipelago and mangroves cover much of the inner islands. The majority of Myanmar coral reefs are found in this area covering an area of about 1686 km². Fringing reefs dominate all sites within this coast (Brown 2007). Commonly, the exposed coasts are wave cut, steep and rocky. Seaward coasts of the outer islands have sandy beaches (Chhibber 1934). Corals in this area have adapted to living under conditions of high turbidity and sediment loads (Brown 2007).

The Tanintharyi continental margin is about 270 km wide with an inner shelf, the Myeik Terrace and the continental slope. The Tanintharyi continental shelf is much wider than the Rakhine coast, with a shelf width of 180–200 km (Rama Rao 1930). North of 13.25°N, the continental shelf is followed by a continental slope. South of 13.25°N, the inner shelf break is at a depth of about 100 m and is followed seawards by the Myeik Terrace, which is 85–140 km wide and has a depth of 200 m at 12°N. The Myeik Terrace is followed by a continental slope and, hence, this region has an inner continental shelf break at 110 m and an outer shelf break at 200 m (Rodolfo 1969*a*). The

Myeik Archipelago lies within the inner shelf, which is less than 100 m deep. At its northern end, the inner shelf is 130 km wide. Southwards, the width decreases and is terminated by a minor slope with 100 m of relief. The Myeik Terrace begins north of 12°N and widens, flattens and deepens southwards. South of 9°N the terrace becomes concave, bifurcating into a western bank and an eastern shelf basin. The Myeik Bank is 70 km broad, 200 km long and crests with occasional pinnacles 1 km below sea level.

Sediments within the protected embayments of the Tanintharyi shelf are mostly carbonate-rich. An isolated patch of fine-grained sediments (sand 5–10%; silt *c.* 46%; clay 45–50%) is present indicating a local source, probably the Dawei (Tavoy) River. Off the Dawei River there is a small patch of silty clay, the rest of the shelf is covered with coralline sands or silty sands. Carbonate content is more than 30%. Satellite images show clear waters, except between the islands of the Archipelago and river mouths of the Tanintharyi (Great Tenasserim) and Ye rivers which have turbid waters due to terrigenous clay from rivers. Tidal currents between the islands are particularly strong and make the waters highly turbid due to resuspension of bottom sediments.

The Myeik Terrace is covered with muddy sands with sub-angular to rounded quartz grains, and mixed with carbonate shells and some fine-grained material. The mineralogy of the terrace sands reflect granitic provenance, and sorting and rounding of the grains indicate reworking over a long period of time. According to Rodolfo (1969*a*), the inner shelf sediments along unprotected coasts are similar to those of the Myeik Terrace. On the Tanintharyi shelf, increased molluscan and coralline activity is reflected in a carbonate content as high as 25%. Over the Myeik Terrace, coralline and foraminiferal sands contribute to over 60% carbonate content, especially on the Myeik Bank (Rodolfo 1969*a*).

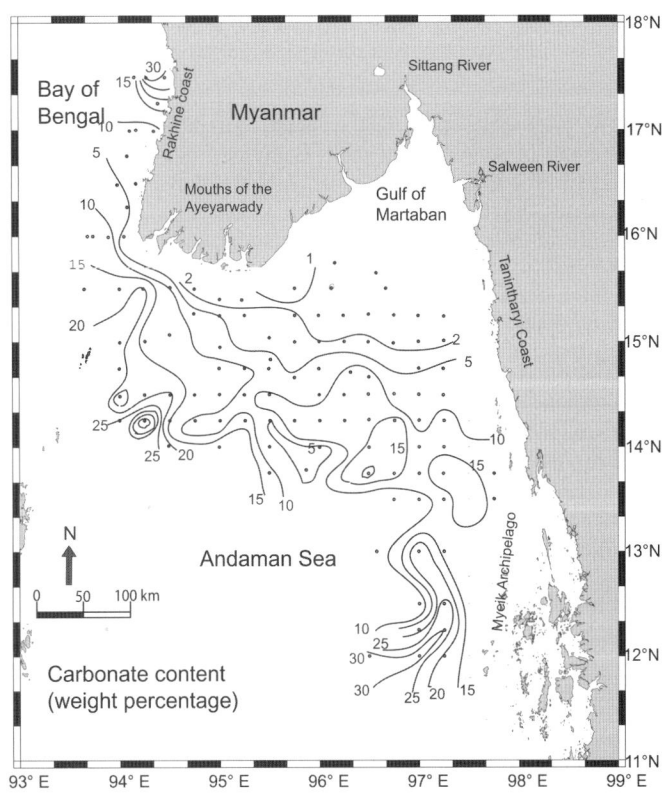

Fig. 17.4. Carbonate content (wt%) of the surficial sediments on the Myanmar continental shelf.

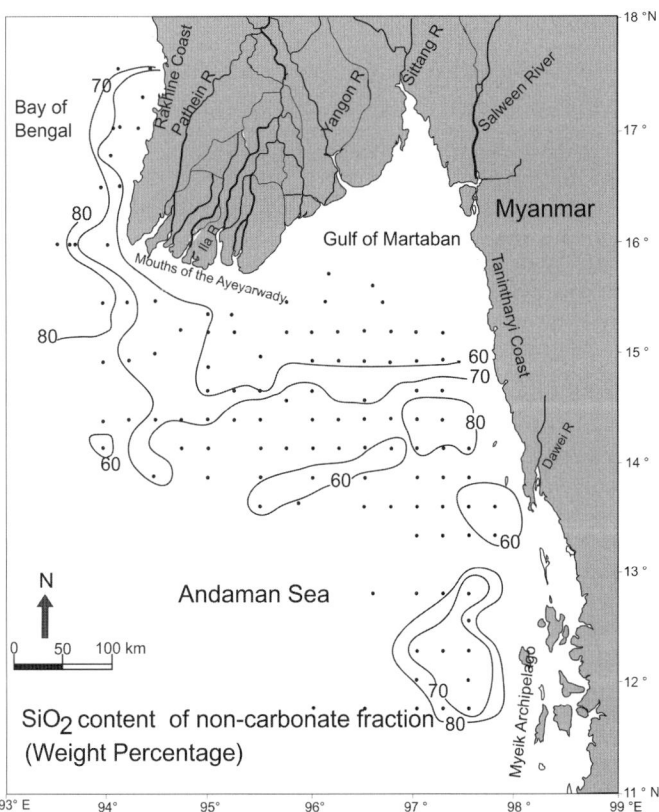

Fig. 17.5. Silica (SiO₂) content (wt%) of the surficial sediments on the Myanmar continental shelf.

Modern sediments on the Ayeyarwady shelf

General composition, texture and grain size

The distribution and sediment texture on the Ayeyarwady shelf shows fine-grained sediments comprising silty-clay and clayey-silt on the Gulf of Martaban and the adjacent inner shelf. The outer shelf is covered with distinctly coarser sediments, mainly sands. A narrow zone of sand–silt–clay is present between fine-grained sediments of the Gulf of Martaban and outer shelf sands. In the western part of the Ayeyarwady shelf, sand and clayey sand sediments extend from a water depth of 30 m to the shelf break. The seafloor in the Martaban Canyon system is covered with sediments varying from silty clays to sands and silty sands (Rao *et al.* 2005).

In the sediments of the Gulf of Martaban and adjacent inner shelf, sand content is negligible (<1%), while the silt and clay content ranges from 33 to 57% and from 42 to 66%, respectively (Rao *et al.* 2005). Outer shelf sediments are predominantly sandy, with sand content varying from 61 to 96%, silt <1–19% and clay 4–23%, while transition zone sediments have near-equal proportions of sand (20–40%), silt (20–50%) and clay (20–35%). The sediments from the Martaban Canyon display greater variation compared to other areas with sand 1–94%, silt 2–59% and clay 4–69%. In the Gulf of Martaban and adjacent inner shelf, silty-clays cover an area of more than 45 000 km^2. From the apex of the Gulf of Martaban, the mud belt is as wide as 250 km and ranks among the largest modern mud belts of the world oceans (Rao *et al.* 2005).

The outer shelf is a zone of non-deposition and is starved of modern fine-grained sediments (Fig. 17.3). The presence of relict foraminifers (Rodolfo 1969*a*; Panchang *et al.* 2008) suggests that the outer shelf sands are relict and probably deposited during the Holocene transgression. Echograms also reveal the relict nature of the outer shelf as topographical irregularities such as pinnacles and undulations that have not been covered by modern deposition. The area covered by relict sands is about 50 000 km^2.

The northern portion of the Martaban Canyon (down to a water depth of 100 m) receives fine-grained sediments from the Martaban mud belt. Beyond a water depth of 100 m, the canyon floor is covered with sands. Redistribution of sediments in the canyon by turbidity currents, gravity-driven mass flows and, perhaps, by tidal currents has resulted in the formation of mixed-type sediments from silty clays in the north, to sands and silty sands in the deeper portion.

Carbonates

The carbonate content of the surficial sediments is low, except in the southern part around the Myeik Archipelago where coralline and shelly sands abound (Fig. 17.4). Sediments of the Martaban province are carbonate-poor, increasing southwards from less than 1% to approximately 5% at the outer delta boundary. Carbonate impoverishment in this area is due to dilution by the tremendous quantities of carbonate-poor silt and clay that pour out of the Ayeyarwady and Salween rivers, and also because strong seasonal salinity changes and extreme turbidity are ecologically unfavourable for foraminifers and molluscs (Ramaswamy & Gaye 2006).

Foraminifers and the remains of molluscs contribute higher percentages of calcium carbonate to the outer and Tanintharyi shelf. The Ayeyarwady outer shelf sediments have a carbonate content of 5–15% as the relict sediments are not diluted by deposition of deltaic sediment. The non-carbonate fraction in the southern Tanintharyi coast is composed of siliceous sand with negligible fine-grained sediment.

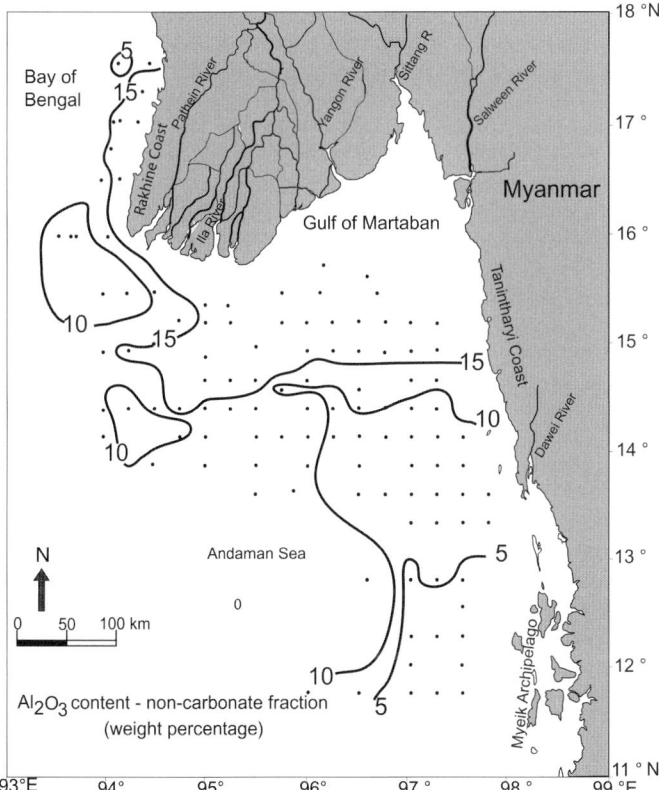

Fig. 17.6. Alumina (Al$_2$O$_3$) content (wt%) of the surficial sediments on the Myanmar continental shelf.

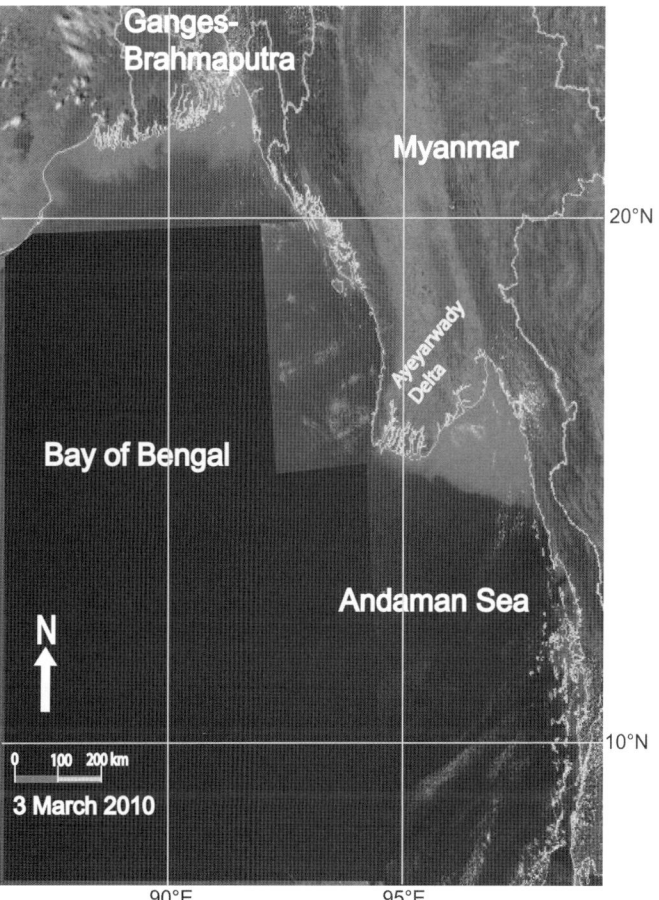

Fig. 17.7. Mosiac of satellite images of 3 March 2010 showing highly turbid zones off the Ayeyarwady river mouths and in the Gulf of Martaban, off the Ganges–Brahmaputra river mouths and in the Myeik Archipelago.

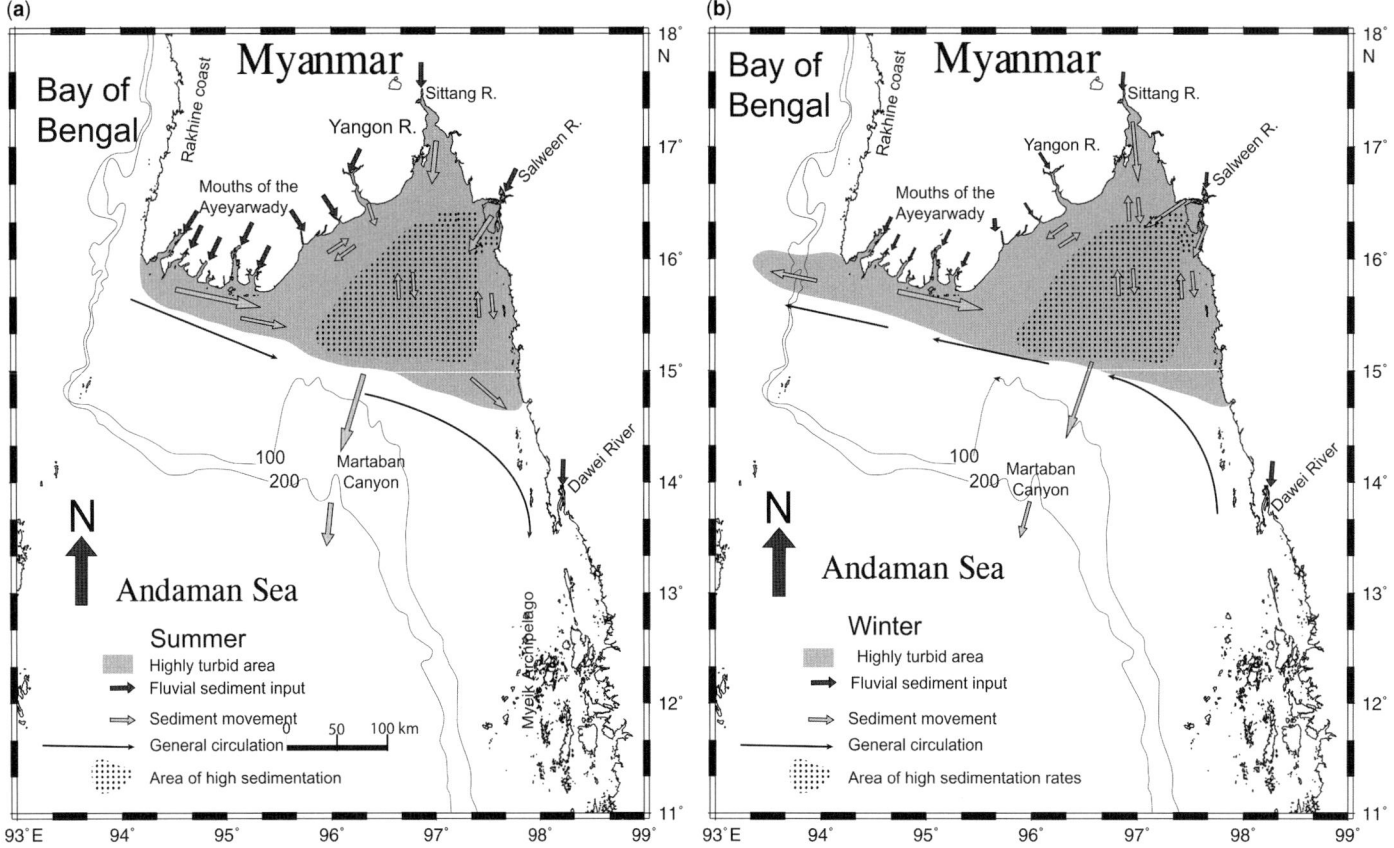

Fig. 17.8. Sediment sources, transport pathways and depositional areas on the Ayeyarwady shelf during (**a**) the summer SW monsoon and (**b**) the winter NE monsoon periods.

Organic matter

The total organic carbon (TOC) content on the Ayeyarwady shelf sediments ranges between 0.07 and 1.4%, and the total nitrogen (TN) content ranges between 0.01 and 0.24% (Ramaswamy *et al.* 2008). Relatively higher TOC and TN concentrations are found in the fine-grained sediments of the inner shelf and continental slope sediments. The outer shelf sediments, composed mostly of relict sands, are low in TOC and TN content. The TOC and TN concentrations are positively correlated with the amount of fine-grained sediments. The TOC:TN ratios in the Gulf of Martaban and Ayeyarwady inner shelf show typical marine values with a range mostly between 6 and 8. However, the $\delta^{13}C$ values of organic matter show a clear terrestrial signature and increase from $-25‰$ in the Gulf of Martaban to about $-22‰$ in the slope regions. Low $\delta^{15}N$ values ranging between $+3.3$ and $+4.8‰$, especially off the Ayeyarwady River mouth, show the influence of freshwater and terrigenous sediment sources. A mixing model indicates that the terrigenous carbon contribution to the inner shelf and Gulf of Martaban sediments is more than 70% as organic matter in rivers is not subject to intensive processing and replacement in the floodplains and deltaic regions and is aided by rapid burial at sea (Ramaswamy *et al.* 2008). Terrigenous organic carbon content decreases gradually offshore, reducing to less than 60% near the continental slope.

Clay-size mineralogy

The clay-size mineralogy of surface sediments of the Ayeyarwady shelf show that illite is the dominant mineral in most of the samples, except for a few samples in the southern part of the Tanintharyi shelf where kaolinite is the dominant clay mineral.

In the Rakhine and Ayeyarwady shelf, chlorite is the dominant mineral after illite. The kaolinite percentages in general are higher compared to those from the Ganges–Brahmaputra river sediments (Segall & Kuehl 1992) as the Ayeyarwady sediments are derived mostly from tropical regions, whereas in the Ganges–Brahmaputra river sediments they are mostly derived from the Himalayas and are dominated by physical weathering (Mallik 1976; Ramaswamy *et al.* 1997). Based on the amount of suspended sediment in various locations on the Ayeyarwady, Stamp (1940) concluded that around 12% of the suspended load of the Ayeyarwady is derived from the Himalayas and 88% is derived from the central dry belt. This is reflected in the Myanmar shelf sediments where illite and chlorite have a broad X-ray diffraction (XRD) peak width and low crystallinity index indicating that the dominant weathering pattern is chemical weathering in an arid tropical environment.

Geochemistry

The major-element distribution of the Ayeyarwady shelf sediments shows that fine-grained sediments in the Gulf of Martaban are homogeneous. Sediment SiO_2 content (Fig. 17.5) has a narrow range of between 54 and 58%, while the Al_2O_3 content ranges between 18 and 20% (Fig. 17.6). The TiO_2 content in the sediments is between 0.86 and 0.89%. The narrow range in major-element content shows that sediments from different sources are mixed well in the Gulf, mainly due to tidal processes. The outer shelf samples generally have a SiO_2 content $>70\%$ as they are composed of relict silica sands. The outer shelf sediments of the Tanintharyi coast have little fine-grained sediment; hence, the silica content of the non-carbonate fraction of the sediments is more than 80%.

Sea-level changes

About 18 ka ago, sea level on the eastern Indian Ocean was about 120 m lower than today (Hanebuth *et al.* 2000). Based on accelerator mass spectrometry (AMS) ^{14}C analysis of relict foraminiferal assemblages, a sea-level curve for the Ayeyarwady coast for the last 16 kyr has been proposed by Panchang *et al.* (2008). At 16 ka ago, the sea level was 140 m below mean sea level (MSL) and, at 13 ka ago, the sea level is comparable to the global sea-level curve. Thereafter there are significant deviations from the global and regional sea-level curves suggesting that the Ayeyarwady shelf underwent major subsidence leading to the destruction of many coral reefs on the shelf. Panchang & Nigam (2012) also reported three increased freshwater pulses on the Ayeyarwady River at about AD 1675, 1750 and 1850 from a study of down-core variations in *Asterorotalia trispinosa*, a benthic foraminifer species with an affinity for low-salinity waters. [210]Pb and calibrated ^{14}C AMS measurements on a gravity core collected at a depth of 37 m off the Ayeyarwady river mouth (15.45°N, 94.5°E) gave an average sedimentation rate of 36 mm a^{-1} (Panchang & Nigam 2012).

Sediment transport processes on the Myanmar shelf

The northern Indian Ocean and Andaman Sea is a highly dynamic environment. High rainfall and river discharge of freshwater and suspended sediments, winds and wind-driven waves, and associated longshore currents, meso- to macrotidal currents, seasonally reversing general circulation, and solitons all play an important role in sediment-dispersal processes in this area (Table 17.1). The combined sediment influx from the Ganges–Brahmaputra and Meghna rivers, about 1.08×10^3 MT a^{-1} (Milliman & Syvitsky 1992), is discharged into the western portion of Bangladesh. Although most of the sediment is transported westwards towards the Swatch-of-No-Ground (Kuehl *et al.* 1989), a small fraction is deposited on the northern portion of the Rakhine coast, and hence this shelf is shallow and deltaic (Gordon & Thwin 1998). Satellite images (Fig. 17.7) show a plume of suspended sediment from the mouths of the Ganges–Brahmaputra river up to the Akyab Bay (18.5°N). South of Sittwe, the waters are very clear and the continental shelf is devoid of sediments, except for the southernmost tip of the Rakhine coast south of 17°N where sediments derived from the Ayeyarwady River are deposited in the inner shelf during the NE monsoon period (Ramaswamy *et al.* 2004).

The Ayeyarwady River sediment is composed mostly of silt and clays with a small amount of sand. As the sediments are debouched into the sea, the sand is the first to be deposited forming beaches, spits and bars. The fine-grained sediment is displaced eastwards by longshore currents aided by the prevailing westerly currents generated by the SW monsoon (Rodolfo 1969a; Rao *et al.* 2005). Tidal currents further push the sediments into the waters of the Gulf of Martaban, where most of the suspended sediment is deposited (Ramaswamy *et al.* 2004). Negligible quantities (<0.1%) of sand-size material in the Gulf of Martaban sediments (Rao *et al.* 2005) indicate that sands are left behind and only fine-grained sediments are transported eastwards. During the SW monsoon, the surface currents flow towards the east and prevent significant quantities of sediment escaping into the Bay of Bengal (Rodolfo 1969a). The general circulation reverses during the NE monsoon period (November–February) and the surface currents flow towards the west. These currents may push some of the suspended sediments westwards into the eastern Bay of Bengal (Fig. 17.8). Satellite images obtained during November and December reveal plumes of suspended sediment heading westwards into the Bay of Bengal (Ramaswamy *et al.* 2004). Rodolfo (1969a) estimated that the sedimentation rates on the eastern and inner Ayeyarwady shelf are about 200 cm per 1000 years, representing about 90% of the river discharge. Sediments of the outer western delta shelf are shelly, muddy sands, relict from low sea stand (Fig. 17.2a), indicating that the western delta shelf is not accumulating significant sediment at present (Rodolfo 1969a).

A very characteristic feature of the suspended sediment concentration (SSC) in the Andaman Sea is the large highly turbid zone in the Gulf of Martaban and off the mouths of the Ayeyarwady River, which appears as a perennial feature throughout the year (Ramaswamy *et al.* 2004). This is one of the largest perennially turbid zones of the world's oceans, covering an area of more than 45 000 km^2 during spring tide, and similar or larger than river plumes of major rivers like the Ganges–Brahmaputra, Amazon and Yangtze. The turbid zone front oscillates nearly 150 km in phase with the tidal cycle (Ramaswamy *et al.* 2004). The maximum aerial extent of the turbid zone (45 000 km^2) is during the spring tide and the minimum ($\sim$17 000 km^2) during the neap tide. Tide-controlled highly turbid waters are also seen in Sittwe Bay, the Myeik Archipelago, the Dawei and Tanintharyi rivers, and also off the Ganges–Brahmaputra rivers. High turbidity is seen simultaneously during the spring tide at all the above locations (Fig. 17.7). The Martaban mud belt and high turbid zone extends partly into the Martaban Canyon where strong tidal forces, gravity flows and possibly hyperpycnal flows assist in transporting sediments directly to the deep Andaman Sea. Water column profiles of suspended sediment concentrations suggest that part of the suspended sediment flows into the Martaban Canyon and is transported towards the deep Andaman Sea as bottom nepheloid layers (Ramaswamy *et al.* 2004). During the fair-weather season, suspended sediment concentrations show an increase with depth due to resuspension of bottom sediments by tidal currents. However, during major storms or cyclones suspended sediment concentrations are more uniform and do not show a pronounced increase with depth (Shi & Wang 2008). The Martaban Canyon acts as a conduit for sediments to the deep Andaman Sea.

We thank A. Chivas for encouragement and for inviting us to contribute this paper. Some of the data used in this paper were collected under the collaborative project 'India–Myanmar Joint Oceanographic Studies in the Andaman Sea' initiated by the Government of India (Ministry of External Affairs), and the Government of Myanmar (Ministry of Education), with support from the Council of Scientific and Industrial Research, New Delhi, and the Ministry of Ocean and Earth Science, New Delhi. We also acknowledge the contribution of S. Thwin and other colleagues from Myanmar for support during the fieldwork. This is NIO contribution number 5454.

References

ALPERS, W., CHEN, H.-W. & HOCK, L. 1997. Observation of internal waves in the Andaman Sea by ERS SAR. *In*: *Proceedings of the 1997 IEEE International Geoscience and Remote Sensing Symposium (IGARSS '97), Singapore, August 3–8, 1997*, Volume **4**. Institute of Electrical and Electronics Engineers, Piscataway, NJ, 1518–1520.

ASHLEY, G. M. 1990. Classification of large-scale sub-aqueous bedforms: a new look at an old problem. *Journal of Sedimentary Petrology*, **60**, 160–172.

BAY OF BENGAL PILOT 1978. *Hydrographer to the Navy*. Ministry of Defence, Taunton, Somerset.

BENDER, F. 1983. *Geology of Burma*. Borntraeger, Berlin.

BIRD, M. I., ROBINSON, R. A. J. ET AL. 2008. A preliminary estimate of organic carbon transport by the Ayeyarwady (Irrawaddy) and Thanlwin (Salween) Rivers of Myanmar. *Quaternary International*, **186**, 113–122.

BROWN, B. E. 2007. Coral reefs of the Andaman Sea – an integrated perspective. *Oceanography and Marine Biology: An Annual Review*, **45**, 173–194.

CHHIBBER, H. L. 1934. *The Geology of Burma*. Macmillan, London.

CURRAY, J. R. 2005. Tectonics and history of the Andaman Sea region. *Journal of the Asian Earth Sciences*, **25**, 187–232.

CURRAY, J. R., MOORE, D. G., LAWVER, L. A., EMMEL, F. J., RAITT, R. W., HENRY, M. & KIECKHEFER, R. 1979. Tectonics of the Andaman Sea and Burma: convergent margins. *In*: WATKINS, J. S., MONTADERT, L. & DICKERSON, P. W. (eds) *Geological and Geophysical Investigations of Continental Margins*. American Association of Petroleum Geologists Memoirs, Tulsa, **29**, 189–198.

FLEMMING, B. W. 1978. Underwater sand dunes along the southeast African continental margin–Observations and implications. *Marine Geology*, **26**, 177–198.

FRITZ, H. M., BLOUNT, C. D., THWIN, S., THU, M. K. & CHAN, N. 2009. Cyclone Nargis storm surge in Myanmar. *Nature Geoscience*, **2**, 448–449.

FURUICHI, T., WIN, Z. & WASSON, R. J. 2009. Discharge and suspended sediment transport in the Ayeyarwady River, Myanmar: centennial and decadal changes. *Hydrological Processes*, **23**, 1631–1641.

GAILLARDET, J., DUPRÉ, B., LOUVAT, P. & ALLÉGRE, C. J. 1999. Global silicate weathering and CO_2 consumption rates deduced from the chemistry of large rivers. *Chemical Geology*, **159**, 3–30.

GORDON, R. 1885. The Irawadi River. *Royal Geographical Society (London) Proceedings (New Series)*, **7**, 292–331.

GORDON, A. L. & THWIN, S. (eds) 1998. *Workshop on the Mixing and Circulation of the Andaman Sea – A Cooperative Myanmar and US Program March, 16–18, 1998, Diamond Jubilee Hall, University of Yangon, Union of Myanmar*. National Science Foundation, Arlington, VA & Ministry of Education, Department of Higher Education, Myanmar.

HANEBUTH, T., STATTEGGER, K. & GROOTES, P. M. 2000. Rapid flooding of the Sunda Shelf: a late-glacial sea-level record. *Science*, **288**, 1033–1035.

HYDER, P., JEANS, D. R. G., CAUQUIL, E. & NERZIC, R. 2005. Observations and predictability of internal solitons in the northern Andaman Sea. *Applied Ocean Research*, **27**, 1–11.

INDIAN TIDE TABLES 2002. *Survey of India*. Dehra Dun, India.

KUEHL, S. A., HARIU, T. M. & MOORE, W. S. 1989. Shelf sedimentation off the Ganges-Brahmaputra River system: evidence for sediment bypassing the Bengal Fan. *Geology*, **17**, 1132–1135.

MALLIK, T. K. 1976. Shelf sediments of the Ganges delta with special emphasis on the mineralogy of the western part, Bay of Bengal, Indian Ocean. *Marine Geology*, **22**, 1–32.

McPHADEN, M. J., FOLTZ, G. R. *ET AL*. 2009 Ocean–atmosphere interactions during Cyclone Nargis. *Eos, Transactions of the American Geophysical Union*, **90**, 53–60.

MILLIMAN, J. D. & SYVITSKI, J. P. M. 1992. Geomorphic/tectonic control of sediment discharge to the ocean: the importance of small mountainous rivers. *Journal of Geology*, **100**, 525–544.

OSBORNE, A. R. 1990. The generation and propagation of internal solitons in the Andaman Sea. *In*: FORDY, A. P. (ed.) *Soliton Theory: A Theory of Results*. Manchester University Press, Manchester, 152–173.

OSBORNE, A. R. & BURCH, T. L. 1980. Internal solitons in the Andaman Sea. *Science*, **208**, 451–460.

PANCHANG, R. & NIGAM, R. 2012. High resolution climatic records of the past ~489 years from Central Asia as derived from benthic foraminiferal species, *Asterorotalia trispinosa*. *Marine Geology*, **307–310**, 88–104.

PANCHANG, R., NIGAM, R., RAVIPRASAD, G. V., RAJAGOPALAN, G., RAY, D. K. & HLA, K. Y. 2008. Relict faunal testimony for sea-level fluctuations off Myanmar (Burma). *Journal of the Paleontological Society of India*, **53**, 185–195.

POTEMRA, J. T., LUTHER, M. E. & O'BRIEN, J. J. 1991. The seasonal circulation of the upper ocean in the Bay of Bengal. *Journal of Geophysical Research*, **96**, 12 667–12 683.

RAJU, K. A. K., RAMPRASAD, T., RAO, P. S., RAO, B. R. & VARGHESE, J. 2004. New insights into the tectonic evolution of the Andaman basin, northeast Indian Ocean. *Earth and Planetary Science Letters*, **221**, 145–162.

RAMA RAO, R. B. S. 1930. *The Geology of the Mergui District*. Geological Survey of India Memoirs, **55**.

RAMASWAMY, V. & GAYE, B. 2006. Regional variations in the fluxes of foraminifera carbonate, coccolithophorid carbonate and biogenic opal in the northern Indian Ocean. *Deep-Sea Research – I*, **53**, 271–293.

RAMASWAMY, V. & RAO, P. S. 2006. Grain size analysis of sediments from the northern Andaman Sea: comparison of laser diffraction and sieve-pipette techniques. *Journal of Coastal Research*, **22**, 1000–1009.

RAMASWAMY, V., VIJAYKUMAR, B., PARTHIBAN, G., ITTEKKOT, V. & NAIR, R. R. 1997. Lithogenic fluxes in the Bay of Bengal measured by sediment traps. *Deep-Sea Research – I*, **44**, 793–810.

RAMASWAMY, V., RAO, P. S., RAO, K. H., THWIN, S., RAO, S. N. & RAIKER, V. 2004. Tidal influence on suspended sediment distribution and dispersal in the northern Andaman Sea and Gulf of Martaban. *Marine Geology*, **208**, 33–42.

RAMASWAMY, V., GAYE, B., SHIRODKAR, P. V., RAO, P. S., CHIVAS, A. R., WHEELER, D. & THWIN, S. 2008. Distribution and sources of organic carbon, nitrogen and their isotopic signatures in sediments from the Ayeyarwady (Irraddy) continental shelf, northern Andaman Sea. *Marine Chemistry*, **111**, 137–150.

RAO, P. S., RAMASWAMY, V. & THWIN, S. 2005. Sediment distribution and transport on the Ayeyarwady continental shelf, Andaman Sea. *Marine Geology*, **216**, 239–247.

ROBINSON, R. A. J., BIRD, M. I. *ET AL*. 2007. The Irrawaddy river sediment flux to the Indian Ocean: the original nineteenth-century data revisited. *Journal of Geology*, **115**, 629–640.

RODOLFO, K. S. 1969a. Sediments of the Andaman basin, northeastern Indian Ocean. *Marine Geology*, **7**, 371–402.

RODOLFO, K. S. 1969b. Bathymetry and marine geology of the Andaman basin, and tectonic implications for Southeast Asia. *Geological Society of America Bulletin*, **80**, 1203–1230.

SEGALL, M. P. & KUEHL, S. A. 1992. Sedimentary processes on the Bengal continental shelf as revealed by clay-sized mineralogy. *Continental Shelf Research*, **12**, 517–541.

SHI, W. & WANG, M. 2008. Three-dimensional observations from MODIS and CALIPSO for ocean responses to cyclone Nargis in the Gulf of Martaban. *Geophysical Research Letters*, **35**, L21603.

STAMP, L. D. 1940. The Irrawaddy River. *The Geographical Journal*, **95**, 329–356.

TOMCZAK, M. & GODFREY, J. S. 1994. *Regional Oceanography: An Introduction*. Pergamon, Oxford.

UNITED KINGDOM HYDROGRAPHIC OFFICE 2004. *Admiralty Tide Tables Volume 3: Indian Ocean & South China Sea (including Tidal Stream Tables, 2004)*. United Kingdom Hydrographic Office, Taunton, Somerset.

VARKEY, M. J., MURTY, V. S. N. & SURYANARAYANA, A. 1996. Physical oceanography of the Bay of Bengal and Andaman Sea. *Oceanography and Marine Biology: An Annual Review*, **34**, 1–70.

WOODROFFE, C. D. 2000. Deltaic and estuarine environments and their Late Quaternary dynamics on the Sunda and Sahul shelves. *Journal of Asian Earth Sciences*, **18**, 393–413.

WYRTKI, K. 1973. Physical oceanography of the Indian Ocean. *In*: ZEITSCHEL, B. & GERLACH, S. A. (eds) *The Biology of the Indian Ocean*. Springer, Berlin, 18–36.

Chapter 18

East Antarctic continental shelf: Prydz Bay and the Mac.Robertson Land Shelf

P. E. O'BRIEN[1]*, P. T. HARRIS[2], A. L. POST[2] & N. YOUNG[3]

[1]*Department of Environment and Geography, Macquarie University, Sydney, New South Wales, Australia*

[2]*Geoscience Australia, GPO 378 Canberra, ACT 2601, Australia*

[3]*Antarctic Climate and Ecosystems CRC and Australian Antarctic Division, Private Bag 80, Hobart, Tasmania 7001, Australia*

*Corresponding author (e-mail: phil.obrien.ant@gmail.com)

Abstract: Prydz Bay and the Mac.Robertson Land Shelf exhibit many of the variations seen on Antarctic continental shelves. The Mac.Robertson shelf is relatively narrow with rugged, inner-shelf topography and shallow outer-shelf banks swept by the west-flowing Antarctic Coastal Current. U-shaped valleys cut across the shelf. It has thin sedimentary cover, deposited and eroded by cycles of glacial advance and retreat through the Neogene and Quaternary. Modern sedimentation is diatom-rich siliceous, muddy ooze in shelf deeps, while, on the banks, phytodetritus, calcareous bioclasts and terrigenous material are mixed by iceberg ploughing. Prydz Bay is a large embayment fed by the Amery Ice Shelf. It has a broad inner-shelf deep area and outer bank, with depths ranging from 2400 m beneath the ice shelf to 100 m on the outer banks. A clockwise gyre flows through the bay. Fine mud and siliceous ooze drape the seafloor; however, banks are scoured by icebergs to depths as great as 500 m. The Mac.Robertson shelf has seen advances to the shelf edge during glacial episodes and retreat during warming and rising sea level. Prydz Bay shows more complexity, with parts of the bay showing partial advance of the ice-grounding zone.

The continental shelf of Antarctica is not uniform but can be divided into major provinces, such as the major embayments of the Weddell, Ross, Bellingshausen and Amundsen seas and Prydz Bay, and the shelves of East Antarctica, the Antarctic Peninsula and West Antarctica. Each has its own climatic and oceanographic characteristics. The understanding of the Antarctic margin has now grown to the point where the entire continent cannot be adequately discussed in one chapter. However, some important variations in shelf type can be illustrated by parts of the margin that are relatively well studied. This chapter seeks to present information on some of the better studied parts of the shelf between 60°E and 80°E as a guide to the possible characteristics of, as yet, unvisited parts of the Antarctic continental shelf, and to illustrate the differences between Antarctica and the rest of the world's continental shelves.

Prydz Bay and the Mac.Robertson Land Shelf are two areas that represent the range of sedimentary environments on the East Antarctic margin (Fig. 18.1; Table 18.1). The Mac.Robertson Land Shelf is relatively narrow and bordered by small coastal glaciers, whereas Prydz Bay is a large embayment that is the downstream end of one of the largest ice drainage basins on the continent. Studies of the two areas, and Prydz Bay in particular, have been important to our developing understanding of the interaction of Antarctic glacial history and the history of life on the continental margin. Data come from marine surveys by five nations, two Ocean Drilling Program (ODP) legs (119 and 188), numerous studies of coastal areas and an extensive study of the adjacent ice shelf using drill holes (Table 18.1).

General setting

The East Antarctic continental shelf forms a broad curve that occupies a narrow latitudinal range from 62°S to 65°S with a major re-entrant, Prydz Bay, extending to 70°S (Fig. 18.1). Almost the entire coast is occupied by glacial ice, mostly of the East Antarctic Ice Sheet, although the region around Cape Poinset is occupied by Law Dome, a separate ice dome (Fig. 18.1). The climate

is cold polar with mean maximum temperatures of −7.3 °C at Davis and −8.3 °C at Mawson, the two Australian stations in the study area (Bureau of Meteorology 2010). Sea ice forms annually and extends over 100 km north of the coast, then mostly melts during spring and summer, although pack ice may persist over summer in some areas and some seasons.

The Polar Low-Pressure Trough tracks the margin for its entire length (Fig. 18.2) (Bromich 1988) and the continent is occupied by a high-pressure region. Precipitation is concentrated in the coastal zone because of rapid cooling of air masses and orographic uplift of air masses on the steep ice-sheet margin (Bromich 1988). The highest accumulation in East Antarctica occurs on Law Dome where easterly moving storm systems run on to the promontory. The polar climate means that precipitation is overwhelmingly snow and that melt rates are very low so there is minimal water run-off. Some surface run-off may occur during exceptionally warm weather. Subglacial water is present in Antarctica in the form of lakes and subglacial stream flow (Kapitsa *et al.* 1996; Fricker *et al.* 2007); however, it has only been observed flowing into the sea once (Goodwin 1987). More cases of subglacial discharge may be observed as awareness of the process grows. Thus, the bulk of drainage from Antarctica crosses the coast as glacial ice, with icebergs and grounding-zone melt being the primary processes of water loss from the continent.

The consistent tracking of extra-tropical lows along the margin results in a predominantly east–west wind regime along the shelf which drives the surface ocean currents in the same direction, with the core of the strong west-flowing current seawards of the shelf break (Fig. 18.3) (Bindoff *et al.* 2000). North of the zone of easterly winds, the strong westerly wind regime drives the Circum-Polar Current from west to east. The zone between the two currents is a zone of upwelling where Circum-Polar Deep Water, derived from North Atlantic Deep Water, rises to the surface. The Antarctic shelf is also an important site for deep-water formation (Rintoul 1998). On the East Antarctic margin outside the Weddell Sea, the major site of deep-water formation is the George V Land Shelf and the Terre Adélie shelf. The process of Adélie Land Bottom Water formation described by Rintoul (1998) involves sea-ice formation through the winter and

From: CHIOCCI, F. L. & CHIVAS, A. R. (eds) 2014. *Continental Shelves of the World: Their Evolution During the Last Glacio-Eustatic Cycle.*
Geological Society, London, Memoirs, **41**, 241–254. http://dx.doi.org/10.1144/M41.18

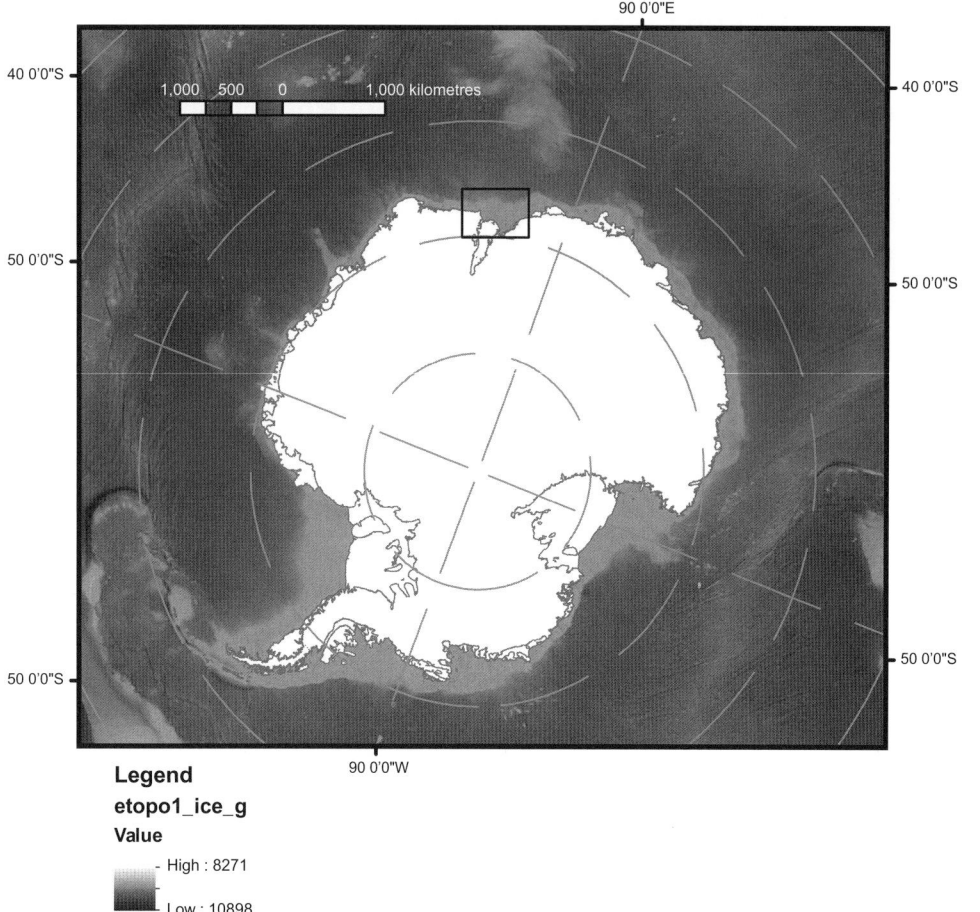

Legend
etopo1_ice_g
Value

- High : 8271

Low : 10898

Fig. 18.1. Location of Prydz Bay and the Mac.Robertson Land Shelf.

spring in a polynya (ice-free area) on the lee side of a floating ice tongue (Mertz Glacier). Salt left in the water forms brine that mixes in shelf depressions with salty Modified Circum-polar Deep Water that intrudes on to the shelf. The resulting 'High Salinity Shelf Water' accumulates in shelf depressions and spills over the shelf break to form Adélie Land Bottom Water, which accounts for about 20% of Antarctic Bottom Water. This process is favoured by the local shelf bathymetry and the presence of a long-lived polynya, and represents a major source of oxygen and nutrients in the deep ocean. In 2010, the Mertz Glacier tongue broke away after impact from a large iceberg. How this will affect long-term bottom water production is yet to be determined but Tamura *et al.* (2012) estimated a 14–20% reduction in sea-ice formation in the Mertz Polynya in the following years.

Table 18.1. *Summary table of the study area*

Length of study area	690 km
Average width	Mac.Robertson shelf – 100 km
	Prydz Bay −200 km
Tides	1–2 m
Dominant process	Currents, iceberg scouring
Average depth of shelf break	500 m
Siliciclastic/biogenic/authigenic/ glacial (%)	20/40/0/40
Modern/relict/palimpsest	60% modern (SMO, compound glacimarine), 40% palimpsest (ice-keel turbates
Tectonics over the last glacial cycle	Stable

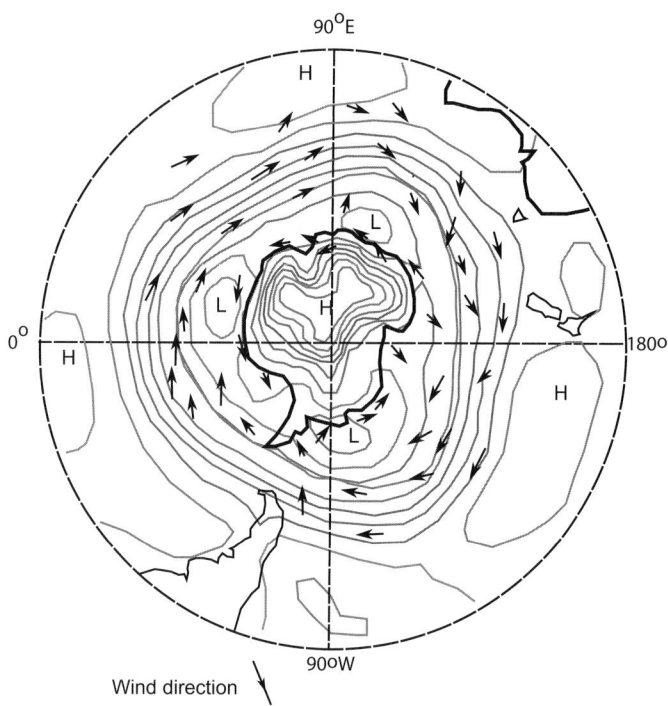

Wind direction

Fig. 18.2. Atmospheric pressure over Antarctica showing the average position of high-pressure (H) and low-pressure (L) systems, and main wind directions based on data presented by the Australian Bureau of Meteorology for Summer 2005 (December–February, www.bom.gov.au/ant). Contour interval is 5 hPa.

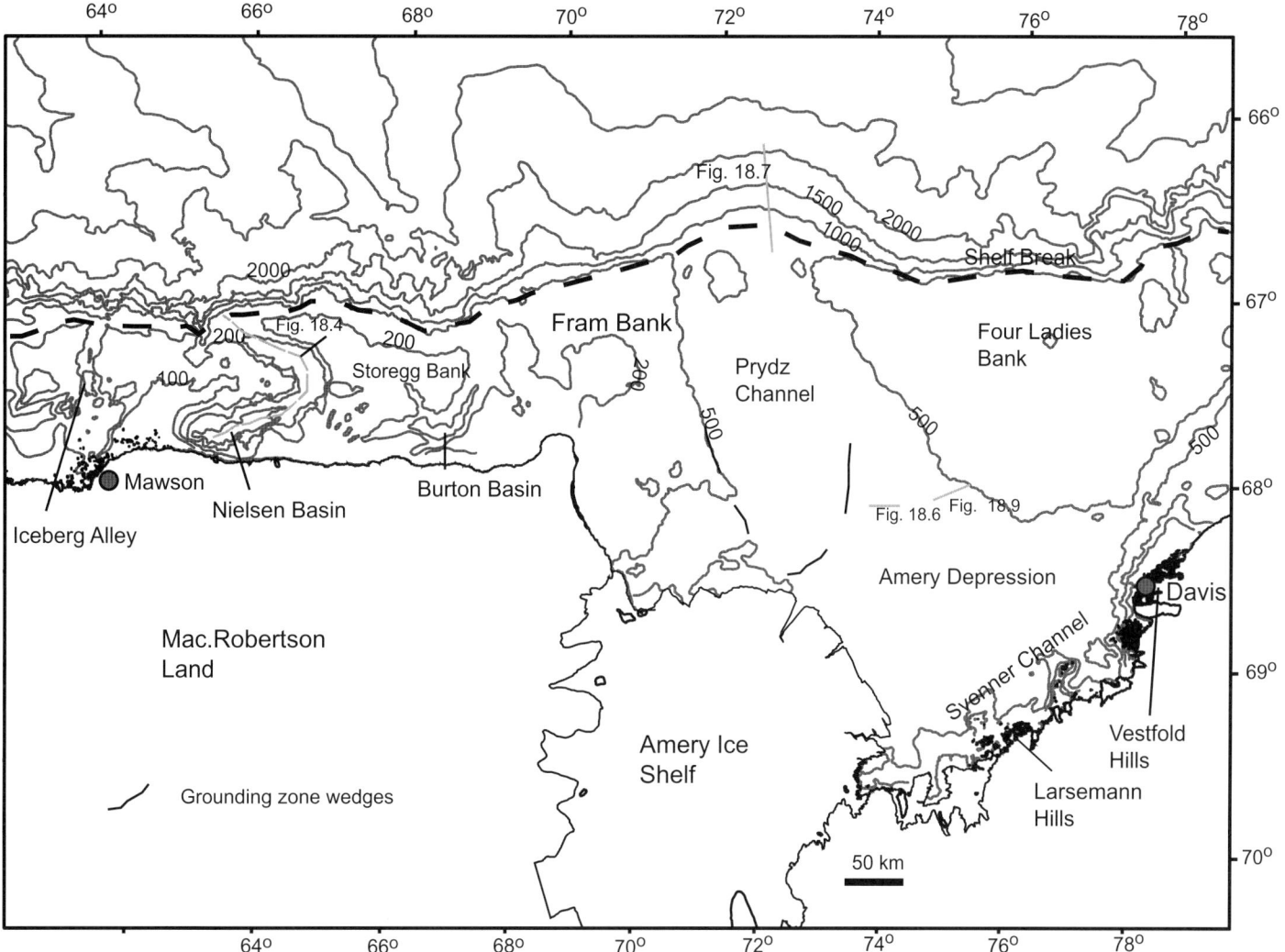

Fig. 18.3. Prydz Bay and Mac.Robertson Land Shelf bathymetry and named features. Bathymetry in metres. The locations of profiles in Figures 18.6, 18.7 and 18.9 are shown.

Modern sediment supply to the shelf comes from two main sources: ice-transported siliciclastic material; and biogenic material. Glaciers transport siliciclastic material as englacial or supraglacial debris when the material is delivered to the surface of the ice, and as subglacial debris when the ice erodes the material at the bed and transports it as a mobile bed driven by shear or frozen in the basal ice (Hambrey 1994). In East Antarctica, supraglacial and englacial debris are rarely observed crossing the coast because outcrops from which they might be delivered to the ice are small compared to the area of ice. Icebergs with bands of probable aeolian sediment are sometimes seen. Aeolian sediment is important in areas adjacent to coastal outcrops, although, where present, it is commonly blown on to sea ice or glacial ice and rafted long distances from the original site of aeolian deposition. Debris bands are visible in basal ice in the few areas where the ice bed is exposed near the coast, although mobile deforming beds, which are thought to be the main mechanism of basal transport in ice streams, have not been observed directly. Flutes and mega-scale glacial lineations (MSGLs) are linear bedforms formed by moulding of deformable subglacial sediment. Their preservation on the modern seafloor indicates that this mechanism was important during past glaciations (O'Brien *et al.* 1999; Mackintosh *et al.* 2011). Deforming subglacial sediment will either be lodged on the glacier bed or deposited at the grounding line where ridges of sediment form (grounding-zone wedges: Powell & Alley 1997). Debris frozen in basal ice will melt out at or downstream of the

grounding zone (e.g. Domack *et al.* 1998; Evans & Pudsey 2002) or be carried further by calving of the ice into icebergs. For icebergs to carry any subglacial debris, some debris has to bypass the grounding zone. This may happen when icebergs break off close to the grounding zone, or where basal freezing takes place close to the grounding zone (Holland *et al.* 2009). In the case of Prydz Bay, zones of basal freezing are many kilometres downstream of the modern grounding zone (Fricker *et al.* 2001). Thus, most ice-rafted debris in the region probably comes from carving of ice cliffs or rapidly retreating ice streams rather than the major ice shelves (Dowdeswell & Bamber 2007). The total amount of glacially delivered siliciclastic material being delivered to the shelf by icebergs is difficult to estimate but the slow flow speed of the Antarctic ice means that it is probably quite low compared to warmer periods in the past.

A major modern sediment source is biological activity. Siliceous phytoplankton dominate primary production south of the Polar Front, with zones of high productivity associated with the interplay of sea ice and open water (Dunbar *et al.* 1989; Anderson 1999). Diatoms are the dominant organisms and they form biosiliceous ooze deposits where sediments accumulate. Even in areas where sediments are not presently accumulating, a layer of diatomaceous material can accumulate during quiet periods, which is then removed by current activity. The other source of biogenic sediment is benthic organisms. Siliceous sponges provide abundant spicules and bryozoans; corals, molluscs and foraminifers provide calcareous sediment in places, although deposits of calcareous sediments seldom reach more than 1 m in thickness (Anderson 1999).

The modern sedimentary regime is dominated by floating ice. Sea ice dampens wave activity (Anderson 1999), while icebergs plough the seafloor to depths as great as 500 m (Barnes 1987; O'Brien & Leitchenkov 1997; Dowdeswell & Bamber 2007). Iceberg scouring is not uniform across the shelf but the factors that control frequency and intensity are poorly known. However, most surface sediment on the shelf shallower than 400 m is probably ice-keel turbate (Anderson 1999); sediment formed by ice keels turning over and ploughing both modern detritus and any older material at the surface.

The East Antarctic shelf is generally tectonically stable with low subsidence rates. Isostatic rebound after the retreat of the ice sheet from its Last Glacial Maximum position may be more important in many places at present (e.g. Zwart *et al.* 1998). Of greater importance in the generation of accommodation space has been long-term glacial erosion (ten Brink *et al.* 1995).

Tectonic framework and geological background

The East Antarctic margin between 50°E and 160°E can be divided into segments based on the style of continent break-up (Stagg *et al.* 2005). The segments differ in the relative role of orthogonal extension v. transtension, and whether Antarctica was separating from Greater India and continental fragments of the Kerguelen Plateau or Australia. Rifting took place in the Mesozoic, with break-up along the Mac.Robertson Land–Prydz Bay segment taking place during the Valanginian (*c.* 140–134 Ma: Stagg *et al.* 2004, 2005). Most of the rift structures occur beneath the continental slope but, beneath the shelf, small half-graben are preserved that are parallel to the major rift structures (Truswell *et al.* 1999). These features are commonly areas of preferential erosion by Cenozoic glacial action but Mesozoic sediments have been sampled in places. The basement of most of these structures is Precambrian metamorphic and igneous rocks.

The complication to this pattern is Prydz Bay, where the Lambert Graben and Prydz Bay Basin intersect the margin. These structures have a history of extension and subsidence since the late Palaeozoic (Stagg 1985). Outcrop studies and ODP drilling have identified sediments from Permian to Cenozoic in the Bay (Barron *et al.* 1991; McLoughlin & Drinnan 1997). Thermochronology studies of the flanks of the Lambert Graben indicate that the last main period of activity was in the Cretaceous (Lisker *et al.* 2007).

Shelf description

Prydz Bay is the largest re-entrant in this segment of the Antarctic margin (Fig. 18.1). It is a roughly rectangular bay between 68°30'E and 80°E. Mac.Robertson Land and Prydz Bay represent about 700 km of shelf and coastline. Prydz Bay is the downstream end of the Amery Ice Shelf drainage basin, which represents about 16% of all Antarctic surface area and is fed by ice originating at the main ice divide in East Antarctica. Prydz Bay represents the ice-free part of a major trough cut by the Amery system that extends 1000 km into the continental interior and which reaches 2560 m below sea level (Damm 2007).

Prydz Bay shows bathymetry that is broadly similar to the rest of the Antarctic shelf in having deeper water near the coast, with depths approaching 1000 m in the SW corner of the bay and a broad topographical basin, the Amery Depression that is around 700 m deep along the front of the Amery Ice Shelf (Fig. 18.3). The Amery Depression shoals gently to outer-shelf banks around 100–200 m deep. The shelf break is at around 400–500 m. The

western side of Prydz Bay features a broad trough crossing from the inner shelf to the shelf edge, Prydz Channel. It is around 100 km wide and is 500 m deep at the shelf break. It is a typical cross-shelf trough of glaciated margins cut by a fast-flowing ice stream (Boulton 1990; O'Brien & Leitchenkov 1997; Passchier *et al.* 2003; O'Brien *et al.* 2007). The SE side of the Bay features a series of troughs and saddles collectively named the Svenner Channel.

The Mac.Robertson Land Shelf extends from 62°E to 68°30'E. Its morphology consists of smooth planar banks that average 100–200 m deep, iceberg scoured and current swept outer banks, an inner-shelf zone of high-relief ridges and valleys, and three deep troughs that reach >1000 m deep near the coast and extend across the shelf to the shelf break where they are 400–500 m deep (O'Brien *et al.* 1994; Harris & O'Brien 1996).

The deep troughs are named informally from west to east, the Iceberg Alley Trough, Nielsen Basin and Burton Basin. The Iceberg Alley Trough extends straight across the shelf from Mawson Station and is a classic U-shaped glacial valley (Fig. 18.4). It is deepest near the coast and shallows to the shelf edge. The Nielsen and Burton basins are arcuate with an east–west segment inshore and a trough extending from their eastern ends in a westerly arc to the shelf edge. The Nielsen Basin is the best mapped and has a long profile that shows depths as great as 1100 m near the coast shoaling to 400 m at the shelf edge (Fig. 18.4a) (Harris & O'Brien 1998). In cross-section, it is clearly U-shaped (Fig. 18.4b). Burton Basin is similar in plan with a maximum depth of 600 m deep.

Trough geometry is a product of underlying structure. Seismic sections show that the inner Nielsen Basin is underlain by a remnant of a Mesozoic half-graben fill with seaward-dipping normal faults and a thin section of Albian non-marine sediments preserved in the deepest part of the basin (Truswell *et al.* 1999). This half-graben has been preferentially eroded by ice during

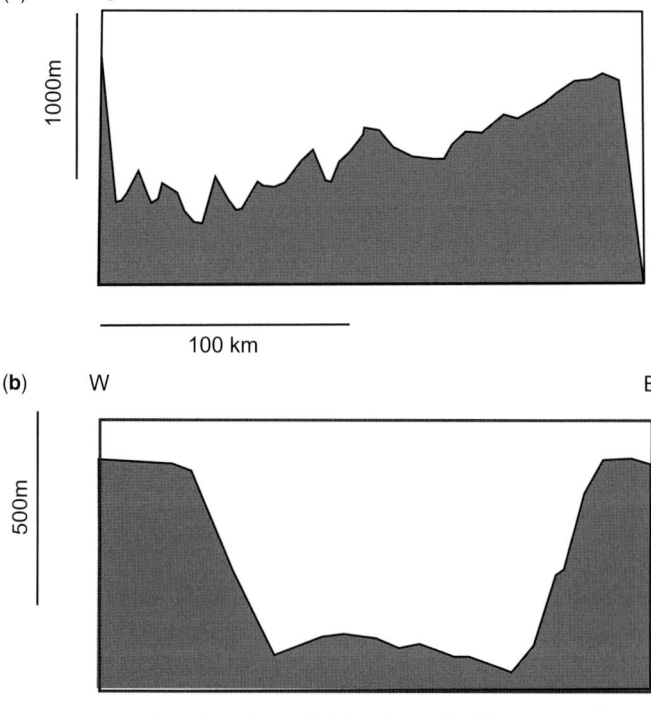

Fig. 18.4. Bathymetric cross-sections of the Mac.Robertson Shelf: (**a**) long profile; Nielsen Basin (Harris & O'Brien 1998); (**b**) cross-section, Nielsen Basin.

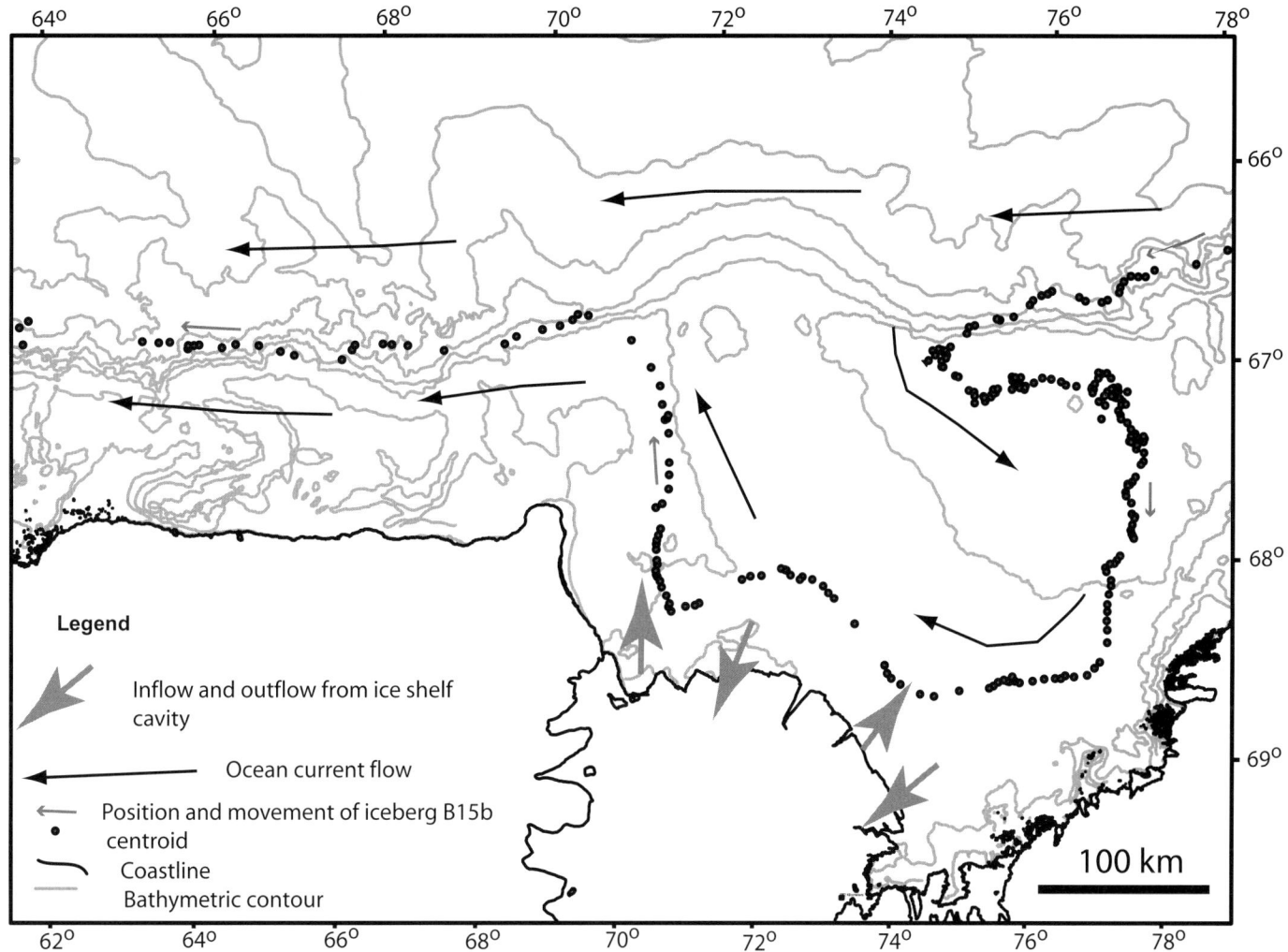

Fig. 18.5. Current flow and iceberg motion through Prydz Bay and the Mac.Robertson Land Shelf based on oceanographic studies and tracking of a large iceberg (B15b). The iceberg track is represented by the position of the iceberg centroid as observed on satellite images between 23 August 2006 and 30 September 2009.

glacial maxima and then the ice flow was diverted seawards where the bounding fault terminates against basement, either by simple reduction in throw or by a transfer fault. The westward curve of the outer part of the Nielsen and Burton basins is harder to explain. This may reflect geological structure or may represent some response of the glacier to westward flowing winds and currents. The Iceberg Alley Trough parallels the orientation of onshore mountains to the south and probably reflects underlying structures in basement.

Oceanography

The dominant feature of ocean circulation is the Antarctic Coastal Current. A clockwise gyre branches off flowing SW along the eastern side of Prydz Bay (Fig. 18.5) (Smith *et al.* 1984; Nunes Vaz *et al.* 1994). The flow becomes more complex along the front of the Amery Ice Shelf where both inflows and outflow have been detected by current meters (Leffanue & Craven 2004). Refrozen seawater has been detected beneath the western side of the ice shelf (Fricker *et al.* 2001), indicating thermohaline flow driven by melting near the grounding zone (Hemer *et al.* 2007). The flow is strong enough to advect phytodetritus into the cavity of the ice shelf for at least 100 km, where mud rich in bio-siliceous material is accumulating and marine organisms have colonized the seafloor (Post *et al.* 2007; Riddle *et al.* 2007).

Evidence from sediment cores (Hemer & Harris 2003; Hemer *et al.* 2007) and sidescan sonar records from part of Prydz Bay formerly covered by the ice shelf (O'Brien *et al.* 1999) show that currents near the grounding zone can be strong enough to transport sand and form large bedforms (Fig. 18.6).

The shelf and upper slope of Mac.Robertson Land are dominated by the east–west-flowing Antarctic Coastal Current (Harris & O'Brien 1998). Current meter and Acoustic Dopler Current Profiler (ADCP) data (Hodgkinson *et al.* 1988, 1991*a*, *b*; Harris & O'Brien 1998) show a general pattern of gentle currents on the inner shelf with stronger flow on the outer shelf and upper slope. Flows are mostly less than 0.20 m s^{-1} but occasional extreme events reach 1.96 m s^{-1}. Flows are generally towards the west and NW, consistent with the primary current being the Antarctic Coastal Current, but with possible episodes where High Salinity Shelf Water (HSSW) spills over the sill where the shelf valleys meet the upper slope, resulting in the strongest flows observed.

Underlying geology

Prydz Bay is underlain by long-lived structures, the Lambert Graben and Prydz Bay Basin (Stagg 1985). These reflect crustal extension dating to the Early Permian or Carboniferous, with reactivation during the Cretaceous (Lisker *et al.* 2007). Up to 10 km of

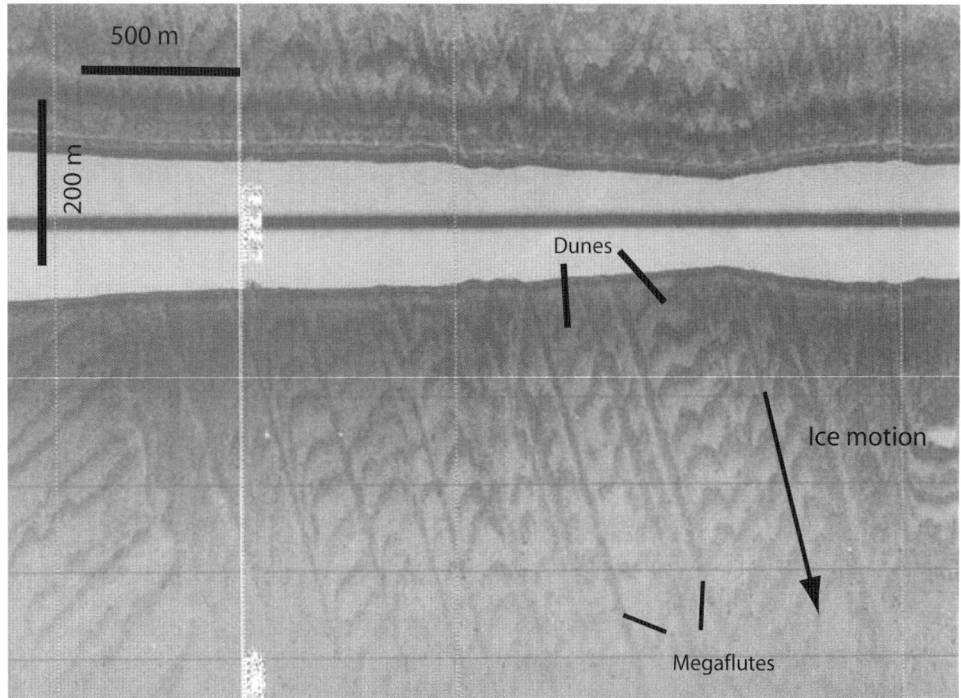

Fig. 18.6. Sidescan sonar records of bedforms from the Amery Depression. The linear ridges are megaflutes formed by the glacier moulding its bed parallel to the flow direction. Between the megaflutes, the arcuate features are dunes formed by water flowing towards the grounding zone. The ice-flow direction indicated by megaflutes is towards slightly west of north (*c.* 355°True).

sediment are present in the thickest areas. The Lambert Graben is an elongate rift that has been mapped by airborne geophysics 1000 km into the continental interior (Damm 2007; McLean *et al.* 2009). The rift flanks are composed of Precambrian metamorphic and igneous basement ranging from Archaean to late Proterozoic age (Mikhalsky *et al.* 2001). Metamorphic grades range from greenschist to granulite facies. The graben itself probably contains sediments that include Permian coal measures and Triassic sandstone overlain in Prydz Bay by undated red beds of probable fluvial origin, non-marine Cretaceous sediments and Palaeogene shallow-marine facies (Cooper & O'Brien 2004). Cooper & O'Brien (2004) found that the first evidence of erosion of the shelf by glaciers and transport of sediment to the lower continental slope was in the middle Miocene, around 14 Ma. Since then, ice has eroded down as far as 2400 m close

to the modern grounding-zone position of the Amery Ice Shelf and to 2560 m inland (Fricker *et al.* 2002).

The Neogene section is dominated by glacial sediments. Whitehead *et al.* (2006) estimated a minimum of 10 Cenozoic ice advances across Prydz Bay. The sections drilled in ODP Leg 119 and Leg 188 are mostly diamicts deposited by subglacial processes or by a mixture of ice rafting and debris flows (Hambrey *et al.* 1991; Passchier *et al.* 2003). There are also thin silicoflagellate and coccolith-bearing muds representing warmer phases (Whitehead & Boharty 2003; O'Brien *et al.* 2007). Advances of glaciers across the shelf are indicated by erosion surfaces, overcompacted intervals (Solheim *et al.* 1991; Forsberg *et al.* 2008) and by prograding wedges of diamicts, and pebbly muddy sands on the upper slope (Fig. 18.7) (O'Brien *et al.* 2007). The best-developed example of such a deposit is a Trough Mouth Fan (Vorren &

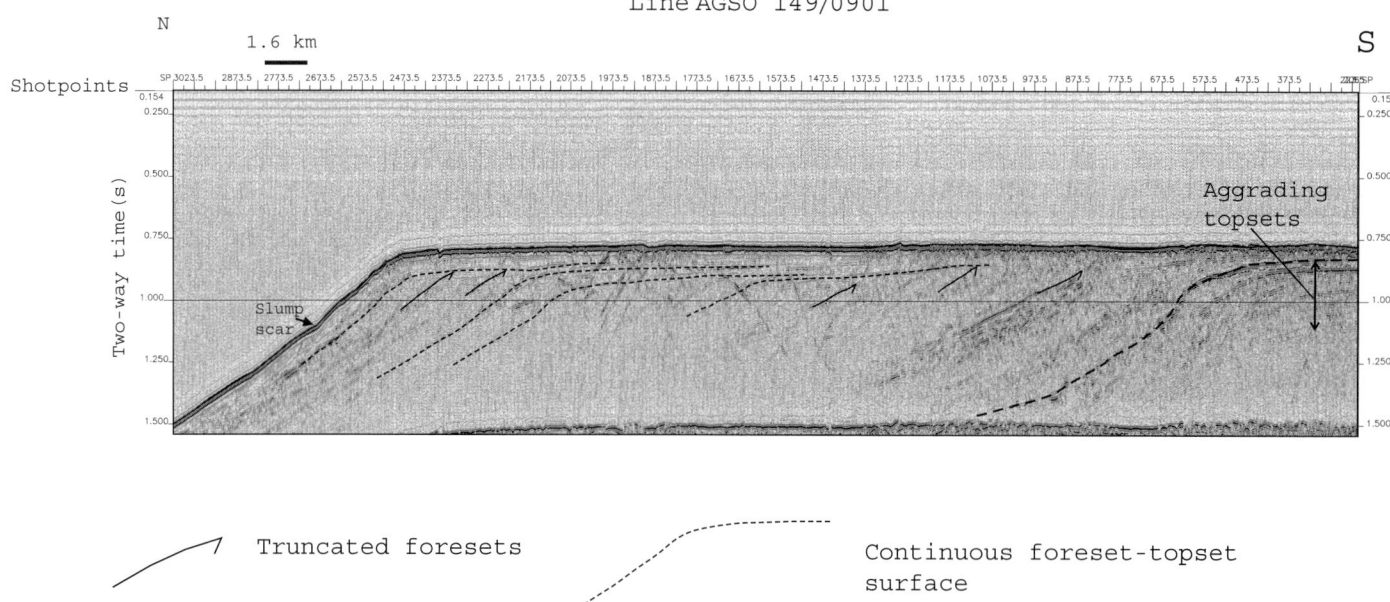

Fig. 18.7. Seismic section across the shelf edge in Prydz Bay showing prograding glacial sediments and topsets (O'Brien *et al.* 2007).

Laberg 1997) where Prydz Channel meets the shelf edge (O'Brien & Leitchenkov 1997; O'Brien et al. 2007). Such upper slope deposits form when rapidly flowing ice streams transport debris at their base to the shelf edge. There it deposits and is remobilized by sediment gravity flows and currents to form the upper slope wedge. The Prydz Channel Fan started accumulating in the Early Pliocene (O'Brien et al. 2004; Whitehead et al. 2006).

Neogene marine sediments crop out near the western side of the Amery Ice Shelf in the northern Prince Charles Mountains (Whitehead et al. 2006). These sediments indicate episodes of open water during warm periods of the Late Miocene and Pliocene. At these times, Prydz Bay extended at least 200 km further south than present, forming a huge fjord with glaciers around its sides but with seasonal open water.

The Mac.Robertson shelf is underlain by Precambrian basement beneath most of the inner shelf. In the Nielsen Basin, seismic sections show a thin remnant of sediment filling a Mesozoic half-graben (Stagg 1985; O'Brien et al. 1994). A wedge of off-lapping sediments underlies the mid to outer shelf (Stagg et al. 2005). The section is much thinner than in Prydz Bay. Dredge and core samples show this wedge to be Late Cretaceous in age in the mid shelf, grading to Eocene beneath the outer shelf and upper slope (Truswell et al. 1999). The relatively great age of sediment at the surface reflects the depth of glacial erosion, which started in the Mid Miocene (Cooper & O'Brien 2004). The rate of tectonic subsidence is probably low compared to the effects of isostatic depression, and rebound and subglacial erosion throughout the Neogene.

Quaternary sediments

The Quaternary sedimentation of Prydz Bay was dominated by glacial–interglacial cycles. Ice advances across the shelf have deposited diamicts or eroded the surface, while interglacials produced mud or diatom ooze. Commonly these fine sediments were eroded by subsequent ice advances. Quaternary sediments form thin sheets that pinch out shorewards and pass into off-lapping wedges at the shelf edge.

ODP Site 1167 on the Prydz Channel Fan (O'Brien et al. 2007) and gravity cores in the Prydz Channel (Domack et al. 1998) indicate that a major change in glaciation took place in the middle Pleistocene. The Amery Ice Shelf grounding zone no longer advanced to the shelf edge in Prydz Channel. Grounding-zone wedges were deposited on the mid to inner shelf (Fig. 18.3) (Domack et al. 1998; O'Brien et al. 1999). The channel accumulated muds, and siliceous mud and ooze (SMO) deposits, in a cavity beneath floating ice (Domack et al. 1998). Iceberg scours are found as deep as 750 m, suggesting that parts of Prydz Bay were not glaciated at the lowest sea level of the last glacial cycle (O'Brien & Leitchenkov 1997). Ice was probably grounded on Four Ladies Bank; however, definitive evidence is not preserved because the bank is so heavily ploughed by icebergs. Evidence from the ice-free areas along the SE edge of the bay suggests that ice had retreated from the Prydz Channel grounding-zone wedges by about 12 ka BP. Sediment cores from beneath the Amery Ice Shelf show that the cavity was open and receiving advected organic matter by 9 ka BP (Hemer et al. 2007; Post et al. 2007).

On the Mac.Robertson shelf, Quaternary sedimentation occurs in contrasting environments. The deepest parts of the shelf valleys contain sequences starting with probable subglacial till passing up into sandy mud passing up into SMO. Sidescan sonar images show subglacial flutes in some of the valleys, indicating till deposition and deformation (Harris & O'Brien 1998). The sandy muds represent sedimentation in the grounding zone of retreating glaciers and beneath the floating ice tongue (Domack & Harris 1998). This facies grades into SMO facies, which may reach 15–20 m thick (Leventer et al. 2006). The SMO section is mostly massive; however, the transition zone between siliciclastic muds and SMO in Iceberg Alley and the Nielsen Basin contain centimetre-scale laminations composed of alternating olive-grey SMO and orange-brown biosiliceous ooze, thought to represent spring blooms of Chaetoceros that occurred during episodes when a strongly stratified water column developed in calving bays during ice retreat (Leventer et al. 2006).

Bank sedimentation is more difficult to detect. Sections recovered by gravity cores tend to be short, recovering compacted tills of unknown age. Horizontal layering imaged by seismic reflection suggests till deposition on the banks, although dating the reflectors has not been possible. The presence of subglacial bedforms, such as flutes, MSGLs in shelf deeps (Harris & O'Brien 1998; Mackintosh et al. 2011) and grounding-zone wedges near the shelf edge (Mackintosh et al. 2011), suggest that the shelf was glaciated almost to the shelf edge during the late Quaternary, including during the Last Glacial Maximum. Dating of sediment cores in shelf deeps (Harris & O'Brien 1998; Leventer et al. 2006; Mackintosh et al. 2011), combined with cosmogenic exposure dating of nunataks in the hinterland (Mackintosh et al. 2007, 2011), indicate that ice retreat started around 14 ka BP. Calving bays formed in the deep valleys, while ice stayed grounded on the banks longer (Harris et al. 1998; Leventer et al. 2006). The present ice-sheet configuration was reached by about 7 ka BP (Mackintosh et al. 2011).

Modern sedimentation

O'Brien & Leitchenkov (1997) described the acoustic character of Prydz Bay surface sediments (Table 18.2) and Harris et al. (1998) described an extensive collection of surface samples. Quilty (1985) described the modern foraminfer assemblage, and Taylor et al. (1997) and Taylor & McMinn (2002) described the diatom assemblages in surface sediments of Prydz Bay. O'Brien & Leitchenkov (1997) recognized seven 'Provinces' on the Prydz Bay shelf (Fig. 18.8). The largest roughly corresponded to Four Ladies Bank and the outer part of the Prydz Channel. It is dominated by small-scale jagged surface roughness 2–8 m high formed by iceberg ploughing, as confirmed by sidescan records (Fig. 18.9) (O'Brien et al. 1999). Iceberg scours occur to depths of >500 m, which suggests that the deeper ones formed during low sea level (O'Brien & Leitchenkov 1997). The next largest Province

Table 18.2. *Seafloor character provinces identified by O'Brien & Leitchenkov (1997) using echo-sounder character and geomorphology*

Province	Seafloor character	Environment
1	Small-scale, jagged seafloor, some smooth patches	Iceberg-scoured shallow banks
2	Smooth or with orientated ridges and swales	Deeper basin with glacially moulded megaflutes
3	Small-scale, jagged seafloor, some smooth patches	Iceberg-scoured shallow bank
4	Steep pinnacles tens of metres high.	Basement outcrop
5	Smooth or with orientated ridges and swales, areas of sub-bottom reflectors	Deep basin with glacially moulded megaflutes and thick Holocene SMO accumulation
6	Steep pinnacles tens of metres high, deep valleys	Basement outcrops and valleys of outlet glaciers
7	Large ridge with small-scale jagged seafloor on crest	Outcrops of Prydz Bay Basin–Lambert Graben sediments
8	Smooth continental slope	Prydz Channel Trough mouth fan
9	Canyons and interfluves on slope	Continental slope

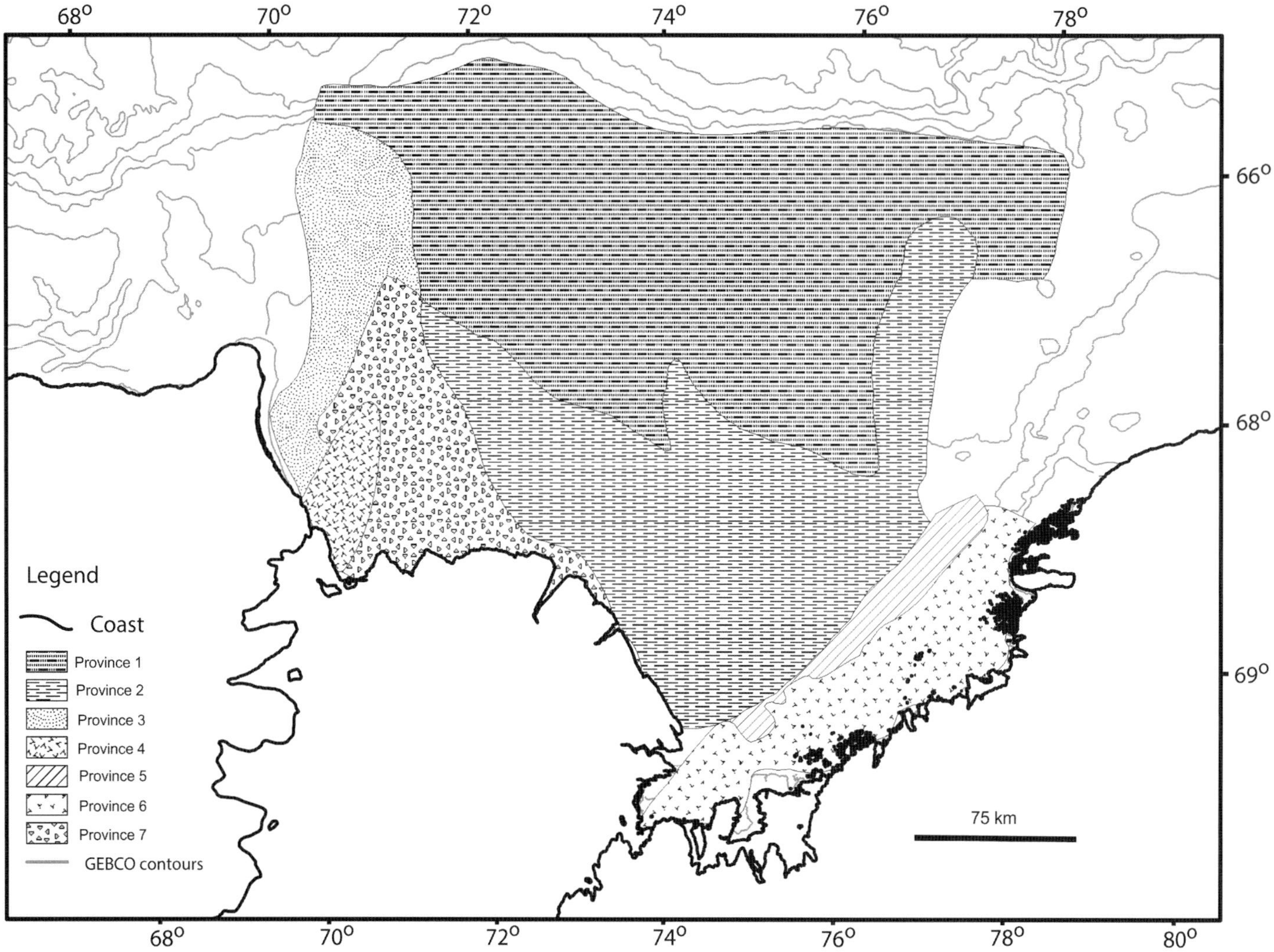

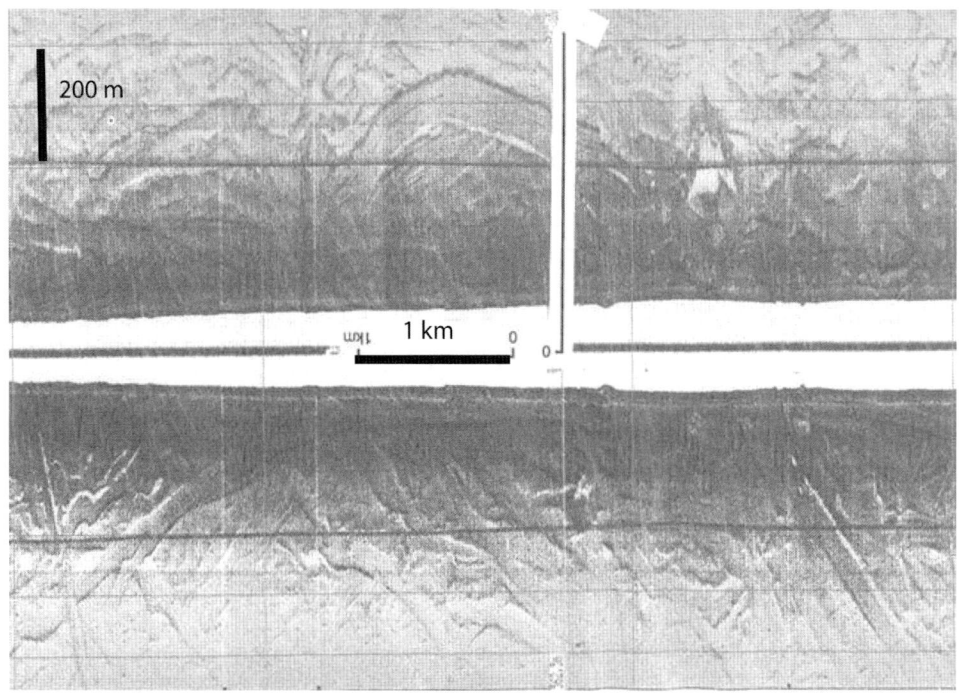

Fig. 18.8. Seabed character provinces from echo-sounder character (O'Brien & Leitchenkov 1997). Province descriptions are presented in Table 18.2.

Fig. 18.9. Iceberg scours on sidescan sonar record, Prydz Bay.

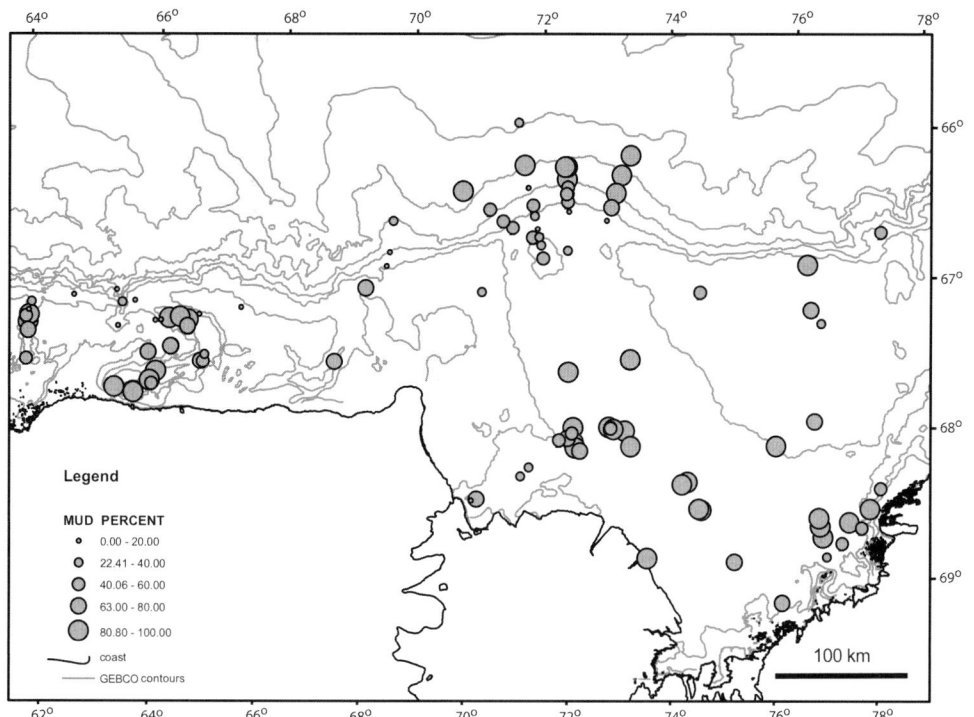

Fig. 18.10. Mud content of surface sediment samples from Prydz Bay and the Mac.Robertson Land Shelf.

corresponds to the Amery Depression and the Prydz Channel deeper than 690 m. This province features ridges and swales up to 10 m high and 2 km across. Sidescan sonar indicates that these features are elongate subglacial bedforms (Fig. 18.6) (O'Brien *et al.* 1999). Their size is similar to megaflutes in the classification of linear subglacial bedforms proposed by Clark (1993).

Harris *et al.* (1998) presented data from surface sediments across the region as contour maps. Here we present the same data as graduated symbols plotted on bathymetric contours (Figs 18.10–18.12). Most samples contain a significant mud fraction (Fig. 18.10) derived from fine biogenic material and fine siliciclastic components derived from reworking of till by iceberg ploughing and currents. The highest mud contents are found in the deep

parts of the shelf, such as the Nielsen Basin, Iceberg Alley and the Amery Depression. Some lower values are found in these deep areas where lower modern sedimentation rates have left glacial sediments near the surface, providing higher sand and gravel components. The lowest mud contents are found in samples on banks where current and iceberg activity have winnowed the surface significantly (Fig. 18.11). Samples in the SE edge of Prydz Bay are also sandy, possibly because of input from wind-blown sand from the Vestfold Hills (Franklin 1996). Only a few samples are gravelly (Fig. 18.12). They are on the outer shelf and have a component of coarse bioclastic material. Some slightly gravelly samples are probably reworked glacial facies on the inner shelf, although ice-rafted pebbles are widely distributed throughout the area.

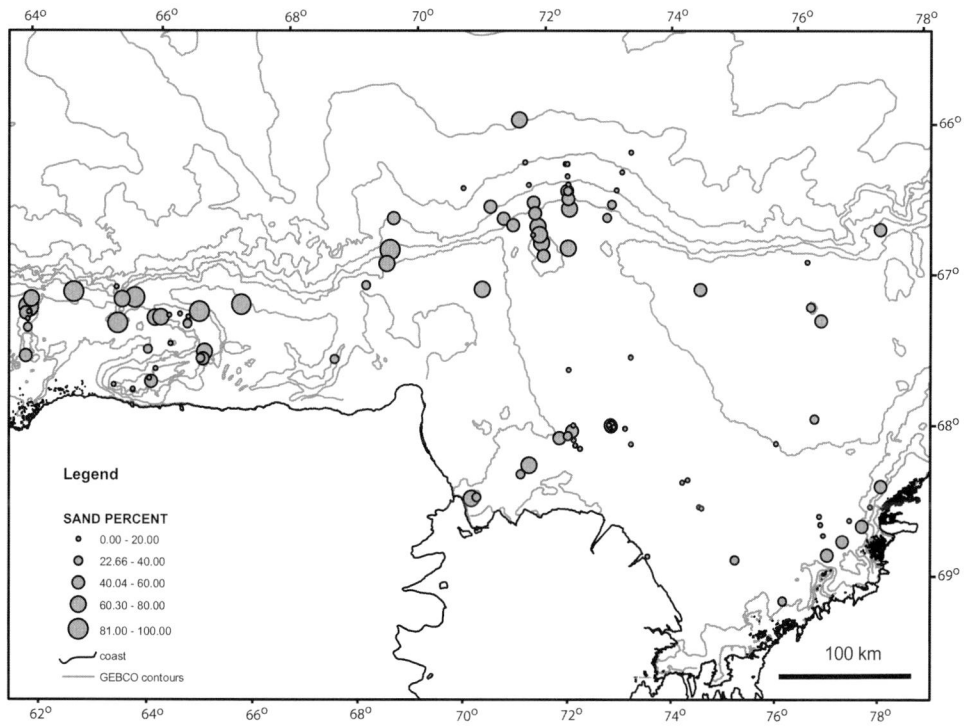

Fig. 18.11. Sand content of surface sediment from Prydz Bay and the Mac.Robertson Land Shelf.

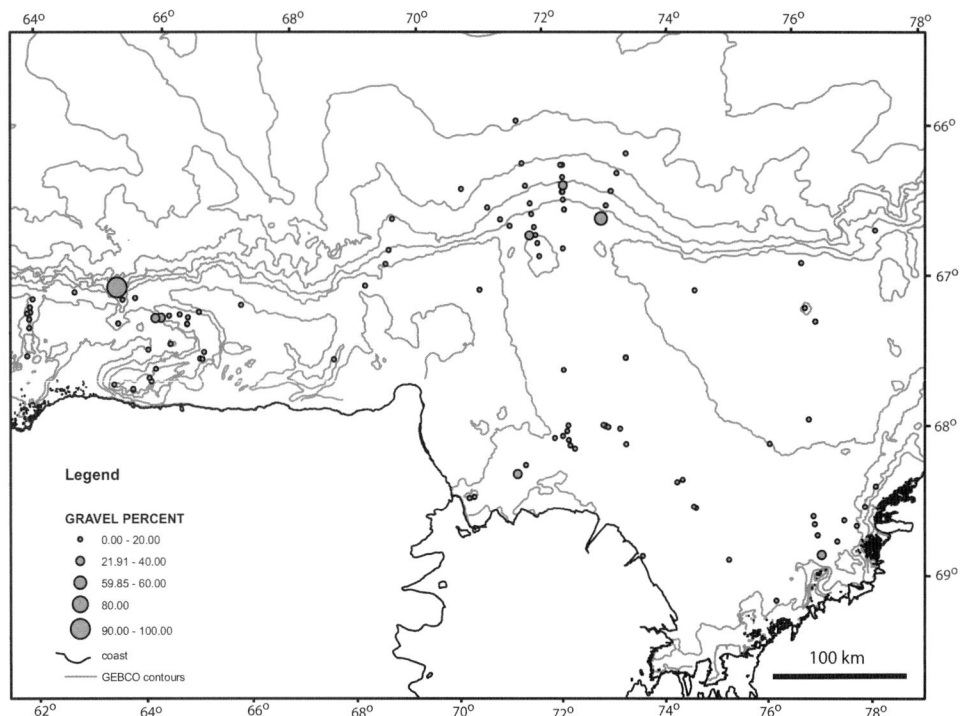

Fig. 18.12. Gravel content of surface sediment from Prydz Bay and the Mac.Robertson Land Shelf.

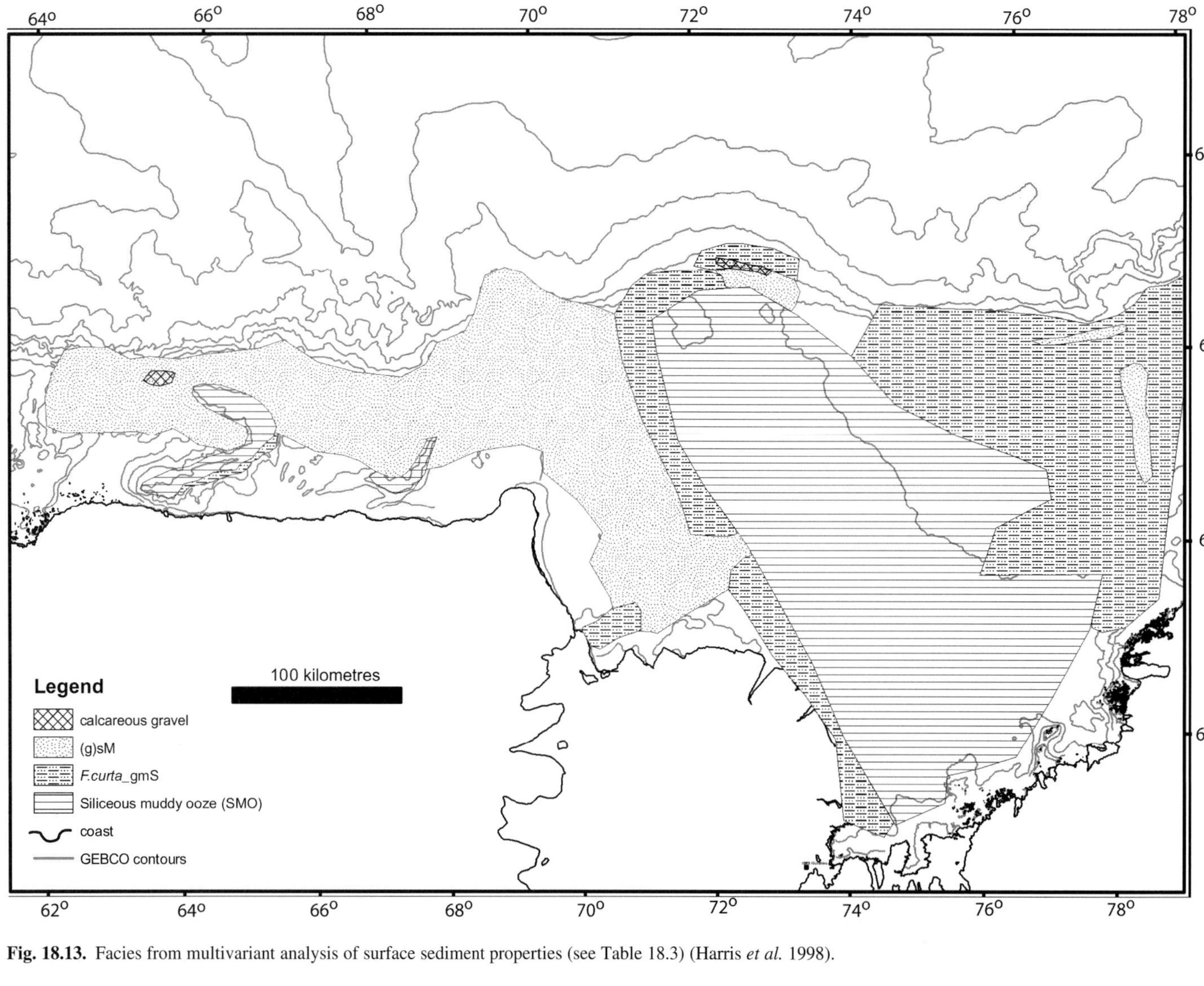

Fig. 18.13. Facies from multivariant analysis of surface sediment properties (see Table 18.3) (Harris *et al.* 1998).

Table 18.3. *Surficial sediment lithofacies for Prydz Bay and the Mac.Robertson shelf based on multivariate analysis of sediment properties and diatom assemblages (Harris* et al. *1998)*

Lithofacies	Description	Interpretation
(g)sM	Slightly gravelly sandy mud	Ice-keel turbates
SMO	Siliceous mud and ooze	Biosiliceous pelagic ooze
F. kergulensis pelagic ooze	*Fragilariopsis kerguelensis*-rich diatom ooze	Pelagic ooze with high open-water diatom species
F. curta mgS	*Fragilariopsis curta*-rich muddy gravelly sand	Muddy, gravelly sand with sea-ice diatoms
Calcareous gravel	Calcareous gravel, mostly bioclasts	Ice-keel turbates and current winnowed facies

Harris *et al.* (1998) used multivariant analysis of grain size, geochemical parameters and diatom assemblages to recognize five facies (Fig. 18.13, Table 18.3). Most of the Bay is covered with sediment comprising more than 20% mud, with slightly gravelly sandy mud ((g)sM) and SMO the most widespread facies. The acoustic character map of O'Brien & Leitchenkov (1997) suggests that the (g)sM, gravelly muddy sand (gmS) and calcareous gravel facies are products of iceberg reworking of the seafloor, mixing modern biogenic sediment and siliciclastic material excavated from older material underlying the modern veneer. Diatom ooze represents a blanket of biogenic ooze draping the seafloor below the depth of iceberg scouring (>500 m). Thick lenses of ooze occupy the NW flank of the Svenner Channel, possibly formed by current redistribution of ooze (O'Brien & Leitchenkov 1997) and similar to the shelf drifts described by Harris *et al.* (2001). The calcareous gravel facies covers small areas near the shelf edge and contains abundant fragments of benthic organisms such as bryozoans and molluscs. Current winnowing probably contributes to the coarse texture of this facies as it has only been recognized on shallow, outer banks and the upper slope.

Modern sedimentation in Prydz Bay is dominated by production of biosiliceous phytodetritus, which accumulates as biosiliceous mud and ooze in the deeper parts of the bay and on shallow banks, although the layer of ooze is highly variable on the banks because it drapes over iceberg scours and is mixed with underlying siliciclastic material by iceberg ploughing, forming the facies described by Harris *et al.* (1998). Modern biosiliceous accumulation is thinnest on the western side of the Amery Ice Shelf, possibly because water flowing from under the ice shelf favours lower productivity. Ice-rafted detritus is carried by icebergs derived from outside the bay or from its eastern coast which float clockwise through the bay.

Although the Antarctic lacks estuaries and coastal waterways common on other shelves, the eastern edge of Prydz Bay features one of the largest areas of shallow, rocky coastline in East Antarctica, the Vestfold Hills (Fig. 18.3). The Vestfold Hills are 400 km^2 of rock and sediment, free of glacial ice. Their coast is highly complex, with fjords extending tens of kilometres inland and numerous islands (Fig. 18.14). A few sandy beaches are present within bays but most of the coast is steep rock outcrop. Water depths reach 130 m in one fjord but most areas around the

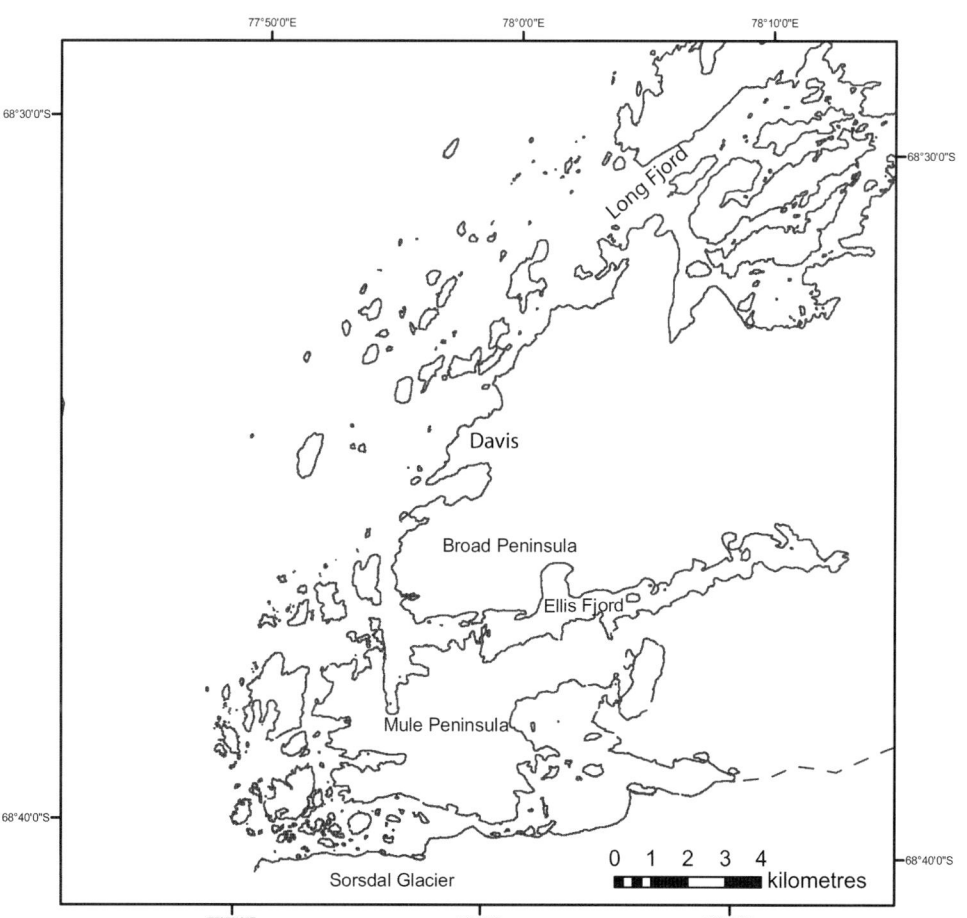

Fig. 18.14. Vestfold Hills coastline showing major fjords.

islands are between 5 and 45 m. Sediments in this coastal zone are mostly sandy mud and muddy sand with common boulders (Franklin 1996). The Vestfold Hills provide a significant amount of wind-blown sand that blows on to sea ice from the ice-free landscape. Ice-keel scours are very common, although they are smaller in width and depth than on the open shelf. Some bays protected by sills, islands and peninsulas lack iceberg scours and feature rippled sand.

The Vestfold Hills coastal zone features areas of rocky seafloor within the photic zone that are colonized by a thick growth of algae, mostly species of the kelp-like *Himanthothallus* and *Iridae*, and attached benthic organisms such as fan worms and sponges. Between the rocky regions, sediment regions are occupied by infauna such as polyceates and sea pens, with mobile forms such as holothurians and regular echinoids common.

The Mac.Robertson shelf surface is not uniformly covered with sediment. Sidescan sonographs show common rocky basement outcrops on the inner shelf (Harris & O'Brien 1998; Harris *et al.* 1998). The deep basins have beds of SMO (Fig. 18.13) (Harris & O'Brien 1996, 1998; Leventer *et al.* 2006) that typically comprise 10–60% angular, fine sand to coarse silt-sized quartz grains and 20–50% biogenic silica derived mostly from diatom frustules. Average total organic carbon content is 1.4%. The samples with highest biogenic silica are found in the deepest inner-shelf basins, with siliciclastic content increasing towards the banks. Ice-rafted detritus is present as sand and pebbles.

The bank tops are covered with poorly sorted sands and gravels, with 5–35% bioclasts comprising benthic and planktonic foraminifers, bivalves, bryozoans, gastropods, echinoids, holothurians, and ice-rafted sand and pebbles to cobbles. Sidescan sonographs show abundant iceberg scours (Harris & O'Brien 1996) and dunes formed by current movement of bedload across the banks.

Modern sedimentation is mostly confined to the valleys where phytodetritus in the form of diatom ooze accumulates with addition of detritus reworked from adjacent banks by icebergs and currents. Net erosion seems more likely than deposition for the Holocene of the Mac.Robertson Land Shelf. Most of the banks have only a thin veneer of ice-keel turbates or mobile sediment composed of coarse detritus and biogenic material being transported by currents (Harris & O'Brien 1996). Biogenic detritus comprises bryozoan fragments, sponge spicules and mollusc fragments.

Response to sea-level change

Deciphering the response of any part of the Antarctic margin to sea-level change requires the unravelling of a complex interplay between external eustatic sea-level forcing, ice-margin retreat, isostatic unloading by ice thinning and isostatic loading of the shelf by rising sea level. In the Mac.Robertson shelf–Prydz Bay region, numerous authors have approached the problem via studies of marine cores from the shelf (Domack *et al.* 1991, 1998; O'Brien & Leitchenkov 1997; Harris & O'Brien 1998; Stickley *et al.* 2005; Leventer *et al.* 2006; Post *et al.* 2007), studies of lake levels and raised marine deposits in coastal areas (e.g. Adamson & Pickard 1986; Zwartz *et al.* 1998), and exposure dating of nunataks (Fink *et al.* 2006; Mackintosh *et al.* 2007; Lilly *et al.* 2010).

For most of the Mac.Robertson shelf and Prydz Bay, rising sea levels saw the retreat of the ice-sheet-grounding zone landwards (Domack *et al.* 1998; Leventer *et al.* 2006; Mackintosh *et al.* 2011). The Mac.Robertson shelf saw retreat from close to the shelf edge, initially at about 14 ka BP in calving bays corresponding to shelf deeps (Leventer *et al.* 2006; Mackintosh *et al.* 2011). Mackintosh *et al.* (2011) combined the marine record with exposure dating of land surfaces in the hinterland and ice-sheet modelling to demonstrate that retreat took place between 14 and 12 ka BP in response to global sea-level rise then from 12 to 7 ka BP in response to warmer ocean waters.

Prydz Bay saw the Amery Ice Shelf grounding-zone retreat, of the order of 450 km, from its mid-shelf position in the Prydz Channel (Domack *et al.* 1998; O'Brien *et al.* 1999) to its present position in the Southern Prince Charles Mountains (Fricker *et al.* 2002). This retreat took place beginning at around 12 ka BP (Domack *et al.* 1998). It is not known when the grounding zone reached its present position but phytodetritus was advecting some 100 km inshore of the present ice-shelf edge by about 9 ka BP (Post *et al.* 2007), indicating a large sub-ice-shelf cavity with well-developed circulation by then. The rock coastal areas around Prydz Bay present a more complex relative sea-level history. Zwart *et al.* (1998) used Vestfold Hills lake cores, which contain a transition from marine to freshwater conditions, to demonstrate a rise in relative sea level to about 8 m above present by 6.2 ka BP followed by a sea-level fall. Verleyen *et al.* (2005) found a similar relative sea-level history in the Larsemann Hills, although the isolation of lakes with similar sill heights in the Larsemann Hills slightly post-dates those in the Vestfold Hills. This pattern reflects initial rising eustatic sea level followed by isostatic rebound following retreat of the ice sheet. They interpreted this pattern as indicating a retreat of about 30–40 km of the ice edge. Evidence from the Vestfold Hills (Colhoun *et al.* 2010) and the nearby Larsemann Hills (Hodgson *et al.* 2001; Kiernan *et al.* 2009) indicates that these areas may have been largely ice free during the Last Glacial Maximum so that this isostatic response is an average response to a complex pattern of advance around the bay.

The authors would like to thank all who have assisted with studies of the area over the years. They would also like to thank Dr R. Larter and the editors for their reviews. P. T. Harris and A. L. Post publish with the permission of the Chief Executive Officer, Geoscience Australia.

References

ADAMSON, D. A. & PICKARD, J. 1986. Cainozoic history of the Vesfold Hills. *In*: PICKARD, J. (ed.) *Antarctic Oasis*. Academic Press, Sydney, 63–97.

ANDERSON, J. B. 1999. *Antarctic Marine Geology*. Cambridge University Press, Cambridge.

BARNES, P. W. 1987. Morphological studies of the Wilkes Land continental shelf, Antarctica – glacial and iceberg effects. *In*: EITTREIM, S. L. & HAMPTON, M. A. (eds) *The Antarctic Continental Margin: Geology and Geophysics of Offshore Wilkes Land*. Circum-Pacific Council for Energy and Mineral Resources Earth Science Series, **5A**, 175–194.

BARRON, J., LARSEN, B. *ET AL.* (eds) 1991. *Proceedings of the Ocean Drilling Program Scientific Results*, **119**. Ocean Drilling Program, College Station, TX.

BINDOFF, N. L., ROSENBERG, M. A. & WARNER, M. J. 2000. On the circulation of water masses over the Antarctic continental slope and rise between 80 and 150°E. *Deep Sea Research Part II: Topical Studies in Oceanography*, **47**, 2299–2326.

BOULTON, G. S. 1990. Sedimentary and sea level changes during glacial cycles and their control on glacimarine facies architecture. *In*: DOWDESWELL, J. A. & SCOURSE, J. D. (eds) *Glacimarine Environments: Processes and Sediments*. Geological Society, London, Special Publications, **53**, 15–52.

BROMICH, D. H. 1988. Snowfall in high southern latitudes. *Reviews of Geophysics*, **26**, 149–168.

BUREAU OF METEOROLOGY. 2010. *Climate Statistics for Australian Sites, Antarctica*. http://www.bom.gov.au/climate/averages/tables/ca_ant_names.shtml

CLARK, C. D. 1993. Mega-scale glacial lineations and cross-cutting ice-flow landforms. *Earth Surface Processes and Landforms*, **18**, 1–29.

COLHOUN, E. A., KIERNAN, K. W., MCCONNELL, A., QUILTY, P. G., FINK, D., MURRAY-WALLACE, C. V. & WHITEHEAD, J. 2010. Late Pliocene age of glacial deposits at Heidemann Valley, East Antarctica:

evidence for the last major glaciation in the Vestfold Hills. *Antarctic Science*, **22**, 53–64.

COOPER, A. K. & O'BRIEN, P. E. 2004. Leg 188 synthesis: transitions in the glacial history of the Prydz Bay region, East Antarctica, from ODP drilling. *In*: COOPER, A. K., O'BRIEN, P. E. & RICHTER, C. (eds) *Proceedings of the Ocean Drilling Program, Scientific Results*, **188**. Ocean Drilling Program, College Station, TX, 1–42.

DAMM, V. 2007. A subglacial topographic model of the southern drainage area of the Lambert Glacier/Amery Ice Shelf system -results from an airborne ice thickness survey south of the Prince Charles Mountains. *Terre Antarctica*, **14**, 85–96.

DOMACK, E. W. & HARRIS, P. T. 1998. A new depositional model for ice shelves, based upon sediment cores from the Ross Sea and the Mac. Robertson shelf, Antarctica. *Annals of Glaciology*, **27**, 281–284.

DOMACK, E. W., JULL, A. J. T. & NAKAO, S. 1991. Advance of East Antarctic outlet glaciers during the Hypsithermal: implications for the volume state of the Antarctic ice sheet under global warming. *Geology*, **19**, 1059–1062.

DOMACK, E., O'BRIEN, P. E., HARRIS, P. T., TAYLOR, F., QUILTY, P. G., DE SANTIS, L. & RAKER, B. 1998. Late Quaternary sediment facies in Prydz Bay, East Antarctica and their relationship to glacial advance onto the continental shelf. *Antarctic Science*, **10**, 236–246.

DOWDESWELL, J. A. & BAMBER, J. L. 2007. Keel depths of modern Antarctic icebergs and implications for sea-floor scouring in the geological record. *Marine Geology*, **243**, 120–131.

DUNBAR, R. B., LEVENTER, A. R. & STOCKTON, W. L. 1989. Biogenic sedimentation in McMurdo Sound, Antarctica. *Marine Geology*, **85**, 155–179.

EVANS, J. & PUDEY, C. J. , 2002. Sedimentation associated with Antarctic Peninsula ice shelves: implications for palaeoenvironmental reconstructions of glacimarine sediments. *Journal of the Geological Society, London*, **159**, 233–237.

FINK, D., MCKELVEY, B., HAMBREY, M. J., FABEL, D. & BROWN, R. 2006. Pleistocene chronology of the Amery Oasis and Radok Lake, northern Prince Charles Mountains, Antarctica. *Earth and Planetary Science Letters*, **243**, 229–243.

FORSBERG, C. F., FLORINDO, F., GRUTZNER, J., VENUTI, A. & SOLHEIM, A. 2008. Sedimentation and aspects of glacial dynamics from physical properties, mineralogy and magnetic properties at ODP Site 1166 and 1167, Prydz Bay, Antarctica. *Palaeogeography, Palaeoclimatology, Palaeoecology*, **260**, 184–201.

FRANKLIN, D. 1996. *The sedimentology of Holocene Prydz Bay: sedimentary patterns and processes and their implications for climate reconstruction*. PhD thesis, Institute of Antarctic and Southern Ocean Studies, University of Tasmania.

FRICKER, H. A., POPOV, S., ALLISON, I. & YOUNG, N. 2001. Distribution of marine ice beneath the Amery Ice Shelf. *Geophysical Research Letters*, **28**, 2241–2244.

FRICKER, H. A., ALLISON, I. *ET AL.* 2002. Redefinition of the Amery Ice Shelf, East Antarctica, grounding zone. *Journal of Geophysical Research Solid Earth*, **107**, CEV1.1–CEV1.9, http://dx.doi.org/10.1029/2001JB000383

FRICKER, H. A., SCAMBOS, T., BINDSCHADLER, R. & PADMAN, L. 2007. An active subglacial water system in West Antarctica mapped from space. *Science*, **315**, 1544–1548.

GOODWIN, I. D. 1987. The nature and origin of a jökulhlaup near Casey Station, Antarctica. *Journal of Glaciology*, **34**, 95–101.

HAMBREY, M. J. 1994. *Glacial Environments*. UCL Press, London.

HAMBREY, M. J., EHRMANN, W. U. & LARSEN, B. 1991. Cenozoic glacial record of the Prydz Bay continental shelf, East Antarctica. *In*: BARRON, J. & LARSON, B. (eds) *Proceedings of the Ocean Drilling Program, Scientific Results*, **119**. Ocean Drilling Program, College Station, TX, 77–132.

HARRIS, P. T. & O'BRIEN, P. E. 1996. Geomorphology and sedimentology of the continental shelf adjacent to Mac.Robertson Land, East Antarctica: a scalped shelf. *Geomarine Letters*, **16**, 287–296.

HARRIS, P. T. & O'BRIEN, P. E. 1998. Bottom currents, sedimentation and ice-sheet retreat facies successions on the Mac.Robertson Shelf, East Antarctica. *Marine Geology*, **151**, 47–72.

HARRIS, P. T., TAYLOR, F., PUSHINA, Z., LEITCHENKOV, G., O'BRIEN, P. E. & SMIROV, V. 1998. Lithofacies distribution in relation to

geomorphic provinces of Prydz Bay, East Antarctica. *Antarctic Science*, **10**, 227–235.

HARRIS, P. T., BRANCOLINI, G. *ET AL.* 2001. Continental shelf drift deposit indicates non-steady state Antarctic bottom water production in the Holocene. *Marine Geology*, **179**, 1–8.

HEMER, M. A. & HARRIS, P. T. 2003. Sediment core from beneath the Amery Ice Shelf, East Antarctica, suggests mid-Holocene ice shelf retreat. *Geology*, **31**, 127–130.

HEMER, M. A., POST, A. L., O'BRIEN, P. E., CRAVEN, M., TRUSWELL, E. M., ROBERTS, D. & HARRIS, P. T. 2007. Sedimentological signatures of the sub-Amery Ice Shelf circulation. *Antarctic Science*, **19**, 497–506.

HODGKINSON, R. P., COLMAN, R. S., KERRY, K. R. & ROBB, M. 1988. Water currents in Prydz Bay during 1985. *ANARE Research Notes*, **81**, 1–127.

HODGKINSON, R. P., COLMAN, R. S., ROBB, M. & WILLIAMS, R. 1991*a*. Current meter moorings in the region of Prydz Bay, Antarctica, 1986. *ANARE Research Notes*, **81**, 1–130.

HODGKINSON, R. P., COLMAN, R. S., ROBB, M. & WILLIAMS, R. 1991*b*. Current meter moorings in the region of Prydz Bay, Antarctica, 1987. *ANARE Research Notes*, **82**, 1–68.

HODGSON, D. A., NOON, P. E. *ET AL.* 2001. Were the Larsemann Hills ice-free through the Last Glacial Maximum? *Antarctic Science*, **13**, 440–454.

HOLLAND, P. R., CORR, H. F. J., VAUGHAN, D. G., JENKINS, A. & SKVARCA, P. 2009. Marine ice in Larsen Ice Shelf. *Geophysical Research Letters*, **36**, L11604, http://dx.doi.org/10.1029/2009GL038162

KAPITSA, A. P., RIDLEY, J. K., ROBIN, G., DE, Q., SIEGERT, M. J. & ZOTIKOV, I. A. 1996. A large deep freshwater lake beneath the ice of central East Antarctica. *Nature*, **381**, 684–686.

KIERNAN, K., GORE, D. B., FINK, D., WHITE, D. A., MCCONNELL, A. & SIGURDSSON, I. A. 2009. Deglaciation and weathering of Larsemann Hills, East Antarctica. *Antarctic Science*, **21**, 373–382.

LEFFANUE, H. & CRAVEN, M. 2004. Circulation and water masses from current meters and T/S measurements at the Amery Ice Shelf. *FRISP Report*, **15**, 73–79.

LEVENTER, A. E., DOMACK, E. *ET AL.* 2006. Marine sediment record from the East Antarctic margin reveals dynamics of ice sheet recession. *GSA Today*, **16**, 4–10.

LILLY, K., FINK, D., FABEL, D. & LAMBECK, K. 2010. Pleistocene dynamics of the interior East Antarctic ice sheet. *Geology*, **38**, 703–706.

LISKER, F., WILSON, C. J. L. & GIBSON, H. J. 2007. Thermal history of the Vestfold Hills (East Antarctica) between Lambert Rifting and Gondwana break-up, evidence from apatite fission track data. *Antarctic Science*, **19**, 97–106.

MACKINTOSH, A., WHITE, D., FINK, D., GORE, D. B., PICKARD, J. & FANNING, P. C. 2007. Exposure ages from mountain dipsticks in Mac.Robertson Land, East Antarctica, indicate little change in ice sheet thickness since the Last Glacial Maximum. *Geology*, **35**, 551–554.

MACKINTOSH, A., GOLLEDGE, N. *ET AL.* 2011. Retreat of the Eastern Antarctic ice sheet during the last glacial termination. *Nature Geoscience*, **4**, 195–202.

MCLEAN, M. A., WILSON, C. J. L., BOGER, S. D., BETTS, P. G., RAWLINGS, T. L. & DAMASKE, D. 2009. Basement interpretation from airborne magnetic and gravity data over the Lambert Rift region of east Antarctica. *Journal of Geophysical Research*, **114**, B06101, http://dx.doi.org/10.1029/2008JB005650

MCLOUGHLIN, S. & DRINNAN, A. N. 1997. Revised stratigraphy of the Permian Bainmedart Coal Measures, northern Prince Charles Mountains, East Antarctica. *Geological Magazine*, **134**, 335–353.

MIKHALSKY, E. V., SHERATON, J. W., LAIBA, A. A., TINGEY, R. J., THOST, D. E., KAMENEV, E. N. & FEDOROV, L. V. 2001. Geology of the Prince Charles Mountains, Antarctica. *AGSO – Geoscience Australia Bulletin*, **247**, 209.

NUNES VAZ, R. A. & LENNON, G. W. 1994. Physical oceanography of the Prydz Bay region of Antarctic waters. *Deep Sea Research Part I: Oceanographic Research Papers*, **43**, 603–641.

O'BRIEN, P. E. & LEITCHENKOV, G. 1997. Deglaciation of Prydz Bay, East Antarctica, based on echo sounding and topographic features. *In*: BARKER, P. F. & COOPER, A. K. (eds) *Geological and Seismic Stratigraphy of the Antarctic Margin, 2*. American Geophysical Union, Antarctic Research Series, **71**, 109–125.

O'BRIEN, P. E., TRUSWELL, E. M. & BURTON, T. 1994. Morphology, seismic stratigraphy and sedimentation history of the Mac.Robertson Shelf. *In*: COOPER, A. K., BARKER, P. F., WEBB, P-N. & BRANCOLINI, G. (eds) *The Antarctic Continental Margin: Geophysical and Geological Stratigraphic Records of Cenozoic Glaciation, Paleoenvironments and Sea-Level Change. Terra Antarctica*, **1**, 407–408.

O'BRIEN, P. E., DE SANTIS, L., HARRIS, P. T., DOMACK, E. & QUILTY, P. G. 1999. Ice shelf grounding zone features of western Prydz Bay, Antarctica: sedimentary processes from seismic and sidescan images. *Antarctic Science*, **11**, 78–91.

O'BRIEN, P. E., COOPER, A. K. ET AL. 2004. Prydz Channel Fan and the history of extreme ice advances in Prydz Bay. *In*: COOPER, A. K., O'BRIEN, P. E. & RICHTER, C. (eds) *Proceedings of the Ocean Drilling Program, Scientific Results*, **188**. Ocean Drilling Program, College Station, TX, 1–32 [CD-ROM].

O'BRIEN, P. E., GOODWIN, I., FORSBERG, C. F., COOPER, A. K. & WHITEHEAD, J. 2007. Late Neogene ice drainage changes in Prydz Bay, East Antarctica and the interaction of the Antarctic ice sheet evolution and climate. *Palaeogeography, Palaeoclimatology, Palaeoecology*, **245**, 390–410.

PASSCHIER, S., O'BRIEN, P. E., DAMUTH, J. E., JANUSZCZAK, N., HANDWERGER, D. A. & WHITEHEAD, J. M. 2003. Pliocene-Pleistocene glaciomarine sedimentation in eastern Prydz Bay and development of the Prydz trough-mouth fan, ODP Sites 1166 and 1167, East Antarctica. *Marine Geology*, **199**, 279–305.

POST, A. L., HEMER, M. A., O'BRIEN, P. E., ROBERTS, D. & CRAVEN, M. 2007. History of benthic colonisation beneath the Amery Ice Shelf, East Antarctica. *Marine Ecology Progress Series*, **344**, 29–37.

POWELL, R. D. & ALLEY, R. B. 1997. Grounding-line systems: processes, glaciological inferences and stratigraphic record. *In*: BARKER, P. F. & COOPER, A. K. (eds) *Geological and Seismic Stratigraphy of the Antarctic Margin, 2*. American Geophysical Union, Antarctic Research Series, **71**, 169–187.

QUILTY, P. G. 1985. Distribution of foraminiferids in sediments of Prydz Bay. *In*: LINDSAY, J. M. (ed.) *Stratigraphy, palaeontology, malacology: papers in honour of Dr Nell Ludbrook*. South Australian Department of Mines and Energy, Special Publications, **5**, 329–340.

RIDDLE, M. J., CRAVEN, M., GOLDSWORTHY, P. M. & CARSEY, F. 2007. A diverse benthic assemblage 100 km from open water under the Amery Ice Shelf, Antarctica. *Paleoceanography*, **22**, PA1204, http://dx.doi.org/10.1029/2006PA001327

RINTOUL, S. R. 1998. On the origin and influence of Adélie Land Bottom Water. *In*: JACOBS, S. & WEISS, R. (eds) *Ocean, Ice and Atmosphere: Interaction at the Antarctic Continental Margin*. American Geophysical Union, Antarctic Research Series, **75**, 151–171.

SMITH, N. R., ZHAOQIAN, D. & WRIGHT, S. 1984. Water masses and circulation in the region of Prydz Bay, Antarctica. *Deep Sea Research Part A. Oceanographic Research Papers*, **31**, 1121–1147.

SOLHEIM, A., FORSBERG, C. F. & PITTENGER, A. 1991. Stepwise consolidation of glacigenic sediments related to the glacial history of Prydz Bay. *In*: BARRON, J. & LARSON, B. (eds) *Proceedings of the Ocean Drilling Program, Scientific Results*, **119**. Ocean Drilling Program, College Station, TX, 169–184.

STAGG, H. M. J. 1985. The structure and origin of Prydz Bay and Mac.Robertson Shelf, East Antarctica. *Tectonophysics*, **114**, 315–340.

STAGG, H. M. J., COLWELL, J. B. ET AL. 2004. Geology of the continental margin of Enderby and Mac.Robertson Lands, East Antarctica: insights from a regional data set. *Marine Geophysical Researches*, **25**, 183–219.

STAGG, H. M. J., COLWELL, J. B. ET AL. 2005. Geological framework of the continental margin in the region of the Australian Antarctic Territory. *Geoscience Australia Record*, **2004/25**, 356.

STICKLEY, C. E., PIKE, J. ET AL. 2005. Deglacial ocean and climate seasonality in laminated diatom sediments, Mac.Robertson Shelf, Antarctica. *Palaeogeography, Palaeoclimatology, Palaeoecology*, **227**, 290–310.

TAMURA, T., WILLIAMS, G. D., FRASER, A. D. & OHSHIMA, K. I. 2012. Potential regime shift in decreased sea ice production after Mertz Glacier calving. *Nature Communications*, **3**, 826, http://dx.doi.org/10.1038/ncomms1820

TAYLOR, F. & MCMINN, A. 2002. Late Quaternary diatom assemblages from Prydz Bay, Eastern Antarctica. *Quaternary Research*, **57**, 151–161.

TAYLOR, F., MCMINN, A. & FRANKLIN, D. 1997. Distribution of diatoms in surficial sediments of Prydz Bay, East Antarctica. *Marine Micropaleontology*, **32**, 231–248.

TEN BRINK, U. S., SCHNEIDER, C. & JOHNSON, A. H. 1995. Morphology and stratal geometry of the Antarctic continental shelf: insights from models. *In*: COOPER, A. K., BARKER, P. F. & BRANCOLINI, G. (eds) *Geology and Seismic Stratigraphy of the Antarctic margin*. American Geophysical Union, Antarctic Research Series, **68**, 1–24.

TRUSWELL, E. M., DETTMANN, M. E. & O'BRIEN, P. E. 1999. Mesozoic palynofloras from the Mac.Robertson shelf, East Antarctica: geological and phytogeographic implications. *Antarctic Science*, **11**, 237–252.

VERLEYEN, E., HODGSON, D. A., MILNE, G. A., SABBE, K. & VYVERMAN, W. 2005. Relative sea-level history from the Lambert Glacier region, East Antarctica, and its relation to deglaciation and Holocene glacier readvance. *Quaternary Research*, **63**, 45–52.

VORREN, T. O. & LABERG, J. S. 1997. Trough mouth fans – palaeoclimate and ice sheet monitors. *Quaternary Science Reviews*, **16**, 865–881.

WHITEHEAD, J. M. & BOHARTY, S. M. 2003. Pliocene summer sea surface temperature reconstruction using silicoflagellates from the Southern Ocean ODP Site 1165. *Paleoceanography*, **18**, 20-1–20-11.

WHITEHEAD, J. M., QUILTY, P. G., MCKELVEY, B. C. & O'BRIEN, P. E. 2006. A review of the Cenozoic stratigraphy and glacial history of the Lambert Graben – Prydz Bay region, East Antarctica. *Antarctic Science*, **18**, 1–17.

ZWART, D., BIRD, M., STONE, J. & LAMBECK, K. 1998. Holocene sea-level change and ice sheet history in the Vestfold Hills, East Antarctica. *Earth and Planetary Science Letters*, **155**, 131–145.

Chapter 19

Carbonate shelf sediments of the western continental margin of Australia

LINDSAY B. COLLINS[1]*, NOEL P. JAMES[2] & YVONNE BONE[3]

[1]*Department of Applied Geology, Curtin University, Perth, WA 6102, Australia*

[2]*Department of Geological Sciences, Queen's University, Kingston, Ontario, Canada K7L 3N6*

[3]*The School of Earth and Environmental Sciences, University of Adelaide, Adelaide, SA 5001, Australia*

*Corresponding author (e-mail: L.Collins@curtin.edu.au)

Abstract: Australia's western margin is adjacent to a low–moderate-relief, semi-arid hinterland extending from northern tropical to southern temperate latitudes. Swell waves occur throughout, and cyclonic storms and tidal influences decline from north to south. The margin is influenced by the poleward-flowing, warm, nutrient-poor Leeuwin Current. There is limited upwelling and localized downwelling of saline water on to the shelf. The North West Shelf (NWS) is an ocean-facing ramp with palimpsest sediments – formed during Marine Isoptope Stage (MIS) 3 and 4; stranded ooids and peloids formed early during the post-Last Glacial Maximum (LGM) sea-level rise – and Holocene particles. Changing oceanography during sea-level rise profoundly affected sediment character.

The SW Shelf (SWS) comprises the subtropical sediment-starved Carnarvon Ramp in the north and the incipiently rimmed, flat-topped, steep-fronted Rottnest Shelf in the south. The inner Carnarvon Ramp includes the Ningaloo Reef and hypersaline Shark Bay. The mid ramp is relict or stranded foraminifer-dominated sand, and represents attenuated carbonate production due to downwelling incursions of Shark Bay water on to the ramp; the outer ramp is planktic foraminiferal sand or spiculitic mud. Rottnest Shelf has coralline algal-encrusted hardgrounds, larger symbiont-bearing foraminifers with abundant cool-water elements including bryzoans, molluscs and smaller foraminifers. The SWS is transitional between warm- and cool-water carbonate realms.

The Australian continent is essentially surrounded by carbonate sedimentation in shelf and slope environments, with cool-water carbonate sediments south of latitude 30°S, and more northerly settings characterized by warm-water, tropical sedimentation. The Great Barrier Reef in NE Australia is known as a globally significant carbonate province, and the southern shelf is a vast area of cool-water carbonate sedimentation (James *et al.* 2001). The western continental margin extends southwards from 16°S to 35°S, and passes from northerly tropical conditions to more cool-water settings in the south (Collins 1988; James *et al.* 1999, 2004). Whilst coral reef systems of the western shelf are relatively poorly known, they are well developed in a variety of settings, from the macrotidal Kimberley Reefs in the north, to the fringing Ningaloo Reef, to the Houtman Abrolhos shelf margin reefs, the southernmost reefs in the Indian Ocean (Fig. 19.1) (Collins 2010; Collins & Testa 2010; Twiggs & Collins 2010). The north–south latitudinal and climatic gradient along the western continental margin is a transitional setting in which to study carbonate sedimentation. A poleward-flowing warm current, the Leeuwin Current, is responsible for the transport of tropical water southwards in winter, the suppression of upwelling, and the maintenance of a broad biotic transition zone between the tropical northern and warm temperate southern biotic zones (Cresswell & Golding 1980; Cresswell 1991). The purpose of this paper is to summarize the sedimentological research completed for various segments of the western shelf, and to provide a regional account of the processes, biota and carbonate facies as part of the late Quaternary history of the region.

General setting

The western continental margin of Australia (Fig. 19.1, Table 19.1) is divisible at latitude 21°S into two components: a NW margin from 16 to 21°S latitude, known as the North West Shelf (NWS); and the South West Shelf (SWS) from 21 to 35°S. The Ningaloo Reef lies close to the boundary of the change in coastal orientation expressed by these two distinct shelf compartments, which are a reflection of the regional geological structure. Most of the NWS is a bathymetrically gentle ramp, up to 250 km across to the 200 m isobath, which then slopes gradually to 1000 m water depth (mwd). A series of isolated, reef-rimmed platforms, the Rowley Shoals, rise to the surface from about 400 mwd on the continental slope of this 600 km long ramp (*sensu* Read 1985).

The SWS has two distinct zones: the Carnarvon Ramp, from 22 to 28°S, up to 100 km wide; and the Rottnest Shelf, about 750 km long, south of 28°S, with a shelf edge close to 200 mwd. The Houtman Abrolhos Reefs (latitude 28–29.5°S) straddle the boundary between the Carnarvon Ramp and the Rottnest Shelf. The Rottnest Shelf has a steep shoreface (0–30 mwd), a wide inner shelf plain (30–50 mwd), a linear ridge complex, which shallows to 40 mwd, and an outer shelf that slopes seawards to the shelf edge at around 200 mwd (Collins 1988; James *et al.* 1999).

Climate and oceanography

As with all southern hemisphere ocean basins, near-surface water circulation in the Indian Ocean is in the form of an anticlockwise gyre, the SE arm of which is the West Australian Current (Fig. 19.1). This situation generally results in an Eastern Boundary Current, a northwards flow of cool water along the coast which, with equatorward winds, leads to coastal upwelling (e.g. the Humboldt Current off Peru and Chile, and the Benguela Current off southern Africa). Instead, off Western Australia, the southward-flowing Leeuwin Current (LC) (Cresswell & Golding 1980; Cresswell 1991) dominates near-surface circulation, with water flowing northwards beneath it as the Leeuwin Undercurrent (LUC). The LC is a narrow (<100 km), shallow (<200 m) stream of comparatively warm, low-salinity, nutrient-depleted oceanic water of tropical origin that flows southwards at relatively high velocity (0.1–0.4 m s^{-1}) along the western continental slope (Pearce 1991), creating strong shear on the outer shelf (Cresswell & Golding 1980). The current, which sheds mesoscale eddies into the Indian Ocean and on to the shelf, is coherent from North West Cape to Cape Leeuwin and into the Great Australian Bight. The LC flows because of strong trade winds in the tropics that

From: CHIOCCI, F. L. & CHIVAS, A. R. (eds) 2014. *Continental Shelves of the World: Their Evolution During the Last Glacio-Eustatic Cycle.*
Geological Society, London, Memoirs, **41**, 255–272. http://dx.doi.org/10.1144/M41.19

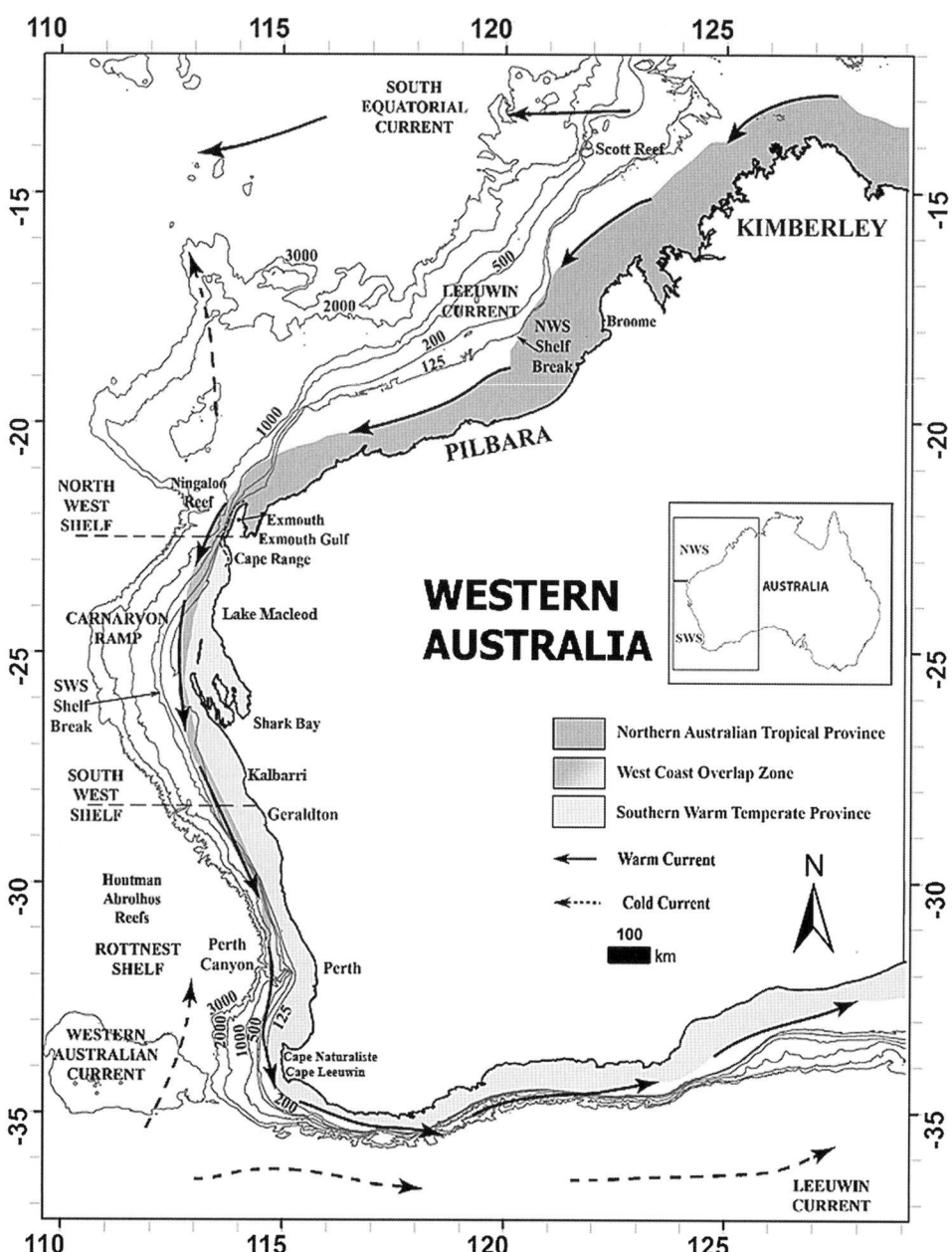

Fig. 19.1. Location map of the western continental margin of Australia showing the North West Shelf and the South West Shelf, including the Carnarvon Ramp and Rottnest Shelf. Coral reefs and biotic zones are also identified.

push the South Equatorial Current westwards through Indonesia. This strong South Equatorial Current flow results in a sea level that is 55 cm higher than that off southern Australia, thus inducing flow to the south (Pattiaratchi & Buchan 1991). This southwards pressure gradient is opposed by winds that blow equatorwards. In general, the LC impedes upwelling of cold, nutrient-rich water from below the outer shelf edge (Gersbach *et al.* 1999). The NWS is macrotidal, while the SWS is mesotidal–microtidal, with tidal range decreasing southwards. Long-period (12–20 s) swell waves influence both the NWS and SWS year round, with cyclonic storms occurring once every three years for the NWS and once every four years for the SWS (see James *et al.* 1999, 2004).

Sedimentation regime

Carbonate sediments dominate throughout the western continental margin shelf settings. The NWS (Fig. 19.1, Table 19.1) sediments are palimpsest, a variable mixture of relict, stranded and Holocene grains. Relict intraclasts, both skeletal and lithic are now localized to the mid-ramp. The most conspicuous stranded particles

are ooids and peloids, which formed in somewhat saline waters during initial stages of post-Last Glacial Maximum (LGM) sea-level rise. Holocene sediment is principally biofragmental, with sedimentation localized to the inner ramp and a ridge of planktic foraminifers offshore. Inner-ramp deposits are a mixture of heterozoan and photozoan elements. Depositional facies reflect episodic environmental perturbation by fluvially derived sediments and nutrients, resulting in a mixed habitat of oligotrophic (coral reefs and large benthic foraminifers) and mesotrophic (macroalgae and bryozoans) indicators (James *et al.* 2004).

Holocene mid-ramp sediment is heterozoan in character, but sparse. Holocene outer-ramp sediment is mainly pelagic, and phosphate accumulations at approximately 200 mwd suggest periodic upwelling or Fe-redox pumping. This ramp depositional system in an arid climate has important applications for the geological record: inner-ramp sediments can contain important heterozoan elements; mid-ramp sediments with bedforms created by internal tides can form in water depths exceeding 50 m, saline outflow can arrest or dramatically slow mid-ramp sedimentation mimicking maximum flooding intervals; and outer-ramp planktic productivity can generate locally important fine-grained carbonate sediment bodies. Coral reefs are developed as isolated platforms at

Table 19.1. *Summary of characteristics of the North West Shelf (NWS) and South West Shelf (SWS), Australia*

	NWS	SWS – RS	SWS – CR
Length of the shelf (km)	600	750	
Average width (km)	250	60	150
Tidal, wave, current ranges	Swell waves ($T = 12–20$ s) macrotidal	Swell waves ($T = 12–20$ s) mesotidal–microtidal	
Dominating process (wave/current/tide)	Swell waves, tides, episodic cyclones	Swell waves, minor tides, episodic cyclones	
Average depth of the shelf break (m)	125	200 – shelf profile, 300 – ramp profile	
Siliciclastic/carbonate/authigenic/ glacial sedimentation	Carbonate-dominated. Siliciclastic <20% inshore	Carbonate-dominated. Siliciclastic <30% inshore	
Modern/relict/palimpsest	>50% modern increasing seawards, 25–50% relict/palimpsest	>70% modern, 0–30% relict/palimpsest	
Tectonic trend over the last glacial cycle (stable/uplifting/subsiding)	Stable: increasing subsidence at the ramp margin	Stable	Stable: subsidence in north
Sea surface temperature (SST)	22–26	15–20	18–26

Abbreviations: NWS, North West Shelf; SWS, South West Shelf; RS, Rottnest Shelf; CR, Carnarvon Ramp.

the ramp margin, except for the fringing Ningaloo Reef (Collins *et al.* 2003; Collins 2010).

The SWS (Fig. 19.1, Table 19.1) is transitional between warm- and cool-water carbonate realms (James *et al.* 1999). The Carnarvon Ramp to the north, outboard of Shark Bay and the Ningaloo Reef, has a euphotic mid ramp with relatively little calcareous benthos, and low numbers of bryozoans, coralline algae and benthic foraminifers, with relict and intraclastic sediments. The outer ramp is pelagic in character, with rippled to burrowed substrates composed largely of pelagic foraminiferal sediments. To the south, the Rottnest Shelf is deep, open and incipiently rimmed, with an inner shelf plain of coralline algae encrusted substrates, rhodolith pavements, bryozoans, sponges and prolific seagrass, and abraded sediments, all within the euphotic zone of wave abrasion. The subphotic outer shelf has bioeroded limestone outcrops with intervening areas of burrowed and rippled sands in which bryozoans are dominant, augmented by benthic foraminifers and molluscs. The upper slope has fine sand and silt composed of bryozoan fragments, and is rich in sponge spicules and pelagic foraminifers. Three isolated platforms, the Houtman Abrolhos coral reefs, are developed from latitudes 28.5–29°S at the southerly limits of reef growth in the Indian Ocean (Collins *et al.* 1993; Collins 2010).

Geological background and tectonic framework

The NWS is presently Australia's premier hydrocarbon province (Purcell & Purcell 1988). The area is anchored by two Precambrian massifs, the Pilbara Block in the SW and the Kimberley Block in the NE, with the Palaeozoic Canning Basin in between. Carbonate sedimentation began in the Late Cretaceous and is in the form of three seaward-prograding sequences totalling 1000–2000 m in thickness (Exon & Colwell 1994). This sediment wedge contains little evidence of outer-shelf reef growth. The hinterland of the Canning Basin is underlain by Cretaceous sedimentary rocks, but more prominent is the Quaternary cover of the Great Sandy Desert (Fig. 19.2). The Pilbara is a moderate relief highland of Archaean and Proterozoic granites, greenstones and sedimentary strata dissected by ephemeral rivers, several of which have substantial drainage basins that discharge into the coastal environment. Palaeoproterozoic rocks comprise igneous intrusions and iron formations, with shales and subordinate carbonates (Trendall 1983).

The Pilbara coast (Fig. 19.2) is a 20–30 km-wide coastal plain of alluvial and tidal flat sediment with nearly continuous Pleistocene and Holocene dune barriers, behind which lie protected embayments, evaporative lagoons and tidal flats (Semeniuk 1993). Pleistocene deposits comprise fan-delta and riverine deposits of terrigenous clastic sediment, including laterite particles that were reworked by aeolian processes to form sand sheets and dunes that are reddened by subaerial diagenetic and pedogenic processes.

The SW continental margin of Australia (Fig. 19.3) lies west of the Archaean (*c.* 2.6 Ga) Yilgarn Craton and is separated from it by the Darling Fault, a 1000 km-long north–south-trending crustal feature, that shows up to 15 km of vertical movement since Permian time (Johnstone *et al.* 1973). The Carnarvon Basin and the Perth Basin lie west of the Darling Fault. The Perth Basin is an intracratonic graben or half-graben filled with post-Ordovician sediments of largely paralic and continental origins but the post-Late Cretaceous sequence consists of prograding carbonates similar to those of the Carnarvon Basin.

The Carnarvon Basin is a continental-margin type basin with some pull-apart structures filled with coeval, mainly marine sediments. The coast of the Carnarvon Basin includes the Ningaloo fringing reef, over 200 km long, backed by Oligo-Miocene anticlines (Collins *et al.* 2006) and aeolianite cliffs of the Shark Bay region to the south. Pleistocene and Holocene aeolian dune barriers dominate the coastal plain of the Perth Basin at more southerly latitudes.

Shallow seismic images show that the outer part of the northern Rottnest Shelf is fault-controlled, with the seaward face formed by numerous listric and normal faults. The onshore Quaternary record (Kendrick *et al.* 1991) indicates relative tectonic stability in the Perth Basin overall. This situation led to reduced accommodation, which maintained the shelf surface in the euphotic, active carbonate-sediment-producing zone during Quaternary sea-level highstands. In the Houtman Abrolhos, with even less recent subsidence, the seafloor was always in shallow water, and so was amenable to growth of coral reefs whenever oceanographic conditions warranted (Collins *et al.* 1993, 1997). The Carnarvon Ramp, which lies atop the more active Carnarvon Basin, was, in contrast, an area of continued subsidence and ever-increasing accommodation.

Pre-Quaternary and Quaternary shelf sedimentation

During the Cenozoic post-rift stage of basin evolution, prograding carbonate sequences characterized the offshore portions of both the Perth and Carnarvon basins (Exon & Colwell 1994). These sequences are best exposed in Oligo-Miocene anticlines in the onshore Carnarvon Basin (Collins *et al.* 2006). Aeolianites of

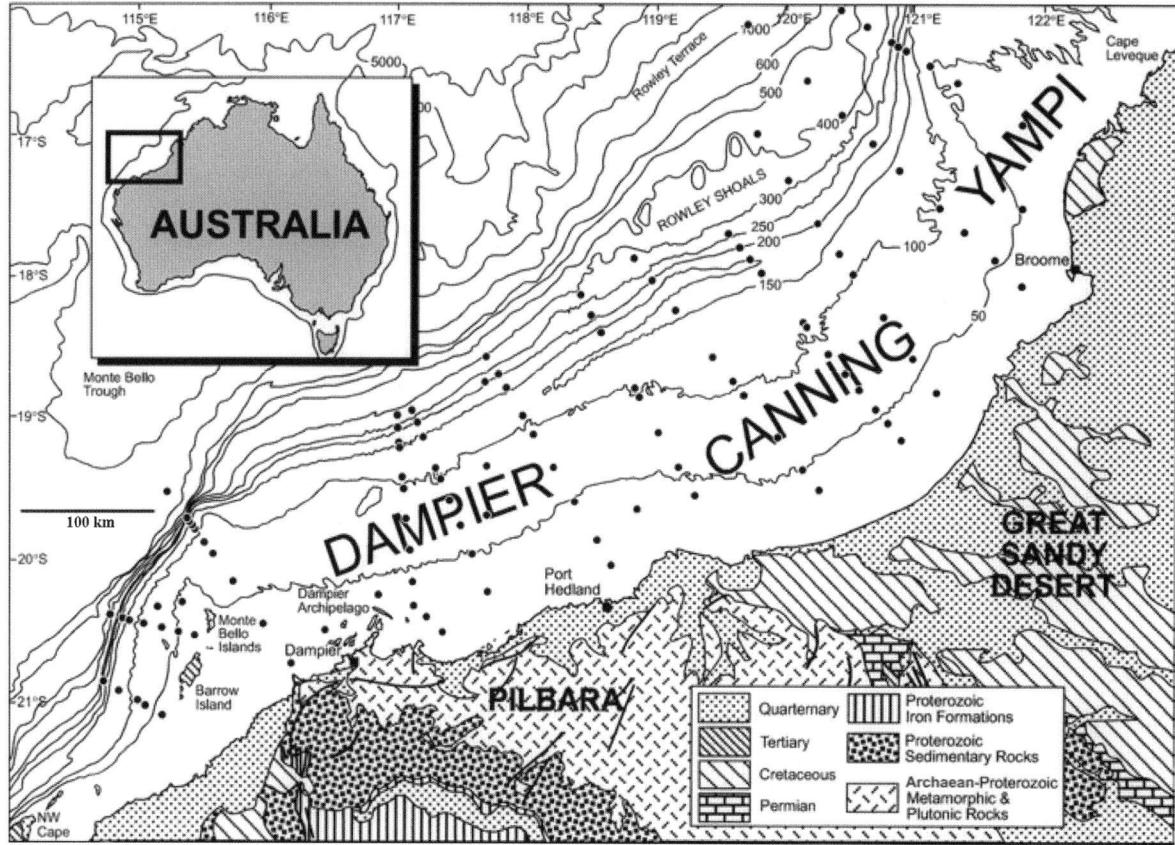

Fig. 19.2. Locality map of the North West Shelf (NWS) area showing offshore bathymetry (in metres) and onshore general geology. Dots are sample sites (after James *et al.* 2004).

the Tamala Limestone, with thin marine intercalations, dominate the Pleistocene stratigraphy and underlie the Holocene sediments on the NWS and SWS (Playford *et al.* 1976).

NWS modern shelf sedimentation

Ramp surficial sediment

Overall, the surface sediments are sand and coarser dominated (Fig. 19.4), approximately 90% carbonate, and contain abundant relict material with authigenic phosphate and glauconite components. Deep-water sediments are planktic muddy sands, while the shelf materials are skeletal, peloidal and oolitic. The coarse fraction (>2 mm) is entirely skeletal, consisting of molluscs, benthic foraminifers, bryozoans, echinoids, calcareous algae and corals. *Amphistegina* and miliolids are the dominant benthic foraminifers, while most of the molluscs are bivalves. The mud-sized fraction is composed of skeletal debris, silicate clays, quartz and micrometre-sized aragonite needles. Terrigenous and high-magnesium calcite (HMC) components are highest near the shoreline, whereas aragonite needles are most abundant on the outer ramp, >100 mwd. Outer-ramp sediment is also rich in planktics, with the rippled sediment boundary between 80 and 120 mwd. A limestone pavement was encountered locally to 282 mwd. The limestone is generally poorly cemented skeletal material and ooids–peloids in micrite matrix/cement.

At the largest scale, aside from the strandline itself, the major break on the seafloor is an escarpment, the base of which lies at 125 mwd. In shallower waters, the sediments are mostly mud free, typically swept into subaqueous dunes or textured by ripples.

Integration of information from Jones (1973) and James *et al.* (2004) permits separation, based largely on sediment composition,

of eight sedimentary facies that are summarized in Table 19.2. Particles in these facies are biofragmental, oolitic and peloidal grains, rod-shaped, hollow carbonate tubes (Fig. 19.5), and terrigenous clastics. These grains are not coeval but a mixture of relict, stranded and Holocene elements. There are two types of relict grains: (1) skeletal intraclasts – typically fragmented to usually well-rounded altered and infilled skeletons; and (2) lithic intraclasts – angular to locally rounded clasts of cemented grainstone, packstone or wackestone of metastable carbonate mineralogy. These particles are a ubiquitous component of Australian continental shelf sediments south of NW Cape and across the cool-water carbonate province of the southern margin (Collins 1988; James *et al.* 1992, 1997a, 1999, 2001). They are interpreted as having formed during former highstands, specifically late Pleistocene MIS 3 and 4 highstands (Fig. 19.6), when sea level was around 50 m lower than today and the shelf was intermittently submerged (James *et al.* 1997b). In contrast, stranded grains are generally lightly stained, mildly abraded, younger than relict particles but clearly out of equilibrium with their modern environment (James *et al.* 1997a, 2001). The most obvious of these grains are coralline algae and ooids in water depths >40 m. Such particles are interpreted as having formed in shallow water and been stranded by rapidly rising sea level during the latest Pleistocene and Holocene.

The following facies are present (see Table 19.2 for facies descriptions). The outer ramp: Facies 1 – Pelagic Sand and Mud; Facies 2 – Pelagic Ridge Sand; Facies 3 – Calcareous Tube Gravel–Pelagic Mud; and Facies 3A – Phosphate Subfacies. The mid ramp (Table 19.2) is dominated by: Facies 4 – Ooid–Peloid Sand; Facies 5 – Relict Intraclast, Ooid–Peloid Sand; and Facies 6 – Relict Intraclast, Biofragment Sand and Gravel. The two main inner-ramp facies (Table 19.2) are: Facies 7 – Benthic Foraminiferal Sand and Gravel; and Facies 8 – Mixed Biofragmental–Terrigenous–Iron Oxide.

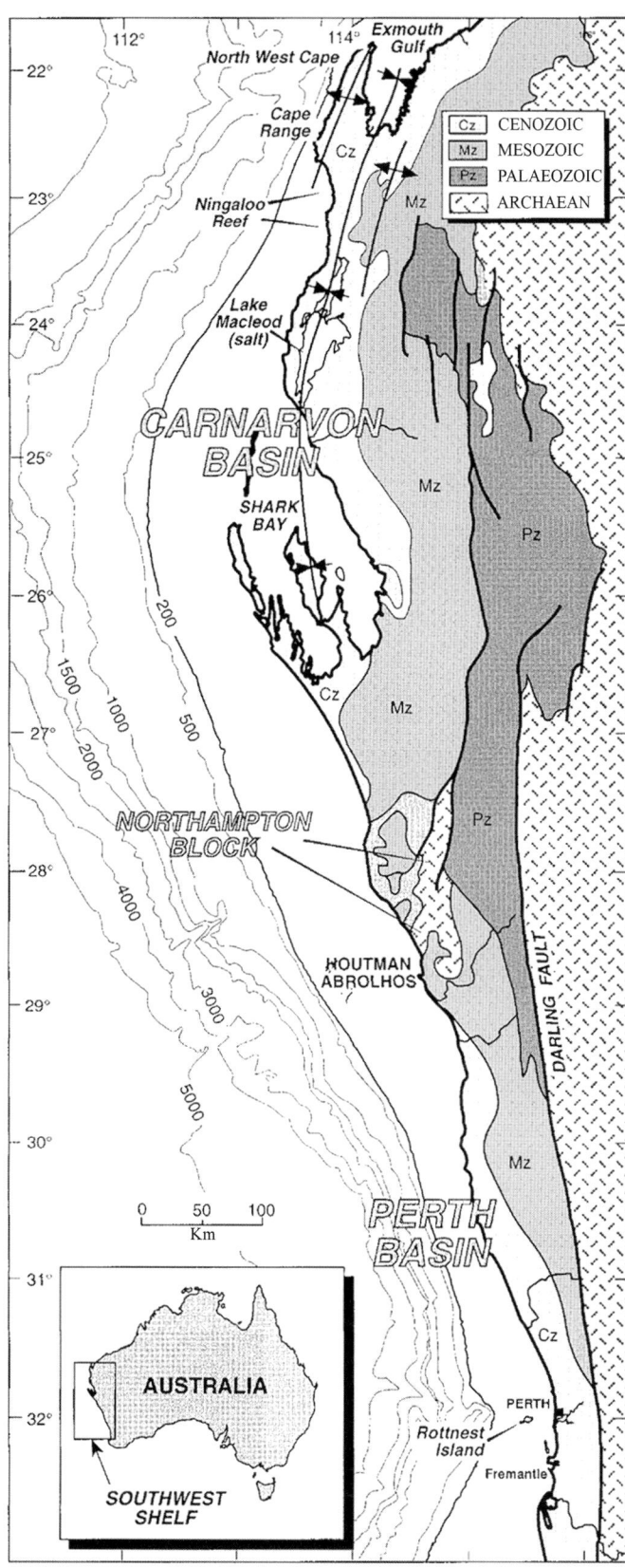

Fig. 19.3. Geological map of the South West Shelf (SWS) region outlining the location of the Perth Basin, Carnarvon Basin and Northampton Block. Bathymetric contours are in metres (after James *et al.* 1999).

Origin of particles

Relict sediment. On the NWS, relict grains are significant components of the sediment on the mid ramp in water depths of

<150 m, and most abundant (>50% of sediment) between 80 and 40 mwd (Fig. 19.7). The brown to orange to yellow-stained particles are clearly separable under a petrographical microscope using a combination of transmitted and reflected light. The brownish material is a complex mixture of clay minerals, iron oxides, aragonite and Mg-calcite manifest as either microcrystalline sediment or alteration of carbonate particles. The material occurs as: (1) microboring fillings along grain peripheries (not common); (2) intraskeletal cavity (pores, zooecia, test holes, stoma) fillings; (3) replacement of the skeletal wall; (4) replacement of micrite between particles in muddy sediment either totally or in clasts that are wackestone or packstone, generally from the periphery inwards; and (5) replacement of microcrystalline grains (peloids).

Stranded sediment. The most obvious of these grains on the NWS are ooids and, by implication, peloids. Some biofragments may also be stranded but there is no clear way of telling which ones have such an origin. The similarity between the modern seafloor biota and biofragmental grains, however, suggests that the stranded bioclastic grains are not numerous (see below).

Ooids are particles that range from unaltered and white to locally beige in colour. Those that are white are typically polished. They are spread across the shelf from 30 to 300 mwd (Fig. 19.8) but are concentrated between 60 and 150 mwd (peloids 60–150 mwd; ooids 60–135 mwd).

Ooids are generally fine- to locally medium-sand size, 0.25–0.75 mm in diameter, in some cases with a well-developed cortex but, most commonly, with only a superficial coating on a peloid core. The nuclei are also locally quartz grains inboard or benthic foraminifers, echinoid grains or bryozoans outboard. As pointed out by Jones (1973), the cortex is aragonite. Scanning electron microscopy (SEM) shows that the cortex comprises layers of tangentially and randomly orientated aragonite needles.

Radiocarbon dating indicates that the ooids formed over a narrow time frame, 15.4–12.7 ka. Incremental dating of the cortex of three ooid populations points to growth occurring over about 2 kyr.

Peloids are generally <2 mm in size, spherical to ovoid in shape, composed of microcrystalline carbonate and of two types (Jones 1973). SEM analysis of these peloids, some of which are ooid cores, indicates that they are composed of aragonite needles, like those forming the ooid cortices. A few are also micritized skeletal grains. Some peloids are stained and so are relict, unlike the ooids. The close association between ooids and peloids implies a common environment of formation and geological age. Open marine ooids with an aragonite cortex typically form in shallow, tidal-dominated settings in water depths of less than 5 m (Simone 1980). More specifically, ooids reflect formation in environments of somewhat elevated salinity: for example, on the Bahama Banks and the Persian Gulf, and in extreme conditions, as at Shark Bay (Logan *et al.* 1970). The rate of NWS ooid growth is similar to that determined from the Bahama Banks (see the discussion in James *et al.* 2004).

The period 15.4–12.7 ka corresponds to the period of rapid sea-level rise during melt water phase 1 A (mwp-1a) associated with the first phase of post-Last Glacial Maximum (LGM) deglaciation (Jensen & Veum 1990), before Younger Dryas cooling and slowed sea-level rise. Comparison with the Barbados curve (Fairbanks 1989) indicates that sea level would have risen from what is now about 105 to approximately 70 mwd during this time. Specifically, it would have rapidly submerged the escarpment and spread across the wide seafloor above the crest of the cliff, creating an expansive, shallow environment. Increased aridity and wind-speed over continental Australia during and immediately after the LGM, with strong offshore winds in the NWS region (Prell *et al.* 1980; Williams 2001), would have increased evaporation, thus promoting ooid formation. Such an interpretation, of evaporative

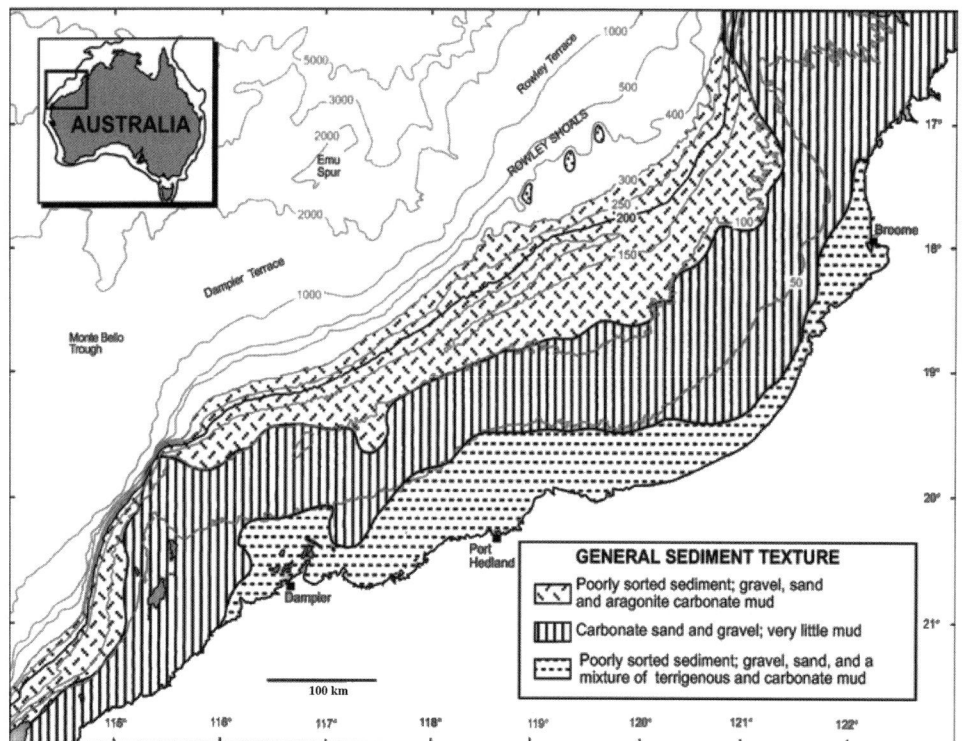

Fig. 19.4. General texture of surficial sediments on the North West Shelf (after James *et al.* 2004).

Bahamian/Persian Gulf-like shallows, also explains, in part, the association of ooids and peloids because, in the Bahamas platform margin, ooid facies also coincide with peloid lithification (Purdy 1963). Oxygen isotopic compositions of ooid cortex aragonite indicate lower $\delta^{18}O$ values with decreasing age, reflecting either warming or, more probably, decreased seawater salinity. Analysis of clays in the region (Gingele *et al.* 2001*a*, *b*) provides evidence that the onset of Leeuwin Current flow, and attendant less saline seawater, began at around 12 ka. Thus, ooid and peloid formation is interpreted as having taken place in shallow evaporative shelf environments during mwp-1a, but as having been arrested by initiation of the less saline waters of the southward-flowing Leeuwin Current during mwp-1b (James *et al.* 2004).

Holocene sediment: biofragmental particles. The association, in general terms, is foramol (foraminifers and molluscs) in character with accessory bryozoans, typical of a heterozoan assemblage (James 1997). Yet, this is too simple a classification. Inshore, to depths of around 30 m in the Barrow and Dampier sectors, the environment is one of coral reefs, bryozoans and macrophytes, together with large (symbiont-bearing) foraminifers. This is an unusual consortium – part photozoan and part heterozoan. There are no ooids here, and the calcareous green algae are not calcified. It is reminiscent of the Houtman Abrolhos reefs far to the south (Collins *et al.* 1997; James *et al.* 1999). Offshore, in waters >40 mwd, the sediment is a heterozoan assemblage with large benthic foraminifers decreasing seawards, a clearly subtropical sediment (James 1997).

Holocene sediment: tube facies and phosphatization. The calcareous tubes (Facies 3) appear to post-date the sediments in which they lie. Foraminifers in the tube-wall sediment are latest Quaternary in age. Radiocarbon dating of bulk sediment from the tube walls (Table 19.2) yields ages of 9.9 and 25.1 ka. Given that the sediment probably contains material of different ages, these results indicate that tubes are <9.9 ka old but the exact age is uncertain.

The sediment in which the tubes are developed has several very shallow-water proxies, particularly the bivalve *Anadara*, the seagrass meadow bivalve *Pinna* and zooxanthellate (hermatypic) corals, all pointing to an inner-ramp environment. This environment, now in water depths >120 m, most probably formed during the LGM lowstand (MIS 2) about 20 ka. The $^{87}Sr/^{86}Sr$ values for seven teeth and bones in the same sediment average 0.709174 (range 0.709143–0.709294), indicating that they are late Pleistocene–Holocene in age (cf. Farrell *et al.* 1995). This age is in accordance with the relatively young age of the tubes upon which phosphate has precipitated. These ages of the tubes imply that arrested sedimentation, and iron and phosphate precipitation and accumulation are continuing today. Thus, this outer-ramp region is an area of weak upwelling and/or phosphatization resulting from Fe-redox cycling within the sediments (cf. Heggie *et al.* 1990).

Holocene sediment: planktic sediment ridge. This elongate feature, first documented by Jones (1971, 1973), is of interest because of its persistence through time. Seafloor samples are always pelagic sediment, particularly pteropods, with a fine-grained benthic component. Surface ripples indicate that sediment movement is to the south. Radiocarbon dating of pteropods and planktic foraminifers from two localities (Fig. 19.8) yields ages of 0.35 and 9.2 ka, confirming the contemporary age of the material.

The structure is interpreted here as a high surface productivity phenomenon. The most likely explanation is that it corresponds to a zone of near-surface shear between the Leeuwin Current and the Indian Ocean Water, thus promoting upwelling and nutrient enhancement.

SWS: modern shelf sedimentation

In the following text, the SWS elements will be described from south to north: the northern Rottnest Shelf; the Houtman Abrolhos reefs; the Transition Zone; and Carnarvon Ramp.

Seafloor sediment components

Most sediments have a distinctive cool-water character (cf. Nelson 1988) but with subtropical attributes (cf. James 1997). The most

Table 19.2. *Composition of surficial sediment facies on the North West Shelf*

Facies	Modern sediment	Relict sediment
1 – Pelagic Sand and Mud (>210 mwd)	Planktic foraminifers and nanofossils	Some planktic foraminifer tests are brown and filled with berthierine, and so interpreted as relict
2 – Pelagic Ridge Sand (130–175 mwd)	Sand, one-quarter pteropod shields and spines + planktic foraminifer tests; minor benthic foraminifers and echinoid pieces. Mud, one-quarter ostracods, sponge spicules, fragmented foraminifers and aragonite needles	None
3 – Calcareous Tube–Pelagic Mud (110–210 mwd)	Sand, one-quarter molluscs with a few bryozoans and benthic foraminifers outboard; but ooid-skeletal sand of Facies 4 adjacent to the 125 mwd escarpment. Numerous benthic foraminifers; conspicuous detritivores (e.g. *Nodosaria*). Mud, one-quarter pelagic foraminifers, small benthic foraminifers (especially *Amphistegina*) and pteropods	Molluscs (50–100%), large gastropods (cones, murex, whelks, olives) + diverse, robust and delicate, infaunal + epifaunal bivalves. Conspicuous whole and fragmented infaunal and epifaunal echinoids (particularly clypeasters). Rare bryozoans (fenestrates > vagrants > delicate branching > flat robust branching). Scarce but ubiquitous pieces of zooxanthellate corals. Numerous (10–60%) *Ophiomorpha* and *Thalassinoides* tubes, locally coated with phosphate and iron oxides. Shark, teleost, tortoise and crocodile teeth. Whale bones. Lithoclasts (10%) = microbioclastic wackestone
4 – Ooid–Peloid Sand (70–130 mwd)	Coarse, one-quarter molluscs, large benthic foraminifers, bryozoans (fenestrate (*Adeona*) > encrusting > vagrant > delicate branching). Sand, one-quarter ooids and peloids (60–90%), molluscs (30%), echinoids (20%) benthic foraminifers (10%) + c. 10% others	Yampi sector = benthic foraminifers (*Amphistegina* but also miliolids and *Marginopora*). Canning sector = same, but more bryozoans, echinoids, molluscs, *Halimeda*, corals + more intraclasts (microbioclastic packestone – wackestone, ooid–peloid skeletal packestone, calcrete). Dampier sector = same + Pleistocene intraclasts and rhodoliths
5 – Relict Intraclast, Ooid–Peloid Sand (40–70 mwd)	Coarse, one-quarter large benthic foraminifers (*Marginopora* and *Heterostegina*), bivalves, bryozoans, gastropods, clypeasters; local rhodoliths and azooxanthellate corals. Sand, one-quarter ooids, peloids (50–70%), benthic foraminifers (15–25%) (*Amphistegina* and *Marginopora*), molluscs (10–15%), equal proportions of bryozoans, corallines and echinoids (each c. 10%). Corals and barnacles present	Skeletal intraclasts = 10–30% bryozoans, 10–40% benthic foraminifers (*Amphistegina, Marginopora* and miliolids). Others = bivalves, encrusting foraminifers, *Halimeda*, coral, gastropods and peloids. Lithic intraclasts = mostly peloid–skeletal wackestone, bioclastic–microbioclastic wackestone, ooid–microbioclastic wackestone, calcrete and laterite. Others are skeletal–peloid and ooid wackestone–packestone
6 – Relict Intraclast, Biofragmental Sand and Gravel (40–70 mwd)	Coarse, one-quarter a full range of bivalves and gastropods (conspicuous pectens). Meagre to abundant bryozoans (fenestrate > delicate branching). Ubiquitous but not abundant solitary corals and echinoids. Barnacles common locally. Sand, one-quarter benthic foraminifers (20–70%), molluscs (20–60%), bryozoans (0–30%) and echinoids (0–30%), usually no or <10% ooids and peloids	Skeletal intraclasts = 25–30% benthic foraminifers (*Amphistegina*, miliolids, *Marginopora* and lesser *Elphidium*), 10% bryozoans (flat robust branching forms) + 10% echinoderms and gastropods. Lithi microbioclastic wackestone – packestone (local quartz and feldspar sand grains), 50% skeletal packestone (locally with *Halimeda* and ooids), ooid–peloid packestone, Pleistocene ooid–peloid–bioclastic grainstone and *Marginopora*–ooid–bryozoan packestone
7 – Benthic Foraminiferal Sand and Gravel (c. <60 mwd)	Coarse, one-quarter similar to living biota; ubiquitous rhodoliths, common bivalves and gastropods, common but few solitary corals, clypeasters in all sediment, hermatypic corals occur < 30 mwd; serpulids locally encrust clasts, and *Dentalium*; barnacles are common. *Halimeda* is conspicuous in Yampi and Barrow sediment. Sand, one-quarter large benthic foraminifers (20–50%), molluscs (20%), bryozoans (10–20%), variable reticulated branching corallines (0–20%) + echinoids (0–20%)	Skeletal intraclasts = most abundant in Dampier sector (80–100%); 20–50% in Yampi sector; 50% *Amphistegina* and *Marginopora*; 20–30% bryozoans; 10% *Halimeda* in the east; zooxanthellate coral locally abundant in the east. Lithic intraclasts = microbioclastic packestone – wackestone, bivalve rudstone, ooid–peloid packestone, peloid–skeletal grainstone, biofragmental wackestone
8 – Mixed Biofragmental–Terrigenous–Iron Oxide (<40 mwd)	Coarse, one-quarter diverse and numerous bivalves, gastropods and bryozoans (fenestrate > encrusting > flat robust branching > vagrants > others), clypeasters, solitary corals, rhodoliths and large benthic foraminifers (*Marginopora* and *Heterostigina*). Local barnacles, hermatypic corals, rhodolites + serpulids. Sand, one-quarter benthic foraminifers (30–70%), bivalves (20–40%), bryozoans (10–30%) + echinoids (10–20%); locally abundant serpulids, barnacles + corallines; a variable number of grains stained grey and black from monosulphides. Conspicuous fresh and living *Amphistegina* + *Marginopora*	Iron-impregnated carbonate grains and terrigenous particles = 50–80% of sediment, with locally up to 20% sand-sized quartz and crystalline rock fragments; laterite + calcrete = 10–40% of particles. Skeletal intraclasts one-quarter 30% benthic foraminifers (*Amphistegina, Marginopora, Pyrgo, Elphidium*), 10% bryozoans, 20% molluscs, trace coral. Lithic intraclasts = highly variable, probable older Pleistocene = quartzose–bioclastic–oolitic grainstone with low-Mg calcite cement; others = microbioclastic packestone – wackestone; ooid–peloid grainstone, *Halimeda*–ooid packestone, biofragmental packestone

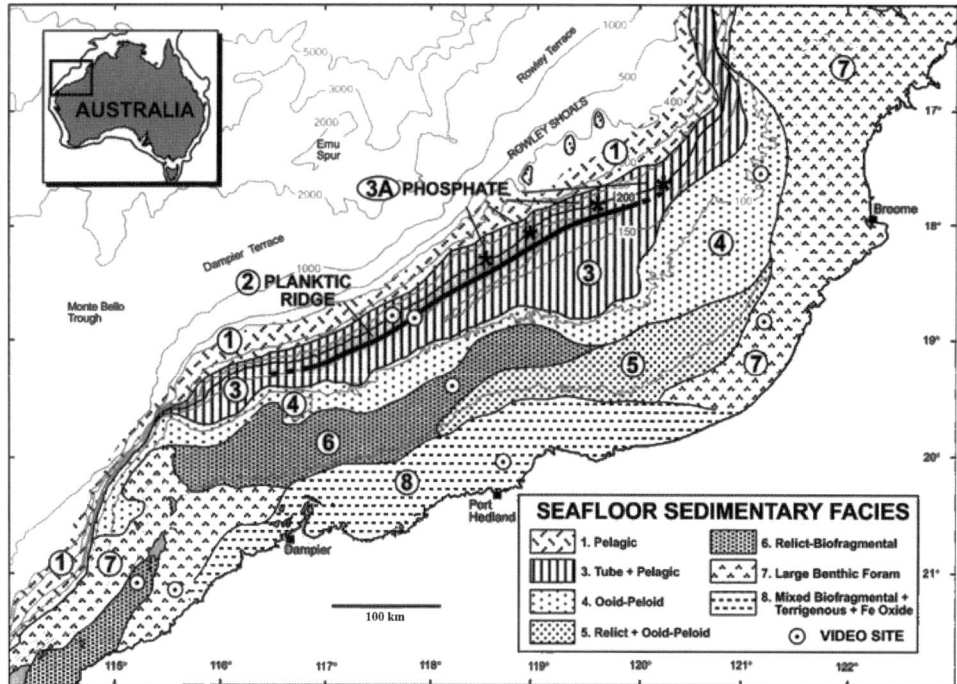

Fig. 19.5. Sedimentary facies of seafloor surface sediments on the North West Shelf (after James *et al.* 2004).

important skeletal elements are coralline algae, bryozoans, molluscs (scaphopods, bivalves and gastropods) and foraminifers (free and encrusting, large and small, benthic and pelagic), and they form particles across the grain-size spectrum (cf. Bone & James 1993; James *et al.* 1997a). The major difference between these sediments and cool-water deposits is the presence of scattered zooxanthellate corals and large, symbiont-bearing foraminifers. The calcareous green alga *Halimeda*, although living in certain facies belts, is poorly calcified and rarely forms sediment particles. Other locally important skeletons are those of serpulid

worms, and epifaunal and infaunal echinoids, together with spicules from sponges, gorgonians and other soft corals. Although most sponges are either free of encrustation or act as a substrate for bryozoans, coralline algae and bivalves, some are impregnated with sediment. Such sand-filled sponges are globular (2–4 cm in diameter) or prone and tabular (2–5 cm thick), and contain 30–40% sediment by volume.

As with most other Australian cool-water carbonate sediments, some particles are variably stained with iron-rich oxides. These materials are present in microborings, replace carbonate skeletons and fill intraskeletal pores. Microprobe analysis of microcrystalline pore-fillings indicates iron-rich clays. Such materials are interpreted as amorphous goethite and poorly crystalline berthierine (chamosite), a 7 Å iron layer-lattice silicate, similar to grains on the subtropical shelf off eastern Australia (Marshall 1983). The particles are referred to as 'stranded' when lightly stained to a buff , and 'relict' when intensely altered to orange and brown (cf. James *et al.* 1997b). True glauconite is abundant locally.

Intraskeletal pores, especially bryozoan zooecia and foraminiferal chambers, can be filled with Mg-calcite cement, generally micrite or micropeloids (40–60 μm) and fine spar (cf. James *et al.* 1976; Macintyre 1985) ranging in composition from 6.0 to 16.0 mol% $MgCO_3$ ($n = 12$). Clasts of various sorts are typically part of the seafloor sediment. These grains can be relict, stranded or modern; they have collectively been called lithoclasts or lithoskels in other studies (Read 1974; Collins 1988). Some sediments also contain blackened, pyrite-impregnated skeletons and clasts, generally associated with the reducing conditions amongst seagrass rhizomes (cf. Maiklem 1967; Davies 1970).

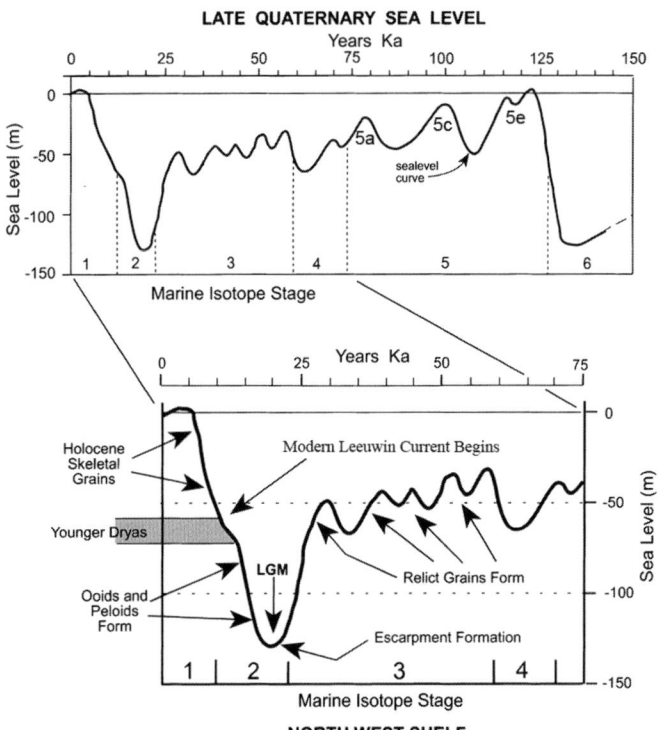

Fig. 19.6. Late Quaternary sea-level record (after Chappell & Shackleton 1986) and expanded portion over the last 75 kyr showing some major events and periods of particle formation on the North West Shelf (after James *et al.* 2004).

Northern Rottnest Shelf

General attributes. This sector of the shelf is the surface expression of the Perth Basin. The region is one of tectonic stability; with Late Cenozoic uplift a little more than a few metres (Kendrick *et al.* 1991). Much of the shelf surface is underlain by aeolianite and skeletal grainstone of the Pleistocene Tamala Limestone. Shelf bathymetry changes from south to north. The shallow depth of the inner-shelf plain and ridge complex means that there is minimum accommodation and that most of the shelf is in the euphotic zone.

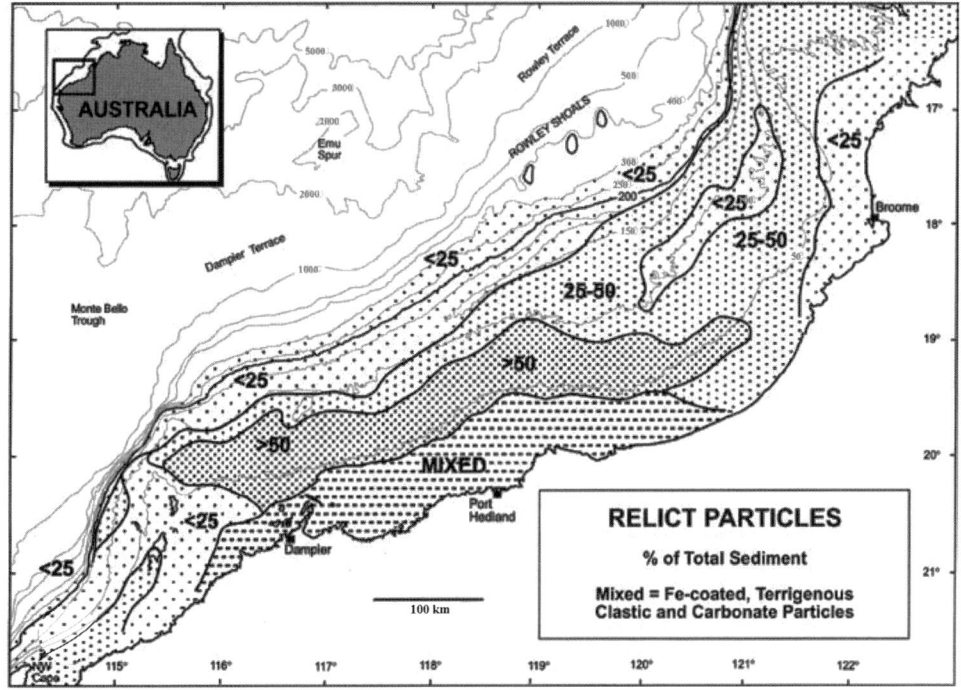

Fig. 19.7. The proportion of relict particles in surface sediment of the NWS (after James *et al.* 2004).

Inner-shelf plain. The steep shoreface extends to about 30 mwd and then merges with a flat to gently oceanward-dipping seafloor locally corrugated by small ridges, probably late Pleistocene beach–dune complexes. Sedimentary facies trends are generally shelf-parallel (Fig. 19.8). Seafloor photographs show: (1) a rocky seafloor almost totally obscured by prolific seagrass, macrophytes and diverse sponges; or (2) rippled sand with clumps of non-calcareous red algae 10–20 cm in diameter; or (3) open rippled sand that is locally degraded by surface traces and burrowing. Some clasts ripped from the seafloor are calcretized Pleistocene calcarenite, generally coralline algal rod–mollusc–benthic foraminifer grainstone–packstone in which most of the particles still consist of metastable carbonate minerals, but with local LMC (low-magnesium calcite) meteoric intergranular cement.

The most abundant organisms are non-calcareous macrophytes, especially *Ecklonia* and non-calcified red algae to 65 mwd, and seagrass (*Thalassodendron*) in waters shallower than 40 mwd. The grasses serve as ephemeral substrates for bryozoans and foraminifers, whereas sponges and ascidians are common amongst the *Ecklonia*.

Coralline red algae form rhodolites around cobbles, boulders and slabs of calcarenite, and encrust the calcretized limestone substrate. Although smooth spherical rhodolites are the norm, corallines are also branching in deep water, approximately 60 mwd, especially leeward of the ridge complex. The coralline algae encrusted surfaces and rhodolites are, in turn, prime substrates for living larger foraminifers (species of *Amphistegina*, *Heterostegina* and *Amphisorus*), and for prolific macrophytes, bryozoans and sponges. *Halimeda* (probably *H. tuna*) is present but sparse,

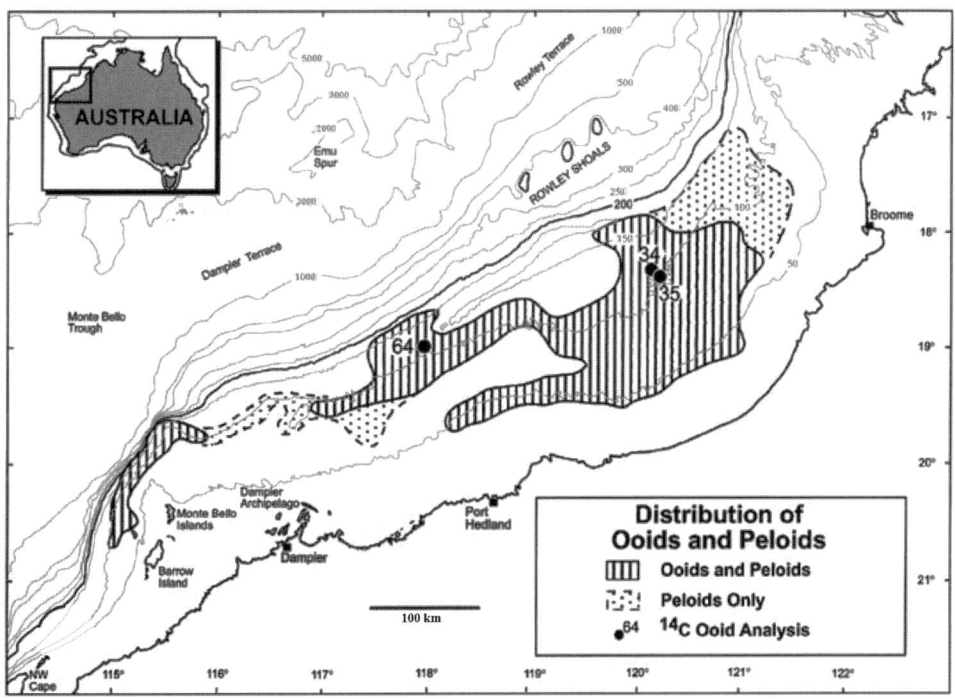

Fig. 19.8. The distribution of ooids and peloids in surface sediment of the NWS (samples used for radiometric dating highlighted) (after James *et al.* 2004).

generally growing attached to hard substrates at depths of less than 50 mwd.

The red encrusting foraminifer *Miniacina* is prolific and usually attached to dead, hard substrates (rocks, rhodolites, bryozoans and molluscs). Limestone cobbles and pebbles are generally bored, with the holes filled by soft or hard lithified sediment.

The dominant inner-shelf plain facies (Fig 19.9, Table 19.2) are: Facies A1 – Skeletal Sand; Facies A2 – Quartzose Skeletal Sand.

Ridge complex. The seafloor here is a zone of irregular topography, 3–8 km wide, that rises from approximately 60 mwd to, at the shallowest measured, about 27 mwd. The inner and outer margins are abrupt, with the front locally terraced (*c.* 50 mwd) but always descending to the inner margin of the outer shelf ramp at around 100–120 mwd. This is the continuation of the 'Offshore Ridges' (Collins 1988) on the southern Rottnest Shelf. Calcarenite hardgrounds are cemented by 11.5 mol% $MgCO_3$ calcite.

Images of the seafloor show a prolific hard substrate biota of intergrown macrophytes, *Halimeda* (locally), *Adeona*, ascidians and hydrozoans, alternating with rhodolite pavements where 90% of the seafloor is covered with nodules and scattered macrophytes (Fig. 19.9). Seagrass grows in the shallowest localities. As on the open shelf plain there are local areas of rippled sand.

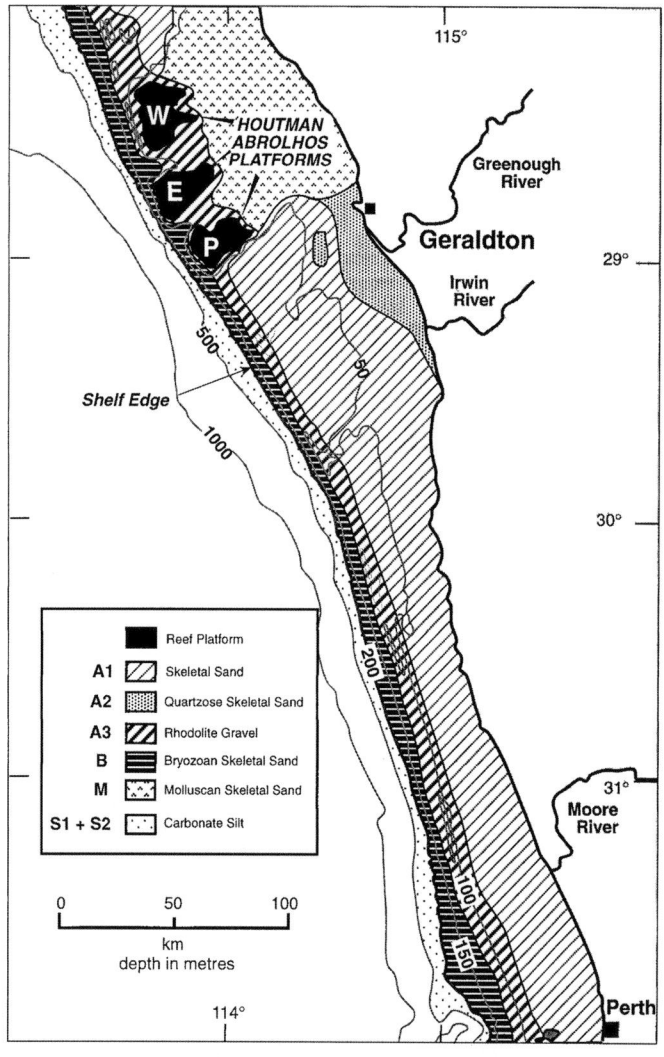

Fig. 19.9. Sediment facies on the northern Rottnest Shelf; see Tables 19.1 and 19.2 for composition of separate facies. Houtman Abrolhos Platforms: W, Wallabi; E, Easter; P, Pelsaert. Bathymetric contours are in metres (after James *et al.* 1999).

The corallines, as crusts and abundant rhodolites, are generally 10–30% alive.

The ridge complex consistently yields the highest numbers of living larger foraminifers. Living hermatypic corals are present but not commonly recovered in bottom samples. Distribution suggests a coral biota of low diversity and abundance, primarily prone, encrusting *Pleisiastrea* and flat *Coscinaraea*. The furthest south that coral was recovered was at 52 mwd at latitude 30°S. Such corals grow sporadically along the coastline south of the Abrolhos but never form reefs (Veron 1995). The dominant ridge complex facies (Fig. 19.9, Table 19.3) is Facies A3 – Coralline Algal Rhodolite Gravel.

Outer shelf. This inclined surface, 10–20 km wide, extends from the base of the ridge complex or reef platforms, where there is a distinct break in slope, to the shelf edge at approximately 200 mwd. It is a zone of rippled sand and rocky outcrop.

The living organisms are a high-diversity sponge–bryozoan community, with accessory ascidians, hydrozoans and bivalves. Rocky substrates support prolific growth of bryozoans (Fig. 19.10), sponges and encrusting bivalves. Bryozoans are most abundant and diverse in shallower parts of the outer shelf, and are intergrown with *Halimeda* (Fig. 19.10) in the euphotic zone. Many bryozoans are attached to sponges. Sponges themselves are especially prolific at the shelf edge. The dominant outer-shelf facies (Fig. 19.9, Table 19.3) is Facies B – Bryozoan Skeletal Sand.

Houtman Abrolhos reefs

General attributes. The shelf here has the same morphology as that to the south except that the ridge complex is topped by three emergent reef platforms (Pelsaert, Easter and Wallabi: Fig. 19.9), separated by 40 m-deep channels. The tabular platforms consist of reefs, reef flats and low islands, and have gentle ocean-facing slopes and steep SE and north margins. The central platforms are Last Interglacial in age, whereas the windward and leeward reefs are Holocene (Collins *et al.* 1993, 1996, 1997). The mean summer sea surface temperature (SST) is 26 °C, falling to 18 °C in winter and staying below 20 °C for 30% of the time. Although the Leeuwin Current generally flows 15–20 km west of the Abrolhos, a large eddy at this latitude spreads warmer water on to the platform (Collins *et al.* 1997).

The reefs contain 184 species and 42 genera of coral, with *Acropora* the dominant coral; it is essentially a tropical biota but with endemic forms. The western, windward margin of the platform has basically no coral; it is instead covered by an unusual mixture of red and brown macrophytes of tropical and temperate (*Ecklonia*) affinities (Hatcher 1991). Leeward margins have a rich and diverse coral community, and seismic profiles and cores indicate up to 40 m of Holocene reef growth (Collins *et al.* 1993, 1996; Collins 2010).

Sediments on the shallow platform tops are coral framestones and sand sheets composed of coral and coralline algae debris, together with lesser bryozoans, benthic foraminifers and mollusc grains. This tropical, warm-water association contains no ooids; although living *Halimeda* is present, it occurs only as trace amounts in sediment.

Inner-shelf plain. A 35–45 m-deep seafloor of open sand and *Posidonia* meadows lies leeward of the platforms. There are many infaunal echinoids, especially species of *Lovenia* (heart urchins) and *Clypeaster* (sand dollars), epifaunal scallops, infaunal bivalves, and gastropods. Numerous tubular sponges with attached bryozoans, mostly fenestrate and articulated branching forms, are locally abundant. Minor coralline algae and agglutinated worm tubes are locally profuse. The dominant inner-shelf plain facies (Fig. 19.9) is Facies M – Molluscan Skeletal Sand.

Table 19.3. *Composition of surficial sediment facies on the South West Shelf*

South West Shelf facies	Colour	Location	Depth (m)	Limestone
Algal Sand and Gravel				
A1 – Skeletal Sand	Pale yellow – 2.5YR7/3 Light grey-pale brown – 10YR7/2, 7/4	RS – inner shelf plain	30–60	Fossiliferous grainstone–rudstone
A2 – Quartzose Skeletal Sand	Light brown – 7.5YR6/4 Pinkish grey – 7.5YR 6/2	RS and HA – innermost shelf		Quartzose fossiliferous grainstone–fossiliferous sandstone
A3 – Rhodolite Gravel	Very pale brown – 10YR 7/4	RS and HA – ridge complex	25–100	Rhodolite rudstone
Molluscan Sand				
M – Molluscan Skeletal Sand	Grey – 10YR5/1 Light brownish grey – 10YR6/2	HA and SB – inner shelf	35–50	Molluscan grainstone–rudstone
Bryozoan Sand				
B – Bryozoan Skeletal Sand	Pale brown – 10YR6/3–6/3 Light yellowish brown – 2.5R6/3 Pale yellow – 2.5Y7/4	RS and HA – outer shelf	120–200	Bryozoan floatstone–rudstone–grainstone
Intraclast Sand				
I1 – Abraded Intraclast–Skeletal Sand	Pale brown – 10YR6/3 Light brownish grey – 10YR6/2	CR – inner mid ramp	20–70	Intraclast, fossiliferous grainstone
I2 – Fragmented Intraclast–Skeletal Sand	Pale brown – 10YR6/3 Light brownish grey – 10YR6/2	CR – outer mid ramp	70–100	Intraclast, fossiliferous packstone–grainstone
Planktic Sand				
P1 – Intraclast–Planktic Sand	Brown – 10YR5/2, 5/3 Light brownish grey – 2.5Y6/2	CR – outer ramp	100–150	Intraclast pelagic fine packstone
P2 – Planktic Sand and Silt	Light.brownish grey – 2.5Y6/2 Light yellowish brown – 2.5Y6/3	CR – outer ramp	120–200	Intraclast pelagic fine packstone–wackestone
P3 Well-Washed Planktic Sand	Light brown – 10YR5/3, 6/3, 6/4, 7/4 Pale brown – 10Y6/3	CR – outer-ramp slope	170–500	Pelagic foraminifer grainstone–packstone
Carbonate Silt				
S1 – Spiculitic Carbonate Silt	Light yellow brown – 2.5Y6/4	RS, HA & CR – slope	>200	Spiculitic wackestone
S2 – Bryozoan Carbonate Silt	Yellow-brown – 2.5Y6/2, 6/3, 6/4	RS – slope	150–200	Bryozoan wackestone

Abbreviations: CR, Carnarvon Ramp; HA, Houtman Abrolhos reefs; RS, Rottnest Shelf.

Rottnest Shelf: Carnarvon Ramp transition zone

The transition zone between the Rottnest Shelf and the Carnarvon Ramp is gradational, with the ridge complex and outer shelf gradually becoming deeper and more subdued northwards until both have disappeared and the seafloor is a gently dipping ramp to a subdued shelf edge. Cross-sections of water temperature and salinity illustrate a different water structure than on the northern Rottnest Shelf. Inner-shelf waters are warm but distinctly saline, locally >36.0‰. This high-salinity water can be traced northwards to the outlet of Shark Bay. Depression of both salinity and temperature contours at the seafloor on the outer shelf indicates overall downwelling. Currents on the inner shelf are weak and multidirectional. This region is one of changing bathymetry and facies (Fig. 19.10).

Carnarvon Ramp

General attributes. The ramp extends from roughly Kalbarri in a curving arc, NW and north, past Shark Bay to NW Cape (Fig. 19.10). The underlying Carnarvon Basin is up to 1000 km long and contains of the order of 7 km in thickness of sedimentary rocks (Hocking 1990). The hinterland is a series of faulted Palaeozoic, Mesozoic and Cenozoic sedimentary rocks of low relief with a surficial soil cover. Detritus is carried from here to coastal areas by sheetfloods and a few intermittent streams. The northern part of the area has undergone Cenozoic tectonism, now reflected by major anticlines and with the syncline between called the

Bullara Sunkland. Uplift is indicated by elevated Pliocene–Pleistocene reef terraces, at a uniform level of 12 m, on the western side of the Cape Range. Pliocene–Pleistocene deposits in the sunklands are mainly dune sands. Thin Quaternary marine carbonate units accumulated when the sea flooded low areas such as Lake Macleod and Shark Bay.

The large evaporite basin of Lake Macleod (Logan 1987) lies in a structural depression and is bounded on the ocean side by the Quobba Ridge, a 3–15 km-wide, 180 km-long horst accentuated by Quaternary aeolianite accretion.

Inner ramp. The Zuytdorp Cliffs extend along the western margin of Shark Bay and consist of nearly vertical escarpments cut into the 300 m-thick, cross-bedded aeolianite. They form a straight, almost unbroken feature for over 220 km, and may be fault controlled. The Murchison River brings terrigenous clastic sediments on to the shelf at Kalbarri (Fig. 19.10). This is reflected in the quartz-rich sediments shallower than 50 mwd that contain minor carbonate, mostly limestone clasts and abraded relict material. Kalbarri is also the northernmost extent of abundant macrophytes, especially *Ecklonia*.

Shark Bay (Fig. 19.10) is a shallow (average 10 mwd) marine embayment, about 14 000 km² in area, composed of a series of gulfs, inlets and basins. The seaward western margin is a string of islands with steep ocean-facing aeolianite cliffs and small, local fringing reefs. It is a large inverse estuary and is characterized by open-ocean salinity at the entrance and hypersalinity at the heads. The sedimentary geology has been documented in detail by Logan *et al.* (1970, 1974) and Playford *et al.* (1976).

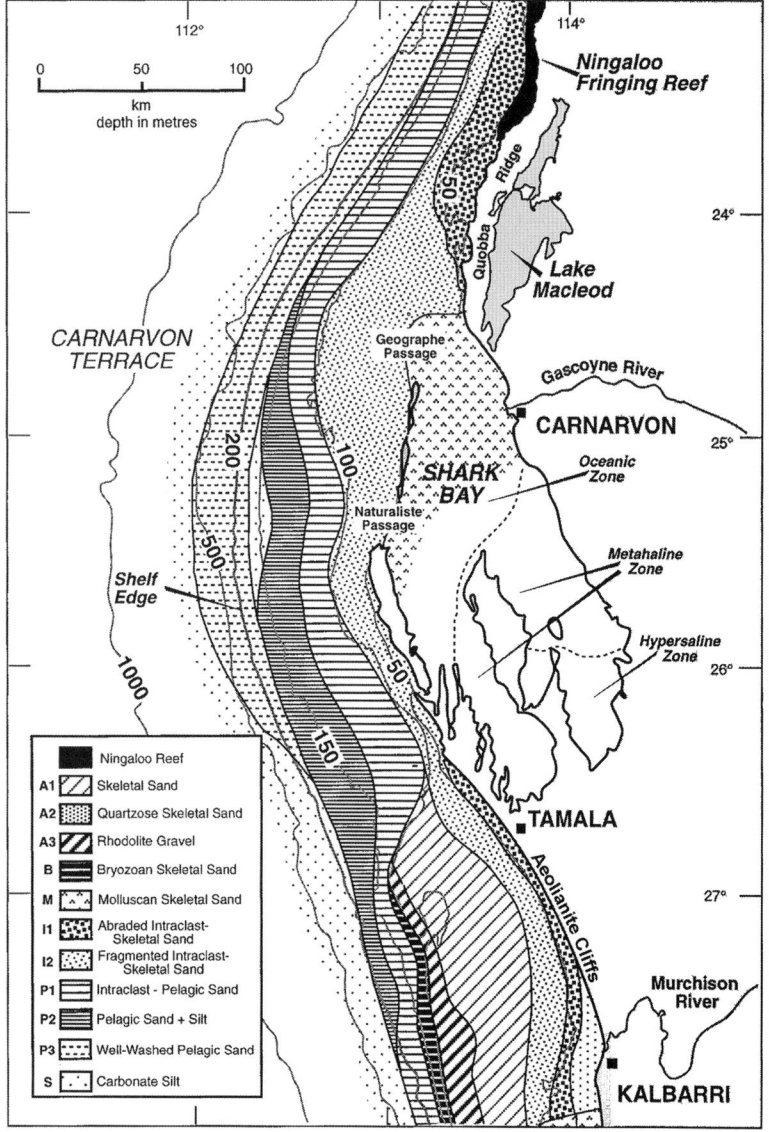

Fig. 19.10. Sediment facies on the Carnarvon Ramp; see Table 19.3 for composition of separate facies. Bathymetric contours are in metres. Note the region of facies change west of Tamala (after James *et al.* 1999).

Bay waters are isohaline, surface to seafloor, but with a strong horizontal gradient (Fig. 19.10). They are well mixed, with high oxygen saturation. Significant exchange of waters through the western openings is probably restricted to winter months, when shelf waters are warmer than Shark Bay waters and southern wind stress is at a minimum (Burling *et al.* 1998). Nutrients are low because river flow is minimal, except in the hypersaline areas, where there are abundant seagrass and algae.

The biota is generally subtropical, reflecting the relatively low winter water temperatures. Constituents, in order of abundance, are coralline algae, molluscs, foraminifers, echinoids, serpulids and bryozoans. Seafloor biotic zonation reflects the salinity gradient. The northern and western part of the bay sampled in this study has oceanic characteristics, and sediments are part of Facies M – Mollusc Skeletal Sand. The seafloor is seagrass-covered, and the coarse fraction (trace–30%) is dominated by bivalves and lithified clasts cemented by aragonite. Because the area is near the outflow of the Gascoyne River (Fig. 19.10), sediments contain 30–60% terrigenous grains. *Amphisorus* is conspicuous in the sand-size fraction throughout.

The Ningaloo Reef is Australia's largest fringing reef system, with reefs separated from the coast by a 0.2–7.0 km-wide lagoon. The reef complex consists of a narrow reef crest, backed by a reef flat, and the shallow (0–4 mwd) lagoon is floored by a skeletal sand sheet interrupted by rocky substrate and corals. Best reef development is along reef passes and in the lagoon;

the outer reef slopes do not have rich coral communities. SST ranges from a high of 28 °C to a low of 22 °C, always tropical. Holocene reefs grew as fringing systems or antecedent topography remnant from a much larger Last Interglacial reef (Collins *et al.* 2004; Collins 2010; Twiggs & Collins 2010).

Mid ramp. This ramp sector (Fig. 19.10) is an open sand plain with few living benthic organisms. The sparse biota is mostly delicate infaunal bivalves but what typifies the region are extensive spreads of vagrant bryozoans. Living rhodolites are common locally, with bryozoans (especially large colonies of *Celleporaria*) and sponges growing on the few hard substrates. *Halimeda* grows on hardgrounds or outcrop (rock or rhodolites) shallower than 50 mwd but is absent in the sediment. The dominant mid-ramp facies (Fig. 19.10, Table 19.3) are Facies I1 – Coarse, Abraded, Intraclast–Skeletal Sand; and Facies I2 – Fragmented, Intraclast–Skeletal Sand.

Outer ramp. As on the mid ramp, there is little alive on the seafloor. Bivalves are the most abundant living organisms but solitary corals, gastropods and echinoids are also present. Vagrant bryozoans litter sand substrates to approximately 140 mwd but not deeper. Brachiopods live in waters beyond 300 mwd. Bryozoans (except for a few delicate cyclostomes) and sponges are rare. The dominant outer-ramp facies (Fig. 19.10, Table 19.3) are Facies P1 – Planktic–Intraclast Sand; Facies P2 – Planktic Sand and Silt; and Facies P3 – Well-Washed Planktic Sand.

Continental slope

Facies S1 – Spiculitic Carbonate Silt. These deposits blanket the seafloor deeper than 300 mwd but are found locally as shallow as 200 mwd (Fig. 19.10). Sediment is green to grey to buff, and mixtures of, approximately, 20% clay-size carbonate, 50% silt and 30% fine sand. Silt is mainly broken fragments of pelagic foraminifers, ostracods and pteropods. Fine sand comprises one-quarter sponge spicules, with the rest comprising variable proportions of faecal pellets and pteropods with trace amounts of molluscs, glaucony and infaunal echinoderm spines. Bryozoans, even the whole and fragmented singlets of articulated zooidal forms, are generally absent. About 20% of the sediment cannot be identified under the binocular microscope.

Facies S2 – Bryozoan-rich Carbonate Silt. These sediments generally lie between 150 and 200 mwd, are localized to the Rottnest Shelf and have a conspicuous bryozoan component. Sediment is approximately one-third clay-size carbonate, one-third silt and one-third fine–medium sand (Table 19.3). Sponge spicules and worm tubes are the other most important constituents. These facies correspond to part of the Rottnest Blanket of Collins (1988).

Synthesis and discussion

NWS surface sediment mosaic

The sediments are palimpsest, with diverse particle types of different ages, variably mixed by biological and hydrodynamic processes to form a complex facies mosaic, some of which contains many particles that are out of equilibrium with the present seafloor setting (Fig. 19.11).

Inner ramp. The seafloor <40 mwd is covered with mainly Holocene carbonate sediment but formed by an unexpected mixture of components exhibiting characteristics of both photozoan and heterozoan associations. This is interpreted as resulting from largely oligotrophic conditions in shallow water that are periodically altered by the fluvial influx of terrestrially derived nutrients during storms from an otherwise arid hinterland. Thus, sediments are either mainly biofragmental (Facies 7), except near the Fortescue and DeGrey rivers, where they contain a significant siliciclastic and iron-particle component (Facies 8). This oceanographic situation is further exacerbated by the fact that turbid waters are trapped against the shoreline for lengthy periods. These waters undergo evaporation and move seawards as a saline bottom flow.

Mid ramp. The modern biota here is largely heterozoan but with numerous large benthic foraminifers, probably because of the deep light penetration. This biota is, however, not flourishing, possibly because, although nutrients can extend across this part of the ramp during upwelling, productivity is attenuated by the flow of saline waters across the seafloor from the inner ramp. Thus, the facies are dominated by relict skeletal and lithic intraclasts, together with stranded ooids and peloids (Fig. 19.11; facies 4, 5 and 6); these sediments are now being reworked and mixed by swell and storm waves, as well as internal tides. The sparsity of Holocene particles implies a relatively low accumulation rate.

Outer ramp. A largely pelagic realm today (Facies 1), the otherwise stranded sediments are burrowed, and upwelling is locally recorded as phosphatization and iron precipitation (Facies 3A),

which again implies a relatively low sedimentation rate. These sediments are partly mixed with stranded sediments produced in shallow water during the LGM lowstand and ooid–peloid particles transported into deeper water (Facies 3). Enhanced pelagic production, possibly linked to an oceanic front development at the margin of the Leeuwin Current, has resulted in the elongate ridge of planktic sediment (Facies 2).

Implications for the understanding of ramp sedimentation. Although the modern environment is not always viewed as a good analogue for ramp sedimentation in the rock record (Burchette & Wright 1992), the NWS is a carbonate ramp on the scale of such carbonate realms as the great Bahama Bank and the southern Persian Gulf, and so has important attributes that are fundamental when reconstructing depositional palaeoenvironments and refining the overall ramp model. Of particular importance from this study is the realization that almost all attributes are determined by regional oceanography, in sometimes unexpected ways.

Inner-ramp facies on the NWS are a complicated mix of photozoan and heterozoan elements, reflecting an oligotrophic oceanic setting repetitively perturbed by the influx of fine sediment and nutrients from land. The high energy of the marine system sequesters these terrigenous clastics adjacent to the shoreline, and they are not transported offshore. Thus, in the rock record, a spectrum of inner-ramp facies comprising photozoan carbonates, local evaporites and terrigenous clastics might be expected to contain a heterozoan, nutrient-dependent component.

Saline water outflow along the seafloor, the result of evaporation on the inner ramp, appears to have subdued or arrested modern benthic carbonate production on much of the mid ramp. This may be why, in part, the mid ramp is not more distally steepened. In the rock record, such suppressed production might result in a thin or absent record that could be mistakenly interpreted as a maximum flooding surface.

Changing late Quaternary oceanography has had a profound effect on NWS carbonate sedimentation. During the last global sea-level lowstand and early parts of the subsequent transgression, the NWS was a site of warm-water carbonate sedimentation, with strong offshore winds probably enhancing upwelling and evaporation. There is clear evidence that, during the initial stages of sea-level rise, ooids and peloids, a classic warm-water facies, developed across many tens of kilometres of the then shallow shelf. Yet, this sedimentation ceased at approximately 12 ka, when the lower salinity Leeuwin Current began to flow strongly. It is this change in oceanography that probably resulted in a shift to the present dominance of skeletal carbonate production. In this situation, carbonate production has not been able to keep pace with rising sea level, implying an overall reduced sedimentation rate that has left the seafloor veneered with a mixture of old and new deposits, many of which are out of equilibrium with modern environments.

Thus, for high-energy ramps, particularly those affected by high-amplitude relatively short-period sea-level change (cf. Read 1985), the sedimentary record, especially on the mid ramp and inner ramp, should be blurred. As the NWS illustrates: (1) the rapid rate of sea-level rise stranded shallow-water sediments; (2) the changing oceanography altered the character of the carbonate factory from photozoan to heterozoan (and mixed heterozoan–photozoan on the inner ramp); and (3) the high energy (waves, swells, cyclones and internal tides) mixed sediments formed at different times during transgression. This situation is most obvious on the NWS because the area is near the temperate–tropical carbonate system boundary and, so, different carbonate sediments from different systems are mixed. It is also likely to be important in similar, but wholly tropical or temperate, high-energy systems but the differences are more subtle.

This large oceanic ramp, located in the tropics yet near the temperate carbonate realm, is one of the largest such systems in the modern world. The surficial carbonate sedimentary facies

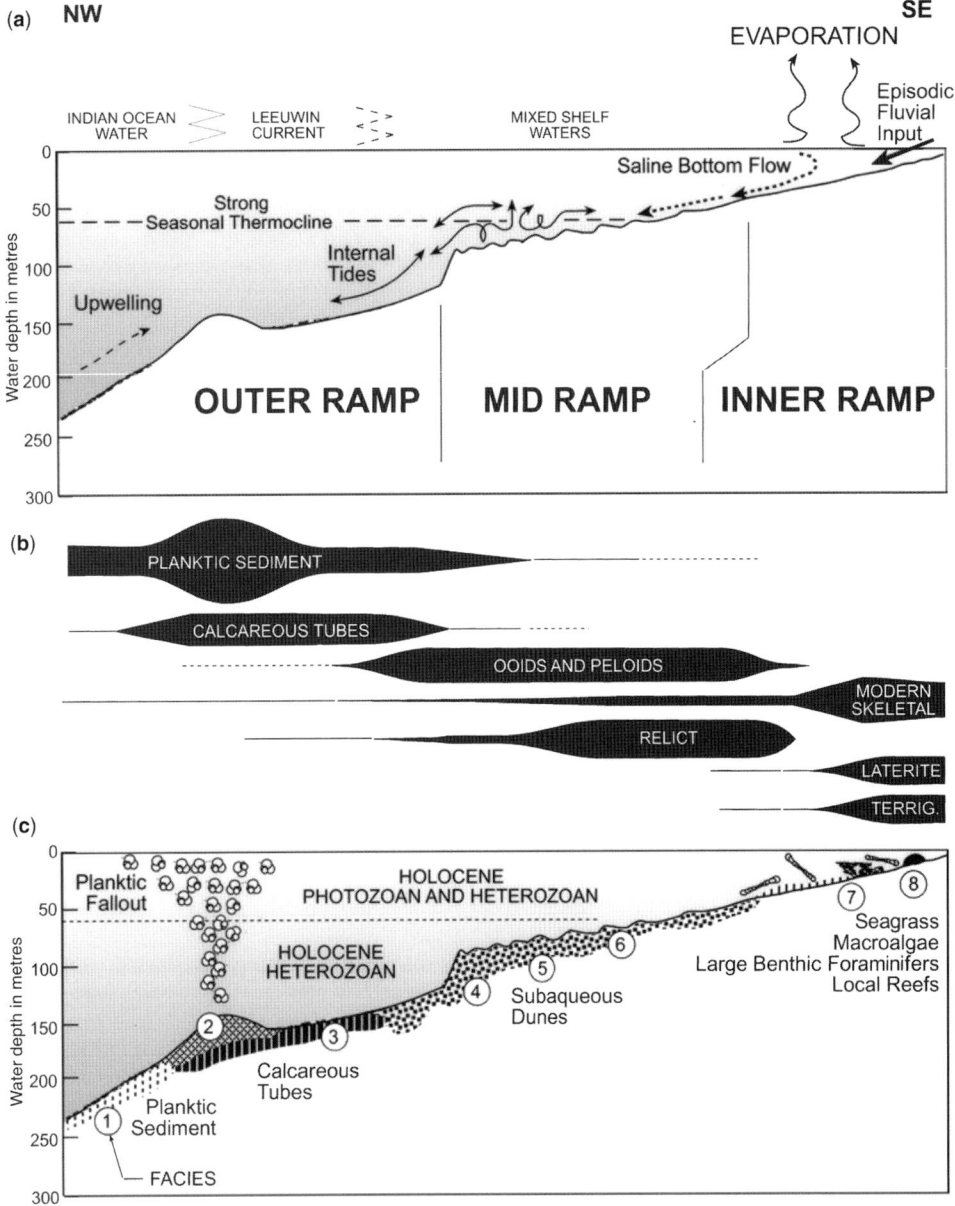

Fig. 19.11. Summary diagrams illustrating: (**a**) the major oceanographic–climatic factors affecting sedimentation on the North West Shelf; (**b**) the spatial distribution of major particle types across the ramp; and (**c**) the sedimentary facies (numbers) (Table 19.2) and major features of the ramp (after James *et al.* 2004).

display complexities that are largely explainable in the context of modern and late Quaternary oceanography. As such, it is an important template to be used when interpreting ramp sedimentation in the rock record.

SWS

The continental margin of SW Australia, facing the Indian Ocean, comprises a southern shelf with a steep seaward face and a northern, gently inclined ramp. The shelf north of Perth is contiguous with the northern part of the Rottnest Shelf described by Collins (1988).

Northern Rottnest Shelf. Throughout its 700 km length from Cape Naturaliste to Kalbarri, the Rottnest Shelf is a deep, open, relatively flat-topped, incipiently rimmed shelf. The rim is at >20 mwd, and only in the Houtman Abrolhos does it support shallow reef platforms. The shelf spans 6° of latitude, with shelf bottom-water temperature in the south and 'subtropical' in the north, and with Perth (latitude 32°S) marking the transition.

In the northern Rottnest Shelf (Fig. 19.12), the wide inner-shelf plain is a sand-and-gravel-veneered limestone substrate with prolific growth of seagrass and soft, non-calcified algae. Coralline algal encrusted rocky substrates are colonized by lush macrophytes, sponges, bivalves, bryozoans and pavements of rhodolites. Intervening seagrass meadows alternate with sweeps of open sand, generally rippled and commonly burrowed by decapods and infaunal bivalves. Grasses are ephemeral substrates for numerous articulated corallines and benthic foraminifers, whereas rhodolites host vibrant growths of larger foraminifers. The outer-ridge complex is similar except that the substrate is rocky and shallower overall, leading to a luxuriant marine forest of supple macrophytes rising from a coralline-algae-encrusted limestone covered by hectares of rhodolites.

Sediment reflects this euphotic, warm (18–22 °C water temperature) carbonate factory, where all environments are within the zone of wave abrasion. The area is one of downwelling with local upwelling of nutrient-depleted waters during the summer. Variably abraded particles are subtropical, a mixture of temperate and tropical elements (cf. James 1997). The bulk of the material is from coralline algae, benthic foraminifers, bryozoans and molluscs. Zooxanthellate corals are scattered, of low diversity and never abundant enough to form reefs. Large benthic foraminifers are conspicuous throughout. *Halimeda* is not sufficiently calcified to produce sediment particles.

(a)

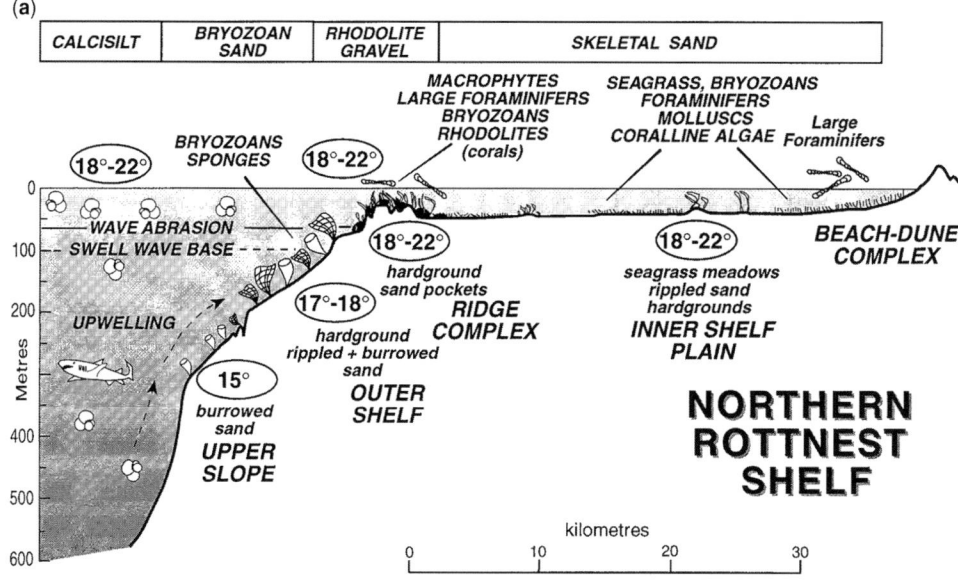

(b)

CARNARVON RAMP

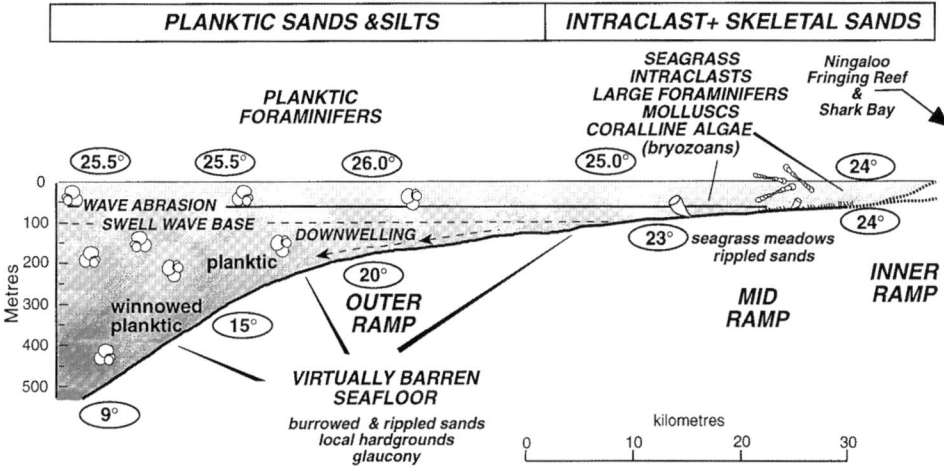

Fig. 19.12. Summary profiles illustrating disposition of major seafloor types, benthic organisms, sediments and water temperatures (in ellipses) of (**a**) the northern Rottnest Shelf and (**b**) the Carnarvon Ramp (after James *et al.* 1999).

The dramatically different subphotic outer shelf is a moderately steep slope of bioeroded limestone outcrop, and intervening burrowed and rippled sand patches in seasonally upwelling waters that range in temperature from 17 to 18 °C. Rocky outcrops of previously exposed Pleistocene limestone and variably cemented Holocene hardgrounds are covered by innumerable sponges with prolific, complex bryozoan communities growing on and between them. Bryozoans, *in situ* molluscs and small benthic foraminifers are poorly sorted deposits episodically disturbed by storm activity and heavily bioturbated by sea stars.

The upper slope is a relatively barren sediment surface with a few sponges, delicate bryozoans and bivalves on a bioturbated muddy seafloor. Bottom waters are generally colder than 15 °C. The fine sand and silt, composed of small bryozoan particles and benthic foraminifers, is exceptionally rich in sponge spicules washed down from shallower environments and pelagic foraminifers that have settled out of the water column.

In short, this is a 'subtropical, incipiently rimmed shelf' whose sediments reflect partly, but not wholly, the living biota. It is similar to other subtropical settings such as the Mediterranean and eastern South America. The sediments do not, however, conform to any of the facies definitions for such shelves because, although they contain abundant corallines and large foraminifers, they lack green calcareous algae.

Carnarvon Ramp. The seafloor of the ramp, a gentle incline throughout (Fig. 19.12), is a modern example of what is visualized as ramp morphology in the geological record (cf. Burchette & Wright 1992). The euphotic mid ramp, strikingly different from the depth-equivalent open-shelf plain and ridge complex of the northern Rottnest Shelf, has no macrophytes, relatively little calcareous benthos and low numbers of bryozoans. Bottom-water temperatures are tropical, 23–24 °C, and there is seasonal downwelling and saline water outflow from Shark Bay. The seafloor is one of seagrass meadows, rippled sand patches and rocky pavements. Sediment, although sandy and extensively reworked, is dominated by relict fragments, numerous intraclasts, coralline algae and benthic foraminifers. Intraclasts are modern and relict: that is, fresh and iron-stained, and cemented by Mg-calcite (12–16 mol% $MgCO_3$). Coralline algae are mainly buff-coloured and so probably stranded. Foraminifers are a mixture of modern, stranded and relict. Relict distributions of *Amphistigina* spp., *Heterostegina depressa* and *Amphisorus* are consistently 20–50 m deeper than their modern occurrences. The overall impression is one of relatively low carbonate production and accumulation.

The outer ramp is pelagic in character. Water temperatures are 20 °C in 100–200 mwd but fall to <15 °C below 300 mwd. The seafloor is a vast, gently sloping surface of burrowed sand that is rippled in shallower waters and virtually devoid of upright

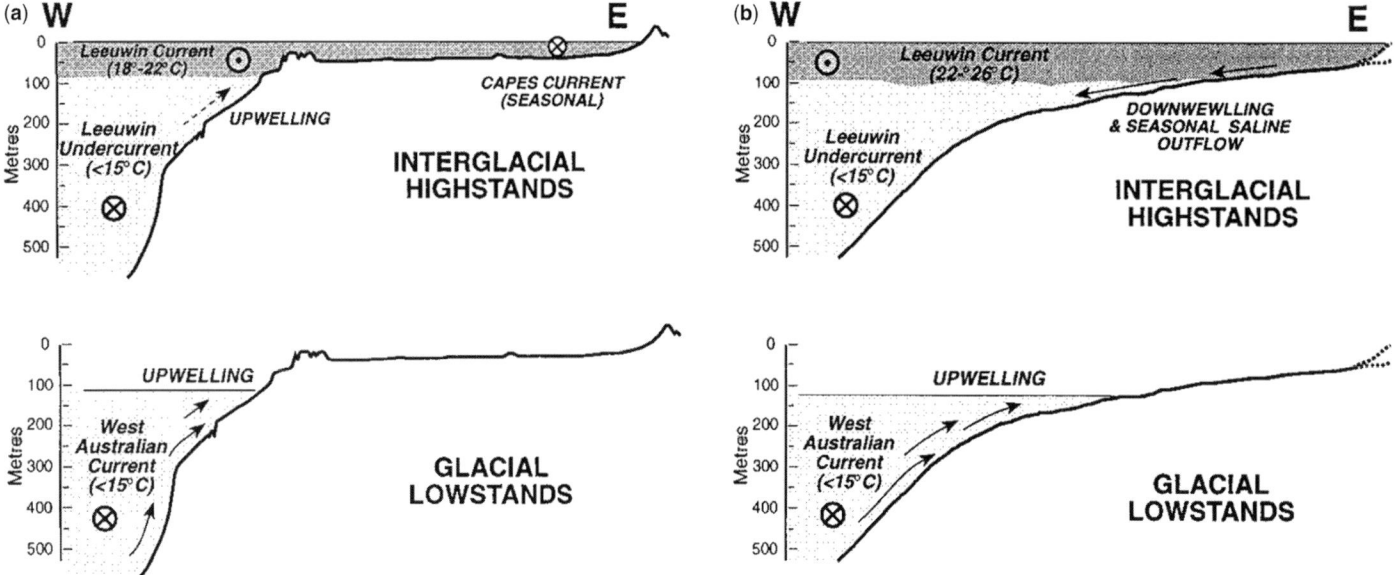

Fig. 19.13. Diagrammatic cross-sections of (**a**) the northern Rottnest Shelf and (**b**) Carnarvon Ramp, illustrating the distribution of major ocean currents during interglacial periods, similar to today, and the interpretation of oceanographic conditions during glacial periods when sea level was significantly lower than it is today (after James *et al.* 1999).

filter-feeding organisms throughout. Seafloor sands comprise mostly pelagic foraminifer shells, mixed with intraclasts in shallower depths, and winnowed into foraminifer test lags between 180 and 500 mwd, especially off Shark Bay and the Ningaloo

Reef. The lack of benthic sediment production is also indicated by the extensively bioeroded nature of carbonate intraclasts, glauconite filling of intraskeletal voids, early seafloor cementation, and local Fe and Mn encrustation.

Northwest Shelf reefs: Ramp margin reefs, high subsidence

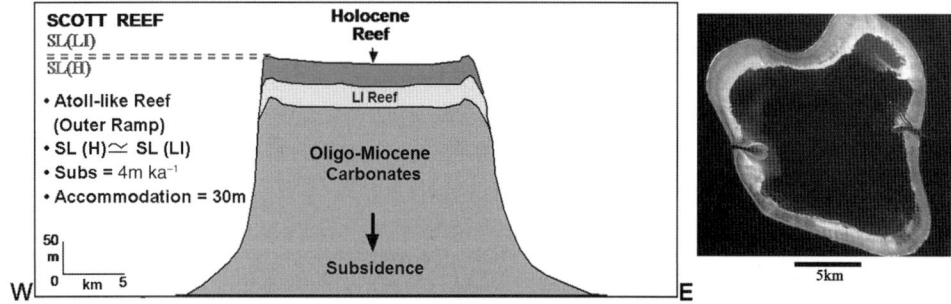

Southwest Shelf reefs: Fringing to outer shelf reefs, low subsidence

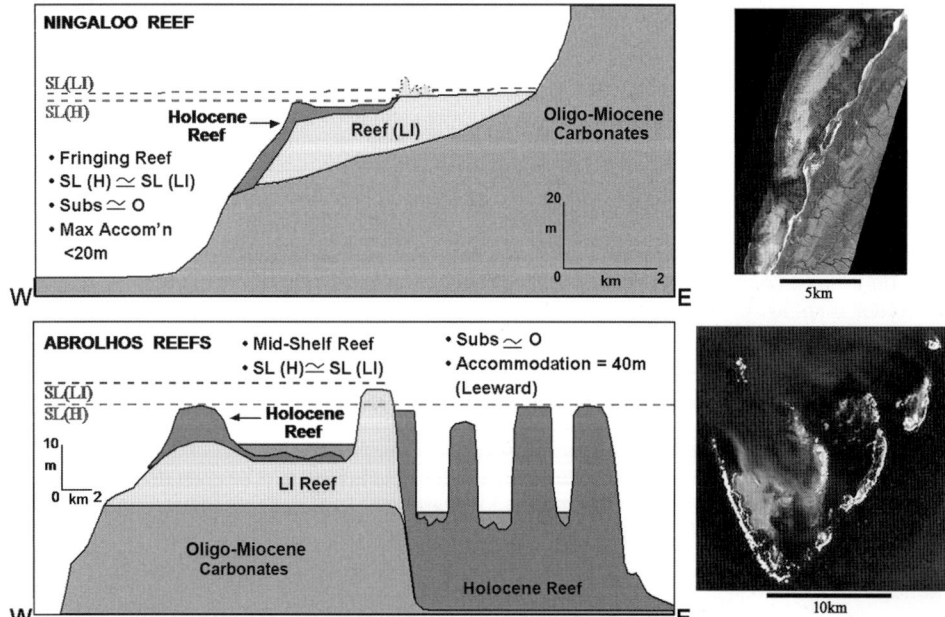

Fig. 19.14. Comparative evolution of NWS and SWS reefs. Note the differences in setting, antecedent topography, subsidence rate and morphology. Scott Reef and Abrolhos imagery sourced from NASA's JSC archive; Ningaloo truecolour Hyperspectral imagery sourced from WAMSI.

This is currently a 'subtropical starved ramp' in the sense that, although mid-ramp bottom-water temperatures are tropical, the biota is largely subtropical and neither the mid ramp nor the outer ramp is a site of active modern carbonate sediment production. Biodegraded sediments and clasts, however, indicate that carbonate production was active in the recent past. The implication is that as sea level rose to its present position, conditions became unfavourable for benthic carbonate production, except by small benthic foraminifers. Such conditions are probably related to the outflow of saline waters from Shark Bay, which are then mixed across the ramp – conditions similar to those that have been described for the Lincoln Shelf, South Australia (James *et al.* 1997*a*). Downwelling seems important, at least for part of the year, thus depriving filter feeders of potential nutrients. Regardless, the ramp is largely unproductive. Finally, strong current flow and sediment winnowing on the outer ramp is similar to the Loop Current along the West Florida Shelf edge (Mullins *et al.* 1987).

Temporal implications. The different sectors reflect, to a first degree, diverse tectonics: the northern Rottnest Shelf is a region of tectonic quiescence and low accommodation; the Carnarvon Ramp is an area of active subsidence and high accommodation. Oceanography appears equally important. Results confirm that the southward-flowing, near-surface Leeuwin Current not only transports warm water into cool regions but also prevents significant upwelling. If the Leeuwin Current was not active during glacial lowstands, as many studies suggest, this whole region would have been one of cool-water carbonate sedimentation, with reefs confined to local shoreline refugia (Fig. 19.13). The margin may be a good modern analogue for 'greenhouse' conditions when equatorial water temperatures were cooler than today and low-latitude shallow-marine environments more 'subtropical'.

Coral reef evolution. Coral reef systems of the NWS and SWS differ in morphology, setting and subsidence history (Collins 2010). The Rowley Shoals and Scott Reef bordering the NWS are isolated reefs overlying carbonate platforms, along a Miocene shoreline distal to the ramp. The combination of high subsidence rate and recurring sea-level highstands caused repeated reef stacking as a sustained pattern of reef growth (Fig. 19.14).

The Ningaloo fringing reef at latitude 20–22°S records Holocene and last interglacial phases of reef growth but the last interglacial reef was widespread here as a fringing system whose elevation was 2–3 m above that of the Holocene transgression. Without subsidence, this coastal setting provided only limited accommodation for Holocene reef development, which thinly mantles the last interglacial reef, the template for Holocene growth.

On the SWS, the Houtman Abrolhos reefs (shelf-edge reefs with atoll-like form: Fig. 19.14) aggraded in a tectonically stable environment with the last interglacial stage of reef growth forming central platform emergent reefs present as islands 2–5 m above sea level today. Rising Holocene sea levels partially drowned the old reef topography which was colonized in eroded portions of the last interglacial substrate, building a 10 m-thick Holocene windward reef, but most Holocene reef development took place by leeward platform growth in wave-protected environments in the lee of the last interglacial platform islands, filling the entire 40 m of available accommodation and building Holocene reefs.

This research grew out of a collaboration initiated by the authors as part of a wider Australian shelf project, and also draws on coral reef research co-ordinated by the first author (L. B. Collins). Funding from the National Sciences and Energy Research Council for N. P. James, and from the Australian Research Council (ARC) for Y. Bone and L. B. Collins is gratefully acknowledged, as is vessel support for Research Vessel (R. V.) *Franklin*, which was funded by the ARC and the Commonwealth Scientific and Industrial Research Organization (CSIRO). The summary provided here draws heavily on R. V. *Franklin* results reported in James *et al.* (1999, 2004). Collaborating researchers from the associated organizations involved, the Captain and crew of R. V. *Franklin*, and research students from Adelaide and Curtin universities are gratefully acknowledged.

References

BONE, Y. & JAMES, N. P. 1993. Bryozoans as carbonate sediment producers on the cool-water Lacepede Shelf, southern Australia. *Sedimentary Geology*, **86**, 247–271.

BURCHETTE, T. P. & WRIGHT, V. P. 1992. Carbonate ramp depositional systems. *Sedimentary Geology*, **79**, 3–57.

BURLING, M. C., PATTIARATCHI, C. B. & IVEY, G. N. 1998. The influence of the Leeuwin Current on the exchange of water between Shark Bay, Western Australia, and the Continental Shelf. Paper presented at the Proceedings of the Ninth Physics of Estuaries and Coastal Seas Conference, Matsuyama, Japan.

CHAPPELL, J. & SHACKLETON, N. J. 1986. Oxygen isotopes and sea level. *Nature*, **324**, 137–140.

COLLINS, L. B. 1988. Sediments and history of the Rottnest Shelf, a swell-dominated, non-tropical carbonate margin. *Sedimentary Geology*, **60**, 15–49.

COLLINS, L. B. 2010. Controls on morphology and growth history of coral reefs of Australia's western margin. *In*: MORGAN, W. A., GEORGE, A. D., HARRIS, P. M., KUPECZ, J. A. & SARG, J. F. (eds) *Cenozoic Carbonate Systems of Australasia*. SEPM Special Publications, Tulsa, **95**, 195–217.

COLLINS, L. B. & TESTA, V. 2010. Quaternary development of resilient reefs on the subsiding Kimberley continental margin, northwest Australia. *Brazilian Journal of Oceanography*, **58**, 67–77.

COLLINS, L. B., ZHU, Z. R. ET AL. 1993. Late Quaternary evolution of coral reefs on a cool-water carbonate margin: the Abrolhos carbonate platforms, southwest Australia. *Marine Geology*, **110**, 203–212.

COLLINS, L. B., ZHU, Z. R. & WYRWOLL, K.-H. 1996. The structure of Easter Platform, Houtman Abrolhos reefs: Pleistocene foundations and Holocene reef growth. *Marine Geology*, **135**, 1–13.

COLLINS, L. B., FRANCE, R. E., ZHU, Z. R. & WYRWOLL, K.-H. 1997. Warm-water platform and cold-water carbonates of the Abrolhos Shelf, southwest Australia. *In*: JAMES, N. P. & CLARKE, J. A. D. (eds) *Cool-Water Carbonates*. SEPM Special Publications, Tulsa, **56**, 23–36.

COLLINS, L. B., ZHU, Z. R., WYRWOLL, K.-H. & EISENHAUER, A. 2003. Late Quaternary structure and development of the northern Ningaloo Reef, Australia. *In*: BLANCHON, P. & MONTAGGIONI, L. (eds) *Late Quaternary Reef Development. Sedimentary Geology*, **159**, 81–94.

COLLINS, L. B., READ, J. F., HOGARTH, J. W. & COFFEY, B. P. 2006. Facies, outcrop gamma ray and C–O isotopic signature of exposed Miocene subtropical continental shelf carbonates, North West Cape, Western Australia. *Sedimentary Geology*, **185**, 1–19.

CRESSWELL, G. R. 1991. The Leeuwin Current – observations and recent models. *In*: PEARCE, A. F. & WALKER, D. I. (eds) *The Leeuwin Current. Journal of the Royal Society of Western Australia*, **74**, 1–14.

CRESSWELL, G. R. & GOLDING, T. J. 1980. Observations of a south-flowing current in the southeastern Indian Ocean. *Deep-Sea Research*, **27A**, 449–466.

DAVIES, G. R. 1970. Carbonate bank sedimentation, eastern Shark Bay, Western Australia. *In*: LOGAN, B. W., DAVIES, G. R., READ, J. F. & CEBULSKI, D. E. (eds) *Carbonate Sedimentation and Environments*. American Association of Petroleum Geologists, Memoirs, **13**, 85–168.

EXON, N. F. & COLWELL, J. B. 1994. Geological history of the outer North West Shelf of Australia: a synthesis. *AGSO Journal of Geology and Geophysics*, **15**, 177–190.

FAIRBANKS, R. G. 1989. A 17 000-year glacio-eustatic sealevel record: influence of glacial melting rates on the Younger Dryas event and deep-ocean circulation. *Nature*, **342**, 637–642.

FARRELL, J. W., CLEMENS, S. C. & GROMET, P. C. 1995. Improved chronostratigraphic reference curve of late Neogene sea-water $^{87}Sr/^{86}Sr$. *Geology*, **23**, 403–406.

GERSBACH, G. H., PATTIARATCHI, C. B., IVEY, G. N. & CRESSWELL, G. R. 1999. Upwelling on the south-west coast of Australia – source of the Capes Current. *Continental Shelf Research*, **19**, 363–400.

GINGELE, F. X., DE DECKKER, P. & HILLENBRAND, C. D. 2001*a*. Clay mineral distribution in surface sediments between Indonesia and NW Australia – source and transport by ocean currents. *Marine Geology*, **179**, 135–146.

GINGELE, F. X., DE DECKKER, P. & HILLENBRAND, C. D. 2001*b*. Late Quaternary fluctuations of the Leeuwin Current and palaeoclimates on the adjacent land masses: clay mineral evidence. *Australian Journal of Earth Science*, **48**, 867–874.

HATCHER, B. G. 1991. Coral reefs in the Leeuwin Current – an ecological perspective. *Journal of the Royal Society of Western Australia*, **74**, 115–128.

HEGGIE, D. T., SKYRING, G. W. ET AL. 1990. Organic carbon recycling and modern phosphorite formation on the East Australian continental margin: a review. *In*: NOTHOLT, A. J. G. & JARVIS, I. (eds) *Phosphorite Research and Development*. Geological Society, London, Special Publications, **52**, 87–117.

HOCKING, R. M. 1990. Carnarvon Basin. *In*: *Geology and Mineral Resources of Western Australia*. Western Australian Geological Survey, Memoirs, **3**, 457–493.

JAMES, N. P. 1997. The cool-water carbonate depositional realm. *In*: JAMES, N. P. & CLARKE, J. A. D. (eds) *Cool-Water Carbonates*. SEPM Special Publications, Tulsa, **56**, 1–22.

JAMES, N. P., GINSBURG, R. N., MARSZALEK, D. M. & CHOQUETTE, P. W. 1976. Facies and fabric specificity of early subsea cements in shallow Belize reefs. *Journal of Sedimentary Petrology*, **46**, 523–544.

JAMES, N. P., BONE, Y., VON DER BORCH, C. C. & GOSTIN, V. A. 1992. Modern carbonate and terrigenous clastic sediments on a cool-water, high-energy, mid-latitude shelf; Lacepede, southern Australia. *Sedimentology*, **39**, 877–903.

JAMES, N. P., BONE, Y., HAGEMAN, S., GOSTIN, V. A. & FEARY, D. A. 1997*a*. Cool-water carbonate sedimentation during the Terminal Quaternary, high-amplitude, sea-level cycle: Lincoln Shelf, Southern Australia. *In*: JAMES, N. P. & CLARKE, J. A. D. (eds) *Cool-Water Carbonates*. SEPM Special Publications, Tulsa, **56**, 53–76.

JAMES, N. P., BONE, Y. & KYSER, T. K. 1997*b*. Brachiopod 18O values do indicate ambient oceanography. *Geology*, **25**, 551–554.

JAMES, N. P., COLLINS, L. B., BONE, Y. & HALLOCK, P. 1999. Subtropical carbonates in a temperate realm: modern sediments on the Southwest Australian shelf. *Journal of Sedimentary Research*, **69**, 1297–1321.

JAMES, N. P., BONE, Y., COLLINS, L. B. & KYSER, T. K. 2001. Surficial sediments of the Great Australian Bight: facies dynamics and oceanography on a vast cool-water carbonate shelf. *Journal of Sedimentary Research*, **71**, 549–567.

JAMES, N. P., BONE, Y., KYSER, T. K., DIX, G. R. & COLLINS, L. B. 2004. The importance of changing oceanography in controlling late Quaternary carbonate sedimentation on a high-energy, tropical, oceanic ramp: north-western Australia. *Sedimentology*, **51**, 1179–1205.

JENSEN, E. & VEUM, T. 1990. Evidence for a two-step deglaciation and its impact on North Atlantic deep water circulation. *Nature*, **343**, 612–616.

JOHNSTONE, M. H., LOWRY, D. C. & QUILTY, P. G. 1973. The geology of southwestern Australia – a review. *Journal of the Royal Society of Western Australia*, **56**, 5–15.

JONES, H. A. 1971. Late Cenozoic sedimentary forms on the northwest Australian continental shelf. *Marine Geology*, **10**, M20–M26.

JONES, H. A. 1973. *Marine Geology of the Northwestern Australian Continental Shelf*. Bureau of Mineral Resources, Geology and Geophysics, Bulletin, **136**.

KENDRICK, G. W., WYRWOLLL, K.-H. & SZABO, B. J. 1991. Pliocene-Pleistocene coastal events and history along the western margin of Australia. *Quaternary Science Reviews*, **10**, 419–439.

LOGAN, B. W. 1987. *The MacLeod Evaporite Basin, Western Australia*. American Association of Petroleum Geologists, Memoirs, **44**.

LOGAN, B. W., DAVIES, G., READ, J. F. & CEBULSKI, D. E. 1970. *Carbonate Sedimentation and Environments, Shark Bay, Western Australia*. American Association of Petroleum Geologists, Memoirs, **13**.

LOGAN, B. W., READ, J. F., HAGAN, G. M., HOFFMAN, P., BROWN, R. G., WOODS, P. J. & GEBELEIN, C. D. 1974. *Evolution and Diagenesis of Quaternary Carbonate Sequences, Shark Bay, Western Australia*. American Association of Petroleum Geologists, Memoirs, **22**.

MACINTYRE, I. G. 1985. Submarine cements – the peloidal question. *In*: SCHNEIDERMANN, N. & HARRIS, P. M. (eds) *Carbonate Cements*. SEPM Special Publications, Tulsa, **36**, 109–116.

MAIKLEM, W. R. 1967. Black and brown speckled foraminiferal sand from the southern part of the Great Barrier Reef. *Journal of Sedimentary Petrology*, **37**, 1023–1030.

MARSHALL, J. F. 1983. Geochemistry of iron-rich sediments on the outer continental shelf off northern New South Wales. *Marine Geology*, **51**, 163–175.

MULLINS, H. T., GARDULSKII, A. F., WISE, S. W. & APPLEGATE, J. 1987. Middle Miocene oceanographic event in the eastern Gulf of Mexico: implications for seismic stratigraphic succession and Loop Current/Gulf Stream circulation. *Geological Society of America, Bulletin*, **98**, 702–713.

NELSON, C. S. 1988. An introductory perspective on non-tropical shelf carbonates. *Sedimentary Geology*, **60**, 3–14.

PATTIARATCHI, C. B. & BUCHAN, S. J. 1991. Implications of long-term change for the Leeuwin Current. *In*: PEARCE, A. F. & WALKER, D. I. (eds) *The Leeuwin Current. Journal of the Royal Society of Western Australia*, **74**, 133–140.

PEARCE, A. F. 1991. Eastern boundary currents of the southern hemisphere. *In*: PEARCE, A. F. & WALKER, D. I. (eds) *The Leeuwin Current. Journal of the Royal Society of Western Australia*, **74**, 35–46.

PLAYFORD, P. E., COCKBAIN, A. E. & LOW, G. H. 1976. Geology of the Perth Basin. *Western Australian Geological Survey, Bulletin*, **124**, 186.

PRELL, W. L., HUTSON, W. H., WILLIAMS, D. F., BE, A. W. H., GEITZENAUER, K. & MOLFINO, B. 1980. Surface circulation of the Indian Ocean during the last glacial maximum, approximately 18,000 years, BP. *Quaternary Research*, **14**, 309–336.

PURCELL, P. G. & PURCELL, R. R. 1988. The North West Shelf of Australia – an introduction. *In*: PURCELL, P. G. & PURCELL, R. R. (eds) *The North West Shelf, Australia. Proceedings of Petroleum Exploration Society of Australia Symposium*. Petroleum Exploration Society of Australia, Perth, 3–15.

PURDY, E. G. 1963. Recent calcium carbonate facies of the Great Bahama Bank. 2. Sedimentary facies. *Journal of Geology*, **71**, 472–497.

READ, J. F. 1974. Calcrete deposits and Quaternary sediments, Edel Province, Shark Bay, Western Australia. *In*: LOGAN, B. W., READ, J. F., HAGAN, G. M., HOFFMAN, P., BROWN, R. G., WOODS, P. J. & GEBELEIN, C. D. (eds) *Evolution and Diagenesis of Quaternary Carbonate Sequences*. American Association of Petroleum Geologists, Memoirs, **22**, 250–282.

READ, J. F. 1985. Carbonate platform facies models. *American Association of Petroleum Geologists Bulletin*, **69**, 1–21.

SEMENIUK, V. 1993. The Pilbara coast: a riverine coastal plain in a tropical arid setting, northwest Australia. *Sedimentary Geology*, **83**, 235–256.

SIMONE, L. 1980. Ooids: a review. *Earth Science Reviews*, **16**, 319–355.

TRENDALL, A. F. 1983. The Hamersley Basin. *In*: TRENDALL, A. F. & MORRIS, R. C. (eds) *Iron Formations – Facts and Problems*. Elsevier, Amsterdam, 69–129.

TWIGGS, E. J. & COLLINS, L. B. 2010. Development and demise of a fringing coral reef during Holocene environmental change, eastern Ningaloo Reef, Western Australia. *Marine Geology*, **27**, 520–36.

VERON, J. E. N. 1995. *Corals in Space and Time*. Comstock/Cornell, Ithaca, NY.

WILLIAMS, M. A. J. 2001. Chapter 1 – Quaternary climate changes in Australia and their environmental effects. *In*: GOSTIN, V. A. (ed.) *Gondwana to Greenhouse*. Geological Society, Australia, Special Publications, **21**, 3–11.

Chapter 20

The continental shelves of SE Australia

COLIN V. MURRAY-WALLACE

School of Earth & Environmental Sciences, University of Wollongong, NSW 2522, Australia
(e-mail: cwallace@uow.edu.au)

Abstract: The continental shelves of SE Australia from the eastern Great Australian Bight to southern New South Wales are situated within a tectonically stable, passive continental margin. The shelves began to develop during Late Cretaceous time and preserve a rich but punctuated record of shallow-marine sedimentation from the Late Paleocene to the present day. The southern margin is the world's largest modern, cool-water temperate carbonate sediment province, a function of the limited terrigenous-clastic sediment being delivered to the coast in response to continental-scale aridity, endorheic drainage and low topographical relief. With the exception of the South Australian gulfs and the protected portions of Bass Strait, the entire shelf sector of SE Australia is a high-energy, storm- and swell-dominated, microtidal province. Along the southern margin, the dominant sedimentary constituents in the modern shelf sediments are skeletal carbonate bioclasts derived from marine invertebrates that include molluscs, foraminifers, bryozoans, coralline algae and echinoids, variably intermixed with quartz sand and heavy minerals (Heterozoan Association). Sediments of Pleistocene and Holocene age are compositionally similar to their modern equivalents, and all show varying degrees of reworking and incorporation of older bioclasts to form the palimpsest sedimentary successions that characterize the entire region. The southern New South Wales sector of the Eastern Australian Shelf is dominated by siliciclastic sediments but preserves relict molluscan-rich carbonate sediments of glacial-age lowstand origin in the outer-shelf region in present water depths exceeding 100 m. The Quaternary stratigraphical record of the SE Australian continental shelves is dominated by successions that formed in the last glacial cycle, but older, inner-shelf successions have been documented from geographically restricted regions such as the South Australian gulfs, Roe and Coorong coastal plains, and the Bass Strait Islands. The outer-shelf successions are not in hydrodynamic equilibrium with modern shelf processes, as reflected in the paucity of mud, and are either largely relict (New South Wales) or dominated by complex modern bryozoan associations.

The continental shelves of SE Australia have resulted from a variety of geological processes operative over a range of time-scales, commencing with the fragmentation of Gondwana in Late Cretaceous time. This chapter outlines the broad physical characteristics of the continental shelves in SE Australia, extending from the eastern sector of the Great Australian Bight through to the southern portion of the Eastern Australian Shelf in New South Wales (NSW). The contrasting modes of sedimentation (the dominance of cool-water skeletal carbonates along much of the southern margin and terrigenous-clastic sediment along the NSW portion of the Eastern Australian Shelf), records of late Quaternary relative sea-level, environmental and climate changes as derived from these shelves, and the significance of geological inheritance in the general development of the continental shelves of this region represent the central themes of this chapter.

Geological background and general environmental setting

The initial development of the continental shelves of SE Australia relates to the early phases of fragmentation of Gondwana. The onset of rifting, giving rise to what is now the coastline of NSW, commenced approximately 83.5 Ma ago with the opening of the Tasman Sea (Veevers 2000) along a broadly NW-trending spreading centre and terminated in the middle Paleocene some 60 Ma ago (Roy & Thom 1991). In contrast, the South Australian coastline and continental shelf were initiated with the onset of rifting around 99 Ma ago. Subsequent drifting and the formation of a well-defined continental shelf began with the onset of more rapid divergence approximately 43 Ma ago (Veevers 2000). Today, the Australian continent is moving slightly in excess of 60 mm NNE per annum. The general configuration of the continental shelves of SE Australia, therefore, reflects these early geological processes and, although representing structurally inherited features, are geologically youthful compared with the overall antiquity of many Australian landscapes (Twidale 2007).

Great Australian Bight

Despite the perceived tectonic quiescence of the Australian continent, the continental shelves of SE Australia traverse a range of geotectonic settings and, although subtle, have influenced the evolution of the continental shelves and coasts of this region. The Great Australian Bight is dominated by a cliffed coastline formed on Palaeogene and Neogene limestones of the Eucla Basin and Nullarbor Plain (Wilson Bluff, Abrakurrie and Nullarbor limestones), represented by the Baxter Cliffs in Western Australia and the Bunda Cliffs in South Australia (SA) that attain heights of between 40 and 90 m (Benbow *et al.* 1995) (Figs 20.1–20.3). Between these features is the Roe Plain, which is backed by the Hampton Range, a relict sea cliff 290 km long (Fig. 20.2). At its widest, the Roe Plain is 35 km across, slopes seawards and is draped by the Roe Calcarenite, a mixed carbonate sand and coquina of latest Neogene–early Pleistocene age (James *et al.* 2006). The Nullarbor Plain is situated landwards of the Roe Plain and, covering an area of approximately 240 000 km^2, is the world's largest karst province (Miller *et al.* 2012). In this region, the modern continental shelf (Eucla Shelf) is up to 230 km wide and across the bight is about 1200 km long (east–west). The inner shelf adjacent the Roe Plain is particularly shallow, with water depths attaining only 10 m up to 5 km off shore. The inner shelf is an erosional feature, having cut through the Abrakurrie and Nullarbor limestones (James *et al.* 2006). The shelf edge is in a present water depth of 200 m (Rollet *et al.* 2001). A marine terrace termed the Ceduna Terrace occurs on the continental slope immediately to the south of the Bunda Cliffs and western Eyre Peninsula (Figs 20.1 & 20.2). The fan-shaped feature has been interpreted as a deltaic succession of Late Cretaceous (Albian–Cenomanian) age (Rollet *et al.* 2001) and is truncated by several submarine canyons. Eyre Terrace is a similar feature on the continental slope south of the Baxter Cliffs and Roe Plain (Fig. 20.1). The Palaeogene shelf margin is accordingly several hundred kilometres landwards of the Late Cretaceous shelf margin. The continental shelf adjacent to the NW Eyre Peninsula (Ceduna sector from Fowlers Bay to Streaky Bay) is a

From: CHIOCCI, F. L. & CHIVAS, A. R. (eds) 2014. *Continental Shelves of the World: Their Evolution During the Last Glacio-Eustatic Cycle.*
Geological Society, London, Memoirs, **41**, 273–291. http://dx.doi.org/10.1144/M41.20

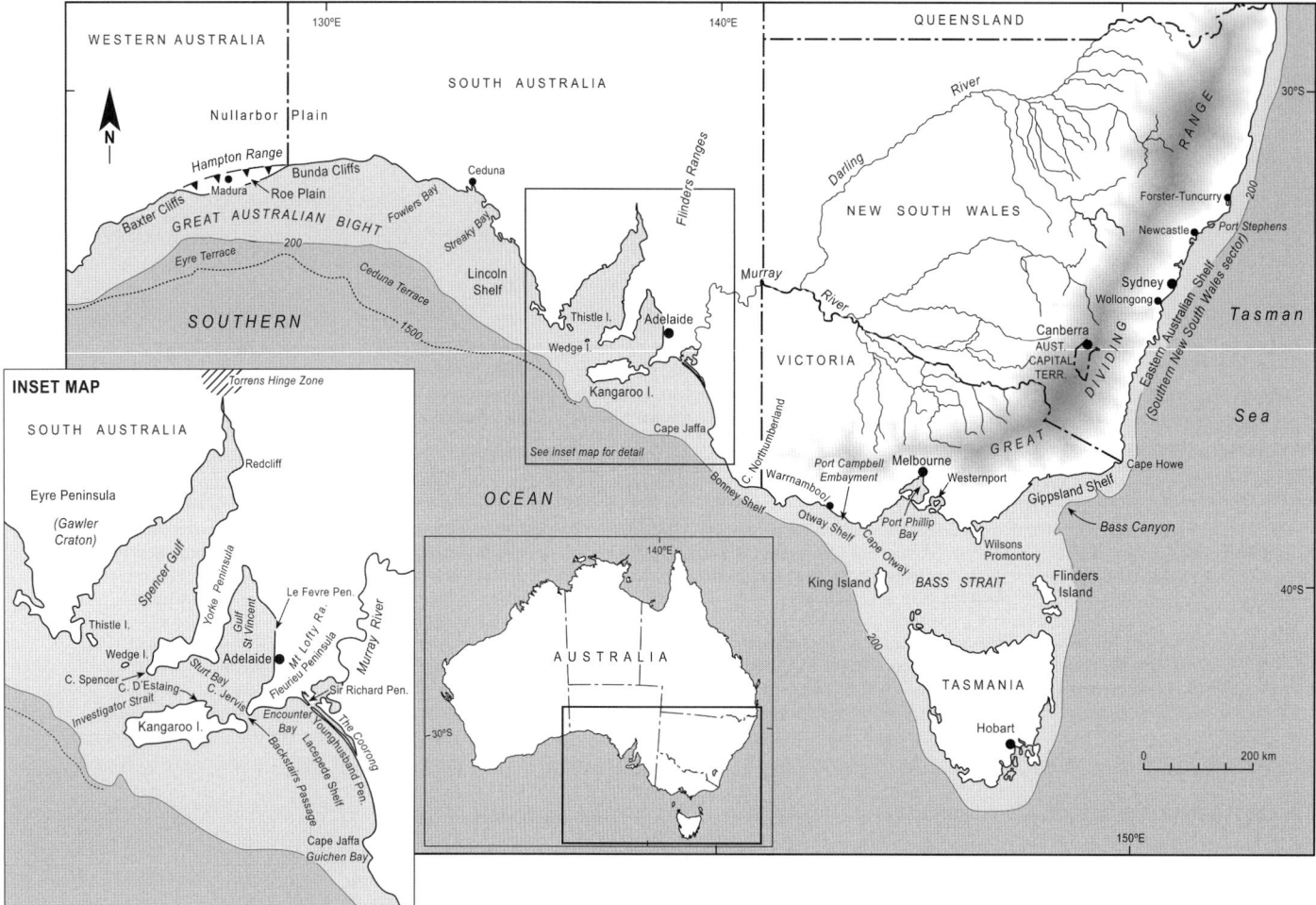

Fig. 20.1. Location map of the continental shelves of SE Australia and associated coastal features.

partially shaved shelf in which the local bedrock (Palaeogene–Neogene and Pleistocene limestones) is eroded and produces a limited supply of sediment that is subsequently transported elsewhere under vigorous current action (James *et al.* 2001; James & Bone 2011).

The Lincoln Shelf (60 000 km^2) is a distally steepened ramp situated to the east of the Eucla Basin in the Great Australian Bight (Fig. 20.1, Table 20.1) and has developed on the southern part of the Gawler Craton on which Eyre Peninsula is situated. The Gawler Craton comprises granites and gneisses that have remained tectonically stable since 1.45 Ga (Parker 1993). Geographically extensive accumulations of aeolianite, large open-coastal embayments with relict barrier–lagoon facies of the last interglacial maximum (Marine Isotope Substage 5e (MIS), 130–118 ka) and bedrock headlands characterize the Eyre Peninsula coastline (Sprigg 1979). The Lincoln Shelf is relatively deep with approximately half the shelf below 100 m of present sea level, and the shelf edge occurs between 150 and 160 m, some 100–180 km offshore (James *et al.* 1997). This shelf sector extends to the south of southern Eyre Peninsula and to the west of Kangaroo Island (Fig. 20.1).

South Australian gulfs

East of the Gawler Craton, Spencer Gulf, Yorke Peninsula, Gulf St Vincent and the Adelaide Fold Belt (Flinders Ranges–Mount Lofty Ranges–Fleurieu Peninsula: Fig. 20.1) define another geotectonic domain. The region is complexly faulted, particularly in the Torrens Hinge Zone adjacent to the Flinders Ranges in northern

Spencer Gulf and on the western margin of the Mount Lofty Ranges adjacent Gulf St Vincent (Fig. 20.1). The gulfs and peninsulas represent well-defined graben and horst structures, respectively (Fenner 1930).

Spencer Gulf (20 000 km^2) and Gulf St Vincent (6800 km^2) represent broadly triangular-shaped, shallow-marine incursions into semi-arid continental Australia. Maximum water depths are less than 45 m (Fig. 20.1). At its mouth, Spencer Gulf is 85 km across, and widens to its maximum width of 145 km between Boston Bay and Hardwicke Bay (Fig. 20.1) in the southern gulf, and extends some 300 km inland. At the entrance to Spencer Gulf, a bedrock high, in part defined by Thistle and Wedge islands, dampens the effect of incoming swell waves (Fig. 20.1). Both gulfs in east–west cross-sections are asymmetrical features. In Spencer Gulf, the seafloor is steeper on its eastern side and the converse occurs in Gulf St Vincent, controlled by the horst structure of Yorke Peninsula. The SE coastline of Eyre Peninsula is largely structurally controlled by the NE-trending Lincoln Fault, and the Ardrossan Fault defines much of the eastern coastline of Yorke Peninsula.

At its mouth between Cape Spencer and Cape Jervis, Gulf St Vincent is 115 km across and extends some 163 km inland (Fig. 20.1). The average water depth of Gulf St Vincent is 21 m (Bye 1976). Investigator Strait (6100 km^2) separates Kangaroo Island from southern Yorke Peninsula, and is up to 55 km wide between Sturt Bay on southern Yorke Peninsula and Cape D'Estaing on Kangaroo Island (Fig. 20.1). Submerged Pleistocene dune structures (aeolianite) have been reported from Investigator Strait (James *et al.* 1997). Backstairs Passage is a flooded erosional depression cutting across rocks of the Adelaide Fold Belt

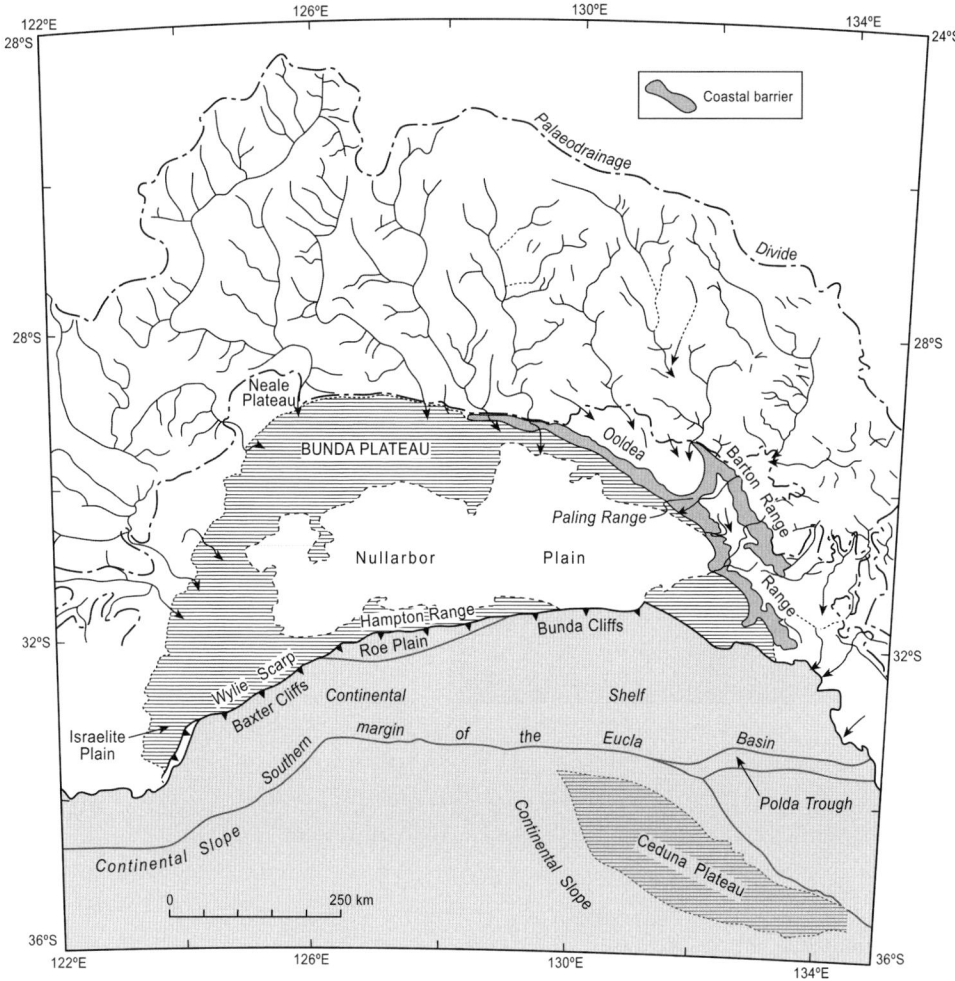

Fig. 20.2. The Eocene Ooldea and Barton Range, coastal barriers of the eastern Eucla Basin, southern Australia, modified after Benbow (1990). Palaeodrainage from the hinterland during Eocene time is also indicated.

(Adelaide Geosyncline) between southern Fleurieu Peninsula and Kangaroo Island. This feature, which is some 15 km wide, has a maximum water depth of 40 m, and represents an oversteepened and exhumed glacial valley formed by westerly flowing ice during Permian time (McGowran & Alley 2008). The St Vincent Basin (cf. Gulf St Vincent) is a larger structure covering some 15 000 km^2 with a Palaeogene–Neogene basin fill. Approximately 60% of the basin is covered by modern Gulf St Vincent (McGowran & Alley 2008) and its northernmost limit is some 60 km further inland than the head of the present gulf.

A record of subsidence since earliest Pleistocene time is evident for the gulfs, with up to 90 m of displacement of marine sediment (early Pleistocene Burnham Limestone) in Gulf St Vincent (Belperio 1995). In contrast, the Mount Lofty Ranges reveal a history of uplift as attested by the Point Ellen Formation at Cape Jervis (Fig. 20.1), a shallow-marine succession some 50 m above present sea level (APSL) and regarded as a correlative of the Burnham Limestone of the St Vincent Basin (Milnes *et al.* 1983). In a similar manner, an embayment fill succession of shelly sands of the last interglacial (MIS 5e, 130–118 ka) Glanville Formation occurs up to 12 m APSL at Normanville on Fleurieu Peninsula (Bourman *et al.* 1999) but on the Adelaide Plains the formation occurs only in subcrop and at Port Adelaide is 3–4.5 m below present sea level (BPSL) (Ludbrook 1976).

The Mount Lofty Ranges comprise sequences of Neoproterozoic–Early Palaeozoic quartzites, shales and limestones and Cambro-Ordovician granites with sequences of Permian glacigene sediments confined to Fleurieu Peninsula (Preiss 1987). The eastern shorelines of the gulfs are represented by well-developed Holocene (<7 ka) peritidal sedimentary successions with extensive sand and mudflats commonly rimmed by stands of mangrove woodlands (*Avecinnia marina*) (Cann *et al.* 2009), whereas the western margins of the gulfs tend to be bedrock dominated coastlines with subdued coastal cliffs (<20 m), narrow coastal plains and pocket-beach embayments. The western margins of the Flinders Ranges adjacent to the northern Spencer Gulf, extending south to the southern Mount Lofty Ranges, are flanked by alluvial fans, the distal portions of which interfinger with shallow-marine facies of the gulfs. South of Adelaide, the alluvial fans have been uplifted and have been deeply eroded on their distal margins during the Late Quaternary. Fleurieu Peninsula is predominantly a cliffed coastline bounded by high cliffs (80–100 m) developed on Neoproterozoic and Cambrian bedrock with small (<1 km wide) pocket beach embayments.

Lacepede and Bonney shelves

The Lacepede Shelf (30 000 km^2) occurs to the SE of Backstairs Passage and represents a large arcuate-shaped embayment that, landwards, is bounded by the Murray Basin, a Palaeogene–Quaternary epicratonic sedimentary basin (300 000 km^2) that has undergone epeirogenic uplift during Quaternary time (McLaren *et al.* 2011) (Fig. 20.1). The coastline of this shelf sector is bounded by the Holocene highstand (<7 ka) coastal barriers, Sir Richard and Younghusband peninsulas, the latter extending for approximately 180 km of coastline on the seaward side of the Coorong Lagoon (Fig. 20.4). From the mouth of the River Murray, the Lacepede Shelf is up to 180 km across to the 100 m isobath. The Lacepede shelf edge is a regular feature bounded to the south and SE of Kangaroo Island by a series of submarine canyons, some of which are related to former courses of the

Fig. 20.3. View looking east along the Nullarbor (Bunda) Cliffs (90 m high) of the Great Australian Bight, southern Australia, and showing in ascending order, the Middle–Late Eocene Wilson Bluff Limestone and the Middle Miocene Nullarbor Limestone. The Abrakurrie Limestone, which occurs between these two formations, is not discernable in this image.

River Murray at times of glacial low sea level (Sprigg 1947; Hill *et al.* 2009). The palaeochannels of the River Murray are 10–20 m deep and 450–1000 m wide (Hill *et al.* 2009), and their sediment infill records differential inputs of fluvially derived clays from the Murray–Darling drainage basin (Gingele *et al.* 2004; Schmidt *et al.* 2010). During glacial maxima, the ancient course of the River Vincent was considered by Sprigg (1947) to have traversed southern Gulf St Vincent and joined the palaeo-River Murray after cutting across Backstairs Passage. Seismic section, sub-bottom profiling of the region, however, failed to find evidence for substantial fluvial incision at times of low sea level, although the presence of low discharge channels in view of glacial age aridity could not be excluded (Hill *et al.* 2009). To the SE, the Lacepede Shelf narrows and merges with the Bonney Shelf, which is a narrow (30–80 km) unrimmed shelf extending SE between Cape Jaffa and Cape Northumberland (Fig. 20.1). In turn, the Bonney Shelf merges with the Otway Shelf, which extends from Cape Northumberland through

to Cape Otway in western Victoria, and is connected with Bass Strait and the narrow shelf of western Tasmania.

Bass Strait

Bass Strait, which separates mainland Australia from the island state of Tasmania, is approximately 500 km long, 300 km wide on average (north–south) and has a maximum water depth of 83 m at its geographical centre. The King Island Rise and Bassian Rise are two prominent metamorphic and crystalline bedrock highs within Bass Strait, represented in part by King Island and Flinders Island, respectively. Collectively, they define an isolation basin (Bass Basin) at times of glacial low sea level (Blom 1988) (Fig. 20.5). The island of Tasmania has a narrow shelf that ranges in width between 20 and 60 km. The western shelf of Tasmania joins with the Otway Shelf to the NW of the island and its eastern shelf merges with the Eastern Australian

Table 20.1. *Summary of the principal attributes of the continental shelves of SE Australia tabulated from west to east*

Shelf attribute	Great Australian Bight	Lincoln Shelf	Investigator Strait	Spencer Gulf	Gulf St Vincent	Backstairs Passage	Lacepede Shelf	Bonney Shelf	Otway Shelf	Bass Strait	Tasmania	Gippsland Shelf	Eastern Australian Shelf
Length (km)	1200	200	150	300	163	30	400	140	400	500	1300	250	700
Average width (km)	230	100	50	530	350	180	180	80	30–80	300	57	70	43
Tidal range (m)	1.2	1	0.9–1.5	2–3.9	1.65–1.8	1.3–1.5	1.7	0.6	0.6	0.6	0.8–3.2	2	1.1–1.3
Dominating process: wave (W); current (C); tide (T)	W	W	W	T	T	W	W	W	W	W > T	W	W	W
Sediment type: carbonate (C); authigenic (A); palimpsest (P); quartz (Q)	C	C, P	Q, C, P	Q, C, P	Q, C, P	Q, C, P	Q, C, P	Q, C, P	Q, C, P	A, C	Q, C, P	Q, P	Q, C, P
Neotectonic trend over the last glacial cycle	Stable	Stable	Stable	Subsiding	Subsiding	Uplifting	Uplifting	Uplifting	Uplifting	Uplifting	Uplifting	Stable	Stable

Fig. 20.4. View looking SSE along the Modern–Holocene coastal barrier Younghusband Peninsula and the back-barrier Coorong Lagoon (in the middle of the image). The coastal barrier represents much of the shoreline of the Lacepede Shelf in southern Australia. The barrier sediments comprise mixed quartz–skeletal carbonate sand, the latter generated on the inner shelf environment from the comminution of marine invertebrates. The sediment has been translated landwards during the post-glacial marine transgression, which culminated in SE Australia approximately 7 ka ago.

Shelf. Bass Canyon is a prominent feature on the NE margin of Bass Strait and south of the East Gippsland coastline (Fig. 20.1). The canyon is 60 km long, 10–15 km wide with canyon head walls 1000 m high, and has eroded through up to 2 km of the eastern Australian continental margin (Hill *et al.* 1998; Mitchell *et al.* 2007*a*, *b*). The canyon is most likely, in part, a relict feature originating from the Gondwanan break-up some 80 Ma ago. Terrigenous-clastic sediment has been transported along the 7–8 km-wide floor of the canyon to the abyssal plain in present water depths of approximately 4000 m (Hill *et al.* 1998).

Within this overall shelf sector of SE Australia, there are several distinct morphotectonic provinces that are largely defined by regionally extensive faults. The underlying bedrock relates to two orogens (Delamerian and Lachlan) also termed the Kanmantoo Trough on the eastern margin of the Adelaide Fold Belt and Lachlan Fold Belt to its east, respectively. The Port Phillip Sunkland near Melbourne is a broad region of subdued topography bounded by prominent faults, the Rowsley and Selwyn faults (Kenley 1976) (Fig. 20.5). The Gippsland area east of Westernport is differentially block-faulted with large embayments and prominences of tectonic origin, and laterally persistent coastal barriers of Holocene and late Pleistocene age (Bird 1993) (Fig. 20.1).

Eastern Australian Shelf (southern sector)

The southern New South Wales (NSW) portion of the Eastern Australian Shelf traverses the Lachlan Fold Belt, Sydney Basin and New England Fold Belt, all of which are tectonically stable. The shelf is significantly narrower than the southern margin. Between Cape Howe and Newcastle (Fig. 20.1), the shelf has a maximum width of 50 km, an average cross-sectional width of only 30 km and gradient of 0.3° (Ferland & Roy 1997). The shelf edge occurs in present water depths of 170–145 m. The present shelf morphology reflects the antecedent conditions of Palaeogene–Neogene sedimentation and concomitant erosion (Roy & Thom 1991). The coastline of NSW is dominated by cliffs developed on Palaeozoic and Mesozoic rocks with siliciclastic estuarine–coastal barrier embayment-fill successions principally of Holocene age (Roy 1998). These sediments infill deeply incised bedrock valleys and as broad coastal plains where unconfined by bedrock. In contrast, during sea-level lowstands, the shelf was rimmed by a low-gradient coastline with fewer embayments. Landwards, the shelf is bounded by the Eastern Highlands (also known as the Great Dividing Range: Fig. 20.1), which rise from a narrow coastal plain, 10–100 km wide, to a height of 500–2000 m APSL.

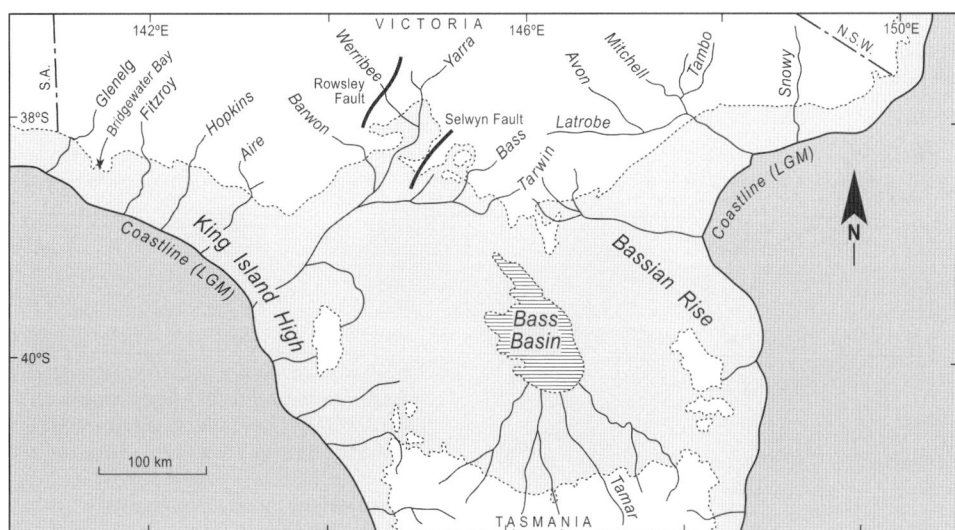

Fig. 20.5. The Tasman Peninsula during the Last Glacial Maximum (*c.* 22 ka) showing the extended drainage and the position of Bass Basin (modified after Jennings 1959; Blom 1988; Bird 1993).

Small river catchments provide sediment to the shelf through complex sediment pathways involving fluvial deposition in incised bedrock valleys and subsequent erosion during sea-level lowstands. During Quaternary time, there has been a progressive loss of sediment from the NSW continental shelf to southern Queensland, forming large sand islands such as Stradbroke (40 km long) and Fraser (124 km long) Islands, as well as sediment ultimately bypassing the shelf to the north of Fraser Island (Boyd *et al.* 2008). The net northward movement of sand along the NSW shelf is due to longshore drift, resulting from the combined effects of southerly storms and long-period swell waves. Over longer timescales, relative sea-level changes have also influenced the net movement of sand, particularly at times of lower sea level, when sand was able to bypass individual coastal compartments defined by bedrock valleys. Accordingly, the southern NSW shelf is sediment-deficient with a steeper gradient, bedrock-dominated coastline, whereas the northern shelf sector towards southern Queensland is a lower gradient, shallower and sediment-rich province.

Physical processes in the modern shelf environments of SE Australia

Several ocean surface boundary currents have an influence on the SE shelf region. The warm southward flowing East Australian Current brings warmer nutrient-rich water south over the NSW Continental Shelf from the Coral Sea (Fig. 20.6). At present, the current extends to about 37°S and then flows back to 34°S, forming large eddies before establishing an easterly path from Australia to New Zealand (Baines 1989; Tomczak & Godfrey 1994). The East Australian Current is strongest and reaches further inshore in summer (December–March). Along the central coastline of NSW, the current is some 30–40 km offshore and is strongest in present water depths of about 150–180 m with a velocity of 0.5–4 m s^{-1} (Ferland & Roy 1997). The path of the current from Australia to New Zealand is termed the Tasman

Front, separating the warm Coral Sea-derived water from the colder water of the Tasman Sea (James & Bone 2011).

In the Southern Ocean, the Antarctic Circumpolar Current (West Wind Drift) is a broad easterly-flowing current that travels across the Southern Ocean between 45°S and 60°S, and to the south of Tasmania. The Flinders Current is a mesoscale subcurrent associated with onshore and offshore transports west of Bass Strait (Bye 1983), which may represent a small bifurcation of the Antarctic Circumpolar Current (Fig. 20.6). In summer, coastal currents associated with the Flinders Current travel in a westerly direction along the Bonney and Lacepede shelves, and the converse occurs in winter.

The southerly flowing Leeuwin Current along the Western Australian coastline brings warm (17–19 °C), low-salinity (35.7–35.8‰) water from near the Indonesian Archipelago and from the Indian Ocean, NE of Exmouth, to Cape Leeuwin, where the current changes direction and moves across the Great Australian Bight, extending to the central bight at about 130°E (Cresswell 1991; Wells & Wells 1994) (Fig. 20.6). The current, which is less than 300 m deep and less than 100 km wide (Pattiaratchi & Woo 2009), is strongest during winter, particularly during La Niña years when additional warm water passes through the Pacific–Indian Ocean gateway in Indonesia (Wells & Wells 1994; Hantoro 1997). The influence of a proto-Leeuwin Current is evident in the Miocene Nullarbor Limestone, with foraminiferal species indicating average inner-shelf temperatures of 20 °C (O'Connell 2011).

In the eastern sector of the Great Australian Bight, the easterly flowing South Australian Current is formed by warm, saline waters during summer (James *et al.* 2001; Cresswell & Domingues 2009). The Bight waters dominantly experience downwelling with upwelling in summer when the influence of the Leeuwin Current weakens. The West Australian Current flows northwards and anticlockwise beneath the Leeuwin Current (Tomczak & Godfrey 1994).

The continental shelf surrounding Tasmania is influenced by the southerly flowing East Australian Current on its eastern margin,

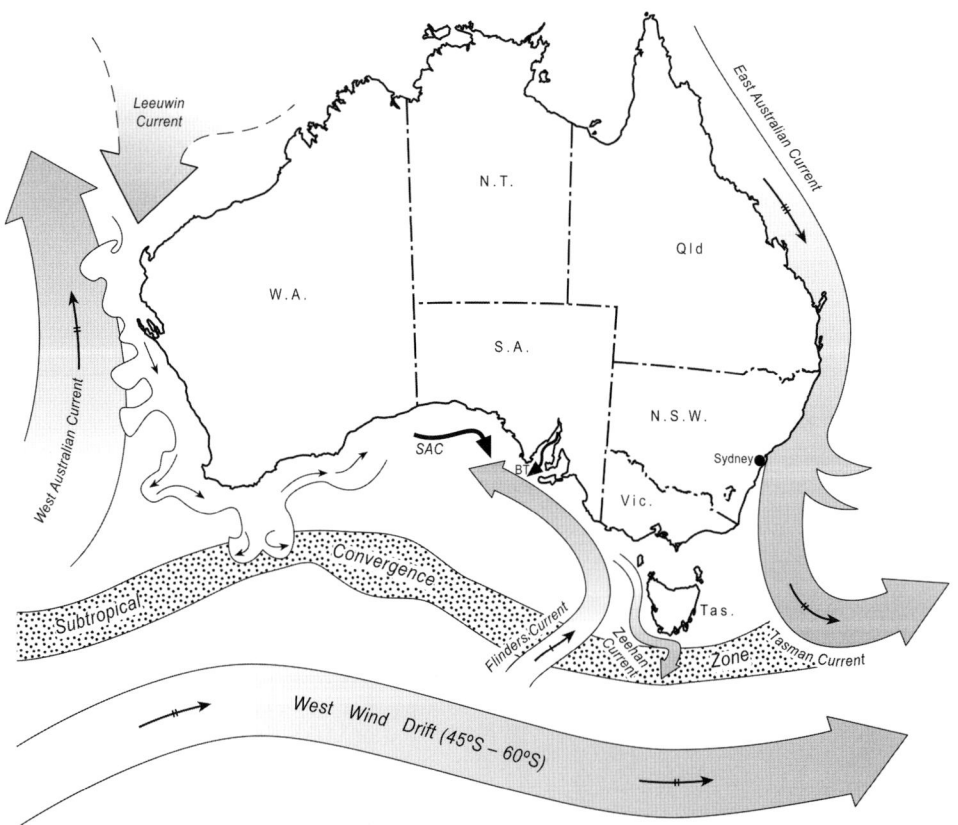

Fig. 20.6. Ocean surface boundary currents that influence continental shelf sedimentation in SE Australia: SAC, South Australian Current; BT, Bonaparte's Tongue. The relative strength of some of the currents is indicated by cross-hatches on the arrows, where three cross-hatches denote the strongest currents.

particularly in the summer months when the current is stronger, and by the Zeehan Current on its western margin (Fig. 20.6). Along the southern sector of the Tasmanian shelf where the currents may converge, their relative influences are affected by the northward movement of the Subtropical Convergence Zone in winter in which sub-Antarctic waters pass northwards beneath subtropical waters (James *et al.* 2008).

With the exception of the South Australian gulfs and portions of Bass Strait, the latter protected from the open ocean by the King Island and Bassian Rises, the entire coastal sector of SE Australia, considered in this review, is a temperate storm- and swell-dominated province (Porter-Smith *et al.* 2004). The Lincoln, Lacepede, Bonney and Otway shelves are high-energy, storm-dominated systems. Modal deep-water wave heights are typically >2.5 m and long-period swell waves (>12 s) with wavelengths of 200 m have been reported (Short & Hesp 1982). For much of southern Australia, significant wave heights exceed 3.5 m for 30–50% of the year and 10–30% per year along the NSW coastline, as determined from satellite altimetry (Porter-Smith *et al.* 2004). On the innermost portion of the Lacepede Shelf, the adjacent Younghusband Peninsula water depths are <20 m and the seafloor gradient is 1:150. Sand is abundant and wave energy is consistently high. More than 75% of the incident wave energy reaches the shore (Short & Hesp 1982) giving rise to dissipative beaches. The magnitude of incident wave energy decreases further south along this coastline due to the steeper gradient of the inner shelf.

The inner continental shelf of NSW is characterized by moderate to high wave energy, with modal deep-water wave heights of 1.5 m and periods of 10 s (Ferland & Roy 1997). More than half of the waves over 1 m are derived from the south, and the remainder are from the SE. Greater than 90% of deep-water wave energy is dampened within 1 km of this shoreline (Thom *et al.* 1992). Deeper-water waves are characteristically long-crested and sinusoidal, with periods of 6–14 s. During high-energy storms, the seabed may be affected to water depths of at least 60 m. The shelves of Tasmania are also characterized by high-energy wave and swell environments. The western shelf is influenced by southwesterly waves, and the eastern shelf by waves from the south and SE. The King Island and Bassian bedrock highs reduce the strength of tidal currents flowing into Bass Strait. Intersecting wave patterns to the east of Wilsons Promontory create characteristically choppy seas (Bird 1993).

Tides in SE Australia are dominantly microtidal, with mesotidal areas restricted to the northern, protected portions of the South Australian Gulfs and northern Tasmania. In Spencer Gulf and Gulf St Vincent, the tidal range increases progressively inland towards the heads of these gulfs. Tidal amplification in Gulf St Vincent is up to 30 cm. In Spencer Gulf the diurnal and semidiurnal tidal components behave as successive waves resulting in a tidal phase lag such that high tide is registered 6 h later at Port Augusta compared with Port Lincoln near the gulf's entrance (Fig. 20.1). In contrast, the influence of a dual opening to Gulf St Vincent via Investigator Strait and Backstairs Passage (Fig. 20.1) results in, with respect to time, essentially uniform tidal changes across the entire gulf but with spatially varying amplitudes (Bye & Kämpf 2008). During the equinoxes, water level within both gulfs remains relatively constant, resulting in a phenomenon termed a 'dodge tide' that may last up to 24 h (Bye & Kämpf 2008).

The NSW open-ocean coastline experiences a microtidal, semidiurnal tidal regime with a diurnal inequality, with tidal range less than 2 m, and a mean spring tidal range of approximately 1.2 m. The tidal pattern is very uniform along this entire coastal sector (Thom *et al.* 1992).

Spencer Gulf and Gulf St Vincent are inverse estuaries due to the high rate of evaporation in the summer months and minimal freshwater input resulting from their arid setting in continental Australia. At the head of Gulf St Vincent, salinities can reach up to 47.0‰ in summer (Bye & Kämpf 2008) and values up to 49.6‰ have been documented at the head of Spencer Gulf (Burne & Colwell 1982; Lennon *et al.* 1987). Clockwise water circulation occurs within both gulfs. In Spencer Gulf, a gravity current of a relatively dense, more saline water mass, termed 'Bonaparte's Tongue', flows from the head of the gulf during winter months across the Lacepede Shelf and over the shelf edge (James *et al.* 1997). The feature, expressed as a single entity is some 100 km long, up to 20 m in vertical profile and several hundred metres in width. The gravity current flows at a rate of approximately 0.1 m s^{-1} and thus has to flow for approximately three months to remove the salt that has accumulated in the northern gulf during the summer months (Lennon *et al.* 1987).

Pre-Quaternary shelf sedimentation

Palaeogene–Neogene marginal marine sedimentation in SE Australia was confined to the Eucla, St Vincent, Murray, Gambier, Otway, Torquay and Gippsland basins, and the region defined by the present continental shelf of NSW (Fig. 20.7). The basin margins reveal that the Eastern Australian Highlands existed as a region of well-defined higher relief before Palaeogene time. The Eucla (*c.* 400 000 km^2) and Murray (300 000 km^2) basins represent the largest marine incursions into southern central Australia post-dating the phase of Cretaceous epicontinental sea sedimentation. Their marine influence extended some 300 and 650 km inland, respectively, from the modern southern Australian coastline. The Palaeogene–Neogene successions crop out extensively along the coastlines of the Eucla, St Vincent, Otway, Torquay and Gippsland basins, and in cliffs along the southern reaches of the River Murray (Alley & Lindsay 1995). Elsewhere, the Palaeogene–Neogene record has been derived from drill core. Along the southern margin, up to 1500–2000 m of temperate carbonate sediments were deposited during Palaeogene–Neogene time in platform to deep-water marine environments. Today, the successions occur within the present continental shelf and slope environments, and extend to varying distances inland of the present coastline. Tectonism and, in particular, regional faulting led to the formation of distinct marginal marine basins in which correlative successions were deposited. A detailed discussion of the Palaeogene–Neogene stratigraphical record including an overview of the principal formations is beyond the scope of this review, and accordingly, only significant trends in continental shelf sedimentation during this interval are considered. Detailed accounts of the lithostratigraphy of these Palaeogene–Neogene basin fills are given by Carter (1985), Alley & Lindsay (1995) and Holdgate & Gallaghar (2003). One of the challenges in assigning ages to these successions has been the presence of extratropical foraminifers, which has made it more difficult to identify biostratigraphically significant P-zones (McGowran 1979).

The earliest evidence for synrift sedimentation in the narrow seaway of southern Australia during Gondwanan fragmentation comes from dredged samples containing Maastrichtian plankton and benthos, and the depositional event has been termed the Ceduna Ingression (McGowran 1991). Varying degrees of marine influence have been described as 'transgressions' and 'ingressions' for episodes of Palaeogene–Neogene sedimentation (McGowran 1991). Ingressions refer to short-lived incursions of marine influence in otherwise marginal marine–paralic sedimentary environments from the Maastrichtian to Early Eocene, whereas transgressions have a more distinctive stratigraphical expression and may relate to an increase in the rate of seafloor spreading post-dating geomagnetic anomaly 19 at 44 Ma ago. The Palaeogene reveals the presence of four principal transgressive episodes that occurred during Eocene time; two in the late Middle Eocene represented by the Wilson Bluff and Tortachilla transgressions; and two in the Late Eocene represented by the Tuketja and Aldinga transgressions. All four transgressions

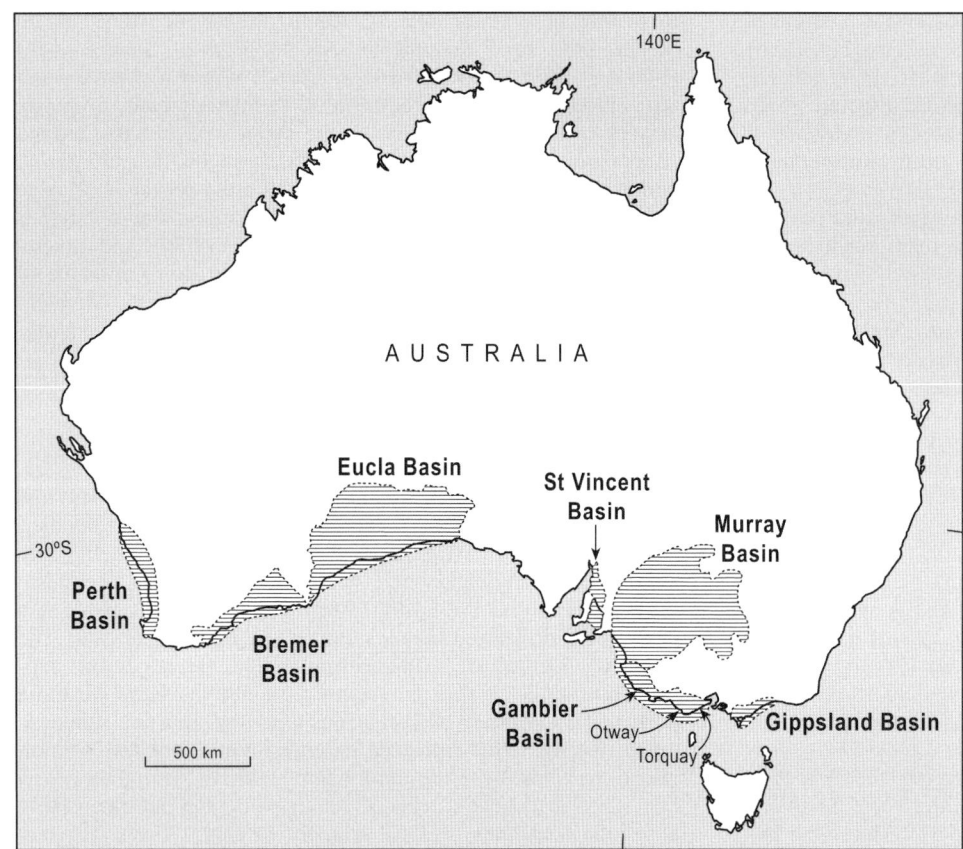

Fig. 20.7. Palaeogene–Neogene sedimentary basins of southern Australia.

are biostratigraphically expressed in the Eucla, St Vincent and Murray basins (McGowran 1989, 1991; Alley & Lindsay 1995). The Tortachilla, Tuketja and, possibly, Aldinga transgressions are also represented in the Otway Basin.

In the southern Australian 'marine basins', the onset of Palaeogene marine sedimentation is diachronous. The earliest record is in the Gambier Embayment of the Otway Basin represented by the Nirranda Group (c. 46 Ma) as a succession of ferruginous and glauconitic sands followed by deposition of the fully marine Gambier Limestone (c. 38 Ma), and in the Eucla Basin (Bunda Plateau) with the onset of deposition of the Wilson Bluff Limestone (c. 43 Ma), a succession of platform carbonates comprising grey wackestone, skeletal mudstone and packstone generally less than 150 m thick (Alley & Lindsay 1995). In the St Vincent Basin, deposition of the South Maslin Sand, a succession of carbonaceous-rich sand with a depauperate marine fauna of bryozoan, molluscs (bivalves and gastropods), echinoid fragments and foraminifers, commenced some 42 Ma ago (Alley & Lindsay 1995). In the Murray Basin, shallow-marine deposition commenced with the Buccleuch Formation some 38 Ma ago. The succession comprises bryozoan and glauconitic limestones with, in places, ferruginous and pyritic clays.

One of the hallmarks of marginal marine sedimentation in Australia during Quaternary time is the presence of laterally persistent coastal barriers, a mode of sedimentation also evident in Palaeogene time. Coastal barrier successions relating to deposition some 37–34 Ma ago have been reported from the NE sector of the Eucla Basin (Benbow 1990). The Ooldea Range, and associated Paling and Barton ranges (Fig. 20.2), are Late Eocene coastal landforms that collectively extend for over 650 km on the eastern margin of the Nullarbor Plain, 25–300 km inland from the present coastline. The Ooldea Range has parallel ridge structures identified as beach ridges (Benbow 1990). The broadly linear accumulations of fluvially sourced frosted quartz sand (0.25–0.50 mm) with marine-sourced sponge spicules represent coastal barriers that formed during the maximum marine incursion of the Eucla Basin, an extensive carbonate platform. At the time of

deposition, active rivers brought abundant supplies of sand to the coastline to be reworked into coastal barrier landforms.

The stratigraphy of the Murray Basin is divisible into three sequence sets comprising: Late Paleocene–Early Oligocene non-marine and marginal marine successions; Late Eocene–Middle Miocene transgressive marine successions; and latest Miocene to Late Pliocene–early Pleistocene marine, coastal and non-marine successions (Alley & Lindsay 1995; Bowler et al. 2006). Lower–Middle Miocene marine sedimentation in the Murray Basin (Gambier Limestone of the Murravian Gulf of Sprigg 1952) records a progressive change in water temperatures with time from cool-water carbonate deposition to warm-temperate to subtropical seas. The sediments were deposited in a low-energy, mesotrophic epeiric ramp environment (Lukasik & James 2006). From the Late Pliocene to towards the end of early Pleistocene time, a major phase of coastal progradation in the Murray Basin resulted in the deposition of the Parilla Sand, a sequence of subdued coastal barriers containing placer, heavy mineral sands (Belperio & Bluck 1990; Bowler et al. 2006).

In southern Victoria, shallow-marine sedimentation during Palaeogene–Neogene time is represented in the two principal basins: Otway (60 000 km^2) and Gippsland (56 000 km^2), as well as in the smaller subsidiary (Port Campbell Embayment, Torquay and Bass) basins. The successions record a transition from siliciclastic, paralic and marginal marine to shallow-water platform and deeper-water carbonate sedimentation. In the Otway Basin, this increasing marine influence is represented by the Wangerrip, Nirranda and Heytesbury groups, and by the Neogene Seaspray Group in the Gippsland Basin. The latter is represented by platform carbonates, which, in the offshore portion of the Gippsland Basin, attain a thickness of up to 2000 m (Holdgate & Gallagher 2003).

The NSW portion of the Eastern Australian Shelf preserves sequences of Palaeogene–Neogene sediments generally less than 500 m thick and confined to the modern shelf region in the form of a prograded wedge (Roy & Thom 1991). On the inner–middle continental shelf, a well-defined marine abrasion surface has formed. The surface has been identified in sidescan surveys extending from the prominent coastal cliffs of the Sydney region for up to 23 km offshore to a water depth of approximately 180 m. It is likely that a substantial portion of the east Australian continental margin preserves a similar marine abrasion surface. The surface was initiated in the Mid-Oligocene and indicates a time-averaged rate of cliff retreat of approximately 1 mm a^{-1} (Thom *et al.* 2010). Only limited Palaeogene–Neogene marine deposition occurred landwards of the modern continental shelf due to the structural barrier presented by the Great Dividing Range, which existed in some form during that time. Seismic stratigraphic analysis of the sedimentary successions has revealed an unconformity that has been interpreted as an erosion surface correlating with the Messinian salinity crisis (Adams *et al.* 1977; Roy & Thom 1991). The unconformity (erosion surface) dips northwards with a gradient of 75 m per degree of latitude, which indicates that at latitude 35°S up to 150 m of uplift has occurred in that region during the past 3 myr (Roy & Thom 1991). The proposed mode of uplift relates to the northward movement of the continent over a stationary 'hot region' that has been responsible for extensive volcanism in eastern Australia (Sutherland 1991).

Quaternary shelf sedimentation and relative sea-level changes

The modes of Palaeogene and Neogene sedimentation evident across the southern margin of the Australian continent continued during Quaternary time but did not penetrate as far inland due to lower sea levels. Collectively, the deposition of Palaeogene–Quaternary marginal-marine successions of southern Australia represents the world's largest cool-water carbonate province (James & Bone 2011). In the following account, the dominant patterns of Quaternary shelf sedimentation for the continental shelves of SE Australia are discussed in an anticlockwise traverse from the Great Australian Bight to the southern New South Wales portion of the Eastern Australian Shelf. Although the discussion primarily focuses on the last glacial cycle, partly because of the greater preservation potential of these successions, some of the older and, in many respects, remarkable continental shelf successions are also briefly considered.

The Roe Plain, Eucla Basin

The Roe Calcarenite, which covers most of the Roe Plain (5800 km^2) in Western Australia, refers to a sheet drape of richly fossiliferous, semi-indurated, medium- to coarse-grained calcarenites and coquinas of latest Neogene–early Pleistocene age (Lowry 1970; Ludbrook 1978; James *et al.* 2006). The formation is, on average, about 1.5–2 m thick and the landward transgressive feather-edge of the formation, which abuts a relict sea cliff (Hampton Range), occurs up to 33 m APSL at Madura, some 42 km inland of the present coastline (Fig. 20.1).

Ludbrook (1978) identified 265 species of fossil molluscs from the Roe Calcarenite. These include species of the bivalve *Katelysia* and the gastropod *Batillaria* (*Zeacumantus*) *diemenensis*, both of which frequent modern intertidal–shallow subtidal environments. Several mollusc species indicate warmer-than-present water temperatures at the time of deposition of the Roe Calcarenite (e.g. the present most southerly occurrence of the gastropod *Angaria tyria* is at Shark Bay on the central Western Australian coastline (S25°25′; E113°35′): Ludbrook 1978). The warmer-water species of foraminifer *Marginopora vertebralis* is also present and is notable for its particularly large size (commonly up to 20 mm in diameter, cf. 5 mm for specimens of last interglacial age from southern Australia: Cann & Clarke 1993). The pelagic janthinid gastropod *Hartungia dennanti chavani*, a biostratigraphically important early Pleistocene species found

elsewhere in southern Australia, is also represented in the fauna (Ludbrook 1984). $^{87}Sr/^{86}Sr$ analyses on fossil molluscs reported by James *et al.* (2006) signify an early Pleistocene age for the Roe Calcarenite given the recently revised position of the Neogene–Quaternary boundary to 2.59 Ma. The high elevation for the more inland portions of the formation relative to present sea level relates to basin inversion and uplift of broad sectors of the southern Australian continental margin. A possible mechanism for this is the subduction of the northern margin of the Indo-Australian plate near Papua New Guinea, so that the northern sector of the Australian continent is sinking while the southern sector is undergoing uplift (i.e. a tilting continent: Murray-Wallace & Belperio 1991; Sandiford 2007).

Spencer Gulf and Gulf St Vincent

Spencer Gulf and Gulf St Vincent preserve a rich record of late Quaternary shallow-marine sedimentation and relative sea-level changes. One of the challenges has been to accurately date the record in view of the paucity of corals suitable for uranium-series dating and, accordingly, a greater reliance has been made on other geochronological methods such as amino acid racemization to constrain the stratigraphical records in this region (Murray-Wallace 2000).

Palaeosols developed on the upper portions of the submarine units (Billing 1984) provide a framework to infer relative changes in sea level and subaerial exposure of the northern Spencer Gulf during the late Quaternary. Sediments of inferred interstadial age (MIS 5c, 105 ka and MIS 5a, 82 ka) have been described from marine vibracores and seismic sections in northern Spencer Gulf (Gostin *et al.* 1984; Hails *et al.* 1984). Holocene and last interglacial successions (MIS 5e) were dated using radiocarbon and amino acid racemization and thermoluminescence, respectively (Belperio *et al.* 1984b). The interstadial successions were named the False Bay Formation (MIS 5a, 105 ka) and the Lowly Point Formation (MIS 5c, 82 ka: Hails *et al.* 1984) (Fig. 20.8). The False Bay Formation comprises bryozoan-rich limestones, as well as poorly consolidated quartz beach sand, oxidized sabkha facies and calcareous estuarine–lagoonal sediments (wackestones). The Lowly Point Formation overlies the False Bay Formation only within the deeper channels of northern Spencer Gulf in present water depths exceeding 14 m. The formation comprises sabkha and laminated lagoonal clays, and shows less pedogenic alteration than the older units (Hails *et al.* 1984). Palaeo-sea levels of −14 and −8 m were inferred from the False Bay (MIS 5a) and Lowly Point (MIS 5c) formations, respectively. However, these values should be regarded as provisional as they were not based on reliable palaeo-sea-level indicators but on the upper bounding surfaces of the shallow-marine formations.

Equivalents of the Glanville Formation (Cann 1978; Ludbrook 1984), a regionally significant stratigraphic marker unit representing shallow-marine and coastal sedimentation during the last interglacial maximum (MIS 5e, c. 130–118 ka ago) of the entire South Australian coastline, were also identified within the submarine cores of northern Spencer Gulf (Hails *et al.* 1984). Hails *et al.* (1984) informally termed the Spencer Gulf succession the Mambray Formation but the original formation name as defined from the Adelaide region (Ludbrook 1984) takes precedence. The Glanville Formation is typically 1–2 m thick, as revealed in the submarine vibracores of northern Spencer Gulf (Hails *et al.* 1984; Cann *et al.* 2000) and in quarries and shallow excavations in Gulf St Vincent on the Adelaide Plains (Cann 1978). The formation contains faunal elements indicating warmer inner-shelf temperatures during the last interglacial, such as the benthic foraminifer *Marginopora vertebralis*, the cockle *Anadara trapezia* and the pearl oyster *Pinctada carchariarum* (Ludbrook 1984). The warmer inner-shelf temperatures may relate to a more vigorous Leeuwin Current during the last interglacial maximum (Cann &

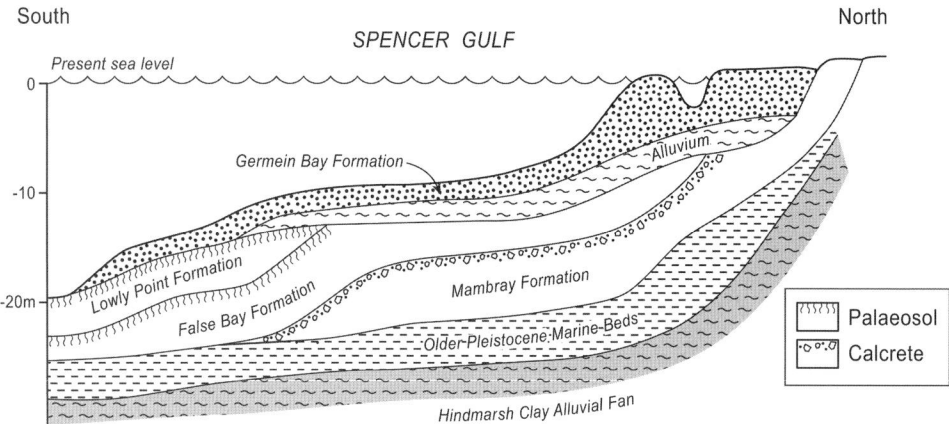

Fig. 20.8. Schematic summary of the middle Pleistocene–Holocene submarine lithostratigraphy of northern Spencer Gulf, South Australia. Width of diagram represents approximately 20 km in a north–south line of section (modified after Hails *et al.* 1984).

Clarke 1993). The extensive distribution of the estuarine mollusc *A. trapezia* in southern Australia, where it no longer lives, reflects the absence today of suitable estuarine habitats, possibly due to increased continental aridity since the last interglacial (Murray-Wallace *et al.* 2000). A higher sea level of 2 m APSL is indicated from the transgressive feather-edge of estuarine–lagoonal successions for over 500 km of coastline from Fowlers Bay to Coffin Bay on western Eyre Peninsula (Murray-Wallace & Belperio 1991).

The older Pleistocene marine beds of northern Spencer Gulf are represented by poorly sorted, mottled and strongly pedogenically modified sandy clays containing *Anadara trapezia* and abraded foraminifera. The succession has been identified in a vibracore transect at Redcliff (Belperio *et al.* 1984*a*) and in submarine vibracores within the northern gulf (Hails *et al.* 1984), and is distinctive for the presence of a well-structured clay-rich B-horizon, a feature not seen in the younger sediments. A correlation with MIS 7 (220 ka) was based on the extent of racemization of several amino acids in *A. trapezia* (Murray-Wallace *et al.* 1988). The marine strata occur up to 2 m APSL and indicate a significantly higher relative sea level in northern Spencer Gulf than otherwise suggested by marine isotope records, which point to an ice-equivalent sea level of some 20 m BPSL during MIS 7 (Shackleton 1987).

Sedimentary successions correlated with the late Pleistocene interstadial MIS 3 have been identified in northern Spencer Gulf (Cann *et al.* 2000) and Gulf St Vincent (Cann *et al.* 1988). In Gulf St Vincent, the sediments are unlithified, yellowish-grey lime mud with, in places, yellow mottles (iron oxide staining). Pedogenic modification (calcrete), characteristic of the last interglacial Glanville Formation, is absent within the interstadial succession. Radiocarbon ages and the extent of amino acid racemization (AAR) revealed a correlation with MIS 3 (Cann *et al.* 1988). An AAR numeric age of 46.5 ± 9.1 ka was based on the extent of valine racemization and isoleucine epimerization in specimens of the cockle *Katelysia rhytiphora* (Murray-Wallace *et al.* 1993). A palaeo-sea level of −30 to −27 m is indicated by the foraminiferal assemblages, in particular the ratio of *Elphidium crispum* to *E. macelliforme* (Cann *et al.* 1988, 2000). The inferred palaeo-sea levels are significantly higher than indicated by the revised Huon Peninsula record of Papua New Guinea (Chappell *et al.* 1996), which suggests that, for the interval 70–30 ka, sea level fell to between −50 and −60 m.

Coastal aeolianites of the southern margin

For much of Quaternary time, along the high wave energy, open-ocean coastal sector from westernmost Eyre Peninsula to western Victoria, Kangaroo, King and Flinders islands, skeletal carbonate sand produced on the shallow-water (<20 m) inner continental shelves has been entrained landwards, resulting in the formation

of extensive aeolianite (calcarenite) successions. The aeolianites mapped as Bridgewater Formation were first formally defined stratigraphically at Bridgewater Bay in western Victoria (Boutakoff 1963) (Fig. 20.5) and refer to aeolian-deposited skeletal carbonate sand with variable proportions of quartz and heavy minerals. Calcium carbonate content ranges between 35 and 90% and, on average, is greater than 50% of the sediment by mass (Murray-Wallace *et al.* 2001). Along the tectonically highly stable coastal sectors of the southern Eyre and Yorke peninsulas, portions of the northern, and much of the southern, coastline of Kangaroo Island, the aeolianite successions, are represented by multiply-stacked blanket-like deposits that extend several kilometres inland from the modern coastline (Sprigg 1979; Wilson 1991; Lachlan 2011) (Fig. 20.9). They are characterized by trough-cross-bedded sand with numerous interbedded palaeosols, rhizoliths and other subaerial-exposure features. The successions, commonly exposed in coastal cliffs of the southern Australian coastline, attain heights of up to 30–40 m APSL and, at Cape Spencer on southern Yorke Peninsula, attain heights of up to 100 m APSL. The Cape Spencer aeolianite section contains up to 20 well-defined calcrete palaeosol units (Sprigg 1979).

On the Coorong Coastal Plain in southern South Australia and extending into western Victoria, epeirogenic uplift has resulted in the physical separation of the aeolianite shoreline deposits to yield morphostratigraphically distinct coastal barrier shorelines with associated back-barrier lagoonal successions (Sprigg 1952; Murray-Wallace *et al.* 1998, 1999). Within the barriers, the aeolianites attain thicknesses of up to 30–40 m above the local basement rocks and in shore-normal cross-section range from 1 to 3 km. Both morphostratigraphical forms (coastal barrier and multiply-stacked sand-sheet successions) of the Bridgewater Formation relate to deposition during interglacial and interstadial sea-level highstands (Huntley *et al.* 1993, 1994; Murray-Wallace *et al.* 1999, 2001, 2010). Correlative (lithologically) aeolianite successions to the Bridgewater Formation have also been described from Warrnambool (Gill 1988) and on the Nepean Peninsula south of Melbourne but the latter relate to deposition during the last glacial cycle based on thermoluminescence dating (Zhou *et al.* 1994).

Bass Basin

Bass Basin is an isolation basin at times of lower sea level during glacial cycles (Fig. 20.5). The basin perimeter is broadly defined by the 60 m isobath and, when glacio-eustatic (ice-equivalent) sea level fell 67 m below present sea level, the basin was completely isolated from the influences of the Indian and Pacific oceans. In contrast, at times of sea level between −55 and −67 m, the bedrock depression formed a marine embayment with an opening to the NW (Blom & Alsop 1988). Bass Basin

Fig. 20.9. Pleistocene aeolianites of the Bridgewater Formation with interbedded palaeosols at Point Reynolds, southern Kangaroo Island, South Australia. The shore platforms and structural benches at the cliff foot and offshore outcrops are represented by Late Eocene and Pliocene limestones. The cliff is approximately 60 m high.

covers an area of 66 000 km^2 and at its deepest is about 83 m, it is some 120 km wide and 400 km long (NW–SE) (Jennings 1959). The basin is unusual in a southern Australian shelf context in that its sediment record is dominated by calcitic mud, rather than skeletal carbonate sediments that otherwise characterize much of the southern Australian shelves in water depths of 70–85 m. The carbonate mud was produced from the accumulation and subsequent comminution of nanoplankton and other bioclastic materials (Blom & Alsop 1988).

The later Pleistocene–Holocene stratigraphy of Bass Basin reveals, in ascending order, a lacustrine facies overlain by an embayment (brackish-water) facies, which is in turn overlain by a fully marine Holocene succession. The base of the Holocene succession was defined by a single radiocarbon measurement (marine reservoir-corrected but not calibrated to sidereal years) on a marine shell (species not reported) of 8700 ± 710 years BP (SUA-2428) (Blom & Alsop 1988) representing the fully marine flooding event. Marine reservoir-corrected radiocarbon ages on the underlying embayment fill facies range between 11 660 ± 300 and 10 290 ± 250 years BP. A succession of highly calcareous (60–92%) poorly sorted muddy bryozoal sands identified towards the bases of several cores described by Blom (1988) most probably relate to deposition during the latter stages of MIS 3 when sea level was at least −60 m.

The margin of Bass Basin was also sensitive to environmental changes during the Holocene. Between approximately 2.8 and 1 ka ago, the land-locked basin of Port Phillip Bay to the south of Melbourne largely dried out (Holdgate *et al.* 2011). This possibly resulted from the constriction of the bay entrance by sand. Evaporation of waters following the constriction subaerially exposed much of the embayment to a depth of −22 m, as revealed from the radiocarbon dating of marine shells from recent sediment infilling the relict fluvial channels.

The Eastern Australian Shelf: southern New South Wales sector

In Newcastle Bight between Newcastle and Port Stephens, the shelf is bounded by a Holocene outer barrier and a late Pleistocene (MIS 5e) inner barrier (Thom 1965). These coastal landforms extend parallel with the modern coastline for some 30 km along Newcastle Bight and also to the north of Port Stephens. The barrier landforms representing regressive successions are capped by transgressive coastal parabolic dunes and preserve estuarine–lagoonal muds in their lee. In shore-normal cross-section, the inner and outer barriers are approximately 4 and 2 km wide, respectively. The sediments comprise largely relict quartzose sands of marine origin with less than 7% shell fragments. The quartz sand has undergone multiple cycles of reworking and the ultimate source of the sand is from fluvial transport to the shelf during sea-level lowstands when much of the shelf was subaerially exposed. The marine reworking of sands extends consistently along the inner shelf to present water depths of 25 m (Roy & Crawford 1980). The active transgressive dunes on the outer barrier may explain the net loss of nearshore sand, with an average shoreline recession of 1–2 m a^{-1} (Roy & Crawford 1980). In many respects, the general morphostratigraphical model of the inner (late Pleistocene) and outer (Holocene) coastal barriers is applicable to much of the sand-dominated coastal systems bordering the inner continental shelves of NSW and the Gippsland coastline in eastern Victoria. Coarse-grained fluvial sands, as well as sediments representing a proto-barrier, have been identified on the inner shelf seawards of the Holocene barrier in Newcastle Bight (Roy & Crawford 1980) and represent relict deposits that formed during the early stages of the post-glacial marine transgression.

Interstadial barriers have been identified in the mid-shelf setting off the Forster–Tuncurry coastline (Fig. 20.1) in water depths of 75–100 m, and correlated with MIS 3 based on thermoluminescence (TL) dating of quartz sand yielding ages of 59–44 ka (Roy *et al.* 1994). Although the barrier deposits show regressive clinoforms in seismic sections, the TL ages indicate a younging trend in a landward direction. However, the TL uncertainty terms reported by Roy *et al.* (1997) reveal that all of the ages are not significantly different when considered in a shoreward transect. The barriers are up to 15 km wide in east–west cross-section, and between 10 and 16 m thick (Roy *et al.* 1997). The sediments comprise fine- to medium-grained, well-sorted quartz-lithic sands with minimal skeletal carbonate sand. The sand may contain up to 26% lithic fragments, 5% feldspar and, although having a lower mineralogical maturity, are texturally very mature. Throughout the barrier profiles, the sediment is mineralogically uniform, suggesting a long period of mixing consistent with a provenance from a remote source and prolonged longshore drift (Roy *et al.* 1997).

Fig. 20.10. Photograph of lowstand glacial age sediments from the continental shelf of New South Wales. Core 130, which retrieved 3.57 m of shelf sediment, was collected in a present water depth of 123 m and contains reworked shells of MIS 6 age intermixed with shells of last glacial age MIS 2, as determined from AAR analyses (Murray-Wallace *et al.* 1996).

On the outer shelf in present water depths exceeding 100 m, successions of lowstand deposits comprise shell-rich, skeletal carbonate sand interbedded with mixed quartz–carbonate sand depleted in macrofossils (Ferland & Roy 1997) (Fig. 20.10). Most of the fossil bivalve molluscs are disarticulated but generally well preserved, retaining much of their original pigmentation. Abraded, pitted and bored shells are also common. Vibracores from the outer continental shelf, in the Newcastle–Wollongong coastal sector, reveal the presence of the shallow subtidal (surf zone and lower shoreface) molluscan species *Placaman placidium*, *Glycymeris* (*radians* and *striatularis*) and *Tawera gallinule*. In the vibracores located in present deeper water, the scallop *Pecten fumatus* is common, consistent with its present-day environmental context in eastern Australia (10–50 m water depth). The shell-rich and depleted horizons correlate with glacial lowstand and interglacial highstand modes of deposition, respectively, based on radiocarbon

and AAR dating (Murray-Wallace *et al.* 1996, 2005). Radiocarbon dating of the fossil molluscs *Glycymeris* sp. and *Pecten fumatus* yielded ages correlating with MIS 2 (28–11 ka) and indicate a palaeo-sea level of approximately 125 m BPSL during the Last Glacial Maximum. Amino acid racemization evidence revealed a higher degree of reworking of fossil molluscs within sediments closer to the inferred location of the palaeoshoreline of Pleistocene glacial maxima. In addition, the degree of racemization for the amino acids valine and leucine indicate the presence of glacial-age deposits on the outer shelf related to MIS 2, 6 and 8 (Murray-Wallace *et al.* 1996, 2005). During glacial maxima, the shelf was a high-energy, wave- and storm-dominated open-shelf environment, and the core of the East Australia Current was most probably located further seawards. The Pleistocene shelf successions are capped by Holocene sediments that are generally <2 m thick (Ferland & Roy 1997).

Holocene sea-level changes and the most recent flooding of the continental shelves

The modern coastal configuration of the continental shelves of SE Australia is the product of deglacial sea-level rise. Sea-level rise accompanying the waning of the Last Glacial continental ice sheets began at about 19 ka ago (Yokoyama *et al.* 2000) and culminated by 7 ka (calendar years) (Thom & Chappell 1975; Thom & Roy 1985; Sloss *et al.* 2007), with an average rate of rise of 15–16 mm a^{-1} (Roy *et al.* 1994; Belperio *et al.* 2002). Australia is located in the far-field of former continental ice sheets and, accordingly, its relative sea-level record is not directly affected by glacio-isostasy. However, the effects of hydro-isostasy are variably expressed in the records of Holocene relative sea-level changes in this region, most noticeably in northern Spencer Gulf where the early Holocene highstand at 7 ka (calendar years) is up to 3 m APSL at the apex of the gulf and progressively falls to about 1 m APSL on the open-ocean coastlines (Belperio *et al.* 2002). The stratigraphical record reveals that since 7 ka relative sea levels fell smoothly to present values, a function of hydro-isostatic adjustment, the magnitude of which is related to the cross-sectional width of the continental shelves. Accordingly, the NSW coastline, bounded by a narrow shelf, reveals a lower amplitude hydro-isostatic signal of about 1–1.5 m from 7.9 to 7.7 ka, and a relative sea-level fall to present from about 2 ka ago (calendar years: Sloss *et al.* 2007). In many respects, the relative sea-level record from this latter region is more ambiguous in terms of the different proxy sea-level indicators used for inferring sea-level change (i.e. sea-level indicators based on sedimentary evidence, such as the elevation of mangrove facies, subtidal seagrass facies or shell beds compared with fixed biological indicators (e.g. serpulid worms)). The fixed biological indicators reveal a potentially more complex record of relative sea-level changes (Baker & Haworth 2000), which may relate to steric changes associated with short-term climate change (e.g. El Niño events) and not just a hydro-isostatic signal.

The transition from the terminal Pleistocene to the Holocene heralded significant changes in the coastal and continental shelf environments of SE Australia. During the Last Glacial Maximum (LGM: 24–22 ka), the Australian continental area was up to 25% larger than today resulting from the subaerial exposure of the shelf when sea level was at least 125 m below present. The rapid rise in sea level accompanying global deglaciation led to profound geomorphological changes. In northern Spencer Gulf and Gulf St Vincent, the post-glacial rise in sea level resulted in the flooding of distal alluvial fans and mixed quartzose–calcareous desert dunes of Glacial age. In southern Gulf St Vincent, the early phase of the post-glacial marine transgression is heralded by the development of a marginal-marine lacustrine system with abundant oogonia (calcified fruiting bodies of charophytes), gypsum and the foraminifer *Miliolinella labiosa* and *Elphidium excavatum*, followed by the development of seagrass meadows inhabited by the foraminifer *Nubecularia lucifuga* and *Discorbis dimidiatus* (Cann *et al.* 2006). Flooding of the Lacepede Shelf terminated the flow of the ancestral River Murray across the continental shelf, as well as rendering the MIS 3 coastal barriers relict features (Hill *et al.* 2009). The Bassian Plain (Fig. 20.5), part of the former Tasman Peninsula with Lake Bass at its geographical centre, which was formerly surrounded by grasslands and steppe with scattered woodlands on the margin of an upland periglacial landscape during the LGM, was also drowned by the post-glacial rise in sea level to create the modern Bass Strait. In Victoria, the onset of the Holocene highstand resulted in renewed coastal erosion along the bedrock-dominated coastlines (e.g. Port Campbell and Otway coasts), and renewed phases of coastal barrier, barrier-island and plain sedimentation (e.g. outer-barrier complex forming Ninety Mile Beach on the East Gippsland coast). Post-glacial sea-level rise on the NSW sector of the Eastern Australian Shelf resulted in the landward translation of large quantities of sand creating barrier shoreline successions, as well as the development of deeply incised bedrock valley, estuarine–lagoonal infill successions.

Modern shelf sedimentation

Mixed quartz–skeletal carbonates represent the dominant modern sediment type extending from the eastern margin of the Lincoln Shelf and throughout the South Australian gulfs, the Lacepede, Bonney and Otway shelves, and across into western Bass Strait (James & Bone 2011). The Lincoln Shelf, however, is one of exclusive carbonate deposition (James *et al.* 1997). The modern shelf sediments of this province are temperate Heterozoan as defined by James *et al.* (1997). This mode of carbonate sedimentation is fostered by: the cool shelf water temperatures, typically <18 °C; limited perennial drainage to deliver terrigenous-clastic sediments along this entire coastline due to regional aridity, low topographical relief; and endorheic drainage, which, affecting some 52% of the Australian continent, has a major influence on the nature of sediment production along the Great Australian Bight (Murray-Wallace 2002; Rivers *et al.* 2007). Skeletal carbonate grains are derived from the fragmentation of molluscs (bivalves and gastropods), bryozoans, planktonic and benthic foraminifers, coralline algae and echinoids, and also include abraded and stained relict carbonate grains and sponge spicules (Belperio *et al.* 1988; James *et al.* 1992). Much of this province is characterized by open-shelf sedimentation, the absence of hermatypic corals, only minor cementation, and extensive bioerosion and maceration of skeletal carbonate grains.

Spatially well-defined zones of modern sediment production, giving rise to distinctive sedimentary facies, have been identified on the continental shelves of SE Australia (James & Bone 2011). The Lacepede Shelf, for example, is characterized by six modern shelf facies that include calcareous quartz sand, quartzose-bryozoan/bivalve sand, bryozoan sand, bryozoan mud, bivalve–coral gravel, and pelagic mud (Foram-nanno ooze) (James *et al.* 1992). In this region, the different modern sedimentary facies are the product of a complex set of processes that relate to: present water depths; proximity to the mouth of the River Murray; partial or complete subaerial exposure of the Lacepede Shelf due to relative sea-level changes; partial erosion of former coastal barrier successions; and reworking of skeletal carbonate grains. The reworking of skeletal carbonate continues today as attested by the erosion of Robe Range, a composite late Pleistocene and Holocene aeolianite coastal barrier that defines the modern coastline of the Bonney Shelf (Fig. 20.1). Robe Range has undergone up to 1 km of lateral erosion and coastal retreat since the culmination of the post-glacial marine transgression some 7 ka ago (calendar years) as attested by the accordant surfaces of shore platforms developed on offshore islands of the late Pleistocene aeolianites of Robe Range. As swell wave base on the Lacepede Shelf extends down to about 70 m, much of the modern shelf sediment is represented by a palimpsest succession, the product of a complex history of sedimentation.

Sediments of Pleistocene and Holocene age are of similar lithological character to modern sediment and contain abundant relict carbonate grains (Rivers *et al.* 2007; Hill *et al.* 2009). The modern skeletal carbonate sands of the southern margin contain varying proportions of relict carbonate grains ranging between 10 and 50% of the sediment of the Lincoln and Lacepede shelves (Li *et al.* 1998). The high degree of racemization of amino acids in modern beach sediments in Encounter Bay and Guichen Bay near Robe on the Coorong Coastal Plain (Fig. 20.1) reveal that a significant component of the relict carbonate grains are of late Pleistocene age (Murray-Wallace *et al.* 2001, 2010).

A particularly distinctive element of the Lincoln and Lacepede shelves is the prolific growth of bryozoans. The dominant sediment producers include delicate branching cyclostome, catencelliform (singlets), vinculariform (rods), reteporiform (fenestrate) and adeoniform (flat branching) forms (James *et al.* 1992, 1997). Although present across the entire shelf, bryozoans grow most prolifically in the outer-shelf to shelf-edge environment in present water depths between −80 and −140 m. In the shelf-edge environment of the Lincoln Shelf, a proportion of the *Celleporaria* sp. bryozoans are relict with uncorrected radiocarbon ages ranging between 13.76 and 21.21 ka BP (James *et al.* 1997).

On the inner shelf of the Great Australian Bight, in water depths of up to 30 m, and in the protected tide-dominated eastern margins of the South Australian gulfs in shallower water (<10 m), the prolific growth of seagrasses plays an important role in promoting sedimentation. The thickly bladed seagrasses *Posidonia australis* and *P. sinuosa* and, in slightly deeper water, *Amphibolis antarctica* (Shepherd & Sprigg 1976), baffle localized water currents resulting in the aggradation of seagrass meadows to produce laterally persistent carbonate banks. Carbonate bank sediments comprise skeletal carbonate grains derived from foraminifers, coralline algae, bivalves and gastropods bound together by partially decayed leaf sheaths to produce a fibrous sediment texture. In tidal-dominated coastal areas, another seagrass, *Zostera muelleri*, which is able to tolerate a degree of intertidal emergence, traps and binds sediment of more muddy character. A closely similar seagrass, *Heterozostrea* sp., grows in the deeper subtidal waters to a depth of about 40 m. At Redcliff in northern Spencer Gulf, the *in situ* growth of seagrass meadows during the past 7 ka has resulted in a blanket-like deposit (sheet-drape) of seagrass facies some 3 m thick and extending over 8 km in shore-normal section, resulting in coastal progradation of that order (Belperio *et al.* 1984a). In the southern gulf, a greater diversity of sedimentary facies is evident, reflecting a wider range of sedimentary processes and proximity to the Lincoln Shelf (Fuller *et al.* 1994).

As with the open-ocean continental shelves of southern Australia, the modern surface sediments of Gulf St Vincent and Spencer Gulf are highly calcareous, and dominated by skeletal carbonate sands with varying proportions of terrigenous sediment. However, in Gulf St Vincent, siliceous sands are more abundant along the western and eastern margins of the gulf (Fuller & Gostin 2008). The northern beaches of metropolitan Adelaide comprise siliceous sand, primarily derived from on-shore sedimentation at the culmination of the post-glacial marine transgression. Persistent, northerly, longshore transport of sand has resulted in the formation of Le Fevre Peninsula, a distinctive barrier spit–beach ridge complex (Fig. 20.1) (Bowman & Harvey 1986). The terrigenous sediment was derived from the erosion of semi-consolidated Neogene–Quaternary alluvial fan successions flanking the southern Mount Lofty Ranges, as well as the older bedrock of the ranges. In addition, the occurrence of fine-grained carbonate sediment plumes under the influence of tidal action represent a further, but unquantified, sediment input to the gulf (Shepherd & Sprigg 1976).

Modern sediments of the Tasmanian open-ocean shelves tend to be relatively coarse grained and comprise three mega-facies that include: (1) quartzose sand; (2) relict sand and gravel; and (3) Holocene biogenic sand and gravel (James *et al.* 2008). Mega-facies 1 comprises at least 50% quartzose sand and lithic fragments, with the skeletal carbonate grains being dominated by molluscs and bryozoans. Megafacies 2 consists of at least 30% relict skeletal carbonate grains derived from molluscs, bryozoans, echinoids, foraminifers and worm tubes. Megafacies 3, which occurs in the deepest-water setting of the three megafacies, consists of fine gravel-size skeletal carbonate fragments with a subordinate sand component (<20%), and extends from water depths of 30 to 400 m. To the north of Tasmania in Bass Strait, minimal terrigenous clastic sediment is being deposited within the basin from the regional denudation of Tasmania and southern Victoria. Accordingly, calcium carbonate content typically exceeds 50% of the modern surface sediment by mass. The sand fraction is dominated by benthic foraminifers, ostracods, and fragments of bryozoans, molluscs, echinoids, brachiopods and worm tubes.

The surface sediments of the inner to mid-continental shelf of NSW are dominantly terrigenous clastic (quartz sand and mud) that have been brought to the shelf during repeated glacial cycles from the erosion of the Eastern Australian Highlands. The mid-shelf surficial sediments are mixed quartz–carbonate, with the mud component ranging between 25 and 50% (Ferland & Roy 1997). The outer continental shelf, however, is dominated by cool-water, palimpsest carbonate sediments comprising slightly muddy, poorly sorted sand with coquina represented by fragmental and entire fossil molluscs (Ferland & Roy 1997). These surface sediments are related to the underlying sediment package and reflect deposition during low sea-level stands during successive glacial maxima (Murray-Wallace *et al.* 1996, 2005). The proportion of calcium carbonate in the sand fraction of the modern shelf sediments increases linearly from about 5% of the total sediment at 50 m BPSL and 4–10 km offshore, up to 80% some 150 m BPSL and 20–40 km offshore (Ferland & Roy 1997). The coarser skeletal carbonate fragments are derived from molluscs, coralline algae, bryozoans, barnacles and echinoderms. Mollusc fragments dominate the bioclastic constituents. The modern sediment zonation essentially mimics the underlying Holocene and Pleistocene sediment record as revealed from vibracores (Ferland & Roy 1997).

On the inner shelf of NSW, extensive sheet-like sand deposits (0.25–1 m thick) occur parallel with the coast and in water depths of 20–70 m, which traverse the shelf for 5–10 km (Roy *et al.* 1994). The sand is dominated by medium-grained quartz with, in places, shelly gravels. The sediments rest on a ravinement surface and appear to be a relict of the post-glacial marine transgression that failed to reach the modern (Holocene) coastline. Evidence for this includes fossil molluscs that would have frequented former estuarine–lagoonal environments in the inner-shelf environment during the transgression. Megaripples and sand ridges are common on the inner shelf but do not display preferred current directions.

A distinctive aspect of the NSW continental shelf is the presence of headland-attached shelf sandbodies (Roy *et al.* 1994). These accumulations of fine- to medium-grained quartzose sand occur on the steeper (>1°) inner continental shelf adjacent prominent headlands. They are typically 20–30 m thick, 4 km wide, may extend parallel with the coastline for up to 30 km and have a convex-upward structure reflecting an early phase of transgression followed by regressive modes of deposition. The deposits formed during the post-glacial marine transgression and have been wave-reworked during the Holocene highstand. The deposits may have represented a source for aeolian cliff-top dune sand on adjacent headlands.

Quaternary shelf evolution

Factors controlling sediment production

Several geological attributes, acting in concert, have influenced the general evolution of the continental shelves and their contrasting sediment properties in south-eastern Australia during the Quaternary. The intraplate setting of the Australian continent has conferred a generally high degree of tectonic stability. Although subtle vertical crustal movements have been discerned, based on the differential elevations of last interglacial (MIS 5e) shoreline successions (Murray-Wallace & Belperio 1991; Murray-Wallace 2002) so that a general neotectonic trend for the continental shelves may be defined (Table 20.1), continental shelf facies of Quaternary age have not been uplifted to positions APSL in SE Australia. An exception is the Plio/Pleistocene Roe Calcarenite, inner continental shelf succession of the Roe Plain which has resulted from the draw-down in Quaternary sea levels associated with the progressive development of the Antarctic Ice Sheet and basin inversion (James *et al.* 2006). The latter may relate to the inferred continental-scale tilting of the Australian landmass (Sandiford 2007).

The dominance of cool-water, temperate carbonate sediments along the entire southern margin has been enhanced by continental aridity, together with the endorheic drainage, paucity of rivers draining to this entire coastal sector and the low topographical relief of continental Australia limiting denudational processes. In contrast, the Eastern Australian Shelf has a significantly higher component of terrigenous-clastic sediment derived from the denudation of the Great Dividing Range.

In the Murray Basin, a significant change in the sediment composition and facies architecture of relict coastal barrier successions coincides with the early–middle Pleistocene transition of approximately 1.3 Ma ago. Although separated by a regional unconformity, a change is evident from the dominantly siliciclastic sediments of the Late Pliocene Parilla Sand (closely spaced coastal ridges of lower relief amplitude) to the carbonate-dominated Bridgewater Formation aeolianite coastal barriers that are widely spaced (typically 5–10 km apart). The transition corresponds with the Mid-Pleistocene transition from the 41 kyr-dominated to the 100 kyr-dominated glacial cycles (Bowler *et al.* 2006). The Parilla Sand and its offshore equivalent, the Bookpurnong Beds (Brown & Stephenson 1991) consisting of marls, appear to have been deposited in shallower water conditions. The Bridgewater Formation, however, was deposited in deeper water and a higher energy wave regime. These contrasting sedimentary regimes may have resulted from the mutual interplay of progressively falling sea levels associated with increasing ice volumes in the Late Pliocene–early Pleistocene, changing bathymetry associated with long-term tectonic changes in the southern portion of the Murray Basin and, at a larger scale, much of the SE margin (Sandiford 2007), and progressive climate changes in the Southern Ocean realm associated with the growth of the Antarctic Ice Sheet (Bowler *et al.* 2006).

Rates of sediment accumulation

Quantifying net carbonate sediment production on the continental shelves of SE Australia, as well as the delivery of terrigenous-clastic sediment to these settings, remains difficult in view of the pervasive reworking of shelf sediments during successive glacial cycles. On the Eastern Australian Shelf, a further complication to quantifying rates of sediment accumulation relates to the net loss of sediment through longshore drift and sediment bypassing down the continental slope to the north of Fraser Island (Boyd *et al.* 2008).

Along the southern Australian margin from western Eyre Peninsula through to western Victoria, extensive aeolianite complexes have formed during the entire Quaternary. These coastal aeolian dune complexes represent major sediment sinks and provide a record of carbonate sediment production within inner continental shelf environments. They also explain the basis for the thin palimpsest sediment cover or shaved shelves that characterize much of this region. The sediment was entrained landwards by high-energy waves and strong SW winds during interglacials and interstadials. The sediments typically comprise variable proportions of abraded and recrystallized relict carbonate grains ($\geq$10%) reworked from older successions (Murray-Wallace *et al.* 2001).

The late Pleistocene shallow-marine formations (MIS 5e, 5c, 5a and 3) of Spencer Gulf and Gulf St Vincent provide insights into the rates of net sediment accumulation, a function of *in situ* skeletal carbonate production, as well as any competing processes that may reduce the thickness of sedimentary successions. The last interglacial Glanville Formation in the South Australian gulfs is commonly 1–1.5 m thick (onshore and as revealed in submarine vibracores: Murray-Wallace & Belperio 1991), indicating an average rate of sediment aggradation of about 0.1–0.15 mm a^{-1} assuming a duration of 10 ka for MIS 5e. The False Bay (MIS 5c) and Lowly Point (MIS 5a) formations, identified in northern Spencer Gulf, are confined to deeper depressions within the gulf (possibly lowstand river channels), favouring the preservation of thicker successions that are atypical for estimations of net sediment accumulation during single marine isotope stages. The False Bay Formation, for example, is documented in isopach maps as exceeding 10 m in thickness (Hails *et al.* 1984). Interstadial, shallow-marine calcitic muds intercepted in vibracores from central Gulf St Vincent reveal up to 3 m of sediment deposited during the latter part of MIS 3 (*c.* 45–30 ka: Cann *et al.* 1988), indicating a net rate of sediment accumulation of 0.2 mm a^{-1}. A higher rate of net sediment accumulation of 0.43 mm a^{-1} is evident for Holocene subtidal seagrass facies from Redcliff, northern Spencer Gulf, for the past 7 kyr (Belperio *et al.* 1984*a*). Submarine vibracores collected from the central portion of Gulf St Vincent (Cann *et al.* 1988) indicate up to 640 mm of sediment aggradation in the past 10 kyr (Murray-Wallace *et al.* 1993) (i.e. 0.064 mm a^{-1}). Collectively, the late Quaternary values for the rate of net sediment accumulation are significantly higher than recorded for Oligocene–Lower Miocene and Lower–Middle Miocene sequences from southern Australia (Lukasik & James 2006). The net accumulation rate values for MIS 5e are some 20 times greater than for the Oligocene–Lower Miocene and five times greater than for the Lower–Middle Miocene. Sedimentary successions from the gulfs indicate that net sediment accumulation rates have been even higher in Holocene time (*c.* up to three times that of MIS 5e). The progressive increase in the rates of net sediment accumulation evident in the Neogene–Quaternary record of southern Australia may be a function of the development of shallow-marine to marginal-marine environments highly conducive to skeletal carbonate production (clear, shallow-marine waters with minimal terrigenous-clastic sediment input, and extensive seagrass meadows in inner shelf and gulf environments).

The lowstand successions of the NSW shelf also provide insights into the rates of net sediment accumulation during glaciations. In vibracores of shelf sediment collected in water depths relatively close to the inferred location of the LGM shoreline, such as Core 130 (present water depth 123 m: Fig. 20.10), the entire 3.57 m sediment package has been reworked and includes shells of MIS 6 age intermixed with shells of MIS 2 age (Murray-Wallace *et al.* 1996). In contrast, vibracores collected in the outer-shelf setting in present water depths of 150 m show minimal evidence for the reworking of shells or sediment. In the outer-shelf setting, the net rate of accumulation for sediments deposited during the LGM, as inferred from several vibracores, is approximately 0.05 mm a^{-1} (Murray-Wallace *et al.* 1996, 2005; Ferland & Roy 1997). This value is, in general, in accord with the net rates of sediment accumulation for the highstand successions from the South Australian gulfs, and provides a first-order approximation of the

probable sediment thicknesses for individual marine isotope stages for many shelf successions from SE Australia, apart from regions characterized by shaved shelves.

The palimpsest character of much of the continental shelf sediments within the entire shelf sector of SE Australia reflects the combined effects of the high storm and swell wave energy of this province, and the additional influence of glacio-eustatic sea-level changes. The latter gives rise to long-term (kyr) cycles of sediment exchange and recycling between coastal and shelf depositional systems. During transgressions and early highstand events, sediment is entrained landwards, forming sand-sheet and coastal-barrier aeolianite successions. Their preservation potential is governed by subsequent sediment availability to blanket and, hence, protect older deposits (e.g. southern Eyre and Yorke peninsulas) or uplift to remove the successions from the influence of coastal erosion (e.g. Coorong Coastal Plain). During regressions, the continental shelves were subaerially exposed by varying magnitudes, depending on the extent of sea-level fall, and some of the successions were subjected to erosion. However, pervasive calcrete development and reduced surface runoff accompanying glacial aridity reduced the susceptibility of these marginal marine successions to denudation. Thus, the on-board carbonate sedimentary environments of the southern margin have been one of net accretion during the Quaternary.

The principal characteristics of the continental shelf of NSW include the dominance of outer-shelf, lowstand mollusc-rich sediments, a lower rate of calcium carbonate bioproductivity during sea-level highstands than evident for the southern margin and a cyclical record of carbonate production on the outer shelf since later middle Pleistocene time. A partitioning in the nature of sediment production accompanies relative sea-level changes on the NSW shelf. During glacial maxima, lowstand carbonate production is favoured on the outer shelf, while sea-level highstands appear to favour siliciclastic sedimentation (Ferland & Roy 1997). These contrasting modes of sedimentation may also reflect contrasting climates with more active rivers during interglacials, delivering greater quantities of terrigenous-clastic sediment to the shelf environments, and more arid glacials heralding a reduction in sediment supply.

As a general observation, sediments of the outer continental shelves of the entire province considered in this review are not in hydrodynamic equilibrium with modern wave and tidal processes, as also observed by Porter-Smith *et al.* (2004). The coarse-grained nature of the skeletal carbonate sediments in the outer-shelf settings is dominantly relict sediment deposited during low sea-level stands. Continental aridity and consequent restricted drainage to the sea has significantly reduced the delivery of mud to the outer-shelf settings.

Conclusions

The continental shelves of SE Australia began to form in Late Cretaceous time along a trailing edge, passive continental margin. The continental shelves in this region are part of the world's largest modern cool-water, temperate carbonate realm, in part a function of regional tectonic stability and slow rates of continental denudation with minimal delivery of terrigenous-clastic sediment to the southern margin. Continental shelf sediments of the southern margin from Palaeogene time to the present are dominantly represented by skeletal carbonates (Heterozoan Association) and are consistently palimpsest in character, reflecting the interplay of sediment reworking due to the high-energy storm and swell wave environment, and over longer timescales the effects of the glacio-eustatic sea-level changes. Although the Eastern Australian Shelf (southern NSW sector) is dominantly covered by siliciclastic sediment, the outer-shelf successions are represented by mollusc-rich skeletal carbonate having formed during sea-level lowstands during glacial maxima. The entire modern shelf and coastline configuration of the SE Australian margin was established after 7 ka ago with the culmination of the post-glacial marine transgression.

I thank J. Cann and P. De Deckker for their constructive reviews of an earlier version of this paper, as well as e-mail discussions with B. McGowran, which have collectively improved this work. I acknowledge, with gratitude, funding over several years from the Australian Research Council which has enabled the production of this review.

References

ADAMS, C. G., BENSON, R. H., KIDD, R. B., RYAN, W. B. F. & WRIGHT, R. C. 1977. The Messinian salinity crisis and evidence of late Miocene eustatic changes in the world ocean. *Nature*, **269**, 383–386.

ALLEY, N. F. & LINDSAY, J. M. 1995. Tertiary. *In*: DREXEL, J. F. & PREISS, W. V. (eds) *The Geology of South Australia, Volume 2, The Phanerozoic*. South Australia Geological Survey Bulletin, **54**, 150–217.

BAINES, P. G. 1989. The physical oceanography of Australian waters – a review. *Australian Meteorological Magazine*, **37**, 155–165.

BAKER, R. G. V. & HAWORTH, R. J. 2000. Smooth or oscillating late Holocene sea-level curve? Evidence from the palaeo-zoology of fixed biological indicators in east Australia and beyond. *Marine Geology*, **163**, 367–386.

BELPERIO, A. P. 1995. Quaternary. *In*: DREXEL, J. F. & PREISS, W. V. (eds) *The Geology of South Australia, Volume 2, The Phanerozoic*. South Australia Geological Survey Bulletin, **54**, 219–280.

BELPERIO, A. P. & BLUCK, R. G. 1990. Coastal palaeogeography and heavy mineral sand exploration targets in the western Murray Basin, South Australia. *Australasian Institute of Mining and Metallurgy, Proceedings*, **295**, 5–10.

BELPERIO, A. P., HAILS, J. R., GOSTIN, V. A. & POLACH, H. A. 1984*a*. The stratigraphy of coastal carbonate banks and Holocene sea levels of northern Spencer Gulf, South Australia. *Marine Geology*, **61**, 297–313.

BELPERIO, A. P., SMITH, B. W. *ET AL.* 1984*b*. Chronological studies of the Quaternary marine sediments of northern Spencer Gulf, South Australia. *Marine Geology*, **61**, 265–296.

BELPERIO, A. P., GOSTIN, V. A., CANN, J. H. & MURRAY-WALLACE, C. V. 1988. Sediment-organism zonation and the evolution of Holocene tidal sequences in southern Australia. *In*: DE BOER, P. L., VAN GELDER, A. & NIO, S. D. (eds) *Tide-influenced Sedimentary Environments and Facies*. Reidel, Dordrecht, 475–497.

BELPERIO, A. P., HARVEY, N. & BOURMAN, R. P. 2002. Spatial and temporal variability in the Holocene sea-level record of the South Australian coastline. *Sedimentary Geology*, **150**, 153–169.

BENBOW, M. C. 1990. Tertiary coastal dunes of the Eucla Basin, Australia. *Geomorphology*, **3**, 9–29.

BENBOW, M. C., LINDSAY, J. M. & ALLEY, N. F. 1995. Eucla Basin and Palaeodrainage. *In*: DREXEL, J. F. & PREISS, W. V. (eds) *The Geology of South Australia, Volume 2, The Phanerozoic*. Geological Survey of South Australia Bulletin, **54**, 178–186.

BILLING, N. B. 1984. Palaeosol development in Quaternary marine sediments and palaeoclimatic interpretations, Spencer Gulf, Australia. *Marine Geology*, **61**, 315–343.

BIRD, E. C. F. 1993. *The Coast of Victoria*. Melbourne University Press, Melbourne.

BLOM, W. D. 1988. Late Quaternary sediments and sea-levels in Bass Basin, southeastern Australia – a preliminary report. *Search*, **19**, 94–96.

BLOM, W. M. & ALSOP, D. B. 1988. Carbonate mud sedimentation on a temperate shelf: Bass Basin, southeastern Australia. *Sedimentary Geology*, **60**, 269–280.

BOURMAN, R. P., BELPERIO, A. P., MURRAY-WALLACE, C. V. & CANN, J. H. 1999. A last interglacial embayment fill at Normanville, South Australia and its neotectonic implications. *Transactions of the Royal Society of South Australia*, **123**, 1–15.

BOUTAKOFF, N. 1963. *The Geology and Geomorphology of the Portland Area*. Victoria Geological Survey, Memoirs, **22**, 172.

BOWLER, J. M., KOTSONIS, A. & LAWRENCE, C. R. 2006. Environmental evolution of the Mallee Region, Western Murray Basin. *Proceedings of the Royal Society of Victoria*, **118**, 161–210.

BOWMAN, G. M. & HARVEY, N. 1986. Geomorphic evolution of a Holocene beach-ridge complex, Le Fevre Peninsula, South Australia. *Journal of Coastal Research*, **2**, 345–362.

BOYD, R., RUMMING, K., GOODWIN, I., SANDSTROM, M & SCHRÖDER-ADAMS, C. 2008. Highstand transport of coastal sand to the deep ocean: a case study from Fraser Island, southeast Australia. *Geology*, **36**, 15–18.

BROWN, C. M. & STEPHENSON, A. E. 1991. *Geology of the Murray Basin, Southeastern Australia*. Bureau of Mineral Resources Bulletin, **235**.

BURNE, R. V. & COLWELL, I. B. 1982. Temperate carbonate sediments of northern Spencer Gulf, South Australia: a high salinity 'foramol' province. *Sedimentology*, **29**, 223–238.

BYE, J. A. T. 1976. Physical oceanography of Gulf St Vincent and Investigator Strait. *In*: TWIDALE, C. R., TYLER, M. J. & WEBB, B. P. (eds) *Natural History of the Adelaide Region*. Royal Society of South Australia, 143–160.

BYE, J. A. T. 1983. Physical oceanography. *In*: TYLER, M. J., TWIDALE, C. R., KING, J. K. & HOLMES, J. W. (eds) *Natural History of the Southeast*. Royal Society of South Australia, Adelaide, 75–84.

BYE, J. A. T. & KÄMPF, J. 2008. Physical oceanography *In*: SHEPHERD, S. A., BRYARS, S., KIRKEGAARD, I., HARBISON, P. & JENNINGS, J. T. (eds) *Natural History of Gulf St Vincent*. Royal Society of South Australia, Adelaide, 56–70.

CANN, J. H. 1978. An exposed reference section for the Glanville Formation. *Quarterly Geological Notes, Geological Survey of South Australia*, **65**, 2–4.

CANN, J. H. & CLARKE, J. D. A. 1993. The significance of *Marginopora vertebralis* (Foraminifera) in surficial sediments at Esperance, Western Australia, and in last interglacial sediments in northern Spencer Gulf, South Australia. *Marine Geology*, **111**, 171–187.

CANN, J. H., BELPERIO, A. P., GOSTIN, V. A. & MURRAY-WALLACE, C. V. 1988. Sea-level history, 45 000 to 30 000 yr B.P., inferred from benthic foraminifera, Gulf St. Vincent, South Australia. *Quaternary Research*, **29**, 153–175.

CANN, J. H., BELPERIO, A. P. & MURRAY-WALLACE, C. V. 2000. Late Quaternary paleosealevels and paleoenvironments inferred from foraminifera, northern Spencer Gulf, South Australia. *Journal of Foraminiferal Research*, **30**, 29–53.

CANN, J. H., MURRAY-WALLACE, C. V., RIGGS, N. J. & BELPERIO, A. P. 2006. Successive foraminiferal faunas and inferred palaeoenvironments associated with the postglacial (Holocene) marine transgression, Gulf St Vincent, South Australia. *The Holocene*, **16**, 224–234.

CANN, J. H., SCARDIGNO, M. F. & JAGO, J. B. 2009. Mangroves as an agent of rapid coastal change in a tidal-dominated environment, Gulf St Vincent, South Australia: implications for coastal management. *Australian Journal of Earth Sciences*, **56**, 927–938.

CARTER, A. N. 1985. A model for depositional sequences in the Late Tertiary of southeastern Australia. *In*: LINDSAY, J. M. (ed.) *Stratigraphy, Palaeontology, Malacology*. South Australian Department of Mines and Energy, Special Publications, **5**, 13–27.

CHAPPELL, J., OMURA, A., ESAT, T., MCCULLOCH, M., PANDOLFI, J., OTA, Y. & PILLANS, B. 1996. Reconciliation of late Quaternary sea levels derived from coral terraces at Huon Peninsula with deep sea oxygen isotope records. *Earth and Planetary Science Letters*, **141**, 227–236.

CRESSWELL, G. R. 1991. The Leeuwin current – observations and recent models. *Journal of the Royal Society of Western Australia*, **74**, 1–14.

CRESSWELL, G. R. & DOMINGUES, C. M. 2009. The Leeuwin current south of Western Australia. *Journal of the Royal Society of Western Australia*, **92**, 83–100.

FENNER, C. 1930. The major structural and physiographic features of South Australia. *Transactions of the Royal Society of South Australia*, **54**, 1–36.

FERLAND, M. A. & ROY, P. S. 1997. Southeastern Australia: a sea-level dependent, cool-water carbonate margin. *In*: JAMES, N. P. & CLARKE, J. A. D. (eds) *Cool-Water Carbonates*. SEPM Special Publications, Tulsa, **56**, 37–52.

FULLER, M. K. & GOSTIN, V. A. 2008. Recent coarse biogenic sediments of Gulf St Vincent. *In*: SHEPHERD, S. A., BRYARS, S., KIRKEGAARD, I., HARBISON, P. & JENNINGS, J. T. (eds) *Natural History of Gulf St Vincent*. Royal Society of South Australia, Adelaide, 29–37.

FULLER, M. K., BONE, Y., GOSTIN, V. A. & VON DER BORCH, C. C. 1994. Holocene cool-water carbonate and terrigenous sediments from southern Spencer Gulf, South Australia. *Australian Journal of Earth Sciences*, **41**, 353–363.

GILL, E. D. 1988. Warrnambool – Port Fairy District. *In*: DOUGLAS, J. G. & FERGUSON, J. A. (eds) *Geology of Victoria*. Geological Society of Australia, Sydney, 374–379.

GINGELE, F. X., DE DECKKER, P. & HILLENBRAND, C.-D. 2004. Late Quaternary terrigenous sediments from the Murray Canyons area, offshore South Australia and their implications for sea level change, palaeoclimate and palaeodrainage of the Murray-Darling Basin. *Marine Geology*, **212**, 183–197.

GOSTIN, V. A., SARGENT, G. E. G. & HAILS, J. R. 1984. Quaternary seismic stratigraphy of northern Spencer Gulf, South Australia. *Marine Geology*, **61**, 167–179.

HAILS, J. R., BELPERIO, A. P., GOSTIN, V. A. & SARGENT, G. E. G. 1984. The submarine Quaternary stratigraphy of northern Spencer Gulf, South Australia. *Marine Geology*, **61**, 345–372.

HANTORO, W. S. 1997. Quaternary sea level variations in the Pacific – Indian Ocean Gateways: response and impact. *Quaternary International*, **37**, 73–80.

HILL, P. J., EXON, N. F., KEENE, J. B. & SMITH, S. M. 1998. The continental margin off east Tasmania and Gippsland: structure and development using new multibeam sonar data. *Exploration Geophysics*, **29**, 410–419.

HILL, P. J., DE DECKKER, P., VON DER BORCH, C. & MURRAY-WALLACE, C. V. 2009. Ancestral Murray River on the Lacepede Shelf, southern Australia: Late Quaternary migrations of a major river outlet and strandline development. *Australian Journal of Earth Sciences*, **56**, 135–157.

HOLDGATE, G. R. & GALLAGHER, S. J. 2003. Tertiary. *In*: BIRCH, W. D. (ed.) *Geology of Victoria*. Geological Society of Australia, Special Publications, **23**, 289–335.

HOLDGATE, G. R., WAGSTAFF, B. & GALLAGHER, S. J. 2011. Did Port Phillip Bay dry up between ~2800 and 1000 cal yr BP? Bay floor channelling evidence, seismic and core dating. *Australian Journal of Earth Sciences*, **58**, 157–175.

HUNTLEY, D. J., HUTTON, J. T. & PRESCOTT, J. R. 1993. The stranded beach-dune sequence of south-east South Australia: a test of thermoluminescence dating, 0–800 ka. *Quaternary Science Reviews*, **12**, 1–20.

HUNTLEY, D. J., HUTTON, J. T. & PRESCOTT, J. R. 1994. Further thermoluminescence dates from the dune sequence in southeast of South Australia. *Quaternary Science Reviews*, **13**, 201–207.

JAMES, N. P. & BONE, Y. 2011. *Neritic Carbonate Sediments in a Temperate Realm*. Springer, Dordrecht.

JAMES, N. P., BONE, Y., VON DER BORCH, C. C. & GOSTIN, V. A. 1992. Modern carbonate and terrigenous clastic sediments on a cool water, high energy, mid-latitude shelf: Lacepede, southern Australia. *Sedimentology*, **39**, 877–903.

JAMES, N. P., BONE, Y., HAGEMAN, S. J., FEARY, D. A. & GOSTIN, V. A. 1997. Cool-water carbonate sedimentation during the terminal Quaternary sea-level cycle: Lincoln Shelf. Southern Australia. *In*: JAMES, N. P. & CLARKE, J. A. D. (eds) *Cool-Water Carbonates*. SEPM Special Publications, Tulsa, **56**, 53–75.

JAMES, N. P., BONE, Y., COLLINS, L. B. & KYSER, T. K. 2001. Surficial sediments of the Great Australian Bight: facies dynamics and oceanography on a vast cool-water carbonate shelf. *Journal of Sedimentary Research*, **71**, 549–567.

JAMES, N. P., BONE, Y., CARTER, R. M. & MURRAY-WALLACE, C. V. 2006. Origin of the Late Neogene Roe Plains and their calcarenite veneer: implications for sedimentology and tectonics in the Great Australian Bight. *Australian Journal of Earth Sciences*, **53**, 407–419.

JAMES, N. P., MARTINDALE, R. C., MALCOLM, I., BONE, Y. & MARSHALL, J. 2008. Surficial sediments on the continental shelf of Tasmania, Australia. *Sedimentary Geology*, **211**, 33–52.

JENNINGS, J. N. 1959. The submarine topography of Bass Strait. *Proceedings of the Royal Society of Victoria*, **71**, 49–72.

KENLEY, P. R. 1976. Otway Basin, Western Part. *In*: DOUGLAS, J. G. & FERGUSON, J. A. (eds) *Geology of Victoria*. Geological Society of Australia, Special Publications, **5**, 147–152.

LACHLAN, T. J. 2011. *Aminostratigraphy and Luminescence Dating of the Pleistocene Bridgewater Formation, Kangaroo Island, South Australia: An Archive of Long Term Climate and Sea-level Change*. PhD thesis, University of Wollongong.

LENNON, G. W., BOWERS, D. G. ET AL. 1987. Gravity currents and the release of salt from an inverse estuary. *Nature*, **327**, 695–697.

LI, Q., JAMES, N. P., MCGOWRAN, B., BONE, Y. & CANN, J. H. 1998. Synergetic influence of water masses and Kangaroo Island barrier on foraminiferal distribution, Lincoln and Lacepede shelves, South Australia: a synthesis. *Alcheringa*, **22**, 153–176.

LOWRY, D. C. 1970. Geology of the Western Australian part of the Eucla Basin. *Geological Survey of Western Australia, Bulletin*, **122**, 201.

LUDBROOK, N. H. 1976. The Glanville Formation at Port Adelaide. *Quarterly Geological Notes, Geological Survey of South Australia*, **57**, 4–7.

LUDBROOK, N. H. 1978. Quaternary molluscs of the western part of the Eucla Basin. *Geological Survey of Western Australia, Bulletin*, **125**, 286.

LUDBROOK, N. H. 1984. *Quaternary Molluscs of South Australia*. South Australia Department of Mines and Energy, Handbook, **9**.

LUKASIK, J. & JAMES, N. P. 2006. Carbonate sedimentation, climate change and stratigraphic completeness on a Miocene cool-water epeiric ramp, Murray Basin, South Australia. *In*: PEDLEY, H. M. & CARANNANTE, G. (eds) *Cool-Water Carbonates: Depositional Systems and Palaeoenvironmental Controls*. Geological Society, London, Special Publications, **255**, 217–244.

MCGOWRAN, B. 1979. The Tertiary of Australia: foraminiferal overview. *Marine Micropaleontology*, **4**, 235–264.

MCGOWRAN, B. 1989. The later Eocene transgressions in southern Australia. *Alcheringa*, **13**, 45–68.

MCGOWRAN, B. 1991. Maastrichtian to early Cainozoic, southern Australia: planktonic foraminiferal biostratigraphy. *In*: WILLIAMS, M. A. J., DE DECKKER, P. & KERSHAW, A. P. (eds) *The Cainozoic in Australia: A Re-Appraisal of the Evidence*. Geological Society of Australia, Special Publications, **18**, 79–98.

MCGOWRAN, B. & ALLEY, N. F. 2008. History of the Cenozoic St Vincent Basin in South Australia. *In*: SHEPHERD, S. A., BRYARS, S., KIRKEGAARD, I., HARBISON, P. & JENNINGS, J. T. (eds) *Natural History of Gulf St Vincent*. Royal Society of South Australia, Adelaide, 13–28.

MCLAREN, S., WALLACE, M. W. ET AL. 2011. Palaeogeographic, climatic and tectonic change in southeastern Australia: the Late Neogene evolution of the Murray Basin. *Quaternary Science Reviews*, **30**, 1086–1111.

MILLER, C. R., JAMES, N. P. & BONE, Y. 2012. Prolonged carbonate diagenesis under an evolving late Cenozoic climate; Nullarbor Plain, southern Australia. *Sedimentary Geology*, **261–262**, 33–49.

MILNES, A. R., LUDBROOK, N. H., LINDSAY, J. M. & COOPER, B. J. 1983. The succession of Cainozoic marine sediments on Kangaroo Island, South Australia. *Transactions of the Royal Society of South Australia*, **107**, 1–35.

MITCHELL, J. K., HOLDGATE, G. R. & WALLACE, M. W. 2007*a*. Pliocene–Pleistocene history of the Gippsland Basin outer shelf and canyon heads, southeast Australia. *Australian Journal of Earth Sciences*, **54**, 49–64.

MITCHELL, J. K., HOLDGATE, G. R., WALLACE, M. W. & GALLAGHER, S. J. 2007*b*. Marine geology of the Quaternary Bass Canyon system, southeast Australia: a cool-water carbonate system. *Marine Geology*, **237**, 71–96.

MURRAY-WALLACE, C. V. 2000. Quaternary coastal aminostratigraphy – Australian data in a global context. *In*: GOODFRIEND, G. A., COLLINS, M. J., FOGEL, M. L., MACKO, S. A. & WEHMILLER, J. F. (eds) *Perspectives in Amino Acid and Protein Geochemistry*. Oxford University Press, New York, 279–300.

MURRAY-WALLACE, C. V. 2002. Pleistocene coastal stratigraphy, sea-level highstands and neotectonism of the southern Australian passive continental margin – a review. *Journal of Quaternary Science*, **17**, 469–489.

MURRAY-WALLACE, C. V. & BELPERIO, A. P. 1991. The last interglacial shoreline in Australia – a review. *Quaternary Science Reviews*, **10**, 441–461.

MURRAY-WALLACE, C. V., KIMBER, R. W. L., GOSTIN, V. A. & BELPERIO, A. P. 1988. Amino acid racemisation dating of the 'Older Pleistocene marine beds', Redcliff, northern Spencer Gulf, South Australia. *Transactions of the Royal Society of South Australia*, **112**, 51–55.

MURRAY-WALLACE, C. V., BELPERIO, A. P., GOSTIN, V. A. & CANN, J. H. 1993. Amino acid racemization and radiocarbon dating of interstadial marine strata (oxygen isotope stage 3), Gulf St. Vincent, South Australia. *Marine Geology*, **110**, 83–92.

MURRAY-WALLACE, C. V., FERLAND, M. A., ROY, P. S. & SOLLAR, A. 1996. Unravelling patterns of reworking in lowstand shelf deposits using amino acid racemisation and radiocarbon dating. *Quaternary Science Reviews (Quaternary Geochronology)*, **15**, 685–697.

MURRAY-WALLACE, C. V., BELPERIO, A. P. & CANN, J. H. 1998. Quaternary neotectonism and intra-plate volcanism: the Coorong to Mount Gambier Coastal Plain, Southeastern Australia: a review. *In*: STEWART, I. S. & VITA-FINZI, C. (eds) *Coastal Tectonics*. Geological Society, London, Special Publications, **146**, 255–267.

MURRAY-WALLACE, C. V., BELPERIO, A. P., BOURMAN, R. P., CANN, J. H. & PRICE, D. M. 1999. Facies architecture of a last interglacial barrier: a model for Quaternary barrier development from the Coorong to Mount Gambier Coastal Plain, southeastern Australia. *Marine Geology*, **158**, 177–195.

MURRAY-WALLACE, C. V., BEU, A. G., KENDRICK, G. W., BROWN, L. J., BELPERIO, A. P. & SHERWOOD, J. E. 2000. Palaeoclimatic implications of the occurrence of the arcoid bivalve *Anadara trapezia* (Deshayes) in the Quaternary of Australasia. *Quaternary Science Reviews*, **19**, 559–590.

MURRAY-WALLACE, C. V., BROOKE, B. P., CANN, J. H., BELPERIO, A. P. & BOURMAN, R. P. 2001. Whole-rock aminostratigraphy of the Coorong Coastal Plain, South Australia: towards a 1 million year record of sea-level highstands. *Journal of the Geological Society, London*, **158**, 111–124.

MURRAY-WALLACE, C. V., FERLAND, M. A. & ROY, P. S. 2005. Further amino acid racemization evidence for glacial age, multiple lowstand deposition on the New South Wales outer continental shelf, southeastern Australia. *Marine Geology*, **214**, 235–250.

MURRAY-WALLACE, C. V., BOURMAN, R. P., PRESCOTT, J. R., WILLIAMS, F., PRICE, D. M. & BELPERIO, A. P. 2010. Aminostratigraphy and thermoluminescence dating of coastal aeolianites and the later Quaternary history of a failed delta: the River Murray mouth region, South Australia. *Quaternary Geochronology*, **5**, 28–49.

O'CONNELL, L. G. 2011. *Sedimentology of the Miocene Nullarbor Limestone; Southern Australia*. Geological Survey of Western Australia Report, **111**.

PARKER, A. J. 1993. Geological Framework. *In*: DREXEL, J. F., PREISS, W. V. & PARKER, A. J. (eds) *Geology of South Australia, Volume 1, The Precambrian*. South Australia Geological Survey Bulletin, **54**, 9–31.

PATTIARATCHI, C. & WOO, M. 2009. The mean state of the Leeuwin Current system between North West Cape and Cape Leeuwin. *Journal of the Royal Society of Western Australia*, **92**, 221–241.

PORTER-SMITH, R., HARRIS, P. T., ANDERSON, O. B., COLEMAN, R., GREENSLADE, D. & JENKINS, C. J. 2004. Classification of the Australian continental shelf based on predicted sediment threshold exceedence from tidal currents and swell waves. *Marine Geology*, **211**, 1–20.

PREISS, W. V. 1987. *The Adelaide Geosyncline – Late Proterozoic Stratigraphy, Sedimentation, Palaeontology and Tectonics*. South Australia Geological Survey Bulletin, **53**.

RIVERS, J. M., JAMES, N. P., KYSER, T. K. & BONE, Y. 2007. Genesis of palimpsest cool-water carbonate sediment on the continental margin of southern Australia. *Journal of Sedimentary Research*, **77**, 480–494.

ROLLET, N., FELLOWS, M., STRUCKMEYER, H. I. M. & BRADSHAW, B. E. 2001. *Seabed character mapping in the Great Australian Bight*. Australian Geological Survey Organisation (AGSO), Record, 2001/42.

ROY, P. S. 1998. Cainozoic geology of the coast and shelf. *In*: SCHEIBNER, E. & BASDEN, H. (eds) *Geology of New South Wales – Synthesis, Volume 2, Geological Evolution*. Geological Survey of New South Wales, Memoirs, **13**, 361–385.

ROY, P. S. & CRAWFORD, E. A. 1980. Quaternary geology of the Newcastle Bight Inner Continental Shelf, New South Wales, Australia. *Records of the Geological Survey of New South Wales*, **19**, 145–188.

ROY, P. S. & THOM, B. G. 1991. Cainozoic shelf sedimentation model for the Tasman Sea margin of southeastern Australia. *In*: WILLIAMS, M. A. J., DE DECKKER, P. & KERSHAW, A. P. (eds) *The Cainozoic in Australia: A Reappraisal of the Evidence*. Geological Society of Australia, Special Publications, **18**, 119–136.

ROY, P. S., COWELL, P. J., FERLAND, M. A. & THOM, B. G. 1994. Wave-dominated coasts. *In*: CARTER, R. W. G. & WOODROFFE, C. D. (eds) *Coastal Evolution: Late Quaternary Shoreline Morphodynamics*. Cambridge University Press, Cambridge, 121–186.

ROY, P. S., ZHUANG, W.-Y., BIRCH, G. F., COWELL, P. J. & CONGXIAN, L. 1997. *Quaternary Geology of the Forster–Tuncurry Coast and Shelf, Southeast Australia*. Geological Survey of New South Wales Report, GS1992/01.

SANDIFORD, M. 2007. The tilting continent: a new constraint on the dynamic topographic field from Australia. *Earth and Planetary Science Letters*, **261**, 152–163.

SCHMIDT, S., DE DECKKER, P., ETCHEBER, H. & CARADEC, S. 2010. Are the Murray Canyons offshore southern Australia still active for sediment transport? *In*: BISHOP, P. & PILLANS, B. (eds) *Australian Landforms*. Geological Society, London, Special Publications, **346**, 43–55.

SHACKLETON, N. J. 1987. Oxygen isotopes, ice volume and sea level. *Quaternary Science Reviews*, **6**, 183–190.

SHEPHERD, S. A. & SPRIGG, R. C. 1976. Substrate sediments and subtidal ecology of Gulf St Vincent and Investigator Strait. *In*: TWIDALE, C. R., TYLER, M. J. & WEBB, B. P. (eds) *Natural History of the Adelaide Region*. Royal Society of South Australia, Adelaide, 161–174.

SHORT, A. D. & HESP, P. A. 1982. Wave, beach and dune interactions in southeastern South Australia. *Marine Geology*, **48**, 259–284.

SLOSS, C. R., MURRAY-WALLACE, C. V. & JONES, B. G. 2007. Holocene sea-level change on the southeast coast of Australia: a review. *The Holocene*, **17**, 999–1014.

SPRIGG, R. C. 1947. Submarine canyons of the New Guinea and South Australian coasts. *Transactions of the Royal Society of South Australia*, **71**, 296–310.

SPRIGG, R. C. 1952. *The Geology of the South-East Province, South Australia, with Special Reference to Quaternary Coast-Line Migrations and Modern Beach Developments*. Geological Survey of South Australia Bulletin, **29**.

SPRIGG, R. C. 1979. Stranded and submerged sea-beach systems of southeast South Australia and the aeolian desert cycle. *Sedimentary Geology*, **22**, 53–96.

SUTHERLAND, F. L. 1991. Cainozoic volcanism, Eastern Australia: a predictive model based on migration over multiple 'hotspot' magma sources. *In*: WILLIAMS, M. A. J., DE DECKKER, P. & KERSHAW, A. P. (eds) *The Cainozoic in Australia: A Reappraisal of the Evidence*. Geological Society of Australia, Special Publications, **18**, 15–43.

THOM, B. G. 1965. Late Quaternary coastal morphology of the Port Stephens – Myall Lakes area, N.S.W. *Journal and Proceedings of the Royal Society of New South Wales*, **98**, 23–36.

THOM, B. G. & CHAPPELL, J. 1975. Holocene sea levels relative to Australia. *Search*, **6**, 90–93.

THOM, B. G. & ROY, P. S. 1985. Relative sea levels and coastal sedimentation in southeast Australia in the Holocene. *Journal of Sedimentary Petrology*, **55**, 257–264.

THOM, B. G., SHEPHERD, M., LY, C. K., ROY, P. S., BOWMAN, G. M. & HESP, P. A. 1992. *Coastal Geomorphology and Quaternary Geology of the Port Stephens–Myall Lakes Area*. Australian National University. Department of Biogeography and Geomorphology, Monograph, **6**.

THOM, B. G., KEENE, J. B., COWELL, P. J. & DALEY, M. 2010. East Australian marine abrasion surface. *In*: BISHOP, P. & PILLANS, B. (eds)

Australian Landscapes. Geological Society, London, Special Publications, **346**, 57–69.

TOMCZAK, M. & GODFREY, J. S. 1994. *Regional Oceanography: An Introduction*. Pergamon Press, Oxford.

TWIDALE, C. R. 2007. *Ancient Australian Landscapes*, Rosenburg, Sydney.

VEEVERS, J. J. (ed.) 2000. *Billion-year Earth History of Australia and Neighbours in Gondwanaland*. Gemoc Press, Sydney.

WELLS, P. E. & WELLS, G. M. 1994. Large-scale reorganization of ocean currents offshore Western Australia during the Late Quaternary. *Marine Micropalaeontology*, **24**, 157–186.

WILSON, C. C. 1991. *Geology of the Quaternary Bridgewater Formation of southwest and central South Australia*. PhD thesis, Flinders University of South Australia.

YOKOYAMA, Y., LAMBECK, K., DE DECKKER, P., JOHNSTON, P. & FIFIELD, L. K. 2000. Timing of the Last Glacial Maximum from observed sea-level minima. *Nature*, **406**, 713–716.

ZHOU, L., WILLIAMS, M. A. J. & PETERSON, J. A. 1994. Late Quaternary aeolianites, palaeosols and depositional environments of the Nepean Peninsula, Victoria, Australia. *Quaternary Science Reviews*, **13**, 225–239.

Chapter 21

A review of sedimentation since the Last Glacial Maximum on the continental shelf of eastern China

SHOUYE YANG[1]*, ZHONGBO WANG[1,2], YANGUANG DOU[1,2] & XUEFA SHI[3]

[1]*State Key Laboratory of Marine Geology, Tongji University, Shanghai 200092, China*

[2]*The Key Laboratory of Marine Hydrocarbon Resources and Environmental Geology, Ministry of Land and Resources, Qingdao Institute of Marine Geology, Qingdao 266071, China*

[3]*First Institute of Oceanography, State Oceanic Administration, Qingdao 266071, China*

*Corresponding author (e-mail: syyang@tongji.edu.cn)

Abstract: The contiguous continental shelf of the Bohai, Yellow and East China seas has a total area of about 0.75×10^6 km^2 and a width of over 600 km. The basement is tectonically stable and the accumulated thickness of Quaternary sediment is up to 500 m. During the post-glacial period, the epicontinental shelf received a large influx of fluvial sediments with complex fluvial–marine interactions. The river-dominated shelf sedimentation is characterized by two types of terrigenous sediment source-to-sink transport processes represented by the two mega-rivers, Changjiang (Yangtze River) and Huanghe (Yellow River), and the small mountainous rivers in Taiwan Island, in total contributing about 10% of the world's total river sediment load. Mud patches and belts, and tidal sand ridges were well developed on the shelf during the post-glacial period. These muds are generally interpreted as eddy sedimentation formed after the highest post-glacial transgression, while the tidal sand ridges with several formation stages developed under variable oceanic regimes. The changes in monsoon climate-induced river flux, sea level and oceanic circulation primarily controlled the stratigraphic framework and sedimentary facies on the shelf during the late Quaternary. Nevertheless, many aspects of sedimentary processes and river–sea interactions at different spatial and temporal scales require more investigation.

General setting of the eastern China continental shelf

The eastern China seas, including the Bohai Sea, Yellow Sea and East China Sea (ECS), are located between the largest continent, Eurasia, and the largest ocean, the Pacific Ocean. The shelf is situated at a mid-latitude of 21–42°N (Fig. 21.1). The general physiographical and oceanographic setting of the shelf is shown in Table 21.1.

The Bohai and Yellow seas are epicontinental seas situated on a uniform and tectonically stable basement overlying continental lithosphere, and the whole basin forms a shelf environment with a maximum water depth of about 140 m. In comparison, the ECS, as a pericontinental sea, consists of a broad shelf about 600 km wide and an incipient intracontinental basin, the Okinawa Trough (Fig. 21.1). The Okinawa Trough is located east of the ECS, behind the Ryukyu arc–trench system. The continental shelves of the Bohai, Yellow and East China seas form a contiguous shelf about 0.75×10^6 km^2 in area, bordered by the Okinawa Trough to the east and connected to the South China Sea (SCS) through the Taiwan Strait to the south.

The Yellow Sea, a typical semi-enclosed epicontinental sea, rests on a flat, broad and tectonically stable seafloor with an average water depth of 44 m and maximum depth of 140 m (Qin *et al.* 1989). The Yellow Sea is separated from the Bohai Sea at its northern extremity by the Shandong Peninsula, and from the ECS to the south by an imaginary line between the Changjiang river mouth and Cheju Island. The Yellow Sea is characterized by a huge delta, wide and open mud flat, and local rocky coast in the western part and, by contrast, numerous ria-type bays, indented islands and a long stretch of tidal flat in the eastern part.

The ECS has an area of 77×10^4 km^2, an average depth of 370 m and almost three-quarters of the area is shallower than 200 m. Geographically, the sea can be divided into three parts: the inner shelf; middle–outer shelf; and the continental slope, which deepens towards the Okinawa Trough. The inner shelf extends from the coastline to about 60 m water depth, and the middle–outer shelf covers the water depth from 60 m to the shelf break at about 200 m, with an average gradient of 1:2000. The Okinawa Trough is a crescent-shaped trough extending in a NE–SW direction parallel to the orientation of the ECS (Qin *et al.* 1987; Qin 1994). From the Changjiang Estuary to the outer shelf, seafloor topography is characterized by undulating sand ridges and intervening swales. Three submarine terraces occur on the shelf that correspond to −110, −60 and −40 m, respectively, which were formed at stillstands during the deglaciation corresponding to 13, 12 and 10 ka, respectively (Feng 1983).

Tidal and current regimes

The eastern China continental shelf is subject to the East Asian monsoon, which causes southerly and southwesterly winds during summer, and northerly and northeasterly winds during winter. Storms are frequent, and wind-induced currents and waves are dominant during winter. Tides are typically semi-diurnal (M2) ranging from <1.5 to about 8 m (Chough *et al.* 2000), and rates of tidal currents vary considerably depending on local topography and seasonal climate.

Uda (1934) first studied the Yellow Sea current regime. The general circulation pattern in the eastern China seas is characterized by several cyclonic gyres, with northward inflow of warm and saline waters along the east side, and southward outflow of cooler and less saline coastal waters along Chinese and Korean coasts (Fig. 21.1). The Kuroshio Current flows NE over the continental slope and outer shelf of the ECS, and enters the Pacific Ocean at the northern extremity of the Okinawa Trough. One branch, the Tsushima Current, enters the Japan Sea through the Tsushima Straits. The Yellow Sea Warm Current is generally regarded as a branch of the Tsushima Current and/or Taiwan Warm Current, with high temperature and salinity, flowing roughly northwards along the Yellow Sea Trough and penetrating to the Bohai Sea, especially in winter. Inshore of the Kuroshio Current, the Taiwan Warm Current also flows NE, and forms anticlockwise gyres in the Yellow Sea and ECS.

From: Chiocci, F. L. & Chivas, A. R. (eds) 2014. *Continental Shelves of the World: Their Evolution During the Last Glacio-Eustatic Cycle.*
Geological Society, London, Memoirs, **41**, 293–303. http://dx.doi.org/10.1144/M41.21

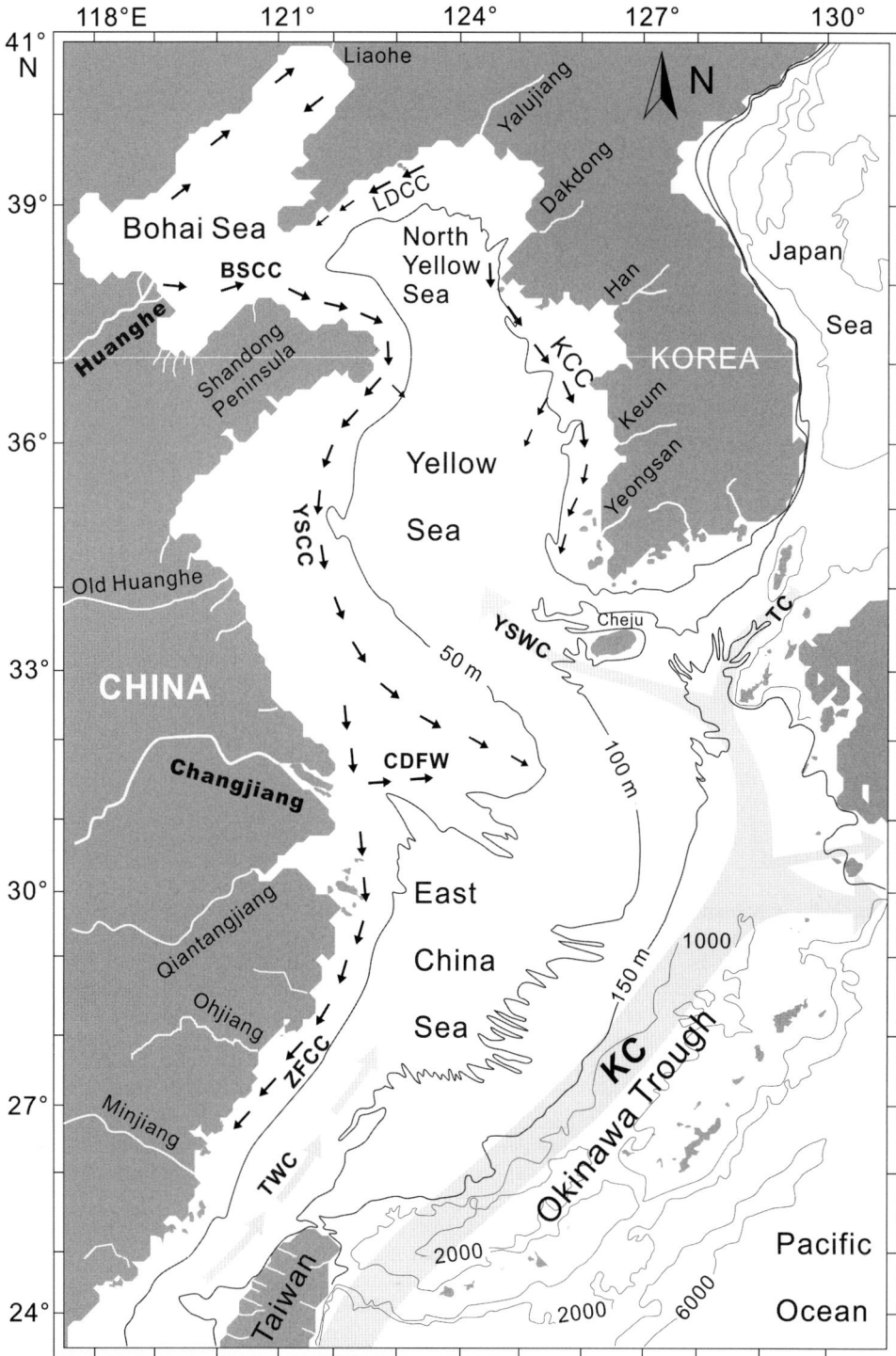

Fig. 21.1. Bathymetry and general circulation of the Bohai, Yellow and East China seas (modified after Beardsley *et al.* 1985; Park & Kim 1992; Hu & Li 1993; Guan 1994; Chough *et al.* 2000). BSCC, Bohai Sea Coastal Current; LDCC, Liaodong Coastal Current; KCC, Korea Coastal Current; YSCC, Yellow Sea Coastal Current; ZFCC, Zhejiang–Fujian Coastal Current; YSWC, Yellow Sea Warm Current; CDFW, Changjiang Diluted Freshwater; TC, Tsushima Current; TWC, Taiwan Warm Current; KC, Kuroshio Current.

The coastal currents, including the Bohai Sea Coastal Current, Liaodong Coastal Current, Yellow Sea Coastal Current and Zhejiang–Fujian Coastal Current, flow south along the Chinese coast, while the Korea Coastal Current flows south along the Korean coast (Fig. 21.1). The coastal water masses are characterized by relatively cold temperatures, low salinity and high suspended-load sediment concentration.

The most significant freshwater discharge into the shelf is from the Changjiang (Yangtze River), which has the largest water discharge in China. The Changjiang freshwater plume predominantly flows SE along the Chinese coast during winter, while it extends as a low-salinity plume to the NE in the direction of Cheju Island during summer (Beardsley *et al.* 1985). Generally, the Changjiang-derived suspended particulate matter does not extend east or NE beyond 123.5°E, based on

physical oceanographic and sediment geochemical observations (Beardsley *et al.* 1985; Hu & Li 1993; Zhang 1999; Wu *et al.* 2003).

Geological and geophysical investigations of the eastern China shelf began in the 1960s, and China, the United States and Japan carried out the early seismic surveys (Niino & Emery 1961; Qin *et al.* 1987, 1989; Qin 1994). Since the 1980s, international cooperation on integrated geological and geophysical investigations has been strengthened in the estuarine and shelf areas, and, in particular, the Yellow and East China seas received more attention. Over the past two decades, the marginal seas in China have increasingly attracted many research subjects because of huge terrigenous inputs from mainland China and Taiwan Island, characteristic land–sea interaction and remarkable palaeoenvironmental changes during the late Quaternary.

Table 21.1. *Selected characteristics of the continental shelves of Bohai, Yellow and East China seas*

	Bohai Sea	**Yellow Sea**	**East China Sea**
Location (longitude, latitude)	117°35′–121°10′E, 37°07′–40°0′N	119°10′–126°50′E, 31°40′–39°50′N	22°–33°N
Maximum length (km)	480	870	1167
Maximum width (km)	300	556	640
Sea area ($\times 10^4$ km^2)	7.7	38	54
Average water depth (m)	18	44	147
Maximum water depth (m)	83	140	2719 (continental slope)
Shelf gradient	1:13 000	1:10 000	1:7000
Annual SST average (°C)	12	16	22
Summer	25	24–27	26–29
Winter	−4 to 0	−2 to 8	2–16
Annual SSS average (‰)	30	31	33–34
Tide	Semi-diurnal dominant, diurnal locally	Semi-diurnal dominant, diurnal locally	Semi-diurnal
Dominant process	Wave, current	Current, tide	Current, tide
Surface sediment character	Siliciclastic dominant, *c.* 10% carbonate	Siliciclastic dominant, *c.* 8% carbonate	Siliciclastic dominant, *c.* 12% carbonate
Surface sediment origin	Modern	70% modern, 10% relict, 20% palimpsest	40% modern, 40% relict, 20% palimpsest
Quaternary sediment thickness (m)	300–500	300–500	200–350

Abbreviations: SSS, sea surface salinity; SST, sea surface temperature.
Data sources: Qin *et al.* (1987, 1989), Yang *et al.* (2003) and Li *et al.* (2005*b*).

Sediment input and modern sediment distribution patterns

The eastern China seas are river-dominated, and the rivers that meet the seas have remarkably different sizes, water and sediment discharges. The Changjiang and the Huanghe (Yellow River), as the two largest rivers in East Asia, contribute about 10% of the world river sediment load (Milliman & Meade 1983), which basically dominates the sedimentation on the eastern China continental shelf. It is noteworthy that, since the late 1970s, the sediment loads of these two large rivers have declined dramatically owing to numerous dam constructions, including the Three Gorges Reservoir, and intense water consumption in the catchment (Yang *et al.* 2003, 2007; Xu & Milliman 2009).

Sediment fluxes of the other rivers from mainland China and the Korea Peninsula are relatively small and vary significantly with season (Fig. 21.2), and may exert influence on nearshore and shelf sedimentation (Yang *et al.* 2003 and references therein; Yang & Youn 2007; Liu *et al.* 2008). Nevertheless, the small mountainous rivers in Taiwan Island can deliver between approximately 230 and >400 Mt a^{-1} of sediment into the Taiwan Strait and Okinawa Trough, mostly by typhoon floods in summer (Fig. 21.2) (Kao & Milliman 2008; Liu *et al.* 2008; Xu *et al.* 2009). Earthquakes, lithology, topography, cyclone-induced rainfall and human disturbance play major roles in the catchment dynamics and in rapid sediment transfer from land to sea. Taiwan Island therefore provides an ideal site for the study of the episodic aspect of the source-to-sink pathway to marine environments (Liu *et al.* 2013).

The sedimentary environments of the eastern China seas comprise coastal, littoral, neritic (open shelf) and semi-pelagic facies. Li *et al.* (2005*b*) suggested four types of sediment origin on the shelf based on mineralogical and chemical composition: terrigenous clastic sediment; biogenic sediment; biogenic–terrigenous sediment; and volcanic debris and hydrothermal chemical sediment. Among these, terrigenous clastic sediment, particularly from the Changjiang and Huanghe, dominates sedimentation on the continental shelf (Yang & Milliman 1983; Milliman *et al.* 1989; Yang *et al.* 2003; Liu J. P. *et al.* 2007). The multi-year observation at Datong gauging station (about 600 km upstream of the river mouth) suggests that about 470–500 Mt a^{-1} of sediment load is transported by the Changjiang into its estuary (Yang *et al.* 2003), while channel aggradation and delta progradation

trap about 70% of the sediment before it reaches the ECS, leaving around 150 Mt a^{-1} of sediments to be transported southwards (Milliman *et al.* 1985*b*; Liu J. P. *et al.* 2007). Only a small part of the Changjiang sediment can be transported to the offshore area up to 124°E over a distance of 250–300 km eastwards (Beardsley *et al.* 1985; Milliman *et al.* 1985*b*; Zhang 1999), and may escape further eastwards to the outer shelf of the ECS, and even to the Okinawa Trough, probably by bottom transport in winter (Yang *et al.* 1992; Guo *et al.* 2002; Katayama & Watanabe 2003; Yang & Liu 2007; Xu *et al.* 2009, 2012). Nevertheless, the quantity and transport mechanism of the river-borne fine particles across the wide shelf to the trough are still unresolved.

Niino & Emery (1961) used echosounder reflection amplitude notations on nautical charts to map the surficial sediments on the eastern China shelf, and defined modern detrital, residual and relict sediment origins. They found that the fine-grained fluvial sediments dominate the inner shelf, while relict sandy sediment is distributed on the mid-outer shelf from which finer sediments are winnowed or prevented from being deposited by the strong Kuroshio Current. Milliman *et al.* (1989) also suggested that the sedimentary environment and history of the eastern China seas have been influenced primarily by the broad and shallow nature of the epicontinental shelf and by the large influx of fluvial sediments from mainland China. In general, the sedimentation on the eastern China shelf is characterized by several mud patches and belts and large sand ridges and sheets (Fig. 21.2). The Bohai Sea is mud-dominated in the estuarine and central parts while tidal sand ridges occupy the Bohai Strait where strong tidal currents occur. Muddy sediments dominate the central part of the Yellow Sea, while sand and muddy sand blanket the eastern and the western parts of the sea (Fig. 21.2). Large-scale sandy deposits occur in the SW Yellow Sea, forming a radial tidal sand-ridge system (Li *et al.* 2001), which is considered a relict deposit (Liu *et al.* 1989; Yang 1989; Li *et al.* 2001). The NE Yellow Sea is floored with coarse-grained transgressive sandy deposits (i.e. another tidal sand ridge system) formed during the last postglacial sea-level rise (Klein *et al.* 1982; Chough *et al.* 2000).

In contrast to the Bohai and Yellow seas, the ECS is characterized by muddy sediments in the estuarine, nearshore and inner shelf areas, and sandy sediments in the form of sand ridges or sand sheets on the middle–outer shelf (Liu 1997; Saito *et al.* 1998; Liu *et al.* 2000, 2007; Wang *et al.* 2013). The sands of the outer continental shelf are part of the 4000 km-long belt of 'relict sediment' that extends almost continuously from Korea to Malaysia (Emery 1968). The sand ridges in the ECS are 8–14 km wide and tens of kilometres long, orientated NW–SE, approximately parallel to the major axis of the tidal ellipse and perpendicular to the continental margin. The tidal sand-ridge system in the ECS can be regarded as one of the largest and deepest sand-ridge systems in the world (Liu Z. X. *et al.* 2007).

Yang *et al.* (1992) and Sun *et al.* (2000) proposed a widely accepted notion that the suspended sediments of the Yellow and East China seas are 'being stored in the summer and transported in the winter'. In summer, a major part of the suspended sediments from the major rivers is trapped in estuarine and coastal areas, and only part escapes to the shelf transported by the river plume. In winter, a large volume of the deposited fine-grained

sediments can be resuspended and transported to the offshore area by the strong coastal current.

Post-glacial sedimentation

Overview

The sedimentary processes on the eastern China continental shelf and their response to environmental change during the late Quaternary have been widely reported over the past 30 years. However, many aspects of the shelf sedimentation are still poorly understood. Milliman *et al.* (1989) suggested that the effect of sea-level fluctuations can be acoustically and sedimentologically

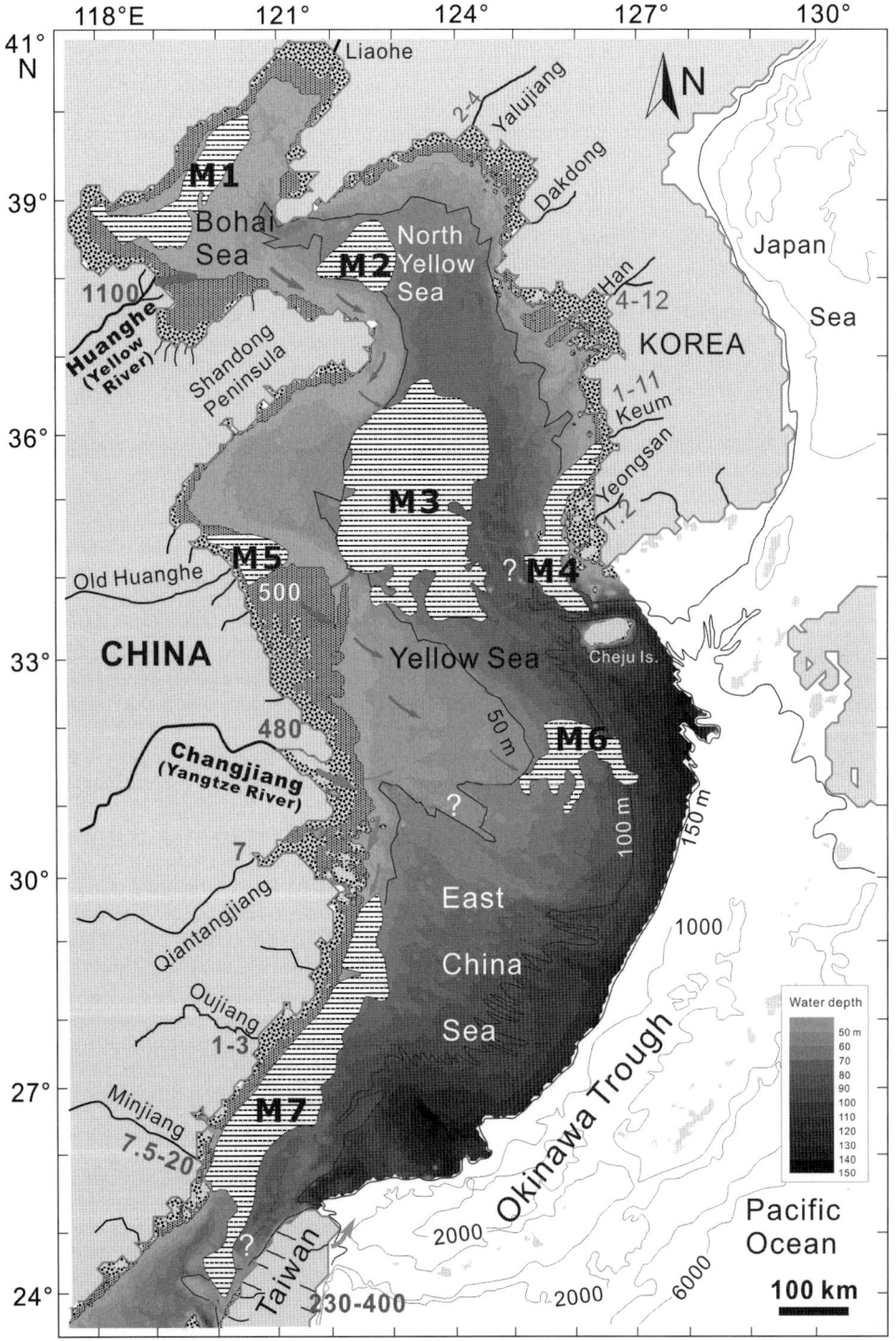

Fig. 21.2. Sediment dispersal and distribution pattern on the eastern China continental shelf (modified after Yang *et al.* 2003; Li *et al.* 2005a, b; Wang *et al.* 2013). The sediment discharges of individual rivers are given in Mt a^{-1}. Main shelf mud depocentres (M1–M7) are highlighted. CBSM, central Bohai Sea mud (M1); NYSM, northern Yellow Sea mud (M2); CYSM, central Yellow Sea mud (M3); SEYSM, the mud belt in the SE Yellow Sea (M4); SWYSM, SW Yellow Sea mud off the northern Jiangsu coast (M5); SWCIM, mud patch in the northern East China Sea or the SW Cheju Island mud (M6); ECSSM, Zhejiang–Fujian offshore mud or inner-shelf mud wedge in the East China Sea (M7). Arrows indicate the pathways of fluvial sediments from modern and old river mouths.

depicted in the Quaternary records on the continental shelf; however, the marine transgression events and corresponding sedimentary records, especially before the last glaciation, have never been investigated in detail. Similarly, the flux and fate of terrigenous sediments on the eastern China continental shelves, particularly those transported by the rivers, remain to be clarified, despite many previous research attempts using various research methods.

During the post-glacial period, most of the terrigenous sediments in the Bohai and Yellow seas have been considered to be derived primarily from the Huanghe, whereas the sediment in the ECS is predominantly sourced from the Changjiang (Qin & Li 1983; Liu et al. 1987; Qin et al. 1987; Milliman et al. 1985a, 1989; Qin 1994). Milliman et al. (1989) suggested that during the Holocene about 90% of the sediment contributed to the Yellow Sea by the Huanghe remained in the deltaic system; the other 10% was eroded and transported SE as far as Cheju Island. Yang & Liu (2007) further suggested that the Huanghe-derived sediment could reach water depths of 80 m in the central South Yellow Sea, about 700 km from the river mouth during the lowstand of sea level, and a very small fraction of the modern riverine sediment can escape the outer shelf or reach the Okinawa Trough.

In comparison, the Changjiang sediment was transported southwards via the coastal current, and cross-shelf transport has been prevented by tidal currents and the northward invasion of the Taiwan Warm Current. However, there is still no quantitative estimation of the spatial and temporal distributions of these fluvial sediments on the shelves. Furthermore, the strong coupling of oceanic circulation evolution and the sediment transport process on the wide shelves is another intractable problem that remains controversial.

Recent work on the clay mineralogy and sediment chemistry of the ECS by Dou et al. (2010, 2012) suggests that the Changjiang-derived sediments dominated the sedimentation on the ECS shelf and the middle Okinawa Trough during the last glaciation and early Holocene (c. 28–8 ka), while the Taiwan-sourced fine-grained sediments might be largely delivered to the outer shelf

and trough during the mid–late Holocene with the strengthening Kuroshio Current. Sedimentological and geochemical work on sediment cores taken from the central south Yellow Sea suggest that muddy Holocene sediments in the central-western part are derived ultimately from Chinese rivers, especially the Huanghe, whereas the eastern sandy sediments primarily came from the Korean rivers during the post-glacial transgression (Yang & Youn 2007).

Although sedimentation on the eastern China continental shelves has been extensively studied from a variety of scientific fields, more detailed studies are needed to better understand the source-to-sink history of fluvial sediments on the shelf and their relationship with changes in sea level, oceanic circulation and monsoon-induced river runoff during the late Quaternary.

Shelf mud deposits: sources and origins

Several mud depocentres are well defined on the eastern China continental shelves: (1) the central Bohai Sea mud (CBSM, M1 in Fig. 21.2): (2) the northern Yellow Sea mud (NYSM, M2); (3) the central Yellow Sea mud (CYSM, M3); (4) the mud belt in the SE Yellow Sea (SEYSM, M4); (5) the SW Yellow Sea mud off the northern Jiangsu coast (SWYSM, M5); (6) the mud patch in the northern ECS or the SW Cheju Island mud (SWCIM, M6); and (7) the Zhejiang–Fujian offshore mud or inner-shelf mud wedge in the ECS (ECSSM, M7 in Fig. 21.2) (Yang et al. 2003; Li et al. 2005b; Liu et al. 2008; Xu et al. 2009, 2012). The origin of these mud deposits has been widely investigated using geophysical, sedimentological and geochemical methods but consensus has rarely been attained. A detailed review of provenance discrimination of Yellow Sea sediments with special emphasis on geochemical approaches was made by Yang et al. (2003).

It is undoubted that the CBSM and NYSM (M1 and M2) are derived predominantly from the Huanghe, and partly from other small rivers such as the Liaohe and Yalujiang (Fig. 21.2). These

Table 21.2. Central position, water depth, area, sediment type and maximum flow velocity of tidal sand-ridge systems on the eastern China continental shelves

Tidal depositional system	Central position (latitude (N), longitude (E))	Water depth (m)	Area (km²)	Sediment type	Maximum flow velocity (knot)
Eastern Yellow Sea			80 000		
West Korean Bay	38°38′, 124°50′	10–50	11 335	FS, MFS	2–3
Kwanghwa Bay	37°20′, 126°32′	10–20		MS, S	2–3
Offshore Keum River	36°48′, 125°37′	30–50	8098	MFS, FS	1–2
Eastern Yellow Sea	37°00′, 125°30′	10–50		FS	
South Yellow Sea (offshore Cheju Island)	34°18′, 124°20′	80–120	17 918	S, T, TY	
Eastern Bohai Sea			11 000		
Laotieshan Channel	38°35′, 121°05′	40–86	3000	FS, GS	4–5
Liaodong Shoal	39°20′, 120°40′	10–36	4000	FS, TS, YS	1.3–2.3
Bozhong Shoal	38°50′, 120°20′	20–40	4000	FS, TS	1.2–1.6
Off Changjiang Estuary			77 719		
Offshore North Jiangsu	33°02′, 121°03′	10–30	23 056	FS, TS	2–3
Yangtze Shoal	30°48′, 123°40′	22–55	54 663	FS, MFS	1.0–2.4
Offshore Taiwan shelves			47 000		
Taiwan Strait	24°30′, 118°30′	50–70	29 000	FS, TS	2–3
Taiwan Shoal	23°00′, 118°30′	20–40	13 000	CS, MS, MFS	±2
Penghu Channel	23°27′, 119°50′	50–100	2000	SG, YS	
Taizhong Shoal	24°15′, 120°00′	9.6–40	36 000	FS	
East China Sea shelf			210 000		
Relict tidal sand ridge	28°10′, 123°00′	60–110	70 000	FS, SS	1.5–2
Relict tidal sand sheet	28°34′, 126°08′	60–110	140 000	FS, FS	

Abbreviations: GS, gravelly sand; SG, sandy gravel; CS, coarse sand; MS, medium sand; MFS, medium–fine sand; FS, fine sand; TS, silty sand; YS, clayey sand; T, silt; SS, shelly sand.
Sources: From Liu et al. (1998), Li et al. (2005b).

two mud deposits constitute the subaqueous deltaic system of the modern Huanghe, and represent the main depocentres of the Huanghe–derived sediments in the Bohai and Yellow seas (Liu *et al.* 2004; Li *et al.* 2005*b*; Yang & Liu 2007).

The CYSM (M3 in Fig. 21.2) and SWYSM (M5 in Fig. 21.2) have been suggested to be derived mostly from the modern and/or ancient Huanghe (Qin & Li 1983; Milliman *et al.* 1987; Lee & Chough 1989; Qin *et al.* 1989; Alexander *et al.* 1991; Park & Khim 1992; Cho *et al.* 1999; Yang *et al.* 2003; Yang & Liu 2007; Liu *et al.* 2010). Yang & Liu (2007) suggested that the alongshore-distributed clinoform muddy deposit around the eastern tip of the Shandong Peninsula in the Yellow Sea can be regarded as the distal subaqueous deltaic lobe of the Huanghe. This distal lobe has formed since the mid-Holocene highstand under a relatively stable sea level. They further deduced that over the past 7 kyr, nearly 30% of the Huanghe-derived sediment has been resuspended and transported out of the Bohai Sea into the Yellow Sea. The SWYSM (M5 in Fig. 21.2) is located immediately offshore of the former Huanghe delta from where

the Huanghe shifted its pathway to the south and directly supplied its sediment to the south Yellow Sea in the Holocene (Milliman *et al.* 1987; Saito 1998; Yang *et al.* 2003; Liu *et al.* 2010).

However, others suggested that the CYSM is a 'relict mud' formed during the last eustatic cycle (Hu 1984) or a 'multi-source mud' with the sediments from the Huanghe, Changjiang and other local rivers, based on mineralogical and geochemical compositions (Zhao *et al.* 1990, 1997). Based on a geochemical study of five cores taken from the central south Yellow Sea, Yang & Youn (2007) argued that the muddy surface sediments in the central-western part are mostly supplied by the modern and/or old Huanghe through transport by the YSCC, whereas the post-glacial muddy sediments were derived from diverse sources varying between the Changjiang and Huanghe origins. In contrast, the muddy sand in the eastern part primarily came from the Keum River during the post-glacial transgression. Overall, Korean rivers, including the Han, Keum and Yeongsan, might not have supplied much fine-grained sediments to the central Yellow Sea during the last glaciation because of their low sediment loads, with most of

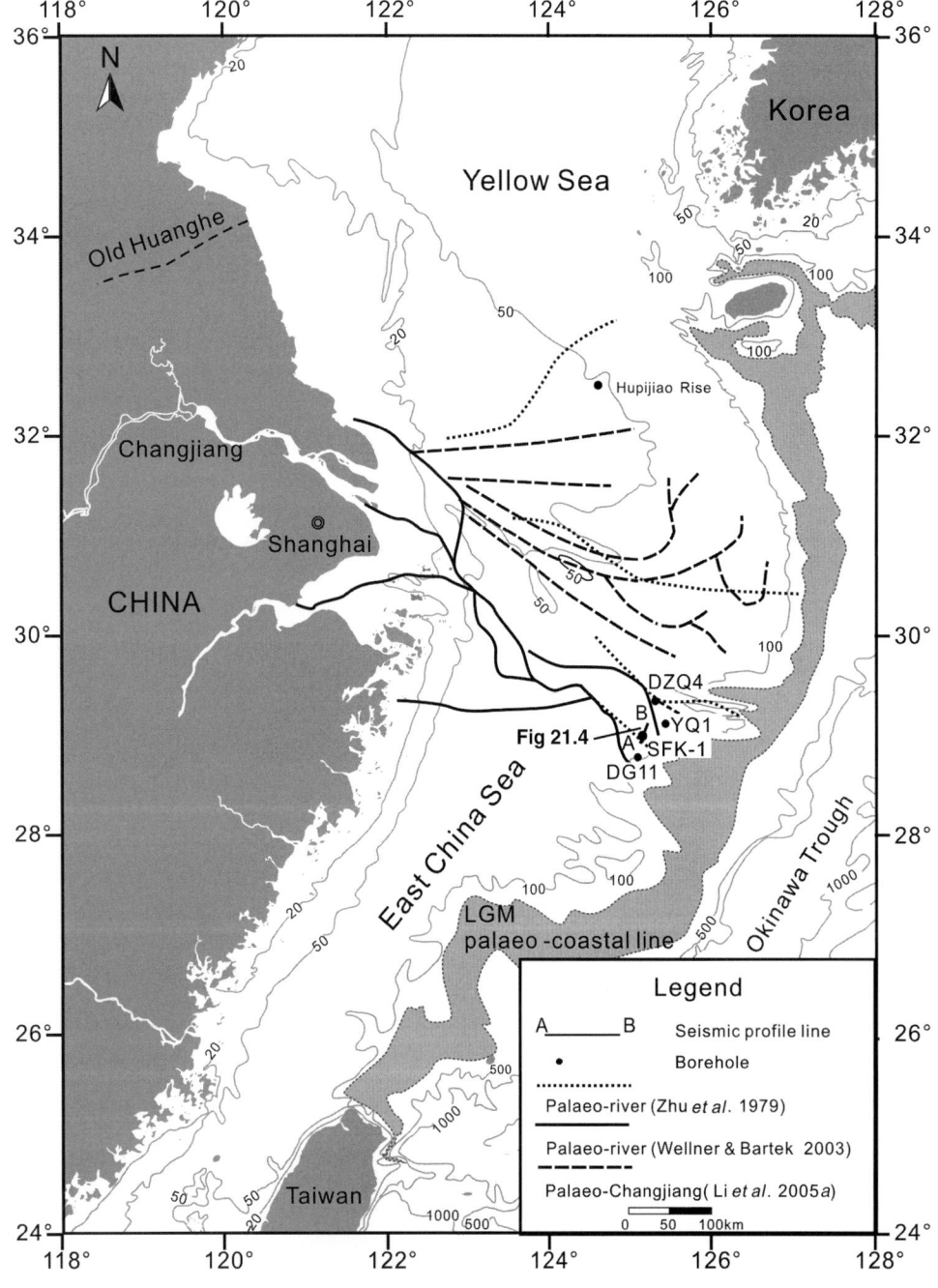

Fig. 21.3. Map of the East China Sea showing the locations of core SFK-1 and other nearby boreholes (Zhu *et al.* 1979; Wellner & Bartek 2003; Li *et al.* 2005*a*). The palaeo-coastal line at the LGM was modified after Saito *et al.* (1998).

their sediments being trapped in estuaries and transported towards the SE Yellow Sea (Cho *et al.* 1999; Chough *et al.* 2000 2004; Lee & Chu 2001; Yang *et al.* 2003; Yang & Youn 2007).

The SEYSM (M4 in Fig. 21.2) is considered to be of mixed sediment sources, including from Korean rivers, the Huanghe and/or the Changjiang, and/or the central Yellow Sea after resuspension (Alexander *et al.* 1991; Cho *et al.* 1999; Chough *et al.* 2000, 2004; Park *et al.* 2000; Zhao *et al.* 2001; Yang *et al.* 2003). The SWCIM (M6 in Fig. 21.2) has even been regarded as the distal mud of the Huanghe dispersal system in the East China Sea (DeMaster *et al.* 1985; Alexander *et al.* 1991) and, accordingly, formed in an anticlockwise cyclonic eddy (Hu & Li 1993).

The ECSSM (i.e. the Zhejiang–Fujian offshore mud) (M7 in Fig. 21.2) has received significant attention over the past decade. It is widely accepted that since about 7 ka, the Changjiang supplied a large volume of fine-grained sediments to form the ECSSM, while the clay from Taiwan rivers was only trapped in a narrow band in the SE part of the Taiwan Strait (Milliman *et al.* 1989; Liu J. P. *et al.* 2007, 2008; Xu *et al.* 2009, 2012). A review by Li *et al.* (2005*a*, *b*) suggested that these several mud deposits started to form when the post-glacial transgression reached the highest stand and that the modern marine environment formed at approximately 8–7 ka. Obviously, the changes in sea level and oceanic circulation exert a significant role on the mud accumulation, which has to be investigated in more detail (Yang *et al.* 2003).

Tidal sand ridge systems: formation mechanism

Tidal sand ridges and sheets are well developed on the continental shelves of the Bohai, Yellow and East China seas, covering a total area of about 425 000 km^2 (Liu *et al.* 1998; Li *et al.* 2005*b*). The location, extent and sediment characters of the typical tidal sand-ridge systems of the eastern China continental shelves are summarized in Table 21.2.

Despite extensive studies over the last two decades, the formation processes of tidal sand-ridge systems on the eastern

China continental shelves have remained controversial and their formation ages vary from the late Quaternary to present (Yang & Sun 1988; Yang 1989; Liu Z. X. *et al.* 1989, 1998, 2000, 2007; Liu 1997; Li *et al.* 2001; Berné *et al.* 2002; Li *et al.* 2005*b*; Wang *et al.* 2013). Overall, two genetic types of tidal sand ridges are recognized. The tidal sand ridges mostly occurring at a water depth of <30–50 m are generally regarded as active at present, such as those in the offshore northern Jiangsu coastal plain in the western Yellow Sea (Liu *et al.* 1989; Li *et al.* 2001; Li *et al.* 2005*a*, *b*), in the west Korean Bay (Off 1963), the east Bohai Sea (Liu *et al.* 1998) and the east Yellow Sea (Jung *et al.* 1998; Park *et al.* 2006). By contrast, the sand ridges on the outer shelf of the ECS are regarded as relict, inactive and moribund (Yang & Sun 1988; Yang 1989; Liu *et al.* 1998; Saito *et al.* 1998; Berné *et al.* 2002; Yoo *et al.* 2002); they are thought to have been formed during the last lowstand of sea level or the early phase of the last transgression (Saito *et al.* 1998; Park *et al.* 2006). For example, Yang & Sun (1988) proposed that the tidal sand ridges in the ECS had formed in the palaeo-Changjiang Estuary at 13.5–11 ka BP, when sea level was lower and the tidal current was stronger than present. The sand-ridge field migrated landwards with sea-level rise during the last transgression and, at present, the sand ridges on the middle–outer shelf of the ECS are moribund and buried, not active (Yang 1989).

Based on more than 5000 km of high-resolution seismic profiles, four tidal sand-ridge sequences developed during marine transgressions at Oxygen Isotopic Stages (IOSs) 7 (or 9), 5, 3 and 1 were recognized on the continental shelf of the ECS, and the ridges were notably influenced by West Pacific tide–wave systems (Liu Z. X. *et al.* 2000, 2007; Berné *et al.* 2002). Based on seismic profiles and sequence stratigraphy with radiocarbon dates, Li *et al.* (2005*a*, *b*) identified three stages of tidal sand-ridge systems on the eastern China continental shelves in response to three phases of sea-level rise at approximately 19–15, 13.8–11 and 9.3–7 ka. Based on bathymetric, seismic and piston-core data, Liu Z. X. *et al.* (2007) further suggested that the tidal sand ridges on the middle–outer shelf of the ECS were developed during the last transgression by SW net transport of regional

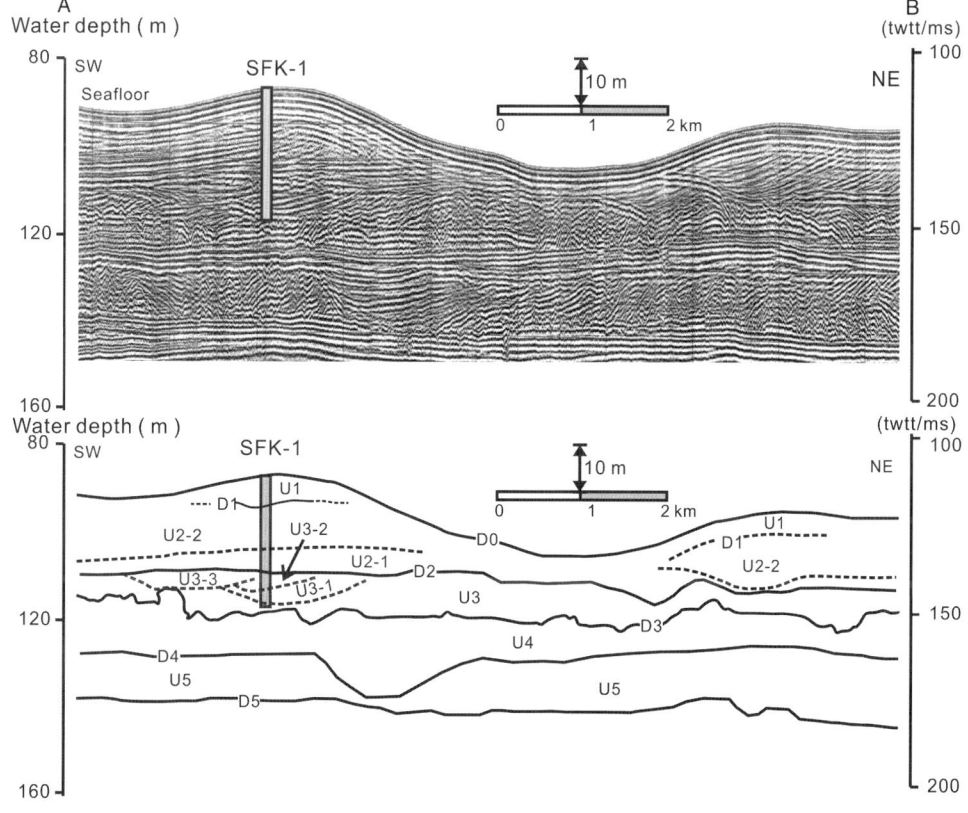

Fig. 21.4. High-resolution seismic profile A–B across borehole SFK-1 (palaeo-Changjiang distributary) and its schematic interpretation (Wang *et al.* 2013). See Figure 21.3 for the location.

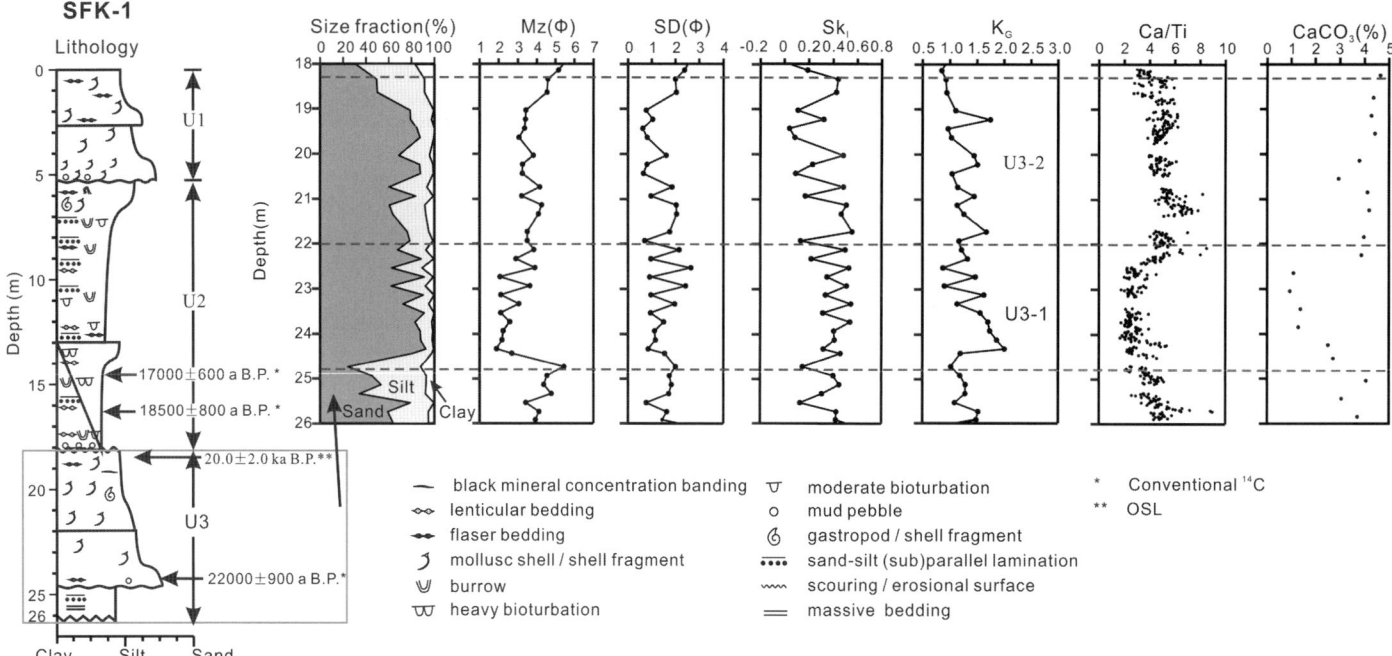

Fig. 21.5. Variations in lithology, particle-size parameters, X-ray fluorescence (XRF)-scanned Ca/Ti ratio and CaCO₃ content in U3 of core SFK-1 (Wang *et al.* 2013), East China Sea. Mz, mean grain size; SD, standard deviation; Sk₁, skewness; K_G, kurtosis.

sediment, so that they gradually migrated towards the SW to form new sand ridges. Therefore, these present-day sand ridges on the shelf are still influenced by modern hydrodynamics and should be referred to as 'quasi-active sand ridges' rather than as moribund or relict sand ridges as suggested by previous studies (Yang & Sun 1988; Yang 1989; Liu *et al.* 1998; Saito *et al.* 1998; Berné *et al.* 2002; Yoo *et al.* 2002; Wang *et al.* 2013). The sediment provenance of the tidal sand ridges and sand sheets on the ECS shelf is thought to be primarily from the palaeo-Changjiang (Liu 1997; Liu Z. X. *et al.* 2000, 2007; Berné *et al.* 2002; Li *et al.* 2005*b*; Wang *et al.* 2013), although there is a lack of direct evidence from the cores.

Apart from the their ages, the tidal sand ridges occurring in nearshore areas and on the middle–outer shelf are characterized by different geometry, morphology, scale and sediment characters, diagnostic of different formation processes in response to variable tidal regimes. Numerical simulations indicate that the tidal

regime during the last transgression was stronger than the modern one, and strong tidal stress on the seafloor resulted in the formation of tidal sand ridges (Uehara *et al.* 2002; Uehara & Saito 2003). Furthermore, the sand ridges on the ECS shelf migrated landwards as the sea level rose and the shoreline retreated. Liu Z. X. *et al.* (2007) suggested that the sand ridges in the ECS show depositional features but those in the Yellow Sea have erosional features, caused by differences in the sediment supply.

Different from the tidal sand-ridge system in the ECS and the southern Yellow Sea, the radial tidal sand ridge in the offshore northern Jiangsu coastal plain, located between the old Huanghe delta and the modern Changjiang, is one of the largest tidal sand-ridge systems in the world, covering an area of about 20 000 km² (Li *et al.* 2001). The sediment of the sand ridge was considered to be derived directly from the Changjiang (Yang 1989; Zhu & An 1993), the old Huanghe (Zhang & Chen 1992) or both rivers (Li *et al.* 2001), based on their morphological,

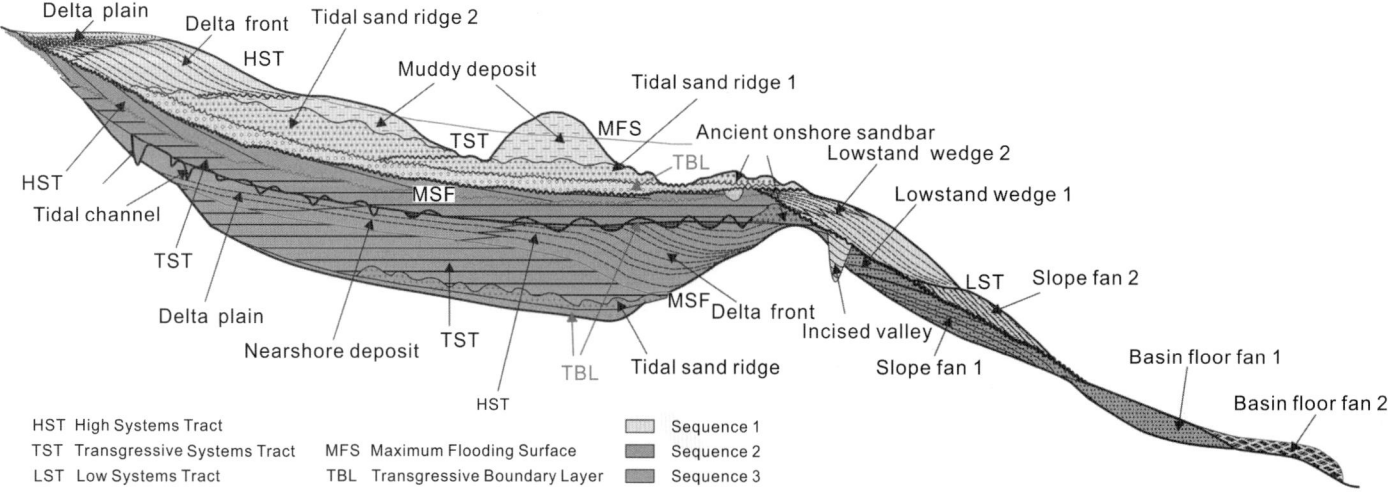

Fig. 21.6. A sequence stratigraphic model for the East China Sea shelf since the late Pleistocene (modified after Yang 2004).

sedimentological and mineralogical characters. The sand ridges are formed after the post-glacial transgression maximum (Li *et al.* 2001; Li *et al.* 2005*b*) or developed during post-glacial sea-level rise (Yang 1989; Zhu & An 1993; Liu *et al.* 1998). In the eastern Yellow Sea, the well-developed tidal sand ridges are composed of recent coastal sands supplied from the Korea Peninsula primarily by strong tidal currents and subsequent shoreface erosion (Klein *et al.* 1982; Lee & Chough 1989; Chough *et al.* 2000, 2004).

Palaeoenvironment and fluvial–marine interaction during the post-glacial period

Alternations of continental and marine strata formed during glacial and interglacial periods are well preserved on the eastern China continental shelf. Post-glacial transgression is obvious, and is readily recognized in coastal areas and on the inner shelf of the Bohai, Yellow and East China seas. Marine strata of variable thickness were developed in the inner shelf, thinning towards the outer shelf. Obviously, the sediment source-to-sink process in the eastern China seas is determined by the broad shallow epicontinental shelf, by the complex oceanic circulation, and by the large influx of fluvial sediments from the Changjiang and Huanghe (Milliman *et al.* 1989; Yang *et al.* 2003). As introduced above, the major rivers in mainland China basically control the sedimentation on the open shelf, while the small mountainous rivers mainly contribute to sedimentation on the coastal and nearshore areas. These two types of terrigenous sediment source-to-sink systems characterize the sedimentation on the eastern China shelf. Nevertheless, the distribution and dispersal patterns of fluvial sediments on the shelf during the late Quaternary remain poorly understood, and the river–sea interaction, such as the coupling between river discharges, sea-level changes and oceanic circulation, needs to be further investigated.

It is noteworthy that at present there is no river with an annual sediment discharge of more than 10^8 t entering the Yellow Sea. The dispersal patterns and transport limits of the modern sediments from the Changjiang and Huanghe in the Yellow and East China seas remain contestable. Alexander *et al.* (1991) suggested that about 15% of the modern Huanghe suspended load can accumulate annually in the Yellow Sea. However, Martin *et al.* (1993), among others, argued that net transport of the modern Huanghe sediment to the Yellow Sea is less than 1% of its sediment discharge. Yang & Liu (2007) estimated that nearly 30% of the Huanghe-derived sediment has been resuspended and transported out of the Bohai Sea into the Yellow Sea during the Holocene. Accordingly, the mechanism and quantity of the modern Changjiang and Huanghe clayey sediments being transported to the outer shelf of the ECS, and even further to the Okinawa Trough, is not known. Similarly, the specific contributions of the ancient Huanghe and Changjiang during the late Quaternary to shelf sedimentation have not yet been revealed. The gap in the estimation of the sediment budgets and dispersal pattern on the shelf is mainly because of the frequent shifts of the river flow pathways, and uncertainty over the past sediment influxes from the ancient Huanghe and Changjiang (Yang *et al.* 2003; Liu J. P. *et al.* 2007; Xu *et al.* 2012). Li *et al.* (2001) and Yang *et al.* (2002) suggested that the Changjiang flowed along its present position, without a significant shift in river pathway, throughout the late Quaternary; however, it has also been proposed that the Changjiang debouched its sediments directly into the south Yellow Sea, forming a palaeo-Changjiang delta therein (Liu *et al.* 1987; Qin *et al.* 1989; Yang 1989; Martin *et al.* 1993; Zhu & An 1993). The late Quaternary changes in monsoonal climate, sea-level and oceanic circulation further complicate the sediment source-to-sink process.

Two hypotheses have been proposed for the route of the Changjiang and Huanghe during the last glaciation (Xiao *et al.* 2004).

First, that the palaeo-Changjiang entered a palaeo-lake north of Jiangsu Province that may have disappeared during the Last Glacial Maximum (LGM) because of the extremely arid climate (Zhao 1984; Xia 1996). A second possibility is that the palaeo-rivers incised into the continental shelf, formed dendritic incised river networks and transported sediments into the Okinawa Trough or the Yellow Sea during the lowest stand of sea level in the LGM (Fig. 21.3) (Qin *et al.* 1989; Li & Zhang 1995; Li *et al.* 2005*a*; Xia & Liu 2001; Warren & Bartek 2002; Wellner & Bartek 2003; Dou *et al.* 2010, 2012; Wang *et al.* 2013). Li *et al.* (2005*a*) suggested that the palaeo-Changjiang incised the continental shelf of the ECS during the last glaciation along the Zhedong and Xihu depression in the SE of the Hupijiao Rise but did not directly enter the Okinawa Trough, despite the surface-water freshening of the northern trough. Six ancient river channel systems on the continental shelf were formed as the main distributaries of the palaeo-Changjiang (Fig. 21.3).

Recent work carried out by Wang *et al.* (2013) confirms that the palaeo-Changjiang incised the middle–outer ECS shelf, and formed a river channel deposit (U3), a tide-influenced estuarine–tidal flat facies (U2), and buried and quasi-active tidal sand ridges (U1) with sea level rising from the lowest stand during the LGM to the highest in the mid-Holocene (Fig. 21.4).

According to the interpretation of the seismic profile (based on the velocity of sound of 1650 m s^{-1}: Yang 1989; Berné *et al.* 2002), the upper portion of core SFK-1 can be divided into three sedimentary units (U1–U3: Figs 21.4 & 21.5). The uppermost 5.2 m (U1) is dominated by fine–very fine sands rich in shell fragments, and U2 (5.2–18.2 m bsf (below seafloor)) is characterized by mud–sand couplets with obvious tidal sedimentary structures and burrowed bioturbation. Unit U3 consists of coarser sand (18.2–24.8 m bsf, the main part of U3: Fig. 21.3), and is the lower part of the tidal deposition with lenticular and flaser bedding (23.8–26.0 m bsf, the lower part of U3: Fig. 21.5). The combination of particle-size composition, the low ratio of Ca/Ti and the low content of $CaCO_3$ of the U3 sediments obviously demonstrate a fluvial sedimentary environment formed during the LGM (Fig. 21.5).

A sequence stratigraphic model was proposed by Yang (2004) to interpret the Last Glacial sedimentation on the eastern China shelf, in which three marine transgressions (Sequence 1–3) and two transgressive–regressive sequences and transgressive boundary layers (TBL) were recognized on the basis of seismic facies interpretation and limited core data (Fig. 21.6). During the lowstand of sea level, the low systems tracts with incised valley, lowstand wedge, relict sandy bodies, and slope and basin floor fans were developed on the open shelf and continental slope. With the rising post-glacial sea level, the transgressive systems tracts were developed with the ancient onshore sandbars considerably reworked to form tidal sand ridges and sheets on the open shelf. During the highstand of sea level in the Holocene, the deltaic sedimentation initiated in the large estuaries, and several typical muddy and sandy systems began to form in the inner and middle–outer shelves, respectively.

In summary, the palaeo-fluvial and marine interaction on the wide eastern China continental shelf was complicated by changes in river flux, sea level and oceanic circulations during the last glacial cycle, which primarily determined the stratigraphic framework and sedimentary facies on the shelf. Undoubtedly, more systematic seismic investigations combined with high-resolution sediment core studies are needed in order to better constrain the source-to-sink process of these fluvial sediments on the epicontinental shelf.

We thank G. X. Li and S. B. Xiao for providing important references that helped to improve the paper. We also thank A. Chivas and F. Chiocci for their editorial work. This work was supported by research funds awarded by the National Natural Science Foundation of China (grants 41225020, 41076018, 41206053 and 41376049) and by China Geologic Survey (GZH201100203).

References

ALEXANDER, C. R., DEMASTER, D. J. & NITTROUER, C. A. 1991. Sediment accumulation in a modern epicontinental shelf setting: the Yellow Sea. *Marine Geology*, **98**, 51–72.

BEARDSLEY, R. C., LIMEBURNER, R., YU, H. & CANNON, G. A. 1985. Discharge of the Changjiang (Yangtze River) into the East China Sea. *Continental Shelf Research*, **4**, 57–76.

BERNÉ, S., VAGNER, P. *ET AL.* 2002. Pleistocene forced regressions and tidal sand ridges in the East China Sea. *Marine Geology*, **188**, 293–315.

CHO, Y. G., LEE, C. B. & CHOI, M. S. 1999. Geochemistry of surface sediments off the southern and western coasts of Korea. *Marine Geology*, **159**, 111–129.

CHOUGH, S. K., LEE, H. J. & YOON, S. H. 2000. *Marine Geology of Korean Sea*. Elsevier, Amsterdam.

CHOUGH, S. K., LEE, H. J., CHUN, S. S. & SHINN, Y. J. 2004. Depositional processes of late Quaternary sediments in the Yellow Sea: a review. *Geoscience Journal*, **8**, 211–264.

DEMASTER, D. J., MCKEE, B. A., NITTROUER, C. A., QIAN, J. & CHENG, G. 1985. Rates of sediment accumulation and particle reworking based on radiochemical measurement from continental shelf deposit in the East China Sea. *Continental Shelf Research*, **4**, 143–158.

DOU, Y. G., YANG, S. Y., LIU, Z. X., CLIFT, P. D., YU, H., BERNÉ, S. & SHI, X. F. 2010. Clay mineral evolution in the central Okinawa Trough since 28 ka: implications for sediment provenance and palaeoenvironmental change. *Palaeogeography, Palaeoclimatology, Palaeoecology*, **288**, 108–117.

DOU, Y. G., YANG, S. Y., LIU, Z. X., LI, J., SHI, X. F., YU, H. & BERNÉ, S. 2012. Sr–Nd isotopic constraints on terrigenous sediment provenances and Kuroshio Current variability in the Okinawa Trough during the late Quaternary. *Palaeogeography, Palaeoclimatology, Palaeoecology*, **365–366**, 38–47.

EMERY, K. O. 1968. Relict sediments on continental shelves of the world. *American Association of Petroleum Geologists Bulletin*, **52**, 445–464.

FENG, Y. 1983. Since 40 ka sea-level change and lowest sea level. *East China Sea*, **1**, 36–42 (in Chinese).

GUAN, B. X. 1994. Pattern and structures of the current in Bohai, Huanghai (Yellow) and East China Seas. *In*: ZHOU, D., LIANG, Y. B. & ZENG, C. K. (eds) *Oceanology of China Seas, Volume 1*. Kluwer Academic, Dordrecht, 3–16.

GUO, Z. G., YANG, Z. S., ZHANG, D. Q., FAN, D. J. & LEI, K. 2002. Seasonal distribution of suspended matter in the northern East China Sea and barrier effect of current circulation on its transport. *Acta Oceanologica Sinica*, **24**, 71–80 (in Chinese, with English abstract).

HU, D. X. 1984. Upwelling and sedimentation dynamics. *Chinese Journal of Oceanology and Limnology*, **2**, 12–19.

HU, D. X. & LI, Y. X. 1993. Study of ocean circulation. *In*: TSENG, C. K., ZHOU, H. O. & LI, B. C. (eds) *Marine Science Study and its Prospect in China*. Qingdao Publishing, Qingdao, 513–516 (in Chinese).

JUNG, W. Y., SUK, B. C., MIN, G. H. & LEE, Y. K. 1998. Sedimentary structure and origin of a mud-cored pseudo-tidal sand ridge, eastern Yellow Sea, Korea. *Marine Geology*, **151**, 73–88.

KAO, S. J. & MILLIMAN, J. D. 2008. Water and sediment discharge from small mountainous rivers, Taiwan: the roles of lithology, episodic events and human activities. *Journal of Geology*, **116**, 431–448.

KATAYAMA, H. & WATANABE, Y. 2003. The Huanghe and Changjiang contribution to seasonal variability in terrigenous particulate load to the Okinawa Trough. *In*: HUVENNE, V. & DAVIES, J. (eds) *Submarine Canyons: Complex Deep-Sea Environments Unravelled by Multidisciplinary Research. Deep Sea Research Part II: Topical Studies in Oceanography*, **50**, 475–485.

KLEIN, G. D., PARK, Y. A., CHANG, J. H. & KIM, C. S. 1982. Sedimentology of a subtidal, tide-dominated sand body in the Yellow Sea, Southwest Korea. *Marine Geology*, **50**, 221–240.

LEE, H. J. & CHOUGH, S. K. 1989. Sediment distribution, dispersal and budget in the Yellow Sea. *Marine Geology*, **87**, 195–205.

LEE, H. J. & CHU, Y. S. 2001. Origin of inner-shelf mud deposit in the southeastern Yellow Sea: Huksan Mud Belt. *Journal of Sedimentary Research*, **71**, 144–154.

LI, C. X. & ZHANG, G. J. 1995. A sea-running Changjiang River during the last glaciation? *Acta Geographica Sinica*, **50**, 459–463 (in Chinese with English abstract).

LI, C. X., ZHANG, J. Q., FAN, D. D. & DENG, B. 2001. Holocene regression and the tidal radial sand ridge system formation in the Jiangsu coastal zone, East China. *Marine Geology*, **173**, 97–120.

LI, G. X., LIU, Y., YANG, Z. G., YUE, S. H. & HAN, X. B. 2005a. Ancient Changjiang channel system in the East China Sea continental shelf during the last glaciation. *Science in China (Series D)*, **48**, 1972–1978.

LI, G. X., YANG, Z. G. & LIU, Y. 2005b. *Genetic Study of Submarine Sedimentary Environments in East China's Seas*. Science Press, Beijing (in Chinese with English abstract).

LIU, J., SAITO, Y. *ET AL.* 2010. Delta development and channel incision during marine isotope 3 and 2 in the western Southern Yellow Sea. *Marine Geology*, **278**, 54–76.

LIU, J. P., MILLIMAN, J. D., GAO, S. & CHENG, P. 2004. Holocene development of the Yellow River's subaqueous delta, North Yellow Sea. *Marine Geology*, **209**, 45–67.

LIU, J. P., XU, K. H., LI, A. C., MILLIMAN, J. D., VELOZZI, D. M., XIAO, S. B. & YANG, Z. S. 2007. Flux and fate of Yangtze River sediment delivered to the East China Sea. *Geomorphology*, **85**, 208–224.

LIU, J. P., LIU, C. S., XU, K. H., MILLIMAN, J. D., CHIU, J. K., KAO, S. J. & LIN, S. W. 2008. Flux and fate of small mountainous rivers derived sediments into the Taiwan Strait. *Marine Geology*, **256**, 65–76.

LIU, J. T., KAO, S. J., HUH, C. A. & HUNG, C. C. 2013. Gravity flows associated with flood events and carbon burial: Taiwan as instructional source area. *Annual Review of Marine Science*, **5**, 12.1–12.22.

LIU, M. H., WU, S. Y. & WANG, Y. J. 1987. *Late Quaternary Sedimentation in the Yellow Sea*. China Ocean Press, Beijing (in Chinese).

LIU, Z. X. 1997. Yangtze Shoal – a modern tidal sand sheet in the northwestern part of the East China Sea. *Marine Geology*, **137**, 321–330.

LIU, Z. X., HUANG, Y. C. & ZHANG, Q. N. 1989. Tidal current ridges in the southwestern Yellow Sea. *Journal of Sedimentary Petrology*, **59**, 432–437.

LIU, Z. X., XIA, D. X., BERNÉ, S., WANG, K. Y., MARSSET, T., TANG, Y. X. & BOURILLET, J. F. 1998. Tidal-depositional systems of China's continental shelf, with special reference to the eastern Bohai Sea. *Marine Geology*, **145**, 225–253.

LIU, Z. X., BERNÉ, S., SAITO, Y., LERICOLAIS, G. & MARSSET, T. 2000. Quaternary seismic stratigraphy and palaeoenvironments on the continental shelf of the East China Sea. *Journal of Asian Earth Science*, **18**, 441–452.

LIU, Z. X., BERNÉ, S. *ET AL.* 2007. Internal architecture and mobility of tidal sand ridges in the East China Sea. *Continental Shelf Research*, **27**, 1820–1834.

MARTIN, J. M., ZHANG, J., SHI, M. C. & ZHOU, Q. 1993. Actual flux of the Huanghe (Yellow River) sediment to the western Pacific Ocean. *Netherlands Journal of Sea Research*, **31**, 243–254.

MILLIMAN, J. D. & MEADE, R. H. 1983. World-wide delivery of river sediment to the oceans. *Journal of Geology*, **91**, 1–21.

MILLIMAN, J. D., BEARDSLEY, R. C., YANG, Z. S. & LIMEBURNER, R. 1985a. Modern Huanghe-derived muds on the outer shelf of the East China Sea: identification and potential transport mechanisms. *Continental Shelf Research*, **4**, 175–188.

MILLIMAN, J. D., SHEN, H. T., YANG, Z. S. & MEADE, R. H. 1985b. Transport and deposition of river sediment in the Changjiang estuary and adjacent continental shelf. *Continental Shelf Research*, **4**, 37–45.

MILLIMAN, J. D., QIN, Y. S., REN, M. E. & SAITO, Y. 1987. Man's influence on the erosion and transport of sediment by Asian rivers: the Yellow River (Huanghe) example. *Journal of Geology*, **95**, 751–762.

MILLIMAN, J. D., QIN, Y. S. & PARK, Y. A. 1989. Sediments and sedimentary processes in the Yellow and East China seas. *In*: TAIRA, A. & MASUDA, F. (eds) *Sedimentary Facies in the Active Plate Margin*. Terry Scientific Publishing, Tokyo, 233–249.

NIINO, H. & EMERY, K. O. 1961. Sediments of shallow portions of the East China Sea and South China Sea. *Geological Society of America Bulletin*, **72**, 731–762.

OFF, T. 1963. Rhythmic linear bodies caused by tidal currents. *American Association of Petroleum Geologists Bulletin*, **47**, 324–341.

PARK, S. C., LEE, H. H., HAN, H. S., LEE, G. H., KIM, D. C. & YOO, D. G. 2000. Evolution of late Quaternary mud deposits and recent sediment budget in the southeastern Yellow Sea. *Marine Geology*, **170**, 271–288.

PARK, S. C., LEE, B. H., HAN, H. S., YOO, D. G. & LEE, C. W. 2006. Late Quaternary stratigraphy and development of tidal sand ridges in the eastern Yellow Sea. *Journal of Sedimentary Research*, **76**, 1093–1105.

PARK, Y. A. & KHIM, B. K. 1992. Origin and dispersal of recent clay mineral in the Yellow Sea. *Marine Geology*, **104**, 205–213.

QIN, Y. S. 1994. Sedimentation in northern China Seas. *In:* ZHOU, D., LIANG, Y. B. & ZENG, C. K. (eds) *Oceanology of China Seas, Volume 2*. Kluwer Academic, Dordrecht, 394–406.

QIN, Y. S. & LI, F. 1983. Study of influence of sediment loads discharged from the Huanghe River on sedimentation in the Bohai Sea and the Yellow Sea. *In: Sedimentation on the Continental Shelf: With Special Reference to the East China Sea, Volume 2*. China Ocean Press, Qingdao, 83–92.

QIN, Y. S., ZHAO, Y. Y., CHEN, L. R. & ZHAO, S. L. 1987. *Geology of the East China Sea*. Science Press, Beijing (in Chinese).

QIN, Y. S., ZHAO, Y. Y., CHEN, L. R. & ZHAO, S. L. 1989. *Geology of the Yellow Sea*. China Ocean Press, Beijing (in Chinese).

SAITO, Y. 1998. Sedimentary environment and budget in the East China Sea. *Bulletin of Coastal Oceanography*, **36**, 43–58 (in Japanese).

SAITO, Y., KATAYAMA, H. *ET AL.* . 1998. Transgressive and highstand systems tracts and post-glacial transgression, the East China Sea. *Marine Geology*, **122**, 217–232.

SUN, X. G., FANG, M. & HUANG, W. 2000. Spatial and temporal variations in suspended particulate matter transport on the Yellow Sea and the East China Sea shelf. *Oceanologia Limnologia Sinica*, **31**, 581–587 (in Chinese with English abstract).

UDA, M. 1934. The results of simultaneous oceanographic investigations in the Japan Sea and its adjacent waters in May and June 1934. *Journal of the Imperial Fisheries Experiment Station*, **5**, 138–190 (in Japanese).

UEHARA, K. & SAITO, Y. 2003. Late Quaternary evolution of the Yellow/East China Sea tidal regime and its impacts on sediment dispersal and seafloor morphology. *Sedimentary Geology*, **162**, 25–38.

UEHARA, K., SAITO, Y. & HORI, K. 2002. Palaeotidal regime in the Changjiang (Yangtze) Estuary, the East China Sea, and the Yellow Sea at 6 ka and 10 ka estimated from a numerical model. *Marine Geology*, **183**, 179–192.

WANG, Z. B., YANG, S. Y., ZHANG, Z. X., LAN, X. H., GU, Z. F. & ZHANG, X. H. 2013. Palaeo-fluvial sedimentation on the outer shelf of the East China Sea during the last glacial maximum. *Chinese Journal of Oceanology and Limnology*, **31**, 886–894.

WARREN, J. D. & BARTEK, L. R. 2002. The sequence stratigraphy of the East China Sea, Where are the incised valleys? *In:* ARMENTROUT, J. M. & ROSEN, N. C. (eds) *Sequence Stratigraphic Models for Exploration and Production: Evolving Methodology, Emerging Models and Application Case Histories*. Proceedings of the 22nd Annual Gulf Coast Section SEPM Foundation Bob F. Perkins Research Conference. Society of Economic Paleontologists and Mineralogists, Gulf Coast Section, Houston, TX, 729–738.

WELLNER, R. W. & BARTEK, L. R. 2003. The effect of sea level, climate, and shelf physiography on the development of incised-valley complexes: a modern example from the East China Sea. *Journal of Sedimentary Research*, **73**, 926–940.

WU, Y., ZHANG, J., LI, D. J., WEI, H. & LU, R. X. 2003. Isotope variability of particulate organic matter at the PN section in the East China Sea. *Biogeochemistry*, **65**, 31–49.

XIA, D. X. 1996. Study on the Yellow River's breaking up during the last glaciation. *Oceanologia Limnologia Sinica*, **27**, 511–517 (in Chinese with English abstract).

XIA, D. X. & LIU, Z. X. 2001. Tracing the Changjiang River's route entering the sea during the last ice age maximum. *Acta Oceanologica Sinica*, **23**, 88–95 (in Chinese with English abstract).

XIAO, S. B., LI, A. C., JIANG, F. Q., LI, T. G., WAN, S. M. & HUANG, P. 2004. The history of the Yangtze River entering sea since the last glacial maximum: a review and look forward. *Journal of Coastal Research*, **20**, 599–604.

XU, K. H. & MILLIMAN, J. D. 2009. Seasonal variations of sediment discharge from the Yangtze River before and after impoundment of the Three Gorges Dam. *Geomorphology*, **104**, 276–283.

XU, K. H., MILLIMAN, J. D., LI, A. C., LIU, J. P., KAO, S. J. & WAN, S. M. 2009. Yangtze- and Taiwan-derived sediments on the inner shelf of East China Sea. *Continental Shelf Research*, **29**, 2240–2256.

XU, K. H., LI, A. C. *ET AL.* . 2012. Provenance, structure, and formation of the mud wedge along inner continental shelf of the East China Sea: a synthesis of the Yangtze dispersal system. *Marine Geology*, **291–294**, 176–191.

YANG, C. S. 1989. Active, moribund and buried tidal sand ridges in the East China Sea and the Southern Yellow Sea. *Marine Geology*, **88**, 97–116.

YANG, C. S. & SUN, J. S. 1988. Tidal sand ridges on the East China Sea shelf. *In:* DE BOER, P. L., VAN GELDER, A. & NIO, S. D. (eds) *Tide-Influenced Sedimentary Environments and Facies*. Reidel, Dordrecht, 23–38.

YANG, S. L., ZHANG, J. & XU, X. J. 2007. Influence of the Three Gorges Dam on downstream delivery of sediment and its environmental implications, Yangtze River. *Geophysical Research Letters*, **34**, L10401, http://dx.doi.org/10.1029/2007GL029472

YANG, S. Y. & YOUN, J. S. 2007. Geochemical compositions and provenance discrimination of the central south Yellow Sea sediments. *Marine Geology*, **243**, 229–241.

YANG, S. Y., LI, C. X., JUNG, H. S. & LEE, H. J. 2002. Discrimination of elemental compositions between the Changjiang and Huanghe sediments and identification of sediment source in northern Jiangsu coast plain, China. *Marine Geology*, **186**, 229–241.

YANG, S. Y., JUNG, H. S., LIM, D. I. & LI, C. X. 2003. A review on the provenance discrimination of the Yellow Sea sediments. *Earth-Science Reviews*, **63**, 93–120.

YANG, Z. G. 2004. *Marine Geology*. Shandong Education Press, Jinan (in Chinese).

YANG, Z. S. & LIU, J. P. 2007. A unique Yellow River-derived distal subaqueous delta in the Yellow Sea. *Marine Geology*, **240**, 169–176.

YANG, Z. S. & MILLIMAN, J. D. 1983. Fine-grained sediments of Changjiang and Huanghe River and sediment sources of the East China Sea. *In: Sedimentation on the Continental Shelf: With Special Reference to the East China Sea, Volume 2*. China Ocean Press, Qingdao, 436–446.

YANG, Z. S., GUO, Z. G., WANG, Z., XU, J. & GAO, W. 1992. The overall pattern of total suspended solids dispersal to the Yellow and East China Sea. *Acta Oceanologica Sinica*, **14**, 81–90 (in Chinese, with English abstract).

YOO, D. G., LEE, C. W., KIM, S. P., JIN, J. H., KIM, J. K. & HAN, H. C. 2002. Late Quaternary transgressive and highstand systems tracts in the northern East China Sea mid-shelf. *Marine Geology*, **187**, 313–328.

ZHANG, J. 1999. Heavy metal compositions of suspended sediments in the Changjiang estuary: significance of riverine transport to the ocean. *Continental Shelf Research*, **19**, 1521–1543.

ZHANG, R. S. & CHEN, C. J. 1992. *Evolution of Sand Bodies in Jiangsu Offshore Zone and Prospect of "Tiaozhimud" Advancing to Land*. Ocean Press, Beijing (in Chinese).

ZHAO, S. L. 1984. Quaternary geological questions of the Yangtze delta. *Marine Sciences*, **5**, 15–20 (in Chinese).

ZHAO, Y. Y., QIN, Z. Y., LI, F. Y. & CHEN, Y. W. 1990. On the source and genesis of the mud in the central area of the south Yellow Sea. *Chin. Journal of Oceanology and Limnology*, **8**, 66–73 (in Chinese).

ZHAO, Y. Y., PARK, Y. A., QIN, Y. S., GAO, S., ZHANG, F. G. & YU, J. J. 1997. Recent development in the southern Yellow Sea sedimentology: China-Korea joint investigation. *The Yellow Sea*, **3**, 47–51 (in Chinese).

ZHAO, Y. Y., PARK, Y. A., QIN, J. Y., GAO, S., LI, F. Y., CHENG, P. & JIANG, R. H. 2001. Material source for the eastern Yellow Sea mud: evidence of mineralogy and geochemistry from China–Korea joint investigations. *The Yellow Sea*, **7**, 22–26 (in Chinese).

ZHU, D. Q. & AN, Z. S. 1993. Formation and evolution of radial sand ridges in Jiangsu offshore zones. *In: Papers Collection of Geography on Celebrating the 80th Birthday of Prof. Ren M. E.* Nanjing University Press, Nanjing (in Chinese).

ZHU, Y. Q., LI, C. Y., ZENG, C. K. & LI, B. G. 1979. Lowest sea level about the continental shelf of the East China Sea during late Pleistocene. *Chinese Science Bulletin*, **7**, 317–320 (in Chinese).

Chapter 22

Pacific margin, Canada shelf physiography: a complex history of glaciation, tectonism, oceanography and sea-level change

J. VAUGHN BARRIE[1]*, RENÉE HETHERINGTON[2] & ROGER MACLEOD[1]

[1]*Natural Resources Canada, Geological Survey of Canada, 9860 West Saanich Road, PO Box 6000, Sidney, B.C., Canada V8L 4B2*

[2]*RITM Corp., 10915 Deep Cove Road, North Saanich, B.C., Canada V8L 5P9*

*Corresponding author (e-mail: renee@ritm.ca)

Abstract: Canada's western continental shelf can be broadly divided into three geographical regions: Salish Sea, Pacific North Coast and Vancouver Island Shelf. Each region's physiography has been uniquely impacted by a history of glaciation, tectonism, oceanography and sea-level change. Retreat of the Fraser Glaciation started from the Pacific towards the mainland of British Columbia first in the north on Haida Gwaii (16 ^{14}C ka BP) then outer Vancouver Island (14 ^{14}C ka BP), and generally progressed east with ice persisting on the Fraser lowland and northern British Columbia mainland, and possibly the central interbasin coast, until after 10 ^{14}C ka BP. Retreat on the Pacific North Coast was rapid, while downwasting and stagnation occurred in the Salish Sea. Sea levels responded to this deglacial history, with submergence of most of the Salish Sea by over 250 m before 12 ^{14}C ka BP then falling to approximately 100 m where it remained until approximately 10 ^{14}C ka BP, and emergence of the Vancouver Island Shelf with the development of a forebulge. In contrast, the Pacific North Coast crustal loading of the British Columbia mainland and development of a peripheral forebulge towards Haida Gwaii resulted in submergence of the mainland by 120 m and emergence of the western basin by more than 120 m, with regional variation throughout. Holocene sedimentation is primarily restricted to the coastal inlets and fjords, except for the southern Salish Sea off the Fraser River.

The western Canadian continental margin comprises the convergent boundary between the Explorer and Juan de Fuca plates and the North America plate along western Vancouver Island, and the transform fault boundary between the Pacific and North America plates along the west coast of Haida Gwaii (formerly named the Queen Charlotte Islands) (Figs 22.1 & 22.2). A ridge-trench transform triple junction lies immediately seawards of the mouth of Queen Charlotte Sound. The western Canadian continental shelf can be broadly divided into three geographical regions: Salish Sea, Pacific North Coast and Vancouver Island Shelf. Each region's physiography has been uniquely impacted by a history of glaciation, tectonism, oceanography and sea-level change (Table 22.1).

Geomorphology, sedimentology, physiography

Salish Sea

The Salish Sea consists of three inland water bodies, the Straits of Georgia and Juan de Fuca and Puget Sound, surrounded by the British Columbia (BC) mainland, Washington State and Vancouver Island, making it one of the world's largest inland seas, encompassing 400 islands and 7500 km of coastline (Figs 22.1 & 22.2). The largest water body, the Strait of Georgia, stretches for approximately 220 km in a NW–SE direction between Vancouver Island and the mainland; the overall width varies from 25 to 55 km. The average depth of the Strait of Georgia is 155 m and the deepest point is 420 m (Thomson 1981). Large areas of the strait reach depths of between 100 and 250 m, although several shallow banks exist down the axis of the basin. The strait connects with the open sea in the south, first through the Gulf Islands and San Juan Islands, and then through the Strait of Juan de Fuca. Bottom topography in the Gulf Islands–San Juan Islands area is complex but mostly shallower than 100 m, except for a few narrow, deep channels. In the north, the Strait of Georgia connects to the open shelf of Queen Charlotte Sound through four narrow channels with sill depths of 90 m or less.

This forearc basin – the Georgia Basin – developed with subsidence that began in the late Cretaceous (85 Ma ago). The three straits consist of a series of structural depressions, overdeepened by Tertiary fluvial erosion and Quaternary glaciation, and partially infilled by glacial and post-glacial sediments. Most of the Salish Sea is presently sediment starved, with sediment capture within the coastal fjords and inlets. The one exception is in the southern Strait of Georgia where sedimentation from the Fraser River dominates the surficial geology, with Holocene sediment thicknesses varying from zero on Pleistocene ridges to more than 300 m within the basin (Mosher & Hamilton 1998). Present-day sedimentation rates vary from 10 cm a^{-1} near the river mouth to less than 3 cm a^{-1} in the distal parts of the prodelta (Hart *et al.* 1998). Bathymetric restriction of the tidally dominated flow between the inland sea and the open Pacific through the restricted channels has resulted in the development of very large subaqueous dune fields, with dune heights up to 28 m in water depths of 170–210 m (Barrie *et al.* 2005, 2009).

Tectonics also impact the region. The North American plate is overriding the oceanic Juan de Fuca plate at a rate of about 45 mm a^{-1} (Riddihough & Hyndman 1991) (Fig. 22.2). Persisting tectonics along this Cascadia Subduction Zone results in ongoing tectonic uplift and volcanism. Regional variations in tectonic regimes, between the Georgia Basin and the more northerly Queen Charlotte Basin, results in variations in heat flow in the two regions. The Georgia Basin heat flow averages 40 mW m^{-2} (Lewis *et al.* 1991; Flück 2003; Flück *et al.* 2003).

Vancouver Island Shelf

The continental shelf west of Vancouver Island ranges from 5 to 75 km wide, and is characterized by an inshore region of complex morphology and a relatively featureless and flat mid and outer shelf. This featureless shelf is disrupted off SW Vancouver Island where large basins bounded by morainal deposits extend approximately two-thirds of the distance across the shelf (Herzer & Bornhold 1982). The shelf break lies in waters between

From: CHIOCCI, F. L. & CHIVAS, A. R. (eds) 2014. *Continental Shelves of the World: Their Evolution During the Last Glacio-Eustatic Cycle.*
Geological Society, London, Memoirs, **41**, 305–313. http://dx.doi.org/10.1144/M41.22

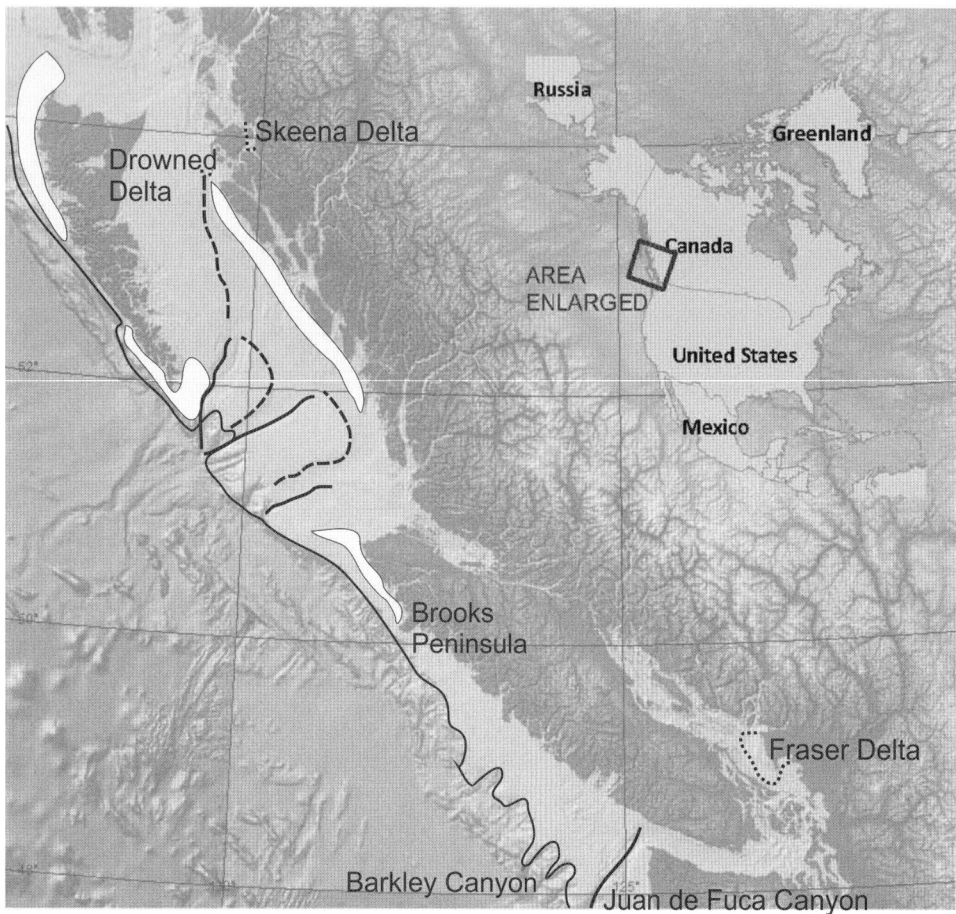

Fig. 22.1. Location map: showing the whole shelf identifying the shelf break (thin black line), the main terrace edges (black dashed lines), the delta front edge (dotted lines), the drowned delta (dotted line identified as 'Drowned Delta'), basement extensive outcrops (shaded white), buried incised valleys (thick black lines) and the canyon heads.

180 and 225 m deep, and is defined by an abrupt change in slope from nearly horizontal to the steep upper continental slope (7–15°). The Vancouver Island Shelf can be divided into three regions from north to south. The NW Vancouver Island Shelf

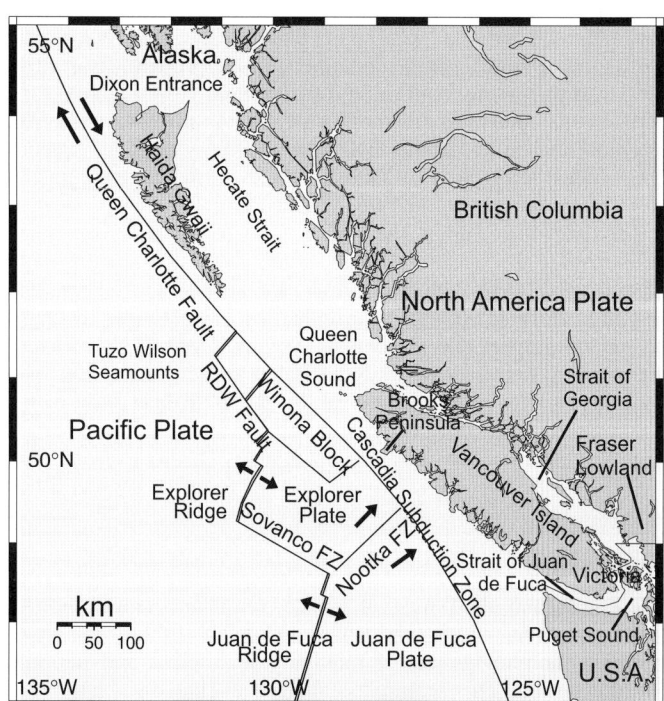

Fig. 22.2. Tectonic setting of Pacific margin of Canada, after Ristau *et al.* (2003); short thick arrows indicate relative plate motions.

is quite narrow (6–35 km wide) and consists of a rugged discontinuous bedrock inner shelf made up of coarse rippled sediments, a mid shelf characterized by coarse to fine well-sorted sand and an outer shelf made up of muddy sand (Bornhold & Barrie 1991). The central shelf, south of Brooks Peninsula (Fig. 22.1), is about 30 km wide and consists of a broad, irregular bank of outcropping Palaeogene sandstone and mudstone with a thin cover of gravel and coarse sand that extends the entire width of the shelf. Southern Vancouver Island Shelf extends from Barkley Sound to the mouth of the Strait of Juan de Fuca (Salish Sea). The inner- and mid-shelf areas off Barkley Sound are characterized by a complex distribution of shallow gravelly banks with small subaqueous dunes and elongate muddy troughs some more than 200 m deep (Herzer & Bornhold 1982; Bornhold & Barrie 1991). Sediments fine to the shelf break where fine sand is dominant and muddy gravel is found at the heads of the canyons, which indent the shelf (Bornhold & Barrie 1991).

Pacific North Coast

The Pacific North Coast is composed of the Queen Charlotte Basin and adjacent inshore areas of mainland BC and northern Vancouver Island. It stretches from northern Vancouver Island to the Alaskan boundary and west to the continental slope of the northern Pacific Ocean. The basin consists of three water bodies: Dixon Entrance separating Haida Gwaii from SE Alaska; Hecate Strait, which separates Haida Gwaii from the BC mainland; and Queen Charlotte Sound, lying south of Haida Gwaii and bounded on the south by Vancouver Island and to the east by the BC mainland (Fig. 22.2). Both Dixon Entrance and Queen Charlotte Sound are open to the Pacific.

The basin stretches for approximately 500 km in a NW–SE direction; the overall width varies from 100 to 225 km. Dixon

Table 22.1. *Shelf characteristics of Canada's Pacific margin*

Shelf characteristics	Salish Sea	Vancouver Island Shelf	Pacific North Coast
Length of shelf (km)	400	400	500
Average width (km)	30	40	175
Tidal, wave, current ranges	3.0 m tide	3.5 m tide	4.5 m tide
Dominating process (wave/current/tide)	Tide/estuarine	Wave/tide	Tide/wave
Average depth of the shelf break (m)	NA	200	220
Modern/relict/palimpsest (%)	70 modern	0 modern	10 modern
Tectonic trend over the last glacial cycle (stable/uplifting/subsiding)	Uplifting	Uplifting	Stable/uplifting
Mean depth (m)	155	100	150
Maximum depth (m)	650	225	440

NA, not applicable.

Entrance is a broad asymmetrical U-shaped trough with the greatest water depths in the north. The rugged topography is a consequence of westward-flowing glaciers (Barrie & Conway 1999). Surface sediments are primarily sand and sandy gravels, a result of the strong tidal flow within Dixon Entrance (Barrie & Conway 1999). Hecate Strait is an asymmetrical channel composed of broad shallow banks on the west (often shallower than 40 m) and a U-shaped southward-opening trough on the east with a narrow bedrock shelf along the BC mainland (Bornhold & Barrie 1991). Most of Hecate Strait is dominated by sand and gravel, with small to large subaqueous dunes. The troughs consist of Holocene muds. In the south, Queen Charlotte Sound consists of three extensive banks less than 100 m deep covered with sands and gravels separated by three broad (10–40 km wide) NE–SW-trending muddy troughs with maximum depths of 300–400 m. The west coast of the Haida Gwaii is extremely narrow, extending offshore less than 5 km in the south and up to 30 km at 54°N latitude to the shelf break, at roughly 200 m water depth (Fig. 22.1). Most of the shelf is devoid of sediment except in areas protected from the severe oceanic energy that dominates this narrow shelf.

The shelf physiography is controlled by the Queen Charlotte Fault, a transform fault boundary between the North American and Pacific lithospheric plates along the west coast of Haida Gwaii (Fig. 22.2). Relative movement along the fault has been calculated at 50–60 mm a^{-1} (Riddihough 1988), with Canada's largest earthquake (magnitude 8.1) occurring off the NW tip of Haida Gwaii in 1949. The Queen Charlotte Basin lithosphere possesses a high heat flow averaging 70 mW m^{-2} (Lewis *et al.* 1991). These data are higher than heat-flow measurements for the Georgia Basin. The Queen Charlotte Basin has a relatively thin elastic lithospheric thickness, whereas that of the Georgia Basin is relatively thicker. This makes the Queen Charlotte Basin very responsive to changes in tectonic movement, ice and water loading, changes in eustatic sea level, sedimentation and erosion processes.

Glacial history

Glaciation affected the Pacific margin of Canada many times, although extensive evidence has been found for only the youngest glacial episode (Barrie & Conway 2002*a*). The Fraser Glaciation, of the last ice age, began approximately 25–30 kyr ago (*c.* 21–25 ^{14}C ka BP: Clague 1977, 1981). Ice sheets up to 2000 m thick depressed the BC mainland, and thinned west and SW towards the continental shelf (Clague *et al.* 1982*b*; Barrie & Conway 1999), generating crustal uplift in areas peripheral to mainland ice where ice was thinner or absent.

Salish Sea

During the early stages of the Fraser Glaciation, thick, well-sorted sand deposits (Quadra Sand) were laid down in front of, and possibly along, the margins of glaciers moving down the Strait of Georgia as distal outwash aprons (Clague 1976, 1977, 1994). Ice moving south from the Coast Mountains of the Canadian Cordillera and Vancouver Island progressively coalesced, overrode and eroded these deposits. Ice flowing SE along the axis of the Strait of Georgia joined ice coming out of the Fraser Valley and then divided into two tongues, one following Juan de Fuca Strait towards the ocean and the other going southwards into Puget Sound. This large glacier reached the south end of the Puget Lowland and the western Juan de Fuca Strait at its maximum extent (Waitt & Thorson 1983; Hewitt & Mosher 2001; Mosher & Johnson 2001; Mosher & Hewitt 2004), at about 14 ^{14}C ka BP (Porter & Swanson 1998). It deposited an ice-contact diamicton (till) of variable thickness throughout most of the basin.

Most of the Strait of Georgia was ice-free by 12–11 ^{14}C ka BP (Hewitt & Mosher 2001; Barrie & Conway 2002*a*; Hetherington & Barrie 2004). A reduced glacial load remained in or proximal to the Fraser Lowland on the mainland until at least 10 ^{14}C ka BP (Clague & James 2002; Hetherington & Barrie 2004). Deglaciation was very rapid with regional downwasting and widespread stagnation (Clague 1981; Guilbault *et al.* 2003). This resulted in a stratigraphy of thick (30–60 m) diamicton, overlain by ice-proximal glaciomarine sediments and a thin, discontinuous ice-distal glaciomarine unit (Barrie & Conway 2002*a*).

Vancouver Island Shelf

The late Wisconsin glacier extended westwards from Vancouver Island across the adjacent shelf, except possibly for the Brooks Peninsula region. The total western extent of the ice sheet is not known and there is limited evidence of the chronology of glacial retreat. The topography of the SW Vancouver Island Shelf is interpreted to have formed through glacial scour and morainal deposition and, as such, is the only unequivocal evidence of continental shelf glaciation beyond the nearshore zone of western Vancouver Island (Bornhold & Barrie 1991). Glaciers, which overtopped Vancouver Island from the Strait of Georgia, flowed through Barkley Sound and joined locally derived glaciers to produce a complex zone of erosional and depositional features seen on the shelf today (Herzer & Bornhold 1982). Evidence from caves at Port Eliza on the NW coast of Vancouver Island suggests that ice inundated the outer coast during the last glacial but this only lasted for 1–2 kyr from 15 to 14 ^{14}C ka BP (Ward *et al.* 2003). Brooks Peninsula on the northern coast of Vancouver Island has been considered not to have been glaciated during the last glacial based on physical and biological characteristics (Hebda *et al.* 1997) but it is unknown whether ice extended around Brooks Peninsula offshore from adjacent shelf areas. Certainly coastal glaciation along Vancouver Island was complex and may not have covered all shelf areas.

Pacific North Coast

In the Queen Charlotte Basin, a glacier from the massive Cordilleran ice sheet extended westwards across northern Hecate Strait and through Dixon Entrance, and coalesced with ice from Haida Gwaii, deflecting it westwards within Dixon Entrance (Sutherland-Brown 1968; Barrie & Conway 1999). This coalescence was probably short-lived (Clague 1989). Ice also moved south down the central trough in Hecate Strait (Barrie & Bornhold 1989) and coalesced with ice flowing through the troughs of Queen Charlotte Sound to the shelf break (Luternauer & Murray 1983; Luternauer *et al.* 1989; Hicock & Fuller 1995; Josenhans *et al.* 1995, 1997). A minimum ice thickness of 400 m is suggested for some shelf areas (Josenhans *et al.* 1995; Barrie & Conway 1999), and approximately 690 m in northern Hecate Strait and Dixon Entrance (Hetherington *et al.* 2004). Glaciation reached its maximum extent some time after 21 ^{14}C ka BP (Blaise *et al.* 1990) and, therefore, earlier than in Georgia Basin.

On Haida Gwaii, small ice caps and piedmont glaciers up to 500 m thick developed that were independent of the Cordilleran Ice Sheet (Clague *et al.* 1982b; Clague 1983). There may have been ice-free areas on the islands and, on the coastal lowlands of Graham Island, glaciation was minimal and of short duration (Clague *et al.* 1982b; Clague 1989). The limited size and extent of the Queen Charlotte Mountain source areas, and the proximity of deep water of the open Pacific Ocean, Dixon Entrance and Queen Charlotte Sound limited expansion of ice on the Haida Gwaii islands (Clague 1981, 1989; Warner *et al.* 1982; Barrie *et al.* 1993; Barrie & Conway 1999).

Based on evidence of low-level cirques along the coast, as well as glacial striae, Sutherland-Brown (1968) suggests that ice moved out on to the shelf, off western Haida Gwaii and formed a small ice shelf. However, Clague (1989) implies that the ice extent was limited and that some areas of the shelf could have been free of ice during the last glaciation. Further, based on recent data, no evidence for diamicton deposition or any identifiable

glacial or deglaciation features exist, such as iceberg scours and boulders, usually common in glaciated areas. It is quite probable, therefore, that the west coast of the Haida Gwaii had little to no ice cover during the last glaciation.

The glaciation along the Pacific North Coast terminated with rapid climatic amelioration, resulting in rapid retreat and melting of the ice. Glaciers had retreated from the lowland areas of Haida Gwaii, beginning around 16 ^{14}C ka BP (Warner *et al.* 1982), but mountain valleys and cirques probably supported remnant ice masses until much later (Clague *et al.* 1982b; Clague 1989). Offshore glacial retreat began some time after 15 ^{14}C ka BP, and ice had largely left the lowlands and offshore of the region by 13–12.6 ^{14}C ka BP (Barrie & Conway 1999; Hetherington *et al.* 2004). This resulted in a stratigraphy of thick (<50 m) diamicton, overlain by thin ice-proximal glaciomarine sediments and an extensive, up to 20 m-thick, ice-distal glaciomarine unit (Barrie & Conway 2002a). By 11.7 ^{14}C ka BP, two coastal plains extended across the 150 km linking the BC mainland to Haida Gwaii (Hetherington *et al.* 2004). Based on crustal uplift observed in Hecate Strait and Queen Charlotte Sound, glacial ice persisted on the northern BC mainland until about 10 ^{14}C ka BP (Clague & James 2002; Hetherington & Barrie 2004). Across NE Vancouver Island, ice thickness that exceeded 1500 m at the height of glaciation (Clague 1983) persisted until after 13.6 ^{14}C ka BP (Howes 1983).

Icebergs were calved from the retreating glacial ice front. These icebergs impacted the seafloor and left characteristic plough features or iceberg furrows that are ubiquitous in the troughs of the basin between 110 and 350 m depth (Luternauer & Murray 1983; Barrie & Bornhold 1989; Barrie & Conway 1999). The curvilinear furrows have incision depths of up to 7 m but are mostly less than 3 m deep. They typically display a preferred orientation in the direction of the trough.

Relative sea-level history, crustal displacement and palaeogeographical reconstructions

The relative sea-level history of the Pacific margin of Canada has been observed, compiled and plotted (Fig. 22.3) (Clague *et al.* 1982a; Josenhans *et al.* 1995, 1997; Barrie & Conway 2002b; Clague & James 2002; James *et al.* 2002, 2005, 2009;

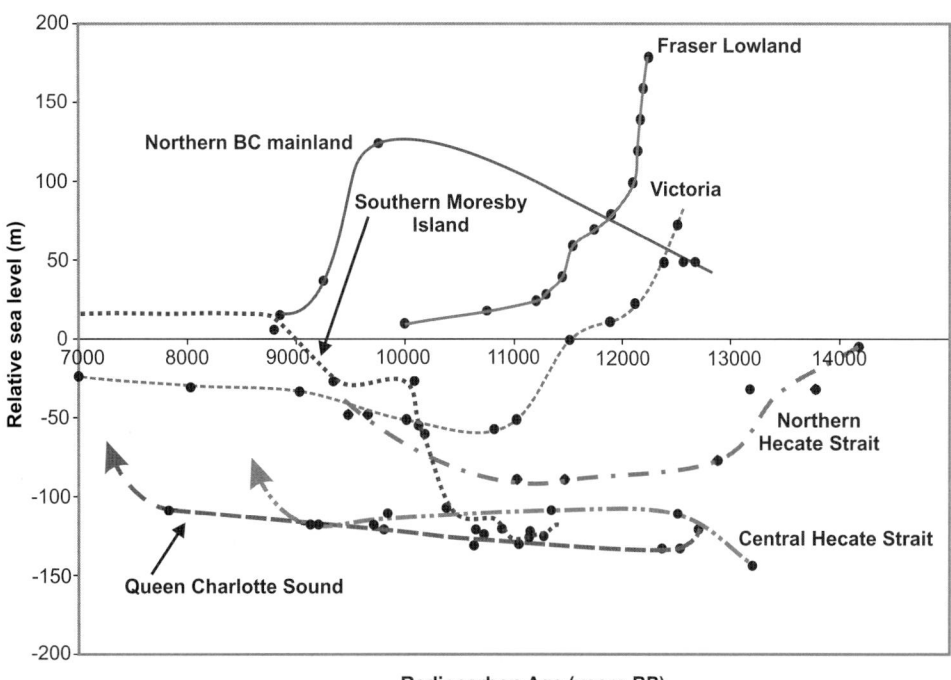

Fig. 22.3. Pacific margin sea-level observations. Queen Charlotte Basin relative sea-level curves include: southern Moresby Island (modified from Josenhans *et al.* 1995, 1997; Hetherington & Barrie 2004), northern BC mainland, northern Hecate Strait, Queen Charlotte Sound and central Hecate Strait (modified from Hetherington & Barrie 2004; Hetherington *et al.* 2004). Georgia Basin sea-level curves include: Fraser Lowland (Clague *et al.* 1982a; James *et al.* 2002) and Victoria (Clague & James 2002; Barrie & Conway 2002b; James *et al.* 2002).

Hetherington & Barrie 2004; Hetherington *et al.* 2004). These observations indicate significant subregional variation in relative sea-level histories as a consequence not only of differences in glacial isostatic history but also of non-glacial tectonic isostatic history (Hetherington & Barrie 2004). We will review these variations by region and their impacts on the palaeogeography of Canada's Pacific margin.

Salish Sea and Vancouver Island Shelf

Sea-level observations for Fraser Lowland on the southern BC mainland and northern Strait of Georgia (James *et al.* 2005) illustrate how the heavy glacial ice load resulted in over 250 m of crustal depression prior to 12 ^{14}C ka BP followed by rapid crustal uplift until about 11 ^{14}C ka BP, after which it stabilized at about 10 ^{14}C ka BP (see Hetherington & Barrie 2004). Local isostatic crustal response combined with glacio-eustatic sea-level change meant that relative sea level fell from nearly 200 m above present before 12 ^{14}C ka BP to near present levels by 10 ^{14}C ka BP.

Across the Strait of Georgia in Victoria on the southern tip of Vancouver Island, ice loading resulted in crustal depression of about 150 m before 13 ^{14}C ka BP until 12.5 ^{14}C ka BP (Hetherington & Barrie 2004; James *et al.* 2009). Slightly more gradual uplift than observed in Fraser Lowland continued until about 11 ^{14}C ka BP when depression reached about 20 m. These findings indicate that ice loading was significantly less on the southern tip of Vancouver Island than it was in the Fraser Lowland and northern Strait of Georgia, and unloading began earlier but occurred more quickly. Further, enough ice persisted in the Fraser Lowland between 11 and 10 ^{14}C ka BP to generate 100 m of depression.

Figure 22.4a provides a geospatial interpolation of crustal displacement in the Georgia Basin at 11.5 ^{14}C ka BP showing crustal depression on the southern BC mainland, and a forebulge to the north and west towards the Vancouver Island Shelf. These crustal changes combined with global eustatic relative sea level about 72 m below present (Fairbanks 1989) resulted in inundation of the Fraser Lowland and other low-lying regions of southern coastal BC and southern Vancouver Island. A palaeogeographical reconstruction of Georgia Basin shows a coastal plain developed on the western and northern edges of Vancouver Island Shelf at 11.5 ^{14}C ka BP (Fig. 22.4b).

Pacific North Coast

Heavy ice loads on the northern BC mainland during the Fraser Glaciation depressed the crust, resulting in relative sea levels at 10 ^{14}C ka BP that were 120 m higher than today (Hetherington

& Barrie 2004; Hetherington *et al.* 2004). Shortly thereafter, by about 8.8 ^{14}C ka BP, relative sea level on the mainland promptly fell to about 13 m above present day. Just 75 km to the west in northern Hecate Strait and Dixon Entrance, a glacial forebulge developed on the continental shelf that caused relative sea levels to fall to almost 90 m below present day by about 11.5 ^{14}C ka BP, rising to 48 m below the present level by at least 9.7 ^{14}C ka BP (Hetherington *et al.* 2004; Hetherington & Barrie 2004). The forebulge extended southwards into Central Hecate Strait and Queen Charlotte Sound, reaching an amplitude of over 80 m and causing relative sea level to fall to between 100 and 120 m below today's level, respectively, by about 12.5 ^{14}C ka BP (Hetherington & Barrie 2004; Hetherington *et al.* 2004). Relative sea level remained consistently low in this southern Queen Charlotte Basin region until the forebulge collapse generated rising relative sea levels after 9 ^{14}C ka BP in Central Hecate Strait and after about 7.8 ^{14}C ka BP in Queen Charlotte Sound (Hetherington & Barrie 2004; Hetherington *et al.* 2004). The forebulge appears to have also extended on to the islands of Haida Gwaii where crustal uplift of over 50 m resulted in relative sea level remaining over 120 m below present on southern Haida Gwaii until about 10.9 ^{14}C ka BP (Hetherington & Barrie 2004; Hetherington *et al.* 2004). However, within 900 years, collapse of the forebulge and subsequent crustal depression on Haida Gwaii resulted in a rapid relative sea-level rise to 27 m below today and within 1000 years to 15 m above today – a rise in relative sea level of 135 m in about 1000 years.

Figure 22.5a provides a geospatial interpolation of crustal displacement in the Queen Charlotte Basin at 11.5 ^{14}C ka BP. It shows crustal depression on the northern BC mainland and a forebulge peripheral to the grounded mainland ice extending across the continental shelf of the Queen Charlotte Basin. As a consequence, the palaeogeography of the region was substantially different from today (see Fig. 22.5b). Two coastal plains developed, one in Hecate Strait and the other in Queen Charlotte Sound, creating a land bridge that extended from Haida Gwaii to the BC mainland. This land bridge closed the marine connection between Hecate Strait and Dixon Entrance, creating the 'Hecate Sea'.

Discussion

Regional comparison of glacial and relative sea-level chronology

The glacial stratigraphy between these three regions of the western Canadian continental shelf is quite different, and results from differing deglacial and corresponding sea-level histories. On the

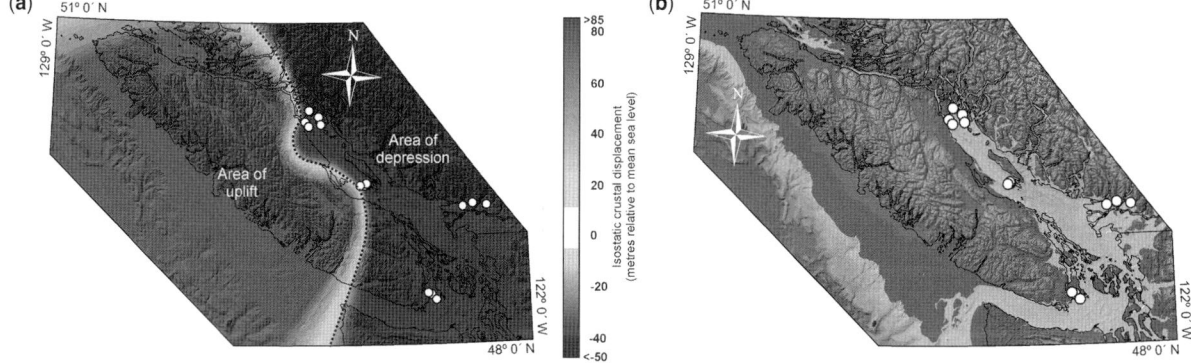

Fig. 22.4. Georgia Basin and surrounding region crustal depression and palaeogeographical maps for 11.5 ^{14}C ka BP. (**a**) Crustal displacement map. Areas portrayed with a black dotted line signify no change (zero isobase); areas to the right of the black dotted line signify depression, where darker regions indicate greater depression; and areas to the left of the black dotted line represent uplift, where darker regions indicate greater uplift. Large dots indicate the location of sample data used to generate the map and more than one data sample may be represented by each dot. Isostatic crustal displacement measured in metres relative to mean sea level. The appearance of reduced crustal uplift on the north coast of Vancouver Island is an artefact of insufficient data across the northern extent of the map area. (**b**) Palaeogeographical reconstruction map. The black line denotes the present coastline.

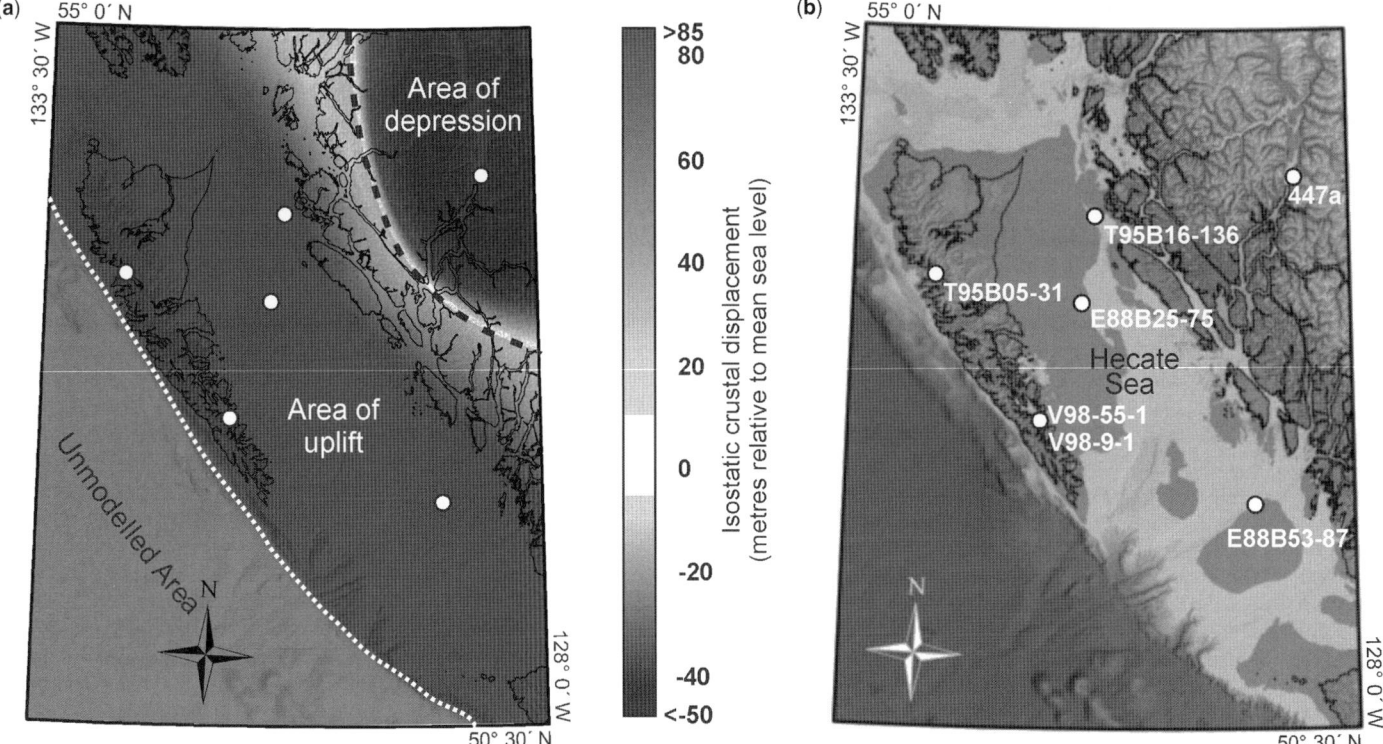

Fig. 22.5. Queen Charlotte Basin and surrounding region crustal depression and palaeogeographical maps for 11.5 ^{14}C ka BP (modified from Hetherington *et al.* 2004). (**a**) Crustal displacement map. Areas portrayed with a black dashed line signify no change (zero isobase); areas depicted to the left of the black dashed line show uplift; and areas illustrated to the right of the black dashed line represent depression; area to the left of the fine white dotted line was not modelled. Large dots indicate the location of sample data used to generate the map and more than one data sample may be represented by each dot. Isostatic crustal displacement measured in metres relative to mean sea level. The appearance of the southern limit to crustal depression on BC mainland is an artefact of insufficient data across the BC mainland. (**b**) Palaeogeographical reconstruction map (modified from Hetherington *et al.* 2004). The black line denotes the present coastline.

Pacific North Coast, isostatic rebound occurred after deglaciation at the western extreme, while, close to the thick ice load of the Cordilleran ice sheet, the crust remained depressed. Retreat appears to have been rapid (Hicock & Fuller 1995; Hetherington & Barrie 2004; Hetherington *et al.* 2004). Sedimentation changes over a very short time period from deposition of till to hemipelagic sedimentation and iceberg rafting (Barrie & Conway 1999). Ice-distal glaciomarine sediments are primarily of ice-rafted origin, with the ice source being well removed from the depositional basin. In southern Haida Gwaii, relative sea level was over 120 m lower than today until about 10.9 ^{14}C ka BP, after which it rose over 90 m in 900 radiocarbon years (Fig. 22.3). Just to the east, relative sea level remained at or below 100 m below present until about 9 ^{14}C ka BP in central Hecate Strait and until 8 ^{14}C ka BP in Queen Charlotte Sound, whereas it rose to 120 m above present by 9.7 ^{14}C ka BP before rapidly dropping to 13 m above present by around 8.8 ^{14}C ka BP on the northern BC mainland (Hetherington & Barrie 2004).

Guilbault *et al.* (1997) suggested that open marine conditions and vigorous circulation occurred shortly after 13 ^{14}C ka BP, when water temperatures changed from glacial to transitional in Dixon Entrance. The intrusion of warmer Pacific waters against the ice front probably enhanced ice melt and the calving of icebergs. Iceberg production must have been significant considering the occurrence of a highly scoured seafloor in water depths greater than 300 m.

In contrast, in the Salish Sea, ice loading resulted in crustal depression of more than 250 m across the Fraser Lowland and northern Strait of Georgia, and more than 150 m at 12.5 ^{14}C ka BP in Victoria. This was followed by gradual uplift until about 11 ^{14}C ka BP across the basin. As a consequence, relative sea level remained high throughout the deglaciation period in the Fraser Lowland and during initial deglaciation in Victoria. By

11.5 ^{14}C ka BP uplift in Victoria was sufficient to overcome lowered eustatic sea level, and relative sea level fell to a lowstand at about 40–50 m below present level (Hetherington & Barrie 2004; James *et al.* 2009). Thick ice in the Strait of Juan de Fuca (Hewitt & Mosher 2001), combined with the restrictive basin physiography, prevented early intrusion of warmer Pacific waters. The high Coast Mountains of the BC mainland, the Olympic Mountains of Washington State and the Vancouver Island Mountains all retarded ice retreat in the basin. The glacial stratigraphy suggests that, once deglaciation began, ice-proximal sedimentation rates were high over a short period of time near 12.4 ^{14}C ka BP. Within approximately 300 years no ice was left within the present marine areas, except to the north in the inter-basin region where ice may have remained longer. Ice-distal glaciomarine sedimentation continued in some areas for the next 800 years. However, the coarse-grained component is far less than that found in the Queen Charlotte Basin, suggesting reduced ice rafting. There is no evidence of any iceberg scouring in the shallowing basin, implying that there was little or no iceberg calving or movement of icebergs across the basin.

Crustal depression on the southern BC mainland and formation of a forebulge to the Vancouver Island Shelf suggests that ice was limited and had retreated to Vancouver Island from the west before 12.5 ^{14}C ka BP and possibly as early as 14 ^{14}C ka BP (Ward *et al.* 2003). Consequently, relative sea levels may have dropped by more than 120 m by 11.5 ^{14}C ka BP (Fig. 22.4b) on the Vancouver Island Shelf.

Differential isostatic crustal response

The variations in coastal deformation along Canada's Pacific margin are a reflection of differences in glacial and non-glacial tectonic isostatic history throughout the region. The Vancouver Island Shelf and Salish Sea have an active subduction regime, whereas the Pacific North Coast has a strike-slip regime. As a result, the Pacific North Coast possesses a much higher average heat flow (70 mW m^{-2}), with values just south of Haida Gwaii reaching between 78 and 159 mW m^{-2} (Lewis *et al.* 1991; Hyndman 1995; Flück 2003), than does the margin of Vancouver in the Georgia Basin (35 mW m^{-2}) (Hyndman 1995) or Victoria on southern Vancouver Island (39 mW m^{-2}) (Lewis *et al.* 1991; Hyndman 1995; Flück 2003). These data suggest a much thinner thermally defined lithosphere in the Pacific North Coast than in the Salish Sea and Vancouver Island Shelf. These findings, combined with crustal displacement reconstructions discussed above, suggest that the flexural wavelengths observed in the Pacific North Coast would be lower than those in the Salish Sea and Vancouver Island Shelf (Hetherington & Barrie 2004). Further, that we would expect to see a slower isostatic crustal response after glacial retreat in the Salish Sea and Vancouver Island relative to that observed in the Pacific North Coast.

Palaeogeographical response

Three distinct palaeo-coastlines from approximately 150 to 40 m water depth with a progressive westward migration towards the present-day Haida Gwaii have been mapped for the Pacific North Coast (Barrie & Conway 2012). The principal terrace would have occurred when relative sea level was at a lowstand. This was the most stable sea-level stillstand in the late Quaternary until present times, and may have lasted for approximately 600 radiocarbon years. Isostatic uplift in Hecate Strait appears to have been greater in the south, assuming the terrace uniquely defines a single sea-level position. At this time Hecate Strait was a northward-narrowing inlet (Hecate Sea) (Fig. 22.5b). The inlet was shallow (50–150 m water depth) and opened to a large embayment within Queen Charlotte Sound that connected to the Pacific Ocean in a SW direction. The width of the inlet facing the wave-cut terrace was approximately 20–60 km, providing a limited fetch for wave exposure. The western coast was primarily straight to gently curved, with smaller embayments, whereas the eastern coastline was more complex as it was bedrock controlled.

In the protected shallow waters of Hecate Sea, the mollusc *Macoma nasuta* overcame the high sediment barrier generated by the glacial meltwater and invaded the laminated muddy sand by 13.2 ^{14}C ka BP (Hetherington & Reid 2003). Fossil remains of herring suggest kelp beds were present, creating a diverse habitat for marine organisms (Hetherington & Reid 2003). Brown bears and, to a lesser extent, black bears, as well as ringed seal, otters, gulls and crows scavenged the beaches (Fedje *et al.* 1996; Heaton *et al.* 1996). Given what we know about modern molluscan species, sea-surface temperatures (SSTs) reached at least 19 °C by between 12.54 and 12.37 ^{14}C ka BP, the upper limit of present average SSTs (Hetherington & Reid 2003).

During this deglacial time the adjacent terrestrial environment would have been treeless tundra, including herb-dominated plant communities with dwarf willow and crowberry shrubs with some highland lodgepole pine (Barrie *et al.* 1993; Lacourse & Mathewes 2005), similar to the present Gulf of Alaska region, from Kodiak Island to the Aleutian Islands (Heusser 1985, 1990). This environment could have provided a habitable landscape for early human populations as early as 13.2 ^{14}C ka BP (Hetherington *et al.* 2003).

As relative sea level rose, a large delta developed at the head of the Hecate Sea inlet (Barrie & Conway 2002b) that was probably the proto-delta of the Skeena River (Fig. 22.1). The glacial and outwash deposits of northern Hecate Strait, Haida Gwaii and the coastal areas of BC provided the sediment source (Barrie & Conway 2002b). With an ameliorating climate, a coniferous forest expanded, primarily consisting of lodgepole pine with some mountain hemlock (Fedje & Josenhans 2000; Lacourse *et al.* 2003). As relative sea level continued to rise, spits, lakes and the delta were drowned and the land bridge connecting the Haida Gwaii was breached, opening up Hecate Strait (Hetherington *et al.* 2004). This would have resulted in a change from an inlet condition to a more open ocean regime with significant tides. The wider inlet also had a greater fetch, resulting in the erosion of coastal landforms. The pine forest changed at this time to an open mixed forest of lodgepole pine, spruce and mountain hemlock with abundant alder and ferns (Lacourse & Mathewes 2005). With the rise of relative sea level to modern conditions, the region became sediment starved with limited terrestrial input (Barrie & Conway 2002b).

In contrast, most of the present Salish Sea was submarine during sea-level change. So, unlike the Pacific North Coast, no subaerial features formed on the seabed except where the Salish Sea meets the Vancouver Island Shelf (see Fig. 22.4b). Consequently, the Salish Sea continental shelf characteristics are primarily glacial in origin with sedimentation from the Fraser River dominating a significant portion of the central basin, and sedimentary bedforms common in many areas where sand and gravel sediments occur (Barrie *et al.* 2005). Figure 22.4b shows that the Vancouver Island Shelf was subaerial at 11.5 ^{14}C ka BP and probably remained so for over 1000 years. However, few data exist at present to confirm the existence of the lowstand or the occurrence of subaerial morphology. The Vancouver Island Shelf, like the west coast of Haida Gwaii, was fully exposed to the Pacific during the early Holocene transgression. Therefore, unlike the more protected waters of the Pacific North Coast, much of the shelf was eroded, leaving a cover of thin palimpsest sediments over bedrock (Bornhold & Barrie 1991).

The authors would like to acknowledge the support of the Geological Survey of Canada. We are grateful to Francesco Chiocci and an anonymous reviewer for their helpful comments on the manuscript. This chapter is a contribution to UNESCO and the International Geological Correlation Program Project 526 'Continental Shelves: Risks, Resources and Record of the Past'.

References

Barrie, J. V. & Bornhold, B. D. 1989. Surficial geology of Hecate Strait, British Columbia continental shelf. *Canadian Journal of Earth Sciences*, **26**, 1241–1254.

Barrie, J. V. & Conway, K. W. 1999. Late Quaternary glaciation and postglacial stratigraphy of the northern Pacific margin of Canada. *Quaternary Research*, **51**, 113–123.

Barrie, J. V. & Conway, K. W. 2002a. Contrasting glacial sedimentation processes and sea-level changes in two adjacent basins on the Pacific margin of Canada. *In*: Dowdeswell, J. & O'Cofaigh, C. (eds) *Glacier-Influenced Sedimentation on High-Latitude Continental Margins*. Geological Society, London, Special Publications, **203**, 181–194.

Barrie, J. V. & Conway, K. W. 2002b. Rapid sea level changes and coastal evolution on the Pacific margin of Canada. *Sedimentary Geology*, **150**, 171–183.

Barrie, J. V. & Conway, K. W. 2012. Paleogeographic reconstruction of Hecate Strait British Columbia: changing sea levels and sedimentary processes reshape a glaciated shelf. *In*: Li, M. Z., Sherwood, C. R. & Hill, P. R. (eds) *Sediments, Morphology and Sedimentary Processes on Continental Shelves*. International Association of Sedimentologists, Special Publications, **44**, 29–46.

Barrie, J. V., Conway, K. W., Mathewes, R. W., Josenhans, H. W. & Johns, M. J. 1993. Submerged Late Quaternary terrestrial deposits and paleoenvironment of northern Hecate Strait, British Columbia continental shelf, Canada. *Quaternary International*, **20**, 123–129.

Barrie, J. V., Hill, P. R., Conway, K. W., Iwanowska, K. & Picard, K. 2005. Georgia Basin: seabed features and marine geohazards. *Geoscience Canada*, **32**, 145–156.

BARRIE, J. V., CONWAY, K. W., PICARD, K. & GREENE, H. G. 2009. Large-scale sedimentary bedforms and sediment dynamics on a glaciated tectonic continental shelf: examples for the Pacific margin of Canada. *Continental Shelf Research*, **29**, 796–806.

BLAISE, B., CLAGUE, J. J. & MATHEWES, R. W. 1990. Time of maximum Late Wisconsin glaciation, west coast of Canada. *Quaternary Research*, **34**, 282–295.

BORNHOLD, B. D. & BARRIE, J. V. 1991. Surficial sediments on the continental shelf off British Columbia. *Continental Shelf Research*, **11**, 685–700.

CLAGUE, J. J. 1976. Quadra Sand and its relation to the late Wisconsin glaciation of southwest British Columbia. *Canadian Journal of Earth Sciences*, **13**, 803–815.

CLAGUE, J. J. 1977. *Quadra Sand: A Study of the Late Pleistocene Geology and Geomorphic History of Coastal Southwest British Columbia.* Geological Survey of Canada, Paper, 77–17.

CLAGUE, J. J. 1981. *Late Quaternary Geology and Geochronology of British Columbia. Part 2: Summary and Discussion of Radiocarbon-dated Quaternary History.* Geological Survey of Canada, Paper, 80–35.

CLAGUE, J. J. 1983. Glacio-isostatic effects of the Cordilleran Ice Sheet, British Columbia, Canada. *In*: SMITH, D. E. & DAWSON, A. G. (eds) *Shorelines and Isostasy.* Institute of British Geographers, Special Publications, **16**, 321–343.

CLAGUE, J. J. 1989. Quaternary geology of the Canadian Cordillera. *In*: FULTON, R. J. (ed.) Chapter 1, *Quaternary Geology of Canada and Greenland.* Geological Survey of Canada, Geology of Canada, **1**, 17–95.

CLAGUE, J. J. 1994. Quaternary stratigraphy and history of south-coastal British Columbia. *In*: MONGER, J. W. H. (ed.) *Geology and Geological Hazards of the Vancouver Region, Southwestern British Columbia.* Geological Survey of Canada, Bulletin, **481**, 181–192.

CLAGUE, J. J. & JAMES, T. S. 2002. History and isostatic effects of the last ice sheet in southern British Columbia. *Quaternary Science Reviews*, **21**, 71–87.

CLAGUE, J. J., HARPER, J. R., HEBDA, R. J. & HOWES, D. E. 1982*a*. Late Quaternary sea levels and crustal movements, coastal British Columbia. *Canadian Journal of Earth Sciences*, **19**, 597–618.

CLAGUE, J. J., MATHEWES, R. W. & WARNER, B. G. 1982*b*. Late Quaternary geology of eastern Graham Island, Queen Charlotte Islands, British Columbia. *Canadian Journal of Earth Sciences*, **19**, 1786–1795.

FAIRBANKS, R. G. 1989. A 17,000-year glacio-eustatic sea level record: influence of glacial melting rates on the Younger Dryas event and deep-ocean circulation. *Nature*, **342**, 637–642.

FEDJE, D. W. & JOSENHANS, H. 2000. Drowned forests and archaeology on the continental shelf of British Columbia, Canada. *Geology*, **28**, 99–102.

FEDJE, D. W., MCSPORRAN, J. B. & MASON, A. R. 1996. Early Holocene archaeology and paleoecology at the Arrow Creek sites in Gwaii Haanas. *Arctic Anthropology*, **33**, 116–142.

FLÜCK, P. 2003. *Contributions to the geodynamics of Western Canada.* PhD thesis. University of Victoria, Victoria, BC.

FLÜCK, P., HYNDMAN, R. D. & LOWE, C. 2003. Effective elastic thickness Te of the lithosphere in Western Canada. *Journal of Geophysical Research*, **108**, 2430.

GUILBAULT, J.-P., PATTERSON, R. T., THOMSON, R. E., BARRIE, J. V. & CONWAY, K. W. 1997. Late Quaternary paleoceanographic changes in Dixon Entrance, northwest British Columbia, Canada. Evidence from the foraminiferal faunal succession. *Journal of Foraminiferal Research*, **27**, 151–174.

GUILBAULT, J.-P., BARRIE, J. V., CONWAY, K. W., LAPOINTE, M. & RADI, T. 2003. Paleoenvironments associated with the deglaciation process in the Strait of Georgia off British Columbia: microfaunal and microfloral evidence. *Quaternary Science Reviews*, **22**, 839–857.

HART, B. S., HAMILTON, T. S. & BARRIE, J. V. 1998. Sedimentation on the Fraser Delta slope and prodelta, Canada, based on high-resolution seismic stratigraphy, lithofacies and ^{137}Cs fallout stratigraphy. *Journal of Sedimentary Research*, **68**, 556–568.

HEATON, T. H., TALBOT, S. L. & SHIELDS, G. F. 1996. An ice age refugium for large mammals in the Alexander Archipelago, southeastern Alaska. *Quaternary Research*, **46**, 186–192.

HEBDA, R. J., HOWES, D. & MAXWELL, B. 1997. Brooks Peninsula as an ice age refugium. *In*: HEBDA, R. J. & HAGGARTY, J. C. (eds) *Brooks Peninsula: An Ice Age Refugium on Vancouver Island.* Occasional Paper, **5**. BC Parks, Ministry of Environment, Lands, and Parks, Victoria, BC, 8.1–8.66.

HERZER, R. H. & BORNHOLD, B. D. 1982. Glaciation and post-glacial history of the continental shelf off southwestern Vancouver Island, British Columiba. *Marine Geology*, **48**, 285–319.

HETHERINGTON, R. & BARRIE, J. V. 2004. Interaction between local tectonics and glacial unloading on the Pacific margin of Canada. *Quaternary International*, **120**, 65–77.

HETHERINGTON, R. & REID, R. G. B. 2003. Malacological insights into the marine ecology and changing climate of the late Pleistocene–early Holocene Queen Charlotte Islands archipelago, western Canada, and implications for early peoples. *Canadian Journal of Zoology*, **81**, 626–661.

HETHERINGTON, R., BARRIE, J. V., REID, R. G. B., MACLEOD, R., SMITH, D. J., JAMES, T. S. & KUNG, R. 2003. Late Pleistocene coastal paleogeography of the Queen Charlotte Islands, British Columbia, Canada, and its implications for terrestrial biogeography and early postglacial human occupation. *Canadian Journal of Earth Sciences*, **40**, 1755–1766.

HETHERINGTON, R., BARRIE, J. V., REID, R. G. B., MACLEOD, R. & SMITH, D. J. 2004. Paleogeography, glacially induced crustal displacement, and Late Quaternary coastlines on the continental shelf of British Columbia, Canada. *Quaternary Science Reviews*, **23**, 295–318.

HEUSSER, C. J. 1985. Quaternary pollen records from the Pacific Northwest coast: Aleutians to the Oregon–California boundary. *In*: BRYANT, V. M., JR & HOLLOWAY, R. G. (eds) *Pollen Records of Late-Quaternary North American Sediments.* American Association of Stratigraphic Palynologists, Dallas, TX, 141–165.

HEUSSER, C. J. 1990. Late Quaternary vegetation of the Aleutian Islands, southwestern Alaska. *Canadian Journal of Botany*, **68**, 1320–1326.

HEWITT, A. T. & MOSHER, D. C. 2001. Late Quaternary stratigraphy and seafloor geology of the eastern Juan de Fuca Strait, British Columbia and Washington. *Marine Geology*, **177**, 295–316.

HICOCK, S. R. & FULLER, E. A. 1995. Lobal interactions, rheologic superposition, and implications for a Pleistocene ice stream on the continental shelf of British Columbia. *Geomorphology*, **14**, 167–184.

HOWES, D. E. 1983. Late Quaternary sediments and geomorphic history of northern Vancouver Island, British Columbia. *Canadian Journal of Earth Sciences*, **20**, 57–65.

HYNDMAN, R. D. 1995. The Lithoprobe corridor across the Vancouver Island continental margin: the structural and tectonic consequences of subduction. *Canadian Journal of Earth Sciences*, **32**, 1777–1802.

JAMES, T. S., HUTCHINSON, I. & CLAGUE, J. J. 2002. *Improved Relative Sea-level Histories for Victoria and Vancouver, British Columbia, From Isolation-basin Coring.* Geological Survey of Canada, Current Research, **2002-A16**.

JAMES, T. S., HUTCHINSON, I., BARRIE, J. V. & CONWAY, K. W. 2005. Relative sea-level change in the northern Strait of Georgia, British Columbia. *Geographie Physique et Quaternaire*, **59**, 113–127.

JAMES, T., GOWAN, E. J., HUTCHINSON, I., CLAGUE, J. J., BARRIE, J. V. & CONWAY, K. W. 2009. Sea-level change and paleogeographic reconstructions, southern Vancouver Island, British Columbia, Canada. *Quaternary Science Reviews*, **28**, 1200–1216.

JOSENHANS, J. W., FEDJE, D. W., CONWAY, K. W. & BARRIE, J. V. 1995. Post glacial sea levels on the western Canadian continental shelf: evidence for rapid change, extensive subaerial exposure, and early human habitation. *Marine Geology*, **125**, 73–94.

JOSENHANS, H. W., FEDJE, D., PIENITZ, R. & SOUTHON, J. 1997. Early humans and rapidly changing Holocene sea levels in the Queen Charlotte Islands–Hecate Strait, British Columbia, Canada. *Science*, **277**, 71–74.

LACOURSE, T. & MATHEWES, R. W. 2005. Terrestrial paleoecology of Haida Gwaii and the continental shelf: Vegetation, climate, and plant resources of the coastal migration route. *In*: FEDJE, D. W. & MATHEWES, R. W. (eds) *Haida Gwaii: Human History and Environment from the Time of Loon to the Time of the Iron People since the Time of Raven.* University of British Columbia Press, Vancouver, 38–58.

LACOURSE, T., MATHEWES, R. W. & FEDJE, D. W. 2003. Paleoecology of the late-glacial terrestrial deposits with in-situ conifers from the submerged continental shelf of western Canada. *Quaternary Research*, **60**, 180–188.

LEWIS, T. J., BENTKOWSKI, W. H. & WRIGHT, J. A. 1991. Thermal state of the Queen Charlotte Basin, British Columbia: warm. *In*: WOODSWORTH, G. J. (ed.) *Evolution and Hydrocarbon Potential of the Queen Charlotte Basin, British Columbia*. Geological Survey of Canada, Paper, **90-10**, 489–506.

LUTERNAUER, J. L. & MURRAY, J. W. 1983. *Late Quaternary Morphologic Development and Sedimentation, Central British Columbia Continental Shelf*. Geological Survey of Canada, Paper, **83-21**.

LUTERNAUER, J. L., CLAGUE, J. J., CONWAY, K. W., BARRIE, J. V., BLAISE, B. & MATHEWES, R. W. 1989. Late Pleistocene terrestrial deposits on the continental shelf of western Canada: evidence for rapid sea–level change at the end of the last glaciation. *Geology*, **17**, 357–360.

MOSHER, D. C. & HAMILTON, T. S. 1998. Morphology, structure and stratigraphy of the offshore Fraser delta and adjacent Strait of Georgia. *In*: CLAGUE, J. J., LUTERNAUER, J. L. & MOSHER, D. C. (eds) *Geology and Natural Hazards of the Fraser River Delta, British Columbia*. Geological Survey of Canada, Bulletin, **525**, 147–160.

MOSHER, D. C. & HEWITT, A. T. 2004. Late Quaternary deglaciation and sea-level history of eastern Juan de Fuca Strait, Cascadia. *Quaternary International*, **121**, 23–29.

MOSHER, D. C. & JOHNSON, S. Y. 2001. *Neotectonic Mapping in the Eastern Strait of Juan de Fuca: Report of Field Activities*. Geological Survey of Canada, Open File, **3868**.

PORTER, S. C. & SWANSON, T. W. 1998. Radiocarbon age constraints on rates of advance and retreat of the Puget lobe of the Cordilleran ice sheet during the last glaciation. *Quaternary Research*, **50**, 205–213.

RIDDIHOUGH, R. P. 1988. The northeast Pacific Ocean and margin. *In*: NAIRN, A. E. M., STEHLI, F. W. & UYEDA, S. (eds) *The Ocean Basins and Margins, The Pacific Ocean*. Plenum, New York, **7B**, 85–118.

RIDDIHOUGH, R. P. & HYNDMAN, R. D. 1991. Modern plate tectonic regime of the continental margin of western Canada. *In*: GABRIELSE, H. & YORATH, C. J. (eds) *Geology of the Cordilleran Orogen in Canada*. Geological Survey of Canada, Geology of Canada, **4**, 435–455.

RISTAU, J., ROGERS, G. & CASSIDY, J. 2003. Moment magnitude-local magnitude calibration for earthquakes off Canada's west coast. *Seismological Research Letters*, **93**, 2296–2300.

SUTHERLAND-BROWN, A. 1968. *Geology of Queen Charlotte Islands, British Columbia*. Department of Mines and Petroleum Resources Bulletin, Victoria, BC.

THOMSON, R. E. 1981. *Oceanography of the British Columbia Coast*. Canadian Special Publications, Canadian Fisheries and Aquatic Sciences, **56**.

WAITT, R. B. & THORSON, R. M. 1983. The Cordilleran ice sheet in Washington, Idaho, and Montana. *In*: PORTER, S. C. (ed.) *Late-Quaternary Environments of the United States, Volume 1. The Late Pleistocene*. University of Minnesota Press, Minneapolis, MN, 53–70.

WARD, B. C., WILSON, M. C., NAGORSEN, D. W., NELSON, D. E., DRIVER, J. C. & WIGEN, R. J. 2003. Port Eliza cave: North American West Coast interstadial environment and implications for human migrations. *Quaternary Science Reviews*, **22**, 1383–1388.

WARNER, B. G., MATHEWES, R. W. & CLAGUE, J. J. 1982. Ice free conditions on the Queen Charlotte Islands, British Columbia, at the height of late Wisconsin glaciation. *Science*, **218**, 678–684.

Chapter 23

Continental shelves as sediment capacitors or conveyors: source-to-sink insights from the tectonically active Oceanside shelf, southern California, USA

JACOB A. COVAULT[1]* & ANDREA FILDANI[2]

[1]*Chevron Energy Technology Company, Clastic Stratigraphy Research and Development, Houston, TX 77002, USA*

[2]*Chevron Energy Technology Company, Clastic Stratigraphy Research and Development, San Ramon, CA 94583, USA*

Corresponding author (e-mail: jcek@chevron.com)

Abstract: Continental shelves are the key interfaces between terrestrial sediment source areas and deep-sea depositional systems, promoting the transfer of sediment across continental margins. Work on shelves in the context of entire continental-margin sediment-routing systems has focused on their importance as capacitors of sediment during several to tens of thousands of years of post-glacial shoreline transgression and sea-level highstand. We demonstrate that the tectonically active Oceanside shelf offshore southern California has served as an efficient conveyor of sediment from land to the deep sea during millennia of significant climatic fluctuations. This conveyance is a result of littoral drift of sediment to canyon heads at narrow segments of the shelf. We compare insights from the Oceanside shelf to other shelves across the tectonically active Pacific margin of the United States, and demonstrate the importance of shelf width, climatic forcings and timescale of observation in assessing the role of shelves as sediment capacitors or conveyors.

Shelves link subaerial and submarine segments of continental-margin sediment-routing systems and act as staging areas for sediment delivery to the deep sea (Posamentier & Kolla 2003). They are located at key positions along continental margins, where subaerial and submarine processes interact to control the offshore distribution of sediment and dissolved materials, which can significantly affect not only carbon budgets but also the distribution of sediment and potentially harmful pollutants to marine ecosystems (Syvitski *et al.* 2005; Galy *et al.* 2008). These dynamic settings receive fluvial and beach-cliff-eroded sediment, which is reworked by ocean currents and stored until submarine sediment-gravity-flow and mass-wasting processes redistribute it into deeper-water depositional environments. Widely applied models of continental-margin stratigraphic evolution highlight shelves as storage areas, or capacitors, of sediment during periods of rising and high sea level (Vail *et al.* 1977; Nardin 1983; Jervey 1988). These models are most appropriate when shoreline transgression creates accommodation on the shelf for appreciable sediment accumulation, such as across broad shelves (hundreds of km) characteristic of tectonically passive margins (Inman & Nordstrom 1971). Recent continental-margin models have recognized the role of shelves as conveyors of sediment from land to the deep sea if canyon heads have been incised across continental shelves to the modern beach or if terrigenous sediment supply is sufficient to push deltas to the edges of shelves (Burgess & Hovius 1998; Covault *et al.* 2007; Boyd *et al.* 2008; Carvajal *et al.* 2009).

Here we focus on the Oceanside shelf of the tectonically active southern California Borderland, where seminal concepts of sedimentological processes of shelves were developed (e.g. Shepard & Emery 1941; Emery & Shepard 1945; Shepard 1963; Inman & Frautschy 1965) (Figs 23.1 & 23.2). The Oceanside shelf is unique in that it is interpreted to have been associated with the distribution of large volumes of sediment to the deep sea regardless of sea level (Covault *et al.* 2007). We outline the source-to-sink physiography and post-glacial (i.e. since *c.* 20 ka: Nardin *et al.* 1981; Lambeck & Chappell 2001) evolution of the linked southern California subaerial drainage basins-to-submarine shelf, slope and basin plain. Our analysis of the Oceanside shelf and its terrestrial and deep-sea counterparts sheds light on to the nature of sediment storage and transfer across narrow shelves. We compare the Oceanside shelf to others across the Pacific margin of the United States, and demonstrate significant variability

in the geomorphology and processes inherent to tectonically active settings. This integrated perspective of hinterland source, staging-area shelf and deep-sea sink provides insights into the physiographical, climatic and temporal controls on continental-margin sediment routing.

Latest Pleistocene–Holocene southern California sediment-routing system

The Oceanside shelf is a component of a southern California sediment-routing system that includes numerous onshore catchments between approximately 34°N and 32.8°N latitude, which drain steep, tectonically active terrain of the Peninsular Ranges, and deep-sea depositional systems of the California Borderland (Lee & Normark 2009) (Fig. 23.2). The California Borderland is a tectonically active region characterized by relatively narrow shelves and complex basin-and-ridge bathymetry (Shepard & Emery 1941) (Figs 23.1 & 23.2). The entire Borderland extends south from Point Conception, offshore the United States, to Bahia Sebatian Vizcaino and Cedros Island, Mexico (Vedder 1987) (Fig. 23.1). The development of the California Borderland began during the Miocene, when subduction associated with the Great Valley forearc basin had ceased and transform tectonism dominated the region (Crouch 1979, 1981; Yeats *et al.* 1981). In the late Miocene, transform tectonism was focused inland at the location of the present San Andreas Fault Zone (Crowell 1979, 1981), and moderate strike-slip fault-related deformation created the present basin-and-ridge bathymetry of the California Borderland (Teng & Gorsline 1991) (Figs 23.1 & 23.2).

The relatively narrow (<9 km) Oceanside shelf extends south from Dana Point (*c.* 33.5°N) to the La Jolla canyon head (*c.* 32.8°N) (Fig. 23.2). The Oceanside shelf edge is bounded by the Holocene-active dextral strike-slip Newport Inglewood–Rose Canyon Fault Zone (Hogarth *et al.* 2007; Ryan *et al.* 2009*a*) (Fig. 23.2). Longshore currents sweep the shelf, and deliver sediment to the Scripps and La Jolla canyon heads at the end of the littoral cell (Inman & Frautschy 1965). The canyon heads filter sediment across steep, fault-bounded slopes to two elliptical seaward basins – the SE Gulf of Santa Catalina and the San Diego Trough, each up to 80 km long and 25 km wide, and trending approximately NW–SE (Covault & Romans 2009).

From: CHIOCCI, F. L. & CHIVAS, A. R. (eds) 2014. *Continental Shelves of the World: Their Evolution During the Last Glacio-Eustatic Cycle.*
Geological Society, London, Memoirs, **41**, 315–326. http://dx.doi.org/10.1144/M41.23

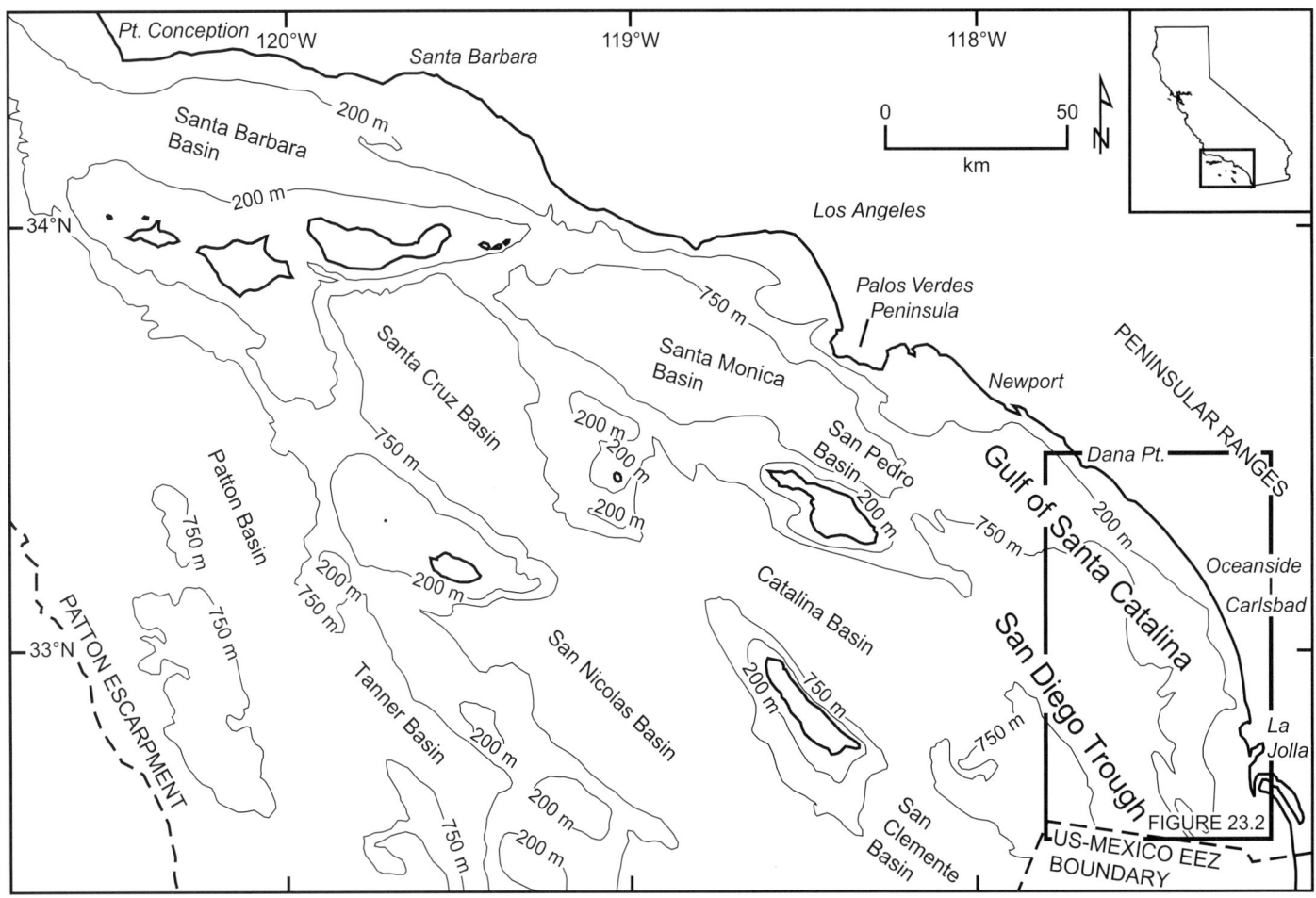

Fig. 23.1. Bathymetric map of the California Borderland (modified from Vedder 1987). The location of the study area in Figure 23.2 is boxed in the SE corner.

These basins are characterized by depths of up to 1200 m below present sea level (mbpsl), and are separated by a series of fault-bounded bathymetric highs, including, from north to south, Crespi Knoll, Carlsbad Ridge and Coronado Bank (Fig. 23.2).

Catchments of the Peninsular Ranges contribute sediment to the Oceanside shelf, and include, from north to south, those of the San Juan Creek, the San Mateo Creek, the Santa Margarita River, the San Luis Rey River and the San Dieguito River (Inman & Jenkins 1999; Warrick & Farnsworth 2009a, b; Covault et al. 2011b) (Fig. 23.2). The steep Peninsular Ranges include resistant Jurassic and Cretaceous plutonic rocks overlain by a veneer of Tertiary and Quaternary sediment and sedimentary rocks (Inman & Jenkins 1999). The mean annual aggregate load of suspended sediment from the catchments to the shelf is approximately 0.2 Mt a^{-1}, estimated for the twentieth century from stream gauges (1943–1998) (Farnsworth & Warrick 2007; Inman 2008). Longer-term, millennial mass loads from catchments, including suspended sediment, bed and dissolved loads, were calculated from cosmogenic radionuclide (^{10}Be) abundances to be approximately 0.1 Mt a^{-1} from the Santa Margarita and San Luis Rey rivers, and around 0.2 Mt a^{-1} delivered to the Oceanside shelf from the entire drainage area (Covault et al. 2011b). The estimated load of suspended sediment from southern California rivers to the Oceanside shelf during the twentieth century is similar to the longer-term, millennial mass load. This suggests that denudation rates of the sediment source area and sediment load to the ocean in this system, and perhaps analogous systems, are consistent over vastly different timescales (Covault et al. 2011b). Mean daily river suspended sediment concentrations during twentieth-century discharge events vary widely,

with most being <20 g l^{-1}, which is less than the threshold for negative buoyancy (>40 g l^{-1} for offshore southern California) and hyperpycnal sediment-gravity-flow initiation (Mulder & Syvitski 1995; Warrick & Farnsworth 2009b). However, the mean daily data can underestimate the significance of sediment discharge at the highest suspended sediment concentrations during extreme events. Warrick & Farnsworth (2009a) noted that twentieth-century sediment load from 12 southern California rivers is punctuated due to the semi-arid climate and the influence of infrequent large storms. They showed that over half of the total post-1900 sediment load of southern California rivers was discharged during El Niño-Southern Oscillation (ENSO)-related events with recurrence intervals >10 years (Inman & Jenkins 1999; Warrick & Farnsworth 2009a). ENSO is the major source of annual–decadal climate variability worldwide, and the coast of California is especially sensitive to the ENSO phase (Masters 2006). During El Niño events, winter storms come from the west and strike California directly with strong winds and high waves and, as a result, precipitation and fluvial discharge are increased (Philander 1990; Masters 2006).

Patsch & Griggs (2006, 2007) calculated that beach cliffs accounted for approximately 29% of the twentieth-century natural sand load to the Oceanside littoral cell and shelf (natural sand yield accounts for reductions in sand contributions behind dams and due to coastal armouring). However, Young & Ashford (2006) showed that during a relatively dry period from April 1998 to April 2004, beach cliffs provided around 67% of the sand and larger-caliber sediment to the Oceanside littoral cell and shelf, thereby demonstrating that beach-cliff sediment contributions might be higher than previous studies indicated.

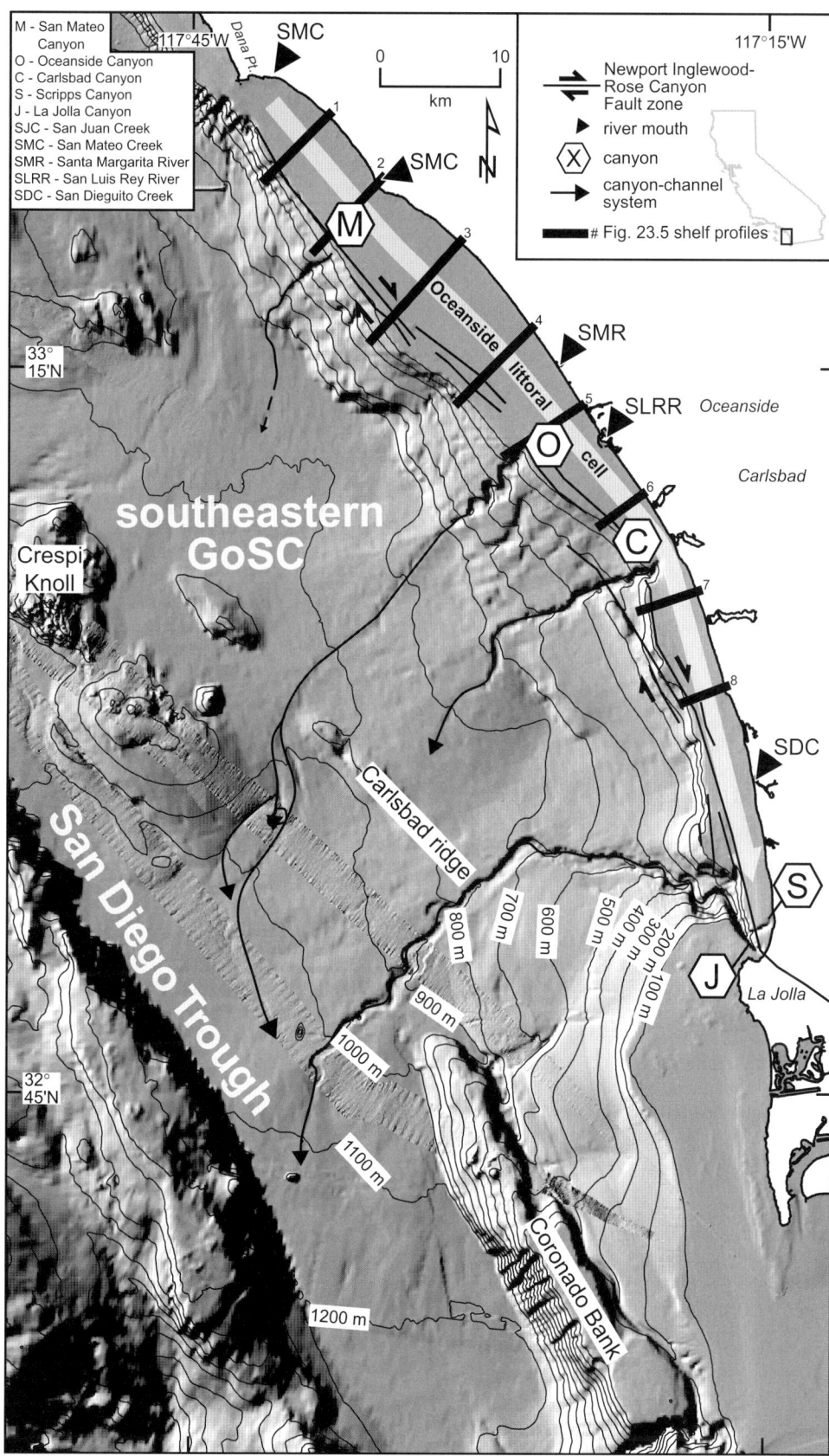

Fig. 23.2. Multibeam bathymetric map of the Oceanside shelf study area in the California Borderland, including the SE Gulf of Santa Catalina (GoSC) and San Diego Trough. Bathymetry from Dartnell *et al.* (2007).

Post-glacial shoreline transgression and climate

Nardin *et al.* (1981) constructed a post-glacial sea-level curve for offshore southern California characterized by a rapid and large-magnitude rise of nearly 130 m, with slowing rates of rise and brief stillstands at approximately 16 ka, at the Younger Dryas at around 11 ka and after about 9 ka, with a longer-term stillstand

occurring since approximately 3 ka (although Masters (2006) interpreted a longer recent stillstand that has occurred since *c.* 6 ka) (Fig. 23.3). During periods of slower rates of transgression, waves are interpreted to have cut shore platforms, thereby providing surfaces for beach development (Masters 2006).

Climate proxy data include magnetic susceptibility, HCl-extractable Al, total inorganic P, total organic matter and $CaCO_3\%$

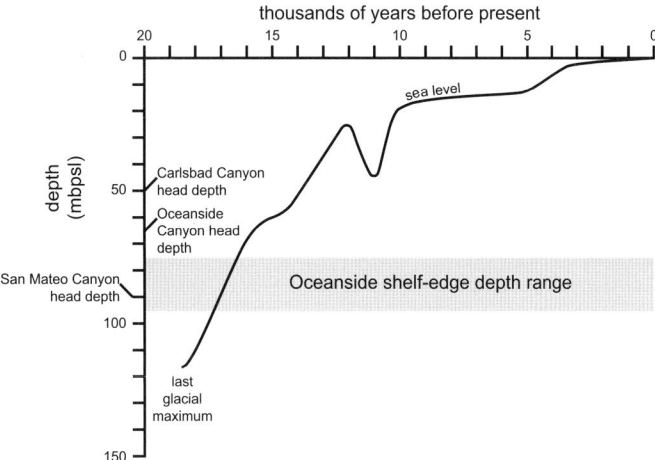

Fig. 23.3. California Borderland sea level since the Last Glacial Maximum based on seismic-stratigraphic analysis of the Santa Monica shelf (see Fig. 23.1) submerged terrace deposits. Modified from Nardin *et al.* (1981).

in drill cores from the bed of Lake Elsinore (33.66°N, 117.35°W) (Kirby *et al.* 2007). These results from the relatively small San Jacinto River drainage basin (<1240 km²) provide the first complete Holocene record of terrestrial climate in southern California (Kirby *et al.* 2007). Figure 23.4a–e shows plots of these climate proxy measurements, which Kirby *et al.* (2007) interpreted to represent onshore southern California variability in precipitation and sediment load into the lake since 9.5 ka (see also Kirby *et al.* 2004). Proxies indicate a wet early Holocene, followed by a long-term drying trend, which is attributed to changes in summer/winter insolation (Wells & Berger 1967; Kirby *et al.* 2005, 2007). Minimum winter, and maximum summer, insolation during the wet early Holocene increased the frequency of winter storms and enhanced the magnitude and spatial extent of the North American monsoon, the frequency of land-falling tropical cyclones and the occurrence of regional convective storms (Kirby *et al.* 2007). The wet early Holocene is interpreted to have been associated with effective erosion and transportation of sediment across the steep terrain of the San Jacinto River drainage basin to Lake Elsinore (Kirby *et al.* 2007). By extension, we posit that these wet conditions in southern California facilitated the transfer of sediment through catchments to the Oceanside shelf (Romans *et al.* 2009; Covault *et al.* 2010).

Marine- and terrestrial-derived climate proxies (diatoms, alkenones, pollen, CaCO₃% and total organic carbon) from Ocean Drilling Program (ODP) Site 1019 offshore northern California (41.7°N, 124.9°W, 980 mbpsl) document a 100-year resolution climatic evolution of coastal California since 16 ka (Barron *et al.* 2003) (Fig. 23.4f). Similar to the Lake Elsinore record, ODP Site 1019 reflects a wet early Holocene, followed by a drying trend through the middle Holocene (11.6–3.2 ka). However, during the late Holocene, after 3.2 ka, a permanent approximately 1 °C increase in alkenone sea surface temperature (SST) signalled a warming of autumn and winter SSTs (Barron *et al.* 2003) (Fig. 23.4). Contemporaneous terrestrial-derived pine pollen, alternating with more alder and redwood pollen, indicates rapid changes in moisture and seasonal temperature, and is evidence of the increased magnitude and frequency of ENSO cycles (Barron *et al.* 2003; for interpretations of enhanced ENSO in the SW United States and southern California, see also Piechota *et al.* 1997; Cayan *et al.* 1999; Kirby *et al.* 2005; Masters 2006; Romans *et al.* 2009; Covault *et al.* 2010).

Between the Last Glacial Maximum and the Holocene epoch (*c.* 20–11 ka), the climate of southern California was relatively cool and wet, with annual precipitation rates as much as three times higher than in the Holocene (Mensing 2001). A hydrological reconstruction for Lake Mojave, Nevada/Arizona, shows high

post-glacial palaeo-lake levels and sedimentation rates that reflect prolonged and enhanced precipitation (i.e. from *c.* 22 to 9 ka: Wells *et al.* 2003) (Fig. 23.4g). By extension, fluvial discharge and sediment delivery to the Oceanside shelf was likely to have been large during these latest Pleistocene pluvial periods. Sommerfield *et al.* (2009) supported these insights with interpretations of large latest Pleistocene fluvial discharge from documentation of buried palaeo-channels incised into southern California shelves (Point Conception–Dana Point).

Oceanside shelf and littoral cell: staging area to the deep sea

The Oceanside shelf is approximately 80 km long, and ranges from 0 to 9 km wide, with an average width of about 5 km. The average depth of the shelf break is around 85 mbpsl (Figs 23.2 & 23.5, Table 23.1). Warrick & Farnsworth (2009b) noted considerable variability in the bathymetry of southern California shelves, with the Oceanside shelf having a moderate slope: that is, between 0.28° and 2.86° (Fig. 23.5). Warrick & Farnsworth (2009b) characterized the floors of shelf-indenting submarine canyons as having steep slopes, >2.86° (e.g. the floor of the Scripps and La Jolla canyon heads). The minimum seabed slope for self-maintaining, fluid-mud (hyperpycnal), gravity-current transport across the shelf was calculated to range between approximately 0.5° and 1.0° (Warrick & Farnsworth 2009b); however, waves and/or currents across the shelf can promote fluid-mud gravity flows along seabeds with lower slopes (Wright & Friedrichs 2006). Thus, the Oceanside shelf has a critical slope necessary for auto-suspending sediment gravity flows, which promote off-shelf sediment transport (Warrick & Farnsworth 2009b). Gradients in the head of La Jolla Canyon, at the southern end of the Oceanside shelf, can be considerably larger (<7°: Covault *et al.* 2011a), which might promote auto-suspending sediment gravity flows with speeds in excess of 100 cm s⁻¹ from littoral supply of sand (cf. the Calleguas Creek to Mugu Fan in Santa Monica Bay of Warrick & Farnsworth 2009b). Slater *et al.* (2002) calculated the volume of unconsolidated sediment of the Oceanside shelf to be 6–7 km³, which is smaller than other Borderland shelves and littoral cells, and categorized the shelf as relatively sediment starved, lacking a thick mud belt (Warrick & Farnsworth 2009b). This type of steep shelf with smaller sediment supply is common to the majority of the southern California Borderland offshore the United States (Warrick & Farnsworth 2009b).

Wave climate, and storm magnitude and frequency are important influences on shelf morphology and sedimentological processes along the western coast of the United States (Inman *et al.* 1976; Wright & Nittrouer 1995). Most of the wave energy for off-shore southern California is generated by mid-latitude winter cyclones in the North Pacific (Adams *et al.* 2008; Xu & Noble 2009) (Fig. 23.6, Table 23.1). Other important sources include waves generated by the northwesterly winds along the California coast during the spring and summer, and swells from winter storms in the Southern Hemisphere, which commonly deliver smaller waves (Adams *et al.* 2008; Xu & Noble 2009). Xu & Noble (2009), in an analysis of the wave climate of the entire Borderland region offshore southern California, documented a statistically significant wave-height minimum and return-period maximum of incipient sediment motion in the area of the Oceanside shelf and littoral cell. This was interpreted to be the result of enhanced sheltering of the Oceanside shelf by the Palos Verdes Peninsula (Fig. 23.1). Winter wave heights of the Oceanside shelf average 1.0–1.3 m; summer heights average 0.9–1.1 m (Xu & Noble 2009) (Table 23.1). Larger wave heights occur during winter storms (November–March); however, the seasonal difference in wave height across the Oceanside shelf is only about 0.1–0.2 m or approximately 11–21% (Xu & Noble

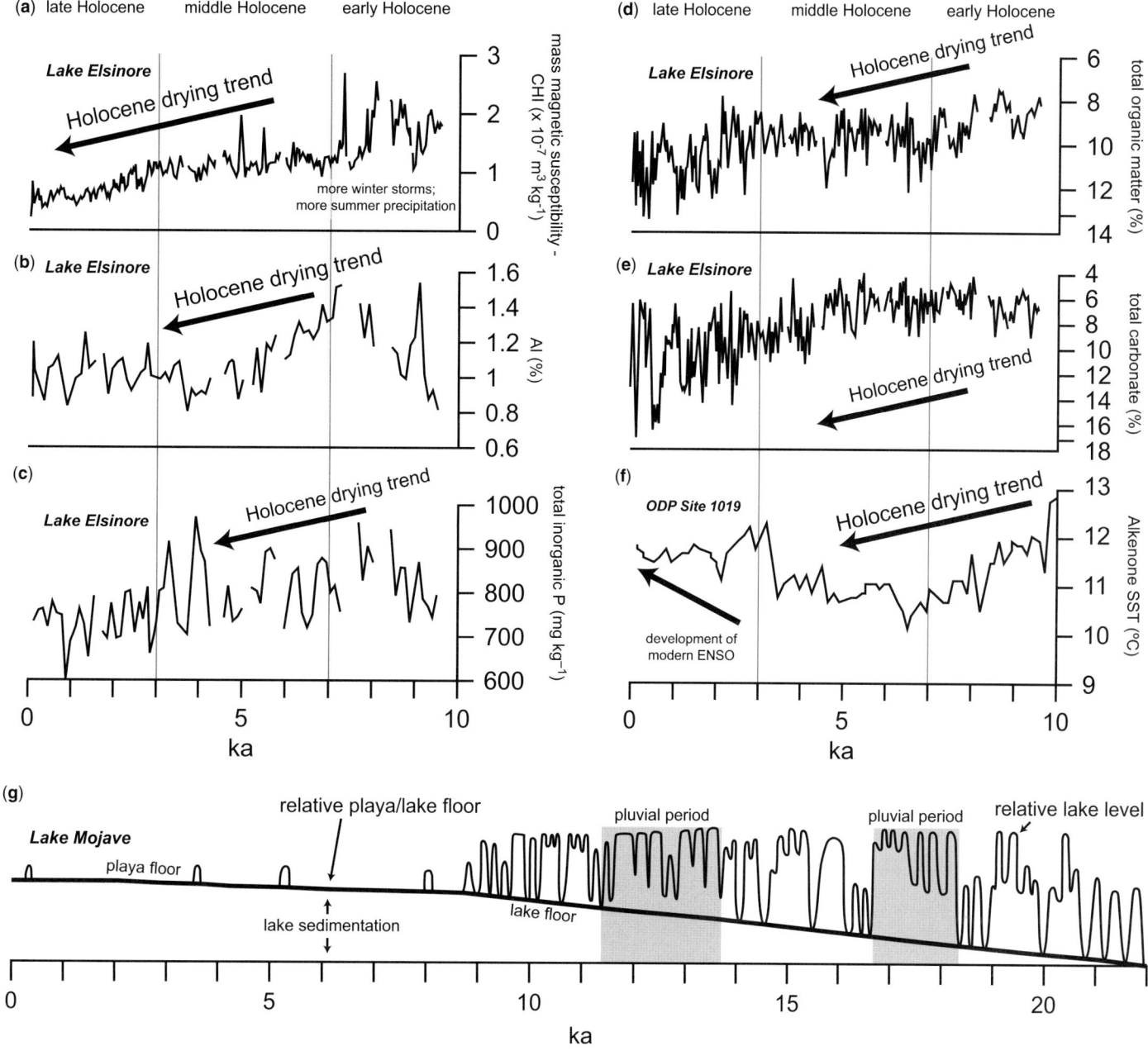

Fig. 23.4. California climate proxies: (**a**)–(**e**) proxies from Lake Elsinore taken from Kirby *et al.* (2007); (**f**) alkenone SST from ODP Site 1019 (Barron *et al.* 2003); and (**g**) simplified post-glacial history of relative water level and basin floor of Lake Mojave and the surrounding Nevada/Arizona lakes (Wells *et al.* 2003).

2009). During the peaks of each of the four strong ENSO events since 1981 (i.e. in 1983, 1986, 1991, and 1998) (Storlazzi & Griggs 2000), the mean wave height was 50–100% larger than that of the same month during non-ENSO years (Xu & Noble 2009). Wiberg *et al.* (2002) observed that on the nearby Santa Monica and San Pedro shelves, storm waves can be as high as 7 m and generate orbital flows sufficient to entrain medium-grained sand across the full width of the shelf (Sommerfield *et al.* 2009).

For most of California, waves from the NW have the greatest influence on longshore currents and littoral drift and, thus, a southward net littoral drift of sand dominates (Patsch & Griggs 2006) (Figs 23.5 & 23.6, Table 23.1). However, during ENSO winters, waves can impact the Oceanside shelf from the west or SW, and the predominance of southward transport can be reduced (Patsch & Griggs 2006). Inman & Frautschy (1965) and Inman *et al.* (1976) noted that the southward littoral drift across the Oceanside shelf delivers sediment to the Scripps and La Jolla canyon heads, where strong, sustained, down-canyon turbidity currents were

documented and interpreted to be related to strong waves coincident with the passage of storm fronts (Normark *et al.* 2009).

Post-glacial Oceanside shelf evolution

The modern Oceanside littoral cell developed < 13 ka, when longshore currents transported sediment to the Scripps and La Jolla canyon heads offshore La Jolla (Covault *et al.* 2007). This interpretation of latest Pleistocene development of the littoral cell is supported by Slater *et al.* (2002), who observed submerged terraces that provided surfaces for beach development on the Oceanside shelf at 20, 40 and 60–80 mbpsl cut during stillstands of sea level at approximately 16, 11 and 9 ka (Nardin *et al.* 1981; Masters 2006) (Fig. 23.3). However, Masters (2006) linked the occurrence of the Pismo clam *Tivela stultorum*, a proxy for an equilibrium beach that includes a sufficient volume of sand to adjust to seasonal changes in wave energy over annual to decadal

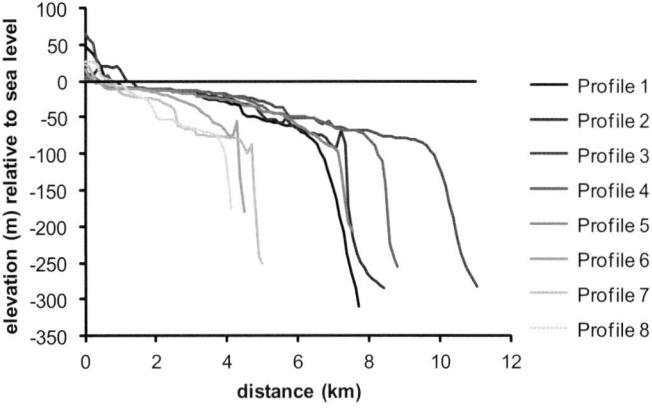

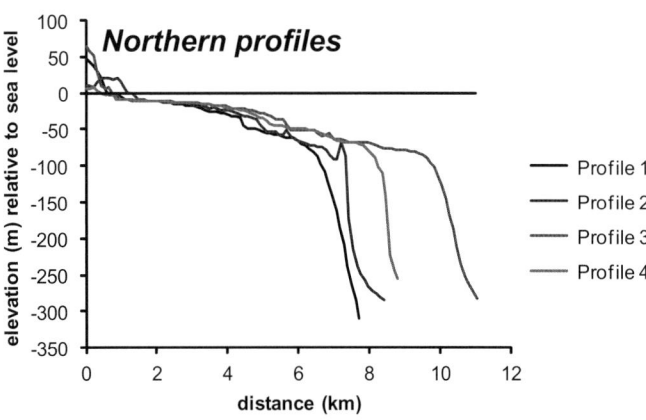

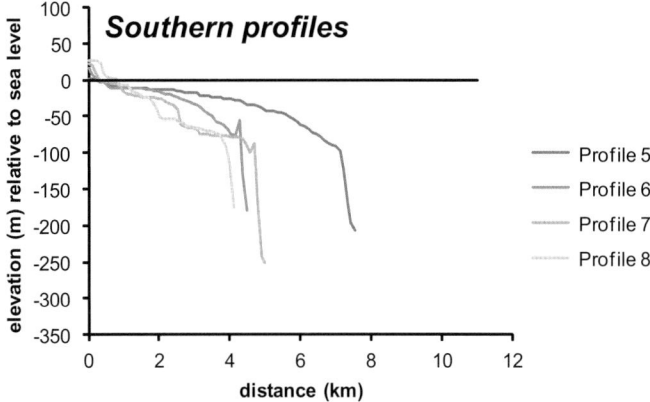

Fig. 23.5. Bathymetric profiles of the Oceanside shelf. See Figure 23.2 for profile locations.

Table 23.1. *Summary of Oceanside shelf characteristics**

Length of the shelf (km)	*c.* 80
Average width (km)	*c.* 5
Tidal, wave, current ranges	Average tide range 1.1 m; winter wave heights average 1.0–1.3 m, summer heights average 0.9–1.1 m; waves from the NW force a southward longshore current
Dominating process	Waves/southward longshore current
Average depth of the shelf break (mbpsl)	*c.* 85
Dominant sediment type	Siliciclastic
Tectonic trend over the last glacial cycle	Uplifting (0.13–2 m ka^{-1}) and strike-slip faulting

*Tide range data from National Ocean Service (2010), http://tidesandcurrents.noaa.gov/.
Wave data from Xu & Noble (2009). Tectonic uplift data from Legg & Kennedy (1979),
Kern & Rockwell (1992) and Hogarth *et al.* (2007).

periods, to the inception of the modern Oceanside littoral cell as late as the middle Holocene (*c.* 5 ka).

Tectonic deformation played a key role in post-glacial Oceanside shelf development. The Newport Inglewood–Rose Canyon Fault Zone, which bounds the Oceanside shelf, limited shelf accretion by initiating submarine mass movements and constrained the location of the La Jolla Canyon head at the southern end of the shelf and littoral cell (Field & Edwards 1980; Hogarth *et al.* 2007; Ryan *et al.* 2009*a*) (Fig. 23.2). A constraining bend of the Rose Canyon Fault offshore La Jolla also created shelfal accommodation for sediment accumulation. Hogarth *et al.* (2007) documented a significant south–north shelf-sediment thickness decrease from >20 m, and shoaling of the latest Pleistocene–Holocene transgressive surface of erosion from the Scripps and La Jolla canyon heads at the southernmost Oceanside shelf. Hogarth *et al.* (2007) proposed that this differential relief is the consequence of tectonic uplift and titling since marine transgression, which yields deformation rates (i.e. >2 m ka^{-1}) of an order of magnitude larger than previous estimates (e.g. 0.13 m ka^{-1}: Legg & Kennedy 1979; Kern & Rockwell 1992).

Slater *et al.* (2002) provided a general model for post-glacial Borderland shelf evolution: as the shoreline regresses across narrow shelves, waves redistribute inner-shelf sediment to shelf-edge and upper-slope staging areas, where sediment is subsequently transported through canyon-and-channel systems to deep-sea fans. A shelf-sediment lens re-establishes as accommodation is created as a result of shoreline transgression. The post-glacial evolution of the Oceanside shelf, however, is more complicated, as sediment was delivered to the deep sea regardless of shoreline position; rather, the pathways of sediment dispersal changed with shoreline migration (Inman & Frautschy 1965; Inman *et al.* 1976; Covault *et al.* 2007) (Fig. 23.7). As sea-level rose following the Last Glacial Maximum, shoreline transgression progressively stranded canyon heads on the outer shelf, and longshore currents transported sediment to the Scripps and La Jolla canyon heads, whose incision kept pace with the transgression (Inman & Frautschy 1965; Inman *et al.* 1976; Covault *et al.* 2007). Thus, the transfer of sediment across the Oceanside shelf as a result of littoral drift inhibited the development of an appreciable sediment lens, as evidenced by the relatively meagre volume of unconsolidated shelf sediment calculated by Slater *et al.* (2002).

Post-glacial shelf–deep sea sediment partitioning

In one of the first comprehensive studies of Quaternary sediment partitioning between Borderland shelf, slope and basin-plain environments, Moore (1969) hypothesized that increased river gradient and discharge during the Last Glacial Maximum was a primary driver of submarine canyon incision and subsequent sediment delivery to the deep sea. Gorsline *et al.* (1968) used radiocarbon ages from the Tanner Basin (Fig. 23.1) to highlight diminished deep-sea sediment supply coincident with marine transgression after about 7.5 ka, which they attributed to reduced river gradients and accommodation creation on drowned Borderland shelves. These models of sea-level lowstand deep-sea sediment delivery and highstand sediment sequestration on the shelf were precursors to seismic- and sequence-stratigraphic concepts, which advocated the importance of accommodation creation and destruction relative to sediment supply in dictating geometries, volumes and facies relationships of stratigraphic sequences (Vail *et al.* 1977; Nardin 1983; Jervey 1988). However, recent studies of land–sea interactions between the Peninsular Ranges catchments, the Oceanside shelf and littoral cell, and the La Jolla Canyon and Fan suggest that the generalized sequence-stratigraphic models of shelf–basin sediment partitioning might not be universally applicable, especially for sediment-routing systems with narrow, sediment-starved shelves dissected by submarine canyons (e.g. Covault *et al.* 2007; Sommerfield *et al.* 2009) (Fig. 23.7).

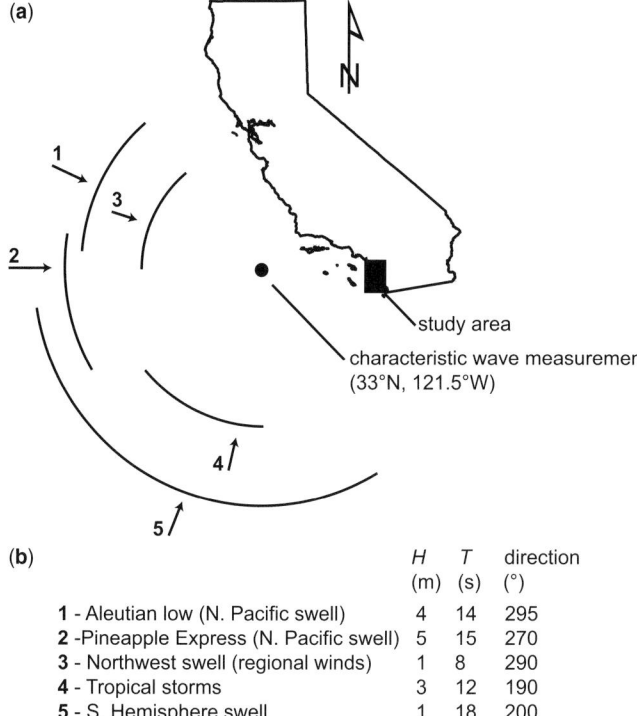

(b)

		H (m)	T (s)	direction (°)
1 - Aleutian low (N. Pacific swell)		4	14	295
2 - Pineapple Express (N. Pacific swell)		5	15	270
3 - Northwest swell (regional winds)		1	8	290
4 - Tropical storms		3	12	190
5 - S. Hemisphere swell		1	18	200

Fig. 23.6. (a) Orientations of five wave types approaching the California Borderland (measured at 33°N, 121.5°W). Arcs represent the range of wave directions; arrows indicate the modal direction. (b) Table of five wave types. Modal values are provided. *H*, modal wave height; *T*, modal wave period. Modified from Adams *et al.* (2008).

Five major canyon heads are incised into the Oceanside shelf: from north to south, the San Mateo, Oceanside, Carlsbad, Scripps and La Jolla canyon heads (Shepard 1963; Inman & Frautschy 1965; Shepard & Buffington 1968; Normark 1970; Piper 1970; Inman *et al.* 1976; Graham & Bachman 1983; Normark *et al.* 2009; Covault & Romans 2009) (Fig. 23.2). The shelf offshore La Jolla actually includes two tributary canyons – the Scripps and La Jolla canyons – that feed into the La Jolla channel and deep-sea fan; therefore, for simplicity, both tributaries will hereafter be referred to as the La Jolla Canyon. These canyons pass from V-shaped conduits at the shelf edge and uppermost slope to U-shaped channels with overbank elements across the lower slope and basin floor, and ultimately contribute sediment to the growth of deep-sea fans (Normark 1970). Shelf widths from the San Mateo, Oceanside and Carlsbad canyon heads to the present shoreline are approximately 7, 6 and 2 km, respectively, which are large relative to the La Jolla Canyon head that has been incised across the shelf nearly to the present shoreline (Fig. 23.2). During late Pleistocene intervals of low sea level, the San Mateo, Oceanside and Carlsbad canyons, along with innumerable ephemeral slope gullies, received terrigenous sediment from segments of the disaggregated Oceanside littoral cell and rivers that were able to extend across the subaerially exposed shelf (Covault *et al.* 2007) (Fig. 23.7). The San Mateo, Oceanside and Carlsbad canyon heads are located in progressively shallower water, from north to south: approximately 90, 65 and 50 mbpsl. As a result, as the shoreline transgressed since the Last Glacial Maximum, the San Mateo Canyon head was drowned first, and ceased to receive terrigenous sediment, followed by the Oceanside and Carlsbad canyons (Covault & Romans 2009). These canyon heads are presently stranded at the outer shelf as a result of the post-glacial sea-level rise, and canyons and channels are inactive today. However, incision of the La Jolla Canyon head kept pace with shoreline transgression, allowing it to receive sediment

from littoral drift (Inman & Frautschy 1965; Inman *et al.* 1976). La Jolla Canyon head lacks a prominent fluvial sediment contributor and was inactive during the period of lower sea level from 40 to 13 ka (Covault *et al.* 2007) (Fig. 23.7).

Covault *et al.* (2007) and Covault & Romans (2009) provided sediment budgets for the deep-sea fans linked to the San Mateo, Oceanside, Carlsbad and La Jolla canyon-and-channel systems. The sea-level lowstand-active systems have a total bulk sediment volume of 70 km^3, whereas the single highstand-active La Jolla system has a volume of 38 km^3. Covault *et al.* (2007) compared the volumetric deposition rates since 40 ka of the Oceanside, Carlsbad and La Jolla fans, and showed that the combined rate of the lowstand-active Oceanside and Carlsbad fans was only half of the rate of the single highstand-active, littoral-drift-fed La Jolla Fan. They concluded that lowstand fans are only part of the turbidite depositional record, and coarse-grained terrigenous sediment has been deposited in California Borderland deep-water basins regardless of sea level.

Efficient sediment transfer from land to deep sea, with the Oceanside shelf serving as the key link, is implied from the chronostratigraphic evidence of consistent deep-sea deposition since the Last Glacial Maximum. Moreover, Covault *et al.* (2011*b*) provided source-to-sink sediment budgets from cosmogenic radionuclide-derived terrestrial denudation rates and submarine-fan deposition rates since 40 ka for the southern California sediment-routing system of this study. They showed that source-area denudation and deep-sea deposition are balanced during the period of generally falling and low sea level from 40 to 13 ka (i.e. mass load from the Santa Margarita and San Luis Rey catchments to the lowstand-active Oceanside deep-sea fan is balanced at *c.* 0.1 Mt a^{-1}) but that deep-sea deposition exceeds terrestrial denudation during the subsequent period of rising and high sea level since 13 ka (i.e. integrated catchment denudation and sediment supply to the Oceanside littoral cell amounts to *c.* 0.2 Mt a^{-1}, whereas the La Jolla Fan deposition rate is *c.* 0.3 Mt a^{-1}). They attributed the additional supply of sediment since 13 ka to enhanced dispersal of sediment across the shelf caused by sea-cliff erosion during post-glacial shoreline transgression and initiation of submarine mass wasting across the tectonically active southern California continental margin. Thus, during periods of both low (40–13 ka) and high (<13 ka) sea level, land and deep-sea sediment loads do not show orders of magnitude imbalances that might be expected in the wake of major sea-level changes. It appears that this southern California sediment-routing system maintains a consistent connection between terrestrial and marine segments, with sediment routing ultimately dependent on characteristics of the Oceanside shelf, particularly submarine canyon physiography (Fig. 23.7). When sea level is low, rivers extend across the subaerially exposed shelf and directly feed numerous canyon-and-channel systems and deep-sea turbidite systems. As sea level rises, sediment discharged from numerous fluvial sources is consolidated into an aggregate transport system, the Oceanside littoral cell, which delivers sediment to a single canyon head. Therefore, despite thousands of years of glacio-eustatic sea-level fluctuations, the Oceanside littoral cell has consistently remained a poor capacitor of sediment. It lacks appreciable accommodation and is characterized by limited accretion of sediment. It is an excellent conveyor of sediment over millennia, since 13 ka, during which the dominant southward-directed longshore current has redistributed terrigenous sediment to the La Jolla Canyon and Fan.

Tectonically active western USA margins: shelves as sediment capacitors or conveyors

In central California, the Monterey Canyon has been incised across the Monterey Bay shelf into the Moss Landing Harbor (Fig. 23.8b).

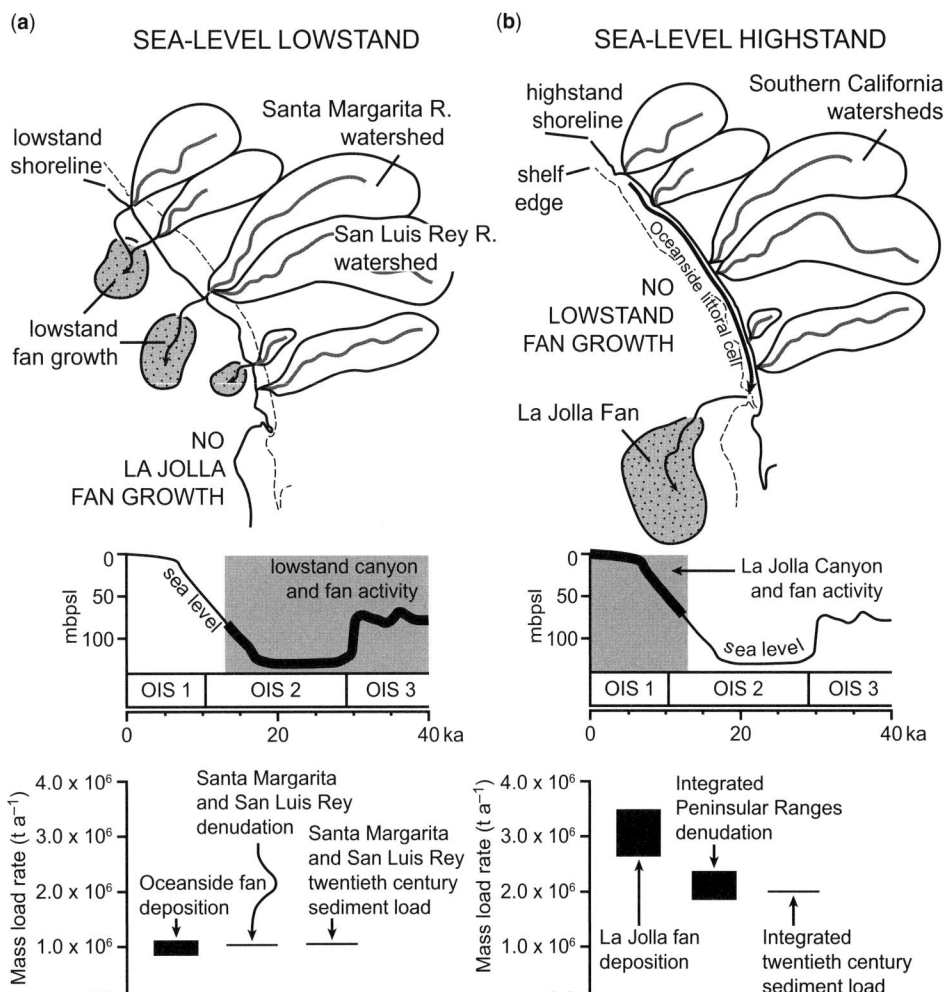

Fig. 23.7. Summary diagram of post-glacial Oceanside shelf evolution at low (**a**) and high (**b**) sea levels, global sea-level and canyon and fan activity since 40 ka, and Oceanside and La Jolla Fan mass-deposition rates, Peninsular Ranges denudation rates, and twentieth-century sediment loads. The shelf width between canyon heads and the littoral zone influences the activity of deep-sea canyon-and-channel systems and fan growth. mbpsl, metres below present sea-level; OIS, Oxygen Isotope Stage; ka, thousands of years ago. Sea level from Lambeck & Chappell (2001). Modified from Covault *et al.* (2011*b*).

This submarine canyon system is one of the largest canyons on the western coast of North America (Greene *et al.* 2002), and an active conduit of sediment to one of the largest deep-sea fans in the world (Fildani & Normark 2004). The Monterey Bay shelf achieves a maximum width of approximately 14 km. Sediment is delivered to the shelf from numerous fluvial systems draining the Santa Cruz Mountains to the north, the Pajaro River and the Salinas River (Farnsworth & Warrick 2007). Sedimentary discharge from the Salinas River is particularly large, nearly 2 Mt a^{-1} of suspended sediment (Farnsworth & Warrick 2007). Similar to the La Jolla Canyon and Fan system in southern California, the Monterey Canyon is presently connected to littoral cells at its head and, as a result, it is estimated that as much as 500 000 m^3 (*c.* 1 Mt) of sand enters the canyon head each year from littoral drift (United States Army Corps of Engineers 1985; Best & Griggs 1991; Paull *et al.* 2005), and that the volume of finer-grained sediment might be even much greater (Eittreim & Noble 2002). Paull *et al.* (2005) postulated that the shelf-bisecting Monterey Canyon acts as a capacitor for sediment over decades–centuries and it is released during enormous gravity-driven mass-movement processes with recurrence intervals of several thousands of years or longer (Piper & Normark 2001; Fildani & Normark 2004). In this way, sediment can be sequestered on the shelf and in the Monterey Canyon over decades–centuries, but, when sediment-dispersal processes are viewed over several thousands of years or longer, forcings with longer-term recurrence intervals can efficiently convey sediment to the deep sea.

The northern California shelf offshore the Eel River is relatively narrow, 10–20 km wide, with the shelf edge at 90–150 mbpsl (Fig. 23.8c). The Eel margin is located in a forearc position directly north of the Mendocino Triple Junction and, as such, it was subjected to complex tectonic processes including folding since about 2 Ma, thrust faulting since approximately 1 Ma and present-day deformation associated with the migrating Mendocino Triple Junction (Burger *et al.* 2002). The Eel Canyon is presently disconnected from a significant fluvial or littoral source of sediment. The Eel River delivers two orders of magnitude more suspended sediment to the shelf (mean annual load *c.* 12 Mt a^{-1}) than all the major catchments feeding the Oceanside shelf combined (mean annual aggregate load *c.* 0.2 Mt a^{-1}: Farnsworth & Warrick 2007; Inman 2008; Covault *et al.* 2011*b*). On an approximately 100-year timescale, fluvial and marine sediment dispersal processes have produced a mid-shelf depocentre on the Eel shelf (Sommerfield & Nittrouer 1999). However, mass-accumulation rates during that approximately 100-year period on the shelf indicate that only about 20% of the mean annual sediment supply to the margin remains on the shelf, with the majority being redistributed across the upper slope and into deeper water (Sommerfield & Nittrouer 1999). Longer-term analyses of the Eel margin (i.e. since *c.* 1 Ma) indicate significant stratigraphic-sequence and -surface developments associated with glacio-eustatic sea-level fluctuations (Burger *et al.* 2001). The separation of the Eel River from the deep-sea canyon during thousands of years of rising and high sea level inhibited the transfer of sediment to the deep sea (Burger *et al.* 2001), and the Eel shelf served as a capacitor of sediment during millennia of shelfal accommodation creation. Therefore, short-term observations of sediment dispersal spanning the last century indicate that the Eel shelf has predominantly been a conveyor of sediment across the shelf edge (Sommerfield & Nittrouer 1999), whereas longer-term observations since about 1 Ma indicate that the Eel shelf has served as a capacitor of sediment (Burger *et al.* 2001).

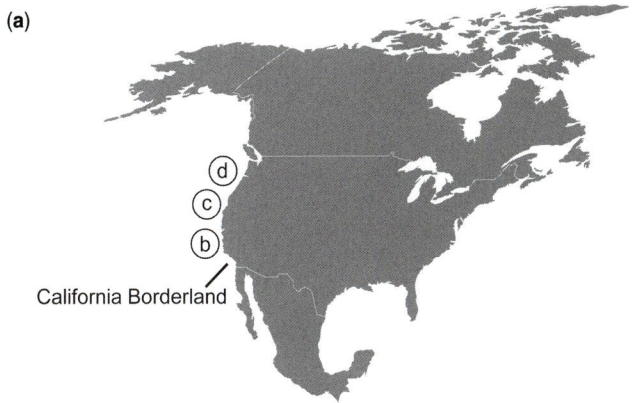

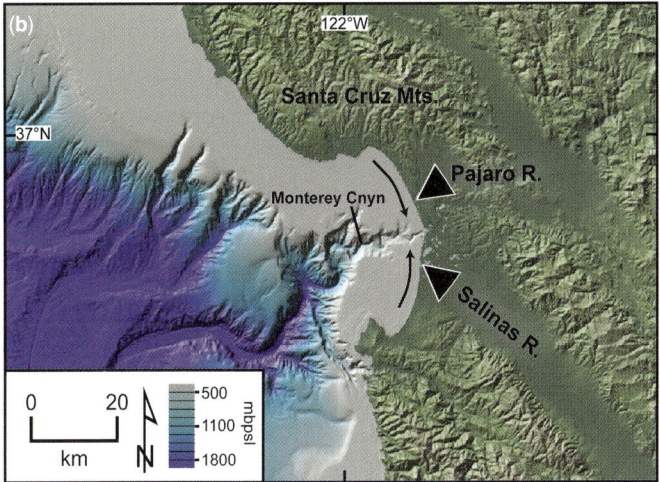

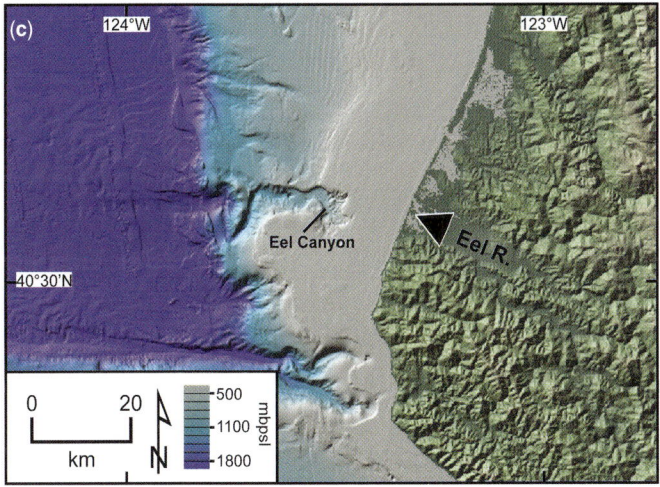

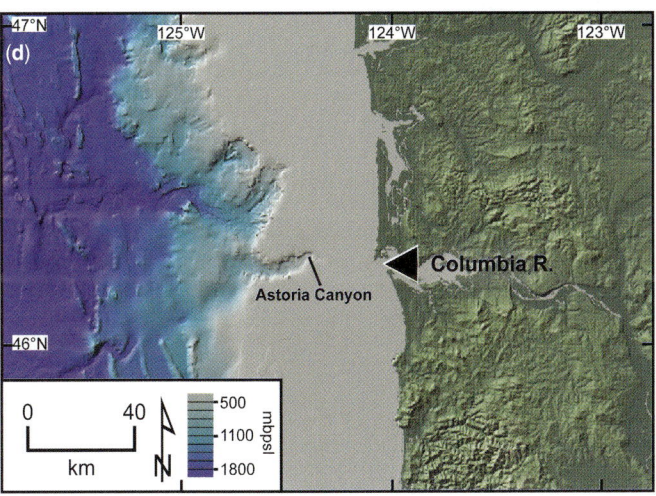

The shelf offshore Washington and Oregon is part of an accretionary prism associated with the subduction of the Juan de Fuca plate beneath North America (Fig. 23.8d). This shelf is wider than the aforementioned California shelves, approximately 25–60 km wide, as a result of Quaternary collisional fold-and-thrust-belt development, which is building westwards at a rate close to the orthogonal component of plate convergence (i.e. $c.$ 40 mm a^{-1}: DeMets $et\ al.$ 1990). The Columbia River is the primary source of sediment to the shelf, with a mean annual sediment load of 5–10 Mt a^{-1} (Milliman & Syvitski 1992; Wise $et\ al.$ 2007); although, catastrophic natural processes inherent to the Pacific Northwest margin can instigate ephemeral periods of extreme sediment discharge from the Columbia River. For example, during the years following the eruption of Mount St Helens (Washington state), the Columbia River discharged an estimated 35 Mt a^{-1} of sediment (Meade & Parker 1985). At least 60% of sediment discharged from the Columbia River is sequestered on the relatively broad Washington and Oregon shelf (Sommerfield & Nittrouer 1999). Sommerfield & Nittrouer (1999) attributed this enhanced sediment sequestration to larger shelf width compared to other shelves offshore the western United States, and to the out-of-phase annual occurrence of maximum fluvial sediment supply and maximum wave energy and current velocity during the twentieth century. That is, a larger shelf width provides greater accommodation for sediment storage, and the delivery of terrigenous sediment to the shelf from the Columbia River and other fluvial sources is out of phase with episodes of intense wave and current reworking of the shelf. Because high-energy wave and current reworking conditions lag behind periods of deposition, consolidation of sediment can develop and increase the erosional yield strength in the seabed (Sommerfield & Nittrouer 1999). Sediment routing from the Columbia River, across the shelf, to the Astoria Canyon and Fan system was particularly efficient during post-glacial marine transgression in spite of shoreline retreat across the wide shelf (Brunner $et\ al.$ 1999; Zuffa $et\ al.$ 2000; Normark & Reid 2003). This is because the Pacific Northwest margin is located in a relatively high-latitude setting sensitive to climatic variability associated with glacial–interglacial transitions, during which catastrophic floods rapidly transported sediment thousands of kilometres from source to sink (Piper & Normark 2009). Radiocarbon age dates from ODP sites 1037 and 1038, which are >1000 km from the Columbia River mouth along a deep-sea conduit, corroborate the synchronicity of deep-sea deposition and terrestrial flooding during post-glacial shoreline transgression (Brunner $et\ al.$ 1999; Zuffa $et\ al.$ 2000). The present Pacific Northwest shelf is a capacitor of sediment; however, during post-glacial melting and consequent sea-level rise, the shelf was a conveyor of sediment to the deep sea.

Summary

This analysis of the Oceanside shelf offshore southern California, when compared to other tectonically active shelves offshore the western United States, highlights the importance of shelf width and submarine canyon dissection, coupled with climatic factors, in land to deep-sea sediment routing. For example, southern and central California shelves that have been incised by canyons across their widths to present shorelines convey sediment through the littoral zone to the deep sea during periods of both high and low sea level (Fig. 23.7). In contrast, wider shelves

Fig. 23.8. Western USA continental shelves. (**a**) Location map. (**b**) Monterey Bay shelf and canyon. Arrows represent dominant longshore-current directions. Monterey Canyon has been incised across the shelf into the Moss Landing Harbor. (**c**) Eel River shelf and canyon. Eel Canyon is stranded on the outer shelf. (**d**) Columbia River shelf and Astoria Canyon. Astoria Canyon is stranded on the outer shelf. Bathymetry is from GeoMapApp (Ryan $et\ al.$ 2009b).

offshore northern California, Oregon and Washington are generally associated with shelfal sediment sequestration as canyons were stranded from terrestrial sources coincident with post-glacial shoreline transgression. The role of shelves as sediment capacitors or conveyors changes according to the timescale of observation, as well as to climatic circumstances. Similar to the Oceanside shelf of southern California (Slater *et al.* 2002; Warrick & Farnsworth 2009*b*), the Monterey Bay shelf and canyon are interpreted to be capacitors of sediment over short, decadal–centennial timescales, but conveyors over millennia as a result of enormous gravity-driven mass-movement processes with thousands of years of recurrence (Paull *et al.* 2005). In contrast, analyses of the Eel margin indicate sediment is conveyed across the shelf over an approximately 100-year time period (Sommerfield & Nittrouer 1999), but longer-term analyses indicate tens to hundreds of thousands of years of appreciable sediment sequestration on the shelf as a result of high sea level (Burger *et al.* 2001). Our analysis of the Oceanside shelf and other tectonically active shelves offshore the western United States demonstrates the importance of the shelf as a filter of sediment *en route* to the deep sea, underscores the influence of shelf-dissecting submarine canyons in sediment routing to the deep sea and highlights the utility of integrated analyses to the understanding of entire continental-margin sediment-routing systems.

We thank the Clastic Stratigraphy R & D team at Chevron Energy Technology Company, and especially B. W. Romans for insightful discussions and insights. We also thank our colleagues at the Stanford Project on Deep-water Depositional Systems, and the US Geological Survey, Coastal and Marine Geology Team. We appreciate thoughtful reviews by S. Hubbard and G. de Alteriis.

References

ADAMS, P. N., INMAN, D. L. & GRAHAM, N. E. 2008. Southern California deep-water wave climate: characterization and application to coastal processes. *Journal of Coastal Research*, **24**, 1022–1035.

BARRON, J. A., HEUSSER, L., HERBERT, T. & LYLE, M. 2003. High-resolution climatic evolution of coastal northern California during the past 16 000 years. *Paleoceanography*, **18**, 1020, http://dx.doi.org/10.1029/2002PA000768

BEST, T. C. & GRIGGS, G. B. 1991. A sediment budget for the Santa Cruz littoral cell, California. *In*: OSBORNE, R. H. (ed.) *From Shoreline to Abyss: Contributions in Marine Geology in Honor of Francis Parker Shepard*. SEPM Special Publications, Tulsa, **46**, 35–50.

BOYD, R., RUMING, K., GOODWIN, I., SANDSTROM, M. & SCHRÖDER-ADAMS, C. 2008. Highstand transport of coastal sand to the deep ocean: a case study from Fraser Island, southeast Australia. *Geology*, **36**, 15–18.

BRUNNER, C. A., NORMARK, W. R., ZUFFA, G. G. & SERRA, F. 1999. Deep-sea sedimentary record from the late Wisconsin cataclysmic floods from the Columbia River. *Geology*, **27**, 463–466.

BURGER, R. L., FULTHORPE, C. S. & AUSTIN, J. A. 2001. Late Pleistocene channel incisions in the southern Eel River Basin, northern California: implications for tectonic vs. eustatic influences on shelf sedimentation patterns. *Marine Geology*, **177**, 317–330.

BURGER, R. L., FULTHORPE, C. S., AUSTIN, J. A. & GULICK, S. P. S. 2002. Lower Pleistocene to present structural deformation and sequence stratigraphy of the continental shelf, offshore Eel River Basin, northern California. *Marine Geology*, **185**, 249–281.

BURGESS, P. M. & HOVIUS, N. 1998. Rates of delta progradation during highstands: Consequences for timing of deposition in deep-marine systems. *Journal of the Geological Society, London*, **155**, 217–222.

CARVAJAL, C., STEEL, R. & PETTER, A. 2009. Sediment supply: the main driver of shelf-margin growth. *Earth-Science Reviews*, **96**, 221–248.

CAYAN, D. R., REDMOND, K. T. & RIDDLE, L. G. 1999. ENSO and hydrologic extremes in the western United States. *Journal of Climate*, **12**, 2881–2893.

COVAULT, J. A. & ROMANS, B. W. 2009. Growth patterns of deep-sea fans revisited: Turbidite-system morphology in confined basins, examples from the California Borderland. *Marine Geology*, **265**, 51–66.

COVAULT, J. A., NORMARK, W. R., ROMANS, B. W. & GRAHAM, S. A. 2007. Highstand fans in the California Borderland: the overlooked deep-water depositional systems. *Geology*, **35**, 783–786.

COVAULT, J. A., ROMANS, B. W., FILDANI, A., MCGANN, M. & GRAHAM, S. A. 2010. Rapid climatic signal propagation from source to sink in a Southern California sediment-routing system. *The Journal of Geology*, **118**, 247–259.

COVAULT, J. A., FILDANI, A., ROMANS, B. W. & MCHARGUE, T. 2011*a*. The natural range of submarine canyon-and-channel longitudinal profiles. *Geosphere*, **7**, 313–332.

COVAULT, J. A., ROMANS, B. W., GRAHAM, S. A., FILDANI, A. & HILLEY, G. E. 2011*b*. Terrestrial source to deep-sea sink sediment budgets at high and low sea levels: insights from tectonically active Southern California. *Geology*, **39**, 619–622.

CROUCH, J. K. 1979. Neogene tectonic evolution of the California Borderland and western Transverse Ranges. *Geological Society of America Bulletin*, **90**, 338–345.

CROUCH, J. K. 1981. Northwest margin of the California Continental Borderland; marine geology and tectonic evolution. *American Association of Petroleum Geologists Bulletin*, **65**, 191–218.

CROWELL, J. C. 1979. The San Andreas fault system through time. *Journal of the Geological Society, London*, **136**, 293–302.

CROWELL, J. C. 1981. Juncture of San Andreas transform system and Gulf of California rift. *Oceanologica Acta, Suppl.*, **4**, 137–142.

DARTNELL, P., NORMARK, W. R., DRISCOLL, N. W., BABCOCK, J. M., GARDNER, J. V., KVITEK, R. G. & IAMPIETRO, P. J. 2007. *Multibeam Bathymetry and Selected Perspective Views Offshore San Diego, California*. United States Geological Survey, Scientific Investigations Map, **2959**.

DEMETS, C., GORDON, R. G., ARGUS, D. F. & STEIN, S. 1990. Current plate motions. *Geophysical Journal International*, **101**, 425–478.

EITTREIM, S. L. & NOBLE, M. 2002. Towards a sediment budget for the Santa Cruz shelf. *Marine Geology*, **181**, 235–248.

EMERY, K. O. & SHEPARD, F. P. 1945. Lithology of the sea floor off southern California. *Geological Society of America*, Bulletin, **56**, 431–478.

FARNSWORTH, K. L. & WARRICK, J. A. 2007. *Sources, Dispersal, and Fate of Fine Sediment Supplied to Coastal California*. United States Geological Survey, Scientific Investigations Report **2007-5254**.

FIELD, M. E. & EDWARDS, B. D. 1980. Slopes of the southern California Continental Borderland: A regime of mass transport. *In*: FIELD, M. E., BOUMA, A. H., COLBOURN, I. P., DOUGLAS, R. G. & INGLE, J. C. (eds) *Proceedings of the Quaternary Depositional Environments of the Pacific Coast: Pacific Coast Paleogeography Symposium No. 4*. Society for Sedimentary Geology, Pacific Section, Tulsa, OK, 169–184.

FILDANI, A. & NORMARK, W. R. 2004. Late Quaternary evolution of channel and lobe complexes of Monterey Fan. *Marine Geology*, **206**, 199–223.

GALY, V., BEYSSAC, O., FRANCE-LANORD, C. & EGLINTON, T. 2008. Recycling of graphite during Himalayan erosion: a geological stabilization of carbon in the crust. *Science*, **322**, 943–945.

GORSLINE, D. S., DRAKE, D. E. & BARNES, P. W. 1968. Holocene sedimentation in Tanner Basin, California continental borderland. *Geological Society of America Bulletin*, **79**, 659–674.

GRAHAM, S. A. & BACHMAN, S. B. 1983. Structural controls on submarine-fan geometry and internal architecture: Upper La Jolla fan system, offshore southern California. *American Association of Petroleum Geologists Bulletin*, **67**, 83–96.

GREENE, H. G., MAHER, N. M. & PAULL, C. K. 2002. Physiography of the Monterey Bay National Marine Sanctuary and implications about continental margin development. *Marine Geology*, **181**, 55–82.

HOGARTH, L. J., BABCOCK, J., DRISCOLL, N. W., LEDANTEC, N., HAAS, J. K., INMAN, D. L. & MASTERS, P. M. 2007. Long-term tectonic control on Holocene shelf sedimentation offshore La Jolla, California. *Geology*, **35**, 275–278.

INMAN, D. L. 2008. Highstand fans in the California borderland: Comment. *Geology*, Online Forum, e166, http://geology.gsapubs.org/content/36/1/e167.full.pdf

INMAN, D. L. & FRAUTSCHY, J. D. 1965. Littoral processes and the development of shorelines. *In*: *Coastal Engineering Proceedings, Special Conference, American Society of Civil Engineers*, Santa Barbara, CA. American Society of Civil Engineers, New York, 511–536.

INMAN, D. L. & JENKINS, S. A. 1999. Climate change and the episodicity of sediment flux of small California Rivers. *The Journal of Geology*, **107**, 251–270.

INMAN, D. L. & NORDSTROM, C. E. 1971. On the tectonic and morphologic classification of coasts. *The Journal of Geology*, **79**, 1–21.

INMAN, D. L., NORDSTROM, C. E. & FLICK, R. E. 1976. Currents in submarine canyons: an air–sea–land interaction. *Annual Reviews of Fluid Mechanics*, **8**, 275–310.

JERVEY, M. T. 1988. Quantitative geological modeling of siliciclastic rock sequences and their seismic expression. *In*: WILGUS, C. K., HASTINGS, B. S., POSAMENTIER, H., VAN WAGONER, J., ROSS, C. A. & KENDALL, C. G. St. C. (eds) *Sea-Level Changes*. SEPM Special Publications, Tulsa, **42**, 47–70.

KERN, J. P. & ROCKWELL, T. K. 1992. Chronology and deformation of Quaternary marine shorelines, San Diego County, California. *In*: FLETCHER, C. H., III & WEHMILLER, J. F. (eds) *Quaternary Coasts of the United States*. SEPM Special Publications, Tulsa, **48**, 377–382.

KIRBY, M. E., POULSEN, S. P., LUND, S. P., PATTERSON, W. P., REIDY, L. & HAMMOND, D. E. 2004. Late Holocene lake-level dynamics inferred from magnetic susceptibility and stable oxygen isotope data: Lake Elsinore, Southern California. *Journal of Paleolimnology*, **31**, 275–293.

KIRBY, M. E., LUND, S. P. & POULSEN, C. J. 2005. Hydrologic variability and the onset of modern El Niño-Southern Oscillation: a 19 250-year record from Lake Elsinore, southern California. *Journal of Quaternary Science*, **20**, 239–254.

KIRBY, M. E., LUND, S. P., ANDERSON, M. A. & BIRD, B. W. 2007. Insolation forcing of Holocene climate change in Southern California: a sediment study from Lake Elsinore. *Journal of Paleolimnology*, **38**, 395–417.

LAMBECK, K. & CHAPPELL, J. 2001. Sea level change through the last glacial cycle. *Science*, **292**, 679–686.

LEE, H. J. & NORMARK, W. R. 2009. *Earth Science in the Urban Ocean: the Southern California Continental Borderland*. Geological Society of America Special Papers, **454**.

LEGG, M. R. & KENNEDY, M. P. 1979. Faulting offshore San Diego and northern Baja, California. *In*: ABBOTT, P. L. & ELLIOTT, W. J. (eds) *Earthquakes and Other Perils, San Diego Region*. San Diego Association of Geologists, San Diego, CA, 29–46.

MASTERS, P. M. 2006. Holocene sand beaches of southern California: ENSO forcing and coastal processes on millennial scales. *Palaeogeography, Palaeoclimatology, Palaeoecology*, **232**, 73–95.

MEADE, R. H. & PARKER, R. S. 1985. *Sediment in Rivers of the United States*. United States Geological Survey, Water-Supply Papers, **2275**, 49–60.

MENSING, S. A. 2001. Late-glacial and early Holocene vegetation and climate change near Owens Lake, eastern California. *Quaternary Research*, **55**, 57–65.

MILLIMAN, J. D. & SYVITSKI, J. P. M. 1992. Geomorphic/tectonic control of sediment discharge to the ocean: the importance of small mountainous rivers. *The Journal of Geology*, **100**, 525–544.

MOORE, D. G. 1969. *Reflection Profiling Studies of the California Continental Borderland: Structure and Quaternary Turbidite Basins*. Geological Society of America Special Papers, **107**.

MULDER, T. & SYVITSKI, J. P. M. 1995. Turbidity currents generated at river mouths during exceptional discharges to the world oceans. *The Journal of Geology*, **103**, 285–299.

NARDIN, T. R. 1983. Late Quaternary depositional systems and sea level change – Santa Monica and San Pedro Basins, California continental borderland. *AAPG Bulletin*, **67**, 1104–1124.

NARDIN, T. R., OSBORNE, R. H., BOTTJER, D. J. & SCHEIDEMANN, R. C. 1981. Holocene sea-level curves for Santa Monica Shelf, California Continental Borderland. *Science*, **213**, 331–333.

NATIONAL OCEAN SERVICE. 2010. *Tide Data Online*. http://tidesandcurrents.noaa.gov/

NORMARK, W. R. 1970. Growth patterns of deep-sea fans. *American Association of Petroleum Geologists Bulletin*, **54**, 2170–2195.

NORMARK, W. R. & REID, J. A. 2003. Extensive deposits on the Pacific Plate from Late Pleistocene North American glacial lake outbursts. *The Journal of Geology*, **111**, 617–637.

NORMARK, W. R., PIPER, D. J. W., ROMANS, B. W., COVAULT, J. A., DARTNELL, P. & SLITER, R. 2009. Submarine canyon and fan systems of the California Continental Borderland. *In*: LEE, H. J. & NORMARK, W. R. (eds) *Earth Science in the Urban Ocean: The Southern California Continental Borderland*. Geological Society of America Special Papers, **454**, 141–168.

PATSCH, K. & GRIGGS, G. 2006. *Littoral Cells, Sand Budgets, and Beaches: Understanding California's Shoreline*. California Coastal Sediment Management WorkGroup, Institute of Marine Sciences, University of California, Santa Cruz, CA.

PATSCH, K. & GRIGGS, G. 2007. *Development of Sand Budgets for California's Major Littoral Cells*. California Coastal Sediment Management WorkGroup, Institute of Marine Sciences, University of California, Santa Cruz, CA.

PAULL, C. K., MITTS, P., USSLER, W., KEATEN, R. & GREENE, H. G. 2005. Trail of sand in upper Monterey Canyon: offshore California. *Geological Society of America Bulletin*, **117**, 1134–1145.

PHILANDER, S. G. 1990. *El Niño, La Niña, and the Southern Oscillation*. Academic Press, San Diego, CA.

PIECHOTA, T. C., DRACUP, J. A. & FOVELL, R. G. 1997. Western US streamflow and atmospheric circulation patterns during El Niño-Southern Oscillation. *Journal of Hydrology*, **201**, 249–271.

PIPER, D. J. W. 1970. Transport and deposition of Holocene sediment on La Jolla deep sea fan, California. *Marine Geology*, **8**, 211–227.

PIPER, D. J. W. & NORMARK, W. R. 2001. Sandy fans-From Amazon to Hueneme and beyond. *American Association of Petroleum Geologists Bulletin*, **85**, 1407–1438.

PIPER, D. J. W. & NORMARK, W. R. 2009. Processes that initiate turbidity currents and their influence on turbidites: a marine geology perspective. *Journal of Sedimentary Research*, **79**, 347–362.

POSAMENTIER, H. W. & KOLLA, V. 2003. Seismic geomorphology and stratigraphy of depositional elements in deep-water settings. *Journal of Sedimentary Research*, **73**, 367–388.

ROMANS, B. W., NORMARK, W. R., McGANN, M. M., COVAULT, J. A. & GRAHAM, S. A. 2009. Coarse-grained sediment delivery and distribution in the Holocene Santa Monica Basin, California: implications for evaluating source-to-sink flux at millennial time scales. *Geological Society of America Bulletin*, **121**, 1394–1408.

RYAN, H. F., LEGG, M. R., CONRAD, J. E. & SLITER, R. W. 2009*a*. Recent faulting in the Gulf of Santa Catalina: San Diego to Dana Point. *In*: LEE, H. J. & NORMARK, W. R. (eds) *Earth Science in the Urban Ocean: The Southern California Continental Borderland*. Geological Society of America Special Papers, **454**, 291–315.

RYAN, W. B. F., CARBOTTEE, S. M. ET AL. 2009*b*. Global multi-resolution topography synthesis. *Geochemistry Geophysics Geosystems*, **10**, Q03014.

SHEPARD, F. P. 1963. *Submarine Geology*. Harper & Row, New York.

SHEPARD, F. P. & BUFFINGTON, E. C. 1968. La Jolla submarine fan-valley. *Marine Geology*, **6**, 107–143.

SHEPARD, F. P. & EMERY, K. O. 1941. *Submarine Topography off the California Coast; Canyon and Tectonic Interpretation*. Geological Society of America Special Papers, **31**.

SLATER, R. A., GORSLINE, D. S., KOLPACK, R. L. & SHILLER, G. I. 2002. Post-glacial sediments of the Californian shelf from Cape San Martin to the US–Mexico border. *Quaternary International*, **92**, 45–61.

SOMMERFIELD, C. K. & NITTROUER, C. A. 1999. Modern accumulation rates and a sediment budget for the Eel shelf: a flood-dominated depositional environment. *Marine Geology*, **154**, 227–241.

SOMMERFIELD, C. K., LEE, H. J. & NORMARK, W. R. 2009. Postglacial sedimentary record of the Southern California continental shelf and slope, Point Conception to Dana Point. *In*: LEE, H. J. & NORMARK, W. R. (eds) *Earth Science in the Urban Ocean: The Southern California Continental Borderland*. Geological Society of America Special Papers, **454**, 89–115.

STORLAZZI, C. D. & GRIGGS, G. B. 2000. Influence of El Niño-Southern Oscillation (ENSO) events on the evolution of central California's shoreline. *Geological Society of America Bulletin*, **112**, 236–249.

SYVITSKI, J. P. M., VÖRÖSMARTY, C. J., KETTNER, A. J. & GREEN, P. 2005. Impact of humans on the flux of terrestrial sediment to the global coastal ocean. *Science*, **308**, 376–380.

TENG, L. T. & GORSLINE, D. S. 1991. Stratigraphic framework of the Continental Borderland basins, southern California. *In*: DAUPHIN, J. P. & SIMOEIT, B. R. T. (eds) *Gulf and Peninsular Province of the Californias*. American Association of Petroleum Geologists, Memoirs, **47**, 127–143.

UNITED STATES ARMY CORPS OF ENGINEERS 1985. *Geomorphology Framework Report, Monterey Bay: Coast of California Storm and Tidal Waves Study*. Los Angeles District Reference CCSTWS, **85-2**.

VAIL, P. R., MITCHUM, R. M., JR & THOMPSON, S. 1977. Seismic stratigraphy and global changes of sea level, Part 4, Global cycles of relative changes of sea level. *In*: PAYTON, C. E. (ed.) *Seismic Stratgraphy; Applications to Hydrocarbon Exploration*. American Association of Petroleum Geologists, Memoirs, **26**, 83–97.

VEDDER, J. G. 1987. Regional geology and petroleum potential of the southern California Borderland. *In*: SCHOLL, D. W., GRANTZ, A. & VEDDER, J. G. (eds) *Geology and Resource Potential of the Continental Margin of Western North America and Adjacent Ocean Basins – Beaufort Sea to Baja California*. Circum-Pacific Council for Energy and Mineral Resources, Earth Science Series, **6**, 403–447.

WARRICK, J. A. & FARNSWORTH, K. L. 2009a. Sources of sediment to the coastal waters of the Southern California Bight. *In*: LEE, H. J. & NORMARK, W. R. (eds) *Earth Science in the Urban Ocean: The Southern California Continental Borderland*. Geological Society of America Special Papers, **454**, 39–52.

WARRICK, J. A. & FARNSWORTH, K. L. 2009b. Dispersal of river sediment in the Southern California Bight. *In*: LEE, H. J. & NORMARK, W. R. (eds) *Earth Science in the Urban Ocean: The Southern California Continental Borderland*. Geological Society of America Special Papers, **454**, 53–67.

WELLS, P. V. & BERGER, R. 1967. Late Pleistocene history of coniferous woodland in the Mohave Desert. *Science*, **155**, 1640–1647.

WELLS, S. G., BROWN, W. J., ENZEL, Y., ANDERSON, R. Y. & McFADDEN, L. D. 2003. Late Quaternary geology and paleohydrology of pluvial Lake Mojave, Southern California. *In*: ENZEL, Y., WELLS, S. G. & LANCASTER, N. (eds) *Paleoenvironments and Paleohydrology of the Mojave and Southern Great Basin Deserts*. Geological Society of America Special Papers, **368**, 79–114.

WIBERG, P. L., DRAKE, D. E., HARRIS, C. K. & NOBLE, M. A. 2002. Sediment transport on the Palos Verdes shelf over seasonal to decadal time scales. *Continental Shelf Research*, **22**, 987–1004.

WISE, D. R., RINELLA, F. A. ET AL. 2007. *Nutrient and Suspended-Sediment Transport and Trends in the Columbia River and Puget Sound Basins, 1993–2003*. United States Geological Survey, Scientific Investigations Report, **2007–5186**.

WRIGHT, L. D. & FRIEDRICHS, C. T. 2006. Gravity-driven sediment transport on continental shelves: a status report. *Continental Shelf Research*, **26**, 2092–2107.

WRIGHT, L. D. & NITTROUER, C. A. 1995. Dispersal of river sediments in coastal seas: six contrasting cases. *Estuaries*, **18**, 494–508.

XU, J. P. & NOBLE, M. A. 2009. Variability of the Southern California wave climate and implications for sediment transport. *In*: LEE, H. J. & NORMARK, W. R. (eds) *Earth Science in the Urban Ocean: The Southern California Continental Borderland*. Geological Society of America Special Papers, **454**, 171–191.

YEATS, R. S., HAQ, B. U. ET AL. 1981. *Initial Reports of the Deep Sea Drilling Project*, **63**, US Government Printing Office, Washington, DC.

YOUNG, A. P. & ASHFORD, S. A. 2006. Application of airborne LIDAR for seacliff volumetric change and beach-sediment budget contributions. *Journal of Coastal Research*, **22**, 307–318.

ZUFFA, G. G., NORMARK, W. R., SERRA, F. & BRUNNER, C. A. 2000. Turbidite megabeds in an oceanic rift valley recording Jökulhlaups of Late Pleistocene glacial lakes of the western United States. *The Journal of Geology*, **108**, 253–274.

Chapter 24

Geometry of the coastline and morphology of the convergent continental margin of Ecuador

JEAN FRANÇOIS DUMONT[1,2]*, ESSY SANTANA[3], MARIE-AUDE BONNARDOT[4,5], NELSON PAZMIÑO[6], KEVIN PEDOJA[7] & BRUNO SCALABRINO[4]

[1]*ESPOL, Campus La Prosperina, Guayaquil, Ecuador*

[2]*IRD Geosciences Azur, UMR 6526, 06235 Villefranche sur mer, France*

[3]*ESPOL, Campus La Prosperina, Guayaquil, Ecuador*

[4]*Géosciences Azur, UMR-CNRS-IRD 6526, Université de Nice-Géoazur, parc Valrose,*
06108 Nice Cedex 02, France

[5]*Present address: FrOG Tech Pty Ltd, Deakin West, Canberra, ACT 2600, Australia*

[6]*INOCAR, Instituto Oceanográfico de la Armada, Base Naval Sur, Av. de la Marina, Guayaquil, Ecuador*

[7]*UMR CNRS 6143 'Morphodynamique Continentale et Côtière' (M2C), Université de Caen, 2–4 rue des Tilleuls,*
14000 Caen, France

Corresponding author (e-mail: dumontjfr@gmail.com)

Abstract: The pacific border of the South American plate presents a more or less symmetrical sinuosity, with a central concave curvature (the Arica Angle located between two side rays along Chile and Peru) and ending in convex arcs (the Patagonian and Talara arcs, respectively). The width of the continental and coastal margins varies significantly according to the geometry of the border. The continental margin of Ecuador corresponds to the northern part of the Talara Arc. Three different segments showing different coastal geomorphology and continental platform characteristics are identified from north to south: the northern segment (Mataje River–Galera Point) shows a wide continental shelf and slope, the upper subducted slab of the subduction plane presents a low dip; the central segment (Galera Point–Santa Elena) stands in front of the Carnegie Ridge, and presents a moderate uplift in the Manta Peninsula, in front of the Carnegie Ridge, and the upper subduction plane is subhorizontal; the southern segment includes the side and inner coasts of the Gulf of Guayaquil, below the gulf the subduction plane shows a low dip. A comparison with published 3D numerical modelling of curved subduction suggests that the geometry of the continental boundary has a significant effect on uplift or subsidence along the continental margin. Also, the subduction of asperities in the trench, such as the Carnegie Ridge, may change the coastal motion from subsiding to uplifting, as is observed in the Esmeraldas area. There is no clear evidence of a shelf developed during the Last Glacial Maximum (LGM) sea-level lowstand, probably due to the vertical motion – uplift or subsidence – observed all along the coastal margin.

The active margin between the South American continent and the Pacific Ocean is among the most continuous on Earth, and is associated with uplift of the Andean Cordillera. This active margin provides the basis for the Chilean-type active margin as opposed to the Marianas-type observed in the western Pacific. The Chilean-type active margin is characterized by a relatively young slab with a low dip subduction angle and a cordillera developing along the subduction zone in the upper plate (Lallemand 1999). The South American subduction system is continuous but made up of 100 km-scale curvature that forms, from north to south, the convex Talara Arc (1°N–6°S), the Arica concave angle (17–21°S) and the convex Patagonian Arc (45–55°S), with relatively straight segments in between them (Fig. 24.1). The main structures of the subducting oceanic plate are the aseismic Carnegie, Nazca and Juan Fernandez ridges (Macharé & Ortlieb 1992; Gutscher *et al.* 1999).

The Ecuadorian continental margin includes the northern half of the Talara Arc, between the bay of Ancón de Sardinas to the north and the Gulf of Guayaquil to the south (Figs 24.1 & 24.2). The Nazca plate carries the Carnegie Ridge, which is subducting in front of the central part of the Ecuadorian margin (Fig. 24.2). North of the Gulf of Guayaquil, the Ecuadorian coastal area belongs to the North Andean Block (Ego *et al.* 1996), a slice of oceanic and continental basement moving northwards between the subduction zone to the west and a complex deformation zone running from the Gulf of Guayaquil to Caracas

to the east. This deformation zone is frequently referred to as the Dolores–Guayaquil Megashear, which, together with the Romeral Fault, defines the suture zone of the accreted oceanic basement against the Andean Cordillera (Case *et al.* 1973; Campbell 1974; Feininger & Seguin 1983). Megard (1987) used the same name for the Early Tertiary sealed suture and the late tertiary reactivated segments of the Romeral–Dolores Megashear, one of these segments turning outside the suture zone towards the Gulf of Guayaquil. Because the present and active deformation zone at the eastern border of the Andean Block is independent of the Romeral–Dolores Megashear, we will define it as the Guayaquil–Caracas Fault Zone that follows and reactivates different segments of strike-slip and reverse faults (Ego *et al.* 1996; Dumont *et al.* 2005*a*).

Based on trends and structures observed along the Ecuadorian coast, three segments are identified: (1) a northern segment from the Mataje River estuary (Colombia–Ecuador boundary) to the Galera Point; (2) a central segment between the Galera and Santa Elena points; and (3) the Gulf of Guayaquil to the south, a large re-entrant that includes different coastal morphologies. The structure and the Quaternary evolution of the Ecuadorian margin is documented for the forearc basins (Deniaud 2000), the evolution of the Gulf of Guayaquil (Deniaud *et al.* 1999; Dumont *et al.* 2005*a, b*; Witt *et al.* 2006), coastal evolution during the Quaternary, including uplifted marine terraces (Pedoja *et al.* 2006*a, b*), erosion and active tectonic processes (Santana *et al.* 2001*a, b*;

From: CHIOCCI, F. L. & CHIVAS, A. R. (eds) 2014. *Continental Shelves of the World: Their Evolution During the Last Glacio-Eustatic Cycle.*
Geological Society, London, Memoirs, **41**, 327–338. http://dx.doi.org/10.1144/M41.24

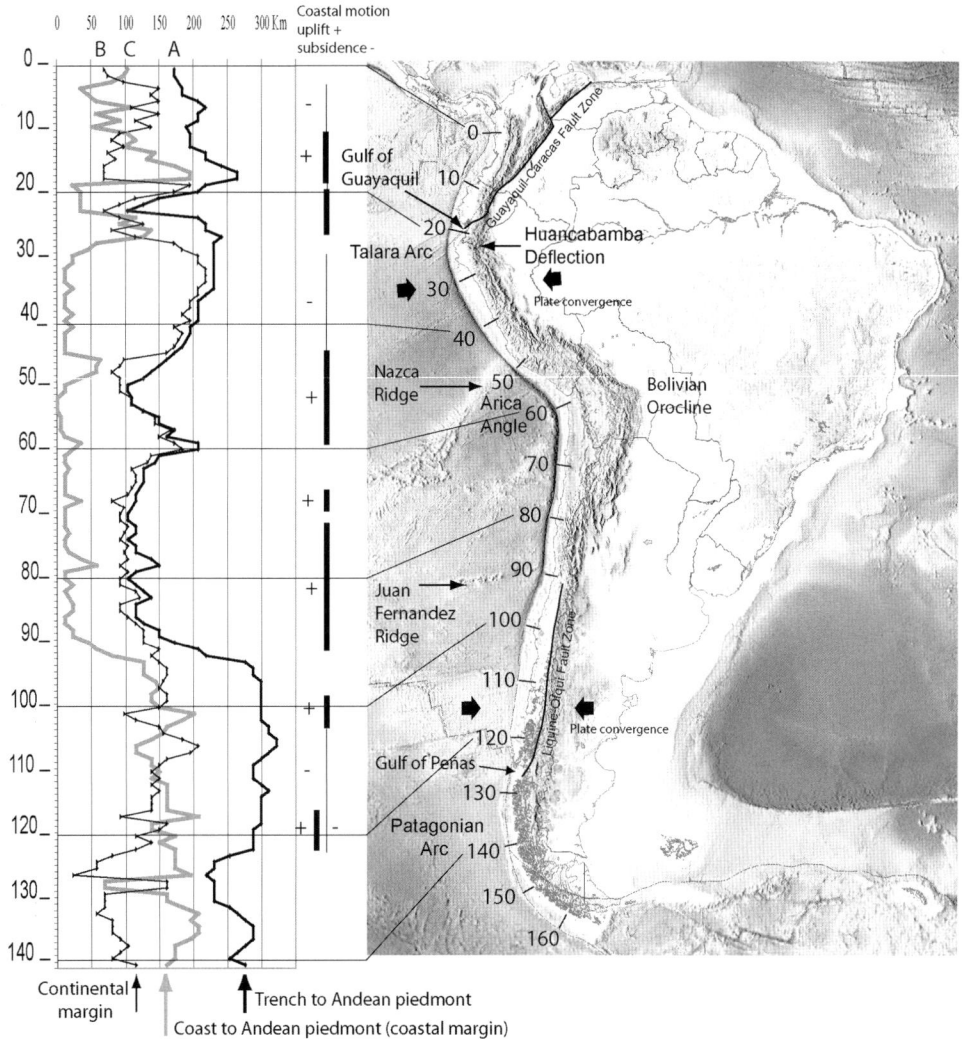

Fig. 24.1. (Right) General view of the South American active margin from a digital model (USGS 2006). The scale along the coast marks the identification points (ID) where the respective widths of (A) the trench to piedmont, (B) the continental margin and (C) coast to piedmont margins have been measured and shown on the scheme to the left. There is an ID point every 60 km. Modified from USGS (2006).

Dumont *et al.* 2006), and the bathymetry and structure of the continental margin (Collot *et al.* 2002, 2004).

The Ecuadorian margin in the framework of the South American active margin

We analysed the variation of trench–coast, coast–piedmont and trench–piedmont distances. Measurements were made at 60 km intervals on the United States Geological Survey digital map (USGS 2006), with a 5 km resolution, in order to determine the main variations of the continental margin relative to the main geometry of the South American active margin (Fig. 24.1, lines A–C). The trench–Andean piedmont distance (Fig. 24.1, line A) outlines the main segmentation of the active margin. This distance varies from about 100 to 320 km, wider along the Talara and Patagonian arcs, and narrower along the Arica Angle. The width of the continental margin (Fig. 24.1, line B) varies from 70 to 200 km, and is close to the trench–piedmont distance all along the Arica Angle and its branches: that is, from the Talara Arc to the north to the Juan Fernandez Ridge to the south. Along the Talara and Patagonian arcs, the width of the coastal margin increases significantly (Fig. 24.1, line C) but this widening, in fact, reflects different situations. A true wide coastal margin appears in central Ecuador, including the forearc ridge (the Coastal Cordillera) and basin (the Guayas Basin). However, the Patagonian Arc presents an outer margin of islands and rias due to the regional pattern of extension faulting and morphological depressions (Diraison *et al.* 1997; Lagabrielle *et al.* 2004; Scalabrino *et al.* 2010). This particular

pattern is related to the subduction of the South Chile Spreading Ridge, and the slab windows that have formed below the Patagonian Andes since about 6 Ma (Lagabrielle *et al.* 2004).

Important short wavelength variations of up to about 10 km are observed in relation to local structures. Along the arcs we observe a large trench–piedmont distance with roughly similar continental margin and coastland widths. The most important variations in the trench–piedmont distances are related to re-entrants (bays) located at the junction of major continental faults with the trench, such as the Gulf of Guayaquil to the north, located at the apex of the Talara Arc and opening tectonically at the junction of the Guayaquil–Caracas Fault Zone with the trench (Deniaud *et al.* 1999; Dumont *et al.* 2005*b*), and the Liquine–Ofqui Fault Zone along the Gulf of Peñas in the Patagonian Arc (Cembrano *et al.* 1996). Both are strike-slip fault zones joining obliquely to the trench, and determining a local offset of the piedmont and trench lines (Villanueva i Bohigas 1997) (Fig. 24.1B, points 20 and 126–127).

Along the Arica Angle and its branches – that is, from the end of the Talara Arc to the north (transition from uplift to stable coast) to the Juan Fernandez Ridge to the south – the coastland is very narrow (Fig. 24.1, line C). Along this segment, the continental margin and the trench–piedmont width are about the same because of the very narrow coastland. However, the width of the continental margin seems to decrease globally from north to south, and we observe, in the same direction, that the age of the subducting Nazca plate increases from around 30 to 50 Ma (Addicott *et al.* 1982). Besides this, two important anomalies are observed: one in front of the subducting Nazca Ridge, and the other in front of the inner part of the Arica Angle. In front of the

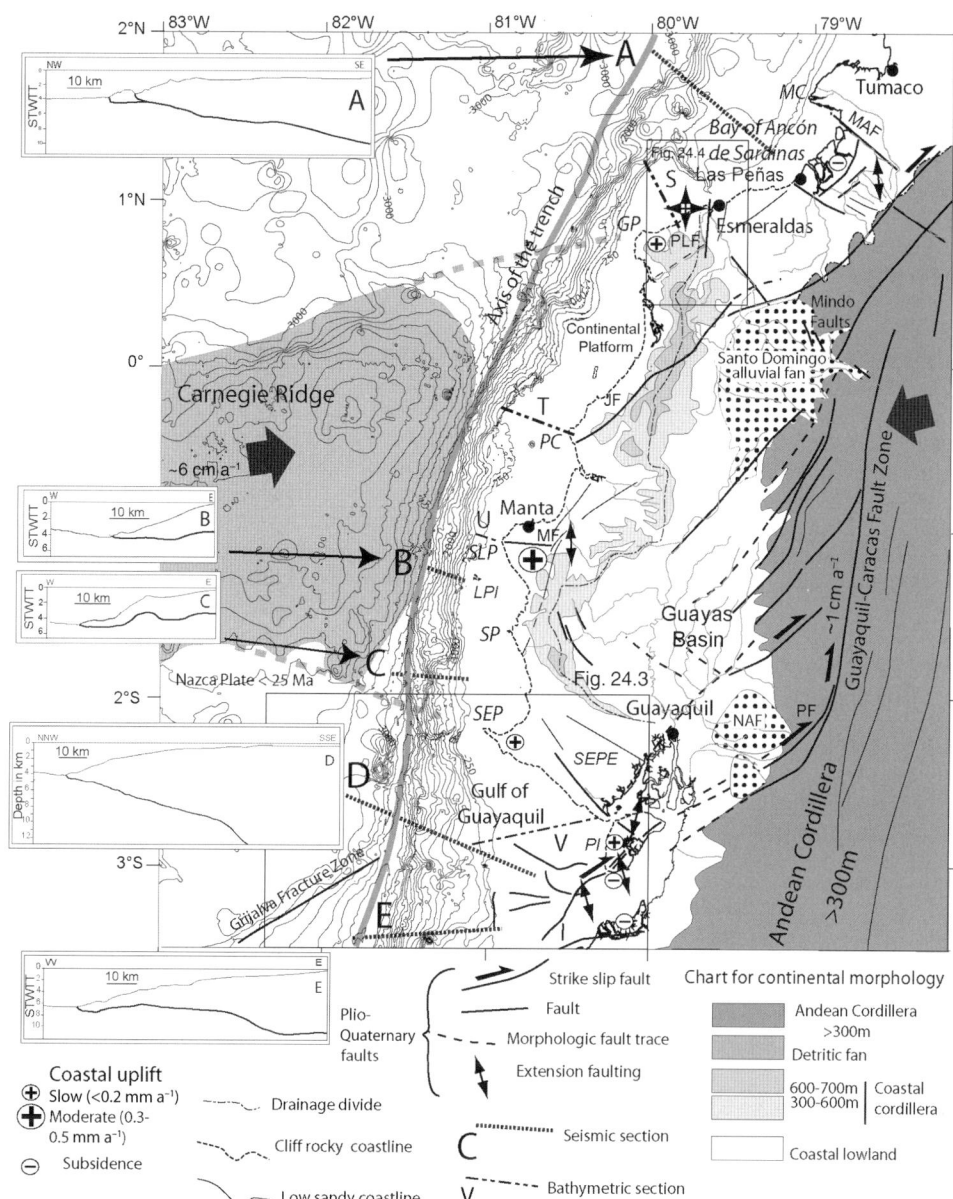

Fig. 24.2. Structural scheme of the active margin of Ecuador. Place names: *MC*, Manglare Point; *GP*, Galera Point; *PC*, Punta Canoa; *SLP*, San Lorenzo Point; *SP*, Salango Point; *SEP*, Santa Elena Point; *SEPE*, Santa Elena Peninsula; *LPI*, La Plata Island; *PI*, Puna Island. Main faults: PLF, Puerto Libre Fault; JF, Jama Fault; MF, Manta Fault; PF, Pallatanga Fault. NAF, Naranjal Alluvial Fan. A–E: positions and simplified profiles of the subduction zone. STWTT, two-way-travel time in seconds. See Figure 24.3 for details. Sections S, T, U and V refer to the profiles in Figure 24.6. Bathymetric chart simplified from Michaud *et al.* (2006).

Nazca Ridge, the width of the coastland increases and the width of the continental margin decreases. The narrowing of the continental margin extends south of the Nazca Ridge in the direction of the migration of the ridge (Macharé & Ortlieb 1992). In the corner of the Arica Angle, the width of the continental margin increases, whereas the coastland remains relatively narrow (Fig. 24.1, line B). This space can be explained by poor geometrical accommodation by the slab of the curvature of the trench along the sharp corner of the Bolivian Orocline.

Geological background

The Talara Arc

The 'Talara Arc' defines the approximately 1000 km-long convex margin extending from northern Peru to northern Ecuador, and includes the trench, the continental shelf, the coastal margin and the adjoining Andean Cordillera (Pedoja *et al.* 2006*b*). The word Talara refers to the geometry and is definitively not related to the notion of a forearc. The structural framework of the Talara Arc results initially from the right angle between the old Tethyian ridge extending across the proto-Caribbean area towards the NW

border of the South American continent and the Late Jurassic–Cretaceous South American subduction zone to the south (Jaillard *et al.* 1990).

The Amotape–Tahuin continental block was accreted against the southern part of the Talara Arc during the Neocomian (Mourier *et al.* 1988) and the Celica oceanic blocks against the north Talara Arc during the Eocene (Jaillard *et al.* 1990, 1995). Post-Oligocene rotations occurred (Mitouard *et al.* 1990, 1992) clockwise and counterclockwise north and south of the Huancabamba deflection, respectively. The limited counterclockwise rotation observed along the south Talara Arc involves the entire Peruvian margin in relation to the uplift and increasing shortening observed SE, towards the Bolivian Orocline. The clockwise rotation observed in the north Talara Arc is interpreted as block rotation in a distributed dextral shear (Mitouard *et al.* 1990). Later, during the Neogene, the opening of the Gulf of Guayaquil along the Guayaquil–Caracas Fault Zone (Deniaud *et al.* 1999; Dumont *et al.* 2005*b*; Winter & Lavenu 1989) reactivated locally the previous suture of the accreted oceanic basement, such as the Romeral Fault (Megard 1987). According to these data, the present shape of the Talara Arc post-dates block accretion and rotation (Oligocene, about 23 Ma), and predates the shortening and rising of the Andean Cordillera of Ecuador (upper Miocene, about 10 Ma) (Steinmann *et al.* 1999).

The Gulf of Guayaquil

The Gulf of Guayaquil is a tectonic structure that opened at the southern tip of the North Andean Block (Winter & Lavenu 1989; Kellogg & Vega 1995; Ego *et al.* 1996) (Fig. 24.1). In front of the Gulf of Guayaquil, the axis of the trench presents a dextral offset of about 60 km relative to the main curve of the Talara Arc (Pedoja 2003). The piedmont of the Andean range presents a similar offset along the Pallatanga Fault (Pedoja 2003; Dumont *et al.* 2005*b*). This is interpreted as resulting from the motion of the North Andean Block since the onset of the uplift of the Andean Cordillera at approximately 9–10 Ma (Steinmann *et al.* 1999). From the Gulf of Guayaquil to the NE, the motion of the North Andean Block is accommodated along the Guayaquil–Caracas Fault Zone. The Gulf of Guayaquil opens on the structural boundary between the continental Amotape–Tahuin Block to the south and the Piñon oceanic block to the north (Calahorrano 2005). However, even on land, it may be difficult to identify clearly the old suture and the new faults (Megard 1987).

Global positioning system (GPS) data suggest a mean offset rate of the North Andean Block of about 10 mm a^{-1} (Kellogg & Vega 1995). Geological and geomorphological data give an offset rate of 4 mm a^{-1} along a segment of the Pallatanga Fault (Fig. 24.2) across the western Andean Cordillera (Winter & Lavenu 1989) and up to 8 mm a^{-1}, including a fault segment, parallel to the Pallatanga Fault to the north (Trenkamp *et al.* 2002). This is consistent with the 6–8 mm a^{-1} of the Zambapala Fault (Fig. 24.3) in the NE part of the Gulf of Guayaquil (Dumont *et al.* 2005*a*). The Banco Peru (Fig. 24.3) located near the southern coast of the Gulf of Guayaquil represents probably the southernmost basin related to the early opening of the Gulf of Guayaquil during the Neogene (Calahorrano 2005).

Structure of the Ecuadorian continental margin

The backbone of the coastal margin of Ecuador includes lower Cretaceous basalts (Piñon Formation), covered by upper Cretaceous and Paleocene volcanic sediments and chert (Cayo Formation) (Bristow & Hoffstetter 1977; Baldock 1982). Forearc basins were filled with marine sediments during the periods of fast convergence (*c.* 200 mm a^{-1} between 48 and 37 Ma), and erosion or continental detritus are registered during periods of slow convergence (*c.* 44 mm a^{-1} between 37 and 20 Ma) (Daly 1989). These basins are located along NE–SW- to NNE–SSW-trending dextral transtension structures or strike-slip fault structures with an en échelon pattern related to the main deformation of the North Andean Block (Daly 1989; E. Jaillard pers. comm. 2006).

The morphology and drainage pattern of the coastal margin of Ecuador (Fig. 24.2) is closely related to the main neotectonic deformations. The three main drainage outlets are the Guayas, Esmeraldas and Cayapas rivers, the Guayas River being the largest river on the Pacific coast of South America. All of the downstream river sections and estuaries are controlled by Plio-Quaternary fault tectonics. The Andean drainage is directed toward the Guayas River to the south and the Esmeraldas River to the north, and there is no drainage transit from the Andes to the coast along the 600 km coastal segment of central Ecuador, which also corresponds to the area uplifted during the Quaternary coastal event. The Guayas River estuary is controlled by tectonic subsidence along the southern segment of the Guayaquil–Caracas Fault Zone (Deniaud *et al.* 1999; Dumont *et al.* 2005*b*). The drainage is gathered along the Andean piedmont in the Guayas Basin before reaching the estuary.

The Guayaquil–Caracas Fault Zone is intersected from Quito to the north by NW–SE-trending lineaments. This structural change matches the volcano-tectonic Pastaza–Esmeraldas Boundary of Hall & Wood (1985). From the western Andean Cordillera to the trench, this structural boundary determines successively the Mindo

Graben, the Santo Domingo and San Miguel alluvial fans, the Esmeraldas River valley (Alvarado 1998), and the Esmeraldas Canyon off the estuary (Collot *et al.* 2005). North of Quito, the NW–SE-trending Mira lineament joins the coast in the Bay of Ancón de Sardinas, with combined NW–SE- and NNE–SSW-trending transtension faults (CODIGEM 1993) determining northward and southward river deviations in the coastal margin (Dumont *et al.* 2006).

Plate motion

The convergence between the Nazca and the South American plates trends nearly east–west, with a mean rate of 70–80 mm a^{-1} according to the NUVEL 1A plate kinematic model (De Mets *et al.* 1989) and 50–70 mm a^{-1} from GPS measurements (Kellogg & Vega 1995; Trenkamp *et al.* 2002). The difference suggests a decrease in plate convergence during recent times (Kendrick *et al.* 2003). The break-up of the Nazca plate at about 25 Ma resulted in a plate younger than 25 Ma to the north of the Grijalva Fracture Zone and a plate older than 32 Ma to the south (Figs 24.2 & 24.3) (Hey 1977; Gutscher *et al.* 1999). The Grijalva Fracture Zone (GFZ) trends N60°E and presents a 3 km-wide morphological ridge with a 700 m-high scarp facing NW. North of the GFZ, the young Nazca plate carries the Carnegie Ridge, which is subducting beneath the South American plate at a constant dip of 25–35° down to a depth of about 200 km (Guillier *et al.* 2001).

The Carnegie Ridge

The Carnegie Ridge is an aseismic volcanic plateau formed during the eastward motion of the Nazca plate over the Galapagos hotspot (Hey 1977; Lonsdale 1978). The ridge extends west–east over more than 1000 km from the Galapagos Islands into the subduction zone of central Ecuador. The ridge has a mean width of 200 km, varying from a few tens of kilometres between the eastern part of the Carnegie Ridge and the Galapagos Islands to 280 km in front of central Ecuador, where diverging borders trend N60°E to the north and west–east to the south (Fig. 24.2). The eastern Carnegie Ridge is a massive plateau at a depth of 1000–1500 m, overlooking the surrounding seafloor at a depth of 4000 m to the south and 2500–3000 m to the north (Fig. 24.2). The subduction of the Carnegie Ridge lifted the bottom of the trench by about 2000 m relative to the depth of the trench in northern Peru (Collot *et al.* 2002) and generated an excess of coastal uplift of 0.2–0.3 mm a^{-1} relative to the southern part of the Talara Arc (Pedoja *et al.* 2006*b*).

The NE-trending Grijalva Fracture Zone determines an oblique front of the Carnegie Ridge in regard to the trend of the trench. This setting suggests a correlation between the onset of subduction of the Carnegie Ridge and Grijalva Scarp beneath central Ecuador during the late Pliocene–early Quaternary and a shortening event registered in the late Miocene–early Pliocene deposits in the Esmeraldas area, and late Pliocene–middle Pleistocene in the Gulf of Guayaquil (Dumont *et al.* 2005*b*). Evidence of faulting and tectonic activity during the upper Pleistocene and Holocene is observed in the form of north–south- and west–east-trending extension in northern Ecuador (Santana *et al.* 2001*b*; Dumont *et al.* 2006) (Fig. 24.2), and approximately north–south extension in front of the Carnegie Ridge (Pedoja *et al.* 2006*a*) and in the Gulf of Guayaquil (Dumont *et al.* 2005*b*).

Description of the Ecuadorian continental margin

The Ecuadorian coast has a length of about 950 km (without the islands), extending from the estuary of the Mataje River to the north at the border with Colombia, and Boca de Capones to

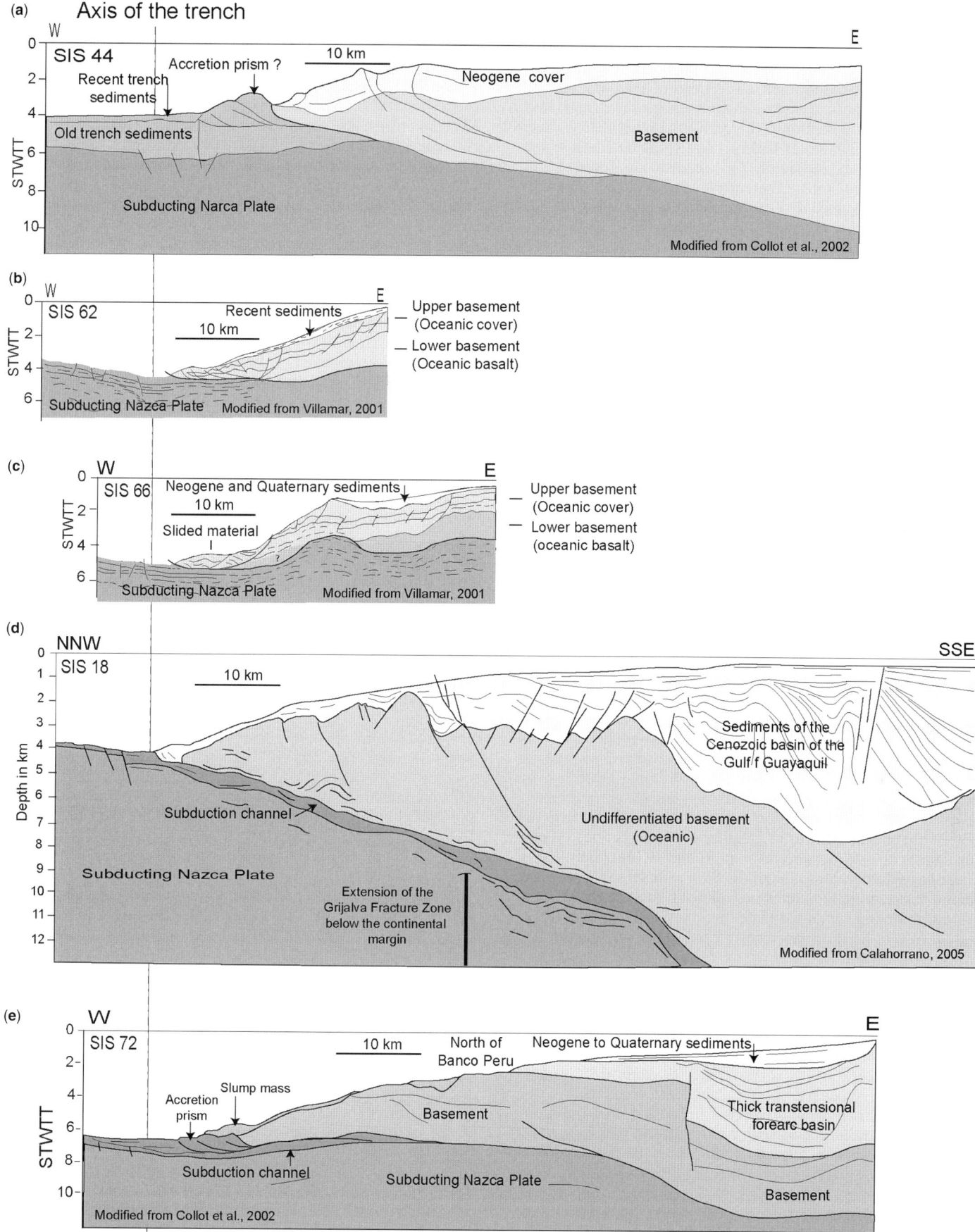

Fig. 24.3. Simplified seismic profiles showing the main structure of the continental margin. See the location in Figure 24.2. STWTT: two way travel time in seconds. The vertical exaggeration is approximately 3.

the south at the border with Peru (De Miro *et al.* 1976). The main trend of the coast of central and northern Ecuador is oblique to the Andean Ranges. The width of the coastal area varies from about 60 km to the north to about 180 km to the south (i.e. measured north of the Gulf of Guayaquil). The trend of the coastline and the structure of the continental margin define three segments, which from north to south are (Fig. 24.2): (1) the Mataje River estuary to Galera Point; (2) Galera Point to Santa Elena Point; and (3) the Gulf of Guayaquil. This segmentation corresponds roughly to the main structural elements: the two north segments are, respectively, located to the north and in front of the Carnegie Ridge, and the southern segment corresponds to the Gulf of Guayaquil at the apex of the Talara Arc.

Northern Ecuador: Mataje River estuary–Galera Point

The northern segment of the Ecuadorian coast between the Galera Point and the Colombian border trends NE–SW, with a trench length (TL) of about 135 km and a coast length (CL) of approximately 170 km, giving a deviation ratio CL/TL of 1.26. Two different coastal morphologies are observed, rocky and cliffy to the west, and low and sandy to the east, changing sharply near Las Peñas.

The coast west of Las Peñas presents marine terraces in Rio Verde and near the Galera Point, allowing a mean uplift rate of about 0.3 mm a^{-1} to be calculated (Pedoja *et al.* 2006*b*). The hinterland is hilly, the Coastal Cordillera joining the coast west of Esmeraldas, and giving a 200 m-high sea cliff in Punta Gorda. The downstream segment of the Esmeraldas River crosses the Coastal Cordillera and the small estuary ends precisely at the head of the canyon (Fig. 24.4). The Esmeraldas Canyon is 95 km long, trending mainly north–south, with a mean slope of 2.5–3%. The upper part of the canyon shows a sinuosity of 1.27 due to channel curves determined principally by fractures and faults (Fig. 24.4). Along the middle part of the Esmeraldas Canyon, a 1000 m-high fault scarp looking towards the NE is defined by bathymetric data (Collot *et al.* 2004). This is consistent with sedimentary dykes filled with sand of probably fluvial origin observed across the coastline near the head of the canyon (Fig. 24.4). These sedimentary dykes and associated faults show a west–east-trending extension that post-dates the Pliocene fan deposit of Punta Gorda (Aalto & Miller 1999), and occurred probably during the late Pleistocene or the Holocene based on the sand filling the dykes. The west–east-trending extension (roughly parallel to the plate motion) alternates in this area with north–south extension (orthogonal to the plate motion). Such a similar deformation pattern has been previously observed in the Aegean Arc (Angelier *et al.* 1982).

The head of the Esmeraldas Canyon cuts across a shallow (*c.* −30 m) and wide (30 km) continental shelf (De Miro *et al.* 1977) eroded in Pliocene turbidite deposits (Aalto & Miller 1999). The high content of terrigenous material and mud turbidites suggests an old marine fan of the Esmeraldas River deposited in the lower part of the continental platform. The Late Miocene–Early Pliocene pre-fan basement crops out in the Galera Point area, showing reverse faults and folds resulting from a west–east shortening (Fig. 24.4A). The present Esmeraldas fan is a deep-sea fan located at the outlet of the canyon to the trench, at a depth of about 3000 m (Collot *et al.* 2002, 2004; Coronel 2002).

From Las Peñas to the border of the upland determined by the Mataje Fault (Fig. 24.2) the coast is low. The intertidal zone includes a 15 km-wide margin of mangrove that constitutes the inner part of the Bay of Ancón de Sardinas. A 5–15 m-high sea cliff cut into weathered Pleistocene detritus limits the intertidal mangrove from the San Lorenzo upland. Active subsidence of the mangrove area is indicated by drowned beach ridges that are limited by striking lineaments interpreted as active faulting inside the

mangrove and along the sea cliff (Dumont *et al.* 2006). The Bay of Ancón de Sardinas opens seawards on a 600 m-deep and 90 km-wide depression corresponding to the present northern Borbon Basin (Deniaud 2000; Collot *et al.* 2002). Off the Bay of Ancón de Sardinas, the continental shelf and slope have widths of 33 and 100 km, respectively, but in front of Galera Point and Manglare Cape the continental shelf narrows to about 10 km. According to Collot *et al.* (2002), the continental slope is tilted trenchwards, eroded at the base, with a small accretion wedge of trench material present at the front (Fig. 24.5A). The Bay of Ancón de Sardinas is subjected to active faulting, represented by NNE–SSW- to NE–SW-trending faults with a 50–100 km spacing (Santana & Dumont 2002; Dumont *et al.* 2006). Kinematic analysis of slickensides on fault planes gives evidence of north–south-trending transtension motion (Dumont *et al.* 2006). The Quaternary state of stress in the northern Andes given both by faults and by subduction earthquakes shows a roughly west–east shortening (Ego *et al.* 1996), which is consistent with the north–south extension observed on the sea cliffs.

Central Ecuador: Galera Point–Santa Elena Point

The coast of central Ecuador has a curved geometry with a succession of points (Pasado Cape, San Lorenzo, Salengo and Santa Elena) and bays. The TL is about 360 km long and the CL about 430 km, giving a deviation ratio CL/TL of 1.19. In the bay areas, the continental shelf and continental slope have a mean width of about 40 and 45 km, respectively. Off the headlands of Galera and San Lorenzo, the width of the continental shelf narrows drastically to less than 10 km, and a little more than 10 km in front of Santa Elena Point (Fig. 24.2). This segment is characterized by the subduction of the Carnegie Ridge, which causes the coastline to bulge out westwards (Pedoja *et al.* 2006*b*) and probably also triggered the uplift that produced the coastal cordillera (Gutscher *et al.* 1999). Marine terraces are observed on the sides of the headlands, the highest being located in the Manta Peninsula at an elevation of about 360 m in front of the axial part of the Carnegie Ridge (Pedoja *et al.* 2006*b*). The main fault zones of the coastal margin (i.e. from north to south, the Puerto Libre, Jama and Manabi faults) determine the Galera, Cabo Pasado and Manta points and peninsulas, respectively (Fig. 24.2). The Cretaceous–Paleocene oceanic basement crops out in the Manta Peninsula, in La Plata Island on the upper edge of the continental slope and at Santa Elena Point (Fig. 24.2).

There are no canyons off the coast of central Ecuador, and the first canyon south of the Esmeraldas Canyon is Santa Elena Canyon, west of Santa Elena Point (Fig. 24.3) (Coronel 2002; Calahorrano 2005). Santa Elena Canyon is 32 km long, beginning at a depth of −400 m and observed downwards towards the trench at −4000 m. It is entrenched by up to 800 m in the upper part of the continental slope (Coronel 2002). There is a single channel with a low sinuosity (1.15) and a mean slope of 12%. A 325 m step located just before the junction to the trench suggests that erosion along the canyon is presently weak or inactive, and there are no fans at the outlet of the canyon to the trench. The origin of this canyon is unclear. On the one hand, there is no evidence of an upslope connection with a fluvial outlet from land, and Santa Elena Peninsula is subjected to arid conditions. However, climatic conditions during glacial periods are unknown. On the other hand, the single channel of Santa Elena Canyon suggests a slope failure canyon (Orange & Breen 1992; Pratson *et al.* 1994) formed from fluids escaping from the oceanic basement of the coastal margin cropping out in the Santa Elena area.

Villamar (2001) describes a seaward tilt and basal erosion of the western part of the continental margin (Fig. 24.5B), with a pattern similar to the section off northern Peru (Fig. 24.5E). The first 30–40 km at least of the subduction plane is nearly horizontal, and even shows a slight upward bulge (Villamar 2001).

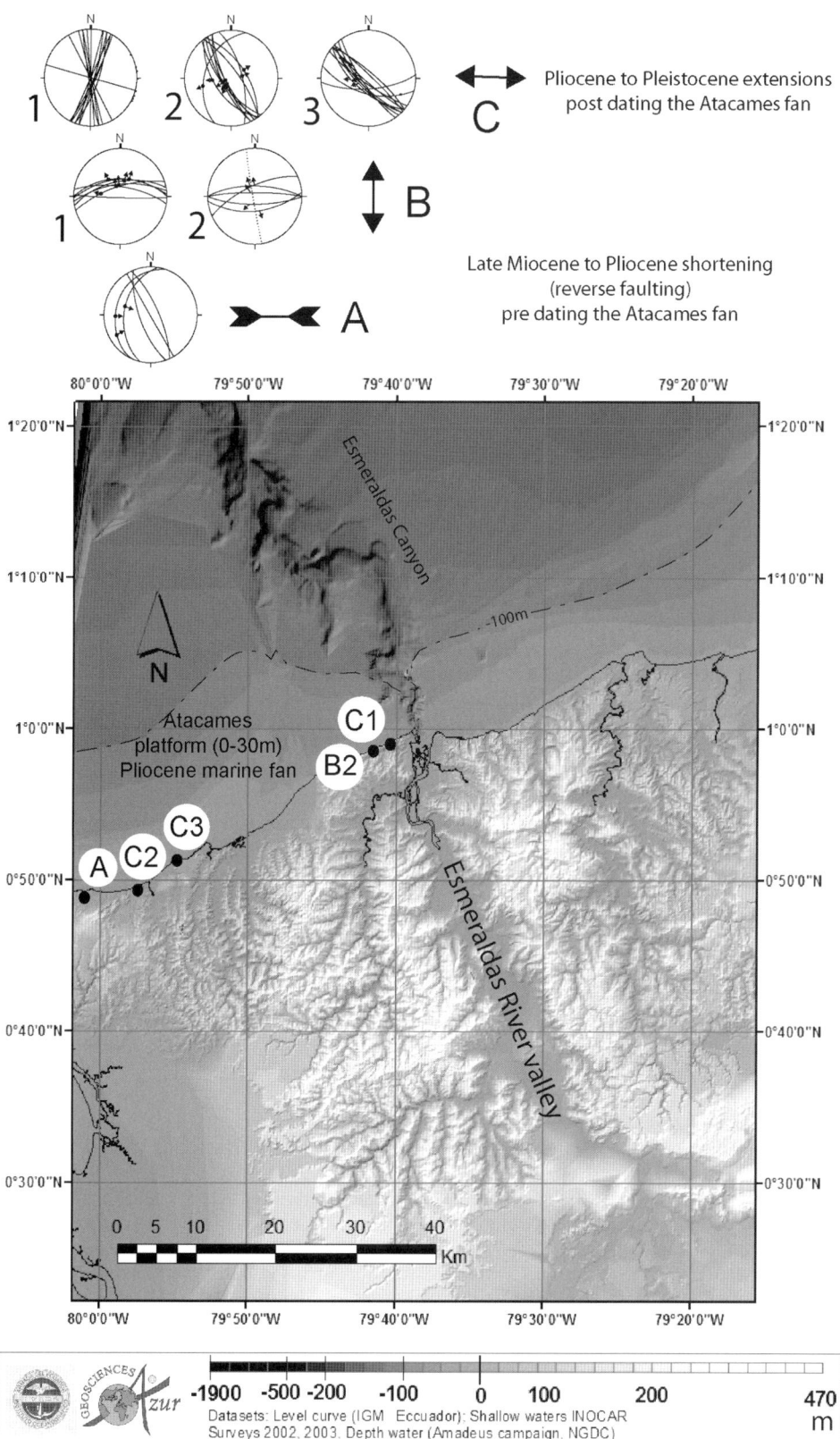

Pliocene to Pleistocene extensions post dating the Atacames fan

Late Miocene to Pliocene shortening (reverse faulting) pre dating the Atacames fan

Fig. 24.4. The digital elevation model shows the transition area between the lower section of the Esmeraldas River valley and the Esmeraldas Canyon. See the position of the Pliocene fan west of the head of the present canyon. Stereograms in the upper part of the figure show the faults planes. A: East–west shortening observed only in pre-Pliocene terrain. B: North–south-trending extension, the point B1 is from Estero el Platano located outside the figure, 10 km west from point A. B2 is from Pliocene deposits. C: East–west-trending extension. C1 shows the sedimentary dykes observed in the Pliocene deposits of the present sea cliff and filled by beach sand. Projections on the lower hemisphere of the Wulff net, the arrows show the slickensides and the observed motion.

South Ecuador: Gulf of Guayaquil

The Gulf of Guayaquil constitutes a 150 km re-entrant on the top of the Talara Arc. The CL from Santa Elena Point to the Peruvian border is about 350 km, corresponding to a TL of about 130 km, and giving a deviation ratio CL/TL of 2.7. This ratio is nearly 2 considering all the coast of the Gulf of Guayaquil from Santa Elena Point to Cabo Blanco in northern Peru (Figs 24.2 & 24.3).

Across the Gulf of Guayaquil, the width of the continental shelf and slope increases to 100 and 80 km, respectively. However, the trench–piedmont distance presents a smaller variation, measuring 200–230 km in northern Ecuador, rising to a maximum of 270 km north of the Gulf of Guayaquil across Guayas Basin, and maintained at 230 km across the gulf and just south of it. However, it narrows to 130 km in northern Peru, in front of the accreted Amotape massif.

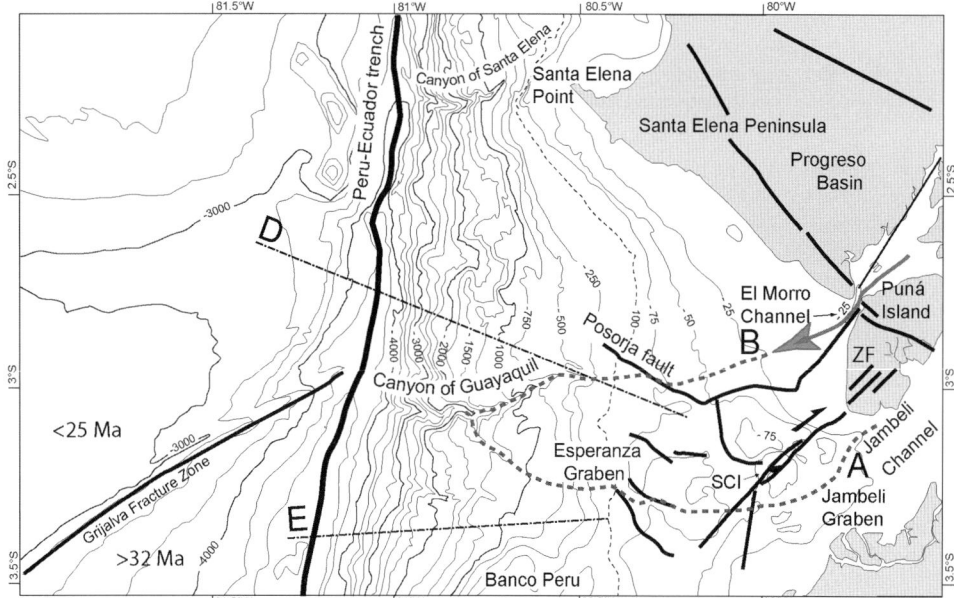

Fig. 24.5. Structural scheme of the Gulf of Guayaquil, combining data from Witt (2002), Calahorrano (2005) and Dumont *et al.* (2005*b*). The lines D and E correspond to the sections in Figure 24.3. The dotted line A shows the main outlet of the Guayas River estuary, and the arrow and dotted line B shows the new channel opened up after the last interglacial period through the El Morro Channel. SCI, Santa Clara Island; ZF, Zambapala Fault. Modified from Villamar (2001), Collot *et al.* (2002) and Calahorrano (2005).

The coast of the Gulf of Guayaquil presents different segments closely related to the structure of the gulf and its Plio-Quaternary tectonic evolution. The northern rocky coast from Santa Elena Point to El Morro channel is nearly straight and trends NW–SE, with up to 50 m-high sea cliffs. It is parallel to the offshore Posorja Fault (Figs 24.2 & 24.3) (Deniaud *et al.* 1999) that controls the northern structural margin of the gulf. Marine terraces are observed all along the coast from Santa Elena Point to Puna Island, allowing a mean uplift rate of 0.2 mm a^{-1} to be calculated (Pedoja *et al.* 2006*b*). This uplifted segment ends sharply to the east on the Zambapala Fault, which constitutes the western boundary of the Guayaquil–Caracas Fault Zone in the Gulf of Guayaquil (Fig. 24.3). The eastern border of the deformation corridor of the Guayaquil–Caracas Fault Zone is close to the Andean piedmont, which also marks the border of the lowland of the Guayas River estuary (Dumont *et al.* 2005*a, b*). The present estuary of the Guayas River is split into two branches at Puna Island, with the narrow El Morro Channel to the west and the wide Jambeli Channel to the east. The continuity of marine terraces from the continent to Puna Island clearly shows that the El Morro Channel (Fig. 24.3) opened up after the last interglacial period (Dumont *et al.* 2005*b*). This channel lines up northwards with the subsidence axis of the Guayas Basin, which is underlain by a basement fault (Fig. 24.2) (CODIGEM 1993). The main estuary of the Guayas River flows east of Puna, joining the continental slope across a wide shallow shelf and the structural depression of Esperanza (Fig. 24.3) (Coronel 2002).

East and south of the Gulf of Guayaquil, the coast trends NE–SW, with a 20 km-wide coastland covered by colluvium issued from the western slopes of the Andean Cordillera. Evidence of coastal uplift begins southwards in the Cancas area of north Peru, on the border of the Amotapes Massif (Pedoja *et al.* 2006*b*).

The Guayaquil Canyon (Fig. 24.3) is the most important canyon south of the Esmeraldas Canyon. It is 70 km long, extending from a depth of about 120 m on the edge of the continental slope to the trench (Coronel 2002). This is a ramified canyon with a main channel trending west–east, and a secondary channel connected from the south. The junction of the secondary channel is suspended over the main channel, suggesting a more recent and/or more active erosion and entrenchment along the main channel (Coronel 2002). The sinuosity of the main channel is low (1.1), with a mean slope of 6.3% and an entrenchment reaching 1000 m. The bathymetry (Michaud *et al.* 2006) suggests a connection between the main channel of the canyon and the El Morro Channel of the Guayas Estuary (Fig. 24.3). The southern secondary channel of the canyon suggests a connection with the main estuary of the Guayas River flowing east of Puna Island (Fig. 24.3). The main channel of the canyon might have been formed since the last interglacial period, and probably during the last glacial periods, because the El Morro Channel was not active before. The lack of a deep-sea fan at the end of the canyon suggests that presently the canyon is not very active. It was probably more active during the Last Glacial Maximum (LGM) sea-level lowstand, when both heads of the canyon were close to the minimum sea level.

Beneath the Gulf of Guayaquil, the initiation of the subduction plane exhibits a low steady dip (Fig. 24.5D) (Calahorrano 2005), similar to the slope observed below the Bay of Acon de Sardinas (Fig. 24.5A). Off the southern part of the gulf, in front of the uplifted Amotapes Massif, the frontal portion of the subduction is approximately flat over the first 45 km from the trench, as it is off the central part of Ecuador (Fig. 24.2).

The continental shelf

The continental shelf presents important variations along the Ecuadorian margin. The larger continental shelf is observed in the Gulf of Guayaquil, with a width of 60 km off the northern margin of the gulf where the multibeam bathymetry is available (Michaud *et al.* 2006) (Fig. 24.6, line A), and more than 100 km measured in an east–west direction south of Puna Island (Fig. 24.5). This wide shelf is due to the opening structure of the Gulf of Guayaquil (Deniaud *et al.* 1999; Witt *et al.* 2006) filling with sediments coming from the Guayas River. The transition from the shelf to the slope of the trench presents two steps: an upper one at a depth of about −100 ± 20 m corresponding to the lower sea level during the last glacial period; and a lower one at a depth of about −200 m.

Off the capes and points of central Ecuador there is no shelf (Fig. 24.6, line D), and only a narrow shelf up to 30 km wide in the bays. The upper edge of the shelf appears close to a depth of −100 m and a lower edge near −200 m. In the area where marine terraces are observed (Pedoja *et al.* 2006*a*), the −100 m shelf can be interpreted as a shelf developed during the LGM sea-level lowstand. However, uplift of the coast should give remnants of intermediate shelves related to substages between −100 m and the present sea level that are not visible on the available bathymetry chart.

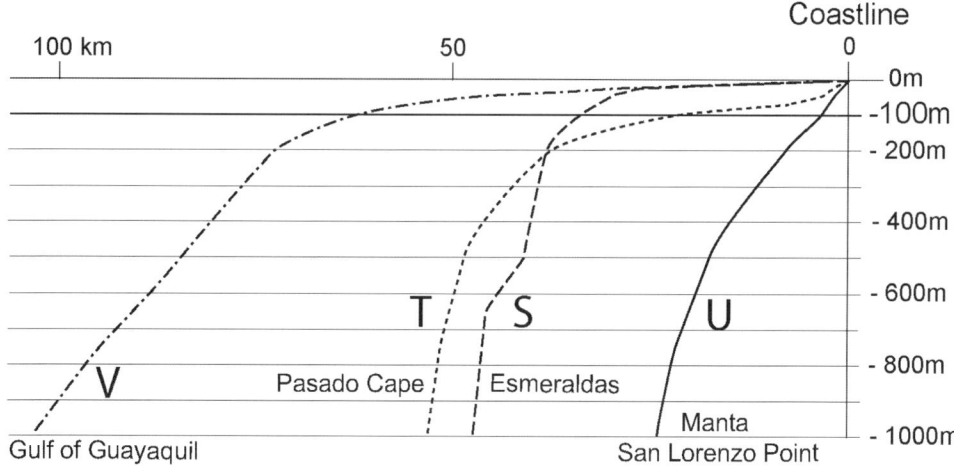

Fig. 24.6. Bathymetric sections of the upper part of the continental margin, located in Figure 24.2.

The Esmeraldas area presents a more complex pattern where only a shallow shelf is observed with an edge at a depth of about 30 m. This shallow shelf is a structural platform cut in the Pliocene fan of the Esmeraldas River (Fig. 24.6, line C), limited by a small step (*c.* 10 m) from the lower edge of the present wave-cut platform (INOCAR 1973). The 30 m-deep shelf is interpreted as having formed by one or several old LGM sea-level lowstands or an early stage of the present interglacial period. Note that we observe the same mean slope for the <30 m shelf observed in the Esmeraldas area and the upper shelf in the Gulf of Guayaquil (Fig. 24.6, lines A and C). There are no bathymetric edges visible at the depth of 100 m NE of Esmeraldas. Active fault tectonics observed in the area (Santana *et al.* 2006), and represented offshore by bathymetry and seismic profiles (Collot *et al.* 2002, 2004), can probably explain the lack of preservation of a shelf developed during the LGM.

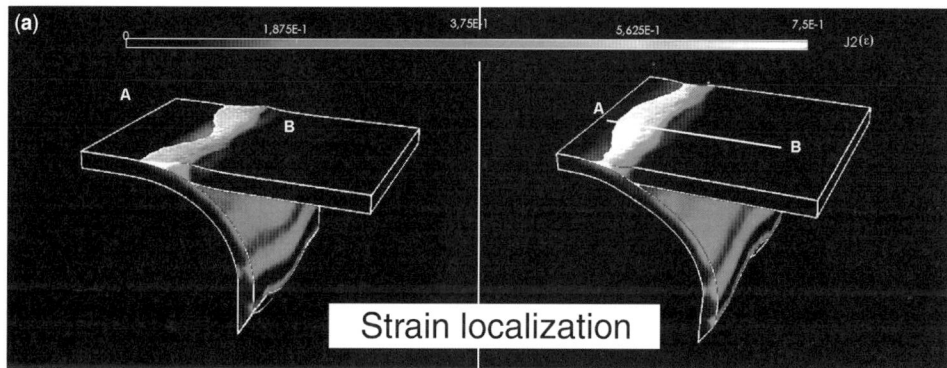

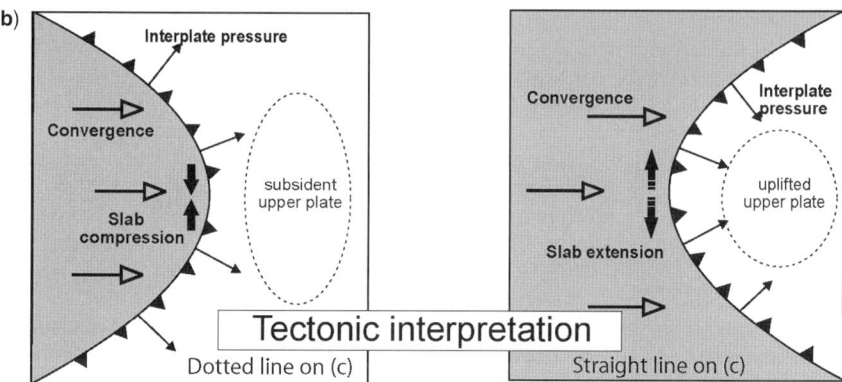

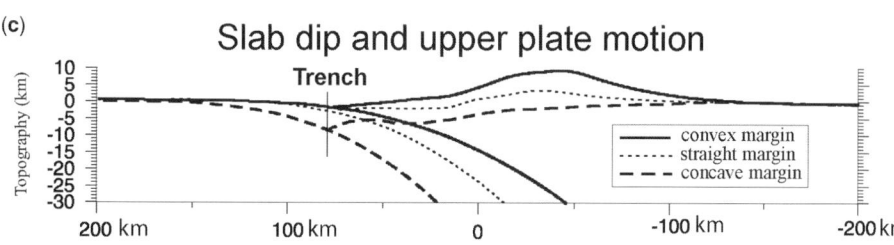

Fig. 24.7. Results of 3D numerical simulations testing the impact of margin curvature on the slab and the overriding plate. (**a**) Strain pattern for convex and concave margins. The dark colour indicates areas of no deformation, and the white denotes the highest strain rate. These models assume that the density contrast between the lithosphere and the asthenosphere is null, and the interplate friction is equal to zero. (**b**) Schematic interpretation of the tectonic regime. An oceanward convex margin induces a strong uplift in the upper plate and is subjected to a strong compression, in contrast to a concave margin, which accommodates an extensional regime (Boutelier 2004; Bonnardot *et al.* 2008). (**c**) Comparison of the interplate slab dip between a straight, convex and concave margin. A very shallow slab dip and a deep slab dip are observed in front of an oceanward convex margin and an oceanward concave margin, respectively. These simulations were performed for a zero interplate friction, as well as for a null density contrast between the lithosphere and the asthenosphere. Modified after Boutelier (2004), Bonnardot (2006) and Bonnardot *et al.* (2008).

Pleistocene–Holocene sedimentation

There are few published data on marine Pleistocene and Holocene deposits of the Ecuadorian continental margin, most of them coming from the Gulf of Guayaquil (Ordóñez 1991; Deniaud 2000). According to Deniaud *et al.* (1999), the main stage of deposition occurred during the Lower Pleistocene with a sedimentation rate of 8600 m Ma^{-1} at the centre of the Gulf of Guayaquil, and 3000 m Ma^{-1} at the centre of the Jambeli Graben (Fig. 24.3). The rate of sedimentation dropped to 700–1400 m Ma^{-1} between 1 Ma and 100 ka BP, but increased up to 4500 m Ma^{-1} from 100 ka BP to the present. Wells and seismic sections give evidence of two different sequences from the Upper Pleistocene to the Holocene period. The lower one (Upper Pleistocene) includes coarsening- and thickening-up layers of regressive deposits, and the upper one (Holocene) shows fining- and thinning-up layers of transgressive deposits (Deniaud *et al.* 1999). The upper Pleistocene regression sequences are linked to a transpression event related to the subduction of the Grijalva Fracture Zone beneath Santa Elena Peninsula, which is considered as a relatively undeformable tectonic block (Deniaud *et al.* 1999; Dumont *et al.* 2005*b*). The transpression tectonics resulted in the rising of shore and beach Pleistocene deposits up to an elevation of 300 m along the pop-up structure of the Zambapala Fault (Deniaud *et al.* 1999; Dumont *et al.* 2005*b*). The transgression sequences during the Holocene are related to the sea-level rise post-glaciation and the increase in tectonic subsidence induced by the present transtensional slip of the Guayaquil Caracas Fault Zone (Winter & Lavenu 1989; Dumont *et al.* 2005*a*) in relation to the main north–south extension observed in about all the coastal margin of Ecuador (Dumont *et al.* 2006).

Comparing the Ecuadorian continental margin with a 3D geometry model of a subduction zone

The structure and recent evolution of the continental and coastal margins of Ecuador underline the importance of geodynamic characteristics (i.e. the Carnegie Ridge, active faults joining the trench) but also the geometry of the subduction zone, and the curvatures and re-entrants at the regional and continental scale (e.g. the Talara Arc and the smaller central Ecuadorian arc associated with uplift, as opposed to large tectonic embayments associated with subsidence). In order to understand how 3D effects may impact on the Ecuadorian active margin, we consider here recent results of 3D numerical modelling of curved subduction zones from Boutelier (2004) and Bonnardot (2006). As numerical modelling was not the scope of this paper, no specific models were designed for the current study. Therefore, we invite the readers to refer to the references cited therein in order to get access to the details of the parametric numerical study. Bonnardot (2006) and Bonnardot *et al.* (2008) used the FEM Adeli to perform 3D mechanical numerical modelling of convergent margins along curved plate boundaries and they showed that the geometry of plate boundary has a fundamental effect on the upper-plate deformation due to variations in the slab dip. Models show that an oceanward convex margin is subjected to an excess of compression, triggering a strong uplift of the upper plate (Fig. 24.7). Conversely, an oceanward concave margin induces an extensional regime within the upper plate resulting in the subsidence of the margin. These results may be explained by the rigidity of the bending subducting plate that prevents the slab deforming along the curved margin. Instead, the slab will tend to bend along a straight interplate plane. As a result and in comparison with a straight margin, a convex margin produces a lower slab dip, whereas a concave margin induces a higher dip angle (Fig 24.7a). These results predict slab dip variations along the margin to accommodate changes in the margin geometry. The upper plate is then likely to deform in response to the slab behaviour. Variation of interplate coupling may, however, impact on slab behaviour. The simulations showed that for a low interplate coefficient of friction, the lithospheric rigidity would prevent the slab deforming, as described above. Conversely, a high coefficient of friction induces a strong deformation within the subducting plate, and results in decreasing the plate–slab coupling (Conrad *et al.* 2004). The weakening zone in the slab allows it to accommodate lateral bending along the curved margin more easily, therefore reducing lateral variations of the slab dip. It is difficult to estimate the interplate coupling to evaluate how the model can fit reality, as data are not well documented for the Ecuadorian margin. The interplate coupling may be variable from south to north, based on the historic record of strong seismic activity: there is apparently no strong seismic activity in the Gulf of Guayaquil, a moderate activity in coastal central Ecuador but strong activity from the Esmeraldas area to Colombia. However, the initial dips of the subducting slab predicted by the model are likely to be correlated with the variations observed along the Ecuadorian margin. For instance, it is worthwhile to note that subsidence probably affects the areas characterized by the re-entrant (the northernmost segment and the Gulf of Guayaquil), in contrast to the uplifted areas mainly observed in the oceanward convex regions (the central segment). In addition, the slight variations in the slab dip observed along the trench from north to south are also coincident with variations in the upper-plate boundary geometry, as the convex rising coast of central Ecuador and the rising Amotapes Massif (south of the Gulf of Guayaquil) are characterized by a flat subducting slab down to the first tens of kilometres in comparison to the 4–6° dip observed below the concave subsiding areas of the Gulf of Guayaquil (Villamar 2001; Collot *et al.* 2002; Calahorrano 2005) and the Bay of Ancón de Sardinas (Collot *et al.* 2002). The comparison between numerical models and the observed data confirms that the 3D geometry of the subduction zone is likely to have had an effect on tectonic behaviour and the vertical motion (i.e. the width) of the continental margin.

Discussion and conclusions

- The continental margin of Ecuador, considered from the Andean piedmont to the trench, varies from approximately 70 to 200 km, and the continental shelf from around 10 to 100 km, according to both local structures and characteristics of the subduction zone. The narrower margin appears in front of the Carnegie Ridge (subducting Nazca plate) and the wider is associated with tectonic embayment, such as the Gulf of Guayaquil and the Bay of Ancón de Sardinas.
- The trench–Andean piedmont distance varies according to the main curvatures of the plate boundary, larger along the Talara and Patagonian arcs, and smaller in the Arica Angle. This parameter is the more reliable to characterize the main curvatures of the active margin at the regional and continental scale. However, the width variation of the continental margin describes only local conditions involving either oceanic or continental structures.
- The curvature of the plate boundary has a fundamental effect on the upper-plate deformation and the slab dip. An oceanward convex margin is subjected to an excess of compression that triggered a strong uplift of the upper plate. Below the rising coast of central Ecuador and south of the Gulf of Guayaquil, the beginning of the subduction is approximately flat, in comparison to the 4–6° dip observed below the subsiding areas of the Gulf of Guayaquil and the Bay of Ancón de Sardinas.
- The Plio-Quaternary tectonic deformation of the continental and coastal margins controls the position and style of the main land–sea sediment transfer. Extension fault tectonics in the Esmeraldas area and the Bay of Ancón de Sardinas,

and the strike-slip fault depression along the southern Guaya-quil Caracas Fault, determine a regional diversion of the drainage transfer. The intermediate area of central Ecuador that receives no drainage from the Andes is subjected to uplift due to the curvature of the plate boundary and the subduction of the Carnegie Ridge.

Change from shortening to extension deformation (reverse and strike-slip faults to normal faulting) observed in the continental and coastal margins of Esmeraldas during the Late Plio-cene–Early Pleistocene resulted in a striking modification of the continental margin. The Pliocene marine fan of the Esmer-aldas River deposited near the present coastline, and previously the lower the edge of the continental shelf, shifted to a deep-sea fan position due to the formation of the Esmeraldas Canyon by extension tectonics.

- The subduction of the Carnegie Ridge has caused a 40–45 km minimum westward shift of the coastline in the Manta Penin-sula since the Late Pliocene, when considering the 300 m contour line (marine terraces observed up to 360 m near San Lorenzo Point). For the same period, the shift in the coastline is about 10 km north of the Carnegie Ridge at Galera Point, and a little more south of the ridge in the Santa Elena area. The coastal relief is about the same in the three areas and, thus, cannot account for the difference. The approximately 30–35 km difference accounts for the *increase* in the width of the continental margin due to the subduction of the Carnegie Ridge, which is the range of the difference between the width of the continental platform observed in front and on the sides of the Carnegie Ridge. During the same period the coastline moved inland at the Bay of Ancón de Sardinas by at least prob-ably 10–15 km.

This synthesis is based on several studies completed in co-operation with IRD, INOCAR (Instituto Oceanográfico de la Armada) and IG-EPN (Instituto Geofísico-Escuela Politécnica Nacional, Quito). This work is part of IGCP 464 'Continental Shelves During the Last Glacial Cycle' and a contribution to IGCP 495 'Land–Ocean Interaction'. It forms part of the publications of Geos-ciences Azur-UMR and of the ESPOL project 'Análisis de la variabilidad de la línea de costa ecuatoriana'. The final manuscript benefited from a review and comments from M. A. Gutscher, for which he is thanked.

References

AALTO, K. R. & MILLER, W. 1999. Sedimentology of the Pliocene Upper Onzole Formation, an inner-trench slope succession in northwestern Ecuador. *Journal of South American Earth Sciences*, **12**, 69–85.

ADDICOTT, W. O., RICHARDS, P. W. & SIDLAUSKAS, F. J. 1982. *Plate-Tectonic Map of the Circum-Pacific Region, Pacific Basin Sheet.* American Association of Petroleum Geologist, Tulsa, OK.

ALVARADO, A. 1998. Variation du champ de contrainte et de déformation et quantification des déformations actives du bloc côtier de l'Equa-teur. Paper presented at the DEA de Géodynamique et Physique de la Terre, Paris XI, Centre d'Orsay.

ANGELIER, J., LYBERIS, N., LE PICHON, X., BARRIER, E. & HUCHON, P. 1982. The tectonic development of the Hellenic Arc and the Sea of Crete: a synthesis. *Tectonophysics*, **86**, 159–196.

BALDOCK, J. W. 1982. *Geología del Ecuador: Boletín de la explicación del Mapa Geológico de la República del Ecuador, Esc. 1:1'000.000.* Min. Rec. Nat. Energ., Quito.

BONNARDOT, M.-A. 2006. *Etude géodynamique de la zone de subduction Tonga-Kermadec par une approche couplée de modélisation numér-ique 3D et de sismotectonique.* PhD thesis, Université Nice-Sofia Antipolis.

BONNARDOT, M.-A., HASSANI, R., TRIC, E., RUELLAN, E. & RÉGNIER, M. 2008. Effect of margin curvature on plate deformation in a 3-D numerical model of subduction. *Geophysical Journal International*, **173**, 1084–1094.

BOUTELIER, D. 2004. *La modélisation expérimentale tridimetionnelle thermo-mécanique de la subduction continentale et l'hexumation*

des roches de ultra haute pression/basse température. PhD thesis, Nice Sofia-Antipolis.

BRISTOW, C. R. & HOFFSTETTER, R. 1977. *Lexique Stratigraphique, Amér-ique Latine, 5–2, Equateur,* CNRS, Paris.

CALAHORRANO, A. 2005. *Structure de la marge du Golfe de Guayaquil (Equateur) et propriétés physiques du chenal de subduction, à partir de données de sismique marine réflexion et réfraction.* PhD thesis, UPMC Paris VI, Paris.

CAMPBELL, C. J. 1974. Ecuadorian Andes. *In*: SPENCER, A. M. (ed.) *Mesozoic–Cenozoic Orogenic Belts, Data for Orogenic Studies.* Geological Society, London, Special Publications, **4**, 725–732.

CASE, J. E., BARNES, J., PARIS, G., GONZALEZ, H. & VINA, A. 1973. Trans-Andean Geophysical Profile, Southern Colombia. *Geological Society of America Bulletin*, **84**, 2895–2904.

CEMBRANO, J., HERVÉ, F. & LAVENU, A. 1996. The Liquine Ofqui fault zone: a long-lived intra-arc fault system in southern Chile. *Tectono-physics*, **259**, 55–66.

CODIGEM 1993. *Mapa tectono-metalogenico de la republica del Ecuador.* British Geological Survey, Keyworth.

COLLOT, J.-Y., CHARVIS, P., GUTSCHER, M.-A. & OPERTO, S. 2002. Exploring the Ecuador–Colombia active margin and interplate seismogenic zone. *Eos, Transactions of the American Geophysical Union*, **83**, 185–190.

COLLOT, J. Y., MARCAILLOU, B. ET AL. 2004. Are rupture zone limits of great subduction earthquakes controlled by upper plate structures? Evidence from multichannel seismic reflection data acquired across the Northern Ecuador–southwest Colombian margin. *Journal of Geophysical Research*, **109**, B11103.

COLLOT, J.-Y., MIGEON, S. ET AL. 2005. Seafloor margin map helps in understanding subduction earthquakes. *Eos, Transactions of the American Geophysical Union*, **86**, 463–465.

CONRAD, C. P., BILEK, S. & LITHGOW-BERTELLONI, C. 2004. Great earth-quake and slab pull: interaction between seismic coupling and plate–slab coupling. *Earth and Planetary Science Letters*, **218**, 109–122.

CORONEL, J. 2002. *Les canyons de la marge équatorienne: approche morphostructurale et évolution.* DEA Dynamique de la lithosphère, Université Pierre et Marie Curie, Villefranche sur mer.

DALY, M. C. 1989. Correlations between Nazca/Farallon plate kinematics and forarc basin evolution in Ecuador. *Tectonics*, **8**, 769–790.

DE METS, C., GORDON, R. G., ARGUS, D. F. & STEIN, S. 1989. Current plate motions. *Geophysical Journal*, **101**, 425–478.

DE MIRO, M., AYON, H. & BENITES, B. 1976. *Morfología y estructura del margen continental del Ecuador. 1.* Instituto Oceanográfico de la Armada (INOCAR), Guayaquil.

DE MIRO, M., CORONEL, V., FRANCO, I. & CUENCA, J. 1977. *Morfología y sedimentos de la plataforma continental de la provincia de Esmeral-das, Ecuador. 1.* Instituto Oceanográfico de la Armada (INOCAR), Guayaquil.

DENIAUD, Y. 2000. *Enregistrements sédimentaire et structural de l'évol-ution géodynamique des Andes Equatoriennes au cours du Néogène: Etude des bassins d'avant arc et bilan de masse.* Laboratoire de géodynamique des chaînes alpines de l'Université Joseph Fourier, Grenoble, Géologie Alpine, Mémoire, **32**.

DENIAUD, Y., BABY, P., BASILE, C., ORDOÑEZ, M., MONTENEGRO, G. & MASCLE, G. 1999. Opening and tectonic and sedimentary evolution of the Gulf of Guayaquil: Neogene and Quaternary fore-arc basin of the south Ecuadorian Andes. *Comptes Rendus de l'Académie des Sciences – Séries II A – Earth and Planetary Science*, **328**, 181–187.

DIRAISON, M., COBBOLD, P. R., GAPAIS, D. & ROSSELLO, E. A. 1997. Magellan strait: part of a Neogene rift system. *Geology*, **25**, 703–706.

DUMONT, J. F., SANTANA, E. & VILEMA, W. 2005a. Morphologic evidence of active motion of the Zambapala Fault, Gulf of Guayaquil (Ecuador). *Geomorphology*, **65**, 223–239.

DUMONT, J. F., SANTANA, E. ET AL. 2005b. Morphological and microtec-tonic analysis of Quaternary deformation from Puna and Santa Clara Islands, Gulf of Guayaquil, Ecuador (South America). *Tectonophy-sics*, **399**, 331–350.

DUMONT, J. F., SANTANA, E., VALDEZ, F., TIHAY, J. P., USSELMANN, P., ITURALDE, D. & NAVARETTE, E. 2006. Fan beheading and drainage diversion as evidence of a 3200–2800 BP earthquake event in the Esmeraldas-Tumaco seismic zone; a case study for the effects of great subduction earthquakes. *Geomorphology*, **74**, 100–123.

EGO, F., SÉBRIER, M., LAVENU, A., YEPES, H. & EGUEZ, A. 1996. Quaternary state of stress in the Northern Andes and the restraining bend model for the Ecuadorian Andes. *Tectonophysics*, **259**, 101–116.

FEININGER, T. & SEGUIN, K. M. 1983. Simple Bouguer gravity anomaly field and the inferred crustal structure of continental Ecuador. *Geology*, **11**, 40–44.

GUILLIER, B., CHATELAIN, J. L., JAILLARD, E., YEPES, H., POUPINET, G. & FELS, J. F. 2001. Seismological evidence on the geometry of the orogenic system in central northern Ecuador (South America). *Geophysical Research Letters*, **28**, 3749–3752.

GUTSCHER, M. A., MALAVIEILLE, J. S. L. & COLLOT, J.-Y. 1999. Tectonic segmentation of the North Andean margin: impact of the Carnegie Ridge collision. *Earth and Planetary Science Letters*, **168**, 255–270.

HALL, M. L. & WOOD, C. A. 1985. Volcano-tectonic segmentation of the Northern Andes. *Geology*, **13**, 203–207.

HEY, R. 1977. Tectonic evolution of the Cocos-Nazca speading center. *Geological Society of America Bulletin*, **88**, 1404–1420.

INOCAR 1973. *Punta Galera y Esmeraldas, mapa de sondeos, mapa I.O.A. 01*. Instituto Oceanografico de la Armada, Guayaquil.

JAILLARD, E., SOLER, P., CARLIER, G. & MOURIER, T. 1990. Geodynamic evolution of the northern and central Andes during Early to Middle Mesozoic times: a Tethyan model. *Journal of the Geological Society, London*, **147**, 1009–1022.

JAILLARD, E., ORDOÑEZ, M., BENÍTEZ, S., BERRONES, G., JIMENEZ, N., MONTENEGRO, G. & ZAMBRANO, I. 1995. Basin development in an accretionary, oceanic-floored fore-arc setting: southern coastal Ecuador during Late Cretaceous–Late Eocene times. *In*: TANKARD, A. J., SUAREZ, S. R. & WELSINK, H. J. (eds) *Petroleum Basins of South America*. American Association of Petroleum Geologists, Memoirs, **62**, 615–631.

KELLOGG, J. N. & VEGA, V. 1995. Tectonic development of Panama, Costa Rica, and the Colombian Andes: constraints from Global Positioning System geodetic studies and gravity. *In*: MANN, P. (ed.) *Geologic and Tectonic Development of the Caribbean Plate Boundary in Southern Central America*. Geological Society of America Special Papers, **295**, 75–90.

KENDRICK, E., BEVIS, M., SMALLEY, R., JR. BROOKS, B., VARGAS, R. B., LAURÍA, E. & FORTES, L. P. S. 2003. The Nazca–South America Euler vector and its rate of change. *Journal of South American Earth Science*, **16**, 125–131.

LAGABRIELLE, Y., SUAREZ, M., ROSESELLO, E. A., HÉRAIL, G., MARTINOD, J., RÉGNIER, M. & DE LA CRUZ, R. 2004. Neogene to Quaternary tectonic evolution of the Patagonian Andes at the latitude of the Chile Triple Junction. *Tectonophysics*, **385**, 211–241.

LALLEMAND, S. 1999. *La Subduction Océanique*. Gordon & Breach Science, Philadelphia, PA.

LONSDALE, P. 1978. Ecuadorian subduction system. *American Association of Petroleum Geologists Bulletin*, **62**, 2454–2477.

MACHARÉ, J. & ORTLIEB, L. 1992. Plio-Quaternary vertical motions and the subduction of the Nazca Ridge, central coast of Peru. *Tectonophysics*, **205**, 97–108.

MEGARD, F. 1987. Cordilleran Andes and marginal Andes; a review of Andean geology north of the Arica Elbow (18°S). *In*: MONGER, J. W. H. & FRANCHETEAU, J. (eds) *Circum-Pacific Orogenic Belts and the Evolution of the Pacific Ocean Basin*. American Geophysical Union, Geodynamics Series, **18**, 71–95.

MICHAUD, F., COLLOT, J.-Y., DE LA TORRE, G., ALVARADO, A., PAZMINO, N. & LOPEZ, E. 2006. *Batimetria y topografia del Ecuador, escala 1:1 000 000*. Instituto Oceanografico de la Armada, Guayaquil.

MITOUARD, P., KISSEL, C. & LAJ, C. 1990. Post-Oligocene rotations in southern Ecuador and northern Peru and the formation of the Huancabamba deflection in the Andean Cordillera. *Earth and Planetary Science Letters*, **98**, 329–339.

MITOUARD, P., LAJ, C., MOURIER, T. & KISSEL, C. 1992. Paleomagnetic study of an arcuate fold belt developed on a marginal orogen: the Cajamarca deflection, northern Peru. *Earth and Planetary Science Letters*, **112**, 41–52.

MOURIER, T., LAJ, C., MÉGARD, F., ROPERCH, P., MITOUARD, P. & MEDRANO, F. 1988. An accreted continental terrane in northwestern Peru. *Earth and Planetary Science Letters*, **88**, 182–192.

ORANGE, D. L. & BREEN, N. A. 1992. The effect of fluid escape on accretionary wedges 2. Seepage forces, slope failure, headless submarine canyons, and vents. *Journal of Geophysical Research*, **97**, 9277–9295.

ORDÓÑEZ, M. 1991. *Bioestratigrafía y Paleoecología del Plio-Pleistoceno del Graben de Jambelí*. Doctora en Geología thesis, Universidad de Guayaquil, Guayaquil.

PEDOJA, K. 2003. *Les terrasses marines de la marge Nord Andine (Equateur et Nord Pérou): relations avec le contexte géodynamique*. PhD thesis, Université Pierre et Marie Curie, Paris.

PEDOJA, K., DUMONT, J. F. *ET AL.* 2006a. Plio-Quaternary uplift of the Manta Peninsula and La Plata Island and the subduction of the Carnegie Ridge, central coast of Ecuador. *Journal of South American Earth Science*, **22**, 1–21.

PEDOJA, K., ORTLIEB, L., DUMONT, J.-F., LAMOTHE, M., GHALEB, B., AUCLAIR, M. & LABROUSSE, B. 2006b. Quaternary coastal uplift along the Talara Arc (Ecuador, Northern Peru) from new marine terrace data. *Marine Geology*, **228**, 73–91.

PRATSON, L. F., RYAN, W. B. F., MOUNTAIN, G. F. & TWICHELL, D. C. 1994. Submarine canyon initiation by downslope-eroding sediment flows: evidence in late Cenozoic strata on the New Jersey continental slope. *Geological Society of America Bulletin*, **106**, 395–412.

SANTANA, E. & DUMONT, J. F. 2002. The San Lorenzo Fault, a new active fault in relation to the Esmeraldas–Tumaco seismic zone. Paper presented at the 5th International Symposium on Andean Geodynamics. IRD, Toulouse.

SANTANA, E., DUMONT, J. F., CRUZ, M. & ORDOÑEZ, M. 2001a. Will Santa Clara Island disappear because of increasing ENSO effects? *In*: *Abstracts of Conference Papers of the Fifth International Conference on Geomorphology. Transactions of the Japanese Morphological Union, Tokyo*, **22**(4), 207.

SANTANA, E., DUMONT, J. F. & KING, A. 2001b. Los efectos del fenomeno El Niño en la ocurencia de una alta tasa de erosion costera en el sector de Punta Gorda, Esmeraldas. *Acta Oceanografica del Pacifico*, **11**, 1–5.

SANTANA, E., DUMONT, J.-F., VALDEZ, F., PAZMINO, A., TIHAY, J. P., USSELMANN, P. & LOPEZ, E. 2006. Método morfo estructural para la identificacion de paleoeventos tecto-sismicos: aplicacion a la zona costera de San Lorenzo, norte de Ecuador. *Acta Oceanografica del Pacifico*, **13**, 224–242.

SCALABRINO, B., LAGABRIELLE, Y. *ET AL.* 2010. A morphotectonic analysis of central Patagonian Cordillera: negative inversion of the Andean belt over a buried spreading center? *Tectonics*, **29**, TC2010.

STEINMANN, M., HUNGERBÜHLER, D., SEWARD, D. & WINKLER, W. 1999. Neogene tectonic evolution and exhumation of the southern Ecuadorian Andes: a combined stratigraphy and fission-track approach. *Tectonophysics*, **307**, 255–276.

TRENKAMP, R., KELLOGG, J. N., FREYMUELLER, J. T. & MORA, H. P. 2002. Wide plate margin deformation, southern Central America and northwestern South America, CASA GPS observations. *Journal of South American Earth Sciences*, **15**, 157–171.

USGS 2006. *USGS CGM InfoBank Atlas: South America Regions*. United States Geological Survey, Reston, VA.

VILLAMAR, R. 2001. *Subduction de la Ride de Carnegie sous la marge d'Equateur: Structure et déformation à partir des données de sismique multitrace*. DEA, Université Pierre et Marie Curie et Géosciences Azur, Paris.

VILLANUEVA I BOHIGAS, G. 1997. *Déformation à grande échelle dans la subduction: réponse de la lithosphère continentale, réponse de la lithosphère océanique, le point triple du Chili*. UBO, Brest.

WINTER, T. & LAVENU, A. 1989. Morphological and microtectonic evidence for a major active right-lateral strike-slip fault across central Ecuador (South America). *Annales Tectonicae*, **III**, 2, 123–139.

WITT, C. 2002. *Evolution du bassin du Golfe de Guayaquil (Equateur) depuis le Quaternaire: Analyse de la déformation à partir de profils de sismique réflexion*. DEA, Villefranche sur Mer, Université Pierre et Marie Curie et UMR Geosciences Azur.

WITT, C., BOURGOIS, J., MICHAUD, F., ORDOÑEZ, M., JIMENEZ, N. & SOSSON, M. 2006. Development of the Gulf of Guayaquil (Ecuador) during the Quaternary as an effect of the North Andean block tectonic escape. *Tectonics*, **25**, 1–22.

Index

Page numbers in *italics* refer to Figures. Page numbers in **bold** refer to Tables.